MICROBIOLOGY
An Introduction

Of Related Interest from the Benjamin/Cummings Series in the Life Sciences

I. E. Alcamo
Fundamentals of Microbiology, Second Edition
(1987)

R. M. Atlas, R. Bartha
*Microbial Ecology: Fundamentals
and Applications*, Second Edition (1987)

W. M. Becker
The World of the Cell (1986)

N. A. Campbell
Biology (1987)

J. Cappuccino and N. Sherman
Microbiology: A Laboratory Manual,
Second Edition (1987)

C. L. Case and T. R. Johnson
Laboratory Experiments in Microbiology (1984)

H. Gest
The World of Microbes (1987)

P. B. Hackett, J. A. Fuchs, and J. W. Messing
*An Introduction to Recombinant DNA Techniques:
Basic Experiments in Gene Manipulation*,
Second Edition (1988)

L. E. Hood, I. L. Weissman, W. B. Wood,
and J. H. Wilson
Immunology, Second Edition (1984)

T. R. Johnson and C. L. Case
*Laboratory Experiments in Microbiology:
Brief Edition*, Second Edition (1989)

E. N. Marieb
Human Anatomy and Physiology (1989)

L. G. Mitchell, J. A. Mutchmor, and W. D. Dolphin
Zoology (1988)

C. F. Norton
Microbiology, Second Edition (1986)

A. P. Spence and E. B. Mason
Human Anatomy and Physiology, Third Edition
(1987)

P. J. VanDemark and B. L. Batzing
*The Microbes: An Introduction to Their Nature
and Importance* (1987)

J. D. Watson, N. H. Hopkins, J. W. Roberts,
J. A. Steitz, and A. M. Weiner
Molecular Biology of the Gene, Fourth Edition
(1987)

MICROBIOLOGY
An Introduction

THIRD EDITION

Gerard J. Tortora
Bergen Community College

Berdell R. Funke
North Dakota State University

Christine L. Case
Skyline College

The Benjamin/Cummings Publishing Company, Inc.
Redwood City, California · Fort Collins, Colorado
Menlo Park, California · Reading, Massachusetts · New York
Don Mills, Ontario · Wokingham, U.K. · Amsterdam · Bonn
Sydney · Singapore · Tokyo · Madrid · San Juan

Sponsoring Editor: Robin J. Williams
Developmental Manager: Jane Reece
Developmental Editors: Camille Cusumano,
    Susan Weisberg, Pat Burner
Production Supervisor: Betsy Dilernia
Art Coordinator: Bruce Lundquist
Designer: Christy Butterfield
Copy Editor: Larry Olsen
Photo Editor: Darcy Lanham
Photo Researcher: Stuart Kenter
Artists: Ben Turner Graphics, Michael Fornalski,
    Linda Harris-Sweezy, Barbara Haynes Cousins,
    Georg Klatt, Elizabeth Morales-Denney,
    Carla Simmons, Carol Verbeeck
Composition and Film: Jonathan Peck Typographers
Printing and Binding: Rand McNally

About the Cover: A researcher inspects a batch of plastic beads coated with HIV antibodies used in the HIV antigen test. Photo courtesy of Abbott Laboratories.

Photograph acknowledgments appear after the Glossary.

Library of Congress Cataloging-in-Publication Data

Tortora, Gerard J.
   Microbiology: an introduction / Gerard J. Tortora, Berdell R. Funke, Christine L. Case.—3rd ed.
      p.   cm.—(Benjamin/Cummings series in the life sciences)
   Includes bibliographies and index.
   ISBN 0-8053-0141-0
   1. Microbiology. I. Funke, Berdell R. II. Case, Christine L., 1948–   . III. Title. IV. Series.
QR41.2.T67 1989   88-28126
576—dc19   CIP

DEFGHIJ-RN-93210

The Benjamin/Cummings Publishing Company, Inc.
390 Bridge Parkway
Redwood City, California 94065

About the Authors

Gerard J. Tortora

Jerry Tortora is a professor of biology and teaches microbiology and human anatomy and physiology at Bergen Community College in Paramus, New Jersey, where he has been a faculty member for the past 21 years. He received his M.A. in biology from Montclair State College in 1965. He belongs to numerous biology/microbiology organizations, such as the American Society of Microbiology (ASM), American Association for the Advancement of Science (AAAS), National Education Association (NEA), New Jersey Education Association (NJEA), and the Metropolitan Association of College and University Biologists (MACUB). Jerry is the author of a number of successful biology textbooks.

Berdell R. Funke

Bert Funke received his Ph.D., M.S., and B.S. in microbiology from Kansas State University. He has been a professor of microbiology for 24 years at North Dakota State University. Currently he is teaching microbiology, pathogenic microbiology, food microbiology, and parasitology. Bert is an active member of the ASM and the AAAS. As a research scientist in the Experiment Station at North Dakota State, he has published numerous papers in soil microbiology and food microbiology.

Christine L. Case

Chris Case is a registered microbiolgist and a professor of microbiology at Skyline College in San Bruno, California, where she has taught for the past 18 years. She received her Ed.D. in curriculum and instruction from Nova University and her M.A. in microbiology from San Francisco State University. Presently she is an active member of the ASM, the Northern California Association of Microbiologists (NCASM), and the Scientific Research Society of North America (Sigma Xi). In addition to teaching, Chris is a regular contributor to the professional literature, develops innovative instructional methodologies, and has maintained a personal and professional commitment to conservation and the importance of science in society. Chris is also an avid photographer, and many of her photographs appear in this volume.

Preface

Microbiology: An Introduction is a comprehensive text for students in a wide variety of programs, including allied-health sciences, biological sciences, environmental studies, animal science, forestry, agriculture, home economics, and liberal arts. It is a beginning text, assuming no previous study of biology or chemistry.

In the seven years since the publication of the first edition, this book has been used at more than 800 colleges and universities, by over 200,000 students. We have been gratified to hear from instructors and students alike that the book has become a favorite among their textbooks—a learning tool that is both effective and enjoyable. Therefore, our goal for the third edition was to maintain our clear coverage of the basic principles of microbiology, while updating the book and integrating the suggestions made by users of the earlier editions.

Features of the Revision

Our primary goal for the third edition of *Microbiology: An Introduction* was to update the book throughout, to make certain it reflected the information and important new discoveries of the last few years. We also wanted to give our book a fresh look—new photomicrographs, new illustrations, and a more inviting design. At the same time, we wanted to keep the book at a manageable length and maintain the emphasis on the fundamental principles significant to the students who use our book. Every page of the book was scrutinized with these goals in mind. The following list highlights the resulting major changes in the third edition:

- Many of the *MMWR* and *Microbiology in the News* boxes have been replaced with more timely articles. *Microbiology Highlights*, a third variety of box, contain interesting information relevant to the study of microbiology that we hope will further motivate students.
- *Microbial Metabolism* (Chapter 5) has been carefully revised to further clarify the basic principles. A new appendix highlighting biochemical pathways (Appendix F) augments the material in Chapter 5 for those interested in further depth.
- *Microbial Genetics* (Chapter 8) has been completely revised to include updated information on molecular biology and recombinant DNA techniques. Many new illustrations will help clarify these difficult topics.
- The *immunology* chapters (Chapters 16–18) have been revised, expanded, and reorganized to reflect the dynamic nature of this important field.
- The three full-color photo essays called "Microviews" have been updated and augmented. *Microview A, Identification of Microorganisms*, covers microbial morphology, bacterial stains, microbial cultures, eucaryotic pathogens, and a brief survey of environmental microbiology. *Microview B, Recombinant DNA at Work*, describes genetic engineering and includes photographs from Cetus and Chiron, leading biotechnology companies. *Microview C, Inside Clinical Laboratories*, is a tour of hospital laboratories at the University of California at San Francisco.

Features Retained from the Earlier Editions

As mentioned earlier, we have retained in this new edition the features that made the previous editions so popular. These include:

- **An appropriate balance between microbiological fundamentals and applications, and between medical applications and other applied areas.** As in previous editions, basic principles are given greater emphasis than applications, and health-related applications are emphasized. Nevertheless—as before—applications are integrated throughout the text, and considerable attention is devoted to microorganisms in habitats outside the human body. We hope that all students will gain an appreciation of the fascinating diversity of microbial life, the central roles of microorganisms in nature, and the importance of microorganisms in our daily lives.
- **Illustration program carefully developed to support the text.** Included are both state-of-the-art electron micrographs that dramatically show microbial structure and good examples of light micrographs that more closely resemble what is usually seen in a microbiology laboratory. As with previous editions, quality line drawings provide significant support to in-text discussions. In the third edition, we have improved many illustrations and added new diagrams to further enhance the visual quality of the text.
- **Straightforward presentation of complex topics.** Although we have endeavored to completely update our text, we have engineered each step of the revision with the student in mind, maintaining the clarity of explanation for which our book has become known.

In-Text Learning Aids

A major goal of writing this text and its revisions has been to create a book that would be an effective tool for learning. Therefore, we have included many student aids in each chapter.

- **Learning Objectives** provide students with guidelines for what they should know after studying the chapter.
- **Tables** summarize, organize, and complement the text discussions.
- **Study Outlines** at the end of each chapter aid review. Each heading is keyed to specific pages in the text.
- **Study Questions** test recall of information presented in the chapter (*Review*) and provide an opportunity to apply knowledge in problem solving and analysis (*Challenge*).
- **Further Reading** suggestions give sources for additional investigation of the topics in the chapter.

Several **appendices** at the end of the book heighten its usefulness. The first appendix is the *Classification of Bacteria According to Bergey's Manual of Systematic Bacteriology*. This is followed by a guide to *Pronunciation of Scientific Names*, which provides basic rules of pronunciation and phonetic pronunciations for genus and species names used in the text. Also included is a guide to *Word Roots Used in Microbiology*, a *Most Probable Numbers (MPN) Table*, a description of *Methods for Taking Clinical Samples*, a new appendix highlighting key *Biochemical Pathways*, and a discussion of *Exponents, Exponential Notation, and Logarithms*. A **Glossary** provides definitions of all important terms used in the text.

Course Sequences

We have organized the book in what we feel is a useful fashion, but we recognize that there are a number of alternative sequences in which the material might be effectively presented. For those who wish to follow another sequence, we have made each chapter as independent as possible and have included numerous cross-references. Thus, the survey of the microbial world, Part Two, could be studied at the beginning of a course, immediately after Chapter 1. Or environmental and industrial microbiology, Part Five, could follow Parts One and Two. Since Chapters 7 and 19 both deal with the control of microbial growth, they could be covered together. The material on microbes and human disease, Part Four, readily lends itself to rearrangement or selective coverage. The various diseases are organized into chapters according to host organ system affected. However, the *Instructor's Guide* provides detailed guidelines for organizing the disease material in several alternative ways.

Supplementary Materials

All the ancillaries to the text have been revised by their original authors.

- The *Study Guide*, by Berdell Funke, will help students master and review major concepts and facts from the text. Each chapter of the study guide begins with a chapter summary organized under the text headings. Important terms are printed in boldface and defined, and important figures and tables from the text are included. Following the summary is an extensive self-testing section containing matching questions, fill-in questions, and an answer key.
- The *Instructor's Guide*, by Christine Case, includes many practical suggestions for using the text in a course. Suggested course outlines and corresponding text pages provide maximum flexibility in organizing your course. Important third edition changes are highlighted for each chapter to facilitate the transition from the second to the third edition. Also included are answers to study questions from the text, clinical issues that ask students to be "medical detectives" and solve the problem presented, and essay questions and multiple-choice tests for each chapter.

- *Benjamin/Cummings Testing Software* is available for the IBM PC, Apple II family, and Macintosh computers.
- **Acetate overhead transparencies** of 100 two-color line drawings from the text are available from the publisher to qualified adopters.
- *Laboratory Experiments in Microbiology: Brief Edition* (Second Edition, 1989), by Ted Johnson and Christine Case, is a very successful lab manual that has been updated throughout with new exercises on genetic engineering, transformation, bacteria of the mouth, and new safety guidelines for the handling of blood and body fluids according to CDC guidelines. The manual has also been reorganized so that lab report forms immediately follow each exercise. A detailed *Instructor's Guide* for the lab manual facilitates preparation for laboratory sessions.
- The *Computerized Student Tutorial*, by Dan Trubovitz, is an easy to use, interactive program for the Apple II family of computers. The tutorial, which is referenced to the text, will assist students by reviewing the chapter, providing practice questions, and assisting them with their incorrect answers.

Acknowledgments

In the preparation of this textbook, we have benefited from the guidance and advice of a large number of microbiology instructors across the country. Rick Bernstein, San Francisco State University, provided an early draft of the microbial genetics chapter. Dedicated teachers participated in focus groups to help plan the revision. Reviewers offered constructive criticism and suggestions at various stages of manuscript preparation. Contributors, focus group attendees, and reviewers are listed on the following page. We gratefully acknowledge our debt to these individuals.

We offer special thanks to the staff at Benjamin/Cummings for their dedication to excellence. Pat Burner's careful attention to detail served to keep information clear. The scientific and editorial contributions of Jane Reece (Gillen) were invaluable. Betsy Dilernia and Bruce Lundquist expertly guided our book through the production process. Robin Williams, our sponsoring editor, and Darcy Lanham, our photo editor, were instrumental throughout the revision process. And, we have enduring appreciation for our students, whose comments and suggestions both provide insight and remind us of their needs. This text is for them.

Gerard J. Tortora
Berdell R. Funke
Christine L. Case

Consultants for and Reviewers of
Microbiology: An Introduction, *Third Edition*

Focus Group Participants

Frank L. Binder, Marshall University
Blanche Griggs, Georgia State University
Rebecca Halyard, Clayton State College
Wallis L. Jones, DeKalb Community College
Michael H. Scholla, E. I. DuPont de Nemours
    & Company
Brian J. Wilkinson, Illinois State University,
    Normal

Reviewers of the Third Edition

Wendall E. Allen, East Carolina University
Kenneth L. Anderson, California State University,
    Los Angeles
Kostia Bergman, Northeastern University
Richard Bernstein, San Francisco State University
Russell F. Bey, University of Minnesota, St. Paul
Russell J. Centanni, Boise State University
William Coleman, University of Hartford
Mary Lynne Perille Collins, University of
    Wisconsin, Milwaukee
Cindy Erwin, City College of San Francisco
Randy Firstman, College of the Sequoias
Joseph J. Gauthier, University of Alabama,
    Birmingham
Ted R. Johnson, St. Olaf College
Diana Kaftan, Diablo Valley College
Ken Keudell, Western Illinois University
Juhee Kim, California State University, Long
    Beach

Robert I. Krasner, Providence College
John Lammert, Gustavus Adolphus College
Thomas Maier, University of Colorado, Denver
Eleanor K. Marr, Dutchess Community College
J. R. Milam, University of Florida, Gainesville
Robert Mitchell, Community College of
    Philadelphia
Helen Oujesky, University of Texas, San Antonio
Robert Pengra, South Dakota State University
Jane Phillips, University of Wisconsin, Madison
Christine Pootjes, Pennsylvania State University,
    University Park
Ralph Rascati, Kennesaw College
Robert Satterfield, College of Du Page
Robert Schoenhoff, Jefferson Community College
Jeffrey J. Sich, Youngstown State University
Josephine Smith, Montgomery County Community
    College
Cynthia V. Sommer, University of Wisconsin,
    Milwaukee
Lisa Steiner, Massachusetts Institute of Technology
Sheldon Steiner, University of Kentucky, Lexington
Teresa Ann Thomas, Southwestern College
Dan Trubovitz, Pasadena City College
Robert Twarog, University of North Carolina,
    Chapel Hill
James E. Urban, Kansas State University
Pat Vary, Northern Illinois University
James White, Prairie State College
Mary K. Wicksten, Texas A & M University
Fred Williams, Iowa State University
Brian J. Wilkinson, Illinois State University,
    Normal
Shanna Yonenaka, San Francisco State University

Brief Table of Contents

Detailed Table of Contents

PART ONE

Fundamentals of Microbiology

The Microbial World and You

In the summer of 1985, newspapers announced that a well-known actor was suffering from AIDS (acquired immune deficiency syndrome). The American public had heard about the disease, but until then AIDS had not been perceived as much of a health threat by the majority of Americans, perhaps because the victims were primarily from a group outside the mainstream of American society—male homosexuals. While the actor sought treatment in hospitals in France, Americans became increasingly aware of the illness that would end the life of one of their screen idols as well as the lives of over 6000 other Americans that same year.

The general public was just beginning to worry about what would become the most frightening epidemic of the century. Yet medical researchers had been gathering information about the devastating disease for several years. The same concepts of microbiology that had helped them identify other diseases, their means of transmission, and treatments were being applied to the study of AIDS.

AIDS had surfaced in 1981, with reports from New York and California that a few young homosexual males had died of a rare form of cancer (Kaposi's sarcoma). All the victims had suffered from problems with their immune systems. As more and more cases appeared, AIDS was also identified among intravenous drug users who had died of a previously rare type of pneumonia known as *Pneumocystis carinii* (nü-mō-sis'-tis kär-i'nē-ē) pneumonia. These victims, too, had experienced immune system problems.

By the middle of 1988, nearly 70,000 people in the United States had been diagnosed as having AIDS, and approximately 50% of them had died as a result of the disease. A great many more people had tested positive for presence of the AIDS virus in their blood. Health officials estimated in 1987 that about 1.5 million Americans carried the virus, and that about 25%

(a)

(b)

Figure 1-1 Human immunodeficiency virus (HIV). (a) Robert Gallo and his colleagues at the National Cancer Institute were one of two research groups that first identified HIV, the virus that causes AIDS. (The other group was headed by Luc Montagnier at the Pasteur Institute in Paris.) **(b)** This scanning electron micrograph shows HIV viruses (the tiny spheres) budding from one of their target cells, a white blood cell called a T4 lymphocyte (×15,000).

to 30% of them would develop the disease within five years. Scientists studying AIDS are still uncertain whether all individuals carrying the virus will eventually develop the disease or whether some may be able to live normally without any symptoms.

Symptoms of AIDS may develop for months or years and include a general feeling of discomfort or uneasiness, low-grade fever or night sweats, extreme fatigue, enlarged lymph nodes in the neck, armpits, and groin, and the blue-violet or brownish spots on the skin (usually the legs) characteristic of Kaposi's sarcoma.

Researchers quickly discovered that the cause of AIDS was a previously unknown virus (Figure 1-1). The virus, now called **human immunodeficiency virus (HIV)**, causes disease in a unique way. Instead of directly causing disease symptoms—as do such viruses as those causing mumps, chickenpox, hepatitis, and smallpox—the AIDS virus destroys certain cells of the immune system, the body's defense system. Sickness and death result from microorganisms or cancerous cells that might otherwise have been defeated by the body's natural defenses. So far, the disease appears to be inevitably fatal.

By studying disease patterns, medical researchers found that the AIDS virus could spread through sexual intercourse, from contaminated needles, from infected mothers to their fetuses before birth, and in blood transfusions—in short, by transmission of bodily fluids from one person to another. Since 1985, blood used for transfusions has been carefully checked for the presence of the AIDS virus, and it is now quite

unlikely for the virus to be spread via this route. So far, means of treating the disease remain elusive, though some experimental drugs show promise. Public health response has focused primarily on prevention.

In the months and years to come, microbiological techniques will be applied intensively to help scientists learn more about the structure of the deadly AIDS virus, how the virus is transmitted, how it grows in cells and causes disease, how drugs can be directed against the virus, and whether an effective vaccine can be developed.

* * *

AIDS poses one of this century's most formidable health threats, but it was not the first serious epidemic of a sexually transmitted disease. Syphilis was also once a fatal epidemic disease. As recently as 1941, syphilis caused an estimated 14,000 deaths in the United States. With few drugs available for treatment and no vaccines to prevent it, efforts to control the disease focused mainly on restraining sexual behavior and on the use of condoms. The eventual development of drugs to treat syphilis contributed significantly to preventing the spread of the disease. Reported cases of syphilis have dropped from a record high of 575,000 in 1943 to 35,398 in 1987.

Just as microbiological techniques helped researchers in the fight against syphilis, in 1976 they helped scientists discover the cause of Legionnaires' disease—a tiny, airborne bacterium not detectable by standard laboratory tests (see Microview A, Photo 3). Though the microorganism was unfamiliar to microbiologists, it had, in fact, caused several unexplained epidemics of pneumonia. Once it had been carefully documented, the bacterium was assigned the scientific name *Legionella pneumophila* (lē-jä-nel'lä nü-mō'fi-lä); the disease itself was later renamed *legionellosis*.

In 1978, toxic shock syndrome (TSS) caught the attention of the public for the first time when a number of otherwise healthy young women suddenly died. Microbiologists were able to show that the cause of death was **toxins** (poisonous substances) produced by the bacterium *Staphylococcus aureus* (staf-i-lō-kok'kus ô'rē-us), a microorganism carried on human skin (see Figure 10-10). In their research on how such a common bacterium could become so toxic, one thing scientists noticed was that all the victims used tampons during menstruation. In 1985, a Harvard research team found that certain tampon fibers absorb magnesium that is normally present in the vagina. When the amount of magnesium in the vagina is reduced, *S. aureus* produces large amounts of toxin. The association of TSS with a particular brand of tampon resulted in the recall of that brand and the publication of cautions about the use of tampons in general.

All the diseases we have mentioned are caused by viruses or bacteria, two classes of microorganisms. This book introduces you to the enormous variety of these microscopic organisms. It shows you how microbiologists use specific techniques and procedures to study the microbes that cause such diseases as AIDS, legionellosis, and TSS. You will also learn about the body's responses to microbial infection and the ways certain drugs combat microbial diseases. Finally, you will learn about the many beneficial roles that microbes play in the world around us.

Microbes in Our Lives

For many people, the words *germ, microbe,* and *microorganism* bring to mind a group of tiny creatures that do not quite fit any of the categories in that old question, "Is it animal, vegetable, or mineral?" Microorganisms are minute living things that individually are too small to be seen with the naked eye. The group includes bacteria (Chapter 10) and fungi (yeasts and molds), protozoans, and microscopic algae (Chapter

---

MICROBIOLOGY
HIGHLIGHTS

What Makes Sourdough Bread Different?

Imagine being a miner during the California gold rush. You've just made bread dough from your last supplies of flour and salt when someone yells, "Gold!" Temporarily forgetting your hunger, you run off to the gold fields. Many hours later, you return. The dough has been rising longer than usual, but you are too cold, tired, and hungry to start a new batch. Later you find that your bread tastes different from previous batches; it is slightly sour. During the gold rush, miners baked so many sour loaves that they were nicknamed "sourdoughs."

Conventional bread is made from flour, water, sugar, salt, shortening, and a living microbe, yeast. The yeast belongs to the Kingdom Fungi and is named *Saccharomyces cerevisiae.* When flour is mixed with water, an enzyme in the flour breaks its starch into two sugars, maltose and glucose. After the ingredients for the bread are mixed, the yeast metabolizes the sugars and produces alcohol (ethanol) and carbon dioxide as waste products. This metabolic process is called fermentation. The dough rises as carbon dioxide bubbles get trapped in the sticky matrix. The alcohol evaporates during baking. It and the carbon dioxide gas form spaces that remain in the bread.

Originally, breads were leavened by wild yeast from the air, which had been trapped in the dough. Later, bakers kept a starter culture of yeast—dough from the last batch of bread—to leaven each new batch of dough. Sourdough bread is made with a special sourdough starter culture that is added to flour, water, and salt. Perhaps the most famous sourdough bread made today comes from San Francisco, where a handful of bakeries have continuously cultivated their starters for more than 100 years and have meticulously maintained the starters to keep out unwanted microbes that can produce different and undesirable flavors. After bakeries in other areas made several unsuccessful attempts to match the unique flavor of San Francisco sourdough, rumors

11). It also includes viruses, those noncellular entities sometimes regarded as being at the border of life and nonlife (Chapter 12). Figure 1-7 on page 17 gives you an idea of the diversity of the microorganisms that you will learn more about in later chapters.

We tend to associate these small organisms only with uncomfortable infections, major diseases such as AIDS, or such common inconveniences as spoiled food. However, the majority of microorganisms make crucial contributions to the welfare of the world's inhabitants by helping to maintain the balance of living organisms and chemicals in our environment. Marine and freshwater microorganisms form the basis of the food chain in oceans, lakes, and rivers. Soil microbes help to break down wastes and incorporate nitrogen gas from the air into organic compounds, thereby recycling chemical elements in the soil, water, and air. Certain bacteria and algae play important roles in photosynthesis, a food- and oxygen-generating process that is critical to life on earth. Humans (and many other animals) depend on the bacteria in our intestines for digestion and synthesis of some vitamins that our bodies require, including some B vitamins for metabolism and vitamin K for blood clotting.

Microorganisms also have commercial applications. They are used in the synthesis of such chemical products as acetone, glycerin, organic acids, enzymes, alcohols, and many drugs. Our food industry frequently uses microbes in producing vinegar, sauerkraut, pickles, alcoholic beverages, green olives, soy sauce, buttermilk, cheese, yogurt, and bread (see the box below).

Though only a minority of microorganisms are **pathogenic** (disease-producing), practical knowledge of microbes is necessary for medicine and the related health sciences. For example, hospital workers must be able to protect patients from bacteria that normally flourish harmlessly on our bodies and around our homes and workplaces but that pose a threat to the sick and injured.

Today, we understand that microorganisms are found almost everywhere. Yet, not long ago, before the invention of the microscope, microbes were unknown to scientists. Thousands of people died in devastating epidemics whose causes were not understood. Food spoilage often could not be controlled, and entire families died because vaccinations and antibiotics were not available to fight infection.

We can get an idea of how our current concepts of microbiology developed by looking at a few of the historic breakthroughs in microbiology that have changed our lives.

---

attributed the taste to a unique local climate or contamination from bakery walls. T. F. Sugihara and L. Kline from the United States Department of Agriculture set out to debunk these rumors and to determine the microbiological reason for the bread's different taste so that it could be made in other areas.

The U.S.D.A. workers found that sourdough is eight to ten times more acidic than conventional bread because of the presence of lactic and acetic acids. These acids account for the sour flavor of the

bread. The workers isolated and identified the yeast in the starter as *Saccharomyces exiguus*, a unique yeast that does not ferment maltose and thrives in the acidic environment of this dough. The sourdough question was not answered, however, because the yeast did not produce the acids and did not use maltose. Sugihara and Kline searched the starter for a second agent capable of fermenting maltose and producing the acids. The bacterium they isolated, so carefully guarded all those years, was classified into the genus *Lactobacillus*. Many

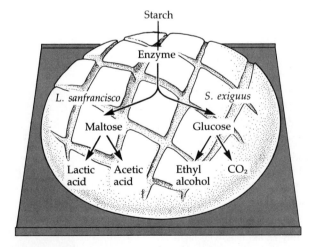

members of this genus are used in dairy fermentations and are part of the normal flora of humans and other mammals. Analyses of cell structure and genetic composition showed that the

sourdough bacterium is genetically different from other previously characterized lactobacilli. It has been given the name *Lactobacillus sanfrancisco*.

A Brief History of Microbiology

The First Observations

One of the most important discoveries in the history of biology occurred in 1665 with the help of a relatively crude microscope. An Englishman, Robert Hooke, reported to the world that life's smallest structural units were "little boxes," or "cells," as he called them. Hooke was able to see individual cells with a microscope that he had developed, an improved version of a compound microscope (one that uses two sets of lenses). Hooke's discovery marked the beginning of the cell theory—the theory that *all living things are composed of cells*. Subsequent investigations into the structure and functions of cells have been based on this theory.

Though Hooke's microscope was capable of showing protozoans and probably bacteria, he lacked the staining techniques (Chapter 3) that would have allowed him to see such small microbes clearly. The Dutch merchant and amateur scientist Anton van Leeuwenhoek was one of the first to actually observe microorganisms through magnifying lenses. Between 1674 and 1723, he wrote a series of letters to the Royal Society of London describing the "animalcules" he saw through his simple, single-lens microscope. Van Leeuwenhoek's detailed drawings of "animalcules" in rain water, in liquid in which peppercorns had soaked, and in material taken from teeth scrapings have since been identified as representations of bacteria and protozoans (Figure 1-2).

The Debate Over Spontaneous Generation

After van Leeuwenhoek discovered the "invisible" world of microorganisms, the scientific community became interested in the origins of these tiny living things. Until the second half of the nineteenth century, many scientists and philosophers believed that some forms of life could arise spontaneously from nonliving matter; this process was called **spontaneous generation**. People thought that toads, snakes, and mice could be born of moist soil, that flies could emerge from manure, and that maggots, the larvae of flies, could arise from decaying corpses.

Evidence Pro and Con

A strong opponent of spontaneous generation, the Italian physician Francesco Redi, set out in 1668 (even before van Leeuwenhoek's later discovery of microscopic life) to demonstrate that maggots do not arise spontaneously from decaying meat, as was commonly believed. Redi filled three jars with decaying meat and sealed them tightly. Then he arranged three other jars similarly but left them open. Maggots appeared in the open vessels after flies entered the jars and laid their eggs. But the sealed containers showed no forms of life (Figure 1-3a). Still, his antagonists were not convinced; they claimed that fresh air was needed for spontaneous generation. So Redi set up a second experiment, in which three jars were covered with a fine net instead of being sealed. No larvae appeared in the gauze-covered jars, even though air was present. Maggots appeared only if flies were allowed to leave their eggs on the meat (Figure 1-3b).

Figure I-2   Van Leeuwenhoek's research with the microscope. (a) A replica of a simple microscope made by Anton van Leeuwenhoek to observe living organisms too small to be seen with the naked eye. The specimen was placed on the tip of the adjustable point and viewed from the other side through the tiny round lens. The highest magnification possible with his lenses was about 300× (times). **(b)** Some of van Leeuwenhoek's drawings of bacteria, made in 1683. He was the first to see the microorganisms we now call bacteria and protozoans.

Location of specimen

Lens

(a)

(b)

Figure 1-3 Redi's experiments demonstrating that maggots do not arise spontaneously from decaying meat. **(a)** In his first experiment, Redi showed that maggots (fly larvae) would appear on uncovered meat but not on the meat in a sealed jar. **(b)** In his second experiment, Redi used gauze to cover half the jars in order to allow the free entrance of air (but not flies). Again, the meat in the covered jars showed no maggots.

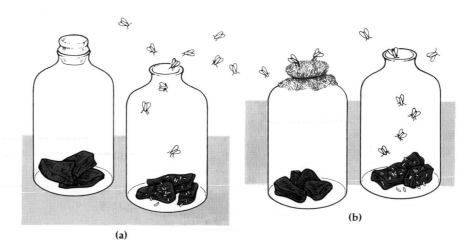

(a)　　　(b)

Redi's results were a serious blow to the long-held belief that large forms of life could arise from nonlife. However, many scientists still believed that small organisms, such as van Leeuwenhoek's "animalcules," were simple enough to be generated from nonliving materials.

The case for spontaneous generation of microorganisms was strengthened in 1745 when John Needham, an Englishman, found that, even after he heated nutrient fluids (chicken broth and corn broth) before pouring them into covered flasks, the cooled solutions were soon teeming with microorganisms. Needham claimed that the microbes developed spontaneously from the fluids. Twenty years later, Lazzaro Spallanzani, an Italian scientist, suggested that microorganisms from the air probably had entered Needham's solutions after they were boiled. Spallanzani showed that nutrient fluids heated *after* being sealed in a flask did not develop microbial growth. Needham responded by claiming that the "vital force" necessary for spontaneous generation had been destroyed by the heat and was kept out of the flasks by the seals.

This intangible "vital force" was given all the more credence shortly after Spallanzani's experiment when Laurent Lavoisier showed the importance of oxygen to life. Spallanzani's observations were criticized on the grounds that there was not enough oxygen in the sealed flasks to support microbial life.

The Theory of Biogenesis

The issue was still unresolved in 1858, when the German scientist Rudolf Virchow challenged spontaneous generation with the concept of **biogenesis**, the claim that living cells can arise only from preexisting living cells. Arguments about spontaneous generation continued until 1861, when the issue was resolved experimentally by Louis Pasteur, the French scientist.

Pasteur demonstrated that microorganisms are indeed present in the air and that they can contaminate seemingly sterile solutions, but air itself does not give rise to microbial life.

Pasteur designed a series of ingenious and persuasive experiments to prove his points. He began by filling several short-necked flasks with beef broth and boiling them. Some were then left open and allowed to cool. In a few days, these flasks were found to be contaminated with microbes. The other flasks, sealed after boiling, remained free of microorganisms. From these results, Pasteur reasoned that microbes in the air were the agents responsible for contaminating nonliving matter such as the broths in Needham's flasks.

Pasteur's next step was to place the boiled broth in open-ended long-necked flasks; he then bent the necks into S-shaped curves (Figure 1-4). The contents of these flasks were then boiled and cooled. The broth in the flasks did not decay and showed no signs of life after days, weeks, and even months. Pasteur's unique design allowed air to pass into the flask, but the curved neck trapped any airborne microorganisms that might contaminate the broth. (Some of these original vessels, which were later sealed, are on display at the Pasteur Institute in Paris. They still show no sign of contamination more than 100 years later.)

Pasteur showed that microorganisms can be present in nonliving matter—on solids, in liquids, and in the air. Furthermore, he demonstrated conclusively that microbial life can be destroyed by heat and that methods can be devised to block the access of airborne microorganisms to nutrient environments. These discoveries form the basis of the **aseptic techniques** (techniques to prevent contamination by unwanted microorganisms) that are now standard practice in laboratory and many medical procedures. Modern

Figure I-4   Pasteur's experiment disproving the theory of spontaneous generation.
(a) Pasteur first poured nutrient broth into a long-necked flask. **(b)** Next he heated the neck of the flask and bent it into an S-shaped curve; then he boiled the solution for several minutes. **(c)** Microorganisms did not appear in the cooled broth, even after long periods of time.

aseptic techniques are among the first and most important things that a beginning microbiologist learns.

Pasteur's work provided evidence that microorganisms cannot originate from mystical forces present in nonliving materials. Rather, any appearance of "spontaneous" life in nonliving solutions can be attributed to microorganisms that were already present in the air or in the fluids themselves. Scientists now believe that a form of spontaneous generation probably did occur on the primitive earth when life first began, but they agree that this does not happen in our present environmental conditions.

The Golden Age of Microbiology

For a period of about 60 years, beginning with the work of Pasteur, there was an explosion of discoveries in microbiology. The period from 1857 to 1914 has been appropriately named the **Golden Age of Microbiology**. During this period, rapid advances, spearheaded mainly by Pasteur and Robert Koch, led to the establishment of microbiology as a science. These years saw discoveries of the agents of many diseases,

the relationships between microorganisms and disease, and the role of immunity in the prevention and cure of disease. During this productive period, microbiologists studied the chemical activities of microorganisms, improved techniques for microscopy and the culturing of microorganisms, and developed vaccines and surgical techniques. Some of the major events that occurred during the Golden Age of Microbiology are listed in Table 1-1 and Figure 1-5, pp. 10–11.

Fermentation and Pasteurization

One of the key steps that established the relationship between microorganisms and disease occurred when a group of French merchants asked Pasteur to find out why wine and beer soured. They hoped to develop a method that would prevent spoiling when those beverages were shipped long distances. At the time, many scientists believed that air converted the sugars in these fluids into alcohol. Pasteur found instead that microorganisms called yeasts convert the sugars to alcohol in the absence of air. This process, called **fermentation** (Chapter 5), is used to make wine or beer.

Souring and spoiling, which occur later, are caused by a different group of microorganisms, which are bacteria. In the presence of air, bacteria change the alcohol in the beverage into vinegar (acetic acid).

Pasteur's solution to the spoilage problem was to heat the alcohol just enough to kill most of the bacteria; the process does not greatly affect the flavor of the beverage. This process, called **pasteurization**, is now commonly used to kill potentially harmful bacteria in milk as well as in some alcoholic drinks. Showing the connection between food spoilage and microorganisms was a key step in establishing the relationship between disease and microbes.

The Germ Theory of Disease

As we have seen, the fact that many kinds of diseases are related to microorganisms was unknown until relatively recently. Before the time of Pasteur, effective treatments for many diseases were discovered by trial and error, but the causes of the diseases were unknown.

The realization that yeasts play a crucial role in fermentation was the first link between a microorganism's activity and physical and chemical changes in organic materials. This discovery alerted scientists to the possibility that microorganisms might have similar relationships with plants and animals—specifically, that microorganisms might cause disease. This idea was called the **germ theory of disease**.

The germ theory was a difficult concept for many people to accept because for centuries it was believed that disease was punishment for an individual's crimes or misdeeds. When the inhabitants of an entire village became ill, people often blamed the disease on foul odors from sewage or on poisonous vapors from swamps. Most people born in Pasteur's time found it inconceivable that "invisible" microbes could travel through the air to infect plants and animals or remain on clothing and bedding to be transmitted from one person to another. But gradually, scientists accumulated the information needed to support the new germ theory.

In 1865, Pasteur was called upon to help fight silkworm disease, which was ruining the silk industry throughout Europe. Years earlier, in 1835, Agostino Bassi, an amateur microscopist, had proved that another silkworm disease was caused by a fungus. Using data provided by Bassi, Pasteur found that the more recent infection was caused by a protozoan, and he developed a method for recognizing afflicted silkworm moths.

In the 1860s, Joseph Lister, an English surgeon, applied the germ theory to medical procedures. Lister was aware that in the 1840s, the Hungarian physician Ignaz Semmelweis had demonstrated that physicians who went from one obstetrical patient to another without disinfecting their hands transmitted puerperal (childbirth) fever. Lister had also heard of Pasteur's work connecting microbes to animal diseases. Disinfectants were unknown at the time, but Lister knew that carbolic acid (now called phenol) kills bacteria, so he began soaking surgical dressings in a mild solution of it. The practice so reduced the incidence of infections and deaths that other surgeons quickly adopted it. Lister's technique was one of the earliest medical attempts to control the infections caused by microorganisms.

The first proof that bacteria actually cause disease came from Robert Koch in 1876. Koch, a brilliant German physician, was Pasteur's young rival in the race to discover the cause of anthrax, a disease that was destroying cattle and sheep in Europe. Koch discovered rod-shaped bacteria now known as *Bacillus anthracis* (bä-sil'lus an-thra'sis) in the blood of cattle that had died of anthrax. He cultured the bacteria on nutrients and then injected samples of the culture into healthy animals. When these animals became sick and died, Koch isolated the bacteria in their blood and compared them with the bacteria originally isolated. He found that the two sets of blood cultures contained the same bacteria.

Koch thus established a sequence of experimental steps for directly relating a specific microbe to a specific disease. These steps are known today as **Koch's postulates** (see Figure 13-3). During the past 100 years, these same criteria have been invaluable in investigations proving that specific microorganisms cause many diseases. Koch's postulates, their limitations, and their application to disease will be discussed in greater detail in Chapter 13.

Vaccination

Often a treatment or preventive procedure is developed before scientists know why it works. The smallpox vaccine is an example of preventive medicine jumping ahead of an understanding of cause. In 1798, almost 70 years before Koch established that a specific microorganism causes anthrax, Edward Jenner, a young British physician, embarked on an experiment to find a protection from smallpox.

Smallpox epidemics were greatly feared. The disease periodically swept through Europe, killing thousands, and it wiped out 90% of the American Indians on the East Coast when European settlers first brought the infection to the New World.

When a young milkmaid told Jenner that she couldn't get smallpox because she already had been sick from cowpox—a much milder disease—he decided to put the girl's story to a test. First Jenner

TABLE 1-1   *The Golden Age of Microbiology (1857–1914)*

Scientist (Country)	Event
Louis Pasteur (France)	Introduced germ theory of disease (**1857**) Showed that yeasts are responsible for fermentation (**1857**) Disproved spontaneous generation (**1861**) Developed pasteurization (**1864**) Developed immunization techniques (**1880**) Discovered *Streptococcus pneumoniae*, the cause of one type of pneumonia (**1880**)
Joseph Lister (England)	Developed aseptic surgery techniques (**1867**)
Ernst Abbé (Germany)	Developed Abbé condenser and oil immersion objective for the compound microscope (**ca. 1870**)
Robert Koch (Germany)	Proved germ theory of disease (**1876**) Discovered *Bacillus anthracis*, the causative agent of anthrax (**1877**) Developed pure culture and staining techniques (**1881**) Discovered *Mycobacterium tuberculosis*, the causative agent of tuberculosis (**1882**) Discovered *Vibrio cholerae*, the causative agent of cholera (**1883**)
Albert Neisser (Germany)	Discovered *Neisseria gonorrhoeae*, the causative agent of gonorrhea (**1879**)
Martinus Beijerinck (Netherlands)	Discovered nitrogen-fixing bacteria in roots of certain plants (**1880**)
Edwin Klebs and Friedrich Loeffler (Germany)	Discovered *Corynebacterium diphtheriae*, the causative agent of diphtheria (**1884**)
Elie Metchnikoff (Russia)	Described phagocytosis and developed a cellular theory of immunity (**1884**)
Hans Christian Gram (Denmark)	Developed a differential staining (dyeing) technique called the Gram stain for observing microorganisms (**1884**)
Arthur Nicolaier (Germany) and Shibasaburo Kitasato (Japan)	Discovered *Clostridium tetani*, the causative agent of tetanus (**1884**)
Theodor Escherich (Germany)	Discovered *Escherichia coli*, the causative agent of urinary tract infections and traveler's diarrhea (**1885**)
Richard Julius Petri (Germany)	Introduced a covered dish for growing microorganisms on a solid medium (**1887**)
Emil von Behring (Germany)	Developed a diphtheria antitoxin (**1890s**)
Dmitri Iwanowski (Russia)	Discovered a filterable organism (virus) that caused tobacco mosaic disease (**1892**)
Sergei Winogradsky (Soviet Union)	Isolated bacteria responsible for the sulfur cycle; isolated nitrogen-fixing clostridia (**1893**)
Kiyoshi Shiga (Japan)	Discovered *Shigella dysenteriae*, the causative agent of bacillary dysentery (**1898**)
Walter Reed (United States)	Proved that mosquitoes transmit yellow fever (**1900**)
Howard T. Ricketts (United States)	Discovered *Rickettsia rickettsii*, the causative agent of Rocky Mountain spotted fever, and proved that ticks transmit the disease (**1909**)
Paul Ehrlich (Germany)	Proposed a theory of immunity in which antibodies are responsible for immunity (**1890s**); synthesized a "magic bullet" for syphilis (**1910**)

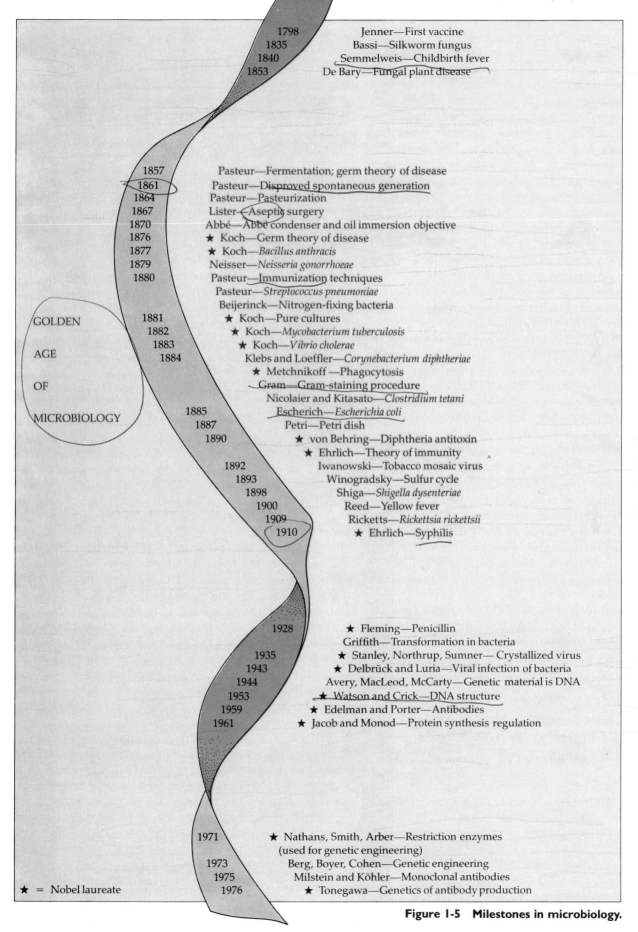

1798	Jenner—First vaccine
1835	Bassi—Silkworm fungus
1840	Semmelweis—Childbirth fever
1853	De Bary—Fungal plant disease
1857	Pasteur—Fermentation; germ theory of disease
1861	Pasteur—Disproved spontaneous generation
1864	Pasteur—Pasteurization
1867	Lister—Aseptic surgery
1870	Abbé—Abbé condenser and oil immersion objective
1876	★ Koch—Germ theory of disease
1877	★ Koch—*Bacillus anthracis*
1879	Neisser—*Neisseria gonorrhoeae*
1880	Pasteur—Immunization techniques
	Pasteur—*Streptococcus pneumoniae*
	Beijerinck—Nitrogen-fixing bacteria
1881	★ Koch—Pure cultures
1882	★ Koch—*Mycobacterium tuberculosis*
1883	★ Koch—*Vibrio cholerae*
1884	Klebs and Loeffler—*Corynebacterium diphtheriae*
	★ Metchnikoff—Phagocytosis
	Gram—Gram-staining procedure
	Nicolaier and Kitasato—*Clostridium tetani*
1885	Escherich—*Escherichia coli*
1887	Petri—Petri dish
1890	★ von Behring—Diphtheria antitoxin
	★ Ehrlich—Theory of immunity
1892	Iwanowski—Tobacco mosaic virus
1893	Winogradsky—Sulfur cycle
1898	Shiga—*Shigella dysenteriae*
1900	Reed—Yellow fever
1909	Ricketts—*Rickettsia rickettsii*
1910	★ Ehrlich—Syphilis
1928	★ Fleming—Penicillin
	Griffith—Transformation in bacteria
1935	★ Stanley, Northrup, Sumner—Crystallized virus
1943	★ Delbrück and Luria—Viral infection of bacteria
1944	Avery, MacLeod, McCarty—Genetic material is DNA
1953	★ Watson and Crick—DNA structure
1959	★ Edelman and Porter—Antibodies
1961	★ Jacob and Monod—Protein synthesis regulation
1971	★ Nathans, Smith, Arber—Restriction enzymes (used for genetic engineering)
1973	Berg, Boyer, Cohen—Genetic engineering
1975	Milstein and Köhler—Monoclonal antibodies
1976	★ Tonegawa—Genetics of antibody production

GOLDEN AGE OF MICROBIOLOGY

★ = Nobel laureate

Figure 1-5 Milestones in microbiology.

collected scrapings from cowpox blisters. Then he made inoculations with the cowpox material by scratching the arm of a healthy volunteer with a pox-contaminated needle. The scratch turned into a raised bump. In a few days, the person became mildly sick but recovered and never again contracted either cowpox or smallpox. The process was called **vaccination**, from the Latin word *vacca*, for cow. The protection from disease provided by vaccination (or by recovery from the disease itself) is called **immunity**. We will discuss the mechanisms of immunity in Chapter 16.

Years after Jenner's experiment, about 1880, Pasteur discovered why vaccinations work. He found that the bacterium that causes a disease called chicken cholera lost its ability to cause disease (lost its *virulence*, or became *avirulent*) after it was grown in the laboratory for long periods. However, it—and other microorganisms with decreased virulence—was able to induce immunity against subsequent infections by its virulent counterparts. The discovery of this phenomenon provided a clue to Jenner's successful experiment with cowpox. Both cowpox and smallpox are caused by viruses. Even though cowpox virus is not a laboratory-produced derivative of smallpox virus, it is so closely related to the smallpox virus that it can induce immunity to both viruses. Pasteur used the term *vaccine* for cultures of avirulent microorganisms used for preventive inoculation.

Jenner's experiment was the first time in a Western culture that a living viral agent—the cowpox virus—was used to produce immunity. Physicians in ancient China had immunized patients by removing scales from drying pustules of a person suffering from a mild case of smallpox, grinding the scales to a fine powder, and inserting the powder into the nose of the person to be protected.

Some vaccines are still produced from avirulent microbial strains that stimulate immunity to the related virulent strain. Other vaccines are made from killed virulent microbes, from isolated components of virulent microorganisms, or by genetic engineering techniques.

The Birth of Modern Chemotherapy: Dreams of a "Magic Bullet"

After the relationship between microorganisms and disease was established, medical microbiologists next focused on the search for substances that could destroy pathogenic microorganisms without damaging the infected animal or human. The treatment of disease by chemical substances is called **chemotherapy**. (The term also commonly refers to chemical treatment of noninfectious diseases, such as cancer.) Chemotherapeutic agents prepared from chemicals in the laboratory are called **synthetic drugs**. Chemicals produced naturally by bacteria and fungi that act against other microorganisms are called **antibiotics**. The success of chemotherapy is based on the fact that some chemicals are more poisonous to microorganisms than to the hosts infected by the microbes.

The First Synthetic Drugs

Paul Ehrlich, a German physician, was the imaginative thinker who fired the first shot in the chemotherapy revolution. As a medical student, Ehrlich speculated about a "magic bullet" that could hunt down and destroy a pathogen without harming the infected host. Ehrlich launched a search for such a bullet. In 1910, after testing hundreds of substances, he found a chemotherapeutic agent called *salvarsan*, an arsenic derivative effective against syphilis. Prior to this discovery, the only known chemical in Europe's medical arsenal was an extract from South American tree bark, *quinine*, which had been used by Spanish conquistadors to treat malaria.

By the late 1930s, researchers had developed several other synthetic drugs that could kill microorganisms. Most of these drugs were derivatives of dyes. The *sulfa drugs* were one important group discovered at that time.

A Fortunate Accident—Antibiotics

In contrast to the sulfa drugs, which were deliberately developed from a series of industrial chemicals, the first antibiotic was discovered by accident. Alexander Fleming, a Scottish physician and bacteriologist, almost tossed out some culture plates that had been contaminated by mold. Fortunately, he took a second look at the curious pattern of growth on the contaminated plates. There was a clear area around the mold where the bacterial culture had stopped growing (Figure 1-6). Fleming was looking at a mold that could inhibit the growth of a bacterium. The mold was later identified as *Penicillium notatum* (pe-ni-sil'lē-um nō-tä'tum), and in 1928 Fleming named the mold's active inhibitor *penicillin*.

Penicillin is an antibiotic produced by a fungus. The enormous usefulness of penicillin was not apparent until the 1940s, when it was finally mass produced and tested clinically. Since then, many other antibiotics have been discovered.

Unfortunately, antibiotics and other chemotherapeutic drugs are not without problems. Many antimicrobial chemicals are too toxic to humans for practical use; they kill the pathogenic microbes, but they also have damaging side effects on the infected host. For reasons we shall discuss later, toxicity for humans is

Figure 1-6   Inhibition of bacterial growth around the mold *Penicillium chrysogenum*. Colonies of the bacteria *Staphylococcus aureus* do not grow in the vicinity of the contaminating *Penicillium* colony near the top of the photo. The reason is that the mold secretes penicillin.

a particular problem in the development of drugs for treating viral diseases; there are as yet very few successful antiviral drugs.

Another major problem associated with antimicrobial drugs is the emergence and spread of new varieties of microorganisms that are resistant to antibiotics. The quest to solve this and other problems requires sophisticated research techniques and correlated studies that were never dreamed of in the days of Koch and Pasteur.

Modern Developments in Microbiology

The groundwork laid during the Golden Age of Microbiology provided the basis for several monumental achievements during the twentieth century (Table 1-2). New branches of microbiology were developed, including immunology and virology. Most recently, the development of a set of new methods called recombinant DNA technology has revolutionized research and practical applications in all areas of microbiology.

Immunology

Immunology, the study of immunity, actually dates back to Jenner's development of the first vaccine in 1798. Since that time, knowledge about the immune system has accumulated steadily, and it expanded rapidly in the twentieth century. Vaccines are now

available for numerous diseases, including measles, mumps, polio, rubella, and hepatitis B. The smallpox vaccine is so effective that not a single case of the disease has been reported worldwide since 1977. In 1960, interferon, a substance generated by the body's own immune system, was discovered. Interferon inhibits replication of viruses and has triggered considerable research related to treatment of viral diseases and cancer. One of this century's biggest challenges for immunologists is learning how the AIDS virus alters the immune system and how the system might be stimulated to ward off the virus.

Virology

The study of viruses, **virology**, actually had its origin during the Golden Age of Microbiology. In 1892, Dmitri Iwanowski reported that the organism that caused mosaic disease of tobacco was so small that it passed through filters fine enough to stop all known bacteria. At the time, Iwanowski was not aware that the organism in question was a virus in the sense that we now understand the term. In 1935, Wendell Stanley showed that the organism, called tobacco mosaic virus (TMV), was fundamentally different from other microbes, and so simple and homogeneous that it could be crystallized like a chemical compound. Stanley's work facilitated the study of viral structure and chemistry. Since the development of the electron microscope in the 1940s, microbiologists have been able to observe the structure of viruses in detail, and today much is known about their structure and activity.

Recombinant DNA Technology

The set of methods called **recombinant DNA technology** had its origins in two related fields. The first, **microbial genetics**, studies the mechanisms by which microorganisms inherit traits. The second, **molecular biology**, specifically studies how genetic information is carried in molecules of DNA and how DNA directs the synthesis of functional proteins.

Although molecular biology encompasses all organisms, much of our knowledge of how genes code for specific traits has been revealed through experiments with bacteria. Until the 1930s, all genetic research was based on the study of plant and animal cells. But in the 1940s, scientists turned to unicellular organisms, primarily bacteria, which have several advantages for genetic and biochemical research. For one thing, bacteria are less complex than plants and animals. For another, the life cycles of many bacteria last less than an hour, so scientists can cultivate very large numbers of individuals for study in a relatively short time.

Once science turned to the study of unicellular life, breakthroughs in genetics began to appear rapidly. DNA was established as the hereditary material in 1944 by Oswald Avery, Colin MacLeod, and Maclyn McCarty. Then, in 1953, James Watson and Francis Crick proposed a model for the structure and replication of DNA. The early 1960s witnessed a further explosion of discoveries relating to the way DNA controls protein synthesis. François Jacob and Jacques Monod discovered messenger RNA in 1961 and also made the first major discoveries about regulation of gene function in bacteria. During this same period, scientists were able to break the genetic code and thus understand the biochemical signals that DNA transmits.

Because of the knowledge gained in such research, microorganisms can now be genetically engineered to manufacture large amounts of human hormones and other badly needed medical substances. With the help of new techniques developed by biochemists, fragments of human or animal DNA that code for important proteins can be attached to bacterial DNA. The resulting hybrid is called **recombinant DNA**. When recombinant DNA is inserted into bacteria, the bacteria can be used to make large quantities of the desired protein.

TABLE 1-2   *Selected Microbiologists Who Have Won the Nobel Prize*

Nobel Laureate	Year of Presentation	Country of Birth	Contribution
Emil A. von Behring	1901	Germany	Developed a diphtheria antitoxin
Robert Koch	1905	Germany	Cultured tuberculosis microbe
Paul Ehrlich	1908	Germany	Developed theories on immunity
Elie Metchnikoff	1908	Russia (work done in Russia and France)	Described phagocytosis, the intake of solid materials by cells
Alexander Fleming, Ernst Chain, and Howard Florey	1945	Scotland (work done in England) England	Discovered penicillin
Wendell Stanley, John Northrop, and James Sumner	1946	United States	Crystallized tobacco mosaic virus
Selman A. Waksman	1952	United States	Discovered streptomycin
Hans A. Krebs	1953	Germany (work done in Germany and England)	Discovered chemical steps of the Krebs cycle in carbohydrate metabolism
John F. Enders, Thomas H. Weller, and Frederick C. Robbins	1954	United States	Cultured polio virus in cell cultures
Joshua Lederberg, George W. Beadle, and Edward L. Tatum	1958	United States	Described genetic control of biochemical reactions in the fungus *Neurospora*
Severo Ochoa and Arthur Kornberg	1959	Spain United States (work done in U.S.)	Discovered enzymes for DNA and RNA synthesis

Naming and Classifying Microorganisms

The system of nomenclature (naming) for organisms in use today was established in 1735 by Carolus Linnaeus. Scientific names are latinized because Latin was the language traditionally used by scholars. Scientific nomenclature assigns each organism two names—the **genus** is the first name and is always capitalized; the **specific epithet (species)** follows and is not capitalized. The species is referred to by both the genus and the specific epithet, and both words are underlined or italicized. By custom, after a scientific name has been mentioned once, it can be abbreviated with the initial of the genus followed by the specific epithet.

Scientific names can, among other things, describe the organism, honor a researcher, or identify the habitat of the species. For example, consider *Staphylococcus aureus*, a bacterium commonly found on the skin of humans and, as we saw earlier, involved in TSS. *Staphylo-* describes the clustered arrangement of the cells; *coccus* indicates that they are shaped like spheres. The specific epithet, *aureus*, is Latin for

TABLE 1-2 *(Continued)*

Nobel Laureate	Year of Presentation	Country of Birth	Contribution
James D. Watson, Frances H. C. Crick, and Maurice A. F. Wilkins	1962	United States England New Zealand (work done in England)	Determined the structure of DNA
François Jacob, Jacques Monod, and André Lwoff	1965	France	Described how protein synthesis is regulated in bacteria
Robert Holley, Har Gobind Khorana, and Marshall W. Nirenberg	1968	United States India United States (work done in U.S.)	Discovered genetic code for amino acids
Max Delbrück, Alfred D. Hershey, and Salvador E. Luria	1969	Germany United States Italy (work done in U.S.)	Described the mechanism of viral infection of bacterial cells
Gerald M. Edelman and Rodney R. Porter	1972	United States England	Described the nature and structure of antibodies
Daniel Nathans, Hamilton Smith, and Werner Arber	1978	United States United States Switzerland	Described the action of restriction enzymes (now used in genetic engineering)
César Milstein and Georges J. F. Köhler	1984	Argentina Germany (Milstein's work done in Argentina and England; Köhler's work done in Germany and Switzerland)	Developed technique for producing monoclonal antibodies
Susumu Tonegawa	1987	Japan (work done in Switzerland and United States)	Described genetics of antibody production

golden, the color of colonies of this bacterium. The genus of the bacterium *Escherichia coli* (esh-er-i'kē-ä kō'lī) is named for a scientist, Theodor Escherich, whereas its specific epithet, *coli*, reminds us that *E. coli* live in the colon, or large intestine.

Before the existence of microbes was known, all organisms were grouped into either the animal kingdom or the plant kingdom. When microscopic organisms with characteristics of animals or plants were discovered late in the seventeenth century, a new system of classification was needed. In 1866, Ernst Heinrich Haeckel, a German zoologist, proposed a classification system with three kingdoms: Animalia, Plantae, and Protista. The Protista consisted of single-celled microorganisms, such as bacteria, some fungi, algae, and protozoans.

In 1969, Robert Whittaker of Cornell University devised a five-kingdom classification system that is based on cellular organization and nutritional patterns of organisms. The five-kingdom system is now widely accepted. It groups all organisms into the following kingdoms:

- Procaryotae or Monera (eubacteria and archaeobacteria)
- Protista (slime molds, protozoans, and some algae)
- Fungi (unicellular yeasts, multicellular molds, and mushrooms)
- Plantae (some algae and all mosses, ferns, conifers, and flowering plants)
- Animalia (including, among others, sponges, worms, insects, and vertebrates)

Classification will be discussed in more detail in Part Two. For now, however, let us simply introduce the major groups of microorganisms studied in this text. These descriptions are only a brief overview; more detailed explanations will follow in later chapters.

The Diversity of Microorganisms

Bacteria

Bacteria are very small, relatively simple, single-celled organisms whose genetic material is not enclosed in a special nuclear membrane (Figure 1-7a). For this reason, bacteria are called **procaryotes**, from Greek words meaning prenucleus. Bacteria make up the kingdom Whittaker calls Monera; others, including microbiologists, call it Procaryotae. (Criteria for classifying bacteria will be discussed in Part Two.)

Bacterial cells generally have one of several shapes. **Bacillus** (rodlike), **coccus** (spherical or ovoid), and **spiral** (or corkscrew) are the most common shapes, but some bacteria are star-shaped or square-shaped (see Figures 4-1 through 4-4 and Microview A, p. A-1). Individual bacteria may form pairs, chains, clusters, or other groupings; such formations are usually characteristic of a particular species.

Bacteria are enclosed in cell walls that are largely composed of a substance called *peptidoglycan*. (By contrast, cellulose is the main substance of plant cell walls.) Bacteria generally reproduce by dividing into two equal daughter cells; this process is called *binary fission*. For nutrition, most bacteria use organic chemicals, which in nature can be derived from other, dead organisms or from a living host. Some bacteria can manufacture their own food by photosynthesis, and some can derive nutrition from inorganic substances. Many bacteria can "swim" by moving appendages called *flagella*.

Fungi

Fungi are **eucaryotes**—organisms whose cells have a distinct nucleus that contains the cell's genetic material and is surrounded by a special envelope called the nuclear membrane. They make up the Kingdom Fungi (see Microview A, pp. A-3 and A-4). Fungi may be unicellular or multicellular. Large multicellular fungi, such as mushrooms, may look somewhat like plants, but they cannot carry out photosynthesis, as most plants can. True fungi have cell walls composed primarily of a substance called *chitin*. The unicellular forms of fungi, yeasts, are oval-shaped microorganisms larger than bacteria. The most typical fungi are *molds*. Molds form *mycelia*, which are long filaments that branch and intertwine (Figure 1-7b). The cottony growths sometimes found on bread and fruit are mold mycelia. Fungi can reproduce sexually or asexually. They obtain nourishment by absorbing solutions of organic material from their environment—whether soil, water, or an animal or plant host.

Protozoans

Protozoans are unicellular, eucaryotic microbes that belong to the Kingdom Protista (see Figure 1-7c and Microview A, p. A-3). Protozoans are classified according to their means of locomotion. Amoebas move using extensions of their cytoplasm called *pseudopods* (false feet). Other protozoans have flagella or numerous shorter appendages called *cilia*. They have a variety

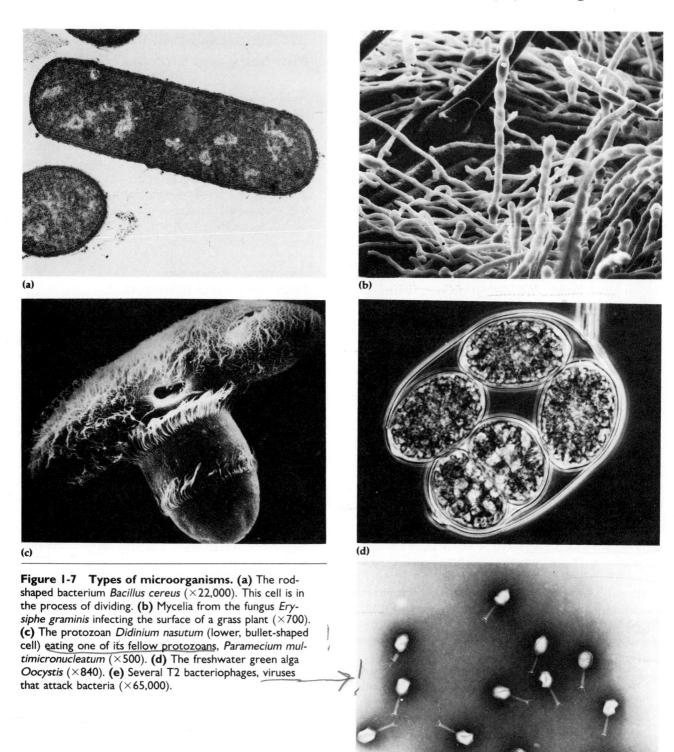

Figure 1-7 Types of microorganisms. (a) The rod-shaped bacterium *Bacillus cereus* (×22,000). This cell is in the process of dividing. **(b)** Mycelia from the fungus *Erysiphe graminis* infecting the surface of a grass plant (×700). **(c)** The protozoan *Didinium nasutum* (lower, bullet-shaped cell) eating one of its fellow protozoans, *Paramecium multimicronucleatum* (×500). **(d)** The freshwater green alga *Oocystis* (×840). **(e)** Several T2 bacteriophages, viruses that attack bacteria (×65,000).

of shapes and live as free entities or parasites (organisms that derive nutrients from living hosts) that absorb or ingest organic compounds from their environment. Protozoans can reproduce sexually or asexually.

Algae

Algae are photosynthetic eucaryotes with a wide variety of shapes and both sexual and asexual reproductive forms (see Figure 1-7d and Microview A, p. A-4). The algae of interest to microbiologists are members of the Kingdom Protista and are usually unicellular. They are abundant in fresh and salt water, soil, and in association with plants. As photosynthesizers, algae need light and air for food production and growth, but they do not generally require organic compounds from the environment. As results of photosynthesis, algae produce oxygen and carbohydrates, which are ingested by other organisms, including animals. Thus they play an important role in the balance of nature.

Viruses

Viruses (Figure 1-7e) are very different from the other microbial groups mentioned here. They are so small that most can be seen only with an electron microscope, and they are not cellular. Structurally very simple, a virus particle contains a core made of only one type of nucleic acid, either DNA (deoxyribonucleic acid) or RNA (ribonucleic acid). This core is surrounded by a protein coat. Sometimes the coat is encased by an additional layer, a lipid membrane called an envelope. All living cells have RNA *and* DNA, can carry out chemical reactions, and can reproduce as self-sufficient units. Viruses can reproduce only inside the cells of other organisms. Thus, all viruses are parasites of other forms of life.

Multicellular Animal Parasites

Although multicellular parasites are not strictly microorganisms, they are of medical importance and so will be discussed in this text. The two major groups of parasites are the flatworms and the roundworms, collectively called **helminths** (see Microview A, p. A-3). During some stages of their life cycle, the helminths are microscopic in size. Laboratory identification of these organisms includes many of the same techniques used for the identification of microbes.

Microbes and Human Welfare

As we mentioned earlier, only a minority of microorganisms are pathogenic. Microbes that cause food spoilage, such as soft spots on fruits and vegetables, decomposition of meats, rancidity of fats and oils, and ropiness of bread, are also a minority. The vast majority of microbes benefit humans as well as other plants and animals in numerous ways. The following sections will outline some of these beneficial activities. In later chapters, we will discuss these activities in greater detail.

Recycling Vital Elements

The discoveries of two microbiologists in the 1880s formed the basis for today's understanding of the biochemical cycles that support life on earth. Martinus Beijerinck and Sergei Winogradsky were the first to show how bacteria help recycle vital elements between the soil and the atmosphere. **Microbial ecology**, the study of the relationship between microorganisms and their environment, and **microbial physiology**, the study of the metabolic activities of microbes, originated with the work of Beijerinck and Winogradsky. Today, microbial ecology has branched out and includes the study of how microbial populations interact in various environments with plants and animals. Among the concerns of microbial ecologists are water pollution and toxic chemicals in the environment.

In the early part of the twentieth century, many advances occurred in microbial physiology as more information became available about chemical reactions in microbes. A major contributor in this field was Albert Kluyver (1888–1956), who showed that, although microorganisms exhibit great diversity, they do have many chemical reactions in common. In fact, he proposed that a similarity of chemical reactions exists among all organisms, not just microbes. Knowledge gained through studies of microbial physiology is being combined with genetic engineering techniques to provide many useful products, such as vaccines, hormones, and interferon.

It is now known that the chemical elements nitrogen, carbon, oxygen, sulfur, and phosphorus are essential for life and are available only in limited, though large, amounts. Nitrogen, oxygen, and carbon (as carbon dioxide) exist as gases in the atmosphere. Sulfur and phosphorus are stored in the Earth's crust. For the most part, it is microorganisms that convert these elements into forms that can be used by plants and animals.

The **nitrogen cycle** is one example of how microorganisms recycle an element (see Figure 26-2). Nitrogen is a major constituent of all living cells. It is present in proteins and in nucleic acids (DNA and RNA). Although 79% of the atmosphere is nitrogen, animals and plants cannot use it in that gaseous form. Certain soil bacteria and cyanobacteria (Monera) in water combine gaseous nitrogen with other elements (carry out "nitrogen-fixation") so that it can be incorporated into living cells and enter the food chain. When other bacteria help to decompose dead plants and animals, nitrogen-containing chemicals are released into the soil; still other microbes then convert the soil nitrogen back into nitrogen gas.

In the **carbon cycle**, green plants and algae remove carbon dioxide from the air and use sunlight to convert the carbon dioxide into food by the process of photosynthesis (see Figure 26-1). When this food is consumed by other organisms, including humans, carbon dioxide is produced in the process called respiration. Microorganisms, mostly bacteria, play another key role in the carbon cycle by also returning carbon dioxide to the atmosphere when they decompose organic wastes and dead plants and animals. Algae, higher plants, and certain bacteria (cyanobacteria) also participate in the **oxygen cycle** by recycling oxygen to the air during photosynthesis. The nitrogen, carbon, and oxygen cycles, as well as those involving sulfur and phosphorus, will be discussed in greater detail in Chapters 5 and 26.

Sewage Treatment: Using Microbes to Recycle Water

With our growing awareness of the need to preserve the environment, we are conscious of our responsibility to recycle precious water and prevent the pollution of rivers and oceans. One major pollutant is sewage, which consists of human and animal excrement, waste water, industrial wastes, and surface run-off. Sewage is about 99.9% water with a few hundredths of one percent suspended solids. The remainder is a variety of dissolved materials.

Sewage plant treatments remove the undesirable materials and harmful microorganisms. Treatments combine various physical and chemical processes with action by beneficial microbes. Large solids such as paper, wood, glass, gravel, and plastic are removed

from sewage; left behind are liquid and organic materials that bacteria convert into such by-products as carbon dioxide, nitrates, phosphates, sulfates, ammonia, hydrogen sulfide, and methane. (We shall discuss sewage treatment in detail in Chapter 26.)

Insect Pest Control by Microorganisms

Insects spread diseases and cause devastating crop damage. Insect pest control is therefore important for agriculture as well as for the prevention of human disease.

Several types of bacteria are now being used to reduce insect infestations. The bacterium *Bacillus thuringiensis* (bä-sil'lus thür-in-jē-en'sis) has been used extensively in the United States to control such pests as alfalfa caterpillars, bollworms, corn borers, cabbageworms, tobacco budworms, and fruit tree leaf rollers. This species of bacteria produces protein crystals that are toxic to the digestive systems of the insects. Enormous quantities of the bacteria are grown, dried, and incorporated into a dusting powder that is applied to the crops these insects eat.

By using microbial rather than chemical insect control, farmers can avoid harming the environment. Many chemical insecticides, such as DDT, are not easily degraded by soil organisms, and they remain in the soil as toxic pollutants. These insecticides are carried to streams and rivers by irrigation and rain water, and are eventually incorporated into the food chain. It has been shown that some insecticides ingested by fish and small animals are poisonous to them. Also, some of these chemicals might be responsible for birth defects in larger animals. To further compound the problem, many chemicals used to control insects are needed in increasingly large concentrations because new generations of resistant insects are no longer killed by the original concentrations sprayed on crops. By contrast, bacteria and other microorganisms used for insect control apparently do not permanently disturb the ecology of the farm, forestland, or public water supply. So far, insects have not developed a resistance to microbial insecticides.

Modern Industrial Microbiology and Genetic Engineering

Earlier in this chapter, we touched on the commercial use of microorganisms to produce some common foods and chemicals. Two recent extensions of industrial microbiology have been attracting considerable attention—**single-cell protein (SCP)**, a microbe-made food substitute, and **genetic engineering**, the use of recombinant DNA technology to expand the potential of bacteria as miniature biochemical factories.

Single-Cell Protein

The demand for food steadily increases along with the world's growing population. In response to this demand, scientists are investigating several new food sources. One such source is SCP—the protein (and other nutrients) produced by microorganisms cultivated on industrial wastes.

The advantage of SCP over plant crops is that microorganisms grow rapidly and can produce a high protein yield, estimated to be 15 times greater than the amount of soybeans (a high-protein food) and 50 times greater than the amount of corn grown in the same time. For example, one-half metric ton of yeast can produce 50 metric tons of protein daily. However, SCP is not very tasty and in some cases might need to be supplemented with essential amino acids to provide a balanced protein source. At present, SCP is used primarily in animal feed.

Genetic Engineering

Genetic engineering also has important potential uses in agriculture, but to date most of its contributions to human welfare have been in the medical area. Recombinant DNA techniques have been used thus far to produce a number of natural proteins that are otherwise expensive and difficult to isolate and purify. Such proteins have great potential for medical use. Among them are insulin, a hormone used to lower blood sugar level in diabetics; a growth hormone required during childhood; interferon, an antiviral (and possibly anticancer) substance; factor VIII, a clotting substance missing in the blood of most hemophiliacs; tissue-plasminogen activator, a substance used to dissolve blood clots in coronary arteries; beta-endorphin, a substance that suppresses pain; hepatitis B protein, used in a vaccine against hepatitis B; and proteins used to treat certain immune defects.

Scientists are using recombinant DNA techniques in attempts to develop vaccines against several microorganisms, including those that cause herpes, AIDS, and malaria. Just recently, they have begun to use recombinant DNA techniques in the diagnosis and treatment of cancer in its early stages. It is possible that certain genetic diseases, such as diabetes and hemophilia, will eventually be eliminated by the insertion of correct DNA sequences directly into unhealthy human cells.

Microbes and Human Disease

We all live in a microbial world, from birth until death. And we all have a variety of microorganisms, or flora, on and inside our bodies (Figure 1-8). Although these **normal flora** usually benefit us, under some circumstances they can cause us illness or infect people we contact.

When is a microbe a welcome part of a healthy human, and when is it a harbinger of disease? The distinction between health and disease is in large part a balance between the natural defenses of the body and the disease-producing properties of microorganisms. Microorganisms can enter our bodies whenever we breathe, swallow, cut our skin, or have close contact with other people and animals. Some microorganisms cause disease by invading body tissues; others attack indirectly by releasing toxins. Usually, large numbers of microorganisms are needed to cause disease.

Whether our bodies overcome the offensive tactics of a particular microbe depends on our **resistance**. Important natural resistance is provided by the barrier of the skin, mucous membranes, cilia, stomach acid, and antimicrobial chemicals such as interferon. Microbes can be destroyed by white blood cells, inflammation, fever, and our powerful immune systems. Sometimes, when our natural defenses are not strong enough to overcome an invader, they have to be supplemented by antibiotics or other drugs.

In the chapters that follow, you will become acquainted with the principles and tools that microbiologists use to understand the many interactions between microbes and their hosts.

Figure 1-8   Several bacteria found as normal flora inside the human mouth (× 3000).

Study Outline

Microbes in Our Lives (pp. 4–5)

1. Living things too small to be seen with the naked eye are called microorganisms.
2. Microorganisms play important roles in the maintenance of an ecological balance on earth.
3. Some microorganisms live in humans and other animals and are needed to maintain the animal's health.
4. Some microorganisms are used to produce foods and chemicals.
5. Some microorganisms cause disease.

A Brief History of Microbiology (pp. 5–14)

The First Observations (p. 6)

1. Robert Hooke observed that plant material was composed of "little boxes"; he introduced the term *cell* (1665).
2. Hooke's observations were the groundwork for development of the cell theory, the concept that all living things are composed of cells.
3. Anton van Leeuwenhoek, using a simple microscope, was the first to observe microorganisms (1673).

The Debate Over Spontaneous Generation (pp. 6–8)

1. Until the mid-1880s, many people believed in spontaneous generation, the idea that living organisms could arise from nonliving matter.
2. Francesco Redi demonstrated that maggots appear on decaying meat only when flies are able to lay eggs on the meat (1668).
3. John Needham claimed that microorganisms could arise spontaneously from heated nutrient broth (1745).
4. Lazzaro Spallanzani repeated Needham's experiments and suggested that Needham's results were due to microorganisms in the air entering his broth (1765).
5. Rudolf Virchow introduced the concept of biogenesis—living cells can arise only from preexisting cells (1858).
6. Louis Pasteur demonstrated that microorganisms are in the air everywhere and offered proof of biogenesis (1861).
7. Pasteur's discoveries led to the development of aseptic techniques used in laboratory and medical procedures to prevent contamination by microorganisms that are in the air.

The Golden Age of Microbiology (p. 8)

1. Rapid advances in the science of microbiology were made between 1857 and 1914.

Fermentation and Pasteurization (pp. 8–9)

1. Pasteur found that yeast ferments sugars to alcohol and that bacteria can oxidize the alcohol to acetic acid.
2. A heating process called pasteurization is used to kill bacteria in some alcoholic beverages and milk.

The Germ Theory of Disease (p. 9)

1. Agostino Bassi (1834) and Pasteur (1865) showed a causal relationship between microorganisms and disease.
2. Joseph Lister introduced the use of a disinfectant to clean surgical dressings in order to control infections in humans (1860s).
3. Robert Koch proved that microorganisms transmit disease. He used a sequence of procedures called Koch's postulates (1876), which are used today to prove that a particular microorganism causes a particular disease.

Vaccination (pp. 9–12)

1. In a vaccination, immunity (resistance to a particular disease) is conferred by inoculation with a vaccine.
2. In 1798, Edward Jenner demonstrated that inoculation with cowpox material provides humans with immunity from smallpox.
3. About 1880, Pasteur discovered that avirulent bacteria could be used as a vaccine for chicken cholera; he coined the word *vaccine*.
4. Modern vaccines are prepared from living avirulent microorganisms or killed pathogens, from isolated components of pathogens, and by recombinant DNA techniques.

The Birth of Modern Chemotherapy: Dreams of a "Magic Bullet" (pp. 12–13)

1. Chemotherapy is the chemical treatment of a disease.
2. Two types of chemotherapeutic agents are synthetic drugs (chemically prepared in the laboratory) and antibiotics (substances produced naturally by bacteria and fungi that inhibit the growth of other microorganisms).
3. Paul Ehrlich introduced an arsenic-containing chemical called salvarsan to treat syphilis (1910).
4. Alexander Fleming observed that the mold (fungus) *Penicillium* inhibited the growth of a bacterial culture. He named the active ingredient penicillin (1928).
5. Penicillin has been used clinically as an antibiotic since the 1940s.
6. Researchers are tackling the problem of drug-resistant microbes.

Modern Developments in Microbiology (pp. 13–14)

1. The study of AIDS, analysis of interferon action, and the development of new vaccines are among the current research interests in immunology.

2. New techniques in molecular biology and electron microscopy have provided tools for advancement of our knowledge of virology.
3. The development of recombinant DNA technology has helped advance all areas of microbiology.

Naming and Classifying Microorganisms   (pp. 15–16)

1. In a nomenclature system designed by Carolus Linnaeus (1735), each living organism is assigned two names.
2. The two names consist of a genus and specific epithet, both of which must be underlined or italicized.
3. All organisms can be classified into five kingdoms: Procaryotae (or Monera), Protista, Fungi, Plantae, and Animalia.

The Diversity of Microorganisms   (pp. 16–18)

Bacteria   (p. 16)

1. Bacteria are one-celled organisms. Because they have no nucleus, the cells are described as procaryotic.
2. The three major basic shapes of bacteria are bacillus, coccus, and spiral.
3. Bacteria have a peptidoglycan cell wall; they divide by binary fission; and they may possess flagella.
4. Bacteria can use a wide range of chemical substances for their nutrition.

Fungi   (p. 16)

1. Fungi (mushrooms, molds, and yeasts) have eucaryotic cells (with a true nucleus). Most fungi are multicellular.
2. Fungi obtain nutrients by absorbing organic material from their environment.

Protozoans   (pp. 16–18)

1. Protozoans are unicellular eucaryotes and are classified according to their means of locomotion.
2. Protozoans obtain nourishment by absorption or ingestion through specialized structures.

Algae   (p. 18)

1. Algae are unicellular or multicellular eucaryotes that obtain nourishment by photosynthesis.
2. Algae produce oxygen and carbohydrates that are used by other organisms.

Viruses   (p. 18)

1. Viruses are noncellular entities that are parasites of cells.
2. Viruses consist of a nucleic acid core (DNA or RNA) surrounded by a protein coat. An envelope may surround the coat.

Multicellular Animal Parasites   (p. 18)

1. The principal groups of multicellular animal parasites are flatworms and roundworms, collectively called helminths.
2. The microscopic stages in the life cycle of helminths are identified by traditional microbiologic procedures.

Microbes and Human Welfare   (pp. 18–20)

1. Microorganisms degrade dead plants and animals and recycle chemical elements to be used by living plants and animals.
2. Bacteria are used to decompose organic matter in sewage.
3. Bacteria that cause diseases in insects are being used as biological controls of insect pests. Biological controls are specific for the pest and do not harm the environment.
4. Microorganisms can be used to help produce foods. They are also food sources (single-cell protein) themselves.
5. Using recombinant DNA, bacteria can produce important human proteins, such as insulin, beta-endorphin, and interferon.

Microbes and Human Disease   (p. 20)

1. Everyone has normal flora in and on the body.
2. The disease-producing properties of a species of microbe and the host's resistance are important factors in determining whether a person will contract a disease.

Study Questions

Review

1. Match the following people to their contribution toward the advancement of microbiology.

_____ Ehrlich
_____ Fleming
_____ Hooke
_____ Koch
_____ Lister
_____ Pasteur
_____ Van Leeuwenhoek

(a) First to observe bacteria
(b) First to observe cells in plant material and name them
(c) Disproved spontaneous generation
(d) Proved that microorganisms can cause disease
(e) Discovered penicillin
(f) Used the first synthetic chemotherapeutic agent
(g) First to employ disinfectants in surgical procedures

2. How did the idea of spontaneous generation come about?

3. Some proponents of spontaneous generation believed that air is necessary for life. They felt that Spallanzani did not really disprove spontaneous generation because he hermetically sealed his flasks to keep air out. How did Pasteur's experiments address the air question without allowing the microbes in the air to ruin his experiment?

4. The genus name of a bacterium is "erwinia" and the specific epithet is "carotovora." Write the scientific name of this organism correctly. Using this name as an example, explain how scientific names are chosen.

5. Match the following microorganisms to their descriptions.

       ____ Algae   (a) Not composed of cells
       ____ Bacteria   (b) Cell walls made of chitin
       ____ Fungi   (c) Cell walls made of peptidoglycan
       ____ Protozoans
       ____ Viruses   (d) Cell walls made of cellulose; photosynthetic
                             (e) Complex cell structure lacking cell walls

6. Into which field of microbiology would the following scientists best fit?

Researcher Who   **Fields**

   ____ Studies biodegradation of toxic wastes.   (a) Immunology
                                        (b) Microbial ecology
   ____ Studies the causative agent of AIDS.   (c) Microbial genetics
                                        (d) Microbial physiology
                                        (e) Molecular biology
                                        (f) Virology

   ____ Studies the production of human proteins by bacteria.
   ____ Studies the symptoms of AIDS.
   ____ Studies the production of TSS toxin by *S. aureus*.
   ____ Studies the life cycle of *Borrelia burgdorferi*.

7. Briefly state the role played by microorganisms in each of the following.
    (a) Biological control of pests
    (b) Recycling of elements
    (c) Normal flora
    (d) Sewage treatment
    (e) Human insulin production
    (f) Vaccine production

Challenge

1. How did the theory of biogenesis lead the way for the germ theory of disease?

2. Since the germ theory of disease was not demonstrated until 1876, why did Semmelweis (1840) and Lister (1867) argue for the use of aseptic techniques?

3. Find at least three supermarket products made by microorganisms. (*Hint:* The label will state the scientific name of the organism or include the word *culture*, *fermented*, or *brewed*.)

Further Reading

Brock, T. D., ed. *Milestones in Microbiology*. Washington, D. C.: American Society for Microbiology, 1971. The writings of Jenner, Lister, Fleming, and others, with instructive editorial comments.

De Kruif, P. *Microbe Hunters*. New York: Harcourt Brace Jovanovich, 1966. The stories of van Leeuwenhoek, Koch, Pasteur, and others, presented in an interesting narrative.

Fraser, D. W., and J. E. McDade. "Legionellosis." *Scientific American* 241(4):82–99, October 1979. A summary of the mysterious epidemic that led to the discovery of a new bacterium.

Gallo, R. C. "The first human retrovirus." *Scientific American* 255(6):88–98, December 1986. A discussion of the spread of this virus from the sixteenth century.

Jaret, P. "Our immune system: the war within." *National Geographic* 169(6):702–734, June 1986. An exciting look at AIDS, malaria, and conception through color-enhanced electron micrographs.

Laurence, J. "The AIDS virus." *Scientific American* 256(1):47–56, January 1987. A summary of the search to identify the cause of AIDS.

Margulis, L., and K. V. Schwartz. *Five Kingdoms: An Illustrated Guide to the Phyla of Life on Earth*, 2nd ed. New York: W. H. Freeman, 1988. A discussion of the classification of living organisms and descriptions of phyla.

"Microbes for Hire." *Science 85* 6(6):30–46, July/August 1985. Three articles that discuss and illustrate technological uses of microbes in large-scale production of foods, vaccines, and drugs.

Rosebury, T. *Life on Man*. New York: Viking Press, 1969. A humorous yet scientific account of the role of microbes on the human body.

CHAPTER 2

Chemical Principles

Learning Objectives

- Discuss the structure of an atom and its relation to the chemical properties of elements.

- Define ionic bond, covalent bond, hydrogen bond, molecular weight, mole.

- Diagram three basic types of chemical reactions.

- Identify the role of enzymes in chemical reactions.

- List several properties of water that are important to living systems.

- Distinguish between inorganic and organic molecules.

- Define acid, base, salt, pH.

- Identify the building blocks of carbohydrates, simple lipids, phospholipids, proteins, and nucleic acids.

- Identify the role of ATP in cellular activities.

We can see a tree rot and smell milk when it sours, but we might not realize what is happening on a microscopic level. In both cases, microbes are conducting chemical operations. The tree rots when microorganisms induce the hydrolysis of wood. (Hydrolysis is a chemical reaction that involves the splitting of water.) Milk turns sour from the production of lactic acid by bacteria. Most of the activities of microorganisms are nothing more than a series of chemical reactions.

Like all organisms, microorganisms must use nutrients to make chemical building blocks that are used for growth and for all the other functions essential to life. For most microorganisms, synthesizing these building blocks requires them to break down nutrient substances and use the energy released to assemble the resulting energy and molecular fragments into new substances. This complicated biochemistry takes place hour by hour in countless microenvironments that in many cases are smaller than the head of a pin.

All matter—whether air, rock, or a living organism—is made up of small units called **atoms**. Atoms interact with each other and in certain combinations to form **molecules**. Living cells are made up of molecules, some of which are very complex. The interaction of atoms and molecules is called **chemistry**.

The chemistry of microbes is the microbiologist's concern. To understand the changes that occur in microorganisms and the changes microbes make in the world around us, you will need to know how molecules are formed and how they interact.

Structure of Atoms

Atoms are the smallest units of matter that enter into chemical reactions. All atoms have a centrally located **nucleus** and particles called **electrons** that move around the nucleus in patterns called electronic configurations (Figure 2-1). The nuclei of most atoms are stable—that is, they do not change spontaneously—and nuclei do not participate in chemical reactions. The nucleus is made up of positively (+) charged particles called **protons** and uncharged (neutral) particles called **neutrons**. The nucleus, therefore, bears a net positive charge. Neutrons and protons have approximately the same weight, which is about 1840 times that of an electron. The charge on electrons is negative (−), and in all atoms the number of protons is equal to the number of electrons. Because the total positive charge of the nucleus equals the total negative charge of the electrons, each atom is electrically neutral.

The number of protons in an atomic nucleus ranges from one (in a hydrogen atom) to more than 100 (in the largest atoms known). Atoms are often listed by their **atomic number**, the number of protons in the nucleus (Table 2-1). The total number of protons and neutrons in an atom is its approximate **atomic weight**.

Chemical Elements

All atoms with the same atomic number behave the same way chemically and are classified as the same **chemical element**. Each element has its own name and a one- or two-letter symbol, usually derived from

TABLE 2-1   *The Elements of Life*

Element	Symbol	Approximate Atomic Weight	Atomic Number
Hydrogen	H	1	1
Carbon	C	12	6
Nitrogen	N	14	7
Oxygen	O	16	8
Fluorine	F	19	9
Sodium	Na	23	11
Magnesium	Mg	24	12
Silicon	Si	28	14
Phosphorus	P	31	15
Sulfur	S	32	16
Chlorine	Cl	35	17
Potassium	K	39	19
Calcium	Ca	40	20
Vanadium	V	51	23
Chromium	Cr	52	24
Manganese	Mn	55	25
Iron	Fe	56	26
Cobalt	Co	59	27
Nickel	Ni	59	28
Copper	Cu	64	29
Zinc	Zn	65	30
Arsenic	As	75	33
Selenium	Se	79	34
Molybdenum	Mo	96	42
Tin	Sn	119	50
Iodine	I	127	53

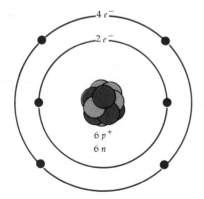

Figure 2-1   Atomic structure. Illustrated here is a very simplified version of a carbon atom (C). It contains six protons (p^+, dark color) and six neutrons (*n*, light color) in its centrally located nucleus. Six electrons (e^-, black) orbit the nucleus.

the English or Latin name for the element. For example, the symbol for the element hydrogen is H, and the symbol for carbon is C. The symbol for sodium is Na—the first two letters of its Latin name, *natrium*—to distinguish it from nitrogen, N, and from sulfur, S. There are 92 naturally occurring elements. However, only about 26 elements are commonly found in living things; these are listed in Table 2-1. The elements most abundant in living matter are hydrogen, carbon, nitrogen, and oxygen.

Most elements have several **isotopes**—atoms with differing numbers of neutrons in their nuclei. All isotopes of an element have the same number of protons in their nuclei, but their atomic weights differ because of the difference in the number of neutrons. For example, in a natural sample of oxygen, all the atoms will contain eight protons. However, 99.76% of the atoms will have eight neutrons, 0.04% will contain nine neutrons, and the remaining 0.2% will contain ten neutrons. Therefore, the three isotopes composing a natural sample of oxygen will have atomic weights of 16, 17, and 18, although all will have the atomic number 8. Atomic numbers are written as a subscript to the left of an element's chemical symbol. Atomic weights are written as a superscript above the atomic number. Thus, natural oxygen isotopes are designated as $^{16}_{8}O$, $^{17}_{8}O$, and $^{18}_{8}O$. Rare isotopes of certain elements are extremely useful in biological research, medical diagnosis, and treatment of some disorders.

Electronic Configurations

In the atom, electrons are arranged in **electron shells**, which are regions corresponding to different **energy levels**. The arrangement is called an **electronic configuration**. Shells are layered outward from the nucleus, and each shell can hold a characteristic maximum number of electrons—two electrons in the innermost shell (lowest energy level), eight electrons in the second shell, and eight electrons in the third shell, if it is the atom's outermost (valence) shell. The fourth, fifth, and sixth electron shells can each accommodate 18 electrons, although there are some exceptions to this generalization. Table 2-2 shows the electronic configurations for atoms of some elements found in living organisms.

There is a tendency for the outermost shell to be filled with the maximum number of electrons. An atom can give up, accept, or share electrons with other atoms to fill this shell. The chemical properties of atoms are largely a function of the number of electrons

in the outermost electron shell. When its outer shell is filled, the atom is chemically stable, or inert: it does not tend to react with other atoms. Helium (atomic number 2) and neon (atomic number 10) are examples of atoms that have filled outer shells. Helium has two electrons in the first and only shell, and neon has two electrons in the first shell and eight electrons in the second shell, its outermost shell.

When an atom's outer electron shell is only partially filled, the atom is chemically unstable. Such an atom reacts with other atoms, and this reaction depends, in part, on the degree to which the outer energy levels are filled. Note the number of electrons in the outer energy levels for the atoms in Table 2-2. We will see later how the number correlates with the chemical reactivity of the elements.

How Atoms Form Molecules: Chemical Bonds

When the outermost energy level of an atom is not completely filled by electrons, it may be thought of as having either unfilled spaces or extra electrons in that energy level. For example, an atom of oxygen, with two electrons in the first energy level and six in the second, has two unfilled spaces in the second electron shell; an atom of potassium has one extra electron in its outermost shell. The most chemically stable configuration for any atom is to have its outermost shell filled, as do the inert gases. Therefore, for these two atoms to attain that state, oxygen has to gain two electrons, and potassium has to lose one electron. All atoms tend to combine so that the extra electrons in the outermost shell of one atom fill the spaces of the outermost shell of the other atom; in the cases of oxygen and potassium, the outermost shell of each atom would have eight electrons when each element combines with another element whose electrons fill the outer shell.

The **valence**, or combining capacity, of an atom is the number of extra or deficient electrons in its outermost electron shell. For example, hydrogen has a valence of 1 (one unfilled space, or one extra electron), oxygen has a valence of 2 (two unfilled spaces), carbon has a valence of 4 (four unfilled spaces, or four extra electrons), and magnesium has a valence of 2 (two extra electrons). Valence may also be viewed as the bonding capacity of an element. Hydrogen can form one chemical bond with another atom, oxygen can form two chemical bonds with other atoms, carbon can form four chemical bonds, and magnesium can form two chemical bonds.

TABLE 2-2   *Electronic Configurations for the Atoms of Some Elements Found in Living Organisms*

Element	Electronic Configuration				Number of Valence (Outermost) Shell Electrons
	First Electron Shell	Second Electron Shell	Third Electron Shell	Diagram	
Hydrogen	1				1
Carbon	2	4			4
Nitrogen	2	5			5
Oxygen	2	6			6
Magnesium	2	8	2		2
Phosphorus	2	8	5		5
Sulfur	2	8	6		6

Basically, atoms gain chemical stability by achieving the full complement of electrons in their outermost energy shells. They do this by combining to form molecules, which can be made up of atoms of one or more elements. A molecule that contains at least two different kinds of atoms, such as H_2O, the water molecule, is called a **compound**. In H_2O, the subscript 2 indicates that there are two atoms of hydrogen to one atom of oxygen. Molecules of a compound hold together because the valence electrons of the combining atoms form attractive forces, called **chemical bonds**, between the atomic nuclei. Because energy is required for chemical bond formation, each chemical bond possesses a certain amount of potential chemical energy. Atoms of a given element can form only a specific number of chemical bonds, based on their electron configurations.

In general, atoms form bonds in one of two ways—by gaining or losing electrons from their outer electron shell or by sharing outer electrons. When atoms have gained or lost outer electrons, the chemical bond is called an **ionic bond**. When outer electrons are shared, the bond is called a **covalent** bond. Although we will discuss ionic and covalent bonds separately, the kinds of bonds actually found in molecules do not entirely belong to either category. Instead, bonds range from the highly ionic to the highly covalent.

Ionic Bonds

Atoms are electrically neutral when the number of positive charges (protons) equals the number of negative charges (electrons). But when an isolated atom gains or loses electrons, this balance is upset. If the atom gains electrons, it acquires an overall negative charge; if the atom loses electrons, it acquires an overall positive charge. Such a negatively or positively charged atom (or group of atoms) is called an **ion**.

Consider the following examples. Sodium (Na) has 11 protons and 11 electrons, with one electron in its outer electron shell. Sodium tends to lose the single outer electron; it is an *electron donor*. When sodium donates an electron to another atom, it is left with 11 protons and only 10 electrons and so has an overall charge of +1. This positively charged sodium atom is called a sodium ion and is written as Na^+. Chlorine (Cl) has a total of 17 electrons, 7 of them in the outer electron shell. Because this outer shell can hold 8 electrons, chlorine tends to pick up an electron that has been lost by another atom; it is an *electron acceptor* (Figure 2-2a). By accepting an electron, chlorine totals 18 electrons. However, it still has only 17 protons in its nucleus. The chloride ion therefore has a charge of −1 and is written as Cl^- (Figure 2-2).

The opposite charges of the sodium ion (Na^+) and chloride ion (Cl^-) attract each other. The attraction, an ionic bond, holds the two atoms together, and a molecule is formed (Figure 2-2b). The formation of this molecule, called sodium chloride (NaCl) or table salt, is a common example of ionic bonding. Thus, an **ionic bond** is an attraction between ions of opposite charge that holds them together to form a stable molecule. Put another way, an ionic bond is an attraction between atoms in which one atom loses electrons and another atom gains electrons. Strong ionic bonds, such as those that hold Na^+ and Cl^- ions together in salt crystals, have only limited importance in living cells. But the weaker ionic bonds formed in aqueous (water) solutions are important in biochemical reactions of microbes and other organisms. For example, weaker ionic bonds assume a role in certain antigen–antibody reactions—that is, reactions in which molecules produced by the immune system (antibodies) combine with foreign substances (antigens) in order to combat infection.

In general, an atom whose outer electron shell is less than half filled will lose electrons and form positively charged ions, called **cations**. Examples of cations are the potassium ion (K^+), calcium ion (Ca^{2+}), and sodium ion (Na^+). The symbol for an ion is the chemical abbreviation followed by the ion's number of positive (+) or negative (−) charges. When an atom's outer electron shell is more than half filled, the atom will gain electrons and form negatively charged ions, called **anions**. Examples are the iodide ion (I^-), chloride ion (Cl^-), and sulfide ion (S^{2-}).

Hydrogen is an example of an atom whose outer level is exactly half filled. The first energy level of a hydrogen atom can hold two electrons, but it contains only one electron. Hydrogen can lose its electron and become a positive ion (H^+), which is precisely what has happened when a hydrogen ion combines with a chloride ion to form hydrochloric acid (HCl). Hydrogen can gain an electron and become a negative ion (H^-), which can form ionic bonds with positive ions, but H^- ions are not very significant in living systems. Hydrogen atoms can also form a more important and altogether different kind of bond, a covalent bond.

Covalent Bonds

A **covalent bond** is a chemical bond formed by two atoms sharing one or more pairs of electrons. Covalent bonds are stronger and far more common in organisms than true ionic bonds. In the hydrogen molecule, H_2, two hydrogen atoms share a pair of electrons. Each hydrogen atom has its own electron plus one electron

Figure 2-2 Ionic bond formation. (a) A sodium atom (Na), left, loses one electron to an electron acceptor and forms a sodium ion (Na⁺). A chlorine atom (Cl), right, accepts one electron from an electron donor to become a chloride ion (Cl⁻). **(b)** The sodium and chloride ions are attracted because of their opposite charges and are held together by an ionic bond to form a molecule of sodium chloride.

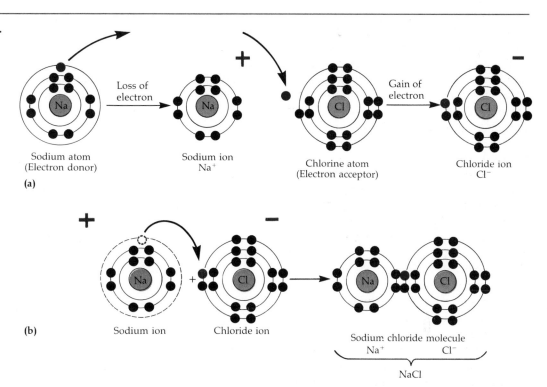

Sodium atom
(Electron donor)

Loss of electron

Sodium ion
Na⁺

Chlorine atom
(Electron acceptor)

Gain of electron

Chloride ion
Cl⁻

(a)

(b)

Sodium ion

Chloride ion

Sodium chloride molecule
Na⁺   Cl⁻

NaCl

from the other atom (Figure 2-3a). The shared pair of electrons actually orbits the nuclei of both atoms. Therefore, the outer electron shells of both atoms are filled. When only one pair of electrons is shared between atoms, a *single covalent bond* is formed. For simplicity, a single covalent bond is expressed as a single line between the atoms (H—H). When two pairs of electrons are shared between atoms, a *double covalent bond* is formed, expressed as two single lines (═) (Figure 2-3b). A *triple covalent bond*, expressed as three single lines (≡), occurs when three pairs of electrons are shared (Figure 2-3c).

The principles of covalent bonding that apply to atoms of the same element also apply to atoms of different elements. Methane (CH_4) is an example of covalent bonding between atoms of different elements (Figure 2-3d). The outer electron shell of the carbon atom can hold eight electrons but has only four. Each hydrogen atom can hold two electrons but has only one. Consequently, in the methane molecule, the carbon atom gains four hydrogen electrons to complete its outer shell, and each hydrogen atom completes its pair by sharing one electron from the carbon atom. Each outer electron of the carbon atom orbits both the carbon nucleus and a hydrogen nucleus. Each hydrogen electron orbits both its own nucleus and the carbon nucleus.

Elements such as hydrogen and carbon, whose outer electron shells are half-filled, form covalent bonds quite easily. In fact, in living organisms, carbon always forms covalent bonds; it never becomes an ion.

However, many atoms whose outer electron shells are almost full will also form covalent bonds. An example is oxygen. It is not necessary for now to know why some atoms tend to form covalent bonds rather than ionic bonds. But it is important to understand the basic principles of bond formation because chemical reactions are nothing more than the making or breaking of bonds between atoms. *Remember*: Covalent bonds are formed by the *sharing* of electrons between atoms. Ionic bonds are formed by *attractions* between atoms that have lost or gained electrons and are therefore positively or negatively charged.

Hydrogen Bonds

Another chemical bond of special importance to all organisms is the **hydrogen bond**, in which a hydrogen atom that is covalently bonded to one oxygen or nitrogen atom is attracted to another oxygen or nitrogen atom. Such bonds are weak and do not bind atoms into molecules. However, they do serve as bridges between different molecules or between various portions of the same molecule.

When hydrogen combines with atoms of oxygen or nitrogen, the relatively large positive nucleus of these larger atoms attracts the hydrogen electron more strongly than does the small hydrogen nucleus. Thus, in a molecule of water (H_2O), all the electrons tend to be closer to the oxygen nucleus than to the

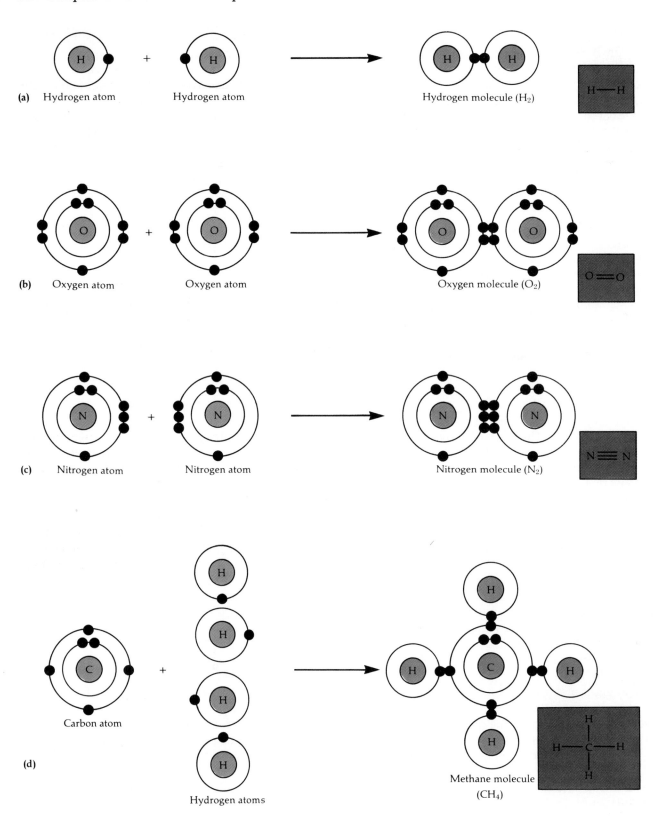

Figure 2-3 **Covalent bond formation.** **(a)** Single covalent bond between two hydrogen atoms. **(b)** Double covalent bond between two oxygen atoms. **(c)** Triple covalent bond between two nitrogen atoms. **(d)** Single covalent bonds between four hydrogen atoms and a carbon atom, forming the methane molecule.

hydrogen nuclei. The oxygen portion of the molecule thus has a slightly negative charge, and the hydrogen portion of the molecule has a slightly positive charge (Figure 2-4a). When the positively charged end of one molecule is attracted to the negative end of another molecule, a hydrogen bond is formed (Figure 2-4b). This attraction can also occur between hydrogen and other atoms of the same molecule, especially in large molecules. Because nitrogen and oxygen have unshared pairs of electrons, they are the elements most frequently involved in hydrogen bonding.

Hydrogen bonds are considerably weaker than either ionic or covalent bonds; they have only about 5% of the strength of covalent bonds. Consequently, hydrogen bonds are formed and broken relatively easily. This property accounts for the temporary bonding that occurs between certain atoms of large and complex molecules, such as proteins and nucleic acids. Even though hydrogen bonds are relatively weak, large molecules containing several hundred of these bonds have considerable strength and stability.

Molecular Weight and Moles

You have seen that bond formation results in the creation of molecules. Molecules are often discussed in terms of units of measure called molecular weight and moles. The **molecular weight** of a molecule is the sum of the atomic weights of all its atoms. To relate the molecular level to the laboratory level, we use a unit called the mole. One **mole** of a substance is its molecular weight expressed in grams. For example, one mole of water weighs 18 grams because the molecular weight of H_2O is 18 ($2 \times 1 + 16$). The word *mole*

can also be applied to atoms or even to ions. Thus, one mole of hydrogen ions is equal to 1 gram.

To determine the number of moles in a certain quantity of a substance, the mass of the substance (in grams) is divided by the molecular weight of the substance. For example, to calculate the number of moles in 500 grams of glucose ($C_6H_{12}O_6$), first calculate the number of grams in one mole ($6 \times 12 + 12 \times 1 + 6 \times 16 = 180$). Then divide 500 grams by 180 grams per mole to get 2.78 moles.

Chemical Reactions

As we said earlier, **chemical reactions** are the making or breaking of bonds between atoms. After a chemical reaction, the total number of atoms remains the same, but there are new molecules with new properties because the atoms are rearranged.

Energy of Chemical Reactions

Some change of energy occurs whenever bonds between atoms are formed or broken during chemical reactions. This energy is called **chemical energy**. When a chemical bond is formed, energy is required. Such a chemical reaction that requires energy is called an **endergonic reaction**, meaning energy is directed inward. When a bond is broken, energy is released. A chemical reaction that releases energy is an **exergonic reaction**, meaning energy is directed outward.

Figure 2-4 Hydrogen bond formation in water. **(a)** In a water molecule, the electrons of the hydrogen atoms are strongly attracted to the oxygen atom. Therefore, the part of the water molecule containing the oxygen atom has a slightly negative charge, and the part containing hydrogen atoms has a slightly positive charge. **(b)** In a hydrogen bond between water molecules, the hydrogen of one water molecule is attracted to the oxygen of another water molecule. Many water molecules may be attracted to each other by hydrogen bonds (black dots).

(a)

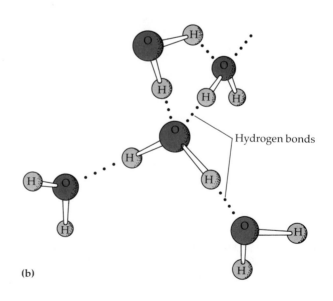

(b)

In this section, we will look at three basic types of chemical reactions common to all living cells. By becoming familiar with these reactions, you will be able to understand the specific chemical reactions we will discuss later, particularly in Chapter 5.

Synthesis Reactions

When two or more atoms, ions, or molecules combine to form new and larger molecules, the reaction is called a **synthesis reaction**. To synthesize means to put together, and a synthesis reaction *forms new bonds*. Synthesis reactions can be expressed in the following way:

$$\text{A} + \text{B} \longrightarrow \text{AB}$$

| Atom, ion, or molecule A | Atom, ion, or molecule B | Combine to form | New molecule AB |

The combining substances, A and B, are called the *reactants*; the substance formed by the combination, AB, is the *product*. The arrow indicates the direction in which the reaction proceeds.

Synthesis reactions in living organisms are collectively called anabolic reactions, or simply **anabolism**. The combining of sugar molecules to form starch, and of amino acids to form proteins, are two examples of anabolism.

Decomposition Reactions

The reverse of a synthesis reaction is a **decomposition reaction**. To decompose means to break down into smaller parts, and in a decomposition reaction, *bonds are broken*. Typically, decomposition reactions split large molecules into smaller molecules, ions, or atoms. A decomposition reaction occurs in this way:

$$\text{AB} \longrightarrow \text{A} + \text{B}$$

| Molecule AB | Breaks down into | Atom, ion, or molecule A | Atom, ion, or molecule B |

Decomposition reactions that occur in living organisms are collectively called catabolic reactions, or simply **catabolism**. The breakdown of food molecules in digestion is an example of catabolism.

Exchange Reactions

All chemical reactions are based on synthesis and decomposition. Many reactions, such as **exchange reactions**, are actually part synthesis and part decomposition. An exchange reaction works in this way:

$$\text{AB} + \text{CD} \longrightarrow \text{AD} + \text{BC}$$

Recombine to form

First, the bonds between A and B and between C and D are broken in a decomposition process. New bonds are then formed between A and D and between B and C in a synthesis process. Exchange reactions are involved in the conversion of organic acids into amino acids.

The Reversibility of Chemical Reactions

All chemical reactions are, in theory, reversible; that is, they can occur in either direction. In practice, however, some reactions do this more easily than others. A chemical reaction that is readily reversible (meaning, the end product can revert to the original molecules) is termed a **reversible reaction**, and is indicated by two arrows, as shown here:

$$\text{A} + \text{B} \underset{\text{Breaks down into}}{\overset{\text{Combines to form}}{\rightleftharpoons}} \text{AB}$$

Some reversible reactions occur because neither the reactants nor the end products are very stable. Other reactions will reverse only under special conditions:

$$\text{A} + \text{B} \underset{\text{Water}}{\overset{\text{Heat}}{\rightleftharpoons}} \text{AB}$$

Whatever is written above or below the arrows indicates the special condition under which the reaction in that direction occurs. In this case, A and B react to produce AB only when heat is applied, and AB breaks down into A and B only in the presence of water.

How Chemical Reactions Occur

The **collision theory** explains how chemical reactions occur and how certain factors affect the rates of those reactions. The basis of the collision theory is that all atoms, ions, and molecules are continuously moving and are thus continuously colliding with one another. The energy transferred by the particles in the collision can disrupt their electron structures enough so that chemical bonds are broken or new ones are formed.

Factors Affecting Chemical Reactions

Several factors determine whether a collision will cause a chemical reaction—the velocities of the colliding particles, their energy, and their specific chemical configurations. Up to a point, the higher the particles' velocities, the greater the probability that their collision will cause a reaction. Also, each chemical reaction requires a specific level of energy. But even if colliding particles possess the minimum energy needed for reaction, no reaction will take place unless the particles are properly oriented toward each other.

Let us assume that molecules of substance X (the reactant) are to be converted to molecules of substance Y (the product). In a given population of molecules of substance X, at a specific temperature, some molecules will possess relatively little energy; the majority of the population will possess an average amount of energy; and a small portion of the population will have high energy. If only the high-energy X molecules are able to react and be converted to Y molecules, then only relatively few molecules at any one time possess enough energy to react in a collision. The collision energy required for a chemical reaction is its **activation energy**, which is the amount of energy needed to disrupt the stable electronic configuration of a specific molecule so that the electrons can be rearranged.

The *reaction rate*, the frequency of collisions containing sufficient energy to bring about a reaction, depends on the number of reactant molecules at or above the activation energy level. One way to increase the reaction rate of a substance is to raise its temperature. By causing the molecules to move faster, heat increases both the frequency of collisions and the number of molecules that attain activation energy. The number of collisions also increases when pressure is increased or the reactants are more concentrated and the distance between molecules is thereby decreased.

Enzymes and Chemical Reactions

Living systems contain large protein molecules called **enzymes** that act as catalysts. A **catalyst** is a substance that changes the rate of a chemical reaction, usually by increasing it. As catalysts, enzymes typically accelerate chemical reactions. The three-dimensional enzyme molecule has an *active site*, a region that will interact with a specific chemical substance (see Figure 5-3). The chemical substance (reactant) an enzyme acts on is referred to as the enzyme's **substrate**.

The enzyme orients the substrate into a position that increases the probability of a reaction. The **enzyme-substrate complex** formed by the temporary binding of enzyme and reactants enables the collisions to be more effective and lowers the activation

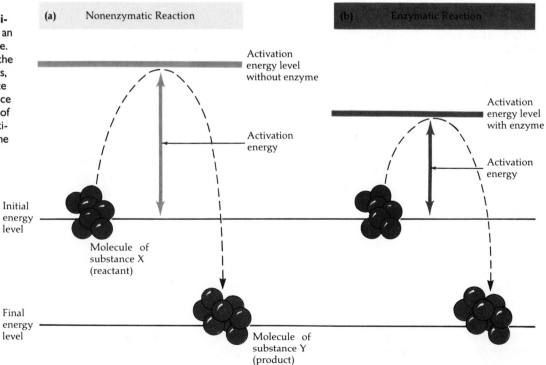

Figure 2-5 Energy requirements of a chemical reaction. (a) Without an enzyme; **(b)** with an enzyme. In **(b)**, the enzyme lowers the activation energy level. Thus, more molecules of substance X are converted to substance Y because more molecules of substance X possess the activation energy needed for the reaction.

energy of the reaction (Figure 2-5). The enzyme therefore speeds up the reaction by increasing the number of X molecules that attain sufficient activation energy to react (Figure 2-5).

An enzyme is able to accelerate a reaction without an increase in temperature. This ability is crucial to living systems because a significant temperature increase would destroy cellular proteins. The crucial function of enzymes, therefore, is to speed up biochemical reactions at a temperature that is compatible with the normal functioning of the cell.

Important Biological Molecules

Biologists and chemists divide compounds into two principal classes, inorganic and organic. **Inorganic compounds** are defined as molecules, usually small, in which ionic bonds may play an important role. Inorganic compounds include water and many salts, acids, and bases.

Organic compounds always contain carbon and hydrogen. Carbon is a unique element in the chemistry of life because it has four electrons in its outer shell and can combine with a variety of atoms, including other carbon atoms, to form straight or branched chains and rings (see Figure 2-9). Carbon chains form the basis of many organic compounds in living cells, including sugars, amino acids, and vitamins. Organic compounds are held together mostly or entirely by covalent bonds. Some organic molecules, such as polysaccharides, proteins, and nucleic acids, are very large and usually contain thousands of atoms. Such giant molecules are called *macromolecules*. In the following section, we will discuss inorganic and organic compounds that are essential for cells.

Inorganic Compounds

Water

All living organisms require a wide variety of inorganic compounds for growth, repair, maintenance, and reproduction. Water is one of the most important, as well as one of the most abundant, of these compounds, and it is particularly vital to microorganisms. Outside the cell, nutrients are dissolved in water, which facilitates their passage through cell membranes. And inside the cell, water is the medium for most chemical reactions. In fact, water is by far the most abundant component of almost all living cells. Water comprises from 5% to 95% or more of each cell, the average being between 65% and 75%. Simply stated, no organism can survive without water.

Water has structural and chemical properties that make it particularly suitable for its role in living cells. As we discussed, the total charge on the water molecule is neutral, but the oxygen region of the molecule has a slightly negative charge, and the hydrogen region has a slightly positive charge (see Figure 2-4a). Any molecule having such an unequal distribution of charges is called a **polar molecule**. The polar nature of water gives it four characteristics that make it a useful medium for living cells.

First, every water molecule is capable of forming four hydrogen bonds with nearby water molecules (see Figure 2-4b). This property results in a strong attraction between water molecules. Because of this strong attraction, a great deal of heat is required to separate water molecules from each other to form water vapor; thus, water has a relatively high boiling point (100°C). Since water has such a high boiling point, it exists in the liquid state on most of the earth's surface. Furthermore, the hydrogen bonding between water molecules affects the density of water, depending on whether it occurs as ice or a liquid. For example, the hydrogen bonds in the crystalline structure of water (ice) make it more spacious. As a result, ice has fewer molecules than an equal volume of liquid water. This makes its crystalline structure less dense than liquid water. It is for this reason that ice floats and can serve as an insulating layer on the surfaces of lakes and streams that harbor living organisms.

Second, the polarity of water makes it an excellent dissolving medium or **solvent**. Many polar substances **dissociate** (separate) into individual molecules in water—that is, dissolve—because the negative part of the water molecule is attracted to the positive part of the molecules in the **solute** (dissolving substance), and the positive part of the water molecules is attracted to the negative part of solute molecules. Substances (such as salts) that are composed of atoms (or groups of atoms) held together by ionic bonds tend to dissociate into separate cations and anions in water. Thus, the polarity of water allows molecules of many different substances to separate and become surrounded by water molecules (Figure 2-6).

Third, polarity accounts for water's characteristic role as a reactant or product in many chemical reactions. Its polarity facilitates the splitting and rejoining of hydrogen (H^+) and hydroxyl (OH^-) groups. Water is a key reactant in the digestive processes of organisms, whereby larger molecules are broken down into smaller ones. Water molecules are also

Figure 2-6 How water acts as a solvent for sodium chloride (NaCl). (a) The positively charged sodium ion (Na⁺) is attracted to the negative part of the water molecule. **(b)** The negatively charged chloride ion (Cl⁻) is attracted to the positive part of the water molecule. In the presence of water molecules, the bonds between the Na⁺ and Cl⁻ ions are disrupted, and the NaCl dissolves in the water.

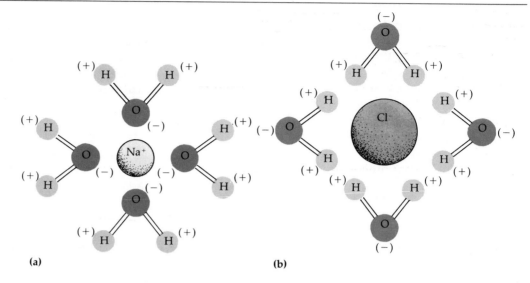

(a) (b)

involved in synthetic reactions—water is an important source of the hydrogen and oxygen that are incorporated into numerous organic compounds in living cells.

Finally, the relatively strong hydrogen bonding between water molecules (see Figure 2-4b) makes water an excellent temperature buffer. A given quantity of water requires a great gain of heat to increase its temperature and a great loss of heat to decrease its temperature, as compared with many other substances. Normally, heat absorption by molecules increases their kinetic energy and thus increases their rate of motion and their reactivity. In water, however, heat absorption first breaks hydrogen bonds rather than increasing the rate of motion. Therefore, much more heat must be applied to raise the temperature of water than to raise the temperature of a nonhydrogen-bonded liquid. The reverse is true as water cools. Thus, water more easily maintains a constant temperature than other solvents and tends to protect a cell from fluctuations in environmental temperatures.

Acids, Bases, and Salts

As we saw in Figure 2-6, when inorganic salts such as sodium chloride (NaCl) are dissolved in water, they undergo **dissociation**, or **ionization**. That is, they break apart into ions. Substances called acids and bases show similar behavior.

An **acid** can be defined as a substance that dissociates into one or more hydrogen ions (H⁺) and one or more negative ions (anions). Thus, an acid can also be defined as a proton (H⁺) donor. A **base** dissociates into one or more positive ions (cations) plus one or more negatively charged hydroxyl ions (OH⁻) that can accept, or combine with, protons. Thus, sodium hydroxide (NaOH) is a base because it dissociates to release hydroxyl ions, which have a strong attraction for protons and are among the most important proton acceptors. A **salt** is a substance that dissociates in water into cations and anions, neither of which is H⁺ or OH⁻. Figure 2-7 shows common examples of each type of compound and how they dissociate in water.

Figure 2-7 Acids, bases, and salts. (a) In water, hydrochloric acid (HCl) dissociates into H⁺ ions and Cl⁻ ions. **(b)** Sodium hydroxide (NaOH), a base, dissociates into OH⁻ ions and Na⁺ ions in water. **(c)** In water, table salt (NaCl) dissociates into positive ions (Na⁺) and negative ions (Cl⁻), neither of which are H⁺ or OH⁻ ions.

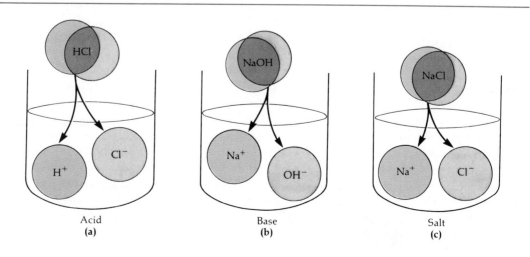

Acid (a)   Base (b)   Salt (c)

Acid–Base Balance

An organism must maintain a fairly constant balance of acids and bases to remain healthy. In the aqueous environment within organisms, acids dissociate into hydrogen ions (H^+) and anions. Bases, on the other hand, dissociate into hydroxyl ions (OH^-) and cations. The more hydrogen ions that are free in a solution, the more acid the solution is. Conversely, the more hydroxyl ions that are free in a solution, the more basic, or alkaline, it is.

Biochemical reactions—that is, reactions in living systems—are extremely sensitive to even small changes in the acidity or alkalinity of the environments in which they occur. In fact, H^+ and OH^- ions are involved in almost all biochemical processes, and the functions of a cell are modified greatly by any deviation from its narrow band of normal H^+ and OH^- concentrations. For this reason, the acids and bases that are continually formed in an organism must be kept in balance.

It is convenient to express the amount of H^+ in a solution by a logarithmic pH scale, which ranges from 0 to 14 (Figure 2-8). The **pH** of a solution is calculated by $-\log_{10}[H^+]$, the negative logarithm to the base 10 of the hydrogen ion concentration, determined in moles per liter $[H^+]$. For example, if the H^+ concentration of a solution is 1.0×10^{-4} moles/liter, or 10^{-4}, its pH equals $-\log_{10}10^{-4} = -(-4) = 4$; this is about the pH of wine. (See Exponents, Exponential Notation, and Logarithms in Appendix G.) In the laboratory, you will usually measure the pH of a solution with a pH meter or with chemical test papers and will not need to do calculations.

Acidic solutions contain more H^+ ions than OH^- ions and have a pH lower than 7. If a solution has more OH^- ions than H^+ ions, it is a **basic**, or **alkaline solution**. In pure water, a small percentage of the molecules are dissociated into H^+ and OH^- ions, so it has a pH of 7. Because the concentrations of H^+ and OH^- are equal, this pH is said to be a **neutral solution**.

The pH scale is logarithmic, so a change of one whole number represents a tenfold change from the previous concentration. Thus, a solution of pH 1 has ten times more H^+ ions than a solution of pH 2 and has 100 times more H^+ ions than a solution of pH 3.

The pH values of some human body fluids and other common substances are shown in Table 2-3.

Keep in mind that the pH of a solution can be changed. We can increase its acidity by adding substances that will increase the concentration of hydrogen ions. As a living organism takes up nutrients, carries out chemical reactions, and excretes wastes,

Figure 2-8   The pH scale. The concentrations of H^+ and OH^- ions are equal at pH 7, which is the neutral point. As the pH values decrease from 14 to 0, the H^+ concentration increases. Thus, the lower the pH, the more acidic the solution; the higher the pH, the more basic the solution. If the pH value of a solution is below 7, the solution is acidic; if the pH is above 7, the solution is basic (alkaline).

its balance of acids and bases tends to change, and the pH fluctuates. Fortunately, organisms possess natural **pH buffers**, compounds that help keep the pH from changing drastically. But the pH in our environment's water and soil can be altered by waste products from organisms, pollutants from industry, or fertilizers used in agricultural fields or gardens. When bacteria are grown in a laboratory medium, they excrete waste products such as acids that can alter the pH of the medium. If this effect were to continue, the pH of the medium would become acidic enough to inhibit bacterial enzymes and cause the death of the bacteria. To prevent this problem, pH buffers are added to the culture medium. One very effective pH buffer for some culture media uses a mixture of K_2HPO_4 and KH_2PO_4 (see Table 6-3).

Different microbes function best within different pH ranges, but most organisms grow best in environments with a pH between 6.5 and 8.5. Fungi are the microbes that are best able to tolerate acid conditions, whereas the procaryotes called cyanobacteria tend to do well in alkaline habitats. *Propionibacterium acnes* (prō-pē-on-ē-bak-ti′rē-um ak′nēz), a bacterium that contributes to acne, has its natural environment on human skin, which tends to be slightly acidic, with a pH of about 4. *Thiobacillus ferrooxidans* (thī-ō-bä-sil′lus fer-rō-oks′i-danz) is a bacterium that grows on elemental sulfur and produces sulfuric acid (H_2SO_4). Its optimum growth pH is from 1 to 3.5. The sulfuric acid produced by this bacterium in mine water is important in dissolving uranium and copper from low-grade ore. (See the box in Chapter 27, p. 708.)

TABLE 2-3 *pH Values of Some Human Body Fluids and Common Substances*

Fluid or Substance	pH
Stomach acid	1.5 to 2.5
Lemon juice	2.2 to 2.4
Carbonated soft drinks	3 to 3.5
Tomato juice	4.2
Urine	5.0 to 7.8
Saliva	6.35 to 6.85
Milk	6.6 to 6.9
Distilled (pure) water	7.0
Blood	7.35 to 7.45
Eggs	7.6 to 8.0
Milk of magnesia	10.0 to 11.0
Limewater	12.3

Organic Compounds

Inorganic compounds, excluding water, constitute about 1% to 1.5% of living cells. These relatively simple components, whose molecules have only a few atoms, cannot be used by cells to perform complicated biological functions. Organic molecules, whose carbon atoms can combine in an enormous variety of ways with other carbon atoms and atoms of other elements, are relatively complex and thus are capable of more complicated biological functions. In the formation of organic molecules, carbon's four outer electrons can participate in up to four covalent bonds, and carbon atoms can bond to each other to form straight-chain, branched-chain, or ring structures (Figure 2-9).

When single bonds hold a carbon atom to four other atoms, those four atoms form a skeleton of a three-dimensional shape called a **tetrahedron** (Figure 2-10). Carbon atoms can also bond with double or triple bonds; the molecular shape then is not tetrahedral.

In addition to carbon, the most frequently found elements in organic compounds are hydrogen (which can form one bond), oxygen (two bonds), and nitrogen (three bonds). Sulfur (two bonds) and phosphorus (five bonds) appear less often. Other elements are found,

but only in a relatively few organic compounds. If you look back at Table 2-1, you will see that the elements most abundant in organic compounds are the same as those most abundant in living cells.

Functional Groups

The chain of carbon atoms in a molecule is called the **carbon skeleton**; there is a huge number of combinations possible for carbon skeletons. Most of these carbons are bonded to hydrogen atoms. The bonding of other elements with carbon and hydrogen forms characteristic **functional groups** (Table 2-4). Functional groups are specific groups of atoms that are

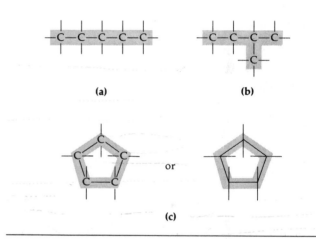

Figure 2-9 Various bonding patterns of carbon. (a) Straight-chain structures. **(b)** Branched-chain structures. **(c)** Ring. In ring structures, the carbon atoms at the corners are often not shown.

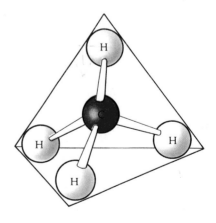

Figure 2-10 A tetrahedral carbon compound. Organic molecules containing carbon atoms with four single bonds have this distinctive shape due to the directions of the four bonds. Here, the carbon atom is bonded to four hydrogen atoms to form the tetrahedral methane molecule (CH_4).

responsible for most of the characteristic chemical properties and many of the physical properties of a particular organic compound. Thus, functional groups help us to classify organic compounds. For example, the —OH group is present in each of the following molecules:

H—C—OH (with H above and below the C) **Methanol**

H—C—C—OH (with H's above and below each C) **Ethanol**

H—C—C—C—H (with H's above, O below middle C) **Isopropyl alcohol**

(ring structure) **Toluol**

Because the characteristic reactivity of the molecules is based on the —OH group, they are grouped together in a class called alcohols. The —OH group is called the *hydroxy group* and is not to be confused with the *hydroxyl ion* (OH⁻) of bases. The hydroxy group of alcohols does not ionize; it is covalently bonded to a carbon atom.

When a class of compounds is characterized by a certain functional group, the letter *R* can be used to stand for the remainder of the molecule. For example, alcohols in general may be written R—OH.

Frequently, more than one functional group is found in a single molecule. For example, an amino acid molecule contains both amino and carboxyl groups. The amino acid glycine is shown below.

Amino group

H—N—C—C (with O and OH) Carboxyl group

Most of the organic compounds found in living organisms are quite complex; a large number of carbon atoms form the skeleton, and many functional groups are attached. In organic molecules, it is important that each of the four bonds of carbon is satisfied (attached to another atom) and each of the attaching atoms has its characteristic number of bonds satisfied. Because of this, such molecules are chemically stable.

TABLE 2-4   *Representative Functional Groups and the Compounds in Which They Are Found*

Functional Group	Name of Group	Class of Compounds
R—O—H	Hydroxy	Alcohol
R—C(=O)—H	Carbonyl (terminal)*	Aldehyde
R—C(=O)—R	Carbonyl (internal)*	Ketone
R—C(H)(H)—NH₂	Amino	Amine
R—C(=O)—O—R'	Ester	Ester
R—C(H)(H)—O—C(H)(H)—R'	Ether	Ether
R—C(H)(H)—SH	Sulfhydryl	Sulfhydryl
R—C(=O)—OH	Carboxyl	Organic acid

*A "terminal" carbonyl group is at an end of a molecule. In contrast, an "internal" carbonyl is at least one C atom removed from both ends.

Macromolecules

Small organic molecules can be combined into the very large molecules called **macromolecules**. Macromolecules are usually **polymers**, large molecules formed by covalent bonding of many repeating small molecules called **subunits** or **monomers**. When two monomers join together, the reaction usually involves the elimination of a hydrogen atom from one monomer and a hydroxy group from the other; these combine to produce water:

$$X\text{—}OH + H\text{—}Y \longrightarrow X\text{—}Y + H_2O$$

This type of reaction is called **dehydration synthesis** or **condensation** because a molecule of water is released. The synthesis of such macromolecules as carbohydrates, lipids, proteins, and nucleic acids by dehydration synthesis is a dominant activity of cell chemistry.

Carbohydrates

The **carbohydrates** are a large and diverse group of organic compounds that includes sugars and starches. The carbohydrates perform a number of major functions in living systems. For instance, one type of sugar (deoxyribose) is a building block of deoxyribonucleic acid (DNA), the molecule that carries hereditary information. Other sugars help to form the cell walls of bacterial cells. Simple carbohydrates are used in the synthesis of amino acids and fats or fatlike substances, which are used to build structures and provide an emergency source of energy. Macromolecular carbohydrates function as food reserves. The principal function of carbohydrates, however, is to fuel cell activities with a ready source of energy.

Carbohydrates are made up of carbon, hydrogen, and oxygen atoms. The ratio of hydrogen to oxygen atoms is always 2:1 in carbohydrates. This ratio can be seen in the formulas for the carbohydrates ribose ($C_5H_{10}O_5$), glucose ($C_6H_{12}O_6$), and sucrose ($C_{12}H_{22}O_{11}$). Although there are exceptions, the general formula for carbohydrates is $(CH_2O)n$, where n indicates that there are three or more CH_2O units. Carbohydrates can be divided into three major groups on the basis of size—monosaccharides, disaccharides, and polysaccharides.

Monosaccharides

Simple sugars are called **monosaccharides**; each molecule contains from three to seven carbon atoms. The number of carbon atoms in the molecule of a simple sugar is indicated by the prefix in its name. For example, simple sugars with three carbons are called trioses. There are also tetroses (four-carbon sugars), pentoses (five-carbon sugars), hexoses (six-carbon sugars), and heptoses (seven-carbon sugars). Pentoses and hexoses are extremely important to living organisms. Deoxyribose is a pentose found in DNA, the genetic material of the cell. Glucose, a common hexose, is the main energy-supplying molecule of living cells.

Disaccharides

Disaccharides are formed when two monosaccharides bond in a dehydration synthesis (condensation) reaction. For example, molecules of two monosaccharides, glucose and fructose, combine to form a molecule of the disaccharide sucrose (table sugar):

$$C_6H_{12}O_6 \quad + \quad C_6H_{12}O_6 \longrightarrow$$
$$\text{Glucose} \qquad\qquad \text{Fructose}$$
$$\text{(monosaccharide)} \qquad \text{(monosaccharide)}$$

$$C_{12}H_{22}O_{11} \quad + \quad H_2O$$
$$\text{Sucrose} \qquad\quad \text{Water}$$
$$\text{(disaccharide)}$$

The formula for sucrose is $C_{12}H_{22}O_{11}$ and not $C_{12}H_{24}O_{12}$ because a molecule of water (H_2O) is lost in the process of disaccharide formation (Figure 2-11a). Similarly, the dehydration synthesis of the monosaccharides glucose and galactose forms the disaccharide lactose (milk sugar).

Glucose $C_6H_{12}O_6$ Fructose $C_6H_{12}O_6$ Sucrose $C_{12}H_{22}O_{11}$ Water

Figure 2-11   Dehydration synthesis and hydrolysis. (a) In dehydration synthesis (left to right), the monosaccharides glucose and fructose combine to form a molecule of the disaccharide sucrose. A molecule of water is lost in the reaction. **(b)** In hydrolysis (right to left), the sucrose molecule breaks down into the smaller molecules glucose and fructose. For the hydrolysis reaction to proceed, water must be added to the sucrose.

It may seem odd that glucose and fructose should have the same chemical formula since they are different monosaccharides. The positions of the oxygens and carbons vary in the two different molecules (see Figure 2-11), and consequently the molecules have different physical and chemical properties. Two molecules with the same chemical formula but different structures and properties are called **isomers**.

Disaccharides can be broken down into smaller, simpler molecules when water is added. This chemical reaction, the reverse of dehydration synthesis, is called **hydrolysis**, which means to split water. A molecule of sucrose, for example, may be hydrolyzed (digested) into its components of glucose and fructose by reacting with the H$^+$ and OH$^-$ of water. This reaction is also represented in Figure 2-11.

Polysaccharides

Carbohydrates in the third major group, the **polysaccharides**, consist of eight or more monosaccharides joined together through dehydration synthesis. Polysaccharides often have side chains branching off the main structure and are classified as macromolecules. Like disaccharides, polysaccharides can be split apart into their constituent sugars through hydrolysis. Unlike monosaccharides and disaccharides, however, they usually lack the characteristic sweetness of sugars such as fructose and sucrose and usually are not soluble in water.

One important polysaccharide is *glycogen*, which is composed of glucose subunits and is synthesized as a storage material by animals and some bacteria. Cel-

lulose, another important glucose polymer, is the main component of the cell walls of plants and most algae. The polysaccharide *dextran*, which is produced as a sugary slime by certain bacteria, is used in a blood plasma substitute. *Starch* is a polymer of glucose produced by plants and used as food by humans.

Lipids

If lipids were suddenly to disappear from the earth, all living cells would collapse in a pool of fluid, because lipids are essential to the structure and function of membranes that separate living cells from their environment. **Lipids** are a second major group of organic compounds found in living matter. Like carbohydrates, they are composed of atoms of carbon, hydrogen, and oxygen, but lipids lack the 2:1 ratio between hydrogen and oxygen atoms. Lipids are a very diverse group of compounds, their common characteristic being that they are very nonpolar molecules. Hence, most are insoluble in water but dissolve readily in nonpolar solvents, such as ether, chloroform, and alcohol. Lipids function in energy storage and provide part of the structure of membranes and cell walls.

Simple Lipids

Simple lipids, or *fats*, contain an alcohol called *glycerol* and a group of compounds known as *fatty acids*. Glycerol molecules have three carbon atoms to which are attached three hydroxy (—OH) groups. The structural formula for glycerol is shown in Figure 2-12a.

Figure 2-12 Structural formulas. (a) Glycerol; **(b)** lauric acid, a fatty acid. In the condensed formula for lauric acid, COOH(CH$_2$)$_{10}$CH$_3$, the subscript 10 means that the molecule has ten CH$_2$ units. **(c)** The chemical combination of a molecule of glycerol with three fatty acid molecules, in this case lauric acid, forms one molecule of fat and three molecules of water in a dehydration synthesis reaction. The addition of three water molecules to a fat forms glycerol and three fatty acid molecules in a hydrolysis reaction.

(a) Glycerol

Carboxyl group Hydrocarbon chain

(b) Lauric acid, COOH(CH$_2$)$_{10}$CH$_3$

Dehydration synthesis

Hydrolysis

Ester linkage

(c) Glycerol + 3 fatty acids

Fat

Fatty acids consist of long hydrocarbon chains (composed only of carbon and hydrogen atoms) ending in a carboxyl (—COOH) (organic acid) group (Figure 2-12b). Most common fatty acids contain an even number of carbon atoms.

A fat molecule is formed when a molecule of glycerol combines with from one to three fatty acid molecules to form a monoglyceride, diglyceride, or triglyceride (Figure 2-12c). In the reaction, from one to three molecules of water are formed (dehydration), depending on the number of fatty acid molecules reacting. The chemical bond formed where the water molecule is removed is an *ester linkage*. In the reverse reaction, hydrolysis, a fat molecule is broken down into its component fatty acid and glycerol molecules.

Because the fatty acids that form lipids have different structures, there is a wide variety of lipids. For example, three molecules of fatty acid A might combine with a glycerol molecule, or one molecule each of fatty acids A, B, and C might unite with a glycerol molecule. Simple lipids serve as energy storage materials.

The terms *saturated fat* and *unsaturated fat* refer to the structure of the hydrocarbon chains of the fatty acids in the fat molecule. If there are no double bonds between carbon atoms in the hydrocarbon chain, then the carbon skeleton contains the maximum number of hydrogen atoms. Such a fat is said to be *saturated*, meaning that it is saturated with hydrogen atoms (Figure 2-13a). Most animal fats are saturated fats. If there is one or more double bonds, which reduces the number of hydrogen atoms that can be attached to the carbon skeleton, then the fat molecule is referred to as *unsaturated* (Figure 2-13a). At each point where a double bond does occur in the hydrocarbon chain of an unsaturated fat, the fatty acid has a kink (bend) in its structure. Plant fats are usually unsaturated. The box on p. 42 describes the roles of saturated and unsaturated fatty acids in plasma membranes.

Complex Lipids

Complex lipids contain such elements as phosphorus, nitrogen, and sulfur in addition to the carbon, hydrogen, and oxygen found in simple lipids. The complex lipids called *phospholipids* are made up of glycerol, two fatty acids, and, in place of a third fatty acid, a phosphate group bonded to one of several organic groups (see Figure 2-13). Phospholipids are the lipids that build membranes; they are thus essential to a cell's survival. Phospholipids have polar as well as nonpolar regions (see Figures 2-13 and 4-12b). When placed in water, phospholipid molecules twist and ori-

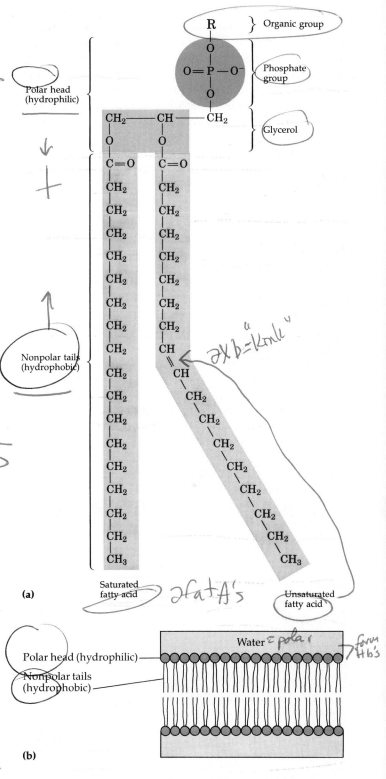

Figure 2-13 Phospholipids. (a) Structure of a phospholipid. The fatty acids and the R group (at top) vary with the particular phospholipid. **(b)** Orientation of phospholipids in water.

ent themselves in such a way that all polar (hydrophilic) portions will orient themselves toward the polar water molecules, with which they then form hydrogen bonds (Figure 2-13b). (*Hydrophilic* means water-loving.) Polar portions consist of a phosphate group and glycerol. In contrast to the polar regions, all nonpolar (hydrophobic) parts of the phospholipid make contact only with the nonpolar portions of neighboring molecules. (*Hydrophobic* means water-fearing.) Nonpolar portions consist of fatty acids. This characteristic behavior makes phospholipids particularly suitable to be a major component of the membranes that enclose cells. Phospholipids enable the membrane to act as a barrier that separates the contents of the cell from the water-based environment in which it lives.

Some complex lipids are useful in identifying certain bacteria. For example, the cell wall of *Mycobacterium tuberculosis* (mī-kō-bak-ti′rē-um tü-ber-kū-lō′sis), the bacterium that causes tuberculosis, is distinguished by its lipid-rich content. The cell wall contains complex lipids such as waxes and glycolipids (lipids with carbohydrates attached) that give the bacterium distinctive staining characteristics. Cell walls rich in such complex lipids are characteristic of all members of the genus *Mycobacterium*.

Steroids

Steroids are structurally very different from the lipids described previously. Figure 2-14 shows the structure of the steroid cholesterol with the characteristic ring structure of steroids. The —OH group in cholesterol makes it a *sterol*. Sterols are important constituents of the plasma membranes of animal cells and of one group of bacteria (mycoplasmas), and they are also found in fungi and plants. Animals synthesize steroid hormones and vitamin D from sterols.

Figure 2-14 Cholesterol, a steroid. Note the four "fused" carbon rings, which are characteristic of steroid molecules. The hydrogen atoms attached to the carbons at the corners of the rings have been omitted.

MICROBIOLOGY
HIGHLIGHTS

Saturation and the Fluidity of Plasma Membranes

No cell could live without a plasma membrane. Plasma membranes hold the cell together, keep harmful substances out of the cell while allowing nutrients to enter it, and prevent the cytoplasm from leaking while allowing waste products to leave the cell. Thus, plasma membranes must be selectively permeable, allowing only certain substances to pass through them. Moreover, plasma membranes must enclose the cell completely and must be self-sealing so that breaks and tears in the membrane will heal themselves without a great expenditure of energy.

Membranes consisting largely of phospholipid molecules meet all these requirements. The phosphate head of each molecule is hydrophilic (attracted to water), and the fatty-acid tail is hydrophobic (repelled by water). Since the fatty-acid tails cling together, phospholipids in the presence of water form a self-sealing phospholipid bilayer, a double layer of phospholipid molecules with the phosphate heads on the outer surfaces of the two layers and the tails facing the inside space between the layers (see figure).

For this bilayer to support the cell and to maintain its selective permeability, the phospholipids must maintain the same viscosity, whatever the temperature of the surroundings may be. The membrane must be about as viscous as olive oil, without getting too fluid when warmed or too thick when cooled. The phospholipid molecules of the membrane, as well as a variety of protein molecules embedded in the membrane, must be able to move about within the membrane freely enough to allow selective permeability but not so freely as to destroy the structure of the membrane.

As everyone who has ever cooked a meal knows, animal fats are usually solid at room temperature, whereas vegetable oils are usually liquid at room temperature. The difference in their respective melting points is caused, among other factors, by the degrees of *saturation* of the fatty-acid chains that comprise the tails of the lipid molecules. The fatty acid is said to be saturated when it has no double bonds — that is, when every carbon atom in the chain has all the hydrogen atoms it can

Proteins

Proteins are organic molecules that contain carbon, hydrogen, oxygen, and nitrogen. Some also contain sulfur. If you were to separate and weigh all the groups of organic compounds in a living cell, the proteins would tip the scale. Hundreds of different proteins can be found in any single cell, and together they make up 50% or more of the cell's dry weight. They are essential ingredients in all aspects of cell structure and function. We have already mentioned enzymes, the proteins that catalyze biochemical reactions. But proteins have other vital functions as well. *Carrier proteins* help to transport certain chemicals into and out of cells. Other proteins, such as the *bacteriocins* produced by many bacteria, kill other bacteria. Toxins produced by certain disease-causing microorganisms are also proteins.

Some proteins play a key role in cell contraction and locomotion. Others, such as the hormones of certain organisms, have regulatory functions. Then there are proteins that play a key role in vertebrate immune systems—antibodies, for instance, are proteins. And some proteins are simply structural, such as the collagen of animal skin and bones.

Amino Acids

Just as the monosaccharides are the building blocks of larger carbohydrate molecules, and fatty acids and glycerol are the building blocks of fats, **amino acids** are the building blocks of proteins. Amino acids contain at least one carboxyl (—COOH) group and one amino (—NH$_2$) group attached to the same carbon atom, called an alpha-carbon (α-carbon) (Figure 2-15). Such amino acids are called *alpha-amino acids*. Also attached to the alpha-carbon is a side group (R group), which is the amino acid's distinguishing factor. The side group can be a hydrogen atom, an unbranched or branched chain of atoms (Figure 2-15b), or a ring structure that is cyclic (all carbon) or heterocyclic (when an atom other than carbon is included in the ring). Figure 2-15b illustrates the structural formula of a branched chain side group—in this case, the amino acid valine. The side group can contain functional groups, such as the sulfhydryl group (—SH), the hydroxy group (—OH), or additional carboxyl or amino groups. These side groups and the carboxyl and alpha-amino groups affect the total structure of a protein, as will be described later. The structures and standard abbreviations of the 20 amino acids found in proteins are shown in Table 2-5, p. 46.

hold. The saturated chains become solid more easily because they are relatively straight (figure, part a) and are thus able to pack together more closely than unsaturated chains. The double bonds of unsaturated chains create kinks in the chain, which keep the chains apart from one

another (figure, part b). Therefore, to maintain the plasma membrane's fluidity, the proper spacing must be maintained between the fatty-acid chains by interspersing unsaturated chains throughout the membrane. In many organisms (including one type of bacterium, Myco-

plasma, which has no cell walls), sterols are dispersed among the lipids. Sterols are alcohols composed of four hydrocarbon rings attached to a hydrocarbon chain. The sterols separate the fatty-acid chains and thus prevent the packing that would harden the plasma membrane at low

temperatures (figure, part c).

In some bacteria, such as *Escherichia coli*, high temperatures stimulate the cells to produce phospholipids that have more saturated fatty acids. The higher degree of saturation prevents the plasma membrane from melting.

(a) Saturated fatty acids (closely packed)

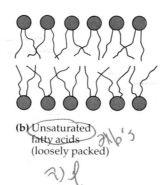

(b) Unsaturated fatty acids (loosely packed)

(c) Sterol molecules separate fatty acid chains

Phospholipid bilayer

Figure 2-15 Amino acids.
(a) General structural formula for an amino acid. The alpha-carbon (α-carbon) is shown in color. The letter R can stand for any of a number of groups of atoms. Different amino acids have different R groups, also called side groups. **(b)** Structural formula for the amino acid valine. Note the side group.

(a)

(b)

Amino acids exist in either of two configurations called *stereoisomers*, designated by D and L. These two configurations are mirror images, corresponding to "right-handed" (D) and "left-handed" (L) three-dimensional shapes (Figure 2-16). The amino acids found in proteins are always the L-isomers (except for glycine,

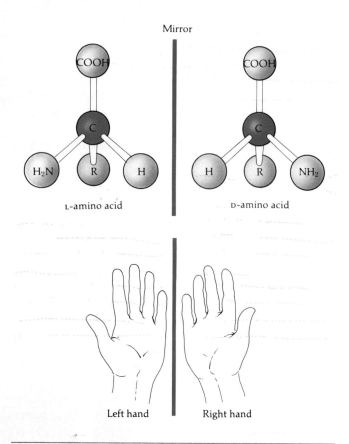

Figure 2-16 The L- and D-isomers of an amino acid, shown with ball-and-stick models. The two isomers, like left and right hands, are mirror images of each other and cannot be superimposed on one another. (Try it!)

the simplest amino acid, which does not have stereoisomers). However, D-amino acids occasionally occur in nature—for example, in certain bacterial cell walls and antibiotics. (Many other kinds of organic molecules also can exist in D and L forms. One example is the sugar glucose, which occurs in nature as D-glucose.)

Although only 20 kinds of amino acids occur naturally in proteins, the total number of amino acids in a single protein molecule can vary from 50 to hundreds or more, and they can be arranged in an infinite number of ways. Proteins have different lengths, different quantities of the various amino acid units, and different specific sequences in which the amino acids are bonded. The number of proteins is practically endless, and every living cell produces many different proteins.

Peptide Bonds

Amino acids bond between the carboxyl (—COOH) group of one amino acid and the amino (—NH$_2$) group of another. For the formation of every bond between two amino acids, one water molecule is released (dehydration). The bonds between amino acids are called **peptide bonds**.

In the example in Figure 2-17, a peptide bond between two amino acids is formed by dehydration synthesis. The carboxyl group of one amino acid supplies an OH$^-$, and the amino group of the other releases an H$^+$ for the formation of water. The peptide bond is between the carbon atom of the carboxyl group of one amino acid and the nitrogen atom of the amino group of another amino acid. The resulting compound is called a *dipeptide* because it consists of two amino acids joined by a peptide bond. Adding another amino acid to a dipeptide would form a *tripeptide*. Further additions of amino acids would produce a long, chain-like molecule called a *polypeptide*, a protein molecule.

Glycine   Alanine   Dipeptide   Water

Figure 2-17 Peptide bond formation by dehydration synthesis. The amino acids glycine and alanine combine to form a dipeptide. The newly formed bond between the nitrogen atom of glycine and the carbon atom of alanine is called a peptide bond.

Levels of Protein Structure

Proteins vary tremendously in structure. Different proteins have different architectures and different three-dimensional shapes. This variation in structure is directly related to their diverse functions.

When a cell makes a protein, the polypeptide chain folds spontaneously to assume a certain shape. One reason for folding of the polypeptide is that some parts of a protein are attracted to water, and other parts are repelled by it. In practically every case, the function of a protein depends on its ability to recognize and bind to some other molecule. As an example, an enzyme binds specifically with its substrate. A hormonal protein binds to a receptor on a cell whose function it will alter. An antibody binds to a foreign substance (antigen) that has invaded the body. The unique shape of a protein permits it to interact with specific other molecules in order to carry out specific functions.

Proteins are described in terms of four levels of organization—primary, secondary, tertiary, and quaternary. The *primary structure* is the unique order (the sequence) in which the amino acids are linked together (Figure 2-18a, p. 48). This sequence is genetically determined. Alterations in sequence can have profound metabolic effects. For example, a single incorrect amino acid in a blood protein can produce the deformed hemoglobin molecule characteristic of sickle-cell anemia. But proteins do not exist as long straight chains. Each polypeptide chain folds and coils in specific ways into a relatively compact structure with a characteristic three-dimensional shape.

A protein's *secondary structure* is the localized, repetitious twisting or folding of the polypeptide chain. This aspect of a protein's shape comes from hydrogen bonds joining the atoms of peptide bonds at different locations along the polypeptide chain. The two types of secondary protein structures are clockwise spirals called helixes and pleated sheets, which form from roughly parallel portions of the chain (Figure 2-18b). Both structures are held together by hydrogen bonds between oxygen and nitrogen atoms that are part of the polypeptide's backbone.

Tertiary structure refers to the overall three-dimensional structure of a polypeptide chain (Figure 2-18c). The folding is not repetitive or predictable, as in secondary structure. Whereas secondary structure involves hydrogen bonding between atoms of the amino and carbonyl groups involved in the peptide bonds, tertiary structure involves interactions between various amino-acid side groups in the polypeptide chain. Thus, the tertiary structure of a protein is elaborately irregular. Hydrogen bonds and other relatively weak interactions play an important role in tertiary structure. In addition, sulfhydryl groups (—SH) on two amino acid subunits can form a covalent, disulfide link (—S—S—) by removal of the hydrogen atoms. This is an oxidation reaction.

Some proteins have a *quaternary structure*, which consists of an aggregation of two or more individual polypeptide chains (subunits) that operate as a single functional unit. Figure 2-18d shows a hypothetical protein consisting of two identical polypeptide chains. Ordinarily, single polypeptide chains of such a protein do not exhibit activity, but they become active when in groups of two. Certain other proteins with quaternary structure have two or more kinds of polypeptide subunits. The bonds that hold a quaternary structure together are basically the same as those that maintain tertiary structure. The overall shape of a protein may be globular (compact and roughly spherical, as in the figure) or fibrous (threadlike).

If a protein encounters a hostile environment in terms of temperature, pH, or salt concentrations, it may unravel and lose its characteristic shape. This

TABLE 2-5   *Twenty Amino Acids Found in Proteins*

Amino Acid (Abbreviation)	Structural Formula	Characteristic of R Group
Glycine (Gly)	$H-\overset{\overset{\displaystyle H}{\mid}}{\underset{\underset{\displaystyle NH_2}{\mid}}{C}}-COOH$	Hydrogen atom
Alanine (Ala)	$CH_3-\overset{\overset{\displaystyle H}{\mid}}{\underset{\underset{\displaystyle NH_2}{\mid}}{C}}-COOH$	Unbranched chain
Valine (Val)	$H_3C-\overset{\displaystyle CH}{\underset{\underset{\displaystyle CH_3}{\mid}}{\mid}}-\overset{\overset{\displaystyle H}{\mid}}{\underset{\underset{\displaystyle NH_2}{\mid}}{C}}-COOH$	Branched chain
Leucine (Leu)	$H_3C-\overset{\displaystyle CH}{\underset{\underset{\displaystyle CH_3}{\mid}}{\mid}}-CH_2-\overset{\overset{\displaystyle H}{\mid}}{\underset{\underset{\displaystyle NH_2}{\mid}}{C}}-COOH$	Branched chain
Isoleucine (Ileu)	$H_3C-CH_2-\overset{\displaystyle CH}{\underset{\underset{\displaystyle CH_3}{\mid}}{\mid}}-\overset{\overset{\displaystyle H}{\mid}}{\underset{\underset{\displaystyle NH_2}{\mid}}{C}}-COOH$	Branched chain
Serine (Ser)	$HO-CH_2-\overset{\overset{\displaystyle H}{\mid}}{\underset{\underset{\displaystyle NH_2}{\mid}}{C}}-COOH$	Hydroxy (—OH) group
Threonine (Thr)	$H_3C-\overset{\displaystyle CH_2}{\underset{\underset{\displaystyle OH}{\mid}}{\mid}}-\overset{\overset{\displaystyle H}{\mid}}{\underset{\underset{\displaystyle NH_2}{\mid}}{C}}-COOH$	Hydroxy (—OH) group
Cysteine (Cys)	$HS-CH_2-\overset{\overset{\displaystyle H}{\mid}}{\underset{\underset{\displaystyle NH_2}{\mid}}{C}}-COOH$	Sulfhydryl (—SH) group
Methionine (Met)	$H_3C-S-CH_2-CH_2-\overset{\overset{\displaystyle H}{\mid}}{\underset{\underset{\displaystyle NH_2}{\mid}}{C}}-COOH$	Sulfhydryl (—SH) group
Glutamic acid (Glu)	$HOOC-CH_2-CH_2-\overset{\overset{\displaystyle H}{\mid}}{\underset{\underset{\displaystyle NH_2}{\mid}}{C}}-COOH$	Additional carboxyl (—COOH) group, acidic

TABLE 2-5   (*Continued*)

Amino Acid (Abbreviation)	Structural Formula	Characteristic of R Group
Aspartic acid (Asp)	$HOOC-CH_2-\overset{\overset{H}{\mid}}{\underset{\underset{NH_2}{\mid}}{C}}-COOH$	Additional carboxyl (—COOH) group, acidic
Lysine (Lys)	$H_2N-CH_2-CH_2-CH_2-CH_2-\overset{\overset{H}{\mid}}{\underset{\underset{NH_2}{\mid}}{C}}-COOH$	Additional amino (—NH₂) group, basic
Arginine (Arg)	$H_2N-\overset{}{\underset{\underset{NH}{\parallel}}{C}}-NH-CH_2-CH_2-CH_2-\overset{\overset{H}{\mid}}{\underset{\underset{NH_2}{\mid}}{C}}-COOH$	Additional amino (—NH₂) group, basic
Asparagine (Asn)	$H_2N-\overset{}{\underset{\underset{O}{\parallel}}{C}}-CH_2-\overset{\overset{H}{\mid}}{\underset{\underset{NH_2}{\mid}}{C}}-COOH$	Additional amino (—NH₂) group, basic
Glutamine (Gln)	$H_2N-\overset{}{\underset{\underset{O}{\parallel}}{C}}-CH_2-CH_2-\overset{\overset{H}{\mid}}{\underset{\underset{NH_2}{\mid}}{C}}-COOH$	Additional amino (—NH₂) group, basic
Phenylalanine (Phe)	C₆H₅—$CH_2-\overset{\overset{H}{\mid}}{\underset{\underset{NH_2}{\mid}}{C}}-COOH$	Cyclic
Tyrosine (Tyr)	$HO-$C₆H₄$-CH_2-\overset{\overset{H}{\mid}}{\underset{\underset{NH_2}{\mid}}{C}}-COOH$	Cyclic
Histidine (His)	imidazole ring$-CH_2-\overset{\overset{H}{\mid}}{\underset{\underset{NH_2}{\mid}}{C}}-COOH$	Heterocyclic
Tryptophan (Trp)	indole ring$-CH_2-\overset{\overset{H}{\mid}}{\underset{\underset{NH_2}{\mid}}{C}}-COOH$	Heterocyclic
Proline (Pro)	pyrrolidine ring$-COOH$	Heterocyclic

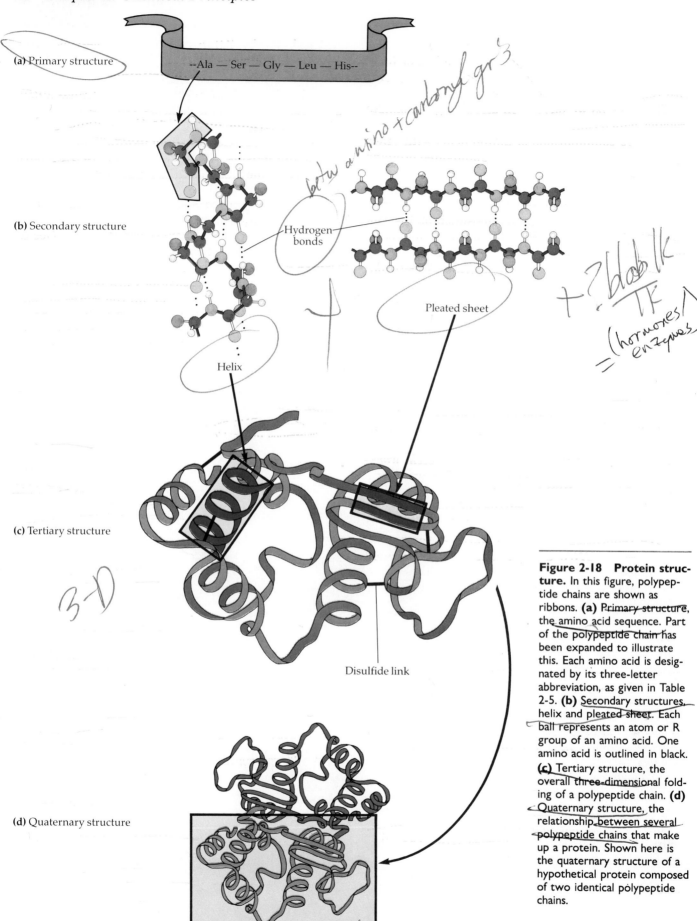

(a) Primary structure

--Ala — Ser — Gly — Leu — His--

(b) Secondary structure

Hydrogen bonds

Pleated sheet

Helix

(c) Tertiary structure

Disulfide link

(d) Quaternary structure

Figure 2-18   Protein structure. In this figure, polypeptide chains are shown as ribbons. **(a)** Primary structure, the amino acid sequence. Part of the polypeptide chain has been expanded to illustrate this. Each amino acid is designated by its three-letter abbreviation, as given in Table 2-5. **(b)** Secondary structures, helix and pleated sheet. Each ball represents an atom or R group of an amino acid. One amino acid is outlined in black. **(c)** Tertiary structure, the overall three-dimensional folding of a polypeptide chain. **(d)** Quaternary structure, the relationship between several polypeptide chains that make up a protein. Shown here is the quaternary structure of a hypothetical protein composed of two identical polypeptide chains.

process is called **denaturation** (see Figure 5-5). As a result of denaturation, the protein is no longer functional. This is discussed in more detail in Chapter 5 with regard to denaturation of enzymes.

The proteins we have been discussing are *simple proteins*, which contain only amino acids. *Conjugated proteins* are combinations of amino acids with other organic or inorganic components. Conjugated proteins are named by their non-amino-acid component. Thus, glycoproteins contain sugars, nucleoproteins contain nucleic acids, metalloproteins contain metal atoms, lipoproteins contain lipids, and phosphoproteins contain phosphate groups. An example of a microbial phosphoprotein is phospholipase C. Synthesized by the bacterium *Clostridium perfringens* (klôs-tri′dē-um pėr-frin′jens), this enzyme breaks down red blood cells and induces some of the symptoms of gas gangrene.

Nucleic Acids

In 1943, three American microbiologists, Oswald Avery, Colin MacLeod, and Maclyn McCarty, discovered that a substance called **deoxyribonucleic acid (DNA)** is the substance of which genes are made. Ten years later, James Watson and Francis Crick, working with molecular models and X-ray information supplied by Maurice Wilkins and Rosalind Franklin, determined the physical structure of DNA. In addition, Watson and Crick also suggested a mechanism for DNA replication and how it works as the hereditary material. DNA, and another substance called **ribonucleic acid (RNA)**, are together referred to as **nucleic acids** because they were first discovered in the nuclei of cells.

DNA

According to the model proposed by Watson and Crick, a DNA molecule consists of two long strands wrapped around each other to form a **double helix** (Figure 2-19). The double helix looks like a twisted ladder. Just as amino acids are the structural units of proteins, so **nucleotides** are the structural units of nucleic acids. Each strand of the DNA double helix is composed of many nucleotides.

Each nucleotide of DNA is composed of three parts—a nitrogen-containing base, a pentose (five-carbon) sugar called deoxyribose, and a phosphate group (phosphoric acid); (see Figure 2-19). The nitrogen-containing bases are cyclic compounds made up of carbon, hydrogen, oxygen, and nitrogen atoms. The bases are named adenine (A), thymine (T), cytosine

(C), and guanine (G). A and G are double-ring structures called **purines**, whereas T and C are single-ring structures referred to as **pyrimidines**.

Nucleotides are named according to their nitrogenous base. Thus, a nucleotide containing thymine is called a **thymine nucleotide**. One containing adenine is called an **adenine nucleotide**, and so on. The term **nucleoside** refers to the combination of a purine or pyrimidine plus a pentose sugar; it does not contain a phosphate group.

Each strand of DNA composing the double helix has a "backbone" consisting of alternating sugar and phosphate groups. The deoxyribose of one nucleotide is joined to the phosphate group of the next. (Refer to Figure 8-4 to see how nucleotides are bonded.) The nitrogen-containing bases make up the rungs of the ladder. Note that the purine A is always paired with the pyrimidine T and that purine G is always paired with the pyrimidine C. The bases are held together by hydrogen bonds; A—T is held by two hydrogen bonds, and G—C is held by three.

Figure 2-19 **Structure of DNA.** Part of a DNA molecule illustrating the double helix.

S = Deoxyribose sugar

P = Phosphate group

A   Adenine

T   Thymine

G   Guanine

C   Cytosine

.... Hydrogen bond

The order in which the nitrogen-base pairs occur along the backbone is extremely specific and in fact contains the genetic instructions for the organism. A certain segment of a nucleotide chain constitutes a gene, and a single DNA molecule may contain thousands of genes. Genes determine all hereditary traits, and they control all the activities that take place within cells.

A very important consequence of nitrogen-containing base pairing is that, if the sequence of bases of one strand is known, then the sequence of the other strand is also known. For example, if one strand has the sequence . . . ATGC . . . , then the other strand has the sequence . . . TACG . . . . Since the sequence of bases of one strand is determined by the sequence of bases of the other, the bases are said to be **complementary**. The actual transfer of information becomes possible because of DNA's unique structure and will be discussed further in Chapter 8.

RNA

RNA, the second principal kind of nucleic acid, differs from DNA in several respects. Whereas DNA is double-stranded, RNA is usually single-stranded. The five-carbon sugar in the RNA nucleotide is **ribose**, which has one more oxygen atom than deoxyribose. Also, one of RNA's bases is uracil (U) instead of thymine. The other three bases (A, G, C) are the same as in DNA. Three major kinds of RNA have been identified in cells. These are referred to as **messenger RNA**, **ribosomal RNA**, and **transfer RNA**. As we shall see in Chapter 8, each type of RNA has a specific role in protein synthesis.

Adenosine Triphosphate (ATP)

Adenosine triphosphate (ATP) is the principal energy-carrying molecule of all cells and is indispensable to the life of the cell. It stores the chemical energy released by some chemical reactions, and it provides the energy for reactions that require energy. ATP consists of an adenosine unit, composed of adenine and ribose, and three phosphate groups (Figure 2-20). In other words, it is an adenine nucleotide (also called adenosine monophosphate, or AMP) with two extra phosphate groups. ATP is called a high-energy molecule because it releases a large amount of usable energy when it loses its terminal phosphate group (abbreviated Ⓟ), to become **adenosine diphosphate (ADP)**. This reaction can be represented as follows:

$$ATP \rightleftharpoons ADP + Ⓟ + Energy$$

| Adenosine triphosphate | Adenosine diphosphate | Inorganic phosphate |

Because the supply of ATP at any particular time is limited, a mechanism exists to replenish it. The addition of a phosphate group to ADP manufactures more ATP. Because energy is required to manufacture ATP, the reaction can be represented as follows:

$$ADP + Ⓟ + Energy \rightleftharpoons ATP$$

| Adenosine diphosphate | Phosphate | | Adenosine triphosphate |

The energy required to attach the terminal phosphate group to ADP is supplied by the cell's various decomposition reactions, particularly the decomposition of glucose. ATP can be stored in every cell, where its potential energy is not released until needed.

Figure 2-20   Structure of ATP. High-energy phosphate bonds are indicated by wavy lines. When ATP breaks down to ADP and inorganic phosphate, a large amount of chemical energy is released for use in other chemical reactions.

Study Outline

Introduction (p. 24)

1. The interaction of atoms and molecules is called chemistry.
2. The metabolic activities of microorganisms involve complex chemical reactions.
3. Nutrients are broken down by microbes to obtain energy and to make new cells.

Structure of Atoms (pp. 25–26)

1. Atoms are the smallest units of chemical elements that enter into chemical reactions.
2. Atoms consist of a nucleus, which contains protons and neutrons, and electrons that move around the nucleus.
3. The atomic number is the number of protons in the nucleus; the total number of protons and neutrons is the atomic weight.

Chemical Elements (pp. 25–26)

1. Atoms with the same atomic number and same chemical behavior are classified as the same chemical element.
2. Chemical elements are designated by letter abbreviations called chemical symbols.
3. There are about 26 elements commonly found in living cells.
4. Atoms that have the same atomic number (are of the same element) but different atomic weights are called isotopes.

Electronic Configurations (p. 26)

1. In an atom, electrons are arranged around the nucleus in electron shells.
2. Each shell can hold a characteristic maximum number of electrons.
3. Chemical properties of an atom are largely due to the number of electrons in its outermost shell.

How Atoms Form Molecules: Chemical Bonds (pp. 26–31)

1. Molecules are made up of two or more atoms; molecules consisting of at least two different kinds of atoms are called compounds.
2. Atoms form molecules in order to fill their outermost electron shells.
3. Attractive forces that bind the atomic nuclei of two atoms together are called chemical bonds.
4. The combining capacity of an atom—the number of chemical bonds the atom can form with other atoms—is its valence.

Ionic Bonds (p. 28)

1. A positively or negatively charged atom or group of atoms is called an ion.
2. A chemical attraction between ions of opposite charge is called an ionic bond.
3. To form an ionic bond, one ion is an electron donor; the other ion is an electron acceptor.

Covalent Bonds (pp. 28–29)

1. In a covalent bond, atoms share pairs of electrons.
2. Covalent bonds are stronger than ionic bonds and are far more common in organisms.

Hydrogen Bonds (pp. 29–31)

1. A hydrogen bond exists when a hydrogen atom covalently bonded to one oxygen or nitrogen atom is attracted to another oxygen or nitrogen atom.
2. Hydrogen bonds form weak links between different molecules or between parts of the same large molecule.

Molecular Weight and Moles (p. 31)

1. The molecular weight is the sum of the atomic weights of all the atoms in a molecule.
2. A mole of an atom, ion, or molecule is equal to its atomic or molecular weight expressed in grams.
3. The number of moles of a substance equals its mass in grams divided by its molecular weight.

Chemical Reactions (pp. 31–34)

1. Chemical reactions are the making or breaking of chemical bonds between atoms.

Energy of Chemical Reactions (pp. 31–32)

1. A change of energy occurs during chemical reactions.
2. Endergonic reactions require energy; exergonic reactions release energy.
3. In a synthesis reaction, atoms, ions, or molecules are combined to form a larger molecule.
4. In a decomposition reaction, a larger molecule is broken down into its component molecules, ions, or atoms.
5. In an exchange reaction, two molecules are decomposed, and their subunits are used to synthesize two new molecules.
6. The products of reversible reactions can readily revert back to form the original reactants.

How Chemical Reactions Occur (pp. 32–34)

1. For a chemical reaction to take place, the reactants must collide with each other.

2. The minimum energy of collision that can produce a chemical reaction is called its activation energy.
3. Specialized proteins called enzymes accelerate chemical reactions in living systems by lowering the activation energy.

Important Biological Molecules   (pp. 34–50)

Inorganic Compounds   (pp. 34–37)

1. Inorganic compounds are usually small, ionically bonded molecules.
2. Water, and many common acids, bases, and salts are examples of inorganic compounds.

Water   (pp. 34–35)

1. Water is the most abundant substance in cells.
2. Because water is a polar molecule, it is an excellent solvent.
3. Water is a reactant in many of the decomposition reactions of digestion.
4. Water is an excellent temperature buffer.

Acids, Bases, and Salts   (p. 35)

1. An acid dissociates into H^+ ions and anions.
2. A base dissociates into OH^- ions and cations.
3. A salt dissociates into negative and positive ions, neither of which is H^+ or OH^-.

Acid-Base Balance   (p. 36)

1. The term pH refers to the concentration of H^+ in a solution.
2. A solution with a pH of 7 is neutral; a pH below 7 indicates acidity; a pH above 7 indicates alkalinity.
3. A pH buffer, which stabilizes the pH inside a cell, can be used in culture media.

Organic Compounds   (pp. 37–50)

1. Organic compounds always contain carbon and hydrogen.
2. Carbon atoms form four bonds with other atoms.
3. Organic compounds are mostly or entirely covalently bonded, and many of them are large molecules.

Functional Groups   (pp. 37–38)

1. A chain of carbon atoms forms a carbon skeleton
2. The letter *R* may be used to denote a particular group of atoms within an organic molecule.
3. Functional groups of atoms are responsible for most of the properties of organic molecules.
4. Frequently encountered classes of molecules are R—OH (alcohols), R—COOH (organic acids), H_2N—R—COOH (amino acids).

Macromolecules   (pp. 38–39)

1. Small organic molecules may combine into very large molecules called macromolecules.
2. Monomers usually bond together by dehydration synthesis or condensation reactions that form water and a polymer.

Carbohydrates   (pp. 39–40)

1. Carbohydrates are compounds consisting of atoms of carbon, hydrogen, and oxygen, with hydrogen and oxygen in a 2:1 ratio.
2. Carbohydrates include sugars and starches.
3. Carbohydrates can be divided into three types, monosaccharides, disaccharides, and polysaccharides.
4. Monosaccharides contain from three to seven carbon atoms.
5. Isomers have the same chemical formula and different structures and properties—for example, glucose ($C_6H_{12}O_6$) and fructose ($C_6H_{12}O_6$).
6. Monosaccharides may form disaccharides and polysaccharides by dehydration synthesis.
7. Polysaccharides and disaccharides may be broken down by hydrolysis, a reaction involving the splitting of water molecules.

Lipids   (pp. 40–43)

1. Lipids are a diverse group of compounds distinguished by their insolubility in water.
2. Simple lipids (fats) consist of a molecule of glycerol and three molecules of fatty acids.
3. A saturated fat has no double bonds between carbon atoms in the fatty acids; an unsaturated fat has one or more double bonds.
4. Phospholipids are complex lipids consisting of glycerol, two fatty acids, and phosphate.
5. Steroids have carbon-ring systems with functional hydroxyl and carbonyl groups.

Proteins   (pp. 43–49)

1. Amino acids are the building blocks of proteins.
2. Amino acids consist of carbon, hydrogen, oxygen, nitrogen, and sometimes sulfur.
3. Twenty amino acids occur naturally.
4. By linking amino acids, peptide bonds (formed by dehydration synthesis) allow the formation of polypeptide chains.
5. Proteins have four levels of structure—primary (sequence of amino acids), secondary (regular coils or pleats), tertiary (overall three-dimensional structure of a polypeptide), and quaternary (two or more polypeptide chains).
6. Conjugated proteins consist of amino acids combined with other organic or inorganic compounds.

Nucleic Acids   (pp. 49–50)

1. Nucleic acids—DNA and RNA—are macromolecules consisting of repeating nucleotides.
2. A nucleotide is composed of a pentose, a phosphate group, and a nitrogenous base. A nucleoside is composed of a pentose and a nitrogenous base.
3. A DNA nucleotide consists of deoxyribose (a pentose) and one of these nitrogenous bases: thymine or cytosine (pyrimidines); or adenine or guanine (purines).

4. DNA consists of two strands of nucleotides wound in a double helix. The strands are held together by hydrogen bonds between purine and pyrimidine nucleotides: A—T and G—C. ~~2~~ 4 b

5. An RNA nucleotide consists of ribose (a pentose) and one of these nitrogenous bases: cytosine, guanine, adenine, or uracil.

Adenosine Triphosphate (ATP) (p. 50)

1. ATP stores chemical energy for various cellular activities.

2. When the bond to ATP's terminal phosphate group is broken, energy is released.

3. The energy from decomposition reactions is used to regenerate ATP from ADP and phosphate.

Study Questions

Review

1. What is a chemical element?

2. Diagram the electronic configuration of a carbon atom.

3. How does $^{14}_{6}C$ differ from $^{12}_{6}C$?

4. What type of bonds will hold the following atoms together?
 (a) Li^+ and Cl^- ions in LiCl
 (b) Carbon and oxygen atoms in CO_2
 (c) Oxygen atoms in O_2
 (d) A hydrogen atom from glutamic acid and a nitrogen atom from lysine in:

$$NH_2$$
$$H-C-COOH$$
$$CH_2$$
$$NH_2 \qquad CH_2$$
$$H-C-COOH \qquad CH_2$$
$$CH_2 \qquad CH_2$$
$$CH_2 \qquad NH_2$$
$$O \diagdown C \diagup OH \cdots\cdots\cdots\cdots$$

Glutamic acid   Lysine

 (e) A hydrogen atom of one nucleotide to nitrogen or oxygen atoms of another nucleotide in:

Phosphate-Deoxyribose   Guanine   Cytosine   Deoxyribose-Phosphate

5. What type of bonding exists between water molecules?

6. Classify the following inorganic molecules as an acid, base, or salt. Their dissociation products are shown to help you.
 (a) $HNO_3 \rightarrow H^+ + NO_3^-$
 (b) $H_2SO_4 \rightarrow 2H^+ + SO_4^{2-}$
 (c) $NaOH \rightarrow Na^+ + OH^-$
 (d) $Mg_2SO_4 \rightarrow 2Mg^{2+} + SO_4^{2-}$

7. Vinegar, pH 3, is how many times more acidic than pure water?

8. Calculate the molecular weight of $C_6H_{12}O_6$.

9. How many moles are in 360 grams of glucose?

10. Classify the following types of chemical reactions.
 (a) Glucose + Fructose $\rightarrow$ Sucrose + H_2O
 (b) Lactose $\rightarrow$ Glucose + Galactose
 (c) $NH_4Cl + H_2O \rightarrow NH_4OH + HCl$
 (d) ATP $\rightarrow$ ADP + Ⓟ

11. The following reaction requires an enzyme. What purpose does the enzyme serve in this reaction? spd up Brk dwn w/H₂O (hydrolysis)

Lactose $\longrightarrow$ Glucose + Galactose

12. Classify the following as subunits of either a carbohydrate, lipid, protein, or nucleic acid.
 (a) $CH_3-(CH_2)_7-CH=CH-(CH_2)_7-COOH$ = lipid unsatur fat A
  Oleic acid
 (b)
$$NH_2$$
$$H-C-COOH$$
$$CH_2$$
$$OH$$
  Serine   protein amino A
 (c) $C_6H_{12}O_6$ carbo (ix sugar)
 (d) Thymine nucleotide Nu cl A

13. Add the appropriate functional group(s) to this ethyl group to produce each of the following compounds: ethyl alcohol, acetic acid, acetaldehyde, ethanolamine, diethylether.

$$H \quad H$$
$$H-C-C-$$
$$H \quad H$$

14. Water plays an important role in these reactions:

$$HO-\overset{\overset{\displaystyle H}{|}}{\underset{\underset{\displaystyle O}{||}}{C}}-\overset{\overset{\displaystyle H}{|}}{\underset{\underset{\displaystyle H}{|}}{C}}-\overset{\overset{\displaystyle H}{|}}{N}-\overset{\overset{\displaystyle H}{|}}{\underset{\underset{\displaystyle O}{||}}{C}}-\overset{\overset{\displaystyle H}{|}}{\underset{\underset{\displaystyle H}{|}}{C}}-\overset{\overset{\displaystyle H}{|}}{N}-H + H_2O \rightleftharpoons$$

$$HO-\overset{\overset{\displaystyle H}{|}}{\underset{\underset{\displaystyle O}{||}}{C}}-\overset{\overset{\displaystyle H}{|}}{\underset{\underset{\displaystyle H}{|}}{C}}-\overset{\overset{\displaystyle H}{|}}{N}-H + HO-\overset{\overset{\displaystyle H}{|}}{\underset{\underset{\displaystyle O}{||}}{C}}-\overset{\overset{\displaystyle H}{|}}{\underset{\underset{\displaystyle H}{|}}{C}}-\overset{\overset{\displaystyle H}{|}}{N}-H$$

(a) What direction is the hydrolysis reaction (left to right or right to left)?
(b) What direction is the dehydration synthesis reaction?
(c) Circle the atoms involved in the formation of water.
(d) Identify the peptide bond.

15. This diagram shows a protein molecule. Indicate the regions of primary, secondary, and tertiary structure. Does this protein have quaternary structure?

16. Draw a simple lipid, and show how it could be modified to a phospholipid.

17. ATP is an energy-storage compound. Where does it get this energy from?

18. The energy-carrying property of the ATP molecule is due to energy dynamics that favor the breaking of bonds between _____ . What type of bonds are these? _____ .

Challenge

1. When you blow bubbles into a glass of water, the following reactions take place:

$$H_2O + CO_2 \xrightarrow{A} H_2CO_3 \xrightarrow{B} H^+ + HCO_3^-$$

(a) What type of reaction is *A*?
(b) What does reaction *B* tell you about the type of molecule H_2CO_3 is?

2. What are the common structural characteristics of ATP and DNA molecules?

3. *Thiobacillus ferrooxidans* was responsible for destroying buildings in the Midwest by causing changes in the earth. The original rock, which contained lime ($CaCO_3$) and pyrite (FeS_2), expanded as the bacteria's metabolism caused gypsum ($CaSO_4$) crystals to form. How did *Thiobacillus* bring about the change from lime to gypsum?

Further Reading

Dickerson, R. E., et al. *Chemical Principles*, 4th ed. Menlo Park, Calif.: Benjamin/Cummings, 1984. An introductory college chemistry textbook with a chapter on the special role of carbon.

Hakomori, S. "Glycosphingolipids." *Scientific American* 254(5):44–53, May 1986. Discusses lipid structure and the role of lipids in cancer.

Howard, F. M. *The Vital Force: A Study of Bioenergetics*. New York: W. H. Freeman, 1986. Chapter 4 covers energy production in bacteria, including photosynthesis, respiration, and translocation. Chapter 5 describes energy utilization such as movement and transport. Chapter 6 uses metabolic examples to illustrate evolution of eucaryotic cells.

Karplus, M., and J. A. McCammon. "The dynamics of proteins." *Scientific American* 254(4):42–51, April 1986. Relates the constant motion of proteins to their biological functions using X-ray crystallography.

Lechtman, M. D., et al. *The Games Cells Play: Basic Concepts of Cellular Metabolism*. Menlo Park, Calif.: Benjamin/Cummings, 1979. An innovative guide to the basics of chemistry for biology students; contains chapter self-tests.

"Molecules of Life." *Scientific American* 253(4); October 1985. Entire issue devoted to biochemistry; includes articles on RNA, DNA, and proteins.

Sharon, N. "Carbohydrates." *Scientific American* 243(5):90–116, November 1980. Illustrates new information on the functions of carbohydrates other than as carbon and energy sources.

Stryer, L. *Biochemistry*, 3rd ed. New York: W. H. Freeman, 1988. A technical survey of biochemistry, with a cellular approach.

Watson, J. D. *The Double Helix*, ed. Gunther S. Stent. New York: W. W. Norton, 1980. Original papers and Watson's account of the struggle to decipher the structure of DNA give insight into creative scientific processes. Includes commentaries and reviews by other scientists.

CHAPTER 3

Observing Microorganisms Through a Microscope

Microorganisms are too small to be seen with the naked eye, so they must be observed with a microscope (*micro* means small, *skopein* means to see). In Anton van Leeuwenhoek's day, looking through a microscope meant looking through rainbow rings and shadows and multiple images. Modern microbiologists, however, have access to microscopes that produce, with great clarity, magnifications from ten to thousands of times greater than those of van Leeuwenhoek's simple single lens (see Figure 1-2a). This chapter will describe how different types of microscopes function and the advantages of each type.

Some microbes are more visible than others. Many have to be escorted through several staining procedures before cell walls, membranes, and structures lose their opaque or colorless natural state. The last part of this chapter will explain the methods of preparing specimens for light microscopy.

You may be wondering how we are going to sort out, measure, and count the specimens we will be studying. We open this chapter with a short discussion of how to use the metric system for measuring microbes.

Learning Objectives

- List the units of measurement used for microorganisms, and know their equivalents.
- Diagram the path of light through a compound microscope.
- Define resolution and total magnification.
- Cite an advantage of darkfield, phase-contrast, and fluorescence microscopy, and compare each with brightfield illumination.
- Explain how electron microscopy differs from light microscopy.
- Differentiate between an acidic dye and a basic dye.
- Compare simple, differential, and special stains.
- List the steps in preparing a Gram stain, and describe the appearance of a gram-positive and gram-negative cell after each step.

Units of Measurement

Because microorganisms and their component parts are so very small, they are measured in units that are unfamiliar to many of us in everyday life. When measuring microorganisms, we use the metric system. The standard unit of length in the metric system is the meter (m). A major advantage of the metric system is that the units are related to each other by factors of 10. Thus, 1 m equals 10 decimeters (dm) or 100 centimeters (cm) or 1000 millimeters (mm). Units in the U.S. system of measure do not have the advantage of conversion by a factor of 10. For example, we have to use 3 feet or 36 inches to equal 1 yard.

Microorganisms and their structural components are measured in even smaller units, such as micrometers, nanometers, and angstroms. A **micrometer (μm)**, formerly known as a micron (μ), is equal to 0.000001 m. The prefix *micro* indicates that the unit following it should be divided by one million, or 10^6 (see Exponential Notation in Appendix G). A **nanometer (nm)**, formerly known as a millimicron (mμ), is equal to 0.000000001 m. The prefix *nano* tells us that the unit after it should be divided by one billion (10^9). An **angstrom (Å)** is equal to 0.0000000001 m, or 10^{-10} m. The angstrom is no longer an official unit of measure; however, because of its widespread presence in scientific literature, you should be familiar with it. Its accepted equivalent is 0.1 nm. Table 3-1 presents the basic metric units of length and some of their U.S. equivalents. In Table 3-1, you can compare the microscopic units of measurement with the commonly known macroscopic units of measurement, such as centimeters, meters, and kilometers.

Microscopy: The Instruments

The simple microscope used by Anton van Leeuwenhoek in the seventeenth century had only one lens and was similar to a magnifying glass. Contemporaries of van Leeuwenhoek, such as Robert Hooke, built microscopes with multiple lenses, called compound microscopes. In fact, history credits a Dutch spectacle-maker, Zaccharias Janssen, with making the first compound microscope around 1600. However, these early compound microscopes were of poor quality. It was not until about 1830 that a significantly better microscope was developed, by Joseph Jackson Lister (the father of Joseph Lister). From Lister's microscope, various improvements were made that resulted in the development of the modern compound microscope, the kind used in microbiology laboratories today.

Compound Light Microscopy

A modern **compound light microscope** has two sets of lenses, objective and ocular, and uses visible light as its source of illumination (Figure 3-1a). Using the compound light microscope, we can examine very small specimens as well as some of their fine detail, or *ultrastructure.* A series of finely ground lenses forms a clearly focused image that is many times larger than the specimen itself. This magnification is achieved when light rays from an *illuminator*, the light source, are passed through a *condenser*, which directs the light rays through the specimen. From here, light rays pass into the *objective lens*, the lens closest to the specimen. The image of the specimen is magnified again by the *ocular lens*, or *eyepiece* (Figure 3-1b).

TABLE 3-1   *Metric Units of Length and U.S. Equivalents*

Metric Unit	Meaning of Prefix	Metric Equivalent	U.S. Equivalent
1 kilometer (km)	*kilo* = 1000	1000 m = 10^3	3280.84 ft or 0.62 mi; 1 mi = 1.61 km
1 meter (m)		Standard unit of length	39.37 inches or 3.28 ft or 1.09 yd
1 decimeter (dm)	*deci* = 1/10	0.1 m = 10^{-1} m	3.94 inches
1 centimeter (cm)	*centi* = 1/100	0.01 m = 10^{-2} m	0.394 inch; 1 inch = 2.54 cm
1 millimeter (mm)	*milli* = 1/1000	0.001 m = 10^{-3} m	
1 micrometer (μm) (formerly micron, μ)	*micro* = 1/1,000,000	0.000001 m = 10^{-6} m	
1 nanometer (nm) (formerly millimicron, mμ)	*nano* = 1/1,000,000,000	0.000000001 m = 10^{-9} m	

Figure 3-1 The compound light microscope. (a) Principal parts and their functions. **(b)** The path of light. **(c)** Refraction. Because the refractive indexes of the glass microscope slide and immersion oil are the same, the light rays do not refract when passing from one to the other.

Ocular (eyepiece) Remagnifies the image formed by the objective lens

Body tube Transmits the image from the objective lens to the ocular

Objective lenses Primary lenses that magnify the specimen

Stage Holds the microscope slide in position

Diaphram Controls the amount of light entering the condenser

Light Source of illumination

Base

Arm

Condenser Focuses light through specimen

Coarse adjustment knob

Fine adjustment knob

(a)

Line of vision

Ocular lens

Path of light

Prism

Body tube

Objective lens

Specimen

Condenser

Illuminator

Base with source of illumination

(b)

Objective

Immersion oil

Air

Unrefracted light

Refracted light without immersion oil

Glass slide

Condenser

Iris diaphragm

Light source

(c)

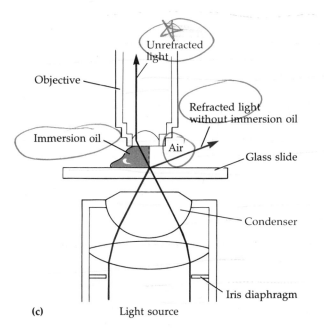

We calculate the total magnification of a specimen by multiplying the objective lens magnification (power) by the ocular lens magnification (power). Most microscopes used in microbiology have several objective lenses, including 10× (low power), 40× (high power), and 100× (oil immersion). Most oculars magnify specimens by 10×. Multiplying the magnification of a specific objective lens with that of the ocular, we see that the total magnifications would be 100× for low power, 400× for high power, and 1000× for oil immersion. Some compound light microscopes can achieve a magnification of 2000× with the oil immersion lens.

A general principle of microscopy is, the shorter the wavelength of light used in the instrument, the greater the resolution. **Resolution**, or **resolving power**, is the ability of the lenses to distinguish fine detail and structure. Specifically, it refers to the ability of the lenses to distinguish between two points at a specified distance apart. For example, if a microscope has a resolving power of 0.4 nm, it is capable of distinguishing two points as separate objects if they are at least 0.4 nm apart. The white light used in a compound light microscope has a relatively long wavelength and cannot resolve structures smaller than 0.2 μm. This fact and other practical considerations limit the magnification achieved by even the best compound light microscopes to about 2000×.

Figure 3-2 shows how the resolution of the human eye, light microscope, and electron microscope relate to various specimens.

To obtain a clear, finely detailed image under a compound light microscope, specimens must be made to contrast sharply with their medium. To attain such contrast, we must change the refractive index of specimens from that of their medium. The **refractive index** is a measure of the relative velocity at which light passes through a material. We change the refractive index of specimens by staining them, a procedure we will discuss shortly. Light rays move in a straight line through a single medium. After staining, when light rays pass through the two materials (the specimen and its medium) with different refractive indexes, the rays change direction (refract) at the boundary between the materials and increase the image's contrast between the specimen and the medium.

Although we want light traveling through the specimen and medium to refract differently, we do not want to lose light rays after they have passed through the stained specimen. To preserve the direction of light rays, immersion oil is used. It is placed between the glass slide and the oil immersion objective lens.

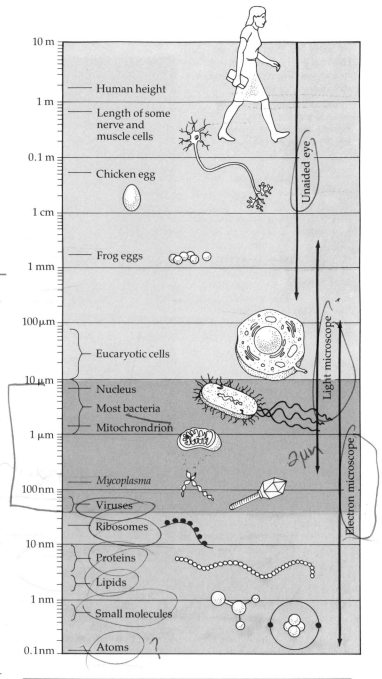

Figure 3-2   Relationships between the sizes of various specimens and the resolution of the human eye, light microscope, and electron microscope. The area in color shows the size range of the organisms we will be studying in this book.

The immersion oil has the same refractive index as glass, so the oil becomes part of the optics of the glass of the microscope. Unless immersion oil is used, light rays are refracted as they enter the air from the slide, and the objective lens would have to be increased in diameter to capture them. The oil has the same effect as increasing the objective diameter; therefore, it improves the resolving power of the lenses. If oil is not used with an oil immersion objective, the image becomes fuzzy, with poor resolution (see Figure 3-1c).

Under usual operating conditions, the field of vision in a compound light microscope is brightly illu-minated. By focusing the light, the condenser pro-duces a **brightfield** illumination (Figure 3-3a).

It is not always desirable to stain a specimen. However, an unstained cell has little contrast with its surroundings and is therefore difficult to see. Unstained cells are more easily observed with the modified compound microscopes described in the next section.

Figure 3-3 Types of images produced by a compound light micro-scope. Shown here are the same cells of a cyanobacterium (also known as a blue-green alga) in the genus *Gloeocapsa* using different microscopic techniques. All cells are shown at a magnification of 1600×. **(a)** Brightfield illumination shows internal structures in the cells and the outline of the transparent sheath that often surrounds two or four recently divided cells. **(b)** Against the black background seen with darkfield micros-copy, edges of the cell are bright, some internal struc-tures seem to sparkle, and the sheath (external covering) is almost invisible. The sheaths that surround individual cells are sharply defined. **(c)** Phase-contrast microscopy shows greater differentiation among the internal structures and also shows the sheath wall. The wide light band around each cell is an artifact resulting from this type of micros-copy—a "phase halo."

(a)

(b)

(c)

Darkfield Microscopy

A **darkfield microscope** is used for examining live microorganisms that either are invisible in the ordinary light microscope, cannot be stained by standard methods, or are so distorted by staining that their characteristics then cannot be identified. Instead of the normal condenser, a darkfield microscope uses a darkfield condenser that contains an opaque disc. The disc blocks light that would enter the objective directly. Only light that is diffracted (bent) by the specimen enters the objective lens (Figure 3-4b). Because there is no direct background light, the specimen appears light against a black background—the dark field (Figure 3-3b). Internal structures of the cell are often not visible. This technique is frequently used to examine unstained microorganisms suspended in liquid. One use for darkfield microscopy is the examination of very thin spirochetes, such as *Treponema pallidum* (tre-pō-nē′mä pal′li-dum), the causative agent of syphilis.

Phase-Contrast Microscopy

Another way to observe microorganisms is with a **phase-contrast microscope**. This technique is especially useful because it permits the detailed examination of internal structures in *living* microorganisms. In addition, it is not necessary to fix (attach the microbes to the microscope slide) or stain the specimen—procedures that could harm the microorganisms. The principle of phase-contrast microscopy is based on slight variations in refractive index. As rays pass from the light source through the specimen, their velocity may be altered by differences in the thickness and physical properties of various portions of the specimens. Light rays passing through the specimen are diffracted (bent) differently and travel different pathways (out of phase with one another) to reach the eye of the viewer (Figure 3-4c). These phase differences are detected by the observer's eye as differences in contrast between the specimen and its background. Details of the internal structures of the specimen also become more sharply defined in phase-contrast microscopy. The internal details of a cell appear as degrees of brightness against a dark background (see Figure 3-3c).

A phase-contrast microscope uses a special condenser that contains an annular (ring-shaped) diaphragm. The diaphragm allows a ring of light to pass

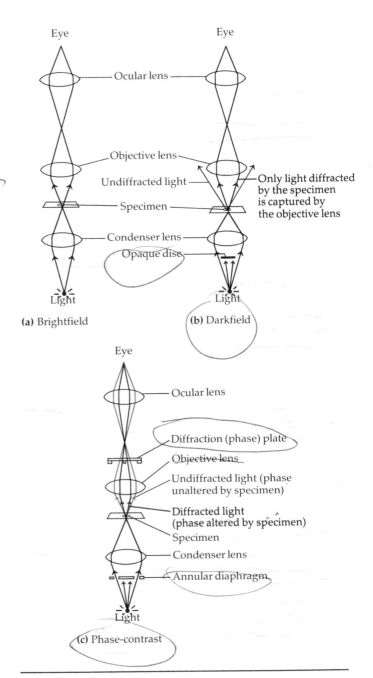

(a) Brightfield

(b) Darkfield

(c) Phase-contrast

Figure 3-4 Brightfield, darkfield, and phase-contrast microscopy. (a) The path of light in brightfield microscopy. **(b)** The path of light in darkfield microscopy. The darkfield microscope is similar to the brightfield microscope, except that a darkfield microscope uses a special condenser that contains an opaque disc that eliminates all light in the center of the beam. The only light that reaches the specimen comes in at an angle; thus, only light diffracted by the specimen (color arrows) reaches the objective lens. **(c)** Phase-contrast microscopy. Light rays are diffracted differently and travel different pathways to reach the eye of the viewer. Diffracted light rays are indicated in color; undiffracted light rays are indicated in black.

through the condenser, focusing light on the specimen and a ring-shaped diffraction (phase) plate in the objective lens. The diffracted and undiffracted rays are then brought in phase with each other to produce the image that meets the eye.

Fluorescence Microscopy

Fluorescence microscopy takes advantage of the fluorescence of substances. Fluorescent substances absorb short wavelengths of light (ultraviolet) and give off light at a longer wavelength that can be seen by the use of special light filters. Some organisms fluoresce naturally under special lighting (Figure 3-5); if the specimen to be viewed does not naturally fluoresce, it is stained with one of a group of fluorescent dyes called *fluorochromes*. When microorganisms stained with a fluorochrome are examined under a fluorescent microscope with an ultraviolet or near-ultraviolet light source, they appear as luminescent, bright objects against a dark background.

Fluorochromes have special attractions for different microorganisms. For example, the fluorochrome auramine O, which glows yellow when exposed to ultraviolet light, is strongly absorbed by *Mycobacterium tuberculosis* (mī-kō-bak-ti′rē-um tü-ber-kū-lō′sis), the bacterium that causes tuberculosis. When the dye is applied to a sample of material suspected of containing the bacterium, the bacterium is detected by the appearance of bright yellow organisms against a dark background. *Bacillus anthracis* (bä-sil′lus an-thrā′sis), the causative agent of anthrax, appears apple green when stained with another fluorochrome, fluorescein isothiocyanate (FITC).

The principal use of fluorescence microscopy is a diagnostic technique called the **fluorescent-antibody technique**, or **immunofluorescence**. **Antibodies** are natural defense molecules that are produced by humans and many animals in reaction to a foreign substance, or **antigen**. Fluorescent antibodies for a particular antigen are obtained as follows: An animal is injected with a specific antigen, such as a bacterium; the animal then begins to produce specific antibodies against that antigen. After a sufficient time, the antibodies are removed from the serum of the animal. Next, a fluorochrome is chemically combined with the antibodies. These fluorescent antibodies are then added to a microscope slide containing an unknown bacterium. If this unknown bacterium is the same bacterium that was injected into the animal, the fluorescent antibodies bind to the surface of the bacterium, causing it to fluoresce (Figure 3-6a).

Figure 3-5 Fluorescence microscopy. Fluorescence microscopy has been used to visualize this triangular-shaped diatom (*Licmorphora*) and strand of an alga (*Spongomorphora*). Under special lighting conditions, the chlorophyll in these organisms fluoresces (×475).

This technique can detect bacteria or other disease-producing microorganisms even within cells, tissues, or other clinical specimens (see Figure 3-6b and Microview C, Photo 19). It is especially useful in diagnosing syphilis and rabies. We will say more about antigen–antibody reactions and immunofluorescence in Chapter 16.

Electron Microscopy

Objects smaller than 0.2 μm, such as viruses or the internal structures of cells, must be examined with an **electron microscope**. Its resolving power is far greater than that of the other microscopes we have mentioned. The better resolution of electron microscopes is due to the shorter wavelengths of electrons; the wavelengths of electrons are about 100,000 times smaller than the wavelengths of visible light. Thus, electron microscopes are used to examine structures too small to be resolved with light microscopes.

In electron microscopy, a beam of electrons is used instead of light. Free electrons travel in waves, just as light photons do. Instead of using glass lenses, an electron microscope uses electromagnetic lenses to focus a beam of electrons through an evacuated tube onto a specimen. There are two types of electron microscopes, the transmission electron microscope and the scanning electron microscope.

(b)

Figure 3-6   The principle of immunofluorescence. (a) A fluorochrome is combined with a specific bacterial antibody. When the preparation is added to bacterial cells on a microscope slide, the antibody attaches to the bacterial cells, and the cells fluoresce when illuminated with ultraviolet light. **(b)** In the fluorescent-antibody technique, the structures to which the specific antibodies attach show up as light regions against a darker background (×3500).

Transmission Electron Microscopy

In the **transmission electron microscope**, a finely focused beam of electrons from an electron gun passes through a specially prepared, ultrathin section of the specimen (Figure 3-7a). The beam is focused on a small area of the specimen by an electromagnetic condenser lens that performs roughly the same function as the condenser of a light microscope—to direct the beam of electrons in a straight line to illuminate the specimen.

Electron microscopes use electromagnetic lenses rather than glass lenses, as in a light microscope. Electromagnetic lenses control illumination, focus, and magnification. Instead of placing the specimen on a glass slide, as in light microscopes, it is usually placed on a copper mesh grid. The beam of electrons passes through the specimen and then through an electromagnetic objective lens, and the image is magnified. Finally, the electrons are focused by an electromagnetic projector lens, rather than by an ocular lens as in a light microscope, onto a fluorescent screen or photographic plate. The final image, called a **transmission electron micrograph**, appears as many light and dark areas, depending on how many electrons are absorbed by different areas of the specimen (Figure 3-8a).

In practice, the transmission electron microscope can resolve objects as close as 2.5 nm, and objects are generally magnified from 10,000 to 100,000 times. Because the density of most microscopic specimens is very low, the contrast between their ultrastructures and the background is weak. Contrast can be greatly enhanced by use of a "stain" that absorbs electrons and produces a darker image in the stained region. Salts of various heavy metals, such as lead, osmium, tungsten, and uranium, are commonly used as stains. These metals can be fixed onto the specimen (positive staining) or used to increase the electron opacity of the surrounding field (negative staining). Negative staining is useful for the study of the very smallest specimens, such as virus particles, bacterial flagella, and protein molecules.

Transmission electron microscopy has the advantage of high resolution and is extremely valuable for examining different layers of specimens. However, it does have certain disadvantages. Because electrons have limited penetrating power, only a very thin section of a specimen (about 100 nm) can be studied effectively. Thus, the specimen has no three-dimensional aspect. In addition, specimens must be fixed, dehydrated, and viewed under a high vacuum. These treatments not only kill the specimen but also cause some shrinkage and distortion.

Figure 3-7 Electron microscopes. (a) In a transmission electron microscope, electrons pass through the specimen and are scattered. Magnetic lenses focus the image onto a fluorescent screen or photographic plate. **(b)** In a scanning electron microscope, the electrons sweep across the specimen and knock electrons from the specimen surface. The electrons are picked up by a collector, amplified, and transmitted onto a viewing screen or photographic plate.

Electron gun

Electron beam

Electromagnetic condenser lens

Specimen

Electromagnetic objective lens

Electromagnetic projector lens

Viewing eyepiece

Fluorescent screen or photographic plate

(a) Transmission

Primary electron beam

Electromagnetic lenses

Viewing screen

Electron collector

Secondary electrons

Specimen

Amplifier

(b) Scanning

(a)

(b)

Figure 3-8 Lymphocyte viewed through a transmission electron microscope and a scanning electron microscope. (a) This transmission electron micrograph shows a thin slice of a lymphocyte, a type of white blood cell. This type of microscopy allows one to see the internal structures present in the slice. **(b)** In a scanning electron micrograph, surface structures can be seen, as demonstrated in this view of a lymphocyte. Note the three-dimensional appearance of this cell, in contrast to the two-dimensional appearance of the cell in (a). Both cells ×18,000.

Scanning Electron Microscopy

The **scanning electron microscope** overcomes the problem of sectioning associated with a transmission electron microscope. A scanning electron microscope provides striking three-dimensional views of specimens (see Figure 3-8b). In scanning electron microscopy, an electron gun produces a finely focused beam of electrons called the primary electron beam. These electrons pass through electromagnetic lenses and are directed over the surface of the specimen. The primary electron beam knocks electrons out of the surface of the specimen, and the secondary electrons thus produced are transmitted to an electron collector, amplified, and used to produce an image on a viewing screen or photographic plate (see Figure 3-7b). The scanning electron microscope is especially useful in studying the surface structures of intact cells and viruses. In practice, the scanning electron microscope can resolve objects as close as 20 nm, and objects are generally magnified from 1000 to 10,000×.

The various types of microscopy just described are summarized in Table 3-2.

Preparation of Specimens for Light Microscopy

Because most microorganisms appear almost colorless when viewed through a standard light microscope, one often must prepare them for observation. One of the ways this can be done is by staining (coloring). Here we will discuss several different staining procedures.

TABLE 3-2   *Summary of Various Types of Microscopes*

Microscope Type	Distinguishing Features	Principal Uses
Brightfield	Uses visible light as a source of illumination; cannot resolve structures smaller than 0.2 μm; specimen appears against a bright background.	Commonly used to observe various stained (killed) specimens and to count microbes; does not resolve very small specimens, such as viruses. Instrument is inexpensive and simple to use.
Darkfield	Uses a special condenser with an opaque disc that blocks light from entering the objective directly; light diffracted by specimen enters objective and specimen appears light against a black background.	Commonly used to examine living microorganisms that are invisible in brightfield microscopy, do not stain easily, or are distorted by staining; frequently used to detect *Treponema pallidum.*
Phase-contrast	Uses a special condenser and diffraction plate to diffract light rays so that they are out of phase with one another; the specimen appears as different degrees of brightness and contrast.	Commonly used to provide detailed examination of the internal structures of living specimens; no staining is required.
Fluorescence	Uses an ultraviolet or near-ultraviolet source of illumination that causes fluorescent compounds in a specimen to emit light.	Principal use is for fluorescent-antibody techniques (immuno-fluorescence) to identify and detect microorganisms in tissues or clinical specimens rapidly.
Electron	Uses a beam of electrons instead of light; because of the shorter wavelength of electrons, structures smaller than 0.2 μm can be resolved.	A transmission electron microscope is used to examine viruses or the internal ultrastructure in thin sections of cells (usually magnified 10,000–100,000×); the image produced is not three dimensional. A scanning electron microscope is used to study the surface features of cells and viruses (usually magnified 1000–10,000×); the image produced is three dimensional.

Preparing Smears and Staining

Most studies of the shapes and cellular arrangements of microorganisms are made from stained preparations. **Staining** simply means coloring the microorganisms with a dye that emphasizes certain structures. Before the microorganisms can be stained, however, they must be attached, or **fixed**, to the microscope slide; otherwise, the stain might wash them from the slide.

When a specimen is fixed, a thin film of material containing the microorganisms is spread over the surface of the slide. This film is called a **smear**. After it is allowed to air dry, the slide is slowly passed through the flame of a Bunsen burner several times, smear side up. Flaming denatures (alters the structure and function of) the microbial proteins and fixes the microorganisms to the slide. This fixing procedure usually kills them. Stain is applied and then washed off with water; then the slide is blotted with absorbent paper. The stained microorganisms are now ready for microscopic examination.

Stains are salts composed of a positive and a negative ion, one of which is colored and is known as the chromophore. The color of so-called **basic dyes** is in the positive ion. In **acidic dyes**, it is in the negative ion. Bacteria are slightly negatively charged at a pH of 7.0. Thus, the colored positive ion in a basic dye is attracted to the negatively charged bacterial cell. Examples of basic dyes are crystal violet, methylene blue, and safranin. Acidic dyes are not attracted to most types of bacteria because the dye's negative ions are repelled by the negatively charged bacterial surface, so the stain colors the background instead. This preparation of colorless bacteria against a colored background is called **negative staining**. It is valuable in the observation of overall cell shapes, sizes, and capsules because the cells are made highly visible against a contrasting dark background. Distortions of cell size and shape are minimized because heat fixing is not necessary, and the cells do not pick up the stain. Examples of acidic dyes are eosin and nigrosin.

To apply acidic or basic dyes, microbiologists use three kinds of staining techniques—simple, differential, and special.

Simple Stains

A **simple stain** is an aqueous or alcohol solution of a single basic dye. Although different dyes bind specifically to different parts of cells, the primary purpose of a simple stain is to highlight the entire microorganism so that cellular shapes and structures are visible. The stain is applied to the fixed smear for a certain amount of time, then washed off, and the slide is dried and examined. Occasionally, a chemical is added to the solution to intensify the stain. Such an additive is called a **mordant**. One function of a mordant is to increase the affinity of a stain for a biological specimen. Another is to coat a structure (such as a flagellum) to make it thicker and easier to see after it is stained with a dye. Some of the simple stains commonly used in the laboratory are methylene blue, carbolfuchsin, crystal violet, and safranin.

Differential Stains

Unlike simple stains, **differential stains** react differently with different kinds of bacteria and thus can be used to distinguish among them. The differential stains most frequently used for bacteria are the Gram stain and the acid-fast stain.

Gram Stain

The **Gram stain** was developed in 1884 by the Danish bacteriologist Hans Christian Gram. It is one of the most useful staining procedures because it divides bacteria into two large groups, *gram-positive* and *gram-negative*.

In this procedure, the heat-fixed smear is covered with a basic purple dye, usually crystal violet (Figure 3-9a). Since the purple stain imparts its color to all cells, it is referred to as a **primary stain**. After a short time, the purple dye is washed off, and the smear is covered with iodine, a mordant (Figure 3-9b). When the iodine is washed off, both gram-positive and gram-negative bacteria appear dark violet or purple. Next, the slide is washed with an ethanol or ethanol–acetone solution (Figure 3-9c). This solution is a **decolorizing agent**, which removes the purple from the cells of some species but not from others. The alcohol is rinsed off, and the slide is then stained with safranin, a basic red dye (Figure 3-9d). The smear is washed again, blotted dry, and examined microscopically.

The purple dye and the iodine combine with each bacterium and color it dark violet or purple. Bacteria that retain this color after the alcohol has attempted to decolorize them are classified as **gram-positive** (see Microview A, Photo 1). Bacteria that lose the dark violet or purple color after decolorization are classified as **gram-negative** (see Microview A, Photo 2). Because gram-negative bacteria are colorless after the alcohol wash, they are no longer visible. It is for this reason that the basic dye safranin is applied; it

Key:
- Purple dye
- Iodine
- Alcohol
- Safranin

(a) Application of crystal violet (purple dye)

(b) Application of iodine (mordant)

(c) Alcohol wash (decolorization)

(d) Application of safranin (counterstain)

Figure 3-9   Gram-staining procedure. Carefully note the key; it indicates which colors replace the bacteria's actual colors. **(a)** A heat-fixed smear of gram-positive cocci and gram-negative rods is first covered with a basic purple dye (primary stain) such as crystal violet, and then the dye is washed off. **(b)** Then the smear is covered with iodine (a mordant) and washed off. At this time, both gram-positive and gram-negative bacteria are purple. **(c)** The slide is washed with ethanol or an alcohol and acetone solution (a decolorizer), and then washed with water. Now the gram-positive cells are purple and the gram-negative cells are colorless. **(d)** In the final step, safranin is added as a counterstain, and the slide is washed, dried, and examined microscopically. Gram-positive bacteria retain the purple dye, even through the alcohol wash. Gram-negative bacteria appear pink because they pick up the safranin counterstain.

turns the gram-negative bacteria pink. Stains such as safranin that have a contrasting color to the primary stain are called **counterstains**. Because gram-positive bacteria retain the original purple stain, they are not affected by the safranin counterstain.

As you will see in the next chapter, different kinds of bacteria react differently to the Gram stain, probably because structural differences in their cell walls affect the retention or escape of a combination of crystal violet and iodine, called the crystal violet–iodine (CV–I) complex. Among other differences, gram-positive bacteria have a thicker cell wall composed of peptidoglycan, whereas gram-negative bacteria have a thinner cell wall, also composed of peptidoglycan (Chapter 4). In addition, gram-negative bacteria contain a layer of lipopolysaccharide external to their cell wall. When applied to both gram-positive and gram-negative cells, crystal violet and then iodine readily enter the cells. Inside the cells, the crystal violet and iodine combine to form CV–I. This complex is larger than the crystal violet molecule that entered the cells, and, because of its size, it cannot be washed out of the intact peptidoglycan layer of gram-positive cells by alcohol. Consequently, gram-positive cells retain the

color of the crystal violet dye. In gram-negative cells, however, the alcohol wash disrupts the outer lipopolysaccharide layer, and the CV–I complex is washed out through the thin layer of peptidoglycan. As a result, gram-negative cells are rendered colorless until counterstained with safranin, after which they are pink. In summary, the gram-positive cells retain the dye and remain purple. The gram-negative cells do not retain the dye. They are colorless until counterstained with a red dye, after which they appear pink.

The Gram method is one of the most important staining techniques in medical microbiology. But Gram staining results are not universally applicable because some bacterial cells stain poorly or not at all. The Gram reaction is most consistent when it is used on young, growing bacteria.

In many cases, the Gram reaction of a bacterium provides valuable information for the treatment of disease. For example, gram-positive bacteria tend to be killed easily by penicillin and sulfonamide drugs. Gram-negative bacteria resist these drugs, but are much more susceptible to such drugs as streptomycin, chloramphenicol, and tetracycline. Thus, Gram-stain identification of a bacterium can help determine which drug will be most effective against a disease.

Bdellovibrio, Predator Extraordinaire

At first glance, the 1-μm-long, curved, gram-negative rods of *Bdellovibrio* do not seem unusual. *Bdellovibrio* exhibit the normal runs and tumbles of flagellated bacteria; however, they move about ten times faster than other bacteria of the same size. Patient observation or use of a microscope equipped with a video camera reveals that *Bdellovibrio*'s behavior is unlike that of any other known bacteria. A unique microbial drama occurs when *Bdellovibrio* are placed in a suspension with other bacteria, such as *E. coli.* Apparently drawn by chemical attractants, *Bdellovibrio* speeds toward and rams the larger, slower *E. coli.* The two cells become attached and rotate rapidly. As the two bacteria whirl about, the *Bdellovibrio* paralyzes the *E. coli* by use of an unknown mechanism that most likely involves one of *Bdellovibrio*'s protein products. *Bdellovibrio* kills *E. coli* without breaking down the cell's contents.

Next, in preparation for moving into its newly acquired home, *Bdellovibrio* adds structural components to the outer membrane of *E. coli.* This activity, which stabilizes the outer membrane, is another ability unique to *Bdellovibrio.* After penetrating *E. coli*'s cell wall and detaching its flagellum, *Bdellovibrio* snuggles into the periplasmic space (that is, the space between the plasma membrane and the cell wall) and prepares for a feast (see figure).

Macromolecules leak from *E. coli*'s cytoplasm into the periplasmic space. *Bdellovibrio* digests the macromolecules and grows into a long, coiled filament. In about two or three hours, the cytoplasmic contents of the host (including its DNA) are digested, and the *Bdellovibrio* filament fragments into as many as 20 individual curved rods. Each rod develops a flagellum and is released when the host cell is broken open by a *Bdellovibrio* enzyme. The new *Bdellovibrio* scurry away to feed on other bacteria.

Bdellovibrio was once regarded as simply a parasite. However, thorough studies have shown that *Bdellovibrio* does not merely siphon off its host's resources and energy, it actually digests the cell's contents as food, so *Bdellovibrio* is now considered to be a bacterial predator. The ecological importance of *Bdellovibrio* is that it, along with protozoans and bacteriophages, helps to control the population growth of soil bacteria.

Some researchers have suggested a use for *Bdellovibrio* in industrial microbiology. Genetically engineered bacteria can be made to produce large quantities of a desired chemical; however, these bacteria tend to store the chemical product rather than release it. This characteristic makes it very difficult to harvest such chemical products in large quantities. Researchers would like to adapt the mechanism used by *Bdellovibrio* to allow large molecules to leave the host cell. If genetically engineered bacteria could be made to secrete rather than to store their metabolic products, the mass production of a desired chemical would be greatly facilitated.

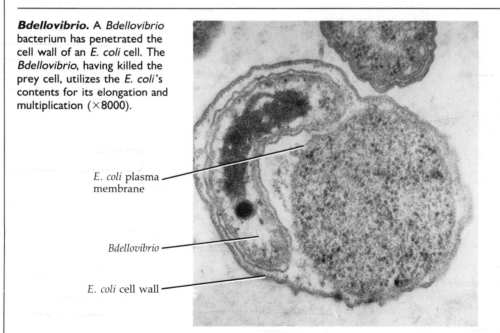

Bdellovibrio. A *Bdellovibrio* bacterium has penetrated the cell wall of an *E. coli* cell. The *Bdellovibrio,* having killed the prey cell, utilizes the *E. coli*'s contents for its elongation and multiplication (×8000).

E. coli plasma membrane

Bdellovibrio

E. coli cell wall

Acid-Fast Stain

Another important differential stain (one that divides bacteria into distinctive groups) is the **acid-fast stain**. Microbiologists use this stain to identify members of bacteria belonging to the genera *Mycobacterium* and *Nocardia* (nō-kär′dē-ä). Although many bacteria in these groups are nonpathogenic, two are important disease producers, *Mycobacterium tuberculosis* and *Mycobacterium leprae* (lep′rī) (see Microview A, Photo 4). The primary difference between *Mycobacterium* and *Nocardia* and other microbes is that these genera contain a waxy material in their cell walls.

In the acid-fast staining procedure, the red dye carbolfuchsin is applied to a fixed smear, and the slide is gently heated for several minutes. Heating enhances penetration and retention of the dye. Then the slide is cooled and washed with water. The smear is next treated with acid-alcohol, a decolorizer, which removes the red stain from bacteria that are not acid-fast. The acid-fast microorganisms retain the red color since the carbolfuchsin is more soluble in the waxes

in the cell wall than in the acid-alcohol. In nonacid-fast bacteria whose cell walls lack the waxy components, the carbolfuchsin is rapidly removed during decolorization, leaving the cells colorless. The smear is then stained with a methylene blue counterstain. Nonacid-fast cells appear blue after application of the counterstain.

Special Stains

Special stains are used to color and isolate specific parts of microorganisms, such as endospores and flagella, and to reveal the presence of capsules.

Negative Staining for Capsules

Many microorganisms contain a gelatinous covering called a **capsule**, which we will discuss in our examination of the procaryotic cell in Chapter 4. In medical microbiology, demonstrating the presence of a capsule is a means of determining the organism's **virulence**, the degree to which a pathogen can cause

(a)   (b)   (c)

Figure 3-10 Negative staining of capsules and staining of endospores and flagella.
(a) India ink provides a dark background, so the capsules of these bacteria, *Thiocapsa floridana*, show up as light areas surrounding the stained cells (×2500). **(b)** Endospores are seen as light areas in the centers of these rod-shaped cells of the bacterium *Bacillus subtilis* (×3000). **(c)** Flagella are shown as wavy extensions from one end of these cells of the bacterium *Pseudomonas aeruginosa* (×3000). In relation to the bodies of the cells, the flagella are much thicker than they normally are because layers of the stain accumulated from treatment of the specimen with a mordant.

Overview of Microorganisms

Bacterial Stains and Morphology

(1) Clusters of *Staphylococcus aureus* demonstrate a gram-positive stain (×1000).

(2) Gram-negative rods of the intestinal bacterium *Escherichia coli* (×1000).

(3) Fluorescent-antibody stain of *Legionella pneumophila* in lung tissue from a fatal case of Legionnnaire's disease (×1500).

(4) Section of a cutaneous nerve showing red, acid-fast bacteria of the leprosy-causing species *Mycobacterium leprae* (×1500).

(5) A capsule stain of *Klebsiella pneumoniae*, clearly showing the light, unstained capsules around the dark cell bodies (×1500).

(6) Numerous intracellular diplococci of *Neisseria gonorrhoeae* are in this urethral discharge smear from a case of gonorrhea. *N. gonorrhoeae* are small red forms; large irregular red bodies are nuclei of white blood cells (×1500).

(7) Short chains of thick rods of *Bacillus anthracis* as seen among tissue cells from a case of anthrax (×1500).

(8) Rickettsias of the species *Rickettsia rickettsii* show up as tiny, red, intracellular rodlike forms in this yolk sac smear (×1500).

(9) Dark, reddish-brown inclusion bodies of *Chlamydia trachomatis* are seen inside three cells of a cultured monolayer (×1500).

Microbial Cultures

(10) Colony growth of *Streptococcus pyogenes* from a throat swab on blood agar demonstrates beta-hemolysis (the light areas around the colonies). Identification of the bacteria is aided by the zone of inhibition around the bacitracin disk (A). (The SXT disk inhibits normal flora, so that inhibition by bacitracin can been seen.)

(11) Colonies of *Pseudomonas aeruginosa* produce a water-soluble pigment that fluoresces under ultraviolet light.

(10)

(11)

(12)

(13)

Vibrio fischeri emits light when energy is released by the transport of electrons to luciferase.

(12) Colonies of *V. fischeri* photographed in daylight.

(13) Colonies of *V. fischeri* photographed in the dark illuminated by their own light.

(14)

(14) Cells of *Proteus mirabilis* are highly motile; *P. mirabilis* colonies exhibit swarming over EMB agar as young cells move away from the colony's center.

(15)

(15) Colonies of *Mycobacterium tuberculosis*; note the rough colony surface characteristic of this species.

(16)

Colonies of lactose-fermenting bacteria are colored by the indicator dyes in EMB agar, while nonlactose fermenters produce colorless colonies.

(17)

(16) Colonies of *E. coli* produce a metallic green sheen.

(17) Colonies of *Shigella boydii*, a nonlactose fermenter, are colorless on EMB agar.

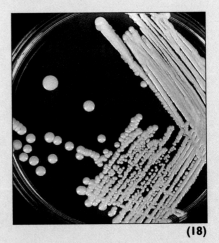

(18)

(18) Colonies of the eucaryotic pathogen, the yeast *Candida albicans*, are raised and mucoid on Sabouraud glucose agar.

Micrographs of Eucaryotic Pathogens

Fungi

(19) Macroconidia of the fungal dermatophyte *Microsporum gypseum* (×600).

(20) Pseudohyphae of the yeast *Candida albicans* in brain tissue (×200).

(21) *Candida albicans;* the large spherical cells are chlamydospores, the much smaller rounded cells are budding cells, and the elongated strands of cells are pseudohyphae (×600).

Protozoans

Cytoplasm
Vacuole
Nucleus

Trypanosome

(22) Blood smear showing the ring stage of the trophozoites of the malarial parasite *Plasmodium vivax* (×2000).

(diamond ring!)

(23) Trypanomastigotes of *Trypanosoma brucei* in blood smear from an experimentally infected animal; note the undulating membrane along the body of each trypanosome (×2000).

Helminths

(24) Scolex of the pork tapeworm *Taenia solium;* note the sucking cups and hooks used to attach to the intestinal wall (×200).

(25) Proglottid of *Taenia solium* filled with eggs (×600).

Environmental Microbiology

(26)

Procaryotes

(26) *Halobacterium* grows in the high-salt concentration of salterns, solar evaporating ponds for manufacturing salt. Its presence causes no negative effect. These bacteria require an environment that is 15% to 25% NaCl.

(27) Green and gray patches of thermophilic cyanobacteria grow on rocks in the hot (60°–75°C) water of Minerva Terrace at Yellowstone National Park. Boiling water is in the white area, upper left.

(27)

Eucaryotes

(28)

(29)

(30)

(28) The brown alga *Laminaria* grows in cold coastal waters (Point Lobos Marine Reserve, California).

(29) A crustose lichen growing on a rock at Crater Lake National Park.

(30) Plasmodium of a yellow acellular slime mold growing on top of a moss.

(31) Two fruiting bodies of the edible basidiomycete (mushroom) *Boletus edulis.*

(32) Bird's nest fungus growing on a twig. The basidiospores seen in one of the cups will pop out when the cup is hit by a raindrop.

(31)

(32)

disease. Capsule staining is more difficult than other types of staining procedures because capsular materials are soluble in water and may be dislodged or removed during rigorous washing. To demonstrate the presence of capsules, a microbiologist can mix the bacteria in a solution containing a fine colloidal suspension of colored particles (usually India ink) to provide a dark background and then can stain the bacteria with a simple stain, such as safranin. Because of their chemical composition, capsules do not accept most biological dyes, such as safranin, and thus appear as halos surrounding each stained bacterial cell. The use of India ink illustrates a negative staining technique; negative stains do not penetrate the cell capsule, and thus they contrast the capsule to the surrounding dark medium (see Figure 3-10a and Microview A, Photo 5).

Endospore (Spore) Stain

Although endospores are relatively uncommon in bacterial cells, they can be formed by six genera of bacteria. When formed within a cell, an **endospore** protects the microorganism from adverse environmental conditions. Endospores cannot be stained by ordinary methods, such as simple staining and Gram staining, because the dyes do not penetrate the wall of the endospore. But because endospores are highly refractive, they are readily detected under the light microscope (Figure 3-10b).

The most commonly used endospore stain is the *Schaeffer–Fulton endospore stain*. Malachite green, the primary stain, is applied to a heat-fixed smear and heated to steaming for about five minutes. The heat helps the stain to penetrate the endospore wall. Then the preparation is washed for about 30 seconds with water to remove the malachite green stain from all of the cells' parts except the endospores. Next, safranin, a counterstain, is applied to the smear to displace any residual malachite green in those cell parts and to stain the cell. In a properly prepared smear, the endospores appear green within red or pink cells.

Flagella Staining

Bacterial **flagella** are structures of locomotion too small to be seen with light microscopes. A tedious and delicate staining procedure uses a mordant and the stain carbolfuchsin to build up the diameters of the flagella until they become visible under the light microscope. Microbiologists use the number and arrangement of flagella as diagnostic aids. Figure 3-10c depicts a flagella stain. A summary of stains is presented in Table 3-3.

TABLE 3-3 *Summary of Various Stains and Their Uses*

Stain	Principal Uses
Simple (methylene blue, carbolfuchsin, crystal violet, safranin)	Aqueous or alcohol solution of a single basic dye. (Sometimes a mordant is added to intensify the stain.) Used to highlight microorganisms to determine cellular shapes and arrangements.
Differential	React differently with different kinds of bacteria in order to distinguish them.
Gram	Divides bacteria into two large groups, gram-positive and gram-negative. Gram-positive bacteria retain the crystal violet stain and appear purple. Gram-negative bacteria do not retain the crystal violet stain and remain colorless until counterstained with safranin and then appear pink.
Acid-fast	Used to distinguish *Mycobacteria* and *Nocardia*. Acid-fast bacteria, once stained with carbolfuchsin and treated with acid-alcohol, remain red because they retain the carbolfuchsin stain. Nonacid-fast bacteria, when stained and treated the same way and then stained with methylene blue, appear blue because they lose the carbolfuchsin stain but retain the methylene blue stain.
Special	Used to color and isolate various structures, such as capsules, endospores, and flagella; sometimes used as a diagnostic acid.
Negative	Used to demonstrate the presence of capsules. Since capsules do not accept most stains, the capsules appear as halos around bacterial cells and stand out against a dark background.
Endospore	Used to detect the presence of endospores in six genera of bacteria. When malachite green is applied to a heat-fixed smear of bacterial cells, the stain penetrates the endospores and stains them green. When safranin (red) is then applied, it stains the remainder of the cells red or pink.
Flagella	Used to demonstrate the presence of flagella. A mordant is used to build up the diameters of flagella until they become visible microscopically when stained with carbolfuchsin.

Study Outline

Units of Measurement (pp. 55–56)

1. The standard unit of length is the meter (m).
2. Microorganisms are measured in micrometers, μm (10^{-6} m), nanometers, nm (10^{-9} m), and angstroms, Å (10^{-10} m).

Microscopy: The Instruments (pp. 56–64)

Compound Light Microscopy (pp. 56–59)

1. The most common microscope used in microbiology is the compound light microscope, which uses two sets of lenses, ocular and objective.
2. We calculate the total magnification of an object by multiplying the magnification of the objective lens by the magnification of the ocular lens.
3. The compound light microscope uses visible light.
4. The maximum resolving power (ability to distinguish two points) of a compound light microscope is 0.2 μm; maximum magnification is $2000\times$.
5. Specimens are stained to increase the difference between the refractive indexes of the specimen and the medium.
6. Immersion oil is used with the oil immersion lens to reduce light loss between the slide and the lens.
7. Brightfield illumination is used for stained smears.
8. Unstained cells are more productively observed using darkfield, phase-contrast, or fluorescence microscopy. These types of microscopy use modified compound microscopes.

Darkfield Microscopy (p. 60)

1. The darkfield microscope shows a light silhouette of an organism against a dark background.
2. It is most useful to detect the presence of extremely small organisms.

Phase-Contrast Microscopy (pp. 60–61)

1. The phase-contrast microscope uses a special condenser to enhance differences in the refractive indexes of the cell's parts and its surroundings.
2. It allows the detailed observation of living organisms.

Fluorescence Microscopy (p. 61)

1. In fluorescence microscopy, specimens are first stained with fluorochromes and then viewed through a compound microscope using an ultraviolet (or near-ultraviolet) light source.
2. The microorganisms appear as bright objects against a dark background.

3. Fluorescence microscopy is primarily used in a diagnostic procedure called fluorescent antibody technique.

Electron Microscopy (pp. 61–64)

1. A beam of electrons, instead of light, is used with an electron microscope.
2. Electromagnets, instead of glass lenses, control focus, illumination, and magnification.
3. Thin sections of organisms can be seen in an electron micrograph produced with a transmission electron microscope. *Resolving power*: 2.5 nm. *Magnification*: 10,000 to 100,000×.
4. Three-dimensional views of the surfaces of whole microorganisms can be obtained with a scanning electron microscope. *Resolving power*: 20 nm. *Magnification*: 1000 to 10,000×.

Preparation of Specimens for Light Microscopy (pp. 64–69)

Preparing Smears and Staining (p. 65)

1. Staining means to color with a dye to make some structures more visible.
2. Fixing uses heat to attach microorganisms to a slide.
3. A smear is a thin film of material used for microscopic examination.
4. Bacteria are negatively charged, and the colored positive ion of a basic dye will stain bacterial cells.
5. The colored negative ion of an acidic dye will stain the background of a bacterial smear; a negative stain is produced.

Simple Stains (p. 65)

1. A simple stain is an aqueous or alcohol solution of a single basic dye.
2. It is used to make cellular shapes and arrangements visible.
3. A mordant may be used to improve bonding between the stain and the specimen.

Differential Stains (pp. 65–68)

1. Differential stains, such as the Gram stain and acid-fast stain, divide bacteria into groups according to their reactions to the stains.
2. The Gram-stain procedure uses a purple stain, iodine as a mordant, an alcohol decolorizer, and a red counterstain.
3. Gram-positive bacteria retain the purple stain after the decolorization step; gram-negative bacteria do not and thus appear red from the counterstain.

4. Acid-fast bacteria, such as members of the genera *Myco-bacterium* and *Nocardia*, retain carbolfuchsin after acid-alcohol-decolorization and appear red; nonacid-fast bacteria take up the methylene blue counterstain and appear blue.

Special Stains (pp. 68–69)

1. Stains such as the endospore stain and flagella stain color only certain parts of microbes.
2. Negative staining is used to make microbial capsules visible.

Study Questions

Review

1. Fill in the following blanks.
1 μm = _____ m
1 _____ = 10^{-9} m
1 Å = _____ m
1 μm = _____ nm = _____ Å

2. Label the parts of the compound light microscope below.

(a) _____
(b) _____
(c) _____
(d) _____
(e) _____

3. Calculate the total magnification of the nucleus of a cell being observed through a compound light microscope using a 10× ocular lens and the oil immersion lens.

4. Which type of microscope would be best to use to observe the following?
(a) A stained bacterial smear
(b) Unstained bacterial cells where the cells are small and no detail is needed
(c) Unstained live tissue where it is desirable to see some intracellular detail
(d) A sample that emits light when illuminated with ultraviolet light
(e) Intracellular detail of a cell that is 1 μm long

5. An electron microscope differs from a light microscope in that _____ focused by _____ is used instead of light, and the image is viewed not through the ocular lenses but on _____ .

6. The maximum magnification of a compound microscope is _____; of an electron microscope, _____ . The maximum resolution of a compound microscope is _____; of an electron microscope, _____ .

7. One advantage of a scanning electron microscope over a transmission electron microscope is _____ .

8. Acidic dyes stain the (cells/background) in a smear and are used for (negative/simple) stains.

9. Basic dyes stain the (cells/background) in a smear and are used for (negative/simple) stains.

10. Why do basic dyes stain bacterial cells? Why don't acidic dyes stain bacterial cells?

11. When is it most appropriate to use:
(a) A simple stain?
(b) A differential stain?
(c) A negative stain?
(d) A flagella stain?

12. Why is a mordant used in the Gram stain? In the flagella stain?

13. What is the purpose of a counterstain in the acid-fast stain?

14. What is the purpose of a decolorizer in the Gram stain? In the acid-fast stain?

15. Choose from the following terms to fill in the blanks: counterstain, decolorizer, mordant, primary stain. In the endospore stain, safranin is the _____ and _____ . In the Gram stain, safranin is the _____ .

16. Fill in the following table regarding the Gram stain.

Steps	Appearance after this step of	
	Gram-positive cells	Gram-negative cells
Crystal violet		
Iodine		
Alcohol–acetone		
Safranin		

Challenge

1. In a Gram stain, one step could be omitted, and you could still differentiate between gram-positive and gram-negative cells. What is that one step?

2. Using a good compound light microscope with a resolving power of 0.3 μm, a 10× ocular lens, and a 100× oil immersion lens, would you be able to discern two objects separated by 3 μm? 0.3 μm? 300 nm? 30,000 Å?

3. Why isn't the Gram stain used on acid-fast bacteria? If you did Gram stain acid-fast bacteria, what would their Gram reaction be? What is the Gram reaction of nonacid-fast bacteria?

Further Reading

Binning, G., and H. Rohrer. "The scanning tunneling microscope." *Scientific American* 253(2):50–56, August 1985. Reports on a new microscope with a resolving power of 10 Å that allows observations of atoms and molecular interactions, such as the attachment of a virus to a cell.

Boatman, E. S., et al. "Today's microscopy." *BioScience* 37(6):384–394, 1987. Provides a background in fundamentals and then describes recent advances in computer enhancement and laser and acoustic microscopy.

Branson, D. *Methods in Clinical Bacteriology.* Springfield, Ill.: Thomas, 1972. Good stepwise directions for staining procedures and preparation of stains. Also see *Procedure Manual for Clinical Bacteriology* (1982), same author and publisher.

Ford, Brian J. *Single Lens: The Story of the Simple Microscope.* New York: Harper & Row, 1985. An interesting history of the simple microscope and its impact on biology.

Gerhardt, P., et al., ed. *Manual of Methods for General Bacteriology.* Washington, D.C.: American Society for Microbiology, 1981. Section II covers microscopy and preparation of specimens.

Gray, P., ed. *Encyclopedia of Microscopy and Microtechnique.* Melbourne, Fla.: Krieger, 1981. An illustrated encyclopedia of terminology and techniques.

Lennette, E. H., et al., ed. *Manual of Clinical Microbiology,* 4th ed. Washington, D.C.: American Society for Microbiology, 1985. Contains a section on microscopy, specimen preparation, and staining.

CHAPTER 4

Functional Anatomy of Procaryotic and Eucaryotic Cells

Despite their complexity and variety, all living cells can be divided into two groups, procaryotes and eucaryotes, based on their ultrastructure as seen with the electron microscope. Plants and animals are entirely composed of eucaryotic cells. In the microbial world, bacteria are procaryotes. Other cellular microbes—fungi (yeasts and molds), protozoans, and true algae—are eucaryotes.

Viruses, as noncellular elements with some cell-like properties, do not fit into any organizational scheme of living cells. They are genetic particles that replicate but are unable to perform the usual chemical activities of living cells. Viral structure and activity will be discussed in Chapter 12. In this chapter, we will concentrate on describing procaryotic and eucaryotic cells.

Procaryotes and eucaryotes are chemically similar, in the sense that they both contain nucleic acids, proteins, lipids, and carbohydrates. They use the same kinds of chemical reactions to metabolize food, build proteins, and store energy. It is primarily the *structure* of cell walls, membranes, and *organelles* (specialized cellular structures that have specific functions) that distinguish procaryotes from eucaryotes.

The chief distinguishing characteristics of **procaryotic cells** (from the Greek for prenucleus) are as follows:

1. Their genetic material (DNA) is not enclosed within a membrane.
2. They lack other membrane-bounded organelles.
3. Their DNA is not associated with histone proteins (special chromosomal proteins found in eucaryotes).
4. Their cell walls almost always contain the complex polysaccharide peptidoglycan.
5. They usually divide by **binary fission**. During this process, the DNA is copied, and the cell splits into two cells. Binary fission involves fewer structures and processes than eucaryotic mitosis and cell division.

Learning Objectives

- Identify the three basic shapes of bacteria.

- Explain the differences between gram-positive and gram-negative cell walls.

- Describe the structure, chemistry, and functions of the procaryotic plasma membrane.

- Define simple diffusion, osmosis, facilitated diffusion, active transport, and group translocation.

- Identify the functions of procaryotic cell structures.

- Compare and contrast the overall cell structure, flagella, nucleus, and ribosomes of procaryotes and eucaryotes.

- Explain what an organelle is.

- Describe the functions of the endoplasmic reticulum, Golgi complex, mitochondria, chloroplasts, and lysosomes.

Eucaryotic cells (from the Greek for true nucleus) have linear structures of DNA called chromosomes; these are found in the cell's nucleus, which is separated from the cytoplasm by a nuclear membrane. The DNA of eucaryotic chromosomes is consistently associated with chromosomal proteins called histones and nonhistones. Eucaryotes also have a mitotic apparatus (various cellular structures that participate in a type of nuclear division called mitosis) and a number of organelles, including mitochondria, endoplasmic reticulum, and sometimes chloroplasts. We will elaborate on the particular characteristics and functions of these organelles later in this chapter.

The Procaryotic Cell

The members of the procaryotic world comprise a vast heterogeneous group of unicellular organisms. This group includes eubacteria, or true bacteria, and archaeobacteria. The thousands of species of bacteria are differentiated by many factors, including morphology (shape), chemical composition (often detected by staining reactions), nutritional requirements, and biochemical activities.

(a)

Diplococci

Streptococci

(b)

Tetrad

(c)

Sarcinae

Staphylococci

Figure 4-1 Arrangements of cocci. The number of planes in which the cell divides determines the arrangement of cells. Shown are diagrams (left) and corresponding photos (right). **(a)** Division in one plane produces diplococci ($\times$15,000) and streptococci ($\times$2500). **(b)** Division in two planes produces tetrads ($\times$25,000). **(c)** Division in three planes produces sarcinae ($\times$16,000) and staphylococci ($\times$2500).

Size, Shape, and Arrangement of Bacterial Cells

There are a great many sizes and shapes among bacteria. Most bacteria fall within a range from 0.20 to 2.0 µm in diameter and from 2 to 8 µm in length. They have a few basic shapes—spherical **coccus** (plural, *cocci*, meaning berries), rod-shaped **bacillus** (plural, *bacilli*, meaning little staffs), and **spiral**.

Cocci are usually round but can be oval, elongated, or flattened on one side. When cocci divide to reproduce, the cells can remain attached to one another. Cocci that remain in pairs after dividing are called **diplococci** (see Figure 4-1a and Microview A, Photo 6). Those that divide and remain attached in chainlike patterns are called **streptococci** (Figure 4-1a). Those that divide in two planes and remain in groups of four are known as **tetrads** (Figure 4-1b). Those that divide in three regular planes and remain attached in cubelike groups of eight are called **sar-** cinae (Figure 4-1c). Those that divide at random planes and form grapelike clusters or broad sheets are called **staphylococci** (Figure 4-1c). These group characteristics are frequently helpful in the identification of certain cocci.

Bacilli divide only across their short axis, so there are fewer groupings of bacilli than of cocci. **Diplobacilli** appear in pairs after division (Figure 4-2b), and **streptobacilli** occur in chains (Figure 4-2c and Microview A, Photo 7). Some bacilli look like straws. Others have tapered ends, like cigars. Still others are oval and look so much like cocci that they are called **coccobacilli** (Figure 4-2d). However, most bacilli appear as single rods (Figure 4-2a).

The term *bacillus* has two meanings in microbiology. As we have just used it, it refers to a bacterial shape. It can also be used as a genus name. For example, the bacterium *Bacillus anthracis* is the causative agent of anthrax.

Spiral bacteria have one or more twists; they are never straight. Curved rods that look like commas are called **vibrios** (Figure 4-3a). Others, called **spirilla**, have a distinctive helical shape, like a corkscrew, and

Figure 4-2 Bacilli. Shown are diagrams (left) and corresponding photos (right). **(a)** Single bacilli. **(b)** Diplobacilli. In the photo, a few joined pairs of bacilli serve as examples of diplobacilli (×2000). **(c)** Streptobacilli (×2000). **(d)** Coccobacilli (×20,000).

(a) Single bacillus

(b) Diplobacillus

(c) Streptobacillus

(d) Coccobacillus

Figure 4-3 Spiral bacteria. Diagrams (left) and corresponding photos (right) of **(a)** vibrio (×4800), **(b)** spirilla (×17,000), **(c)** spirochete (×4000).

(a) Vibrio

(b) Spirilla

(c) Spirochete

(a)

(b)

Figure 4-4 Star-shaped and square-shaped cells. (a) *Stella* (star-shaped) (×36,000). **(b)** *Arcula* (square-shaped) (×1700).

fairly rigid bodies (Figure 4-3b). Yet another group of spirals are called **spirochetes** (Figure 4-3c). Unlike the spirilla, which have flagella, spirochetes move by means of an axial filament, which resembles a flagellum but is contained under an external flexible sheath. As you will see later, some genera of spiral bacteria have more complex shapes and arrangements.

In addition to the three basic shapes, there are also star-shaped cells (genus *Stella*) and recently discovered square-shaped, flat cells (proposed genus *Arcula*) (Figure 4-4).

The shape of a bacterium is determined by heredity. However, a number of environmental conditions can alter that shape. If the shape is altered, identification becomes even more difficult. Moreover, some bacteria, such as *Rhizobium* (rī-zō'bē-um) and *Corynebacterium* (kô-rī-nē-bak-ti'rē-um), are genetically **pleomorphic**, which means they can have many shapes, not just one.

Figure 4-5 shows the structure of a typical procaryotic (bacterial) cell. We will discuss its components according to the following organization: (1) structures external to the cell wall, (2) the cell wall, and (3) structures internal to the cell wall.

Structures External to the Cell Wall

Among the structures external to the procaryotic cell wall are the glycocalyx, which can take the form of a capsule or a slime layer, flagella, axial filaments, and pili.

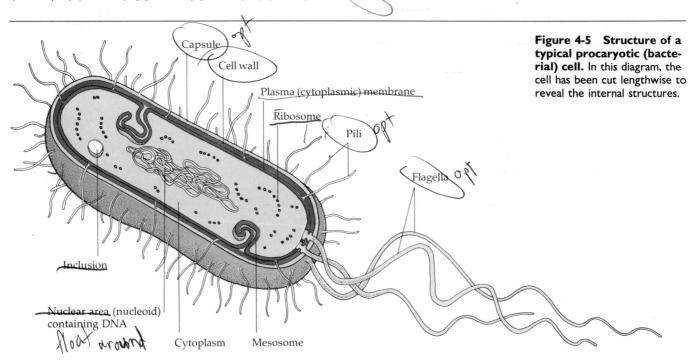

Figure 4-5 Structure of a typical procaryotic (bacterial) cell. In this diagram, the cell has been cut lengthwise to reveal the internal structures.

Capsule

Cell wall

Plasma (cytoplasmic) membrane

Ribosome

Pili

Flagella

Inclusion

Nuclear area (nucleoid) containing DNA

Cytoplasm

Mesosome

Glycocalyx

Glycocalyx is the general term used for the substances that surround bacterial cells. (An alternative term for the glycocalyx is **extracellular polymeric substance**, or **EPS**.) The glycocalyx is a gelatinous polymer composed of polysaccharide, polypeptide, or both. The chemical complexity of the glycocalyx of different species of bacteria varies widely. The material of the glycocalyx is viscous (sticky). For the most part, it is made inside the cell and excreted to the cell surface. If the substance is organized and is firmly attached to the cell wall, the glycocalyx is described as a **capsule** (see Figure 14-2).

The presence of a capsule can be determined by use of negative staining, such as the India ink method mentioned in Chapter 3 (see Figure 3-10a and Microview A, Photo 5). If the substance is unorganized and only loosely attached to the cell wall, the glycocalyx is described as a **slime layer**.

Capsules are important in contributing to bacterial virulence (the degree to which a pathogen causes disease) in certain species. Capsules often protect pathogenic bacteria from phagocytosis by the cells of the host. Phagocytosis, which is discussed in Chapter 15, is a process by which certain white blood cells engulf and destroy microbes (see Figure 15-8). *Streptococcus pneumoniae* (strep-tō-kok'kus nü-mō'nē-ī) causes pneumonia when the cells are protected by a polysaccharide capsule. Unencapsulated *S. pneumoniae* are readily phagocytized and cannot cause pneumonia. *Bacillus anthracis* (bä-sil'lus an-thrā'sis) produces a capsule of D-glutamic acid. Recall from Chapter 2 that the D forms of amino acids are unusual. Since only encapsulated *B. anthracis* cause anthrax, it appears that the capsule prevents phagocytosis.

Another function of the sticky glycocalyx is to attach a bacterium to various surfaces in order to survive in its natural environment. Through attachment, bacteria can grow on diverse surfaces, such as rocks in fast-moving streams, plant roots, human teeth and tissues, and even other bacteria. *Streptococcus mutans* (mū'tans), an important cause of dental caries, attaches itself to the surface of teeth by a glycocalyx. The capsule of *Klebsiella pneumoniae* (kleb-sē-el'lä nü-mō'nē-ī) prevents phagocytosis and allows this bacterium to adhere to and colonize the respiratory tract.

S. mutans may use its capsule as a source of nutrition by breaking it down and utilizing the sugars when energy stores are low. A glycocalyx can protect a cell against dehydration. Also, its viscosity may inhibit the movement of nutrients from the cell.

Flagella

A second structural component of some procaryotic cells is flagella (see Figure 4-5). **Flagella** (singular, *flagellum*, meaning whip) are long filamentous appendages that propel bacteria.

Bacterial cells have four arrangements of flagella (Figure 4-6a)—**monotrichous** (single polar flagellum), **amphitrichous** (single flagella at both ends of the cell), **lophotrichous** (two or more flagella at one or both poles of the cell), and **peritrichous** (flagella distributed over the entire cell).

A flagellum has three basic parts (Figure 4-6b). The long outermost region, the *filament*, is constant in diameter and contains the globular (roughly spherical) protein *flagellin* arranged in several chains that intertwine and form a helix around a hollow core. Flagellar proteins serve to identify certain pathogenic bacteria. In most bacteria, filaments are not covered by a membrane or sheath, as in eucaryotic cells. The filament is attached to a slightly wider *hook*, which consists of a different protein. The third portion of a flagellum is the *basal body*, which anchors the flagellum to the cell wall and plasma membrane.

The basal body is composed of a small central rod inserted into a series of rings. Gram-negative bacteria contain two pairs of rings. The outer pair of rings is anchored to various portions of the cell wall, and the inner pair of rings is anchored to the plasma membrane. In gram-positive bacteria, only the inner pair is present. As you will see later in the chapter, the flagella (and cilia) of eucaryotic cells are more complex than those of procaryotic cells.

Bacteria with flagella are motile. That is, they have the ability to move on their own. Each procaryotic flagellum is a semirigid, helical rotor that moves the cell by rotating from the basal body. The rotation of a flagellum is either clockwise or counterclockwise around its long axis. Eucaryotic flagella, by contrast, undulate in a wavelike motion. The movement of a procaryotic flagellum results from rotation of the flagella's basal body and is similar to the movement of the shaft of an electric motor. As the flagella rotate, they form a bundle that pushes against the water and propels the bacterium (Figure 4-6c). Although the exact mechano-chemical basis for this biological "motor" is not completely understood, we know that it depends on the cell's continuous generation of energy.

Bacterial cells have the ability to alter the speed and direction of rotation of flagella and thus are capa-

Figure 4-6   Flagella. (a) Four basic types of flagellar arrangements: amphitrichous (*Spirillum ostreae*, ×3500); monotrichous (marine vibrio, ×21,000); lophotrichous (*Ectothiorhodospira mobilis*, ×2000); peritrichous (*Proteus vulgaris*, ×4200). **(b)** Parts and attachment of a flagellum of a gram-negative bacterium. **(c)** Types of bacterial motility, showing a "run" and a "tumble."

Amphitrichous

(a)

Monotrichous

Lophotrichous

Peritrichous

Knowns →

Filament — *outermost*

Hook

Gram⁻ [inbrant only hv) !

Outer pair of rings

Outer membrane

Peptidoglycan layer — Cell wall

Inner pair of rings

anchor

Basal body

Plasma membrane

Rod

Cytoplasm

(b)

Run

Tumble

(c)

ble of various patterns of motility (Figure 4-6c). When a bacterium moves in one direction for a period of time, the movement is called a "run" or "swim." "Runs" are interrupted by periodic, abrupt, random changes in direction called "tumbles." Then, a "run" resumes again. "Tumbles" are caused by a reversal of flagellar rotation. Some species of bacteria endowed with large numbers of flagella (*Proteus*, for example) can "swarm," or show rapid wavelike growth across a solid culture medium (see Microview A, Photo 14).

One advantage of motility is that it enables a bacterium to move toward a favorable environment or away from an adverse one. The movement of a bacterium toward or away from a particular stimulus is called **taxis**. Such stimuli include chemicals (**chemotaxis**) and light (**phototaxis**). Motile bacteria contain receptors in various locations, such as the cell wall or just under the cell wall. These receptors pick up chemical stimuli, such as oxygen, ribose, and galactose. In response to the stimuli, information is passed to the flagella. If the chemotactic signal is positive, called an *attractant*, the bacteria move toward the stimulus with many swimming movements and few tumbles. If the chemotactic signal is negative, called a *repellent*, the frequency of tumbles increases as the bacterium moves away from the stimulus (see the box on p. 80).

Axial Filaments

Spirochetes are a group of bacteria that have unique structure and motility. One of the best-known spirochetes is *Treponema pallidum* (tre-pō-nē'mä pal'li-dum), the causative agent of syphilis. Spirochetes move by means of **axial filaments**, bundles of fibrils that arise at the ends of the cell beneath the outer sheath and spiral around the cell (Figure 4-7). Axial filaments, which are anchored at one end of the spirochete, have a structure similar to that of flagella. The rotation of the filaments causes the rigid helical cell to rotate in the opposite direction and move as a corkscrew. Eucaryotic cells do not contain axial filaments.

Pili

Pili are hairlike appendages attached to bacterial cells in much the same way as flagella are. But pili are considerably shorter and thinner than flagella (see Figure 4-5). Like flagella, pili consist of a protein (called *pilin*) arranged helically around a central core. Pili can occur at the poles of the bacterial cell, or they can be evenly distributed over the entire surface of

Figure 4-7 Axial filaments. (a) Photomicrograph of the spirochete *Leptospira*, showing an axial filament (×14,000). **(b)** Diagram of axial filaments wrapping around part of a spirochete.

(a)

Axial filament   Cell wall   Outer sheath

(b)

the cell. There can be anywhere from a few to several hundred pili per cell (Figure 4-8). Many gram-negative bacteria have pili. Eucaryotic cells do not have pili.

Pili are of two types, and each has a different function. The first type, called **common pili**, allows a cell to adhere to surfaces, including the surfaces of other cells. This function is similar in the pili and the glycocalyx. Pili associated with the bacterium *Neisseria gonorrhoeae* (nī-se′rē-ä go-nôr-rē′ī), the causative agent of gonorrhea, help the microbe to colonize mucous membranes. Once colonization occurs, the bacteria are capable of causing the disease. When pili are absent (because of genetic mutation), colonization cannot occur, and disease does not occur.

The second type of pili, referred to as **sex pili**, functions to join bacterial cells prior to the transfer of DNA from one cell to another. These pili are discussed in detail in Chapter 8.

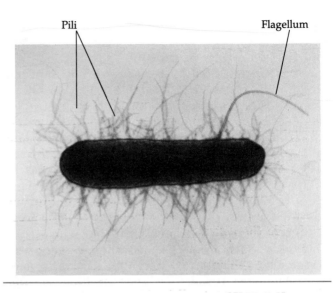

Figure 4-8   Pili. Common pili seem to bristle from this cell of *E. coli* (×22,000). Note the flagellum.

MICROBIOLOGY HIGHLIGHTS

Microbial Magnets

For nearly 100 years, biologists have suspected that birds can use the earth's magnetic field for orientation during flight. During the 1970s, magnetotaxis (response to a magnetic field) was demonstrated in some birds as well as in honeybees and some bacteria. In 1975, R. P. Blakemore made an observation that has led to a more complete understanding of the possible mechanism behind magnetotaxis. He reported highly motile bacteria in sulfide-rich mud that ". . . swam in the same geographic direction even when the microscope was turned around, moved to another location, or covered with a pasteboard box. To my astonishment, when a magnet was

brought near the microscope, the hundreds of swimming cells instantly turned and rushed away from the end of the magnet! They were always attracted by the end that also attracted the north-seeking end of a compass needle and they were repelled by its opposite end. Their swimming speed was very fast, on the order of 100 μm per second, and the entire population swerved in unison as the magnet was moved about nearby."*

Magnetotactic bacteria synthesize magnetite, iron oxide (Fe_3O_4), and store it in inclusions called magnetosomes, as shown in the figure. The magnetosomes act like magnets and cause the bacterial cells to align on the earth's geomagnet-

ic field. All bacteria with magnetosomes are gram-negative, have pili or a polysaccharide glycocalyx to provide attachment, and may be either rods or cocci. These bacteria are found in sediments throughout the world. In the Northern Hemisphere, north-seeking bacteria predominate, and in the Southern Hemisphere, south-seeking bacteria predominate.

Aquaspirillum magnetotacticum (see figure) metabolizes organic molecules for carbon and energy and requires a microaerophilic environment. It was found that when *A. magnetotacticum* are separated from an attachment site they swim along what appear to be geomagnetic

lines. They move downward, toward either the north or south pole, until reaching a suitable attachment site.

How these bacteria synthesize magnetite is presently being studied, and the function of magnetite in the cell's metabolism is not yet clear. The formation of ferric iron (Fe^{3+}) from ferrous iron (Fe^{2+}) could provide adenosine triphosphate (ATP) for the cell, but the estimated amount of ATP is so small that the magnetite must serve another function. In vitro, magnetosomes can decompose hydrogen peroxide, which forms in cells in the presence of oxygen and is usually degraded by the enzyme catalase before it reaches toxic levels.

The Cell Wall

The **cell wall** of the bacterial cell is a complex, semi-rigid structure that is responsible for the characteristic shape of the cell. The cell wall surrounds the underlying, fragile plasma (cytoplasmic) membrane and protects it and the internal parts of the cell from adverse changes in the surrounding environment (see Figure 4-5). Almost all procaryotes have cell walls.

The major function of the cell wall is to prevent bacterial cells from rupturing when the osmotic pressure inside the cell is greater than that outside the cell. It also helps maintain the shape of a bacterium, serves as a point of anchorage for flagella, and produces symptoms of disease in some species. As the volume of a bacterial cell increases, there is a corresponding extension of the plasma membrane and cell wall. Clinically, the cell wall is important because it is the site of action of some antibiotics.

Although some eucaryotes, including plants, algae, and fungi, contain cell walls, their cell walls differ chemically from those of procaryotes, are simpler in structure, and are less rigid.

Composition and Characteristics

The bacterial cell wall is composed of a macromolecular network called *peptidoglycan* (*murein*). Peptidoglycan is a mucopolysaccharide consisting of a repeating disaccharide attached to chains of four or five amino acids. The component monosaccharides, called N-acetylglucosamine (NAG) and N-acetylmuramic acid (NAM) (from *murus*, meaning wall), are

---

Researchers speculate that magnetosomes protect the cell against hydrogen peroxide accumulation in vivo.

The synthesis and function of magnetosomes are of interest to microbiologists to further our understanding of cells and may provide a model to explain the formation of similar iron oxides in birds, insects, and other animals. The ecology of magnetic microbes may also provide a tool for scientists studying the earth's magnetic field. Evidence suggests that the earth's magnetic field may have reversed or shifted several times. Examination of the orientation of fossils of magnetotactic bacteria preserved in rock or sediment may shed light on the movement of the continents and magnetic poles in relation to one another. Magnetotactic bacteria might also help us learn about evolution. Normally, the earth's magnetic field repels a portion of the ionized gas that continually emanates from the sun. During a magnetic reversal, more of this radiation strikes the earth, which may lead to an increase in mutations. The orientation of fossilized magnetotactic bacteria in relation to mutant fossils might help us learn whether magnetic reversals affect evolution.

*Reproduced, with permission, from the *Annual Review of Microbiology*, Volume 36, © 1982 by Annual Reviews Inc. and Richard P. Blakemore. (Richard P. Blakemore, "Magnetotactic bacteria." *Annual Review of Microbiology* 36:217–238, 1982.)

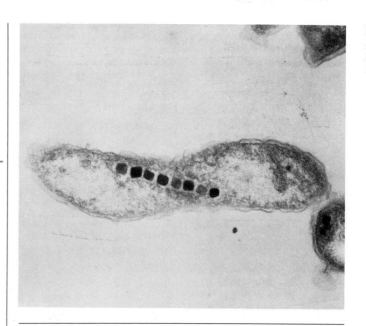

Aquaspirillum magnetotacticum, showing chain of magnetosomes. Outer membrane of gram-negative wall is also visible ($\times 30,000$).

related to glucose and have amino acid groups attached. The structural formulas for NAG and NAM are shown in Figure 4-9.

The various components of peptidoglycan are assembled in the cell wall as follows. The NAG and NAM alternate in rows, each row containing from 10 to 65 sugars and forming a carbohydrate "backbone" (Figure 4-10). To each molecule of NAM is attached

a *tetrapeptide side chain,* consisting of four amino acids. The amino acids occur in an alternating pattern of D and L forms (see Figure 2-16). This is unique since the amino acids found in proteins, except for peptidoglycan, are L forms. Parallel tetrapeptide side chains may be directly bonded to each other or linked by a *peptide cross bridge,* consisting of from one to five amino acids.

In most gram-positive bacteria, the cell wall consists of several layers of peptidoglycan connected by peptide side chains and cross bridges, the structure shown in Figure 4-10. This arrangement provides a very rigid framework. The layers of peptidoglycan are considerably thicker in gram-positive bacteria than in gram-negative bacteria (Figure 4-11a). The cell walls of many gram-positive bacteria contain *teichoic acids.* These acids make it possible for the bacteria to be identified by serological means, which will be described in Chapter 9. Teichoic acids consist primarily of an alcohol (such as glycerol or ribitol) and phosphate. The acids are bonded to the peptidoglycan layers or to the plasma membrane. Because of their negative charge (from the phosphate groups), teichoic acids may bind and regulate the movement of cations (positive ions) into and out of the cell. They may also assume a role in cell growth. When the cell wall grows, enzymes called *autolysins* are needed. These enzymes separate components of the wall to allow insertion of new subunits. Teichoic acids regulate the activity of autolysins, thus preventing extensive wall breakdown and possible wall *lysis* (rupturing). Finally, teichoic acids may be involved in storing phosphorus. The cell

Figure 4-9   N-acetylglucosamine (NAG) and N-acetylmuramic acid (NAM) joined together as in peptidoglycan. The colored areas show the differences between the two molecules. The linkage between them is called a β 1,4 linkage.

Figure 4-10   Chemical structure and organization of peptidoglycan of *Staphylococcus aureus*. The frequency of peptide cross bridges and the number of amino acids in these bridges vary with the species of bacterium. In addition to the bridges between chains in each peptidoglycan sheet, there are bridges between different sheets. Thus, a peptidoglycan macromolecule might be quite thick.

N-acetylglucosamine (NAG)

N-acetylmuramic acid (NAM)

Amino acid

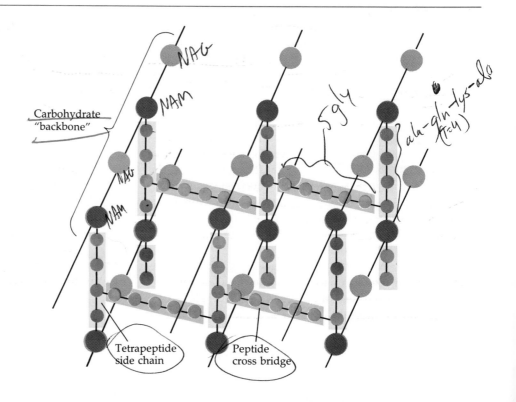

Carbohydrate "backbone"

Tetrapeptide side chain

Peptide cross bridge

Figure 4-11 Comparison of the structures and chemical components of (a) gram-positive and (b) gram-negative cell walls.

walls of acid-fast bacteria, such as *Mycobacterium* (mī-kō-bak-tī'rē-um), consist of peptidoglycan, and as much as 60% of the wall is lipids.

Gram-negative bacteria also contain peptidoglycan, but in very small amounts, and they contain no teichoic acids at all. The peptidoglycan is found in the periplasmic space, a space between the plasma membrane and outer membrane, and is covalently bonded to lipoproteins in the outer membrane. The periplasmic space is a gel-like substance that contains a high concentration of degradative enzymes and transport proteins. Because the cell walls of gram-negative bacteria contain only a small amount of peptidoglycan, they are more susceptible to mechanical breakage.

The peptidoglycan layer of the gram-negative cell is surrounded by an *outer membrane*, which consists of lipoproteins, lipopolysaccharides (LPS), and phospholipids (Figure 4-11b). Whereas lipoproteins of the plasma membrane consist of protein bound to lipid covalently, lipoproteins of the outer membrane of the cell wall are composed of protein bound to lipid noncovalently. The outer membrane has several specialized functions. Its strong negative charge is an important factor in evading phagocytosis and the action of complement, two components of the defenses of the host (discussed in detail in Chapter 15).

The outer membrane also provides a barrier to

such substances as certain antibiotics (for example, penicillin), lysozyme (an enzyme that can break down the cell walls of all gram-positive bacteria but of only a few gram-negative bacteria), detergents, heavy metals, bile salts, digestive enzymes, and certain dyes.

However, the outer membrane does not provide a barrier to all substances in the environment since nutrients must pass through to sustain the metabolism of the cell. Part of the permeability of the outer membrane is due to proteins in the membrane called **porins** that form channels. Porins are nonspecific and permit the passage of small molecules up to a molecular weight of about 800. The permeability of the outer membrane is also related to membrane proteins called **specific channel proteins**. These proteins permit the passage of specific substances only (for example, vitamin B_{12}, iron, nucleotides, and maltose) and also function as attachment sites for viruses and bacteriocins (proteins produced by some bacteria that inhibit or kill closely related species).

The lipopolysaccharide component of the outer membrane provides two important characteristics of gram-negative bacteria. The polysaccharide portion of the lipopolysaccharide is composed of sugars, called *O polysaccharides*, that function as antigens and are useful for distinguishing species of gram-negative bacteria (for example, *Salmonella* species) by serological means. This role is comparable with that of teichoic acids in gram-positive cells. The lipid portion of the lipopolysaccharide, called *lipid A*, is referred to as *endotoxin* and is toxic when in the host's bloodstream. It causes fever and shock. The nature and importance of endotoxins and other bacterial toxins will be discussed in Chapter 14.

A comparison of some characteristics of gram-positive and gram-negative bacteria is presented in Table 4-1.

Atypical Cell Walls

Among procaryotes, there are cells that naturally have no walls or have very little wall material. These include members of the genus *Mycoplasma* (mī-kō-

plaz'mä) and related organisms. Mycoplasmas are the smallest known bacteria that can grow and reproduce outside living host cells. Because they have no cell walls, they pass through most bacterial filters and were first mistaken for viruses. Their plasma membranes are unique among bacteria in having lipids called *sterols*, which are thought to help protect them from osmotic lysis. Archaeobacteria have walls composed of sugars and proteins and never have peptidoglycan.

Other atypical bacterial cells are the L forms (named after the Lister Institute, where they were discovered). These are tiny mutant bacteria with defective cell walls. Certain chemicals and antibiotics like penicillin induce many bacteria to produce L forms. Although some L forms can revert to the original bacterial form, others are stable. L forms tend to

TABLE 4-1   *Some Comparative Characteristics of Gram-Positive and Gram-Negative Bacteria*

Characteristic	Gram-Positive	Gram-Negative
Gram reaction	Retain crystal violet dye and stain dark violet or purple	Can be decolorized to accept counterstain (safranin) and stain red
Peptidoglycan layer	Thick (multilayered)	Thin (single-layered)
Lipopolysaccharide (LPS) content	Virtually none	High
Lipid and lipoprotein content	Low (acid-fast bacteria have lipids linked to peptidoglycan)	High (due to presence of outer membrane)
Teichoic acids	Present in many	Absent
Periplasmic space	Absent	Present
Outer membrane	Absent	Present
Flagellar structure	2 rings in basal body	4 rings in basal body
Toxins produced	Primarily exotoxins	Primarily endotoxins
Resistance to physical disruption	High	Low
Cell wall disruption by lysozyme	High	Low (requires pretreatment to destabilize the outer membrane)
Susceptibility to penicillin and sulfonamide	Marked	Much less marked
Susceptibility to streptomycin, chloramphenicol, and tetracycline	Much less marked	Marked
Inhibition by basic dyes	Marked	Much less marked
Susceptibility to anionic detergents	Marked	Much less marked
Resistance to sodium azide	Marked	Much less marked
Resistance to drying	High	Low

contain just enough cell wall material to prevent lysis from occurring in dilute solutions.

Recall from Chapter 3 that the mechanism for the Gram stain is related to the structure and chemical composition of the cell wall. It might serve you well to review that mechanism before reading the next section.

Damage to the Cell Wall

Chemicals that damage bacterial cell walls, or interfere with their synthesis, often do not harm the cells of an animal host because the wall is made of chemicals unlike those in eucaryotic cells. Thus, cell wall synthesis is the target for some antimicrobial drugs. One way that the cell wall can be damaged is by exposure to the enzyme *lysozyme*. This enzyme occurs naturally in some eucaryotic cells and is a constituent of tears, mucus, and saliva. Lysozyme is particularly active on the major cell wall components of most gram-positive bacteria, making them vulnerable to rupture or lysis. The lysozyme catalyzes the hydrolysis of the bonds between the sugars in the polysaccharide chain of peptidoglycan. This act is analogous to cutting the steel supports of a bridge with a cutting torch. The gram-positive bacterium's cell wall is almost completely destroyed by lysozyme. The cellular contents that remain surrounded by the plasma membrane may remain intact if osmotic lysis does not occur (defined below); this wall-less cell is termed a **protoplast**. Typically, the protoplast is spherical and is still capable of carrying on metabolism.

When lysozyme is similarly applied to gram-negative cells, usually the wall is not destroyed to the same extent as in gram-positive cells; some of the outer membrane also remains. In this case, the cellular contents, plasma membrane, and remaining outer wall layer are called a **spheroplast**, also a spherical structure. For lysozyme to exert its effect on gram-negative cells, the cells are first treated with ethylenediamine tetraacetic acid (EDTA), a substance that weakens ionic bonds in the outer membrane and damages it, giving the lysozyme access to the peptidoglycan layer.

Protoplasts and spheroplasts burst in pure water or very dilute salt or sugar solutions because the water molecules from the surrounding fluid rapidly move into and enlarge the cell, which has a much lower internal concentration of water. This rupturing is called **osmotic lysis**, and will soon be discussed in detail.

Certain antibiotics, such as penicillin, destroy bacteria by interfering with the formation of the peptide cross bridges of peptidoglycan, thus preventing the formation of a functional cell wall. Most gram-negative bacteria are not as susceptible to penicillin as gram-positive bacteria are because the outer membrane of gram-negative bacteria forms a barrier that inhibits the entry of this and other substances. Organisms that lack cell walls, such as mycoplasmas, are also resistant to antibiotics that interfere with cell wall synthesis. Gram-negative bacteria are quite susceptible to β-lactam antibiotics, other than penicillin. Antibiotic drugs will be discussed in more detail in Chapter 19.

Structures Internal to the Cell Wall

Thus far, we have discussed the procaryotic cell wall and structures external to it. We will now look inside the procaryotic cell and discuss the structures and functions of the plasma membrane and other components within the cytoplasm of the cell.

Plasma (Cytoplasmic) Membrane

The **plasma (cytoplasmic) membrane** is a thin structure lying inside the cell wall and enclosing the cytoplasm of the cell (see Figure 4-5). The plasma membrane of procaryotes consists primarily of phospholipids (see Figure 2-13), which are the most abundant chemicals in the membrane, and proteins. Eucaryotic plasma membranes also contain carbohydrates and sterols, such as cholesterol. Because they lack sterols, procaryotic plasma membranes are less rigid than eucaryotic membranes. One exception to this is the wall-less procaryote called mycoplasma, which contains membrane sterols.

Structure

In electron micrographs, procaryotic and eucaryotic cell membranes both look like two-layered structures; there are two dark lines with a light space between the lines (Figure 4-12a). The phospholipid molecules are arranged in two parallel rows, called a *phospholipid bilayer* (Figure 4-12b and c). Each phospholipid molecule contains a polar head, composed of a phosphate group and glycerol that is hydrophilic (water-loving) and soluble in water, and nonpolar tails, composed of fatty acids that are hydrophobic (water-fearing) and are insoluble in water (Figure 4-12a).

(a)

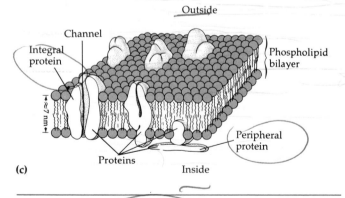

(b)

(c)

Figure 4-12   Plasma membrane. (a) Electron micrograph showing the phospholipid bilayer forming the plasma membrane of the bacterium *Bacillus brevis* (×175,000). Layers of the cell wall can be seen outside the plasma membrane. **(b)** Space-filling models of several molecules as they are arranged in the phospholipid bilayer. **(c)** Drawing of a membrane showing phospholipid bilayer and proteins.

The polar heads are on the two surfaces of the phospholipid bilayer, and the nonpolar tails are in the interior of the bilayer.

The protein molecules in the membrane can be arranged in a variety of ways (Figure 4-12c). Some, called *peripheral proteins*, are easily removed from the membrane by chemical treatment and are thought to lie at the inner or outer surface of the membrane. They may function as enzymes that catalyze chemical reactions, as a "scaffold" for support, and as mediators of changes in membrane shape during movement. Other proteins, called *integral proteins*, can be removed from the membrane only after disrupting the bilayer. Integral proteins are believed to penetrate the membrane completely or possibly partly. Some of these proteins contain channels through which substances enter and exit the cell. Studies by physicists and biologists have demonstrated that the phospholipid and protein molecules in membranes are not static but seem to move quite freely within the membrane surface. This movement is most likely associated with the many functions performed by the plasma membrane. This dynamic arrangement of phospholipid and protein is referred to as the *fluid mosaic model* (see the box in Chapter 2, p. 42).

Functions

The most important function of the plasma membrane is to serve as a selective barrier through which materials enter and exit the cell. In this function, plasma membranes are **selectively permeable** (sometimes called **semipermeable**). This term indicates that certain molecules and ions pass through the membrane, but others are restricted. The permeability of the membrane depends on several factors. Large molecules (such as proteins) cannot pass through the plasma membrane. This may be because these molecules are larger than the channels in integral proteins. But smaller molecules (such as water, amino acids, and some simple sugars) usually pass through easily. Ions penetrate the membrane only very slowly. Substances that dissolve easily in lipids (such as oxygen, carbon dioxide, and nonpolar organic molecules) enter and exit more easily than do other substances because the membrane consists mostly of phospholipids. The movement of materials across plasma membranes also depends on carrier molecules, which are described shortly.

Plasma membranes are also important to the breakdown of nutrients and the production of energy. The plasma membranes of bacteria contain enzymes capable of catalyzing the chemical reactions that break down nutrients and produce ATP. In some bacteria, pigments and enzymes involved in photo-synthesis (the conversion of light energy into chemical energy) are found in infoldings of the plasma membrane that extend into the cytoplasm. These membrane layers are called **chromatophores** or **thylakoids** (Figure 4-13).

Mesosomes

The plasma membranes of bacteria often contain one or more large, irregular folds called **mesosomes** (see Figure 4-5). Generally, mesosomes are seen in association with the nuclear area at or near the site of cell division. Mesosomes are not present in eucaryotic cells.

Although the exact role of mesosomes is unknown, they might play a role in reproduction and metabolism. When a bacterial cell divides (binary fission), a cross wall called a *transverse septum* forms and partitions the genetic material of the parent cell into each of the two identical daughter cells (Figure 4-14). Mesosomes might begin the formation of the transverse septum and attach bacterial DNA to the plasma membrane. Mesosomes might also help to separate DNA into each daughter cell following binary fission. Some scientists also believe that mesosomes may play a role in the generation of ATP and in secretion of enzymes from the cell. And the ability of cells to concentrate nutrients might be enhanced by the presence of mesosomes because their folds increase the surface area of the plasma membrane.

There is considerable debate regarding the structure and function of mesosomes. It has even been suggested that mesosomes are artifacts (structures that appear in microscopic preparations due to the method of preparation). In other words, mesosomes may not be actual structures in the living cell.

Destruction of the Plasma Membrane by Antimicrobial Agents

Because the plasma membrane is vital to the bacterial cell, it is not surprising that it is the site at which several antimicrobial agents exert their effect. In addition to the chemicals that damage the cell wall and hence indirectly expose the membrane to injury, many compounds specifically damage plasma membranes. These compounds include certain alcohols and

Transverse septum

Plasma membrane

Cell wall

Cytoplasm

Nuclear area (nucleoid)

Ribosomes

Mesosomes

Figure 4-14 Transverse septum in a dividing cell of *Sporosarcina ureae.* Two mesosomes can be seen extending into the cytoplasm from the innermost edges of the septa (×72,000).

Figure 4-13 Chromatophores. Membranous structures that are sites of photosynthesis can be seen in this transmission electron micrograph of *Rhodospirillum rubrum,* a purple sulfur bacterium (×30,000).

quaternary ammonium compounds, which are used as disinfectants. By disrupting the membrane's phospholipids, a group of antibiotics known as the *polymyxins* cause leakage of intracellular contents and subsequent cell death. This mechanism is discussed in Chapter 19.

Movement of Materials Across Membranes

When the concentration of a substance is stronger on one side of a membrane than on another, a **concentration gradient** (difference) exists. If the substance can cross the membrane, it will move to the more dilute side until the concentrations are equal or until other forces stop its movement.

Materials move across plasma membranes of both procaryotic and eucaryotic cells by two kinds of processes, passive and active. In **passive processes**, substances cross the membrane from an area of high concentration to an area of low concentration (with the concentration gradient), without any expenditure of energy (ATP) by the cell. Examples of passive processes are simple diffusion, osmosis, and facilitated diffusion, to be discussed shortly. In **active processes**, the cell must use energy (ATP) to move substances from areas of low concentration to areas of high concentration (against the concentration gradient). Examples of active processes are active transport and group translocation, which we will also discuss shortly.

Simple Diffusion

Simple diffusion is the net (overall) movement of molecules or ions from an area of high concentration to an area of low concentration (Figure 4-15). The movement from areas of high to low concentration continues until the molecules or ions are evenly distributed. The point of even distribution is called *equilibrium*. Cells rely on diffusion to transport certain small molecules, such as oxygen and carbon dioxide, across their cell membranes.

Facilitated Diffusion

When **facilitated diffusion** occurs, the substance (glucose, for example) to be transported combines with a *carrier protein* in the plasma membrane. Such carriers are sometimes called *permeases.* The carrier can transport a substance across the membrane from an area of high concentration to one of low concentration. Carrier proteins bind to the substance they transport on the outer surface of the plasma membrane and, by

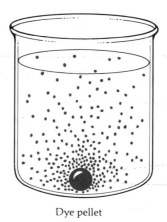

Dye pellet

Figure 4-15   The principle of simple diffusion. The molecules of dye in the pellet are diffusing into the water from an area of high dye concentration to areas of low dye concentration.

a mechanism still poorly understood, transport the substance through the membrane to the inner surface and release it into the cytoplasm. One proposed mechanism for facilitated diffusion is that the carrier protein remains in place in the membrane but undergoes a change in shape that transports substances from one side of the membrane to the other (Figure 4-16). Facilitated diffusion is similar to simple diffusion, in that the cell does not need to expend energy because the substance moves from a high to a low concentration. The process differs from simple diffusion in its use of carriers.

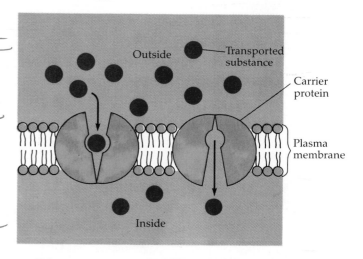

Figure 4-16   Facilitated diffusion. Carrier proteins in the plasma membrane transport molecules across the membrane from an area of high concentration to one of low concentration (with the concentration gradient). The carrier molecule probably undergoes a change in shape to transport the substance. The process does not require ATP.

In some cases, molecules that bacteria need are too large to be transported into the cells by the methods just described. Most bacteria, however, produce enzymes that can break down large molecules, such as proteins into amino acids, and polysaccharides into simple sugars. Such enzymes, which are released by the bacteria into the surrounding medium, are appropriately called *extracellular enzymes*. Once the enzymes degrade the large molecules, the subunits are transported by permeases into the cell. For example, specific carriers retrieve DNA bases, such as the purine guanine, from extracellular media and bring them into the cell's cytoplasm.

Osmosis

Osmosis is the net movement of solvent molecules across a selectively permeable membrane from an area in which the solvent molecules are highly concentrated to an area of low concentration. In living systems, the chief solvent is water.

Osmosis may be demonstrated with the apparatus shown in Figure 4-17a and b. A sack constructed from cellophane, which is a selectively permeable membrane, is filled with a solution of 20% sucrose (table sugar). The opening of the cellophane sack is plugged with a rubber stopper, through which a glass tubing is fitted. The cellophane sack is placed into a beaker containing distilled water. Initially, the concentrations of water on either side of the membrane are different. Because of the sucrose molecules, the concentration of water is lower inside the cellophane sack than outside. Therefore, water moves from the beaker (where its concentration is higher) into the cellophane sack (where its concentration is lower).

There is no movement of sugar out of the cellophane sack into the beaker, however, because the cellophane is impermeable to molecules of sugar—the sugar molecules are too large to go through the pores of the membrane. As water moves into the cellophane sack, the sugar solution becomes increasingly dilute, and, because the cellophane sack is at maximum

Figure 4-17 The principle of osmosis. (a) Setup at the beginning of the experiment. Water molecules start to move from the beaker into the sack along the concentration gradient. **(b)** Setup at equilibrium. The osmotic pressure exerted by the solution in the sack pushes water molecules from the sack back into the beaker to balance the rate of water entry into the sack. The final height of solution in the glass tube in (b) is a measure of the osmotic pressure. **(c)–(e)** Effects of various solutions on bacterial cells.

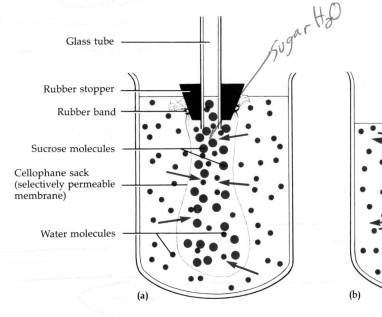

Glass tube

Rubber stopper

Rubber band

Sucrose molecules

Cellophane sack (selectively permeable membrane)

Water molecules

(a)

(b)

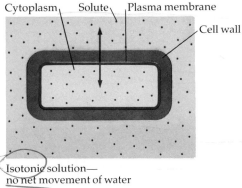

Cytoplasm Solute Plasma membrane

Cell wall

Isotonic solution— no net movement of water

(c)

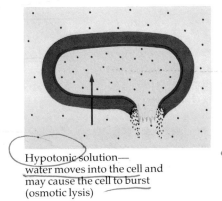

Hypotonic solution— water moves into the cell and may cause the cell to burst (osmotic lysis)

(d)

Hypertonic solution— water moves out of the cell, causing it to shrink (plasmolysis)

(e)

expansion due to its increased volume of water, water begins to move up the glass tubing. In time, the water that has accumulated in the cellophane sack and the glass tubing exerts a downward pressure that forces water molecules out of the cellophane sack and back into the beaker. This movement of water through a selectively permeable membrane produces a pressure called osmotic pressure. **Osmotic pressure** is the pressure required to prevent the movement of pure water (water with no solutes) into a solution containing some solutes. In other words, osmotic pressure is the pressure needed to stop the flow of water across the selectively permeable membrane (cellophane). When water molecules leave and enter the cellophane sack at the same rate, equilibrium is reached.

A bacterial cell may be subjected to three kinds of osmotic solutions, isotonic, hypotonic, or hypertonic. An **isotonic solution** (*iso* means equal) is one in which the overall concentrations of solutes are the same on both sides of the membrane. Water leaves and enters the cell at the same rate (no net change); the cell's contents are in equilibrium with the solution outside the cell wall (Figure 4-17c).

Earlier we mentioned that lysozyme and certain antibiotics damage bacterial cell walls by causing the cells to rupture, or lyse. Such rupturing occurs because bacterial cytoplasm usually contains such a strong concentration of solutes that, when the wall is weakened or removed, additional water enters the cell by osmosis. The damaged (or removed) cell wall cannot constrain the swelling of the cytoplasmic membrane, and the membrane bursts. This is an example of osmotic lysis caused by immersion in a hypotonic solution. A **hypotonic solution** (*hypo* means under or less) outside the cell is a medium whose concentration of solutes is lower than that inside the cell. Most bacteria live in hypotonic solutions, and swelling is contained by the cell wall. Cells with weak cell walls, such as gram-negative bacteria, may burst or undergo osmotic lysis due to excessive water intake (Figure 4-17d).

A **hypertonic solution** (*hyper* means above) is a medium having a higher concentration of solutes than the cell has. Most bacterial cells placed in a hypertonic solution shrink and collapse or plasmolyze because water leaves the cells by osmosis (Figure 4-17e).

Keep in mind that the terms *isotonic*, *hypotonic*, and *hypertonic* describe the concentration of solutions outside the cell *relative* to the concentration inside the cell.

Active Transport

In performing **active transport**, the cell uses energy in the form of adenosine triphosphate (ATP) to move substances across the plasma membrane. The move-

ment of a substance in active transport is usually from outside to inside, even though the concentration might be much higher inside the cell. Like facilitated diffusion, active transport depends on carrier proteins in the plasma membrane (see Figure 4-16). There appears to be a different carrier for each transported substance or group of closely related transported substances.

Group Translocation

In active transport, the substance that crosses the membrane is not altered by transport across the membrane. In **group translocation**, a special form of active transport that occurs exclusively in procaryotes, the substance is chemically altered during transport across the membrane. Once the substance is altered and inside the cell, the plasma membrane becomes impermeable to it, so it remains inside the cell. This is important because a cell can accumulate various substances even though they may be in low concentrations outside the cell. Group translocation requires energy, supplied by phosphoenolpyruvate (PEP), a high-energy phosphate compound.

One example of group translocation is the transport of the sugar glucose, which is often used in growth media for bacteria. During transport of the glucose molecule across the membrane by a specific carrier protein, a phosphate group is added to the sugar. The energy needed for the phosphate addition is provided by PEP. The phosphorylated form of glucose, which cannot be transported out, can then be used in the cell's metabolic pathways.

Simple diffusion and facilitated diffusion are useful mechanisms for transporting substances into cells when the concentration of the substances is greater outside the cell. However, when a bacterial cell is in an environment in which nutrients are in low concentration, the cell must utilize active transport and group translocation to accumulate the needed substances.

Eucaryotic cells can use two additional active transport processes called phagocytosis and pinocytosis. Both processes are described in later sections.

Cytoplasm

For a procaryotic cell, the term **cytoplasm** refers to the internal matrix of the cell contained inside the plasma membrane (see Figure 4-5). Cytoplasm is about 80% water and contains primarily proteins (enzymes), carbohydrates, lipids, inorganic ions, and many low-molecular-weight compounds. Inorganic ions are present in much higher concentrations in cytoplasm than in most media. Cytoplasm is thick, semitransparent, and elastic. The major structures in the cytoplasm are DNA, particles called ribosomes, and reserve deposits called inclusions.

Procaryotic cytoplasm lacks certain features of eucaryotic cytoplasm, such as a cytoskeleton and cytoplasmic streaming. These features will be described later.

Nuclear Area

The **nuclear area**, or **nucleoid**, of a bacterial cell (see Figure 4-5) contains a single long circular molecule of double-stranded DNA, the **bacterial "chromosome"** (see Figure 8-2a). This is the cell's genetic information, its DNA, carrying all the information required for the cell's structures and functions. Unlike the chromosomes of eucaryotic cells, bacterial chromosomes do not include histones and are not surrounded by a nuclear envelope. The nuclear area can be spherical, elongated, or dumbbell-shaped.

Bacteria often contain, in addition to the bacterial chromosome, small circular, double-stranded DNA molecules called **plasmids** (see the F factor in Figure 8-24a). These are extrachromosomal genetic elements; that is, they are not connected to the main bacterial chromosome, and they replicate independently of chromosomal DNA. Plasmids usually contain from five to 100 genes, which are generally not crucial for the survival of the bacterium under the usual environmental conditions, and plasmids may be gained or lost without harming the cell. Under certain conditions, however, plasmids are an advantage to cells. Plasmids may carry genes for such activities as antibiotic resistance, tolerance to toxic metals, production of toxins, and synthesis of enzymes. Plasmids can be transferred from one bacterium to another.

Ribosomes

All eucaryotic and procaryotic cells contain **ribosomes**, which function as the sites of protein synthesis. Cells that have high rates of protein synthesis have a great number of ribosomes. The cytoplasm of a procaryotic cell contains thousands of these very small structures, which give the cytoplasm a granular appearance (see Figure 4-5).

Ribosomes are composed of two subunits, each subunit being composed of protein and a type of RNA called ribosomal RNA (rRNA). Procaryotic ribosomes

differ from eucaryotic ribosomes in the number of proteins and rRNA molecules they contain and are somewhat smaller and less dense than ribosomes of eucaryotic cells. Accordingly, procaryotic ribosomes are called 70S ribosomes, and those of eucaryotic cells are known as 80S ribosomes. The letter *S* refers to Svedberg units, which indicate the relative rate of sedimentation during ultra-high-speed centrifugation. Sedimentation rate is a function of the size, weight, and shape of a particle. The subunits of a 70S ribosome are a small 30S subunit containing one molecule of rRNA and a larger 50S subunit containing two molecules of rRNA.

Several antibiotics, such as streptomycin, neomycin, and tetracyclines, exert their antimicrobial effects by inhibiting protein synthesis on the ribosomes. Because of differences in procaryotic and eucaryotic ribosomes, the microbial cell can be killed by the antibiotic while the eucaryotic host cell is unaffected.

Inclusions

Within the cytoplasm of procaryotic (and eucaryotic) cells are several kinds of reserve deposits, known as **inclusions**. Some inclusions are common to a wide variety of bacteria, whereas others are limited to a small number of species and therefore serve as a basis for identification. Among the more prominent bacterial inclusions are the following.

Metachromatic Granules

These inclusions sometimes stain red with certain blue dyes, such as methylene blue, and are collectively known as **volutin**. Volutin represents a reserve of inorganic phosphate (polyphosphate) that can be used in the synthesis of ATP. It is generally formed by cells that grow in phosphate-rich environments. Metachromatic granules are found in algae, fungi, and protozoans, as well as in bacteria. These granules are quite large and are characteristic of *Corynebacterium diphtheriae* (kô-rī-nē-bak-ti′rē-um dif-thi′rē-ī), the causative agent of diphtheria; thus, they have diagnostic significance.

Polysaccharide Granules

These inclusions typically consist of glycogen and starch, and their presence can be demonstrated when iodine is applied to the cells. In the presence of iodine, glycogen granules appear reddish brown, and starch granules appear blue.

Lipid Inclusions

Lipid inclusions appear in various species of *Mycobacterium*, *Bacillus*, *Azotobacter* (ä-zō-tō-bak'tėr), *Spirillum* (spī-ril'lum), and other genera. (A lipid commonly found as a storage material and unique to bacteria is the polymer *poly-β-hydroxybutyric acid*.) Lipid inclusions are revealed by use of fat-soluble dyes, such as Sudan dyes.

Sulfur Granules

Certain bacteria, known as the "sulfur bacteria," which belong to the genus *Thiobacillus*, derive energy by oxidizing sulfur and sulfur-containing compounds. These bacteria may deposit sulfur granules in the cell, where they serve as an energy reserve.

Carboxysomes

These are polyhedral and hexagonal inclusions that contain the enzyme ribulose 1,5-diphosphate carboxylase. Bacteria that use carbon dioxide as their sole source of carbon require this enzyme for photosynthesis. Among the bacteria containing carboxysomes are nitrifying bacteria, cyanobacteria, and thiobacilli.

Gas Vacuoles

These are hollow cavities found in many aquatic procaryotes, including cyanobacteria, anoxygenic photosynthetic bacteria, and halobacteria. Each vacuole consists of rows of several individual *gas vesicles*, which are hollow cylinders covered by protein. The function of gas vacuoles is to maintain buoyancy so that the cells can remain at the depth in the water appropriate for them to receive sufficient amounts of oxygen, light, and nutrients.

For a discussion of inclusions in magnetotactic bacteria, see the box on p. 80.

Endospores

When essential nutrients are depleted, or when water is unavailable, certain gram-positive bacteria, such as those of the genera *Clostridium* (klos-tri'dē-um) and *Bacillus*, form specialized "resting" cells called **endospores**. Unique to bacteria, endospores are highly durable dehydrated bodies with thick walls or additional layers. They are formed *inside* the bacterial cell membrane. Released into the environment, they can survive extreme heat, lack of water, and exposure to many toxic chemicals and radiation.

The process of endospore formation within a vegetative (parent) cell is known as **sporulation** or **sporogenesis** (Figure 4-18a and b). It is not clear what biochemical events trigger this process. In the first observable stage of sporogenesis, a newly replicated bacterial chromosome and a small portion of cytoplasm are isolated by an ingrowth of the plasma membrane called a *spore septum*. The spore septum becomes a double-layered membrane that surrounds the chromosome and cytoplasm. This structure, entirely enclosed within the original cell, is called a *forespore*. Thick layers of peptidoglycan are laid down between the two membrane layers. Then a thick *spore coat* of protein forms around the outside membrane. It is this coat that is responsible for the resistance of endospores to many harsh chemicals.

The diameter of the endospore may be the same as, smaller than, or larger than the diameter of the vegetative cell. Depending on the species, the endospore might be located *terminally* (at one end), *subterminally* (near one end), or *centrally* inside the vegetative cell (Figure 4-18c). When the endospore matures, the vegetative cell wall dissolves (lyses), and the endospore is freed.

Most of the water present in the forespore cytoplasm is eliminated by the time sporogenesis is complete, and endospores do not carry out metabolic reactions. The highly dehydrated endospore core contains only DNA, small amounts of RNA, ribosomes, enzymes, and a few important small molecules. The latter include a strikingly large amount of an organic acid called *dipicolinic acid* (found in the cytoplasm), which is accompanied by a large amount of calcium ion. These cellular components will be essential for resuming metabolism later.

Endospores can remain dormant for a long time, even hundreds of years. An endospore returns to its vegetative state by a process called **germination**. Germination is triggered by physical or chemical damage to the endospore's coat. The endospore's enzymes then break down the extra layers surrounding the endospore, water enters, and metabolism resumes. Because one vegetative cell forms a single endospore, which after germination remains one cell, sporogenesis in bacteria is *not* a means of reproduction. There is no increase in the number of cells.

Endospores are important from a clinical viewpoint and in the food industry because they are resistant to processes that normally kill vegetative cells. Such processes include heating, freezing, desiccation, use of chemicals, and radiation. Whereas most vegetative cells are killed by temperatures above 70°C, endospores can survive in boiling water for an hour or more. Endospore-forming bacteria are a problem in the food industry because they are likely to survive

Figure 4-18 Endospores.
(a) Sporogenesis, the process of endospore formation. **(b)** Electron micrograph of an endospore in *Bacillus sphaericus*. **(c)** Various locations of endospores. *Left*: Central endospore in *Bacillus megaterium* (×40,500). *Center*: Subterminal endospore in *Bacillus anthracis* (×38,000). *Right*: Round, terminal endospore at the ends of a cell of *Clostridium tetani* (approx. ×14,000).

(a)

(b)

(c)

any underprocessing, and if conditions for growth occur, some species produce toxins and disease. Special methods used to control organisms that produce endospores are discussed in Chapter 7.

As noted in Chapter 3, endospores are difficult to stain for detection. Thus, a specially prepared stain must be used along with heat. (The Schaeffer–Fulton endospore stain is commonly used.)

Having examined the functional anatomy of the procaryotic cell, we will now look at the functional anatomy of the eucaryotic cell.

The Eucaryotic Cell

As mentioned earlier, eucaryotic organisms include algae, protozoans, fungi, higher plants, and animals. The eucaryotic cell (Figure 4-19) is typically larger and structurally more complex than the procaryotic

cell. By comparing the structure of the procaryotic cell in Figure 4-5 with that of the eucaryotic cell, we can see the differences between the two types of cells.

To review briefly, eucaryotic cells contain membrane-bounded organelles; procaryotic cells do not. **Organelles** are specialized structures that perform specific functions. Although both types of cells carry on the same basic functions, procaryotic cells do not localize these functions in specific organelles. The genetic material (DNA) of procaryotic cells is not membrane-bounded or associated in a regular way with protein. The genetic material of eucaryotic cells *is* membrane-bounded, organized into chromosomes, and closely associated with histones and other proteins.

The following discussion of eucaryotic cells will parallel our discussion of procaryotic cells by starting with structures that extend external to the cell. At the end of the discussion, the principal differences between procaryotic and eucaryotic cells will be summarized in Table 4-2.

Figure 4-19 Highly schematic diagram of a composite eucaryotic cell.

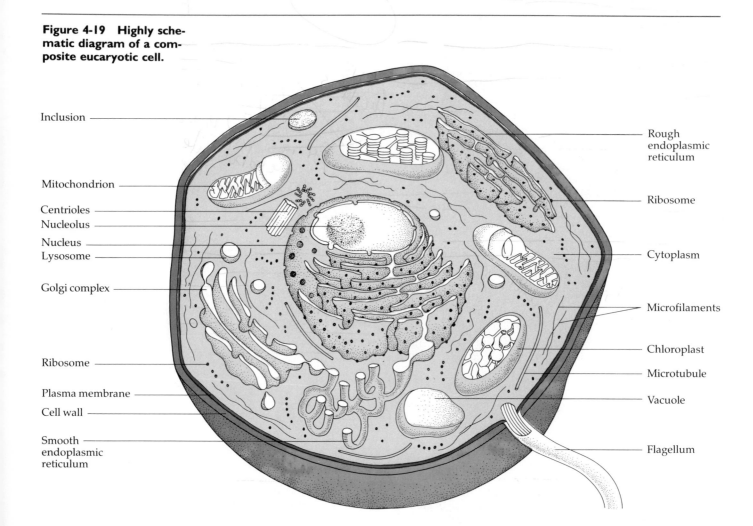

Flagella and Cilia

Many types of eucaryotic cells have projections that are used for cellular locomotion or for moving substances along the surface of the cell. These projections contain cytoplasm and are enclosed by the plasma membrane. If the projections are few and are long in relation to the size of the cell, they are called **flagella**. If the projections are numerous and short, resembling hairs, they are called **cilia**. Both flagella and cilia consist of nine pairs of microtubules arranged in a ring, plus two single microtubules in the center of the ring. Euglenoid algae use a flagellum for locomotion, whereas protozoans, such as *Paramecium* (pār-ä-mē′sē-um), utilize cilia for locomotion (Figure 4-20a and b). A procaryotic flagellum rotates, but a eucaryotic flagellum moves in a wavelike manner (Figure

4-20c). To help keep foreign material out of the lungs, ciliated cells of the human respiratory system move the material along the surface of the cells in the bronchial tubes and trachea toward the throat and mouth (see Figure 15-3).

The Cell Wall

In general, the eucaryotic **cell wall** is considerably simpler than that of the procaryotic cell. Most algae have cell walls consisting of the polysaccharide *cellulose* (as do all plants). Cell walls of some fungi also contain cellulose, but in most fungi, the principal structural component of the cell wall is the polysaccharide *chitin*, a polymer of N-acetylglucosamine (NAG) units. (Chitin is also the main structural component of the exoskeleton of crustaceans and insects.)

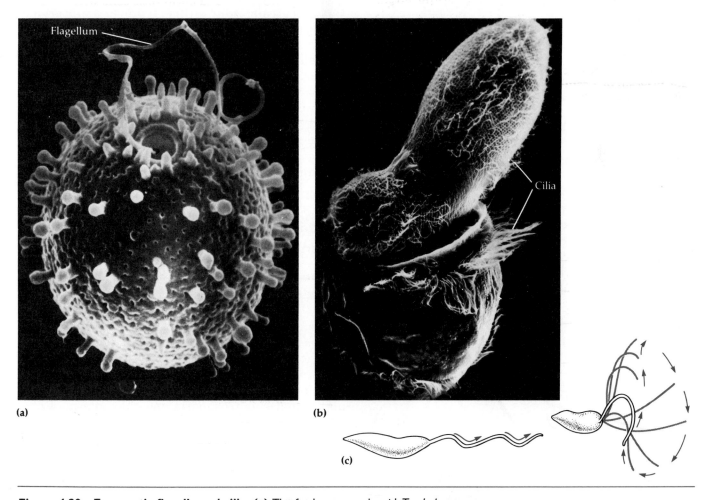

(a)   (b)   (c)

Figure 4-20 Eucaryotic flagella and cilia. (a) The freshwater euglenoid *Trachelomonas* sp. (×5000). Its long, whiplike flagellum is emerging from the cell body at the top of this scanning electron micrograph. **(b)** Ciliate eats ciliate. *Didinium nasutum*, the lower cell with two rows of cilia around its body, has begun to engulf a cilia-covered cell of *Paramecium multimicronucleatum* (×1500). **(c)** Movement of a eucaryotic flagellum.

The cell walls of yeasts contain the polysaccharides *glucan* and *mannan*. In eucaryotes that lack a cell wall, the plasma membrane may be the outer covering; however, cells such as slime molds and amoebas, which have direct contact with the environment, may have coatings outside the plasma membrane (see Microview A, Photo 30). Protozoans do not have a typical cell wall; instead, they have a flexible outer covering called a *pellicle*.

An important clinical consideration is that eucaryotic cells do not contain peptidoglycan, the framework of the procaryotic cell wall. This is significant medically because antibiotics, such as penicillins and cephalosporins, act against peptidoglycan and therefore do not affect human eucaryotic cells.

The Plasma (Cytoplasmic) Membrane

In eucaryotic cells that lack a cell wall, the **plasma membrane** is the external covering of the cell. In function and basic structure, the eucaryotic and procaryotic plasma membranes are very similar. There are, however, differences in the proteins. In eucaryotes, carbohydrates serve as receptor sites that assume a role in such functions as cell-to-cell recognition. These carbohydrates also provide adherence for bacteria. Eucaryotic plasma membranes also contain *sterols*, complex lipids not found in procaryotic plasma membranes (with the exception of the mycoplasmas). Sterols seem to be associated with the ability of the membranes to resist lysis due to increased osmotic pressure.

Substances can cross eucaryotic and procaryotic plasma membranes by simple diffusion, facilitated diffusion, osmosis, or active transport. Group translocation does not occur in eucaryotic cells. However, eucaryotic cells can utilize an additional mechanism called **endocytosis**. This occurs when a segment of the plasma membrane surrounds a particle or large molecule, encloses it, and brings it into the cell. Endocytosis is one of the ways that viruses can enter animal cells.

Two very important types of endocytosis are phagocytosis and pinocytosis. During phagocytosis, cellular projections called pseudopods engulf particles and bring them into the cell. Phagocytosis is used by white blood cells to destroy bacteria and foreign substances (see Figure 15-8 and further discussion in Chapter 15). In pinocytosis, the plasma membrane folds inward, bringing extracellular fluid into the cell along with whatever substances are dissolved in the fluid.

Cytoplasm

The **cytoplasm** of the eucaryotic cell encompasses the matrix inside the cell's membrane and outside the nucleus (see Figure 4-19). The cytoplasm is the substance in which various cellular components are found. A major difference between eucaryotic and procaryotic cytoplasm is that eucaryotic cytoplasm has a complex internal structure, consisting of exceedingly small rods called *microfilaments* and *intermediate filaments* and cylinders called *microtubules*. Together, they form the **cytoskeleton**. The cytoskeleton and other cytoplasmic components are held together by a three-dimensional membrane of fine filaments called the **microtrabecular lattice**. It provides support and shape, organizes chemical reactions that occur in cytoplasm, and assists in transporting substances through the cell and even moving the entire cell, as in phagocytosis. The movement of eucaryotic cytoplasm from one part of the cell to another, which helps to distribute nutrients and move the cell over a surface, is called *cytoplasmic streaming*. Another difference between procaryotic and eucaryotic cytoplasm is that many of the important enzymes found in the cytoplasmic fluid of procaryotes are sequestered in the organelles of eucaryotes.

Organelles

Nucleus

The most characteristic eucaryotic organelle is the nucleus (see Figure 4-19). The **nucleus** is usually spherical or oval (Figure 4-21), is frequently the largest structure in the cell, and contains almost all of the cell's hereditary information (DNA). Some DNA is also found in mitochondria and chloroplasts of photosynthetic organisms.

The nucleus is separated from the cytoplasm by a double membrane called the *nuclear envelope*. Each of the two membranes resembles the plasma membrane in structure. Minute *pores* in the nuclear membrane allow the nucleus to communicate with the membranous network in the cytoplasm, called the endoplasmic reticulum (see next section). Substances entering and exiting the nucleus are believed to pass

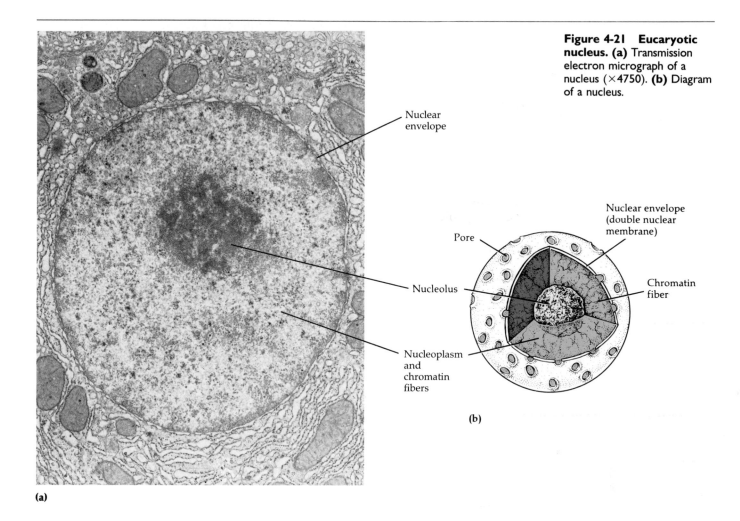

Figure 4-21 Eucaryotic nucleus. (a) Transmission electron micrograph of a nucleus (×4750). **(b)** Diagram of a nucleus.

Nuclear envelope

Nuclear envelope (double nuclear membrane)

Pore

Nucleolus

Chromatin fiber

Nucleoplasm and chromatin fibers

(b)

(a)

through the tiny pores. Within the nuclear envelope is a gelatinous fluid called *nucleoplasm*. Spherical bodies called *nucleoli* are also present. These structures are a center for the synthesis of ribosomal RNA, an essential constituent of ribosomes (see next section).

Finally, there is the DNA, which is combined with a number of proteins, including several basic proteins called *histones* and *nonhistones*. The combination of about 165 base pairs of DNA and 9 molecules of histones is referred to as a *nucleosome*. When the cell is not reproducing, the DNA and its associated proteins appear as a threadlike mass called **chromatin**. Prior to nuclear division, the chromatin coils into shorter and thicker rodlike bodies called **chromosomes**. Nucleosomes are the basic structural units of chromosomes. Procaryotic "chromosomes" do not undergo this process, do not have histones, are not formed of nucleosomes, and are not enclosed in a nuclear envelope.

Eucaryotic cells divide by two elaborate mechanisms called mitosis and meiosis (Chapter 8). Neither occurs in procaryotic cells.

Endoplasmic Reticulum

Within the cytoplasm, there is a system consisting of pairs of parallel membranes that enclose narrow vesicles or sacs of varying shapes. This system, not present in procaryotes, is known as the **endoplasmic reticulum**, or **ER** (Figure 4-22).

The ER is a network of tubes running throughout the cytoplasm that is continuous with both the plasma membrane and the nuclear membrane (see Figure 4-19). It is thought that the ER provides a surface area for chemical reactions, a pathway for the transportation of molecules within the cell, and a storage area for synthesized molecules. The ER plays a role in both lipid synthesis and protein synthesis.

Endoplasmic reticulum
covered with ribosomes

Ribosomes free
in cytoplasm

(a)

Endoplasmic
reticulum

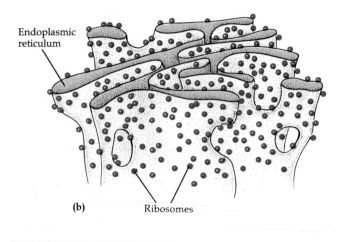

(b)   Ribosomes

Figure 4-22   Endoplasmic reticulum and ribosomes.
(a) Transmission electron micrograph of endoplasmic reticulum
and ribosomes in cross section (×24,000). **(b)** Three-dimensional
drawing of endoplasmic reticulum and ribosomes.

Ribosomes

Attached to the outer surface of some of the ER are
ribosomes, which are also found free in the cytoplasm
(Figure 4-22). As in procaryotes, ribosomes are the
sites of protein synthesis in the cell.

The ribosomes of the eucaryotic ER and cytoplasm
are somewhat larger and denser than those of procaryotic cells. These eucaryotic ribosomes are 80S
ribosomes. An 80S ribosome consists of a large 60S
subunit containing three molecules of rRNA and a
smaller 40S subunit with one molecule of rRNA. Chloroplasts and mitochondria contain 70S ribosomes,
which may indicate their evolution from procaryotes.
This hypothesis is discussed later in this chapter. The
role of ribosomes in protein synthesis will be discussed
in more detail in Chapter 8.

Golgi Complex

Another organelle found in the cytoplasm of eucaryotic cells is the **Golgi complex**. This structure usually
consists of from four to eight flattened sacs (cisternae)
with expanded ends (vesicles). The cisternae are
stacked one on the other (Figure 4-23). The Golgi complex is sometimes connected to the ER. One function
of the Golgi complex is to package and secrete (release
from the cell) certain proteins and lipids. Another
function of the Golgi complex is the synthesis of carbohydrates and the combining of carbohydrates and
proteins to form complexes called *glycoproteins*. Carbohydrates and glycoproteins are also secreted from
the cell in vesicles.

Mitochondria

Spherical or rod-shaped organelles called **mitochondria** appear throughout the cytoplasm of eucaryotic
cells (see Figure 4-19). A mitochondrion consists of a
double membrane similar in structure to the plasma
membrane (Figure 4-24). The outer mitochondrial
membrane is smooth, but the inner mitochondrial
membrane is arranged in a series of folds called *cristae*. The center of the mitochondrion is a semifluid
substance called the *matrix*. Because of the nature
and arrangement of the cristae, the inner membrane
provides an enormous surface area on which chemical
reactions can occur. Some proteins that function in
cellular respiration, including the enzyme that makes
ATP, are located on the cristae of the inner mitochondrial membrane, and many of the metabolic steps
involved in cellular respiration are concentrated in
the matrix (Chapter 5). Mitochondria are frequently
called the "powerhouses of the cell" because of their
central role in the production of ATP.

Mitochondria contain 70S ribosomes and some
DNA of their own, as well as the machinery necessary
to replicate, transcribe, and translate the information
encoded by their DNA. In addition, mitochondria can
reproduce more or less on their own by growing and
dividing in two.

Chloroplasts

Algae and green plants contain a unique organelle
called a **chloroplast** (Figure 4-25). A chloroplast is
a membrane-bounded structure that contains both

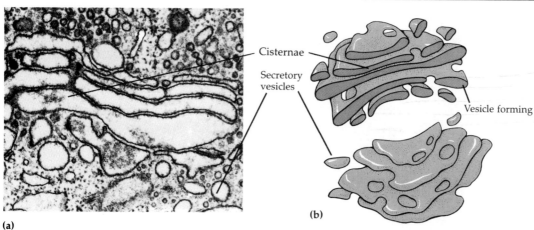

Figure 4-23 **Golgi complex. (a)** Transmission electron micrograph of a Golgi complex in cross section (×26,000). **(b)** Three-dimensional drawing of a Golgi complex.

Cisternae

Secretory vesicles

Vesicle forming

(a)

(b)

the pigment chlorophyll and the enzymes required for the light-gathering phases of photosynthesis (Chapter 5). The chlorophyll is contained in flattened membrane sacs called *thylakoids*. Stacks of thylakoids are called *grana*. Like mitochondria, chloroplasts contain 70S ribosomes, DNA, and enzymes involved in protein synthesis. They are capable of multiplying on their own within the cell. Interestingly, their way of multiplying—by increasing in size and then dividing in two—is strikingly reminiscent of bacterial multiplication. This same similarity exists between mitochondrial multiplication and bacteria, as well.

Lysosomes

As viewed under the electron microscope, **lysosomes**, which are formed from Golgi complexes, appear to be membrane-enclosed spheres in eucaryotic cytoplasm. Unlike mitochondria, lysosomes have only a single membrane and lack detailed structure (see Figure 4-19). But they contain powerful digestive enzymes capable of breaking down many kinds of molecules. Moreover, these enzymes can also digest bacteria that enter the cell. Human white blood cells, which use phagocytosis to ingest bacteria, contain large numbers of lysosomes.

Centrioles

Located near the nucleus are a pair of cylindrical structures, the **centrioles** (see Figure 4-19). Each centriole is a ring of nine evenly spaced bundles, each of which consists of three microtubules. The two centrioles are situated so that the long axis of one is at

Matrix   Cristae

Inner membrane   Outer membrane

(a)

Matrix

Cristae

Inner membrane

(b)

Figure 4-24 **Mitochondria. (a)** Transmission electron micrograph of a mitochondrion in longitudinal section from a rat pancreas cell (×34,000). **(b)** Three-dimensional drawing of a mitochondrion.

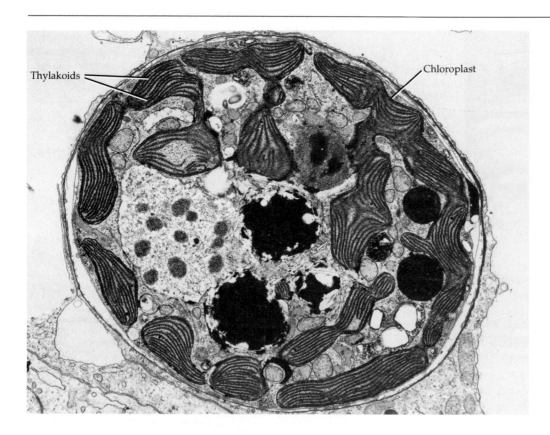

Figure 4-25 Chloroplasts. Transmission electron micrograph of chloroplasts in a cell of the alga *Gymnodynium microadriaticum*, which lives within cells of a coral (×11,200). Photosynthesis occurs in the chloroplasts, providing nutrients for both the alga and the coral. The light-trapping pigments are located on the thylakoids.

right angles to the long axis of the other. Centrioles play a role in eucaryotic cell division and as basal bodies in the formation of cilia and flagella.

The principal differences between procaryotic and eucaryotic cells are presented in Table 4-2. Most of the differences have been discussed in this chapter; a few will be treated in greater detail in subsequent chapters.

Evolution of Eucaryotes

At present, there are two major views on the origin of eucaryotes. According to the **autogenous hypothesis**, specialized internal membranes that derived from a procaryotic plasma membrane evolved into organelles characteristic of eucaryotic cells. According to this view, the nuclear envelope, endoplasmic reticulum, Golgi complex, and organelles bounded by single membranes (such as lysosomes) evolved this way. These organelles are sometimes referred to as the *endomembrane system*. The other view, the **endosymbiotic hypothesis**, proposes that the forerunners of eucaryotic cells were associations of small, symbiotic (living-together) procaryotic cells living within larger procaryotic cells. This hypothesis

focuses primarily on the origin of mitochondria and chloroplasts (which multiply similarly to bacteria). The proposed ancestors of mitochondria are believed to have been oxygen-requiring heterotrophic bacteria. Heterotrophs obtain food molecules by eating other organisms or their by-products. Chloroplasts are believed to be descendants of photosynthetic procaryotes. Photosynthetic organisms use light energy to produce food molecules. It has been suggested that oxygen-requiring heterotrophs and photosynthetic procaryotes both gained entry into the large procaryotic cell as undigested prey or internal parasites. It is possible that the origin of eucaryotes from procaryotes took place by a combination of both mechanisms. The nuclear envelope and other cellular membranes may have evolved as outlined in the autogenous hypothesis, whereas mitochondria and chloroplasts may have evolved by the endosymbiotic mechanism.

* * *

Our next concern is to examine microbial metabolism. In Chapter 5, you will learn about the importance of enzymes to microbes and the ways microbes produce and utilize energy.

TABLE 4.2 / *Principal Differences Between Procaryotic and Eucaryotic Cells*

Characteristic	Procaryotic	Eucaryotic
Size of cell	Typically 1–10 μm in diameter	Typically 10–100 μm in diameter
Nucleus	No nuclear membrane or nucleoli	True nucleus, consisting of nuclear membrane and nucleoli
Membrane-bounded organelles	Absent	Present; examples include lysosomes, Golgi complex, endoplasmic reticulum, mitochondria, and chloroplasts
Flagella	Consist of two protein building blocks	Complex; consist of multiple microtubules
Glycocalyx	Extracellular polymeric capsule or slime layer	Absent
Cell wall	Usually present; chemically complex	When present, chemically simple
Plasma membrane	No carbohydrates and generally lacks sterols	Sterols and carbohydrates that serve as receptors present
Cytoplasm	No cytoskeleton or cytoplasmic streaming	Cytoskeleton; cytoplasmic streaming
Ribosomes	Small size (70S)	Large (80S); small (70S) sizes in organelles
Chromosome (DNA) arrangement	Single circular chromosome; lacks histones	Several or many linear chromosomes with histones
Cell division	Binary fission	Mitosis
Sexual reproduction	No meiosis; exchange of DNA fragments only	Involves meiosis

Study Outline

Introduction   (pp. 73–74)
1. Procaryotic and eucaryotic cells are similar in their chemical composition and chemical reactions.
2. Procaryotic cells lack membrane-bounded organelles (including a nucleus).
3. Peptidoglycan is found in procaryotic cell walls and not in eucaryotic cell walls.

The Procaryotic Cell   (pp. 74–94)
1. Bacteria are unicellular, and most of them multiply by binary fission.

2. Bacterial species are differentiated by morphology, chemical composition, nutritional requirements, and biochemical activities.

Size, Shape, and Arrangement of Bacterial Cells   (pp. 75–76)
1. Most bacteria are from 0.20 to 2.0 μm in diameter and from 2 to 8 μm in length.
2. The three basic bacterial shapes are coccus (spheres), bacillus (rods), and spiral (twisted).
3. Pleomorphic bacteria can assume several shapes.

Structures External to the Cell Wall   (pp. 76–80)

Glycocalyx   (p. 77)

1. The glycocalyx (capsule or slime layer) is a gelatinous polysaccharide and/or polypeptide covering.
2. Capsules protect pathogens from phagocytosis.
3. Capsules provide adherence to surfaces, prevent desiccation, and may prevent loss of nutrients.

Flagella   (pp. 77–79)

1. Flagella are relatively long filamentous appendages consisting of a filament, hook, and basal body.
2. Procaryotic flagella rotate to push the cell.
3. Motile bacteria exhibit taxis—positive taxis is movement toward an attractant and negative taxis is movement away from a repellent.

Axial Filaments   (p. 79)

1. Spiral cells that move by means of an axial filament are called spirochetes.
2. Axial filaments are similar to flagella, except that they wrap around the cell.

Pili   (pp. 79–80)

1. Pili are short, thin appendages.
2. Common pili help cells adhere to surfaces.
3. Sex pili are used to bring cells together for the transfer of DNA from one cell to another.

The Cell Wall   (pp. 81–85)

Composition and Characteristics   (pp. 81–84)

1. The cell wall surrounds the plasma membrane and protects the cell from changes in osmotic pressure.
2. The bacterial cell wall consists of peptidoglycan, a polymer consisting of NAG and NAM and short chains of amino acids.
3. Many gram-positive bacteria also contain teichoic acids.
4. Gram-negative bacteria have a lipoprotein–lipopolysaccharide–phospholipid outer membrane surrounding a thin peptidoglycan layer.
5. The outer membrane protects the cell from phagocytosis and penicillin, lysozyme, and other chemicals.
6. Porins are proteins that permit small molecules to pass through the outer membrane; specific channel proteins allow other molecules to move through the outer membrane.
7. The lipopolysaccharide component of the outer membrane consists of sugars that function as antigens and lipid A, which is an endotoxin.

Atypical Cells Walls   (pp. 84–85)

1. *Mycoplasma* are bacteria that naturally lack cell walls.
2. Archaeobacteria do not have peptidoglycan.
3. L forms are mutant bacteria with defective cell walls.

Damage to the Cell Wall   (p. 85)

1. In the presence of lysozyme, gram-positive bacteria's cell walls are destroyed, and the remaining cellular contents are referred to as a protoplast.
2. In the presence of lysozyme, gram-negative bacteria's cell walls are not completely destroyed, and the remaining cellular contents are referred to as a spheroplast.
3. Protoplasts and spheroplasts are subject to osmotic lysis.
4. Antibiotics such as penicillin interfere with cell-wall synthesis.

Structures Internal to the Cell Wall   (pp. 85–94)

Plasma (Cytoplasmic) Membrane   (pp. 85–88)

1. The plasma membrane encloses the cytoplasm and is a phospholipid bilayer with protein (fluid mosaic).
2. The plasma membrane is selectively permeable.
3. Plasma membranes carry enzymes for metabolic reactions, such as nutrient breakdown, energy production, and photosynthesis.
4. Mesosomes—irregular infoldings of the plasma membrane—may function in cell division and metabolism.
5. Plasma membranes can be destroyed by alcohols and polymyxins.

Movement of Materials Across Membranes   (pp. 88–90)

1. Movement across the membrane may be by passive processes, in which materials move from higher to lower concentration and no energy is expended by the cell.
2. In simple diffusion, molecules and ions move until equilibrium is reached.
3. In facilitated diffusion, substances are transported by permeases across membranes from high to low concentration.
4. Osmosis is the movement of water from high to low concentration across a selectively semipermeable membrane until equilibrium is reached.
5. In active transport, materials move from low to high concentrations by permeases, and the cell must expend energy.
6. In group translocation, energy is expended to modify chemicals and transport them across the membrane.

Cytoplasm   (pp. 90–91)

1. Cytoplasm is the fluid component inside the plasma membrane.
2. The cytoplasm is mostly water, with inorganic and organic molecules, DNA, ribosomes, and inclusions.

Nuclear Area   (p. 91)

1. The nuclear area contains the DNA of the bacterial chromosome. Bacteria can also contain plasmids, which are extrachromosomal DNA circles.

Ribosomes (p. 91)

1. The cytoplasm of a procaryote contains numerous 70S ribosomes; ribosomes consist of rRNA and protein.
2. Protein synthesis occurs at ribosomes; this can be inhibited by certain antibiotics.

Inclusions (pp. 91–92)

1. Inclusions are reserve deposits found in procaryotic and eucaryotic cells.
2. Among the inclusions found in bacteria are metachromatic granules (inorganic phosphate), polysaccharide granules (usually glycogen or starch), lipid inclusions, sulfur granules, carboxysomes (ribulose 1,5-diphosphate carboxylase), and gas vacuoles.

Endospores (pp. 92–94)

1. Endospores are resting structures formed by some bacteria for survival during adverse environmental conditions.
2. The process of endospore formation is called sporulation; the return of an endospore to its vegetative state is called germination.

The Eucaryotic Cell (pp. 94–101)

Flagella and Cilia (p. 95)

1. Flagella are few and long in relation to cell size; cilia are numerous and short.
2. Flagella and cilia are used for motility, and cilia also move substances along the surface of the cells.
3. Both flagella and cilia contain a "nine pairs + two pairs" arrangement of microtubules.

The Cell Wall (pp. 95–96)

1. The cell walls of most algae and some fungi consist of cellulose.
2. The main material of fungal cell walls is chitin.
3. Yeast cell walls consist of glucan and mannan.

The Plasma (Cytoplasmic) Membrane (p. 96)

1. Like the procaryotic plasma membrane, the eucaryotic plasma membrane is a phospholipid bilayer containing proteins.
2. Eucaryotic plasma membranes contain carbohydrates attached to the proteins and sterols not found in procaryotic cells (except *Mycoplasma*).
3. Eucaryotic cells can move materials across the plasma membrane by passive processes used by procaryotes, active transport, and endocytoses (phagocytosis and pinocytosis).

Cytoplasm (p. 96)

1. The cytoplasm of eucaryotic cells includes everything inside the plasma membrane and external to the nucleus.
2. The chemical characteristics of the cytoplasm of eucaryotic cells resemble that of the cytoplasm of procaryotic cells.
3. Eucaryotic cytoplasm has a cytoskeleton and exhibits cytoplasmic streaming.

Organelles (pp. 96–100)

1. Organelles are specialized membrane-bounded structures in the cytoplasm.
2. Organelles are characteristic of eucaryotic cells.
3. The nucleus, which contains DNA in the form of chromosomes, is the most characteristic eucaryotic organelle.
4. The nuclear membrane is connected to a system of parallel membranes in the cytoplasm, called the endoplasmic reticulum.
5. The endoplasmic reticulum provides a surface for chemical reactions, serves as a transporting network, and stores synthesized molecules.
6. 80S ribosomes are found in the cytoplasm or attached to the endoplasmic reticulum.
7. The Golgi complex consists of cisternae. It functions in secretion, carbohydrate synthesis, and glycoprotein formation.
8. Mitochondria are the primary sites of ATP production. They contain small 70S ribosomes and DNA, and they multiply by fission.
9. Chloroplasts contain chlorophyll and enzymes for photosynthesis. Like mitochondria, they contain 70S ribosomes and DNA and multiply by fission.
10. Lysosomes are formed from Golgi complexes. They store powerful digestive enzymes.
11. A pair of cylindrical structures called centrioles that are involved in cell division are near the nucleus.

Evolution of Eucaryotes (p. 100)

1. Two hypotheses offer explanations of the origin of eucaryotes from procaryotic ancestors.
2. The autogenous hypothesis states that organelles (for example, the nucleus and Golgi complex) evolved from internal membranes derived from the plasma membrane.
3. According to the endosymbiotic hypothesis, organelles (for example, mitochondria and chloroplasts) evolved from symbiotic procaryotes living inside other procaryotic cells.

Study Questions

Review

1. Draw the bacterial shapes listed in a, b, and c. Show how d, e, and f are special conditions of a, b, and c, respectively.
   (a) Spiral (d) Spirochete
   (b) Bacillus (e) Streptobacilli
   (c) Coccus (f) Staphylococci

2. List three differences between procaryotic and eucaryotic cells.

3. Diagram each of the following flagellar arrangements:
   (a) Lophotrichous
   (b) Monotrichous
   (c) Peritrichous

4. Match the structures to their functions.
   ____ Cell wall   (a) Attachment to
   ____ Endospore   surfaces
   ____ Flagella   (b) Cell wall formation
   ____ Glycocalyx   (c) Motility
   ____ Pili   (d) Protection from
   ____ Plasma membrane   osmotic lysis
   ____ Ribosomes   (e) Protection from
                                phagocytes
                           (f) Resting
                           (g) Protein synthesis
                           (h) Selectively
                                permeable
                           (i) Transfer of genetic
                                material

5. Of what value is each of the following to the cell?
   (a) Metachromatic granules
   (b) Polysaccharide granules
   (c) Lipid inclusions
   (d) Sulfur granules
   (e) Carboxysomes
   (f) Gas vacuoles

6. Endospore formation is called _____ . It is initiated by _____ . Formation of a new cell from an endospore is called _____ . This process is triggered by _____ .

7. Why is an endospore called a resting structure? Of what advantage is an endospore to a bacterial cell?

8. Explain what would happen in the following experiments.
   (a) A suspension of bacteria is placed in distilled water.
   (b) A suspension of bacteria is placed in distilled water with lysozyme.
   (c) A suspension of bacteria is placed in an aqueous solution of lysozyme and 10% sucrose.

9. Compare and contrast the following:
   (a) Simple diffusion and facilitated diffusion
   (b) Active transport and facilitated diffusion
   (c) Active transport and group translocation

10. Why are mycoplasmas resistant to antibiotics that interfere with cell wall synthesis?

11. Answer the following questions using the diagrams below, which represent cross sections of bacterial cell walls.
    (a) Which diagram represents a gram-positive bacterium? How can you tell?
    (b) Explain how the Gram stain works to distinguish between these two types of cell walls.
    (c) Why does penicillin have no effect on most gram-negative cells?

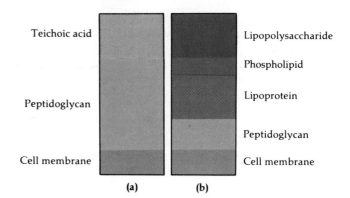

12. Starch is readily metabolized by many cells, but a starch molecule is too large to cross the plasma membrane. How does a cell get the glucose molecules from a starch polymer? How does the cell get these glucose molecules across the plasma membrane?

13. Match the following characteristics of eucaryotic cells with their functions.
    ____ Chloroplasts   (a) Intracellular
    ____ Endoplasmic reticulum   transport
    ____ Golgi complex   (b) Photosynthesis
    ____ Lysosomes   (c) ATP production
    ____ Mitochondria   (d) Digestive
                                    enzyme storage
                               (e) Secretion

14. Eucaryotic cells might have evolved from early procaryotic cells living in close association. What do you know about eucaryotic organelles that would support this theory?

15. What process would a eucaryotic cell use to ingest a procaryotic cell? To ingest a virus?

Challenge

1. Why can procaryotic cells be smaller than eucaryotic cells and still carry on all the functions for life?

2. Within a three-day period at a large hospital, five patients undergoing hemodialysis developed fever and chills. *Pseudomonas aeruginosa* and *Klebsiella pneumoniae* were isolated from three of the patients. *P. aeruginosa*, *K. pneumoniae*, and *Enterobacter agglomerans* were isolated from the dialysis system. Why do all three bacteria cause similar symptoms?

3. Two types of procaryotic cells have been distinguished: eubacterial and archaeobacterial. How do these cells differ from each other? How are they similar?

Further Reading

Adler, J. "The sensing of chemicals by bacteria." *Scientific American* 235(4):40–47, April 1976. Describes the chemotactic response of bacteria using molecules that detect the presence of chemicals in the environment.

Allen, M. M. "Cyanobacterial cell inclusions." *Annual Review of Microbiology* 38:1–25, 1984. Discusses the effects of starvation and feeding on carboxysomes, gas vacuoles, lipid inclusions, and other cell inclusions.

Costerton, J. W., R. T. Irvin, and K.-J. Cheng. "The bacterial glycocalyx in nature and disease." *Annual Review of Microbiology* 35:299–324, 1981. A thorough discussion of the functions of the glycocalyx.

Doetsch, R. N., and R. D. Sjoblad. "Flagella structure and function in eubacteria." *Annual Review of Microbiology* 34:69–108, 1980. A detailed description of procaryotic flagellar ultrastructure and mechanics.

Ferris, F. G., and T. J. Beveridge. "Functions of bacterial cell surface structure." *BioScience* 35:172–177, 1985. Includes ultrastructures and functions of the cell wall, capsules, flagella, pili, and a recently discovered structure, spinae.

Henrichsen, J. "Twitching motility." *Annual Review of Microbiology* 37:81–93, 1983. Describes a newly observed locomotion that is not related to flagella.

Thomas, L. *The Lives of a Cell: Notes of a Biology Watcher.* New York: Viking Press, 1974. Thoughtful and entertaining essays by a physician and researcher, including essays on evolution of eucaryotic cells and the importance of microbes.

Unwin, N., and R. Henderson. "The structure of proteins in biological membranes." *Scientific American* 250(2):78–94, February 1984. Illustrated with electron micrographs of membrane-bounded proteins, including bacteriorhodopsin.

CHAPTER 5

Microbial Metabolism

Now that you are familiar with the structure of procaryotic cells, including the internal and external features, we can discuss the activities that allow these microbes to thrive. The life-support activity of even the most structurally simple organism involves a large number of complex biochemical reactions. Most, though not all, of the biochemical processes of bacteria also occur in eucaryotic microbes and in the cells of multicellular organisms, including humans. However, the reactions that are unique to microbes are fascinating because they allow microorganisms to do things we cannot do. For example, some bacteria (the chemoautotrophs) can grow on a diet of strictly inorganic compounds!

This chapter examines some representative chemical reactions that either produce energy (catabolic reactions) or use energy (anabolic reactions) in microorganisms. We will also look at how these various reactions are integrated within the cell.

Catabolic and Anabolic Reactions

We use the term **metabolism** to refer to the sum of all chemical reactions within a living organism. Because chemical reactions either release or require energy, metabolism can be viewed as an energy balancing act. Accordingly, metabolic activity can be divided into two classes of chemical reactions—**catabolic (degradative) reactions**, releasing energy, and **anabolic (synthetic) reactions**, requiring energy.

In living cells, the chemical reactions that combine simple substances into more complex molecules are collectively known as **anabolism**. Anabolic processes often involve dehydration synthesis reactions (reactions that release water) and require energy to form new chemical bonds. Examples of anabolic processes are the formation of proteins from amino acids, nucleic acids from nucleotides, and polysaccharides from simple sugars. These biosynthetic reactions generate the materials for cell growth.

The chemical reactions that break down complex organic compounds into simple ones are collectively known as **catabolism**. Catabolic reactions are generally hydrolytic reactions (reactions that use water to break chemical bonds) that release chemical energy stored in organic molecules. An example of catabolism occurs when cells break down unneeded proteins into individual amino acids for further use.

Whereas anabolic reactions are mostly energy-requiring reactions, catabolic reactions furnish the energy to drive anabolic reactions. This coupling of energy-requiring and energy-releasing reactions is achieved through adenosine triphosphate (ATP). (You can review its structure in Figure 2-20.) ATP stores energy derived from catabolic reactions and releases it later to drive anabolic reactions. Recall from Chapter 2 that a molecule of ATP consists of an adenine molecule, a ribose molecule, and three phosphate groups. When the terminal phosphate group, $\textcircled{P}$, is split from ATP, adenosine diphosphate (ADP) is formed, and the energy released is used to drive anabolic reactions (ATP → ADP + $\textcircled{P}$ + energy). Then,

the energy from catabolic reactions is used to combine ADP and a phosphate group to resynthesize ATP (ADP + $\textcircled{P}$ + energy → ATP). This concept of coupled reactions is very important. We will see why at the end of this chapter. For now, you should know that the chemical composition of a living cell is constantly changing; some molecules are being broken down while others are being synthesized. This balanced flow of chemicals and energy maintains the life of a cell.

The role of ATP in linking anabolic and catabolic reactions is shown in Figure 5-1. Only part of the energy released in catabolism is actually available for cellular functions. Part of the energy is lost to the environment as heat. Thus, there is a continuous need for new external sources of energy for the cell.

Before we discuss how cells produce energy, let us first consider the principal properties of a group of proteins involved in almost all biologically important chemical reactions. These proteins, the enzymes, were described briefly in Chapter 2. Although it is beyond the scope of this text to name and discuss in detail the actions of individual enzymes, you should be aware of the central role of enzymes in metabolic reactions. It is important to understand that a cell's metabolic pathways are determined by its enzymes, which are in turn determined by the cell's genetic makeup.

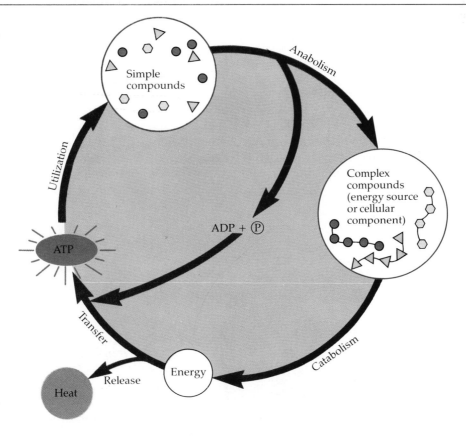

Figure 5-1 Relationship between anabolism, catabolism, and ATP. When simple compounds are combined to form complex compounds (anabolism), ATP provides the energy for synthesis. When large compounds are split apart (catabolism), much of the energy is given off as heat. Some energy is transferred to and trapped in ATP and then is used to drive anabolic reactions.

Enzymes

We indicated in Chapter 2 that chemical reactions occur when chemical bonds are made or broken. In order for reactions to take place, atoms, ions, or molecules must collide with each other. The effectiveness of the collision depends on the velocity of the particles, the amount of energy required for the reaction to occur (**activation energy**), and the specific configuration of the particles. The physiological temperature and pressure of living organisms are too low for chemical reactions to occur quickly enough to maintain the life of the organism. Raising the temperature, pressure, and the amounts of the reacting molecules can increase the frequency of collisions and also increase the rate of chemical reaction. However, such changes would damage or kill the organism. The living cell's solution to this problem is enzymes. Enzymes speed up chemical reactions by increasing the frequency of collisions, lowering the activation energy, and properly orienting the colliding molecules. And they do this without increasing the temperature or pressure—in other words, without disrupting or killing the living cell.

Substances that can speed up a chemical reaction by increasing the frequency of collisions or lowering the activation energy requirement, without themselves being altered, are called **catalysts**. In living cells, **enzymes** serve as biological catalysts.

As catalysts, enzymes are specific. Each particular enzyme will affect only specific **substrates**—reactant molecules in one particular chemical reaction. The specificity of enzymes is made possible by their structures. Enzymes are generally large globular proteins that range in molecular weight from about 10,000 to somewhere in the millions. Of the thousand or more known enzymes, each has a characteristic three-dimensional shape with a specific surface configuration due to its primary, secondary, and tertiary structures (see Figure 2-18).

Enzymes are extremely efficient. Under optimum conditions, they can catalyze reactions at rates that are from 10^8 to 10^{10} times (up to 10 billion times) higher than those of comparable reactions occurring without enzymes. The **turnover number** (number of substrate molecules converted to product per enzyme molecule per second) is generally between 1 and 10,000 and can be as high as 500,000.

Yet, as previously mentioned, enzymes are specific in the reactions they catalyze as well as in the substrates they act upon. From the large number of diverse molecules in the cell, an enzyme must "find" the correct substrate. Moreover, the enzyme-catalyzed reactions usually take place in aqueous solution and at relatively low temperatures—conditions that otherwise would not favor rapid movement of molecules or rapid chemical reactions.

Enzymes are subject to various cellular controls. Their rate of synthesis and their concentration at any given time are under the control of the cell's genes and are influenced by various other molecules in the cell, as will be discussed in Chapter 8. Many enzymes exist in the cell in both active and inactive forms. The rate at which the inactive form becomes active or the active form becomes inactive is determined by the cellular environment.

Enzyme Components

Some enzymes consist entirely of proteins. Many enzymes, however, consist of both a protein portion called an **apoenzyme** and a nonprotein component called the **cofactor**. Together, the apoenzyme and cofactor form a **holoenzyme**, or whole enzyme (Figure 5-2). If the cofactor is removed, the apoenzyme will not function. The cofactor can be a metal ion or a complex organic molecule called a **coenzyme**.

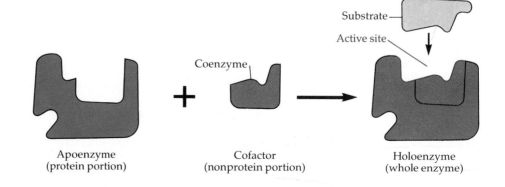

Figure 5-2 Components of a holoenzyme. Many enzymes require both an apoenzyme (protein portion) and a cofactor (nonprotein portion) to become active. The cofactor can be a metal ion or an organic molecule, called a coenzyme (as shown here). The apoenzyme and cofactor together make up the holoenzyme, or complete enzyme. The substrate is the reactant acted upon by the enzyme.

Apoenzyme (protein portion)

Coenzyme

Cofactor (nonprotein portion)

Substrate

Active site

Holoenzyme (whole enzyme)

Coenzymes assist the enzyme by accepting atoms removed from the substrate or donating atoms required by the substrate. Many coenzymes are derived from vitamins (Table 5-1). Two of the coenzymes most important in cellular metabolism are **NAD$^+$** (nicotinamide adenine dinucleotide) and **NADP$^+$** (nicotinamide adenine dinucleotide phosphate). Both compounds contain derivatives of the B vitamin nicotinic acid (niacin), and both function with various enzymes in removing hydrogen ions and electrons (which together form hydrogen atoms) from substrates and transferring them in later reactions to other molecules. Enzymes that remove hydrogen atoms from a molecule are called *dehydrogenases*, and the process is called *dehydrogenation*.

The flavin coenzymes, such as **FMN** (flavin mononucleotide) and **FAD** (flavin adenine dinucleotide), contain derivatives of the B vitamin riboflavin and are important in the photosynthetic reactions. Like NAD$^+$ and NADP$^+$, these coenzymes also function in hydrogen transfer reactions with dehydrogenase enzymes.

Another important coenzyme, *coenzyme A (CoA)*, contains a derivative of pantothenic acid, another B vitamin. This coenzyme plays an important role in the synthesis and breakdown of fats and in a series of oxidizing reactions called the Krebs cycle (discussed later in this chapter). Coenzyme A is used in decarboxylation reactions (removal of CO_2) and is associated with a useful fragment of cellular catabolism, the acetyl group:

$$CH_3 - \overset{\displaystyle O}{\overset{\|}{C}} -$$

When a coenzyme is tightly bonded to its apoenzyme, the coenzyme is called a **prosthetic group**. One example is the heme (iron-containing) group of a protein called *cytochrome c*. **Cytochromes** are proteins that function as electron carriers in respiration and photosynthesis, as we shall see later in this chapter. These compounds contain pigments and metal ions and are structurally related to hemoglobin and chlorophyll.

TABLE 5-1   *Vitamins and Their Functions*

Vitamin	Function
Water-Soluble Vitamins	
Vitamin B$_1$ (thiamine)	Part of coenzyme cocarboxylase; has many functions, including the metabolism of pyruvic acid
Vitamin B$_2$ (riboflavin)	Coenzyme in flavoproteins; active in electron transfers
Niacin (nicotinic acid)	Part of NAD molecule; active in electron transfers
Vitamin B$_6$ (pyridoxine)	Coenzyme in amino acid metabolism
Vitamin B$_{12}$ (cyanocobalamin)	Coenzyme (methyl cyanocobalamide) involved in transfer of methyl groups; active in red-blood-cell formation, amino acid metabolism
Pantothenic acid	Part of coenzyme A molecule; involved in metabolism of pyruvic acid and lipids
Biotin	Involved in carbon dioxide fixation reactions, fatty acid synthesis
Folic acid	Coenzyme used in synthesis of purines and pyrimidines
Vitamin C (ascorbic acid)	Involved in collagen deposition in connective tissue
Lipid-Soluble Vitamins	
Vitamin A	Active in formation of visual pigment, bone and teeth growth
Vitamin D	Involved in intestinal absorption of calcium and phosphorus
Vitamin E	Needed for cellular and macromolecular syntheses
Vitamin K	Coenzyme used in electron transport (napthoquinones and quinones) and formation of blood-clotting proteins

As noted earlier, some cofactors are metal ions, including iron, copper, magnesium, manganese, zinc, calcium, and cobalt. It is believed that such cofactors bridge the enzyme and the substrate, thus binding them to facilitate substrate transformation. For example, magnesium (Mg^{2+}) is required by many phosphorylating enzymes (enzymes that attach a phosphate group to another compound) that act together with ATP. The Mg^{2+} can form a link between the enzyme and the ATP molecule. Most trace elements required by living cells are probably used to activate cellular enzymes.

Mechanism of Enzymatic Action

Although scientists do not completely understand how enzymes lower activation energy, the general sequence of events is believed to be as follows (Figure 5-3):

1. The surface of the substrate contacts a specific region on the surface of the enzyme molecule, called the **active site**.
2. A temporary intermediate compound forms, called an **enzyme-substrate complex**.
3. The substrate molecule is transformed by rearrangement of existing atoms, a breakdown of the substrate molecule, or combining with another substrate molecule.
4. The transformed substrate molecules, the products of the reaction, are released from the enzyme molecule.
5. The unchanged enzyme, now freed, reacts with other substrate molecules.

As mentioned earlier, enzymes have *specificity* for particular substrates. For example, a specific enzyme may be capable of hydrolyzing a peptide bond only between two specific amino acids. Other enzymes are capable of hydrolyzing starch but not cellulose; even though both starch and cellulose are polysaccharides composed of glucose subunits, the orientations of the subunits in the two polysaccharides differ. Enzymes have this specificity because the three-dimensional shape of the active site fits the substrate somewhat as a lock fits with its key. The substrate is usually much smaller than the enzyme, and relatively few of the enzyme's amino acids make up the active site.

A certain compound can be a substrate for a number of different enzymes that catalyze different reactions, so the fate of a compound depends on the enzyme that acts upon it. For example, glucose-6-phosphate, a molecule important in cell metabolism, can be acted upon by at least four different enzymes, and each reaction will give a different product.

Classifying Enzymes

The names of enzymes usually end in *-ase*. All enzymes can be grouped into six classes, according to the type of chemical reaction they catalyze (Table 5-2). Enzymes within each of the major classes are named according to the more specific types of reactions they carry out. For example, the class called oxidoreductases is involved with oxidation-reduction reactions (described shortly). During some oxidation reactions, a substrate loses hydrogen; in others, a substrate gains oxygen. Enzymes in the oxidoreductase

Figure 5-3   Mechanism of enzymatic action. The substrate contacts the active site on the enzyme to form an enzyme–substrate complex. The substrate is then transformed into products, and the enzyme is recovered unchanged.

TABLE 5-2 *Enzyme Classification Based on Type of Chemical Reaction*

Class	Type of Chemical Reaction	Examples
Oxidoreductase	Oxidation-reduction in which oxygen and hydrogen are gained or lost	Cytochrome oxidase, D-lactate dehydrogenase
Transferase	Transfer of functional groups, such as an amino group, acetyl group, or phosphate group	Acetate kinase, alanine deaminase
Hydrolase	Hydrolysis (addition of water)	Lipase, sucrase
Lyase	Without hydrolysis, removal of groups of atoms	Oxalate decarboxylase, isocitrate lyase
Isomerase	Rearrangement of atoms within a molecule	Glucose-phosphate isomerase, alanine racemase
Ligase	Joining of two molecules (using energy derived from breakdown of ATP)	Acetyl-CoA synthetase, methionyl-tRNA synthetase

class that remove hydrogen are called *dehydrogenases*; those that add oxygen are called *oxidases*. As you will see later, even dehydrogenase and oxidase enzymes have more specific names, such as D-lactate dehydrogenase and cytochrome oxidase, depending on the specific substrates on which they act.

Factors Influencing Enzymatic Activity

Several factors influence the activity of an enzyme. Among the more important are temperature, pH, substrate concentration, and inhibitors.

Temperature

The speed of most chemical reactions increases as the temperature rises. At low temperatures, molecules move more slowly than at higher temperatures and do not always have enough energy to cause a chemical reaction. For enzymatic reactions, however, elevation beyond a certain temperature drastically reduces the reaction's rate (Figure 5-4a). This decrease in the reaction's rate is due to the enzyme's denaturation, a phenomenon common to all proteins. **Denaturation** (Figure 5-5) usually involves breakage of the hydrogen bonds and other noncovalent bonds that hold the

(a)

(b)

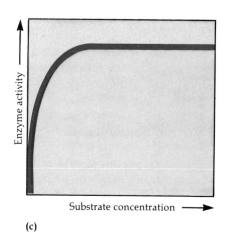

(c)

Figure 5-4   Factors that influence enzymatic activity, plotted for a hypothetical enzyme.
(a) Temperature. The enzymatic activity (rate of reaction catalyzed by the enzyme) increases with increasing temperature until the enzyme, a protein, is denatured by heat and inactivated. At this point, the reaction rate falls steeply. **(b)** pH. The level at which the enzyme illustrated is most active is about pH 6. **(c)** Substrate concentration. With increasing concentration of substrate molecules, the rate of reaction increases until the active sites on all the enzyme molecules are filled, at which point the maximum rate of reaction is reached.

Figure 5-5   Denaturation of a protein. Breaking the noncovalent bonds (such as hydrogen bonds) that hold the active protein in its three-dimensional shape makes the denatured protein no longer functional.

Active protein   Denatured protein

enzyme in its characteristic three-dimensional structure (tertiary configuration).

A common example of denaturation is the transformation of uncooked egg white (albumin) to a hardened state after it is heated. As might be expected, this alteration in structure changes the arrangement of the amino acids in the active site, altering its shape and causing the enzyme to lose its catalytic ability and vital biological activity. In some cases, denaturation is partially or fully reversible. But if denaturation continues to the point of the enzyme losing its solubility and coagulating (as with cooked albumin), denaturation becomes irreversible and the enzyme cannot regain its original properties. Enzymes can also be denatured by concentrated acids, bases, heavy metal ions (such as copper, zinc, silver, arsenic, or mercury), alcohol, and ultraviolet radiation.

pH

Most enzymes have a pH at which their activity is characteristically maximal; this level is the **pH optimum**. Above or below this pH value, enzyme activity, and therefore reaction rates, decline (Figure 5-4b). When the H^+ concentration (pH) in the medium is changed, the many ionizable groups essential for stabilizing the enzymatic three-dimensional structure are affected. Extreme changes in pH can cause denaturation.

Substrate Concentration

There is a maximum rate at which a certain amount of enzyme can catalyze a specific reaction. Only when the concentration of substrate(s) is extremely high can this maximum rate be attained. Under conditions of high substrate concentration, the enzyme is said to be **saturated**; that is, its active site is always occupied by substrate or product molecules. In this condition, a further increase in substrate concentration will not affect the reaction rate because all active sites are already in use (Figure 5-4c). If a substrate's concentration exceeds a cell's saturation level for a particular

enzyme, the rate of reaction can be increased only if the cell produces additional enzyme molecules. But under normal cellular conditions, enzymes are not saturated with substrate(s). At any given time, many of the enzyme molecules are inactive for lack of substrate; thus, the rate of reaction is likely to be determined (restricted) by the substrate concentration.

Inhibitors

As you will see in later chapters, an effective way to control the growth of bacteria is to control their enzymes. Certain poisons, such as cyanide, arsenic, mercury, and nerve gas, combine with enzymes and prevent their functioning. The result is that the cells stop functioning and die. Some antimicrobial drugs used in medicine are enzyme inhibitors.

Enzyme inhibitors are classified according to their mechanism of action as competitive inhibitors and noncompetitive inhibitors. **Competitive inhibitors** compete with the normal substrate for the active site of the enzyme. The competitive inhibitor is able to do this because its shape and chemical structure are very similar to those of the normal substrate (Figure 5-6a and b). Competitive inhibitors bind reversibly with their enzymes and slow down the reaction rate. One good example of a competitive inhibitor is sulfanilamide (a sulfa drug), which inhibits the enzyme whose normal substrate is para-aminobenzoic acid (PABA).

Sulfanilamide   PABA

PABA is an essential nutrient used by many bacteria in the synthesis of folic acid, a vitamin that functions as a coenzyme. When sulfanilamide is administered to bacteria, the enzyme that normally converts PABA to folic acid combines instead with the

Figure 5-6 Enzyme inhibitors. (a) Uninhibited enzyme and its normal substrate. **(b)** Competitive inhibitor. **(c)** One type of noncompetitive inhibitor, causing allosteric inhibition.

(a) (b) (c)

[handwritten: inhibitor bind @ site ≠ bind site]

[handwritten: am activators tool.]

sulfanilamide. The folic acid is not synthesized, and the bacteria cannot grow. Human cells do not use PABA to make their folic acid, so sulfanilamide selectively kills bacteria but does not harm human cells.

Noncompetitive inhibitors do not compete with the substrate for the enzyme's active site; instead, they act on a part of the enzyme other than the active site, altering it to prevent binding of enzyme and substrate. The site of the noncompetitive inhibitor's binding is called an **allosteric site** ("other space") and is specific for the inhibitor (Figure 5-6c). This process is an **allosteric inhibition**—an enzyme's activity is altered due to a change in shape caused by selective binding of an inhibitor at a site that does not overlap with the substrate's binding site. The change to the enzyme can be either reversible or irreversible. In some cases, allosteric interactions can activate an enzyme rather than inhibit it.

Another type of noncompetitive inhibition can operate on enzymes that require metal ions for their activity. Certain chemicals can bind or tie up the metal-ion activators and thus prevent an enzymatic reaction. Cyanide can bind the iron in iron-containing enzymes, and fluoride can bind the calcium or magnesium in enzymes that contain those ions.

Feedback Inhibition

Allosteric noncompetitive inhibitors play a role in a biochemical control called **feedback inhibition** (or **end-product inhibition**). In some metabolic reactions, several steps are required for the synthesis of a particular chemical compound, called the end-product. The cell makes this end-product by converting one intermediate substance to another in the presence of specific enzymes (Figure 5-7). In many biosynthetic pathways, the final product can inhibit activity of one of the enzymes earlier in the pathway. This phenomenon is feedback inhibition (Figure 5-7). This regulatory mechanism prevents the cell from making excessive end-product.

Feedback inhibition acts on enzymes that have already been synthesized; it does not affect the synthesis of these enzymes. A single enzyme, usually the first one in a metabolic pathway, is allosterically inhibited by the end-product in a reversible reaction. Because the enzyme is inhibited, the product of the first enzymatic reaction in the pathway is not synthesized. Because that unsynthesized product would normally be the substrate for the second enzyme in the pathway, the second reaction stops immediately as well. Thus, even though only the first enzyme in the pathway is inhibited, the entire pathway shuts down, and no new end-product is formed. As the exist-

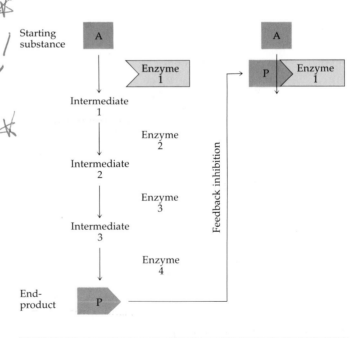

Figure 5-7 Feedback inhibition. In some biochemical reactions, a cell makes an end-product by converting one intermediate substance into another through the sequential action of a number of different enzymes (left). In feedback inhibition (right), the end-product of a biochemical reaction can inhibit an enzyme in the sequence and prevent the production of excess end-product.

Figure 5-8 Feedback inhibition in *E. coli*. Shown here is a sequence of reactions in which threonine is converted to isoleucine. If isoleucine is added to the culture medium, *E. coli* no longer synthesizes isoleucine because the isoleucine combines with the enzyme that catalyzes the first step of the synthesis. Thus, isoleucine inhibits the synthesis of more isoleucine.

ing end-product is used up by the cell, the first enzyme's allosteric site will more often remain unbound, and the pathway will resume activity.

The bacterium *Escherichia coli* (esh-ėr-i'kē-ä kō'lī) can be used to demonstrate feedback inhibition in the synthesis of the amino acid isoleucine. In this metabolic pathway, five steps are taken to enzymatically convert the amino acid threonine to isoleucine (Figure 5-8). If isoleucine is added to the *E. coli* culture medium, the bacteria no longer synthesize isoleucine since the isoleucine inhibits the first enzyme. This condition is maintained until the supply of isoleucine is depleted. This type of feedback inhibition is also involved in regulating the cell's production of other amino acids, vitamins, purines, and pyrimidines.

In the following section on energy production methods, you will see the key roles enzymes play in both catabolic and anabolic reactions.

Energy Production Methods

Nutrient molecules, like all molecules, have energy stored in the bonds between their atoms. When this energy is spread throughout the molecule, it is difficult for the cell to use. Various reactions in catabolic pathways, however, concentrate the energy as it is released into the high-energy bonds of ATP, which serves as a convenient energy carrier. Before discussing the catabolic pathways, we will consider three general aspects of energy production—the concept of oxidation-reduction, mechanisms of ATP generation, and the categorization of organisms according to the energy-producing schemes they use.

Oxidation-Reduction

Oxidation is the removal of electrons (e^-) from a molecule. In many cellular oxidations, two electrons and two hydrogen ions (H^+) are removed at the same time; this is equivalent to the removal of two hydrogen atoms. Because most biological oxidations involve the loss of hydrogen atoms, they are called **dehydrogenation** reactions. For example, when lactic acid is oxidized to form pyruvic acid, the lactic acid loses two hydrogen atoms:

$$\begin{array}{ccc} \text{COOH} & & \text{COOH} \\ | & & | \\ \text{CHOH} & \xrightarrow{-2\text{H (oxidation)}} & \text{C}=\text{O} \\ | & & | \\ \text{CH}_3 & & \text{CH}_3 \\ \text{Lactic acid} & & \text{Pyruvic acid} \end{array}$$

When a compound has picked up electrons or hydrogen atoms, it is said to be reduced. Thus, **reduction** is a gain of electrons and is the opposite of oxidation. Consider the reduction of pyruvic acid to form lactic acid:

$$\begin{array}{ccc} \text{COOH} & & \text{COOH} \\ | & & | \\ \text{C}=\text{O} & \xrightarrow{+2\text{H (reduction)}} & \text{CHOH} \\ | & & | \\ \text{CH}_3 & & \text{CH}_3 \\ \text{Pyruvic acid} & & \text{Lactic acid} \end{array}$$

Oxidation and reduction reactions are always coupled. In other words, each time one substance is oxidized, another is almost simultaneously reduced. The pairing of these reactions is **oxidation-reduction**.

When a substance is oxidized in this way, the freed hydrogen atoms do not remain free in the cell but are transferred immediately by coenzymes to another compound. Two coenzymes commonly used by living cells to carry hydrogen atoms are derivatives of the vitamin niacin. They are, as mentioned earlier, nicotinamide adenine dinucleotide (NAD^+) and nicotinamide adenine dinucleotide phosphate ($NADP^+$). The oxidation and reduction states of NAD^+ and $NADP^+$ can be symbolized as follows:

$$\begin{array}{lcl} NAD^+ & \underset{-2\text{H (2H}^+ + 2e^-)}{\overset{+2\text{H (2H}^+ + 2e^-)}{\rightleftharpoons}} & NADH + H^+ \\ \text{Oxidized} & & \text{Reduced} \end{array}$$

$$\begin{array}{lcl} NADP^+ & \underset{-2\text{H (2H}^+ + 2e^-)}{\overset{+2\text{H (2H}^+ + 2e^-)}{\rightleftharpoons}} & NADPH + H^+ \\ \text{Oxidized} & & \text{Reduced} \end{array}$$

In the preceding equations, $2\text{H} (2\text{H}^+ + 2e^-)$ means that two neutral hydrogen atoms (2H) are equivalent to two hydrogen ions (2H^+) and two electrons ($2e^-$).

When NAD^+ is reduced to NADH, two hydrogen ions ($2H^+$) and two electrons ($2e^-$) are gained. (Actually, one H^+ ion and two electrons are delivered to NAD^+, and the other H^+ ion is released into the surrounding solution.) The addition of a H^+ ion and $2e^-$ to NAD^+ neutralizes the charge on NAD^+ and adds a hydrogen atom so that the reduced form is NADH. The same holds true for the reduction of $NADP^+$ to NADPH.

Conversely, when NADH is oxidized to NAD^+, two hydrogen ions ($2H^+$) and two electrons ($2e^-$) are lost by NADH. (Actually, one H^+ ion and two electrons are lost from NADH, and the other H^+ ion is picked up from the surrounding medium.) The loss of a H^+ ion and $2e^-$ by NADH results in one less hydrogen atom and an additional positive charge, so that the oxidized form is NAD^+. NADPH is oxidized to $NADP^+$ by the same mechanism.

Thus, when lactic acid is *oxidized* to form pyruvic acid, the two hydrogen atoms removed in the reaction are used to *reduce* NAD^+. This coupled oxidation-reduction reaction may be written as follows:

$$\underset{\text{Lactic acid}}{\begin{array}{c} COOH \\ | \\ CHOH \\ | \\ CH_3 \end{array}} \quad \underset{\xrightarrow{\hspace{2cm}}}{\overset{NAD^+ \quad \underset{+H^+}{NADH}}{\overbrace{\hspace{2cm}}}} \quad \underset{\text{Pyruvic acid}}{\begin{array}{c} COOH \\ | \\ C=O \\ | \\ CH_3 \end{array}}$$

An important point to remember about oxidation-reduction reactions is that oxidation is usually an energy-producing reaction. Cells take nutrients, some of which serve as energy sources, and degrade them from highly reduced compounds (with many hydrogen atoms) to highly oxidized compounds (with many oxygen atoms or multiple bonds). For example, when a cell oxidizes a molecule of glucose ($C_6H_{12}O_6$) to CO_2 and H_2O, the energy in the glucose molecule is removed in a stepwise manner and ultimately trapped by ATP, which is then an energy source for energy-requiring reactions. Compounds such as glucose that have many hydrogen atoms are highly reduced compounds, containing more potential energy than oxidized compounds. Thus, glucose is a valuable nutrient for organisms.

Generation of ATP: Types of Phosphorylation

The energy released during oxidation reactions is trapped within the cell by the formation of ATP. To summarize what was discussed at the beginning of this chapter, a phosphate group (symbolized $\textcircled{P}$) is added to ADP, with the input of energy, to form ATP.

$$\underbrace{\text{Adenosine}—\textcircled{P} \sim \textcircled{P}}_{\text{ADP}} + \text{Energy} + \textcircled{P} \longrightarrow$$
$$\underbrace{\text{Adenosine}—\textcircled{P} \sim \textcircled{P} \sim \textcircled{P}}_{\text{ATP}}$$

The symbol $\sim$ designates a high-energy bond, one that can readily be broken to release usable energy. The high-energy bond that attaches the third phosphate contains the energy stored in this reaction. The addition of a phosphate to a chemical compound is called **phosphorylation**. Organisms use three mechanisms of phosphorylation to generate ATP from ADP.

In **substrate-level phosphorylation**, ATP is generated when a high-energy phosphate group is directly transferred from an intermediate phosphorylated metabolic compound to ADP. (In eucaryotic cells, this occurs in the cytoplasm of the cell.) The following example shows only the carbon skeleton and the phosphate group of the metabolic compound:

$$C—C—C \sim \textcircled{P} + ADP \longrightarrow C—C—C + ATP$$

When **oxidative phosphorylation** occurs, electrons are removed from organic compounds (usually by NAD^+) and passed through a series of electron acceptors to molecules of oxygen (O_2) or other inorganic molecules. This process occurs in the plasma membrane of procaryotes and in the inner mitochondrial membrane of eucaryotes. The series of electron acceptors (or carriers) used in oxidative phosphorylation is called an **electron transport chain** (see Figure 5-17). The transfer of electrons from one electron acceptor to the next releases energy, which is used to generate ATP from ADP through a process called chemiosmosis, to be described shortly.

The third mechanism of phosphorylation, **photophosphorylation**, occurs only in photosynthetic cells, which contain light-trapping pigments such as chlorophylls. In this mechanism, light energy activates chlorophyll to transfer an electron to a series of carrier molecules, an electron transport chain. As the electron is passed along the series of carriers, each transfer releases energy used in chemiosmosis to convert ADP to ATP. In **cyclic phosphorylation**, the electron eventually returns to chlorophyll, and ATP is formed using the energy released by electron transport. In **noncyclic photophosphorylation**, the electrons released from chlorophyll do not return to chlorophyll but become incorporated into the coenzyme nicotinamide adenine dinucleotide phosphate ($NADP^+$). The electrons lost from chlorophyll are replaced by electrons from water. The products of noncyclic phosphorylation are ATP (formed using the energy of electron transfer), O_2 (from water molecules), and NADPH + H^+ (from $NADP^+$, electrons from chlorophyll, and H^+ from water) (Figure 5-9). Note that both oxidative phosphorylation and photophosphorylation involve the passing of electrons to electron acceptors in an electron transport chain.

The significance of these three forms of phosphorylation will become clearer later in this chapter, as we discuss their role in energy production in more detail.

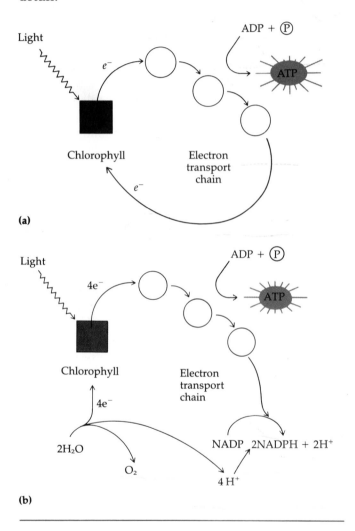

(a)

(b)

Figure 5-9   Photophosphorylation. (a) In cyclic photophosphorylation, electrons released from chlorophyll by light return to chlorophyll after passage along the electron transport chain. The energy from electron transfer is converted to ATP. **(b)** In noncyclic photophosphorylation, electrons released from chlorophyll are replaced by electrons from water. The chlorophyll electrons are passed along the electron transport chain to the electron acceptor $NADP^+$.

Chemiosmotic Mechanism of ATP Generation

The chemiosmotic mechanism of ATP synthesis—**chemiosmosis**—was first proposed by British biochemist Peter Mitchell in 1961. Although it took a number of years to gain full acceptance, it is now recognized as a major milestone in the history of biochemistry.

What is chemiosmosis, and how does it drive the energy-requiring synthesis of ATP from ADP? To understand the answers to these questions, we need to recall several concepts that were introduced in Chapter 4 as part of the section Movement of Materials Across Membranes (pp. 88–90). Recall that substances diffuse passively across membranes from regions of high concentration to regions of low concentration; this diffusion yields energy. Recall also that the movement of substances *against* such a concentration gradient *requires* energy and that, in such an active transport of molecules or ions across biological membranes, the required energy is usually provided by ATP. In chemiosmosis, the energy released when a substance moves along a gradient is used to *synthesize* ATP. The "substance" in this case refers to protons (H^+). The basic mechanism of chemiosmosis is the same for both oxidative phosphorylation and photophosphorylation. The steps of chemiosmosis are as follows (Figure 5-10):

1. Energy from sunlight (in photophosphorylation) or from NADH (in oxidative phosphorylation) powers some of the carriers of the electron transport chain to pump (actively transport) hydrogen ions from one side of the membrane to the other. Because hydrogen ions are in fact protons, such carrier molecules are called *proton pumps*.
2. The phospholipid membrane is normally impermeable to protons, so this one-directional pumping establishes a proton gradient (a difference in concentration of protons on the two sides of the membrane). Thus, the fluid on one side of the membrane acquires a positive charge, while the fluid on the other side is left with a negative charge. The proton gradient has potential energy, called the **proton motive force**.
3. The protons on the side of the membrane with the higher proton concentration can diffuse across the membrane with the help of an enzyme in the membrane called **adenosine triphosphatase (ATPase)**. When this movement occurs, energy is released and is used by the enzyme to synthesize ATP from ADP.

② High H⁺ concentration

H⁺   H⁺

Membrane

Electron transport chain (includes proton pumps)

①

ATPase

③

ADP + Ⓟ → ATP

Energy from sunlight or NADH

Low H⁺ concentration

Figure 5-10 Chemiosmosis. This process is described on p. 116.

Both procaryotic and eucaryotic cells use the chemiosmotic mechanism to generate energy for ATP production in both oxidative phosphorylation and photophosphorylation. The membrane that contains the electron transport carriers and the ATPase may be the plasma membrane of procaryotes, the inner mitochondrial membrane in eucaryotes, or the thylakoid membrane in cyanobacteria and eucaryotic chloroplasts. An unusual and very simple proton pump is described in the box on p. 120.

Nutritional Patterns Among Organisms

To acquire the energy for various metabolic activities, microorganisms need raw materials—mainly a carbon source for new organic compounds. Nutritional patterns among organisms can be distinguished on the basis of two criteria, energy source and principal source of carbon. First considering the energy source, we can generally classify organisms as phototrophs or chemotrophs. **Phototrophs** use light as their primary energy source, whereas **chemotrophs** depend on oxidation-reduction reactions of inorganic or organic compounds for energy. For their principal carbon source, **autotrophs** (self-feeders) use carbon dioxide, and **heterotrophs** (feeders on others) require an organic carbon source. Autotrophs may also be referred to as *lithotrophs* (rock-eating), and heterotrophs may also be referred to as *organotrophs*.

If we combine the energy and carbon sources, we derive the following nutritional classifications for organisms—*photoautotrophs, photoheterotrophs, chemoautotrophs,* and *chemoheterotrophs*. This nutritional classification of microbes is summarized in Table 5-3. Almost all of the medically important microorganisms discussed in this book are chemoheterotrophs.

Photoautotrophs

Photoautotrophs use light as a source of energy and carbon dioxide as their chief source of carbon. They include photosynthetic bacteria (green sulfur and purple sulfur bacteria and cyanobacteria), algae, and green plants. The process by which photoautotrophs transform carbon dioxide and water into carbohydrates and oxygen gas is called **photosynthesis**. Essentially, photosynthesis is the conversion of light energy into chemical energy; chlorophyll molecules trap the light. In the process, CO_2 from the atmosphere is converted to more reduced carbon compounds, primarily sugars. This is called **carbon fixation**. Continuation of life on Earth depends on the recycling of carbon in this way, as well as on the

TABLE 5-3 *Nutritional Classification of Organisms*

Nutritional Type	Energy Source	Carbon Source	Examples
Photoautotroph	Light	Carbon dioxide (CO_2)	Photosynthetic bacteria (green sulfur and purple sulfur bacteria), cyanobacteria, algae, plants (some green and purple sulfur bacteria may grow chemotrophically)
Photoheterotroph	Light	Organic compounds	Purple nonsulfur and green nonsulfur bacteria (some purple nonsulfur bacteria may grow photoautotrophically and chemotrophically)
Chemoautotroph	Inorganic compounds	Carbon dioxide (CO_2)	Hydrogen, sulfur, iron, and nitrifying bacteria
Chemoheterotroph	Organic compounds	Organic compounds	Most bacteria and all fungi, protozoans, and animals

production of O_2 for use in aerobic respiration. Cyanobacteria, algae, and green plants all contribute to this vital recycling with photosynthesis, which can be represented as follows:

$$6CO_2 + 12H_2O \xrightarrow[\text{Chlorophyll, enzymes}]{\text{Light}}$$

Carbon dioxide | Water

Raw materials | Necessary conditions

$$C_6H_{12}O_6 + 6O_2 + 6H_2O$$

Sugar | Oxygen | Water

Products

In the synthesized sugar, the carbon (C) and the oxygen (O) come from carbon dioxide (CO_2), and the hydrogen (H) comes from water (H_2O). The oxygen (O) in water (H_2O) is eventually given off as oxygen gas (O_2). The energy for the photosynthetic production of sugar is derived from ATP, which was generated by photophosphorylation, using the chemiosmotic mechanism. Ultimately, the energy for photosynthesis comes from light.

The fixation of carbon requires a large amount of energy from ATP and a source of electrons to carry out the reduction of CO_2 to sugar. Both ATP production and the release of electrons are part of a series of reactions driven by light and collectively called the **light reactions**. The electrons that reduce CO_2 come from NADPH. In the next sequence of reactions, called the **dark reactions**, the energy from ATP and electrons from NADPH are used to incorporate CO_2 into sugar.

The dark reactions are so called because no light is directly required for them to occur. They make up a complex cyclic pathway called the **Calvin–Benson cycle**, in which CO_2 is fixed (Figure 5-11; see also Appendix F). For the net synthesis of one molecule of a three-carbon sugar called glyceraldehyde 3-phosphate, the cycle of reactions must occur three times, fixing three molecules of CO_2. These three turns of the cycle require the input of nine molecules of ATP and six molecules of NADPH. To make glucose, a six-carbon sugar, three more turns of the cycle are required. Thus, to make one molecule of glucose, the cycle must turn six times and requires an investment of 6 CO_2, 18 ATP, and 12 NADPH (+ 12H^+).

In the photosynthetic reactions of cyanobacteria, algae, and green plants, the hydrogen atoms of water are used to reduce carbon dioxide, and oxygen gas is given off. Because this photosynthetic process pro-

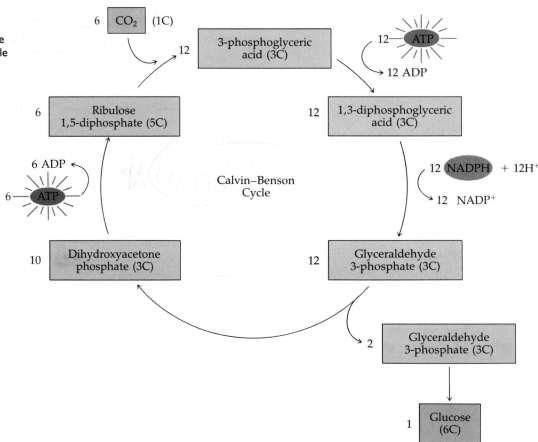

Figure 5-11 Simplified version of the Calvin–Benson cycle. To make one molecule of glucose, the cycle must turn six times so that there is a total investment of 6 CO_2, 18 ATP, and 12 NADPH. A more detailed version is presented in Appendix F.

duces O_2, it is sometimes called **oxygenic**. The light-trapping pigment principally used by green plants, algae, and cyanobacteria is *chlorophyll a*. It is located in the membranous thylakoids of chloroplasts in algae and green plants (see Figure 4-25), and in the thylakoids found in the photosynthetic structures of cyanobacteria.

In addition to the cyanobacteria (see Microview A, Photo 27), there are several other families of photosynthetic procaryotes. Each is classified according to the way it reduces carbon dioxide. These bacteria cannot use water to reduce carbon dioxide and cannot carry on photosynthesis when oxygen is present (they must have an *anaerobic* environment). Consequently, their photosynthetic process does not produce oxygen gas and is called **anoxygenic**. Two of the families are photoautotrophs—the green sulfur and purple sulfur bacteria. The **green sulfur bacteria**, such as *Chlorobium* (klô-rō'bē-um), use sulfur, sulfur compounds, or hydrogen gas to reduce carbon dioxide and form organic compounds. Applying the energy from light and the appropriate enzymes, these bacteria oxidize sulfur to sulfuric acid, hydrogen sulfide to sulfur, or hydrogen gas to water. The **purple sulfur bacteria**, such as *Chromatium* (krō-mā'tē-um), also use sulfur, sulfur compounds, or hydrogen gas to reduce carbon dioxide. They are distinguished from the green sulfur bacteria biochemically and morphologically.

The chlorophylls used by these photosynthetic bacteria are called *bacteriochlorophylls*, and they absorb light at longer wavelengths than light absorbed by chlorophyll *a*. Bacteriochlorophylls of green sulfur bacteria are found in vesicles called *chlorosomes* (or *chlorobium vesicles*) underlying and attached to the plasma membrane. In the purple sulfur bacteria, the bacteriochlorophylls are located in invaginations of the plasma membrane (intracytoplasmic membranes).

Several characteristics that distinguish eucaryotic photosynthesis from procaryotic photosynthesis are presented in Table 5-4. See the box on p. 120 for a discussion of an exceptional photosynthetic system that exists in *Halobacterium* (see also Microview A, Photo 26). The system does not use chlorophyll.

Photoheterotrophs

Photoheterotrophs use light as a source of energy but cannot convert carbon dioxide to sugar; rather, they use organic compounds as sources of carbon. These organic compounds include alcohols, fatty acids, other organic acids, and carbohydrates. Photoheterotrophs are anoxygenic. Among the photoheterotrophs are **green nonsulfur bacteria** such as *Chloroflexus* (klô-rō-flex'us) and **purple nonsulfur bacteria** such as *Rhodopseudomonas* (rō-dō-sū-dō-mō'nas).

Chemoautotrophs

Chemoautotrophs use inorganic compounds as a source of energy, and they use carbon dioxide as their principal source of carbon (see Figure 26-1, the carbon cycle). Inorganic sources of energy for these organisms include hydrogen sulfide (H_2S) for *Beggiatoa* (bej-jē-ä-tō'ä); elemental sulfur (S) for *Thiobacillus*; ammonia (NH_3) for *Nitrosomonas* (nī-trō-sō-mō'näs);

TABLE 5-4 *Comparison of Eucaryotic and Selected Procaryotic Photosynthesis*

	Eucaryotes	Procaryotes		
Characteristic	**Algae, Plants**	**Cyanobacteria**	**Green Sulfur Bacteria**	**Purple Sulfur Bacteria**
Substance that reduces CO_2	H atoms of H_2O	H atoms of H_2O	Sulfur, sulfur compounds, H_2 gas	Sulfur, sulfur compounds, H_2 gas
Oxygen production	Oxygenic	Oxygenic (and anoxygenic)	Anoxygenic	Anoxygenic
Light-trapping pigment	Chlorophyll *a*	Chlorophyll *a*	Bacteriochlorophyll *a*	Bacteriochlorophyll *a* or *b*
Site of photosynthesis	Chloroplasts with thylakoids	Thylakoids	Chlorosomes	Intracytoplasmic membrane
Environment	Aerobic	Aerobic (and anaerobic)	Anaerobic	Anaerobic

nitrites (NO$_2$) for *Nitrobacter* (nī-trō-bak'tėr); hydrogen gas (H$_2$) for *Hydrogenomonas* (hī-drō-je-nō-mō'näs); and iron (Fe^{2+}) for *Thiobacillus ferrooxidans* (fer-rō-oks'i-danz)(see the box in Chapter 27). The energy derived from the oxidation of these inorganic compounds is eventually stored in ATP, which is produced by oxidative phosphorylation.

Chemoheterotrophs

When we discuss photoautotrophs, photoheterotrophs, and chemoautotrophs, it is easy to categorize energy source and carbon source because they occur as separate entities. However, in **chemoheterotrophs**, the distinction is not so clear because both energy source

and carbon source are usually the same organic compound—glucose, for example.

Heterotrophs are further classified according to their source of organic molecules—**saprophytes** live on dead organic matter, and **parasites** derive nutrients from a living host. The vast majority of bacteria, and all fungi, protozoans, and animals, are chemoheterotrophs.

Summary of Energy Production Methods

In the living world, energy is passed from the sun to photosynthetic organisms and then to other organisms in the form of the potential energy contained in

MICROBIOLOGY HIGHLIGHTS

Photosynthesis Without Chlorophyll

The interesting archaeobacterium *Halobacterium* lives where very little else can grow. This bacterium is found in salt lakes, in the salt licks on ranches, in salt flats, or in any other environment with a concentration of salt that is from five to seven times greater than that of the ocean. Halobacteria are easy to detect because they turn their environment red. These bacteria cannot ferment carbohydrates and do not contain chlorophyll, so it was assumed that all their energy comes from oxidative phosphorylation. The exciting discovery of a new system of photophosphorylation arose through the study of the plasma membrane of *Halobacterium halobium*.

Researchers found that the plasma membrane of *H. halobium* fragments into two fractions (red and purple) when the cell is broken down and its com-

ponents are sorted. The red fraction, which comprises most of the membrane, contains cytochromes, flavoproteins, and other parts of the electron transport chain that carries out oxidative phosphorylation. The purple fraction is more interesting. This purple membrane occurs in distinct patches of hexagonal lattices within the plasma membrane. The purple color comes from a protein that makes up 75% of the purple membrane. This protein is similar to the retinal pigment in the rod cells of the human eye, rhodopsin, so the protein was named bacteriorhodopsin. At the time it was discovered, its function was not known.

In the 1970s, further study of *H. halobium* suggested some startling explanations. When starved

cells in anaerobic, dark environments were exposed to either light or oxygen, ATP synthesis increased. The table compares the effects of light and oxygen on anaerobic photosynthetic bacteria, photosynthetic green algae, a facultative anaerobe (a bacterium that can grow with or without O$_2$), *E. coli*, and *H. halobium*.

E. coli can grow in any environment because it requires neither oxygen nor light. The photosynthetic bacteria and green

algae synthesize ATP only when they are in the appropriate concentration of oxygen and only in the presence of light. *H. halobium* can grow in the presence of either light or oxygen but cannot grow when neither is present.

This unexpected result suggested that *Halobacterium* can obtain energy by using either of two systems, one that operates in the presence of oxygen (oxidative phosphorylation) and one that operates in the presence of light (some

| | Synthesis of ATP | | | |
| | Oxygen | | No Oxygen | |
Microorganism	Light	Dark	Light	Dark
Photosynthetic bacteria	–	–	+	–
Green algae	+	–	–	–
E. coli	+	+	+	+
H. halobium	+	+	+	+

the bonds of chemical compounds. To obtain energy in a usable form, a cell must first have an electron (or hydrogen) donor, which serves as an initial energy source within the cell. Electron donors can be as diverse as photosynthetic pigments, glucose or other organic compounds, elemental sulfur, ammonia, or hydrogen gas (Figure 5-12). Next, electrons removed from the chemical energy sources are transferred to electron carriers, such as the coenzymes NAD^+, $NADP^+$, and FAD (Figure 5-12b). This transfer is an oxidation-reduction reaction; the initial energy source is oxidized as this first electron carrier is reduced. During this phase, some ATP is produced. In the third stage, electrons are transferred from electron carriers to their final electron acceptors in further oxidation-reduction reactions with the additional production of ATP by chemiosmosis (Figure 5-12c).

As we will discuss in later sections, when oxygen (O_2) serves as the final electron acceptor, the process is called aerobic respiration. In the process called anaerobic respiration, inorganic substances other than oxygen, such as nitrate ions (NO_3^-) or sulfate ions (SO_4^{2-}), serve as the final electron acceptors. In fermentation, organic compounds serve as the final electron acceptors. In aerobic and anaerobic respiration, a series of electron carriers called an electron transport chain releases energy that is used by the mechanism of chemiosmosis to synthesize ATP. Regardless of their energy sources, all organisms use similar oxidation-reduction reactions to transfer elec-

---

kind of photophosphorylation). It was found that the rate of ATP synthesis by *H. halobium* is highest when the cells receive light that is between 550 and 600 nm in wavelength; this range exactly corresponds to the absorption spectrum of bacteriorhodopsin. This and other clues suggested that the mechanism of photophosphorylation was tied to the pigment bacteriorhodopsin.

Researchers hypothesized that bacteriorhodopsin, like chlorophyll-containing systems, acts as a proton pump to create a proton gradient across a cell membrane; in this case, the gradient is created across the purple membrane. To test this theory, researchers isolated pure purple membrane and used a special technique to form the membrane into vesicles. Illumination caused the pH

level of the medium inside and surrounding the vesicles to change. This change indicated that protons were being transported across the purple membrane. According to the chemiosmotic theory, this transport of protons in intact cells creates a proton gradient that can do cellular work—drive the synthesis of ATP (see figure) or transport solutes.

Halobacterium synthesizes the purple membrane only when it is in low oxygen concentrations and in the presence of light. It is clear that the membrane provides the bacterium with energy when oxygen concentrations in its highly salinic environment become too low to support oxidative phosphorylation; that condition occurs frequently. Cells can accumulate solutes by group translocation to prevent water loss.

Until the discovery of the halobacterial system,

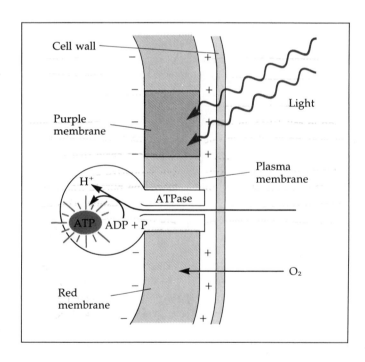

chlorophyll-containing systems were the only ones known to generate ATP using sunlight. Use of the pigment bacteriorhodopsin by *Halobacterium* is the simplest form of photophosphorylation known, and it provides a useful model for the study of

chemiosmosis. Bacteriorhodopsin is itself an intriguing discovery. Its similarity to rhodopsin is the basis of much conjecture about both the mechanism and evolution of sight (the perception of light) and its relationship to the generation of cellular energy.

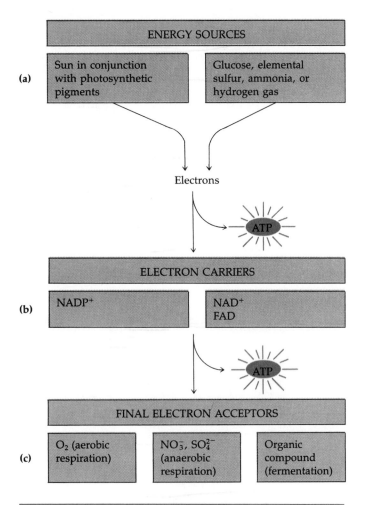

Figure 5-12 ATP production. The production of ATP requires **(a)** an energy source, **(b)** the transfer of electrons to an electron carrier during an oxidation-reduction reaction, and **(c)** the transfer of electrons to a final electron acceptor.

trons and similar mechanisms to use the energy released to produce ATP.

We will next consider the specific biochemical pathways that organisms use to produce energy.

Biochemical Pathways of Energy Production

Organisms produce and store energy from organic molecules by a series of controlled reactions rather than in a single burst. If the energy were released all at once, it would be released as a large amount of heat, which could not be readily used to drive chemical reactions and would, in fact, damage the cell. To extract energy from organic compounds and store it

in cells, organisms pass electrons from one compound to another through a series of oxidation-reduction reactions.

A sequence of enzymatically catalyzed chemical reactions occurring in a cell is called a **biochemical pathway**. Pathways are usually written as follows:

$$ \overset{H_2O \qquad ATP \qquad ADP + \textcircled{P} \qquad CO_2}{A \longrightarrow B \longrightarrow C \Longrightarrow D \rightleftharpoons E} $$

Note that pathways shown in this way are not usually written as balanced chemical equations. This hypothetical pathway converts starting material A into the end product E in a series of four steps. (Of course, if this were an actual pathway, chemical structures or names would replace letters A through E.)

The curved arrow originating at H_2O indicates that water is a reactant in the reaction converting A to B. Similarly, the curved arrow pointing to CO_2 indicates that carbon dioxide is a product of the reaction converting C to D. To drive the reaction converting B to C, a molecule of ATP must be broken down to ADP and a phosphate in a *coupled reaction* (that is, the two reactions occur almost simultaneously, each depending on the other). The reaction converting D to E is readily reversible, as indicated by the double arrow. Keep in mind that almost every reaction in a biochemical pathway is catalyzed by a specific enzyme; sometimes the name of the enzyme is printed near the arrow.

Carbohydrate Catabolism

Most microorganisms oxidize carbohydrates to provide most of the cell's energy. Carbohydrate catabolism, the breakdown of carbohydrate molecules to produce energy, is therefore of great importance in cell metabolism. Glucose is the carbohydrate energy source most used by cells. Microorganisms can also catabolize various lipids and proteins for energy production, as you will see later.

To produce energy from glucose, microorganisms use two general processes, respiration and fermentation. In both processes, electrons produced from the oxidation of glucose are passed to NAD^+. However, the ultimate fate of the electrons differs for the two processes. The first stage of both respiration and fermentation is the oxidation of glucose to pyruvic acid. This is most commonly accomplished by a process called glycolysis.

Glycolysis

Glycolysis, the oxidation of glucose to pyruvic acid, is usually the first stage in carbohydrate catabolism. Most microorganisms use this pathway; it is characteristic of *E. coli*, the streptococci, and most other microbial heterotrophs as well as eucaryotes.

Glycolysis is also called the **Embden–Meyerhof pathway**. The word *glycolysis* means splitting of sugar, and this is exactly what happens during the glycolytic pathway. Glycolysis splits glucose, a six-carbon sugar, into two three-carbon sugars, releasing energy in the process. These sugars are then oxidized and rearranged to form two molecules of pyruvic acid. In addition, during glycolysis, NAD^+ is reduced to NADH, and there is a net production of two ATPs by substrate-level phosphorylation. Glycolysis does not require oxygen; it can occur under either aerobic or anaerobic conditions. This pathway is a series of ten chemical reactions, each catalyzed by a different enzyme. The steps can be outlined as follows (Figure 5-13; see also Appendix F for a more detailed representation of glycolysis):

Step 1. Glucose enters the cell and is phosphorylated. A molecule of ATP is invested. The product of this reaction is glucose 6-phosphate.

Step 2. Glucose 6-phosphate is rearranged to form fructose 6-phosphate.

Step 3. The phosphate from another ATP is required to produce fructose 1,6-diphosphate, still a 6-carbon compound. (Note the total investment of 2 ATPs up to this point.)

Step 4. An enzyme cleaves (splits) the sugar into two three-carbon molecules, dihydroxyacetone phosphate and glyceraldehyde 3-phosphate.

Step 5. Dihydroxyacetate phosphate and glyceraldehyde 3-phosphate are interconvertible.

Step 6. The next enzyme converts glyceraldehyde phosphate to another three-carbon compound, 1,3-diphosphoglyceric acid. Since every dihydroxyacetone phosphate molecule can be converted to glyceraldehyde 3-phosphate, and each glyceraldehyde 3-phosphate to 1,3-diphosphoglyceric acid, the result is two molecules of 1,3-diphosphoglyceric acid for each initial molecule of glucose. Glyceraldehyde 3-phosphate is oxidized by the

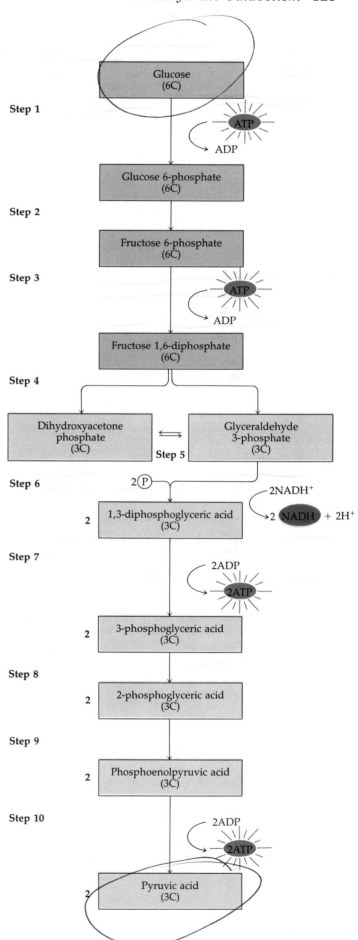

Figure 5-13   Outline of the reactions of glycolysis. A more detailed version is presented in Appendix F.

transfer of two hydrogen atoms to NAD^+ to form $NADH + H^+$. The enzyme couples this reaction with the creation of a high-energy bond between the sugar and a phosphate group. The three-carbon sugar now has two phosphate groups.

Step 7. The high-energy phosphate group is moved to ADP, forming ATP, the first ATP production of glycolysis. (Since the sugar-splitting in step 4, all products are doubled. Therefore, this step actually repays the earlier investment of two ATPs.)

Step 8. An enzyme relocates the remaining phosphate group of 3-phosphoglyceric acid in preparation for the next step.

Step 9. The sugar is rearranged in such a way that the remaining phosphate bond becomes a high-energy bond.

Step 10. The high-energy phosphate group is transferred to ADP, forming ATP. For each initial glucose molecule, the result of this step is *two* molecules of ATP and two molecules of a three-carbon compound called pyruvic acid.

In summary, three things have happened to glucose during glycolysis:

1. First, two molecules of ATP are used when a six-carbon glucose molecule is phosphorylated, rearranged, and split into two three-carbon compounds—glyceraldehyde 3-phosphate and dihydroxyacetone phosphate. These two compounds are interconvertible.

2. Next, the two three-carbon molecules are oxidized in several stages to two molecules of pyruvic acid. In these reactions, two molecules of $NADH + H^+$ are produced, and four molecules of ATP are formed by substrate-level phosphorylation.

3. Because two molecules of ATP were needed to get glycolysis started, and four molecules of ATP have been generated when the process is completed, *there is a net gain of two molecules of ATP for each molecule of glucose that is oxidized.*

Pentose Phosphate Pathway

Many bacteria have an alternate pathway in addition to glycolysis for the oxidation of glucose. This is called the **pentose phosphate pathway** (or **hexose monophosphate shunt**), and it operates simultaneously with glycolysis. The pentose phosphate pathway is a cyclic pathway that provides a means for the breakdown of five-carbon sugars (pentoses) as well as glucose (see the detailed figure in Appendix F). A key feature of this pathway is that it produces important

intermediate pentoses that act as precursors in the synthesis of (1) nucleic acids, (2) glucose from carbon dioxide in photosynthesis, and (3) certain amino acids. The pathway is an important producer of NADPH from $NADP^+$; 12 molecules of $NADPH + 12H^+$ can be produced from each molecule of glucose if the glucose is completely oxidized. NADPH is then used in various biosynthetic reactions in the cell. Unlike glycolysis, the pentose phosphate pathway yields a net gain of only one molecule of ATP for each molecule of glucose oxidized. Bacteria that use the pentose phosphate pathway include *Bacillus subtilis* (su'til-us), *E. coli*, *Leuconostoc mesenteroides* (lü-kō-nos'tok mes-en-ter-oi'dēz), and *Streptococcus faecalis* (fē-kāl'is).

Entner-Doudoroff Pathway

The **Entner–Doudoroff pathway** is still another pathway for the oxidation of glucose to pyruvic acid. From each molecule of glucose, two molecules of NADPH and one molecule of ATP are produced for use in cellular biosynthetic reactions (Figure 5-14; see also Appendix F for a more detailed representation). Bacteria that have the enzymes for the Entner–Doudoroff pathway can metabolize without either glycolysis or the pentose phosphate pathway. The Entner–Doudoroff pathway is found in some gram-negative bacteria, including *Rhizobium* (rī-zō'bē-um), *Pseudomonas* (sū-dō-mō'nas), and *Agrobacterium* (ag-rō-bak-ti'rē-um). It is generally not found among gram-positive bacteria. Tests for the ability to oxidize glucose by this pathway are sometimes used to identify *Pseudomonas* in the clinical laboratory.

Respiration

After glucose has been broken down to pyruvic acid, the pyruvic acid can undergo further fermentation (to be described later), or it can undergo respiration. **Respiration** is an ATP-generating process in which chemical compounds are oxidized, and the final electron acceptor is almost always an inorganic molecule, most commonly O_2. Pyruvic acid produced by glycolysis is split, and a fragment of it is attached to a coenzyme molecule called *coenzyme A* (or *CoA*); that combination is acetyl CoA. Acetyl CoA next enters the Krebs cycle, a series of reactions that release electrons and protons (H^+). Electrons released during glycolysis, the formation of acetyl CoA, and the Krebs cycle are passed to NAD^+ or FAD molecules. These molecules carry the electrons to an electron transport chain, where ATP is formed. A major feature of respiratory processes is the operation of an electron transport chain. As we discussed before, electrons

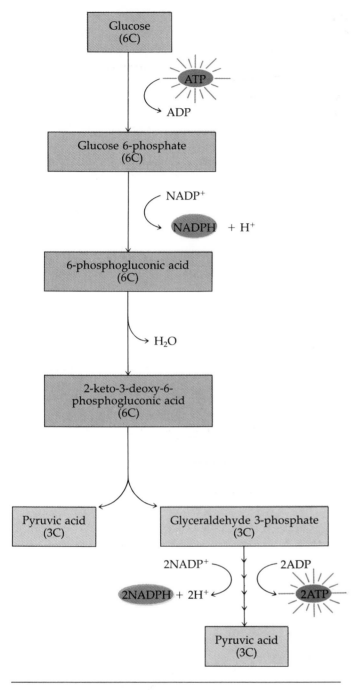

Figure 5-14 Simplified version of the Entner–Doudoroff pathway. A more detailed version is presented in Appendix F.

move through the chain from reduced coenzymes to an inorganic molecule (such as O_2), losing energy in the process, and making possible the generation of ATP by chemiosmosis. This is the process called oxidative phosphorylation.

Any naturally occurring organic molecule can be degraded by some microbe during respiration. The organic molecule is usually oxidized completely to carbon dioxide and H_2O. Consequently, the yield of ATP per molecule of substrate is much greater in respiration than in fermentation.

In most respirations, the final hydrogen (electron) acceptor is molecular oxygen (O_2). This type of respiration is called **aerobic respiration**. A few bacteria carry on **anaerobic respiration**, in which the final hydrogen (electron) acceptor is an inorganic molecule other than molecular oxygen. Examples of these inorganic acceptors are such ions as nitrate (NO_3^-), sulfate (SO_4^{2-}), and carbonate (CO_3^{2-}). Anaerobic respiration can produce an energy yield almost as high as that of aerobic respiration. Anaerobic respiration by bacteria using nitrate and sulfate as final acceptors is essential for the nitrogen and sulfur cycles that occur in nature.

Aerobic Respiration

Before pyruvic acid can participate in biochemical reactions that completely oxidize it to carbon dioxide and water, it must lose one molecule of carbon dioxide and become a two-carbon compound. This process is called **decarboxylation**. The two-carbon compound, called an *acetyl group*, then attaches to coenzyme A (CoA); the resulting complex is known as *acetyl coenzyme A (acetyl CoA)* (Figure 5-15.) During this reaction, NAD^+ is reduced to NADH + H^+. Remember that the oxidation of one glucose molecule produces two molecules of pyruvic acid, so for each molecule of glucose, two molecules of carbon dioxide are lost, and 2 NADH + $2H^+$ are produced. Each molecule of NADH will later yield three molecules of ATP in the electron transport chain. Once the pyruvic acid has undergone decarboxylation and its derivative (the acetyl group) has attached to CoA, the resulting acetyl CoA is ready to enter the Krebs cycle.

Figure 5-15 Formation of acetyl CoA from pyruvic acid and coenzyme A. A high-energy bond is formed between the acetyl group and CoA.

Krebs cycle The **Krebs cycle**, also called the **tricarboxylic acid cycle (TCA)** or **citric acid cycle**, is a series of biochemical reactions in which the large amount of potential chemical energy stored in intermediate substances ultimately derived from pyruvic acid is released step by step. In this cycle, a series of oxidations and reductions transfers that potential energy, in the form of electrons, to a number of coenzyme molecules. The pyruvic acid derivatives are oxidized; the coenzymes are reduced.

As acetyl CoA enters the Krebs cycle, CoA detaches from the acetyl group and can pick up more acetyl groups for the next round of the Krebs cycle. Meanwhile, the acetyl group combines with a substance called oxaloacetic acid to form citric acid. This synthesis reaction requires energy, which is provided by the cleavage of the high-energy bond between the acetyl group and CoA. The Krebs cycle begins with the formation of citric acid. The major chemical reactions of this cycle are outlined in Figure 5-16; a more detailed representation of the Krebs cycle is provided in Appendix F. Keep in mind that each reaction is

Figure 5-16 The Krebs cycle. Note the points where CO_2 is produced, NADH and $FADH_2$ are formed, and guanosine triphosphate (GTP) is generated. GTP is used directly in certain cell processes and can also be used to convert ADP to ATP. A more detailed version is presented in Appendix F.

catalyzed by a specific enzyme.

The chemical reactions of the Krebs cycle fall into several general categories. One of these is decarboxylation. Isocitric acid, a six-carbon compound, is decarboxylated to the five-carbon compound called α-ketoglutaric acid. By losing a molecule of CO_2 and picking up a molecule of CoA, α-ketoglutaric acid is then decarboxylated to form the four-carbon compound called succinyl CoA. Once again, remember that the oxidation of one molecule of glucose produces two molecules of acetyl CoA that enter the Krebs cycle. This means that from each molecule of glucose, two molecules of CO_2 are produced when pyruvic acid is decarboxylated, and four molecules of CO_2 are produced in the Krebs cycle.

Another general category of Krebs-cycle chemical reactions is oxidation-reduction. In the conversion of the six-carbon isocitric acid to a five-carbon compound, two hydrogen atoms are lost. In other words, the six-carbon compound is oxidized. The hydrogen atoms released in the Krebs cycle are picked up by the coenzymes NAD^+ and FAD. When NAD^+ picks up hydrogen atoms, it is reduced and may be represented as $NADH + H^+$: similarly, FAD is reduced to $FADH_2$. Note the points in the synthesis of acetyl CoA and in the Krebs cycle where coenzymes pick up hydrogen atoms (Figure 5-16).

If we look at the Krebs cycle as a whole, we see that, for each two molecules of acetyl CoA that enter the cycle, four molecules of carbon dioxide are liberated by decarboxylation, $6 \ NADH + 6H^+$ and $2 \ FADH_2$ are produced by oxidation-reduction reactions, and two molecules of guanosine triphosphate (GTP, the equivalent of ATP) are generated by substrate level phosphorylation. Many of the intermediates in the Krebs cycle also play a role in other pathways, especially in amino acid biosynthesis (discussed later in this chapter). In the electron transport chain, the $6 \ NADH + 6H^+$ will later yield 18 ATP molecules, and the $2 \ FADH_2$ will yield 4 ATP molecules.

The carbon dioxide produced in the Krebs cycle is liberated into the atmosphere as a gaseous by-product of aerobic respiration. (Humans produce carbon dioxide from the Krebs cycle in each cell of the body and discharge it through the lungs during exhalation.) The reduced coenzymes ($NADH$ and $FADH_2$) are the most important outcome of the Krebs cycle because they contain the energy originally stored in glucose and then stored in pyruvic acid. During the next phase of aerobic respiration, a series of reductions indirectly transfers the energy stored in the coenzymes to ATP. These reactions are collectively called the electron transport chain.

Electron transport chain The **electron transport chain** consists of a sequence of carrier molecules that are capable of oxidation and reduction. As electrons are passed through the chain, there is a stepwise release of energy, which is used to drive the chemiosmotic generation of ATP. In aerobic respiration, the terminal electron acceptor of the chain is molecular oxygen (O_2); in anaerobic respiration, the final acceptor is an inorganic molecule other than O_2 (or, rarely, an organic compound). The final oxidation is irreversible in both respirations. In eucaryotic cells, the electron transport chain is contained in the inner mitochondrial membrane; in procaryotic cells, it is found in the plasma membrane.

The respiration of organic compounds in electron transport chains involves three classes of carrier molecules. The first are called **flavoproteins**. These proteins contain flavin, a coenzyme derived from riboflavin (vitamin B_2), and are capable of performing alternating oxidations and reductions. One important flavin coenzyme is called **flavin mononucleotide (FMN)**. The second class of carrier molecules is called **cytochromes**, proteins that contain an iron-containing prosthetic group (heme) capable of existing alternately as a reduced form (Fe^{2+}) and an oxidized form (Fe^{3+}). The several cytochromes involved in the electron transport chain described shortly are cytochrome b (cyt b), cytochrome c_1 (cyt c_1), cytochrome c (cyt c), cytochrome a (cyt a), and cytochrome a_3 (cyt a_3). The third class of carriers is known as **ubiquinones** (or **coenzyme Q**), symbolized **Q**, nonprotein carriers of low molecular weight.

The electron transport chains of different bacterial systems are somewhat diverse, in that the particular carriers used and the order in which they act may differ from those of other bacteria and from those of eucaryotic mitochondrial systems. However, keep in mind that all electron transport chains achieve the same basic goal. Much is known about the electron transport chain in the mitochondria of eucaryotic cells, so this is the one we will describe.

The first step in the mitochondrial electron transport chain involves the transfer of high-energy electrons from $NADH + H^+$ to FMN, the first carrier in the chain (Figure 5-17a). (Two $NADH + 2H^+$ are generated from glycolysis, two from the formation of acetyl CoA, and six from the Krebs cycle.) This transfer actually involves the passage of a hydrogen atom with two electrons to FMN, which then picks up an additional H^+ from the surrounding aqueous medium. As a result of the first transfer, $NADH + H^+$ is oxidized to NAD^+, and FMN is reduced to $FMNH_2$.

In the second step in the electron transport chain, $FMNH_2$ passes $2H^+$ to the other side of the mitochondrial membrane and passes two electrons to Q.

Figure 5-17   An electron transport chain. (a) The electrons pass along the chain in a gradual and stepwise fashion, so energy is released in manageable quantities that can be used to pump protons (H⁺) across a membrane. **(b)** Organization of electron carriers into three complexes. Note the three points at which protons (H⁺) are pumped across the membrane. In a eucaryotic cell, they are pumped from the matrix side of the mitochondrial membrane to the opposite side. The flow of electrons is shown in color.

(a)

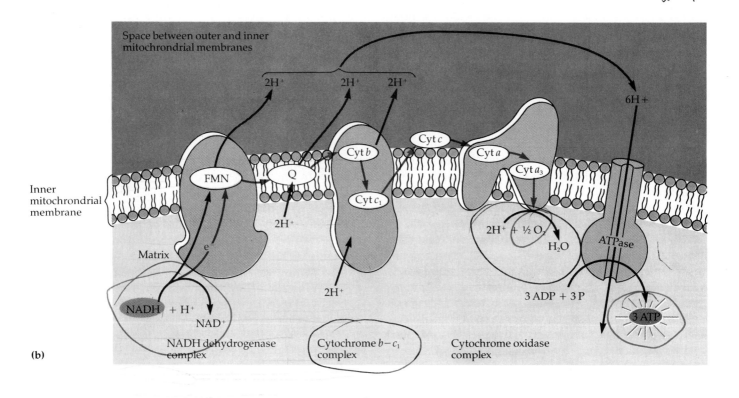

(b)

As a result, FMNH$_2$ is oxidized to FMN. Q also picks up an additional 2H$^+$ from the surrounding aqueous medium.

The next part of the electron transport chain involves the cytochromes. Electrons are passed successively from Q to cyt b, cyt c_1, cyt c, cyt a, and cyt a_3. Each cytochrome in the chain is reduced as it picks up electrons and is oxidized as it gives up electrons. The last cytochrome, cyt a_3, passes its electrons to molecular oxygen (O$_2$), which becomes negatively charged and then picks up 2H$^+$ from the surrounding medium to form H$_2$O.

Note in Figure 5-17a that FADH$_2$, derived from the Krebs cycle, is another source of electrons. However, FADH$_2$ adds its electrons to the electron transport chain at a lower level than NADH. Because of this, the electron transport chain produces about one third less energy for ATP generation when FADH$_2$ donates electrons than when NADH is involved.

An important feature of the electron transport chain is the presence of some carriers, such as FMN and Q, that accept H$^+$ as well as electrons, and other carriers, such as cytochromes, that transfer electrons only. As a result, electron flow down the chain is accompanied at several points by the active transport (pumping) of protons (H$^+$) from the matrix side of the inner mitochondrial membrane to the opposite side of the membrane. The result is the proton gradient described earlier, which is responsible for the chemiosmotic generation of ATP (Figure 5-17b).

Within the mitochondrial membrane, the carriers of the electron transport chain are organized into three complexes: (1) NADH dehydrogenase complex, which contains FMN, (2) cytochrome b-c_1 complex, which contains cytochromes b and c_1, and (3) cytochrome oxidase complex, which contains cytochromes a and a_3 (Figure 5-17b). Q transfers electrons between the first and second complexes, and cytochrome c transfers electrons between the second and third complexes. Three components of the electron transport chain pump protons (H$^+$)—the NADH dehydrogenase complex, the carrier Q, and the cytochrome b-c_1 complex.

Recall that the potential energy stored in the proton gradient is the energy source for the conversion of ADP to ATP by the adenosine triphosphatase (ATPase) complexes in the membrane. By coupling electron transport to the generation of ATP, the proton gradient drives oxidative phosphorylation. The chemiosmotic generation of ATP is responsible for most of the ATP produced during cellular respiration.

The various electron transfers in the electron transport chain generate 34 ATP molecules from each molecule of glucose oxidized: three from each of the ten molecules of NADH + 10H$^+$ (30), and two from each of the two molecules of FADH$_2$(4). During aerobic respiration, a total of 38 ATPs can be generated from one molecule of glucose. Note that four of those ATPs come from substrate-level phosphorylation in glycolysis and the Krebs cycle. Table 5-5 provides a detailed accounting of the ATP yield during aerobic respiration.

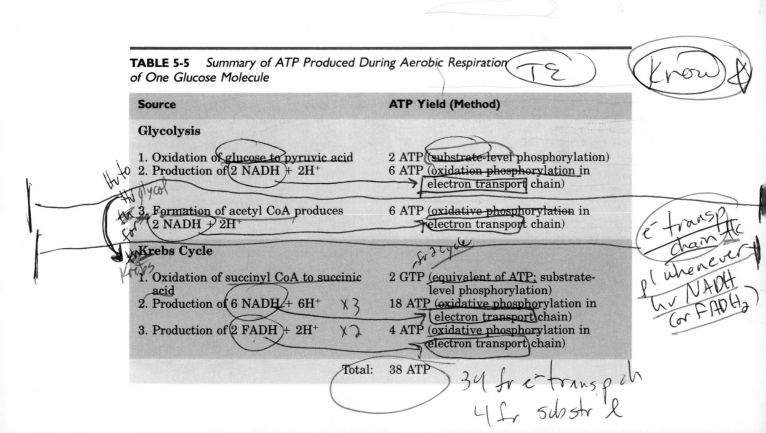

TABLE 5-5 *Summary of ATP Produced During Aerobic Respiration of One Glucose Molecule*

Source	ATP Yield (Method)
Glycolysis	
1. Oxidation of glucose to pyruvic acid	2 ATP (substrate-level phosphorylation)
2. Production of 2 NADH + 2H$^+$	6 ATP (oxidation phosphorylation in electron transport chain)
3. Formation of acetyl CoA produces 2 NADH + 2H$^+$	6 ATP (oxidative phosphorylation in electron transport chain)
Krebs Cycle	
1. Oxidation of succinyl CoA to succinic acid	2 GTP (equivalent of ATP; substrate-level phosphorylation)
2. Production of 6 NADH + 6H$^+$	18 ATP (oxidative phosphorylation in electron transport chain)
3. Production of 2 FADH + 2H$^+$	4 ATP (oxidative phosphorylation in electron transport chain)
Total:	38 ATP

We can now summarize the overall reaction for aerobic respiration as follows:

$$C_6H_{12}O_6 + 6O_2 + 38\,ADP + 38\,\textcircled{P} \rightarrow$$

Glucose   Oxygen

$$6CO_2 + 6H_2O + 38\,ATP$$

Carbon   Water
dioxide

A summary of the various stages of aerobic respiration is presented in Figure 5-18.

Anaerobic Respiration

In the process called anaerobic respiration, the final electron acceptor is an inorganic substance other than oxygen (O_2). Some bacteria, such as *Pseudomonas* and *Bacillus*, can use a nitrate ion (NO_3^-) as a final electron acceptor; it is reduced to nitrite ion (NO_2^-), nitrous oxide (N_2O), or nitrogen gas (N_2). Other bacteria, such as *Desulfovibrio* (dē-sul-fō-vib′rē-ō), use sulfate (SO_4^{2-}) as the final electron acceptor, to form hydrogen sulfide (H_2S). Still other bacteria use carbonate (CO_3^{2-}) to form methane (CH_4). (See the box in Chapter 26.) Anaerobic respiration yields varying amounts of ATP, depending on the organism and the pathway.

Fermentation

After glucose is broken down to pyruvic acid, the pyruvic acid can undergo respiration, as previously described, or it can undergo further fermentation. Fermentation can be defined in several ways (see the box on p. 000), but we will define it here as a process that:

1. Releases energy from sugars or other organic molecules, such as amino acids, organic acids, purines, and pyrimidines.
2. Does not require oxygen (but sometimes can occur in its presence).
3. Does not require use of the Krebs cycle or an electron transport chain.
4. Uses an organic molecule as the final electron acceptor.
5. Produces only small amounts of ATP (only one or two ATP molecules for each molecule of starting material) because much of the energy remains in the chemical bonds of the end products, such as lactic acid or ethanol.

During fermentation, hydrogen atoms are transferred from $NADH + H^+$ to pyruvic acid, which is reduced to various end-products; substrate-level phos-

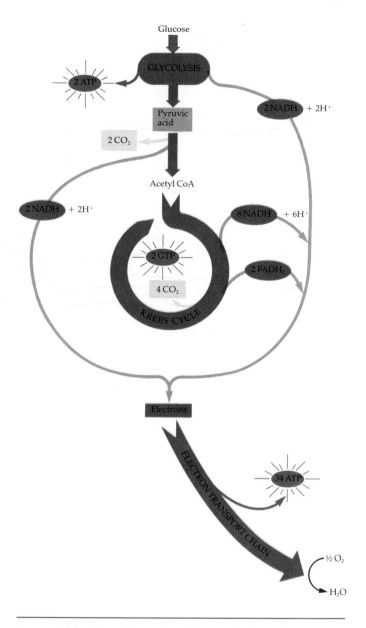

Figure 5-18 Overview of aerobic respiration. Glucose is broken down completely to carbon dioxide and water, and ATP is generated. This process has three major phases—glycolysis, the Krebs cycle, and the electron transport chain. The key event in the process is that electrons are picked up from intermediates of glycolysis and the Krebs cycle by NAD^+ or FAD and are carried by NADH or $FADH_2$ to the electron transport chain. NADH is also produced in the conversion of pyruvic acid to acetyl CoA. Most of the ATP generated by aerobic respiration is made by the chemiosmotic mechanism during the electron-transport-chain phase; this is called oxidative phosphorylation. The process of fermentation uses only the glycolysis portion of this sequence and generates much less ATP, by substrate-level phosphorylation.

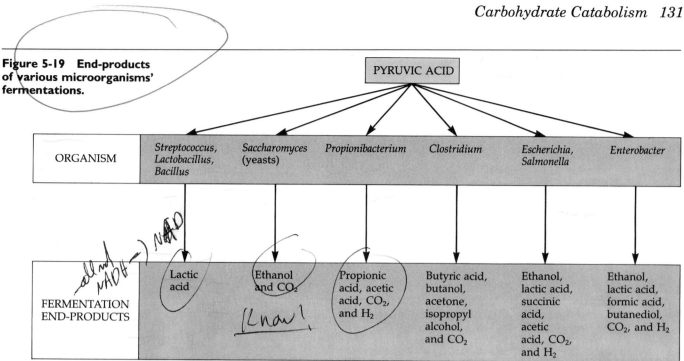

Figure 5-19 End-products of various microorganisms' fermentations.

phorylation converts ADP to ATP. Various microorganisms are able to ferment various substrates; the end-products depend on the particular microorganism, the substrate, and the enzymes that are present and active. Chemical analyses of these end-products are useful in the identification of microorganisms. The products of various microorganisms' fermentations of pyruvic acid are shown in Figure 5-19.

In discussing the process of fermentation, we will consider two of the more important processes, lactic acid fermentation and alcoholic fermentation.

Lactic Acid Fermentation

During the first phase of **lactic acid fermentation**, glycolysis, a molecule of glucose is oxidized to two molecules of pyruvic acid (Figure 5-20a; see also Figure 5-13). This oxidation generates the energy with which the two molecules of ATP are formed. In the next step, the two molecules of pyruvic acid are reduced by $2 NADH + 2H^+$ to form two molecules of lactic acid. Because lactic acid is the end product of the reaction, it undergoes no further oxidation, and

---

MICROBIOLOGY HIGHLIGHTS

What Is Fermentation?

To many people, fermentation simply means the production of alcohol. Grains and fruits are fermented to produce beer and wine. If a food soured, you might say it was "off" or fermented. Here are some definitions of fermentation. They range from informal, general usage to more scientific definitions that you will need to know for this course.

Fermentation is:

1. Any process that produces alcoholic beverages (general use).

2. Any spoilage of food by microorganisms (general use).

3. Any large-scale microbial process occurring with or without air (common definition used in industry).

4. Any energy-releasing metabolic process that takes place only under anaerobic conditions (becoming more scientific).

5. All metabolic processes that release energy from a sugar or other organic molecule, do not require oxygen or an electron transport system, and use an organic molecule as the final electron acceptor. This is the definition we will use in this book.

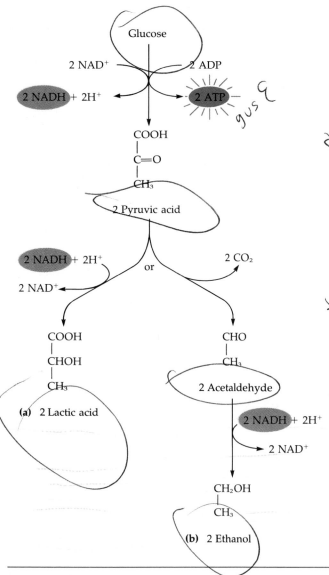

Figure 5-20   Fermentation. (a) Lactic acid fermentation.
(b) Alcoholic fermentation.

most of the energy produced by the reaction remains stored in the lactic acid. Thus, this fermentation yields only a small amount of energy. The overall reaction for lactic acid fermentation is:

$$\text{Glucose} + 2\ \text{ADP} + 2\ \textcircled{P} \longrightarrow 2\ \text{Lactic acid} + 2\ \text{ATP}$$

Two important genera of lactic acid bacteria are *Streptococcus* and *Lactobacillus* (lak-tō-bä-sil′lus). Because these microbes produce only lactic acid, they are referred to as **homolactic** (or *homofermentative*). Lactic acid fermentation can result in food spoilage. However, the process is also important in food production. Lactic acid fermentation can produce yogurt from milk, sauerkraut from fresh cabbage, and pickles from cucumbers.

Alcohol Fermentation

Alcohol fermentation also begins with the glycolysis of a molecule of glucose and the yield of two molecules of pyruvic acid and two molecules of ATP. In the next reaction, the two molecules of pyruvic acid are converted to two molecules of acetaldehyde and two molecules of CO_2 (Figure 5-20b). The two molecules of acetaldehyde are next reduced by 2 NADH + 2H⁺ to form two molecules of ethanol (ethyl alcohol). Again, alcohol fermentation is a low-energy-yield process because most of the energy contained in the initial glucose molecule remains in the ethanol, the end-product. Alcohol fermentation is carried out by a number of bacteria and yeast.

The ethanol and carbon dioxide produced by the yeast *Saccharomyces* (sak-ă-rō-mī′sēs) are waste products for yeast cells but are useful to humans. Ethanol made by yeast is the alcohol in alcoholic beverages, and carbon dioxide made by yeast causes bread dough to rise. (See the box in Chapter 1, p. 4.)

Organisms that produce lactic acid as well as other acids or alcohols are known as **heterolactic** (or **heterofermentative**) and often use the pentose phosphate pathway. The overall reaction for a typical heterolactic fermentation is:

$$\text{Glucose} + \text{ADP} + \textcircled{P} \longrightarrow$$
$$\text{Lactic acid} + \text{Ethanol} + CO_2 + \text{ATP}$$

Table 5-6 lists some of the various microbial fermentations used by industry to convert inexpensive raw materials into useful end-products. A summary of aerobic respiration, anaerobic respiration, and fermentation is given in Table 5-7.

Lipid Catabolism

Thus far in our discussion of energy production, we have emphasized the oxidation of glucose, the principal energy-supplying carbohydrate. But microbes also oxidize lipids and proteins, and the oxidations of all these nutrients are related.

Fats, as you recall, are lipids consisting of fatty acids and glycerol. Microbes produce extracellular enzymes called *lipases* that break fats down into their fatty acid and glycerol components. Each component is then metabolized separately.

Many types of microbes can convert glycerol into dihydroxyacetone phosphate, one of the three-carbon intermediates formed during glycolysis (see Figure 5-13). The dihydroxyacetone phosphate is then metabolized by glycolysis and the Krebs cycle, as shown in Figure 5-21.

TABLE 5-6 *Some Industrial Fermentations*

Fermentation Product	Commercial Use	Starting Material	Microorganism
Ethanol	Beer	Malt extract	*Saccharomyces cerevisiae* (yeast, a fungus)
	Wine	Grape or other fruit juices	*Saccharomyces ellipsoideus* (yeast, a fungus)
Acetic acid	Vinegar	Alcohol	*Acetobacter* (bacterium)
Lactic acid	Cheese, yogurt	Milk	*Lactobacillus, Streptococcus* (bacteria)
	Rye bread	Grain, sugar	*Lactobacillus bulgaricus* (bacterium)
	Summer sausage	Meat	*Pediococcus* (bacterium)
Propionic acid and carbon dioxide	Swiss cheese	Milk	*Propionibacterium freudenreichii* (bacterium)
Acetone and butanol	Pharmaceutical, industrial uses	Molasses	*Clostridium acetobutylicum* (bacterium)
Glycerol	Pharmaceutical, industrial uses	Molasses	*Saccharomyces cerevisiae* (yeast, a fungus)
Citric acid	Flavoring	Molasses	*Aspergillus* (fungus)
Methanol	Fuel	Agricultural wastes	*Clostridium* (bacterium)

TABLE 5-7 *Comparison of Aerobic Respiration, Anaerobic Respiration, and Fermentation*

Energy-Producing Process	Growth Conditions	Final Hydrogen (Electron) Acceptor	Type of Phosphorylation Used to Build ATP	ATP Molecules Produced per Glucose Molecule
Aerobic respiration	Aerobic	Molecular oxygen (O_2)	Substrate-level and oxidative	38
Anaerobic respiration	Anaerobic	Usually an inorganic substance (such as NO_3^-, SO_4^{2-}, or CO_3^{2-}), but not molecular oxygen (O_2)	Substrate-level and oxidative	Variable
Fermentation	Aerobic or anaerobic	An organic molecule	Substrate-level	2

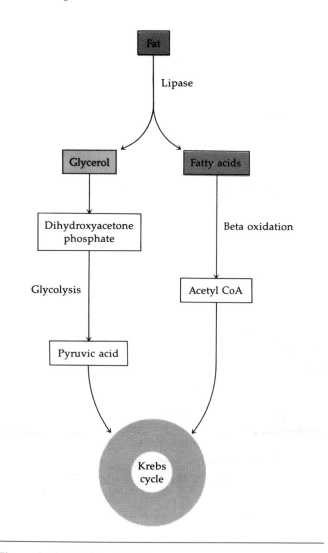

Figure 5-21   Lipid catabolism.

lized, they must be converted to substances that can enter the Krebs cycle. In one such conversion, called **deamination**, the amino group of an amino acid is removed and converted to an ammonium ion (NH_4^+), which can be excreted from the cell. The remaining organic acid can enter the Krebs cycle. Other conversions involve decarboxylation (removal of —COOH) and dehydrogenation. These conversions are complex; the important thing to remember is that various methods can prepare amino acids to enter the Krebs cycle.

Biochemical Pathways of Energy Utilization (Anabolism)

Up to now we have been considering energy production. Through the oxidation of organic molecules, organisms produce energy by aerobic respiration, anaerobic respiration, and fermentation. Much of this energy is given off as heat. The complete metabolic oxidation of glucose to carbon dioxide and water is considered a very efficient process, but about 60% of the energy of glucose is lost as heat. Cells use the remaining energy, which is trapped in molecules of ATP, in a variety of ways. Microbes use ATP to provide energy for the transport of substances across plasma membranes—the process called active transport that we discussed in Chapter 4. Certain microbes can also use some of their ATP for flagellar motion (also discussed in Chapter 4). Most of the ATP, however, is used in the biosynthesis of new cellular components. Biosynthesis is a continuous process in cells, and, in general, it occurs at a faster rate in procaryotic cells than in eucaryotic cells.

Autotrophs build their organic compounds by fixing carbon dioxide in the series of reactions called the Calvin–Benson cycle (see Figure 5-11). This requires both energy (ATP) and electrons (from the oxidation of NADPH). Heterotrophs, by contrast, must have a ready source of organic compounds for biosynthesis. These compounds are used as both carbon source and energy source for the cell. We will next consider the biosynthesis of a few representative classes of biological molecules—carbohydrates, lipids, amino acids, purines, and pyrimidines. As we do so, keep in mind that synthesis reactions require a net input of energy.

Fatty acids are catabolized somewhat differently. The mechanism of fatty acid oxidation is called **beta oxidation**. In this process, carbon fragments of a long chain of fatty acid are removed two at a time, and acetyl CoA is formed from them. As the molecules of acetyl CoA form, they enter the Krebs cycle, as do the acetyl CoA molecules formed by the oxidation of pyruvic acid (see Figure 5-16). Thus, the oxidation of glycerol and fatty acids is directly linked to the oxidation of glucose by the Krebs cycle.

Protein Catabolism

Proteins are too large to pass unaided through plasma membranes. Microbes produce extracellular *proteases* and *peptidases*, which break down proteins into their component amino acids, which can cross the membrane. However, before amino acids can be catabo-

Biosynthesis of Polysaccharides

Microorganisms synthesize sugars and polysaccharides. The carbon atoms required to synthesize glucose are derived from the intermediates produced during glycolysis and the Krebs cycle and from lipids or amino acids. Following the synthesis of glucose (or other simple sugars), bacteria may assemble it into the more complex polysaccharides. One such example is glycogen. In order for bacteria to build glucose into glycogen, glucose units must be phosphorylated and linked together. The product of glucose phosphorylation is called glucose 6-phosphate. Such a process involves the expenditure of energy, usually in the form of ATP, and dehydration (loss of water). The key intermediate in the synthesis of glycogen in bacteria

is called *adenosine diphosphoglucose (ADPG)*. It is formed when a molecule of ATP is added to glucose 6-phosphate (Figure 5-22). Once the building block is synthesized, it is linked with similar units through dehydration to form glycogen.

Using uridine triphosphate (UTP) as a source of energy, animals synthesize glycogen and many other carbohydrates from *uridine diphosphoglucose, UDPG* (Figure 5-22). A compound related to UDPG called *UDP-N-acetyl glucosamine (UDPNAG)* is a key starting material in the biosynthesis of peptidoglycan, the substance that forms bacterial cell walls. UDPNAG is formed from fructose 6-phosphate, and the reaction also uses UTP.

GLYCOLYSIS

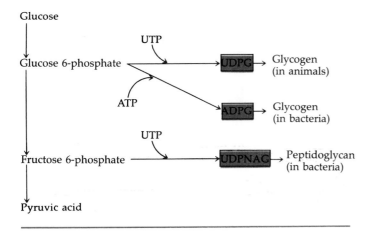

Figure 5-22 Biosynthesis of polysaccharides.

Biosynthesis of Lipids

Because lipids vary in chemical composition, they are synthesized by a variety of routes. Cells synthesize fats by uniting glycerol and fatty acids. The glycerol portion of the fat is derived from dihydroxyacetone phosphate, an intermediate formed during glycolysis. Fatty acids, which are long-chain hydrocarbons (hydrogen linked to carbon), are built up when two-carbon fragments of acetyl CoA are successively added to each other (Figure 5-23). As with polysaccharide synthesis, the building units of fats and other lipids are linked via dehydration reactions that require

Figure 5-23 Biosynthesis of simple lipids.

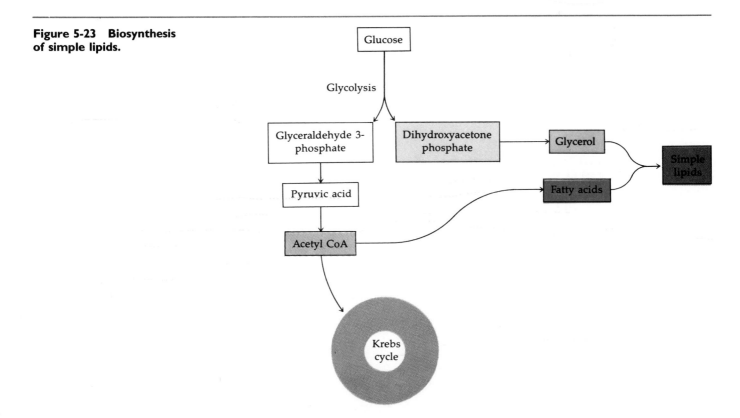

energy, not always in the form of ATP. In the synthesis of phospholipids, for example, the energy is derived from the hydrolysis of a nucleotide called cytidine triphosphate (CTP).

As mentioned previously, the most important role of lipids is as structural components of biological membranes, and most membrane lipids are phospholipids. A lipid of a very different structure, cholesterol, is also found in plasma membranes of eucaryotic cells. Other lipids, such as carotenoids, provide the red, orange, and yellow pigments of some microorganisms. Some lipids form portions of chlorophyll molecules. Lipids also function in energy storage. The lipids called waxes are important components of the cell wall of acid-fast bacteria. Recall that the breakdown products of lipids after biological oxidation feed into the Krebs cycle.

Biosynthesis of Amino Acids

Amino acids are required for protein biosynthesis. Some microbes, such as *E. coli*, contain the enzymes necessary to use starting materials, such as glucose

and inorganic salts, for the syntheses of all the amino acids they need. Organisms with the necessary enzymes can synthesize all amino acids directly or indirectly from intermediates of carbohydrate metabolism (Figure 5-24a). Other microbes require that the environment provide some preformed amino acids.

One important source of these *precursors* (intermediates) used in amino acid synthesis is the Krebs cycle. Adding an amine group to pyruvic acid or to an appropriate organic acid of the Krebs cycle converts the acid into an amino acid. This process is called *amination*. If the amine group comes from a preexisting amino acid, the process is called *transamination* (Figure 5-24b). Other precursors for amino acid biosynthesis are derived from the pentose phosphate pathway and from the Entner–Doudoroff pathway.

Most amino acids within cells are destined to be building blocks for protein synthesis. Proteins play major roles in the cell as enzymes, structural components, and toxins, just to name a few uses. The joining of amino acids to form proteins involves dehydration synthesis and requires energy in the form of ATP. The mechanism of protein synthesis is elaborate and will be discussed in Chapter 8.

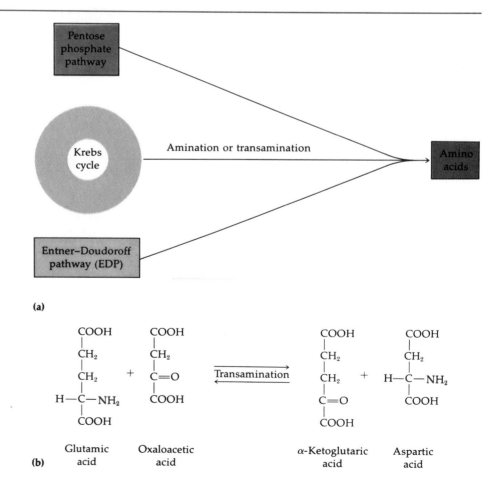

Figure 5-24 Biosynthesis of amino acids. (a) Pathways of amino acid biosynthesis through amination or transamination of intermediates of carbohydrate metabolism. **(b)** Transamination, a process by which new amino acids are made with the amine groups from old amino acids. Glutamic acid and aspartic acid are both amino acids. The other two compounds are intermediates in the Krebs cycle.

Biosynthesis of Purines and Pyrimidines

As you may recall from Chapter 2, the informational molecules DNA and RNA consist of repeating units called *nucleotides*, each of which consists of a purine or pyrimidine, a pentose (five-carbon sugar), and a phosphate group. The five-carbon sugars of nucleotides are derived from either the pentose phosphate pathway or the Entner–Doudoroff pathway. Certain amino acids—aspartic acid, glycine, and glutamine—produced during glycolysis and in the Krebs cycle participate in the biosyntheses of purines and pyrimidines (Figure 5-25). The carbon and nitrogen atoms derived from these amino acids form the purine and pyrimidine rings, and the energy for synthesis is provided by ATP. DNA contains all the information necessary to determine the specific structures and functions of cells. Both RNA and DNA are required for protein synthesis. In addition, such nucleotides as ATP, NAD$^+$, and NADP$^+$ assume roles in stimulating and inhibiting the rate of cellular metabolism. The synthesis of DNA and RNA from nucleotides will be discussed in Chapter 8.

Integration of Metabolism

We have seen thus far that the metabolic processes of microbes produce energy from light, inorganic compounds, and organic compounds. Reactions also occur in which energy is utilized for biosynthesis. With such a variety of activity, it might be imagined that anabolic and catabolic reactions occur independently of each other in space and time. Actually, anabolic and catabolic reactions are joined through a group of common intermediates (identified as key intermediates in Figure 5-26). Both anabolic and catabolic reactions also share some biochemical pathways, such as the Krebs cycle. For example, reactions in the Krebs cycle not only participate in the oxidation of glucose, they also produce intermediates that can be converted to amino acids. Biochemical pathways that function in both anabolism and catabolism are called **amphibolic pathways**, meaning that they are dual-purpose.

Amphibolic pathways bridge the reactions that lead to the breakdown and synthesis of carbohydrates,

Figure 5-25 Biosynthesis of purine and pyrimidine nucleotides.

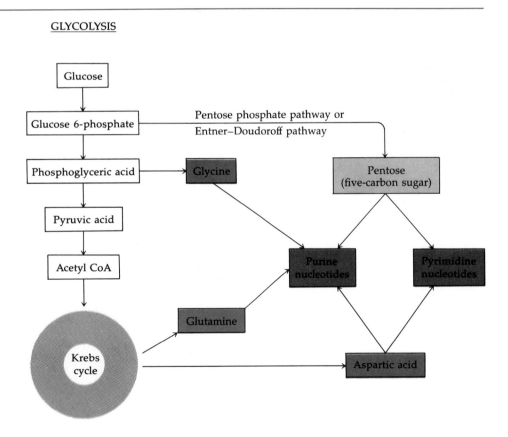

Figure 5-26   Integration of metabolism. Key intermediates are outlined in color. Although not indicated in the figure, amino acids and ribose are used in the synthesis of purine and pyrimidine nucleotides (see Figure 5-25). The double arrows indicate the amphibolic pathways.

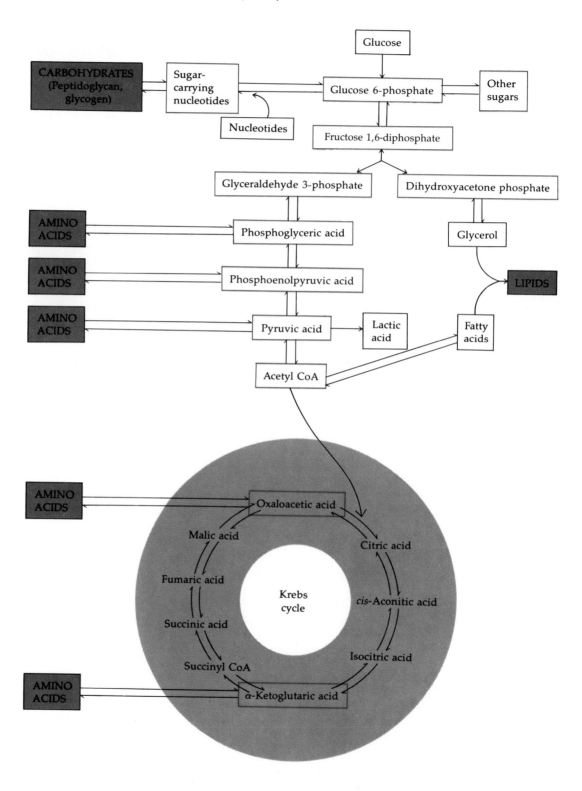

lipids, proteins, and nucleotides. Such pathways enable simultaneous reactions to occur in which the breakdown product formed in one reaction is used in another reaction to synthesize a different compound, and vice versa. Since various intermediates are common to both anabolic and catabolic reactions, mechanisms exist that regulate synthesis and breakdown pathways and allow these reactions to occur simultaneously. One such mechanism involves the use of different coenzymes for opposite pathways. For example, NAD^+ is involved in catabolic reactions, whereas $NADP^+$ is involved in anabolic reactions.

Enzymes can also coordinate anabolic and catabolic reactions by accelerating or inhibiting the rates of biochemical reactions. The energy stores of a cell can also affect the rates of biochemical reactions. For example, if ATP begins to accumulate, an enzyme shuts down glycolysis. This control helps to synchronize the rates of glycolysis and the Krebs cycle. Thus, if citric acid consumption increases, either because of a demand for more ATP or because anabolic pathways are draining off intermediates of the citric acid cycle, glycolysis accelerates and meets the demand.

* * *

In the next chapter, we will turn our attention to microbial growth and apply some of these concepts of microbial metabolism.

Study Outline

Catabolic and Anabolic Reactions   (pp. 106–107)

1. The sum of all chemical reactions within a living organism is known as metabolism.
2. Anabolism refers to chemical reactions in which simpler substances are combined to form more complex molecules. Anabolic reactions usually require energy.
3. Catabolism refers to chemical reactions that result in the breakdown of more complex organic molecules into simpler substances. Catabolic reactions usually release energy.
4. The energy of catabolic reactions is used to drive anabolic reactions.
5. The energy for chemical reactions is stored in ATP.

Enzymes   (p. 108)

1. Enzymes are proteins produced by living cells that catalyze chemical reactions.
2. Enzymes are generally globular proteins with characteristic three-dimensional shapes.
3. Enzymes catalyze chemical reactions by lowering the activation energy.
4. Enzymes are efficient, are able to operate at relatively low temperatures, and are subject to various cellular controls.

Enzyme Components   (pp. 108–110)

1. Most enzymes are holoenzymes, consisting of a protein portion (apoenzyme) and a nonprotein portion (cofactor).
2. The cofactor can be a metal ion (iron, copper, magnesium, manganese, zinc, calcium, or cobalt) or a complex organic molecule known as a coenzyme (NAD, NADP, FMN, FAD, and coenzyme A).

Mechanism of Enzymatic Action   (p. 110)

1. When an enzyme and substrate combine, the substrate is transformed, and the enzyme is recovered.
2. Enzymes are characterized by specificity, which is a function of their active sites.

Classifying Enzymes   (pp. 110–111)

1. Enzymes' names usually end in -ase.
2. The six classes of enzymes are defined on the basis of the types of reactions they catalyze.

Factors Influencing Enzymatic Activity   (pp. 111–113)

1. At high temperatures, enzymes undergo denaturation and lose their catalytic properties; at low temperatures, the reaction rate decreases.
2. The pH at which enzymatic activity is maximal is known as the pH optimum.
3. Within limits, enzymatic activity increases as substrate concentration increases.
4. Competitive inhibitors compete with the normal substrate for the active site of the enzyme. Noncompetitive inhibitors act on other parts of the apoenzyme or on the cofactor and decrease the enzyme's ability to combine with the normal substrate.

Feedback Inhibition   (pp. 113–114)

1. Feedback inhibition occurs when the end product of a pathway inhibits an enzyme's activity in the pathway.

Energy Production Methods   (pp. 114–122)

Oxidation-Reduction   (pp. 114–115)

1. Biological oxidation of a substance means either the addition of oxygen or the removal of a pair of electrons from it. Hydrogen ions are usually removed with the electrons.
2. Reduction of a substance refers to its gain of a pair of electrons or hydrogen atoms.
3. Each time a substance is oxidized, another is simultaneously reduced.
4. NAD^+ and $NADP^+$ are oxidized forms; NADH and NADPH are reduced forms.

Generation of ATP: Types of Phosphorylation   (pp. 115–116)

1. Energy released during certain metabolic reactions can be trapped to form ATP from ADP and Ⓟ (phosphate). Addition of a phosphate group to a molecule is called phosphorylation.
2. During substrate-level phosphorylation, a high-energy phosphate group from an intermediate in catabolism is added to ADP.
3. During oxidative phosphorylation, energy is released as electrons are passed to a series of electron acceptors (an electron transport chain) and finally to oxygen or another inorganic compound.
4. During photophosphorylation, energy from light is trapped by chlorophyll, and an electron that passes to a series of electron acceptors is liberated. The electron transfer releases energy used for the synthesis of ATP.
5. In cyclic photophosphorylation, the electron returns to chlorophyll. In noncyclic photophosphorylation, the electron is picked up by $NADP^+$.

Chemiosmotic Mechanism of ATP Generation   (pp. 116–117)

1. Protons being pumped across the membrane generate a proton motive force, as electrons move through a series of acceptors or carriers.
2. Energy produced from movement of the protons back across the membrane is used by ATPase to make ATP from ADP and Ⓟ.

Nutritional Patterns Among Organisms   (pp. 117–120)

1. Photoautotrophs obtain energy by photophosphorylation and fix carbon from CO_2 via the Calvin–Benson cycle to synthesize organic compounds.
2. Cyanobacteria are oxygenic phototrophs. Green sulfur bacteria, purple sulfur bacteria, and purple nonsulfur bacteria are anoxygenic phototrophs.
3. Photoheterotrophs use light as an energy source and an organic compound for their carbon source or electron donor.
4. Chemoautotrophs use inorganic compounds as their energy source and carbon dioxide as their carbon source.
5. Chemoheterotrophs use complex organic molecules as their carbon and energy sources.

Summary of Energy Production Methods   (pp. 120–122)

1. In oxidation-reduction reactions, energy is derived from the transfer of electrons.
2. To produce energy, a cell needs an electron donor (organic or inorganic), a system of electron carriers, and a final electron acceptor (organic or inorganic).

Biochemical Pathways of Energy Production   (p. 122)

1. Series of enzymatically catalyzed chemical reactions called biochemical pathways store energy in and release energy from organic molecules.

Carbohydrate Catabolism   (p. 122)

1. Most of a chemoheterotrophic cell's energy is produced from the oxidation of carbohydrates.
2. Glucose is the most commonly used carbohydrate.

Glycolysis   (pp. 123–124)

1. The most common pathway for the oxidation of glucose is glycolysis. Pyruvic acid is the end product.
2. Two ATP and two NADH molecules are produced from one glucose molecule.
3. Some organisms also use the pentose phosphate pathway or the Entner–Doudoroff pathway to oxidize glucose.

Respiration   (pp. 124–130)

1. During respiration, organic molecules are oxidized. Energy is generated from the electron transport chain.
2. In aerobic respiration, oxygen functions as the final electron acceptor.
3. In anaerobic respiration, the final electron acceptor can be nitrate, sulfate, or carbonate.
4. Two-carbon acetyl groups are oxidized in the Krebs cycle. Electrons are picked up by NAD^+ and FAD for the electron transport chain.
5. Thirty-eight ATPs can be produced from complete oxidation of a glucose molecule in glycolysis, the Krebs cycle, and the electron transport chain.
6. The electron transport chain consists of carriers, including NADH dehydrogenase, cytochromes, and cytochrome oxidase.

Fermentation   (pp. 130–132)

1. Fermentation releases energy from sugars or from other organic molecules.
2. Oxygen is not required in fermentation.
3. ATP is produced by substrate-level phosphorylation; the final electron acceptor is an organic molecule (pyruvic acid).
4. Reduction of pyruvic acid can produce lactic acid or ethanol.

Lipid Catabolism   (pp. 132–134)

1. Lipases hydrolyze lipids into glycerol and fatty acids.
2. Glycerol is catabolized by conversion to dihydroxyacetone phosphate, and fatty acids are catabolized by beta oxidation.
3. Catabolic products can be further broken down in glycolysis and the Krebs cycle.

Protein Catabolism   (p. 134)

1. Before amino acids can be catabolized, they must be converted to various substances that enter the Krebs cycle.
2. Transamination, decarboxylation, and dehydrogenation reactions convert the amino acids to be catabolized.

**Biochemical Pathways
of Energy Utilization (Anabolism)**   (pp. 134–137)

Biosynthesis of Polysaccharides   (pp. 134–135)

1. Glycogen is formed from ADPG.
2. UDPNAG is the starting material for the biosynthesis of peptidoglycan.

Biosynthesis of Lipids   (pp. 135–136)

1. Lipids are synthesized from fatty acids and glycerol.
2. Glycerol is derived from dihydroxyacetone phosphate, and fatty acids are built from acetyl CoA.

Biosynthesis of Amino Acids   (p. 136)

1. Amino acids are required for protein biosynthesis.
2. All amino acids can be synthesized either directly or indirectly from intermediates of carbohydrate metabolism, particularly from the Krebs cycle.

Biosynthesis of Purines and Pyrimidines   (p. 137)

1. The sugars composing nucleotides are derived from either the pentose phosphate pathway or the Entner–Doudoroff pathway.
2. Carbon and nitrogen atoms from certain amino acids form the backbones of the purines and pyrimidines.

Integration of Metabolism   (pp. 137–139)

1. Anabolic and catabolic reactions are integrated through a group of common intermediates.
2. Such integrated pathways are referred to as amphibolic pathways.

Study Questions

Review

1. Define metabolism.
2. Distinguish between catabolism and anabolism. How are these processes related?
3. Using the diagrams below, show
   (a) Where the substrate will bind.
   (b) Where the competitive inhibitor will bind.
   (c) Where the noncompetitive inhibitor will bind.

Enzyme   Substrate   Competitive   Noncompetitive
                      inhibitor   inhibitor

   (d) Which of those four elements could be the inhibitor in feedback inhibition?

4. What will the effect of the reactions in Question 3 be?

5. Why are most enzymes active at one particular temperature? Why are enzymes less active below this temperature? What happens above this temperature?

6. Which substances in each of the following reactions are being oxidized? Which are being reduced?

$$
\begin{array}{c}
\text{H} \\
| \\
\text{C}{=}\text{O} + \text{NADH} + \text{H}^+ \\
| \\
\text{CH}_3
\end{array}
\longrightarrow
\begin{array}{c}
\text{H} \\
| \\
\text{H}-\text{C}-\text{OH} + \text{NAD}^+ \\
| \\
\text{CH}_3
\end{array}
$$

(a) Acetaldehyde   Ethanol

$$
\begin{array}{c}
\text{COOH} \\
| \\
\text{H}-\text{C}-\text{H} \\
| \\
\text{H}-\text{C}-\text{H} \\
| \\
\text{COOH}
\end{array}
+ \text{FAD} \longrightarrow
\begin{array}{c}
\text{COOH} \\
| \\
\text{C}-\text{H} \\
\| \\
\text{C}-\text{H} \\
| \\
\text{COOH}
\end{array}
+ \text{FADH}_2
$$

(b) Succinic acid   Fumaric acid

7. There are three mechanisms for the phosphorylation of ADP to produce ATP. Write the name of the mechanism that describes each of the following reactions in the table below.

ATP Generated by	Reaction
	An electron, liberated from chlorophyll by light, is passed down an electron transport chain.
	Cytochrome *c* passes two electrons to cytochrome *a*.
	$$\begin{array}{c}\text{CH}_2 \\ \| \\ \text{C}-\text{O}\sim\text{\textcircled{P}} \\ | \\ \text{COOH}\end{array} \longrightarrow \begin{array}{c}\text{CH}_3 \\ | \\ \text{C}{=}\text{O} \\ | \\ \text{CH}_3\end{array}$$ Phosphoenolpyruvic acid Pyruvic acid

8. What is the fate of pyruvic acid in an organism that uses respiration? Fermentation?

9. List four compounds that can be made from pyruvic acid by an organism that uses fermentation only.

10. Fill in the table below with the carbon source and energy source of each type of organism.

Organism	Carbon Source	Energy Source
Photoautotroph		
Photoheterotroph		
Chemoautotroph		
Chemoheterotroph		

11. Define *oxidation-reduction*, and differentiate between the following terms:
    (a) Aerobic and anaerobic respiration
    (b) Respiration and fermentation
    (c) Cyclic and noncyclic photophosphorylation

12. An enzyme and substrate are combined. The rate of reaction begins as shown in this graph. To complete the graph, show the effect of increased substrate concentration on a constant enzyme concentration. Show the effect of increased temperature.

Use the following diagrams for Questions 13 through 20.

(a)

(b)

(c)

13. Name the pathways diagrammed in a, b, and c.

14. Show where glycerol is catabolized and where fatty acids are catabolized.

15. Show where the amino acid glutamic acid is catabolized:

$$HOOC—CH_2—CH_2—\underset{NH_2}{\overset{H}{C}}—COOH$$

16. Show how these pathways are related.

17. Where is ATP required in pathways a and b?

18. Where is CO_2 released in pathways a and b?

19. Where is NADH (or FADH$_2$ or NADPH) used and produced in these pathways?

20. Identify four places where anabolic and catabolic pathways are integrated.

21. The pentose phosphate pathway produces only one ATP. List four advantages of this pathway for the cell.

22. Explain how ATP is a key intermediate in metabolism.

Challenge

1. Write your own definition of the chemiosmotic mechanism of ATP generation. On Figure 5-10, mark the following:
   (a) The acidic side of the membrane
   (b) The side with a positive electrical charge
   (c) Potential energy
   (d) Kinetic energy

2. Explain why, even under ideal conditions, *Streptococcus* grows slowly.

3. Why must NADH be reoxidized? How does this happen in an organism that uses respiration? Fermentation?

4. The graph below shows the normal rate of reaction of an enzyme and its substrate (black) and the rate when an excess of competitive inhibitor is present (color). Explain why the graph appears as it does.

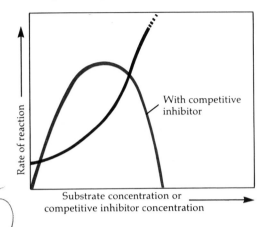

5. Compare and contrast carbohydrate catabolism and energy production in the following bacteria:
   (a) *Pseudomonas*, an aerobic chemoheterotroph
   (b) *Spirulina*, an oxygenic photoautotroph
   (c) *Ectothiorhodospira*, an anoxygenic photoautotroph

Further Reading

Blakemore, R. P., and R. B. Frankel. "Magnetic navigation in bacteria." *Scientific American* 245(6):58–65, December 1981. Describes synthesis and function of magnets in bacterial cells.

Gottschalk, G. *Bacterial Metabolism*, 2nd ed. New York: Springer-Verlag, 1986. A detailed description of metabolic pathways in bacteria.

Lehninger, A. L. *Bioenergetics: The Molecular Basis of Biological Energy Transformation*, 2nd ed. Menlo Park, Calif.: Benjamin/Cummings, 1971. A classic reference that provides clear explanations of bioenergetics; predates the chemiosmotic theory, however.

Nicholls, D. G. *Bioenergetics: An Introduction to the Chemiosmotic Theory*. New York: Academic Press, 1982. An introductory text on this theory.

Oliver, D. "Protein secretion in *Escherichia coli*." *Annual Review of Microbiology* 39:615–648, 1985. A summary of the processes by which membrane proteins and extracellular proteins are synthesized and excreted from cells.

Proton Pumping. *BioScience* 35:14–48, January 1985. Includes seven articles, with excellent illustrations, describing proton chemistry and photosynthetic, halobacterial, and eucaryotic membrane systems.

Stryer, L. *Biochemistry*, 3rd ed. New York: W. F. Freeman, 1988. The latest edition of a clearly written and widely used textbook.

Youvan, D. C., and B. L. Marrs. "Molecular mechanisms of photosynthesis." *Scientific American* 256(6):42–48, June 1987. Describes modern techniques from physics and molecular genetics used to identify the electron transport chain in bacterial (cyclic) photophosphorylation.

Microbial Growth

When we talk about microbial growth, we are really referring to the *number* of cells, not the *size* of the cells. Microbes that are "growing" are increasing in number, accumulating into *clumps* of hundreds, *colonies* of hundreds of thousands, or *populations* of billions. For the most part, we are not concerned with the size of an individual cell because it does not vary much during the cell's lifetime. Thus, the total mass of a bacterial population is usually roughly proportional to the number of cells in the population; the rate of increase in a population's mass is often a measure of its reproductive rate.

By understanding the conditions necessary for microbial growth, we can predict how quickly microorganisms will grow in various situations, and we can determine how to control their growth. Microbial populations can become very large in a very short time, and unchecked microbial growth can cause serious disease and food spoilage.

We will examine in this chapter the physical and chemical requirements for microbial growth, the various kinds of culture media, bacterial division, the phases of microbial growth, and the methods of measuring microbial growth. Once you understand these concepts, you will have a better idea of how unwanted microbial growth can be retarded or stopped.

Requirements for Growth

The requirements for microbial growth can be divided into two categories, physical and chemical. Physical aspects include temperature, pH, and osmotic pressure. Chemical requirements include water, sources of carbon and nitrogen, minerals, oxygen, and organic growth factors.

Physical Requirements

Temperature

Microorganisms grow well at the normal temperatures favored by higher animals. However, certain bacteria are capable of growing in extreme cold or extreme heat, at temperatures that would certainly hinder the survival of most higher organisms.

Microorganisms are divided into three groups on the basis of their preferred range of temperature. These are the **psychrophiles** (cold-loving microbes), **mesophiles** (moderate-temperature-loving microbes), and **thermophiles** (heat-loving microbes). Most bacteria grow only within a limited range of temperatures, and their maximum and minimum growth temperatures are about 30°C apart. They grow poorly at some temperatures within their range and well at others. Each bacterial species grows at particular minimum, optimum, and maximum temperatures. The **minimum growth temperature** is the lowest temperature at which the species will grow. The **optimum growth temperature** is the temperature at which the species grows best. The **maximum growth temperature** is the highest temperature at which growth is possible. By graphing the growth response over a temperature range, we can see that the optimum growth temperature is usually near the top of the range; above that temperature, the rate of growth drops off rapidly (Figure 6-1). This presumably happens because the high temperature has inactivated necessary enzymatic systems of the cell.

The ranges and maximum growth temperatures that define bacteria as psychrophile, mesophile, or thermophile are not rigidly defined and vary from reference to reference. Psychrophiles, for example, were originally considered simply as organisms capable of growing at 0°C. However, there seem to be two fairly distinct groups of organisms capable of growth at that temperature. It has therefore been suggested that the category of psychrophile be reserved for organisms with an optimum growth temperature of about 15°C and a maximum growth temperature of about 20°C; these requirements would indicate that they cannot grow in reasonably warm rooms (20°C = 68°F). Such organisms are found mostly in the oceans' depths or in certain arctic regions, and they seldom present concerns to humans such as in the preservation of food.

A common and often troublesome group of psychrophiles will grow at refrigerator temperatures, or even below 0°C, but can also grow at temperatures above 20°C. However, few will grow well at temperatures above 30°C, and their optimum growth tem-

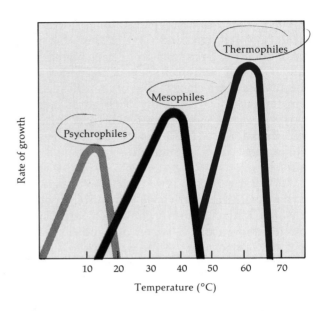

Figure 6-1 Typical growth responses to temperature by psychrophiles, mesophiles, and thermophiles. The optimum growth (fastest reproduction) is represented by the peak of the curve. Note that the reproductive rate drops off very quickly at temperatures only a little above the optimum. At either extreme of the temperature range, the reproductive rate is much less than the rate at the optimum temperature.

peratures tend to be relatively low (often in the upper teens or low twenties). It has been proposed that these organisms be called **psychrotrophs**, and many sources use this term to describe bacteria capable of growth at a common refrigerator temperature (4°C). Other sources apply the term *psychrophile* to both types of organisms, those that grow best between 15°C and 20°C and those whose range is between 0°C and 30°C.

Refrigeration is the most common method of preserving household food supplies. It is based on the principle that microbial reproductive rates decrease at low temperatures. Although microbes usually survive even subfreezing temperatures (they might become entirely dormant), they gradually decline in numbers. Some species decline faster than others. Psychrophiles, or psychrotrophs, actually do not grow well at low temperatures, except in comparison with other organisms, but given time they are able to slowly deteriorate the quality of food. This deterioration might take the form of mold mycelium, slime on food surfaces, or off tastes or colors in foods. The temperature inside a properly set refrigerator will prevent the growth of most spoilage organisms and will prevent the growth of almost all the pathogenic bacteria.

The mesophile, with an optimum growth temperature of between 25° and 40°C, is the most common type of microbe. Organisms that have adapted to live in the bodies of higher animals usually have an optimum temperature close to that of their host. The optimum temperature for many pathogenic bacteria is about 37°C, and incubators for clinical cultures are usually set at about this temperature. The mesophiles include most of our common spoilage and disease organisms.

A very interesting group of microorganisms are the thermophiles, those capable of growth at high temperatures. Many of these organisms have an optimum growth temperature between 50° and 60°C. This is about the temperature of the water from a hot-water tap. Such temperatures can also be reached in soil on which sunlight falls and in thermal waters, such as the hot springs in Yellowstone National Park. (See Microview A, Photo 27.) Some of these organisms grow at temperatures well above 90°C, near the boiling point of water. Remarkably, many thermophiles are incapable of growth at temperatures below about 45°C. (Thermophilic fungi usually have an upper limit of 55° to 60°C, somewhat lower than that of most thermophilic bacteria.) Endospores formed by thermophilic bacteria are unusually heat-resistant. They survive the usual heat treatment given canned goods but will not grow at normal storage temperatures. Although elevated temperatures may cause surviving endospores to germinate and grow, these thermophilic bacteria are not a public health problem. Thermophiles are also important in organic compost piles, in which the temperature can rise rapidly to 50° or 60°C. An unusual group of thermophiles is described in the box below.

pH

As you recall from Chapter 2, pH refers to the acidity or alkalinity of a solution. Most bacteria grow best in a narrow range of pH near neutrality, between pH 6.5 and 7.5. Very few bacteria grow at an acid pH

MICROBIOLOGY HIGHLIGHTS

Hydrothermal Bacteria

Until humans explored the ocean floor in the late 1970s, scientists believed that only a few forms of life could survive in that extremely cold, completely dark, oxygen-poor environment. Then, in 1977, the first manned vehicle able to penetrate to the bottom of the deepest oceans carried two scientists 2600 meters below the surface at the Galápagos Rift, about 350 km northeast of the Galápagos Islands. There, amid the vast expanse of barren basalt pillows, the scientists found unexpectedly rich oases of life, including mollusks, crustaceans, and worms. How do such creatures survive in these harsh conditions?

Ecosystem of the Hydrothermal Vents

Life at the surface of the world's oceans depends on photosynthetic organisms, such as bacteria, plankton, and green plants, to harness the sun's energy and make carbohydrates. At the ocean floor, where no light penetrates, photosynthesis is not possible. The scientists found that the so-called primary producers at the ocean floor are chemoautotrophic bacteria. Using such inorganic compounds as hydrogen sulfide (H_2S) as a source of energy and carbon dioxide (CO_2) as a source of carbon, the chemoautotrophs create an environment that supports higher life forms. Hydro-thermal vents in the sea floor supply the H_2S and CO_2. As superheated water from within the earth rises through fractures in the earth's crust called vents, it reacts with surrounding rock and dissolves metal ions, sulfides, and CO_2. The ecosystems of these vents depend on an abundance of sulfur compounds in the hot water. The concentration of sulfide (S^{2-}) is three times greater than the concentration of molecular oxygen in and around the vents. Such high sulfide concentrations are toxic to many organisms, but not to the creatures that inhabit this exotic environment.

Mats of bacteria grow along the sides of the vents, where temperatures can be as high as 350°C. This is the highest temperature that any organism is known to tolerate. Above the vent, where temperatures are about 20°C, the concentration of bacteria is about four times greater than that in water farther from the vents, and the growth rate of bacteria is equal to that found in productive, sunlit coastal waters. The bacteria in and around the vents create an environment in which bacteria consumers such as clams, mussels, and worms and decomposers can exist.

A Rare Relationship
One of the most astonishing discoveries in the hydrothermal ecosystem is

below about 4.0. That is the reason why a number of foods, such as sauerkraut, pickles, and many cheeses, are preserved by the acids of bacterial fermentation. Nonetheless, some bacteria, **acidophiles**, are remarkably tolerant of acidity. One autotrophic bacterium, which is found in the drainage water from coal mines and which oxidizes sulfur to form sulfuric acid, can survive at a pH of 1 (see the box in Chapter 27). Molds and yeasts will grow over a greater pH range than bacteria will, but the optimum pH of molds and yeasts is generally below that of bacteria, usually about pH 5 to 6. Alkalinity also inhibits microbial growth, but is rarely used to preserve foods.

When bacteria are cultured in the laboratory, they produce acids that can interfere with desired bacterial growth. To neutralize the acids and maintain the proper pH, chemicals called **buffers** are included in the growth medium. The peptones and amino acids in some media act as buffers, and many media also contain phosphate salts. Phosphate salts have the advantage of exhibiting their buffering effect in the pH growth range of most bacteria. They are also nontoxic; in fact, they provide phosphorus, an essential nutrient element.

Osmotic Pressure

Microbes obtain almost all of their nutrients in solution from the surrounding water. They therefore require water for growth and are actually about 80% to 90% water. High osmotic pressures have the effect of removing necessary water from a cell. When a microbial cell is in a solution that has a higher concentration of solutes than in the cell (hypertonic), the cellular water passes out through the cytoplasmic membrane to the high salt concentration. (See the discussion of osmosis in Chapter 4, and review Figure 4-17 for the three types of solution a cell may encounter). This osmotic loss of water causes **plasmolysis**, or shrinkage of the cell (Figure 6-2). What is so important about this phenomenon is that the growth of the cell is inhibited as the cytoplasmic membrane pulls

of animals that live with some species of chemoautotrophic bacteria in mutually beneficial relationships called *endosymbioses*, in which the bacteria live inside the animal. One such relationship involves *Riftia*, a tubeworm that can grow up to one meter in length and has no mouth or gut. The photograph shows hundreds of such tubeworms at a hydrothermal vent on the Galápagos Rift. *Riftia* exists in an unusual endosymbiotic relationship, as it actually receives food from the parasitic bacteria it hosts. *Riftia*'s tentacles absorb metal ions, sulfides, CO_2, and O_2 from the plume of hot water. These elements are transported through the worm's circulatory system, so that the bacteria throughout its body are supplied with CO_2 and H_2S. (Although hydrogen sulfide is usually toxic to animals, *Riftia* appears to have in its blood a protein that binds the sulfide so that it cannot inhibit metabolism.) The bacteria fix the carbon from CO_2 into organic compounds using energy obtained from oxidation of sulfide to sulfate: $CO_2 + H_2S \rightarrow$ carbohydrate $+ SO_4^{2-}$. These carbohydrates are food for *Riftia*. Thus, it is ultimately a chemical reaction—the oxidation of H_2S by its parasite bacteria—and not the sun that gives *Riftia* its energy.

Normal cell

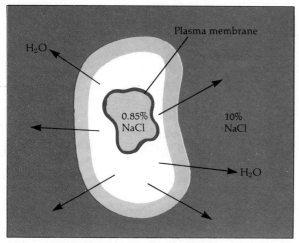

Plasmolyzed cell

Figure 6-2   Plasmolysis. If the concentration of solutes, such as sodium chloride, is higher in the surrounding medium (hypertonic), then water tends to leave the cell. The cell membrane shrinks away from the cell wall (an action called plasmolysis) and cell growth is inhibited.

away from the cell wall. Thus, the addition of salts (or other solutes) to a solution, and the resulting increase in osmotic pressure, can be used to preserve foods. Salted fish, honey, and sweetened condensed milk are preserved largely by this mechanism; the high salt or sugar concentrations draw water out of any microbial cells that are present and thus prevent their growth.

Some bacteria have adapted so well to an environment of high salt concentrations that they actually require them for growth. These bacteria are called **extreme halophiles**. To grow, bacteria from such saline waters as the Dead Sea often require nearly 30% salt, and the inoculating loop (a device for han-

dling bacteria in the laboratory) used to transfer them must first be dipped into a saturated salt solution. More common are **facultative halophiles**, which do not require high salt concentrations but are able to grow at salt concentrations up to 2%, a concentration that inhibits the growth of many other bacteria. A few species of facultative halophiles can even tolerate 15% salt. A halophile is described in the box in Chapter 5, p. 120.

The concentration of agar (a complex polysaccharide isolated from algae) used to solidify microbial growth media is usually about 1.5%. If markedly higher concentrations are used, the growth of some bacteria can be inhibited by the increased osmotic pressure. Because the effects of osmotic pressure are roughly related to the number of molecules in a volume of solution, a low-molecular-weight compound such as sodium chloride has a greater antimicrobial effect per gram than does, for example, sucrose, which has a somewhat higher molecular weight.

If the osmotic pressure is unusually low (hypotonic), such as in distilled water, for example, water tends to enter the cell rather than leave it. Some microbes that have a relatively weak cell wall may actually be lysed by such treatment. Red blood cells, which do not have the tough cell wall of bacteria, can only be handled in weak saline solutions that approximate the osmotic pressures of their normal habitats in the body.

Chemical Requirements

Carbon

Besides water, one of the most important requirements for microbial growth is a source of carbon. Carbon is the structural backbone of living matter; it is needed for all the organic compounds that make up a living cell. As we saw in Chapter 2, its four valences allow it to be used in constructing extended and complicated organic molecules (see Figures 2-9 and 2-10). One half of the dry weight of a typical bacterial cell is carbon.

Chemoheterotrophs get most of their carbon from the source of their energy, namely, organic materials such as proteins, carbohydrates, and lipids. Chemoautotrophs and photoautotrophs derive their carbon from carbon dioxide.

Because the utilization of carbon is such a basic trait of life on Earth, one way to test for life on another planet is to look for a sign that carbon is being metabolized. If carbon compounds are either being formed or broken down, then living organisms might be present.

Nitrogen, Sulfur, and Phosphorus

In addition to carbon, other elements are needed by microbes for the synthesis of cellular material. For example, protein synthesis requires considerable amounts of nitrogen as well as some sulfur. The syntheses of DNA and RNA also require nitrogen and some phosphorus, as does the synthesis of ATP, the molecule so important for storage and transfer of chemical energy within the cell. Together, nitrogen, sulfur, and phosphorus constitute about 18% of the dry weight of the cell; of that amount, 15% is nitrogen.

Nitrogen is used primarily to form the amino group of the amino acids of proteins. Many bacteria meet this requirement by decomposing protein-containing material and reincorporating the amino acids into newly synthesized proteins and other nitrogen-containing compounds. Other bacteria use nitrogen from ammonium ions (NH_4^+), which are already in the reduced form and are usually found in organic cellular material. Still other bacteria are able to derive nitrogen from nitrates (compounds that dissociate to give the nitrate ion, NO_3^-, in solution).

Some important bacteria, including the cyanobacteria (blue-green algae), use gaseous nitrogen (N_2) directly from the atmosphere. The process by which they absorb N_2 is called **nitrogen fixation**. Some organisms that can use this method are free-living, mostly in the soil, but others live in symbiosis with the roots of certain plants. *Symbiosis* is a close association of two or more different species of organisms. The most important symbiotic nitrogen-fixing bacteria belong to the genera *Rhizobium* (rī-zō'bē-um) and *Bradyrhizobium* (brad-ē-rī-zō'bē-um) and are associated with such legumes as clover, soybeans, alfalfa, beans, and peas. The nitrogen fixed in the symbiosis is used by both the plant and the bacterium. This reaction greatly increases soil fertility and is therefore very important in agriculture.

Sulfur is used to synthesize sulfur-containing amino acids and vitamins such as thiamine and biotin. The box on p. 146 describes an unusual ecosystem that is based on a supply of H_2S that would be toxic to most organisms. Important natural sources of sulfur include the sulfate ion (SO_4^{2-}), hydrogen sulfide (H_2S), and the sulfur-containing amino acids. Phosphorus is essential to synthesize nucleic acids and the phospholipids of cell membranes. Among other places, it is also found in the energy bonds of ATP. An important source of phosphorus is the phosphate ion (PO_4^{3-}). Potassium, magnesium, and calcium are also examples of elements that microorganisms require, often as cofactors for enzymes (Chapter 5).

Trace Elements

Microbes require very small amounts of other mineral elements, such as iron, copper, molybdenum, and zinc; these are referred to as **trace elements**. Most are essential for activity of certain enzymes, usually as cofactors. Although these components are sometimes added to a laboratory medium, they are usually assumed to be naturally present in tap water and other components of media. Even most distilled waters contain adequate amounts, but tap water is sometimes specified to ensure that these trace minerals will be present in culture media.

Oxygen

We are accustomed to thinking of molecular oxygen (O_2) as a necessity of life, but it is actually a rather corrosive gas. Molecular oxygen did not exist in the atmosphere during most of our planet's history. It is speculated that life could not have arisen if oxygen had been present. However, many current forms of life have metabolic systems that require oxygen for aerobic respiration. As we have seen, oxygen combines with hydrogen atoms that have been stripped from organic compounds and forms water. This process yields a great deal of energy while neutralizing a potentially toxic gas—a very neat solution, all in all. When oxygen is not available, the energy yield is much lower because the organic material is only partially degraded to compounds such as acids or alcohols. These compounds still contain considerable amounts of unreleased chemical energy.

Microbes that use molecular oxygen, called **aerobes**, produce more energy from nutrients than do microbes that do not use oxygen. Organisms that require oxygen to live are called **obligate aerobes** (Figure 6-3a).

Obligate aerobes are at a disadvantage, in that oxygen is poorly soluble in water, and much of the environment is oxygen poor. Therefore, many of the aerobic bacteria have developed, or retained, the ability to continue growing in the absence of oxygen. Such organisms, including some that exist in your intestinal tract, are called **facultative anaerobes** (Figure 6-3b). In other words, facultative anaerobes can use oxygen when it is present but are able to continue growth by using fermentation or anaerobic respiration when oxygen is not available. However, their efficiency in producing energy decreases in the absence of oxygen. Examples of facultative anaerobes include the familiar *E. coli* and many yeasts.

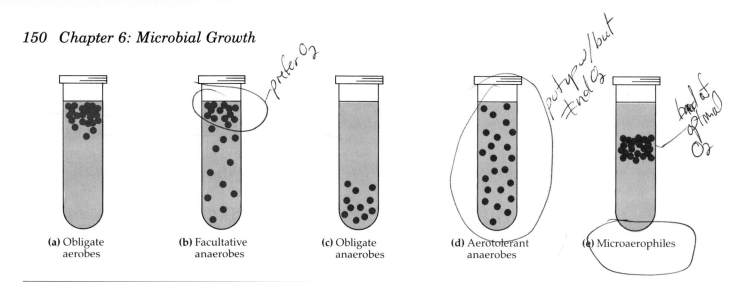

(handwritten annotations: "prefer O₂" above tube b; "put type w/ but fnd O₂" above tube d; "bad of optimal O₂" pointing to tube e)

(a) Obligate aerobes   **(b)** Facultative anaerobes   **(c)** Obligate anaerobes   **(d)** Aerotolerant anaerobes   **(e)** Microaerophiles

Figure 6-3   Effect of oxygen concentration on the growth of various bacteria in tubes of solid medium. (a) Obligate aerobes—growth occurs only in the short distance to which the oxygen diffuses into the medium. **(b)** Facultative anaerobes—growth is best near the surface, where oxygen is available, but occurs throughout the tube. **(c)** Obligate anaerobes—oxygen is toxic, and there is no growth near the surface. **(d)** Aerotolerant anaerobes—growth occurs evenly throughout the tube but is not better at the surface because the organisms do not use oxygen. **(e)** Microaerophiles, aerobic organisms, do not tolerate atmospheric concentrations of oxygen—growth occurs only in a narrow band of optimal oxygen concentration.

In laboratory identification of microbes, use is often made of the ability, or inability, of the organisms to ferment different sugars. Also important in identification of microbes may be differing abilities to substitute other electron acceptors, such as nitrates, for oxygen (anaerobic respiration).

Obligate anaerobes (Figure 6-3c) are bacteria that are unable to use molecular oxygen for energy-yielding reactions. In fact, most are harmed by it. When such bacteria have an electron transport chain, they cannot use molecular oxygen as a final electron acceptor. These bacteria do use oxygen atoms in cellular materials; the atoms are usually obtained from water.

Understanding how organisms can be harmed by oxygen requires a brief discussion of the toxic forms of oxygen.

1. **Singlet oxygen** is normal molecular oxygen (O_2) that has been boosted into a higher energy state and is extremely reactive. It is formed most commonly by the action of visible light. Present in phagocytic cells (Chapter 15), it plays a role in the destruction of foreign cells after they are ingested by the phagocytes.

2. **Superoxide free radicals** (O_2^-) are formed in small amounts during the normal respiration of organisms that use oxygen as a final electron acceptor, forming water (see Figure 5-17). In the presence of oxygen, obligate anaerobes, such as bacteria of the genus *Clostridium* (klos-tri′dē-um), also appear to form some superoxide free radicals. These free radicals are so toxic to cellular components that all organisms attempting to grow in atmospheric oxygen must produce an enzyme, **superoxide dismutase,** to neutralize them. Their toxicity is due to the fact that the radical lacks an electron and avidly steals it from a neighboring molecule, which in turn becomes a radical and steals an electron, and so on. Unless this reaction series is quickly neutralized, extreme damage to the cell results. Aerobic bacteria, facultative anaerobes growing aerobically, and aerotolerant anaerobes (discussed later) produce this enzyme (Table 6-1). Using superoxide dismutase, they convert the superoxide free radical into molecular oxygen (O_2) and hydrogen peroxide (H_2O_2):

$$O_2^- + O_2^- + 2H^+ \longrightarrow H_2O_2 + O_2$$

The H_2O_2 is then immediately converted into oxygen and water by another enzyme, **catalase:**

$$2H_2O_2 \longrightarrow 2H_2O + O_2$$

Catalase is easily detected by its action on hydrogen peroxide. When a drop of hydrogen peroxide is added to a colony of bacterial cells producing catalase, oxygen bubbles are released. This colony is designated as a catalase-positive organism. Anyone who has put hydrogen peroxide on a wound

TABLE 6-1 *Oxygen-Related Enzymes of Bacteria*

Type of Microbe	Catalase	Superoxide Dismutase	Comments
Aerobes and facultative anaerobes	Present	Present	Oxygen tolerated, all harmful forms neutralized
Aerotolerant anaerobes	Absent	Present	Oxygen tolerated because harmful forms partially neutralized
Obligate anaerobes	Absent	Absent	Oxygen is toxic, harmful forms not neutralized

will recognize that cells in human tissue also contain catalase. The actions of the two enzymes, catalase and superoxide dismutase, neutralize toxic forms of oxygen and permit the cell to grow while using oxygen as a final electron acceptor. Another enzyme that breaks down hydrogen peroxide is **peroxidase**; it differs from catalase in that its reaction does not produce oxygen:

$$H_2O_2 + 2H^+ \longrightarrow 2H_2O$$

3. **Peroxide** (O^{2-}) is a toxic form of oxygen that is probably generated in small amounts during normal respiration. But it is better known as the active principle of the antimicrobials hydrogen peroxide and benzoyl peroxide; these are discussed in the next chapter.

4. The **hydroxyl free radical** $(OH \cdot)$ is another intermediate form of oxygen and probably the most reactive. It is formed in the cellular cytoplasm by ionizing radiation. It is also produced by a reaction between superoxide free radicals and peroxide, so most aerobic respiration produces some hydroxyl radicals.

Obligate anaerobes, such as most members of the genus *Clostridium*, produce neither superoxide dismutase nor catalase. Since aerobic conditions probably lead to an accumulation of superoxide free radicals in the cell's cytoplasm, they are extremely sensitive to oxygen. The ability to form endospores has probably aided the survival of these oxygen-sensitive bacteria.

Aerotolerant anaerobes cannot use oxygen for growth, but they tolerate it fairly well. On the surface of a solid medium, they will grow without the special techniques (discussed later) required by obligate anaerobes (Figure 6-3d). Many of the aerotolerant bacteria characteristically ferment carbohydrates to lactic acid. As lactic acid accumulates, it inhibits the growth of aerobic competitors and establishes an eco-

logical niche for the lactic-acid producers. A common example of lactic-acid-producing aerotolerant anaerobes are the lactobacilli used in the production of many acidic fermented foods, such as pickles and cheese. In the laboratory, they are handled and grown much like any other bacteria, but they make no use of the oxygen in the air. These bacteria can tolerate oxygen because they possess superoxide dismutase (Table 6-1) or an equivalent system that neutralizes the toxic forms of oxygen discussed above.

A few bacteria are **microaerophilic**, meaning that they grow only in oxygen concentrations lower than those of air. However, they are aerobic; they require oxygen. In a test tube of solid nutrient medium, they grow only at a depth where small amounts of oxygen have diffused into the medium; they do not grow at the oxygen-rich surface (Figure 6-3e). This limited tolerance is probably due to their sensitivity to superoxide free radicals and peroxides, which they produce in lethal concentrations under oxygen-rich conditions.

Organic Growth Factors

Organic growth factors are essential organic compounds that the organism is unable to synthesize; they must be directly obtained from the environment. A group of organic growth factors for humans is vitamins. Most vitamins function as coenzymes, the accessories required by certain enzymes in order to function. Many bacteria can synthesize all their own vitamins and are not dependent on outside sources. However, some bacteria lack the enzymes needed for the synthesis of certain vitamins, and for them those vitamins are organic growth factors. Other organic growth factors required by some bacteria are amino acids, purines, and pyrimidines.

Culture Media

A nutrient material prepared for the growth of microorganisms in a laboratory is called a **culture medium**. Some bacteria can grow well on just about any culture medium; others require special media, and still others cannot grow on any nonliving medium yet developed. The microbes that grow and multiply in or on a culture medium are referred to as a **culture**.

Suppose we want to grow a culture of a particular microorganism, perhaps the microbes from a particular clinical specimen. What criteria must the culture medium meet? First, it must contain the right nutrients for the particular microorganism we want to grow. It should also contain sufficient moisture, oxygen, and a properly adjusted pH. The medium must initially be **sterile**—that is, it must initially contain no living microorganisms—so that the culture will contain only the microorganisms we add to the medium (and their offspring). Finally, the growing culture should be incubated at the proper temperature.

A wide variety of media are available for the growth of microorganisms in the laboratory. Most of these, which are available from commercial sources, have premixed components and require only the addition of water and sterilization. Media are constantly being developed or revised for use in the isolation and identification of bacteria that are of interest to researchers in such areas as food, water, and clinical microbiology.

When it is desirable to grow bacteria on a solid medium, a solidifying agent such as agar is added to the medium. **Agar** is a complex polysaccharide derived from a marine alga, and it has long been used as a thickener in foods such as jellies, soups, and ice cream.

Agar has some very important properties that make it valuable to microbiology, and no really satisfactory substitute has ever been found. Few microbes can degrade agar, so it remains a solid. Also important is the fact that it melts at about the boiling point of water but remains in a liquid state until the temperature drops to about 40°C. For laboratory use, agar is therefore held in water baths at about 50°C. At this temperature, it does not injure most bacteria when it is poured over a bacterial inoculum. Bacterial suspensions can also be mixed with the melted agar in the water bath so that a uniform suspension of bacteria is made; this can be used for tests of antibiotic susceptibility and for other purposes. Once the agar has solidified, it can be incubated at temperatures approaching 100°C before it liquefies; this property is particularly useful when thermophilic bacteria are being grown.

Agar media are usually contained in test tubes or Petri plates. The test tubes are called slants when they are allowed to solidify with the tube held at an angle so a large surface for growth is available. When the agar solidifies in a vertical tube, it is called a deep. Petri plates, named for their inventor, are shallow dishes with a lid that nests over the bottom to prevent contamination.

Chemically Defined Media

When a medium is being prepared for microbial growth, consideration must be given to providing an energy source as well as sources of carbon, nitrogen, sulfur, phosphorus, and any necessary growth factors that the organism is unable to synthesize. A **chemically defined medium** is one in which the exact chemical composition is known. Table 6-2 shows a chemically defined medium used for the growth of an autotrophic organism capable of extracting energy from the oxidation of ammonium ions to nitrite ions.

As Table 6-3 demonstrates, many organic growth factors must be provided in the chemically defined medium used to cultivate species of *Lactobacillus*.

TABLE 6-2 *Chemically Defined Medium for the Growth of a Chemoautotrophic Bacterium That Is Able To Use Ammonium Ions for Energy*

Constituent	Concentration (Grams/Liter)
Ammonium sulfate [$(NH_4)_2SO_4$]	0.5
Sodium bicarbonate [$NaHCO_3$]*	0.5
Sodium phosphate, dibasic [Na_2HPO_4]	13.5
Potassium phosphate, monobasic [KH_2PO_4]	0.7
Magnesium sulfate [$MgSO_4 \cdot 7H_2O$]	0.1
Ferric chloride [$FeCl_3 \cdot 6H_2O$]	0.014
Calcium chloride [$CaCl_2 \cdot 2H_2O$]	0.18
Water	1 L

*The $NaHCO_3$ is a source of carbon dioxide (the carbon source) in solution.
Source: Pramer, D., and E. L. Schmidt, *Experimental Soil Microbiology*. Minneapolis: Burgess, 1964.

selectv

organ comp

TABLE 6-3 *Chemically Defined Medium for the Growth of a Fastidious Heterotrophic Bacterium, Such as* Lactobacillus *Species*

Constituent	Concentration (Grams/Liter)
Carbon and Energy Sources	
Glucose	10.0
Sodium acetate	6.0
Salts (added as two solutions of mixed salts)	
Ammonium chloride [NH_4Cl]	2.5
Ammonium sulfate [$(NH_4)_2SO_4$]	2.5
Potassium phosphate monobase [KH_2PO_4]	0.50
Potassium phosphate, dibasic [K_2HPO_4]	0.50
Sodium chloride [$NaCl$], Ferrous sulfate [$FeSO_4$], and Manganese sulfate [$MnSO_4$], each	0.005
Amino Acids	
Arginine, cystine, glycine, histidine, hydroxyproline, proline, tryptophan, and tyrosine, each	0.20
Alanine, aspartic and glutamic acids, isoleucine, leucine, lysine, methionine, norleucine, phenylalanine, serine, threonine, and valine, each	0.10
Glutamine	0.025
Purines	
Adenine, guanine, and uracil, each	0.010
Vitamins	
Pyridoxamine-HCl	0.0004
Riboflavin, thiamine, niacin, and pantothenic acid, each	0.0002
Para-aminobenzoic acid	0.00004
Folic acid	0.00002
Biotin	0.0000002
Water	1 L

Source: Thimann, K., *The Life of Bacteria*, 2nd ed. New York: Macmillan, 1963.

TABLE 6-4 *Chemically Defined Medium for the Growth of a Less Fastidious Heterotroph, Such as* E. coli

Constituent	Concentration (Grams/Liter)
Glucose	5.0
Ammonium phosphate, monobasic [$NH_4H_2PO_4$]	1.0
Sodium chloride [$NaCl$]	5.0
Magnesium sulfate [$MgSO_4 \cdot 7H_2O$]	0.2
Potassium phosphate, dibasic [K_2HPO_4]	1.0
Water	1 L

Organisms that require many growth factors are described as *fastidious*. Organisms of this type, such as *Lactobacillus*, are sometimes used in tests that determine the concentration of a particular vitamin in a substance. For such *microbiological assays* (tests), the growth medium to which the test substance and the bacterium will be added contains all the growth requirements of the bacterium except the vitamin being assayed. When the medium, test substance, and bacterium are combined, the growth of bacteria will be proportional to the amount of vitamin in the test substance. This growth is reflected by the amount of lactic acid produced. The more lactic acid, the more the *Lactobacillus* has been able to grow, so the more vitamin is present.

A much simpler chemically defined medium is required for the growth of *E. coli*, a less fastidious bacterium. The composition of this medium is shown in Table 6-4. The simple composition reflects the fact that *E. coli* has a greater synthesizing ability than *Lactobacillus*.

Complex Media

Most heterotrophic bacteria and fungi are routinely grown on **complex media**, for which the exact chemical composition varies slightly from batch to batch. These complex media are made up of nutrients such as extracts from yeasts, beef, or plants, or digests of proteins from these and other sources. Table 6-5 gives one widely used recipe.

In complex media such as this one, the energy, carbon, nitrogen, and sulfur requirements of the growing microorganisms are met largely by protein. Protein is a large, relatively insoluble molecule that few

TABLE 6-5   *Composition of Nutrient Agar, a Complex Medium for the Growth of Heterotrophic Bacteria*

Constituent	Concentration (Grams/Liter)
Peptone	5.0
Beef extract	3.0
Sodium chloride	8.0
Agar	15.0
Water	1 L

microorganisms can utilize directly, but a partial digestion by acids or enzymes reduces the protein to shorter chains of amino acids called peptones. These small, soluble fragments can be digested by the bacteria.

The vitamins and other organic growth factors are provided by meat extracts or yeast extracts. The soluble vitamins and minerals from the meats or yeasts are dissolved in the extracting water, which is then evaporated so that these factors are concentrated. (These extracts also supplement the organic nitrogen and carbon compounds.) Yeast extracts are particularly rich in the B vitamins. If this type of medium is in liquid form, it is called **nutrient broth**. When agar is added, it is called **nutrient agar**. (This terminology sometimes confuses students, who may think that agar itself is a nutrient—it is not.)

Anaerobic Growth Media and Methods

The cultivation of anaerobic bacteria poses a special problem. Because anaerobes might be killed by exposure to oxygen, special media called **reducing media** must be used. These media contain ingredients, such as sodium thioglycolate, that chemically combine with dissolved oxygen and deplete the oxygen in the culture medium. To routinely grow and maintain pure cultures of obligate anaerobes, microbiologists use reducing media stored in ordinary, tightly capped test tubes. These media are heated shortly before use, so that any absorbed oxygen is driven off.

When the culture must be grown in Petri plates so that individual colonies can be observed, special jars are used that hold several Petri plates in an oxygen-free atmosphere. The oxygen is removed by the following process: A packet of chemicals (sodium bicarbonate and sodium borohydride) in the jar is moistened with a few milliliters of water, and the jar

is sealed. Hydrogen and carbon dioxide are produced by the reaction of the chemicals with the water. A palladium catalyst in the jar combines the oxygen in the jar with the hydrogen produced by the chemical reaction, and water is formed. As a result, the oxygen disappears in a short time. Moreover, the carbon dioxide that is produced aids the growth of many anaerobic bacteria (Figure 6-4).

Researchers working with anaerobes on a regular basis use anaerobic glove boxes, which are small, transparent chambers equipped with air locks and filled with inert gases. The technicians are able to manipulate the equipment by inserting their hands into airtight rubber gloves fitted to the wall of the chamber (see Figure 6-5 and Microview C, Photo 13).

Colonies of anaerobes can also be grown in deep test tubes instead of in Petri plates. The natural atmosphere in the test tube is replaced with an inert gas, such as nitrogen. A bacterial inoculum can be mixed with melted nutrient medium in the tube, and the tube, tightly capped, is rolled on horizontal rollers.

Figure 6-4   Anaerobic container used in cultivation of anaerobic bacteria on Petri plates. When water is mixed with the chemical packet containing sodium bicarbonate and sodium borohydride, hydrogen and carbon dioxide are generated. Reacting on the surface of a palladium catalyst in a screened reaction chamber, the hydrogen and the atmospheric oxygen in the jar combine to form water. The oxygen is thus removed. An anaerobic indicator is also in the container. It contains methylene blue, which is blue when oxidized and turns colorless when the oxygen is removed from the container.

Labels in figure:
Clamp with clamp screw
Palladium catalyst pellets
Lid with O-ring gasket
Envelope containing sodium bicarbonate and sodium borohydride
CO_2
H_2
Anaerobic indicator (methylene blue)
Petri plates

Figure 6-5 Anaerobic chamber. Microbiologists who work regularly with oxygen-sensitive anaerobes often use an anaerobic chamber. The transparent enclosure is filled with an inert, oxygen-free gas. Organisms and materials enter and leave through an air lock, and manipulations are made by means of glove ports.

The nutrient medium solidifies against the interior walls of the test tube (usually referred to as **roll tubes**), and colonies that appear there can be counted and picked, as on a shallow layer of agar in a Petri plate.

Special Culture Techniques

Some bacteria have never been successfully grown on artificial laboratory media. *Mycobacterium leprae* (lep′rī), the leprosy bacterium, was once grown on the foot pads of mice and is now usually grown in armadillos, which have a relatively low body temperature. Another example is the syphilis spirochete, although certain specialized strains of the latter have been grown on laboratory media. With few exceptions, the obligate intracellular parasites, such as the rickettsias and the chlamydia bacteria, do not grow on artificial media. They, like viruses, can reproduce only in a living host cell.

Many clinical laboratories have special *carbon dioxide incubators* in which to grow aerobic bacteria that require concentrations of CO_2 higher or lower than that found in the atmosphere. Desired carbon dioxide levels are maintained by electronic controls. High levels of carbon dioxide are also obtained with simple *candle jars* (Figure 6-6). Petri dish cultures

Figure 6-6 Candle jar. Plates and tubes inoculated with, for example, *Neisseria meningitidis* are placed in a jar with a lighted candle, and the jar is sealed. The candle will be extinguished when the atmosphere contains approximately 10% carbon dioxide.

are placed in a large sealed jar containing a lighted candle. The candle stops burning when the atmosphere in the jar has a low concentration of oxygen (but one that is still adequate for the growth of aerobic bacteria) and the high CO_2 concentration required by certain clinically important bacteria, such as those causing brucellosis, gonorrhea, and meningococcal meningitis.

Selective and Differential Media

In clinical and public health microbiology, it is frequently necessary to detect the presence of specific microorganisms associated with disease or poor sanitation. For this task, selective and differential media are used. **Selective media** are designed to suppress the growth of unwanted bacteria and encourage the growth of the desired microbes. For example, bismuth sulfite agar is used to isolate the typhoid bacterium, the gram-negative *Salmonella typhi* (sal-mōn-el′lä tī′fē) from feces. Bismuth sulfite inhibits gram-positive bacteria, and the most numerous gram-negative intestinal bacteria as well. Sabouraud's glucose agar, which has a pH of 5.6, is used to isolate fungi that outgrow bacteria at this pH. Dyes such as brilliant green selectively inhibit gram-positive bacteria, and this dye is the basis of a medium called Brilliant Green agar that is used to isolate the gram-negative *Salmonella*.

Differential media make it easier to distinguish colonies of the desired organism from other colonies growing on the same plate. Similarly, pure cultures of microorganisms have identifiable reactions with differential media in tubes or plates. Blood agar (which contains red blood cells) is a dark, reddish brown medium that microbiologists often use to identify bacterial species that destroy red blood cells.

These species, such as *Streptococcus pyogenes* (streptō-kok′kus pī-äj′en-ēz), the bacterium that causes strep throat, show a clear ring around their colonies where they have lysed the surrounding blood cells (see Figure 6-7 and Microview A, Photo 10).

Sometimes, selective and differential media are used together. Suppose we want to isolate the common bacterium *Staphylococcus aureus* (staf-i-lō-kok′kus ô′rē-us), found in the nasal passages. One of the characteristics of this organism is its tolerance for high concentrations of sodium chloride; another characteristic is its ability to ferment the carbohydrate mannitol to form acid. There is an isolation medium called mannitol salt agar medium that combines features of both selective and differential media. Mannitol salt agar contains 7.5% sodium chloride, which will discourage the growth of competing organisms and thus *select for* (favor the growth of) *Staphylococcus*. This salty medium also contains a pH indicator that changes color if the mannitol in the medium is fermented to acid; the mannitol-fermenting colonies of *S. aureus* are thus differentiated from colonies of bacteria that do not ferment mannitol. Bacteria that grow at the high salt concentration *and* ferment mannitol to acid can be readily identified by the color change. These are probably colonies of *S. aureus*, and their identification can be confirmed with additional tests.

Another medium that is both selective and differential is MacConkey agar. This medium contains bile salts and crystal violet, which inhibit the growth of gram-positive bacteria. Because this medium also contains lactose, gram-negative bacteria that can grow on this disaccharide can be differentiated from similar bacteria that cannot. The ability to distinguish between lactose fermenters (red or pink colonies) and nonfermenters (colorless colonies) is useful in distinguishing between the pathogenic *Salmonella* bacteria and other related bacteria.

Figure 6-7 Differential medium containing red blood cells. The organism has lysed the red blood cells (hemolysis), so a clear area has formed around the colonies.

Colonies

Hemolysis

Enrichment Culture

Because bacteria present in small numbers can be missed, or because the bacterium to be isolated might be an unusual physiological type, it is sometimes necessary to resort to an **enrichment culture**. The medium for such a culture is usually liquid and provides nutrients and environmental conditions that favor the growth of a particular microbe but are not suitable for the growth of other types of microbes. In this sense, it is also a selective medium, but it is designed to increase very low numbers of desired microbes to detectable levels.

Let us assume that we want to isolate from a soil sample a microbe capable of growth on phenol. If the soil sample is placed in a liquid enrichment medium, in which phenol is the only source of carbon and energy, microbes unable to metabolize phenol will not grow. The culture medium is allowed to incubate for a few days, and then a small amount of it is transferred into another flask of the same medium. After a series of such transfers, the surviving population will be bacteria capable of metabolizing phenol. The bacteria are given time to grow in the medium between transfers; this is the enrichment stage. Any nutrients in the original inoculum are rapidly diluted out with the successive transfers. When the last dilution is streaked onto a solid medium of the same composition, only those colonies of organisms that are capable of using phenol should grow.

Table 6-6 summarizes the purposes of the main types of culture media.

Obtaining Pure Cultures

Most infectious materials, such as pus, sputum, or urine, contain several kinds of bacteria; so do samples of soil, water, or foods. If these materials are plated out onto the surface of a solid medium, colonies will form that are clones of the same organism. A **colony** theoretically arises from a single spore or vegetative cell or from a group of the same microorganisms attached to one another in clumps or chains. Microbial colonies often have a distinctive appearance that distinguishes one microbe from another (Figure 6-8). The bacteria must, of course, be distributed widely enough so that the colonies that grow are visibly separate from each other.

Most bacteriological work requires pure cultures, or clones, of bacteria. The isolation method most commonly used to get pure cultures is the **streak plate method** (Figure 6-9). A sterile inoculating loop is

Figure 6-8 Bacterial colonies with a variety of morphologies. (a) *Klebsiella pneumoniae* on MacConkey agar. **(b)** *Bacillus anthracis* on chemically defined medium. **(c)** *Mycobacterium smegmatis* on Penassay agar. **(d)** *Micrococcus luteus* on blood agar. These colonies are shown somewhat larger than actual life size.

(a)

(b)

(c)

(d)

Colonies

Figure 6-9   Streak plate method for isolation of pure cultures of bacteria. The direction of streaking is indicated by arrows. Streak series 1 is made from the original bacterial mixture. The inoculating loop is sterilized following each streak series, which dilutes out the number of cells in each succeeding series. In series 3 of this example, note that well-isolated colonies of two different types of bacteria have been obtained.

dipped into a mixed culture that contains more than one type of microbe and is streaked in a pattern over the surface of the nutrient medium. As the pattern is traced, bacteria are rubbed off the loop onto the medium in paths of fewer and fewer cells. The last cells to be rubbed off the loop are far enough apart to grow into isolated colonies. These colonies can be

TABLE 6-6   *Culture Media*

Type of Media	Purpose
Chemically defined	Grow chemoautotrophs, photoautotrophs, and for microbiological assays
Complex	Most chemoheterotrophic organisms
Reducing	Obligate anaerobes
Selective	Suppress unwanted microbes, encourage desired microbes
Differential	Distinguish colonies of desired microbes from others
Enrichment	Similar to selective media, but designed to increase numbers of desired microbes to detectable levels

picked up with an inoculating loop and transferred to a test tube of nutrient medium to form a pure culture containing only one type of bacterium.

The streak plate method works well when the organism to be isolated is present in large enough numbers. However, sometimes the microbe to be isolated is present only in very small numbers. The soil microbe that is capable of using the disinfectant phenol as a nutrient is one such example. Its numbers must be greatly increased by enrichment before streak plate isolation.

Preserving Bacterial Cultures

Refrigeration can be used for short-term storage of bacterial cultures. To preserve microbial cultures for long periods of time, two common methods are deep-freezing and lyophilization. **Deep-freezing** is a process in which a pure culture of microbes is placed in a suspending liquid and quick-frozen at temperatures ranging from $-50°$ to $-95°C$. The culture can usually be thawed and used up to several years later. During **lyophilization (freeze-drying)**, a suspension of microbes is quickly frozen at temperatures ranging from $-54°$ to $-72°C$, and the water is removed by a high vacuum (sublimation). While under vacuum, the container is sealed by a high-temperature torch. The remaining powderlike residue that contains the surviving microbes can be stored for years. The microbes can be revived at any time by hydration with suitable liquid nutrient medium.

Growth of Bacterial Cultures

Bacterial Division

As we mentioned at the beginning of this chapter, bacterial growth is an increase in bacterial numbers, not an increase in the size of the individual cells. Bacteria normally reproduce by binary fission (Figure 6-10). The first step in division is cell elongation and duplication of the chromosomal DNA. The cell wall and cell membrane then begin to grow inward from all sides at a point between the two regions of the chromosomal DNA. Eventually, the ingrowing cell walls meet, and two individual cells are formed, each of which is essentially identical to the parent cell.

A few bacterial species **bud**; that is, they form a small initial outgrowth that enlarges until its size approaches that of the parent cell, and then it separates. Some filamentous bacteria (certain actinomycetes) reproduce by producing chains of spores carried externally at the tips of the filaments. A few filamentous species simply fragment, and the fragments initiate the growth of new cells.

Generation Time

For purposes of calculating generation time, we will consider only reproduction by binary fission, which is by far the most common method. As can be seen in Figure 6-11, one cell's division produces two cells, two cells' division produces four cells, and so on. When the arithmetic number of cells in each generation is expressed as a power of 2, the exponent tells the number of doublings that have occurred. (You might at this point wish to refer to the discussion of exponential notation in Appendix G.)

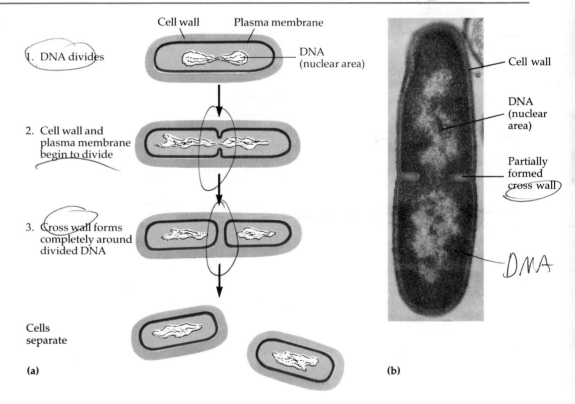

Figure 6-10 Binary fission in bacteria. (a) Diagram of sequence of cell division. **(b)** Electron micrograph of a thin section of a cell of *Bacillus licheniformis* starting to divide ($\times 34,000$).

1. DNA divides
2. Cell wall and plasma membrane begin to divide
3. Cross wall forms completely around divided DNA

Cells separate

(a)

Cell wall   Plasma membrane
DNA (nuclear area)

Cell wall
DNA (nuclear area)
Partially formed cross wall
DNA

(b)

Figure 6-11 Cell division. When the arithmetic number of cells in each generation is expressed as a power of 2, the exponent tells the number of doublings that have occurred.

Arithmetic numbers of cells	Expressed as the power of 2	
1	2^0	●
2	2^1	●●
4	2^2	●●●●
8	2^3	●●●●●●●●
16	2^4	●●●●●●●●●●●●●●●●
32	2^5	●●●●●●●●●●●●●●●●●●●●●●●●●●●●●●●●

The time required for a cell to divide and its population to therefore double is called the **generation (doubling) time**. It varies considerably among organisms. Most bacteria have a generation time from one to three hours; others require over 24 hours per generation. If binary fission continues unchecked, an enormous number of cells will be produced. If a doubling occurred every 20 minutes—which *E. coli* would do under favorable conditions—in 20 generations a single initial cell would increase to over a million cells. This would require a little less than 7 hours. In 30 generations, or 10 hours, the population would be a billion, and in 24 hours it would be a number trailed by 21 zeroes. It is difficult to graph population changes of such enormous magnitude using arithmetic numbers. This is why logarithmic scales rather than arithmetic scales are generally used to graph bacterial growth. Understanding logarithmic representations of bacterial populations requires a bit of study of mathematics, and it is necessary for anyone studying microbiology to be able to do this.

Logarithmic Representation of Bacterial Populations

To illustrate the difference between logarithmic and arithmetic graphing of bacterial populations, let us graph 20 bacterial generations both logarithmically and arithmetically. In 5 generations (2^5), there would be 64 cells, in 10 generations (2^{10}), there would be 1024 cells, and so on. (If your calculator has a y^x key and a *log* key, you can duplicate the calculations in the third column.)

Generation Number	Arithmetic Number of Cells	Log$_{10}$ of Arithmetic Number of Cells
0	1	0
5 (2^5) =	32	1.5
10 (2^{10}) =	1024	3.0
15 (2^{15}) =	32768	4.5
16 (2^{16}) =	65536	4.8
17 (2^{17}) =	131072	5.1
18 (2^{18}) =	262144	5.4
19 (2^{19}) =	524288	5.7
20 (2^{20}) =	1048576	6.0

In Figure 6-12, note that the arithmetically plotted line (solid) does not clearly show the population changes in the early stages of the growth curve at this

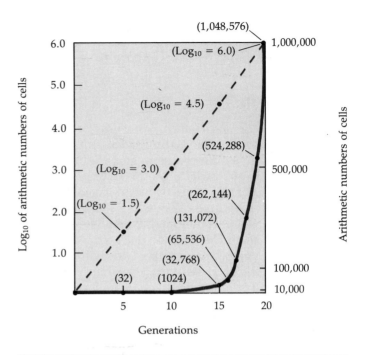

Figure 6-12 Growth curve for an exponentially increasing population, plotted logarithmically and arithmetically.

scale. In fact, the first 10 generations essentially do not leave the baseline. Furthermore, another arithmetic generation graphed to the same scale would double the height of the graph, and two more generations would take the line off the page.

The dashed line in Figure 6-12 shows how these plotting problems can be avoided by graphing the log$_{10}$ of the population numbers. The log$_{10}$ of the population is plotted at 5, 10, 15, and 20 generations. Note that a straight line is formed, and that a thousand times this population (1,000,000,000, or log$_{10}$ 9.0) could be accommodated in relatively little extra space. However, this advantage is obtained at the cost of distorting our "common sense" perception of the actual situation. We are not accustomed to thinking in logarithmic relationships, but it is necessary in order to properly understand graphs of microbial populations.

Phases of Growth

When a few bacteria are inoculated into a liquid growth medium and the population is counted at intervals, it is possible to plot a typical **bacterial growth curve** that shows the growth of cells over time (Figure 6-13). There are four basic phases of growth.

Lag Phase

For a while, there is little or no change in the number of cells because they do not immediately reproduce in a new medium. This period of little or no cell division is called the **lag phase**, and it can last for an hour or several days. But during this time, the cells are not dormant. The microbial population has intense metabolic activity, in particular, enzyme synthesis. (The situation is analogous to a factory newly equipped to produce automobiles—there is considerable initial activity but no immediate increase in the automobile population.) Near the end of the lag phase, some cells will double or triple in size in preparation for reproduction.

Log Phase

Eventually, the cells begin to divide and enter a period of growth, or logarithmic increase, called the **log phase** or **exponential growth phase**. Cellular reproduction is most active during this period, and generation time reaches a constant minimum. There is apparently a characteristic minimum generation time—a maximum rate of doubling—genetically determined for each species. Because the doubling time is constant, a logarithmic plot of growth during the log phase is a straight line. During the log phase, cells begin to show their visible characteristics—the shape, color, density, and groupings of their colonies. The log phase is also the time when cells are most active metabolically. (In industrial production, this is the peak period of activity and efficiency.)

However, during their log phase of growth, microorganisms are more sensitive to adverse conditions than they usually are. Radiation and many antimicrobial drugs—for example, the antibiotic penicillin—exert their effect by interfering with some important step in the growth process and are therefore most harmful to cells during this phase.

Stationary Phase

If exponential growth continues unchecked, startlingly high numbers of cells could arise. For example, a single bacterium at a weight of 9.5×10^{-13} grams per cell dividing every 20 minutes for only 25.5 hours can theoretically produce an enormously large population, equivalent in weight to that of an 80,000 ton aircraft carrier. But this does not happen. Eventually, the growth rate slows, the number of microbial deaths balances the number of new cells, and the population stabilizes. The metabolic activities of individual surviving cells also slow at this stage. This period of equilibrium is called the **stationary phase**.

The reasons for cessation of exponential growth are not always clear. Waste products that are toxic to cells accumulate and certain required nutrients become exhausted. Changes occur in pH and temperature. In a specialized apparatus called a *chemostat*, it is possible to keep a population in the exponential growth phase indefinitely by continually draining off spent medium and adding fresh medium.

Death Phase

Usually, the number of deaths soon exceeds the number of new cells formed, and the population enters the **death phase**, or **logarithmic decline phase**. This continues until the population is diminished to a tiny fraction of the more resistant cells, or the population might die out entirely. Some species pass through the entire series of phases in only a few days; others retain some surviving cells almost indefinitely. Microbial death will be discussed further in the next chapter.

Measurement of Microbial Growth

There are a number of ways to measure bacterial growth. Some methods measure cell numbers; other methods measure the population's total mass, which is often directly proportional to cell numbers. Population numbers are usually recorded as the number of cells in a milliliter of liquid or in a gram of solid material. Because bacterial populations are usually

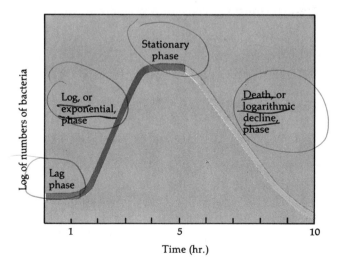

Figure 6-13   Bacterial growth curve, showing the four typical phases of growth.

very large, most methods of counting them are based on direct or indirect counts of very small samples; calculations then determine the size of the total population. Assume, for example, that a millionth of a milliliter (10^{-6}ml) of sour milk is found to contain 70 bacterial cells. Then there must be 70 times 1 million cells, or 70 million cells, per milliliter.

The problem in this method is that it is not practical to measure out a millionth of a milliliter of liquid or a millionth of a gram of food. Therefore, the procedure is done indirectly, in a series of dilutions. For example, if we add one milliliter of milk to 99 milliliters of water, each milliliter of this dilution now has one-hundredth as many bacteria as each milliliter of the original sample had. By making a series of such dilutions, we can readily estimate the number of bacteria in our original sample. To count microbial populations in foods (such as hamburger), a homogenate of one part food to nine parts water is finely ground in a food blender. Samples can then be transferred with a pipette for further dilutions or cell counts.

Plate Counts

The most frequently used method for the measurement of bacterial populations is the **plate count**. An important advantage of this method is that it measures the number of viable cells. One disadvantage may be that it takes some time, usually 24 hours or more, for visible colonies to form. This can be a serious problem in some applications, such as quality control of milk, where it is not possible to hold a particular lot for this amount of time.

The plate count is based on the assumption that each bacterium grows and divides to produce a single colony. It is also assumed that the original inoculum is homogeneous and that no aggregates of cells are present.

When a plate count is performed, it is important that only a limited number of colonies develop in the plate. When too many colonies are present, some cells are overcrowded and do not develop; these conditions cause inaccuracies in the count. Generally, only those plates with from 30 to 300 colonies are counted. To ensure that some colony counts will be within this range, the original inoculum is diluted several times in a process called **serial dilution** (Figure 6-14a).

Serial dilutions   Let us say, for example, that a milk sample has 10,000 bacteria per milliliter. If 1 ml of this sample were plated out, there would theoretically be 10,000 colonies formed in the Petri plate of medium. Obviously, this would not be countable. If 1 ml of this sample were transferred to a tube contain-

ing 9 ml of sterile water, each ml of fluid in this tube would now contain 1000 bacteria. If 1 ml of this sample were inoculated into a Petri plate, there would still be too many potential colonies to count on a plate. Therefore, another serial dilution could be made. One ml containing 1000 bacteria would be transferred to a second tube of 9 ml of water. Each milliliter of this tube would now contain only 100 bacteria, and if 1 ml of the contents of this tube were plated out, potentially 100 colonies would be formed.

Pour plates and spread plates   A plate count is done by either the pour plate or the spread plate method. The **pour plate method** follows the procedure shown in Figure 6-14b. Dilutions of the bacterial suspension in the amounts of 1.0 ml or 0.1 ml are introduced into a Petri dish. The nutrient medium, in which the agar is kept liquid by holding it in a water bath at about 50°C, is poured over the sample, which is then mixed into the medium by gentle agitation of the plate. When the agar solidifies, the plate is incubated. With the pour plate technique, colonies will grow within the nutrient agar (from cells suspended in the nutrient medium as the agar solidifies) as well as on the surface of the agar plate. This technique has some drawbacks since some relatively heat-sensitive microorganisms may be damaged by the melted agar, preventing them from forming colonies. Also, when using certain differential media, the distinctive appearance of the colony on the surface is essential for diagnostic purposes. Colonies that form beneath the surface of a pour plate are not satisfactory for such tests. To avoid these problems, the **spread plate method** is frequently used instead (Figure 6-14c). A 0.1 ml inoculum is added to the surface of a prepoured, solidified agar medium. The inoculum is then spread uniformly over the surface of the medium with a specially shaped sterilized glass rod. This method positions all the colonies on the surface and avoids contact of the cells with melted agar.

Filtration

Where the quantity of bacteria is very low, as in lakes or relatively pure streams, bacteria can be counted by **filtration methods**. One hundred milliliters or more of water are passed through a thin membrane filter whose pores are too small to pass bacteria. Thus, the bacteria are sieved out and retained on the surface of the filter. This filter is then transferred to a plate of nutrient agar, where colonies arise from the bacteria on the filter's surface (Figure 6-15). This method is applied frequently to coliform bacteria, which are indicators of fecal pollution of food or water (Chapter 26). The colonies formed by these bacteria are distinctive when a differential nutrient medium is used.

Calculation: Number of colonies on plate × dilution of sample = number of bacteria /ml

(a) Serial dilutions

(b) Pour plate

(c) Spread plate

Figure 6-14 Plate counts and serial dilutions. (a) In serial dilutions, the original inoculum is diluted out in a series of dilution tubes. In our example, each succeeding dilution tube will have only one tenth of the number of microbial cells as the preceding tube. **(b)** Pour plate. In making a pour plate, the inoculum is placed in a Petri plate. The melted agar medium is poured over it and allowed to solidify. **(c)** Spread plate. In making a spread plate, 0.1 ml of inoculum is placed on the surface of a prepoured plate of solidified agar medium and then is spread evenly.

Figure 6-15   Counting bacteria by filtration. (a) The bacteria in 100 ml of water were sieved out onto the surface of a membrane filter (×2000). **(b)** The filter was placed on a pad saturated with liquid nutrient media, and the individual bacteria grew into visible colonies. There are 65 colonies visible, so we would record 65 bacteria per 100 ml of water sample.

(a)   (b)

Most Probable Number Method

Another method for the determination of the number of bacteria in a sample is the **most probable number (MPN) method**. This statistical estimating method is based on the fact that, the greater the number of bacteria in a sample, the more dilution is needed to reduce density to the point where no bacteria are left to grow in tubes in a dilution series. The MPN method is most useful in cases where the microbes being counted will not grow on solid media (such as the autotrophic nitrifying bacteria). It is also useful when bacterial growth in a liquid differential medium is used to identify them (such as coliform bacteria in water testing). The MPN is only a statement that there is a 95% chance that the bacterial population falls within a certain range and that the MPN is statistically the most probable number. Figure 6-16 illustrates the MPN technique.

Direct Microscopic Count

In the method known as the **direct microscopic count**, a measured volume of a bacterial suspension must be placed inside a defined area on a microscope slide. For example, in the *Breed count method* for milk bacteria, a 0.01 ml sample is spread over a marked square centimeter of slide and stained. By using an oil immersion objective, you can inspect many microscope fields and average the number of bacteria seen in each field.

Since the area of the viewing field of the objective can be determined, the average number of bacteria per viewing field can be obtained. By solving the following equation,

$$\frac{\text{Average number of bacteria per viewing field}}{\text{Area of the viewing field}} = \frac{\text{Number of bacteria applied to the slide}}{1 \text{ square millimeter}}$$

you can determine the number of bacteria in the marked square centimeter on the slide, and you can then determine the number of bacteria in the 0.01 ml sample applied to the slide.

A specially designed slide called a *Petroff–Hausser counter* is also used in direct microscopic counts (Figure 6-17). A shallow well of known volume is indented into the surface of a microscope slide and covered with a thin glass inscribed with squares of known area. The well is filled with the microbial suspension. The average number of bacteria in each of a series of these squares is calculated, and then multiplied by the factor that produces the count per milliliter. Motile bacteria are difficult to count by this method, and, as happens with other microscopic methods, dead cells are about as likely to be counted as live ones. In addition to these disadvantages, a rather high concentration of cells is required, about 10 million bacteria per milliliter. The chief advantage of microscopic counts is that no incubation time is required, and they are usually reserved for applications having time as the primary consideration. This advantage also holds for *electronic cell counters*, which automatically count the number of cells in a measured volume of liquid. These instruments are available in some research laboratories and hospitals.

Figure 6-16 Most probable number (MPN) determination. (a) In this example, there are 3 sets of tubes and 5 tubes in each set. The first set of 5 tubes each receives 10 ml of the inoculum, such as a sample of water. The second set of 5 tubes each receives 1 ml of the sample, and the third set, 0.1 ml, each. In our example, there were enough bacteria in the sample so that all 5 tubes of the set receiving 10 ml showed bacterial growth and were recorded as positive. In the next set of tubes, which received only one tenth as much inoculum, only 3 tubes were positive. In the third set of tubes, which received one hundredth as much inoculum, only 1 tube was positive. **(b)** Tables of MPN allow us to calculate the microbial numbers in a sample that are statistically likely to lead to such a result. The number of positive tubes is recorded for each set: 5, 3, and 1. If we look up this combination in an MPN table (see Appendix D for complete MPN table), we find that the MPN index per 100 ml is 110. Statistically, this means that we would get the results we did if each 100 ml of our water sample contained 110 bacteria.

Combination of Positives	MPN Index/ 100 mL	95% Confidence Limits	
		Lower	Upper
4-2-0	22	9.0	56
4-2-1	26	12	65
4-3-0	27	12	67
4-3-1	33	15	77
4-4-0	34	16	80
5-0-0	23	9.0	86
5-0-1	30	10	110
5-0-2	40	20	140
5-1-0	30	10	120
5-1-1	50	20	150
5-1-2	60	30	180
5-2-0	50	20	170
5-2-1	70	30	210
5-2-2	90	40	250
5-3-0	80	30	250
5-3-1	110	40	300
5-3-2	140	60	360

Volume of inoculum for each set of 5 tubes	Tubes of nutrient medium (sets of 5 tubes)	Number of positive tubes in set
10 ml		5
1 ml		3
0.1 ml		1

(a)

(b)

Figure 6-17 Direct microscopic count of bacteria with a Petroff–Hausser cell counter. The average number of cells within a large square multiplied by a factor of 1,250,000 gives the number of bacteria per milliliter.

Grid with 25 large squares
Cover glass
Slide

(a) Bacterial suspension is added here and fills the shallow well under the squares by capillary action.

Bacterial suspension
Location of squares on cover glass
Cover glass
Slide

(b) Cross section of a cell counter. The depth under the cover glass is known, and the area of the squares is known, so the volume of the bacterial suspension under the squares can be calculated (depth × area).

(c) Microscopic count—all cells in several large squares are counted, and the numbers are averaged. The large square shown here has 14 bacterial cells.

(d) The volume of fluid under the large square is 1/1,250,000th of a milliliter. If it contains 14 cells, as shown here, then there are 14 times 1,250,000 cells in a milliliter.

Estimations of Bacterial Numbers by Indirect Methods

It is not always necessary to count microbial cells to estimate their numbers. Science and industry determine microbial numbers or activity by some of the following indirect means as well.

Turbidity

For some types of experimental work, estimating turbidity is a practical way of following bacterial growth. As bacteria multiply in a liquid medium, the medium becomes **turbid**, or cloudy with cells.

The instrument used to measure turbidity is a spectrophotometer, or colorimeter. In the spectrophotometer, a beam of light is transmitted through a bacterial suspension to a photoelectric cell (Figure 6-18). As bacteria increase, less light will reach the photoelectric cell. This change of light will register on the instrument's scale as the *percentage of transmission*. Also printed on the instrument's scale is a logarithmic expression called the *absorbance* (sometimes called *optical density* or *OD*), a value derived from the percentage of transmission. The absorbance is used to plot bacterial growth. When the bacteria are in logarithmic growth or decline, the graph of the absorbance versus time will form an approximately straight line. If absorbance readings are matched with plate counts of the same culture, this correlation can be used in future estimations of bacterial numbers made directly by turbidity.

More than a million cells per milliliter must be present for the first traces of turbidity to be visible. About 10 to 100 million cells per milliliter are needed to make a suspension turbid enough to be read on a spectrophotometer. Therefore, turbidity is not a useful measure of contamination of liquids by relatively small numbers of bacteria.

Metabolic Activity

Another indirect way to estimate bacterial numbers is by the measurement of the **metabolic activity** of the population. Rather than attempting to count bacterial colonies or estimate turbidity, we measure the amount of a certain metabolic product and assume it to be in direct proportion to the number of bacteria present. The metabolic product might be acid or CO_2. Bacterial numbers can also be estimated with a reduction test, which measures oxygen uptake directly or indirectly. To a medium such as milk, a dye that changes color in the presence or absence of oxygen is added, and the filled tube of milk is tightly capped. Methylene blue, for example, is blue in the presence of oxygen and colorless in its absence. The bacteria use the oxygen as they metabolize the milk. The faster the dye loses color, the faster the oxygen is being depleted and the more bacteria are presumed to be present in the milk. Reduction tests are frequently

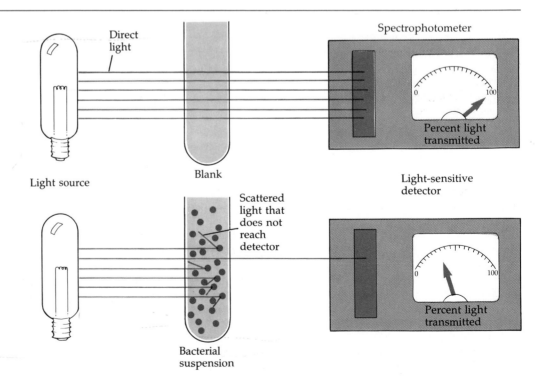

Figure 6-18 Turbidity estimation of bacterial numbers. The amount of light striking the light-sensitive detector is inversely proportional to the number of bacteria under standardized conditions. The less light transmitted, the more bacteria in the sample.

Direct light

Light source

Blank

Spectrophotometer

Percent light transmitted

Light-sensitive detector

Scattered light that does not reach detector

Bacterial suspension

Percent light transmitted

used in microbiology teaching laboratories, but they lack accuracy and are seldom used in commercial applications.

Dry Weight   ? underst

For filamentous organisms, such as molds, the usual measuring methods are less satisfactory. A plate count would not measure this filamentous increase in biomass. In plate counts of molds, the number of asexual spores is counted instead, and this is often not a good measure of growth. One of the better measure-

ments of growth of filamentous organisms is by *dry weight*. In this procedure, the fungus is removed from the growth medium, filtered to remove extraneous material, placed in a weighing bottle, and dried in a desiccator. For bacteria, the same basic procedure is followed. It is customary to remove the bacteria from the culture medium by centrifugation.

You now have a basic understanding of the requirements for, and measurements of, microbial growth. In Chapter 7, we will look at ways in which this growth is controlled in laboratories, hospitals, industry, and our homes.

Study Outline

Requirements for Growth   (pp. 144–148)

1. The growth of a population is an increase in the number of cells or in mass.
2. Requirements for microbial growth are divided into physical and chemical needs.

Physical Requirements   (pp. 145–148)

1. On the basis of growth range of temperature, microbes are classified as psychrophiles (cold-loving), mesophiles (moderate-temperature-loving), and thermophiles (heat-loving).
2. The minimum growth temperature is the lowest temperature at which a species will grow; the optimum growth temperature is the temperature at which it grows best; and the maximum growth temperature is the highest temperature at which growth is possible.
3. Most bacteria grow best at a pH between 6.5 and 7.5.
4. In a hypertonic solution, microbes undergo plasmolysis; halophiles can tolerate high salt concentrations.

Chemical Requirements   (pp. 148–151)

1. All organisms require a carbon source; chemoheterotrophs use an organic molecule, and autotrophs typically use carbon dioxide.
2. Nitrogen is needed for protein and nucleic acid synthesis. Nitrogen can be obtained from decomposition of proteins or from NH_4^+ or NO_3^-; a few bacteria are able to fix N_2.
3. On the basis of oxygen requirements, organisms are classified as obligate aerobes, facultative anaerobes, obligate anaerobes, aerotolerant anaerobes, and microaerophiles.
4. Aerobes, facultative anaerobes, and aerotolerant anaerobes must have the enzymes superoxide dismutase ($2O_2^{2-}$ + $2H^+ \rightarrow O_2 + H_2O_2$) and either catalase ($2H_2O_2 \rightarrow 2H_2O + O_2$) or peroxidase ($H_2O_2 + 2H^+ \rightarrow 2H_2O$).

5. Other chemicals required for microbial growth include sulfur, phosphorus, trace elements, and, for some microorganisms, organic growth factors.

Culture Media   (pp. 152–158)

1. A culture medium is any material prepared for the growth of bacteria in a laboratory.
2. Microbes that grow and multiply in or on a culture medium are known as a culture.
3. Agar is a common solidifying agent for a culture medium.

Chemically Defined Media   (pp. 152–153)

1. A chemically defined medium is one in which the exact chemical composition is known.

Complex Media   (pp. 153–154)

1. A complex medium is one in which the exact chemical composition is not known.

Anaerobic Growth Media and Methods   (pp. 154–155)

1. Reducing media chemically remove molecular oxygen (O_2) that might interfere with the growth of anaerobes.
2. Petri plates can be incubated in an anaerobic jar or anaerobic glove box.

Special Culture Techniques   (pp. 155–156)

1. Some parasitic and fastidious bacteria must be cultured in living animals or in cell cultures.
2. CO_2 incubators or candle jars are used to grow bacteria requiring an increased CO_2 concentration.

Selective and Differential Media   (p. 156)

1. By inhibiting unwanted organisms with salts, dyes, or other chemicals, selective media allow growth of only the desired microbe.

2. Differential media are used to distinguish between different organisms.

Enrichment Culture (p. 157)

1. An enrichment culture is used to encourage the growth of a particular microorganism in a mixed culture.

Obtaining Pure Cultures (pp. 157–158)

1. Colonies are the ~~progeny of a single cell~~ that have grown in or on a culture medium.
2. Pure cultures are usually obtained by the streak plate method.

Preserving Bacterial Cultures (p. 158)

1. Microbes can be preserved for long periods of time by deep-freezing or lyophilization (freeze-drying).

Growth of Bacterial Cultures (pp. 159–167)

Bacterial Division (p. 159)

1. The normal reproductive method of bacteria is binary fission, in which a single cell divides into two identical cells.
2. Some bacteria ~~reproduce~~ by budding, aerial spore formation, or fragmentation.

Generation Time (pp. 159–160)

1. The time required for a cell or population to divide is known as the generation time.

Logarithmic Representation of Bacterial Populations (p. 160)

1. Bacterial division occurs according to a logarithmic progression (2 cells, 4 cells, 8 cells, etc.).

Phases of Growth (pp. 160–161)

1. During the lag phase, there is little or no change in the number of cells, but metabolic activity is high.
2. During the log phase, the bacteria multiply at the fastest rate possible under the conditions provided.
3. During the stationary phase, there is an equilibrium between cell division and death.
4. During the death phase, the number of deaths exceeds the number of new cells formed.

Measurement of Microbial Growth (pp. 161–165)

1. A standard plate count reflects the number of viable microbes and assumes that each bacterium grows into a single colony.
2. A plate count may be done by either the pour plate or spread plate methods.
3. In filtration, bacteria are retained on the surface of a membrane filter and then are transferred to a culture medium to grow and subsequently be counted.
4. The most probable number method (MPN) can be used for microbes that will grow in a liquid medium; it is a statistical estimation.
5. In a direct microscopic count, the microbes in a measured volume of a bacterial suspension are counted with the use of a specially designed slide.

Estimations of Bacterial Numbers by Indirect Methods (pp. 166–167)

1. A spectrophotometer is used to determine turbidity by measuring the amount of light that passes through a suspension of cells.
2. An indirect way of estimating bacterial numbers is by measurement of the metabolic activity of the population, for example, acid production or oxygen consumption.
3. For filamentous organisms such as fungi, measuring dry weight is a convenient method of growth measurement.

Study Questions

Review

1. Describe binary fission.

2. What physical and chemical properties of agar make it useful in culture media?

3. Draw a typical bacterial growth curve. Label and define each of the four phases.

4. Macronutrients (needed in relatively large amounts) are often listed as CHONPS. What does each of these letters indicate, and why are they needed by the cell?

5. Most bacteria grow best at pH _____ .

6. Why can high concentrations of salt or sugar be used to preserve food?

7. Explain five bacterial categories, based on their requirements for oxygen.

8. Which of these curves best depicts the log phase of:
(a) A thermophile incubated at room temperature?
(b) *Listeria monocytogenes* growing in a human?
(c) A psychrophile when incubated at 9°C?

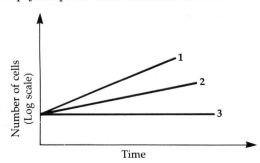

9. Define the following terms and explain the importance of each enzyme.

(a) Catalase (d) Superoxide free radical
(b) Hydrogen peroxide (e) Superoxide dismutase
(c) Peroxidase

10. *Clostridium* can be cultured in an anaerobic incubator or in the presence of atmospheric oxygen if thioglycolate is added to the nutrient broth. Compare these two techniques. Using terms from Question 9, explain why elaborate culture techniques are used for *Clostridium*.

11. A pastry chef inoculated a cream pie with six *S. aureus* cells. If *S. aureus* has a generation time of 60 minutes, how many cells would be in the cream pie after seven hours?

12. Seven methods of measuring microbial growth were explained in this chapter. Categorize each as a direct or indirect method.

13. Flask A contains yeast in glucose-minimal salts broth, incubated at 30°C with aeration. Flask B contains yeast in glucose-minimal salts broth, incubated at 30°C in an anaerobic jar.
(a) Which culture produced the most ATP?
(b) Which culture produced the most alcohol?
(c) Which culture had the shortest generation time?
(d) Which culture had the greatest cell mass?
(e) Which culture had the greatest absorbance?

14. A normal fecal sample has a high concentration of coliform bacteria. A patient with typhoid fever will excrete *Salmonella* along with normal bacteria. Feces from a suspected case of typhoid fever are inoculated on an S-S agar plate. After proper incubation, coliform bacteria form red colonies, and *Salmonella* form colorless colonies with a black center. Is S-S agar selective? Differential? How can you tell?

15. Differentiate between complex and chemically defined media.

16. By deep-freezing, bacteria can be stored without harm for extended periods of time. Why then do refrigeration and freezing preserve foods?

17. Draw the following growth curves for *E. coli*, starting with 100 cells with a generation time of 30 min at 35°C.
(a) Incubated for 5 hours at 35°C.
(b) After 5 hours, the temperature is changed to 20°C for 2 hours.
(c) After 5 hours at 35°C, the temperature is changed to 5°C for 2 hours followed by 35°C for 5 hours.

Challenge

1. *E. coli* was incubated with aeration in a nutrient medium containing two carbon sources, and the growth curve below was made from this culture.
(a) Explain what happened at the time marked *x*.
(b) Which substrate provided "better" growth conditions for the bacteria? How can you tell?

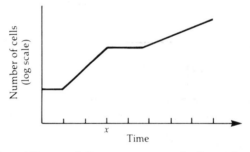

2. *Clostridium* and *Streptococcus* are both catalase-negative. *Streptococcus* grows by fermentation. Why is *Clostridium* killed by oxygen, while *Streptococcus* is not?

3. Design an enrichment medium and procedure for growing an endospore-forming bacterium that fixes nitrogen and uses cellulose for its carbon source.

4. Most laboratory media contain a fermentable carbohydrate and peptone because the majority of bacteria require carbon, nitrogen, and energy sources in these forms. How are these three needs met by glucose-minimal salts medium? (See Table 6-4.)

Further Reading

Atlas, R. M., and R. Bartha. *Microbial Ecology: Fundamentals and Applications*, 2nd ed. Menlo Park, Calif.: Benjamin/Cummings, 1986. Discusses the interrelationships of microbes with their environments; includes enumeration and isolation techniques.

Costerson, J. W., et al. "Bacterial biofilms in nature and disease." *Annual Review of Microbiology* 41:435–464, 1987. Describes methods for examining the growth of mixed cultures in natural environments and in disease. Line drawings illustrate growth from the cell's view.

Difco Manual: Dehydrated Culture Media and Reagents for Microbiology, 10th ed. Detroit: Difco Laboratories, 1984. Formulas and applications of media from the manufacturer.

Dworkin, M. *Developmental Biology of the Bacteria*. Menlo Park, Calif.: Benjamin/Cummings, 1985. Describes life cycles, cell structure, and ecology of selected bacteria, including endospore-formers, cyanobacteria, and actinomycetes.

Ingraham, J. L., O. Maaloe, and F. C. Niedhardt. *Growth of the Bacterial Cell*. Sunderland, Mass.: Sinauer Associates, 1983. Provides comprehensive coverage of cell growth.

Krieg, N. R., and P. S. Hoffman. "Microaerophily and oxygen toxicity." *Annual Review of Microbiology* 40:107–130, 1986. Offers seven explanations for microaerophily.

Sutter, V., et al. *Wadsworth Anaerobic Bacteriology Manual*, 4th ed. Belmont, Calif.: Starr Publishing Co., 1986. Describes methods for isolation and cultivation of anaerobes.

Control of Microbial Growth

The scientific control of microbial growth began only about 100 years ago. Prior to that time, epidemics commonly killed thousands of people. In some hospitals, 25% of delivering mothers died of infections carried by the hands and instruments of attending nurses and physicians. Ignorance of microbes was such that, during the American Civil War, a surgeon might have cleaned his scalpel on his bootsole between incisions.

Ignatz Semmelweis (1816–1865), a Hungarian physician working in Vienna, and Joseph Lister (1827–1912), an English physician, first introduced the concept of microbial control. At the obstetrics ward in the Vienna General Hospital, Semmelweis required all personnel to wash their hands in chlorinated lime; that procedure significantly lowered the infection rate. Lister, meanwhile, read about Pasteur's work with microbes and concluded that the number of infected surgical wounds (sepsis) could be decreased through procedures that prevented the access of microbes to the wound. This system, known as **aseptic surgery**, included the heat sterilization of surgical instruments and, following surgery, the application of phenol (carbolic acid) to wounds.

We have come a long way in controlling microbial growth since the time of Semmelweis and Lister. Today's procedures are far more sophisticated and effective. They are used not only to control disease organisms but also to curb microbial growth that results in food spoilage. This chapter will discuss how microbial growth can be controlled by physical methods and chemical agents. Physical methods include the use of heat, low temperatures, desiccation, osmotic pressure, filtration, and radiation. Chemical agents include several groups of substances that destroy or limit microbial growth on body surfaces or on inanimate objects. In this chapter, discussion will be limited to agents that prevent microbial infection. In Chapter 19, we will discuss methods for the control of microbes once infection has occurred. Chemotherapy is one such method.

Before we begin our survey of physical and chemical methods used to control microbial growth, it would be useful first to define the important terms related to control. Microbial control is needed to prevent the transmission of infection, contamination, and spoilage. Microbial control does not always mean killing the microbes; some situations require that they simply be inhibited or removed. A variety of agents and procedures have been developed; for different situations, each has its own application and range of controlling effects. Table 7-1 contains a list of terms related to the destruction or suppression of microbial growth.

TABLE 7-1 *Terminology Related to the Control of Microbial Growth*

Term	Definition	Comments
Terms Related to Destruction of Organisms		
Sterilization	The process of destroying all forms of microbial life on an object or in a material. This includes the destruction of endospores—the most resistant form of microbial life. *Sterilization* is absolute; there are no degrees of sterilization.	Typical temperatures required are: moist heat, 121°C for 15 minutes; dry heat, 170°C for 120 minutes. Other methods are ionizing radiation and gases such as ethylene oxide.
Disinfection	The process of destroying vegetative pathogens, but not necessarily endospores or viruses. Usually, a *disinfectant* is a chemical applied to an object or a material. Disinfectants tend to reduce or inhibit growth; they usually do not sterilize.	Term usually applied to use of liquid chemical solutions on surfaces or to elimination of pathogens in water—for example, by chlorination.
Antisepsis	Chemical disinfection of the skin, mucous membranes, or other living tissues.	Term especially applied to treatment of wounds. *Antisepsis* is a specific kind of *disinfection*.
Germicide (*cide*, kill)	A chemical agent that rapidly kills microbes but not necessarily their endospores.	A *bactericide* kills bacteria, a *sporicide* kills endospores, a *fungicide* kills fungi, a *virucide* kills viruses, and an *amoebicide* kills amoebas.
Terms Related to Suppression of Organisms		
Bacteriostasis (*stasis*, halt)	A condition in which bacterial growth and multiplication are inhibited, but the bacteria are not killed. If the bacteriostatic agent is removed, bacterial growth and multiplication may resume. *Fungistasis* refers to the inhibition of fungal growth.	Refrigeration is bacteriostatic; many chemicals such as dyes are bacteriostatic rather than bacteriocidal.
Asepsis (*asepsis*, without infection)	The absence of pathogens from an object or area. Aseptic techniques are designed to prevent the entry of pathogens into the body. Whereas *surgical asepsis* is designed to exclude all microbes, *medical asepsis* is designed to exclude microbes associated with communicable diseases.	Air filtration, ultraviolet lights, personnel masks, gloves, and gowns, and instrument sterilization are all factors in achieving *asepsis*.
Degerming	The removal of transient microbes from the skin by mechanical cleansing or by the use of an antiseptic.	For routine injections, alcohol swabs are often used; before surgery, iodine-containing products are often used.
Sanitization	The reduction of pathogens to safe public health levels on eating utensils by mechanical cleansing or chemicals.	Any chemicals must be compatible with safety and palatability of foods.

Conditions Influencing Microbial Control

Usually, if we want to sterilize an object, such as a laboratory flask, we do not bother to identify the species of microbes on it. Instead, we launch a general attack calculated to be strong enough to kill the most resistant microbial forms that could possibly be present—endospores. At any time, the flask could carry a multitude of microorganisms in different phases of their growth cycles, with different metabolic states and even with different microenvironments. Thus, it is possible that an antimicrobial agent will not kill all the microbial population immediately. The time it takes to reduce or eliminate microbial growth on any object depends, in part, on the number and species of microbes on the object. It also depends on the following other factors.

Temperature

Everyone is familiar with the use of ice or mechanical refrigeration to control microbial growth. Like all chemical reactions, the biochemical reactions required for growth are slowed considerably by low temperatures. Chemical disinfectants are also often inhibited by low temperatures. Because their activity is due to temperature-dependent chemical reactions, disinfectants tend to work somewhat better in warm solutions. You might have noticed that directions on disinfectant containers frequently specify that a warm solution be used.

Types of Microbes

Many disinfectants and antiseptics tend to have a greater effect on gram-positive bacteria, as a group, than on gram-negative bacteria, although this differentiation is by no means as clear cut as it is with antibiotic activity. For example, a certain group of gram-negative bacteria, the pseudomonads (genus *Pseudomonas*), is unusually resistant to chemical activity and will even grow actively *in* some disinfectants and antiseptics. These bacteria are able to maintain themselves in such unlikely substrates as simple saline (salt) solutions, and they are also resistant to many antibiotics. This resistance to antimicrobials is probably related to the size of the *porins* (structural openings in the wall) of pseudomonads (see Chapter 4). Porins are much smaller in this group of organisms. These properties have made these common and normally harmless bacteria very troublesome to hospitals. Pseudomonads are **opportunistic pathogens**—microbes that do not ordinarily cause disease, but become pathogenic in the absence of normal competitive flora, as when antibiotics suppress the growth of other microbes but do not adversely affect the pathogen.

The mycobacteria are another group of nonendospore-forming organisms that exhibit greater than normal resistance to disinfectants. This group includes *Mycobacterium tuberculosis*, the microbe that causes tuberculosis, and a number of opportunistically pathogenic mycobacteria that infect persons with depressed immune systems. Special tuberculocidal tests have been developed to evaluate the effectiveness of chemical antimicrobials acting on the mycobacteria.

The endospores of bacteria and cysts of protozoans are also highly resistant to chemical agents, as are some viruses. Their resistance to chlorine disinfection in water treatment is a particularly important factor affecting public health efforts.

Physiological State of the Microbe

Microorganisms that are actively growing tend to be more susceptible to chemical agents than older cells. This might be because an actively metabolizing and reproducing cell has more points of vulnerability than does an older, possibly dormant cell. When microorganisms have formed endospores, the endospores are generally more resistant than the vegetative cells. The endospores of the resistant strains of *Clostridium botulinum* (bo-tū-lī′num), for example, can withstand boiling for several hours.

Environment

Organic matter frequently interferes with the action of chemical agents. In hospitals, the presence of organic matter in vomit and feces influences the selection of disinfectants because some disinfectants are more effective than others under these conditions. Even bacteria found in food tend to be protected by the organic matter of the food. For example, bacteria in cream are protected by proteins and fats and can survive more heat than bacteria in skim milk. Heat is much more effective in killing bacteria under acidic conditions than at a neutral pH. This is also true of a few chemicals.

Let us now examine the mechanisms of action of control agents, that is, the way various agents actually kill or inhibit microbes.

Actions of Microbial Control Agents

Alteration of Membrane Permeability

The plasma membrane, located just inside the cell wall, is the target of many control agents. This membrane actively controls the passage of nutrients into the cell and the elimination of wastes from the cell. Damage to the lipids or proteins of the plasma membrane by antimicrobial agents, such as quaternary ammonium compounds, typically causes cellular contents to leak into the surrounding medium and interferes with the growth of the cell. Several types of chemical agents and antibiotics work at least partially in this manner.

Damage to Proteins and Nucleic Acids

You might recall that the functional properties of proteins are due to their three-dimensional shape. This shape is maintained by chemical bonds that hold together adjoining portions of the amino acid chain as it folds back and forth upon itself. Some of those bonds are hydrogen bonds, which are susceptible to breakage by heat or certain chemicals; breakage results in denaturation of the protein. Covalent bonds, which are stronger, are also subject to attack. For example, disulfide bridges, which play an important role in protein structure by joining amino acids with exposed sulfhydryl (—SH) groups, can be broken by certain chemicals or sufficient heat.

Because DNA and RNA carry the genetic message, damage to them by radiation or chemicals is frequently lethal to the cell; this damage prevents both replication and normal functioning.

Rate of Microbial Death

When bacterial populations are heated or treated with antimicrobial chemicals, they usually die at a constant rate. For example, say that a population of 1 million microbes has been treated for one minute, and

TABLE 7-2 *Example of Microbial Death Rate*

Time (minutes)	Deaths per Minute	Number of Survivors
0	0	1,000,000
1	900,000	100,000
2	90,000	10,000
3	9,000	1,000
4	900	100
5	90	10
6	9	1

Source: O. Rahn, *Physiology of Bacteria*. New York: McGraw-Hill, 1932.

90% of the population has died. We are now left with 100,000 microbes. If the population is treated for another minute, 90% of *those* microbes die, and we are left with 10,000 survivors. In other words, for each minute that the treatment is applied, 90% of the remaining population is killed (Table 7-2). If the death curve is plotted logarithmically, the death rate is seen to be constant, as shown by the straight line in Figure 7-1. Obviously, the more microbes there are at the beginning, the longer it will take to kill them all.

Now that you have an understanding of the basic principles of microbial control, we will examine some representative physical and chemical methods.

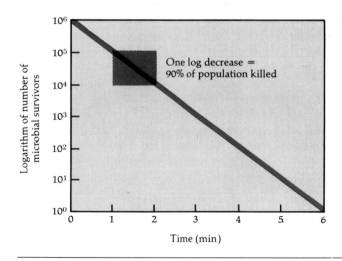

Figure 7-1 Microbial death curve plotted logarithmically and arithmetically. In this case the cells are dying at the rate of 90% each minute.

Physical Methods of Microbial Control

Heat

Probably the most common method by which microbes are killed is heat. A visit to any supermarket will demonstrate that canned goods—in which microorganisms have been killed by moist heat—easily outnumber the foods preserved by other methods. Laboratory media and glassware and hospital instruments are also usually sterilized by heat.

Heat is not only the most widely applicable and effective agent for sterilization, it is also the most economical and easily controlled. Heat appears to kill microbes by denaturing their enzymes. In sterilization by heat, the degree of the bacteria's heat resistance must be considered. Heat resistance varies among different microbes; these differences can be expressed through the concept of thermal death point. **Thermal death point (TDP)** is the lowest *temperature* required to kill all of the microorganisms in a liquid suspension in 10 minutes.

Another factor to be considered in sterilization is the length of time required for the material to be rendered sterile. This is expressed as **thermal death time (TDT)**, the minimal length of *time* in which all bacteria in a liquid culture will be killed at a given temperature. Both TDP and TDT are useful guidelines that indicate the severity of treatment required to kill a given population of bacteria.

Decimal reduction time (DRT or **D value)** is a third concept related to the bacteria's degree of heat resistance. It is the time, in minutes, in which 90% of the population of bacteria at a given temperature will be killed (Figure 7-1). DRT is especially useful in the canning industry.

The heat used in sterilization can be applied in the form of moist heat or dry heat. Dry heat kills by oxidation effects. A simple analogy is the slow charring of paper in a heated oven, when the temperature is below the ignition point of paper. Moist heat kills microorganisms more quickly because the water hastens the breaking of hydrogen bonds that hold proteins in their three-dimensional structure.

Moist Heat

One type of moist heat sterilization is boiling (100°C at sea level), which kills vegetative forms of bacterial pathogens, many viruses, and fungi and their spores within about 10 minutes. Free-flowing (unpressur-ized) steam is equivalent in temperature to boiling water. Endospores and some viruses, however, are not destroyed so quickly. The hepatitis virus, for example, can survive up to 30 minutes of boiling, and some bacterial endospores have resisted boiling temperatures for more than 20 hours. Boiling is therefore not always a reliable sterilization procedure. However, a few minutes of boiling temperatures will kill most pathogens and generally make food or water safe to eat or drink.

Reliable sterilization with moist heat requires temperatures above that of boiling water. These high temperatures are most commonly achieved by steam under pressure in an **autoclave** (Figure 7-2). As the preferred method of sterilization, autoclaving is used unless the material to be sterilized can be damaged by heat or moisture. The higher the pressure in the autoclave, the higher the temperature. For example, when free-flowing steam at a temperature of 100°C is placed under a pressure of 1 atmosphere above sea level pressure, that is, about 15 pounds of pressure per square inch (psi), the temperature rises to 121°C. Increasing the pressure to 20 pounds per square inch raises the temperature to 126°C. The relationship between temperature and pressure is shown in Table 7-3. In an autoclave, steam at a pressure of about 15 pounds per square inch (121°C) will kill *all* organisms and their endospores in about 15 minutes, or perhaps a bit longer, depending on the type and volume of material being sterilized.

Autoclaving is used to sterilize culture media, instruments, dressings, intravenous equipment, applicators, solutions, syringes, transfusion equipment,

TABLE 7-3 *Relationship Between Pressure and Temperature of Steam at Sea Level**

Pressure (psi in excess of atmospheric pressure)	Temperature (°C)
0 psi	100
5 psi	110
10 psi	116
15 psi	121
20 psi	126
30 psi	135

*At higher altitudes, the gauge pressure would be higher for a particular temperature.

Exhaust valve (to remove steam after sterilization)

Steam to chamber

Safety valve

Pressure gauge

Operating valve (controls steam from jacket to chamber)

Steam

Air

Door

Sediment screen

Thermometer

Pressure regulator for steam supply

Steam jacket

Automatic ejector valve

To waste line

Steam supply

Figure 7-2 Autoclave. The entering steam forces the air out of the bottom (black arrows). The automatic ejector valve remains open as long as an air–steam mixture is passing out of the waste line. When all the air has been ejected, the higher temperature of the pure steam closes the valve, and the pressure in the chamber increases.

and numerous other items that can withstand high temperature and pressure. Large industrial autoclaves are called *retorts*, but the same principle is used for the common household pressure cooker in the home canning of foods.

Heat requires extra time to reach the center of solid materials, such as canned meats, in which heat is conducted without the efficient convection currents in liquids. Heating large containers also requires extra time. Table 7-4 shows the different time requirements for various container sizes. It is important to remember that, to sterilize a surface, steam has to actually contact it. This is not necessary when sterilizing liquids, but with dry glassware, bandages, and the like, care must be taken to ensure that contact is made. For example, aluminum foil is impervious to steam and should not be used to wrap containers that are to be sterilized; paper should be used instead. Care should also be taken to avoid trapping air in the bottom of a dry container because trapped air will not be replaced by steam, which is lighter than air. The

TABLE 7-4 *Effect of Container Size on Sterilization Times for Liquid Solutions**

Container Size	Liquid Volume	Sterilization Time (minutes)
Test tube: 18 × 150 mm	10 ml	15
Erlenmeyer flask: 125 ml	95 ml	15
Erlenmeyer flask: 2000 ml	1500 ml	30
Fermentation bottle: 9000 ml	6750 ml	70

*Larger containers require additional time to reach sterilizing temperatures. For a test tube, this additional time might be only 5 minutes, but for a 9000-ml bottle it might be 60 minutes. A container is usually not filled past 75% of its capacity.

trapped air is the equivalent of a small hot-air oven, which, as we shall see shortly, requires a higher temperature and longer time. Air-trapping containers should be placed in a tipped position so that the steam will force out the air. Products that do not permit penetration by moisture, such as mineral oil or petroleum jelly, are not sterilized by the same treatments that would sterilize aqueous solutions.

Steam under pressure fails to sterilize when the air is not completely exhausted. This is usually due to the premature closing of the autoclave's automatic ejector valve (Figure 7-2). Anyone familiar with home canning knows that the steam must flow vigorously out of the valve in the lid for several minutes to entrain and remove all the air before the pressure cooker is sealed. If the air is not completely exhausted, the container will not reach the temperature expected for a given pressure.

Pasteurization

You may recall from Chapter 1 that Louis Pasteur, in the early days of microbiology, found a practical method of preventing the spoilage of beer and wine. Pasteur used mild heating, which was sufficient to kill the organisms that caused the particular spoilage problem without seriously damaging the taste of the product. The same principle was later applied to milk to produce what we now call pasteurized milk. Milk was first pasteurized to eliminate the tuberculosis bacterium. Many relatively heat-resistant (*thermoduric*) organisms survive pasteurization, but these are unlikely to cause disease or cause refrigerated milk to spoil in a short time.

In the classic pasteurization treatment of milk, the milk was exposed to a temperature of about 63°C for 30 minutes. Most milk pasteurization today uses higher temperatures, at least 72°C, but for only about 15 seconds. This treatment, known as **high-temperature, short-time (HTST) pasteurization**, is applied as the milk flows continuously past a heat exchanger. Although the temperature is higher than needed to kill pathogens, it lowers total bacterial counts, so the milk keeps well under refrigeration. Milk can also be sterilized—something quite different from pasteurization—by **ultra high temperature (UHT)** treatments so that it can be stored without refrigeration. This is more useful in parts of the world with poor refrigeration facilities. In the United States, sterilization is sometimes used on the small containers of coffee creamers found in restaurants. To avoid imparting a cooked taste to the milk, it was necessary to

devise a UHT system in which the liquid milk never touches a surface hotter than the milk itself while being heated by steam. The milk falls in a thin film through a chamber of superheated steam and reaches 140°C in less than a second. It is held for 3 seconds in a holding tube and then cooled in a vacuum chamber, where the steam flashes off. With this process, in less than 5 seconds the milk temperature rises from 74°C to 140°C and drops back to 74°C.

The heat treatments we have just discussed illustrate the concept of **equivalent treatments**—as the temperature is increased, much less time is needed to kill the same number of microbes. For example, the destruction of highly resistant endospores might take 70 minutes at 115°C, whereas only 7 minutes would be needed at 125°C. Both treatments yield the same result. The concept of equivalent treatments also explains why classic pasteurization temperatures of 63°C for 30 minutes, HTST temperatures of 72°C for 15 seconds, and UHT temperatures of 140°C for less than a second can have similar effects.

Dry Heat Sterilization

One of the simplest methods of dry heat sterilization is **direct flaming**. You will use this procedure many times in the microbiology laboratory when you sterilize inoculating loops. To sterilize the inoculating loop, you heat the wire to a red glow, which is 100% effective. A similar principle is used in **incineration**, an effective way to sterilize and dispose of contaminated paper cups, bags, and dressings.

Another form of dry heat sterilization is **hot air sterilization**. Items to be sterilized by this procedure are placed in an oven. Generally, a temperature of about 170°C maintained for nearly two hours ensures sterilization. The longer period of time and higher temperature (relative to moist heat) are required because the heat in water is more readily transferred to a cool body than the heat in air. For example, imagine the different effects of immersing your hand in boiling water at 100°C (212°F) and of holding it in a hot-air oven at the same temperature for the same time.

Filtration

Filtration is the passage of a liquid or gas through a screenlike material with pores small enough to retain microorganisms (this is often the same apparatus used for counting; see Figure 6-15). A vacuum that is created in the receiving flask aids gravity in pulling the liquid through the filter. Filtration is used to sterilize heat-sensitive materials, such as some culture media, enzymes, vaccines, and antibiotic solutions.

Some operating theaters and rooms occupied by burn patients receive filtered air. **High-efficiency particulate air (HEPA) filters** remove almost all microorganisms larger than about 0.3 μm in diameter.

In the early days of microbiology, hollow cylindrical candle-shaped filters of unglazed porcelain were used to filter liquids. The long and indirect passageways through the walls of the filter adsorbed the bacteria. Unseen pathogens that passed through such filters (causing such diseases as rabies) were called *filterable viruses*.

In recent years, **membrane filters**, composed of such substances as cellulose esters or plastic polymers, have become popular for industrial and laboratory use (see Figures 6-15 and 7-3). These filters are only 0.1 mm thick. The pores of membrane filters are closely matched in size. In some brands, plastic film is irradiated so that very uniform holes, where the radiation particles have passed, are etched in the plastic. The pores of membrane filters include 0.22 and 0.45 μm sizes, intended for bacteria, and range down to only 0.01 μm, a size that will retain viruses and even some large protein molecules.

Low Temperature

The effect of low temperatures on microbes depends on the particular microbe and the intensity of the application. For example, at temperatures of ordinary refrigerators (0° to 7°C), the metabolic rate of most microbes is so reduced that they cannot reproduce or elaborate products. In other words, ordinary refrigeration has a bacteriostatic effect. Yet, psychrophilic (cold-loving) species do grow slowly at refrigerator temperatures and will alter the appearance and taste of foods after a considerable time. Surprisingly, some bacteria can grow at temperatures several degrees below freezing. One third of the population of some vegetative bacteria might survive a year of freezing; some other psychrophilic species might have very few survivors. Rapidly attained subfreezing temperatures render microbes dormant but do not necessarily kill them. Freezing is most harmful to bacteria when done slowly; the ice crystals that form and grow disrupt the cellular and molecular structure of the bacteria.

Desiccation

To grow and multiply, microbes require water. In the absence of water—a condition known as **desiccation**—microbes are not capable of growth or reproduction but can remain viable for years. Then, when water is made available to them, they can resume their growth and division. This ability is used in the laboratory when microbes are preserved by lyophilization, a process described in Chapter 6.

Vegetative cells' resistance to desiccation varies with the species and the organism's environment. For example, the gonorrhea bacterium can withstand dryness for only about an hour, but the tuberculosis bacterium can remain viable for months. A normally susceptible bacterium is much more resistant if it is embedded in mucus, pus, or feces. Viruses are generally resistant to desiccation, but not as resistant as endospores. This ability of certain dried microbes and endospores to remain viable is important in a hospital setting. Dust, clothing, bedding, and dressings might contain infectious microbes in dried mucus, urine, pus, and feces.

Figure 7-3 Filter sterilization with a disposable, presterilized plastic unit. The sample is placed into the upper chamber and forced through the membrane filter by a vacuum in the lower chamber. Pores in the membrane filter are smaller than the bacteria, so bacteria are retained on the filter. The sterilized sample can then be decanted from the lower chamber. Similar equipment with removable filter disks is used to count bacteria in samples (see Figure 6-15).

Nomadic peoples such as some Native Americans used desiccation to preserve foods. They found that if meat is sliced thin and sun dried (made into jerky), it can be preserved for long periods of time. Other examples include raisins and similarly dried fruits.

Osmotic Pressure

The use of high concentrations of salts and sugars to preserve food is based on the effects of **osmotic pressure**. High concentrations of these substances create a hypertonic environment that causes water to leave the microbial cell (see Chapters 4 and 6). This process resembles preservation by desiccation, in that both methods deny the cell the moisture it needs for growth. As water leaves the microbial cell, the plasma membrane shrinks away from the cell wall (plasmolysis), and the cell stops growing, although it might not immediately die. The principle of osmotic pressure is used in the preservation of foods. For example, concentrated salt solutions are used to cure meats, and thick sugar solutions are used to preserve fruits.

As a general rule, molds and yeasts are much more capable than bacteria of growing in materials with low moisture or high osmotic pressures. This property of molds, sometimes combined with their ability to grow under acidic conditions, is the reason fruits and grains are spoiled by molds rather than by bacteria. It is also part of the reason why molds are able to form "mildew" growth on a damp wall or a shower curtain.

Radiation

Radiation has various effects on cells, depending on its wavelength, intensity, and duration. There are two types of sterilizing radiation, ionizing and nonionizing.

Ionizing radiation, such as **gamma rays** or **high-energy electron beams**, has a wavelength shorter than that of nonionizing radiation, less than about 1 nm. Therefore, it carries much more energy (Figure 7-4). Gamma rays are emitted by radioactive cobalt, and electron beams are produced by accelerating electrons to high energies in special machines. Gamma rays penetrate deeply but may require hours for sterilization of large masses; high-energy electron beams have much lower penetrating power but usually require only a few seconds of exposure. The principal effect of ionizing radiation is the ionization of water, which forms highly reactive hydroxyl radicals (see the discussion of toxic forms of oxygen in Chapter 6).

The food industry has recently renewed its interest in the use of radiation for food preservation. (This subject will be discussed more fully in Chapter 27.) Low-level radiation, used for years in many countries, has been approved recently in the United States for certain purposes. However, public apprehension about the safety of radiation has limited its use.

Ionizing radiation, especially high-energy electron beams, is increasingly used for the sterilization of pharmaceuticals and disposable dental and medical supplies, such as plastic syringes, surgical gloves, suturing materials, and catheters. Radiation has been replacing gases for sterilizing these items. Sterilizing gases will be discussed later in this chapter.

Nonionizing radiation has a wavelength longer than that of ionizing radiation, usually greater than about 1 nm. A good example of nonionizing radiation is ultraviolet (UV) light. UV light damages the DNA of exposed cells. It causes bonds to form between adjacent thymines in DNA chains (see Figure 8-17). These *thymine dimers* inhibit correct replication of the DNA during reproduction of the cell. The UV wavelengths most effective for killing microorganisms are about 260 nm; these wavelengths are specifically absorbed by cellular DNA. UV radiation is also used to control microbes in the air. A UV or "germicidal" lamp is commonly found in hospital rooms, nurseries, operating rooms, and cafeterias. UV light is also used to sterilize vaccines, sera and toxins, and occasionally municipal waste waters and drinking waters. A major drawback of UV light as a sterilizer is that it has low

Figure 7-4   Radiant energy spectrum.

penetrability, so the organisms to be killed must be directly exposed to the rays. Organisms protected by solids and such coverings as paper, glass, and textiles are not affected. Another potential problem is that UV light can damage the eyes, and prolonged exposure to UV light can cause burns and skin cancer.

Sunlight contains some UV light, but the most effective (shorter) wavelengths are screened out by the atmosphere. The antimicrobial effect of sunlight is due almost entirely to its forming singlet oxygen in the cytoplasm (see Chapter 6). Many pigments produced by bacteria provide protection from sunlight.

Microwaves do not have much direct effect on microorganisms but kill them indirectly by heating the medium, such as food. Table 7-5 (pp. 180–181) summarizes the physical methods of microbial control.

Chemical Methods of Microbial Control

Chemical agents are used to control microbes on living tissue and inanimate objects. Unfortunately, few chemical agents achieve sterility; most of them merely reduce microbial populations to safe levels or remove vegetative forms of pathogens from objects. A common problem in disinfection is the selection of an agent that will kill all organisms in the shortest period of time without damaging the contaminated material. Just as there is no single physical method of microbial control that can be used in every situation, so also there is no one disinfectant that will be appropriate for all circumstances.

Principles of Effective Disinfection

To select a disinfectant for a particular job, we must first understand its action. For example, what are the properties of the disinfectant? What is it designed to do? Simply by reading the label, we can learn a great deal about a disinfectant's properties. We should also remember that the concentration of a disinfectant will affect its action. A disinfectant should always be diluted exactly as suggested by the manufacturer. Solutions that are too weak may be ineffective or be bacteriostatic instead of bactericidal. Solutions that are too strong can be dangerous to humans who come in contact with them.

Also to be considered is the nature of the material being disinfected. Are organic materials present that might interfere with the action of the disinfectant? (See the box on p. 180.) The presence of organic mate-

rials and the pH of the medium in which the microbes are present might determine whether a chemical control agent is only inhibitory or is lethal to the microbes.

Another very important consideration is whether the disinfectant will easily make contact with the microbes. An area might have to be scrubbed and rinsed before the disinfectant is applied. In general, disinfection is a gradual process. Thus, to be effective, a disinfectant might have to be left on a surface for several hours.

A final factor to keep in mind is, the higher the temperature at which the disinfectant is applied, the more effective it usually is.

Evaluating a Disinfectant

Phenol Coefficient

There is an obvious need to compare the effectiveness of disinfectants and antiseptics. Phenol, which was once a widely used disinfectant, is still used as a standard for these comparisons. However, some antimicrobials cannot be compared with phenol, particularly if their action is bacteriostatic or if they have a prolonged residual activity on the skin. Nonetheless, the **phenol coefficient test** is still used and serves to illustrate the general principles involved in all such comparisons. For this test, three test organisms are used: *Staphylococcus aureus* (a gram-positive bacterium), *Salmonella typhi* (a gram-negative bacterium), and *Pseudomonas aeruginosa* (ā-rü-ji-nō′sä) (a gram-negative organism often resistant to antimicrobials). In broth cultures under standard conditions, all three test organisms are exposed to the test chemical for a specified time. If the test chemical must be used at a greater concentration, or for a longer period of time, than phenol is used to achieve comparable effects, the test chemical's phenol coefficient will be less than 1. This rating indicates that the test chemical is weaker than phenol. A coefficient greater than 1 indicates that the chemical is more active than phenol. For example, if Brand X at a dilution concentration of 1:200 is effective under the specified conditions, and if phenol is equally effective at a concentration of 1:100, then Brand X is more efficient than phenol by a factor of 200/100 and has a phenol coefficient of 2. The result of a phenol coefficient test is usually confirmed by the use-dilution test.

The **use-dilution test** is sometimes considered the official evaluation of a disinfectant. In this test, standardized preparations of several test bacteria are

(Text continues on p. 182)

KNOW

TABLE 7-5 *Summary of Physical Methods Used to Control Microbial Growth*

Method	Mechanism of Action	Comment	Preferred Use
Heat			
1. Moist heat sterilization			
a. Boiling or flowing steam	Denaturation	Kills vegetative bacterial and fungal pathogens and many viruses within 10 minutes; less effective on endospores.	Dishes, basins, pitchers, various equipment.
b. Autoclaving	Denaturation	Very effective method of sterilization; at about 15 pounds of pressure (121°C), all vegetative cells and their endospores are killed in about 15 minutes.	Microbiological media, solutions, linens, utensils, dressings, equipment, and other items that can withstand temperature and pressure.
2. Dry heat sterilization			
a. Direct flaming	Burning to ashes	Very effective method of sterilization.	Inoculating loops.
b. Incineration	Burning to ashes	Very effective method of sterilization.	Paper cups, dressings, animal carcasses, bags, and wipes.
c. Hot-air sterilization	Oxidation	Very effective method of sterilization, but requires temperature of 170°C for about 2 hours.	Empty glassware, instruments, needles, and syringes.
3. Pasteurization	Denaturation	Heat treatment for milk (72°C for about 15 seconds) that kills all pathogens and some nonpathogens.	Milk, cream, and certain alcoholic beverages (beer and wine).

MMWR   *Centers for Disease Control*
Morbidity and Mortality Weekly Report

A CASE OF CONTAMINATED DISINFECTANT

During a six-week period in 1962, eleven cases of septic arthritis, a painful joint inflammation, were diagnosed in a town in Pennsylvania. The infection, caused by the presence of the bacterium *Serratia marcescens* in the joint fluid, required prolonged hospital treatment, including surgical drainage and antibiotic injections into the joints.

The outbreak drew the attention of the Centers for Disease Control (CDC). Septic arthritis due to *Serratia* is uncommon. However, nosocomial (hospital-acquired) epidemics caused by *Serratia* have been traced to various sources, including contaminated disinfectants, intravenous fluids, respirators, catheters, and hospital personnel.

An investigation by the CDC revealed that the five women and six men affected by the disease all received treatment at the same orthopedic surgery office. During the course of routine treatment, all the patients had received injections of methylprednisolone and lidocaine, drugs given for painful, inflamed joints. The CDC found that the methylprednisolone had been stored according to the manufacturer's directions. A jar in which cotton balls were stored was washed about once a month with detergent and tap water. Both the methylprednisolone and the jar of cotton balls contained *S. marcescens*. But the drug and the cotton balls were not the ultimate culprits. The disinfectant—a commercial benzalkonium chloride product used to clean the cotton ball jar, drug vials, and injection area—was actually the source of contamination.

However, it was the routine procedure that contaminated the disinfectant that transmitted the infection. The

TABLE 7-5 (Continued)

Method	Mechanism of Action	Comment	Preferred Use
Filtration	Separation of bacteria from suspending liquid	Passage of a liquid or gas through a screenlike material that traps microbes; most filters in use consist of cellulose acetate or nitrocellulose.	Useful for sterilizing liquids (toxins, enzymes, vaccines) that are destroyed by heat.
Low Temperature			
1. Refrigeration	Decreased chemical reactions and possible changes in proteins	Has a bacteriostatic effect.	Food, drug, and culture preservation.
2. Deep-freezing (see Chapter 6)	Decreased chemical reactions and possible changes in proteins	An effective method for *preserving* microbial cultures, in which cultures are quick-frozen between $-50°$ and $-95°C$.	Food, drug, and culture preservation.
3. Lyophilization (see Chapter 6)	Decreased chemical reactions and possible changes in proteins	Most effective method for long-term preservation of microbial cultures; water removed by high vacuum at low temperature.	Food, drug, and culture preservation.
Desiccation	Disruption of metabolism	Involves removing water from microbes; primarily bacteriostatic.	Food preservation.
Osmotic pressure	Plasmolysis	Results in microbial cells losing water.	Food preservation.
Radiation			
1. Ionizing	Destruction of DNA by gamma rays and high-energy electron beams	Not widespread in routine sterilization.	Used for sterilizing pharmaceuticals and medical and dental supplies.
2. Nonionizing	Damage to DNA by UV light	Radiation not very penetrating.	Practical application is the UV (germicidal) lamp.

cotton balls, which had been soaked in the disinfectant, were used to wipe out the multiple-use vials. Then, the medication was drawn into disposable syringes from the vials. The site of injection on each patient was also wiped with a soaked cotton ball.

The CDC investigators knew that the contamination did not originate with the methylprednisolone because unopened bottles of the drug from the same lot numbers tested sterile. But they also knew that, as a quaternary ammonium compound (quat), benzalkonium chloride could pose several problems. For one thing, organic matter such as cotton fibers can deactivate the disinfecting action of quats. And, an environmental contaminant, such as the bacteria, can gain access to the quat through the repeated opening of the jar to remove cotton or to add fresh cotton balls. Then, during exposure to the low concentrations of the disinfectant in the jar, the resistant bacteria can grow.

The CDC tests confirmed that the epidemic strain was indeed resistant to low concentrations of benzalkonium chloride and that the bacteria could grow in the methylprednisolone, which, by sheer coincidence, was also preserved with a quat. This quat was just as ineffective at preventing growth of these gram-negative bacteria.

Although quats are strongly bactericidal against gram-positive bacteria, they are not as effective against gram-negative bacteria (such as *S. marcescens*). Therefore, one way this epidemic might have been prevented was by the use of a different disinfectant. The CDC made no recommendation.

Source: A. K. Nakashima et al. *Journal of Clinical Microbiology* 25:1014–1021, 1987.

Figure 7-5 Evaluation of disinfectants by the filter paper method. The paper disks (lettered) are soaked in a solution of the disinfectant and are placed on the surface of a nutrient medium, on which a culture of test bacteria has been streaked. Note that the gram-positive *Staphylococcus aureus* was the only one susceptible to the inhibitory effects of all the disinfectants. The two gram-negative bacteria were relatively less affected; only sodium hypochlorite (chlorine bleach) (C) was effective against them, but the clear zones were smaller than in *Staphylococcus*. If higher concentrations of O-phenylphenol (P) and benzalkonium chloride (Q) (a quaternary ammonium compound) had been used, they would have shown some effectiveness against *E. coli*. Hexachlorophene (H) is effective against only gram-positive bacteria.

Staphylococcus aureus (+)

Escherichia coli (−)

Pseudomonas aeruginosa (−)

added to a series of tubes containing increasingly strong concentrations of the test disinfectant. The tubes are then incubated, and growth, or lack of it, is recorded. The more highly diluted the chemical can be and still be effective, the higher its rating.

Many specialized tests can be used to determine specific effects of disinfectants, such as their sporicidal, tuberculocidal, or fungicidal effectiveness.

Filter Paper Method

The **filter paper method** is often used in teaching laboratories to evaluate the efficacy of a chemical agent. A disk of filter paper is soaked with a chemical and placed on the surface of an agar plate that has been previously inoculated and incubated with the test organism. After incubation, if the chemical agent is effective, a clear zone representing inhibition of growth can be observed around the disk (Figure 7-5).

Types of Disinfectants

Phenol and Phenolics

Phenol (carbolic acid), the substance first used by Joseph Lister in his operating room, is rarely used as an antiseptic and disinfectant because it irritates the skin and has a disagreeable odor. It is often used in

throat lozenges for its local anesthetic effect but has no antimicrobial effect at the low concentrations used. In concentrations above 1%, however, phenol has an antibacterial effect. One commercial throat spray with a 1.4% phenol concentration has a significant effect on recovery from sore throats. The structure of a phenol molecule is shown in Figure 7-6a.

Derivatives of phenol, called **phenolics**, contain a molecule of phenol that has been chemically altered to reduce its irritating qualities or increase its antibacterial activity in combination with a soap or detergent. Phenolics exert antimicrobial activity by injuring plasma membranes, inactivating enzymes, and denaturing proteins. Phenolics are frequently used as disinfectants because they remain active in

Figure 7-6 Structure of phenol and phenolics. (a) Phenol. **(b)** O-phenylphenol. **(c)** Hexachlorophene.

the presence of organic compounds, they are stable, and they persist for long periods of time after application. For these reasons, phenolics are suitable agents for disinfecting pus, saliva, and feces. The addition of halogens such as chlorines to phenolics usually increases their antimicrobial activity.

One of the most frequently used phenolics is derived from coal tar, a group of chemicals called *cresols*. A very important cresol is *O-phenylphenol* (Figures 7-5 and 7-6b), the main ingredient in most formulations of Lysol®. Cresols are very good environmental disinfectants.

Another phenolic, used in the past much more widely than now, is *hexachlorophene*, consisting of two molecules of phenol (a bisphenol) joined together (Figure 7-6c). This phenolic was originally an ingredient in soaps and lotions (such as pHisoHex®) used for surgical scrubs and hospital microbial control procedures, cosmetic soaps, deodorants, feminine hygiene sprays, and even toothpaste. As a bacteriostatic agent, hexachlorophene is especially effective against gram-positive staphylococci and streptococci (Figure 7-5), which can cause infections of the skin. Moreover, the hexachlorophene persists on the skin for long periods of time. For these reasons, it was once heavily used to control staphylococcal and streptococcal infections in hospital nurseries. However, in 1972, it was found that *excessive* use of hexachlorophene, such as bathing infants with it several times a day, could lead to neurological damage. Hexachlorophene is still used in hospitals to control nosocomial (hospital-acquired) infections and as a scrub for hospital personnel. However, it is no longer casually used by the general public. At present, a prescription is required for the purchase of a 3% or stronger solution of hexachlorophene.

Chlorhexidine

Chlorhexidine, which is not a phenol, although its structure and application resemble hexachlorophene, is frequently used for disinfection of skin and mucous membranes as an alternative to hexachlorophene. It is combined with a detergent or alcohol for surgical hand scrubs and preoperative skin preparation in patients. In such applications, it works more rapidly than hexachlorophene and is equally persistent on the skin. The skin does not absorb it, and no toxicity has been reported. Its killing effect is related to damage to the plasma membrane. It is effective against most vegetative bacteria, but is not sporicidal. Only certain enveloped (lipophilic) viruses are affected by it.

Halogens

The **halogen elements**, particularly iodine and chlorine, are effective antimicrobial agents, both alone (as I_2 or Cl_2 in solution) and as constituents of inorganic or organic compounds. *Iodine* (I_2) is one of the oldest and most effective antiseptics. Iodine is effective against all kinds of bacteria, many endospores, various fungi, and some viruses. One proposed mechanism for the activity of iodine is that it combines with the amino acid tyrosine, a common component of many enzymes and other cellular proteins (Figure 7-7). As a result, microbial protein function is inhibited. Iodine also reacts with other amino acids and apparently with the fatty acids of the plasma membranes.

Iodine is available as a *tincture*, that is, in solution in aqueous alcohol, and as an iodophor. An *iodophor* is a combination of iodine and an organic molecule, usually a detergent in which the iodine is released slowly. Iodophors have the antimicrobial activity of iodine, but they do not stain and are less irritating than iodine. The preparations most commonly seen commercially are Betadine® and Isodine®. These are *povidone-iodines*; povidone is a surface-active iodophor that improves the wetting action. The main use of iodines is for skin disinfection and wound treatment.

Chlorine, as a gas or in combination with other chemicals, is another widely used disinfectant. Its ger-

Figure 7-7 A proposed explanation of the mechanism for the activity of iodine. Iodine reacts with tyrosine, an amino acid found in protein. The diiodotyrosine formed inhibits the function of the protein in the cell. Iodine probably also reacts with other cellular constituents.

micidal action is caused by the hypochlorous acid ($HOCl$) that forms when chlorine is added to water.

(1) Cl_2 + H_2O $\rightleftharpoons$
        Chlorine   Water

        H^+ + Cl^- + $HOCl$
      Hydrogen   Chloride   Hypochlorous
         ion   ion   acid

(2) $HOCl$ $\rightleftharpoons$ H^+ + OCl^-
    Hypochlorous   Hydrogen   Hypochlorite
        acid   ion   ion

Exactly how hypochlorous acid exerts its killing power is not completely known. It is a strong oxidizing agent that prevents functioning of much of the cellular enzyme system. Hypochlorous acid is the most effective form of chlorine because it is neutral in electrical charge and diffuses as rapidly as water through the cell wall. Because of its negative charge, the hypochlorite ion (OCl^-) cannot enter the cell freely.

A liquid form of compressed chlorine gas is used extensively for disinfecting municipal drinking water, water in swimming pools, and sewage. Several compounds of chlorine are also effective disinfectants. For example, solutions of *calcium hypochlorite* ($Ca(OCl)_2$) are used to disinfect dairies, barns, slaughterhouses, and restaurant eating utensils. Another chlorine compound, *sodium hypochlorite* ($NaOCl$), is used as a household disinfectant and a bleach (Clorox®) as well as a disinfectant in dairies, food-processing establishments, and hemodialysis systems (Figure 7-5). When the quality of drinking water is in question, household bleach can achieve a rough equivalent of municipal chlorination. After two drops of bleach are added to a liter of water (four drops if the water is cloudy) and the mixture has sat for 30 minutes, the water is considered safe for drinking under emergency conditions.

Another group of chlorine compounds, the *chloramines*, consist of chlorine and ammonia. They are used as disinfectants, antiseptics, or sanitizing agents. Chloramines are very stable compounds that release chlorine over long periods of time. They are relatively effective in organic matter, but they have the disadvantages of acting more slowly and being less effective purifiers than many other chlorine compounds. Chloramines are used to sanitize glassware and eating utensils and to treat dairy and food-manufacturing equipment. Ammonia is usually mixed with chlorine in municipal water treatment systems to form chloramines. The chloramines control taste and odor problems caused by the reaction of chlorine with other nitrogenous compounds in the water. Because chloramines are less effective as germicides, sufficient chlorine must be added to ensure a residual of chlorine in the form of $HOCl$. (Chloramines are toxic to aquarium fish, but pet shops sell chemicals to neutralize them.)

Alcohols

Alcohols effectively kill bacteria and fungi but not endospores and nonenveloped viruses. The mechanism of action of alcohol is usually protein denaturation, but alcohol can also disrupt membranes and dissolve many lipids, including the lipid component of enveloped viruses. Alcohols have the advantage of acting and then evaporating rapidly and leaving no residue. In a quick swabbing of the skin (degerming) before an injection, most of the microbial control activity comes from a simple *wiping away* of dirt and microorganisms. However, alcohols are unsatisfactory when applied to wounds. They cause coagulation of a layer of protein under which bacteria continue to grow.

Two of the most commonly used alcohols are ethanol and isopropanol. The recommended optimum concentration of *ethanol* is 70%, but concentrations between 60% and 95% seem to kill as fast (Table 7-6). Pure alcohol is less effective than aqueous solutions (alcohol mixed with water) because denatura-

TABLE 7-6   *Germicidal Action of Various Concentrations of Ethanol Against* Streptococcus pyogenes*

Percentage of Ethanol	Seconds				
	10	20	30	40	50
100	+	+	+	+	+
95	−	−	−	−	−
90	−	−	−	−	−
80	−	−	−	−	−
70	−	−	−	−	−
60	−	−	−	−	−
50	+	+	−	−	−
40	+	+	+	+	+

* + = growth, microbes not killed; − = no growth.
Source: H. E. Morton, "Relationship of concentration and germicidal efficiency of ethyl alcohol." *Annals of the New York Academy of Sciences* 53:191–196, 1950.

tion requires water. *Isopropanol*, often sold as rubbing alcohol, is slightly superior to ethanol as an antiseptic and disinfectant. Moreover, it is less volatile, less expensive, and more easily obtained than ethanol.

Alcohols are also used to enhance the effectiveness of other chemical agents. For example, an aqueous solution of Zephiran™ (described shortly) kills about 40% of the population of a test organism in two minutes, whereas a tincture of Zephiran™ kills about 85% in the same period of time. You can compare the effectiveness of tinctures and aqueous solutions in Figure 7-10 (see page 187).

Heavy Metals and Their Compounds

Several heavy metals, such as silver, mercury, and copper, can be germicidal or antiseptic. The ability of very small amounts of heavy metals, especially silver and copper, to exert antimicrobial activity is referred to as **oligodynamic action** (*oligo* means few). This action can be seen when we place a coin or other clean piece of metal containing silver or copper on an inoculated Petri plate. Extremely small amounts of metal diffuse from the coin and inhibit the growth of bacteria for some distance around the coin (Figure 7-8). This effect is produced by the action of heavy metal ions on microbes. When the metal ions combine with the —SH groups on cellular proteins, denaturation results.

Silver is used as an antiseptic in a 1% *silver nitrate* solution. The solution is bactericidal for most organisms. Many states require that the eyes of newborns be treated with a few drops of silver nitrate to guard against a gonococcal infection of the eyes called gonorrheal ophthalmia neonatorum, which the infants might have contracted as they passed through the birth canal. In recent years, antibiotics have been replacing silver nitrate for this purpose.

Inorganic mercury compounds, such as *mercuric chloride*, probably have the longest history of use as disinfectants. They have a very broad spectrum of activity; their effect is primarily bacteriostatic. However, their use is now limited because of their toxicity, corrosiveness, and ineffectiveness in organic matter. Organic mercury compounds, such as Mercurochrome and Merthiolate, are less irritating and less toxic than inorganic mercury compounds. Mercurochrome and Merthiolate are antiseptics used on skin and mucous membranes. Unfortunately, their effects are reversed if they are removed (washed away).

Copper in the form of *copper sulfate* is used chiefly to destroy green algae (algicide) that grow in reservoirs, swimming pools, and fish tanks. If the water does not contain excessive organic matter, copper sul-

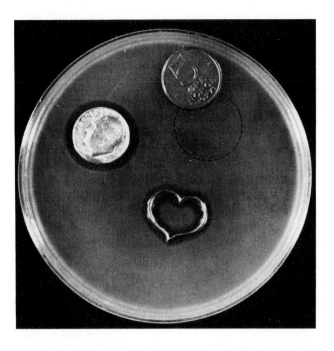

Figure 7-8 Oligodynamic action of silver. Clear zones where bacterial growth has been inhibited are seen around the dime and the heart, both of which contain silver. The nonsilver coin has been pushed aside to show the growth beneath it.

fate is effective in concentrations of one part per million of water. To prevent mildew, copper compounds are sometimes included in paint.

Another metal used as an antimicrobial is zinc. *Zinc chloride* is a common ingredient in mouthwashes, and *zinc oxide* is probably the most widely used antifungal in paints.

Surface-Active Agents

Surface-active agents, or **surfactants**, can decrease surface tension among molecules of a liquid. Such agents include soaps and detergents. Soap has little value as an antiseptic, but it does have an important function in the mechanical removal of microbes through scrubbing. The skin normally contains dead cells, dust, dried sweat, microbes, and oily secretions from oil glands. Soap breaks the oily film into tiny droplets, a process called **emulsification**, and the water and soap together lift up the emulsified oil and debris and float them away as the lather is washed off. In this sense, soaps are good degerming and emulsifying agents. Many so-called deodorant soaps contain compounds such as *triclocarban* that strongly inhibit gram-positive bacteria.

Acid-anionic surface-active sanitizers are very important in the dairy industry's cleaning of utensils and equipment. Their cleansing ability is related to the negatively charged portion (anion) of the molecule. These sanitizers act on a wide spectrum of microbes, including troublesome thermoduric bacteria. They are nontoxic, noncorrosive, and fast acting.

Quaternary Ammonium Compounds

The most widely used surface-active agents are the cationic detergents, especially the **quaternary ammonium compounds (quats)**. Their cleansing ability is related to the positively charged portion—the cation—of the molecule. Their name comes from the fact that they are modifications of the four-valence ammonium ion, NH_4^+ (Figure 7-9). Quaternary ammonium compounds are strongly bactericidal against gram-positive bacteria, and they are somewhat less strong against gram-negative bacteria.

Quats are also fungicidal, amoebicidal, and virucidal against enveloped viruses. They do not kill endospores or tuberculosis bacteria, but they are bacteriostatic. Quats' mode of action is unknown, but they most likely affect the plasma membrane. They change the cell's permeability and cause the loss of essential cytoplasmic constituents, such as potassium.

Two popular quats are Zephiran™, benzalkonium chloride (see Figure 7-5), and Cepacol®, cetylpyridinium chloride. They are strongly antimicrobial, colorless, odorless, tasteless, stable, easily diluted, and nontoxic, except at high concentrations. If your mouthwash bottle fills with foam when shaken, the mouthwash probably contains a quat. However, organic matter interferes with their activity, and they are rapidly neutralized by soaps and anionic detergents.

Anyone associated with medical applications of quats should remember that certain bacteria, such as some species of *Pseudomonas*, not only survive in quaternary ammonium compounds but actively grow in them (see the box on p. 180). This resistance occurs not only to the disinfectant solution, but also to the

moistened gauze and bandages, whose fibers tend to neutralize the quats. This resistance is another example of the peculiar and troublesome characteristics of the pseudomonads. Their ability to metabolize unusual substrates for growth permits *Pseudomonas* to grow in unlikely places, such as in simple saline solutions and even distilled water. They are using as nutrients such things as traces of soap residues and cap-liner adhesives.

Before we move on to the next group of chemical agents, refer to Figure 7-10, which compares the effectiveness of some of the antimicrobials we have discussed so far.

Organic Acids and Derivatives

A number of organic acids are used as preservatives to control mold growth. *Sorbic acid* (often as the salt *potassium sorbate*) is used to inhibit mold growth in acidic foods, such as cheese. *Benzoic acid* (or its salt *sodium benzoate*) is an antifungal that is effective at low pH levels and has wide use in soft drinks and other acidic foods. The *parabens*, such as *methylparaben* and *propylparaben*, are often used to inhibit mold growth in liquid cosmetics and shampoos. The parabens are derivatives of benzoic acid but work at a neutral pH. *Calcium propionate* prevents mold growth in bread.

The activity of these organic acids is not related to their acidity but to their ability to inhibit enzymatic and metabolic activity. Organic acids are listed on the labels of many foods and a few cosmetic preparations. By and large, the body metabolizes these organic acids readily, so their use is considered quite safe.

Aldehydes

Aldehydes are among the most effective antimicrobials. Two examples are formaldehyde and glutaraldehyde. They inactivate proteins by forming covalent crosslinks with a number of organic functional groups on proteins ($-NH_2$, $-OH$, $-COOH$, and $-SH$). *Formaldehyde gas* is an excellent disinfectant. However, it is more commonly used as *formalin*, a 37% aqueous solution of formaldehyde gas. (Methanol added to formalin helps maintain stability of formaldehyde and enhances antimicrobial effectiveness.) Formalin was once used extensively to preserve biological specimens, embalm corpses, and inactivate bacteria and viruses in vaccines. Formalin is sporicidal at high concentrations. However, its sterilizing and disinfecting applications are limited by its tissue-irritating qualities, poor penetration, slow action, unpleasant odor, and its property of leaving a white residue on treated materials.

Figure 7-9 The ammonium ion and a quaternary ammonium compound. Note how other groups replace the hydrogens of the ammonium ion, benzalkonium chloride (Zephiran™).

Figure 7-10 Comparison of the effectiveness of various antiseptics. The steeper the downward slope of the killing curve of the antiseptic, the more effective it is. A 1% iodine in 70% ethanol solution is the most effective; soap and water are the least effective. Note that tinctures are more effective than aqueous solutions of the same antiseptic.

Glutaraldehyde is a chemical relative of formaldehyde that is less irritating and more pleasant to handle than formaldehyde. Glutaraldehyde is used to sterilize hospital instruments, including respiratory-therapy equipment. When used in a 2% solution (Cidex™), it is bactericidal, tuberculocidal, and virucidal in 10 minutes and sporicidal in from 3 to 10 hours. It is probably the only liquid chemical disinfectant that can be considered a possible sterilizing agent. Since such a long exposure time is required for sporicidal activity, chemical agents cannot be relied on as sterilants.

Gaseous Chemosterilizers

Gaseous chemosterilizers are chemicals that sterilize in a closed chamber (similar to an autoclave). A gas suitable for this method is *ethylene oxide*:

$$H_2C — CH_2$$
$$\diagdown \diagup$$
$$O$$

Its activity depends on the denaturation of proteins: The proteins' labile hydrogens, such as —SH, —COOH, or —OH, are replaced by alkyl groups (*alkylation*), such as —CH$_2$CH$_2$OH. Ethylene oxide kills all microbes and endospores. It is toxic and explosive in its pure form, so it is usually mixed with a nonflammable gas, such as carbon dioxide or nitrogen. One of its principal advantages is that it is highly penetrative. Although materials to be sterilized with ethylene oxide must be exposed to it for from 4 to 18 hours, its remarkable penetrating power is one reason why ethylene oxide was chosen to sterilize spacecraft sent to land on the Moon and other planets. Using heat to sterilize the electronic gear on these vehicles was not practical.

Because of their ability to sterilize without heat, gases like ethylene oxide are also widely used on medical supplies and equipment. Examples include disposable sterile plasticware, such as syringes and Petri plates, textiles, sutures, lensed instruments, artificial heart valves, heart–lung machines, and mattresses. Many large hospitals have ethylene oxide chambers, some large enough to sterilize mattresses, as part of their sterilizing equipment. Propylene oxide and beta-propiolactone are also used for gaseous sterilization:

$$H_3C — CH — CH_2 \qquad H_2C — CH_2$$
$$\diagdown \diagup \qquad\qquad |\qquad\quad |$$
$$O \qquad\qquad\quad O —— C = O$$

Propylene oxide   Beta-propiolactone

All these gases may be hazardous by causing mutations and are suspected carcinogens, especially beta-propiolactone. For this reason, there has been concern about exposure of hospital workers to ethylene oxide from such sterilizers; certainly, it is wise to minimize exposures.

Oxidizing Agents

Oxidizing agents exert antimicrobial activity by oxidizing cellular components of the treated microbes. Examples of oxidizing agents are ozone and hydrogen peroxide. *Ozone* (O_3) is a highly reactive form of oxygen that is generated by high-voltage electrical discharges. It is responsible for the air's rather fresh odor after a lightning storm, in the vicinity of electric sparking, or around an ultraviolet light. There is considerable interest in using ozone to replace chlorine in the disinfection of water. Although ozone is a more effective killing agent, its residual activity is difficult to maintain in water, and it is more expensive than chlorine.

Hydrogen peroxide is an antiseptic found in many household medicine cabinets and in hospital supply rooms. It is not a good antiseptic for open wounds because it is quickly broken down to water and gaseous oxygen by the action of the enzyme catalase, which is present in human cells (Chapter 6). Although its usefulness as an antiseptic is limited, hydrogen peroxide is effectively used to disinfect inanimate objects, an application where it is even sporicidal. On a nonliving surface, the normally protective enzymes of aerobic bacteria and facultative anaerobes are overwhelmed by the high concentrations of peroxide used. The food industry is increasing its use of hydrogen

peroxide for aseptic packaging (Chapter 27). The packaging materials pass through a hot solution of the chemical before being assembled into a container. Many wearers of contact lenses are familiar with disinfection by hydrogen peroxide. A platinum catalyst in the lens disinfecting kit destroys residual hydrogen peroxide so it does not persist on the lens, where it might be an irritant.

TABLE 7-7   *Summary of Chemical Agents Used to Control Microbial Growth*

Chemical Agent	Mechanism of Action	Preferred Use	Comment
Phenol and Phenolics			
1. Phenol	Disruption of plasma membrane, denaturation, inactivation of enzymes.	Still used as a standard for the effectiveness of other disinfectants (phenol coefficient).	Seldom used as a disinfectant or antiseptic because of its irritating qualities and disagreeable odor.
2. Phenolics	Disruption of plasma membrane, denaturation, inactivation of enzymes.	Environmental surfaces, instruments, skin surfaces, and mucous membranes.	Derivatives of phenol that are reactive even in the presence of organic material; examples include O-phenylphenol and hexachlorophene.
Chlorhexidine	Disruption of plasma membrane.	Skin degerming, especially for surgical scrubs.	Bactericidal to gram-positives and gram-negatives. Nontoxic, persistent.
Halogens	Iodine inhibits protein function and is a strong oxidizing agent; chlorine forms the strong oxidizing agent hypochlorous acid, which alters cellular components.	Effective antiseptic available as a tincture and an iodophor; chlorine gas is used to disinfect water; chlorine compounds are used to disinfect dairy equipment, eating utensils, household items, and glassware.	Iodine and chlorine may act alone or as components of inorganic and organic compounds.
Alcohols	Denaturation and lipid dissolution.	Thermometers and other instruments; in a quick swabbing of the skin with alcohol before an injection, most of the disinfecting action probably comes from a simple wiping away (cleansing) of dirt and some microbes; usually another antiseptic is used in addition.	Bactericidal and fungicidal, but not effective against endospores, nonenveloped viruses; commonly used alcohols are ethanol and isopropanol.
Heavy Metals and Their Compounds	Denature enzymes and other essential proteins.	Silver nitrate is used to prevent gonococcal eye infections; Mercurochrome and Merthiolate disinfect skin and mucous membranes; copper sulfate is an algicide.	Heavy metals such as silver and mercury are germicidal or antiseptic.

Zinc peroxide and hydrogen peroxide are useful in irrigation of deep wounds, where the oxygen released makes an environment that inhibits the growth of anaerobic bacteria.

Benzoyl peroxide is another useful treatment for wounds infected by anaerobic pathogens, but it is probably much more familiar as the main ingredient in over-the-counter medications for treating acne,

which is caused by an anaerobic bacterium.

A summary of chemical agents that control microbial growth is presented in Table 7-7.

The compounds discussed in this chapter are not generally useful in the treatment of diseases. Because antibiotics are used in chemotherapy, antibiotics and the pathogens against which they are active will be discussed together, in Chapter 19.

TABLE 7-7 *(Continued)*

Chemical Agent	Mechanism of Action	Preferred Use	Comment
Surface-Active Agents			
1. Soaps and acid-anionic detergents	Mechanical removal of microbes through scrubbing.	Skin degerming and emulsification of debris.	Many cosmetic soaps contain antimicrobials such as triclocarban.
2. Acid-anionic detergents	Not certain; may involve enzyme inactivation or disruption.	Sanitizers in dairy and food-processing industry.	Wide spectrum of activity, nontoxic, noncorrosive, fast acting.
3. Cationic detergents (quaternary ammonium compounds)	Enzyme inhibition, protein denaturation, and disruption of plasma membranes.	Antiseptic for skin, instruments, utensils, rubber goods.	Bactericidal, bacteriostatic, fungicidal, and virucidal against enveloped viruses; examples of quats are Zephiran™ and Cepacol®.
Organic Acids	Metabolic inhibitors, mostly affecting molds; action not related to their acidity.	Sorbic acid and benzoic acid effective at low pH; parabens much used in cosmetics, shampoos; calcium propionate used in bread; all are mainly antifungals.	Widely used to control molds and some bacteria in foods and cosmetics.
Gaseous Sterilants	Denaturation	Excellent sterilizing agent, especially for objects that would be damaged by heat.	Ethylene oxide is most commonly used.
Aldehydes	Protein inactivation	Formalin (37% aqueous solution of formaldehyde) preserve specimens and inactivate microbes for vaccines; glutaraldehyde is less irritating than formaldehyde and is used for sterilization of medical equipment (Cidex™) and for embalming corpses.	Very effective antimicrobials; glutaraldehyde is considered a liquid sterilant.
Oxidizing Agents	Oxidation	Contaminated surfaces; some deep wounds, in which they are very effective against oxygen-sensitive anaerobes.	Ozone is gaining attention as a potential replacement for chlorination; hydrogen peroxide is a poor antiseptic but a good disinfectant.

Study Outline

Introduction (pp. 170–172)

1. Ignatz Semmelweis introduced handwashing with chlorinated lime by hospital personnel to reduce infections.
2. Joseph Lister introduced aseptic surgery, including heat sterilization of instruments and disinfection of wounds.
3. Microbial growth can be controlled by physical and chemical methods.
4. Control of microbial growth can prevent infections and food spoilage.
5. Sterilization is the process of destroying all microbial life on an object.
6. Disinfection is the process of reducing or inhibiting microbial growth.

Conditions Influencing Microbial Control (pp. 172–173)

1. A general rule of disinfection is to try to kill the most resistant microbes that can be found on the object to be disinfected.
2. The number and species of microorganisms present will affect the rate of disinfection.

Temperature (p. 172)

1. Biochemical reactions occur more rapidly at warm temperatures.
2. Disinfectant activity is enhanced by warm temperatures.

Types of Microbes (p. 172)

1. Gram-positive bacteria are generally more susceptible to disinfectants than gram-negative bacteria.
2. Pseudomonads can even grow in some disinfectants and antiseptics.

Physiological State of the Microbe (p. 172)

1. An actively growing microorganism tends to be less resistant to chemical agents than an older microorganism.
2. Endospores are resistant to chemical agents and physical methods.

Environment (pp. 172–173)

1. Organic matter and pH level frequently interfere with the actions of chemical control agents.
2. Examples of such organic matter include vomit, feces, pus, and food.

Actions of Microbial Control Agents (p. 173)

Alteration of Membrane Permeability (p. 173)

1. The susceptibility of the plasma membrane is due to its lipid and protein components.
2. Certain chemical control agents damage the plasma membrane by altering its permeability.

Damage to Proteins and Nucleic Acids (p. 173)

1. Some microbial control agents damage cellular proteins by breaking hydrogen and covalent bonds.
2. Other agents interfere with DNA and RNA replication and protein synthesis.

Rate of Microbial Death (p. 173)

1. Bacterial populations subjected to heat or antimicrobial chemicals die at a constant rate.
2. Such a death curve can be plotted logarithmically.

Physical Methods of Microbial Control (pp. 174–179)

Heat (pp. 174–176)

1. Heat is frequently used to eliminate microorganisms; it is economical and easily controlled.
2. The mechanism or action of heat is denaturation of enzymes.
3. Thermal death point (TDP) is the lowest temperature at which all the bacteria in a liquid culture will be killed in 10 minutes.
4. Thermal death time (TDT) is the length of time required to kill all bacteria in a liquid culture at a given temperature.
5. Decimal reduction time (DRT) is the length of time in which 90% of a bacterial population will be killed at a given temperature.
6. Boiling (100°C) kills many vegetative cells and viruses within 10 minutes.
7. Autoclaving (steam under pressure) is the most effective method of moist heat sterilization. The steam must directly contact the material to be sterilized.
8. In pasteurization, a high temperature is used for a short time (72°C for 15 seconds) to destroy pathogens without altering the flavor of the food.
9. Methods of dry heat sterilization include direct flaming, incineration, and hot-air sterilization.
10. Different methods that produce the same effect (reduction in microbial growth) are called equivalent treatments.

Filtration (pp. 176–177)

1. Filtration is the passage of a liquid or gas through a filter with pores small enough to retain microbes.
2. Microbes can be removed from air by high-efficiency particulate air filters.
3. Membrane filters composed of nitrocellulose or cellulose acetate are commonly used to filter out bacteria, viruses, and even large proteins.

Low Temperature (p. 177)

1. The effectiveness of low temperatures depends on the particular microorganism and the intensity of the application.
2. Most microorganisms do not reproduce at ordinary refrigerator temperatures (0° to 7°C).
3. Many microbes survive (but do not grow) at the subzero temperatures used to store foods.

Desiccation (pp. 177–178)

1. In the absence of water, microorganisms cannot grow but can remain viable.
2. Viruses and endospores can resist desiccation.

Osmotic Pressure (p. 178)

1. Microorganisms in high concentrations of salts and sugars undergo plasmolysis.
2. Molds and yeasts are more capable of growing in materials with low moisture or high osmotic pressure than bacteria are.

Radiation (pp. 178–179)

1. The effects of radiation depend on its wavelength, intensity, and duration.
2. Ionizing radiation (gamma rays and high-energy electron beams) has a high degree of penetration and exerts its effect primarily by ionizing water and forming highly reactive hydroxyl radicals.
3. Ultraviolet radiation, a form of nonionizing radiation, has a low degree of penetration and causes cell damage by making DNA thymine dimers that interfere with the DNA replication; the most effective germicidal wavelength is 260 nm.
4. Microwaves can kill microbes indirectly as materials get hot.

Chemical Methods of Microbial Control (pp. 179–189)

1. Chemical agents are used on living tissue (as antiseptics) and on inanimate objects (as disinfectants).
2. Few chemical agents achieve sterility.

Principles of Effective Disinfection (p. 179)

1. Careful attention should be paid to the properties and concentration of the disinfectant to be used.
2. The presence of organic matter, degree of contact with microorganisms, and temperature should also be considered.

Evaluating a Disinfectant (pp. 179–182)

1. The phenol coefficient is the comparison of one chemical's disinfecting action with that of phenol, applied for the same length of time on the same organism under identical conditions.
2. In the use-dilution test, a series of tubes contain increasing concentrations of disinfectant; the more the chemical can be diluted and still be effective, the higher its rating.
3. In the filter paper method, a disk of filter paper is soaked with a chemical and placed on an inoculated agar plate; a clear zone of inhibition indicates effectiveness.

Types of Disinfectants (pp. 182–189)

Phenol and Phenolics (pp. 182–183)
1. Phenolics exert their action by injuring plasma membranes, inactivating enzymes, and denaturing proteins.
2. Common phenolics are cresols and hexachlorophene.

Chlorhexidine (p. 183)
1. Chlorhexidine damages plasma membranes of vegetative cells.

Halogens (pp. 183–184)
1. Some halogens (iodine and chlorine) are used alone or as components of inorganic or organic solutions.
2. To inactivate enzymes and other cellular proteins, iodine combines with the amino acid tyrosine.
3. Iodine is available as a tincture (in solution with alcohol) or as an iodophor (combined with an organic molecule).
4. The germicidal action of chlorine is based on the formation of hypochlorous acid when chlorine is added to water.
5. Chlorine is used as a disinfectant in gaseous form (Cl_2) or in the form of a compound, such as calcium hypochlorite, sodium hypochlorite, and chloramines.

Alcohols (pp. 184–185)
1. Alcohols exert their action by denaturing proteins and dissolving lipids.
2. In tinctures, they enhance the effectiveness of other antimicrobial chemicals.
3. Aqueous ethanol (60% to 95%) and isopropanol are used as disinfectants.

Heavy Metals and Their Compounds (p. 185)
1. Silver, mercury, copper, and zinc are used as germicidals or antiseptics.
2. They exert their antimicrobial action through oligodynamic action. When heavy metal ions combine with —SH groups, proteins are denatured.

Surface-Active Agents (pp. 185–186)
1. Surface-active agents decrease the tension between molecules that lie on the surface of a liquid; soaps and detergents are examples.

2. Soaps have limited germicidal action but assist in the removal of microorganisms through scrubbing.
3. Acid-anionic detergents are used to clean dairy equipment.

Quaternary Ammonium Compounds (p. 186)
1. Quats are cationic detergents attached to NH_4^+.
2. By disrupting plasma membranes, they allow cytoplasmic constituents to leak out of the cell.
3. They are most effective against gram-positive bacteria.

Organic Acids and Derivatives (pp. 186–187)
1. Sorbic acid, benzoic acid, and propionic acid inhibit fungal metabolism. They are used as food preservatives.

Aldehydes (p. 187)
1. Aldehydes such as formaldehyde and glutaraldehyde exert their antimicrobial effect by inactivating proteins.
2. They are among the most effective chemical disinfectants.

Gaseous Chemosterilizers (p. 187)
1. Ethylene oxide is the gas most frequently used for sterilization.
2. It penetrates most materials and kills all microorganisms by protein denaturation.

Oxidizing Agents (pp. 187–189)
1. Ozone and peroxide are used as antimicrobial agents.
2. They exert their effect by oxidizing molecules inside cells.

Study Questions

Review

1. Name the cause of cell death resulting from damage to the following:
(a) Cell wall
(b) Plasma membrane
(c) Proteins
(d) Nucleic acids

2. A bacterial culture was in log phase. At time x, a bactericidal compound was added to the culture medium. Explain why the viable count does not immediately drop to zero.

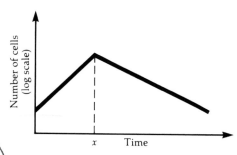

3. Fill in the following table.

Method of Sterilization	Temp.	Time	Type of Heat	Pref'd Use	Type of Action
Autoclaving					
Hot air					
Pasteurization					

4. How do the examples in Question 3 illustrate the concept of equivalent treatments?

5. The thermal death time for a suspension of *B. subtilis* endospores in dry heat is 30 minutes, and, in an autoclave, less than 10 minutes. Which type of heat is more effective? Why?

6. If pasteurization does not achieve sterilization, why is food treated by pasteurization?

7. Heat-labile solutions such as glucose-minimal salts broth can be sterilized by _____ .

8. Thermal death point is not considered an accurate measure of the effectiveness of heat sterilization. List three factors that can alter the thermal death point.

9. The antimicrobial effect of gamma radiation is due to _____ . The antimicrobial effect of ultraviolet radiation is due to _____ .

10. How do salts and sugars preserve foods? Why are these considered physical rather than chemical methods of control? Name one food that is preserved with sugar, and one preserved with salt. How do you account for the occasional growth of *Penicillium* mold in jelly, which is 50% sucrose?

11. List five factors to consider before selecting a disinfectant.

12. Give the method of action and at least one standard use of each of the following types of disinfectants:
(a) Phenolics   (b) Iodine
(c) Chlorine   (d) Alcohol
(e) Heavy metals   (f) Aldehydes
(g) Ethylene oxide   (h) Oxidizing agents

13. The phenol coefficients against *S. aureus* for ethanol and isopropanol are 0.039 and 0.054, respectively. Which is the more effective antimicrobial agent? What is a phenol coefficient?

14. The use-dilution values for two disinfectants tested under the same conditions are, Disinfectant A—1:2; Disinfectant B—1:10,000. If both disinfectants are designed for the same purpose, which would you select?

15. A large hospital washes burned patients in a stainless steel tub. After each patient, the tub is cleaned with a quat. It was noticed that 14 out of 20 burn patients acquired *Pseudomonas* infections after being bathed. Provide an explanation for this high rate of infection.

Challenge

1. The filter paper method was used to evaluate three disinfectants. The results were as follows:

Disinfectant	Zone of Inhibition
X	0 mm
Y	5 mm
Z	10 mm

(a) Which disinfectant was the most effective against the organism?

(b) Can you determine whether compound Y was bactericidal or bacteriostatic?

2. A bacterial culture was in log phase. At time *x*, a bacteriostatic compound was added to the culture. Explain why this graph of the number of viable organisms does not look like the graph in Review Question 2.

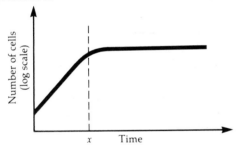

3. *Entamoeba histolytica* and *Giardia lamblia* were isolated from the stool sample of a 45-year-old teacher, and *Shigella sonnei* was isolated from the stool sample of an 18-year-old woman. Both patients experienced diarrhea and severe abdominal cramps, and, prior to onset of digestive symptoms, both had been treated by the same chiropractor.

The chiropractor had administered colonic irrigations (enemas) to these patients. The device used for this treatment was a gravity-dependent apparatus using 12 liters of tap water. There were no check valves to prevent backflow, so all parts of the apparatus could have become contaminated with feces during each colonic treatment. The chiropractor provided colonic treatment to four or five patients per day. Between patients, the adaptor piece that is inserted into the rectum was placed in a "hot water sterilizer."

What two errors did the chiropractor routinely make?

4. Between March 9 and April 12, five chronic peritoneal dialysis patients at one hospital became infected with *Pseudomonas aeruginosa*. Four patients developed peritonitis (inflammation of the abdominal cavity), and one developed a skin infection at the catheter insertion site. All patients with peritonitis had low-grade fever, cloudy peritoneal fluid, and abdominal pain. All patients had permanent indwelling peritoneal catheters, which were wiped with gauze soaked with an iodophor (Prepodyne®) solution each time the catheter was connected to or disconnected from the machine tubing. Aliquots of the iodophor were transferred from stock bottles to small in-use bottles.

Cultures from the dialysate concentrate and the internal areas of the dialysis machines were negative; iodophor from a small in-use plastic container yielded a pure culture of *P. aeruginosa*.

What improper technique led to this infection?

Further Reading

American Society for Microbiology, Board of Education and Training. *Infection Control in the Hospital Environment*. Washington, D.C.: American Society for Microbiology (Proceedings), 1976. Lectures and articles on disinfection and environmental monitoring in the hospital.

Block, S. S., ed. *Disinfection, Sterilization and Preservation*, 3rd ed. Philadelphia: Lea and Febiger, 1983. A comprehensive reference on principles and procedures for disinfection and sterilization.

Borick, P. M., ed. *Chemical Sterilization*. New York: Van Nostrand Reinhold, 1973. A reference on applications and theory of common chemical disinfectants and sterilizers.

Brewere, J. H., ed. *Lectures on Sterilization*. Durham, N.C.: Duke University Press, 1973. Lectures addressed to hospital personnel who use supplies and are responsible for maintaining sterility and asepsis.

Castle, M. *Hospital Infection Control*. New York: Wiley, 1980. An excellent reference for practical surveillance and maintenance of asepsis.

Perkins, J. J. *Principles and Methods of Sterilization in Health Sciences*, 2nd ed. Springfield, Ill.: Thomas, 1983. A comprehensive reference on theories and applications of sterilization methods.

Phillips, G. B., and W. S. Miller, eds. *Industrial Sterilization*. Durham, N.C.: Duke University Press, 1973. Methods and applications for the use of gas, formaldehyde, and radiation sterilization.

Russell, A. *The Destruction of Bacterial Spores*. New York: Academic Press, 1982. A discussion of physical and chemical control of endospores in industry and hospitals.

CHAPTER 8

Microbial Genetics

All the characteristics of microorganisms are inherited. These characteristics include shape and structural features (morphology), the biochemical reactions the microbes carry out (metabolism), their ability to move or behave in other ways, and their ability to interact with other organisms—perhaps causing disease. Individual organisms transmit these characteristics to their offspring through genes, which are the units of the hereditary material (deoxyribonucleic acid, DNA) that contain the information that determines these characteristics.

Chromosomes are the cellular structures containing genes that physically carry hereditary information. **Genetics**, the science of heredity, is the study of what genes are, how they carry information, how they are replicated and passed to future generations of cells or passed between organisms, and how the expression of their information within an organism determines the particular characteristics of that organism.

Learning Objectives

- Define genetics, chromosome, gene, genetic code, genotype, phenotype, mutagen, genetic recombination.

- Describe how DNA serves as genetic information.

- Describe DNA replication.

- Describe protein synthesis.

- Explain the regulation of gene expression in bacteria by induction, repression, attenuation, and catabolite repression.

- Classify mutations by type.

- Outline methods for direct and indirect selection of mutants.

- Compare mechanisms of genetic transfer in bacteria.

- Define plasmid and transposon and discuss their functions.

- Outline how recombinant DNA is produced in vitro and list its potential uses.

Structure and Function of the Genetic Material

Genes consist of DNA. We saw in Chapter 2 that DNA is a macromolecule composed of repeating units called *nucleotides*. You may recall that each nucleotide consists of a nitrogenous base—adenine (A), thymine (T), cytosine (C), or guanine (G)—deoxyribose (a pentose sugar), and a phosphate group. Review the general arrangement of the nucleotides in a DNA molecule in Figure 2-19. The DNA within a cell exists as long strands of nucleotides twisted together in pairs to form a double helix. Each strand has a string of alternating sugar and phosphate groups, its *sugar-phosphate backbone*, and to each sugar is attached a nitrogenous base. The two strands are held together by hydrogen bonds between their nitrogenous bases. The bases are paired in a specific way. Adenine always pairs with thymine, and cytosine always pairs with guanine. Because of the specific base pairing, if the base sequence of one DNA strand is known, then the sequence of the other strand is also known. The two strands of DNA are thus *complementary*. A good analogy to complementary sequences in DNA strands is the relationship between a positive photograph and its negative. The complementary structure of DNA helps explain how DNA stores and transmits genetic information, as we shall see later.

The genetic language uses an alphabet with only four letters—the bases A, T, G, and C. But 1000 of these four bases (the number contained in an average gene) can be arranged in 4^{1000} different ways, which explains how genes can be varied enough to provide all the information a cell needs to grow and perform its functions. The genetic code, which determines how a nucleotide sequence is converted into the amino acid sequence of a protein, is discussed in more detail later in this chapter.

The numerous hydrogen bonds along even a short stretch of DNA provide significant bonding energy to make the double helix a stable structure. In spite of this stability, the two strands of DNA do separate under the influence of special unwinding proteins. This allows the two separated strands of DNA to serve as templates for replication, which leads to the production of two daughter DNA molecules.

The specific sequence of nucleotides contained in the DNA is usually duplicated accurately by enzymes each time the cell divides. This accurate duplication allows DNA to carry genetic information from cell to cell and from generation to generation. The information in the genetic code determines the characteristics of the cell and transfers these characteristics to subsequent generations of cells. Later in this chapter, we will describe several important experiments that clearly identified DNA as the genetic material of cells.

A **gene** can be defined as a segment of DNA (a sequence of nucleotides in DNA) that codes for a functional product. The final product can be a molecule of ribosomal RNA (rRNA) or transfer RNA (tRNA). Usually, however, the final product is a protein. Genes that code for proteins are called **structural genes**, and each one has a unique base sequence that codes for a unique protein. The DNA of a cell therefore specifies the complete collection of proteins to be found in the cell.

Proteins are so important that much of a cell's machinery is concerned with translating the DNA of

structural genes into specific proteins. The DNA sequence is *transcribed* (copied) to produce a specific molecule of RNA called messenger RNA (mRNA). The information encoded in the mRNA is *translated* into a specific sequence of amino acids that forms a protein.

When the ultimate molecule that a gene codes for (a protein, rRNA, or tRNA) has been produced, we say that the gene has been *expressed*. Thus, gene expression involves the transcription of DNA to produce specific molecules of RNA. If the RNA is mRNA, the information encoded in it is then translated into protein. This process can be symbolized as follows:

$$DNA \xrightarrow{\text{Transcription}} mRNA \xrightarrow{\text{Translation}} Protein$$

DNA can *change*, or *mutate*. Although some genetic mutations kill the cell or cause harm, others create new characteristics that give the offspring inheriting the mutation a better chance of survival in new environments. Thus, DNA's ability to mutate is in the long run an advantage that contributes to the successful evolution of species. We will discuss various types of mutation later in this chapter.

Genotype and Phenotype

The **genotype** of an organism is its genetic makeup, the information that codes for all the particular characteristics of the organism. The genotype represents the *potential* properties but not the properties themselves. **Phenotype** refers to the *actual*, expressed properties, such as the organism's ability to perform a particular chemical reaction. The phenotype, then, is the manifestation of the genotype.

In molecular terms, an organism's genotype is its collection of genes, its entire DNA. What constitutes the organism's phenotype in molecular terms? Put simply, an organism's phenotype is its collection of proteins. Most of a cell's properties derive from the structures and functions of its proteins. In microbes, most proteins are either enzymatic (catalyzing particular reactions) or structural (participating in large functional complexes such as membranes or ribosomes). Even those phenotypes that depend on structural macromolecules other than protein (such as lipids or polysaccharides) rely indirectly on proteins. For instance, the structure of a complex lipid or a polysaccharide molecule results from the catalytic activities of enzymes that synthesize, process, and degrade them. Thus, although it is not completely accurate to say that phenotypes are due only to proteins, it is a useful simplification.

DNA and Chromosomes

Although the structure of DNA is fairly well understood, the packaging of DNA in chromosomes is not. Evidence available so far suggests that the DNA in each chromosome—even in the large, complex chromosomes of eucaryotes—is one long double helix to which is attached various proteins that function in DNA replication and expression.

The DNA of *E. coli*, the most studied bacterium, is contained in a typical procaryotic chromosome, a single circular molecule of DNA (Figure 8-1a). The chromosome is attached at one or several points to the plasma membrane. The DNA of *E. coli* has about 4 million base pairs and is about 1 mm long—1000 times longer than the entire cell. However, DNA is very thin and is tightly packed inside the cell, so that this twisted, coiled macromolecule takes up only about 10% of the cell's volume. Some proteins are normally attached to bacterial DNA.

Eucaryotic chromosomes (Figure 8-1b) contain DNA that is even more highly coiled (condensed) than procaryotic DNA. Eucaryotic chromosomes contain much more protein than procaryotic chromosomes do. In eucaryotic cells, a group of proteins known as histones form complexes around which DNA is wound. The structure and function of a eucaryotic chromosome are also influenced by diverse nonhistone chromosomal proteins, which help determine many tissue-specific and species-specific phenotypes. The detailed structure of the eucaryotic chromosome, and the precise arrangement of DNA with proteins, is still under investigation. Researchers believe that an understanding of the detailed structure is likely to reveal how the cell turns genes on and off to produce crucial proteins when needed. This regulation of gene expression governs the differentiation of eucaryotic cells into the different types of cells found in multicellular organisms as well as an individual cell's ongoing activities.

We will next examine more closely how DNA functions as the genetic material in replication and protein synthesis.

DNA Replication

Replication of DNA results in the synthesis of a duplicate double-stranded DNA molecule. Thus, from one parental DNA molecule, two daughter DNA mole-

(a)

(b)

Figure 8-1   Chromosomes. (a) Electron micrograph of a procaryotic chromosome. The tangled mass and looping strands of DNA emerging from this disrupted cell of *E. coli* are part of its single chromosome (×6000). **(b)** Electron micrograph of a eucaryotic chromosome. This human chromosome is one of 46 chromosomes found in a normal human cell. The individual fibers are strands of DNA with attached proteins (×28,000).

cules are produced. The complementary structure of the nitrogenous base sequences in the DNA molecule provides the key to understanding DNA replication. Because the bases that comprise the two strands of double-helical DNA are complementary, one strand can act as a template for the production of the other strand.

When DNA replicates, the two strands of parental DNA unwind and separate from each other in one small DNA segment after another. Free nucleotides present in the cytoplasm of the cell match up to the exposed bases of the single-stranded parental DNA. Where thymine is present on the original strand, only adenine can fit into place on the new strand; where guanine is present on the original strand, only cytosine can fit into place, and so on. Once aligned, the newly added nucleotide is joined to the growing DNA strand by an enzyme called DNA polymerase. Then the parental DNA unwinds a bit further to allow the addition of the next nucleotides. The point at which replication is occurring is called the **replication fork** (Figure 8-2).

Figure 8-2 DNA replication. The double helix of the parent DNA separates as weak hydrogen bonds between the nucleotides on opposite strands break due to the action of specific replication enzymes. Next, hydrogen bonds join new complementary nucleotides with each strand of the parental template to form new base pairs. Enzymes catalyze the formation of sugar-phosphate bonds between sequential nucleotides on each resulting daughter strand.

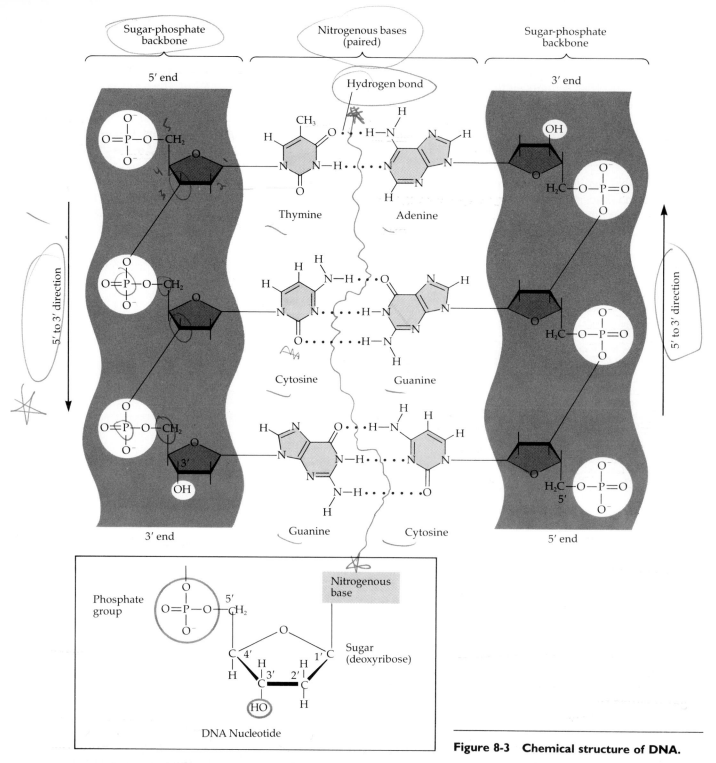

Figure 8-3 **Chemical structure of DNA.**

As the replication fork moves along the parental DNA, each of the unwound single strands combines with new nucleotides. The original strand and this newly synthesized daughter strand then rewind. Because each new double-stranded DNA molecule contains one original strand (conserved) and one new strand, the process of replication is referred to as **semiconservative replication**.

Before discussing DNA replication in more detail, let us take a closer look at the structure of DNA (Figure 8-3). Note in the inset of Figure 8-3 that the carbon atoms of the sugar component of each nucleotide are numbered 1' to 5'. Although the two strands of DNA are complementary, their backbones have different chemical senses of direction in terms of the orientation of the sugar groups. Each phosphate group in the sugar-phosphate backbone of DNA attaches the 5' carbon of one sugar to the 3' carbon of the next sugar. (The nitrogenous base attaches to the 1' carbon.) If the sugar-phosphate backbone is broken by the action of an enzyme or the DNA is a linear molecule, a hydroxyl group is found at one end of each DNA strand. The end of the DNA strand that has the

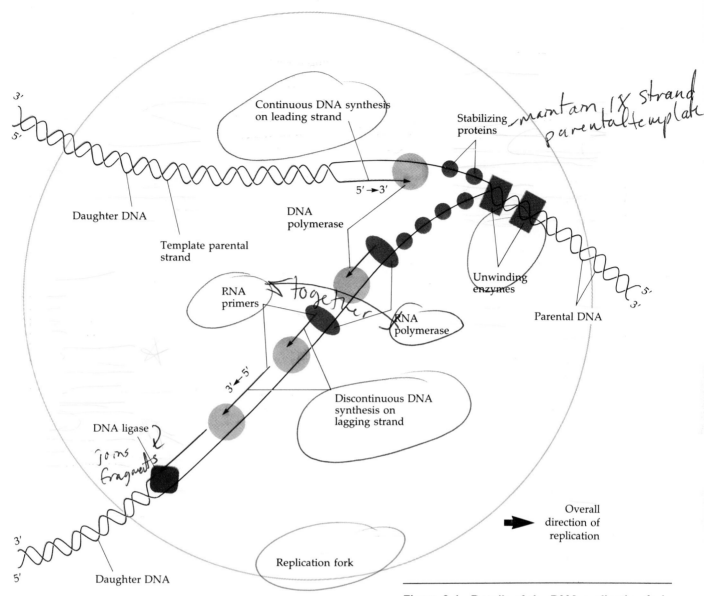

Figure 8-4 Details of the DNA replication fork.

Handwritten annotations: maintain IX strand parental template · together · joins fragments

hydroxyl attached to the 3′ carbon of the sugar is called the 3′ end of the DNA strand; the end having a phosphate group attached to the 5′ carbon of the sugar is called the 5′ end. Because of the way in which the two strands of DNA fit together, if one strand has the chemical sense of direction 5′→3′, the other must be reversed, 3′→5′. The two strands of the DNA double helix are said to be *antiparallel.*

The antiparallel structure of DNA affects the replication process. DNA polymerases can join new nucleotides to the parental strand only by adding them to the 3′ end. That is, a DNA strand grows only in the 5′→3′ direction. Therefore, as the site of the replication fork moves along the parental DNA, the two new DNA strands must grow in slightly different fashions. One new DNA strand, called the *leading strand,* is synthesized continuously in the 5′→3′ direction (from a template parental strand running 3′→5′). In contrast, the *lagging strand* of new DNA is synthesized *discontinuously* in fragments of about 1000

nucleotides, which must later be joined to make one continuous strand (Figure 8-4).

The process of DNA synthesis requires many proteins and enzymes, including several types of DNA polymerase molecules. At the site of the replication fork, the parental double helix is unwound by enzymes. Stabilizing proteins help maintain the single-stranded parental template DNA. DNA polymerase cannot initiate synthesis of a DNA strand entirely on its own, but it can attach additional nucleotides to the free 3′ end of an existing DNA or RNA strand to elongate the strand. While the leading strand of DNA is being synthesized by DNA polymerase, an RNA polymerase starts the synthesis of each DNA fragment of the lagging strand with a stretch of about 20 RNA nucleotides called an *RNA primer.* DNA synthesis catalyzed by DNA polymerase can then proceed from the RNA primers. The RNA is later digested away by the 5′→3′ *exonuclease* activity that is part

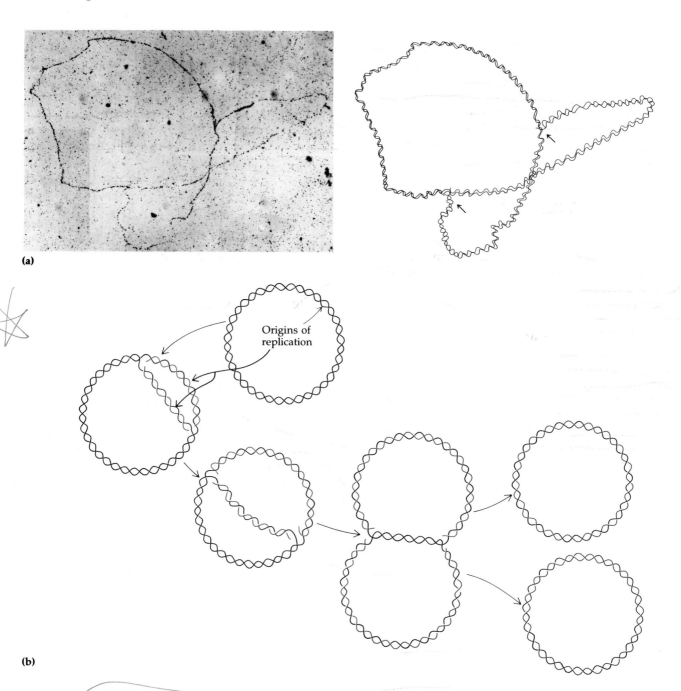

(a)

(b)

Origins of
replication

Figure 8-5   Replication of bacterial DNA. (a) Autoradiograph of the replication of an *E. coli* chromosome (corresponding diagram at right) (×170). The arrows point to the two replication forks. The chromosome is about one-third replicated. Note that one of the new helices is crossed over the other one. **(b)** Diagrammatic representation of the bidirectional replication of a circular DNA molecule. The new strand is shown in color.

of the function of DNA polymerase. Another enzyme, DNA ligase, joins the discontinuous fragments together with covalent bonds to make a continuous lagging strand. Each newly synthesized DNA strand forms a double helix with its parental strand; the two daughter DNA double helices are indistinguishable from each other and from the original parental DNA.

In bacteria, the replication process begins at a unique site on the bacterial chromosome. It is known that DNA replication of some bacteria, such as *E. coli*, goes *bidirectionally* around the chromosome (Figure 8-5). Two replication forks move in opposite directions away from the origin of replication. Because the bacterial chromosome is a closed loop, the replication forks eventually meet when the replication of the chromosome is completed. Much evidence exists to show an association between the bacterial cell membrane and each replication fork, with its unwinding proteins, stabilizing proteins, DNA polymerases, and other enzymes. The DNA–membrane attachment in procaryotes might serve to ensure that in cell division each daughter cell receives one copy of the DNA molecule, that is, one complete chromosome.

The Rate of DNA Replication

DNA synthesis is a surprisingly fast process, about 1000 nucleotides per second in *E. coli* growing at 37°C. At first glance, this speed seems improbable, considering that nucleotide substrates must be synthesized and then must diffuse to the replication fork. Furthermore, several attempts are probably made by wrong nucleotides to pair at each position before the correct bases pair up. Nevertheless, the speed and specificity of DNA replication are governed by chemical principles only.

Under some conditions, namely log phase growth in a rich nutrient medium (see Figure 6-12), *E. coli* can grow faster than the two replication forks can complete the circular chromosome. Under these conditions, the cell initiates *multiple* replication forks at the origin on the chromosome; a new pair of forks

begins before the last pair has finished. In this way, the overall rate of DNA synthesis matches the rate at which the cell divides. Similarly, when the cell's growth greatly slows, the initiation of DNA synthesis at the origin of replication may be delayed. The rate at which DNA nucleotides are polymerized is generally constant (at a stable temperature), but the cell regulates its overall rate of DNA replication to match its rate of growth and cell division. The cell does not regulate the rate of movement of each replication fork but instead regulates the rate at which replication forks on the chromosome are initiated. By regulating the initiation of DNA synthesis, the cell conserves energy because no energy is wasted on the synthesis of incomplete products that have no function.

The Flow of Genetic Information

DNA replication makes possible the flow of genetic information from one generation to the next. As shown in Figure 8-6, the DNA of a cell replicates before cell division, so that each daughter cell receives a chromosome identical to the parent's. Within each metabolizing cell, the genetic information contained in DNA also flows in another way: It is transcribed into RNA and then translated into protein. We describe the processes of transcription and translation in the next section.

Parent cell

DNA replication and cell division

Metabolizing daughter cells

Figure 8-6 Flow of genetic information. Genetic information can be transferred between generations of cells and within a cell. The cell represented in this diagram is a bacterium, which has a single, circular chromosome.

RNA and Protein Synthesis

How is the information in DNA translated into the proteins that control cell activities? In a process called transcription, genetic information in DNA is copied, or transcribed, into a complementary base sequence of RNA. The cell then uses the information encoded in this RNA message to synthesize specific proteins through the process of translation. We will now take a closer look at these two processes as they occur in a bacterial cell.

Transcription

Transcription can be defined as the synthesis of a complementary strand of RNA from a DNA template. As we discussed earlier, there are three kinds of RNA in microbial cells—messenger RNA (mRNA), ribosomal RNA (rRNA), and transfer RNA (tRNA). Ribosomal RNA forms an integral part of ribosomes, the cellular machinery for protein synthesis. Transfer RNA is also involved in protein synthesis, as we will see later. Messenger RNA codes for proteins. We will concentrate our discussion of transcription on

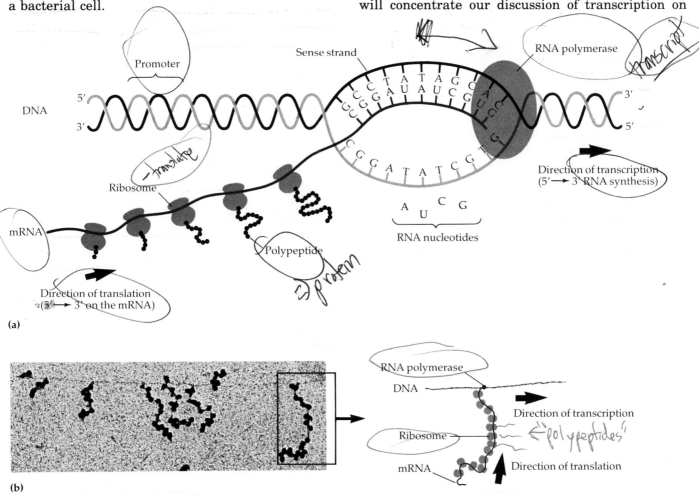

Figure 8-7 Transcription of mRNA in bacteria. (a) Diagram of a partly uncoiled DNA helix acting as a template for RNA synthesis. A molecule of RNA polymerase binds to a promoter, a nucleotide sequence that signals the start of a gene. Transcription takes place on the temporarily unwound region of DNA, where specific basepairing between the sense strand of DNA and RNA nucleotides determines the correct sequence of mRNA. The free end of completed mRNA has begun to be translated by ribosomes. In procaryotes, transcription and translation can be coupled in this way. **(b)** Electron micrograph of transcription from a single gene (×69,000). Many molecules of mRNA are being synthesized simultaneously, starting at the promoter site. The longest mRNA molecules were the first to begin at the promoter. Note the many RNA polymerase molecules along the DNA, and the ribosomes attached to the newly forming mRNA. The newly synthesized polypeptides are not visible here.

mRNA and then describe its translation into protein products.

A strand of mRNA is synthesized during transcription, using a specific gene, a portion of the cell's DNA, as a template (Figure 8-7). In other words, the genetic information stored in the sequence of nitrogenous bases of DNA is rewritten so that the same information appears in the base sequence of mRNA. As in DNA replication, a G in the DNA template dictates a C in the mRNA being made; a C in the DNA template dictates a G in the mRNA; and a T in the DNA template dictates an A in the mRNA. However, an A in the DNA template dictates a uracil (U) in the mRNA because RNA contains U instead of T. (U has a chemical structure slightly different from T, but it base-pairs in the same way.) If, for example, the template portion of DNA has the base sequence ATGCAT, then the newly synthesized mRNA strand will have the complementary base sequence UACGUA.

The process of transcription requires an enzyme called RNA polymerase and a supply of RNA nucleotides. Only one of the two DNA strands serves as the template for RNA synthesis. This is called the *sense strand*. (The other strand, the *antisense strand*, is the complement of the sense strand.) The region where RNA polymerase initially binds to DNA is known as the **promoter site**; this is also where transcription begins. Like DNA, RNA is synthesized in the $5' \rightarrow 3'$ direction. The region of DNA that signals the end point for transcription is the **terminator site**; at this site, the RNA polymerase and newly formed single-stranded mRNA are released from the DNA. The DNA double helix then reforms. The base sequence that specifies the promoter site on DNA is not translated, nor is the terminator.

In summary, the genetic information for protein synthesis is stored in DNA and passed to mRNA during transcription.

Translation

The process in which the nitrogenous-base sequence of mRNA dictates the amino acid sequence of a protein is called **translation**. In bacteria, which lack a membrane-enclosed nucleus, translation of an mRNA molecule can begin even before transcription is completed (see Figure 8-7). The events involved in translation are described in this section and shown in Figure 8-8, p. 204, and Figure 8-9, p. 205.

First, the 5' end of the mRNA molecule becomes associated with a ribosome, the site of protein synthesis (Figure 8-8a). Ribosomes consist of a special type of RNA called **ribosomal RNA (rRNA)** and protein. Each ribosome consists of two subunits. At the start of translation, the two subunits come together with the mRNA.

In solution in the cytoplasm are 20 different amino acids that may participate in protein synthesis (see Table 2-5). These amino acids can be synthesized by the cell or taken up from the external medium. However, before the appropriate amino acids can be joined together to form a protein, they must be *activated* by attaching to **transfer RNA (tRNA)** (Figure 8-9a). For each different amino acid, there is a different type of tRNA. During amino acid activation, a specific amino acid attaches to its specific tRNA. This attachment is accomplished with an *amino acid activating enzyme* and energy from ATP (Figure 8-9b).

Now we are ready to see how the mRNA actually determines the order in which amino acids are linked together to form a protein. Each set of three nucleotides of mRNA, called a **codon**, specifies a single amino acid. For example, the sequence

AUGCCAGGCAAA

contains four codons, specifying the amino acids methionine (AUG) – proline (CCA) – glycine (GGC) – lysine (AAA).

The tRNA molecules "read" the coded message on the mRNA. On one special part of each tRNA molecule is a set of three nucleotides, called an **anticodon**, that is complementary to the codon for the amino acid carried by the tRNA (Figure 8-9c). During translation, the anticodon of a molecule of tRNA attaches to its complementary codon on mRNA. For example, a tRNA with anticodon AGC pairs with the mRNA codon UCG (Figure 8-9c). The pairing of codon and anticodon occurs only where mRNA is attached to a ribosome.

After the first tRNA, with its amino acid, attaches to mRNA at the *start codon* (Figure 8-8b), the ribosome moves along the mRNA strand, and the second tRNA molecule, with its amino acid, is moved into position on the second codon. The two amino acids are joined by a peptide bond, and the first tRNA molecule detaches from the mRNA strand (Figure 8-8c). (The sites on the ribosome where the tRNA molecules base-pair with the mRNA are called the *A site* and the *P site*.) The detached tRNA can now pick up another molecule of its specific amino acid. Each tRNA can be used over and over again. As the proper amino acids are brought into line one by one, peptide bonds form between the amino acids, and a polypeptide chain is formed (Figure 8-8d).

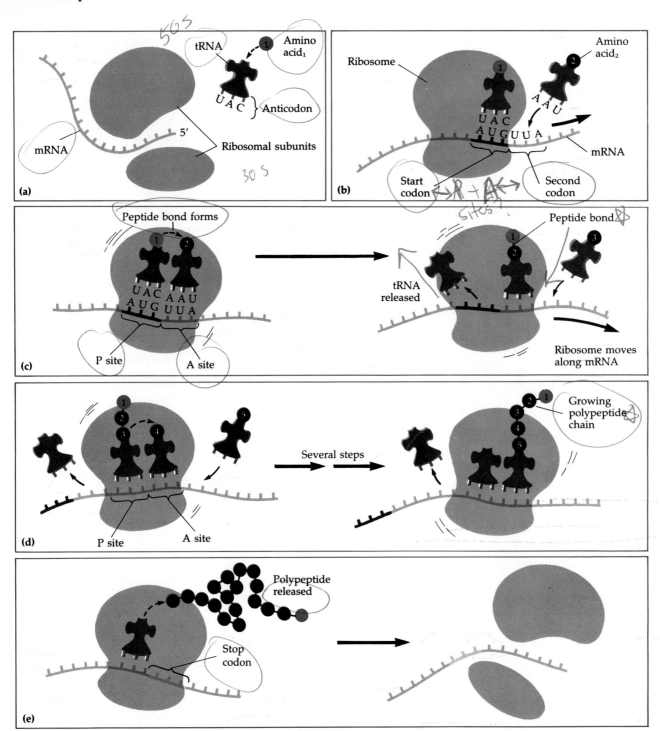

Figure 8-8   Translation. (a) Components needed to begin translation. **(b)** A tRNA carrying the first amino acid pairs with the start codon on the mRNA; this takes place at the A site on the ribosome. **(c)** The ribosome then moves along mRNA until the first tRNA is in the P site on the ribosome. In the A site, the second codon of the mRNA pairs with the tRNA carrying the second amino acid. The first amino acid is joined to the second by a peptide bond, and the first tRNA is released. **(d)** The ribosome moves along the mRNA until the next tRNA is in the P site, and the process continues. (Nucleotide bases are shown only for the first two codons.) The chain of amino acids (polypeptide) grows. **(e)** When the ribosome reaches the stop codon, the polypeptide is released, the last tRNA is released, and the ribosome comes apart.

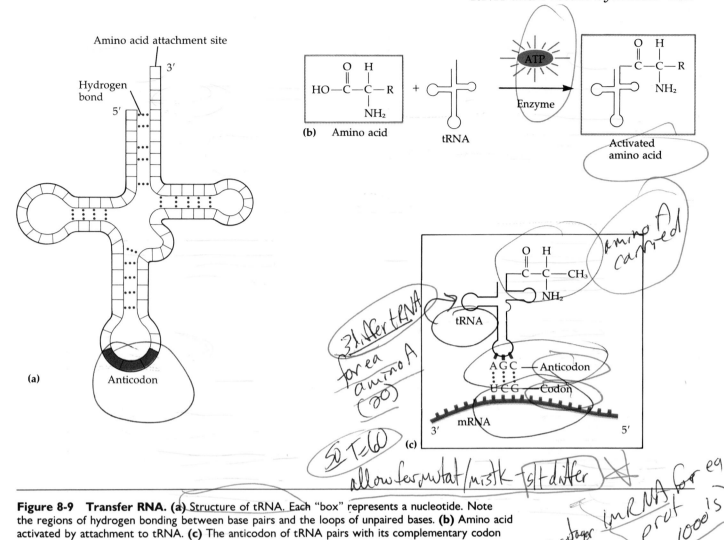

Figure 8-9 Transfer RNA. (a) Structure of tRNA. Each "box" represents a nucleotide. Note the regions of hydrogen bonding between base pairs and the loops of unpaired bases. **(b)** Amino acid activated by attachment to tRNA. **(c)** The anticodon of tRNA pairs with its complementary codon on an mRNA strand. The tRNA shown here carries the amino acid alanine. Codons are always read in the 5′ → 3′ direction; hence the codon for alanine is read G–C–U.

A special *stop codon* in the mRNA (called a *nonsense codon* because it does not specify any amino acid) signals the end of a polypeptide chain and the chain's release from the ribosome (Figure 8-8e). The ribosome then comes apart into its two subunits.

Before the ribosome moving along the mRNA completes translation of that gene, another ribosome, and then another, can attach and begin translation of the same mRNA molecule; there can be a number of ribosomes attached at different positions to the same mRNA molecule (see Figure 8-7b). In this way, a single mRNA strand can be translated simultaneously into several identical protein molecules. The combination of an mRNA strand with several attached ribosomes is called a **polyribosome**.

In summary, the synthesis of a protein requires that the genetic information in DNA first be transferred to a molecule of mRNA by a process called transcription. Then, in a process called translation, the mRNA directs the assembly of amino acids into a polypeptide. The mRNA attaches to a ribosome, and tRNA delivers the amino acids to mRNA according to the complementary base pairings of codons and anticodons. The amino acids are then linked to form polypeptides. Proteins can be composed of one or more polypeptide chains, each of which may range in length from 50 to several hundred amino acids.

Because a typical polypeptide has about 300 amino acids, the typical mRNA must have about 300

TABLE 8-1   *Major Components of Protein Synthesis*

	Transcription	Translation
Messenger RNA (mRNA)	Synthesized as copy of information in DNA	Acts as coded message specifying amino acid sequence of proteins
Monomer subunits	RNA nucleotides (4)	Amino acids (20)
Proteins		
Enzymes	RNA polymerase	Amino acid activating enzymes and peptidyl transferase (responsible for peptide bonds)
Structural proteins	None	Ribosomal proteins
Protein factors	Initiation and termination factors	Initiation, elongation, and termination factors
Other RNAs		
Transfer RNA (tRNA) Ribosomal RNA (rRNA)	tRNA and rRNA synthesized from DNA template	tRNAs with amino acids attached rRNAs, structural components of ribosomes
Chemical energy	From RNA nucleotide triphosphates (ATP, GTP, CTP, UTP)	From ATP and GTP

codons and thus be at least 900 nucleotides long. The DNA of *E. coli* is long enough to contain about 4000 genes and can therefore specify about 4000 different kinds of proteins. There are only a few copies of some proteins, but there are several thousand copies of other proteins. Table 8-1 shows the components of protein synthesis involved in transcription and translation.

The Genetic Code

The **genetic code** refers to the relationship between the nitrogenous base sequence of DNA, the corresponding codons of mRNA, and the amino acids that the codons stand for (Figure 8-10). Codons are written in terms of their base sequence in mRNA. Note that

there are 64 possible codons but only 20 amino acids. This means that most amino acids are signaled by several alternative codons; this situation is referred to as the *degeneracy* of the code. For example, leucine has 6 codons, and alanine has 4 codons.

Of the 64 codons, 61 are sense codons, and 3 are nonsense codons. **Sense codons** code for amino acids, and **nonsense codons** do not. Rather, they signal the end of the protein molecule's synthesis. The nonsense codons are UAA, UAG, and UGA. The initiator codon that starts the synthesis of the protein molecule is AUG, which is also the sense codon for methionine in eucaryotes and archaeobacteria, and for a variant of methionine called formylmethionine in eubacteria. The initiating methionine is usually removed later, so all proteins do not begin with methionine.

Figure 8-10 The genetic code. The three nucleotides in an mRNA codon are designated, respectively, as the first position, second position, and third position of the codon, going in the 5′ → 3′ direction on the mRNA. Each set of three nucleotides specifies a particular amino acid, represented by a three-letter abbreviation (see Table 2-5). The codon AUG (which specifies the amino acid methionine) is the start for protein synthesis. The word *End* stands for the nonsense codons that serve as signals to terminate protein synthesis.

[handwritten annotations: "usually 3rd Bvaries", "so can survive minor mutat (dmg)", "=start!"]

First Position	Second Position				Third Position
	U	C	A	G	
U	UUU, UUC — Phe; UUA, UUG — Leu	UCU, UCC, UCA, UCG — Ser	UAU, UAC — Tyr; UAA End, UAG End	UGU, UGC — Cys; UGA End, UGG — Trp	U, C, A, G
C	CUU, CUC, CUA, CUG — Leu	CCU, CCC, CCA, CCG — Pro	CAU, CAC — His; CAA, CAG — Gln	CGU, CGC, CGA, CGG — Arg	U, C, A, G
A	AUU, AUC, AUA — Ile; AUG — Met	ACU, ACC, ACA, ACG — Thr	AAU, AAC — Asn; AAA, AAG — Lys	AGU, AGC — Ser; AGA, AGG — Arg	U, C, A, G
G	GUU, GUC, GUA, GUG — Val	GCU, GCC, GCA, GCG — Ala	GAU, GAC — Asp; GAA, GAG — Glu	GGU, GGC, GGA, GGG — Gly	U, C, A, G

Regulation of Gene Expression in Bacteria

In Chapter 5 we learned that the bacterial cell is involved in an enormously large number of metabolic reactions. Some of these reactions are concerned with biosynthesis and are known as anabolic reactions. Others are concerned with degradation and are known as catabolic reactions. The common feature of all metabolic reactions is that they are catalyzed by enzymes.

We have seen that genes, through transcription and translation, direct the synthesis of proteins, many of which serve as enzymes. These are the very enzymes used for cellular metabolism. Therefore, the genetic machinery and metabolic machinery of a cell are integrated and interdependent. We will next look at how gene expression, and therefore metabolism, is regulated. Because protein (enzyme) synthesis requires a tremendous expenditure of energy (ATP), the regulation of protein synthesis is important to the cell's energy economy. Synthesizing only the proteins needed at a particular time conserves energy.

Enzymes that are present at all times in a cell are called *constitutive.* They are produced from genes that are effectively turned on all the time. The production of other enzymes is regulated so that they are present only when needed.

Repression, Induction, and Attenuation

Mechanisms called repression, induction, and attenuation regulate the transcription of messenger RNAs and consequently the synthesis of enzymes from the mRNAs. These genetic control mechanisms regulate the formation and amounts of enzymes in the cell, not their activities.

Repression and Induction

The regulatory mechanism that inhibits gene expression and decreases the synthesis of enzymes is called **repression**. Repression usually is a response to the presence of an end-product of a metabolic pathway;

there is a decrease in the rate of synthesis of the enzymes leading to the formation of that product. Repression is mediated by regulatory proteins called *repressors*, which block the ability of RNA polymerase to initiate transcription from the repressed genes.

Induction is the process that turns on the transcription of a gene or genes. A substance that acts to induce transcription of a gene is called an *inducer*, and enzymes that are synthesized in the presence of inducers are *inducible enzymes*. The genes required for lactose metabolism in *E. coli* are a well-known example of an inducible system. One of these genes codes for the enzyme β-galactosidase, which splits the substrate lactose into two simple sugars, glucose and galactose. (β is the Greek letter "beta" and refers to the type of linkage that joins the glucose and galactose.) If *E. coli* is placed in a medium in which no lactose is present, the organism is found to contain almost no β-galactosidase enzyme. However, when lactose is added to the medium, the bacterial cells produce a large quantity of the enzyme. Lactose is converted in the cell to the related compound allolactose, which is the inducer for these genes; the presence of lactose thus indirectly induces the cells to synthesize more enzyme. This response, which is under genetic control, is termed **enzyme induction**.

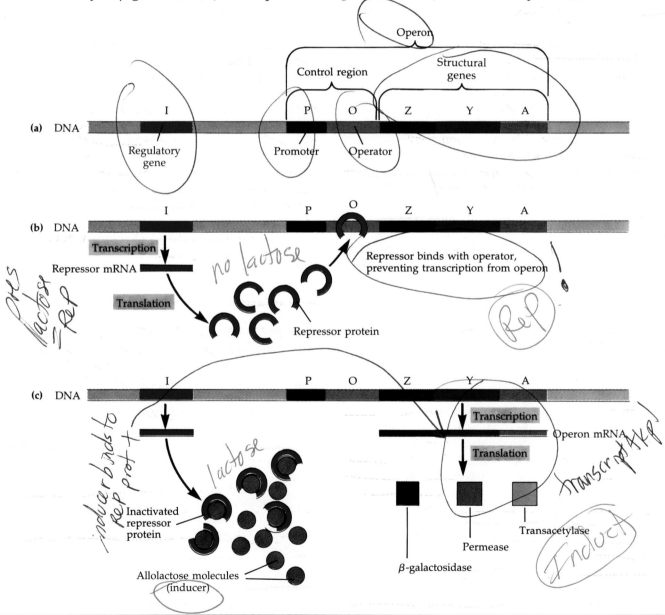

Figure 8-11   The *lac* operon. (a) This segment of the DNA molecule of *E. coli* shows the genes involved in lactose catabolism. In addition to the promoter and operator sites, the lactose operon contains three structural genes, which code for the proteins β-galactosidase, β-galactoside permease, and thiogalactoside transacetylase. **(b)** Control of the operon in the absence of lactose (repression). **(c)** Control of the operon in the presence of lactose (induction). When the inducer binds to the repressor protein, the repressor can no longer block transcription.

Operon Model

Details of the control of gene expression by induction and repression are described by the operon model. In 1961, François Jacob and Jacques Monod formulated this general model to account for the regulation of protein synthesis. They based their model on studies of the induction of the enzymes of lactose catabolism in *E. coli*. These enzymes include β-galactosidase, which breaks the disaccharide lactose into glucose and galactose, β-galactoside permease, which is involved in the transport of lactose into the cell, and thiogalactoside transacetylase, which metabolizes certain disaccharides other than lactose.

The structural genes for the three enzymes involved in lactose uptake and utilization are next to each other on the bacterial chromosome and are regulated together (Figure 8-11a). These genes are transcribed and translated rapidly and simultaneously when lactose is introduced into the culture medium. We will now see how this regulation occurs.

Next to the lactose operon on the bacterial chromosome is a regulatory gene called the *I* gene, which codes for the synthesis of a *repressor protein*. Another region of bacterial DNA, the *operator*, is a regulatory site located next to the structural genes. The operator controls the transcription of the structural genes. Recall that the promoter is the region of DNA where RNA polymerase binds to initiate transcription. (The operator and promoter sites and the structural genes they control are referred to as an *operon*.) When lactose is absent (Figure 8-11b), the repressor protein binds tightly to the operator site. This binding prevents the RNA polymerase enzyme from transcribing the adjacent structural genes; consequently, no mRNA is made, and no enzymes are synthesized. But when lactose is present (Figure 8-11c), some of it diffuses into the cells and is converted into the inducer allolactose. The inducer binds the repressor protein and alters it so that it cannot bind to the operator site. In the absence of an operator-bound repressor, the RNA polymerase enzyme can transcribe the structural genes into mRNA, which is then translated into enzymes. This is why, in the presence of lactose, enzymes are produced. Lactose is said to induce enzyme synthesis, and the *lac* operon is called an inducible operon.

In repressible operons, the repressor requires the presence of a small molecule called a **corepressor**, which helps it bind to the operator gene. Without the corepressor, the repressor is unable to bind. The enzymes involved in several biosynthetic pathways are regulated in this manner.

Remember that many genes, perhaps from 60% to 80%, are not regulated but are constitutive. Usually these genes code for enzymes that the cell needs in fairly constant amounts for its major life processes. The enzymes of glycolysis are examples.

Further regulation of the lactose operon depends on the level of glucose in the medium, which in turn controls the intracellular level of the small molecule **cyclic AMP**, a substance derived from ATP that serves as a cellular alarm signal. The cell prefers glucose to lactose as a carbon source because glucose is more directly metabolized as a carbon and energy source. Enzymes that metabolize glucose are constitutive, and cells grow at their maximal rate using glucose as their carbon source (Figure 8-12a). Cells have evolved alarm signals to respond to environmental and nutritional stress, such as the lack of glucose in the medium. When glucose is no longer transported across the cell membrane of *E. coli*, the cell responds by producing cyclic AMP. Increasing the level of cyclic AMP in the cell triggers a series of events that result in transcription of the lactose operon, enabling the cell to grow on lactose (Figure 8-12b). Similar mechanisms allow the cell to grow on other sugars. Inhibition of the metabolism of alternative carbon sources by glucose is termed **catabolite repression** or the **glucose effect**.

Table 8-2 summarizes the components of regulation of the lactose operon.

Attenuation

A control mechanism that allows mRNA synthesis to begin but then prematurely terminates it is called **attenuation**. For example, cells of *E. coli* growing in a medium lacking any amino acids contain the enzymes necessary for the synthesis of all 20 amino acids needed to make proteins. But introducing an amino acid such as tryptophan to the culture greatly decreases the synthesis of enzymes needed to make that amino acid. This response is adaptive and conserves resources because the enzymes are not needed when the end product of the biosynthetic pathway (tryptophan, in the case just described) is available to the cell.

The mechanism of attenuation involves the coupled processes of transcription and translation that were shown in Figure 8-7a. RNA polymerase begins transcribing the genes that code for the biosynthetic enzymes needed to make tryptophan, for example, but in the presence of tryptophan soon reaches a site on the newly made RNA called the *attenuator*, where it stops transcribing.

The attenuator only serves as a termination site when tryptophan is present. Why is that? It turns out

Figure 8-12 Growth rate of *E. coli* on glucose and lactose. The steeper the straight line, the faster the growth rate. **(a)** Bacteria growing on glucose as a sole carbon source grow faster than on lactose. **(b)** Bacteria growing in a medium containing glucose and lactose first consume the glucose and then, after a short lag time, the lactose. During the lag time, the intracellular cyclic AMP increases, the *lac* operon is transcribed, more lactose is transported into the cell, and β-galactosidase is synthesized to break down lactose.

TABLE 8-2 *Components of Gene Expression of the Lactose Operon*

	Component	Function
Proteins	RNA polymerase	Binds to the promoter site on DNA and catalyzes the synthesis of mRNA
	lac repressor	Binds to the operator site on DNA and the inducer allolactose
	CAP (catabolite activator protein)*	Binds to the promoter site and cyclic AMP
Small molecules	Lactose	Forms the inducer allolactose inside the cell
	Cyclic AMP	Binds to CAP
Sites in DNA	Promoter site	Attachment site for RNA polymerase and CAP
	Operator site	Attachment site for repressor protein (when not bound by inducer)

*CAP is a protein that is always present in small amounts in the cytoplasm; it is active only when bound to cyclic AMP.

that there are repeated codons that specify tryptophan in the leading portion of mRNA synthesized before RNA polymerase reaches the attenuator. If enough tryptophan is present in the cell, translation of these codons takes place, and transcription and translation both proceed smoothly until RNA polymerase reaches the attenuator region, where it terminates transcription. In the absence of tryptophan, translation stalls, and the partial mRNA folds up in such a way that the attenuator cannot function. In that case, synthesis of the complete mRNA for the tryptophan operon can be completed, and all the tryptophan biosynthetic enzymes are made. With these enzymes available, tryptophan is restored to the cell in large amounts. Attenuation is an elegant form of repression that regulates the genes that specify amino acid biosynthesis in bacteria.

Mutation: Change in the Genetic Material

A **mutation** is a change in the base sequence of DNA. We can reasonably expect that a change in the base sequence of a gene will sometimes cause a change in the product coded by that gene. When the gene for an enzyme mutates, for example, the enzyme coded by the gene may become inactive or less active because its amino acid sequence has changed. Such a change in genotype may be disadvantageous or even lethal if the cell loses a phenotypic trait it needs. Yet, a mutation can be beneficial if, for instance, the altered enzyme coded by the mutant gene has a new activity that benefits the cell.

Many simple mutations are neutral: the change in DNA base sequence causes no change in the activity of the product coded by the gene. Neutral mutations commonly occur when one nucleotide is substituted for another in the DNA, especially at a location corresponding to the third position of the mRNA codon. Because of the degeneracy of the genetic code, the resulting new codon might still code for the same amino acid. Even if the amino acid is changed, there may be no change in protein function if the amino acid is in a nonvital portion of the protein or is chemically very similar to the original amino acid.

Types of Mutations

The most common type of mutation involving single base pairs is **base substitution** or **point mutation**, in which a single base at one point in the DNA is replaced with a different one. Then, when the DNA replicates, the result is a substituted base pair (Figure 8-13). For example, A-T might be substituted for G-C, or C-G for G-C. If a base substitution occurs in a portion of the DNA molecule that codes for a protein,

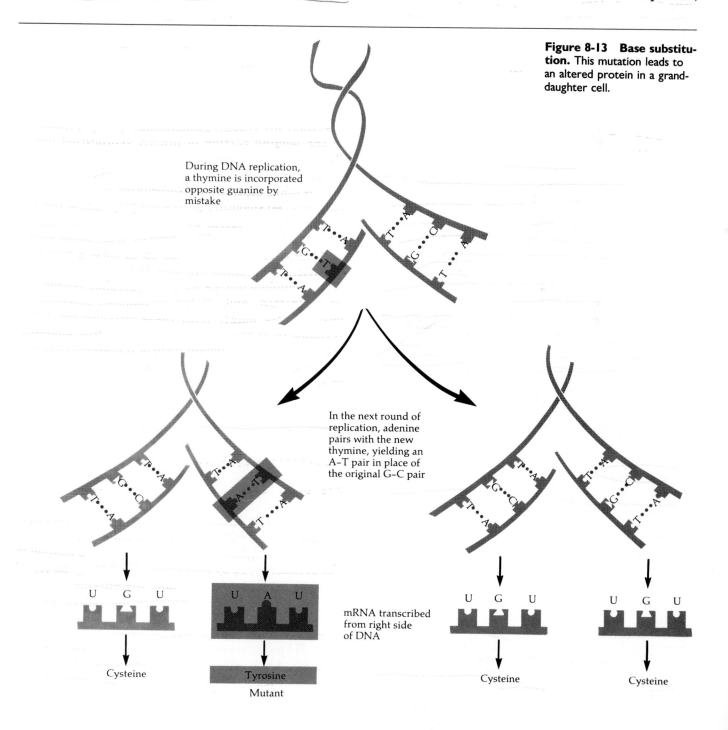

Figure 8-13 Base substitution. This mutation leads to an altered protein in a granddaughter cell.

During DNA replication, a thymine is incorporated opposite guanine by mistake

In the next round of replication, adenine pairs with the new thymine, yielding an A–T pair in place of the original G–C pair

mRNA transcribed from right side of DNA

Cysteine

Tyrosine
Mutant

Cysteine

Cysteine

then the mRNA transcribed from the gene will carry an incorrect base at some position. When the mRNA is translated into protein, the incorrect base can cause the insertion of an incorrect amino acid in the protein. Thus, the base substitution in DNA can result in an amino acid substitution in the synthesized protein. This is known as a **missense mutation** (Figure 8-14a and b).

By creating a stop (nonsense) codon in the middle of an mRNA molecule, some base substitutions effectively prevent the synthesis of a functional protein. Only a fragment of the protein is synthesized. A base substitution thus resulting in a nonsense codon is called a **nonsense mutation** (Figure 8-14c).

Besides base pair mutations, there are also changes in DNA called **frameshift mutations**. Here, one or a few nucleotide pairs are deleted or inserted in the DNA (Figure 8-14d). This can shift the "translational reading frame," that is, the three-by-three grouping of nucleotides recognized as codons by the tRNAs during translation. For example, inserting one nucleotide pair in the middle of a gene causes many amino acids downstream from the site of the original mutation to change. Frameshift mutations almost always result in a long stretch of missense and an inactive protein produced for the mutated gene. In most cases, a nonsense codon will eventually be generated and thereby terminate translation.

Base substitutions and frameshift mutations may occur spontaneously because of occasional mistakes made during DNA replication. These **spontaneous mutations** are mutations that occur without known intervention of mutation-causing agents. Agents in the environment, such as certain chemicals and radiation, that directly or indirectly bring about mutations are called **mutagens**.

Chemical Mutagens

Among the many chemicals known to be mutagenic is nitrous acid. Figure 8-15 shows how exposure of DNA to nitrous acid can convert the base adenine (A) to a form that no longer pairs with thymine (T) but instead pairs with cytosine (C). When DNA containing such modified adenines replicates, one daughter DNA molecule will have a base-pair sequence different from that of the parent DNA. Eventually, some A-T base pairs of the parent will have been changed to G-C base pairs in a granddaughter cell. Nitrous acid is thus an effective **base pair mutagen** that makes a specific mutational change in DNA. Like all mutagens, it alters DNA at random locations.

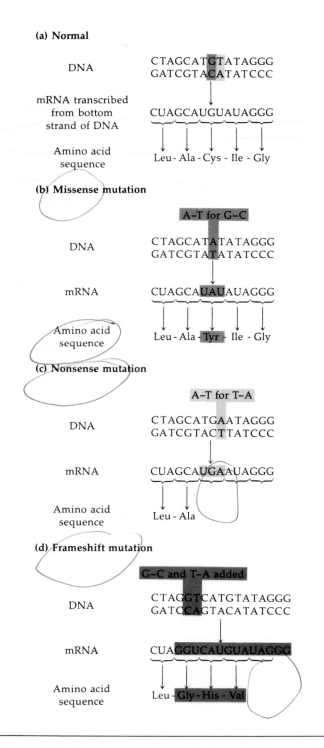

Figure 8-14   Types of mutations. (a) Normal DNA molecule. **(b)** Missense mutation. **(c)** Nonsense mutation. **(d)** Frameshift mutation.

Recombinant DNA at Work

Since ancient times people have harnessed the metabolic abilities of microorganisms for human use—primarily for the making of foods such as bread, cheese, yogurt, wine, and beer. In the past half century, the range of microbial products has expanded to include chemicals for medical use (antibiotics) and industrial use. Within the past 10 years, however, biotechnology has entered a whole new phase. Laboratory techniques for making recombinant DNA—DNA containing genes from two or more sources—have made it possible to "engineer" microorganisms to make products that were previously unique to the cells of animals and plants.

These micrographs show four types of cells used in the production of recombinant-DNA products. The micrographs on this page are from Genentech, a biotechnology company.

(1) *E. coli* (×1500).

(2) Cultured mammalian cells (×1500).

(3) Yeast (×1500).

(4) *Bacillus* (×1500).

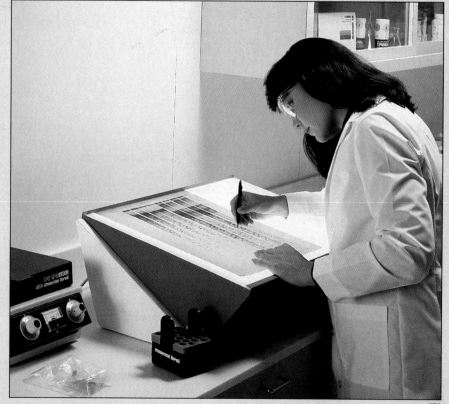

(5) One of the techniques used in the making of recombinant DNA is slab gel electrophoresis. This researcher is studying a photograph of a slab gel. The gel was treated with a DNA-binding dye; the dark bands on the photo represent DNA molecules of different sizes. On the next three pages, you will learn more about this and other processes of the new biotechnology.

High Tech: Tour of a Biotechnology Company

On these two pages, we walk through the labs of Cetus, a biotechnology company headquartered in Emeryville, California. Using recombinant-DNA techniques, Cetus is developing a number of products for medical use, including beta-interferon and interleukin-2. Here we will look at a few of the steps in preparing a hypothetical mammalian protein product, "Protein X."

(6)

(6) Preparing plasmid DNA. A research scientist inoculates a flask of liquid medium with plasmid-carrying *E. coli*. When the culture has grown, she will collect the cells isolate their DNA, and separate out the small, circular plasmids.

(7) Isolating the gene for Protein X. The bottles contain cul-

(7)

tures of mammalian cells, about to be "harvested." These cells grow on the bottle surface, nourished by the red liquid medium. From the harvested mammalian cells, cell biologists isolate mRNA for Protein X. Then they incubate the mRNA with the enzyme reverse transcriptase and the four deoxyribonucleotides to make complementary DNA (cDNA), which is the gene for Protein X (see Chapter 8).

(8) Using sterile techniques and working under a hood, a technician pipettes fresh medium into a culture bottle to feed a growing mammalian cell culture.

(9) Constructing the recombinant DNA. To make the recombinant DNA, the plasmid DNA is incubated with a restriction enzyme to cut it open, and then in a second step, incubated with the gene for Protein X and the enzyme ligase, which joins the two DNA molecules together (see diagram). As the photo shows, no high-technology tools are needed for these steps—only an ice bath, a water bath, small test tubes, and a pipette for measuring microliter amounts.

(9a)

(9b)

(8)

(10)

(10) Cloning the recombinant DNA. The recombinant plasmids are introduced into microbial cells by transformation and reproduce themselves as the cells multiply in culture. Individual clones of cells are tested for the presence of the gene for Protein X or for Protein X itself. The type of microbe used for production of Protein X may be *E. coli*, as in the above micrograph, which shows localized deposits of interleukin-2 (bubble-like areas) within the cells (×2000).

(11) Producing Protein X. The procedures for growing the microbial strain that makes Protein X must be scaled up for the eventual manufacturing process. This photo shows a 1500-liter fermenter. For some products, fermenters as large as 150,000 liters would be used.

(12) Purifying the product. Column chromatography is being used to purify Protein X. In the foreground, a computer is controlling the collection of the liquid flowing out of the column.

(11)

(12)

Some Tools of the Trade

(13) Ultracentrifuge. Spinning at speeds high enough to generate forces of 100,000× gravity or more, the ultracentrifuge can sediment particles as small as protein or nucleic acid molecules. It is used in many of the steps that lead to the construction of the recombinant microbial strain—for example, the isolation of mammalian mRNA and the purification of plasmid DNA.

(13)

(14) Amino acid analyzer. This machine is used to determine the amino acid composition of proteins. The polypeptide is broken to amino acids, which the machine identifies by chromatography. Because it is different for every protein, amino acid composition is an excellent criterion for identifying a protein sample or determining its purity. Thus the amino acid analyzer is used at many stages in the development and production of a protein product. It is also used to determine amino acid sequences.

(14)

(15)

(15) Slab gel electrophoresis. Electrophoresis on slabs of polyacrylamide gel can determine the size and/or electrical charges of DNA and protein molecules. It is based on the fact that molecules of different charges migrate differently in an electrical field. If a molecule is uniformly charged, as is DNA, the rate of migration is a measure of the size of the molecule. This method is useful for identifying a particular DNA by the pattern of fragments produced after treatment with a restriction enzyme (see Photo 9), and it is an essential tool in DNA sequencing.

(16)

(16) Biochemical assay devices. At virtually every stage in the development and manufacture of a protein product, the activity of the protein in various samples must be determined by some sort of biological or biochemical assay. The device shown here is a computerized pipetting system that can quickly carry out assays on a large number of samples in microliter quantities.

Benefits of Biotechnology

Medicine

(17) At Genentech, bottles of human growth hormone produced by recombinant-DNA techniques are in the final stages of manufacture.

(18) Children with growth-hormome deficiencies are already benefiting from this new product.

(17)

(18)

(19) Researchers at Chiron, another biotechnology company, are developing new vaccines against the HIV virus. These scientists are developing a rapid assay for the virus, for use in evaluating immune responses to recombinant AIDS vaccines.

(20) Chiron is also working on new diagnostic tests for AIDS. The test strip seen in this photo is used to determine what kinds of AIDS antibodies are being made by a patient infected with the virus.

(19)

(20)

Agriculture

(21) At Cetus, cotton is being genetically engineered to improve fiber quality and to achieve insect resistance.

(21)

Figure 8-15 Mutagenesis by nitrous acid (HNO₂). The nitrous acid alters an adenine (asterisk), with the result that it pairs with cytosine instead of thymine.

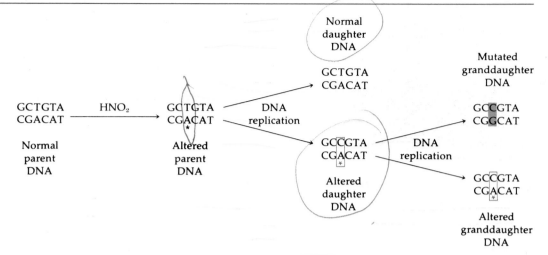

Other chemical mutagens are **base analogs**, like 2-aminopurine and 5-bromouracil. These molecules are structurally similar to normal nitrogenous bases, but they have slightly altered basepairing properties. The 2-aminopurine molecule (Figure 8-16a) is incorporated into DNA in place of adenine (A) but can sometimes pair with cytosine (C). The 5-bromouracil molecule (Figure 8-16b) is incorporated into DNA in place of thymine (T) but often pairs with guanine (G). When base analogs are given to growing cells, the analogs are randomly incorporated into cellular DNA in place of the normal bases. Then, during DNA replication, the analogs cause mistakes in basepairing. The wrongly paired bases will be copied during subsequent replication of the DNA, resulting in base pair substitutions in the progeny cells. Some antiviral and antitumor drugs are base analogs.

Still other mutagens can cause small deletions or insertions instead of substitutions. For instance, under certain conditions, benzpyrene, which is present in smoke and soot, is an effective **frameshift mutagen**. Likewise, aflatoxin, produced by *Aspergillus flavus* (a-sper-jil´lus flā´vus), a mold that grows on peanuts and grain, is a frameshift mutagen, as are the acridine dyes used experimentally against herpesvirus infections. Frameshift mutagens usually have the right size and chemical properties to slip between the stacked base pairs of the DNA double helix. It is believed that frameshift mutagens work by slightly offsetting the two strands of DNA, leaving a gap in one strand or a bulge in the other. When the staggered DNA strands are copied during DNA synthesis, one or more base pairs can be inserted or deleted in the new double-stranded DNA. Interestingly, frameshift mutagens are often potent carcinogens.

Figure 8-16 Base analogs and the nitrogenous bases they replace. (a) 2-aminopurine and adenine. **(b)** 5-bromouracil and thymine.

Radiation

X rays, gamma rays, and other forms of ionizing radiation are potent mutagens. The penetrating rays of ionizing radiation cause electrons to pop out of their usual shells. The affected electrons bombard other molecules and cause more damage, and many of the resulting ions and free radicals (molecular fragments with unpaired electrons) are very reactive. Some of these ions can combine with bases in the DNA. Such a combination results in mistakes in base pairing during replication and leads to base substitutions. An even more serious outcome is the breaking of covalent bonds in the sugar-phosphate backbone of DNA. These cause physical breaks in chromosomes.

Another form of mutagenic radiation is ultraviolet light (UV), a nonionizing component of ordinary sunlight. The most important effect of UV light on DNA is the formation of harmful covalent bonds among the bases. Adjacent thymines in a DNA strand can crosslink to form thymine dimers (Figure 8-17a). Such dimers, unless repaired, may cause serious damage or death to the cell because the cell cannot properly transcribe or replicate such DNA.

Bacteria and other organisms can repair radiation damage with processes involving enzymes. Some components of the repair processes also participate in recombination. Some repair processes depend on visible light, and some do not. One type of light-independent repair of UV damage is illustrated in Figure 8-17b. Enzymes cut out the distorted crosslinked thymines, widen the gap, and then fill in the gap with newly synthesized DNA that is complementary to the undamaged strand, so that the original base pair sequence is restored. The last step is the covalent sealing of the new DNA backbone to the old by the enzyme DNA ligase. Occasionally, such a repair process makes an error, and the original base pair sequence is not properly restored. The result of this error is a mutation.

Frequency of Mutation

The **mutation rate** is the probability that a gene will mutate when a cell divides. The rate is usually stated as a power of 10, and because mutations are very rare, the exponent is always a negative number. For example, if there is 1 chance in 10,000 that a gene will mutate when the cell divides, the mutation rate is 1/10,000, which is expressed as 10^{-4}. Spontaneous mistakes in DNA replication occur at a very low rate because the replication machinery is remarkably faithful. Perhaps only once in 10^9 replicated base pairs does an error occur. Because the average gene has about 10^3 base pairs, the spontaneous rate of mutation is about once in 10^6 (a million) replicated genes. Mutations usually occur more or less randomly along a chromosome. Having random mistakes at low frequency is an essential aspect in the design of living organisms, for evolution requires that genetic diversity be generated randomly and at a low rate. Typically, a mutagen will increase the spontaneous rate from 10 to 1000 times; that is, a mutagen will produce a mutation rate of 10^{-5} to 10^{-3} per gene per cell per generation.

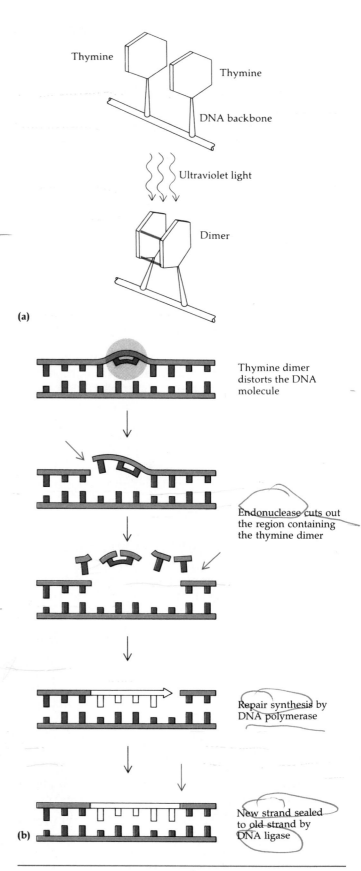

Figure 8-17 Mutagenesis by ultraviolet light. (a) Ultraviolet light crosslinks adjacent thymines. **(b)** Mechanism of repair of crosslinked thymines. This mechanism does not require light.

Identifying Mutants

Mutants can be detected by selecting or testing for an altered phenotype. Whether or not a mutagen is used, mutant cells with specific mutations are always rare as compared with other cells in the population. The problem is one of detecting a rare event.

Experiments are done more easily with bacteria than with other organisms because bacteria reproduce rapidly, and large numbers of organisms (more than 10^9 per milliliter of broth) can easily be used. Furthermore, because bacteria generally have only one copy of each gene per cell, the effects of a mutated gene are not masked by the presence of a normal version of the gene, as in many eucaryotic organisms.

Positive (direct) selection involves the detection of mutant cells by rejection of the unmutated parent cells. For example, suppose we were trying to find mutant bacteria that are resistant to penicillin. When the bacterial cells are plated on a medium containing penicillin, the mutant can be identified directly. The few cells in the population that are resistant (mutants) will grow and form colonies. The normal, penicillin-sensitive, parental cells cannot grow.

For identifying mutations in other kinds of genes, it may be necessary to use **negative (indirect) selection**. This is made possible by the **replica-plating technique**. In replica plating, about 100 bacterial cells are first inoculated onto an agar plate. This plate, called the master plate, contains a rich, nonselective medium on which all cells will grow. After several hours of incubation, each cell reproduces to form a colony. Then a pad of sterile velvet is pressed over the master plate (Figure 8-18). Some of the cells from each colony adhere to the velvet. Next, the velvet is pressed down onto two (or more) sterile plates. One plate contains a rich medium, and one contains a minimal medium on which the original, nonmutant bacteria

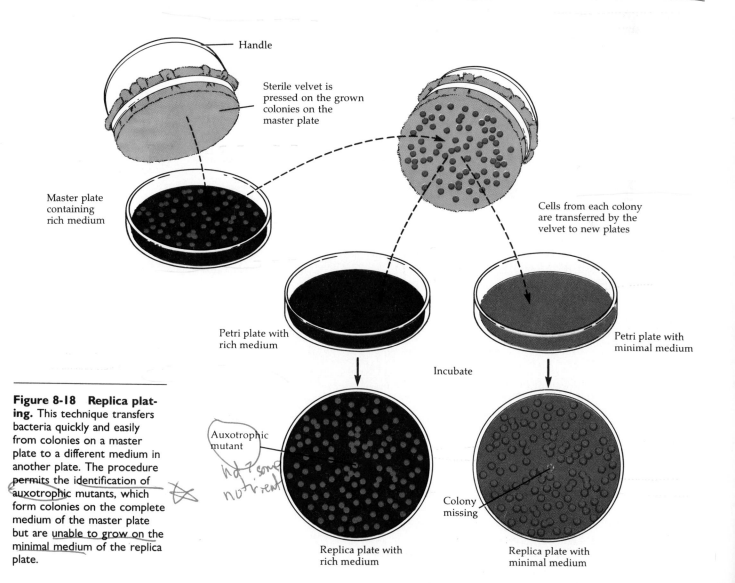

Handle

Sterile velvet is pressed on the grown colonies on the master plate

Master plate containing rich medium

Cells from each colony are transferred by the velvet to new plates

Petri plate with rich medium

Petri plate with minimal medium

Incubate

Auxotrophic mutant

Colony missing

Replica plate with rich medium

Replica plate with minimal medium

Figure 8-18   Replica plating. This technique transfers bacteria quickly and easily from colonies on a master plate to a different medium in another plate. The procedure permits the identification of auxotrophic mutants, which form colonies on the complete medium of the master plate but are unable to grow on the minimal medium of the replica plate.

can grow (for example, glucose and inorganic salts). A mutant colony that grew on the rich medium of the master plate but that has a new requirement for a growth factor will not grow on the minimal medium, which lacks that growth factor. Cells taken from the mutant colony on the master plate can then be tested on minimal media supplemented with various growth factors. These tests determine precisely which factor is required.

The replica-plating technique is a very effective means of isolating mutants that require one or more new growth factors. Any mutant microorganism possessing a nutritional requirement not possessed by the parent is known as an **auxotroph**. For example, an auxotroph may be lacking an enzyme needed for synthesizing a particular amino acid and therefore require that amino acid as a growth factor in its nutrient medium.

Figure 8-19   Ames test for suspected carcinogen. (a) Mutant *Salmonella* bacteria unable to synthesize histidine (His⁻) are mixed with a potentially mutagenic chemical and rat liver extract, which provides a source of mammalian activating enzymes. The test bacteria are then plated on growth medium with a very small amount of histidine, to start the bacteria growing. (Some growth is necessary for mutagenesis.) Control His⁻ bacteria, which have not been treated with the mutagen, are inoculated onto another plate containing the same medium. Only bacteria whose His⁻ phenotype has mutated back (reverted) to His⁺ (able to synthesize histidine) can grow into colonies. The control plate may show a few spontaneous His⁺ revertants. The test plates will show an increase in the number of His⁺ revertants if the test chemical is indeed a mutagen and potential carcinogen. **(b)** Quantitative results can be expressed as a mutational dose-response curve. This graph shows the effect of increasing concentrations of a mutagenic mycotoxin.

Identifying Chemical Carcinogens

Many known mutagens have been found to be carcinogens. A **carcinogen** is any substance that causes cancer in animals, including humans. In recent years, chemicals in the environment, the work place, and the diet have been implicated as causes of cancer in humans. The usual subjects of tests to determine potential carcinogens are animals, and the testing procedures are both time consuming and expensive. Now there are faster and less expensive procedures for the preliminary screening of potential carcinogens. One of these, called the **Ames test**, uses bacteria as carcinogen indicators.

The Ames test is based on the observation that exposure of mutant bacteria to mutagenic substances may cause new mutations that reverse the effect (the phenotype) of the original mutation. Specifically, the test measures the reversion of histidine auxotrophs of *Salmonella* (His⁻ cells—mutants that have lost the ability to synthesize the amino acid histidine) to histidine-synthesizing cells (His⁺) after treatment with a mutagen. Bacteria are incubated in both the presence and absence of the substance being tested (Figure 8-19). Since many chemicals must be activated (transformed chemically) by animal enzymes for mutagenic or carcinogenic activity to be exhibited, the chemical to be tested and the mutant bacteria are incubated together with rat liver extract, a rich source of activation enzymes. If the substance being tested is mutagenic, it will cause reversion of His⁻ bacteria to His⁺ at a rate higher than the spontaneous reversion rate. The number of observed revertants provides an indication of the degree to which a substance is mutagenic and therefore possibly carcinogenic.

The test can be used in many ways. Several potential mutagens can be qualitatively tested by spotting the individual chemicals on small paper disks on a single plate inoculated with bacteria. Not just pure chemicals but also mixtures such as wine, blood, smoke condensates, and extracts of foods can be tested for mutagenic substances. (An example of the use of the Ames test is given in the box in Chapter 14, p. 390.)

About 90% of the substances found by the Ames test to be mutagenic have also been shown to be carcinogenic in animals. Furthermore, the more mutagenic substances have generally been found to be more carcinogenic.

* * *

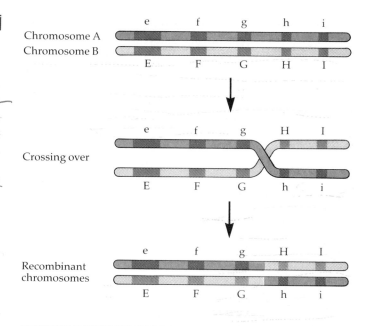

Figure 8-20 Genetic recombination between two related chromosomes. Chromosomes A and B each carry one copy of genes E through I. The chromosomes cross over by breaking and rejoining, sometimes in more than one location. The result is two recombinant chromosomes, each of which carries genes originating from both chromosomes.

Genetic Transfer and Recombination in Procaryotes

Genetic recombination refers to the exchange of genes between two DNA molecules to form new combinations of genes on a chromosome. Figure 8-20 shows one type of recombinational event occurring between two pieces of DNA, which we shall regard as chromosomes for the sake of simplicity. We have called one chromosome A, and the other, B. If these two chromosomes break and rejoin as shown (a process called **crossing over**), some of the genes carried by these chromosomes are shuffled. Thus, the original chromosomes have been *recombined* so that each now carries a portion of the other chromosome's genes.

If A and B represent DNA from different individuals, how are they brought close enough to recombine? Genes of higher organisms are recombined during meiosis, when chromosomes from the two parents pair (a process called synapsis). Thus, the gametes resulting from meiosis contain recombined DNA. For eucaryotes, genetic recombination is an ordered process that usually occurs as part of the sexual cycle of the organism. In bacteria, genetic recombination can happen in a number of ways, which we will discuss in the following sections.

Like mutation, genetic recombination contributes to a population's genetic diversity, which is the source of variation in evolution. Recombination is more likely than mutation to be beneficial because recombination is less likely to destroy the function of a gene and may bring together combinations of genes that enable the organism to carry out a valuable new function.

Genetic material can be transferred between bacteria in several ways. In all of the mechanisms, the transfer involves a **donor cell** that gives only a portion of its total DNA to a **recipient cell**. Once transferred, part of the donor's DNA is usually incorporated into the recipient's DNA; the remainder is degraded by cellular enzymes. The recipient cell that incorporates donor DNA into its own DNA is called a **recombinant**. The transfer of genetic material between bacteria is by no means a frequent event. It may occur in only 1% or less of an entire population. Let us now examine the specific types of genetic transfer in detail.

Transformation in Bacteria

During the process of **transformation**, genes are transferred from one bacterium to another as "naked" DNA in solution. This process was first demonstrated more than 50 years ago, although it was not understood at that time. Not only did transformation show that genetic material could be transferred from one bacterial cell to another, but study of this phenomenon eventually led to the conclusion that DNA is the genetic material. The initial experiment on transformation was performed by Frederick Griffith in England in 1928 while he was working with two strains of *Streptococcus pneumoniae*. One, a virulent strain, has a polysaccharide capsule and causes pneumonia; the second, an avirulent strain, lacks the capsule and does not cause disease.

Griffith was interested in determining whether

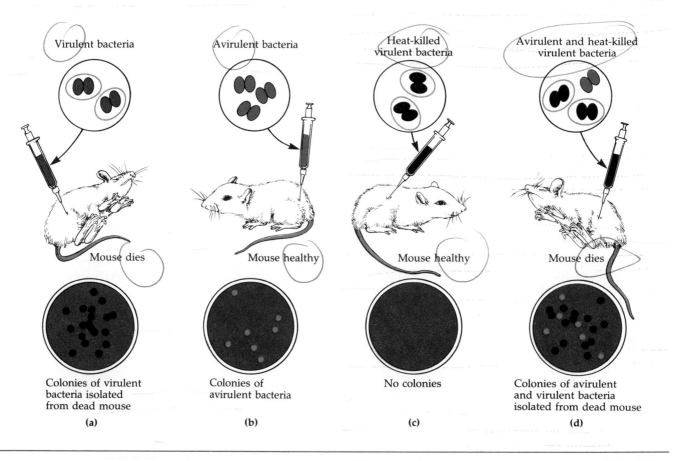

Virulent bacteria

Avirulent bacteria

Heat-killed virulent bacteria

Avirulent and heat-killed virulent bacteria

Mouse dies

Mouse healthy

Mouse healthy

Mouse dies

Colonies of virulent bacteria isolated from dead mouse

Colonies of avirulent bacteria

No colonies

Colonies of avirulent and virulent bacteria isolated from dead mouse

(a)

(b)

(c)

(d)

Figure 8-21 **Transformation.** Griffith's experiment demonstrating genetic transformation. Some material from the heat-killed virulent bacteria transformed the living avirulent bacteria into virulent bacteria, which killed the mouse. Avirulent bacteria lack capsules and are readily destroyed by the host; therefore, few show up as colonies on the medium in (b) and (d).

injections of heat-killed (at 60°C) virulent pneumococci could be used to vaccinate mice against pneumonia. He found that the injected mice did not become ill from the heat-killed strain. However, when these dead cells were mixed with live avirulent cells and injected into the mice, many of the mice died. In the blood of the dead mice, Griffith found living, encapsulated, virulent bacteria. It was inferred that hereditary material from the dead bacteria had entered the live cells and changed them genetically so that their offspring were encapsulated, virulent forms (Figure 8-21).

Subsequent investigations based on Griffith's research revealed that bacterial transformation could be carried out using standard microbiological culturing techniques. A broth (liquid culture medium) was inoculated with live bacteria without capsules. Dead bacteria with capsules were then added to the broth, which was incubated for some hours. A drop of the culture was smeared over agar culture medium in Petri plates. After a period of incubation, it was found that significantly more of the bacterial colonies than would be expected from spontaneous reversion had the smooth appearance typical of encapsulated bacteria—that is, they had been transformed. The bacteria had acquired a new hereditary trait.

The next step was to extract various chemical components from the killed cells and determine which of these chemicals caused the transformation. These crucial experiments were performed by Oswald T. Avery and his associates at the Rockefeller Institute in the United States. After years of research, they announced in 1944 that the chemical responsible for transforming harmless pneumococci into virulent strains was DNA.

Since the time of Griffith's experiment, considerable information has been gathered about transformation. In nature, some bacteria, perhaps after death and cell lysis, release their DNA into the environment. Other bacteria can then encounter the DNA, and, depending on the particular species and growth conditions, take up fragments of DNA into their cytoplasm and recombine the DNA into their own chromosomes. A recipient cell having this new combination of genes is a kind of hybrid, or recombinant cell (Figure 8-22). All the descendants of such a recombinant cell will be identical to it. Transformation occurs naturally among very few genera of bacteria. These include *Bacillus*, *Hemophilus* (hē-mä′fi-lus), *Neisseria*, *Acinetobacter* (ā-si-ne′tō-bak-tėr), *Rhizobium*, and certain strains of *Streptococcus* and *Staphylococcus*.

Transformation works best when the donor and recipient cells are very closely related and when the recipient cells are in the late log phase of growth. Even though only a small portion of a cell's DNA is transferred to the recipient, it is still a very large molecule that must pass through the recipient cell wall and membrane. When a recipient cell is in a physiological state in which it can take up the donor DNA, it is said to be **competent**. Competence may be related to alterations in the cell wall that make it permeable to the large DNA molecule, or competence may involve the synthesis of specific receptor sites on the bacterial cell's surface. Among the traits whose transformation has been studied are capsule formation (already noted), nutritional needs, pathogenicity, pigment formation, and drug resistance.

E. coli is not naturally competent for transformation. However, a simple laboratory treatment enables *E. coli* to readily take up DNA from other *E. coli* cells or even from unrelated cells. The discovery of this treatment, along with our recently developed ability to make recombinant DNA in the laboratory, has brought about the genetic engineering revolution.

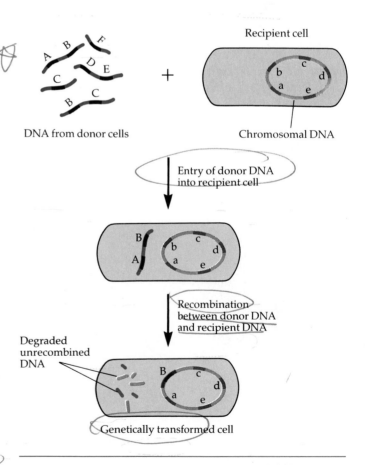

Figure 8-22   Mechanism of genetic transformation.

Conjugation in Bacteria

Conjugation is another mechanism by which genetic material is transferred from one bacterium to another.

Conjugation is mediated by *plasmids*, which are circular pieces of DNA that replicate independently from the cell's chromosome. They are like minichromosomes, but they differ from bacterial chromosomes in that the genes they carry are usually not essential for the growth of the cell. The plasmids responsible for conjugation are transmissible between cells during conjugation.

Conjugation differs from transformation in two major ways. First, conjugation requires that there be direct cell-to-cell contact. Second, the conjugating cells must be of opposite "mating type"—donor cells must carry the plasmid, and recipient cells must not. The plasmid carries genes that code for the synthesis of *sex pili*, projections from the donor's cell surface that contact the recipient and help to bring the two cells into direct contact. In the process of conjugation, the plasmid is replicated during transfer of a single-stranded copy of the plasmid DNA to the recipient, where the complementary strand is synthesized.

Because most experimental work on conjugation has been done with *E. coli*, we will describe the process in this organism. In *E. coli*, the *F factor* was the first plasmid observed to transfer between cells during conjugation (Figure 8-23). Donors carrying F (F⁺ cells) transfer the plasmid to recipients (F⁻ cells), which become F⁺ cells as a result (Figure 8-24a). In some cells carrying F, the F factor integrates into the chromosome, forming an **Hfr (high frequency of recombination) cell** (Figure 8-24b). During conjugation between an Hfr and an F⁻ cell, the Hfr cell's chromosome, with its integrated F factor, replicates, and the new copy of the chromosome is transferred to the recipient cell (Figure 8-24c). Replication of the Hfr chromosome begins within the F factor, and a small piece of the F factor leads the chromosomal genes into the F⁻ cell. Most of the integrated F factor enters the recipient cell last, if at all. Usually, the chromosome breaks before it is completely transferred. Once within the recipient cell, donor DNA can recombine with the recipient's DNA, as occurs in transformation. Donor DNA that is not integrated is degraded.

By the process of conjugation between an Hfr cell and an F⁻ cell, the F⁻ cell can acquire new versions of chromosomal genes (just as in transformation).

Transduction in Bacteria

A third mechanism of genetic transfer between bacteria is **transduction**. In this process, "naked" DNA is not passed, as it is in transformation or during cell-to-cell contact in conjugation. Rather, DNA is transferred inside a bacterial virus, called a **bacteriophage** (or simply, **phage**), from the donor cell to the recipient cell. (Phages will be discussed further in Chapter 12.)

To understand how transduction works, we shall consider the life cycle of one type of transducing phage of *E. coli*; this phage carries out a process called **generalized transduction**. In the process of infection, the phage attaches to the bacterial cell wall and injects its DNA into the bacterium (Figure 8-25a). The phage DNA acts as a template for the synthesis of new phage DNA and also directs the synthesis of phage protein coats (Figure 8-25b). During phage development inside the infected cell, the bacterial chromosome breaks apart, and at least some fragments of the bacterial chromosome are accidentally packaged inside phage protein coats (Figure 8-25c). (Even plasmid DNA or DNA of another virus that happens to be inside the cell can be given phage protein coats.) The resulting phage particles then carry bacterial DNA instead of phage DNA. When the released phage particles later infect a new population of bacteria, bacterial genes will be transferred to the newly infected cells at low frequency (Figure 8-25d).

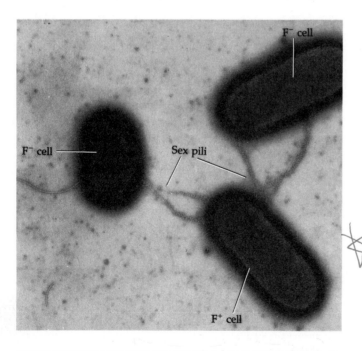

Figure 8-23 Bacterial conjugation. The DNA-donating *E. coli* cell extends sex pili to a recipient. Later the two cells will come into direct contact, and DNA will be transferred from one to the other (×25,000).

Transduction of cellular DNA by a virus can lead to recombination between the DNA of the first host cell and the DNA of the second host cell (Figure 8-25e). Transduction is thus another way that bacteria can acquire new genotypes, in addition to mutation, transformation, and conjugation. The process of generalized transduction just described is typical of bacteriophages such as phage P1 of *E. coli* and phage P22 of *Salmonella*.

Figure 8-24   Conjugation in *E. coli*. (a) When an F factor (a plasmid) is transferred from donor (F⁺) to recipient (F⁻), the F⁻ cell is converted into an F⁺ cell. **(b)** Integration of an F factor into the bacterial chromosome makes the cell an Hfr. **(c)** An Hfr donor passes a portion of its chromosome into an F⁻ cell. A recombinant F⁻ cell results.

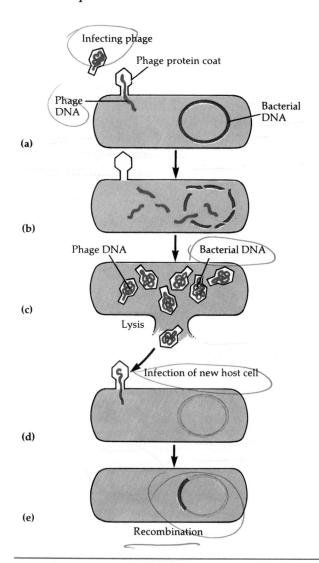

(a)

(b)

(c)

(d)

(e)

Infecting phage

Phage protein coat

Phage DNA

Bacterial DNA

Phage DNA   Bacterial DNA

Lysis

Infection of new host cell

Recombination

Figure 8-25   Transduction by bacteriophage. Shown here is generalized transduction, in which any bacterial DNA can be transferred from one cell to another.

All genes contained within a bacterium infected by a generalized transducting phage are equally likely to be packaged in a phage coat and transferred. There is another type of transduction, called **specialized (restricted) transduction**, in which only certain bacterial genes are transferred. Specialized transduction will be discussed in Chapter 12.

Recombination in Eucaryotes

In transformation, conjugation, and transduction in bacteria, only a small segment of the DNA of the donor cell enters and recombines with the DNA of the recipient cell. In eucaryotes, genetic recombination is the result of a sexual reproductive process. In contrast to

the processes in bacteria, sexual reproduction is a regular event for most eucaryotes and is necessary for the survival of many species.

Each species of eucaryotic organism has a characteristic **chromosome number**, that is, the number of chromosomes in each ordinary body cell (called *somatic cells*). The human chromosome number is 46 (23 pairs). Each parent contributes one set of 23 chromosomes, which carries genes for virtually all the activities of human cells. The other set of 23 chromosomes is more or less a duplicate set; it contains the same or a different version of each gene.

Sexual reproduction in eucaryotes involves the fusion of the haploid nuclei of two parent cells (Figure 8-26). A **haploid cell** is one that contains only one of each chromosome; the number of chromosomes in such a cell is called the **haploid number**, symbolized n. In humans, the haploid number of chromosomes is 23. The haploid cells that fuse are called **gametes**. (In animals, the male gametes are sperm, and the female gametes are ova.) Fusion of the gametes produces a fertilized ovum, or **zygote**, which has received half its chromosomes from one parent and half from the other. Procaryotic organisms (and many eucaryotic microorganisms) are typically haploid and regularly reproduce asexually from a single parent cell.

A **diploid cell** contains two of each type of chromosome; the number of chromosomes in such a cell is called the **diploid number**, symbolized $2n$. In humans, the diploid number of chromosomes is 46. The paired chromosomes are known as **homologous chromosomes**.

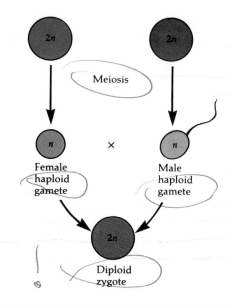

$2n$ $2n$

Meiosis

n $\times$ n

Female haploid gamete   Male haploid gamete

$2n$

Diploid zygote

Figure 8-26   Sexual reproduction in eucaryotes.

For eucaryotes to reproduce sexually by fusing their gametes to form zygotes, the chromosome number must be reduced from diploid in the somatic cells to haploid in the gametes. This is accomplished by a special cell division process called **meiosis**. In the process of meiosis, portions of homologous chromosomes can be exchanged with one another by **crossing over** (Figure 8-20). Crossing over is in fact the process of genetic recombination; and in eucaryotic cells it can often be observed microscopically. Once haploid gametes are produced by meiosis, the gamete's nucleus can fuse with another gamete's nucleus and form a diploid zygote. Subsequent growth and development of a zygote produce a mature organism capable of sexual reproduction. The regular events of crossing over in meiosis, segregation of chromosomes into gametes, and diploid formation in zygotes generate enormous genetic diversity in every generation.

Genetic Engineering

Recombination is a natural biological process in virtually all organisms. In recent years, however, the terms *recombination* and *recombinant DNA* have taken on special, somewhat different meanings. Our detailed knowledge about plasmids and about enzymes involved in DNA metabolism has made it possible for us to isolate genes from various organisms and incorporate them into other bacterial or eucaryotic cells. This manipulation of genetic material in the laboratory is referred to as **genetic engineering** (see Microview B). Further explanation of plasmids will help your understanding of recombinant DNA, the hallmark of genetic engineering.

Plasmids

Plasmids are found in some eucaryotic microorganisms, such as the yeast *Saccharomyces cerevisiae* (se-ri-vis′ē-ī), as well as in bacteria. As discussed before in connection with the F plasmid, plasmids may be **conjugative plasmids**; that is, they may have genes for carrying out their transfer to another cell (and for transferring the bacterial chromosome or other plasmids). Plasmids are usually dispensable—the cell can survive under most conditions without the use of the proteins coded by plasmid genes.

However, under certain conditions, genes carried by the plasmids can be crucial to the survival and growth of the cell. **Dissimilation plasmids** code for enzymes that catalyze the catabolism of certain unusual sugars and hydrocarbons. Some species of *Pseu-*

domonas can actually use such substances as toluene, camphor, and the high-molecular-weight hydrocarbons of petroleum as primary carbon and energy sources due to genes carried on plasmids.

Bacteriocinogenic plasmids contain genes for the synthesis of *bacteriocins*, toxic proteins that kill other bacteria. These plasmids have been found in many genera, and they are useful markers for the identification of certain bacteria in clinical laboratories.

Resistance factors (R factors) are plasmids of great medical importance. R factors carry genes that confer upon their host cell resistance to antibiotics, heavy metals, or cellular toxins. Many R factors are conjugative plasmids that contain two groups of genes. One group is called the **resistance transfer factor (RTF)** and includes genes for plasmid replication and conjugation. The other group, the **r-determinant**, has the resistance genes; it codes for the production of enzymes that inactivate certain drugs or toxic substances (Figure 8-27). Different R factors,

(a)

R-determinant (genes for resistance to various antibiotics)

RTF (genes for replication and conjugation)

(b)

Figure 8-27   R factors, a type of plasmid. (a) Electron micrograph showing two different resistance (R) factors (×15,000). The larger loop of DNA is a plasmid isolated from the bacterium *Bacteroides fragilis* and encodes resistance to the antibiotic clindamycin. The smaller loop of DNA is a plasmid from *E. coli* and provides resistance to tetracycline. **(b)** Diagram of an R factor with multiple resistance determinants.

when present in the same cell, readily recombine to produce R factors with new combinations of genes on their r-determinants.

R factors present very serious problems for the treatment of infectious diseases using antibiotics. The widespread use of antibiotics in medicine and agriculture (many animal feeds contain antibiotics) has led to the preferential survival (selection) of bacteria that have R factors. So populations of resistant bacteria grow larger and larger. The transfer of resistance between bacterial cells of a population, and even between bacteria of different genera, also contributes to the problem. (See the boxes in Chapter 19, p. 494, and Chapter 25, p. 646.)

Transposons

Transposons (also called *transposable genetic elements*) are small segments of DNA that can move (be "transposed") from one region of the DNA molecule to another. In the popular press, they are often called "jumping genes." These pieces of DNA measure from 700 to 40,000 base pairs long.

There is now strong evidence that transposons occur in all organisms, but they have been studied most thoroughly in microorganisms. They are found in the main chromosomes of organisms, in plasmids, and in the genetic material of viruses. This wide

MICROBIOLOGY HIGHLIGHTS

Agrobacterium: A Natural Genetic Engineer

A pathogen that causes growth of tumors in plants is helping researchers find a way to insert desired genes into plants. Several species of soil bacteria of the genus *Agrobacterium* cause tumor growth in a wide variety of broad-leafed (dicotyledonous) plants. The disease, called *crown gall* (reportedly the third most important plant disease in the United States), causes a lump of tumor tissue (see figure) to grow at the site of infection. The cells of crown gall are similar in many ways to cancerous cells in animals. They exhibit uncontrolled growth, and once they take hold of plants they continue to grow, even if the bacteria are killed by antibiotics.

The Ti Plasmid

Researchers have found that the tumor-inducing agent in strains of *Agrobacterium* is actually a plasmid, called Ti (tumor inducing), which carries the genes for crown gall disease. Like all plasmids, Ti plasmids exist in the bacterial cells as independently replicating circles of DNA. But once a plant is infected, the bacterium inserts a portion of its Ti plasmid, called T-DNA, into the host plant cell genome. T-DNA genes direct the plant cell to produce unusual amino acid derivatives called opines, which the bacteria can use as a source of carbon and nitrogen, although the plant cells cannot. In short, *Agrobacterium* commandeers plant cells and makes them produce chemicals for its own growth. What is more, T-DNA has several genes that suppress shoot and root formation by tumor cells in the infected plant. The incorporated T-DNA is replicated along with the plant's chromosomes every time a plant cell divides. It is thought that some of the T-DNA genes direct synthesis of substances that are similar to plant hormones that

Crown gall disease in a pecan tree. Wartlike growths from a few inches to a foot in diameter form on the base of the tree trunk and on larger roots.

occurrence is not surprising because transposons can jump from plasmid to plasmid and from genome to plasmid and vice versa. They may move from one site on the chromosome to another site on the same chromosome or to another chromosome. Transposons can insert themselves into almost any stretch of DNA. As you might imagine, frequent movement of transposons could wreak havoc within a cell. For example, as transposons move about on chromosomes, they may insert themselves *inside* genes, inactivating them. Fortunately, transposition occurs relatively rarely. The frequency of transposition is comparable with the spontaneous mutation rate that occurs in bacteria, that is, from 10^{-5} to 10^{-7} per generation.

All transposons contain the information for their own transposition. The simplest transposons, also called **insertion sequences (IS)**, contain only a gene that codes for an enzyme (transposase) that catalyzes the cutting and ligating of DNA that occurs in transposition and recognition sites. Recognition sites are short regions of DNA the enzyme recognizes as recombination sites between the transposon and the chromosome.

Complex transposons also carry other genes not connected with the transposition process. For example, many bacterial transposons contain genes that confer antibiotic resistance. R factors are frequently made up of a collection of transposons (Figure 8-28).

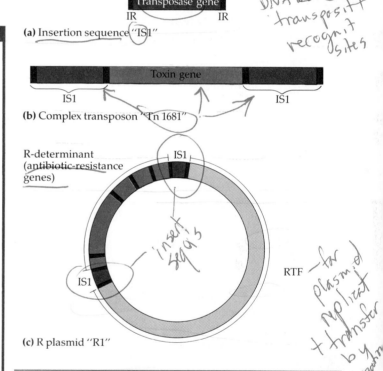

(a) Insertion sequence "IS1"

(b) Complex transposon "Tn 1681"

(c) R plasmid "R1"

Figure 8-28   Transposons and insertion sequences. (a) The simplest type of transposon is the insertion sequence (IS)—a segment of DNA that contains a gene for transposase (the enzyme that catalyzes transposition) and is bounded at each end by nucleotide sequences that the enzyme recognizes as sites of recombination. IS1 is one example of an insertion sequence. **(b)** Complex transposons carry other genetic material in addition to transposase genes. The example shown here, Tn 1681, carries the gene for a bacterial toxin and has complete copies of the insertion sequence IS1 at each end. **(c)** A large transposon makes up a major portion of the plasmid R1, which has two parts. One part, the RTF, contains the genes needed for plasmid replication and transfer of the plasmid by conjugation; the other, the r-determinant, carries genes for resistance to five different antibiotics. The presence of an insertion sequence at each end of the group of antibiotic-resistance genes makes this part of the plasmid a transposon.

regulate normal growth. These hormone analogs interfere with normal regulation of plant growth, allowing plant cells to reproduce in an uncontrolled manner.

Agrobacterium preys on wounded plants, attaching its cell-wall lipopolysaccharide to receptor sites on the plant cell walls. Within the next few hours, the bacterium produces cellulose fibers to secure its attachment. A single-stranded molecule of T-DNA is then produced in the bacterium. It is speculated that the single-stranded T-DNA is then transferred to the plant cell through its damaged wall by a process not unlike bacterial conjugation.

Potential Benefits
Research groups in the United States and Europe have been closely investigating the possible use of the Ti plasmid as a *genetic vector*, a means for transferring genes from one organism to another for

the genetic engineering of plants. Altered Ti plasmids that do not induce tumors might be used as vectors to insert desirable genes into plants. The genes could be inserted into plant cells that, when cloned, produce plants that carry the new gene in their seeds. Work is in progress to incorporate genes for amino acid production and nitrogen fixation into T DNA so *Agrobacterium* can insert the genes into crop plants. These new genes would make the crops more nutritious and able to grow with less fertilizer. One obstacle is that many important crops, including grasses such as corn and wheat, are monocots, which are normally resistant to infection by *Agrobacterium*. Recently, a strain of *Agrobacterium* was used to insert a foreign gene into corn. This success offers the hope that *Agrobacterium* may eventually be used to improve a wide variety of crops.

Transposons with antibiotic-resistance genes are of particular practical interest, but there is no limitation on the kinds of genes that transposons can have. Thus, transposons provide a powerful natural mechanism for the movement of genes from one chromosome to another. Furthermore, since they may be carried between cells on plasmids or viruses, they can also spread from one organism—or even species—to another.

Recombinant DNA

Recombinant DNA results from the joining together of DNA from different sources in vitro. (*In vitro* means "in glass," that is, not within a living organism.) Often the DNA sources are different species. When the recombinant DNA is introduced into an appropriate microbe, the microbe can often synthesize the products of the new gene it has acquired.

Special bacterial enzymes called *restriction enzymes* are used to make recombinant DNA. Restriction enzymes recognize unique sites in DNA that are short symmetrical sequences of bases called *palindromes*. Restriction enzymes often cut double-stranded DNA within the palindromic recognition site at staggered positions (cut sites) on the two DNA strands, leaving single-stranded self-complementary ends called sticky ends. For instance, the *E. coli* restriction enzyme *Eco* RI acts as follows:

Note that the recognition site is six nucleotide pairs long and has the same sequence on both strands. The result is that pieces of DNA cut by *Eco* RI will have sticky ends with the same sequence:

```
A A T T C                 G
                   
          G                 C T T A A
```

As a consequence, the single-stranded ends of any two of such pieces of DNA can basepair and join the two pieces together in a linear or circular form. The linear form has this arrangement:

```
A A T T C          G A A T T C          G

       G          C T T A A G          C T T A A
```

The circular form looks like this:

Junctions are covalently sealed by *DNA ligase*, another enzyme that can be used in vitro.

The construction of a recombinant plasmid using this technique is illustrated in Figure 8-29. Once constructed, the recombinant plasmid is introduced into bacterial cells by transformation. Chemical treatment of the cell walls of the recipient bacteria removes structural barriers and increases their permeability to DNA during transformation. The transformed cells are hybrids with genotypes that never before existed in nature. Each time the hybrid bacteria divide, so do the plasmids carrying the recombinant DNA. Often such recombinant bacteria can produce an enormous amount of whatever substance is encoded in their new genes.

Most bacterial cells do not readily secrete the substances produced into the culture medium. After cultivation, the cells must be collected and lysed to harvest their product. Current efforts in genetic engineering are focused on the culturing of recombinant cells that secrete their product naturally and on the use of eucaryotic cells as recipients of recombinant DNA and as producers of recombinant DNA products.

Special techniques are required to construct recombinant plasmids containing eucaryotic genes if a functional protein product from the gene is needed because many eucaryotic genes are made up of *exons* and *introns* (Figure 8-30a). **Exons** are regions of a gene that code for the protein product of the gene; **introns** are intervening regions within the gene that do not code for the protein. The eucaryotic cell transcribes the DNA of the entire gene, including both exons and introns, removes the intron RNA, and finally splices together the exons to form the mRNA that is translated into the functional protein.

Bacterial cells do not contain the enzymes to remove the intron RNA; therefore, a bacterial cell cannot make the functional mRNA or protein from such a eucaryotic gene. The DNA of a eucaryotic gene that contains exons and introns is not used directly for making recombinant DNA; instead, mRNA is collected. A DNA copy of the mRNA is then made in

Figure 8-29 Method for in vitro construction of recombinant DNA. (a) A plasmid, such as pBR322, is isolated from *E. coli* and treated with a restriction enzyme that cuts the plasmid at a single known site, which is not within vital plasmid genes or origin of replication (ori). The palindromic recognition site of the restriction enzyme *Eco* RI is illustrated. **(b)** A segment of foreign DNA is cut with the same restriction enzyme to generate many fragments of different lengths. Many restriction enzymes make staggered cuts, shown here on both plasmid and foreign DNA, resulting in short single-stranded complementary ends called sticky ends on each DNA fragment. **(c)** Cut foreign DNA is mixed with the cut plasmid, into which it inserts via sticky ends. The enzyme DNA ligase seals the gaps to make a hybrid plasmid, a recombinant DNA molecule made in vitro. The recombinant plasmid is introduced into bacteria by transformation. Cells with the recombinant plasmid may be selected because of the new gene introduced. Successive generations of hybrid bacteria maintain the recombinant plasmid.

DNA ligase seals the gaps (■)

Figure 8-30   Special features of eucaryotic genes and their use in recombinant DNA. (a) Transcription and processing of a gene containing exons and introns in the eucaryotic cell. **(b)** Reverse transcription of mRNA into DNA in vitro is brought about by the viral enzyme reverse transcriptase. The cDNA contains only the sequence of the exons and can now be properly transcribed and translated by a bacterium.

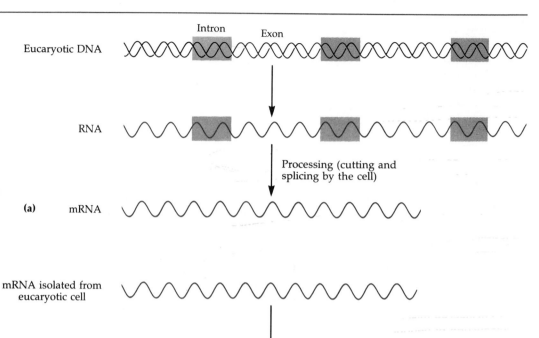

vitro, with viral reverse transcriptase and other enzymes as catalysts (Figure 8-30b; see also Figure 12-21). This **cDNA** (*complementary DNA*) represents the entire coding sequence of DNA for the protein of interest. That is, it contains all the exons of the gene in the correct sequence but does not contain the introns. When the cDNA is introduced into a bacterial cell in a recombinant plasmid, the bacterium can then transcribe and translate the eucaryotic cDNA.

Recombinant DNA research not only increases our understanding of how genes are regulated but also has widespread practical applications. Animal genes can be inserted into bacterial DNA so that bacteria produce substances of therapeutic importance to humans. For example, bacteria have already been used to synthesize *somatostatin,* a hormone related to growth; *insulin,* a hormone used by diabetics; *human growth hormone,* a hormone that promotes growth in children; and *interferon,* an important antiviral (and possible anticancer) substance that will be discussed in Chapter 15.

Recombinant DNA techniques are also being used to produce certain proteins that appear on the surface of pathogenic microorganisms. Such proteins can be given to animals as a vaccine, thereby causing the animals to produce natural antibodies against the pathogenic organism and protecting them from infection.

A third application of recombinant DNA research is to improve existing pharmaceutical processes for making microbial products such as antibiotics. A fourth application is its use by the chemical and energy industries—for example, using bacteria to convert ethylene to ethylene glycol or to ferment plants into ethanol for use as fuel. The box on p. 229 lists some uses of microbes in biotechnological research and development. In virtually all of the categories shown, recombinant microorganisms are playing a major role.

Genes and Evolution

We have now seen how genes can be controlled by the cell's internal regulatory mechanisms and how genes can be altered by mutation and recombination. All these processes provide diversity in the descendants of cells, and diversity is the driving force of evolution. Without genetic diversity, there can be no natural selection by environmental factors. The different kinds of microorganisms that exist today, all with relatively stable properties, are the result of a long history of evolution. During that history, microorganisms have continuously changed by altering their genetic properties and adapting to many different habitats.

MICROBIOLOGY HIGHLIGHTS

Microbial Products in Biotechnology

In Human Medicine and Nutrition

1. Production of vaccines, such as hepatitis vaccine, by genetically engineered microbes.
2. Production of hormones, such as insulin and human growth hormone, by genetically engineered microbes.
3. Production of peptide factors, such as streptokinase for dissolving blood clots, by genetically engineered microbes.
4. Production of vitamins for food supplements by microbial fermentation.
5. Production of amino acids for food supplements by microbial fermentation.
6. Production of antibiotics by fungi and bacteria.

In Agriculture and Animal Husbandry

1. Production of vaccines, such as for hoof-and-mouth disease, by genetically engineered microbes.
2. Production of single-cell protein, microbial biomass, as animal feed.
3. Production of animal feed supplements, antibiotics, and hormones by microbial fermentations.
4. Genetic engineering of crop plants using microbial genes or enhancing plant–microbe associations.
5. Genetic engineering of bacteria to protect crop plants, such as "ice-minus" bacteria (*Pseudomonas syringae*) to protect against frost.
6. Production of insect toxins, such as anticaterpillar toxin, by bacteria (*Bacillus thuringiensis*).

In Manufacturing

1. Production of energy from plant biomass by microbial fermentation.
2. Production of chemicals, such as solvents and raw materials for plastics.
3. Production of enzymes, such as invertase from yeast (used in candy making) and rennin, a calf enzyme made in genetically engineered yeast (used in cheese making).
4. Processing of low quality ores for the mining industry by bacteria (*Thiobacillus*—see the box in Chapter 27).
5. Manufacture of *biochips*, semiconductors made using engineered bacteria (under development).
6. Production of fuel cells with membranes of photosynthetic bacteria (under development).

In Environmental Quality

1. Sewage treatment.
2. Water processing and recycling.
3. Degradation of xenobiotics, such as halogenated insecticides, plastics, and toxic pollutants, by genetically engineered bacteria.

Study Outline

Structure and Function of the Genetic Material (pp. 194–207)

1. Genetics is the study of what genes are, how they carry information, how their information is expressed, and how they are replicated and passed to subsequent generations or other organisms.
2. DNA in cells exists as a double-stranded helix; the two strands are held together by hydrogen bonds between specific nitrogenous base pairs: A-T and C-G.
3. A gene is a segment of DNA, a sequence of nucleotides, that codes for a functional product, usually a protein.
4. The genetic information in a region of DNA is transcribed to produce RNA; mRNA is translated into proteins.

Genotype and Phenotype (p. 195)

1. Genotype is the genetic composition of an organism—its entire DNA.
2. Phenotype is the expression of the genes—the proteins of the cell and their ability to perform particular chemical reactions.

DNA and Chromosomes (p. 196)

1. The DNA in a chromosome exists as one long double helix associated with various proteins that regulate genetic activity.
2. Bacterial DNA is circular; the chromosome of *E. coli*, for example, contains about 4 million base pairs and is approximately 1000 times longer than the cell.
3. The DNA of eucaryotic chromosomes is more condensed and is associated with much protein, including histones.

DNA Replication (pp. 196–201)

1. During DNA replication, the two strands of the double helix separate at the replication fork, and each strand is used as a template by DNA polymerases to synthesize two new strands of DNA according to the rules of nitrogenous basepairing.

2. The result of DNA replication is two new strands of DNA, each having a base sequence complementary to one of the original strands.

3. Because each double-stranded DNA molecule contains one original and one new strand, the replication process is called semiconservative.

4. DNA is synthesized in one chemical direction called 5'→3'. At the replication fork, one daughter strand is synthesized continuously, the other, discontinuously.

5. DNA synthesis of new fragments on the lagging or discontinuous strand moves in the opposite direction from the movement of the replication fork. Completion of the lagging strand requires RNA primers, RNA polymerase, and DNA ligase as well as DNA polymerase.

RNA and Protein Synthesis (pp. 202–206)

Transcription (pp. 202–203)

1. During transcription, the enzyme RNA polymerase synthesizes a strand of RNA from one strand of double-stranded DNA, which serves as a template.

2. RNA is synthesized from nucleotides containing the bases A, C, G, and U, which pair with the bases of the DNA sense strand.

3. The starting point for transcription, where RNA polymerase binds to DNA, is the promoter site; the region of DNA that is the end point of transcription is the terminator site.

4. The product of transcription is RNA. This RNA, called mRNA, directs protein synthesis.

Translation (pp. 203–206)

1. The process in which the information in the nucleotide-base sequence of mRNA is used to dictate the amino acid sequence of a protein is called translation.

2. The mRNA associates with ribosomes, which consist of rRNA and protein.

3. Three-base segments of mRNA that specify amino acids are called codons.

4. Specific amino acids are attached to molecules of tRNA. Another portion of the tRNA has a base triplet called an anticodon.

5. The basepairing of codon and anticodon at the ribosome results in specific amino acids being brought to the site of protein synthesis.

6. The ribosome moves along the mRNA strand as amino acids are joined to form a growing polypeptide.

7. A polyribosome is an mRNA with several ribosomes attached to it.

The Genetic Code (p. 206)

1. The genetic code refers to the relationship between the nucleotide-base sequence of DNA, the corresponding codons of mRNA, and the amino acids for which the codons code.

2. The genetic code is degenerate; that is, most amino acids are coded for by more than one codon.

3. Of the 64 codons, 61 are sense codons (which code for amino acids), and 3 are nonsense codons (which do not code for amino acids and are stop signals for translation).

4. The start codon, AUG, codes for formylmethionine in eubacteria and for methionine in eucaryotes and archaeobacteria.

Regulation of Gene Expression in Bacteria (pp. 207–210)

1. Gene regulation of protein synthesis is energy efficient because some proteins (enzymes) are only synthesized as they are needed.

2. Constitutive enzymes are always present in a cell.

Repression, Induction, and Attenuation (pp. 207–210)

1. For these gene regulatory mechanisms, the control is aimed at mRNA synthesis.

Repression and Induction (pp. 207–208)

1. Repression controls the synthesis of one or several (repressible) enzymes.

2. When cells are exposed to a particular end product, the synthesis of enzymes related to that product decreases.

3. In the presence of certain chemicals (inducers), cells synthesize more enzymes. This process is called induction.

4. An example of induction is the production of β-galactosidase by *E. coli* in the presence of lactose, so lactose can be metabolized.

Operon Model (p. 209)

1. The formation of enzymes is controlled by structural genes. In bacteria, a group of coordinately regulated structural genes with related metabolic functions and the promoter and operator sites that control their transcription are called an operon.

2. In the operon model, a regulatory gene codes for repressor protein.

3. When the inducer is absent, the repressor binds to the operator, and no mRNA is synthesized.

4. When the inducer is present, it binds to the repressor so that it cannot bind to the operator; thus mRNA is made, and enzyme synthesis is induced.

5. In repressible systems, the repressor requires a corepressor in order to bind to the operator site; thus, the corepressor controls enzyme synthesis.

6. Constitutive genes produce proteins regardless of how much substrate is present. Examples are genes for most of the enzymes of glycolysis.

7. In catabolite repression, transcription of structural genes for catabolic enzymes (such as β-galactosidase) is induced by the absence of glucose and the presence of an alternative substrate (such as lactose).

Attenuation (pp. 209–210)

1. Attenuation is a regulatory mechanism that acts by causing premature termination of mRNA synthesis, it was discovered in studies in the operon for tryptophan synthesis.

2. Near the start of the operon for tryptophan synthesis are a number of tryptophan codons.

3. When tryptophan is present at a high level, the coupling of transcription and translation that occurs in bacteria facilitates premature termination of transcription at the attenuator site.

4. When tryptophan is at a low level, the attenuator does not function and the mRNA is completed and translated into enzymes for making more tryptophan.

Mutation: Change in the Genetic Material (pp. 210–217)

1. A mutation is a change in the nitrogeneous base sequence of DNA; that change causes a change in the gene product coded for by the mutated gene.

2. Many mutations are neutral, some are disadvantageous, and some are beneficial.

Types of Mutations (pp. 211–212)

1. A base substitution occurs when one base pair in DNA is replaced with a different base pair.

2. In a frameshift mutation, one or a few base pairs are deleted or added to DNA.

3. Alterations in DNA can result in missense mutations (amino acid substitutions) or nonsense mutations (which create stop codons).

4. Mutagens are agents in the environment that cause permanent changes in DNA.

5. Some mutations occur without the presence of a mutagen; these are called spontaneous mutations.

Chemical Mutagens (pp. 212–213)

1. Chemical mutagens include base-pair mutagens (for example, nitrous acid), base analogs (for example, 2-aminopurine and 5-bromouracil), and frameshift mutagens (for example, benzpyrene).

Radiation (pp. 213–214)

1. Ionizing radiation causes the formation of ions and free radicals that react with DNA; base substitutions or breakage of the sugar-phosphate backbone result.

2. Ultraviolet radiation is nonionizing; it causes bonding between adjacent thymines.

3. In the presence of light, radiation-induced damage can be repaired by photoreactivation; it can be repaired in the dark by enzymes that cut out and replace the damaged portion of DNA. Errors in the repair process result in mutations.

Frequency of Mutation (p. 214)

1. Mutation rate is the probability that a gene will mutate when a cell divides; the rate is expressed as 10 to a negative power.

2. Mutations usually occur randomly along a chromosome.

3. A low rate of spontaneous mutations is beneficial in providing the genetic diversity needed for evolution.

Identifying Mutants (pp. 215–216)

1. Mutants can be detected by selecting or testing for an altered phenotype.

2. Positive selection involves the selection of mutant cells and rejection of nonmutated cells.

3. Replica plating is used for negative selection, to detect, for example, auxotrophs that have nutritional requirements not possessed by the parent (nonmutated) cell.

Identifying Chemical Carcinogens (p. 217)

1. The Ames test is a relatively inexpensive and rapid test for identifying possible chemical carcinogens.

2. The test assumes that a mutant cell can revert to a normal cell in the presence of a mutagen and that many mutagens are carcinogens.

3. Histidine auxotrophs of *Salmonella* are exposed to an enzymatically treated potential carcinogen, and reversions to the nonmutant are selected.

Genetic Transfer and Recombination in Procaryotes (pp. 217–229)

1. Genetic recombination, which refers to the rearrangement of genes from separate groups of genes, usually involves DNA from different organisms; it contributes to genetic diversity.

2. In crossing over, genes from two chromosomes are recombined into one chromosome containing some genes from each original chromosome.

3. Mechanisms for genetic transfer in bacteria involve a portion of the cell's DNA being transferred from donor to recipient.

4. When some of the donor's DNA has been integrated into the recipient's DNA, the resultant cell is called a recombinant.

Transformation in Bacteria (pp. 218–219)

1. During this process, genes are transferred from one bacterium to another as "naked" DNA in solution.

2. This process was first demonstrated in *Streptococcus pneumoniae.*

3. It occurs naturally among a few genera of bacteria.

Conjugation in Bacteria   (p. 220)

1. This process requires contact between living cells.

2. One type of genetic donor cell is an F^+; recipient cells are F^-. F^+ cells contain plasmids called F factors; these are transferred to the F^- cell during conjugation.

3. When the plasmid becomes incorporated into the chromosome, the cell is called an Hfr (high-frequency recombinant).

4. During conjugation, an Hfr can transfer chromosomal DNA to an F^-. Usually, the Hfr chromosome breaks before it is fully transferred.

Transduction in Bacteria   (pp. 220–222)

1. In this process, DNA is passed from one bacterium to another in a bacteriophage, and is then incorporated into the recipient's DNA.

2. In generalized transduction, any bacterial genes can be transferred.

Recombination in Eucaryotes   (pp. 222–223)

1. In eucaryotes, genetic recombination is associated with sexual reproduction.

2. Sexual reproduction involves the fusion of haploid gametes to form a diploid zygote.

3. Haploid gametes are produced through meiosis, in which portions of chromosomes can be exchanged in the recombination process of crossing over.

Genetic Engineering   (pp. 223–228)

1. The procedure of manipulating genetic material in the laboratory is called genetic engineering.

Plasmids   (pp. 223–224)

1. Plasmids are self-replicating circular molecules of DNA carrying genes that are not usually essential for survival of the cell.

2. There are several types of plasmids, including conjugative, dissimilative, bacteriocinogenic, and resistance plasmids.

Transposons   (pp. 224–226)

1. Transposons are small segments of DNA that can move from one region of a chromosome to another.

2. Transposons, found in the main chromosomes of organisms, in plasmids, and in the genetic material of viruses, vary from simple (insertion sequences) to complex.

3. Transposons can carry any type of gene, including antibiotic-resistance genes, and are thus a natural mechanism for moving genes from one chromosome to another.

Recombinant DNA   (pp. 226–228)

1. Recombinant DNA results from the union of DNA from different sources, often different species; its production in vitro uses restriction enzymes and DNA ligase.

2. Foreign DNA is inserted into a bacterial plasmid, and the process of transformation introduces the plasmid into bacterial cells.

3. cDNA, made from mRNA by reverse transcription, is used to insert eucaryotic genes into bacteria.

4. Through genetic engineering, bacteria can produce human substances of therapeutic importance, carry out improved fermentation processes, and synthesize chemicals for industry.

Genes and Evolution   (p. 228)

1. Diversity is the precondition of evolution.

2. Genetic mutation and recombination provide a diversity of organisms, and the process of natural selection allows the growth of those best adapted for a given environment.

Study Questions

Review

1. Briefly describe the components of DNA, and explain its functional relationship to RNA and protein.

2. Draw a diagram showing a portion of a chromosome undergoing replication.
   (a) Identify the replication fork.
   (b) What is the role of DNA polymerase? Of RNA polymerase?
   (c) What is the role of DNA ligase?
   (d) Why is the lagging strand synthesized discontinuously?
   (e) How does this process represent semiconservative replication?

3. Below is a code for a strand of DNA.

(3')   (5')
A T A T T A C T T T G C T T G G A C T
1 2 3 4 5 6 7 8 9 10 11 12 13 14 15 16 17 18 19
TATA = Promoter sequence

   (a) Write the code for the complementary strand of DNA.
   (b) Using the genetic code provided in Figure 8-10, identify the sequence of amino acids coded for by

mRNA transcribed from this strand of DNA (the complementary strand is the antisense strand).

(c) What would be the effect if C was substituted for T at base 10?

(d) What would be the effect if A was substituted for G at base 11?

(e) What would be the effect if G was substituted for T at base 14?

(f) What would be the effect if C was inserted between bases 9 and 10?

(g) How would ultraviolet radiation affect this strand of DNA?

(h) Identify a nonsense sequence in this strand of DNA.

4. Describe translation, and be sure to include the following terms: ribosome, rRNA, amino acid activation, tRNA, anticodon, and codon.

5. Contrast the structures of a bacterial chromosome and a eucaryotic chromosome.

6. Explain the selection of an antibiotic-resistant mutant by direct selection and the selection of an antibiotic-sensitive mutant by indirect selection.

7. Match the following examples of mutagens.

_____ A mutagen that is incorporated into DNA in place of a normal base

_____ A mutagen that causes the formation of highly reactive ions

_____ A mutagen that alters adenine so that it basepairs with cytosine

_____ A mutagen that causes insertions

_____ A mutagen that causes the formation of pyrimidine dimers

(a) Frameshift mutagen
(b) Base analog
(c) Base-pair mutagen
(d) Ionizing radiation
(e) Nonionizing radiation

8. Describe the principle of the Ames test for identifying chemical carcinogens.

9. Differentiate between transformation and transduction.

10. Define plasmids, and explain the relationship between F plasmids and conjugation.

11. Outline the production of recombinant DNA in vitro. How is cDNA made? What are the purposes of DNA ligase and restriction enzymes?

12. List three practical applications of genetic engineering. What new product have you heard about that is made by genetic engineering?

13. Use this metabolic pathway to answer the questions below:

$$\text{Substrate } A \xrightarrow[a]{\text{enzyme}} \text{Intermediate } B \xrightarrow[b]{\text{enzyme}} \text{End-product } C$$

(a) If enzyme *a* is inducible and is not being synthesized at present, a _____ protein must be bound tightly to the _____ site. When the inducer is present, it will bind to the _____ so that _____ can occur.

(b) If enzyme *a* is repressible, end produce *C*, called a _____, causes the _____ to bind to the _____. What causes derepression?

(c) If enzyme *a* is constitutive, what effect, if any, will the presence of *A* or *C* have on it?

14. How does tRNA^{trp} stop transcription of tryptophan structural genes?

15. How does a decrease in glucose result in transcription of new catabolic genes?

16. Define the following terms:
(a) Genotype
(b) Phenotype
(c) Recombination

17. Why are mutation and recombination important in the process of natural selection and the evolution of organisms?

18. You are provided with cultures with the following characteristics.

Culture 1: F$^+$, genotype $A^+ B^+ C^+$
Culture 2: F$^-$, genotype $A^- B^- C^-$

(a) Indicate the possible genotypes of a recombinant cell resulting from the conjugation of cultures 1 and 2.

(b) Indicate the possible genotypes of a recombinant cell resulting from conjugation of the two cultures after the F$^+$ has become an Hfr.

Challenge

1. Base analogs and ionizing radiation are used in the treatment of cancer. These are mutagens, so how do you suppose they are used to treat diseases?

2. Why are semiconservative replication and degeneracy of the genetic code advantageous to the survival of species?

3. Replication of *E. coli*'s chromosome takes from 40 to 45 minutes, and the cell has a generation time of 26 minutes. How does the cell have complete chromosomes for each daughter cell? For each granddaughter cell?

4. Why are Ames test cultures incubated in the dark?

Further Reading

Ayala, F. J., and J. A. Kiger, Jr. *Modern Genetics*, 2nd ed. Menlo Park, Calif: Benjamin/Cummings, 1984. A genetics textbook with excellent coverage of bacterial and viral genetics.

Cech, T. R. "RNA as an enzyme." *Scientific American* 255(5):64–75, November 1986. Describes how eucaryotic RNA works to remove exons to produce an RNA molecule. Poses the question of whether proteins are the only enzymes.

Devoret, R. "Bacterial tests for potential carcinogens." *Scientific American* 241(2):40–49, August 1979. Describes short tests using bacteria instead of animals to detect carcinogenic chemicals and the effects of carcinogens on DNA.

Freifelder, D. *Molecular Biology: A Comprehensive Introduction to Procaryotes and Eucaryotes*, 2nd ed. Boston: Jones and Bartlett, 1987. Includes fairly detailed discussions of bacterial and viral genetics.

Gilbert, W., and L. Villa-Komaroff. "Useful proteins from recombinant bacteria." *Scientific American* 242(4):74–94, April 1980. Describes how genes are spliced into bacteria so the bacteria can produce nonbacterial proteins, such as insulin and interferon.

Howard-Flanders, P. "Inducible repair of DNA." *Scientific American* 245(5):72–80, November 1981. Describes the roles of two enzymes that respond to an "SOS" for DNA damage.

Stahl, F. W. "Genetic recombination." *Scientific American* 256(2):91–101, February 1987. Describes current studies in genetic recombination using a lambda-phage–*E. coli* model.

Steitz, J. A. " 'Snurps.' " *Scientific American* 258(6):56–63, June 1988. Describes how introns are removed from eucaryotic RNA by *small nuclear ribonucleoproteins*.

Stewart, G. J., and C. A. Carlson. "The biology of natural transformation." *Annual Review of Microbiology* 40:211–235, 1986. An overview of information on transformation acquired from laboratory studies and a discussion of the importance of transformation in nature.

Watson, J. D., N. H. Hopkins, J. W. Roberts, J. A. Steitz, and A. M. Weiner. *Molecular Biology of the Gene*, 4th ed. 2 volumes. Menlo Park, Calif.: Benjamin/Cummings, 1987. A recent, authoritative textbook that discusses topics of microbial genetics in great detail; very clear explanations.

Watson, J. D., J. Tooze, and D. T. Kurtz. *Recombinant DNA: A Short Course*. New York: W. H. Freeman, 1983. Describes laboratory techniques used to identify, isolate, and transfer genes in procaryotic and eucaryotic cells; includes a chapter on the recombinant DNA industry.

PART TWO

Survey of the Microbial World

CHAPTER 9

Classification of Microorganisms

The science of classification, especially the classification of living forms, is called **taxonomy**, from the Greek words for law and order. The objective of taxonomy is to classify living organisms, that is, to establish the relationship between one group of organisms and another and to differentiate between them. A taxonomic system enables us to classify organisms that have not yet been studied in detail. That is, a heretofore unknown organism can be identified, then grouped or classified with other organisms that have similar characteristics. For example, very little is known about bacteria that live at great depths in the ocean, but data suggest these bacteria are related to the known archaeobacteria.

The practical value of taxonomy is to provide a method of identifying organisms already classified. For example, when a bacterium suspected of causing a specific disease is isolated from a patient, characteristics of that isolate are matched to lists of characteristics of previously classified bacteria to identify the isolate. After the bacterium has been identified, drugs can be selected that affect that bacterium.

Taxonomy is a basic but necessary tool for scientists. It provides a universal language of communication and a common reference for identification. Modern taxonomy is an exciting and dynamic field; new techniques in molecular biology and genetics are providing new insights into classification and evolution.

In this chapter, you will learn the various classification systems, the different criteria used for classification, and tests that are used to identify bacteria that are already classified.

Learning Objectives

- Define taxonomy and binomial nomenclature.
- List the major characteristics used to differentiate among kingdoms in the five-kingdom system and the three-kingdom system.
- Identify the characteristics of the Kingdom Procaryotae that differentiate it from other kingdoms.
- Describe the divisions used in *Bergey's Manual*.
- Compare and contrast classification and identification.
- Describe ten methods of identification and classification of microorganisms.

Phylogenetic Relationships

Charles Darwin's (1809–1882) theory of evolution explains that the many similarities among organisms are a result of their descent from common ancestors. The arrangement of organisms into taxonomic categories, collectively called **taxa**, reflects degrees of relatedness among organisms. That is, the hierarchy of taxa shows evolutionary or **phylogenetic** (from a common ancestor) relationships.

The most fundamental taxonomic or phylogenetic division is that between procaryotes and eucaryotes. One widely held theory (discussed in Chapter 4) is that eucaryotic cells evolved from procaryotic cells living inside one another (as endosymbionts). In fact, the similarities between procaryotic cells and eucaryotic organelles (Table 9-1) provide striking evidence for this endosymbiotic relationship. (Recall that procaryotes do not have a nucleus separated from the cytoplasm by a membrane, as eucaryotes do. Instead, they have a nuclear region in which the genetic material is more or less localized.) The oldest known fossils are the remains of organisms that lived more than 3.5 billion years ago. They are procaryotes. Eucaryotic cells evolved more recently, about 1.4 billion years ago.

A Five-Kingdom System

In 1969, Robert H. Whittaker proposed a **five-kingdom system** of biological classification that shows procaryotes to be the ancestors to all eucaryotes (Figure 9-1). In this system, all procaryotes are included in the Kingdom **Procaryotae** (or Monera). The four eucaryotic kingdoms are distinguished according to nutritional requirements, patterns of development, tissue differentiation, and possession of 9 + 2 flagella. (Recall that eucaryotic flagella and cilia are composed of nine outer pairs plus an inner pair of microtubules, whereas procaryotic flagella are composed of repeating flagellin subunits.)

Simple eucaryotic organisms, mostly unicellular, are grouped as the **Protista**. Protists generally have flagella at some time during their life cycle. The multicellular protists lack tissue organization. The Kingdom Protista includes funguslike water molds, slime molds, protozoans, and the primitive eucaryotic algae.

Fungi, plants, and animals comprise the three kingdoms of more complex eucaryotic organisms, most of which are multicellular.

Fungi include the unicellular yeasts, multicellular molds, and macroscopic species such as mushrooms. To obtain raw materials for vital functions, a fungus absorbs dissolved organic matter through its plasma membranes. The cells of a multicellular fungus are commonly joined together to form thin tubes called hyphae. These cells lack complete walls, so they form continuous, multinucleated cytoplasm rather than discrete cells. Most fungi lack flagella. Fungi develop from spores or from fragments of hyphae.

The Kingdom **Plantae** (plants) includes some algae and all mosses, ferns, conifers, and flowering plants. All members of this kingdom are multicellular. To obtain energy, a plant uses photosynthesis; this process converts carbon dioxide and water into organic molecules used by the cell.

The kingdom of multicellular organisms called **Animalia** (animals) includes sponges, various worms, insects, and animals with backbones (vertebrates). Animals obtain nutrients and energy by ingesting organic matter through a mouth of some kind.

A Three-Kingdom System

In 1978, Carl R. Woese proposed a three-kingdom system of classification (Figure 9-2). The grouping of all bacteria into Procaryotae in the five-kingdom system had been based on microscopic observations. Woese's

TABLE 9-1   *A Comparison of Procaryotic and Eucaryotic Cells and Organelles*

	Procaryotic Cells	Eucaryotic Cells	Mitochondria and Chloroplasts (Eucaryotic Organelles)
DNA	Circular	Linear	Circular
Histones	No	Yes	No
Ribosomes	70S	80S	70S
Growth	Binary fission	Mitosis	Binary fission

Figure 9-1 Five-kingdom system. This is a commonly accepted system of classification.

FUNGI
(absorption)

PLANTS
(photosynthesis)

ANIMALS
(ingestion)

Arthropods   Chordates

(Eucaryotic;
mostly multicellular)

Yeasts

Conjugation
fungi (molds)

Club fungi
(mushrooms)

Green algae

Sac fungi
(molds)

Brown algae

Other
protists

Plasmodial
slime molds

Water
molds

Red algae

Flagellates

Euglenoids

Ciliates

PROTISTA
(Eucaryotic;
mostly unicellular)

Foraminifera

Cellular
slime molds

Dinoflagellates

Amoeba

Other
bacteria

Cyanobacteria

PROCARYOTAE (MONERA)
(procaryotic; unicellular)

proposal was based on modern techniques in molecular biology and biochemistry that revealed that there are actually two types of procaryotic cells. This discovery was based on the observation that ribosomes are not the same in all cells (as you will recall from Chapter 4). In particular, comparing the sequences of nucleotides in ribosomal RNA (rRNA) from different kinds of cells shows that there are three distinctly different cell groups, the eucaryotes and two different

types of procaryotes—the **eubacteria**, or "true bacteria," and the **archaeobacteria** (from *archaios*, meaning ancient).*

Woese believed, therefore, that the archaeobacteria and the eubacteria (the Procaryotae in the five-kingdom system), while similar in appearance, should

*Debate over the proper spelling of the term *archaeobacteria* currently exists. *Archaeobacteria* is used in *Bergey's Manual of Systematic Bacteriology*; other references use *archaebacteria.*

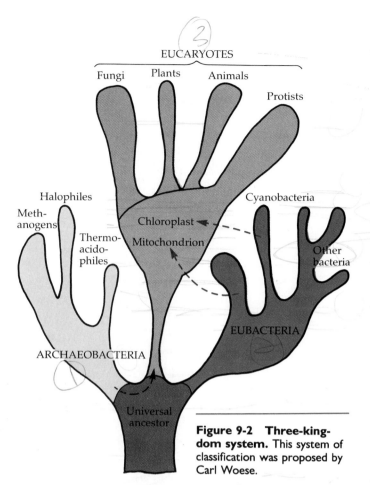

Figure 9-2 Three-kingdom system. This system of classification was proposed by Carl Woese.

form their own separate branches on the evolutionary tree. All eucaryotes (divided into Protista, Fungi, Plantae, and Animalia in Whittaker's system), he believed, should be grouped into a third kingdom. Differences in membrane lipid structure, transfer RNA molecules, and sensitivity to antibiotics also support classification of the three cell types into separate kingdoms (Table 9-2).

Archaeobacteria differ from eubacteria in a number of other ways. For example, their cell walls never contain peptidoglycan, they often live in extreme environments, and they carry out unusual metabolic processes. They include three groups:

1. The methanogens, strict anaerobes that produce methane (CH_4) from carbon dioxide and hydrogen.
2. Extreme halophiles, which require high concentrations of salt for survival.
3. Thermoacidophiles, which normally grow in hot, acidic environments.

Future research may reveal whether eubacteria actually evolved from archaeobacteria or from a different ancestor. (An archaeobacterium may have provided the original host in which endosymbiotic eubacteria developed into organelles.) Biochemical similarities between archaeobacteria and eucaryotes

TABLE 9-2   *Some Characteristics of Archaeobacteria, Eubacteria, and Eucaryotes*

	Archaeobacteria	Eubacteria	Eucaryotes
Cell type	Procaryotic	Procaryotic	Eucaryotic
Cell wall	Varies in composition; contains no peptidoglycan	Contains peptidoglycan	Varies in composition; contains carbohydrates
Membrane lipids	Composed of branched carbon chains attached to glycerol by ether linkage	Composed of straight carbon chains attached to glycerol by ester linkage	Composed of straight carbon chains attached to glycerol by ester linkage
Start signal for protein synthesis	Methionine	Formylmethionine	Methionine
Antibiotic sensitivity	No	Yes	No
rRNA	Procaryotic loop lacking*	Procaryotic loop present*	Procaryotic loop lacking*
Common arm of tRNA**	Lacking	Present	Present

*Binds to ribosomal protein; found in all eubacteria.
**A sequence of bases on tRNA found in all eucaryotes and eubacteria: guanine-thymine-pseudouridine-cytosine-guanine.

(Table 9-2) suggest that the original eucaryotic cell inhabited by symbiotic eubacteria was an archaeobacterium such as *Thermoplasma* (thėr-mō-plaz′mä), whose chromosome contains proteins similar to eucaryotic histones. An ancestor of the modern eubacteria *Bdellovibrio* (del-lō-vib′rē-ō) or *Daptobacter* (dap-tō-bak′tėr), which can invade and reproduce within other bacteria, may have been the original symbiont. Acceptance of the three-kingdom system depends, at least in part, on answering the question about the origin of eucaryotes. At present, the five-kingdom system is accepted by most biologists as the most useful taxonomic tool.

Taxonomists are searching for a natural classification system that depicts phylogenetic relationships. As you can see, taxonomy has developed into a tool for clarifying the evolution of organisms as well as their relationships to each other.

Classification of Organisms

Scientific Nomenclature

In a world inhabited by millions of living organisms, biologists must be sure they know exactly which organism is being discussed. We cannot use common names because the same name is often used for many different organisms in different regions. For example, there are two different plants with the common name Spanish moss, and neither one is a moss. Three different animals are referred to as a gopher. Because common names are rarely specific and can often be misleading, a system of scientific names, referred to as scientific nomenclature, was developed in the eighteenth century by Carolus Linnaeus.

This system gives every organism two names, usually derived from either Latin or Greek. These are the **genus** name and **specific epithet (species)**, and both names are printed underlined or italicized. The genus name is always capitalized and is always a noun. The species name is written in lowercase and is usually an adjective. Because this system gives two names to each organism, the system is called **binomial nomenclature**.

Let us consider a few examples. Your own genus and specific epithet are *Homo sapiens* (hō′mō sā′pē-ens); one bacterium that causes pneumonia is called *Klebsiella pneumoniae* (kleb-sē-el′lä nü-mō′nē-ī); and a mold that contaminates bread is called *Rhizopus*

Figure 9-3   Classification of living organisms.

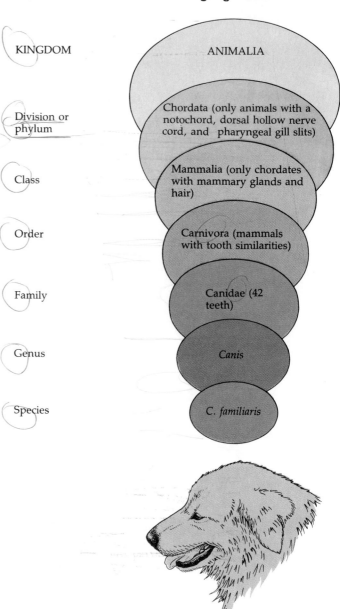

KINGDOM — ANIMALIA

Division or phylum — Chordata (only animals with a notochord, dorsal hollow nerve cord, and pharyngeal gill slits)

Class — Mammalia (only chordates with mammary glands and hair)

Order — Carnivora (mammals with tooth similarities)

Family — Canidae (42 teeth)

Genus — *Canis*

Species — *C. familiaris*

nigricans (rī-zō′pus nī′gri-kans). The genus and specific epithet are spelled out the first time they are used (*Klebsiella pneumoniae*), but thereafter the genus name may be abbreviated (*K. pneumoniae*).

Rules for assigning names to newly classified bacteria and for assigning bacteria to taxa are established by the International Committee on Systematic Bacteriology and are published in the *Bacteriological Code*. Descriptions of bacteria and evidence for their classifications are published in the *International Journal of Systematic Bacteriology* before being incorporated into a reference called *Bergey's Manual*.

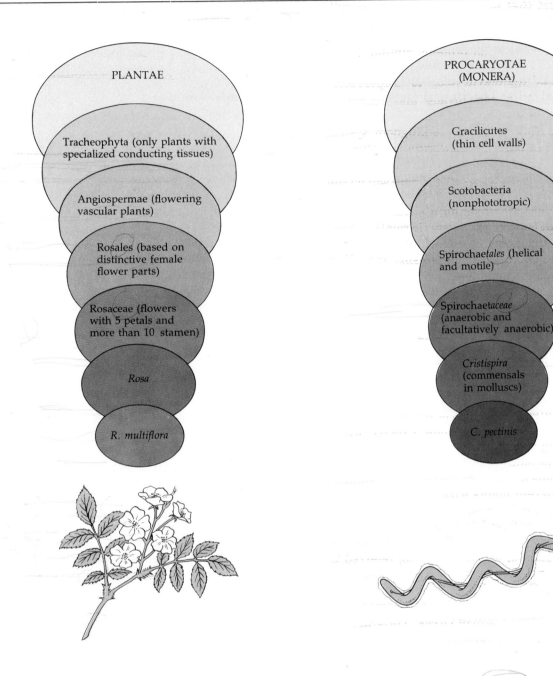

According to the *Bacteriological Code*, scientific names are to be taken from Latin (a genus name can be taken from Greek) or latinized by the addition of the appropriate suffix. Suffixes for order and family are specified as *-ales* and *-aceae*, respectively (Figure 9-3).

The Taxonomic Hierarchy

All organisms can be grouped into a series of subdivisions that make up the taxonomic hierarchy. For eucaryotic organisms, a species is a group of closely related organisms that breed among themselves. (Bacterial species will be discussed shortly.) A genus (plural, *genera*) consists of species that differ from each other in certain ways but are related by descent. For example, *Quercus* (kwer'kus), the genus name for oak, consists of all types of oak trees (white oak, red oak, bur oak, velvet oak, and so on). Even though each species of oak differs from every other species, they are all related genetically. Just as a number of species make up a genus, related genera make up a **family**. A group of similar families constitutes an **order**, and a group of similar orders makes up a **class**. Related classes, in turn, comprise a **division**. (In zoology, the comparable term **phylum** is used.) Thus, a particular organism (or species) has a genus name and specific epithet and belongs to a family, order, class, and phylum or division. All phyla or divisions that are related to each other make up a **kingdom** (Figure 9-3).

A Phylogenetic Hierarchy

Grouping organisms according to common properties implies that a group of organisms evolved from a common ancestor. Each species retains some of the characteristics of the ancestor. Some of the information used to classify and to determine evolutionary relationships in higher organisms comes from fossils. Bones, shells, or stems that contain mineral matter or have left imprints in rock that was once mud are examples of fossils.

The structures of most microorganisms are not readily fossilized. Some notable exceptions are:

1. A marine protist whose fossilized colonies form the White Cliffs of Dover, England.
2. Stromatolites, the fossilized remains of microbial communities that flourished between 0.5 and 2 billion years ago.
3. The fossilized cyanobacteria found in 3- to 3.5-billion-year-old rocks in Western Australia. These are the oldest known fossils. Some fossils of procaryotes are shown in Figure 9-4.

Since fossil evidence is not available for the majority of procaryotes, their phylogeny must be based on other kinds of evidence. Conclusions from DNA hybridization studies (discussed shortly) of selected orders and families of eucaryotes are in agreement with the fossil records. This has encouraged workers to use DNA and RNA hybridization and RNA sequencing to gain an understanding of the evolutionary relationships among procaryotic groups.

Classification of Bacteria

The taxonomic classification scheme for bacteria may be found in *Bergey's Manual of Systematic Bacteriology*. In *Bergey's Manual*, bacteria are divided into four divisions (archaeobacteria comprise their own division). Each division is divided into classes; in all, there are seven classes of bacteria. The divisions and classes are listed in Table 9-3. Classes are divided into orders; orders, into families; families, into genera; and genera, into species.

A bacterial species is defined somewhat differently than a eucaryotic species, which is a group of closely related organisms that can interbreed. Unlike reproduction in eucaryotic organisms, cell division in bacteria is not directly tied to sexual conjugation, which is infrequent and does not always need to be species-specific. A **bacterial species**, therefore, is defined simply as a population of cells with similar characteristics. (The types of characteristics will be discussed later in this chapter). The members of a bacterial species are essentially indistinguishable from each other but are distinguishable from members of other species, usually on the basis of several features. In some cases, pure cultures of the same species are not identical in all ways. Each such group is called a **strain**, which is a group of cells all derived from a single cell. Strains are identified by numbers, letters, or names that follow the specific epithet.

Figure 9-4 Fossilized procaryotes. (a) Filamentous procaryotes from the Early Precambrian (ca. 3500 million years ago) of Western Australia (×1000). **(b)** Coccoid cyanobacteria from the Late Precambrian (ca. 850 million years ago) of central Australia (×1800). **(c)** A filamentous cyanobacterium from the Late Precambrian (ca. 700 million years ago) from China (×5000).

(a)

(b)

(c)

TABLE 9-3 *Divisions and Classes in the Kingdom Procaryotae (Monera) Identified by Common Names*

Division	Class
Typical gram-negative cell wall	Nonphotosynthetic bacteria
	Anaerobic photosynthetic bacteria
	Cyanobacteria
Typical gram-positive cell wall	Rods and cocci
	Actinomycetes and related organisms
Wall-less procaryotes	Mycoplasmas
Unusual walls	Archaeobacteria

Bergey's Manual provides a reference for identifying bacteria in the laboratory as well as a classification scheme for bacteria. Suggested evolutionary relationships are given for most groups. Information used to build (and modify) the phylogenic models comes from analyses of nucleotide sequences in DNA and RNA, from DNA hybridizations, and from chemical analyses of cellular components. (All these techniques will be described later in this chapter.) A scheme of the suggested evolutionary relationships of bacteria is shown in Figure 9-5.

Not all bacteria have yet been surveyed, and those needing further investigation are temporarily placed in descriptive categories or sections, such as "Endosymbionts" and "Dissimilatory Sulfate- or Sulfur-Reducing Bacteria."

Classification of Viruses

Viruses are not classified as part of any of the five kingdoms discussed earlier in this chapter. Viruses are not composed of cells, and they use the anabolic machinery of host cells to multiply inside living cells. Because a viral genome (genetic material) can direct biosynthesis inside a host cell, and because some viral genomes can become incorporated into the host genome, it is possible that viruses might be more closely related to their hosts than to other viruses.

Viruses are obligatory intracellular parasites, so they must have evolved after a suitable host cell had evolved. There are two hypotheses on the origin of viruses: (1) they arose from independently replicating strands of nucleic acids (such as plasmids), and (2) they developed from degenerative cells that, through many generations, gradually lost the ability to survive independently but could survive when associated with another cell. Viruses will be discussed in Chapter 12.

Figure 9-5 Phylogenetic relationships of procaryotes. Pathways indicate possible major lines of descent of bacterial groups.

Criteria for Classification and Identification of Microorganisms

A classification scheme provides a list of characteristics and a means for comparison to aid in the identification of an organism. Once an organism is identified, it can be placed into a previously devised classification scheme. Microorganisms are *identified* for practical purposes—for example, to determine an appropriate treatment for an infection. They are not necessarily identified by the same techniques by which they are classified. Most identification procedures are easily performed in a laboratory and require as few procedures or tests as possible.

Bergey's Manual of Determinative Bacteriology has been a widely used reference for identification of bacteria since publication of the first edition in 1923. *Bergey's Manual of Systematic Bacteriology* includes identifying characteristics for some bacteria. Volume 1 includes wall-less eubacteria and some gram-negative eubacteria. Volume 2 describes gram-positive eubacteria. Volume 3 includes some gram-negative eubacteria (photosynthetic bacteria, chemolithotrophic bacteria, budding, appendaged, and sheathed bacteria, gliding and fruiting bacteria) and the archaeobacteria. Volume 4 describes the actinomycetes.

At the time this chapter is being written, only Volumes 1 and 2 of *Bergey's Manual of Systematic Bacteriology* are yet published. Until the remaining volumes are published, *Bergey's Manual of Determinative Bacteriology*, 8th edition, can be used for laboratory identification. The latter does not classify bacteria according to evolutionary relatedness but provides identification schemes based on such criteria as cell wall composition, morphology, differential staining, oxygen requirements, and biochemical testing.

In recent years, microbiologists have been acquiring information on relationships among bacteria and are constructing identification schemes that reflect the evolutionary relationships used in *Bergey's Manual of Systematic Bacteriology*.

Medical microbiology (the branch of microbiology dealing with human pathogens) has dominated interest in microbes, and this interest is reflected in many identification schemes. However, to put the pathogenic properties of bacteria in perspective, of the more than 1800 species listed in the *Approved Lists of Bacterial Names*, fewer than 200 of these are human pathogens.

We will next discuss several criteria and methods for the classification of microorganisms and the routine identification of some. In addition to these methods, the source and habitat of a bacterial isolate will be considered as part of the classification and identification processes.

Morphological Characteristics

Higher organisms are frequently classified according to observed anatomical detail. However, many microorganisms appear too similar to be classified by their structures. Through a microscope, there is a great deal of similarity between organisms that might differ in metabolic or physiological properties. There are literally hundreds of bacterial species that are small rods or small cocci.

Cell morphology tells us little about phylogenetic relationships. However, morphological characteristics are still useful in identifying bacteria. For example, differences in such structures as endospores or flagella can be helpful.

Differential Staining

One of the first steps in the identification of bacteria is differential staining. As we have seen from earlier chapters, most bacteria are either gram-positive or gram-negative. Other differential stains, such as the acid-fast stain, can be useful for a more limited group of microorganisms. Recall that these stains are based on the chemical composition of the cell walls and are therefore not useful in identifying either the wall-less bacteria or the archaeobacteria with unusual walls.

Biochemical Tests

Enzymatic activities are widely used to differentiate among bacteria. Even closely related bacteria can usually be separated into distinct species by subjecting them to biochemical tests, such as their ability to ferment an assortment of selected carbohydrates.

Enteric, gram-negative bacteria are a large heterogeneous group of microbes whose natural habitat is the intestinal tract of humans and animals. Among the enteric bacteria are members of the genera *Escherichia* (esh-ėr-i′kē-ä), *Enterobacter* (en-te-rō-bak′tėr), *Shigella* (shi-gel′lä), and *Salmonella* (sal-mōn-el′lä). *Escherichia* and *Enterobacter* can be distinguished from *Salmonella* and *Shigella* according to their ability to ferment lactose—the former ferment lactose to produce acid and gas, whereas the latter do not. Further biochemical testing, as shown in Figure 9-6, can differentiate between the genera.

The time needed to identify bacteria can be reduced considerably by the use of selective and differential media or by rapid identification methods. Recall from Chapter 6 that selective media contain ingredients that suppress the growth of competing organisms and encourage the growth of desired ones, and differential media allow the desired organism to form a colony that is somehow distinctive.

Rapid identification tools are manufactured for groups of medically important bacteria, such as the enterics. Such tools are designed to perform several biochemical tests simultaneously and can identify bacteria within 4 to 24 hours. The results of each test are assigned a number (which varies with the tool being used) based on the relative reliability and importance of each test. (See Microview C.)

In the example shown in Figure 9-7, an unknown enteric bacterium is inoculated into a tube that is designed to perform 15 biochemical tests. After incubation, results in each compartment are recorded. Positive results are marked on the scoring form. Note that each test is assigned a value; the number derived from all the tests is called the I.D. value. Fermentation of glucose is important, and a positive reaction

is valued at 4, as compared with the production of acetoin (V-P test), where a positive reaction is assigned the value of 2. A computerized interpretation of the simultaneous test results is essential and is provided by the manufacturer. In this example, the I.D. value indicates that the bacterium is *K. pneumoniae*, whether the strain is V-P-positive or V-P-negative and citrate-positive or citrate-negative. *Bergey's Manual* does not evaluate the relative importance of each biochemical test and does not always describe strains. In clinical diagnosis of a disease, it is important that a particular species and even a particular strain be identified in order to proceed with proper treatment. Use of another rapid identification test is described in Microview C.

Serology

Serology is the science that deals with the blood serum and particularly with immune responses evident in serum (see Chapter 17). Microorganisms are antigenic; that is, injected microorganisms are capable of stimulating antibody formation in animals. For example, the immune system of a rabbit injected with killed typhoid bacteria responds by producing antibodies against typhoid bacteria. The antibodies are

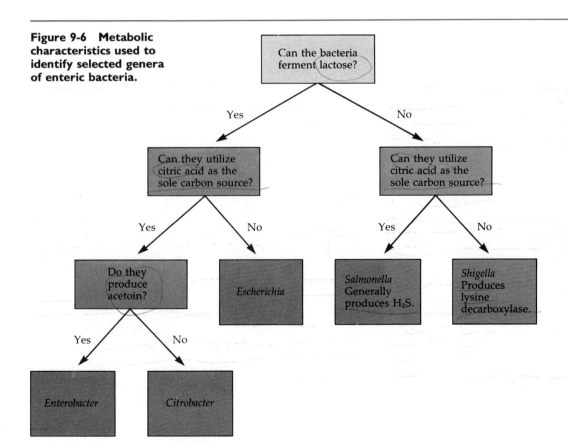

Figure 9-6 Metabolic characteristics used to identify selected genera of enteric bacteria.

Can the bacteria ferment lactose?

Yes → Can they utilize citric acid as the sole carbon source?

No → Can they utilize citric acid as the sole carbon source?

(Yes branch of first left) Can they utilize citric acid as the sole carbon source?
- Yes → Do they produce acetoin?
  - Yes → *Enterobacter*
  - No → *Citrobacter*
- No → *Escherichia*

(No branch / second box) Can they utilize citric acid as the sole carbon source?
- Yes → *Salmonella* Generally produces H₂S.
- No → *Shigella* Produces lysine decarboxylase.

Figure 9-7   One type of rapid identification method for bacteria. Enterotube® II from Roche Diagnostics. **(a)** One tube containing media for 15 tests is inoculated. **(b)** After incubation, the tube is observed for positive results (colored boxes). **(c)** The value for each positive test is circled, and the numbers from each group of tests are added to give the I.D. value. **(d)** Comparing the I.D. value with a computerized listing shows that the organism in the tube is *Klebsiella pneumoniae*. Different strains of *K. pneumoniae* may produce different test results, which are listed in the Atypical Test Results column.

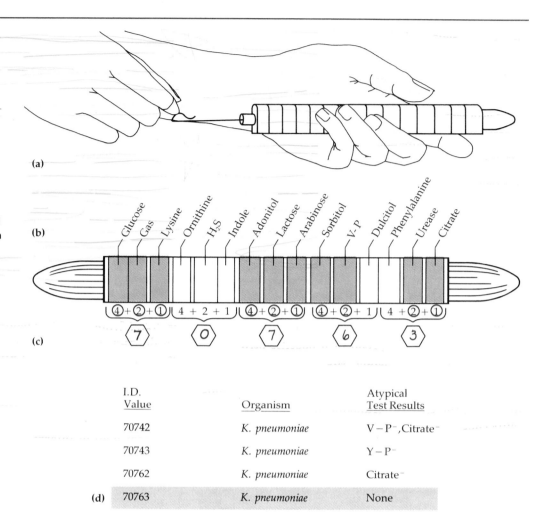

I.D. Value	Organism	Atypical Test Results
70742	*K. pneumoniae*	V – P⁻, Citrate⁻
70743	*K. pneumoniae*	Y – P⁻
70762	*K. pneumoniae*	Citrate⁻
70763	*K. pneumoniae*	None

proteins that circulate in the blood and combine in a highly specific way with the bacteria that caused their production. **Antisera** (singular, **antiserum**) are commercially available solutions of such antibodies used in the identification of many medically important bacteria. If an unknown bacterium is isolated from a patient, it can be tested against known antisera and often identified quickly.

In a procedure called a **slide agglutination test**, the unknown bacterium is placed in a drop of saline on a slide and mixed with a drop of known antiserum. The same bacterium is mixed with other different antisera on different slides. The bacteria agglutinate (clump) when mixed with antibodies that are produced to combine with that strain or species. A positive test is therefore determined by agglutination. A slide agglutination test is illustrated in Figure 9-8.

These serological techniques can differentiate not only among microbial species but also among strains within species. Because closely related bacteria produce some of the same antigens, serological testing is also useful in screening bacterial isolates for possible similarities. If an antiserum reacts with proteins from different bacterial strains or species, these bacteria can be tested further for relatedness.

Phage Typing

Phage typing, like serological testing, looks for similarities among bacteria. Both techniques are useful in tracing the origin and course of a disease outbreak. **Phage typing** is a test for determining which phages a bacterium is susceptible to. Recall from Chapter 8 that bacteriophages (phages) are bacterial viruses and that they usually cause lysis of the bacterial cells they infect. They are highly specialized, in that they usually infect only members of a particular species or even particular strains within a species. One bacterial strain might be susceptible to two different phages, whereas another strain of the same species might be susceptible to those two phages plus a third phage. Bacteriophages will be discussed further in Chapter 12.

Figure 9-8 Slide agglutination test. (a) Positive test. The grainy appearance is due to the clumping (agglutination) of the bacteria. Agglutination results when the bacteria are mixed with antibodies that are produced to react with the same strain. **(b)** Negative test. The bacteria are still evenly distributed in the saline and antiserum.

(a)   (b)

To see how phage typing can help researchers trace the origin and spread of a specific strain of a bacterium, let us consider the following example. Years ago, there was a very high incidence of hospital-associated (nosocomial) staphylococcal infections. Moreover, the chain of transmission was unknown in many cases. Although such outbreaks still occur, they are less frequent because their sources can be traced by phage typing. In one version of this procedure, a plate of staphylococci growing on an agar medium is marked off in small squares, and a drop of each different phage type to be used in the test is placed on the bacteria (Figure 9-9). In squares in which the phages are able to infect and lyse the bacterial cells, clearings in the bacterial growth (called plaques) appear. Such a test might show that staphylococci isolated from a surgical wound have the same pattern of phage sensitivity as those isolated from the operating surgeon or surgical nurses.

Amino Acid Sequencing

A great deal of information can be derived from the study of the amino acid sequences of proteins. The sequence of amino acids in a protein directly reflects the base sequence of the DNA that encodes the protein. The longer the time between the evolution of two organisms, the more changes in the DNA sequence (mutations) encoding the same protein in the two organisms. Thus, the comparison of amino acid sequences from proteins of two different organisms can help determine the similarity in the DNA sequence and therefore the evolutionary relatedness of the species. The more similar the proteins, the more closely related are the organisms. The phylogenetic tree derived for animals from traditional methods, such as fossils and comparative anatomy, is being reinforced today by such a comparison of amino acid sequences.

The sequencing of amino acids in several important proteins, such as cytochrome *c*, has been useful in the study of phylogenetic relationships among

Figure 9-9 Phage typing of a strain of *Staphylococcus aureus*. The tested strain was grown over the entire plate. Plaques, or areas of lysis, were produced by bacteriophages 6, 42E, 47, 54, 75, and 85, indicating that the strain was sensitive to infection by these phages. Accordingly, the strain is identified as phage type 6/42E/47/54/75/85. (DM is a dye marker.)

Figure 9-10   Amino acid sequences for a fragment of cytochrome c from tuna and from two species of bacteria. Three-letter abbreviations indicate the amino acids. Amino acids of the same chemical type are shown in the same color (for example, Glu and Asp are both "acidic amino acids," having an extra —COOH group). The similarity among the three sequences is striking, although the two bacteria are clearly more closely related to each other than to tuna.

many different organisms (Figure 9-10). Amino acid sequencing is limited to use with groups of organisms that share a common protein. Also, it cannot be assumed that all organisms that share a protein are related. Although it is unlikely, it is not impossible for two unrelated organisms to evolve the same solution to a problem—for example, the production of a certain enzyme to transport a needed nutrient.

Protein Analysis

Amino acid sequencing focuses on the sequence of one or a few known proteins in different organisms. Protein analysis compares the pattern of all proteins found in the cells of different organisms. Since a cell's proteins are products of its genes, each species is expected to produce a unique set of proteins. This concept provides the basis for making profiles of the proteins in a cell.

Protein profiles are obtained by **polyacrylamide gel electrophoresis (PAGE)**. In this process, proteins from lysed cells are dissolved in a detergent, and a sample of the resulting solution is placed on a thin slab of polyacrylamide gel. In a process called **electrophoresis,** an electric current is then passed through the gel. While the charge is applied, different proteins migrate through the gel at different rates, which depend on the size and charge of each protein. After the gel is stained, it shows a characteristic pattern of the cells' proteins (Figure 9-11).

We assume that identical organisms have identical protein profiles, that the profiles of closely related organisms show only minor differences, and that profiles of distantly related organisms show major differences. As procedures and materials are developed further, this method may provide a useful tool for

Figure 9-11   PAGE patterns. Twelve strains of *Bifidobacterium* were isolated from humans and other mammals. All 12 strains produced identical results from biochemical tests. The PAGE patterns, however, reveal that there are two species. Lanes 1–8 are a new species, *B. globosum;* lanes 9–12 are *B. pseudolongum.*

identification as well as classification. By comparing the protein profile of the unknown cell with a set of standard profiles, a laboratory technician might be able to identify an organism within 24 hours of its isolation.

PAGE analysis is now being used to compare one group of proteins, the cytochromes, in different organisms. Although the presence of certain cytochromes is affected by the environment, analysis of bacterial cytochromes has disclosed some patterns that may be useful for classifying organisms. The following table shows the cytochromes that are common to three metabolic categories of organisms. The letters correspond to the cytochromes that are present in these cells.

Organisms	Cytochromes
Gram-positive chemoheterotrophs	b, c, a, a_3, o
Gram-negative chemoheterotrophs	b, c, d, o, a
Gram-negative chemoautotrophs	b, c, a, a_3, o, a

A limitation of the PAGE process is that different proteins may not be resolved from one another because they migrate to the same position. However, Patrick O'Farrell has developed a more sensitive version called **two-dimensional PAGE**, shown in Figure 9-12. This process differs from the PAGE process because the protein solution is exposed to electrophoresis twice, once to separate proteins by charge and again to separate them by size. Proteins from lysed cells are mixed with a special pH buffer, and a sample of the mixture is placed on a polyacrylamide gel (Figure 9-12a). When exposed to the electric field, the buffer molecules arrange themselves in a pH gradient (basic to acidic), and the proteins migrate through the gel until each reaches its **isoelectric point**—the pH at which the protein is no longer charged and therefore no longer migrates (Figure 9-12b). This process is called **isoelectric focusing**. The gel is then placed on a larger polyacrylamide slab gel and exposed to an electric field in a detergent buffer at a 90° angle to the electric field in the isoelectric focusing gel. This process separates the proteins at each pH by size (Figure 9-12c). The gel is stained to make the proteins visible. It is unlikely that two proteins will have both the same isoelectric point and the same size. Comparison of protein patterns from two bacteria shows differences between the two species (Figures 9-12d and 9-12e). If enough research had been done to constitute a library of these protein patterns, researchers would be able to identify unknown bacteria by comparing their two-dimensional PAGE protein patterns with those of known bacteria.

Base Composition of Nucleic Acids

A classification technique that has come into wide use among taxonomists because it has a good possibility of at least suggesting evolutionary relationships is the

Figure 9-12 Two-dimensional polyacrylamide gel electrophoresis. (a) A sample of cell extract is placed at one end of a gel strip. **(b)** During electrophoresis, the buffer establishes a pH gradient in the gel, and charged proteins migrate through the gel to their isoelectric points. **(c)** The gel strip is then placed on a larger gel slab and submitted to a second electrophoresis at right angles to the first. In this case, a detergent is present that eliminates the differences in charge, so the migration rates of the proteins depend only on their sizes. **(d)** *Vibrio* and **(e)** *Enterobacter* protein patterns. Proteins shown in boxes are not found in both species. (Note that proteins are present in different amounts as shown by the sizes of spots.)

Acidic (+)

Current

Basic (−)

(a)

(b)

Current

(c)

Acidic

1st Dimension

Larger proteins   Smaller proteins

Basic   2nd Dimension

(d) *Vibrio*

Acidic

1st Dimension

Larger proteins   Smaller proteins

Basic   2nd Dimension

(e) *Enterobacter*

determination of the nitrogenous-base composition of DNA. This base composition is usually expressed as percentage of guanine plus cytosine (G + C). The base composition of a single species is theoretically a fixed property; thus a comparison of the G + C content in different species can reveal the degree of species relatedness. As we saw in Chapter 8, each guanine (G) in DNA has a complementary cytosine (C). Similarly, each adenine (A) in the DNA has a complementary thymine (T). Therefore, the percentage of DNA bases that are G-C pairs also tells us the percentage that are A-T pairs (G-C + A-T = 100%). Two organisms that are closely related and hence have many identical or similar genes will have similar amounts of the various bases in their DNA. However, if there is a difference of more than 10% in their percentage of G-C pairs (for example, if one bacterium's DNA contains 40% G-C and another bacterium has 60% G-C), then these two organisms are probably not related. Of course, two organisms that have the same percentage of G-C are not necessarily closely related; other supporting data are needed to draw conclusions about evolutionary relationships.

Actually, determining the entire *sequence* of bases in an organism's DNA is theoretically possible with modern biochemical methods, but this is currently impractical for all but the smallest microorganisms (viruses). However, the use of restriction enzymes allows the comparison of base sequences of different organisms in two ways. Restriction enzymes cut a molecule of DNA everywhere a specific base sequence occurs, producing restriction fragments (as discussed in Chapter 8). For example, the enzyme *Eco* R1 cuts DNA at the arrows in the sequence

$$\downarrow$$
$$...\, G\, AATTC\, ...$$
$$...\, CTTAA\, G\, ...$$
$$\uparrow$$

In the first technique, using restriction enzymes, the DNA from two microorganisms is treated with the

.. Microbiology in the News ....

Mass Deaths of Dolphins Worry Experts

During a normal year, ten or twelve dead dolphins are found washed ashore on beaches between New Jersey and Virginia. By mid August of 1987, however, the annual total had already surpassed 200, and scientists believe that hundreds more may have died offshore. Since this represents a large part of the Atlantic dolphin population, some experts are worried that this population may ultimately be destroyed. Moreover, alarmed swimmers have asked whether they themselves are also in danger from whatever is killing the dolphins.

Marine mammal researchers examine a Commerson's dolphin.

Since large numbers of *Streptococcus* and *Vibrio* have been found in the dead dolphins, the first indications are that they may be dying from septicemia. However, these bacteria are a normal part of the flora of dolphins and of coastal waters. If these bacteria are the cause of death, why are the dolphins suddenly susceptible to bacteria that normally do not harm them? Are chemical spills weakening the dolphins, or are viruses at fault?

Information Is Scarce
These questions are the concern of veterinary microbiology, which until recently has been a neglected branch of medical microbiology. Although the diseases of such animals as cattle, chickens, and mink have been studied—partly because of their availability to researchers—the microbiology of wild animals, especially marine mammals, is a newly emerging field. There is still very little information, for example, about the normal flora of healthy dolphins.

A team of scientists at the University of California at Berkeley is trying to identify bacteria in a wide variety of marine mammals, including the normal flora in healthy specimens. However, gathering samples of animals that live in the open ocean and performing bacteriological analyses on them are very difficult. Currently, the animals being same restriction enzyme, and the restriction fragments that are produced are separated by electrophoresis on a slab gel (see Microview B, Photo 5). A comparison of the number and sizes of restriction fragments that are produced from different organisms provides information about their genetic similarities and differences—the more similar the patterns, the more closely related the organisms are expected to be.

Restriction enzymes can also be used to facilitate the determination of the base sequence of specific genes. Although the techniques are still somewhat tedious, base sequences can be determined for entire genes or gene fragments cut from the organism's DNA by restriction enzymes. As mentioned earlier, the sequencing of bases in the genes that code for ribosomal RNA has provided useful information about evolutionary relationships among bacteria. We assume that closely related genera have a similar base sequence that differs from that of distantly related genera. An indirect determination of the degree of similarity between the DNA nucleotide sequences of two organisms can be made with the method of nucleic acid hybridization.

Nucleic Acid Hybridization

If a double-stranded molecule of DNA is subjected to heat, the complementary strands will separate as the hydrogen bonds between the bases break. If the single strands are then cooled slowly, they will reunite to form a double-stranded molecule identical to the original double strand. (This reunion occurs because the single strands have complementary sequences.) When this technique is applied to separated DNA strands from two different organisms, it is possible to determine the extent of similarity between the base sequences of the two organisms. This method is known as **nucleic acid hybridization**. The procedure

· ·

studied are those that live in captivity and those that come onto the shore to breed, such as the northern fur sea lion.

The scientists are identifying bacteria in marine mammals by using conventional test batteries and rapid-identification test kits developed for human microbiology. The bacteria are compared with species described in *Bergey's Manual* to assign names or to identify them, although they are not always a perfect match. Perhaps new species of bacteria will be found in the marine mammals.

From Dolphin to Human or from Human to Dolphin?

Results indicate that some microbes found in dolphins can be transmitted to humans, and vice versa. Two dolphins in captivity died from a strain of *Staphylococcus aureus* that produces β-lactamase (penicillinase), the same strain that was isolated from the trainer's nares. However, the human pathogens isolated from dolphins (such as *Aeromonas* and *Plesiomonas*) are also found in water as well as on a variety of domestic animals. As information becomes more complete, pathogens involved in future epidemics of marine mammals can be more readily differentiated from normal flora.

The possibility of humans contracting a respiratory infection from dolphins is low because humans are not normally near a dolphin's nostril when it exhales. There may be a moderate chance of transmission of intestinal pathogens (such as *Plesiomonas* and *Yersinia*) through ingesting water contaminated with feces, but the greatest chance of transmitting infections from dolphins to humans is by bacteria (such as *Aeromonas* and *Erysipelothrix*) entering through skin wounds.

Veterinary microbiologists hope that increased study of the microbiology of wild animals, including marine mammals, will not only promote improved wildlife management but also will provide models for the study of human diseases.

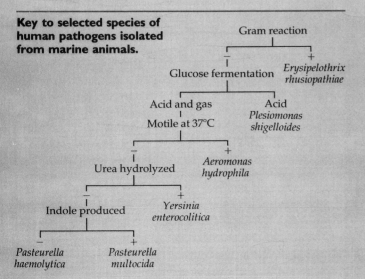

Key to selected species of human pathogens isolated from marine animals.

Figure 9-13   DNA hybridization. The greater the amount of pairing between DNA strands from different organisms (hybridization), the more closely the organisms are related.

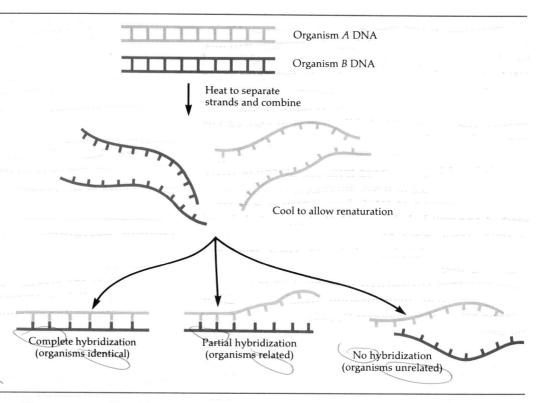

Figure 9-14   DNA probes used to identify bacteria. (a), (b) Method of producing *Salmonella* DNA probes. **(c), (d)** DNA probes are used to identify *Salmonella* in a sample of food.

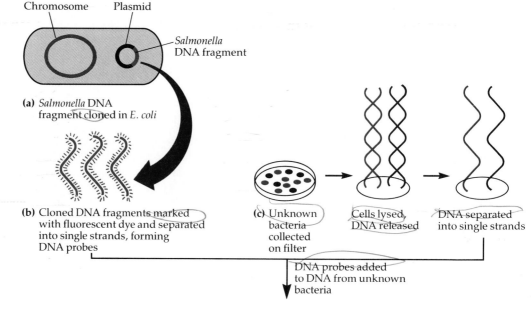

(a) *Salmonella* DNA fragment cloned in *E. coli*

(b) Cloned DNA fragments marked with fluorescent dye and separated into single strands, forming DNA probes

(c) Unknown bacteria collected on filter

Cells lysed, DNA released

DNA separated into single strands

DNA probes added to DNA from unknown bacteria

(d) DNA probes hybridize with *Salmonella* DNA from sample; excess probe is washed off; fluorescence indicates presence of *Salmonella* assumes that, if two species are similar or related, a major portion of their nucleic acid sequences will also be similar. The procedure measures the ability of DNA strands from one organism to hybridize (bind through complementary base pairing) with the DNA strands of another organism (Figure 9-13). The greater the degree of hybridization, the greater the degree of relatedness.

As we learned in Chapter 8, RNA is single-stranded and is transcribed from one strand of DNA; a particular strand of RNA, therefore, is complementary to the strand of DNA from which it was transcribed and will hybridize with that separated strand of DNA. DNA–RNA hybridization can thus be used to determine relatedness between DNA from one organism and RNA from another organism in the same way that DNA–DNA hybridization is used.

Nucleic acid hybridization tools called **DNA probes** are being developed for rapid identification of bacteria. For example, DNA extracted from *Salmonella* is broken into fragments using a restriction enzyme, and a specific fragment is selected as the probe for *Salmonella*. This fragment must be able to hybridize with the DNA of all *Salmonella* strains but not with the DNA of closely related enteric bacteria. The chosen DNA fragment is cloned in a plasmid in

E. coli (Figure 9-14a), producing hundreds of specific *Salmonella* DNA fragments. These fragments are tagged with radioactive isotopes or a fluorescent dye, and are separated into single strands of DNA (Figure 9-14b). The resulting DNA probes can then be mixed with single-stranded DNA from a food sample suspected of containing *Salmonella* (Figure 9-14c). If *Salmonella* are present, the DNA probes will hybridize with the *Salmonella*'s DNA (Figure 9-14d), and hybridization can be detected by the radioactivity or fluorescence of the probes.

Genetic Recombination

Genetic recombination also might provide information useful to the classification and identification of bacteria. Recall from Chapter 8 that genetic recombination (that is, the rearrangement of genes to form new combinations) can occur by transformation, transduction, or conjugation. The use of transformation in identification of a microorganism (Figure 9-15) is based on the fact that recombination between bacterial chromosomes is likely to occur only between closely related organisms. This technique is limited to those organisms that can take up a strand of DNA.

DNA from unknown (suspected *N. gonorrhoeae*)

N. gonorrhoeae pro⁻ (auxotrophic mutant obtained by exposing *N. gonorrhoeae* to a chemical mutagen)

Growth of transformed auxotroph, pro⁺ **(a)**

(b)   **(c)**

Figure 9-15   Using transformation to identify bacteria. If the unknown species and the auxotroph are the same species, transformation can occur, and **(a)** the resultant recombinant cell will grow. Lack of growth of the DNA alone **(b)** and the auxotroph **(c)** serves as a control.

A test to identify *Neisseria gonorrhoeae* (nī-se′rē-ä go-nôr-ré′ä) uses transformation. DNA is extracted from bacteria suspected to be *N. gonorrhoeae*. The DNA is then incubated with an auxotrophic *N. gonorrhoeae* mutant. After incubation, samples are inoculated onto a minimal medium that will not support growth of the mutant (Figure 9-15). Only transformed cells—those that have exchanged genetic material with the DNA of the normal *N. gonorrhoeae* and have acquired the necessary wild-type gene—will grow. A relatively high amount of growth suggests that the unknown bacterium is *N. gonorrhoeae*.

The use of genetic recombination for identification is desirable because few media are used and a pure culture of the unknown organism is not necessary.

Numerical Taxonomy

Numerical taxonomy (Figure 9-16) has been used to infer evolutionary relationships between organisms. Many microbial characteristics are listed, such as the ability to use a certain carbohydrate, to form a pigment, or to produce a certain acid, and the presence

(a)

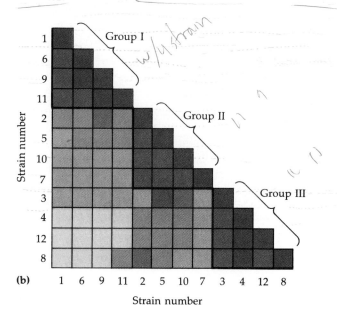

(b)

Figure 9-16 Numerical taxonomy. With the aid of a computer, a similarity matrix (array) can be constructed to show relationships among different organisms. As many as 100 different tests may be run on each organism. Each organism is then given a numerical score, and the score is shown on a matrix **(a)**. The closer the scores of two organisms, the more similar they are. **(b)** In another matrix, organisms with similar test results are grouped together. Groups I, II, and III represent three species, each containing four strains. **(c)** A dendrogram constructed from the similarity values shows species 1, 6, 9, and 11 are closely related to each other but not to the other organisms tested. The remaining strains are related to each other; they form a family that is divided into two genera, each consisting of four species.

(c)

or absence of each characteristic is scored for each organism. Amino acid sequences, the percentage of G-C pairs, and the percentage of hybridization observed between two nucleic acid strands can be listed. Tests must be selected carefully because each test is weighted equally. That is, a positive reaction is assigned a value of 1, and a negative result, 0, unlike the rapid identification methods, in which a relative weighting can be assigned to each test.

A computer then matches the characteristics of each organism against those of other organisms. The greater the number of characteristics shared by two organisms, the closer the taxonomic relationship is assumed to be. A match of 90% similarity or higher usually indicates a single taxonomic unit, or species. The computer can also generate a dendrogram (Figure 9-16c) that illustrates phylogenetic relationships derived from similarities from hybridization and other genetic experiments.

Morphological characteristics, differential staining, and biochemical testing were the only identification tools available just a few years ago. Technological advancements are making it possible to use nucleic acid and protein analysis techniques, once reserved for classification, for routine identification. (A practical example is described in the box in Chapter 19, p. 494.) A summary of taxonomic criteria and their uses is provided in Table 9-4.

TABLE 9-4 *A Summary of Classification and Identification Criteria*

Criterion	Used for	
	Classification	Identification
Morphological characteristics	No	Yes
Differential staining	Yes (for cell wall type)	Yes
Biochemical testing	No	Yes
Serology	No	Yes
Phage typing	No	Yes
Amino acid sequencing	Yes	No
Protein analysis	Yes	No
Base composition of nucleic acids	Yes	No
Nucleic acid hybridization	Yes	Yes (DNA probes)
Genetic recombination	Yes	Yes (being developed)
Numerical taxonomy	Yes	No

Study Outline

Phylogenetic Relationships (pp. 237–240)

1. Taxonomy is the science of the classification of organisms, with the goal of showing relationships among organisms.
2. Taxonomy provides a means to identify organisms.

A Five-Kingdom System (p. 237)

1. Living organisms can be classified into five kingdoms.
2. Procaryotic organisms are placed in the Kingdom Monera, or Procaryotae.
3. Eucaryotic organisms may be classified into the Kingdoms Protista, Fungi, Plantae, or Animalia.
4. Protista are mostly unicellular organisms.
5. Fungi are absorptive heterotrophs that develop from spores.
6. Multicellular photoautotrophs are placed in the Kingdom Plantae.
7. Multicellular ingestive heterotrophs are classified as Animalia.

A Three-Kingdom System (pp. 237–240)

1. Another proposal divides living organisms into three kingdoms.
2. In this system, plants, animals, fungi, and protists belong to a single kingdom.
3. Eubacteria (with peptidoglycan) form a second kingdom.

4. Archaeobacteria (with unusual cell walls) form a third kingdom.

Classification of Organisms   (pp. 240–243)

Scientific Nomenclature   (pp. 240–241)

I. According to scientific nomenclature, each organism is assigned two names (binomial nomenclature), a genus and a specific epithet, or species.
2. Rules for the assignment of names to bacteria are published in the *Bacteriological Code.*

The Taxonomic Hierarchy   (pp. 241–242)

I. A eucaryotic species is a group of organisms that interbreeds but does not breed with individuals of another species.
2. Similar species are grouped into a genus; similar genera are grouped into a family; families, into an order; orders, into a class; classes, into a division or phylum; and phyla, into a kingdom.

A Phylogenetic Hierarchy   (p. 242)

I. Organisms are grouped into taxa according to phylogenetic (from a common ancestor) relationships.
2. Some of the information for eucaryotic relationships is obtained from the fossil record.
3. Procaryotic relationships are determined by molecular biology techniques.

Classification of Bacteria   (pp. 242–243)

I. *Bergey's Manual of Systematic Bacteriology* is the standard reference on bacterial classification.
2. In *Bergey's Manual,* bacteria are divided into four divisions.
3. A division is divided into classes; classes are composed of orders; orders consist of families; a family is divided into genera; and a genus is composed of species.
4. A group of bacteria derived from a single cell is called a strain.
5. Closely related strains constitute a species.

Classification of Viruses   (p. 243)

I. Viruses are not placed in a kingdom. They are not composed of cells and are incapable of growth without a host cell.

Criteria for Classification and Identification of Microorganisms   (pp. 244–255)

I. *Bergey's Manual of Systematic Bacteriology* and, until the remaining volumes are published, *Bergey's Manual of Determinative Bacteriology,* 8th edition, are references for laboratory identification of bacteria.
2. Morphological characteristics are useful in the identification of microorganisms, especially when aided by differential staining techniques.
3. The possession of various enzymes as determined by biochemical tests is used in the identification of microorganisms.
4. Serological tests, involving the reactions of microorganisms with specific antibodies, are useful in determining the identity of strains and species as well as relationships between organisms.
5. Phage typing is the identification of bacterial species and strains by the determination of their susceptibility to various phages.
6. The sequences of amino acids in proteins of related organisms are similar.
7. Related organisms have identical proteins; this characteristic can be ascertained by PAGE "fingerprints."
8. The percentage of G-C pairs in the nucleic acid of cells can be used in the classification of organisms.
9. Single strands of DNA, or of DNA and RNA, from related organisms will hydrogen-bond to form a double-stranded molecule; this bonding is called nucleic acid hybridization.
10. Transformation of an auxotrophic mutant by the DNA of an unknown bacterium can be used in identification of the unknown bacterium.
II. Grouping many different characteristics to show relatedness is called numerical taxonomy.

Study Questions

Review

I. What is taxonomy?
2. List and define the five kingdoms used in the five-kingdom system of classification.
3. What characterizes the Kingdom Procaryotae, and what organisms are placed in this kingdom?
4. Discuss the evidence that supports a three-kingdom system of classification.

5. What is binomial nomenclature?
6. Why is binomial nomenclature preferable to the use of common names?
7. Put the following terms in the correct sequence from the most general to the most specific: order, class, genus, kingdom, species, division, family.
8. Define species.

9. List the eleven bases discussed in this chapter for the classification of microorganisms. Separate your list into those tests used primarily for taxonomic classification and those used primarily for identification of microorganisms already classified.

(handwritten: see below ↓)

10. Higher organisms are arranged into taxonomic groups on the basis of evolutionary relationships. Why is this type of classification only just being developed for bacteria? *(handwritten: level technol to explore/xper —fossil record eukaryote > prokaryote)*

11. Describe the use of a DNA probe and transformation for:
    (a) Rapid identification of an unknown bacterium.
    (b) Determining which of a group of bacteria are most closely related.

12. Can you tell which of these organisms are most closely related?

Organism	G-C moles %
A	48–52
B	49–53
C	50–52
D	46–48

13. Using *Bergey's Manual*, identify the organism with the following characteristics: gram-negative rod; facultative anaerobe; motile; oxidase-negative; fermentative; acid and gas from lactose; indole produced; methyl red +; Voges–Proskauer −; citrate −; H₂S not produced; does not grow in KCN broth. *(handwritten: Pasturella multocida? p. 251)*

Challenge

1. Here is some additional information on the organisms in Question 12:

Organisms	% DNA Hybridization
A and B	70–100
A and C	40–50
A and D	20–30

   Which of these organisms are most closely related? Compare this answer with your response to Question 12.

2. In *Bergey's Manual*, the gram-positive cocci includes the genera *Micrococcus* and *Staphylococcus*. The reported G-C content of *Micrococcus* is 66 to 75 moles %, and of *Staphylococcus*, 30 to 40 moles %. According to this information, would you conclude that these two genera are closely related?

3. Compare and contrast *Bergey's Manual of Systematic Bacteriology* and *Bergey's Manual of Determinative Bacteriology*.

Further Reading

Goodfellow, M., and R. G. Board., eds. *Microbiological Classification and Identification*. New York: Academic Press, 1980. Comprehensive coverage of modern identification techniques, including genetics, immunology, and protein electrophoresis.

Holt, J. G., ed. *Bergey's Manual of Systematic Bacteriology*, 1st ed., Vol. 1. Baltimore: Williams and Wilkins, 1984. A discussion of bacterial classification that provides a perspective on the relative importance of classification and identification.

Knoll, A. H. "The distribution and evolution of microbial life in the late Proterozoic Era." *Annual Review of Microbiology* 39:391–417, 1985. Describes fossil and chemical evidence for early life and provides living models for early life.

Lennette, E. H., ed. *Manual of Clinical Microbiology*. Washington, D.C.: American Society for Microbiology, 1985. Provides useful methods for standard bacteriological tests for identification and includes new rapid identification and molecular biology techniques.

Margulis, L. *Early Life*. Boston: Science Books International, 1982. A good book that explains hypotheses and experiments pertaining to the evolution of procaryotes, eucaryotes, fermentation, and respiration.

Olsen, G. J., D. J. Lane, S. J. Giovannoni, and N. R. Pace. "Microbial ecology and evolution: a ribosomal RNA approach." *Annual Review of Microbiology* 40:337–365, 1986. Demonstrates construction of phylogenetic trees from RNA and DNA–RNA hybridization; includes a discussion of the use of DNA probes to identify bacteria in natural samples.

Schleifer, K. H., and E. Stackebrandt. "Molecular systematics of procaryotes." *Annual Review of Microbiology* 37:143–187, 1983. Describes modern methods of taxonomy for rapid identification and for classifying bacteria.

Woese, C. R. "Archaebacteria." *Scientific American* 244(6):98–122, June 1981. Gives detailed descriptions of the archaeobacteria and comparisons with eubacteria and eucaryotes.

(handwritten table: taxonom class / ID those alrdy class — Morphol char: No / Yes; D. After Stain: Yes-all wall type / Yes; Biochem Test: No / Yes; Serol: No / Yes; Phage Type: No / No; Amino Sequ: Yes / No; Prot Analys: Yes / No; B comp Nucl A: Yes / Yes - DNA probes; Nucl A Hybrdizat: Yes / Yes - be devel; Genet Recomb: Yes / ; Numer Taxon: Yes / No)

CHAPTER 10

Bacteria

Learning Objectives

- List at least six characteristics used to classify and identify bacteria according to *Bergey's Manual of Systematic Bacteriology*.
- List two major characteristics of each section of *Bergey's Manual of Systematic Bacteriology* described in this chapter.
- Identify the sections that contain species of medical importance.

Bacteria are the most metabolically diverse of all microorganisms. Relatively few species of bacteria cause disease in humans or in other organisms, and without the activities of the many other kinds of bacteria, all life on Earth would cease. Indeed, one widely accepted theory suggests that eucaryotic organisms probably evolved from bacterialike organisms. Because bacteria are relatively simple in structure and many are readily cultured and controlled in the laboratory, microbiologists have devoted intensive study to their life processes. From such studies, we have learned that the basic processes of life are the same for all organisms.

Bacterial Groups

This chapter will introduce you to many of the more common bacteria. All bacteria share the property of procaryotic cellular organization, but it is difficult to correlate structural and functional characteristics to show evolutionary relationships. As we mentioned in the previous chapter, the most widely accepted taxonomic classification for bacteria is *Bergey's Manual of Systematic Bacteriology*. The most recent revision of *Bergey's Manual* divides bacteria into four divisions (or phyla) according to the characteristics of cell walls (see Table 9-3). Each division is subdivided into *sections* according to such characteristics as Gram-stain reaction, cell shape, cell arrangements, oxygen requirements, motility, and nutritional and metabolic properties. Groups are named according to these characteristics—for example, *spirochetes, gram-negative aerobic rods and cocci, anoxygenic photosynthetic bacteria*, and *endospore-forming rods and cocci*. Each section consists of a number of genera. In some sections, genera are grouped into families and orders; in other sections, they are not (see Table 10-1 and Appendix A).

This chapter briefly describes the principal bacterial groups according to the basic organization of *Bergey's Manual*. Our discussion emphasizes bacteria considered to be of practical importance, those important in medicine, or those that illustrate biologically unusual or interesting principles.

Spirochetes

The spirochetes are found in contaminated water, in sewage, soil, and decaying organic matter, and within the bodies of humans and animals. One of the first microorganisms described by Anton van Leeuwenhoek in the 1600s was a large spirochete taken from saliva and tooth scrapings. These bacteria are typically coiled, like a metal spring. Some are tightly coiled; others resemble a spring that is stretched.

All spirochetes are actively motile and achieve their motility by means of two or more *axial filaments*. The axial filaments are enclosed in the space between an outer sheath and the body of the cell.

One end of each axial filament is attached near a pole of the cell (Figure 10-1a and b). By stretching and relaxing its axial filament, the cell moves through liquids, rotating like a corkscrew. This mechanism is especially well adapted for movement through liquids that are too viscous to allow motility by flagella, the method used by most other motile bacteria. This adaptation might explain why several species of spirochete parasites are restricted to animal tissue and fluids.

The spirochetes can be aerobic, facultatively anaerobic, or anaerobic, and they do not have flagella or endospores. All divide by binary fission.

The spirochetes include a number of important pathogenic bacteria. The best known is the genus *Treponema*, which includes *Treponema pallidum* (tre-pō-nē'mä pal'li-dum), the cause of syphilis (Figure 10-1c). A number of other species of *Treponema* commonly inhabit the mouth. Members of the genus *Borrelia* (bôr-rel'ē-ä) cause relapsing fever and Lyme disease, serious diseases that are usually transmitted by either ticks or lice.

Leptospirosis is a disease usually spread to humans by water contaminated by *Leptospira* (lep-tō-spī'rä) species. The bacteria are excreted in the urine of such animals as dogs, rats, and swine, so domestic dogs and cats are routinely immunized against it. The cells of *Leptospira* typically have hooked ends (see Figure 25-4).

(a)

(b)

(c)

Figure 10-1 Spirochetes. These helically shaped bacteria move by means of an axial filament. **(a)** These two axial filaments are anchored to one end of the cell, but not to the other end. By flexing, they cause the cell to rotate in a corkscrew fashion and propel it through fluids. **(b)** A cross section of a spirochete showing its multiple axial filaments (f) located under the outer sheath (e). The plasma membrane of the cell is indicated (m) (×161,000). **(c)** The spirochete *Treponema pallidum*, the causative agent of syphilis, attached to another cell (×5000).

TABLE 10-1 *Summary of Selected Characteristics of Bacterial Groups,
from* Bergey's Manual of Systematic Bacteriology, *1st Edition*

Name of Section	Important Genera	Habitat	Special Features
Spirochetes	*Treponema, Borrelia, Leptospira*	Aquatic; animal parasites	Helical morphology; motility by axial filaments; several important pathogens; gram-negative
Aerobic/micro-aerophilic, motile, helical/vibrioid, gram-negative bacteria	*Spirillum, Campylobacter, Azospirillum*	Soil and aquatic environments; human intestinal tract and oral cavity	Helical morphology; motility by flagella, not axial filaments; vibrioids do not have a complete turn
Nonmotile (or rarely motile), gram-negative, curved bacteria	*Spirosoma, Meniscus*	Aquatic and sedimentary environments	Uncommon; mostly aquatic, not pathogenic; form S-shapes, C-shapes, rings
Gram-negative; aerobic rods and cocci	*Pseudomonas, Neisseria, Branhamella, Brucella, Bordetella, Francisella, Legionella, Agrobacterium*	Soil; water; animal parasites	Organisms of medical, industrial, and environmental importance
Facultatively anaerobic, gram-negative rods	*Salmonella, Shigella, Klebsiella, Yersinia, Vibrio, Hemophilus, Gardnerella, Pasteurella*	Soil; plants; animal intestinal tracts	Many important pathogens
Anaerobic, gram-negative straight, curved, or helical rods	*Bacteroides, Fusobacterium*	Animals and insects	Obligate anaerobes, mostly of intestinal tract; some common in mouth and genital tract
Dissimilatory sulfate-reducing or sulfur-reducing bacteria	*Desulfovibrio*	Anaerobic sediments	Reduce oxidized forms of sulfur to H_2S; gram-negative
Anaerobic gram-negative cocci	*Veillonella*	Mostly animal intestinal tracts	Nonmotile anaerobes
Rickettsias and chlamydias	*Rickettsia, Coxiella, Chlamydia*	Parasites of arthropods and animals	Obligate intracellular bacteria; many important pathogens; gram-negative
Mycoplasmas	*Mycoplasma*	Parasites of animals, plants, insects	Pleomorphic; lack cell walls; some pathogens; gram-negative
Endosymbionts	*Holospora, Blattabacterium*	Intracellular symbionts of protozoa, insects, helminths, plants, etc.	Assorted bacteria that live symbiotically in protozoa, insects, and fungi; gram-negative
Gram-positive cocci	*Staphylococus, Micrococcus, Streptococcus*	Soils; skin and mucous membranes of animals	Some important pathogens

TABLE 10-1 (*Continued*)

Name of Section	Important Genera	Habitat	Special Features
Endospore-forming rods and cocci	*Bacillus, Clostridium*	Soil; animal intestinal tract	*Bacillus,* aerobic or facultative anaerobes; *Clostridium,* anaerobic; gram-positive
Regular, nonsporing, gram-positive rods	*Lactobacillus, Listeria*	Dairy products; genital and oral cavities, animal feces	*Lactobacillus* forms lactic acid from carbohydrates, important industrially; *Listeria,* an animal pathogen
Irregular, nonsporing, gram-positive rods	*Gardnerella, Corynebacterium, Propionibacterium, Actinomyces*	Human pathogens, soil organisms	Pleomorphic morphology; several important pathogens
Mycobacteria	*Mycobacterium*	Soil; plants; animals	*Mycobacterium* is an important pathogen; acid-fast; gram-positive
Nocardioforms	*Nocardia*	Soil and animals	Form branched filaments; reproduce by fragmentation; often acid-fast; some pathogens
Budding and/or appendaged bacteria	*Hyphomicrobium, Caulobacter*	Mostly aquatic, some soil	Possess prosthecae; *Hyphomicrobium* reproduces by budding; *Caulobacter* is stalked
Sheathed bacteria	*Sphaerotilus*	Aquatic cause of sewage treatment problems	Cells encased in hollow sheath
Nonphotosynthetic, nonfruiting, gliding bacteria	*Cytophaga, Beggiatoa*	Aquatic	*Cytophaga* degrade cellulose; *Beggiatoa* oxidize H_2S
Gliding, fruiting bacteria	*Myxococcus*	Dung, soil	Cells aggregate to form a fruiting body
Aerobic chemolithotrophic bacteria	*Nitrosomonas, Nitrobacter, Thiobacillus*	Soil	Nitrifying and sulfur-oxidizing bacteria; agriculturally and environmentally important
Archaeobacteria	*Methanobacterium, Halobacterium, Sulfolobus*	Anaerobic sediments; environments of extreme temperature and osmotic pressure	Not related to other bacterial groups; no peptidoglycan in cell walls; methane producers useful in sewage treatment
Anoxygenic photosynthetic bacteria	*Chromatium, Rhodospirillum, Chlorobium*	Anaerobic sediments	Includes the green and purple sulfur and nonsulfur bacteria; green and purple sulfur bacteria use H_2S as an electron donor and release sulfur
Oxygenic photosynthetic bacteria	*Chroococcus, Anabaena*	Aquatic	Produce oxygen during photosynthesis; many species fix atmospheric nitrogen
Actinomycetes	*Streptomyces, Frankia, Micromonospora*	Soil; some aquatic	Common in soil; branching filaments with reproductive conidiospores; *Frankia* involved in nitrogen-fixing symbiosis with plants

Aerobic/Microaerophilic, Motile, Helical/Vibrioid Gram-Negative Bacteria

A distinctly helical shape does not serve as the universal criterion for classification of a bacterium as a spirochete. Some helical bacteria are not included with the spirochetes because they lack an axial filament. They use flagella instead. These bacteria possess a single flagellum at one or both poles or sometimes tufts of flagella at these locations (Figure 10-2a). Spiral bacteria, unlike spirochetes, are rigid helices or curved rods. These bacteria tend to be microaerophilic.

Most spiral bacteria are harmless aquatic organisms, such as *Spirillum volutans* (spī-ril′um vol′ü-tans). This is an unusually large bacterium, as much as 60 μm in length. The box in Chapter 4, p. 80, describes a member of the genus *Aquaspirillum* that is able to respond to magnetic fields.

Agricultural microbiologists have been interested in members of the genus *Azospirillum*, a soil bacterium that grows in close association with the roots of many plants, especially tropical grasses. It utilizes nutrients excreted by the plants and in return fixes nitrogen from the atmosphere. This nitrogen fixation is most significant in some tropical grasses and in sugar cane, although it can be isolated from the root system of many temperate climate plants, such as corn.

The spiral bacteria also include pathogenic organisms. One species, *Campylobacter fetus* (kam-pī-lō-bak′tėr fē′tus), causes abortion in domestic animals. Another species, *C. jejuni* (jē-jū′nē), causes outbreaks of foodborne intestinal disease (Chapter 24).

MICROBIOLOGY HIGHLIGHTS

Why Microbiologists Study Termites

Although termites are famous for their ability to eat wood, causing damage to wooden structures and recycling cellulose in the soil, they are unable to digest the wood that they eat. To break down the cellulose, termites enlist the help of a variety of microorganisms. Some termites, for example, dig tunnels in the wood, then inoculate the tunnels with fungi that grow on the wood. These termites then eat the fungi, not the wood itself.

What microbiologists find most interesting are the termites that contain, within their digestive tracts, symbiotic microorganisms that digest the cellulose that the termites chew and swallow. Even more fascinating to microbiologists is the fact that these microorganisms themselves can survive only because of even smaller symbionts that live on and within them, without which they would not even be able to move. By studying how a single termite survives, microbiologists have begun to gain an entirely new understanding of symbiosis.

The termite's dependence on nitrogen-fixing bacteria to supply its nitrogen and on protozoans such as *Trichonympha sphaerica* to digest cellulose is an example of endosymbiosis, a symbiotic relationship with an organism that lives inside the body of the host organism (in this case, within the hindgut of the termite).

Theoretically, the protozoan produces the cellulolytic enzymes that digest the cellulose. The picture is more complicated than this, however, for *T. sphaerica* itself is unable to digest cellulose without the aid of bacteria that live within its body; in other words, the protozoan has its own endosymbionts.

Certain hindgut ciliates such as *T. sphaerica* also demonstrate another form of symbiosis—ectosymbiosis, a symbiotic relationship with organisms that live outside its body. Recent advances in microscopy have shown that these ciliates are covered by precise rows consisting of thousands of bacteria, either bacilli or spirochetes. If these bacteria are killed, the protozoan is unable to move. Evidently, the protozoan does not use its cilia to move; instead, the rows of bacteria move the protozoan like rows of oarsmen in a boat.

The protozoan *Mixotricha*, for example, has rows of spirochetes on its surface. As shown in part (a) of the figure, the end of each spirochete abuts against a swelling known as a bracket. The spirochetes undulate in unison, thereby creating waves of motion along *Mixotricha*'s surface.

In 1978, S. L. Tamm observed rod-shaped bacteria aligned in grooves that covered the surface of devescovinids, another group of termite-hindgut protozoa. Each rod has twelve flagella that overlap the flagella of the adjacent bacteria to form a continuous filament along the

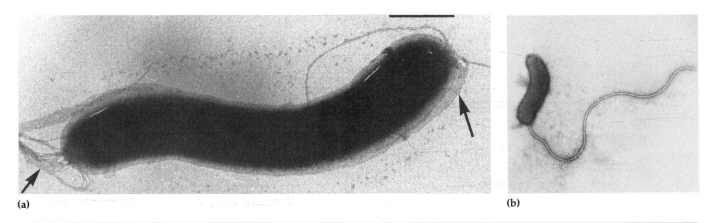

(a)   (b)

Figure 10-2   Helical/vibrioid bacteria. (a) Spiral cell of *Aquaspirillum bengal*, which is motile by means of flagella. The arrows point to the tufts of flagella at each end of the cell (×17,000; bar = 1 μm). **(b)** *Bdellovibrio bacteriovorus*, an example of a helical bacterium that does not make a complete twist, resulting in a comma shape. Note the polar flagellum (×22,000).

groove (see part (b) of the figure). The bacteria beat their flagella to create coordinated waves along all these rows of filaments, thus propelling the protozoan.

An important question remains: Who controls whom? Do the bacteria detect a food source and move the protozoan toward it, or does the protozoan force the bacteria to respond to its commands?

Protozoan flagella

Spirochetes

Bracket

Protozoan surface

(a)

Filament composed of overlapping flagella

Rod-shaped bacteria in grooves

Protozoan surface

(b)

Arrangements of bacteria on the surfaces of two protozoans. (a) Spirochetes (color) attached to brackets over the surface of a *Mixotricha* protozoan align themselves and move in unison. **(b)** On this devescovinid protozoan, the flagella from one rod-shaped bacterium (color) overlap the next to form a continuous filament.

Sometimes, helical bacteria do not make a complete turn or twist; these are **vibrioids**. A particularly interesting example is the comma-shaped bacterium *Bdellovibrio*, which attacks other bacteria (see the box in Chapter 3, p. 67, and Figure 10-2b). It reproduces within the periplasmic space of the gram-negative bacteria that it preys upon. However, if there are no prey bacteria available, a different reproductive cycle occurs. The cell elongates into a tight spiral, which then fragments almost simultaneously into a number of individual flagellated cells.

For a discussion of an especially interesting group of spirochetes, see the box on p. 262.

Gram-Negative Aerobic Rods and Cocci

This group contains many microorganisms of medical, industrial, and environmental importance. Some of the more interesting are those of the genus *Pseudomonas* (Figure 10-3). The common name for these bacteria is *pseudomonads*. These organisms are rod-shaped and have polar flagella. Many species excrete extracellular, water-soluble pigments that diffuse into their media. One species, *Pseudomonas aeruginosa* (sū-dō-mō′nas ā-rü-ji-nō′sä), produces a characteristically soluble, blue-green pigmentation. Under the right conditions, particularly in weakened hosts,

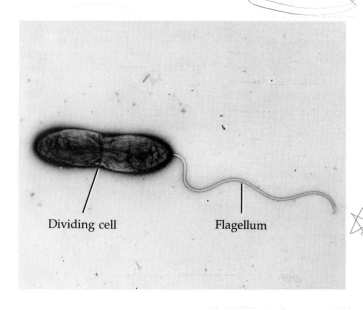

Figure 10-3 *Pseudomonas andropogonis.* Note the single polar flagellum characteristic of this species (×13,600).

this organism can infect the urinary tract, burns, and wounds and can cause septicemia, abscesses, and meningitis. Other pseudomonads produce soluble fluorescent pigments that glow when illuminated by ultraviolet light (see Microview A, Photo 11).

Pseudomonads are very common in soil and other natural environments and are generally no threat to a healthy individual. These bacteria are less efficient than some other heterotrophic bacteria in utilizing many of the common nutrients, but pseudomonads have compensating characteristics. For example, many are psychrophilic (or psychrotrophic) and grow at refrigerator temperatures; these impart off-tastes and colors to foods. Pseudomonads are also capable of synthesizing an unusually large number of enzymes and probably contribute significantly to the decomposition of chemicals, such as pesticides, that are added to soils.

In hospitals and other places where pharmaceuticals are prepared, the ability of pseudomonads to grow on minute traces of unusual carbon sources, such as soap residues or cap-liner adhesives found in a solution, has been unexpectedly troublesome. They are even capable of growth in some antiseptics, such as quarternary ammonium compounds. Their resistance to most antibiotics has also been a source of medical concern. This resistance is related to the relatively small size of the cell wall porins (Chapter 4). Some of these considerations will be discussed in more detail later. Many pseudomonad genes that code for unusual metabolic characteristics, including resistance to antibiotics, are carried on plasmids (Chapter 8).

Although pseudomonads are classified as aerobic, like many other bacteria they are capable of substituting nitrate for oxygen as a terminal electron acceptor. This process, anaerobic respiration, yields almost as much energy as aerobic respiration yields (Chapter 5). Pseudomonads are one of the bacteria that cause important losses of valuable nitrogen in fertilizer and soil. Nitrate is the form of fertilizer nitrogen most easily used by plants. Under anaerobic conditions, as in waterlogged soil, pseudomonads eventually convert nitrate into nitrogen gas (N_2), which is lost to the atmosphere (Chapter 26).

Several genera of gram-negative aerobic rods and cocci are of medical importance.

Legionella (lē-jä-nel′lä) is a recently discovered genus, which now contains six species. Originally isolated during a search for the cause of a pneumonia now known as legionellosis, these bacteria do not grow on usual laboratory isolation media, and they do not stain with the usual histological staining techniques. After intensive effort, they were isolated through tests of guinea pigs inoculated with infected tissue taken from patients. A specially buffered charcoal–yeast extract agar has since been developed for use in their isolation and growth. It has been demonstrated that this is a relatively common organism in streams, and it colonizes such habitats as warm-water supply lines in hospitals and water in cooling towers of air conditioning systems.

Neisseria is a genus of nonendospore-forming diplococci. These bacteria are aerobic or facultatively anaerobic, parasitic on human mucous membranes, and able to grow well only at temperatures near body temperature. Pathogenic *Neisseria* include the gonococcus bacterium *Neisseria gonorrhoeae* (nī-se′rē-ä go-nôr-rē′ī), the causative agent of gonorrhea (Figure 10-4), and *N. meningitidis* (me-nin-ji′ti-dis), the agent of meningococcal meningitis (Chapter 21).

Members of the genus *Moraxella* are strictly aerobic coccobacilli. Coccobacilli are egg-shaped, a structure between cocci and rods. *Moraxella lacunata* (mô-raks-el′lä la-kü-nä′tä) is implicated in conjunctivitis (pink eye), an inflammation of the conjunctiva, the membrane that covers the eye and lines the eyelids.

Brucella (brü-sel′lä) is a small nonmotile coccobacillus. All species of *Brucella* are obligate parasites of mammals and cause brucellosis. Of medical interest is the unusual ability of *Brucella* to survive phagocytosis, an important element of our defense against bacteria (Chapter 15).

Bordetella (bôr-de-tel′lä) is a nonmotile rod. Virulent forms have capsules. Found only in humans, *Bordetella pertussis* (bôr-de-tel′lä pėr-tus′sis) is the primary cause of whooping cough.

Francisella is a genus of small, pleomorphic bacteria that grow only on complex media enriched with blood or tissue extracts. *Francisella tularensis* (fran-sis-el′lä tü-lä-ren′sis) causes tularemia.

Thus far in our discussion of this group of bacteria, most genera have been largely of medical importance. There are also several genera that are of industrial and environmental importance.

Rhizobium (rī-zō′bē-um) and *Bradyrhizobium* are especially important to agriculture. These organisms cause formation of nodules on the roots of leguminous plants, such as beans or clover. Living within these nodules in a symbiotic relationship with the plants, they fix nitrogen from the air (Chapter 26).

Some nitrogen-fixing bacteria, such as *Azotobacter* (ä-zō-tō-bak′tėr) and *Azomonas* (ä-zō-mō′nas), are free-living in the soil. They are frequently used in laboratory demonstrations of nitrogen fixation. However, to fix agriculturally significant amounts of nitrogen, they would require energy sources, such as carbohydrates, that are in limited supply in soil.

Agrobacterium tumefaciens (ag-rō-bak-ti′re-um tü-mē-fā′shens) is a plant pathogen that causes a dis-

Figure 10-4   The gram-negative coccus *Neisseria gonorrheae.* Note the paired arrangement (diplococcus) (×15,000). The pili and the capsule contribute to the pathogenicity of the organism, which causes gonorrhea.

ease of plants called crown gall. The tumorlike gall grows when *A. tumefaciens* inserts a plasmid that contains bacterial genetic information into the plant's chromosomal DNA (see the box in Chapter 8, p. 224). For this reason, microbial geneticists are deeply interested in this organism. They hope that eventually it may be a vehicle for inserting useful genetic information into plants—perhaps even the ability to fix nitrogen into plants, such as wheat and corn, that do not have that capability.

Acetobacter (ä-sē-tō-bak′tėr) and *Gluconobacter* (glü-kon-ō-bak′tėr) are industrially important aerobic organisms that convert ethanol to vinegar. *Zoogloea* (zō-ō-glē′ä) will be discussed in Chapter 26 in the context of aerobic sewage treatment processes, such as the activated sludge system. As they grow, *Zoogloea* form unusual fluffy, slimy masses that are essential to the proper operation of such systems.

Facultatively Anaerobic Gram-Negative Rods

From a medical viewpoint, facultatively anaerobic gram-negative rods are a very important group of bacteria. Many of them cause diseases of the gastrointestinal tract as well as diseases of other organs. Discussed here are three families, Enterobacteriaceae, Vibrionaceae, and Pasteurellaceae.

Enterobacteriaceae (Enterics)

The family **Enterobacteriaceae**, or **enterics**, as they are commonly called, includes a group of bacteria that inhabit the gastrointestinal tracts of humans and other animals. Some species are permanent residents, others are found in only a fraction of the population, and still others are present only under disease conditions. Most enterics are active fermenters of glucose and other carbohydrates.

Because of the clinical importance of enterics, there are many techniques for their isolation and identification (see Microview C, Photo 9). An identification key to some enterics is shown in Figure 9-6, and a modern tool using 15 biochemical tests is shown in Figure 9-7. Biochemical tests are especially important in the detection of foodborne contamination by *Salmonella*. Enterics can also be distinguished from each other according to antigens present on their surfaces (Chapter 17).

Enterics include motile as well as nonmotile species; those that are motile have peritrichous flagella. Many enterics have pili (see Figure 4-8) that help them adhere to surfaces of mucous membranes. Specialized sex pili aid in the exchange of genetic information between cells, which often includes antibiotic resistance (see Figure 8-23).

Enterics, like most bacteria, produce proteins called *bacteriocins* that cause the lysis of closely related species of bacteria. Bacteriocins produced by some enterics have been most thoroughly studied. Bacteriocins might help maintain the ecological balance of various enterics in the small intestine.

Among the important genera included as enterics are *Escherichia, Salmonella, Shigella, Klebsiella, Serratia, Proteus, Yersinia, Erwinia,* and *Enterobacter.*

Escherichia   The facultative anaerobe *E. coli* is one of the most common inhabitants of the intestinal tract and is probably the most familiar organism of microbiology (see Microview A, Photo 2). As you may remember from earlier chapters, a great deal is known about the biochemistry and genetics of *E. coli*, and it continues to be an important tool for basic biological research. Its presence in water or food is also important as an indication of fecal contamination (Chapter 26). *E. coli* is not usually considered to be pathogenic. However, it can be a common cause of urinary tract infections, and certain strains produce enterotoxins that commonly cause traveler's diarrhea (Chapter 24).

Salmonella   Almost all members of this genus are potentially pathogenic. Because of this, there are extensive biochemical and serological tests to clinically isolate and identify salmonellae. They are common inhabitants of the intestinal tracts of many animals, especially poultry and cattle. Under unsanitary conditions, they can contaminate food.

Typhoid fever, caused by *Salmonella typhi* (sal-mōn-el′lä ti′fē), is the most severe illness caused by a member of the genus *Salmonella*. Gastrointestinal diseases less severe than typhoid fever caused by other salmonellae are called *salmonellosis*. Salmonellosis is one of the most common forms of foodborne illness.

Although many members of the genus *Salmonella* have specieslike names, such as *S. typhimurium* (tī-fi-mùr′ē-um) and *S. dublin* (dub′lin), no individual species are recognized. Instead, this genus is taxonomically divided into about 2000 **serovars** (serotypes). That is, they are differentiated by serological means. Serovars can be further differentiated by special biochemical or physiological properties into **biovars**, or biotypes.

When the salmonellae are injected into appropriate animals, their flagella, capsules, and cell walls cause the formation of antibodies that are specific for each of these structures. (This topic is more fully discussed in Chapter 16.) These specific antibodies, which are available commercially, can be used to differentiate *Salmonella* serovars by a system known as the Kauffmann–White scheme. The Kauffmann–White scheme designates an organism by numbers and letters that correspond to specific antigens on the organism's capsule, body, and flagella. For example, the antigenic formula of *S. typhimurium* is 1,4,[5],12,i,1,2. Some salmonellae are named only by their antigenic formulas.

Shigella   Species of *Shigella* are responsible for a disease called bacillary dysentery or shigellosis. These organisms are second only to *E. coli* as a cause of traveler's diarrhea. Some strains of *Shigella* can cause a life-threatening dysentery.

Klebsiella   The organism *Klebsiella pneumoniae* (kleb-së-el′lä nü-mō′nē-ī) is a major cause of septicemia in pediatric wards and is also a cause of one form of pneumonia, which is contracted especially by persons with upper respiratory tract infections or chronic alcoholism (see Microview A, Photo 5).

Serratia   *Serratia marcescens* (ser-rä′tē-ä mär-ses′sens), which is distinguished by its production of red pigment, has been used as an experimental organism for tests of air dispersal of bacteria for biological warfare. *S. marcescens* has become increasingly important in recent years because of its connection with nosocomial infections, that is, infections acquired in hospitals (see the box in Chapter 7, p. 180). The organism causes urinary and respiratory tract infections and has been found on catheters, in saline irrigation solutions, and in other supposedly sterile solutions.

Proteus *Proteus* (prō'tē-us) bacteria are very actively motile. They are implicated in infections of the urinary tract and infections of wounds and in infant diarrhea (see Figure 10-5 and Microview A, Photo 14).

Yersinia *Yersinia pestis* (yėr-sin'ē-ä pes'tis) is the causative agent of bubonic plague ("black death"). Urban rats in some parts of the world and ground squirrels in the American Southwest are carriers of these organisms. Fleas usually transmit the organisms among animals and to humans, although contact with animals and respiratory droplets from infected persons can be involved with transmission.

Erwinia *Erwinia* (ėr-wi'nē-ä) species are primarily plant pathogens; some cause plant soft-rot diseases. These species produce enzymes that hydrolyze the pectin between individual plant cells and cause the plants to rot.

Enterobacter Two *Enterobacter* species, *E. cloacae* (klō-ā'kī) and *E. aerogenes* (ā-rä'jen-ēz), can cause urinary tract infections and nosocomial infections. They are widely distributed in man and animals as well as in water, sewage, and soil (Chapter 26).

Vibrionaceae

This is a family of gram-negative, facultatively anaerobic rods, many of which are slightly curved. They are found mostly in aquatic habitats. *Vibrio* is the most important genus in the family.

Vibrio Members of the genus *Vibrio* are rods that are frequently slightly curved (Figure 10-6). Although most are nonpathogenic, one important pathogen is *Vibrio cholerae* (vib'rē-ō kol'ėr-ī), the causative agent of cholera. The disease is characterized by a profuse and watery diarrhea. *V. parahaemolyticus* (pa-rä-hē-mō-li'ti-kus) causes a less serious form of gastroenteritis. *V. parahaemolyticus* usually inhabits coastal salt waters and is transmitted to humans, mostly by raw or undercooked shellfish.

Pasteurellaceae

The third medically important family among the gram-negative facultatively anaerobic rods includes *Pasteurella*, *Hemophilus*, and *Actinobacillus* (see the box in Chapter 15, p. 401).

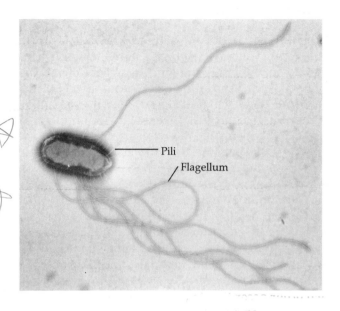

Figure 10-5 *Proteus mirabilis*, **a gram-negative rod.** Note the flagella distributed about the body, and the short pili (×14,000).

Figure 10-6 *Vibrio cholerae.* Numerous cells of *Vibrio cholerae* are visible here on the surface of the small intestine of a mouse. Note the characteristic slight curvature of the rods (×4000).

Pasteurella This genus is primarily known as a pathogen of domestic animals. It causes septicemia in cattle, fowl cholera in chickens and other fowl, and pneumonias in several types of animals. The best known species is *Pasteurella multocida* (pas-tyėr-el'lä mul-tō'si-dä), which can be transmitted to humans by dog and cat bites.

Hemophilus   *Hemophilus*, also spelled *Haemophilus*, is a very important group of pathogenic bacteria that receives less attention than it should. These organisms commonly inhabit the mucous membranes of the upper respiratory tract, mouth, vagina, and intestinal tract. The best-known species that affects humans is *Hemophilus influenzae* (hē-mä′fi-lus in-flü-en′zī), named long ago after the erroneous conclusion that it was responsible for influenza.

The name *Hemophilus* is derived from the bacteria's requirement for blood in their culture medium. They are unable to synthesize important parts of the cytochrome system needed for respiration. They obtain these substances from the heme fraction, known as the X factor, of blood hemoglobin. In some formulations, such as chocolate agar, the blood is heated, which causes it to turn a dark brown. The culture medium must also supply the cofactor nicotinamide adenine dinucleotide (NAD^+ or $NADP^+$), which is known as V factor. Clinical laboratories use tests for the requirement of X and V factors to identify isolates as *Hemophilus* species.

Hemophilus is responsible for several important diseases. It is the most common cause of meningitis in young children and is one of the common causes of earaches. Other clinical conditions caused by *H. influenzae* include epiglottitis (a sore throat complication in which the epiglottis becomes infected), septic arthritis in children, bronchitis, and pneumonia. It can also be involved in endocarditis and pericarditis (infections of the heart). If it enters the bloodstream during childbirth, it can cause puerperal sepsis, a complication that threatens the mother's life.

Gardnerella

Gardnerella vaginalis (gärd-nė-rel′lä va-jin-al′is), a bacterium that causes one of the most common forms of vaginitis, is not presently assigned to any family. In 1955, researchers named it *Hemophilus vaginalis*, and it was placed with *Hemophilus* in *Bergey's Manual*, 8th edition, because it did not seem to fit anywhere else. Its pleomorphic (highly irregular) form and occasional gram-positive reaction have caused it to be also classified at times with the genus *Corynebacterium*, which will be discussed later in this chapter. The organism has a cell wall that is structurally similar to that of a gram-positive cell. However, the wall is quite thin, which leads to a gram-negative staining reaction. The present edition of *Bergey's Manual* places this genus in the group of gram-negative rods that we are now discussing. However

Bergey's Manual also includes *Gardnerella* in a group of irregular, nonsporing gram-positive rods, which we will discuss later in this chapter.

There are at least five genera in *Bergey's Manual* that have a Gram-staining reaction that does not match their cell-wall anatomy, and they are similarly grouped in two different taxonomic sections. Such dual classification is unheard of in botany or zoology today. It illustrates the difficulty of applying taxonomy to bacteria. The types of evolutionary criteria used to classify plants and animals are not so evident among bacteria. For example, although a bat and a bird superficially seem closely related, it can be determined that the bat is a mammal and the bird occupies another evolutionary branch more closely related to reptiles. Such relationships between bacteria may eventually become clearer as we learn to detect similarities and differences in ribosomal RNA and chromosomal DNA of bacteria. A footnote in Appendix A lists the other bacterial genera that, for one reason or another, have been placed in two different sections.

Anaerobic, Gram-Negative, Straight, Curved, and Helical Rods

Among the anaerobic gram-negative bacteria, the genus *Bacteroides* (bak-tė-roi′dēz) is a large group of microbes that live in the human intestinal tract (Figure 10-7a). Some species of this genus also reside in the oral cavity, genital tract, and upper respiratory tract. *Bacteroides* are nonendospore-forming and nonmotile. Infections due to *Bacteroides* often result from puncture wounds or surgery and are a frequent cause of peritonitis.

Another genus of gram-negative anaerobic bacteria is *Fusobacterium* (fü-sō-bak-ti′rē-um). These microbes are long and slender with pointed rather than blunt ends (Figure 10-7b). In humans, they are found most often in the gums and are responsible for dental abcesses.

Dissimilatory Sulfate-Reducing or Sulfur-Reducing Bacteria

The sulfur-reducing bacteria are obligately anaerobic organisms that, in the absence of oxygen, use oxidized forms of sulfur, such as sulfates (SO_4^{2-}) or elemental sulfur (S), as electron acceptors. The product of this reduction is hydrogen sulfide (H_2S). The activity of these bacteria releases millions of tons of H_2S into the atmosphere every year. H_2S is an essential element of the sulfur cycle and is discussed in Chapter 26.

Figure 10-7 Gram-negative anaerobes. (a) Chains of *Bacteroides hypermegas* (×1000). **(b)** *Fusobacterium* (×2200). Note the slender shape with pointed ends.

These bacteria are found in anaerobic muds and sediments and also in the intestinal tracts of humans and animals. *Desulfovibrio* (dē-sul-fō-vib'rē-ō) is the best-known genus. Still other sulfur bacteria, which we will discuss later in this chapter, are able to use the H_2S either as part of photosynthesis or as an autotrophic energy source. Altogether, the sulfur bacteria of different types are some of the most interesting microorganisms known; their means of energy production are physiologically unique and ecologically crucial.

Anaerobic Gram-Negative Cocci

The cells of anaerobic gram-negative cocci typically occur in pairs, but they can occur singly, in clusters, or in chains. They are nonmotile and nonendospore-forming. Bacteria of the genus *Veillonella* (vi-lo-nel'lä) are found as part of the normal flora of the mouth and are components of dental plaque.

Rickettsias and Chlamydias

Both rickettsias and chlamydias are obligate intracellular parasites, which means that they can reproduce only within a host cell. In this respect, they are similar to viruses. In fact, they are even smaller than some of the largest viruses. However, in morphological and biochemical aspects, they resemble bacteria and are therefore classified as such. A comparison of rickettsias and chlamydias appears in Table 10-2.

TABLE 10-2 *Comparison of Biochemical Properties of Rickettsias and Chlamydias*

Property	Rickettsias	Chlamydias
Genetic material	RNA and DNA	RNA and DNA
Ribosomes	Present	Present
Structural integrity maintained during multiplication	Yes	Yes
Peptidoglycan in cell wall	Present	Absent
Macromolecular synthesis	Carried out	Carried out
ATP-generating system	Present	Absent
Sensitivity to antibacterial antibiotics	Sensitive	Sensitive

Source: Thomas D. Brock, *Biology of Microorganisms*, 3rd ed., © 1979, p. 720. Reprinted by permission of Prentice-Hall, Inc., Englewood Cliffs, N.J.

Rickettsias (Figure 10-8a) are rod-shaped bacteria or coccobacilli that have a high degree of pleomorphism. They are gram-negative and nonmotile and divide by binary fission. Rickettsias range in length from 1 to 2 μm. One distinguishing feature of most rickettsias is that they are transmitted to humans by insects and ticks. The one exception is *Coxiella burnetii* (käks-ē-el'lä bėr-ne'tē-ē), which causes Q fever; it can be transmitted by airborne or foodborne routes.

Recently, a sporulation cycle has been reported in *C. burnetii*. This cycle might explain the bacterium's relatively high resistance to pasteurization temperatures and antimicrobial chemicals.

Diseases caused by rickettsias include epidemic typhus, caused by *Rickettsia prowazekii* (ri-ket'sē-ä prou-wä-ze'kē-ē) and transmitted by lice; endemic murine typhus (not to be confused with typhoid fever), caused by *R. typhi* (tī'fē) and transmitted by rat fleas;

Figure 10-8 Rickettsias and chlamydias. (a) Electron micrograph of a slice of infected tick tissue (×42,000), showing cross sections of several rickettsias in the tissue. **(b)** Electron micrograph of *Chlamydia psittaci* in a slice of cytoplasm of a host cell (×25,000). The dense, dark, relatively small elementary bodies have thin walls similar to those of other gram-negative bacteria. One reticulate body is shown dividing, which it can do only in a host cell such as this one. The intermediate bodies shown are transitional stages between elementary and reticulate forms.

(a)

(b)

and Rocky Mountain spotted fever, caused by *R. rickettsii* (ri-ket'sē-ē) and transmitted by ticks (see Microview A, Photo 8). In humans, rickettsial infections damage the permeability of capillaries; severe infections can cause the cardiovascular system to collapse. Rickettsias are usually cultivated in the yolk sac of chicken embryos.

Chlamydias (Figure 10-8b) are coccoid bacteria that range in size from 0.2 to 1.5 μm. They are gram-negative and nonmotile and, unlike most rickettsias, do not require insects for transmission. They are transmitted by interpersonal contact or by airborne respiratory routes. The developmental cycle of chlamydias is perhaps their most distinguishing characteristic. First, the microbe's infectious form, called the **elementary body**, attaches to a host cell. The host cell phagocytizes the elementary body and houses it in a vacuole. (This cycle differs from that of the rickettsias, which usually multiply in the cytoplasm of the host cell.) Within the host cell, the chlamydia's elementary body reorganizes and develops into a larger, less infective **reticulate body**. This body then divides successively and eventually condenses to produce the smaller infectious elementary bodies. Finally, these bodies are released from the host cell and spread to infect surrounding host cells.

There are only two species of chlamydias. *Chlamydia trachomatis* (kla-mi'dē-ä trä-kō'mä-tis) is the causative agent of trachoma, the most common cause of blindness in humans (see Microview A, Photo 9). *C. trachomatis* seems to be the primary causative agent of nongonococcal urethritis (NGU), which might now be the most common venereal disease in the United States, and lymphogranuloma venereum, another venereal disease. *C. psittaci* (sit'tä-sē) is the causative agent of psittacosis (ornithosis). Chlamydias can be cultivated in laboratory animals, cell cultures, or the yolk sac of chicken embryos.

Mycoplasmas

Mycoplasmas are bacteria that do not form cell walls. They should not be confused with L forms, which are mutants of wall-forming bacteria that fail to form normal cell walls. Mycoplasmas cannot revert to normal cell-wall-containing types, as L forms can. Moreover, mycoplasmas do not require high concentrations of salt to maintain their cellular integrity, whereas L forms do. Also, the plasma membranes of mycoplasmas have a high sterol content, whereas those of L forms do not. Finally, the growth of mycoplasmas is not inhibited by penicillin (which inhibits synthesis of the cell wall). These differences between mycoplasmas and L forms are summarized in Table 10-3.

TABLE 10-3 *Comparison Between Mycoplasmas and L Forms*

Property	Mycoplasmas	L Forms
Stability	Cannot revert to cell-wall-containing type	May revert to cell-wall-containing type
Growth medium	Do not require high salt medium to maintain cellular integrity	Require high salt medium to maintain cellular integrity
Plasma membrane	High sterol content	No sterols
Reaction to penicillin	No adverse effect	Reproduction is inhibited

The majority of *Mycoplasma* (mī-kō-plaz'mä) species are aerobes or facultative anaerobes. Because they lack cell walls, they are highly pleomorphic (Figure 10-9). They can produce filaments that resemble fungi; hence their name (*myco* means fungus). *Mycoplasma* cells are very small, ranging in size from 0.1 to 0.25 μm. The most significant human pathogen among mycoplasmas is *M. pneumoniae*, the causative agent of primary atypical pneumonia (commonly called "walking" pneumonia).

Figure 10-9 *Mycoplasma pneumoniae.* Bacteria such as *Mycoplasma pneumoniae* have no cell wall, and their morphology is irregular (pleomorphic). The filament in the center shows a series of bulges; the organism reproduces by fragmentation of the filaments at these sites (×20,000).

Also included in this group of bacteria are the *Spiroplasma*, cells with a tight corkscrew morphology that are serious plant pathogens and are common parasites of plant-feeding insects. The *Ureaplasma*, (ū-rē-ä-plas′mä), so called because they can enzymatically split the urea in urine, are occasionally associated with urinary tract infections. The *Thermoplasma*, which were originally isolated from spontaneously heating coal-refuse piles, are harmless organisms occasionally found in household hot water systems. This genus is taxonomically grouped with the archaeobacteria as well as with the mycoplasmas.

Mycoplasmas can be grown on artificial media that provide them with sterols, if necessary, and other special nutritional or physical requirements. Colonies are less than 1 mm in diameter and have a characteristic "fried egg" appearance when viewed through a microscope. Due to the small colony size on artificial media, cell culture methods are often more satisfactory.

Gram-Positive Cocci

Most gram-positive cocci of medical importance are members of the genera *Staphylococcus* or *Streptococcus*.

Staphylococcus

Staphylococci typically occur in grapelike clusters (Figure 10-10 and Microview A, Photo 1). The most important staphylococcal species is *Staphylococcus aureus* (staf-i-lō-kok′kus ô′rē-us), named for its yellow-pigmented colonies (*aureus* means golden). Members of this species are aerobes or facultative anaerobes.

Figure 10-10 *Staphylococcus aureus.* Note the grapelike clusters of these gram-positive cocci (×2500).

Understanding the characteristics of the staphylococci may help in understanding the reasons for their pathogenicity, which takes many forms. The spherical shape, combined with the strength of the gram-positive cell wall, enables them to survive and grow at high osmotic pressures. This ability partially explains why they can grow and survive in nasal secretions (many of us carry the bacteria in our noses) and on the skin. This ability also explains how *S. aureus* can grow in certain foods with high osmotic pressures (such as ham and other cured meats) or in low-moisture foods that tend to inhibit the growth of other organisms. Furthermore, the yellow pigment probably confers some protection from the antimicrobial effects of sunlight.

S. aureus produces many toxins that contribute to the bacterium's pathogenicity by increasing its ability to invade the body or damage tissue. The infection of surgical wounds by *S. aureus* is a common problem in hospitals, as we will see in Chapter 13. The bacterium's ability to develop resistance quickly to such antibiotics as penicillin contributes to its danger in hospitals. *S. aureus* is the agent of toxic shock syndrome, a severe infection causing high fever and vomiting and sometimes death. *S. aureus* also produces an *enterotoxin* that causes vomiting and nausea when ingested, one of the most common types of food poisoning.

Streptococcus

Members of the genus *Streptococcus* are spherical gram-positive bacteria. They are probably responsible for more illnesses and cause a greater variety of diseases than any other group of bacteria. Among the diseases caused by streptococci are scarlet fever, pharyngitis (sore throat), and pneumococcal pneumonia. They are also important as lactic acid producers in many dairy products.

Streptococci typically appear in chains that can contain as few as four to six cocci or as many as 50 or more (Figure 10-11). One species, *Streptococcus pneumoniae*, is usually found only in pairs (see Figure 23-8). Streptococci do not use oxygen, although most are aerotolerant. A few are obligately anaerobic.

One basis for the classification of streptococci is their action on blood agar. *Alpha-hemolytic* species produce a substance called alpha hemolysin that reduces hemoglobin (red) to methemoglobin (green). This reduction causes a greenish zone to surround the colony. *Beta-hemolytic* species produce a hemolysin that forms a clear zone of hemolysis on blood agar (see Microview A, Photo 10). Some species have no appar-

Figure 10-11 *Streptococcus mutans.* Most streptococci form chains, as shown in this scanning electron micrograph (×11,000).

ent effect on red blood cells. These are referred to as *nonhemolytic* or, sometimes less logically, as gamma-hemolytic.

Like staphylococci, pathogenic streptococci produce a number of extracellular substances that contribute to disease. Among them are products that destroy phagocytic cells (see Figure 15-8) that are essential to the body's defenses, enzymes that digest connective tissue of the host and spread the infection, and enzymes that lyse fibrin. Fibrin is a fibrous protein deposited when blood clots that limits movement of pathogens in infected areas.

Endospore-Forming Gram-Positive Rods and Cocci

The formation of endospores by bacteria is important both to medicine and to the food industry because of the endospores' resistance to heat and many chemicals. The majority of endospore-forming rods and cocci are gram-positive. With respect to oxygen requirements, they can be strict aerobes, facultative anaerobes, obligate anaerobes, or microaerophiles. The two important genera are *Bacillus* and *Clostridium* (Figure 10-12).

(a)

(b)

(c)

Figure 10-12 Representative gram-positive, endospore-forming rods. (a) A germinating cell of *Bacillus cereus* (×18,000). **(b)** *Bacillus thuringiensis*; the diamond-shaped crystal shown next to the endospore is toxic to insects that ingest it (×6600). **(c)** *Clostridium botulinum*, with terminal spore (×23,000).

In figure (b): Toxic crystal · Endospore · Collapsed *B. thuringiensis* cell

Bacillus anthracis (bä-sil'lus an-thrā'sis) causes anthrax, a disease of cattle, sheep, and horses that can be transmitted to humans. The anthrax bacillus is a nonmotile facultative anaerobe ranging in length from 4 to 8 μm. It is one of the largest bacterial pathogens. The endospores of *B. anthracis* are centrally located, and smears prepared from tissue show a capsule.

Bacillus thuringiensis (thür-in-jē-en'sis) is probably the best-known microbial insect pathogen (Figure 10-12b). *B. thuringiensis* produces intracellular crystals of toxic glycoproteins when it sporulates. Commercial preparations containing endospores and crystalline toxin of this bacterium are sold in gardening supply shops to be sprayed on plants. When ingested by an insect, the toxin quickly causes paralysis of the insect's gut, and the insect ceases to feed. In some insects the ingested endospores germinate and bacterial growth is a factor in the death of the insect. Attempts to use this bacterium in mosquito control are also promising.

Members of the genus *Clostridium* are obligate anaerobes. They vary in length from 3 to 8 μm, and in most species the cells containing endospores appear swollen (Figure 10-12c). Some clostridial endospores can withstand temperatures of 120°C for 15 minutes. Diseases associated with clostridia include tetanus, or lockjaw, caused by *C. tetani* (te'tan-ē); botulism, caused by *C. botulinum*; and gas gangrene, caused by *C. perfringens* (per-frin'jens) and other clostridia. *C. perfringens* is also the cause of a common form of foodborne diarrhea.

Regular Nonsporing Gram-Positive Rods

A chief representative of nonsporing gram-positive rods is the genus *Lactobacillus* (lak-tō-bä-sil'lus). Lactobacilli lack a cytochrome system and are unable to use oxygen as an electron acceptor. Instead, these aerotolerant rods produce lactic acid from simple carbohydrates. The acidity inhibits competing bacteria, but lactobacilli grow well in acidic environments. In humans, lactobacilli are located in the vagina, gastrointestinal tract, and oral cavity. Common industrial uses of lactobacilli are in the production of sauerkraut, pickles, buttermilk, and yogurt. Typically, a succession of lactobacilli, each more acid-tolerant than the other, participates in these lactic acid fermentations.

A pathogen in this group, *Listeria monocytogenes* (lis-te'rē-ä mo-nō-sī-tô'je-nēz), can contaminate foods; this problem has led to recalls of dairy products. *L. monocytogenes* survives within phagocytic cells and is capable of growth at refrigeration temperatures.

The organism poses the threat of stillbirth or serious damage to the fetus if it infects a pregnant woman.

Irregular Nonsporing Gram-Positive Rods

The organisms in this group are often grouped under the general term **corynebacteria** (*coryne* means club-shaped). They tend to be pleomorphic (highly irregular in morphology), which often varies with the age of the cells (Figure 10-13). They may be aerobic, anaerobic, or microaerophilic. The best-known and most widely studied species is *Corynebacterium diphtheriae* (kō-rī-nē-bak-ti'-rē-um dif-thi'rē-ī), the causative agent of diphtheria. One related species, *Propionibacterium acnes* (prō-pē-on-ē-bak-ti'rē-um ak'nēz), is commonly found on human skin and is implicated in acne.

The genus *Actinomyces* consists of anaerobes that are found in the mouth and throat of humans and animals. They occasionally form filaments that can fragment into coryneform cells. One species, *Actinomyces israelii* (ak-tin-ō-mī'sēs is-rā'lē-ē), shown in Figure 10-14, causes actinomycosis, a tissue-destroying disease usually affecting the head, neck, or lungs.

Figure 10-13 *Corynebacterium xerosis.* Note the irregular, club-shaped morphology (×1000).

Figure 10-14 *Actinomyces israelii.* Note the branched, filamentous morphology (×2000).

Mycobacteria

Mycobacteria are aerobic, nonendospore-forming, nonmotile, rod-shaped organisms. Their name (*myco* means fungus), was suggested by their occasional exhibition of filamentous growth. Most of the pathogenic species are acid-fast. A number of species of *Mycobacterium* are found in the soil. Others are important pathogens: *Mycobacterium tuberculosis* (mī-kō-bak-ti′rē-um tü-ber-kū-lō′sis) causes tuberculosis, and *Mycobacterium leprae* (lep′rī) causes leprosy (see Microview A, Photos 4 and 15).

Nocardioforms

The best-known genus in this group is *Nocardia*. They morphologically resemble *Actinomyces*; however, *Nocardia* are aerobic. To reproduce, they form rudimentary filaments, which fragment into short rods. The structure of their cell wall resembles that of the mycobacteria; therefore, they are often acid-fast. *Nocardia* are common in soil. Some species, such as *Nocardia asteroides* (nō-kär′dē-ä as-ter-oi′dēz) (Figure 10-15), occasionally cause chronic, difficult-to-treat pulmonary nocardiosis (Chapter 23). *N. asteroides* is also one of the causative agents of mycetoma, a localized destructive infection of the feet or hands.

Gliding, Sheathed, and Budding and/or Appendaged Bacteria

Appendaged Bacteria

This is an unusual group of bacteria that is linked taxonomically by the presence of **prosthecae**. Prosthecae are protrusions, such as stalks and buds. Most of these organisms are found in low-nutrient aquatic environments such as lakes. One of the best studied is the genus *Caulobacter* (kô-lō-bak′tėr), which has stalks that anchor the organisms to surfaces (Figure 10-16a). This tends to increase their nutrient uptake because they are exposed to a continuously changing flow of water and because the stalk increases the surface-to-volume ratio of the cell. Also, if the surface to which they anchor is a living host, these bacteria can use the host's excretions as nutrients. When nutrient concentration is exceptionally low, the size of the stalk increases, evidently to provide an even greater surface area for nutrient absorption.

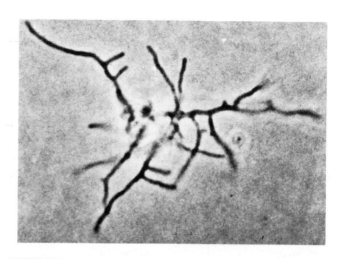

Figure 10-15 *Nocardia asteroides.* Note the branched, filamentous morphology (×2000).

Another unusual characteristic of *Caulobacter* is that reproduction by fission does not result in two essentially identical cells. When a stalked *Caulobacter* cell divides, the result is one stalked cell and one flagellated swarmer cell (Figure 10-16a). The swarmer cell is highly motile and confers the advantage of movement to new locations. After a period of motility, the swarmer cell loses its flagellum and becomes another stalked cell. Because of this unusual reproductive cycle, biologists studying the mechanics and genetics of cellular differentiation have found *Caulobacter* to be a useful research tool.

Gliding, Nonfruiting Bacteria

Gliding bacteria are classified as such on the basis of their method of motility. Embedded in a layer of slime, they glide over surfaces, often leaving a visible trail. Members of the genus *Cytophaga* (sī-täf′äg-ä) are important cellulose-degraders in the soil. Another gliding bacterium is *Beggiatoa* (bej-jē-ä-tō′ä). This organism morphologically resembles certain cyanobacteria, but it is not photosynthetic (Figure 10-16b). Nutritionally, it uses hydrogen sulfide (H_2S) as an autotrophic energy source and accumulates granules of sulfur within the cytoplasm. *Beggiatoa*'s affinity for H_2S allows it to have a symbiotic relationship with rice. Rice grows in muds in which bacteria such as *Desulfovibrio* produce H_2S, which the plant is sensitive to. The *Beggiatoa* lower the amounts of toxic H_2S, and in return the bacteria benefit from low levels of oxygen that the plant transports to the root area.

(a)

(b)

(c)

**Figure 10-16   Gliding, sheathed, and budding, and/or
appendaged bacteria. (a)** *Caulobacter* spp. (×20,000). The
stalked cell is dividing, and a flagellated swarmer cell is being
formed. **(b)** *Beggiatoa alba*, an example of a gliding, nonfruiting
bacterium (×200). Sulfur granules are visible in the cell. **(c)** *Stigmatella aurantiaca*, an example of a gliding, fruiting bacterium
(×9000). Numerous vegetative cells have aggregated to form the
stalk shown here, on which four spores have formed. When
growth conditions improve, these spores can germinate into new
vegetative cells with gliding motility.

Gliding, Fruiting Bacteria

This group of bacteria, the *myxobacteria* (*myxa* means
slime), have a remarkable life cycle. Large numbers
of vegetative, gram-negative cells converge upon a
single point. They move by gliding, leaving a slime
trail. An important source of nutrition is the lysis of
bacteria they encounter. Where the moving cells
aggregate, they differentiate and form a stalked fruiting body that contains resting cells called *myxospores*
(Figure 10-16c). Under proper conditions, the myxospores germinate and form new vegetative gliding
cells. This is probably the most complex life cycle
among procaryotic cells.

Budding Bacteria

Budding bacteria do not divide by fission into nearly
identical cells. Members of the genus *Hyphomicrobium* (hī-fō-mī-krō′bē-um) and a few other genera
instead form buds (Figure 10-17). The process in principle resembles the asexual reproductive processes of
many yeasts. The parent cell retains its identity,
while the bud increases in size until it separates as a
complete new cell.

Figure 10-17   The budding bacterium *Hyphomicrobium.*

Sheathed Bacteria

Sheathed bacteria, such as *Sphaerotilus natans* (sfe-rä′ti-lus nä′tans), are found in fresh water and in sewage. These gram-negative, polarly flagellated bacteria form a hollow, filamentous sheath to live in (Figure 10-18). They are the cause of bulking, an important problem in sewage treatment (Chapter 26).

Chemoautotrophic Bacteria

These organisms are of great importance to the environment and to agriculture. They are autotrophs capable of using inorganic chemicals as energy sources and carbon dioxide as the only source of carbon. They possess great synthesizing ability. Especially important for agriculture are the nitrifying bacteria. The energy sources of the genera *Nitrobacter* (nī-trō-bak′tėr) and *Nitrosomonas* (nī-trō-sō-mō′näs) are reduced nitrogenous compounds, such as ammonium (NH_4^+) and nitrite (NO_2^-); the bacteria convert these compounds into nitrates (NO_3^-). Nitrate is a nitrogen form that is mobile in soil and is therefore likely to be encountered and used by plants. *Thiobacillus* (thī-ō-bä-sil′lus) and other sulfur-oxidizing bacteria are parts of the important sulfur cycle. These bacteria are capable of obtaining energy by oxidizing the reduced forms of sulfur, such as hydrogen sulfide (H_2S) or elemental sulfur (S), into sulfates (SO_4^{2-}).

Archaeobacteria

This is an exceptionally interesting group of bacteria. They include extreme halophiles, such as *Halobacterium* (hal-ō-bak-te′rē-um) (see the box in Chapter 5, p. 120, and Microview A, Photo 26) and *Halococcus* (hal-ō-kok′kus), which live in high concentrations of sodium chloride (NaCl), and actually require such environmental conditions for growth. When an inoculating loop is used to transfer these bacteria from their normal habitat, the loop must first be dipped into a concentrated NaCl solution so that the cells do not lyse from exposure to low concentrations of salt. Other archaeobacteria thrive in acidic, sulfur-rich hot springs. Such an organism is *Sulfolobus* (sul-fō-lō′bus), which has a pH optimum of about 2 and a temperature optimum of more than 70°C.

Of considerable practical importance are the methane-producing bacteria (methanogens) that are used in sewage treatment processes (Figure 10-19; see also Chapter 26). These archaeobacteria derive energy from combining hydrogen (H_2) with CO_2 to

Figure 10-18 *Sphaerotilus natans.* This organism, which is found in sewage and aquatic environments, forms elongated sheaths that the bacteria live in. The bacteria have flagella (not visible here) and are capable of swimming free of the sheath eventually (×1500).

(a)

(b)

Figure 10-19 Methanogens. Transmission electron micrographs of **(a)** *Methanobacterium* (×20,000) and **(b)** *Methanosarcina* (×6000). Cells of *Methanosarcina* accumulate in irregular "sandlike" aggregates.

form methane (CH_4). An essential part of the treatment of sewage sludge is encouraging the growth of these organisms, in anaerobic digestion tanks, to convert sewage sludge into CH_4.

Phototrophic Bacteria

There are three groups of procaryotes that use light as an energy source. These **phototrophic** bacteria are the purple bacteria, the green bacteria, and the blue-green bacteria (cyanobacteria). The energetics of phototrophic bacteria were discussed in Chapter 5.

Purple and Green Phototrophic Bacteria

The purple and green bacteria (which are not necessarily these colors) are generally anaerobic (Figure 10-20). Their habitat is usually the deep sediments of lakes and ponds. Like plants, algae, and the cyanobacteria, purple and green bacteria carry out photosynthesis. However, unlike plantlike photosynthesis, the photosynthesis of purple or green phototrophic bacteria is **anoxygenic**—it does not produce oxygen. Plants, algae, and the cyanobacteria produce oxygen (O_2) from water (H_2O) as they carry out photosynthesis, as in equation 1:

$$(1) \quad H_2O + CO_2 \xrightarrow{\text{light}} (CH_2O) + H_2O + O_2$$

The **purple sulfur** and **green sulfur bacteria** use reduced sulfur compounds, such as hydrogen sulfide (H_2S), instead of water, and they produce granules of sulfur (S) rather than oxygen, as shown in equation 2:

$$(2) \quad H_2S + CO_2 \xrightarrow{\text{light}} (CH_2O) + H_2O + 2S$$

At one time, an important question in biology concerned the source of the oxygen produced by plant photosynthesis; was it from the CO_2 or from the H_2O? Until the advent of radioisotope tracers, which traced the oxygen in water and carbon dioxide and finally settled the question, the comparison of equations (1) and (2) above was the only evidence that the oxygen source was indeed H_2O.

Other phototrophs, the **purple nonsulfur** and **green nonsulfur bacteria**, use organic compounds, such as acids and carbohydrates, for the photosynthetic reduction of carbon dioxide. Morphologically, the photosynthetic bacteria are very diverse, with spirals, rods, cocci, and even budding forms.

(a)　　　　　　　　　　　　　(b)

Figure 10-20　Intracellular sulfur granules within two different bacteria. (a) *Thiocapsa floridana*, a photosynthetic purple sulfur bacterium. **(b)** *Chromatium vinosum* ($\times 2000$).

Cyanobacteria

The **cyanobacteria** are essentially aerobic. They carry out oxygen-producing photosynthesis, much as plants and the eucaryotic algae do (Chapter 11). Some, however, are capable of using reduced sulfur compounds for anoxygenic photosynthesis. A considerable number of cyanobacteria are capable of fixing nitrogen. Specialized cells called *heterocysts* contain enzymes that fix nitrogen gas (N_2) into ammonium (NH_4^+) for use by the growing cell (Figure 10-21a). *Gas vacuoles* are found in many species that grow in water. The gas vacuole is a series of chambers or gas vesicles surrounded by a protein wall that is permeable to air but not to water. The gas vacuoles provide buoyancy that helps the cell to move to favorable light. Cyanobacteria that are motile move about by gliding.

The cyanobacteria are morphologically varied. They have unicellular forms that divide by simple fission (Figure 10-21b), colonial forms that divide by multiple fission, and filamentous forms (Figure 10-21c) that reproduce by fragmentation of the filaments. The filamentous forms usually exhibit some differentiation of cells, which are often bound together within an envelope or sheath.

The cyanobacteria, especially those that fix nitrogen, are extremely important to the environment. They occupy environmental niches similar to those occupied by the eucaryotic algae described in Chapter 11 (see Roles of Algae in Nature, p. 296), but the ability of cyanobacteria to fix nitrogen makes them even more adaptable (see Microview A, Photo 27). The environmental role of the cyanobacteria is presented more fully in Chapter 26, in the discussion of eutrophication.

Actinomycetes

Actinomycetes are filamentous bacteria. Superficially, their morphology resembles that of the filamentous fungi; however, the filaments of actinomycetes consist of procaryotic cells with diameters much smaller than those of molds. Some actinomycetes further resemble molds by using externally carried asexual spores for reproduction.

Actinomycetes are very common inhabitants of soil, where mycelia (a filamentous habit of growth) have advantages. The organism can bridge water-free gaps between soil particles to move to new nutritional sites. This morphology also gives the organism a much higher surface-area-to-volume ratio and improves its nutritional efficiency in the highly competitive soil environment.

One genus, *Frankia* (frank'ē-ä), causes nitrogen-fixing nodules to form in alder tree roots, much as *Rhizobium* cause nodules on the roots of legumes (Chapter 26). The best-known genus of actinomycetes is *Streptomyces* (strep-tō-mī'sēs), which is one of the bacteria most commonly isolated from soil (Figure 10-22). The reproductive asexual spores of *Streptomyces*, conidiospores, are formed at the ends of aerial filaments. If each conidiospore lands on a suitable substrate, it is capable of germinating into a new colony. These organisms are strict aerobes. They often produce extracellular enzymes that enable them to utilize proteins, polysaccharides such as starch or cellulose, and many other organic materials found in soil. *Streptomyces* produce a gaseous compound called *geosmin* that gives fresh soil its typical musty odor. Species of *Streptomyces* are valuable because they produce most of our commercial antibiotics (Chapter 19).

* * *

In the next chapter, we will turn our attention to algae, fungi, protozoans, and helminths.

Figure 10-21 Cyanobacteria. (a) Light micrograph of a filamentous cyanobacterium (×2000). Note the heterocysts in which nitrogen-fixing activity is located. **(b)** A nonfilamentous cyanobacterium, *Chroococcus turgidus*, as seen with a phase-contrast microscope (×930). The single cell has just divided and is still contained within the surrounding sheath. **(c)** Scanning electron micrograph of *Phormidium luridum*, showing long chains of rod-shaped cells (×4400).

Figure 10-22 *Streptomyces* spp. Scanning electron micrograph of a typical *Streptomyces* (×6500). Reproduction is by means of the conidiospores carried in long spiral chains at the end of branched filaments of the bacterium.

Study Outline

Bacterial Groups   (pp. 258–279)

1. *Bergey's Manual* divides bacteria into sections based on Gram-stain reaction, cellular morphology, oxygen requirements, and nutritional properties.

2. In some cases, the sections include families and orders, and some bacteria are included as genera of uncertain affiliation.

3. Spirochetes are long, thin, helical cells that move by means of an axial filament.

4. Aerobic/microaerophilic, motile, helical/vibrioid gram-negative bacteria move by means of one or more polar flagella.

5. Gram-negative aerobic rods and cocci have polar flagella, if flagellated, and can utilize a wide variety of organic compounds.

6. Facultatively anaerobic gram-negative rods have flagella, if motile, and include the enterics, Vibrionaceae, Pasteurellaceae, and the genus *Gardnerella*.

7. Members of the anaerobic gram-negative straight, curved, and helical rods can be found in humans.

8. Dissimilatory sulfate-reducing or sulfur-reducing bacteria are anaerobes that are important in the sulfur cycle.

9. Anaerobic gram-negative cocci are normal flora of the human mouth.

10. Rickettsias and chlamydias are obligate intracellular parasites.

11. Mycoplasmas are bacteria that lack cell walls.

12. Gram-positive cocci include the catalase-positive *Staphylococcus* and catalase-negative *Streptococcus*.

13. Endospore-forming gram-positive rods and cocci may be aerobic, facultatively anaerobic, or anaerobic.

14. The diverse group of regular, nonsporing, gram-positive rods includes *Lactobacillus* and *Listeria*.

15. Irregular, nonsporing, gram-positive rods include the irregular-staining corynebacteria.

16. Pathogenic species of mycobacteria are acid-fast.

17. Nocardioform bacteria may be acid-fast; they form short filaments.

18. Bacteria with unusual morphologies are discussed in this book in terms of the following groups: budding and/or appendaged; gliding, nonfruiting; gliding, fruiting; budding; sheathed.

19. The chemoautotrophic bacteria play important roles in the cycles of elements in the environment.

20. Extreme halophiles, acidophiles, thermophiles, and methane-producing bacteria are included in the archaeobacteria.

21. Photosynthetic purple and green bacteria are included in the group of anoxygenic phototrophic bacteria; they do not produce molecular oxygen.

22. Cyanobacteria produce molecular oxygen during photosynthesis; they are oxygenic phototrophs.

23. Actinomycetes produce mycelia and reproduce by external spores.

Study Questions

Review

1. The following outline is a key that can be used to identify the medically important groups of bacteria. Fill in the name of the group indicated by the key.

Name of Group and Representative Genus

I. Gram-positive
   A. Endospore-forming   _____
   B. Nonsporing
      1. Cocci   _____
      2. Rods
         a. Regular   _____
         b. Irregular   _____
         c. Acid-fast   _____
   C. Mycelium produced
      1. Acid-fast   _____
      2. Produce chains
         of conidia   _____
II. Gram-negative
   A. Cells are helical or curved
      1. Axial filament   _____
      2. No axial filament
         a. Aerobic   _____
         b. Anaerobic   _____
   B. Cells are rods or cocci
      1. Aerobic,
         nonfermenting   _____
      2. Facultatively
         anaerobic   _____
      3. Anaerobic   _____
   C. Intracellular
      parasites   _____
III. Lacking cell wall   _____

2. Compare and contrast each of the following:
(a) Cyanobacteria and algae
(b) Actinomycetes and fungi
(c) *Bacillus* and *Lactobacillus*
(d) *Pseudomonas* and *Escherichia*
(e) *Leptospira* and *Spirillum*
(f) *Veillonella* and *Bacteroides*
(g) *Rickettsia* and *Chlamydia*
(h) *Thermoplasma* and *Mycoplasma*

3. Matching:
I. Gram-positive
  A. Nitrogen-fixing
II. Gram-negative
  A. Phototrophic
    1. Anoxygenic
    2. Oxygenic
  B. Chemoautotrophic
    1. Oxidize inorganics, such as NO_2^-
    2. Reduce CO_2 to CH_4
  C. Chemoheterotrophic
    1. Move via a slime layer
    2. Form myxospores

(a) Archaeobacteria
(b) Cyanobacteria
(c) *Cytophaga*
(d) *Desulfovibrio*
(e) *Frankia*
(f) *Hyphomicrobium*
(g) Methanogenic bacteria
(h) Myxobacteria
(i) *Nitrobacter*
(j) Purple bacteria
(k) *Sphaerotilus*
(l) *Sulfolobus*

3. Reduce sulfate to H_2S
  a. Anaerobic
  b. Thermophilic
4. Long filaments, found in sewage
5. Form projections from the cell
III. Unusual cell wall (lacking peptidoglycan)

Challenge

I. Place each section listed in Table 10-1 in the appropriate division:
(a) Typical gram-positive
(b) Typical gram-negative
(c) Wall-less
(d) Unusual walls

2. Where are each of the following classified in *Bergey's Manual*? Why are they described more than once?
(a) *Gardnerella*
(b) Nocardioforms

Further Reading

Carr, N.G., and B. A. Whitton. *The Biology of Cyanobacteria. Botanical Monographs*, Vol. 19. Berkeley: University of California Press, 1982. A comprehensive reference on morphology, metabolism, and reproduction of cyanobacteria.

Finegold, S. M., and E. J. Baron. *Diagnostic Microbiology*, 7th ed. St. Louis: Mosby, 1986. A good reference for procedures used in the clinical microbiology laboratory.

Goodfellow, M., M. Mordarski, and S. Williams. *Biology of the Actinomycetes*. New York: Academic Press, 1984. A survey of the biology, ecology, and pathogenicity of this diverse group of bacteria.

Harwood, C. S., and E. Canale-Parola. "Ecology of spirochetes." *Annual Review of Microbiology* 38:161–192, 1984. Describes the role of spirochetes in decomposition and in their associations with humans and other organisms.

Holt, J. G., ed. *Bergey's Manual of Systematic Bacteriology*, 1st ed. Baltimore: Williams and Wilkins; Vol. 1, 1984; Vol. 2, 1986; Vols. 3 and 4, forthcoming, 1989. These four volumes are the standard reference for identification and classification of bacteria.

Kloos, W. E. "Natural populations of the genus *Staphylococcus.*" *Annual Review of Microbiology* 34:559–592; 1980. Illustrates the use of modern technology to determine a species; includes a discussion of staphylococci that live on birds and mammals.

Maniloff, J. "Evolution of wall-less prokaryotes." *Annual Review of Microbiology* 37:477–499, 1983. Mycoplasmas and archaeobacteria are examples in this discussion of evolution; information comes from advances in molecular biology.

Shapiro, J. A. "Bacteria as multicellular organisms." *Scientific American* 258(6):82–89, June 1988. Describes bacterial colonies that function as multicellular organisms.

Stanier, R. Y., J. L. Ingraham, M. L. Wheelis, and P. R. Painter. *The Microbial World*, 5th ed. Englewood Cliffs, N.J.: Prentice-Hall, 1986. An especially good, concise discussion of bacterial groups, with an emphasis on physiology.

Starr, M. P., H. Stolp, H. G. Truper, A. Balows, and H. G. Schlegel. *The Prokaryotes: A Handbook on Habitats, Isolation and Identification of Bacteria*. 2 volumes. New York: Springer-Verlag, 1981. Discusses characteristics and provides information on cultivation, isolation, and identification of bacteria.

CHAPTER 11

Fungi, Algae, Protozoans, and Multicellular Parasites

In this chapter, we will examine the eucaryotic micro-organisms—fungi, algae, protozoans, and parasitic helminths, or worms. (See Table 11-1 for a comparison of their characteristics.) Although most adult helminths are not microscopic, they do have microscopic stages in their development, and some of these stages cause disease; therefore, microbiologists study helminths. Microbiologists also study arthropods (jointed-legged animals) because some of them transmit microbial diseases.

Fungi

The study of fungi is called **mycology**. The fungi include yeasts, molds, and fleshy fungi. Yeasts are unicellular organisms. Molds are multicellular filamentous organisms, such as mildews, rusts, and smuts. Fleshy fungi include multicellular mushrooms, puffballs, and coral fungi.

All fungi are chemoheterotrophs, requiring organic compounds for energy and carbon. Fungi are aerobic or facultatively anaerobic; no strictly anaerobic fungi are known. The majority of fungi are saprophytes in soil and water; there they primarily decompose plant material. Like bacteria, fungi contribute significantly to the decomposition of matter and recycling of nutrients. By using extracellular enzymes, such as cellulases and pectinases, fungi are the primary decomposers of the hard parts of plants that cannot be digested by most animals. Table 11-2 lists the basic differences between fungi and bacteria. Of the more than 100,000 species of fungi, only about 100 are pathogenic for humans and other animals. However, thousands of fungi are pathogenic to plants. Virtually every economically important plant is attacked by one or more of the fungi.

Learning Objectives

- Differentiate between asexual and sexual reproduction, and use as examples the processes of a fungus and a protozoan.

- List the defining characteristics of the four phyla of fungi described in this chapter.

- Describe the outstanding characteristics of the six phyla of algae discussed in this chapter.

- List the distinguishing characteristics of lichens, and describe their nutritional needs.

- Compare and contrast cellular slime molds and plasmodial slime molds.

- Describe the outstanding characteristics of the four phyla of protozoans discussed in this chapter.

- List the characteristics of the three classes of parasitic helminths, and give an example of each class.

- Define intermediate host, definitive host, and arthropod vector.

TABLE 11-1 *Major Differences Among Fungi, Algae, Protozoans, and Helminths (All Are Eucaryotes)*

	Fungi	Algae	Protozoans	Helminths
Kingdom	Fungi	Protista and Plantae	Protista	Animalia
Nutritional type	Chemoheterotroph	Photoautotroph	Chemoheterotroph	Chemoheterotroph
Multicellular	All, except yeasts	Some	None	All
Cellular arrangement	Unicellular, filamentous, fleshy (such as mushrooms)	Unicellular, colonial, filamentous; tissues	Unicellular	Tissues and organs
Food acquisition	Absorptive	Absorptive	Absorptive; cytostome	Mouth (ingestive); absorptive
Characteristic features	Sexual and asexual spores	Pigments	Motility; some form cysts	Many have elaborate life cycles, including egg, larva, and adult
Embryo	None	Some	None	All

TABLE 11-2 *Comparision of Selected Features of Fungi and Eubacteria*

	Fungi	Eubacteria
Cell type	Eucaryotic with well-defined nuclear membrane	Procaryotic
Cell membrane	Sterols present	Sterols absent, except in *Mycoplasma*
Cell wall	Glucans; mannans; chitin (no peptidoglycan)	Peptidoglycan
Spores	Produce a wide variety of sexual and asexual reproductive spores	Endospores (not for reproduction); some asexual reproductive spores
Metabolism	Limited to heterotrophic; aerobic, facultatively anaerobic	Heterotrophic, chemoautotrophic, photoautotrophic; aerobic, facultatively anaerobic, anaerobic

Source: After B. D. Davis et al. *Microbiology*, 3rd ed. New York: Harper & Row, 1980, p. 829.

Characteristics of Fungi

Yeast identification, like bacterial identification, uses biochemical tests. However, multicellular fungi are identified on the basis of physical appearance, including colony characteristics and spore formation. Colonies are described as **vegetative** structures because they are composed of the cells involved in catabolism and growth. Fungal reproduction is usually by spores.

Vegetative Structures

Molds and Fleshy Fungi

A **thallus** (body) of a mold or fleshy fungus consists of long filaments of cells joined together; these filaments are called **hyphae** (singular, **hypha**). In most molds, the hyphae contain crosswalls called **septa** (singular, **septum**), which divide the hyphae into distinct, uninucleate (one-nucleus) cell-like units. These hyphae are called **septate hyphae** (Figure 11-1a). In a few classes of fungi, however, the hyphae contain

Figure 11-1   Vegetative structures of fungi. (a) Septate hyphae have crosswalls dividing the hyphae into cell-like units. **(b)** Coenocytic hyphae lack crosswalls. **(c)** Hyphae grow by elongating at the tips. **(d)** Pseudohyphae are short chains of cells formed by some yeasts.

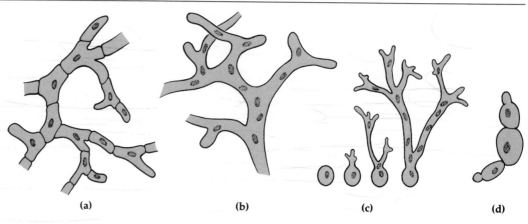

(a)   (b)   (c)   (d)

no septa and appear as long continuous cells with many nuclei. These hyphae are called **coenocytic hyphae** (Figure 11-1b). Even in fungi with septate hyphae, there are usually openings in the septa that make the cytoplasm of adjacent "cells" continuous; so these fungi are actually coenocytic organisms, too. The hyphae of a thallus grow by elongating at the tips (Figure 11-1c). Each part of a hypha is capable of growth, and, when a fragment breaks off, it can elongate to form a new hypha. In the laboratory, fungi are usually grown from fragments obtained from a fungal thallus.

When environmental conditions are suitable, the hyphae grow, intertwine, and form a mass called a **mycelium**, which is visible to the naked eye. The portion of the mycelium concerned with obtaining nutrients is called the **vegetative mycelium**; the portion concerned with reproduction is the **reproductive** or **aerial mycelium**, so called because it projects above the surface of the medium on which the fungus is growing, as shown in Figure 11-2. The aerial mycelium often bears reproductive spores, which we will discuss later.

Yeasts

Yeasts are nonfilamentous, unicellular fungi that are typically spherical or oval in shape. Like molds, yeasts are widely distributed in nature; yeasts are frequently found as a white powdery coating on fruits and leaves. Because most yeasts are colonies of unicellular organisms, they do not reproduce as a unit. Instead, the colony grows as the number of yeast cells increases. This increase usually happens by **budding**. In budding, the parent cell forms a protuberance (bud) on its outer surface. As the bud elongates, the parent cell's nucleus divides, and one nucleus migrates into the bud. Cell-wall material is then laid down between the bud and parent cell, and the bud eventually breaks away (Figure 11-3).

One yeast cell can in time produce up to 24 daughter cells by budding. Some yeasts produce buds that fail to detach themselves; these buds form a short chain of cells called a **pseudohypha** (see Figure 11-1d). A few types of yeast grow by fission. During fission, the parent cell itself elongates, its nucleus divides, and two daughter cells are produced.

(a)

(b)

Figure 11-2   *Aspergillus niger* grown on agar. (a) Vegetative mycelium. **(b)** Aerial mycelia bearing reproductive spores (×780).

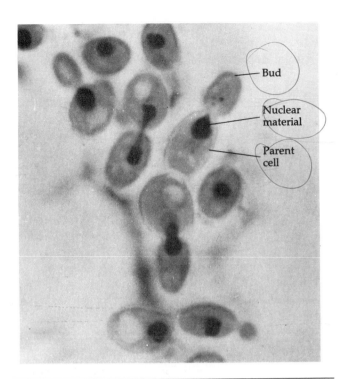

Figure 11-3 Bakers' yeast, *Saccharomyces cerevisiae*. Various stages of budding. Note the distribution of the darkly stained nuclear material from some of the parent cells into the buds (×1570).

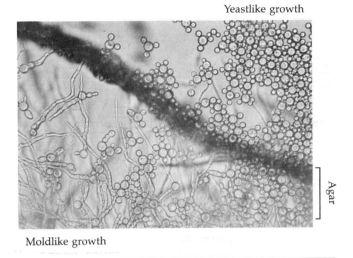

Figure 11-4 Dimorphism. Dimorphism in the fungus *Mucor rouxii* depends on CO_2 concentration. On the agar surface, *Mucor* exhibits yeastlike growth, but in the agar it is moldlike (×960).

Increases in the number of yeast cells on a solid medium produce a colony similar to a bacterial colony (see Microview A, Photo 18).

Yeasts are capable of facultative anaerobic growth. As you may recall from Chapter 5, yeasts can use oxygen or an organic compound as the final electron acceptor; this is a valuable attribute. If given access to oxygen, yeasts perform aerobic respiration to metabolize carbohydrates to carbon dioxide and water. Denied oxygen, they ferment carbohydrates and produce ethanol and carbon dioxide. This fermentation is the basis of the brewing, wine-making, and baking industries. Species of *Saccharomyces* (sak-ä-rō-mī'sēs) produce ethanol, used for brewing beverages, and carbon dioxide, used for raising dough.

Dimorphic Fungi

Some fungi, most notably the pathogenic species, exhibit **dimorphism**, that is, two forms of growth. Such fungi can grow either as a mold or as a yeast. Frequently, dimorphism is temperature dependent: At 37°C, the fungus is yeastlike, and at 25°C, it is moldlike. The appearance of the dimorphic fungus shown in Figure 11-4 changes with CO_2 concentration.

Reproductive Structures

Viewed with a microscope, the hyphae of fungi often look alike. When fungi are identified, the reproductive structures, or **spores**, must be examined.

Reproduction in fungi occurs by spore formation. These spores, however, are quite different from bacterial endospores. Bacterial endospores (Chapter 4) are formed so the bacterial cell will survive adverse environmental conditions. A single vegetative bacterial cell forms one endospore, which eventually germinates to produce a single vegetative bacterial cell. Formation of an endospore by bacteria is not reproduction because it does not increase the total number of bacterial cells. But after a mold forms a spore, the spore detaches from the parent and germinates into a new mold (Figure 11-1c). Unlike the bacterial endospore, this is a true reproductive spore—one organism produces many spores. Spores are formed from the aerial mycelium in a great variety of ways, depending on the species.

Fungal spores can be asexual or sexual. **Asexual spores** are formed by the aerial mycelium of one organism. When these spores germinate, they become organisms that are genetically identical to the parent. **Sexual spores** result from the fusion of nuclei from two opposite mating strains of the same species of fungus. Organisms that grow from sexual spores will have genetic characteristics of both parental strains. Because spores are of key importance in the identification of fungi, we will next look at some of the various types of asexual and sexual spores.

Asexual Spores

Asexual spores are produced by an individual fungus through mitosis and subsequent cell division; there is no fusion of the nuclei of cells. Several types of asexual spores are produced by fungi. One type, an **arthrospore**, is formed by the fragmentation of a septate hypha into single, slightly thickened cells (see Figures 11-5a and 23-14). One species that produces such spores is *Coccidioides immitis* (kok-sid-ē-oi'dēz im'mitis). Another type of asexual spore is a **chlamydospore**, a thick-walled spore formed by segmentation within a hypha (Figure 11-5b). A fungus that produces chlamydospores is *Candida albicans* (kan'did-ä al'bi-kans) (see Microview A, Photo 21). A **sporangiospore** is an asexual spore formed within a sac (**sporangium**) at the end of an aerial hypha called a **sporangiophore**. The sporangium can contain hundreds of sporangiospores (Figure 11-5c). Such spores are produced by *Rhizopus*. A fourth principal type of asexual spore is a **conidiospore**, which is a unicellular or multicellular spore that is not enclosed in a sac (Figure 11-5d). Conidiospores are produced in a chain at the end of a **conidiophore**. Such spores are produced by *Penicillium*. A fifth type of asexual spore, a **blastospore**, consists of a bud coming off the parent cell (Figure 11-5e). Such spores are found in some yeasts.

Sexual Spores

A fungal sexual spore results from sexual reproduction, consisting of three phases:

1. A haploid nucleus of a donor cell (+) penetrates the cytoplasm of a recipient cell (−).
2. The (+) and (−) nuclei fuse to form a diploid zygote nucleus.
3. By meiosis, the diploid nucleus gives rise to haploid nuclei (sexual spores), some of which may be genetic recombinants.

Fungi produce sexual spores less frequently than asexual spores. Often, the sexual spores are produced only under special circumstances. The plant-pathogenic ascomycetes only produce ascospores when host plants are at the end of a growing season. This suggests that the ascospores are initiated by changes in moisture, temperature, or nutrient availability. The ascospores survive until the plant hosts are growing again. The sexual spores produced by fungi are a criterion used to group the fungi into several phyla.

One kind of sexual spore is a **zygospore**, a large spore enclosed in a thick wall (Figure 11-6a). This type of spore results when the nuclei of two cells that are morphologically similar to each other fuse. Such spores are produced by fungi of the Phylum Zygomycota. A second type of sexual spore is an **asco-**

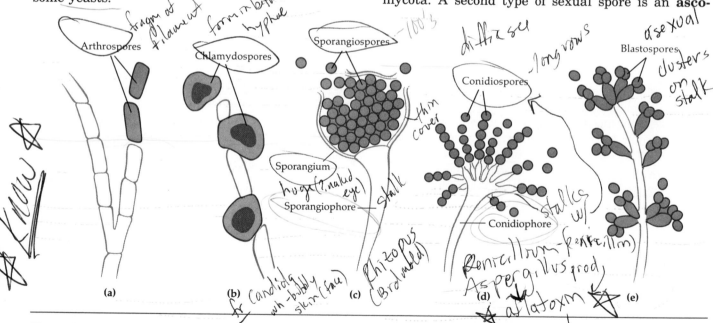

Figure 11-5   Representative asexual spores. (a) Fragmentation of hyphae results in formation of arthrospores. **(b)** Chlamydospores are thick-walled cells within the hyphae. **(c)** Sporangiospores are formed within a sporangium (spore sac). **(d)** Conidiospores are arranged in chains at the end of a conidiophore. **(e)** Blastospores are formed from buds of the parent cell.

Handwritten notes at top:
≠⟩phylum ④ Deuteromycote — no sex form yet dscvrd
↳ cause most hum infed

(a)

Handwritten: Zygomycota

(b)

Handwritten: Ascomycota (penicilum)

Handwritten: Basidiomycota

(c)

Figure 11-6 Sexual spores. (a) Zygospores, characteristic of the Phylum Zygomycota, are produced from the fusion of two cells that are morphologically alike. **(b)** Ascospores, produced by the Ascomycota, are formed within an ascus. **(c)** Basidiospores, produced only by Basidiomycota, are formed on the tip of a pedestal called a basidium.

spore, a spore resulting from the fusion of the nuclei of two cells that can be either morphologically similar or dissimilar. These spores are produced in a saclike structure called an **ascus** (Figure 11-6b). There are usually from two to eight ascospores in an ascus. Such spores are produced by the Phylum Ascomycota. Another type of sexual spore is the **basidiospore**, a spore formed externally on a base pedestal called a **basidium** (Figure 11-5c). There are usually four basidiospores per basidium. Basidiospores are produced only by the Phylum Basidiomycota, which includes the mushrooms. Mature basidiospores are often discharged explosively when a raindrop or an animal touches the mushroom (see Microview A, Photo 32).

Nutritional Adaptations

Fungi are generally adapted to environments that would be adverse to bacteria; thus, they can survive in places where microbial growth is not expected. Fungi are chemoheterotrophs, and, like bacteria, they absorb nutrients rather than ingesting them, as animals do. However, fungi differ from bacteria in certain environmental requirements and in the following nutritional characteristics.

1. Fungi usually grow better in an acidic pH (5.0), which is too acidic for the growth of most common bacteria.
2. Most molds are aerobic, so they grow on surfaces rather than throughout a substrate. Yeasts are facultative anaerobes.
3. Most fungi are more resistant to osmotic pressures than bacteria are; most fungi are therefore able to grow in high sugar or salt concentrations.
4. Fungi are capable of growing on substances with a very low moisture content, generally too low to support the growth of bacteria.
5. Fungi require somewhat less nitrogen for growth than bacteria.
6. Fungi are capable of using complex carbohydrates, such as lignin (wood), that most bacteria cannot metabolize.

These characteristics allow fungi to grow on such unlikely substrates as painted walls or shoe leather. The undesirable effects of these adaptations will be discussed shortly. Some of the beneficial uses of fungi are discussed in Chapter 27.

Medically Important Phyla of Fungi

This section provides an overview of medically important phyla of fungi. The actual diseases they cause will be studied in Part Four, Chapters 20–25. Note that not all fungi cause disease. There are other phyla of fungi not discussed here, because they contain no pathogens.

The genera named in the following phyla are readily found as contaminants in foods and in laboratory bacterial cultures. Although these genera are not of primary medical importance, they are typical examples of their respective groups. See Table 11-3 for a specific list of human pathogens.

Deuteromycota

The Deuteromycota are also known as the Fungi Imperfecti. These fungi are "imperfect" because they have not yet been found to produce sexual spores. Members of this phylum produce the asexual chlamydospores, arthrospores, and conidiospores; budding also occurs. Deuteromycota have septate hyphae.

Most of the pathogenic fungi are, or once were, classified as Deuteromycota. This phylum might be described as a "holding category" in which fungi are placed until sexual spores are observed and the fungus can be properly classified. Table 11-4 lists some well-known Fungi Imperfecti that have only recently been classified. Note that the generic names are changed with reclassification. When different species with a deuteromycete genus are observed to have morpho-

TABLE 11-3   *Characteristics of Some Parasitic Fungi*

Phylum	Growth Characteristics	Asexual Spore Types	Human Pathogens
Zygomycota	Nonseptate hyphae	Sporangiospores	Rhizopus
			Mucor
Ascomycota	Septate hyphae	Conidiospores	Allescheria boydii
	Dimorphic	Conidiospores	Aspergillus*
			Blastomyces dermatitidis*
			Histoplasma capsulatum*
	Septate hyphae, strong affinity for keratin	Conidiospores	Microsporum*
		Arthrospores	Trichophyton*
Basidiomycota	Septate hyphae include the fleshy fungi (mushrooms), rust and smuts, and plant pathogens; yeastlike encapsulated cells	Conidiospores	Cryptococcus neoformans*
Deuteromycota	Septate hyphae	Conidiospores	Epidermophyton
		Chlamydospores	Cladosporium werneckii
	Dimorphic	Conidiospores	Sporothrix schenkii
		Arthrospores	Coccidioides immitis
	Yeastlike, pseudohyphae	Chlamydospores	Candida albicans*
	Dimorphic	Chlamydospores Arthrospores Conidiospores	Paracoccidioides brasiliensis
	Septate hyphae, produce melaninlike pigments	Conidiospores	Phialophora Fonsecaea
	Yeastlike pseudohyphae	Arthrospores	Trichosporon

*Imperfect name. Refer to Table 11-4 for current classification and names.

logically different sexual spores, the species might be reclassified into two or more different genera, as was done with *Penicillium*.

Zygomycota

The **Zygomycota**, or conjugation fungi, are saprophytic molds that have coenocytic hyphae. A common example is *Rhizopus nigricans* (rī-zō'pùs nī'gri-kans), the common black bread mold (Figure 11-7). The asexual spores of *Rhizopus* are sporangiospores. The dark sporangiospores inside the sporangium give *Rhizopus* its descriptive common name. When the sporangium breaks open, the sporangiospores are dispersed. If they fall on a suitable medium, they will germinate into a new mold thallus. The sexual spores are zygospores.

TABLE 11-3 *(Continued)*

Clinical Notes	Type of Mycosis	Habitat	Refer to Chapter
Opportunistic pathogen	Systemic	Ubiquitous	23
Opportunistic pathogen	Systemic	Ubiquitous	23
Primary cause of maduromycosis	Subcutaneous	Soil	20
Opportunistic pathogen	Systemic	Ubiquitous	23
Inhalation	Systemic	Unknown	23
Inhalation	Systemic	Soil	23
Tinea capitis (Ringworm)	Cutaneous	Soil, animals	20
Tinea pedis (Athlete's foot)	Cutaneous	Soil, animals	20
Inhalation	Systemic	Soil, bird feces	21
Tinea cruris (jock itch), tinea unguium (of fingernails or toenails)	Cutaneous	Soil, humans	20
Tinea nigra	Superficial	Ubiquitous	20
Puncture wound	Subcutaneous	Soil	—
Inhalation	Systemic	Soil	23
Opportunistic pathogen	Cutaneous, systemic, mucocutaneous	Human normal flora	20, 25
Inhalation	Systemic	Presumably soil	—
Chromomycosis	Subcutaneous	Soil, plant debris	20
Chromomycosis	Subcutaneous	Soil, plant debris	20
White piedra	Superficial	Soil (?), humans	20

TABLE 11-4 *Several Genera of Reclassified Imperfect Fungi*

Imperfect Name	Reclassified to Phylum	Perfect Name
Aspergillus	Ascomycota	*Sartorya, Eurotium, Emericella*
Blastomyces	Ascomycota	*Ajellomyces*
Candida	Ascomycota	*Pichia*
Cryptococcus	Basidiomycota	*Filobasidiella*
Histoplasma	Ascomycota	*Emmonsiella, Gymnoascus*
Microsporum	Ascomycota	*Nannizia*
Penicillium	Ascomycota	*Talaromyces, Carpenteles*
Petriellidium	Ascomycota	*Allescheria*
Trichophyton	Ascomycota	*Arthroderma*

Ascomycota

The **Ascomycota**, or sac fungi, include molds with septate hyphae and yeasts. They are called sac fungi because their sexual spores are ascospores produced in an ascus. Their asexual spores are usually conidiospores produced in long chains from the conidiophore. The arrangements of conidiospores in *Penicillium* (pen-i-sil'lē-um) and *Aspergillus* (a-spėr-jil'lus) are shown in Figure 11-8. *Conidia* means dust, and these spores freely detach from the chain at the slightest disturbance and float in the air like dust.

(a)

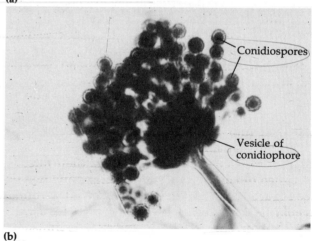

(b)

Figure 11-8 Conidiospores. The arrangement of conidiospores is useful in identification of fungi. **(a)** *Penicillium marneffei* produces conidiospores from a branched conidiophore (×800). **(b)** Conidiospores of *Aspergillus tamarii* are produced from the enlarged terminal end (vesicle) of the conidiophore (×2000).

Figure 11-7 *Rhizopus nigricans.* Note the sporangia at the tops of the sporangiophores (×1050).

Basidiomycota

The **Basidiomycota**, or club fungi, also possess septate hyphae. The common name is derived from the shape of the basidium that bears the sexual basidiospores. Some of the basidiomycota produce asexual conidiospores. Representative basidiomycetes are shown in Figure 11-9 (see also Microview A, Photos 31 and 32).

Fungal Diseases

Any fungal infection is called a **mycosis**. Mycoses are classified as one of five groups, according to the level of infected tissue and mode of entry into the host. Mycoses are classified as systemic, subcutaneous, cutaneous, superficial, or opportunistic. Characteristics of the medically important fungi are summarized in Table 11-3.

Systemic mycoses are fungal infections deep within the body. They are not restricted to any particular region of the body; they can affect a number of tissues and organs. Systemic, or deep, mycoses are usually caused by saprophytic fungi that live in the soil. Inhalation of spores is the route of transmission; these infections typically begin in the lungs and then spread to other body tissues. They are not contagious from animal to human or from human to human. Two systemic mycoses, histoplasmosis and coccidioidomycosis, are discussed in Chapter 23.

Subcutaneous mycoses are fungal infections beneath the skin, and they are caused by saprophytic fungi that live in soil and on vegetation. Infection occurs by direct implantation of spores or mycelial fragments into a puncture wound in the skin.

Fungi that infect only the epidermis, hair, and nails are called **dermatophytes**, and their infections are called **dermatomycoses** or **cutaneous mycoses** (see Figure 20-14). Dermatophytes secrete *keratinase*, an enzyme that degrades keratin. Keratin is a protein found in hair, skin, and nails. Infection is transmitted from human to human or from animal to human by direct contact or by contact with infected hairs and epidermal cells (as from barber shop clippers or shower room floors). (See Microview A, Photo 19.)

The fungi that cause **superficial mycoses** are localized along hair shafts and in superficial (surface) epidermal cells. These infections are prevalent in the tropics.

(a)

(b)

(c)

Figure 11-9 Representative basidiomycota. (a) *Calvulina cristata*, one kind of coral fungus. **(b)** *Agaricus augustus* is a popular produce item. **(c)** Puffballs of the species *Lycoperdon periatum* (about one-half actual size).

An **opportunistic pathogen** is generally harmless in its normal habitat but can become pathogenic in a host who is seriously debilitated or traumatized or who is under treatment with broad spectrum antibiotics or immunosuppressive drugs. A person's normal flora can become opportunistic pathogens under these conditions.

Mucormycosis is an opportunistic mycosis caused by *Rhizopus* and *Mucor* (mū-kôr); the infection occurs mostly in patients with ketoacidosis resulting from diabetes mellitus, leukemia, or treatment with immunosuppressive drugs. **Aspergillosis** is also an opportunistic mycosis; it is caused by *Aspergillus* (see Figure 11-2). This disease occurs in persons who have debilitating lung diseases or cancer and have inhaled *Aspergillus* spores. The mycosis **candidiasis** (see Figure 20-15) is most frequently caused by *Candida albicans* and may occur as vulvovaginal candidiasis during pregnancy (see Microview A, Photos 20 and 21). **Thrush**, a mucocutaneous candidiasis, is an inflammation of the mouth and throat; it frequently occurs in newborns.

Some fungi cause disease by producing toxins. These toxins will be discussed in Chapter 14.

Economic Effects of Fungi

Fungi can have undesirable effects for industry and agriculture because of their nutritional adaptations. As most of us have observed, mold spoilage of fruits, grains, and vegetables is relatively common, but bacterial spoilage of such foods is not. There is little moisture on the unbroken surfaces of such foods, and the interiors of fruits are too acidic for many bacteria to be able to grow there. Jams and jellies also tend to be acidic, and they have a high osmotic pressure from the sugars they contain. These factors all discourage bacterial growth but readily support the growth of mold. A paraffin layer on top of a jar of homemade jelly helps deter mold growth because molds are aerobic and the paraffin layer keeps out the oxygen. However, fresh meats and certain other foods are such good substrates for bacterial growth that bacteria will not only outgrow molds but will actively suppress mold growth in these foods.

The ability of fungi to grow at low moisture levels is of particular importance in their role as plant pathogens. Bacterial plant pathogens are far less common than fungal pathogens. In fact, the fungus that caused the great potato blight in Ireland during the early 1800s (*Phytophthora infestans*) was one of the first microorganisms to be associated with a disease.

The spreading chestnut tree, of which Longfellow wrote, no longer grows in this country except in a few widely isolated sites. An important fungal blight killed virtually all the trees. First seen in the United States in 1904, the ascomycete *Endothia parasitica* (en-dō′thē-ä par-ä-si′ti-kä) was introduced from China. The tree roots live and put forth shoots regularly, but the shoots are just as regularly killed by the fungus. Another devastating fungal plant disease that was also imported to this country is Dutch elm disease, caused by *Ceratocystis ulmi* (sē-rä-tō-sis′tis ul′mē). Carried from tree to tree by a bark beetle, the fungus blocks the afflicted tree's circulation.

Algae

Algae are familiar as the large brown kelp in coastal waters, the green scum in a puddle, and the green stains on soil or on a rock. Algae are mostly aquatic, although some are found in soil or on trees when sufficient moisture is available there. Unusual algal habitats include the hair of both the sedentary South American sloth and the polar bear. Water is necessary for algae's physical support, reproduction, and diffusion of nutrients. Generally, algae are found in cool temperate waters, although the large floating mats of the brown alga *Sargassum* (sär-gas′sum) are found in the subtropical Sargasso Sea. Some species of brown algae grow in antarctic waters.

Algae are photoautotrophs; hence, they may be found throughout the photic (light) zone of water. Their locations depend on the availability of nutrients, wavelengths of light, and type of substratum. Red and brown algae require rocks, snails, or other algae to which their holdfasts can attach. Probable locations for representative marine algae are shown in Figure 11-10.

Chlorophyll *a* (light-trapping pigment) and accessory pigments involved in photosynthesis are responsible for the distinctive colors of many algae. Some algae exist as single cells; others exist as colonies of hundreds or thousands of cells. Algae are classified as protists or plants according to their structures, pigments, and other qualities (Table 11-5, p. 295). Following are descriptions of some phyla of algae.

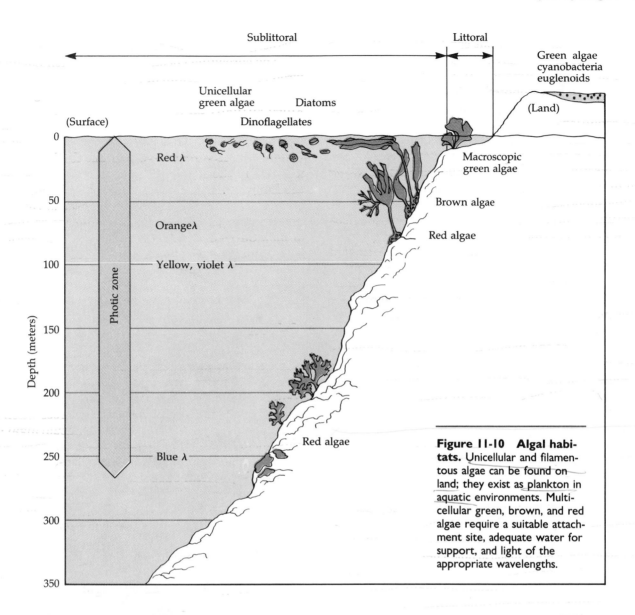

Figure 11-10 Algal habitats. Unicellular and filamentous algae can be found on land; they exist as plankton in aquatic environments. Multicellular green, brown, and red algae require a suitable attachment site, adequate water for support, and light of the appropriate wavelengths.

Phyla of Algae

Dinoflagellates (Figure 11-11a) are unicellular **planktonic**, or free-floating, **algae**. Their cell walls consist of many individual plates made of cellulose and silica. Marine dinoflagellates in the genus *Gonyaulax* (gō-nē-ō′laks) produce neurotoxins that cause **paralytic shellfish poisoning**. This disease spreads to humans when large numbers of dinoflagellates are eaten by mollusks—such as mussels or clams—that in turn are eaten by humans. Large concentrations of *Gonyaulax* give the ocean a deep red color, from which the name *red tide* originates. Mollusks should not be harvested for human consumption during a red tide. A similar disease called **ciguatera** occurs when the dinoflagellate *Gambierdiscus toxicus* (gam-bir′dis-

kus toks′i-kus) passes up the food chain and is concentrated in large fish. Ciguatera is endemic (constantly present) in the South Pacific Ocean and Caribbean Sea.

Euglenoids (Figure 11-11b) are unicellular, flagellated algae. Euglenoids have a semirigid plasma membrane called a *pellicle,* and they move by means of an anterior flagellum. Some euglenoids are facultative chemoheterotrophs. In the dark, they ingest organic matter through the *cytostome*. Euglenoids are frequently studied with protozoans because they lack a cell wall and can move and feed using the flagellum.

All euglenoids except one have a red *eyespot* at the anterior end. This carotenoid-containing organelle senses light and directs the cell in the appropriate direction using the *preemergent flagellum.*

(a)

(b)

(c)

(d)

Figure 11-11   Algae. (a) *Peridinium*, a dinoflagellate. Some dinoflagellates have two flagella in perpendicular, opposing grooves. When the two flagella beat simultaneously, they cause the cell to spin. **(b)** *Euglena*. Semirigid rings supporting the plasma membrane allow *Euglena* to change shape. **(c)** *Nereocystis*, a brown alga. The hollow stipe and gas-filled pneumatocyst hold the thallus upright to ensure that an adequate amount of sunlight is received for growth. **(d)** *Microcladia*, a red alga. The delicately branched red algae get their color from phycobiliprotein accessory pigments. **(e)** *Chlamydomonas*, a green alga. Two whiplike flagella propel this cell.

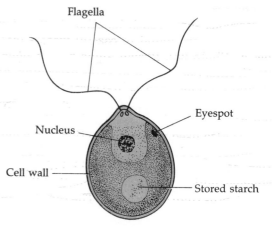

(e)

TABLE 11-5 *Characteristics of Selected Algae Phyla*

	Dinoflagellates	Euglenoids	Diatoms	Brown Algae	Red Algae	Green Algae
Classification	Protist	Protist	Protist	Protist	Protist	Plant
Color	Brownish	Green	Brownish	Brownish	Reddish	Green
Cell wall	Cellulose and silica	None; protein pellicle	Pectin and silica	Cellulose and alginic acid	Cellulose	Cellulose
Flagella	2	1	In gametes	In gametes	No	In unicellular forms and in gametes
Cell arrangement	Unicellular	Unicellular	Unicellular	Multicellular	Most are multicellular	Unicellular and multicellular
Nutrition	Autotrophic; some heterotrophic	Autotrophic; some heterotrophic	Autotrophic	Autotrophic	Autotrophic	Autotrophic
Pigments	Chlorophyll a and c, carotene, xanthins	Chlorophyll a and b, carotene	Chlorophyll a and c, carotene, xanthophylls	Chlorophyll a and c, xanthophylls	Chlorophyll a and d, phycobiliproteins	Chlorophyll a and b
Sexual reproduction	In a few (?)	No	Yes	Yes	Yes	Yes
Storage material	Starch	Glucose polymer	Oil	Carbohydrate (laminarin)	Starchlike	Starch

Diatoms (Figure 11-12) are unicellular or filamentous algae with complex cell walls that consist of pectin and a layer of silica. The two parts of the wall fit together like the halves of a Petri plate. The distinctive patterns of the walls are a useful tool in diatom identification. Diatoms store energy captured through photosynthesis in the form of oil. Much of the world's petroleum was formed from diatoms that lived over 300 million years ago.

The **brown algae** or **kelp** (see Figure 11-11c) are macroscopic; some reach lengths of 50 meters. Most brown algae are found in coastal waters (see Microview A, Photo 28). Brown algae have a phenomenal growth rate. Some grow at rates exceeding 20 cm per day and therefore can be harvested regularly. *Algin*, a thickener used in many foods, such as ice cream and cake decorations, is extracted from their cell walls. Algin is also used in the production of a wide variety of nonfood goods, including rubber tires and hand lotion. Most **red algae** (Figure 11-11d) are delicately branched, multicellular organisms that live at greater ocean depths than other algae. The red pigments allow

(a)

(b)

Figure 11-12 Diatoms. (a) In this scanning electron micrograph of *Achinanthes minutissima*, note how the two parts of the cell wall fit together (×2000). **(b)** Asexual reproduction. During mitosis, each daughter cell retains one half of the cell wall from the parent (color) and must synthesize the remaining half (gray).

red algae to absorb the blue light that penetrates deepest into the ocean. Agar is extracted from the red alga *Gelidium* (jel-id´ē-um). Another gelatinous material called carrageen comes from a species of red algae called Irish moss. Carrageen can be a thickening ingredient in evaporated milk, ice cream, and pharmaceuticals. **Green algae** (see Figure 11-11e) are often classified as plants because they have cellulose cell walls, contain chlorophyll *a* and *b*, and store starch, as plants do. Green algae are believed to have given rise to terrestrial plants. Most green algae are microscopic. Some filamentous kinds form the grass-green scum in ponds.

Structure and Reproduction

The body of a multicellular alga is called a **thallus**. Thalli of the larger multicellular algae consist of branched **holdfasts** (which anchor the alga to a rock), stemlike and often hollow **stipes**, and leaflike **blades** (Figure 11-11c). The cells covering the thallus can carry on photosynthesis. The thallus lacks conductive tissue (xylem and phloem) characteristic of vascular plants; algae absorb nutrients from the water over their entire surface. The stipe is not lignified or woody, so it does not offer the support of a plant's stem. Instead, the pressure of surrounding water supports the algal thallus; some algae are also buoyed by a floating, gas-filled bladder.

All algae can reproduce asexually. The multicellular and filamentous forms fragment; each piece is capable of forming a new thallus. When a unicellular alga divides, its nucleus first divides (mitosis), then the two nuclei move to opposite parts of the cell. The cell then divides into two complete cells (cytokinesis).

Sexual reproduction occurs in some algae. In some species, asexual reproduction may occur for several generations; then, under different conditions, the same species reproduce sexually. Other species alternate generations so that the offspring resulting from sexual reproduction reproduce asexually, and the next generation then reproduces sexually.

Roles of Algae in Nature

Algae are an important part of any aquatic food chain because algae fix carbon dioxide into organic molecules that can be consumed by chemoheterotrophs. Using the energy produced in photophosphorylation, algae convert the carbon dioxide in the atmosphere into carbohydrates. Molecular oxygen (O_2) is a by-

product of their photosynthesis. The top few meters of any body of water contain planktonic algae. As three-fourths of the Earth is covered with water, it is estimated that 80% of the Earth's photosynthesis is carried on by planktonic algae.

Seasonal changes in nutrients, light, and temperature cause fluctuations in algal populations; periodic increases in numbers of planktonic algae are called *blooms*. Blooms of dinoflagellates are responsible for seasonal red tides. Blooms of a few species indicate that the water in which they grow is polluted because these algae thrive in high concentrations of organic materials that exist in sewage or industrial wastes. When algae die, the decomposition of the large numbers of cells associated with an algal bloom depletes the level of dissolved oxygen in the water. (This phenomenon is discussed in Chapter 26.)

When diatoms and other planktonic organisms that grew millions of years ago died and were buried by sediments, the organic molecules they contained did not decompose but returned to the carbon cycle as CO_2. Heat and pressure resulting from the Earth's geologic movements altered the oil stored in the cells as well as the cell membranes. Oxygen and other elements were eliminated, leaving a residue of hydrocarbons in the form of petroleum and natural gas deposits.

Many unicellular algae are symbionts in animals. The giant clam (*Tridacna*) has evolved special organs that host dinoflagellates. As the clam sits in shallow water, the algae proliferate in these organs when they are exposed to the sun. The algae release glycerol into the clam's bloodstream, thus supplying the clam's carbohydrate requirement. Evidence suggests that the clam gets needed proteins by phagocytizing old algae.

Lichens

A **lichen** is a combination of a green alga (or cyanobacterium) and a fungus. The two organisms exist in a *mutualistic* relationship, in which each partner benefits. The lichen is very different from either the alga or fungus growing alone, and, if the partners are separated, the lichen no longer exists. There are approximately 20,000 species of lichens occupying quite diverse habitats. Because lichens can inhabit areas in which neither fungi nor algae could survive alone, lichens are often the first life forms to colonize on newly exposed soil or rock. Lichens secrete organic acids that chemically weather rock, and they accumulate nutrients needed for plant growth. Lichens are some of the slowest-growing organisms on Earth.

Lichens can be grouped into three morphologic categories. **Crustose lichens** (see Figure 11-13a and Microview A, Photo 29) grow flush or encrusting onto the substratum; **foliose lichens** (Figure 11-13b) are more leaflike; and **fruticose lichens** (Figure 11-13c) have fingerlike projections. The lichen's thallus, or body, forms when fungal hyphae (**medulla**) grow around algal cells. Fungal hyphae project below the lichen body to form **rhizoids**, or holdfasts. Fungal hyphae form a **cortex**, or protective covering, over the algal layer and sometimes under it as well. After incorporation into a lichen thallus, the alga continues to grow by mitosis in the algal layer of the lichen (Figure 11-13d), and the growing hyphae can incorporate new algal cells. In a lichen, the fungus reproduces sexually.

When the algal partner is cultured separately in vitro, about 1% of the carbohydrates produced during photosynthesis are released into the culture medium; however, when the alga is associated with a fungus, the algal plasma membrane is more permeable, and up to 60% of the products of photosynthesis are released to the fungus or are found as end-products of fungus metabolism. It is clear that the fungus benefits from its association. The alga, which appears to be giving up valuable nutrients, is compensated by the protection from desiccation and the holdfast that the fungus provides.

Populations of lichens readily incorporate cations (positively charged ions) into their thalli. The concentrations and types of cations in the atmosphere can be determined by chemical analyses of lichen thalli. The presence or absence of species that are quite sensitive to pollutants can be used to ascertain air quality. A 1985 study in the Cuyahoga Valley (Ohio) revealed that 81% of the 172 lichen species that were present in 1917 are gone. Since this area is severely affected by air pollution, the inference is that air pollutants, primarily sulfur dioxide (the major contributor to acid rain), caused the death of sensitive species.

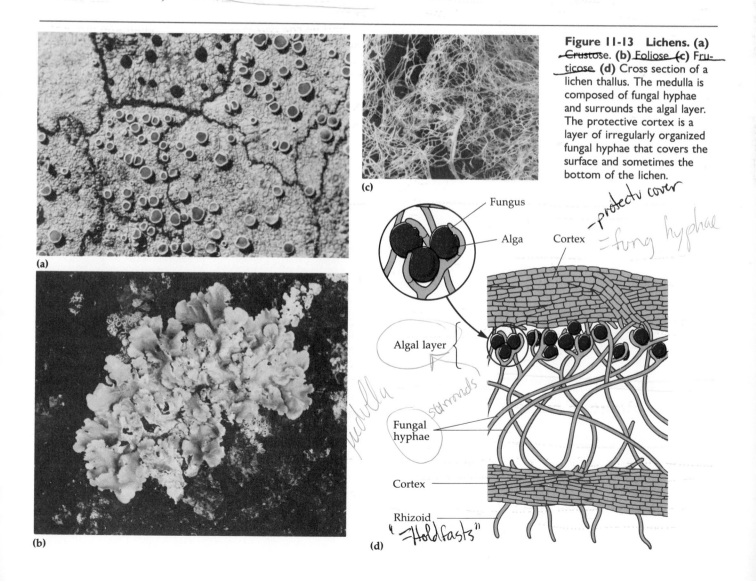

Figure 11-13 Lichens. (a) Crustose. **(b)** Foliose **(c)** Fruticose. **(d)** Cross section of a lichen thallus. The medulla is composed of fungal hyphae and surrounds the algal layer. The protective cortex is a layer of irregularly organized fungal hyphae that covers the surface and sometimes the bottom of the lichen.

(a)

(b)

(c)

(d)

Fungus

Alga

Cortex

Algal layer

Fungal hyphae

Cortex

Rhizoid

Slime Molds

Slime molds have both fungal and animal characteristics and have membrane-bounded nuclei. Thus, they are classified as protists. **Cellular slime molds** (Figure 11-14) are typical eucaryotic cells that resemble amoebas. The amoeboid cells ingest other microoganisms and bacteria by phagocytosis. Cellular slime molds are of interest to biologists who study cellular migration and aggregation. When conditions are unfavorable, large numbers of amoeboid cells aggregate to form a single structure. This aggregation happens because some individual amoebas produce the chemical cyclic AMP (cAMP), toward which the other amoebas migrate. The aggregated amoebas are enclosed in a slimy sheath called a *slug*. The slug migrates as a unit toward light. After a period of hours, the slug ceases to migrate and becomes vertically oriented. Some of the amoeboid cells form a stalk; others swarm up the stalk to form a spore cap,

and most of these differentiate into spores. When spores are released under favorable conditions, they germinate to form single amoebas.

In 1973, a Dallas resident discovered a pulsating red blob in his backyard. The news media claimed that a "new life form" had been found. For some people, the "creature" evoked a spine-chilling recollection of an old science fiction movie. Before imaginations got carried away too far, biologists calmed everyone's worst fears (or highest hopes). The amorphous mass was merely a plasmodial slime mold, they explained. Its unusually large size (46 cm in diameter), however, startled even scientists.

Plasmodial (acellular) slime molds were first scientifically reported in 1729. They belong to a separate phylum. A plasmodial slime mold exists as a mass of protoplasm with many nuclei (it is multinucleated). This mass of protoplasm is called a **plasmodium** (Figure 11-15). The entire plasmodium moves as a giant amoeba (see Microview A, Photo 30). It engulfs organic debris and bacteria. Biologists have found that musclelike proteins forming structures called

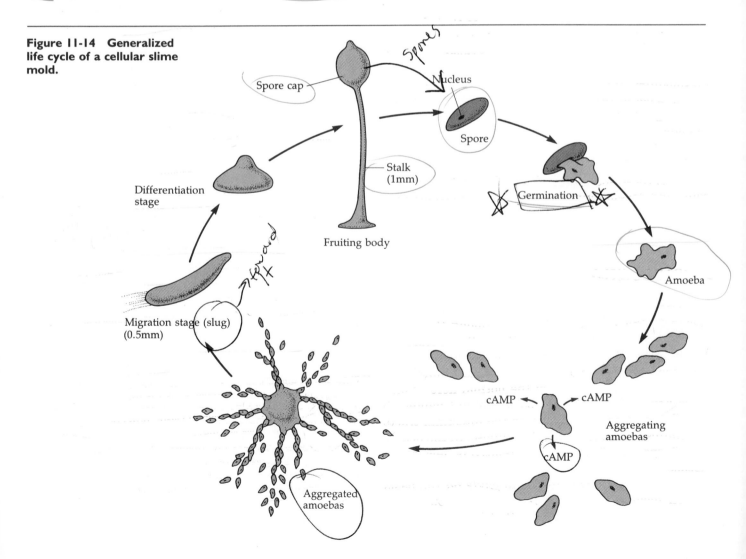

Figure 11-14   Generalized life cycle of a cellular slime mold.

Spore cap

Nucleus

Spore

Germination

Amoeba

Differentiation stage

Stalk (1mm)

Fruiting body

cAMP   cAMP

Migration stage (slug) (0.5mm)

Aggregating amoebas

cAMP

Aggregated amoebas

Figure 11-15 Acellular slime mold. Plasmodium of an acellular plasmodial slime mold (×30).

microfilaments account for the movement of the plasmodium. When plasmodial slime molds are grown in laboratories, a phenomenon called **cytoplasmic streaming** is observed. During cytoplasmic streaming, the protoplasm within the plasmodium moves and changes both its speed and direction so that the oxygen and nutrients are evenly distributed.

The plasmodium grows as long as there is enough food and moisture. When either is in short supply, the plasmodium separates into many groups of protoplasm; each of these groups forms a stalked sporangium, in which spores (a resistant, resting form of the slime mold) develop (see Figure 11-13b). Nuclei within these spores undergo meiosis and form uninucleate haploid amoebalike cells. When conditions improve, these spores germinate, fuse to form diploid cells, and develop into a multinucleate plasmodium.

Protozoans

Protozoans are one-celled, eucaryotic organisms that belong to the Kingdom Protista. Among the protozoans are many variations on this cell structure, as we shall see. Protozoans inhabit water and soil and feed upon bacteria and small particulate nutrients. Some protozoans are part of the normal flora of animals. For example, the box on p. 262 in Chapter 10 describes protozoans that live in the intestinal tracts of termites to help them digest cellulose. Of the nearly 20,000 species of protozoans, relatively few cause disease.

Protozoans are classified into phyla on the basis of their means of motility. We will restrict our study to the four phyla of protozoans that include disease-

causing species. One such phylum, the **Sarcodina**, consists of amoebas (Figure 11-16a). The amoebas move by extending usually blunt, lobelike projections of the cytoplasm called pseudopods. Any number of pseudopods can flow from one side of the amoeba cell, and the rest of the cell will flow toward the pseudopods.

Another phylum, the **Mastigophora**, or flagellates, possess flagella. Flagella are capable of whiplike movements that pull the cell through the medium (Figure 11-19). Most flagellates have one or two flagella, but some of the parasitic species have as many as eight. Species in the phylum **Ciliata** have projections called cilia that are similar to but shorter than flagella. The cilia are in a precise arrangement over the cell (Figure 11-16b) and move to propel the cell through its medium. The species in the fourth phylum we will study, the **Sporozoa**, are incapable of independent movement.

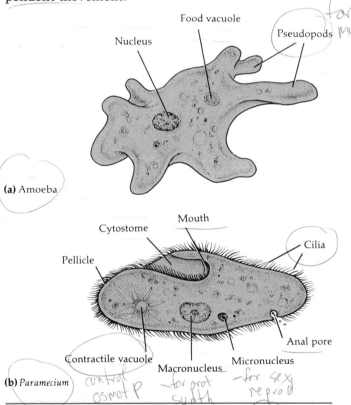

Figure 11-16 Examples of protozoans. (a) Amoeba. To move and to engulf food, amoebas extend cytoplasmic portions called pseudopods. Once surrounded by the pseudopods, the food is in a food vacuole. **(b)** *Paramecium*, a ciliate. Rows of cilia cover the cells of ciliates. The cilia are moved in unison for locomotion and to bring food particles to the protozoan. *Paramecium* has specialized structures for ingestion (mouth), elimination of wastes (anal pore), and regulation of osmotic pressure (contractile vacuoles). The macronucleus is involved with protein synthesis and other ongoing cellular activities. The micronucleus functions in sexual reproduction.

Protozoan Biology

Nutrition

Protozoans are mostly aerobic heterotrophs, although many intestinal protozoans are capable of anaerobic growth. One chlorophyll-containing flagellate, *Euglena* (ū-glē′nä) (Figure 11-11b), can also grow in the dark as a heterotroph and is often included with the protozoans.

All protozoans live in areas with a large supply of water. Some protozoans transport food across the plasma membrane. However, some have a protective covering called the **pellicle** and require specialized structures to take in food. The ciliates take in food by waving their cilia toward a mouthlike opening called the **cytostome**. The amoebas engulf food by surrounding it with pseudopods and phagocytizing it. In the protozoans, digestion takes place in membrane-bounded **vacuoles**, and waste may be eliminated through the plasma membrane or through a specialized **anal pore**.

Reproduction

Protozoans reproduce asexually by fission, budding, or schizogony. **Schizogony** is *multiple fission*. The nucleus undergoes multiple divisions before the cell divides. After many nuclei are formed, a small portion of cytoplasm concentrates around each nucleus, and separation of the single cell into daughter cells follows.

Sexual reproduction has been observed in some protozoans. The ciliates reproduce sexually by **conjugation** (Figure 11-17), which is very different from the bacterial process of the same name. During protozoan conjugation, two cells fuse, and a haploid nucleus (the micronucleus) from each cell migrates to the other cell. That haploid nucleus fuses with the haploid nucleus within the cell. The parent cells separate, each now a fertilized cell. When the cells later divide, they produce daughter cells with recombined DNA. Some protozoans produce **gametes** or **gametocytes**, haploid sex cells. In reproduction, two gametes fuse to form a diploid zygote.

Encystment

Under certain adverse conditions, some protozoans are capable of producing a protective capsule called a **cyst**. A cyst permits the organism to survive when food, moisture, or oxygen are lacking, when temper-

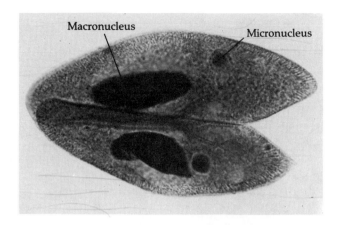

Figure 11-17   *Paramecium* conjugation. Sexual reproduction in ciliates is by conjugation. Each cell has two nuclei, a micronucleus and a macronucleus. The micronucleus is haploid and is specialized for conjugation. One micronucleus from each cell will migrate to the other cell during conjugation (×200).

atures are not suitable, or when toxic chemicals are present. A cyst also enables a parasitic species to survive outside a host. This is important because parasitic protozoans can have life cycles that involve more than one host.

Medically Important Phyla of Protozoans

Sarcodina

A well-known parasitic amoeba is *Entamoeba histolytica* (en-tä-mē′bä his-tō-li′ti-kä), the causative agent of amoebic dysentery (Figure 11-18). The primary food of *E. histolytica* is red blood cells. *E. histolytica* is transmitted from human to human through cysts that pass out with feces and are ingested by the next host.

Mastigophora

As we mentioned previously, members of the Phylum Mastigophora move by means of flagella. Their outer membrane is a tough, flexible pellicle. Flagellates are typically spindle-shaped, with flagella projecting from the front end. Food is ingested through a cytostome. Some flagellates have an **undulating membrane**, which appears to consist of highly modified flagella (Figure 11-19).

An example of a flagellate that is a human parasite is *Giardia lamblia* (jē-är′dē-ä lam′lē-ä). (See Figures 11-20 and 24-15, which show the vegetative form,

Figure 11-18 *Entamoeba histolytica* **in the vegetative form.** The dark circles inside the amoeba are engulfed red blood cells from its human host (×800).

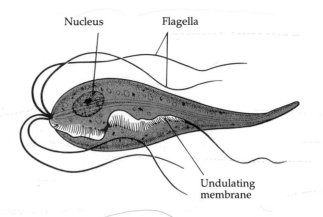

Figure 11-19 *Trichomonas vaginalis.* This flagellate is the cause of urinary and genital tract infections. It has a small undulating membrane. This flagellate does not have a cyst stage.

or **trophozoite**, as well as the cyst stage of *G. lamblia.*) The parasite is found in the small intestine of humans and other mammals. It is passed out of the intestine and survives as a cyst before being ingested by the next host. Diagnosis of giardiasis, the disease caused by *G. lamblia*, is often based on identification of cysts in feces.

Another parasitic flagellate is *Trichomonas vaginalis* (trik-ō-mōn′as va-jin-al′is), shown in Figure 11-19. *T. vaginalis* does not have a cyst stage and must be transferred from host to host quickly before desiccation occurs. *T. vaginalis* is found in the vagina and in the male urinary tract. It is usually transmitted by sexual intercourse but can be transmitted by toilet facilities or towels.

The hemoflagellates are transmitted by the bites of blood-sucking insects and are found in the circulatory system of the bitten host. To survive in this viscous fluid, hemoflagellates have long slender bodies and an undulating membrane. The genus *Trypanosoma* (tri-pa-nō-sō′mä) includes the species that cause African sleeping sickness (*T. brucei gambiense*, which is transmitted by the tsetse fly; see the box on p. 302 and Microview A, Photo 23). *T. cruzi* (see Figure

(a) Trophozoite

Cyst wall

(b) Cyst

Figure 11-20 *Giardia lamblia.* **(a)** The trophozoite of this intestinal parasite has eight flagella and two prominent nuclei, giving it a distinctive appearance. **(b)** The cyst provides protection in the environment before being ingested by a new host.

Nuclei

21-16), the causative agent of Chagas' disease, is transmitted by the "kissing bug," so-called because it bites on the face. After entering the insect, the trypanosome rapidly multiplies by fission. If the insect then defecates while biting a human, the insect can release trypanosomes that can contaminate the bite wound.

Ciliata

The only ciliate that is a human parasite is *Balantidium coli* (bal-an-tid′ē-um kō′lī), the causative agent of a severe though rare type of dysentery. When cysts are ingested by the host, they enter the colon, into which the trophozoites are released. The trophozoites feed on bacteria and fecal debris as they multiply, and cysts are passed out with feces (see Figure 24-16).

Sporozoa

Sporozoans are not motile in their mature forms and are obligate intracellular parasites. They have complex life cycles that ensure their survival and transmission from host to host. An example of a sporozoan is *Plasmodium* (plaz-mō′dē-um), the causative agent of malaria. A vaccine against malaria is now being

.. Microbiology in the News ....

Parasite Outwits the Immune System, Intrigues Scientists

"A leopard," the old saying goes, "can't change its spots." The fearsome hunter of the African forest could never hide its identity by altering its coat. But a tiny yet far more menacing native of Africa can do just that. The trypanosome, a protozoan parasite that causes the devastating disease known as sleeping sickness (see the photo, ×1000), can rapidly change its protein coat to disguise itself from the immune system of its host. In the past few years, this strange creature has attracted a great deal of attention from molecular biologists, who hope their research will ultimately lead to a vaccine or drug against sleeping sickness. Scientists also find the parasite fascinating in its

own right. Its remarkable defense mechanisms may shed light on one of the most pressing questions of molecular genetics—how individual genes express themselves.

An Unflagging Invader
Sleeping sickness, or African trypanosomiasis, is spread by the tsetse fly, which hosts the African trypanosome (*Trypanosoma* spp.). When an infected fly bites a human or animal, trypanosomes enter the mammalian host's bloodstream, multiply, and eventually invade the central nervous system. The resulting disease is one of the most terrible known. The victim suffers increasing fever, lethargy,

and mental deterioration, finally lapsing into a coma. Death is nearly certain without treatment. Although existing drug therapies improve the prognosis somewhat, they are not highly effective and have severe side effects.

In the broad belt across central Africa that is home to the tsetse fly, the impact of sleeping sickness has been enormous. During one epidemic around the turn of the century, two-thirds of the population of the north shore of Lake Victoria died. Today the disease affects about a million people, with 20,000 new cases appearing each year. No livestock can be kept throughout this vast

region, which encompasses over a third of the African continent.

What makes the trypanosome so deadly, scientists have found, is its ability to rebound from an attack by the host's immune system. When a foreign organism invades a human or animal, the host's immune system makes antibodies against it. These antibodies are produced in response to specific antigens (large, antibody-provoking molecules) on the invader's surface, which they recognize and attack. In the case of trypanosomes, the antigens are the protein molecules that form its coat.

In the early stages of sleeping sickness, when trypanosomes are circulating in the blood, the host produces enormous numbers of antibodies, destroying over 99% of the invaders within a week. With most pathogens this response would end the disease before it became serious. But such is not tested on humans (see the box in Chapter 17, p. 438).

Plasmodium grows by schizogony in human liver cells and in red blood cells (Figure 11-21), in which it completes its asexual cycle. When an *Anopheles* (an-of'el-ēz) mosquito carrying the infective stage of *Plasmodium* (called a **sporozoite**) bites a human, sporozoites can be injected into the human (Figure 11-21a). The sporozoites are carried by the blood to the liver. Here the sporozoites undergo schizogony in liver cells and produce thousands of progeny called **merozoites**. Merozoites enter the bloodstream and infect red blood cells (Figure 11-21b). The young trophozoite looks like a ring in which the nucleus and cytoplasm are visible. This is called a **ring stage** (see Microview A, Photo

22). The ring stage enlarges and divides repeatedly, and the red blood cells eventually rupture and release more merozoites. Upon release of the merozoites, their waste products, which cause fever and chills, are also released. Most of the merozoites infect new red blood cells and perpetuate their cycle of asexual reproduction. However, some develop into male and female sexual forms called **gametocytes**. Even though the gametocytes themselves cause no further damage, they can be picked up by the bite of another *Anopheles* mosquito (Figure 11-21c); they then enter the mosquito's intestine and begin their sexual cycle. Here the male and female gametocytes unite to form a zygote. The zygote forms an oocyst, in which cell division

the case with trypanosomes, for the remaining 1% have changed their coat proteins so that the antibodies no longer recognize them. These survivors multiply in the blood while the body manufactures new antibodies against them, and eventually they, too, are destroyed—except the 1% that have changed coats again. Each time the host produces new antibodies, the parasite produces new antigens, until finally the parasite penetrates the nervous system and kills its host.

Probing the Mystery
Molecular biologists are intrigued by the trypanosome's quick-change artistry, and have been probing the mysteries of its gene expression. The trypanosome's coat is made of a single protein, called *variable surface glycoprotein*, or VSG, so it can be entirely transformed by the alteration of a single gene. The trypanosome has genes for many variants of VSG, and these

genes are switched on and off in quick succession.

An individual trypanosome probably has from several hundred to several thousand VSG genes, but only those on the ends of chromosomes are expressed. Most VSG genes are normally located in the interior portions of chromosomes, so when these genes are to be transcribed, they are first duplicated and moved to an "expression site" at a chromosome end. No one yet knows what determines which gene is to be transcribed or how the moving is accomplished. Complicating the picture further, genes normally found at ends of chromosomes are sometimes not expressed, which means that additional on-and-off mechanisms are at work.

Researchers are excited by the possibility that understanding these mechanisms will lead to a better understanding of gene expression in all organisms. The very fact that trypanosomes alter so

rapidly makes them excellent subjects for study. Most organisms are so stable at the molecular level that there is little to observe, but trypanosomes turn their genes on and off before the observers' eyes.

A Grim Future?
Will understanding how trypanosomes outwit the immune system lead to development of a vaccine for sleeping sickness? The prospects seem poor. To eliminate an organism that can interchange hundreds or thousands of surface antigens, most of which appear in seemingly random order, vaccination might have to produce hundreds or thousands of kinds of antibodies. The discovery that the same 12 to 14 antigens always appear in the first five days or so of an infection provides a slim ray of hope; perhaps producing antibodies against all of these simultaneously could halt the disease in its earliest stage. But it seems quite possible that some of the trypanosomes would

simply produce other antigens and survive the onslaught.

What about new drug therapies? Trypanosomes are remarkable, not only for their mechanisms of gene expression but also for a host of unique metabolic pathways. They are very ancient organisms, and their relationship to most other living things is very distant, so it is not surprising that they are distinctive in many ways. Because they are unique, there is enormous interest in studying trypanosomes, and metabolic research could prompt the development of specific antitrypanosomal drugs. Pharmaceutical companies, however, have shown little interest in developing drugs against parasites since parasitic infections are common only where most people are very poor. The scourge of a continent is worthy of basic research—but the end of the affliction is not in sight.

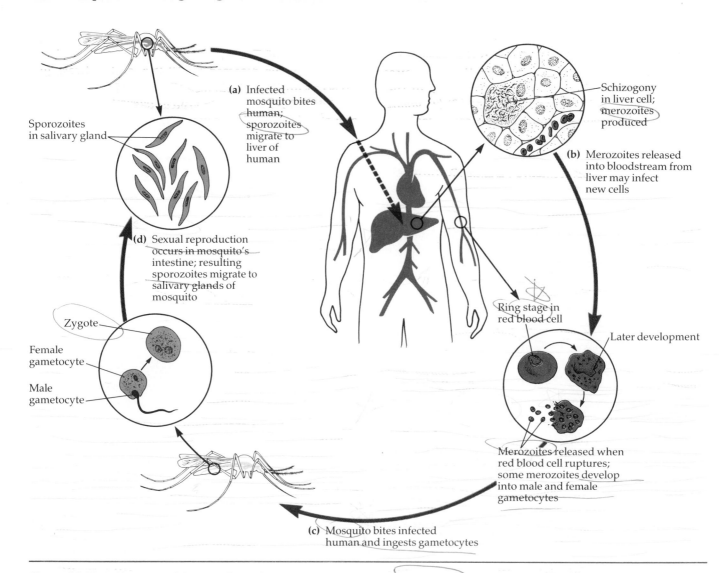

Figure 11-21   Life cycle of *Plasmodium vivax*. Asexual reproduction (schizogony) of the parasite takes place in the liver and in the red blood cells of a human host. Sexual reproduction occurs in the intestine of an *Anopheles* mosquito after the mosquito has ingested gametocytes.

occurs, and asexual sporozoites are formed. When the cyst ruptures, the sporozoites migrate to the salivary glands of the mosquito (Figure 11-21d). They can then be injected into a new human host by the biting mosquito.

Malaria is diagnosed in the laboratory by microscopic observation of thick blood smears for the presence of *Plasmodium* (see Figure 22-19). A peculiar characteristic of malaria is that the interval between periods of fever caused by release of merozoites is always the same for a given species of *Plasmodium*, and the interval is always a multiple of 24 hours. The

reason and mechanism for such precision has intrigued scientists. After all, why should a parasite need a biological clock? Frank Hawking and his coworkers have demonstrated that *Plasmodium*'s development is regulated by the host's body temperature, which normally fluctuates over a 24-hour period. The parasite's careful timing ensures that gametocytes are mature at night, when mosquitoes are feeding, to enable transmission of the parasite to a new host.

Toxoplasma gondii (toks-ō-plaz′mä gon′dē-ē) is another intracellular parasite of humans. The life cycle of this sporozoan involves domestic cats. The trophozoites reproduce sexually and asexually in an infected cat, and oocysts, each containing eight sporozoites, are passed out with feces. If the oocysts are ingested by humans or other animals, the sporozoites emerge as trophozoites, which can reproduce in the tissues of the new host (see Figure 22-16). *T. gondii* is dangerous to pregnant women, as it can cause congenital infections in utero. Tissue examination and observation of *T. gondii* is used for diagnosis. Antibodies may be detected by ELISA and indirect fluorescent antibody tests (see Chapter 17).

Table 11-6 lists some typical parasitic protozoans and the diseases they cause.

TABLE 11-6   *Some Representative Parasitic Protozoans*

Phylum	Human Pathogens	Distinguishing Features	Disease	Source of Human Infections	See Figure	Refer to Chapter
Sarcodina (amoebas)	*Acanthamoeba*	Pseudopods	Keratitis	Water	—	20
	Entamoeba histolytica	Pseudopods	Amoebic dysentery	Fecal contamination of drinking water	11-18	24
	Naegleria fowleri	Some flagellated forms	Microencephalitis	Water in which people swim	—	21
Mastigophora (flagellates)	*Giardia lamblia*	Two nuclei, eight flagella	Giardial dysentery	Fecal contamination of drinking water	11-20	24
	Trichomonas vaginalis	No encysting stage	Urethritis; vaginitis	Contact with vaginal–urethral discharge	11-19	25
	T.b. gambiense, T.b. rhodesiense	Undulating membrane	African trypanosomiasis	Bite of tsetse fly	—	21
	Trypanosoma cruzi		Chagas' disease	Bite of *Triatoma* (kissing bug)	—	22
Ciliata	*Balantidium coli*	Only parasitic ciliate of humans	Balantidial dysentery	Fecal contamination of drinking water	23-16	24
Sporozoa	*Babesia microti*		Babesiosis	Domestic animals, ticks	—	23
	Cryptosporidium	Complex life cycles may require more than one host	Diarrhea	Cows, other animals	—	23
	Isospora		Coccidiosis	Domestic animals	—	Box, Chapter 12; 18
	Plasmodium		Malaria	Bite of *Anopheles* mosquito	11-21	22
	Pneumocystis carinii		Pneumonia	Other animals	—	23
	Toxoplasma gondii		Toxoplasmosis	Cats, other animals; congenital	21-15	22

Helminths

There are a number of parasitic animals that spend part or all of their lives in humans. Most of these animals belong to two phyla, **Platyhelminthes** (flatworms) and **Aschelminthes** (roundworms). These worms are commonly called **helminths**. There are also free-living species in these phyla, but we will limit our discussion to the parasitic species.

Helminth Biology

Helminths are multicellular eucaryotic animals, generally possessing digestive, circulatory, nervous, excretory, and reproductive systems. Parasitic helminths must be highly specialized to live inside their hosts. The following generalizations about parasitic helminths distinguish them from their free-living relatives:

1. They either *lack* a digestive system or have a greatly simplified one. They can absorb nutrients from the host's food, body fluids, and tissues.
2. Their nervous system is *reduced*. They do not need an extensive nervous system because they do not have to search for food or respond to their environment. The environment within a host is fairly constant.
3. Their means of locomotion is either *reduced* or *completely lacking*. Because they are transferred from host to host, they do not have to search for a suitable habitat.
4. Their reproductive system is often *more* complex; it produces more fertilized eggs, by which a suitable host is infected.

Reproduction

Adult helminths may be **dioecious**—male reproductive organs are in one individual, and female reproductive organs are in another. In those species, reproduction occurs only when two adults of the opposite sex are in the same host.

Adult helminths may also be **hermaphroditic**—one animal has both male and female reproductive organs. Two hermaphrodites may copulate and simultaneously fertilize each other. A few types of hermaphrodites fertilize themselves.

Life Cycle

The life cycle of parasitic helminths can be extremely complex, involving a succession of hosts. The term **definitive host** is given to the organism that harbors the adult, sexually mature helminth. One or more **intermediate hosts** might be necessary for completion of each *larval* (developmental) stage of the parasite.

Platyhelminthes

Members of the Phylum Platyhelminthes (flatworms) are flattened from top to bottom. They have what is called an *incomplete digestive system*. This type of digestive system has only one opening (mouth), through which food enters and wastes leave. The classes of parasitic flatworms include the trematodes and cestodes.

Trematodes

Trematodes, or flukes, have flat, leaf-shaped bodies with a ventral sucker and an oral sucker (Figure 11-22). The suckers hold the organism in place and suck fluids from the host. Flukes can also obtain food by absorbing it through their nonliving outer covering, called the cuticle. Flukes are named according to the tissue of the definitive host in which the adults live (for example, lung fluke, liver fluke, blood fluke).

To exemplify a fluke's life cycle, let us look at the

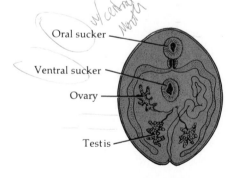

Figure 11-22   Adult fluke. This generalized diagram of the anatomy of an adult fluke shows the oral and ventral suckers. The suckers attach the fluke to the host. The mouth is located in the center of the oral sucker. Flukes are hermaphroditic; each animal contains both testes and ovaries.

lung fluke, *Paragonimus westermanni* (pār-ä-gŏn′e-mus we-ster-man′nē), shown in Figure 11-23. The adult lung fluke lives in the bronchioles of humans and other mammals. The adults are approximately 6 mm wide by 12 mm long. The hermaphroditic adults liberate eggs into the bronchi. Because sputum that contains eggs is frequently swallowed, the eggs are usually excreted in feces from the host.

To continue the life cycle, the eggs must be excreted into a body of water. Inside the egg, a **miracidial larva**, or **miracidium**, then develops. When the egg hatches, the larva enters a suitable snail. Only certain species of aquatic snails can be the intermediate host. Inside the snail, the lung fluke undergoes asexual reproduction and produces **rediae**. Each redia develops into a **cercaria** that bores out of the snail and penetrates the cuticle of a crayfish. The parasite encysts as a **metacercaria** in the muscles and other tissues of the crayfish. When the crayfish is eaten by a human, the metacercaria is freed in the human's small intestine. It bores out and wanders around until it penetrates the lungs, enters the bronchioles, and develops into an adult lung fluke.

In a laboratory diagnosis, sputum and feces are examined microscopically for eggs. Infection results from eating undercooked crayfish, and the disease could be prevented by thoroughly cooking crayfish.

The cercariae of the blood fluke *Schistosoma* (shis-tō-so′mä) are not ingested. Instead, they burrow through the skin of the human host and enter the circulatory system. The adults are found in certain abdominal and pelvic veins. The disease schistosomiasis is a major world health problem; it will be discussed further in Chapter 22.

Figure 11-23 Life cycle of *Paragonimus westermanni*. The free-swimming miracidium invades the first intermediate host, a snail. Two generations of rediae develop within the snail, and the rediae give rise to cercariae. The cercariae leave the snail and penetrate the second intermediate host, a crayfish. Here they encyst as metacercariae. When a human eats raw or undercooked crayfish, the metacercariae are released to develop into adults.

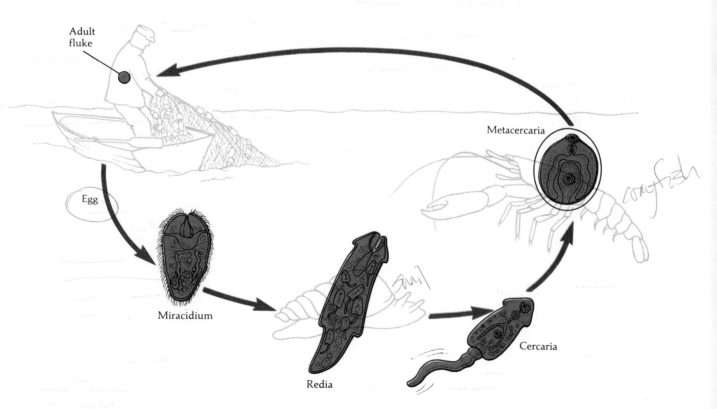

Adult fluke

Egg

Miracidium

Redia

Cercaria

Metacercaria

Cestodes

Cestodes, or tapeworms, are intestinal parasites. Their structure is shown in Figure 11-24. The head, or **scolex**, has suckers and might have small hooks for attaching to the intestinal mucosa of the host (see Microview A, Photo 24). Tapeworms do not ingest the tissues of their hosts; in fact, they completely lack a digestive system. To obtain nutrients from the small intestine, they absorb food through their cuticle. The body consists of segments called **proglottids** (see Microview A, Photo 25). Proglottids are continually

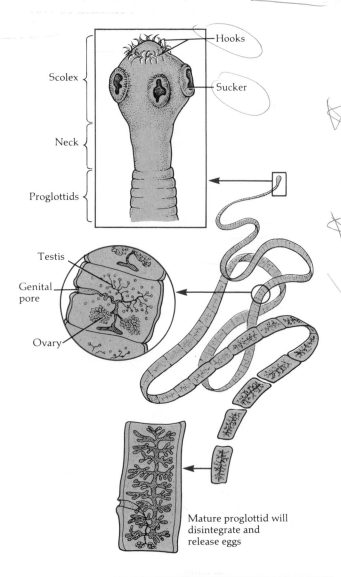

Figure 11-24 General anatomy of an adult tapeworm.
The scolex consists of suckers and hooks that attach to the host's tissues. The body lengthens as new proglottids form at the neck. Each proglottid contains both testes and ovaries.

produced by the neck region of the scolex, as long as the scolex is attached and alive. Each proglottid contains both male and female reproductive organs. The proglottids farthest away from the scolex are the mature ones containing fertilized eggs.

Humans as Definitive Hosts

The adults of *Taenia saginata* (te′nē-ä sa-ji-nä′tä), beef tapeworm, live in humans and can reach a length of 6 meters. The scolex is about 2 mm in length and is followed by a thousand or more proglottids. The life cycle of the beef tapeworm is illustrated in Figure 24-19. The feces of an infected human contain mature proglottids, and each proglottid contains thousands of eggs. As the proglottids wriggle away from the fecal material, they increase their chances of being ingested by an animal that is grazing. Upon ingestion by cattle, the larvae hatch from the eggs and bore through the intestinal wall. The larvae migrate to muscle (meat), in which they encyst as **cysticerci**. When the cysticerci are ingested by humans, all but the scolex is digested. The scolex anchors itself in the small intestine and begins producing proglottids.

Diagnosis is based on the presence of mature proglottids and eggs in feces. Cysticerci can be seen macroscopically in meat; their presence is referred to as "measly beef." Inspecting beef that is intended for human consumption for "measly" appearance is one way to prevent infections by beef tapeworm. Another method of prevention is to avoid the use of untreated human sewage as fertilizer in grazing pastures.

Humans as Intermediate Hosts

Humans are the intermediate hosts for *Echinococcus granulosus* (e-kīn-ō-kok′kus gra-nū-lō′sus), shown in Figure 11-25. Dogs and cats, both wild and domestic, are the definitive hosts for this minute (2 to 8 mm) tapeworm. Eggs are excreted with feces and are ingested by deer or humans. Humans can become infected by contaminating their hands with dog feces or saliva from a dog's tongue. The eggs hatch in the human's small intestine, and the larvae migrate to the liver or lungs. The host forms a cyst around the larvae in these organs. This cyst, called a **hydatid cyst**, contains "brood capsules," in which thousands of scoleces (singular, *scolex*) might be produced. In the wild, the cysts might be in a deer that is eaten by a wolf. The scoleces would be able to attach themselves in the wolf's intestine and produce proglottids.

Diagnosis of hydatid cysts is frequently made only on autopsy, although X rays can detect the cysts (Figure 24-20).

(a) Adult

Hydatid cyst

Scolex

Brood capsule

prod 1000's scolex

Egg

Intermediate host

Definitive host

(b)

Hydatid cyst

Intermediate host

Figure 11-25 *Echinococcus granulosus.* This tiny tapeworm is found in the intestines of dogs, cats, wolves, and foxes. **(a)** Adult. **(b)** Life cycle. Eggs are excreted from the definitive host and ingested by an intermediate host, such as a deer. In the intermediate host, the eggs hatch, and the larvae form hydatid cysts in the host's tissues. A hydatid cyst is a fluid-filled sac containing many scoleces. For the cycle to be complete, the cysts must be ingested by a definitive host that eats the intermediate host. A human serving as the intermediate host is a dead end for the parasite unless the human is eaten by an animal.

Aschelminthes

Members of the Phylum Aschelminthes, the roundworms, are cylindrical in shape and tapered at each end. Roundworms have a *complete* digestive system, consisting of a mouth, an intestine, and an anus. Most species are dioecious. Males are smaller than females and have one or two hardened *spicules* on their posterior ends. Spicules are used to guide sperm to the vagina. Parasites of humans belong only to the class consisting of nematodes.

Nematodes

Some species of nematodes are free-living in soil and water, and others are parasites on plants and animals. Parasitic nematodes do not have the succession of larval stages exhibited by flatworms. Some nematodes pass their entire life cycle, from egg to mature adult, in a single host.

Nematode infections of humans can be divided into two categories—those in which the egg is infective and those in which the larva is infective.

Eggs Infective for Humans

The pinworm *Enterobius vermicularis* (en-te-rō′bē-us ver-mi-kū-lar′is) spends its entire life in a human host (Figure 11-26). Adult pinworms are found in the large intestine. From there, the female pinworm migrates to the anus to deposit her eggs in the perianal skin. The eggs can be ingested by the host or by another person exposed through contaminated clothing. Pinworm infections are diagnosed by the Graham sticky-tape method. A piece of tape is placed on the perianal skin in such a way that the sticky side is exposed to pick up eggs that were deposited earlier. The tape is microscopically examined for the presence of eggs adhering to it.

Ascaris lumbricoides (as′kar-is lum-bri-koi′dēz) is a large nematode (30 cm in length). It is dioecious with **sexual dimorphism**; that is, the male and female worms look distinctly different, the male being smaller with a curled tail. The adult *Ascaris* lives in the small intestines of humans and domestic animals (such as pigs and horses); it feeds primarily on semi-digested food. Eggs, excreted with feces, can survive in the soil for long periods until accidentally ingested by another host. The eggs hatch in the intestine of the host, mature in the lungs, and from there migrate to the intestines.

Diagnosis is frequently made when the adult worms are excreted with feces. Prevention of the disease in humans is managed by proper sanitary habits. The life cycle of *Ascaris* in pigs can be interrupted by keeping pigs in areas free of fecal material.

Larvae Infective for Humans

Adult hookworms, *Necator americanus* (ne-kā′tôr ä-me-ri-ka′nus), live in the small intestines of humans (Figure 11-27); the eggs are excreted with feces. The

TABLE 11-7   *Representative Parasitic Helminths*

Phylum	Class	Human Parasites	Intermediate Host	Definitive Host	Stage Passed to Humans	Disease	Location in Humans	See Figure
Platyhelminthes	Trematodes	*Paragonimus westermanni*	Freshwater snails and crayfish	Humans; lungs	Metacercaria in crayfish	Paragoni-miasis (lung fluke)	Lungs	11-23
		Schistosoma	Freshwater snails	Humans	Cercariae through skin	Schisto-somiasis	Veins	22-20, 22-21, 22-22
	Cestodes	*Taenia saginata*	Cattle	Humans; small intestine	Cysticerci in beef	Tapeworm	Small intestine	11-25, 24-19
		Echinococcus granulosus	Humans	Dogs and other animals; intestines	Eggs from other animals	Hydatid cyst	Lungs	11-25
Aschelminthes	Nematodes	*Ascaris lumbricoides*	—	Humans; small intestine	Ingestion of eggs	Ascariasis	Small intestine	—
		Enterobius vermicularis	—	Humans; large intestine	Ingestion of eggs	Pinworm	Large intestine	11-26a
		Necator americanus	—	Humans; small intestine	Penetration of larvae through skin	Hookworm	Small intestine	11-27
		Trichinella spiralis	—	Humans, swine, and other mammals; small intestine	Ingestion of larvae in meat, especially pork	Trichinosis	Muscles	24-21

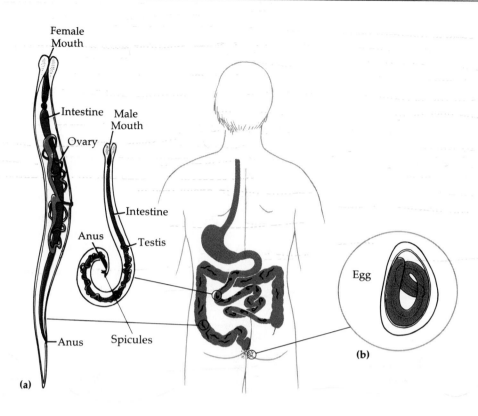

Female Mouth

Intestine

Male Mouth

Ovary

Intestine

Anus

Testis

Anus

Spicules

(a)

Egg

(b)

Figure 11-26 *Enterobius vermicularis.* **(a)** Adult pinworms, which live in the intestines of humans. Roundworms have a complete digestive system with a mouth, intestine, and anus. Most roundworms are dioecious, and the female (left) is often distinctly larger than the male (right). **(b)** Eggs, which the female deposits on the perianal skin at night.

larvae hatch in the soil and feed on bacteria there. A larva enters its host by penetrating the host's skin. The larva then enters a blood or lymph vessel, which carries it to the lungs. The larva is coughed up in sputum, swallowed, and finally carried to the small intestine. Diagnosis is based on the presence of eggs in feces. People can avoid hookworm infections by wearing shoes.

Trichinella spiralis (tri-kin-el′lä spī-ra′lis) infections, called *trichinosis*, are usually acquired by eating encysted larvae in pork or bear meat. In the human digestive tract, the larvae are freed from the cysts. They mature into adults in the intestine and sexually reproduce there. Eggs develop in the female, and she gives birth to live nematodes. The larvae enter lymph and blood vessels in the intestines and migrate from there throughout the body. They encyst in muscles and other tissues and remain there until ingested by another host (see Figure 24-21).

Diagnosis of trichinosis is made by microscopic examination for larvae in a muscle biopsy. Trichinosis can be prevented by thoroughly cooking meat prior to consumption.

Table 11-7 lists representative parasitic helminths of each phylum and class and the diseases they cause.

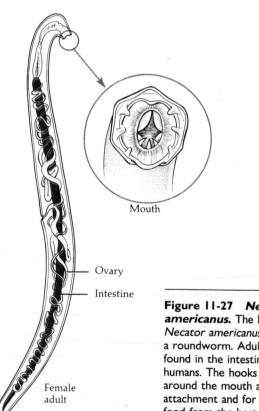

Mouth

Larva

Ovary

Intestine

Female adult

Figure 11-27 *Necator americanus.* The hookworm *Necator americanus* is actually a roundworm. Adults are found in the intestine of humans. The hooks located around the mouth are used for attachment and for removing food from the host tissue. The free-living larvae are in the soil and penetrate the skin of the definitive host.

Arthropods as Vectors

Arthropods are jointed-legged animals. With nearly 1 million species, this is the largest phylum in the animal kingdom. We will briefly describe arthropods here because a few suck the blood of humans and other animals and can transmit microbial diseases while doing so. Arthropods that carry disease-causing microorganisms are called **vectors**. Scabies, a disease caused by an arthropod, is described in the box in Chapter 20, p. 513.

Representative classes of arthropods include the following:

1. Arachnida (eight legs): spiders, mites, ticks
2. Crustacea (four antennae): crabs, crayfish
3. Chilopoda (two legs per segment): centipedes
4. Diplopoda (four legs per segment): millipedes
5. Insecta (six legs): bees, flies

Table 11-8 lists those arthropods that are important vectors, and Figures 11-28 and 11-29 illustrate some of them. These insects and ticks reside on an animal only when they are feeding. An exception to this is the louse, which spends its entire life on its host and cannot survive long away from a host. Among insect vectors, only the adult female flies (including mosquitoes) bite.

Some vectors are just a mechanical means of transport for a pathogen. For example, houseflies are attracted to decaying organic matter in which to lay their eggs. A housefly can then pick up a pathogen on

TABLE 11-8   *Important Arthropod Vectors of Human Diseases*

Class	Order	Vector	See Figure	Disease	Refer to Chapter
Arachnida	Mites and ticks	*Dermacentor* (tick)	11-29a	Rocky Mountain spotted fever	22
		Ixodes (tick)	—	Lyme disease, babesiosis	22
Insecta		*Ornithodorus*	—	Relapsing fever	22
	Sucking lice	*Pediculus* (human louse)	11-29b	Epidemic typhus	22
	Fleas	*Xenopsylla* (rat flea)	11-29c	Endemic murine typhus, plague	22
	True flies	*Chrysops* (deer fly)	11-29d	Tularemia	22
		Aedes (mosquito)	11-28a	Dengue fever, yellow fever	22
		Anopheles (mosquito)	11-28b	Malaria	22
		Culex (mosquito)	—	Arboviral encephalitis	21
		Glossina (tsetse fly)	—	African trypanosomiasis	21
	True bugs	*Triatoma* (kissing bug)	11-28e	Chagas' disease	22

its feet or body from feces and transport the pathogen to food.

Some parasites multiply in their vectors. When this happens, the parasites can accumulate in the vector's feces or saliva. Large numbers of parasites can then be deposited on the skin of the host while the vector is feeding there.

Plasmodium is an example of a parasite that requires that its vector also be its host. *Plasmodium* can sexually reproduce only in the gut of an *Anopheles* mosquito (Figure 11-28b). *Plasmodium* is introduced into a human host with the mosquito's saliva, which acts as an anticoagulant to keep blood flowing.

To eliminate vector-borne diseases (such as African sleeping sickness), health workers focus on eradicating the vectors.

Figure 11-29 Arthropod vectors. (a) The tick, *Dermacentor.* **(b)** The human louse, *Pediculus.* **(c)** The rat flea, *Xenopsylla.* **(d)** The deer fly, *Chrysops.* **(e)** The kissing bug, *Triatoma.* See Table 11-8 for the diseases these vectors carry.

(a) Tick

(b) Louse

(c) Rat flea

(d) Deer fly

(e) Kissing bug

(a)

(b)

Figure 11-28 Mosquitoes. (a) *Aedes aegypti* (female), sucking blood from human skin. This insect vector transmits several diseases from person to person, including the viral diseases yellow fever and dengue fever. **(b)** *Anopheles* transmits malaria.

Study Outline

Fungi (pp. 282–292)

1. The Kingdom Fungi includes yeasts, molds, and fleshy fungi (mushrooms).
2. Fungi are aerobic or facultatively anaerobic chemoheterotrophs.
3. Most fungi are decomposers, and a few are parasites of plants and animals.

Characteristics of Fungi (pp. 283–287)

Vegetative Structures (pp. 283–285)

1. A fungal thallus consists of filaments of cells called hyphae; a mass of hyphae is called a mycelium.
2. Yeasts are unicellular fungi that reproduce by budding and occasionally by fission.
3. Buds that do not separate from the mother cell form pseudohyphae.
4. Dimorphic fungi are yeastlike at 37° C and moldlike at 25° C.

Reproductive Structures (pp. 285–287)

1. The following spores can be produced asexually: arthrospores, chlamydospores, sporangiospores, conidiospores, and blastospores.
2. Fungi are classified according to the type of sexual spore that they form.
3. Sexual spores include zygospores, ascospores, and basidiospores.

Nutritional Adaptations (p. 287)

1. Fungi can grow in acidic, low-moisture, aerobic environments.
2. They are able to metabolize complex carbohydrates.

Medically Important Phyla of Fungi (pp. 287–291)

1. Deuteromycota reproduce asexually by one or more of the following: chlamydospores, arthrospores, conidiospores, or budding.
2. Deuteromycota are called the Fungi Imperfecti because sexual spores have not yet been seen.
3. The Zygomycota have coenocytic hyphae and produce sporangiospores and zygospores.
4. The Ascomycota have septate hyphae and produce ascospores and frequently conidiospores.
5. Basidiomycetes have septate hyphae and produce basidiospores; some produce conidiospores.

Fungal Diseases (pp. 291–292)

1. Systemic mycoses are fungal infections deep within the body and affect many tissues and organs.
2. Subcutaneous mycoses are fungal infections beneath the skin.
3. Cutaneous mycoses affect keratin-containing tissues such as hair, nails, and skin.
4. Superficial mycoses are localized on hair shafts and superficial skin cells.
5. Opportunistic mycoses are caused by normal flora or fungi that are not usually pathogenic.
6. Opportunistic mycoses include mucormycosis, caused by some Zygomycetes; aspergillosis, caused by *Aspergillus*; and candidiasis, caused by *Candida*.
7. Opportunistic mycoses can infect any tissues. However, they are usually systemic.

Economic Effects of Fungi (p. 292)

1. Mold spoilage of fruits, grains, and vegetables is more common than bacterial spoilage of these products.
2. Some fungi cause diseases in plants (for example, in potatoes, chestnuts, and elms).

Algae (pp. 292–296)

1. Algae are photoautotrophs that produce oxygen.
2. Algae are classified as plants or protists according to their structures and pigments.

Phyla of Algae (pp. 293–296)

1. Unicellular dinoflagellates have cellulose and silica cell walls; some produce a neurotoxin that causes paralytic shellfish poisoning and ciguatera.
2. Englenoids have a semirigid cell membrane and one flagellum.
3. Diatoms have pectin and silica cell walls.
4. Multicellular algae include the brown algae, red algae, and green algae.
5. Brown algae (kelp) are harvested for algin.
6. Red algae grow deeper in the ocean than other algae because their red pigments can absorb the blue light that penetrates to deeper levels.
7. Green algae have cellulose and chlorophyll *a* and *b* and store starch.

Structure and Reproduction (p. 296)

1. The thallus (or body) of multicellular algae usually consists of a stipe, holdfast, and blades.

2. Algae reproduce asexually by cell division and fragmentation.
3. Many algae reproduce sexually.

Roles of Algae in Nature (p. 296)

1. Algae are the primary producers in aquatic food chains.
2. Planktonic algae produce most of the molecular oxygen in the Earth's atmosphere.
3. Petroleum is the fossil remains of planktonic algae.
4. Unicellular algae are symbionts in such animals as *Tridacna*.

Lichens (pp. 296–297)

1. A lichen is a symbiotic relationship between an alga and a fungus.
2. The alga photosynthesizes, providing carbohydrates for the lichen; the fungus provides a holdfast.
3. Lichens colonize habitats that are unsuitable for either the alga or fungus alone.
4. Lichens may be classified on the basis of morphology as crustose, foliose, or fruticose.
5. In the lichen thallus, the alga reproduces by cell division, and the fungus forms sexual spores.

Slime Molds (pp. 298–299)

1. Cellular slime molds resemble amoebas and ingest bacteria by phagocytosis.
2. A plasmodial (acellular) slime mold is a multinucleated mass of protoplasm that engulfs organic debris and bacteria as it moves.

Protozoans (pp. 299–305)

1. Protozoans are unicellular eucaryotes in the Kingdom Protista.
2. Protozoans are found in soil and water and as normal flora in animals.
3. Protozoans are classified by their means of locomotion: members of the Phylum Sarcodina move by amoeboid motion; the Mastigophora (flagellates) use flagella for motility; the Ciliata possess cilia; and the Sporozoa lack a means of locomotion and are obligate parasites.

Protozoan Biology (p. 300)

1. Most protozoans are heterotrophs and feed on bacteria and particulate organic material.
2. Protozoans are complex cells that have a pellicle, a cytostome, and an anal pore.
3. Asexual reproduction is by fission, budding, or schizogony.
4. Sexual reproduction is by conjugation.
5. During protozoan conjugation, two haploid nuclei fuse to produce a zygote.
6. Some protozoans can produce a cyst for protection during adverse environmental conditions.

Medically Important Phyla of Protozoans (pp. 301–305)

1. Parasitic sarcodinae that cause amoebic dysentery are found in the genus *Entamoeba*.
2. Parasitic flagellates (Mastigophora) include the following: *Giardia lamblia*, causing an intestinal infection called giardiasis; *Trichomonas vaginalis*, causing genitourinary infections that may be transmitted by coitus; and *Trypanosoma*, which are found in the blood of human hosts and are transmitted by blood-sucking insects.
3. The only ciliate that is a parasite of humans is *Balantidium coli*, the cause of one form of dysentery.
4. *Plasmodium* is the sporozoan that causes malaria.
5. Asexual reproduction of *Plasmodium* occurs in red blood cells and the liver of humans.
6. Sexual reproduction of *Plasmodium* takes place in the intestine of the female *Anopheles* mosquito.
7. *Toxoplasma gondii* is a sporozoan that infects humans; it can be transmitted to a fetus in utero.

Helminths (pp. 306–311)

1. Some parasitic flatworms belong to the Phylum Platyhelminthes.
2. Parasitic roundworms belong to the Phylum Aschelminthes.

Helminth Biology (p. 306)

1. Helminths are multicellular animals; a few are parasites of humans.
2. The anatomy and life cycles of parasitic helminths are modified for parasitism.
3. Helminths can be hermaphroditic or dioecious.
4. The adult stage of a parasitic helminth is found in the definitive host.
5. Each larval stage of a parasitic helminth requires an intermediate host.

Platyhelminthes (pp. 306–309)

1. Flatworms are dorsoventrally flattened animals that have an incomplete digestive system.
2. Adult trematodes or flukes have an oral and ventral sucker with which they attach to and feed on host tissue.
3. Eggs of trematodes hatch into free-swimming miracidia that enter the first intermediate host; two generations of rediae develop in the first intermediate host; the rediae become cercariae that bore out of the first intermediate host and penetrate the second intermediate host; cercariae encyst as metacercariae in the second intermediate host; after they are ingested by the definitive host, the metacercariae develop into adults.
4. A cestode, or tapeworm, consists of a scolex (head) and proglottids.
5. Humans serve as the definitive host for the beef tapeworm, and cattle are the intermediate host.

6. Humans serve as the intermediate host for *Echinococcus granulosus*; the definitive hosts are dogs, wolves, and foxes.

Aschelminthes   (pp. 309–311)

1. Members of the Phylum Aschelminthes are roundworms with a complete digestive system.
2. Parasitic members of Aschelminthes are in the class consisting of nematodes.
3. The Aschelminthes that infect humans with their eggs are *Ascaris lumbricoides* and *Enterobius vermicularis* (pinworm).

4. The Aschelminthes that infect humans with their larvae are *Necator americanus* (hookworm) and *Trichinella spiralis*.

Arthropods as Vectors   (pp. 312–313)

1. Jointed-legged animals, including ticks and insects, belong to the Phylum Arthropoda.
2. Arthropods that carry diseases are called vectors.
3. Elimination of vector-borne diseases is best done by the control or eradication of the vectors.

Study Questions

Review

1. Contrast the mechanism of conidiospore and ascospore formation by *Penicillium*.

2. Fill in the following table.

Phylum	Spore Type(s)	
	Sexual	Asexual
Zygomycota		
Ascomycota		
Basidiomycota		
Deuteromycota		

3. Fungi are classified into phyla on the basis of _____ .

4. The following is a list of fungi, their methods of entry into the body, and sites of infections they cause. Categorize each type of mycosis as cutaneous, opportunistic, subcutaneous, superficial, or systemic.

Genus	Method of Entry	Site of Infection	Mycosis
Blastomyces	Inhalation	Lungs	
Sporothrix	Puncture	Ulcerative lesions	
Microsporum	Contact	Fingernails	
Trichosporon	Contact	Hair shafts	
Aspergillus	Inhalation	Lungs	

5. A mixed culture of *Escherichia coli* and *Penicillium chrysogenum* is inoculated onto the following culture media. On which media would you expect each to grow? Why?
(a) 0.5% peptone in tap water
(b) 10% glucose in tap water

6. What is the role of the alga in a lichen? What is the role of the fungus?

7. Briefly discuss the importance of lichens in nature. Briefly discuss the importance of algae.

8. Complete the following table.

Phylum	Cell-Wall Composition	Special Features/ Importance
Dinoflagellates		
Euglenoids		
Diatoms		
Red algae		
Brown algae		
Green algae		

Indicate which phyla consist primarily of unicellular forms. Which phylum would you include in the plant kingdom? Why? Into which kingdom would you place the others?

9. Differentiate between cellular and plasmodial slime molds. How does each survive adverse environmental conditions?

10. Why is it significant that *Trichomonas* does not have a cyst stage? Name a protozoan parasite that does have a cyst stage.

11. Protozoans are classified on the basis of _____ .

12. Recall the life cycle of *Plasmodium*. Where does asexual reproduction occur? Where does sexual reproduction occur? Identify the definitive host. Identify the vector.

13. Transmission of helminthic parasites to humans usually occurs by _____ .

14. To what phylum and class does this animal belong?

List two characteristics that put it in this phylum. Name the body parts. What is the name of the encysted larva of this animal?

15. Most nematodes are dioecious. What does this term mean? To what phylum do nematodes belong?

16. Vectors can be divided into three major types, according to the roles they play for the parasite. List the three types of vectors and a disease transmitted by each.

Challenge

1. A generalized life cycle of the liver fluke (*Clonorchis sinensis*) is shown below. Identify the intermediate host(s). Identify the definitive host(s). To what phylum and class does this animal belong?

(a)

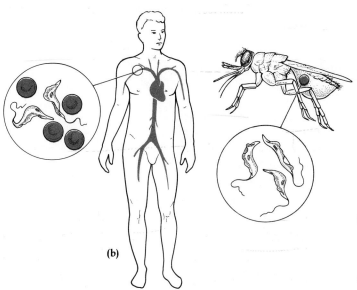

(b)

2. *T. b. gambiense* (part a, at right) is the causative agent of African sleeping sickness. To what phylum and class does it belong? Part b shows a generalized life cycle for *T. b. gambiense*. Identify the host and vector of this parasite.

3. The size of a cell is limited by its surface-to-volume ratio; that is, if the volume gets too great, internal heat cannot be dissipated, and nutrients and wastes cannot be efficiently transported. How do plasmodial slime molds manage to circumvent the surface-to-volume rule?

Further Reading

Alexopoulos, C. J., and C. W. Mims. *Introductory Mycology*, 3rd ed. New York: Macmillan, 1979. A classic textbook on the biology of fungi.

Barnett, H. L., and B. B. Hunter. *Illustrated Genera of Imperfect Fungi*, 4th ed. New York: Macmillan, 1987. Good line drawings facilitate identification of saprophytic molds.

Barnett, J. A., R. W. Payne, and D. Yarrow. *Yeasts: Characteristics and Identification*. New York: Cambridge University Press, 1984. A complete key to yeasts, based on morphologic and biochemical characteristics.

Beaver, P. C., and R. C. Jung. *Animal Agents and Vectors of Human Disease*, 5th ed. Philadelphia: Lea and Febiger, 1985. Includes a discussion of arthropod vectors and culture techniques for protozoans and helminths.

Bold, H. C., and M. J. Wynne. *Introduction to the Algae*, 2nd ed. Englewood Cliffs, N.J.: Prentice-Hall, 1985. Covers biology of freshwater and marine unicellular and multicellular algae.

Bonner, J. T. "Hormones in social amoebae and mammals." *Scientific American* 220(6):78–91, June 1969. Description of "slug" formation and function of cAMP in slime molds.

Donelson, J. E., and M. J. Turner. "How the trypanosome changes its coat." *Scientific American* 252(2):44–51, February 1985. In studying the trypanosome's variable surface proteins, scientists may have a model for how gene expression is controlled.

Godson, G. N. "Molecular approaches to malaria vaccines." *Scientific American* 242(5):51–59, May 1985. Illustrates the use of molecular biology techniques to isolate genes and produce proteins that are unique to *Plasmodium*.

Hale, M. E. *Biology of Lichens*, 3rd ed. Baltimore: Edward Arnold, 1983. An authoritative work on the structure, physiology, and ecology of lichens; includes some discussion of identification and culture techniques.

Stark, A-A. "Mutagenicity and carcinogenicity of mycotoxins." *Annual Review of Microbiology* 34:235–262, 1980. A description of fungal toxins and their effects on eucaryotic cells. Possible binding of DNA is discussed as a method of action.

CHAPTER 12

Viruses

In 1892, in an experiment to isolate the cause of tobacco mosaic disease, Dmitri Iwanowski filtered the sap of diseased tobacco plants through a porcelain filter that was designed to retain bacteria. He expected to find the microbe trapped in the filter. Instead, he found that the infectious agent had passed through the minute pores of the filter. When he injected healthy plants with the filtered fluid, they contracted tobacco mosaic disease.

Iwanowski still believed the infectious agent to be a bacterium that was small enough to pass through the filter. But later, other scientists, led by Martinus Beijerinck, observed that the behavior of this agent was different from that of bacteria. In the early 1900s, a distinction was finally made between bacteria and viruses, the filterable agents that cause tobacco mosaic disease and many other diseases. (*Virus* is the Latin word for poison.) In 1935, an American chemist, Wendell M. Stanley, isolated the tobacco mosaic virus (TMV), making it possible for the first time to carry out chemical and structural studies on a purified virus. At about the same time, the invention of the electron microscope made it possible to see viruses for the first time.

We now know that viruses are found as parasites in all types of organisms. They can only reproduce within cells. Many diseases of humans are known to be caused by viruses; some diseases of agriculturally important animals and plants are also known to be caused by viruses. Viruses infect fungi, bacteria, and protists as well.

General Characteristics of Viruses

The question of whether or not viruses are living organisms has an ambiguous answer. Life can be defined as a complex set of processes resulting from the actions of proteins specified by nucleic acids. The nucleic acids of living cells are in action all the time. Because viruses are inert outside of living host cells, in this sense they are not considered to be living organisms. However, once viruses enter a host cell, the viral nucleic acids become active, and viral multiplication results. In this sense, viruses are alive when they multiply in the host cells they infect. From a clinical point of view, viruses can be considered alive because they cause infection and disease, just as pathogenic bacteria, fungi, and protozoans do. Depending on one's viewpoint, a virus may be regarded as an exceptionally complex aggregation of nonliving chemicals or as an exceptionally simple living microorganism.

How, then, do we define a virus? Viruses were originally distinguished from other infectious agents because they are especially small (filterable) and because they are **obligatory intracellular parasites**—that is, they absolutely require living host cells in order to multiply. However, both of these properties are shared by certain small bacteria, such as some rickettsias. The truly distinctive features of viruses are now known to relate to their simple structural organization and composition and their mechanism of multiplication. Accordingly, **viruses** are entities that:

1. Contain a single type of nucleic acid, either DNA or RNA.
2. Contain a protein coat (sometimes itself enclosed by an envelope of lipids, proteins, and carbohydrates) that surrounds the nucleic acid.
3. Multiply inside living cells using the synthesizing machinery of the cell.
4. Cause the synthesis of specialized structures that can transfer the viral nucleic acid to other cells.

Viruses have few or no enzymes of their own for metabolism—for example, viruses lack enzymes for protein synthesis and ATP generation. To multiply, viruses must take over the metabolic machinery of the host cell. This fact has considerable medical significance for the development of antiviral drugs because most drugs that would interfere with viral multiplication would also interfere with the functioning of the host cell and hence are too toxic for clinical use. (Antiviral drugs are discussed in Chapter 19.)

However, the presence of lipids in the coverings of some viruses makes these viruses sensitive to disinfection (outside of a host cell). For example, a 1:10 dilution of household chlorine bleach can inactivate HIV (AIDS virus) outside the body. Viral lipids may also be damaged by lipid solvents, such as ether, and by emulsifying agents, such as bile salts and detergents.

Host Range

The **host range** of a virus is the spectrum of host cells that the virus can infect. Viruses multiply only in cells of particular species and thus are divided into three main classes—*animal viruses*, *bacterial viruses* (bacteriophages), and *plant viruses*. (Protists and fungi can also be hosts for viruses, but in this chapter we are concerned mainly with animal viruses and bacterial viruses.) Within each class, each virus is usually only able to infect cells of certain species.

The particular host range of a virus is determined by the virus's requirements for its specific attachment to the host cell and the availability within the potential host of cellular factors required for viral multiplication. For the virus to infect the host cell, the outer surface of the virus must chemically interact with specific receptor sites on the surface of the cell. The two complementary components are held together by weak bonds, such as hydrogen bonds. For some bacteriophages, the receptor site is part of the cell wall of the host; in other cases, it is part of pili or flagella. For animal viruses, the receptor sites are on the plasma membranes of the host cells.

Size

The sizes of viruses were first estimated by filtration through membranes of known pore diameters. Viral sizes are determined today by ultracentrifugation and by electron microscopy, which seems to produce the most accurate results. Viruses vary considerably in size. Although most are quite a bit smaller than bacteria, some of the larger viruses (such as the smallpox virus) are about the same size as some very small bacteria (such as the mycoplasmas, rickettsias, and chlamydias). Viruses range from 20 to 300 nm in diameter. The comparative sizes of several viruses and bacteria are shown in Figure 12-1.

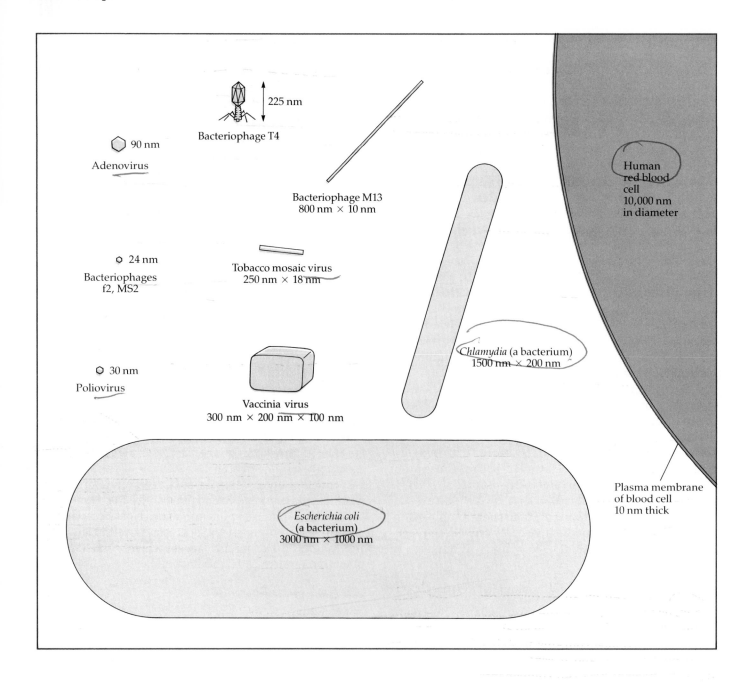

Figure 12-1 **Virus sizes.** Sizes of several viruses (light color) and bacteria (gray) are compared with a human red blood cell, shown to the right of the microbes. Dimensions are given in nanometers (nm) and are either diameters or length by width.

Viral Structure

A **virion** is a complete, fully developed viral particle composed of nucleic acid surrounded by a coat that protects it from the environment and serves as a vehicle of transmission from one host cell to another.

Nucleic Acid

As we noted earlier, the core of a virus contains a single kind of nucleic acid, either DNA or RNA, which is the genetic material. The percentage of nucleic acid in relation to protein is about 1% for the influenza virus and about 50% for certain bacteriophages. The total amount of nucleic acid varies from a few thousand nucleotides (or pairs) to as many as 250,000 nucleotides.

In contrast to procaryotic and eucaryotic cells, in which DNA is always the primary genetic material (and RNA plays an auxiliary role), a virus can have either DNA or RNA but never both. The nucleic acid of a virus can be single-stranded or double-stranded. Thus, there are viruses with the familiar double-stranded DNA, with single-stranded DNA, with double-stranded RNA, and with single-stranded RNA. Depending on the virus, the nucleic acid can be linear or circular. In some viruses (such as the influenza virus), the nucleic acid is in several separate segments.

Capsid and Envelope

The nucleic acid of a virus is surrounded by a protein coat called the **capsid** (Figure 12-2a). The capsid, whose architecture is ultimately determined by the viral nucleic acid, accounts for most of the mass of a virus, especially of small ones. Each capsid is composed of protein subunits called **capsomeres**. In some viruses, the proteins composing the capsomeres are of a single type; in other viruses, several types of protein may be present. Individual capsomeres are often visible in electron micrographs (see Figure 12-4b, for example). The arrangement of capsomeres is characteristic of a particular virus.

In some viruses, the capsid is covered by an **envelope** (Figure 12-2b), which usually consists of some combination of lipids, proteins, and carbohydrates. The molecular organization of these envelopes is generally unknown. Some animal viruses are released from the host cell by an extrusion process that coats the virus with a layer of the host cell's plasma membrane; that layer becomes the viral envelope. In many cases, the envelope contains proteins determined by viral nucleic acid, and it contains materials derived from the normal host cell components.

Depending on the virus, envelopes may or may not be covered by **spikes**, which are carbohydrate–protein complexes that project from the surface of the envelope. Some viruses attach to host cells by means of spikes. Spikes are such a reliable characteristic of some viruses that they can be used as a means of identification. The ability of certain viruses, such as the influenza virus, to clump red blood cells is associated with spikes. Such viruses bind to red blood cells and form bridges between them. The resulting clumping is called **hemagglutination** and is the basis for several useful laboratory tests.

Viruses whose capsids are not covered by an envelope are known as **naked viruses** or **nonenveloped**

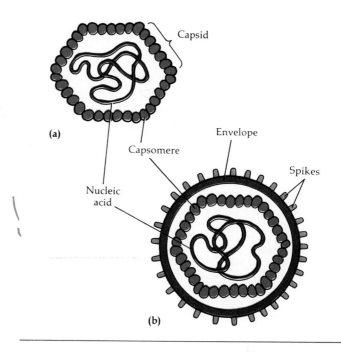

Figure 12-2   General structure of two types of viruses. **(a)** Naked virus. **(b)** Enveloped virus with spikes.

viruses (Figure 12-2a). The capsid of a naked virus protects the nucleic acid from nuclease enzymes in biological fluids and promotes the virus's attachment to susceptible host cells.

When infected by a virus, the immune system produces antibodies (proteins that react with the virus) to inactivate that virus and stop the infection. Some viruses can escape antibodies because regions of the genes that code for these viruses' surface proteins are susceptible to mutations. Mutant viruses alter their surface proteins so that the antibodies are not able to react with them. Influenza virus frequently undergoes such changes in its spikes. This is the reason why you can get influenza each year. Although you may have produced antibodies to one influenza virus, the virus will mutate and be able to infect you again.

General Morphology

Viruses may be classified into several morphological types, based on their capsid architecture as revealed by electron microscopy and a technique called x-ray crystallography.

Helical viruses   Helical viruses resemble long rods that may be rigid or flexible. Surrounding the nucleic acid, their capsid is a hollow cylinder with a helical structure. An example of a helical virus that is a rigid rod is the tobacco mosaic virus (Figure 12-3). Another is bacteriophage M13.

Polyhedral viruses   Many animal, plant, and bacterial viruses are polyhedral viruses; that is, they are many-sided. The capsid of most polyhedral viruses is in the shape of an *icosahedron*, a regular polyhedron with 20 triangular faces and 12 corners (Figure 12-4a). The capsomeres of each face form an equilateral triangle. An example of a polyhedral virus in the shape of an icosahedron is the adenovirus (Figure 12-4b). Another icosahedral virus is the poliovirus.

Enveloped viruses   As we noted earlier, the capsid of some viruses is covered by an envelope. Enveloped viruses are roughly spherical but highly pleomorphic (variable in shape) because the envelope is not rigid. When helical or polyhedral viruses are enclosed by

Figure 12-3   Morphology of a helical virus. (a) Diagram of a portion of a tobacco mosaic virus. Several rows of capsomeres have been removed to reveal the nucleic acid. **(b)** Electron micrograph of a tobacco mosaic virus showing helical rods (×147,000).

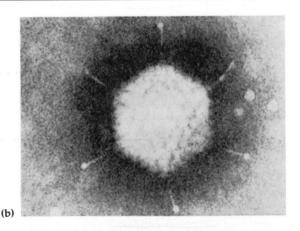

Figure 12-4   Morphology of a polyhedral virus in the shape of an icosahedron. (a) Diagram of an icosahedron. **(b)** Electron micrograph of an adenovirus (×174,000). Individual capsomeres in the protein coat are visible.

Figure 12-5   Morphology of an enveloped helical virus. (a) Diagram of an enveloped helical virus. **(b)** Electron micrograph of influenza viruses (×119,000). Note the halo of spikes projecting from the outer surface of each envelope (see Chapter 23).

Nucleic acid Capsomere

Envelope

Spikes

(a)

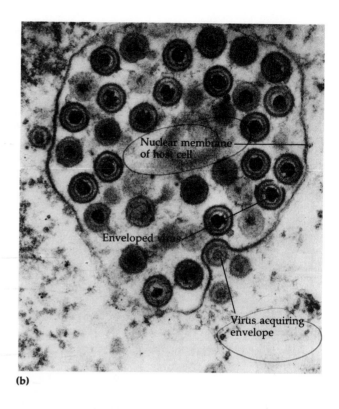

Nuclear membrane of host cell

Enveloped virus

Virus acquiring envelope

(b)

Figure 12-6 Morphology of an enveloped polyhedral (icosahedral) virus. (a) Diagram of an enveloped polyhedral virus. **(b)** Electron micrograph of a group of herpes simplex viruses (×48,000). To the lower right, a virus particle is acquiring its envelope as it buds out through the nuclear membrane of a host cell.

envelopes, they are called *enveloped helical* and *enveloped polyhedral viruses.* An example of an enveloped helical virus is the influenza virus (Figure 12-5). An example of an enveloped polyhedral (icosahedral) virus is the herpes simplex virus (Figure 12-6).

Complex viruses Some viruses, particularly bacterial viruses, have very complicated structures and are called complex viruses. Examples of complex viruses are poxviruses, which do not contain clearly

identifiable capsids but have several coats around the nucleic acid (Figure 12-7a). Certain bacteriophages have capsids to which additional structures are attached (Figure 12-7b and c). If you take a close look at the bacteriophages shown in Figure 12-7b and c, you will note that the capsid (head) is polyhedral and the tail is helical. The head contains the nucleic acid. Later in this chapter, we will discuss the functions of the other structures, such as the tail sheath, tail fibers, plate, and pin.

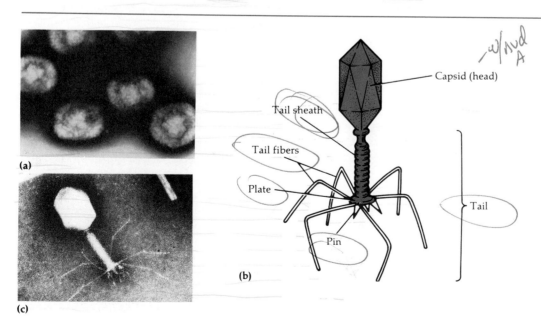

(a)

(c)

Capsid (head)

Tail sheath

Tail fibers

Plate

Pin

Tail

(b)

Figure 12-7 Morphology of complex viruses. (a) Electron micrograph of vaccinia virus, a strain of poxvirus used to vaccinate humans against smallpox (×42,000). **(b)** Diagram of a T-even bacteriophage (×110,000). **(c)** T4 bacteriophage (×110,000).

Taxonomy of Viruses

For our purposes in this chapter, we are grouping viruses as animal viruses, bacterial viruses, or plant viruses, according to host range. This classification is convenient, but it is not scientifically acceptable.

In the oldest system, an animal virus was classified according to the organs that were affected by the diseases it caused. This is a classification by symptomatology. In this system, poliovirus might be considered a nervous system virus; herpesvirus, a skin virus; and hepatitis virus, a liver virus. However, as we learned that the same virus can cause more than one disease, this scheme became unsatisfactory.

In the past few decades, hundreds of viruses have been isolated from plants, animals, and humans. As the list grows, the problem of viral classification becomes more and more complex. Present classification systems are based on such factors as the type of nucleic acid, the mechanism of nucleic acid replication, morphological class, presence of an envelope, size of capsid, and number of capsomeres. A summary of the classification of animal viruses based on these morphological, biochemical, and physical properties is presented in Table 12-1 (pp. 326–327). Other classification schemes also take into account susceptibility to physical and chemical agents, immunologic properties, site of multiplication (nucleus or cytoplasm), and natural methods of transmission.

Figure 12-8 shows electron micrographs of several DNA-containing animal viruses, and Figure 12-9 shows electron micrographs of several RNA-containing animal viruses.

Figure 12-8   DNA-containing viruses. (a) Negatively stained capsid of a herpesvirus (×400,000). The individual capsomeres are clearly visible. **(b)** Negatively stained viruses that have been concentrated in a centrifuge gradient (×198,000). The larger capsids are an adenovirus, and the smaller "adeno-associated particles" are a parvovirus.

(a)

(b)

Figure 12-9   RNA-containing viruses. (a) Capsids of A2 influenza virus (×92,000). The spikes projecting from each capsid are visible. **(b)** Particles of vesicular stomatitis virus (VSV). This is a rhabdovirus, a rod-shaped virus similar to rabies (×37,000). **(c)** Mouse mammary tumor virus, a B-type retrovirus (×160,000).

(a)

(b)

(c)

Isolation, Cultivation, and Identification of Viruses

The fact that viruses cannot multiply outside a living host cell complicates their detection, enumeration, and identification. It is necessary to provide viruses with living cells instead of a fairly simple chemical medium. Living plants and animals are difficult and expensive to maintain, and disease-causing viruses that grow only in higher primates and human hosts cause additional complications. However, viruses that use bacterial cells as a host (bacteriophages) are rather easily grown on bacterial cultures. This is one reason why so much of our understanding of viral multiplication has come from bacteriophages.

Growth of Bacteriophages in the Laboratory

Bacteriophages can be grown either in suspensions of bacteria in liquid media or in bacterial cultures on solid media. The use of solid media makes possible the **plaque method** for detecting and counting viruses. Only simple materials and equipment are needed for this procedure, which was first developed for use with bacteriophages. A sample of bacteriophage is mixed with host bacteria and melted agar. The agar containing bacteriophage and host bacteria is then poured into a Petri plate containing a hardened layer of agar growth medium. The virus–bacteria mixture solidifies into a thin top layer that contains a layer of bacteria approximately one cell thick. Each virus infects a bacterium, multiplies, and releases several hundred new viruses. These newly produced viruses infect other bacteria in the immediate vicinity, and more new viruses are produced. Following several viral-multiplication cycles, all the bacteria in the area surrounding the original virus are destroyed. This produces a number of clearings, or **plaques**, visible against a "lawn" of bacterial growth on the surface of the agar (Figure 12-10). While the plaques form, uninfected bacteria elsewhere in the Petri plate multiply rapidly by fission and produce a turbid background.

Each plaque theoretically corresponds to a single virus in the initial suspension. Because a single plaque can arise from more than one virion, and because some virions may not be infectious, the concentrations of viral suspensions measured by the number of plaques are usually given in terms of **plaque-forming units (pfu)**.

Figure 12-10 Viral plaque formed by bacteriophage. In the Petri plate on the left, clear viral plaques of varying sizes have been formed by the bacteriophage λ on a lawn of *E. coli.* For comparison, the Petri plate on the right contains a bacterial culture without phage.

TABLE 12-1   *Classification of Animal Viruses by Nucleic Acid Replication*

Nucleic Acid Category	Viral Group and Specific Examples	Morphological Class	Dimensions of Capsid (Diameter in nm)	Clinical or Special Features
Single-stranded DNA viruses	Parvoviruses (adenosatellite)	Naked polyhedral	18–25	Very small viruses, most of which depend on coinfection with adenoviruses for growth; probably infect only rats, mice, and hamsters.
Double-stranded DNA viruses	Papovaviruses (papilloma, polyoma, simian)	Naked polyhedral	40–57	Small viruses that induce tumors; the human wart virus (papilloma) and certain viruses that produce cancer in animals (polyoma and simian) belong to this family. Refer to Chapters 20 and 25.
	Adenoviruses	Naked polyhedral	70–80	Medium-sized viruses that cause various respiratory infections in humans; some cause tumors in animals.
	Herpesviruses (herpes simplex, herpes zoster)	Enveloped polyhedral	150–250	Medium-sized viruses that cause various human diseases, such as fever blisters, chickenpox, shingles, and infectious mononucleosis; implicated in a type of human cancer called Burkitt's lymphoma. Refer to Chapters 20, 22, and 25.
	Poxviruses (variola, cowpox, vaccinia)	Enveloped complex	200–350	Very large, complex, brick-shaped viruses that cause diseases such as smallpox (variola), molluscum contagiosum (wartlike skin lesion), cowpox, and vaccinia; vaccinia virus gives immunity to smallpox. Refer to Chapter 20.
Sense strand RNA viruses	Picornaviruses (poliovirus, rhinovirus, hepatitis A virus)	Naked polyhedral	18–38	Smallest RNA-containing viruses; at least 70 human enteroviruses are known, including the polio-, coxsackie-, and echoviruses; more than 100 rhinoviruses exist and are the most common cause of colds. Refer to Chapters 21, 22, 23, and 24.
	Togaviruses (arboviruses, alpha virus, flavivirus)	Enveloped polyhedral	40–60	Included are many viruses transmitted by arthropods; diseases include eastern equine encephalitis (EEE), Venezuelan equine encephalitis (VEE), St. Louis encephalitis (SLE), yellow fever, and dengue. Refer to Chapters 21 and 22.

TABLE 12-1 (*Continued*)

Nucleic Acid Category	Viral Group and Specific Examples	Morphological Class	Dimensions of Capsid (Diameter in nm)	Clinical or Special Features
Antisense strand RNA viruses	Orthomyxoviruses (influenza A, B, C)	Enveloped helical	80–200	Medium-sized viruses with a spiked envelope; have the ability to agglutinate red blood cells; cause influenza. Refer to Chapter 23.
	Paramyxoviruses (measles, mumps)	Enveloped helical	150–300	Morphologically similar to orthomyxoviruses, but generally larger; cause parainfluenza, measles, mumps. Refer to Chapters 20, 23, and 24.
	Coronaviruses	Enveloped helical	80–130	Associated with upper respiratory tract infections and the common cold. Refer to Chapter 23.
	Rhabdoviruses (rabies)	Enveloped helical	70–180	Bullet-shaped viruses with a spiked envelope; cause rabies and Newcastle disease of chickens. Refer to Chapter 21.
	Arenaviruses (lassa)	Enveloped helical	50–300	Viruses contain RNA-containing granules; some members produce slow viral infections.
Double-stranded RNA viruses	Reoviruses	Naked polyhedral	60–80	Involved in mild respiratory infections and infantile gastroenteritis; causes Colorado tick fever.
Reverse transcription viruses	Retroviruses	Enveloped helical	100–120	Includes all RNA tumor viruses. Double-stranded RNA viruses; cause leukemia and tumors in animals; cause AIDS. Refer to Chapter 18.
	Hepatitis B	Enveloped polyhedral	42	Double-stranded DNA virus; cause hepatitis; cause liver tumors. Refer to Chapter 24.

Growth of Animal Viruses in the Laboratory

In Living Animals

Some animal viruses can only be cultured in *living animals*, such as mice, rabbits, and guinea pigs. Most experiments to study the immune system's response to viral infections must also be performed in virally infected live animals. Animal inoculation may be used as a diagnostic procedure for identifying and isolating a virus from a clinical specimen. After the animal is inoculated with the specimen, the animal is observed for signs of disease or is killed so that infected tissues can be examined for the virus.

In Embryonated Eggs

If the virus will grow in an *embryonated egg*, this can be a fairly convenient and inexpensive form of animal host for many animal viruses (Figure 12-11). A hole is drilled in the shell of the embryonated egg, and a viral suspension or suspected virus-containing tissue is injected into the fluid of the egg. There are several membranes in the egg on which the virus can be made to grow, and the virus is injected into the proper location in the egg. Viral growth is signaled by the death of the embryo, by embryo cell damage, or by the formation on the membranes of the egg of typical pocks or lesions that result from viral growth. This method was once the most widely used method of viral isolation and growth and is still used to grow viruses for some vaccines. You may be asked if you are allergic to eggs before receiving a vaccination since egg proteins may be present in the viral vaccine preparations. (Allergy is discussed in Chapter 18.)

In Cell Culture

Cell cultures (sometimes called *tissue cultures*, although that is not the best term) have replaced embryonated eggs as growth media for many viruses. Cell cultures consist of cells grown in culture media in the laboratory. Because these cultures are generally rather homogeneous collections of cells and can be propagated and handled much like bacterial cultures, they are more convenient to work with than whole animals or embryonated eggs.

Cell culture lines are readily started by treatment of a slice of animal tissue with enzymes that separate the individual cells (Figure 12-12). These cells are suspended in a solution that provides the osmotic pressure, nutrients, and growth factors needed for the cell to grow. The cells tend to adhere to the glass or plastic container and reproduce to form a monolayer. Viruses

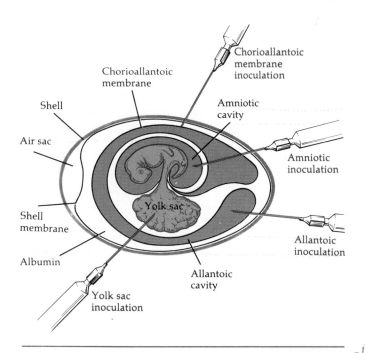

Figure 12-11   Inoculation of embryonated eggs. The injection site determines the membrane on which the viruses will grow.

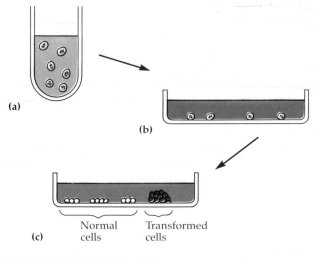

Figure 12-12   Cell cultures. (a) A tissue is treated with enzymes to separate the cells. **(b)** Cells are suspended in culture medium. **(c)** Normal cells or primary cell lines grow in a monolayer across the glass or plastic container. Transformed cells or continuous cell cultures do not grow in a monolayer.

(a) **(b)**

Figure 12-13 Cytopathic effect of viruses. (a) A monolayer of uninfected mouse L cells. **(b)** The same cells 24 hours after infection with VSV (see Figure 12-9). Note the separation and "rounding up" of the cells. (Both photos, ×800.)

infecting such a monolayer sometimes cause the cells of the monolayer to deteriorate as they multiply. This tissue deterioration is called the **cytopathic effect (CPE)** and is illustrated in Figure 12-13. The CPE can be detected and counted in much the same way as plaques caused by bacteriophages on a lawn of bacteria.

Primary cell lines, derived from tissue slices, tend to die out after only a few generations. Certain cell lines called **diploid cell lines** developed from human embryos can be maintained for about a hundred generations and are widely used for diagnostic work concerning human diseases. Cell lines developed from embryonic human cells are used to culture rabiesvirus for a rabies vaccine called human diploid culture vaccine (Chapter 21).

When viruses are routinely grown in a laboratory, **continuous cell lines** are used. These are "transformed" cells that can be maintained through an indefinite number of generations, and they are sometimes called "immortal" cell lines (see the discussion of transformation at the end of this chapter). One of these, the HeLa cell line, was isolated from the cancer of a woman who died in 1951. After years of laboratory cultivation, many such cell lines have lost almost all the original characteristics of the cell, but these changes have not interfered with the use of the cells for viral propagation. In spite of the success of cell culture in viral isolation and growth, there are still some viruses that have never been successfully cultivated in cell culture.

The idea of cell culture dates back to the end of the last century, but it was not a practical laboratory technique until the development of antibiotics in the years following World War II. A major problem with cell culture is that the cell lines must be kept free of microbial contamination. The maintenance of cell culture lines requires trained technicians with consid-

erable experience working on a full-time basis. Because of these difficulties, most hospital laboratories and many state health laboratories do not isolate and identify viruses in clinical work. Instead, the tissue or serum samples are sent to central laboratories that specialize in this work.

Viral Identification

The identification of viral isolates is not an easy task. For one thing, viruses cannot be seen at all without the use of an electron microscope. Serological methods are the most commonly used means of identification. In these tests, the virus is detected and identified by its reaction with antibodies. Antibodies, which are proteins produced by animals in response to exposure to a virus, are highly specific for the virus that causes their formation. We will discuss antibodies in detail in Chapter 16 and a number of immunological tests for identifying viruses in Chapter 17. Later in this chapter, we will describe how viruses can be identified by observation of their effects on host cells.

Using modern biochemical methods, molecular biologists can identify and characterize purified viruses by isolating the particular nitrogenous base sequences of the virus's nucleic acid. For DNA viruses, restriction enzymes are used to cut DNA into specific nitrogenous base sequences, or fragments (Chapter 9). Fragments of closely related viruses have similar electrophoretic patterns. Using more sophisticated methods, researchers can even determine detailed nitrogenous base sequences of entire viral genomes. However, these laboratory research methods are not practical for the routine identification of clinical samples.

Viral Multiplication

The nucleic acid in a virion contains only a few of the genes needed for the synthesis of new viruses. These include genes for the virion's structural components, such as the capsid proteins, and genes for a few of the enzymes used in the viral life cycle. Most **viral enzymes** (or **virus-specific enzymes**)—the enzymes encoded in viral nucleic acid—are not part of the virion. They are synthesized and functional only when the virus is within the host cell. Viral enzymes are almost entirely concerned with replicating or processing viral nucleic acid. They are almost never concerned with the machinery of protein synthesis or energy production. Although the smallest naked virions do not contain any preformed enzymes, the larger virions may contain one or a few enzymes, which usually function in helping the virus to penetrate the host cell or to replicate its own nucleic acid.

Thus, for a virus to multiply, it must invade a host cell and take over the host's metabolic machinery. A single virus can give rise to several or even thousands of similar viruses in a single host cell. This process drastically changes the host cell and often causes its death.

Multiplication of Bacteriophages

Although the means by which a virus enters and exits a host cell may vary, the basic mechanism of viral multiplication is similar for animal viruses, plant viruses, and bacteriophages. The best-understood viral life cycles are those of the bacteriophages (phages). Because the so-called *T-even bacteriophages* (T2, T4, and T6) have been studied most extensively, we will first describe the multiplication of T-even bacteriophages in their host, *E. coli*.

T-Even Bacteriophages

The T-even bacteriophages have large, complex, naked virions, whose characteristic head-and-tail structure is shown in Figure 12-13. The length of DNA contained in these bacteriophages is only about 6% of that contained in *E. coli*; the bacteriophage has enough DNA for over 100 genes. The multiplication cycle of these phages, like that of all viruses, can be divided into several distinct stages—attachment, penetration, biosynthesis of viral components, maturation, and release.

Attachment of phage to host cell   After a chance collision between phage particles and bacteria, **attachment**, or **adsorption**, occurs. During this process, an attachment site on the virus attaches to a complementary receptor site on the bacterial cell. This attachment is a chemical interaction in which weak bonds are formed between the attachment and receptor sites. T-even bacteriophages use fibers at the end of the tail as attachment sites. The complementary receptor sites are on the bacterial cell wall (Figure 12-14a). Other phages attach to flagella or pili.

Penetration   After attachment, the T-even bacteriophage injects its DNA (nucleic acid) into the bacterium. To do this, the bacteriophage's tail releases an enzyme, *phage lysozyme*, which breaks down a portion of the bacterial cell wall. During the process of **penetration**, the tail sheath of the phage contracts, and the tail core is driven through the cell wall. When the tip of the core reaches the plasma membrane, the DNA from the bacteriophage's head passes through the tail core and through the plasma membrane and enters the bacterial cell. The capsid of most bacteriophages remains outside the bacterial cell (Figure 12-14b).

Biosynthesis of viral components   Once the bacteriophage DNA has reached the cytoplasm of the host cell, biosynthesis of viral nucleic acid and protein occurs.

Initially, the phage uses the host cell's nucleotides and several of its enzymes to synthesize many copies of phage DNA. Soon after, biosynthesis of viral proteins begins. Any RNA transcribed in the cell is mRNA transcribed from phage DNA for biosynthesis of phage enzymes and capsid protein. The host cell's ribosomes, enzymes, and amino acids are used for translation (Figure 12-14c).

For several minutes following infection, complete phages cannot be found in the host cell. Only separate components can be detected—DNA and proteins. The time period during viral multiplication when complete, infective virions are not yet present is called the **eclipse period**.

Maturation   In the next sequence of events, **maturation** occurs. In this process, bacteriophage DNA and capsids are assembled into complete virions. The assembly process is guided by the products of certain viral genes in a step-by-step sequence. The phage heads and tails are separately assembled from protein subunits, the head is packaged with phage DNA, and the tail is attached (Figure 12-14d). (For many simpler viruses, the nucleic acid and capsid proteins assemble spontaneously to form virions, without the intervention of other phage gene products.)

Figure 12-14 Multiplication cycle of T-even bacteriophage.

Release   The final stage of viral multiplication is the **release** of virions from the host cell. The term **lysis** is generally used for this stage in the multiplication of T-even phages because in this case the plasma membrane actually breaks open (lyses). Lysozyme, whose code is provided by a phage gene, is synthesized within the cell. This enzyme causes a breakdown of the bacterial cell wall, and the newly produced bacteriophages are released from the host cell (Figure 12-14e). The released bacteriophages infect other susceptible cells in the area, and the viral multiplication cycle is repeated within those cells.

Lysozyme is not synthesized early in the multiplication cycle because it would cause lysis of the host cell before complete phages could be assembled. Accordingly, there are genetic controls that regulate when different regions of phage DNA are transcribed into mRNA during the multiplication cycle. For example, there are early messages that are translated into early phage proteins, the enzymes used in the synthesis of phage DNA. Also, there are late messages that are translated into late phage proteins for the synthesis of capsid proteins and lysozyme. This control mechanism is mediated by RNA polymerase.

The time elapsed from phage attachment to release is known as **burst time** and averages between 20 to 40 minutes. The number of newly synthesized phage particles released from a single cell is referred to as **burst size** and usually ranges from about 50 to 200.

The various stages involved in the multiplication of phages can be demonstrated experimentally in what is known as a **one-step growth experiment** (Figure 12-15). In this procedure, a phage suspension is diluted until a sample containing only a few phage particles can be obtained. These particles are then introduced into a culture of host cells. Periodically, samples of phage particles are removed from the culture and inoculated onto a plate-culture of susceptible host cells; the plaque method is used to determine the number of infective phage particles on this culture (see Figure 12-10). A few minutes after attachment, there are no infective particles present. However, phage nucleic acid is found inside the infected cells, and capsid proteins will be synthesized. The time required for maturation is the interval between the appearance of phage nucleic acid and the synthesis of mature phages. After a few minutes, the number of infective phage particles found in subcultures begins to rise. The burst size is determined once the number of infective phage particles remains constant, indicating that no further phage multiplication will occur.

Lysogeny

In the sequence of events just described for the multiplication of T-even bacteriophages, the release of the phages causes lysis and death of the host cell. Thus, such a sequence is called a **lytic cycle**. Some phages can either proceed through a lytic cycle or incorporate their DNA into the host cell's DNA. In the latter state, the phage remains latent and does not cause lysis of the host cell. Such a state is called **lysogeny**, and such phages are called **lysogenic phages** or **temperate phages**. The participating bacterial host cells are known as **lysogenic cells**.

Let us go through the steps in lysogeny for bacteriophage λ (lambda), a well-studied temperate phage (Figure 12-16). Upon penetration into an *E. coli* cell, the linear phage DNA forms a circle. This circle can multiply and be transcribed, leading to the production of new phage and to cell lysis (lytic cycle). Or, the circle can recombine with and become part of the circular bacterial DNA (lysogenic cycle). The inserted

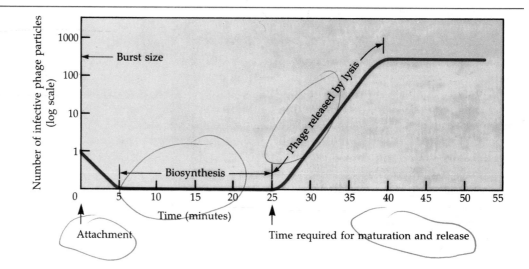

Figure 12-15   A bacteriophage one-step growth curve. No new infective phage particles are found in a culture until after biosynthesis and maturation have taken place.

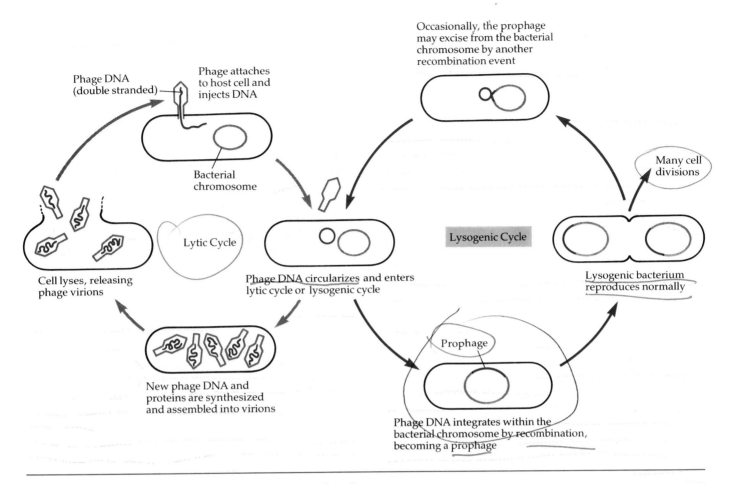

Figure 12-16 Lysogenic cycle of bacteriophage λ in *E. coli*.

phage DNA is now called a **prophage**. Most of the prophage genes are repressed by two repressor proteins that are the products of phage genes. These repressors stop transcription of all the other phage genes by binding to operators. Thus, the phage genes that would otherwise direct the synthesis and release of new virions are turned off, in much the same way that the genes of the *E. coli lac* operon are turned off by the *lac* repressor (Chapter 8).

Every time the host cell's machinery replicates the bacterial chromosome, it also replicates the prophage DNA. The prophage remains latent within the progeny cells. However, a rare spontaneous event or the action of ultraviolet light or certain chemicals can lead to the excision (popping-out) of the phage DNA—and initiation of the lytic cycle.

There are three important outcomes of lysogeny. First, the lysogenic cells are immune to reinfection by the same phage. (However, the host cell is not immune to infection by other phage types.) The second outcome of lysogeny is that the host cell may exhibit new properties. For example, the bacterium *Coryne-*

bacterium diphtheriae, which causes diphtheria, is a pathogen whose disease-producing properties are related to the synthesis of a toxin. The organism is capable of producing toxin only when it carries a temperate phage because the prophage carries the gene coding for the toxin. As another example, only streptococci carrying a temperate phage are capable of producing the toxin associated with scarlet fever. The toxin produced by *Clostridium botulinum*, which causes botulism, is coded by a prophage gene.

The third possible outcome of lysogeny is **specialized transduction**. Recall from Chapter 8 that bacterial genes can be picked up in a phage coat and transferred to another bacterium in a process called generalized transduction (see Figure 8-25). Any bacterial genes can be transferred by generalized transduction because the host chromosome is broken down into fragments, any of which can be packaged into a phage coat. In specialized transduction, however, only certain bacterial genes can be transferred.

[handwritten margin notes: "only the cert / spec genes w/th = close / to phage genes / (unlike the / any part / @ all)"]

(a) *gal*⁺ host   **(b)**   **(c)** *gal*⁻ host   **(d)**   **(e)** *gal*⁺ cell

Bacterial
DNA

Figure 12-17   Specialized transduction. When a prophage is excised from its host chromosome, it can take with it a bit of the adjacent DNA from the bacterial chromosome. **(a)** Prophage in galactose-utilizing *E. coli.* **(b)** Excised phage genome carries the adjacent host gene. **(c)** Phage carries the *gal*⁺ gene from the original host's chromosome to its new (*gal*⁻) host. **(d)** Along with the prophage, the bacterial *gal*⁺ gene becomes integrated into the new host's DNA. **(e)** Lysogenic cell can now metabolize galactose.

Specialized transduction is mediated by a lysogenic phage, which packages bacterial DNA *along with* its own DNA in the same capsid. When a prophage is excised from the host chromosome, adjacent genes from either side may remain attached to the phage DNA. The genes that remain attached to bacteriophage λ are genes for galactose utilization and biotin synthesis. A phage particle newly synthesized from this DNA may thus contain one or more of the bacterial genes along with phage DNA. Upon lysogenizing a new host cell, the phage can confer new characteristics to the cell; a cell infected by bacteriophage λ, for example, could gain the ability to metabolize galactose or synthesize biotin. In Figure 12-17, bacteriophage λ has picked up the *gal* gene for galactose fermentation from its galactose-positive host. The phage carries this gene to a galactose-negative cell, which can become galactose-positive.

Certain animal viruses are able to undergo processes very similar to lysogeny. Animal viruses that can remain latent in cells for long periods without multiplying or causing disease may become inserted in a host chromosome or remain separate from host DNA in a repressed state (as some lysogenic phages). Cancer-causing viruses may also be lysogenic, as will be discussed later in this chapter.

Multiplication of Animal Viruses

The multiplication of animal viruses follows the basic pattern of bacteriophage multiplication but has several notable differences. Animal viruses differ from phages in their mechanisms of entering the host cell. Also, once the virus is inside, the synthesis and assembly of the new viral components are somewhat different, partly because of the differences between procaryotic cells and eucaryotic cells. Animal viruses may have certain types of enzymes not found in phages. Finally, the mechanisms of maturation and release, and the effects on the host cell, differ in animal viruses and phages.

In the following discussion of the multiplication of animal viruses, we will first consider the processes that are shared by both DNA- and RNA-containing animal viruses. These processes are attachment, penetration, and uncoating. Then we will examine how DNA- and RNA-containing viruses differ with respect to their processes of biosynthesis and release from host cells.

Attachment

Like bacteriophages, animal viruses have attachment sites that attach to complementary receptor sites on the host cell's surface. However, the receptor sites of animal cells are proteins and glycoproteins of the plasma membrane (Figure 12-18a). Moreover, animal viruses do not possess appendages like the tail fibers of some bacteriophages. The attachment sites of animal viruses are distributed over the surface of the virus. The sites themselves vary from one group of viruses to another. In adenoviruses, which are naked icosahedral viruses, the attachment sites are small fibers at the corners of the icosahedron. In many of

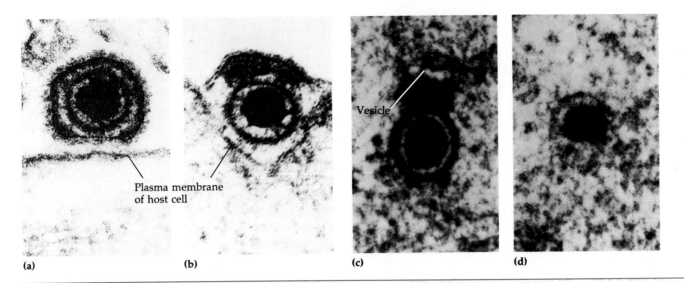

Figure 12-18 Entry of herpes simplex virus into an animal cell. (a) Attachment of the viral envelope to the plasma membrane (×180,000). **(b)** Formation of a vesicle around the virus, which results in loss of the envelope (×180,000). **(c)** Unenveloped capsid entering the cytoplasm of the cell from the vesicle (×180,000). **(d)** Digestion of the capsid leaves only the nucleic acid core (×180,000).

the enveloped viruses, such as the myxoviruses, the attachment sites are spikes located on the surface of the envelope (see Figure 12-5a). As soon as one spike attaches to a host receptor, additional receptor sites on the same cell migrate to the virus. Attachment is completed when many sites are bound. Moreover, attachment is enhanced by the presence of cations.

Receptor sites are inherited characteristics of the host. Consequently, the receptor for a particular virus can vary from person to person. This could account for the individual differences in susceptibility to a particular virus. Understanding the nature of attachment can lead to the development of drugs that prevent viral infections. Monoclonal antibodies (discussed in Chapter 16) that combine with a virus's attachment site or the cell's receptor site may soon be used to treat some viral infections.

Penetration

Following attachment, penetration occurs. In enveloped animal viruses, penetration occurs by **endocytosis**. Endocytosis is an active cellular process by which nutrients and other molecules are brought into a cell. A cell's plasma membrane continuously folds inward to form vesicles. These vesicles contain elements that originate outside the cell and are brought into the interior of the cell to be digested. If a virion attaches to a small outfolding (called a microvillus) on the plasma membrane of a potential host cell, the host cell will enfold the virion into a fold of plasma membrane, forming a vesicle (Figure 12-18b). Once the virion is enclosed within the vesicle, its viral envelope is destroyed. The capsid is digested when the cell attempts to digest the vesicle's contents. The viral nucleic acid is then released into the cytoplasm of the host cell (Figure 12-18c).

An alternative method of penetration called **fusion** has been hypothesized—the viral envelope fuses with the plasma membrane and releases the capsid into the host cell's cytoplasm.

Uncoating

Uncoating is the separation of the viral nucleic acid from its protein coat. It is a poorly understood process, and one that apparently varies with the type of virus. Uncoating occurs only in animal viruses; recall that phage DNA is inserted directly into the host cell. Some animal viruses accomplish uncoating by the action of lysosomal enzymes contained inside phagocytic vacuoles and coated vesicles of the host cell. These enzymes degrade the proteins of the viral capsid. The uncoating of poxviruses is completed by a specific enzyme coded by the viral DNA and synthesized soon after infection. For other viruses, uncoating appears to be exclusively caused by enzymes in the host cell cytoplasm (Figure 12-18d). For at least one virus, the poliovirus, uncoating seems to begin while the virus is still attached to the host cell's plasma membrane.

Biosynthesis for DNA-Containing Viruses

When a poxvirus multiplies, the viral components are synthesized and assembled in the cytoplasm of the host cell. Other DNA-containing viruses (parvoviruses, papovaviruses, adenoviruses, and herpesviruses) synthesize their components differently from poxviruses. They replicate their DNA in the nucleus of the host cell and synthesize their proteins in the cytoplasm. Then, this stage is followed by the migration of the proteins into the nucleus of the host cell; there they are assembled into complete viruses. To show an example of the multiplication of a DNA virus, we will follow the sequence of events in adenovirus (Figure 12-19).

After attachment, penetration, and uncoating, the viral DNA is released into the nucleus of the host cell. Next occurs the transcription of a portion of the viral DNA—the "early" genes. Translation follows. The products of these genes are enzymes required for the multiplication of viral DNA. In most DNA viruses, early transcription is carried out with the host's transcriptase (RNA polymerase); poxviruses, however, contain their own transcriptase. Sometime after the initiation of DNA multiplication, transcription and translation of the remaining, "late," viral genes occur. This leads to the synthesis of capsid proteins,

which occurs in the cytoplasm of the host cell. After the capsid proteins migrate into the nucleus of the host cell, maturation occurs—the viral DNA and capsid proteins assemble themselves to form complete viruses.

Biosynthesis for RNA-Containing Viruses

The multiplication of RNA viruses is essentially the same as that of DNA viruses, except that several different mechanisms of mRNA formation occur among different groups of RNA viruses. Although the details of these mechanisms are beyond the scope of this text, for comparative purposes we will trace the multiplication cycles of some representative RNA viruses (Figure 12-21, pp. 338–339, summarizes the multiplication cycles of the four major types of RNA viruses.) Multiplication of RNA viruses takes place in the host cell's cytoplasm. The major differences among the multiplication processes of these viruses lie in how mRNA and viral RNA are produced. Once viral RNA and viral proteins are synthesized, maturation occurs by similar means among all animal viruses. This will be discussed shortly.

Picornavirus   Picornaviruses, such as poliovirus, are single-stranded RNA viruses. They are the smallest

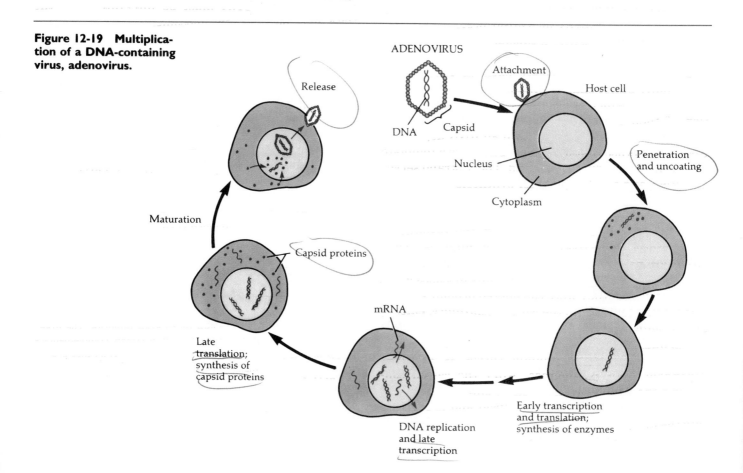

Figure 12-19   Multiplication of a DNA-containing virus, adenovirus.

viruses, hence, the prefix *pico* (small) plus *RNA* gives these viruses their name. The RNA within the virion is identified as a **+** or **sense strand** because it can act as mRNA. After attachment, penetration, and uncoating are completed, the single-stranded viral RNA is translated into two principal proteins, which inhibit the host cell's synthesis of RNA and protein and which form an enzyme called *RNA-dependent RNA polymerase*. This enzyme catalyzes the synthesis of another strand of RNA, which is complementary in base sequence to the original infecting strand. This new strand is called a **−** or **antisense strand**. It serves as a template to produce additional **+** strands. The **+** strands may serve as mRNA for the translation of capsid protein, may become incorporated into capsid protein to form a new virus, or may serve as a template for continued RNA multiplication. Once viral RNA and viral protein are synthesized, maturation occurs.

Togavirus Togaviruses, which include arthropod-borne viruses (Chapter 21), also contain a single **+** strand of RNA. Togaviruses are enveloped viruses; their name is from the Latin word for covering, *toga*. Keep in mind that these are not the only enveloped viruses. After a **−** or antisense strand is made from the **+** strand, two types of mRNA are transcribed from the **−** strand. One type of mRNA is a short strand that codes for envelope proteins. The other, longer strand serves as mRNA for capsid proteins and can become incorporated into a capsid.

Rhabdovirus Rhabdoviruses, such as rabiesvirus, are usually bullet-shaped. *Rhabdo* is from the Greek word for rod, which is not really an accurate description of their morphology. They contain a single **−** strand of RNA. They also contain an RNA-dependent RNA polymerase that uses the **−** strand as a template from which to produce a **+** strand. The **+** strand serves as mRNA and as a template for synthesis of new viral RNA.

Reovirus Reoviruses are found in the respiratory and enteric (digestive) systems of humans. They either are not pathogenic or are of such low virulence that they were not associated with any diseases when first discovered; they were considered orphan viruses. Their name comes from the words *respiratory*, *enteric*, and *orphan*.

In addition to its structural role for the virus, one of the reovirus capsid proteins serves as an RNA-dependent RNA polymerase. After the capsid containing the double-stranded RNA enters a host cell, mRNA is produced inside the capsid and is released into the cytoplasm. There it is used to synthesize more viral proteins. One of the newly synthesized viral pro-

teins acts as RNA-dependent RNA polymerase to produce more **−** strands of RNA. The mRNA **+** and **−** strands form the double-stranded RNA that is then surrounded by capsid proteins.

Retrovirus Many retroviruses infect vertebrates. One group of retroviruses, the human immunodeficiency viruses (HIV-1 and HIV-2), causes AIDS (see the boxes on p. 341 and in Chapter 18, p. 468). The retroviruses that cause cancer will be discussed later in this chapter.

The formation of mRNA and RNA for new retrovirus virions is quite interesting. These viruses carry their own polymerase, which uses the RNA of the virus to synthesize a complementary strand of DNA, which in turn is replicated to form double-stranded DNA. This enzyme also degrades the original viral RNA. The enzyme is an RNA-dependent DNA polymerase called *reverse transcriptase*, so called because it carries out a reaction (RNA → DNA) that is exactly the reverse of the familiar transcription of DNA → RNA (Figure 12-20). The name *retrovirus* is derived from the enzyme name *reverse transcriptase*. The formation of complete viruses requires that DNA be transcribed back into the RNA that will serve as mRNA for viral-protein synthesis and be incorporated into new virions. However, before transcription can take place, the viral DNA must be integrated into the DNA of a host cell chromosome. In this integrated state, the viral DNA is called a **provirus**. Unlike a prophage, the provirus never comes out of the chromosome.

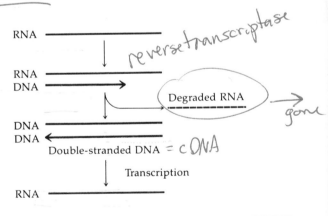

Figure 12-20 Reverse transcriptase. This enzyme catalyzes the synthesis of a strand of DNA on an RNA template. The RNA strand is then degraded, and another DNA strand, complementary to the first, is synthesized. The double-stranded DNA thus produced serves as a template for transcription to form both mRNA and RNA for new virions.

Figure 12-21 Pathways of multiplication used by various RNA-containing viruses. After uncoating, ssRNA viruses with a + strand genome are able to synthesize proteins directly from their + strand. Using the + strand as a template, they transcribe − strands to produce additional + strands to serve as mRNA and be incorporated into capsid protein as viral genome. The ssRNA viruses with a − strand genome must transcribe a + strand to serve as mRNA before they begin synthesizing proteins. The mRNA transcribes additional − strands for incorporation into capsid protein. Both ssRNA and dsRNA must use mRNA (+ strand) to code for proteins, including capsid protein.

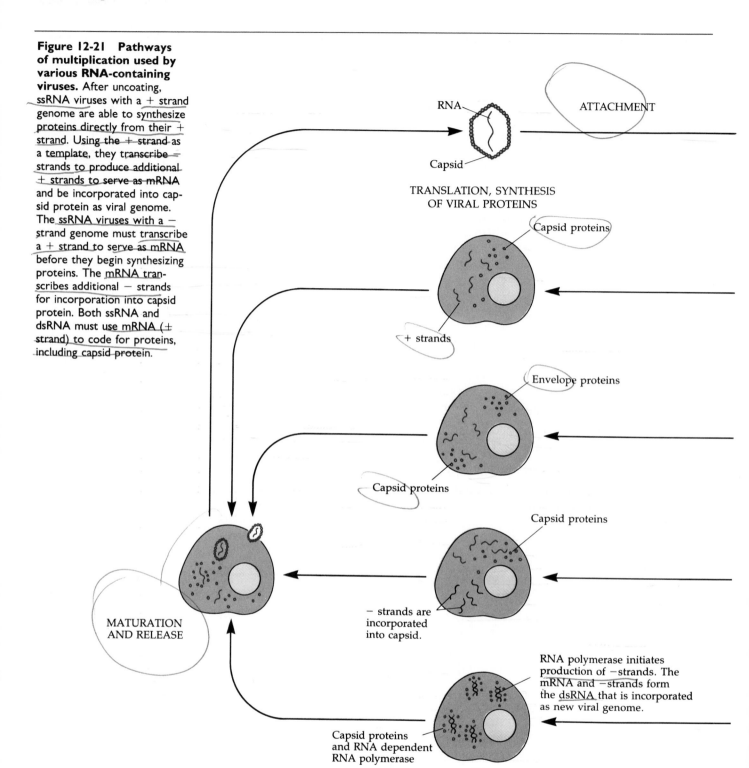

RNA

Capsid

ATTACHMENT

TRANSLATION, SYNTHESIS
OF VIRAL PROTEINS

Capsid proteins

+ strands

Envelope proteins

Capsid proteins

Capsid proteins

− strands are
incorporated
into capsid.

MATURATION
AND RELEASE

RNA polymerase initiates
production of −strands. The
mRNA and −strands form
the dsRNA that is incorporated
as new viral genome.

Capsid proteins
and RNA dependent
RNA polymerase

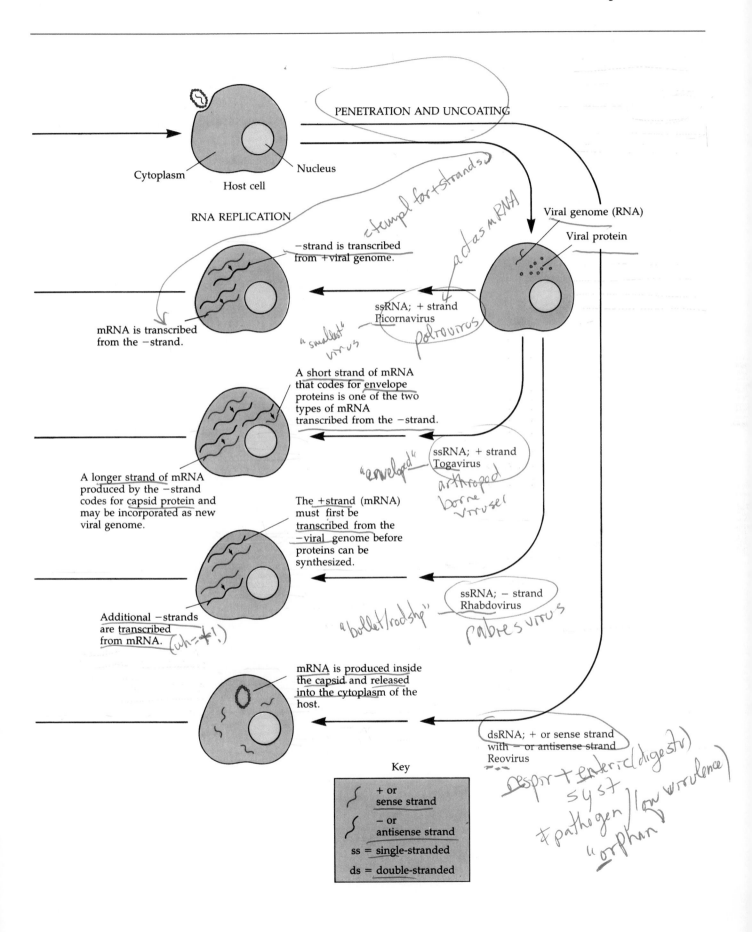

Once the provirus is integrated into the host cell's DNA, several things can happen. Sometimes the provirus simply remains in a latent state and replicates when the DNA of the host cell replicates. In other cases, the provirus becomes transcribed and produces new viruses, which may infect adjacent cells. The provirus can also convert the host cell into a tumor cell; possible mechanisms will be discussed later. How a retrovirus infects and multiplies inside a host cell is summarized in Figure 12-22.

Maturation and Release

The first step in viral maturation is the assembly of the protein capsid; this assembly is usually a spontaneous process. The capsids of many of the RNA-containing animal viruses are enclosed by an envelope consisting of protein, lipid, and carbohydrate, as noted earlier. Examples of such viruses include myxoviruses and paramyxoviruses. The envelope protein is synthesized by the virus and is incorporated into the

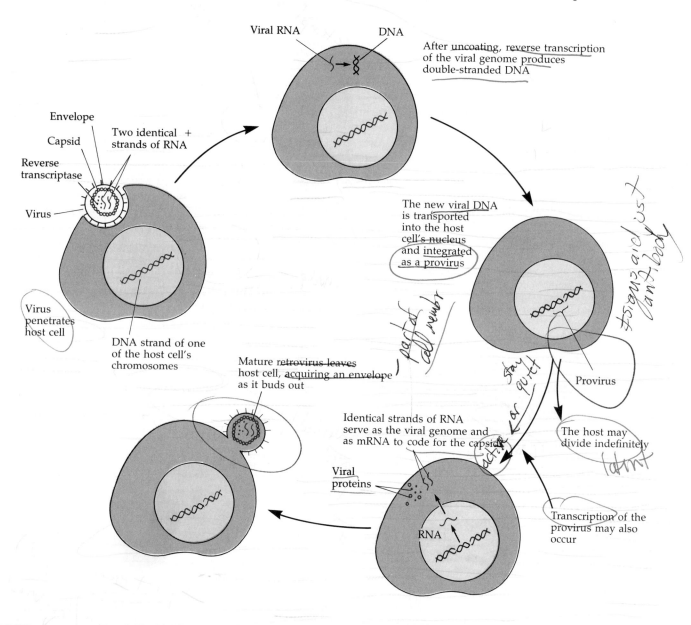

Figure 12-22 Multiplication and inheritance processes of a retrovirus. Retrovirus may become a provirus that replicates in a latent state, and it may produce new retroviruses.

∧∧∧∧∧ MMWR   *Centers for Disease Control*
Morbidity and Mortality Weekly Report

AIDS DIAGNOSIS UPDATE:
THE RISK TO HEALTH-CARE WORKERS

Aids Surveillance and Diagnosis

Since AIDS was identified in 1981, the number of cases has increased markedly every six months (see Figure 18-11). The primary victims are homosexual men, intravenous drug users, and hemophilia patients. Other high-risk groups are the heterosexual partners of AIDS sufferers and transfusion patients. Ninety percent of AIDS victims in the United States are between the ages of 20 and 49 years, and 94 percent are males. AIDS has occurred in all 50 states. The average incubation period exceeds seven years. Forty-eight percent of known AIDS cases have died within two years of diagnosis. Although AIDS is underreported worldwide because of lack of recognition and lack of surveillance, the disease pattern reported from Western Europe, Canada, Australia, New Zealand, and Latin America is similar to that of the United States.

Although the AIDS virus has been isolated from both semen and saliva, there is no proof of transmission by saliva. Thousands of family members, coworkers, and friends of AIDS patients do not have AIDS. The virus is transmitted by sexual contact or the parenteral route (chiefly by contaminated blood) and not by casual contacts.

What Is AIDS?

AIDS is an illness that is characterized by a positive HIV-antibody test and by the presence of *indicator* diseases. Indicator diseases are diseases that are found in AIDS sufferers much more frequently than in others and to which the AIDS sufferer is more than usually susceptible. One example is the cancer called Kaposi's sarcoma. Indicator diseases also include opportunistic parasitic infections and neurological conditions.

The multiple opportunistic infections (OI) that accompany AIDS are often caused by protozoans, many of which were either previously considered nonpathogenic or occurred rarely. These OI include *Pneumocystis carinii*, *Toxoplasma, Cryptosporidium,* and *Isospora.*

Neurological conditions are also among the indicator diseases. In 1985, the isolation of HIV from the central nervous system (CNS) led scientists to speculate that CNS involvement was an early stage of HIV infection. However, since then, the paralytic disease tropical spastic paraparesis has been attributed to another retrovirus.

The following CNS symptoms are included in the case definition of AIDS: *HIV encephalopathy* (a disabling cognitive and/or motor dysfunction that interferes with the patient's ability to function) and *HIV wasting disease* (profound involuntary weight loss).

The Risk to Health-Care Workers

Many nurses and other health-care workers have been understandably concerned about the risk of contracting AIDS after being exposed to the blood of infected patients through needle-stick wounds or the contamination of open wounds or mucous membranes. So far, nine health-care workers who denied other risk factors and who were exposed to blood from patients infected with HIV have been reported to have HIV infection, four of these following needle-stick injuries. The remaining five cases had extensive contact with blood or body fluids of infected patients and did not observe routinely recommended barrier precautions. The exact route of transmission in these cases is not known. Additionally, two workers in different laboratories producing large quantities of highly concentrated HIV have been reported to have laboratory-acquired HIV infections. One worker's infection was presumed to be caused by "undetected skin contact with virus culture supernatant." The other worker's infection followed "an injury with a potentially contaminated needle."

Routine precautions vary, according to the extent of the exposure, from gloves alone to gloves, masks, gowns, and eye coverings. Hands and other skin surfaces accidentally contaminated with blood should be washed thoroughly and immediately.

Three ongoing studies have supplied data on the magnitude of the risk. According to a National Institutes of Health study, none of 103 workers with percutaneous exposures to AIDS patients' blood or body fluids, and none of 229 workers with mucous membrane exposures, was seropositive. A study at the University of California showed no seropositive results among 63 workers with exposure through open wounds or mucous membranes. A Centers for Disease Control study showed that one worker out of 298 became seropositive after needle-stick injuries.

These studies indicate that the risk of transmission during exposure of open wounds or mucous membranes to contaminated blood must be very low, although the precise risk has not been defined. Nonetheless, the increasing prevalence of HIV infection increases the potential for such exposures, especially when health-care workers do not follow routine precautions.

Adapted from *MMWR* 36:75 (12/25/87); *MMWR* 36:19 (5/22/87); *MMWR* 37/S-4 (4/1/88); *MMWR* 37:286 (5/13/88).

plasma membrane of the host cell. The envelope lipid and carbohydrate are synthesized by host cell enzymes and are present in the plasma membrane. The envelope actually develops around the capsid by a process called **budding** (Figure 12-23).

Figure 12-23 Budding of an enveloped virus. (a) Diagram of the budding process. **(b)** The small "bumps" seen on this freeze-fractured plasma membrane are Sindbis virus particles caught in the act of budding out from an infected cell (×38,000).

After the sequence of attachment, penetration, uncoating, and biosynthesis of viral nucleic acid and protein, the assembled capsid-containing nucleic acid pushes through the plasma membrane. As a result, a portion of the plasma membrane, now the envelope, adheres to the virus. This extrusion of a virus from a host cell is one method of release. Budding does not immediately kill the host cell, and in some cases the host cell survives.

Naked viruses are released through ruptures in the host cell plasma membrane. In contrast to budding, this type of release usually results in the death of the host cell.

Effects of Animal Viral Infection on Host Cells

Infection of a host cell by an animal virus usually kills the host cell. Death can be caused by the accumulation of large numbers of multiplying viruses, by the effects of viral proteins on the permeability of the host cell's plasma membrane, or by inhibition of host DNA, RNA, or protein synthesis. The different abnormalities that can lead to damage or death of a host cell are known as **cytopathic effects**, or **CPE** (see Table 14-4). These effects are frequently used as tools for the diagnosis of many viral infections.

One type of CPE is an **inclusion body**, some abnormal body of material in a cell. Some inclusion bodies arise as a result of the accumulation of assembled or unassembled viruses in the nucleus or cytoplasm of the host cell. Other inclusion bodies arise at sites of earlier viral synthesis but do not contain assembled viruses or their components. Inclusion bodies are important because their presence can help to identify the virus causing an infection. For example, rabiesvirus produces inclusion bodies (Negri bodies) in the cytoplasm of nerve cells, and their presence in the brain tissue of animals suspected of being rabid has been used as a diagnostic tool for rabies. Diagnostic inclusion bodies are also associated with the measles virus, vaccinia virus, smallpox virus, herpesvirus, and adenoviruses (Figure 12-24).

As a result of viral infection, some host cells exhibit another type of CPE. At times, several adjacent infected cells fuse to form giant cells called **polykaryocytes**. These are produced from infections from myxoviruses. Other viral infections result in chromosomal damage to the host cell, most often chromosomal breakage. Some viral infections result in changes in the host cell's functions with no visible changes in the infected cells. For example, if a virus

Figure 12-24 Viral inclusion bodies. The dark spots in the sections of two mouse muscle cells shown here are intranuclear bodies of coxsackie A4 virus (×5600). Notice the large crystalline array of virus particles in the lower left.

changes the production level of a hormone, the cells normally affected by that hormone will not function properly.

One very significant response of some virus-infected cells is their production of a substance called **interferon**. Viral infection induces cells to produce interferon, but it is the host cell's DNA that codes for the interferon. This substance protects neighboring uninfected cells from viral infection. More will be said about interferon in Chapter 15.

Viruses and Cancer

A number of types of cancer are now known to be caused by viruses. Research on cancer-causing viruses has led to breakthroughs in our general understanding of cancer. Molecular biological research shows that the mechanisms of the diseases are similar, even when a virus is not the cause of the cancer.

Cancer is not a single disease, but many diseases. The human body contains more than 100 different kinds of cells, each of which can malfunction in its own distinctive way to cause cancer. When cells multiply in an uncontrolled way, the excess tissue that develops is called a tumor. A cancerous tumor is known as a malignant tumor, whereas a noncancerous tumor is called a benign tumor. (Tumors are generally

named by attachment of the suffix *-oma* to the name of the tissue from which the tumor arises.) Most persons who die from cancer are not killed by the primary tumor that develops but by **metastasis**, the spread of the cancer to other parts of the body. Metastatic tumors are harder to detect and eliminate than primary tumors.

Not all cancerous cells form solid tumors. When the malignant cells are those that give rise to white blood cells, the resulting cancer, characterized by an excess of white cells in the circulatory system, is called leukemia.

The relationship between cancers and viruses was first demonstrated in 1908 when Wilhelm Ellerman and Olaf Bang were trying to isolate the causative agent of chicken leukemia. They found that leukemia could be transferred to healthy chickens by cell-free filtrates that contained viruses. Three years later, Francis Rous found that a chicken **sarcoma** (cancer of connective tissue) can be similarly transmitted. Virus-induced **adenocarcinomas** (cancers of glandular epithelial tissue) in mice were discovered in 1936. At that time, it was clearly shown that mouse mammary gland tumors are transmitted from mother to offspring through the mother's milk.

The viral cause of cancer can often go unrecognized, for several reasons. First, most of the particles of some viruses infect cells but do not induce cancer. Second, cancer might not develop until long after viral infection. Third, cancers do not seem to be contagious, as viral diseases usually are.

Transformation of Normal Cells into Tumor Cells

Almost anything that can alter the genetic material of a cell has the potential to make a normal cell cancerous. It is now believed that these cancer-causing alterations to cellular DNA affect parts of the genome called **oncogenes**. All nucleated cells have these oncogenes, which are capable of transforming normal cells into cancerous cells. Essentially the same oncogenes have been isolated from a wide variety of eucaryotes, from yeasts to humans, and from oncogenic viruses as well. More than 20 oncogenes have been isolated from animal cells; some of those associated with human cancers are listed in Table 12-2. Note that most of the oncogenes listed are also found in viruses. Because the same oncogenes are isolated from different types of cancer, it is thought that the total number of different oncogenes will be less than 100. Furthermore, oncogenes can be grouped by nucleotide sequence and protein-product to form a much smaller number of classes.

TABLE 12-2   *Representative Cellular Oncogenes Isolated in Human Cells*

	Found in		Location in Human Chromosome (Chromosome Number)
Oncogene	Cancer Cells	Virus	
abl	Chronic myelogenous leukemia	Abelson murine leukemia virus	9
fes	Acute promyelocytic leukemia	Fujinami (feline) sarcoma virus	15
fos	Burkitt's lymphoma	FBJ osteosarcoma virus	2
mos	Acute myeloblastic leukemia	Moloney murine sarcoma virus	8
myb	Acute lymphocytic leukemia	Avian myeloblastosis virus	6
myc	Burkitt's lymphoma	Avian myelocytomatosis virus	8
ras^H-1	Wilms' tumor	Harvey murine sarcoma virus	11
ras^K-2	Acute lymphocytic leukemia	Kristen murine sarcoma virus	12
ras^N	Myeloid tumors	Not associated with viruses	1
sis	Burkitt's lymphoma; chronic myelogenous leukemia	Simian sarcoma virus	22

Source: P. D'Eustachio, "Gene mapping and oncogenes." *American Scientist* 72:32–40, 1984; and L. Fink, "Unraveling the molecular biology of cancer." *Bioscience* 34:75–77, 1984.

The wide occurrence of oncogenes suggests that they are not only normal but vital for cell functions. Under normal conditions, these genes probably code for proteins that are necessary to the cell's growth, but mutations can interfere with the normal regulation of oncogenes or cause oncogenes to code for abnormal proteins. (Oncogenes that function normally are sometimes called proto-oncogenes to differentiate them from their abnormally activated counterparts.)

The most common types of oncogene products isolated so far are protein kinases. Normal **protein kinases** remove a phosphate group from ATP and add it to the amino acids *serine* or *threonine* within a protein. Phosphorylation modifies a protein's function. The abnormal protein kinases isolated from cancerous cells attach a phosphate group to the amino acid *tyrosine*. The exact role of the phosphorylated proteins in transforming normal cells into cancerous cells is unknown. However, hypotheses being investigated relate their activity to loss of normal controls on cell shape and cell growth.

Oncogenes can be activated by a variety of agents, including mutagenic chemicals, high-energy radiation, and viruses. Viruses capable of producing tumors in animals are called **oncogenic viruses**.

Both DNA- and RNA-containing viruses are capable of inducing tumors in animals. When this occurs, the tumor cells are **transformed** in such a way that they acquire properties that are distinct from properties of uninfected cells or from infected cells that do not form tumors. An outstanding feature of all oncogenic viruses is that their genetic material integrates into the host cell's DNA and replicates along with the host cell's chromosome. This mechanism is similar to the phenomenon of lysogeny in bacteria, and it can alter the host cell's characteristics in the same way.

Transformed cells also lose a property called **contact inhibition.** Normal animal cells in tissue culture move about randomly by amoeboid movement and divide repeatedly until they come into contact with each other. Then, both movement and cell division stops. This phenomenon is known as contact inhibition. Transformed cells in cell culture do not exhibit contact inhibition but instead form tumorlike cell masses. Transformed cells sometimes produce tumors when injected into susceptible animals.

After being transformed by viruses, many tumor cells contain a virus-specific antigen on their cell surface, called **tumor-specific transplantation antigen (TSTA)**, or an antigen in their nucleus, called the **T antigen.** Transformed cells tend to be rounder than normal cells and tend to exhibit certain chromosomal abnormalities, such as unusual numbers of chromosomes and fragmented chromosomes.

DNA-Containing Oncogenic Viruses

Oncogenic viruses are found within several groups of DNA-containing viruses. These groups include adenoviruses, herpesviruses, poxviruses, and papovaviruses. Among the papovaviruses are the papilloma viruses that cause benign warts in humans and other animals, polyoma viruses that cause several kinds of tumors when injected into newborn mice, and simian virus 40 (SV40), which was originally isolated from cell cultures being used to cultivate polioviruses for vaccine production. Papilloma viruses are frequently found in cancerous cervical cells, providing evidence that they may cause uterine (cervical) cancer.

During polyoma and SV40 infection, there is an increase in the host cell's synthesis of DNA. This increase is followed by the appearance of TSTA and T antigen. When the host cell is transformed, viral DNA is integrated into the host cell's DNA as a provirus. This mechanism, similar to lysogeny in bacteria, results in the transformed cells' distinctive properties noted earlier.

Among the herpesviruses that affect humans are the herpes simplex virus, the herpes zoster virus, and the Epstein–Barr (EB) virus. The EB virus has an attraction for lymphocytes (a type of white blood cell) and has the potential for transforming lymphocytes into highly proliferating cells. This EB virus, in addition to being the cause of infectious mononucleosis, has been implicated as the causative agent of two human cancers, Burkitt's lymphoma and nasopharyngeal carcinoma. Burkitt's lymphoma is a rare cancer of the lymphatic system; it affects mostly children in certain areas of Africa. Nasopharyngeal carcinoma, a cancer of the nose and throat, is found worldwide. Some researchers have also suggested that EB virus is involved in Hodgkin's disease, a cancer of the lymphatic system.

It is estimated that about 80% of the population of the United States carries the latent stage of the EB virus in their lymphocytes but have no disease. This latent stage is indicated by the presence of antibodies to EB virus in blood serum. Although the EB virus leads to no apparent symptoms in healthy persons, it can cause infectious mononucleosis, mostly in teenagers. TSTA and T antigen specific for the EB virus have been found in specimens from persons with either Burkitt's lymphoma or nasopharyngeal cancer. Moreover, DNA from EB virus is always found in malignant cells from these cancers.

The role of EB virus in human cancer was accidentally demonstrated in 1985 when a 12-year-old boy received a bone marrow transplant. David was born without a functioning immune system and was kept from all microbes inside a sterile "bubble" his entire life—hence he was known as "the bubble boy." (The lymphoctyes that provide immunity are produced in bone marrow.) Several months after the transplant, he developed signs of infectious mononucleosis (caused by EB virus), and a few months after that he died of cancer. An autopsy revealed that the EB virus had been unwittingly introduced into the boy with the bone marrow transplant. The case confirmed suspicions that EB virus can induce cancer in immunosuppressed individuals.

One type of herpes simplex virus (HSV 1) produces cold sores. Another type, HSV 2, is associated with more than 90% of genital herpes infections. Women with cervical cancer have more HSV 2 antibodies than asymptomatic patients have, so it has been suggested that HSV 2 might be associated with cervical cancer.

Another virus implicated in cancer is hepatitis B virus (HBV). HBV is an enveloped DNA virus with a small, circular genome. The DNA from HBV can become incorporated into the host genome. Its presence has been associated with liver cancer.

RNA-Containing Oncogenic Viruses

Among the RNA-containing viruses, only the retroviruses seem to be oncogenic. Human T cell leukemia viruses (HTLV 1 and HTLV 2) are retroviruses that are associated with leukemia and lymphoma in humans. (T cells are a type of white blood cell involved in immune responses to disease.) HTLV 1 has been isolated from adult T cell leukemia (ATL) patients. Antibodies against the virus are found in people living in regions where the disease is endemic (constantly present), but antibodies are not found in people outside these areas. The presence of antibodies shows that HTLV infections occur in the same geographic areas as ATL, which suggests that the viral infection induces the cancer.

Sarcoma viruses of cats, chickens, and rodents and the mammary tumor viruses of mice are also retroviruses. Another retrovirus, feline leukemia virus (FeLV), causes leukemia in cats and is transmissible between cats. A test to detect the virus in cats' saliva is being developed.

The ability of retroviruses to induce tumors is related to their production of a reverse transcriptase by the mechanism described earlier (see Figures 12-21 and 12-22). The provirus, which is the double-stranded DNA molecule synthesized from the viral

RNA, becomes integrated into the host cell's DNA; new genetic material is thereby introduced into the host's genome, and this is the key reason why retroviruses can contribute to cancer. Some retroviruses contain oncogenes; others contain promoters that turn on oncogenes or other cancer-causing factors.

Activation of Oncogenes

Cancer appears to be a *multistep* process. At least two oncogenes must be activated to abnormal functioning for malignancy to result. At present, four mechanisms for the activation of oncogenes are proposed.

1. A **single mutation** such as a nitrogenous base pair substitution resulting from exposure to a chemical mutagen or radiation could alter the amino acid sequence of a gene product. For example, the *ras* oncogene in malignant cells differs from the normal *ras* protooncogene in only one nucleotide. It is thought that the mutated *ras* gene increases a cell's response to growth-regulating hormones.

   A new protein resulting from a mutation may contribute directly to transformation. Alternatively, the mutation might affect a gene whose product normally regulates an oncogene. If that oncogene codes for a cell growth factor and is no longer properly regulated, the growth factor might be produced in an uncontrolled way and lead to cancer.

2. **Transduction** of oncogenes by a virus could remove the oncogenes from normal cellular controls and place them under control of viral regulatory proteins. The viral proteins could cause an oncogene product to be produced at an abnormal time or in abnormal amounts. Some human oncogenes that have been isolated from viruses are listed in Table 12-2.

3. **Translocation** of oncogenes from one chromosomal locus to another can occur during normal cellular activities and could place the oncogenes into a site in which normal controls are not active. In Burkitt's lymphoma, the *myc* oncogene is moved from its normal position on chromosome 8 to chromosome 14, and its product is increased because it is now regulated by an antibody-regulating gene. It is thought that the continual presence of the *myc* gene product might allow cells to replicate indefinitely.

4. **Gene amplification** could cause the cell to produce unusually large amounts of oncogenic products. A gene is amplified when an unknown mechanism causes the gene to be duplicated several times. In one central nervous system cancer (glioma), more than 50 copies of a particular gene were found in a malignant cell.

Of these four major mechanisms of oncogene activation, viruses are or may be involved in the last three.

Latent Viral Infections

A virus can remain in equilibrium with the host and not actually produce disease for a long period, often many years. The classic example of such a **latent viral infection** is the infection of the skin by herpes simplex virus, which produces cold sores. This virus can inhabit the host's nerve cells but cause no damage until it is activated by such stimuli as fever and sunburn—hence the term *fever blister.*

It is interesting that in some individuals viral production occurs, but the symptoms never appear. Even though a large percentage of the human population carries the herpes simplex virus, most people never exhibit the disease. The virus of some latent infections can exist in a lysogenic state within host cells.

The varicella-zoster virus (also called herpes zoster virus) can also exist in a latent state. Chickenpox (varicella) is a childhood disease in which the viral agent is indistinguishable from the viral agent (herpes zoster) that causes shingles, an adult neurological disorder. Shingles occurs only in a small fraction of the population who have had chickenpox. Many researchers believe that shingles may be caused by reactivation of the virus that has remained latent in certain nerve cells for years.

Slow Viral Infections

The term **slow viral infection** was coined in the 1950s to refer to a disease process that occurs gradually over a long period of time and was thought to be caused by a virus. (The term does not imply that viral multiplication is unusually slow.) Typically, slow viral infections are fatal.

A number of diseases called slow viral infections have in fact been shown to be caused by conventional viruses. For example, several years after causing mea-

TABLE 12-3 *Examples of Slow Viral Infections Caused by Conventional Viruses*

Disease	Host	Organ Primarily Affected	Virus
Subacute sclerosing panencephalitis	Human	Brain	Measles (paramyxovirus)
Progressive encephalitis	Human	Brain	Rubella (togavirus)
Progressive multifocal leukoencephalopathy	Human	Brain	Papovavirus
Progressive pneumonia	Sheep	Lung	Retrovirus
Lymphocytic choriomeningitis	Mouse	Kidney, brain, liver	Arenavirus
Aleutian mink disease	Mink	Reticuloendothelial system	Parvovirus
Canine demyelinating encephalomyelitis	Dog	Brain, spinal cord	Distemper (paramyxovirus)

Source: B. D. Davis, et al. *Microbiology*, 3rd ed. New York: Harper & Row, 1980, p. 1043.

sles, the measles virus can be responsible for a rare form of encephalitis called *subacute sclerosing panencephalitis (SSPE)*. Slow viral infection is apparently different from latent viral infection in that, in most slow viral infection, detectable infectious virus gradually builds up over a long period, rather than appearing suddenly.

Several examples of slow viral infections caused by conventional viruses are listed in Table 12-3.

Unconventional Agents of Disease

Other diseases called slow viral infections have not been found to have a viral cause and are now thought to be caused by unconventional agents. Ironically, many scientists have abandoned the term "slow viral infection" for truly viral diseases such as SSPE (instead using terms such as "persistent viral infection" for SSPE); some medical scientists now reserve "slow viral infection" for diseases that are apparently *not* caused by conventional viruses. The prototype of these latter diseases is scrapie, a neurologic disease of sheep, for which the infectious agent appears to be pure protein. The mystery of how an infectious agent can lack nucleic acid has prompted several hypotheses, including the following:

1. The infectious agent might actually be a conventional virus, but one whose nucleic acid is extraordinarily difficult to detect.

2. The infectious agent might be an undetectably small nucleic acid enclosed in a *host* protein. The nucleic acid might not encode protein at all but might induce disease by somehow directly disrupting the host cell, perhaps by interacting with the host DNA. Although the agent might physically resemble a virus, it would be protected from the host immune system because its coat of host protein would not be recognized as foreign. This hypothetical infectious agent has been called a **virino**.

3. The infectious agent might really be pure protein. If so, how would it replicate? In the past, the unlikely possibility of protein self-replication was raised. Now, however, there is evidence that, at least for scrapie, a major portion of the infectious agent is a protein whose gene is found in normal host DNA. Questions remain, including the questions of how the gene is turned on, how its product acts as an infectious disease agent, and whether other genes and proteins are involved. The postulated infectious protein agent has been named a **prion**, for **pro**teinaceous **in**fectious particle. (The name would more logically be "proin," but the coiner thought that "prion" sounded better.)

All diseases thought to be caused by unconventional agents are neurologic diseases. In addition to scrapie, these include Creutzfeldt–Jakob disease, kuru, and Gerstmann–Sträussler syndrome. (Neurologic diseases are discussed in Chapter 21.)

Plant Viruses and Viroids

Plant viruses resemble animal viruses in many respects; plant viruses are morphologically similar to animal viruses, and they have similar types of nucleic acid (Table 12-4). In fact, some plant viruses can multiply inside insect cells. Plant viruses cause many diseases of economically important crops, including tomatoes (tomato spotted wilt virus), corn and sugarcane (wound tumor virus), and potatoes (potato yellow dwarf virus). Viruses can cause color change, deformed growth, wilting, and stunted growth in their plant hosts. Some hosts, however, remain symptomless and only serve as reservoirs of infection.

Plant cells are generally protected from disease by an impermeable cell wall. Viruses must enter through wounds or be assisted by other plant parasites, including nematodes, fungi, and, most often, insects that suck the plant's sap. Once one plant is infected, it can spread infection to other plants in its pollen and seeds.

In laboratories, plant viruses are cultured in protoplasts (plant cells with the cell walls removed) and in insect cell cultures.

Some plant diseases are caused by **viroids**, short pieces of naked RNA, only 300 to 400 nucleotides long, with no protein coat. The nucleotides are often internally paired, so the molecule has a closed, folded, three-dimensional structure that presumably helps protect it from attack by cellular enzymes. The RNA does not code for any proteins. Thus far, viroids have been conclusively identified as pathogens only of plants. Infections by viroids, such as potato spindle tuber viroid (Figure 12-25), result in losses of millions of dollars in crop damage.

Current research on viroids shows a similarity in base sequences with introns. Recall from Chapter 8 that introns are sequences of genetic material that do not code for polypeptides. This observation has led to the hypothesis that viroids evolved from introns—which leads to speculation that future work may discover animal viroids.

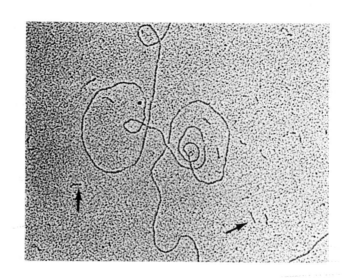

Figure 12-25   Potato spindle tuber viroid (PSTV). The arrows point to the viroids, which are adjacent to much longer viral nucleic acid from bacteriophage T7, for comparison in this electron micrograph ($\times$56,300).

TABLE 12-4   *Classification of Some Major Plant Viruses According to Morphological, Chemical, and Physical Properties*

Virus	Morphological Class	Nucleic Acid*	Dimensions of Capsid (nm)	Method of Transmission	Shows Similarities to
Tobacco mosaic virus	Helical	ss RNA	300 (length)	Wounds	Picornavirus
Brome grass mosaic virus	Polyhedral	ss RNA	23 (diameter)	Pollen	Picornavirus
Potato yellow dwarf virus	Bullet-shaped	ss RNA	380 (length)	Leafhoppers and aphids	Rhabdovirus
Wound tumor virus	Polyhedral	ds segmented DNA	70 (diameter)	Leafhoppers	Reovirus
Cauliflower mosaic virus	Polyhedral	ds DNA	50 (diameter)	Aphids	Papilloma

*ss = single-stranded; ds = double-stranded.

Study Outline

General Characteristics of Viruses (pp. 319–320)

1. Depending on one's viewpoint, viruses may be regarded as exceptionally complex aggregations of nonliving chemicals or as exceptionally simple living microbes.
2. Viruses contain a single type of nucleic acid (DNA or RNA) and a protein coat, sometimes enclosed by an envelope composed of lipids, proteins, and carbohydrates.
3. Viruses are obligatory intracellular parasites. They multiply by using the host cell's synthesizing machinery to cause the synthesis of specialized elements that can transfer the viral nucleic acid to other cells.
4. A virion is a complete, fully developed viral particle composed of nucleic acid surrounded by a coat.

Host Range (p. 319)

1. Host range refers to the spectrum of host cells in which a virus can multiply.
2. Depending on its host range, a virus is generally classified as an animal virus, bacterial virus (bacteriophage), or plant virus. A virus can infect only certain species within each class.
3. Host range is determined by the specific attachment site on the host cell's surface and the availability of host cellular factors.

Size (pp. 319–320)

1. Viral size is determined by filtration through membrane filters, ultracentrifugation, and electron microscopy.
2. Viruses range from 20 to 300 nm in diameter.

Viral Structure (pp. 320–323)

Nucleic Acid (pp. 320–321)

1. The proportion of nucleic acid in relation to protein in viruses ranges from about 1% to about 50%.
2. Viruses contain either DNA or RNA, never both, and the nucleic acid may be single- or double-stranded, linear or circular, or divided into several separate molecules.

Capsid and Envelope (p. 321)

1. The protein coat surrounding the nucleic acid of a virus is called the capsid.
2. The capsid is composed of subunits, the capsomeres, which can be a single type of protein or several types.
3. The capsid of some viruses is enclosed by an envelope consisting of lipids, proteins, and carbohydrates.

4. Some envelopes are covered with carbohydrate–protein complexes called spikes.
5. Viruses without envelopes are called naked viruses.

General Morphology (pp. 321–323)

1. Helical viruses (for example, tobacco mosaic virus) resemble long rods, and their capsids are hollow cylinders surrounding the nucleic acid.
2. Polyhedral viruses (for example, adenovirus) are many-sided. Usually the capsid is an icosahedron.
3. Enveloped viruses are covered by an envelope and are roughly spherical but highly pleomorphic. There are also enveloped helical viruses (for example, influenza) and enveloped polyhedral viruses (for example, herpes simplex).
4. Complex viruses have complex structures. For example, many bacteriophages have a polyhedral capsid with a helical tail attached.

Taxonomy of Viruses (p. 324)

1. Classification of viruses is based on type of nucleic acid, morphological class, size of capsid, and number of capsomeres.
2. Other classification schemes take into account the virus's susceptibility to microbial control agents, immunological properties, site of multiplication, and method of transmission.

Isolation, Cultivation, and Identification of Viruses (pp. 325–329)

1. Viruses must be grown in living cells.
2. The easiest viruses to grow are bacteriophages.

Growth of Bacteriophages in the Laboratory (p. 325)

1. The plaque method mixes bacteriophages with host bacteria and nutrient agar.
2. After several viral multiplication cycles, the bacteria in the area surrounding the original virus are destroyed; the area of lysis is called a plaque.
3. Each plaque can originate with a single viral particle or more than one; the number of viral particles required to initiate a plaque is called a plaque-forming unit.

Growth of Animal Viruses in the Laboratory (pp. 328–329)

1. Cultivation of some animal viruses requires whole animals.
2. Some animal viruses can be cultivated in embryonated eggs.

3. Cell cultures are cells growing in culture media in the laboratory.
4. Primary cell lines and embryonic diploid cell lines grow for a short time in vitro.
5. Continuous cell lines can be maintained in vitro indefinitely.
6. Viral growth can cause cytopathic effects in the cell culture.

Viral Identification   (p. 329)

1. Serological tests are used most often to identify viruses.
2. Viruses may be identified by restriction enzyme fragments and nucleic acid base sequencing.

Viral Multiplication   (pp. 330–342)

1. Viruses do not contain enzymes for energy production or protein synthesis.
2. For a virus to multiply, it must invade a host cell and direct the host's metabolic machinery to produce viral enzymes and components.

Multiplication of Bacteriophages   (pp. 330–334)

T-Even Bacteriophages   (pp. 330–332)
1. The T-even bacteriophages that infect *E. coli* have been studied extensively.
2. In attachment, sites on the phage's tail fibers attach to complementary receptor sites on the bacterial cell.
3. In penetration, phage lysozyme opens a portion of the bacterial cell wall, the tail sheath contracts to force the tail core through the cell wall, and phage DNA enters the bacterial cell. The capsid remains outside.
4. In biosynthesis, transcription of phage DNA produces mRNA coding for proteins necessary for phage multiplication. Phage DNA is replicated, and capsid proteins are produced. During the eclipse period, separate phage DNA and protein can be found.
5. During maturation, phage DNA and capsids are assembled into complete viruses.
6. During release, phage lysozyme breaks down the bacterial cell wall, and the multiplied phages are released.
7. The time from phage adsorption to release is called burst time (20 to 40 minutes). Burst size, the number of newly synthesized phages produced from a single infected cell, ranges from 50 to 200.

Lysogeny   (pp. 332–334)
1. During a lytic cycle, a phage causes the lysis and death of a host cell.
2. Some viruses can either cause lysis or have their DNA incorporated as a prophage into the DNA of the host cell. The latter situation is called lysogeny.
3. Prophage genes are regulated by a repressor coded for by the prophage. The prophage is replicated each time the cell divides.

4. Exposure to certain mutagens can lead to excision of the prophage and initiation of the lytic cycle.
5. Because of lysogeny, lysogenic cells become immune to reinfection with the same phage, and the host cell can exhibit new properties.
6. A lysogenic phage can transfer bacterial genes from one cell to another through transduction. Any genes can be transferred in generalized transduction, and specific genes can be transferred in specialized transduction.

Multiplication of Animal Viruses   (pp. 334–342)

1. Animal viruses attach to the plasma membrane of the host cell.
2. Penetration of enveloped viruses occurs by endocytosis.
3. Animal viruses are uncoated by viral or host-cell enzymes.
4. The DNA of most DNA viruses is released into the nucleus of the host cell. Transcription of viral DNA and translation produce viral DNA and, later, capsid protein. Capsid protein is synthesized in the cytoplasm of the host cell.
5. Multiplication of RNA viruses occurs in the cytoplasm of the host cell. RNA-dependent RNA polymerase synthesizes a double-stranded RNA.
6. Picornavirus + strand RNA acts as mRNA and directs the synthesis of RNA-dependent RNA polymerase.
7. Togavirus + strand RNA acts as a template for RNA-dependent RNA polymerase, and mRNA is transcribed from a new − RNA strand.
8. Rhabdovirus − strand RNA is a template for viral RNA-dependent RNA polymerase, which transcribes mRNA.
9. mRNA is produced inside the capsid of reoviruses.
10. Retroviruses carry reverse transcriptase (RNA-dependent DNA polymerase), which transcribes DNA from RNA.
11. After maturation, viruses are released. One method of release (and envelope formation) is budding. Naked viruses are released through ruptures in the host cell membrane.

Effects of Animal Viral Infection
on Host Cells   (pp. 342–343)

1. Cytopathic effects (CPE) are abnormalities that lead to damage or death of a host cell.
2. Cytopathic effects include inclusion bodies, polykaryocytes, and altered function.
3. Interferon is produced by virus-infected cells and protects neighboring cells from viral infection.

Viruses and Cancer   (pp. 343–346)

1. An excess of tissue due to unusually rapid cell multiplication is called a tumor. Tumors are malignant (cancerous) or benign (noncancerous). Metastasis refers to the spread of cancer to other parts of the body.
2. Tumors are usually named by attachment of the suffix *-oma* to the name of the tissue from which the tumor arises.
3. The earliest relationship between cancer and viruses was demonstrated in the early 1900s, when chicken leukemia

and chicken sarcoma were transferred to healthy animals by cell-free filtrates.

Transformation of Normal Cells into Tumor Cells (pp. 343–344)

1. Eucaryotic cells have protooncogenes that code for proteins necessary for the cells' normal growth. When activated to oncogenes, these genes transform normal cells into cancerous cells.
2. Mutagenic chemicals, radiation, and viruses can activate oncogenes.
3. Viruses capable of producing tumors are called oncogenic viruses.
4. Several DNA viruses and retroviruses can be oncogenic.
5. The genetic material of oncogenic viruses becomes integrated into the host cell's DNA.
6. Transformed cells lose contact inhibition, contain virus-specific antigens (TSTA and T antigen), exhibit chromosomal abnormalities, and can produce tumors when injected into susceptible animals.

DNA-Containing Oncogenic Viruses (p. 345)

1. Oncogenic viruses are found among adenoviruses, herpesviruses, poxviruses, and papovaviruses.
2. The EB virus, a herpesvirus, causes infectious mononucleosis and has been implicated in Burkitt's lymphoma and nasopharyngeal carcinoma. One type of herpes simplex (HSV 2), associated with over 90% of genital herpes infections, might be implicated in cervical cancer.

RNA-Containing Oncogenic Viruses (pp. 345–346)

1. Among the RNA viruses, only retroviruses seem to be oncogenic.
2. HTLV 1 and HTLV 2 have been associated with human leukemia and lymphomas.
3. The virus's ability to produce tumors is related to the production of reverse transcriptase. The DNA synthesized from the viral RNA becomes incorporated as a provirus into the host cell's DNA.
4. A provirus can remain latent, can be multiplied, or can transform the host cell.

Activation of Oncogenes (p. 346)

1. A single mutation can result in the production of a protein required for transformation.
2. Transduction of oncogenes could result in oncogene products being made in abnormal amounts or at the wrong time.
3. Translocation of oncogenes could remove normal controls.
4. Gene amplification causes unusually large amounts of oncogene products.

Latent Viral Infections (p. 346)

1. A latent viral infection is one in which the virus remains in the host cell for long periods of time without producing an infection.
2. Examples are cold sores and shingles.

Slow Viral Infections (pp. 346–347)

1. Slow viral infections are disease processes that occur over a long period of time and are generally fatal.
2. Some of these diseases are caused by conventional viruses; others apparently are not.

Unconventional Agents of Disease (p. 347)

1. The "slow viral infections" not caused by conventional viruses—such as scrapie, Creutzfeldt–Jakob disease, and certain other neurologic diseases—may be caused by unconventional infectious agents that are pure protein, without nucleic acid (prions).
2. Another hypothesized agent of such diseases is the virino, a tiny piece of nucleic acid associated with host protein.

Plant Viruses and Viroids (p. 348)

1. Plant viruses must enter plant hosts through wounds or with invasive parasites, such as insects.
2. Some plant viruses also multiply in insect (vector) cells.
3. Viroids are infectious pieces of RNA that cause some plant diseases, such as potato spindle tuber viroid disease.

Study Questions

Review

1. Viruses were first detected because they are filterable. What do we mean by the term *filterable*, and how could this property have helped their detection before invention of the electron microscope?
2. Why do we classify viruses as obligatory intracellular parasites?
3. List the four properties that define a virus. What is a virion?
4. Describe the four morphological classes of viruses, then diagram and give an example of each.
5. Describe how bacteriophages are detected and enumerated by the plaque method.
6. Explain how animal viruses are cultured in each of the following:
(a) Whole animals
(b) Embryonated eggs
(c) Cell cultures

7. Why are continuous cell lines of more practical use than primary cell lines for culturing viruses? What is unique about continuous cell lines?

8. Describe the multiplication of a T-even bacteriophage. Be sure to include the essential features of attachment, penetration, biosynthesis, maturation, and release.

9. *Streptococcus pyogenes* produces erythrogenic toxin and is capable of causing scarlet fever only when it is lysogenic. What does this mean?

10. Describe the principal events of attachment, penetration, uncoating, biosynthesis, maturation, and release of an enveloped DNA-containing virus.

11. Assume that this strand of RNA is the nucleic acid for an RNA-containing animal virus: UAGUCAAGGU.
    (a) Describe the steps of RNA replication for a virus that contains a + strand of RNA.
    (b) Describe the steps of RNA replication for a virus that contains a − strand of RNA.
    (c) Describe the steps of RNA replication for a virus that contains double-stranded RNA.
    (d) Describe the steps of RNA replication for a virus that contains reverse transcriptase.

12. Provide an explanation for the chronic recurrence of cold sores in some people. (*Note:* The cold sores almost always return at the same site.)

13. Slow viral infections such as _____ might be caused by _____ that are _____.

14. What are cytopathic effects?

15. Recall from Chapter 1 that Koch's postulates are used to determine the etiology of a disease. Why is it difficult to determine the etiology of
    (a) a viral infection such as influenza?
    (b) cancer?

16. The DNA of DNA-containing oncogenic viruses can become integrated into the host DNA. When integrated, the DNA is called a _____. How does this process result in transformation of the cell? Describe the changes of transformation. How can an RNA-containing virus be oncogenic?

17. Contrast viroids and unconventional agents. Name a disease thought to be caused by each.

18. Plant viruses cannot penetrate intact plant cells because _____; therefore, they enter cells by _____. Plant viruses can be cultured in _____.

Challenge

1. Discuss the arguments for and against the classification of viruses as living organisms.

2. In some viruses, capsomeres function as enzymes as well as structural supports. Of what advantage is this to the virus?

3. Discuss four mechanisms of oncogene activation, and discuss how viruses may be implicated.

4. Prophages and proviruses have been described as being similar to bacterial plasmids. What similar properties do they exhibit? How are they different?

Further Reading

Croce, C. M., and G. Klein. "Chromosome translocations and human cancer." *Scientific American* 252(3):54–60, March 1985. Describes one method of oncogene activation.

Diener, T. O. "Viroids and their interactions with host cells." *Annual Review of Microbiology* 36:239–258, 1982. Comprehensive coverage of the structure, pathogenesis, and possible origin of viroids by their discoverer.

Fucillo, D. A., J. E. Kurent, and J. L. Sever. "Slow virus diseases." *Annual Review of Microbiology* 28:231–264, 1974. A summary of information about viral diseases caused by conventional and unconventional agents.

Hogle, J. M., M. Chow, and D. J. Filman. "The structure of poliovirus." *Scientific American* 256(3):42–49, March 1987. Describes virus assembly and provides a mechanism for protein coat changes that allow a virus to escape the host's immune response.

Hunter, T. "The proteins of oncogenes." *Scientific American* 251(2):70–79, August 1984. Explains what the products of oncogenes do.

Karpas, A. "Viruses and leukemia." *American Scientist* 70:277–285, 1982. A report on cell transformation and transmission of leukemia.

Prusiner, S. B. "Prions." *Scientific American* 251(4):50–59, October 1984. The scientist who discovered and named prions poses intriguing and as yet unanswered questions regarding the origin and nature of the prion and its genome.

Roizman, B., and A. E. Sears. "An inquiry into the mechanisms of herpes simplex virus latency." *Annual Review of Microbiology* 41:543–571, 1987. Describes host cell changes and provides models for establishment and maintenance of latency.

Varmus, H. "Reverse transcription." *Scientific American* 257(3):56–64, September 1987. Presents the action of this enzyme and its role in retroviruses, hepatitis B virus, and evolution.

Watson, J. D., N. H. Hopkins, J. W. Roberts, J. A. Steitz, and A. M. Weiner. *Molecular Biology of the Gene*, 4th ed. Menlo Park, Calif.: Benjamin/Cummings, 1987. Chapter 7 discusses phages and Chapter 24 gives an excellent overview of the molecular biology of eucaryotic viruses.

Weinberg, R. A. "A molecular basis of cancer." *Scientific American* 249(5):126–141, November 1983. Describes two methods of oncogene activation, by point mutation and by viruses.

White, D. O., and F. J. Fenner. *Medical Virology*, 3rd ed. New York: Academic Press, 1986. A comprehensive textbook on the biology of medically important viruses.

PART THREE

Interaction Between Microbe and Host

CHAPTER 13

Principles of Disease and Epidemiology

Learning Objectives

- Define normal flora; compare commensalism, mutualism, and parasitism, and give one example of each.

- List Koch's postulates.

- Define a reservoir of infection; contrast human, animal, and nonliving reservoirs, and give one example of each.

- Define nosocomial infections and explain their importance.

- Explain four methods of transmission of disease.

- Categorize diseases according to incidence.

- Define epidemiology, and describe three methods of epidemiologic investigation.

- Differentiate between a communicable and a noncommunicable disease.

- Define pathogen, etiology, infection, host, disease, acute, chronic, subacute, and latent.

- Identify four predisposing factors for disease.

- Put the following terms in proper sequence in terms of the pattern of disease: period of decline, period of convalescence, period of illness, crisis, prodromal period, period of incubation.

Now that you have a basic understanding of the structures and functions of microorganisms and some idea of the variety of microorganisms that exist, we can consider how the human body and various microorganisms interact in terms of health and disease.

We all have defense mechanisms to keep us healthy. For instance, unbroken skin and mucous membranes are effective barriers against microbial invasion. Within the body, certain cells and certain specialized proteins called antibodies can work together to destroy microbes. In spite of our defenses, however, we are still susceptible to **pathogenic** (disease-causing) microorganisms. Some bacteria can invade our tissues and resist our defenses by producing capsules or enzymes. Other bacteria release toxins that can seriously affect our health. A rather delicate balance exists between our defenses and the disease-producing mechanisms of microorganisms. When our defenses resist these disease-producing capabilities, we maintain our health. But when the disease-producing capability overcomes our defenses, disease results. After the disease has become established, an infected person may recover completely, suffer temporary or permanent damage, or die. The outcome depends on many factors.

In Part Three, we examine some of the principles of infection and disease, the mechanisms of pathogenicity, the body's defenses against disease, and the ways that microbial diseases can be prevented by immunization and controlled by drugs. This first chapter discusses the general principles of disease, starting with a discussion of the meaning and scope of pathology. In the last section of this chapter, Epidemiology, you will learn how these principles are useful in studying and controlling disease.

Pathology, Infection, and Disease

Pathology is the scientific study of disease (*pathos* means suffering; *logos* means science). Pathology is first concerned with the cause, or **etiology**, of disease. Second, it deals with **pathogenesis**, the manner in which a disease develops. Third, pathology is concerned with the structural and functional changes brought about by disease and with its final effects on the body.

Although the terms *infection* and *disease* are sometimes used interchangeably, they differ somewhat in meaning. **Infection** is the invasion or colonization of the body by pathogenic microorganisms. **Disease** occurs when an infection results in any change from a state of health. Disease is an abnormal state in which part or all of the body is not properly adjusted or is not capable of carrying on its normal functions. An infection may exist in the absence of detectable disease. For example, the body may be infected with the virus that causes AIDS, but there may be no symptoms of the disease.

The presence of a particular type of microorganism in a part of the body where it is not normally found is also called an infection—and may lead to disease. For example, although large numbers of *E. coli* are normally present in the healthy intestine, their infection of the urinary tract usually results in disease.

Not all microorganisms are pathogenic. In fact, the presence of some microorganisms can even benefit the host. Therefore, before we discuss the role of microorganisms in causing disease, let us examine the relationship of normal flora to the human body.

Normal Flora

Animals, including humans, are germ-free in utero. At birth, however, normal and characteristic microbial populations begin to establish themselves. Just before a woman gives birth, lactobacilli in her vagina multiply rapidly. The newborn's first contact with microorganisms is usually with these lactobacilli, and they become the predominant organisms in the newborn's intestine. More microorganisms are introduced to the newborn's body from the environment when breathing begins and feeding starts. After birth, *E. coli* and other bacteria acquired from foods begin to inhabit the large intestine. These microorganisms remain there throughout life and, in response to altered environmental conditions, may increase or decrease in number and contribute to disease.

Many other harmless microorganisms establish themselves inside other parts of the normal adult body and on its surface. It is estimated that a typical human body contains 1×10^{13} body cells, yet harbors 1×10^{14} bacterial cells. This gives you an idea of the abundance of normal organisms residing in the human body. The microorganisms that establish more or less permanent residence (colonize) but that do not produce disease under normal conditions are known as **normal flora** or **normal microbiota** (Figure 13-1).

(a) (b) (c)

Figure 13-1 Representative normal flora for different regions of the body. (a) Lactic acid bacteria on the surface of the stomach epithelium ($\times$3900). **(b)** Plaque on enamel near the gums ($\times$1900). **(c)** Mucous membrane of large intestine showing bacteria in crypts (depressions) ($\times$1000).

Others, called **transient flora**, may be present for several days, weeks, or months and then disappear. Microorganisms are not found throughout the entire human body but are localized in certain regions, as shown in Figure 13-2.

The principal normal flora in different regions of the body and some distinctive features of each region are listed in Table 13-1. Normal flora are also discussed more specifically in Part Four.

Relationships Between Normal Flora and Host

Once established, the normal flora can benefit the host by preventing the overgrowth of harmful microorganisms, a phenomenon called **microbial antagonism**. When this balance is upset, disease can result from normal flora. For example, the normal bacterial flora of the adult human vagina maintain a local pH of from 3.5 to 4.5. The presence of these bacteria inhibits overgrowth of the yeast *Candida albicans* (kan′did-ä al′bi-kans), which cannot grow under acidic conditions and can normally be present in small amounts in the vagina (see Microview A, Photo 21). If the bacterial population is eliminated by antibiotics, excessive douching, or deodorants, the pH of the vagina reverts to nearly neutral, and *C. albicans* can flourish and become the dominant microorganism there. This condition can lead to a form of vaginitis (vaginal infection).

The relationship between normal flora and their host is called **symbiosis** (living together). In the symbiotic relationship called **commensalism**, one of the organisms is benefited and the other is unaffected. Many of the microorganisms that make up our normal flora are commensals; these include the corynebacteria that inhabit the surface of the eye and certain saprophytic mycobacteria that inhabit the ear and external genitals. These bacteria live on secretions and sloughed-off cells, and they bring no apparent benefit or harm to the host.

Mutualism is a type of symbiosis that benefits both organisms. For example, the large intestine contains bacteria, such as *E. coli*, that synthesize vitamin K and some B vitamins. These vitamins are absorbed by the blood and distributed for use by body cells. In exchange, the large intestine provides nutrients for the bacteria so that they can survive. In still another kind of symbiosis, one organism is benefited at the expense of the other; this relationship is called **parasitism**. Many disease-causing bacteria are parasites.

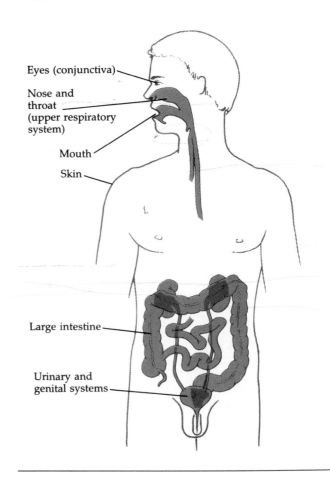

Eyes (conjunctiva)

Nose and throat (upper respiratory system)

Mouth

Skin

Large intestine

Urinary and genital systems

Figure 13-2   Locations of normal flora on and in the human body. Various species of microbes are found in all the named regions.

Opportunistic Organisms

Although it is convenient to categorize symbiotic relationships by type, we must keep in mind that under certain conditions the relationship can change. For example, given the proper circumstances, a mutualistic organism, such as *E. coli*, can become harmful. *E. coli* is generally harmless as long as it remains in the large intestine. But if it gains access to other body sites, such as the urinary bladder, lungs, spinal cord, or wounds, it may cause urinary tract infections, pulmonary infections, meningitis, or abscesses.

Opportunists are potentially pathogenic organisms that ordinarily do not cause disease in their normal habitat in a healthy person. For example, organisms that gain access through broken skin or mucous membranes can cause opportunistic infections. Or, if the host is already weakened or compromised by infection, microbes that are usually harmless can cause disease. The disease AIDS is often accompanied by a common opportunistic infection, *Pneumocystis carinii* pneumonia, caused by the oppor-

Kaow → 4 /organism frea one

TABLE 13-1 *Representative Normal Flora by Body Region**

Region	Principal Flora	Comment
Skin	*Propionibacterium acnes, Staphylococcus epidermidis, Staphylococcus aureus, Corynebacterium xerosis, Pityrosporum* spp., *Candida* spp.	Most of the microbes in direct contact with skin do not become residents because secretions from sweat and oil glands have antimicrobial properties.
Eyes (conjunctiva)	*S. epidermidis, S. aureus,* diphtheroids	The conjunctiva, a continuation of the skin or mucous membrane, contains basically the flora found on the skin.
Nose and throat (upper respiratory system)	*S. aureus, S. epidermidis,* and aerobic diphtheroids in the nose; *S. epidermidis, S. aureus,* diphtheroids, *Streptococcus pneumoniae, Hemophilus,* and *Neisseria* in the throat	Although some of the normal flora are potential pathogens, their ability to cause disease is reduced by microbial antagonism.
Mouth	Various species of *Streptococcus, Lactobacillus, Actinomyces, Bacteroides, Fusobacterium, Treponema, Corynebacterium, Candida*	Abundant moisture, warmth, and constant presence of food make the mouth an ideal environment that supports very large and diverse microbial populations on the tongue, cheeks, teeth, and gums.
Large intestine	*Bacteroides, Fusobacterium, Lactobacillus, Streptococcus, Bifidobacterium, Escherichia coli, Enterobacter, Citrobacter, Proteus, Klebsiella, Shigella, Candida*	The large intestine contains the largest numbers of resident flora in the body because of its available moisture and available nutrients.
Urinary and genital systems	*Staphylococcus epidermidis,* aerobic micrococci, *Streptococcus, Lactobacillus,* aerobic diphtheroids, *Pseudomonas, Klebsiella,* and *Proteus* in urethra; lactobacilli in vagina; aerobic diphtheroids, *Streptococcus, Staphylococcus, Bacteroides, Clostridium, Candida albicans,* and *Trichomonas vaginalis* in vagina	The lower urethra in both sexes has a resident population; the vagina has its acid-tolerant population of microbes because of the nature of its secretions.

*This table lists some flora that are not discussed in this chapter; they will be discussed in Part Four.

tunistic protozoan *Pneumocystis carinii*. This secondary infection can develop in AIDS sufferers because their immune systems are suppressed (see the box in Chapter 12, p. 341). Prior to the AIDS epidemic, this type of pneumonia was rare.

In addition to the usual symbionts, many people carry other microorganisms that are generally regarded as pathogenic but that may not cause disease in those people. Among the pathogens that are frequently carried in normal individuals are **enteric cytopathogenic human orphan (ECHO) viruses**, which can cause intestinal diseases, and **adenoviruses**, which can cause respiratory diseases. *Neisseria meningitidis* (nī-se′rē-ä me-nin-ji′ti-dis), which often resides benignly in the respiratory tract, can cause meningitis, a disease that inflames the coverings of the brain and spinal cord. *Streptococcus pneumoniae*, a normal resident of the nose and throat, can cause a type of pneumonia.

Etiology of Infectious Diseases

Some diseases, such as polio, have a well-known etiology. Some have an etiology that is not completely understood—for example, the relationship between herpes zoster (chickenpox) and shingles. For still others, such as Creutzfeldt–Jakob disease, the etiology is unknown. Of course, not all diseases are caused by microorganisms. For example, the disease hemophilia is an *inherited disease*; osteoarthritis and cirrhosis are considered *degenerative diseases*. There are several other categories of disease, but here we will discuss only *infectious diseases*, those caused by microorganisms. In fact, half of all human diseases are infectious. To see how microbiologists determine the etiology of an infectious disease, we will discuss in greater detail the work of Robert Koch that was introduced in Chapter 1.

Koch's Postulates

In the historical overview of microbiology presented in Chapter 1, we briefly discussed Koch's famous postulates. Koch, you might recall, was a German physician who played a major role in establishing that microorganisms cause specific diseases. In 1877, Koch published some early papers on anthrax, a disease of cattle that can also occur in humans. Koch demonstrated that certain bacteria, today known as *Bacillus anthracis*, were always present in the blood of animals that had the disease and were not present in healthy animals. He knew that the mere presence of the bacteria did not prove that they caused the disease; the bacteria could have been there as a result of the disease. Thus he experimented further. Koch took a sample of blood from a diseased animal and injected it into a healthy animal. The second animal developed the same disease and died. He repeated this procedure many times, always with the same results. (A key criterion in the validity of any scientific proof is that experimental results be repeatable.) Koch also cultivated the microorganism in fluids outside the animal's body, and he demonstrated that the bacterium would cause anthrax even after many culture transfers. In short, Koch showed that a specific infectious disease (anthrax) is caused by a specific microorganism (*B. anthracis*) that can be isolated and cultured on artificial media.

Koch later used the same methods to show that the bacterium *Mycobacterium tuberculosis* is the causative agent of tuberculosis. As part of this work, he also developed a specific stain for use in microscopic examination of the bacteria.

Koch's research provides a framework for the study of the etiology of any infectious disease. Today, we refer to Koch's experimental requirements as **Koch's postulates** (Figure 13-3). They are summarized as follows:

1. The same pathogen must be present in every case of the disease.
2. The pathogen must be isolated from the diseased host and grown in pure culture.
3. The pathogen from the pure culture must cause the disease when it is inoculated into a healthy, susceptible laboratory animal. In exclusively human diseases, this step is not morally acceptable, but it sometimes happens accidentally in a laboratory or through other circumstances. The case of the "bubble boy" (p. 345) is an example where a pathogen accidentally introduced into a human confirmed the cause of a disease.

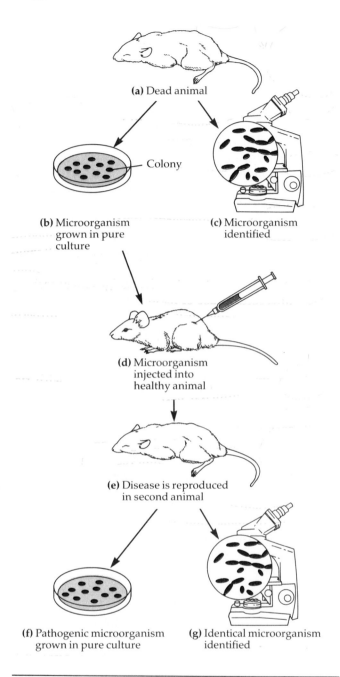

(a) Dead animal
(b) Microorganism grown in pure culture — Colony
(c) Microorganism identified
(d) Microorganism injected into healthy animal
(e) Disease is reproduced in second animal
(f) Pathogenic microorganism grown in pure culture
(g) Identical microorganism identified

Figure 13-3   Koch's postulates. (a) Microorganisms from a dead animal are **(b)** grown in pure culture and **(c)** identified. **(d)** A sample of the pure culture is inoculated into a healthy animal, and **(e)** the disease is reproduced in the animal. **(f)** Pathogenic microorganisms are again grown in pure culture and **(g)** identified, confirming that they are identical with the original pathogen.

4. The pathogen must again be isolated from the inoculated animal and must be shown to be the original organism.

Exceptions to Koch's Postulates

Although Koch's postulates are useful in determining the causative agent of most bacterial diseases, there are some exceptions. For instance, it is known that the bacterium *Treponema pallidum* (trē-pō-nē′mä pal′li-dum) is the causative agent for syphilis, but virulent strains have never been cultured on artificial media. Moreover, many rickettsial and all viral pathogens cannot be cultured on artificial media because they multiply only within cells.

The discovery of microorganisms that cannot grow on artificial media has made necessary some modifications of Koch's postulates and the use of alternative methods of culturing and detecting particular microbes. For example, when investigators searching for the microbial cause of legionellosis (Legionnaires' disease) were unable to isolate the microbe directly from a victim, they took the alternative step of inoculating a victim's lung tissue into guinea pigs. These guinea pigs developed the disease's pneumonialike symptoms, whereas guinea pigs inoculated with tissue from an unafflicted person did not. Then tissue samples from the diseased guinea pigs were cultured in yolk sacs of chick embryos—a method, discussed in Chapter 12, that reveals the growth of extremely small microbes. After the embryos were incubated, electron microscopy revealed rod-shaped bacteria in the chick embryos. Finally, modern immunologic techniques (which will be discussed in Chapter 17) were used to show that the bacteria in the chick embryos were the same bacteria as those in the guinea pigs and in afflicted humans.

In a number of situations, a human host exhibits certain signs and symptoms that are associated only with a certain pathogen and its disease. For example, the pathogens responsible for diphtheria and tetanus give rise to distinguishing signs and symptoms that can be produced by no other microbe. They are unequivocally the only organisms that produce their respective diseases. But some infectious diseases are not quite so clear-cut. For example, nephritis (inflammation of the kidneys) can involve any of several different pathogens, all of which give rise to the same signs and symptoms. Thus, it is often difficult to know which particular microorganism is causing a disease. Other infectious diseases that have poorly defined etiologies are pneumonia, meningitis, and peritonitis (inflammation of the peritoneum, the membrane that lines the abdomen and pelvis and covers the organs within them).

Some pathogens can cause several pathologies. *Mycobacterium tuberculosis*, for example, is implicated in diseases of the lungs, skin, bones, and internal organs. *Streptococcus pyogenes* (pī-äj′en-ēz) can cause sore throat, scarlet fever, skin infections (erysipelas), puerperal fever, and osteomyelitis (inflammation of bone), among other diseases. When clinical signs and symptoms are used together with laboratory methods, these infections can usually be distinguished from infections of the same organs caused by other pathogens.

Classifying Infectious Diseases

Each disease affecting the body alters body structures and functions in particular ways, and these alterations are usually indicated by several kinds of evidence. For example, the patient may experience certain *symptoms*—changes in body function—such as pain and malaise (a vague feeling of body discomfort). These subjective changes are not apparent to the observer. The patient can also exhibit *signs*, which are objective changes that the physician can observe and measure. Frequently evaluated signs include lesions (changes produced in tissues by disease), swelling, fever, and paralysis. Sometimes a specific group of symptoms or signs always accompanies a particular disease. Such a group is called a **syndrome**. The diagnosis of a disease is achieved by evaluation of the signs and symptoms together with the results of laboratory tests.

Diseases are often classified in terms of how they behave within a host and within a given population. Any disease that spreads from one host to another, either directly or indirectly, is said to be a *communicable disease*. Chickenpox, measles, genital herpes, typhoid fever, and tuberculosis are examples. Chickenpox and measles are also examples of *contagious diseases* that are easily spread from one person to another. A *noncommunicable disease* is not spread from one host to another. These diseases are caused by microorganisms that normally inhabit the body and only occasionally produce disease or by microorganisms that reside outside the body and produce disease only when introduced into the body. An example is tetanus. *Clostridium tetani* (klô-stri-dē-um te′tan-ē) produces disease only when it is introduced into the body via abrasions or wounds.

Occurrence of Disease

To understand the full scope of a disease, we should know something about its occurrence. The **incidence** of a disease is the fraction of a population that contracts it during a particular period of time. The **prevalence** of a disease is the fraction of the population having the disease at a specified time. Knowing the incidence and prevalence of a disease in different populations (for example, in populations representing different geographical regions or different racial groups) shows epidemiologists the range of the disease's occurrence and its tendency to affect some groups of people more than others.

Frequency of occurrence is another criterion that is used in the classification of diseases. If a particular disease occurs only occasionally, it is called a **sporadic disease**. Typhoid fever in the United States is such a disease. A disease constantly present in a population is called an **endemic disease**; an example of such a disease is the common cold. If many people in a given area acquire a certain disease in a relatively short period of time, it is called an **epidemic disease**. Influenza is an example of a disease that often achieves epidemic status. Public health authorities consider AIDS near or at epidemic status at this time (see Figure 18-11). Some authorities consider gonorrhea and certain other sexually transmitted diseases as epidemic at this time as well. An epidemic disease that occurs worldwide is called a **pandemic disease**. We experienced a pandemic of influenza in the late 1960s.

Severity or Duration

Another useful way of defining the scope of a disease is in terms of its severity or duration. An **acute disease** is one that develops rapidly but lasts only a short time. A good example is influenza. A **chronic disease** develops more slowly, and the body's reactions may be less severe, but the disease is likely to be continual or recurrent for long periods of time. Infectious mononucleosis, tuberculosis, syphilis, and leprosy fall into this category. A disease that is intermediate between acute and chronic is described as a **subacute disease**. An example is subacute sclerosing panencephalitis, a rare brain disease characterized by diminished intellectual function and loss of nervous function. A **latent disease** is one in which the causative agent remains inactive for a period of time but then becomes active

to produce symptoms of the disease. An example is shingles, one of the diseases caused by the varicella-zoster virus.

Extent of Host Involvement

Infections can also be classified according to the extent to which the host's body is affected. A **local infection** is one in which the invading microorganisms are limited to a relatively small area of the body. Examples of local infections are boils and abscesses. In a **systemic**, or **generalized**, **infection**, microorganisms or their products are spread throughout the body by the blood or lymphatic system. Measles is an example of a systemic infection. Very frequently, agents of a local infection enter a blood or lymph vessel and spread to other parts of the body. We refer to this condition as a **focal infection**. Focal infections can arise from infections in the teeth, tonsils, or sinuses. The presence of bacteria in the blood is known as **bacteremia**, and if the bacteria actually multiply in the blood, the condition is called **septicemia** (see Microview C, p. C-2). **Toxemia** refers to the presence of toxins in blood, and **viremia** refers to the presence of viruses in blood.

The state of host resistance also determines the extent of infections. A **primary infection** is an acute infection that causes the initial illness. A **secondary infection** is one caused by an opportunist microbe after the primary infection has weakened the body's defenses. Secondary infections of the skin and respiratory tract are common and are sometimes more dangerous than the primary infections. *Pneumocystis* pneumonia as a consequence of AIDS is an example of a secondary infection that is more serious than the primary one. Streptococcal bronchopneumonia following influenza is another example. An **inapparent**, or **subclinical**, **infection** is one that does not cause any noticeable illness. Polio and hepatitis A, for example, can be carried by persons who show no illness.

Spread of Infection

Reservoirs

For a disease to perpetuate itself, there must be a continual source of the disease organisms. This source can be either a living organism or an inanimate object that provides a pathogen with adequate conditions for

(a)

(b)

(c)

Figure 13-4 Reservoirs of disease. (a) Human. **(b)** Animal. **(c)** Nonliving.

survival and multiplication and an opportunity for transmission. Such a source is called a **reservoir of infection**. These reservoirs may be human reservoirs, animal reservoirs, or nonliving reservoirs (Figure 13-4).

Human Reservoirs

The principal living reservoir of human disease is the human body itself. Many people harbor pathogens and transmit them directly or indirectly to others. People with signs and symptoms of a disease are obviously transmitters of the disease. However, some people can harbor pathogens and transmit them to others without exhibiting any signs of illness. These people, called carriers, are important living reservoirs of infection. Some carriers have infections for which no signs or symptoms are ever exhibited. Other people carry a disease during its symptom-free stages—during the incubation period (before symptoms appear) or during the convalescent period (recovery). Human carriers play an important role in the spread of such diseases as AIDS, diphtheria, typhoid fever, hepatitis, gonorrhea, amoebic dysentery, and streptococcal infections.

Animal Reservoirs

Both sylvatic (wild) and domestic animals are living reservoirs of microorganisms that can cause human diseases. Diseases that occur primarily in wild and domestic animals and can be transmitted to humans are called **zoonoses**. Some strains of influenza (transmitted by swine), rabies (transmitted by bats, skunks, foxes, dogs, and cats), and Rocky Mountain spotted fever (transmitted by ticks) are examples of zoonoses. Other representative zoonoses are presented in Table 13-2.

About 150 zoonoses are known. The transmission of zoonoses to humans can occur via one of many routes—by direct contact with infected animals, by

contamination of food and water, by contact with contaminated hides, fur, or feathers, by consumption of infected animal products, or by insect vectors (insects that transmit pathogens).

Nonliving Reservoirs

The two major nonliving reservoirs of infectious disease are soil and water. Soil harbors such pathogens as fungi, which cause mycoses, and *Clostridium botulinum* (klô-stri′dē-um bo-tū-lĭ′num), the bacterial agent that causes botulism. Water that has been contaminated by the feces of humans and other animals is a reservoir for several pathogens, most notably those responsible for gastrointestinal diseases.

Transmission of Disease

The causative agents of disease can be transmitted from the reservoir of infection to a susceptible host by four routes: (1) contact, (2) a common vehicle, (3) the airborne route, and (4) vectors.

Contact Transmission

Contact transmission is the spread of the agent of disease by direct contact, indirect contact, or by droplet (Figure 13-5a). *Direct contact transmission* refers to the direct transmission of an agent from its source to a susceptible host; an intermediate object is not involved. This is also known as *person-to-person transmission*. The most common forms of direct contact transmission are touching, kissing, or sexual intercourse. Among the diseases transmitted from person to person by direct contact are viral respiratory tract diseases (common cold and influenza), hepatitis A, measles, scarlet fever, smallpox, and sexually transmitted diseases (syphilis, gonorrhea, and genital herpes). Direct contact is also one way of spreading AIDS and infectious mononucleosis. Potential pathogens can be transmitted by direct contact from animals (or animal products) to humans. Examples are

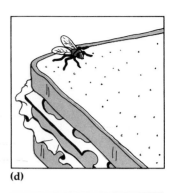

(a)   (b)   (c)   (d)

Figure 13-5   Means of transmission of disease.
(a) Contact transmission. (b) Common vehicle transmission. (c) Airborne transmission.
(d) Vector transmission (mechanical).

TABLE 13-2   *Selected Zoonoses That Can Be Transmitted to Humans*

Disease	Etiology	Reservoir	Method of Transmission	Refer to Chapter
Viral				
Influenza (some strains)	Myxovirus	Swine	Direct contact	23
Rabies	Rhabdovirus	Bats, skunks, foxes, dogs, cats	Direct contact (bite)	21
Western equine encephalitis	Arbovirus	Horses, birds	*Culex* mosquito bite	21
Yellow fever	Arbovirus	Monkeys	*Aedes* mosquito bite	22
Bacterial				
Anthrax	*Bacillus anthracis*	Domestic livestock	Direct contact with contaminated hides or animals; air; food	22
Brucellosis	*Brucella* spp.	Domestic livestock	Direct contact with contaminated milk, meat, or animals	22
Bubonic plague	*Yersinia pestis*	Rodents	Flea bites	22
Cat-scratch fever	Gram-negative bacillus	Domestic cats	Direct contact	22
Leptospirosis	*Leptospira*	Wild mammals, domestic dogs and cats	Direct contact with urine, soil, water	25
Lyme disease	*Borrelia burgdorferi*	Birds, deer	Tick bites	22
Pneumonic plague	*Yersinia pestis*	Rodents	Direct contact	22
Psittacosis (Ornithosis)	*Chlamydia psittaci*	Birds, especially parrots	Direct contact	23
Q fever	*Coxiella burnetii*	Domestic livestock	Inhalation, tick bites (between animals)	23

the pathogens causing rabies and anthrax.

Indirect contact transmission occurs when the agent of disease is transmitted from its reservoir to a susceptible host by means of a nonliving object. Any nonliving object involved in the spread of an infection is called a **fomite**. Examples of fomites are tissues, handkerchiefs, towels, bedding, diapers, drinking cups, eating utensils, and thermometers. Contaminated syringes serve as fomites in the transmission of AIDS and hepatitis B.

Droplet infection is a third type of contact transmission. Agents of disease can be spread by droplets of saliva or mucus discharged into the air by coughing, sneezing, laughing, or talking. Such droplets travel less than one meter through the air from the reservoir

to the host. Disease agents that travel such short distances are not regarded as airborne. An example of a disease spread by droplet infection is influenza.

Common Vehicle Transmission

Common vehicle transmission refers to the transmission of disease agents by a common inanimate reservoir (food, water, drugs, blood) to an individual (Figure 13-5b). Examples of diseases transmitted in this way are water-borne shigellosis, food-borne salmonellosis, and various kinds of nosocomial (hospital-acquired) infections (discussed later in this chapter). Such a disease outbreak at a college football game is described in the box in Chapter 24, p. 624.

TABLE 13-2 (*Continued*)

Disease	Etiology	Reservoir	Method of Transmission	Refer to Chapter
Rocky Mountain spotted fever	*Rickettsia rickettsii*	Rodents	Tick bites	22
Salmonellosis	*Salmonella* spp.	Poultry, rats, turtles	Ingestion of contaminated food, water	24
Tularemia	*Francisella tularensis*	Wild and domestic mammals, especially wild rabbits	Direct contact with infected animals; deer-fly bites	22
Typhus fever	*Rickettsia typhi*	Rodents	Flea bites	22
Fungal				
Ringworms	*Trichophyton Microsporum Epidermophyton*	Domestic mammals	Direct contact; fomites	20
Protozoan				
Chagas' disease	*Trypanosoma cruzi*	Wild mammals	"Kissing bug" bite	22
Malaria	*Plasmodium* spp.	Monkeys	*Anopheles* mosquito bite	22
Toxoplasmosis	*Toxoplasma gondii*	Cats and other mammals	Direct contact with infected tissues or fecal material	22
Helminthic				
Hydatid cyst	*Echinococcus granulosus*	Dogs	Direct contact with fecal material	11
Tapeworm (beef)	*Taenia saginata*	Cattle	Ingestion of contaminated beef	24
Trichinosis	*Trichinella spiralis*	Pigs, bears	Ingestion of contaminated meat	24

Airborne Transmission

Airborne transmission refers to the spread of agents of infection by droplet nuclei or dust at a distance of more than one meter from the reservoir to the host (Figure 13-5c). Some microbes are carried in small drops of mucus discharged in a fine spray from the mouth and nose during coughing, sneezing (Figure 13-6), laughing, and even talking. These mucus droplets that contain microbes can be so small that they evaporate, travel more than one meter from the source, and remain airborne for prolonged periods of time. Such droplets are called **droplet nuclei**. The virus that causes measles and the bacterium that causes tuberculosis are transmitted via droplet nuclei. Infectious agents contained in dust particles are also carried in airborne transmission. Q fever is one disease transmitted by the airborne route. Spores produced by certain fungi are also transmitted by the airborne route and can cause such diseases as histoplasmosis, coccidioidomycosis, and blastomycosis (see Chapter 23).

Vectors

Arthropods are the most important group of disease **vectors**—animals that carry pathogens from one host to another. (Insects and other arthropod vectors were discussed in Chapter 11.) Arthropod vectors transmit disease by two general methods. *Mechanical transmission* is the transport of the pathogens on the insect's feet or other body parts (Figure 13-5d). If the insect makes contact with a host's food, pathogens can be transferred to the food and later swallowed by the host. Houseflies, for instance, are capable of transferring the pathogens of typhoid fever and bacillary dysentery (shigellosis) from the feces of infected persons to food.

Biological transmission is more complex. The arthropod bites an infected person or animal and ingests some of the infected blood. The pathogens then reproduce in the vector, and the increase in the number of pathogens increases the possibility that they will be transmitted to another host. Some parasites reproduce in the gut of the arthropod; these are passed with feces. If the arthropod defecates while biting a potential host, the parasite can enter the wound. Other parasites reproduce in the vector's gut and migrate to the salivary gland. These are directly injected into a bite. Some protozoan and helminthic parasites use the vector as a host for a developmental stage in their life cycle.

Table 13-3 lists a few important arthropod vectors and the diseases they transmit.

Portals of Exit

In the next chapter, you will learn how microorganisms enter the body through a preferred route or portal of entry, such as mucous membrane or skin. For now, since we are mainly concerned with the spread of disease throughout a population, it is important to know that pathogens also have definite routes of exit, called **portals of exit**. In general, portals of exit are related to the part of the body that has been infected.

The most common portals of exit are the respiratory and gastrointestinal tracts. For example, many pathogens living in the respiratory tract exit in discharges from the mouth and nose; these discharges are expelled during coughing or sneezing. These microorganisms are found in droplets formed from mucus. Pathogens that cause tuberculosis, whooping cough, pneumonia, scarlet fever, meningococcal meningitis, measles, mumps, smallpox, and influenza are discharged through the respiratory route. Other pathogens exit from the gastrointestinal tract in feces or saliva. Feces may be contaminated with pathogens associated with salmonellosis, cholera, typhoid fever, paratyphoid fever, shigellosis, amoebic dysentery, and poliomyelitis. Saliva can also contain pathogens, such as the rabies virus.

Another important route of exit is the urogenital tract. Microbes responsible for sexually transmitted diseases are found in secretions from the penis and vagina. Urine can also contain the pathogens respon-

Figure 13-6   Droplet transmission. This high-speed photograph shows the spray of small droplets that come from the mouth during a sneeze.

TABLE 13-3 *Representative Arthropod Vectors and the Diseases They Transmit*

Disease	Causative Agent	Arthropod Vector	Refer to Chapter
Malaria	*Plasmodium* spp.	*Anopheles* (mosquito)	22
African trypanosomiasis	*Trypanosoma brucei*	*Glossina* sp. (tsetse fly)	21
American trypanosomiasis	*Trypanosoma cruzi*	Kissing bug	22
Yellow fever	Arbovirus (yellow fever virus)	*Aedes* (mosquito)	22
Dengue	Arbovirus (dengue fever virus)	*Aedes aegypti* (mosquito)	22
Arthropod-borne encephalitis	Arbovirus (encephalitis virus)	*Culex* (mosquito)	21
Epidemic typhus	*Rickettsia prowazekii*	*Pediculus humanus* (louse)	22
Endemic murine typhus	*Rickettsia typhi*	*Xenopsylla cheopis* (rat flea)	22
Rocky Mountain spotted fever	*Rickettsia rickettsii*	*Dermacentor andersoni* and other species (tick)	22
Plague	*Yersinia pestis*	*Xenopsylla cheopis* (rat flea)	22
Relapsing fever	*Borrelia* sp.	*Ornithodorus* spp. (soft ticks)	22
Lyme disease	*Borrelia burgdorferi*	*Ixodes dammini* (tick)	22

sible for typhoid fever and brucellosis, which can exit via the urinary tract. Skin or wound infections are other portals of exit. Drainage from wounds can spread infections to another person directly or by contact with a contaminated fomite. Infected blood can be removed and reinjected by biting insects and contaminated needles and syringes to spread infection within the population. Examples of diseases so transmitted are AIDS, yellow fever, Rocky Mountain spotted fever, tularemia, malaria, and hepatitis B.

Nosocomial (Hospital-Acquired) Infections

Infections acquired as a result of a hospital stay are called **nosocomial infections**. (The word *nosocomial* is derived from the Greek word for hospital; the term also includes infections acquired in nursing homes and other health-related facilities.) The Centers for Disease Control (a public health service) estimates that from 5% to 15% of all hospital patients acquire some type of nosocomial infection. It is estimated that about 20,000 people die of nosocomial infections each year.

Nosocomial infections result from the interaction of several factors: (1) microorganisms in the hospital environment, (2) the compromised (or weakened) status of the host, and (3) the chain of transmission in the hospital. Figure 13-7 illustrates how it is not simply the presence of any single one of these factors but rather is the interaction of all three factors that poses a significant risk of nosocomial infection.

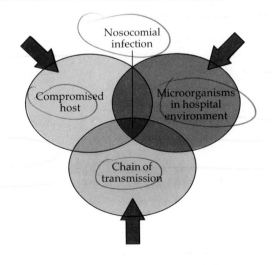

Figure 13-7 Nosocomial infections. Interaction of three principal factors that contribute to nosocomial infections.

Microorganisms in the Hospital

Although every effort is made to disinfect and check the growth of microorganisms in the hospital, the hospital environment is a major reservoir for a variety of pathogens. One reason is that the normal flora from the human body are opportunists and present a particularly strong danger to hospital patients. In fact, most of the bacteria that cause nosocomial infections do not cause disease in healthy people, but are pathogenic only for individuals whose defenses have been weakened by illness or therapy.

In the past, most nosocomial infections were caused by gram-positive microbes. At one time, gram-positive *Staphylococcus aureus* (staf-i-lō-kok′kus ô′rē-us) was the primary cause of nosocomial infections. Although antibiotic-resistant strains of that organism are still significant factors, the major causes today are gram-negative bacteria, such as *E. coli* and *Pseudomonas aeruginosa* (sū-dō-mō′nas ā-rü-ji-nō′sä), especially the antibiotic-resistant strains. *P. aeruginosa* has the ability to cause opportunistic skin infections, especially in surgical and burn patients. The major bacteria involved in nosocomial infections are summarized in Table 13-4.

In addition to the threat of opportunists in the hospital, some microorganisms become resistant to antimicrobial drugs. For example, *P. aeruginosa* and other gram-negative bacteria tend to be difficult to control with antibiotics because of their R factors, which carry genes that determine resistance to antibiotics (Chapter 8). As the R factors recombine, new and multiple resistance factors are produced. These strains become part of the flora of patients and hospital personnel and become progressively more resistant to antibiotic therapy. In this way, people become part of the reservoir (and chain of transmission) for antibiotic-resistant strains of bacteria. Usually, if the host's resistance is high, the new strains are not much of a problem. But if disease, surgery, or trauma has weakened the host's defenses, secondary infections may be difficult to treat.

Compromised Host

A **compromised host** is one whose resistance to infection is impaired by disease, therapy, or burns. Three principal conditions can compromise the host: (1) broken skin or mucous membranes, (2) a suppressed immune system, and (3) impaired cell activity.

TABLE 13-4   *Bacteria Involved in More Than 70% of All Nosocomial Infections*

Bacteria	Percentage of Total Infections	Infections Caused
Enterobacteria (*E. coli, Klebsiella* species, *Proteus* species, *Enterobacter* species, and *Serratia marcescens**)	Over 40	Urinary tract infections, peritonitis, bacteremia, pneumonia, septicemia, gastrointestinal inflammations, and neonatal (in newborns) meningitis
Staphylococcus aureus	11	Urinary and respiratory tract infections and endocarditis (inflammation of the lining of the heart)
Enterococci	10	Urinary tract and surgical wound infections
Pseudomonas aeruginosa	9	Burn and surgical wound infections, septicemia, and pneumonia

*See the box in Chapter 7, p. 180.

As long as the skin and mucous membranes remain intact, they provide formidable physical barriers against most pathogens. Burns, surgical wounds, trauma (such as accidental wounds), injections, invasive diagnostic procedures, intravenous therapy, and urinary catheters (used to drain urine) can all break the first line of defense and make an individual more susceptible to disease in hospitals. Burn patients are especially susceptible to nosocomial infections because their skin is no longer an effective barrier to microorganisms. A burn patient can have tissue damage caused by heat, electricity, radiation, or chemicals, all of which destroy proteins and cells in affected tissues.

The risk of infection is also related to invasive procedures, such as administering anesthesia, which may alter respiration and contribute to pneumonia, and tracheotomy, in which an incision is made into the trachea to assist breathing.

In healthy individuals, white blood cells called T lymphocytes provide resistance to disease by killing pathogens directly, mobilizing phagocytes, and secreting chemicals that kill pathogens. Drugs, radiation therapy, steroid therapy, burns, diabetes, leukemia, kidney disease, and malnutrition can all adversely affect the actions of T lymphocytes and compromise the host.

A final factor that may compromise host defenses is impairment of the development or activities of white blood cells called B lymphocytes, which develop into antibody-producing cells. Antibodies provide immunity by such actions as neutralizing toxins, inhibiting attachment of a pathogen to host cells, and helping to lyse pathogens. Conditions that alter the actions of B lymphocytes include malnutrition and cirrhosis of the liver.

A summary of the principal kinds of nosocomial infections is presented in Table 13-5.

Chain of Transmission

Given the variety of pathogens (and potential pathogens) in the hospital and the compromised state of the host, routes of transmission are a constant concern. The principal routes of transmission of nosocomial infections are direct contact transmission from hospital staff to patient and from patient to patient, and indirect contact transmission through fomites and the hospital's ventilation system (airborne transmission).

Because hospital personnel are often in direct contact with patients, they can often transmit disease. For example, a physician or nurse may transmit her or his flora to a patient when changing a dressing, or a kitchen worker who carries *Salmonella* can infect a food supply.

Certain areas of a hospital are reserved for specialized care. These include the burn, hemodialysis, recovery, intensive care, and oncology units. Unfortunately, these same units also group patients together and provide environments for the epidemic spread of nosocomial infections from patient to patient.

Many diagnostic and therapeutic hospital procedures provide a fomite route of transmission. The urinary catheter used to drain urine from the bladder is a fomite in many nosocomial infections. Intravenous catheters, which pass into a vein through the skin to provide fluids, nutrients, or medication, can also transmit nosocomial infections. Respiratory aids can introduce contaminated fluids into the lungs. Needles may introduce pathogens into muscle or blood, and surgical dressings can become infected and promote disease.

TABLE 13-5   *The Principal Kinds of Nosocomial Infections*

Type of Infection	Comment
Urinary tract infection	Most common, usually accounting for about 50% of all nosocomial infections. Typically related to urinary catheterization.
Surgical wound infection	Ranks second in incidence (about 25%). It is estimated that from 5% to 12% of all surgical patients develop postoperative infections; the percentage can reach 30% for certain surgeries, such as colon surgery and amputations.
Lower respiratory infection	Nosocomial pneumonias rank third in incidence (about 12%) and have high mortality rates. Most of these pneumonias are related to respiratory devices that aid breathing or administer medications.
Bacteremia	Bacteremias account for about 6% of hospital infections. Intravenous catheterization is implicated in nosocomial infections of the bloodstream, particularly infections caused by bacteria and fungi.
Cutaneous infection	Among the least common of all nosocomial infections. However, newborns have a high rate of susceptibility to skin and eye infections.

Control of Nosocomial Infections

Control measures aimed at preventing nosocomial infections vary from one institution to another, but certain procedures are generally implemented. It is important to reduce the number of pathogens to which patients are exposed by using aseptic techniques, handling contaminated materials carefully, insisting on frequent and conscientious handwashing, educating personnel with basic infection control measures (see the box below), and using isolation rooms and wards. Tubs used to bathe patients should be disinfected between uses so that bacteria from the previous patient will not contaminate the next one. Respirators and humidifiers provide both a suitable growth environment for some bacteria and a method of airborne transmission. These sources of nosocomial infections must be kept scrupulously clean and disinfected, and materials used for bandages and intubation (insertion of tubes into organs, such as the trachea) should be sterilized prior to use. Packaging used to maintain sterility should be removed aseptically. Physicians can help improve patients' resistance to infection by prescribing antibiotics only when necessary, avoiding invasive procedures if possible, and minimizing the use of immunosuppressive drugs.

Accredited hospitals should have an infection control committee. Most hospitals at least have an infection control nurse or epidemiologist (an individual who studies disease in populations). The role of these personnel is to identify problem sources, such as antibiotic-resistant strains of bacteria and improper sterilization techniques. The infection control officer should make periodic examinations of equipment to determine the amount of microbial contamination. Samples should be taken from tubing, catheters, respirator reservoirs, and other equipment.

Patterns of Disease

A definite sequence of events usually occurs during infection and disease. As you already know, there must be a reservoir of infection as a source of pathogens for an infectious disease to occur. Next, there must be transmission of the pathogen either by direct contact, indirect contact, or vectors to a susceptible

MMWR   *Centers for Disease Control*
Morbidity and Mortality Weekly Report

OUTBREAK OF HEPATITIS B ASSOCIATED WITH AN ORAL SURGEON—NEW HAMPSHIRE

During the first six months of 1986, four clinical cases of hepatitis B were reported in a city in New Hampshire. Each case was serologically confirmed. All of the patients had been seen by the same oral surgeon, and all of them had undergone tooth extractions from three to five months before becoming ill. All four patients denied other risk factors for hepatitis B virus infection. One patient developed inflammation of the outer layer of arteries with severe complications, including colonic perforation, diseases of nerves with paralysis, and ulceration into the joint space of one ankle.

Of the four patients, one remained seropositive for hepatitis B surface antigen (HBsAg) and became a chronic hepatitis B carrier. He was tested and found to have HBsAg subtype ad, the same subtype that the oral surgeon carried.

The oral surgeon had been practicing in the city for 25 years. His practice was limited to dental extractions. He had never had any symptoms suggestive of hepatitis B, and had never received hepatitis B vaccine. He had never been tested for hepatitis B serologic markers prior to the outbreak. In July 1986, he was seropositive for HBsAg, indicating that he was probably a hepatitis B carrier. He was not aware of having had any skin lesions on his hands in the past year. Although he was careful to scrub his hands between surgical procedures, he did not wear gloves.

Editorial Note: Eight other outbreaks of hepatitis B traceable to dentists or oral surgeons have been reported since 1974. The number of clinically infected patients in each outbreak ranged from three to 55. Two of the nine clinically ill patients in one outbreak died of fulminant hepatitis B. In each outbreak, the implicated dentist or oral surgeon was seropositive for HBsAg and did not use gloves during dental or surgical procedures. None of the dentists who were hepatitis B carriers was aware of their chronic infections. Traumatic procedures (surgery, extractions) have been associated with a higher infection risk than nontraumatic procedures (fillings, denture fittings, etc.). Transmission has been thought to occur through apparent or inapparent lesions on the dentist's hands.

The repeated occurrence of outbreaks associated with dentists or oral surgeons is especially disturbing because measures to prevent them are easily available and widely recommended. A safe, effective vaccine against hepatitis B became available in 1982, and, since the late 1970s, national dental authorities have urged dental practitioners to wear gloves during all procedures involving hand contact with patients' mouths. In March 1986, a national random telephone survey revealed that 44% of nonfederal practicing dentists and oral surgeons in the United States had been vaccinated against hepatitis B. Only 15% of respondents used gloves routinely for all procedures.

Recurrent, avoidable outbreaks such as this one should prompt dentists and oral surgeons to seek hepatitis B vaccination and to use gloves routinely when treating patients.

Source: *MMWR* 36:132–133 (3/13/87).

host. This event is followed by invasion, in which the microorganism enters the host and multiplies. Following invasion, the microorganism injures the host through a process called pathogenesis (which you will learn more about in the next chapter). The extent of injury depends on the extent to which host cells are damaged, either directly or by toxins. Despite the effects of all these factors, the occurrence of disease ultimately depends on the resistance of the host to the offensive weapons of the pathogen.

Predisposing Factors

Certain predisposing factors also affect the occurrence of disease. A *predisposing factor* is one that makes the body more susceptible to a disease and may alter the course of the disease. Gender is sometimes a predisposing factor. For example, females have a higher incidence of urinary tract infections than males. Males have higher rates of pneumonia and meningitis. Other aspects of genetic background may play a role as well. Individuals with sickle-cell anemia, for instance, are actually more resistant to malaria than others.

Climate and weather seem to have some effect on the incidence of infectious diseases. In temperate regions, the incidence of respiratory diseases increases during the winter. This increase correlates with people staying indoors and having closer contact with each other, which facilitates the spread of respiratory pathogens. Other predisposing factors include inadequate nutrition, fatigue, age, environment, habits, life-style or occupation, preexisting illness, chemotherapy, and emotional disturbances. It is often difficult to know the exact relative importance of different predisposing factors.

Development of Disease

Once a microorganism does overcome the defenses of the host, development of the disease follows a certain sequence of steps that tends to be similar whether the disease be acute or chronic.

Period of Incubation

The period of incubation is the time interval between the actual infection and the first appearance of any signs or symptoms. In some diseases, the incubation period is always the same; in others, it is quite variable. The time of incubation depends on the specific microorganism involved, its virulence (degree of path-

ogenicity), the number of infecting microorganisms, and the resistance of the host. Table 14-1 in the next chapter lists the incubation periods for a number of microbial diseases.

Prodromal Period

The prodromal period is a relatively short period of time that follows the period of incubation in some diseases. The prodromal period is characterized by early, mild symptoms of disease, such as general aches and malaise, and certain very specific symptoms, such as Koplik's spots in measles.

Period of Illness

During the period of illness, the disease is most acute. The person exhibits overt signs and symptoms of disease, such as fever, chills, muscle pain (myalgia), sensitivity to light (photophobia), sore throat (pharyngitis), lymph node enlargement (lymphadenomegaly), and gastrointestinal disturbances. During the period of illness, the number of white blood cells increases or decreases. Generally, the patient's immune response and other defense mechanisms overcome the pathogen, and the period of illness ends. When the disease is not successfully overcome (or treated medically), the patient usually dies during this period.

Period of Decline

During the period of decline, the signs and symptoms subside. The fever decreases, and the feeling of malaise diminishes. If the period of decline is short—less than 24 hours, for instance—it is said to occur by *crisis.* If, instead, it takes several days, with the fever decreasing a little each day until body temperature returns to normal, then the period of decline is said to occur by *lysis* (not to be confused with the disintegration of cells). During this phase, the patient is vulnerable to secondary infections.

Period of Convalescence

During the period of convalescence, the person regains strength, and the body returns to its prediseased state. Recovery has occurred.

We all know that, during the period of illness, people serve as reservoirs of disease and can easily spread infections to other persons. However, you should also know that persons can spread infection during incubation and convalescence. This is especially true of diseases such as typhoid fever and cholera, where the convalescing person carries the pathogenic microorganism for months or even years.

Epidemiology

In today's crowded, overpopulated world, where frequent travel and mass production and distribution of foods and other goods are a way of life, diseases can spread rapidly. A contaminated food or water supply, for example, can affect many thousands of people very quickly. It is desirable to identify the causative agent so that a disease can be effectively controlled and treated. It is also desirable to understand the mode of transmission and geographical distribution of the disease. The science that deals with when and where diseases occur and how they are transmitted in the human population is called **epidemiology**.

Modern epidemiology began in 1854 when John Snow, a British physician, reasoned that cholera was transmitted by contaminated water. He questioned some victims of a cholera epidemic in London and found that most of them obtained drinking water from the Broad Street well. He also determined that the incidence of cholera was higher among people who drank from the London section of the Thames River than among the population using cleaner water upstream. Even before Pasteur's work, Snow used epidemiologic investigation to improve public health. Today, epidemiologists study specified groups of individuals to discover the causes and prevent the spread of diseases.

An epidemiologist not only determines the etiology of a disease, he or she also identifies other possibly important factors and patterns concerning the persons affected. An important part of the epidemiologist's work is assembling and analyzing such data as age, sex, occupation, personal habits, socioeconomic status, history of immunization, the presence of any other diseases, and the common history of affected individuals (such as eating the same food or visiting the same doctor's office). Also important for prevention of future outbreaks is knowledge of the site at which a susceptible host came in contact with the agent of infection. In addition, the epidemiologist considers the time period during which the disease occurs, either on a seasonal basis (to indicate whether the disease is prevalent during the summer or winter) or on a yearly basis (to indicate the effects of hygiene or immunization).

Figure 13-8 shows graphs indicating the incidence of selected diseases. Such graphs provide information about whether disease outbreaks are sporadic or epidemic and, if epidemic, how the disease might have spread. By establishing the frequency of a disease in a population and identifying the factors responsible for its transmission, an epidemiologist provides phy-

sicians with information that is important in determining the prognosis and treatment of a disease. Epidemiologists also evaluate how effectively a disease is being controlled in a community—by a vaccination program, for example. Finally, epidemiologists can provide data to help in evaluating and planning overall health care for a community.

Epidemiologists use three basic types of investigations when analyzing the occurrence of a disease—descriptive, analytical, and experimental.

Descriptive Epidemiology

Descriptive epidemiology entails collecting all data that describe the occurrence of the disease under study. Relevant information usually includes information about the affected persons and the place and time period in which the disease occurred.

Such a study is generally *retrospective* (looking backward after the episode has ended). In other words, the epidemiologist backtracks to the cause and source of the disease (see the box in Chapter 7, p. 180). The search for the cause of toxic shock syndrome is an example of a fairly recent retrospective study. In the initial phase of an epidemiological study, retrospective studies are more common than *prospective* (looking forward) studies, in which an epidemiologist chooses a group of persons who are free of a particular disease to study. The group's subsequent disease experiences are then recorded for a given period of time. This type of study was used to test the Salk polio vaccine in 1954 and 1955.

Analytical Epidemiology

Analytical epidemiology analyzes a particular disease to determine its probable cause. This study can be done in two ways. With the *case control method*, the epidemiologist looks for factors that might have preceded the disease. A group of persons who have the disease is compared with another group who are free of the disease. For example, one group with meningitis and one without the disease might be matched by age, sex, socioeconomic status, and location. These statistics are compared to determine which of all the possible factors—genetic, environmental, nutritional, and so forth—might be responsible for the meningitis. Using the *cohort method*, the epidemiologist studies two populations, one that has had contact with the agent causing a disease and another that has not (both groups are called *cohort groups*). For example, a comparison of one group composed of people who have received blood transfusions and one composed of people who have not could reveal an association between blood transfusions and the incidence of hepatitis B virus.

Experimental Epidemiology

Experimental epidemiology begins with a hypothesis about a particular disease; experiments to test the hypothesis are then conducted with a group of people. One such hypothesis could be the assumed effectiveness of a drug. A group of infected individuals is selected and divided randomly so that some receive the drug and others receive a placebo, a substance that has no effect. If all other factors are kept constant between the two groups, and if it is determined that those people who received the drug recovered more rapidly than those who received the placebo, it can be concluded that the drug was the experimental factor (variable) that made the difference.

Case Reporting

We noted earlier in this chapter that establishing the chain of transmission for a disease is extremely important. Once known, the chain can be interrupted in order to slow down or stop the spread of the disease.

An effective way to establish the chain of transmission is case reporting, a procedure that requires health personnel to report specified diseases to local, state, and national health officials (see the box in Chapter 7, p. 180). Examples of such diseases include AIDS, measles, gonorrhea, tetanus, and typhoid fever. Case reporting provides epidemiologists with an approximation of the incidence and prevalence of a disease. This information helps officials decide whether or not to investigate a given disease.

Case reporting provided epidemiologists with valuable leads regarding the origin and spread of AIDS. In fact, the first clue about AIDS came from reports of young males with Kaposi's sarcoma, formerly a dis-

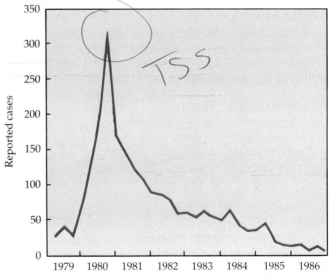

Figure 13-8 Epidemiological graphs. Epidemiologists use various types of graphs to describe patterns of disease. **(a)** This graph of the number of legionellosis cases reported from 1976 to 1987 simply shows the yearly occurrence of the disease during that time period. **(b)** This bar graph looks at legionellosis from a different perspective, allowing epidemiologists to begin drawing some conclusions about the disease. It records the incidence by month, for one year (1979). Over a period of years, researchers saw that the incidence was consistently highest during the hot months. These data helped them realize that the disease was transmitted by an airborne pathogen through air-conditioning cooling towers in use during the hot weather. Note that this graph records the number of cases per 100,000 people, rather than the total number of cases. **(c)** Annual incidence of toxic shock syndrome among menstruating women, 1979–1986. This graph shows how public education can help in disease control. Once women learned how tampons were involved in the disease and decreased use of them, incidence of the disease fell sharply.

ease of older males. Using these reports, epidemiologists began various studies of the patients. If an epidemiological study shows that a large enough segment of the population is affected by a disease, an attempt is then made to isolate and identify its causative agent. Identification is accomplished by a number of different microbiological methods. Very frequently, identification of the causative agent provides valuable information regarding the reservoir for the disease.

Once the chain of transmission is discovered, it is possible to apply control measures to stop the disease from spreading. These might include elimination of the source of infection, isolation and segregation of contaminated persons, development of vaccines, and, as in the case of AIDS, education.

Centers for Disease Control (CDC)

Epidemiology is a major concern of state and federal public health departments. The Centers for Disease Control (CDC), a U.S. Public Health Service branch located in Atlanta, Georgia, is a central source of epidemiological information in the United States.

The CDC issues a publication called *Morbidity and Mortality Weekly Report (MMWR)*, which is read by microbiologists, physicians, and other hospital and public health personnel. The *MMWR* contains data on the incidence of specific notifiable diseases (morbidity) and on the deaths from these diseases (mortality); these data are usually organized by state. (Notifiable diseases are those for which physicians are required by law to report cases to the Public Health Service.) Publication articles include reports of disease outbreaks, case histories of special interest, and summaries of the status of particular diseases during a recent period. These articles often include recommendations for procedures for diagnosis, immunization, and treatment. A number of graphs and other data in this book are from *MMWR*, and some of the boxes are direct excerpts from this publication.

*   *   *

In the next chapter, we consider the mechanisms of pathogenicity. We will discuss in more detail the methods by which microorganisms enter the body and cause disease, the effects of disease on the body, and the means by which pathogens leave the body.

Study Outline

Introduction   (p. 354)
1. Pathogenic microorganisms have special properties that allow them to invade the human body or produce toxins.
2. When the microorganism overcomes the body's defenses, a state of disease results.

Pathology, Infection, and Disease   (p. 355)
1. Pathology is the scientific study of disease.
2. Pathology is concerned with the etiology (cause) of disease, pathogenesis (development), and effects of disease.
3. Infection is the invasion and growth of pathogens in the body.
4. A host is an organism that shelters and supports the growth of pathogens.
5. Disease is an abnormal state in which part or all of the body is not properly adjusted or is not capable of carrying on normal functions.

Normal Flora   (pp. 355–357)
1. Animals, including humans, are germ-free in utero.
2. Microorganisms begin colonization in and on the surface of the body soon after birth.
3. Microorganisms that establish permanent colonies inside or on the body without producing disease are called normal flora.
4. Transient flora are present for varying periods of time and then disappear.

Relationships Between Normal Flora and Host   (p. 356)
1. Normal flora can prevent pathogens from causing an infection by microbial antagonism.
2. Normal flora and humans exist in a symbiosis (living together).
3. The three types of symbiosis are commensalism (one organism benefits and the other is unaffected), mutualism (both organisms benefit), and parasitism (one organism benefits and one is harmed).

Opportunistic Organisms (pp. 356–357)

1. Opportunists (opportunistic pathogens) do not cause disease under normal conditions but cause disease under special conditions.

Etiology of Infectious Diseases (pp. 357–359)

Koch's Postulates (p. 358)

1. Koch's postulates are a method for establishing that specific microbes cause specific diseases.
2. Koch's postulates state the following requirements: (1) the same pathogen must be present in every case of the disease; (2) the pathogen must be isolated in pure culture; (3) the pathogen isolated from pure culture must cause the same disease in a healthy, susceptible laboratory animal; and (4) the pathogen must be reisolated from the inoculated laboratory animal.

Exceptions to Koch's Postulates (p. 359)

1. Koch's postulates are modified to establish etiologies of diseases caused by viruses and some bacteria, which cannot be grown on artificial media.
2. Some diseases, such as tetanus, have unequivocal signs and symptoms.
3. Some pathologies, such as pneumonia and nephritis, may be caused by a variety of microbes.
4. Some pathogens, such as *S. pyogenes*, cause several different diseases.

Classifying Infectious Diseases (pp. 359–360)

1. A patient may exhibit symptoms (subjective changes in body functions) and signs (measurable changes), which are used by a physician to make a diagnosis (identification of the disease).
2. A specific group of symptoms or signs that always accompanies a specific disease is called a syndrome.
3. Communicable diseases are transmitted directly or indirectly from one host to another.
4. Noncommunicable diseases are caused by microorganisms that normally grow outside the human body and are not transmitted from one host to another.
5. A contagious disease is one that is easily spread from one person to another.

Occurrence of Disease (p. 360)

1. Disease occurrence is reported by incidence (number of people with the disease) and prevalence (incidence at a particular time).
2. Diseases are classified by frequency of occurrence—sporadic, endemic, epidemic, and pandemic.

Severity or Duration (p. 360)

1. The scope of a disease may be defined as acute, chronic, subacute, or latent.

Extent of Host Involvement (p. 360)

1. A local infection affects a small area of the body; a systemic infection is spread throughout the body through the circulatory system.
2. A secondary infection can occur after the host is weakened from a primary infection.
3. An inapparent, or subclinical, infection does not cause any signs of disease in the host.

Spread of Infection (pp. 360–365)

Reservoirs (pp. 360–361)

1. A continual source of infection is called a reservoir of infection.
2. People who have a disease or are carriers of pathogenic microorganisms are human reservoirs of infection.
3. Zoonoses (diseases that occur in wild and domestic animals) can be transmitted to humans from animal reservoirs of infection.
4. Some pathogenic microorganisms grow in nonliving reservoirs, such as soil and water.

Transmission of Disease (pp. 361–364)

1. Transmission by direct contact involves close physical contact between the source of the disease and a susceptible host.
2. Transmission by fomites (inanimate objects), food, or water constitutes indirect contact.
3. Transmission by saliva or mucus is called droplet infection.
4. Transmission to a large number of people from the same inanimate reservoir is called common vehicle transmission.
5. Airborne transmission refers to pathogens carried on water droplets or dust for a distance greater than one meter.
6. Arthropod vectors carry pathogens from one host to another by both mechanical and biological transmission.

Portals of Exit (pp. 364–365)

1. Just as pathogens have preferred portals of entry, they also have definite portals of exit.
2. Three common portals of exit are the respiratory tract via coughing or sneezing, the gastrointestinal tract via saliva or feces, the genital tract via secretions from the vagina and penis.
3. Arthropods and syringes provide a portal of exit for microbes in blood.

Nosocomial (Hospital-Acquired) Infections (pp. 365–368)

1. Nosocomial infections are acquired during the course of stay in a hospital.
2. From 5% to 15% of all hospitalized patients acquire nosocomial infections.

Microorganisms in the Hospital   (p. 366)

1. Normal flora are often responsible for nosocomial infections when they are introduced into the body through such medical procedures as surgery and catheterization.
2. Opportunistic, drug-resistant gram-negative bacteria are the most frequent causes of nosocomial infections.

Compromised Host   (pp. 366–367)

1. Patients with burns, surgical wounds, and suppressed immune systems are most susceptible to nosocomial infections.

Chain of Transmission   (p. 367)

1. Nosocomial infections are transmitted by direct contact between staff and patient and between patients.
2. Fomites such as catheters, syringes, and respiratory aids can transmit nosocomial infections.

Control of Nosocomial Infections   (pp. 367–368)

1. Aseptic techniques can prevent nosocomial infections.
2. Hospital infection-control personnel are responsible for overseeing proper storage and handling of equipment and supplies.

Patterns of Disease   (pp. 368–369)

Predisposing Factors   (p. 369)

1. A predisposing factor is one that makes the body more susceptible to disease or alters the course of a disease.
2. Examples include climate, age, fatigue, and inadequate nutrition.

Development of Disease   (p. 369)

1. The period of incubation is the time interval between the actual infection and the first appearance of signs and symptoms.
2. The prodromal period is characterized by the appearance of the first mild signs and symptoms.
3. During the period of illness, the disease is at its height, and all disease signs and symptoms are apparent.
4. During the period of decline, the signs and symptoms subside.
5. During the period of convalescence, the body returns to its prediseased state, and health is restored.

Epidemiology   (pp. 370–372)

1. The science of epidemiology is the study of the transmission, incidence, and frequency of disease.
2. Data about infected persons are collected and analyzed in descriptive epidemiology.
3. In analytical epidemiology, a group of infected persons is compared with an uninfected group.
4. Controlled experiments designed to test hypotheses are performed in experimental epidemiology.
5. Case reporting provides data on incidence and prevalence to local, state, and national health officials.
6. The Centers for Disease Control (CDC) is the main source of epidemiologic information in the United States.
7. The CDC publishes *Morbidity and Mortality Weekly Report* to provide information on morbidity (incidence) and deaths (mortality).

Study Questions

Review

1. Differentiate between the following pairs of terms.
   (a) Etiology and pathogenesis
   (b) Infection and disease
   (c) Communicable disease and noncommunicable disease

2. What is meant by normal flora? How do they differ from transient flora?

3. Define symbiosis. Differentiate between commensalism, mutualism, and parasitism, and give an example of each.

4. What is a reservoir of infection? Match the following diseases with their reservoirs.
   ____ Influenza   (a) Nonliving
   ____ Rabies   (b) Human
   ____ Botulism   (c) Animal

5. Describe how Koch's postulates establish the etiology of many infectious diseases. Why don't Koch's postulates apply to all infectious diseases?

6. Describe the various ways diseases can be transmitted in each of the following categories.
   (a) Transmission by direct contact
   (b) Transmission by indirect contact
   (c) Transmission by arthropod vectors
   (d) Droplet transmission
   (e) Common vehicle transmission
   (f) Airborne transmission

7. Indicate whether each of the conditions described is typical of subacute, chronic, or acute infections.
   (a) Patient experiences rapid onset of malaise; symptoms last five days.
   (b) Patient experiences cough and breathing difficulty for months.
   (c) Patient has no apparent symptoms and is a known carrier.

8. Of all the hospital patients with infections, one third do not enter the hospital with an infection. How do they acquire these infections? What is the method of transmission of these infections? What is the reservoir of infection?

9. Differentiate between an endemic and epidemic state of infectious disease.

10. What is epidemiology? What is the role of the Centers for Disease Control (CDC)?

11. Distinguish between symptoms and signs as signals of disease.

12. How can a local infection become a systemic infection?

13. Why are some organisms that constitute normal flora described as commensals, whereas others are described as mutualistic?

14. Put the following in the correct order to describe the pattern of disease: period of convalescence, crisis, prodromal period, period of decline, period of incubation, period of illness.

15. List four predisposing factors to disease.

Challenge

1. Ten years before Robert Koch published his work on anthrax, Anton De Bary showed that potato blight was caused by the fungus *Phytophthora infestans.* Why do you suppose we use Koch's postulates instead of something called "De Bary's postulates"?

2. Florence Nightingale gathered the following data in 1855.

Population sampled	Deaths due to contagious diseases
Englishmen	0.2%
English soldiers (in England)	18.7%
English soldiers (in Crimean War)	42.7%
English soldiers (in Crimean War) after Nightingale's sanitary reforms	2.2%

Discuss how Nightingale employed the three basic types of epidemiologic investigation. The contagious diseases were primarily cholera and typhus; how are these diseases transmitted and prevented?

3. Name the method of transmission of the following diseases.
   (a) Malaria
   (b) Tuberculosis
   (c) Nosocomial infections
   (d) Salmonellosis
   (e) Streptococcal pharyngitis
   (f) Mononucleosis
   (g) Measles
   (h) Hepatitis A
   (i) Tetanus
   (j) Hepatitis B
   (k) Chlamydial urethritis

4. This graph shows the incidence of typhoid fever in the United States from 1955 to 1987. Mark the graph to show when this disease occurred sporadically and epidemically. What appears to be the endemic level? What would have to be shown to indicate a pandemic of this disease? How is typhoid fever transmitted?

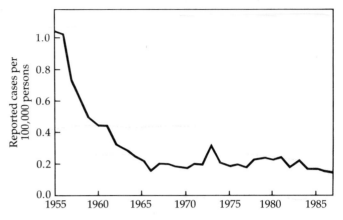

5. Three patients in a large hospital acquired infections of *Pseudomonas cepacia* during their stay. All three patients received cryoprecipitate, which is prepared from blood that has been frozen in a standard plastic blood transfer pack. The transfer pack is then placed in a water bath to thaw. What is the probable origin of the nosocomial infections? What characteristics of *Pseudomonas* would allow it to be involved in this type of infection?

6. Three days before a nurse developed meningococcemia, she assisted with intubation of a patient with an *N. meningitidis* infection. Of the 24 medical personnel involved, only this nurse became ill. The nurse recalled that she had exposure to nasopharyngeal secretions and did not receive antibiotic prophylaxis. What two mistakes did the nurse make? How is meningitis transmitted?

7. The following is a case history of a 49-year-old man. Identify each period in the pattern of disease that he experienced. On February 7, he handled a parakeet with a respiratory illness. On March 9, he experienced intense pain in the legs, followed by severe chills and headaches. On March 16, he had chest pains, cough, and diarrhea, and his temperature was 40°C. Appropriate antibiotics were administered on March 17, and his fever subsided within 12 hours. He continued taking antibiotics for 14 days. (*Note*: The disease is psittacosis. Can you find the etiology?)

Further Reading

Ayliffe, G. A., and L. J. Taylor. "Hospital-acquired infections: principles and prevention." Littleton, Mass: John Wright-PSG, 1982. Discusses nosocomial infections and how to prevent them.

Chesney, P. J., M. S. Bergdoll, and J. P. Davis. "The disease spectrum, epidemiology, and etiology of toxic-shock syndrome." *Annual Review of Microbiology* 38:315–338, 1984. Includes the history, pathogenesis, and method of action of toxic-shock toxin.

Epidemiology Issue. *Science* 234:921, 951–979, November 1986. Three articles (on AIDS, salmonellosis, leishmaniasis, and malaria) about the tools and techniques used to identify and control an epidemic.

Mandell, G. L., R. G. Douglas, and J. E. Bennett, eds. "Nosocomial infections." In *Principles and Practices of Infectious Disease*, 2nd ed. New York: Wiley, 1985. Discusses specific nosocomial infections, such as pneumonia.

Mausner, J. S. and S. Kramer. *Mausner and Bahn Epidemiology*, 2nd ed. Philadelphia: Saunders, 1984. A comprehensive reference on the collection and interpretation of epidemiological data.

McEvedy, C. "The bubonic plague." *Scientific American* 258(2):118–123, February 1988. A historical account of plague with hypotheses about the occurrence of epidemics.

Murray, J. D. "Modeling the spread of rabies." *American Scientist* 75:280–284; May–June 1987. Provides a way to predict an epidemic while addressing the current European rabies epidemic.

Riley, L. W. "The epidemiologic, clinical, and microbiologic features of hemorrhagic colitis." *Annual Review of Microbiology* 41:383–407, 1987. A complete epidemiologic report on the unique pathogen *E. coli* 0157:H7; includes descriptive epidemiology and case-control studies.

Roueche, B. *The Medical Detectives*, 2 vols. New York: Truman Talley Books. Vol. 1 by Times Books, Volume 2 by Washington Square Press, 1984. True accounts describing the scientific investigations of epidemic diseases. Written so the reader identifies the etiology with the epidemiologic investigator.

CHAPTER 14

Mechanisms of Pathogenicity

Now that you have a basic understanding of how microorganisms cause disease, we will take a look at some of the specific properties of microorganisms that contribute to pathogenicity, the ability to cause disease in a host, and **virulence**, the degree of pathogenicity. Keep in mind that many of the properties contributing to microbial pathogenicity and virulence are unclear or unknown. We do know, however, that if the microbes' attack overpowers the host defenses, disease results.

Entry of a Microorganism into the Host

Most pathogens must gain access to the host, adhere to host tissues, penetrate host defenses, and damage the host tissue to cause disease. Some microbes, such as those that cause dental caries and acne, can cause disease without penetrating the body. Pathogens can gain entrance to the human body and other hosts through several avenues, which we call **portals of entry**.

Mucous Membranes

Many bacteria and viruses gain access to the body by penetrating mucous membranes lining the respiratory tract, gastrointestinal tract, genitourinary tract, and conjunctiva, a delicate membrane that covers the eyeball and lines the eyelids. Most pathogens enter through the mucous membranes of the gastrointestinal and respiratory tracts. The respiratory tract is the easiest and most frequently traveled portal of entry for infectious microorganisms. Microbes are inhaled into the nose or mouth in drops of moisture

Learning Objectives

- Define portal of entry, pathogenicity, virulence, and LD_{50}.

- Explain how adherence, capsules, cell-wall components, and enzymes contribute to pathogenicity.

- Compare the effects of hemolysins, leukocidins, coagulase, kinases, hyaluronidase, and collagenase.

- Contrast the nature and effects of exotoxins and endotoxins.

- Outline the mechanisms of action of diphtherotoxin, botulinum toxin, tetanus toxin, choleragen, and lipid A.

- List seven cytopathic effects of viral infections.

- Discuss the causes of symptoms in fungal, protozoan, and helminthic diseases.

and dust particles. Some diseases that are commonly contracted via the respiratory tract include the common cold, pneumonia, tuberculosis, influenza, measles, and smallpox.

Microorganisms can gain access to the gastrointestinal tract in food and water. Bringing contaminated fingers to the mouth is another common mode of entry. Most microbes that enter the body in these ways are destroyed by hydrochloric acid (HCl) and enzymes in the stomach or by bile and enzymes in the small intestine. Those that survive can cause disease. Microbes in the gastrointestinal tract can cause poliomyelitis, hepatitis A, typhoid fever, amoebic dysentery, bacillary dysentery (shigellosis), and cholera. These pathogens are then eliminated with feces and can be transmitted to other hosts via water, food, or contaminated fingers.

An important pathogen capable of penetrating mucous membranes of the genitourinary tract is *Treponema pallidum*, the causative agent of syphilis.

Skin

Although most microorganisms cannot penetrate unbroken skin, the skin is sometimes a vulnerable barrier. Some microbes gain access to the body through openings in the skin, such as hair follicles and sweat gland ducts. It has been demonstrated that the hookworm *Necator americanus* (ne-kā′tôr ä-me-ri-ka′nus) actually bores through intact skin, and some fungi grow on the keratin in skin or infect the skin itself.

Parenteral Route

Other microorganisms gain access to the body when they are deposited directly into the tissues beneath the skin or into mucous membranes when these barriers are penetrated or injured. This route is called the **parenteral route**. Punctures, injections, bites, cuts, wounds, surgery, and splitting due to swelling or drying can all establish parenteral routes.

Preferred Portal of Entry

Even after microorganisms have entered the body, they do not necessarily cause disease. Occurrence of disease depends on several factors, only one of which is the portal of entry. Many pathogens have a preferred portal of entry, and using that portal is a prerequisite to their being able to cause disease. If they

gain access to the body by another portal, disease might not occur. For example, the bacteria of typhoid fever, *Salmonella typhi* (sal-mōn-el′lä tī′fē) produce all the signs and symptoms of the disease when swallowed (preferred route), but if the same bacteria are rubbed on the skin no reaction (or only a slight inflammation) occurs. Streptococci that are inhaled (preferred route) can cause pneumonia; those that are swallowed generally do not produce signs or symptoms. Some pathogens, such as the microorganism that causes plague, can initiate disease from more than one portal of entry. The portals of entry for some common pathogens are given in Table 14-1, pp. 380–381.

Numbers of Invading Microbes

If only a few microbes enter the body, it is likely that they will be overcome by the host's defenses. However, if large numbers of microbes gain entry, then the stage is probably set for disease. Thus, the likelihood of disease increases as the number of pathogens increases.

The virulence of a microbe or the potency of its toxin is often expressed as the LD_{50} (**lethal dose**), the number of microbes in a dose that will kill 50% of inoculated test animals under normal conditions. The dose required to produce a demonstrable infection in 50% of the test animals is called the ID_{50} (**infectious dose**).

Adherence

Once pathogens gain entry to a host, almost all of them have some means of attaching themselves to host tissues. For most pathogens, this attachment, called **adherence**, is a necessary step in pathogenicity. (Of course, nonpathogens also have structures for attachment.) The attachment between pathogen and host takes place by means of surface molecules on the pathogen called **adhesins** or **ligands** that bind specifically to complementary surface receptors on the cells of certain host tissues (Figure 14-1). (Nonpathogenic microbes that colonize the body also have adhesins. The presence or absence of complementary cell receptors in various host tissues determines which tissues are colonized.) Adhesins may be located on a microbe's glycocalyx or on other microbial surface structures, such as pili (see Chapter 4).

The majority of adhesins on the microorganisms we have studied so far are glycoproteins or lipoproteins. The receptors on host cells are typically sugars, such as mannose. Adhesins on different strains of the

(a)

(b)

Figure 14-1 **Adherence. (a)** Selective attachment of a pathogenic strain of *E. coli* to the intestinal tissue of a rabbit (×17,000). **(b)** Scanning electron micrograph of bacteria adhering to the skin of a salamander (×1260).

same species of pathogen can vary in structure. Different cells of the same host can also have different receptors that vary in structure. If adhesins, receptors, or both can be altered to interfere with adherence, it is possible to prevent or control infection.

The following examples illustrate the diversity of adhesins. *Streptococcus mutans*, a bacterium that plays a key role in tooth decay, attaches to the surface of teeth by its glycocalyx. An enzyme (*dextransucrase*) produced by *S. mutans* converts glucose into a sticky polysaccharide called dextran that forms the glycocalyx and facilitates the organism's attachment to teeth. Enteropathogenic strains of *E. coli* (those

responsible for gastrointestinal disease) have adhesins on pili that adhere only to specific kinds of cells in certain regions of the small intestine. *Neisseria gonorrhoeae* (go-nôr-rē'ī), the causative agent of gonorrhea, also has pili-containing adhesins, which in this case permit attachment to cells with appropriate receptors in the genitourinary tract, eye, and pharynx. *Staphylococcus aureus*, which can cause skin infections, binds to skin with a mechanism of adherence that resembles viral attachment (discussed in Chapter 12).

How Pathogens Penetrate Host Defenses

Although some pathogens can cause damage on the surface of tissues, most must penetrate tissues to cause disease. Here we will consider several factors that contribute to a microorganism's ability to invade a host.

Capsules

We noted in Chapter 4 that some bacteria make glycocalyx material that forms capsules around their cell walls (Figure 14-2); this property increases the vir-

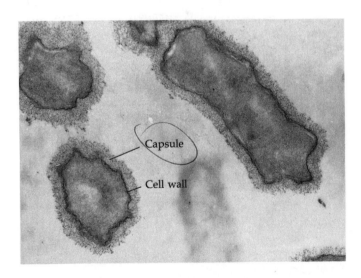

Capsule

Cell wall

Figure 14-2 **A bacterial capsule.** Electron micrograph of a thin section through cells of an encapsulated strain of *Klebsiella aerogenes* (×23,000). The capsule appears as a thick gray layer of material outside the dark cell wall. Capsules play an important role in the resistance of some pathogenic bacteria to phagocytosis.

ulence of the species. The capsule resists the host's defenses by impairing phagocytosis, a process by which certain cells of the body engulf and destroy microbes (to be discussed in Chapter 15). The chemical nature of the capsule appears to prevent the phagocytic cell from adhering to the bacterium. However, the human body is capable of producing antibodies against the capsule, and when these antibodies are present on the capsule surface, the encapsulated bacteria are easily destroyed by phagocytosis.

One bacterium that owes its virulence to the presence of a polysaccharide capsule is *Streptococcus pneumoniae*, the causative agent of lobar pneumonia (see Figure 4-5). Some strains of this organism have cap-

sules, and others do not. Strains with capsules are virulent, but strains without capsules are avirulent because they are susceptible to phagocytosis. Other bacteria that produce capsules related to virulence are *Klebsiella pneumoniae* (kleb-sē-el′lä nü-mō′nē-ī), a causative agent of bacterial pneumonia; *Hemophilus influenzae* (hē-mä′fi-lus in-flü-en′zī), a cause of pneumonia and meningitis in children; *Bacillus anthracis*, the cause of anthrax; and *Yersinia pestis*, the causative agent of bubonic plague. Keep in mind that capsules are not the only cause of virulence. Many nonpathogenic bacteria produce capsules, and the virulence of some pathogens is not related to the presence of a capsule.

TABLE 14-1 *Causative Agents for Some Common Diseases, Arranged by Portal of Entry*

Portal of Entry	Causative Agent*	Disease	Incubation Period
Respiratory tract	*Corynebacterium diphtheriae*	Diphtheria	2–5 days
	Neisseria meningitidis	Meningococcal meningitis	1–7 days
	Streptococcus pneumoniae	Pneumococcal pneumonia	Variable
	*Mycobacterium tuberculosis***	Tuberculosis	Variable
	Bordetella pertussis	Whooping cough (pertussis)	12–20 days
	Myxovirus	Influenza	18–36 hours
	Paramyxovirus	Measles (rubeola)	11–14 days
	Togavirus	German measles (rubella)	2–3 weeks
	Epstein–Barr virus (herpesvirus)	Infectious mononucleosis	2–6 weeks
	Varicella–Zoster virus (herpesvirus)	Chickenpox (varicella)	14–16 days
	Poxvirus	Smallpox (variola)	12 days
	Coccidioides immitis (fungus)	Coccidioidomycosis (primary infection)	1–3 weeks
	Histoplasma capsulatum (fungus)	Histoplasmosis	5–18 days
Gastrointestinal tract	*Shigella* species	Bacillary dysentery (shigellosis)	1–2 days
	Brucella melitensis	Brucellosis (undulant fever)	6–14 days
	Vibrio cholerae	Cholera	1–3 days
	Salmonella enteritidis, Salmonella typhimurium, Salmonella choleraesuis	Salmonellosis	7–22 hours
	Salmonella typhi	Paratyphoid fever	7–24 days

Components of the Cell Wall

The cell walls of certain bacteria contain chemical substances that contribute to virulence. For example, *Streptococcus pyogenes* contains a heat-resistant and acid-resistant protein called **M protein**. This protein is found on both the fimbriae and the cell surface of *S. pyogenes*. The M protein mediates attachment of the bacterium to epithelial cells of the host and helps the bacterium resist phagocytosis by white blood cells. The protein thereby increases the virulence of the microorganism. Immunity to *S. pyogenes* depends on the body's production of an antibody specific to M protein.

The waxes that make up the cell wall of *Mycobacterium tuberculosis* also increase virulence by resisting phagocytosis. In fact, *M. tuberculosis* can even multiply inside phagocytes.

Enzymes

The virulence of some bacteria is thought to be aided by the production of extracellular enzymes (exoenzymes) and related substances. These chemicals can break cells open, dissolve materials between cells, and form or dissolve blood clots, among other functions.

TABLE 14-1 *(Continued)*

Portal of Entry	Causative Agent*	Disease	Incubation Period
	Salmonella typhi	Typhoid fever	5–14 days
	Hepatitis A virus (picornavirus)	Hepatitis A	15–50 days
	Paramyxovirus	Mumps	2–3 weeks
	Poliovirus	Poliomyelitis	4–7 days
	Trichinella spiralis (helminth)	Trichinosis	2–28 days
Genitourinary tract	*Neisseria gonorrhoeae*	Gonorrhea	3–8 days
	Treponema pallidum	Syphilis	9–90 days
Skin or parenteral route	*Clostridium perfringens*	Gas gangrene	1–5 days
	Clostridium tetani	Tetanus	3–21 days
	Leptospira interrogans	Leptospirosis	2–20 days
	Yersinia pestis	Plague	2–6 days
	Rickettsia rickettsii	Rocky Mountain spotted fever	3–12 days
	Hepatitis B virus**	Hepatitis B	6 weeks–6 months
	Rhabdovirus	Rabies	10 days–1 year
	Togavirus	Yellow fever	3–6 days
	Plasmodium species (protozoan)	Malaria	2 weeks

*All causative agents are bacteria, unless indicated otherwise. For viruses, only the viral group is given, except where the virus has a name different from the disease it causes.
**These pathogens can also cause disease after entering the body via the gastrointestinal tract.

However, the importance of some of these enzymes to bacterial virulence has not been proved conclusively.

Substances called **leukocidins**, produced by some bacteria, can destroy neutrophils, a type of white blood cell (leukocyte) that is very active in phagocytosis. Leukocidins are also active against macrophages (phagocytic cells) present in tissue. Among the bacteria that secrete leukocidins are staphylococci and streptococci. Leukocidins produced by streptococci degrade lysosomes within leukocytes and thereby cause the death of the white blood cell (see the box in Chapter 15, p. 401). Hydrolytic enzymes released from the leukocytic lysosomes can damage other cellular structures and thus intensify streptococcal lesions. This type of damage to white blood cells decreases host resistance.

Hemolysins are another group of enzymes that are produced by bacteria and might contribute to their virulence. Hemolysins cause the lysis of red blood cells. Bacteria produce a number of hemolysins that differ in their ability to lyse different kinds of red blood cells (in humans, sheep, and rabbits, for example) and the type of lysis they cause. Important products of hemolysins are staphylococci, streptococci, and *Clostridium perfringens* (klô-stri′dē-um pėr-frin′ jens), the most common causative agent of gas gangrene.

Coagulases are bacterial enzymes that coagulate (clot) the fibrinogen in blood. Fibrinogen, a plasma protein produced by the liver, is converted by coagulases into fibrin, the threads that form a blood clot. The fibrin clot caused by coagulases may protect the bacterium from phagocytosis and isolate it from other defenses of the host. Coagulases are produced by some members of the genus *Staphylococcus*; they may be involved in the walling-off process in boils produced by staphylococci. However, some staphylococci that do not produce coagulases are still virulent. A capsule, and not coagulases, may be the more important factor in the virulence of these bacteria.

Bacterial kinases are another group of enzymes that may contribute to bacterial virulence. By breaking down fibrin, the kinases dissolve clots formed by the body to isolate the infection. One of the better-known kinases is **streptokinase (fibrinolysin)**, which is produced by such streptococci as *Streptococcus pyogenes*. Another kinase, **staphylokinase**, is produced by such staphylococci as *Staphylococcus aureus*. Injected directly into the blood, streptokinase has been used successfully to dissolve some types of blood clots in cases of heart attacks due to obstructed coronary arteries.

Hyaluronidase is yet another enzyme secreted by certain bacteria, such as streptococci, and possibly related to microbial virulence. It dissolves hyaluronic acid, a type of polysaccharide that holds together certain cells of the body, particularly cells in connective tissue. It is believed that this dissolving action is involved in the tissue blackening of infected wounds and helps the microorganism spread from its initial site of infection. Hyaluronidase is also produced by some clostridia that cause gas gangrene. For therapeutic use, hyaluronidase may be mixed with a drug to promote the spread of the drug through a body tissue. Another enzyme, **collagenase**, produced by several species of *Clostridium*, facilitates the spread of gas gangrene. Collagenase breaks down the protein collagen, which forms the connective tissue of muscles and other body organs and tissues.

Other bacterial substances believed to contribute to virulence are *necrotizing factors*, which cause the death of body cells; *hypothermic factors*, which decrease body temperature; *DNase* and *streptodornase*, which improve bacterial motility by reducing the viscosity of purulent (pus) discharges; *lecithinase*, which destroys plasma membrane, especially around red blood cells; and *protease*, which breaks down proteins, especially in muscle tissue.

Damage to Host Cells

When a microorganism invades a body tissue, it initially encounters phagocytes of the host. If the phagocytes are successful in destroying the invader, no further damage is done to the host. But if the pathogen overcomes the host defense, then the microorganism can damage host cells in three basic ways: (1) by causing direct damage in the *immediate vicinity* of the invasion; (2) by producing toxins, transported by blood and lymph, that damage sites *far removed* from the original site of invasion; and (3) by inducing hypersensitivity reactions. This latter mechanism is considered in detail in Chapter 18. For now, we will discuss only the first two mechanisms.

Direct Damage

Once a pathogen is attached to host cells, it can pass through them to invade other tissues. During this invasion, the pathogens metabolize and multiply to kill host cells. Some bacteria, such as *E. coli*, *Shigella* (shi-gel′lä), *Salmonella*, and *Neisseria gonorrhoeae*, can induce host epithelial cells to engulf them by a process that resembles phagocytosis. These pathogens can also be extruded from the host cells by a reverse phagocytosis process in order to enter other host cells. Some bacteria can also penetrate host cells by excreting enzymes and by their own motility; such penetration can itself damage the host cell. Most damage by bacteria, however, is done by toxins.

Toxins

Toxins are poisonous substances that are produced by certain microorganisms. They are frequently the primary factor that contributes to the pathogenic properties of those microbes. The capacity of microorganisms to produce toxins is called **toxigenicity**. Toxins transported by the blood or lymph can cause serious, and sometimes fatal, effects. Some toxins produce fever, cardiovascular disturbances, diarrhea, and shock. Toxins can also inhibit protein synthesis, destroy blood cells and blood vessels, and disrupt the nervous system by causing spasms. Of the 220 known bacterial toxins, nearly 40% cause disease by damaging eucaryotic cell membranes. The term **toxemia** refers to symptoms caused by toxins in the blood. Toxins are of two types, exotoxins and endotoxins.

Exotoxins

Exotoxins are produced inside some bacteria as part of their growth and metabolism and are released into the surrounding medium. Exotoxins are proteins, and many are enzymes that catalyze only certain biochemical reactions. Most bacteria that produce exotoxins are gram-positive. The genes for most of them (perhaps all) are carried on bacterial plasmids or phages. Because exotoxins are soluble in body fluids, they can easily diffuse into the blood and are rapidly transported throughout the body.

Exotoxins work by destroying particular parts of the host's cells or by inhibiting certain metabolic functions. Exotoxins are among the most lethal substances known. Only 1 mg of the exotoxin botulinum is enough to kill 1 million guinea pigs. Fortunately, only a few bacteria produce such potent exotoxins.

Diseases caused by bacteria that produce exotoxins are often caused by minute amounts of exotoxins, not by the bacteria themselves. It is the exotoxins that produce the specific signs and symptoms of the disease. Thus, exotoxins are disease-specific. For example, the infection of a wound by *Clostridium tetani* need be no larger or more painful than a pin prick, yet the organisms in a wound that size can produce enough tetanus toxin to kill an unvaccinated human.

The body produces antibodies called **antitoxins** that provide immunity to exotoxins. When exotoxins are inactivated by heat, formaldehyde, iodine, or other chemicals, they no longer cause the disease but are still able to stimulate the body to produce antitoxins. Such altered exotoxins are called **toxoids**. When toxoids are injected into the body as a vaccine, they stimulate antitoxin production so that immunity is produced to disease. Diphtheria and tetanus are common diseases that can be prevented by toxoid vaccination.

Here we will briefly describe a few of the more notable exotoxins (antitoxins will be discussed further in Chapter 17). Exotoxins may be grouped into three principal types, based on their mode of action: (1) **cytotoxins**, which kill host cells or affect their functions; (2) **neurotoxins**, which interfere with normal nerve impulse transmission; and (3) **enterotoxins**, which affect cells lining the gastrointestinal tract.

Diphtheria toxin　*Corynebacterium diphtheriae* (kô-rī-nē-bak-ti'rē-um dif-thi'rē-ī) produces the diphtheria toxin (diphtherotoxin) only when it is infected by a lysogenic phage carrying the *tox* gene. This cytotoxin inhibits protein synthesis in eucaryotic cells. The mechanism by which it does this is an excellent example of how an exotoxin interacts with host cells (Figure 14-3). Diphtherotoxin is a protein consisting of two different polypeptides, designated A (active) and B (binding). (Recall from Chapter 2 that polypeptides are long, chainlike molecules of amino acids.) Although only polypeptide A is the active component that causes symptoms in the host, polypeptide B is required for A to be active. Polypeptide B binds to surface receptors on the host cell and causes the transport of the entire protein across the plasma membrane into the cell. Once this is accomplished, polypeptide A inhibits protein synthesis within the target cell. (This model also applies to the cholera exotoxin, which is described shortly.)

Erythrogenic toxins　*Streptococcus pyogenes* has the genetic material to synthesize three types of cytotoxins, designated as A, B, and C. These erythrogenic toxins damage blood capillaries under the skin and produce a red skin rash. Scarlet fever, produced by *S. pyogenes*, is named for this characteristic rash.

Botulinum toxin　Botulinum toxin, produced by *Clostridium botulinum*, is not a typical exotoxin. It is made within the clostridial cell and is not released into the medium until cell death occurs. Botulinum

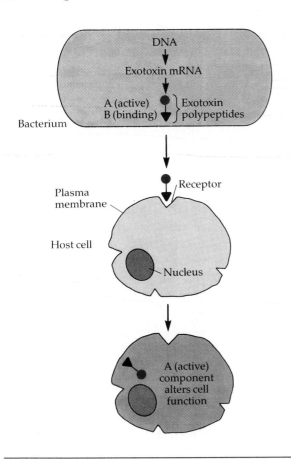

Figure 14-3   Exotoxin. Proposed model for mechanism of action of diphtherotoxin.

toxin is a neurotoxin. It acts at the neuromuscular junction (the junction between nerve cell and muscle cell) and prevents the transmission of impulses from the nerve cell to muscle. The toxin accomplishes this by binding to the nerve cell and inhibiting the release of a neurotransmitter called acetylcholine. As a result, botulinum toxin causes paralysis in which muscle tone is lacking (flaccid paralysis). *C. botulinum* produces eight different types of botulinum toxin, and each possesses a different potency.

Tetanus toxin   *Clostridium tetani* produces tetanus neurotoxin, also known as tetanospasmin. This toxin reaches the central nervous system and binds to nerve cells that control contraction of various skeletal muscles. These nerve cells normally send inhibiting impulses that prevent random contractions and terminate completed contractions. The binding of tetanospasmin blocks the inhibitory nerve impulses to one skeletal muscle while the opposing muscle is contracting. The result is uncontrollable muscle contrac-

tion, producing the convulsive symptoms (spasmodic contractions) of tetanus, or "lockjaw."

Vibrio enterotoxin   *Vibrio cholerae* (vib′rē-ō kol′-ėr-ī) produces an enterotoxin called **choleragen.** Like diphtherotoxin, choleragen consists of two polypeptides, A (active) and B (binding). The B component binds to plasma membranes of epithelial cells lining the small intestine, and the A component induces the formation of cyclic AMP from ATP in the cytoplasm (Chapter 8). As a result, epithelial cells discharge large amounts of fluids and electrolytes (ions). Normal muscular contractions are disturbed, leading to severe diarrhea that may be accompanied by vomiting. **Heat-labile enterotoxin** (so called because it is more sensitive to heat than most toxins) produced by some strains of *E. coli* has an action identical to that of choleragen.

Staphylococcal enterotoxin   *Staphylococcus aureus* produces an enterotoxin that affects the intestines in the same way as choleragen. A strain of *S. aureus* also produces enterotoxins that result in the symptoms associated with toxic shock syndrome, discussed in Chapter 1.

A summary of diseases produced by exotoxins is presented in Table 14-2.

Endotoxins

Endotoxins differ from exotoxins in several ways. Endotoxins are part of the outer portion of the cell wall of most gram-negative bacteria. Recall from Chapter 4 that gram-negative bacteria have an outer membrane surrounding the peptidoglycan layer of the cell wall. This outer membrane consists of lipoproteins, phospholipids, and lipopolysaccharides, LPS (see Figure 4-11). The lipid portion of LPS, called *lipid A*, is the endotoxin. Thus, endotoxins are lipopolysaccharides, whereas exotoxins are proteins.

Endotoxins exert their effects when gram-negative bacteria die and their cell walls undergo lysis, thus liberating the endotoxin. Antibiotics used to treat diseases caused by gram-negative bacteria can lyse the bacterial cells; this reaction releases endotoxin and may lead to an immediate worsening of the symptoms, but the condition improves as the antibiotic begins to exert its effect. All endotoxins produce the same signs and symptoms, regardless of the species of microorganism, although not to the same degree. Responses by the host include fever, weakness, generalized aches, and, in some cases, shock. Endotoxins can also induce miscarriage and prevent blood from clotting.

TABLE 14-2 *Diseases Caused by Exotoxins*

Disease	Bacterium	Mechanism
Botulism	*Clostridium botulinum*	Neurotoxin prevents transmission of nerve impulses; flaccid paralysis results
Tetanus	*Clostridium tetani*	Neurotoxin blocks nerve impulse that permits relaxation of one skeletal muscle while the opposing muscle is contracting
Gas gangrene and food poisoning	*Clostridium perfringens* and other species of *Clostridium*	One exotoxin (cytotoxin) causes massive red blood cell destruction (hemolysis); another exotoxin (enterotoxin) is related to food poisoning and causes diarrhea
Diphtheria	*Corynebacterium diphtheriae*	Cytotoxin inhibits protein synthesis, especially in nerve, heart, and kidney cells
Scalded skin syndrome, food poisoning, and toxic shock syndrome (TSS)	*Staphylococcus aureus*	One exotoxin causes skin layers to separate and slough off (scalded skin); another exotoxin (enterotoxin) produces diarrhea and vomiting; still another exotoxin produces symptoms associated with TSS
Cholera	*Vibrio cholerae*	Enterotoxin induces diarrhea
Scarlet fever	*Streptococcus pyogenes*	Cytotoxins cause vasodilation that results in the characteristic rash
Traveler's diarrhea	Enterotoxigenic *Escherichia coli* and *Campylobacter jejuni*	Enterotoxin causes excessive secretion of ions and water; diarrhea results

The fever (pyrogenic response) caused by endotoxins is believed to occur as depicted in Figure 14-4. When gram-negative bacteria are ingested by phagocytes and degraded in vacuoles, the lipopolysaccharides (LPS) of the bacterial cell wall are released. The lipopolysaccharides cause macrophages to produce small protein molecules called **interleukin-1 (IL-1)**, formerly called **endogenous pyrogen**, which are carried via the blood to the hypothalamus,

a temperature control center in the brain. IL-1 induces the hypothalamus to release lipids called prostaglandins that reset the thermostat in the hypothalamus at a higher temperature. The result is a fever. Bacterial cell death caused by lysis or antibiotics can also produce fever by this mechanism. Both aspirin and acetaminophen reduce fever by inhibiting the synthesis of prostaglandins. The function of fever in the body is discussed in Chapter 15.

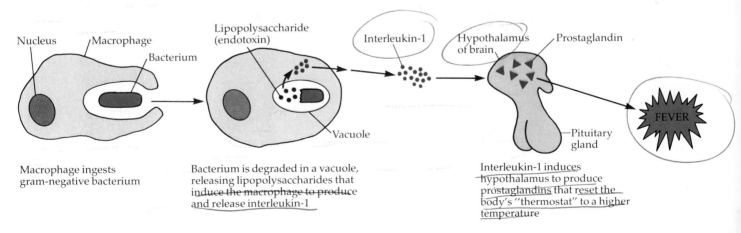

Macrophage ingests gram-negative bacterium

Bacterium is degraded in a vacuole, releasing lipopolysaccharides that induce the macrophage to produce and release interleukin-1

Interleukin-1 induces hypothalamus to produce prostaglandins that reset the body's "thermostat" to a higher temperature

Figure 14-4 Pyrogenic response. The postulated mechanism by which endotoxins cause fever.

Shock caused by gram-negative bacteria is called septic or endotoxin shock. The shock produced by endotoxins, like fever, is related to secretion of a substance by macrophages. Phagocytosis of gram-negative bacteria causes the phagocytes to secrete a polypeptide called **tumor necrotizing factor (TNF)**, or **cachectin**. This substance binds to many tissues in the body and alters their metabolism in a number of ways. One effect of TNF is damage to blood capillaries; their permeability is increased, and they lose large amounts of fluid. The result is a drop in blood pressure that results in shock. Low blood pressure has serious effects on the kidneys, lungs, and gastrointestinal tract. TNF also alters fat metabolism so that blood glucose level increases (hyperglycemia).

Endotoxins do not promote the formation of effective antitoxins. Antibodies are produced, but they tend not to counter the effect of the toxin; sometimes, in fact, they actually enhance its effect.

Representative microorganisms that produce endotoxins are *Salmonella typhi* (the causative agent of typhoid fever), *Proteus* species (frequently the causative agents of urinary tract infections), and *Neisseria meningitidis* (the causative agent of meningococcal meningitis). A comparison of exotoxins and endotoxins appears in Table 14-3.

Plasmids, Lysogeny, and Pathogenicity

Recall from Chapters 4 and 8 that plasmids are small circular DNA molecules that are not connected to the main bacterial chromosome and are capable of autonomous replication. One group of plasmids, called resistance (R) factors, is responsible for the resistance of some microorganisms to antibiotics. In addition, a plasmid may carry the information that determines a microbe's pathogenicity. Some examples of virulence factors encoded by plasmid genes include tetanospasmin, the neurotoxin produced by *Clostridium tetani*; heat-labile enterotoxin, produced by certain strains of *E. coli*; and staphylococcal enterotoxin, produced by *Staphylococcus aureus*. Other examples include dextransucrase, an enzyme produced by *Strep-*

TABLE 14-3 *Comparison of Exotoxins and Endotoxins*

Property	Exotoxin	Endotoxin
Bacterial source	Mostly from gram-positive bacteria	Almost exclusively from gram-negative bacteria
Relation to microorganism	Metabolic product of growing cell	Present in lipopolysaccharide (LPS) of outer membrane of cell wall and released only with destruction of cell
Chemistry	Protein or short peptide	Lipid portion (lipid A) of LPS of outer membrane
Heat stability	Unstable; can usually be destroyed at 60° to 80°C (except staphylococcal enterotoxin)	Stable; can withstand autoclaving (121°C for one hour)
Toxicity (power to cause disease)	High	Low
Immunology (relation to antibodies)	Can be converted to toxoids; neutralized by antitoxin	Not easily neutralized by antitoxin; therefore, effective toxoids cannot be made
Pharmacology (effect on body)	Specific for a particular cell structure or function in the host (mainly affects cell functions, nerves, and gastrointestinal tract)	General, such as fever, weaknesses, aches, and shock; all produce the same effects
Lethal dose	Small	Considerably larger
Representative diseases	Gas gangrene, tetanus, botulism, diphtheria, scarlet fever	Salmonellosis, epidemic meningitis, and tularemia

tococcus mutans that is involved in tooth decay; adhesins and coagulase produced by *Staphylococcus aureus*; and a type of pilus specific to enteropathogenic strains of *E. coli*.

In Chapter 12, we noted that some bacteriophages (viruses that infect bacteria) can incorporate their DNA into the bacterial chromosome and in that way remain latent and not cause lysis of the bacterium. Such a state is called lysogeny, and the bacteriophage DNA that has been incorporated into the chromosome of its host bacterial cell is called a prophage. One outcome of lysogeny is that the host bacterial cell and its progeny may exhibit new properties coded for by the bacteriophage DNA. Among the bacteriophage genes that contribute to pathogenicity are the genes for diphtherotoxin, a cytotoxin produced by *Corynebacterium diphtheriae*; erythrogenic toxins, produced by strains of *Streptococcus pyogenes*; staphylococcal enterotoxin and pyrogenic toxin, produced by *Staphylococcus aureus*; botulinum neurotoxin, produced by *Clostridium botulinum*; and the capsule produced by *Streptococcus pneumoniae*.

Pathogenic Properties of Nonbacterial Microorganisms

Viruses

Viruses use the host cell's metabolism to interfere with normal cell reactions or to produce substances that can cause observable changes in infected cells. These observable changes, called **cytopathic effects (CPE)**, vary with the virus. One difference is the point in the viral infection cycle at which the effects occur. Some viral infections result in early changes in the host cell; in other infections, changes are not seen until a much later stage. Moreover, some viruses cause **cytocidal effects** (changes resulting in cell death), whereas other viruses are noncytocidal. Some viruses are cell-specific; that is, they attack only particular cells of the body. For example, the AIDS virus attacks specific cells of the immune system called T lymphocytes. A virus can produce one or more of the cytopathic effects listed below. (Some of these effects have already been discussed in Chapter 12; others will be discussed in more detail in later chapters.)

1. At some stage in their multiplication, cytocidal viruses cause the macromolecular synthesis within the host cell to stop. Some viruses, such as herpes simplex, irreversibly stop mitosis.

2. Infection by a cytocidal virus can destroy cells, releasing enzymes from lysosomes. These enzymes cause autolysis of the cell.

3. Inclusion bodies (Figure 14-5a) are granules found in the cytoplasm or nucleus of some infected cells. These granules are viral parts—nucleic acids or proteins in the process of being assembled into virions. The granules vary in size, shape, and staining properties, according to the virus. Inclusion bodies are characterized by their ability to stain with an acidic stain (acidophilic) or with a basic stain (basophilic).

(a)

(b)

Figure 14-5 Effects of viruses on cells. (a) Cytoplasmic inclusion body (arrow) in a human brain cell from a person who died of rabies (×8000). **(b)** The relatively clear area in the center of this micrograph is a giant cell formed in a culture of cells infected with measles virus. The numerous dark, oval bodies around the cell's edges are its multiple nuclei (×500).

4. Some viral infections cause host cells to fuse, producing multinucleated "giant" cells (Figure 14-5b).
5. Many viral infections induce antigenic changes on the surface of the infected cells. These antigenic changes elicit a host antibody response against the infected cell, and thus they target the cell for destruction by the host's immune system.
6. Some viruses induce chromosomal changes in the host cell.
7. Normal cells cease growing when they come close to another cell, a phenomenon known as *contact inhibition*. Viruses capable of causing cancer *transform* host cells, as discussed in Chapter 12. Transformation results in an abnormal, spindle-shaped cell that does not recognize contact inhibition (Figure 14-6). Loss of contact inhibition results in unregulated cell growth.

Some representative viruses that cause cytopathic effects are presented in Table 14-4. In subsequent chapters, we will discuss the pathological properties of viruses in more detail.

TABLE 14-4 *Cytopathic Effects (CPE) of Selected Viruses*

Virus	CPE
Poliovirus	Cell death
Papovavirus	Acidophilic inclusion bodies in nucleus
Adenovirus	Basophilic inclusion bodies in nucleus
Rhabdovirus	Acidophilic inclusion bodies in cytoplasm
Measles, cytomegalovirus	Acidophilic inclusion bodies in nucleus and cytoplasm
Measles	Cell fusion
Polyoma	Transformation
AIDS virus (HIV)	Destruction of T lymphocytes

Figure 14-6 Transformed cells in culture. In the center of this micrograph is a cluster of chick embryo cells transformed by Rous sarcoma virus. Such a concentration of transformed cells in a cell culture is called a *focus*; it results from multiplication of a single transforming virus that originally infected a cell at the site. Note how the transformed cells of the focus appear dark in contrast to the monolayer of light, flat, normal cells around them. This appearance is caused by their spindle shapes and uninhibited growth on top of one another; there is no contact inhibition (×800).

Fungi, Protozoans, Helminths, and Algae

Although fungi cause disease, they do not have a well-defined set of virulence factors. For example, *Cryptococcus neoformans* (krip-tō-kok'kus nē-ō-fôr'manz) is a fungus that causes a type of meningitis; it produces a capsule that helps it to resist phagocytosis. Other fungi have metabolic products that are toxic to human hosts. In such cases, however, the toxin is only an indirect cause of disease, as the fungus is already growing in or on the host. Fungal infection can also provoke an allergic response in the host.

The human disease called *ergotism*, which was common in Europe during the Middle Ages, is caused by a toxin produced by an ascomycete plant pathogen, *Claviceps purpurea* (kla'vi-seps pùr-pù-rē'ä), that grows on grains. The toxin is contained in *sclerotia* (Figure 14-7), highly resistant portions of the fungus's mycelia that can detach. The toxin itself, **ergot**, is an alkaloid that can cause hallucinations resembling those produced by LSD; in fact, ergot is a natural source of LSD. Ergot also constricts capillaries and can cause gangrene of the extremities through prevention of proper blood circulation. Although *Claviceps* still occasionally occurs on grains, modern milling usually removes the sclerotia.

Figure 14-7 *Claviceps purpurea*. The sclerotia of *Claviceps purpurea* are visible on this rye flower. *Claviceps* is a natural source of the drug LSD.

Sclerotia on rye
= LSD

A number of other toxins are produced by fungi that grow on grains or other plants. For example, peanut butter is occasionally recalled because of excessive amounts of **aflatoxin**, a toxin that has been found to cause cancer of the liver in laboratory animals (no link to human cancer has yet been found). Aflatoxin is produced by the growth of the mold *Aspergillus flavus* (a-spėr-jil′lus flā′vus). It is believed that the toxin might be altered in the animal's body to a mutagenic compound. (See the box on p. 390 for a discussion of how the product of an organism may be the indirect cause of a disease.)

A few mushrooms produce toxins that exert their effects after they have been ingested. Perhaps the most notable **mycotoxins** (toxins produced by fungi) are **phalloidin** and **amanitin**, produced by *Amanita phalloides* (am-an-ī′ta fal-loi′dēz), commonly known as the death angel. These neurotoxins are so potent that ingestion of the *Amanita* mushroom may result in death.

The presence of protozoans or helminths often produces disease symptoms in the host. Some of these organisms actually use host tissues for their own growth or produce large parasitic masses that block the flow of lymph; the resulting cellular damage evokes the symptoms. An example is the roundworm *Wuchereria bancrofti* (vū-kėr-ār′ē-ä ban-krof′tē), the causative agent of elephantiasis. Waste products of the metabolism of these parasites can also contribute to the symptoms of a disease.

A few species of algae produce neurotoxins. For example, some genera of dinoflagellates, such as *Gonyaulax* (gä-nē-ä′luks), are important because they produce a neurotoxin called **saxitoxin**. Although mollusks that feed on the dinoflagellates that produce saxitoxin show no symptoms of disease, people who eat the mollusks develop symptoms similar to botulism. Recall from Chapter 11 that extensive dinoflagellate growth is responsible for a situation called red tide. Public health agencies frequently prohibit human consumption of mollusks during red tides.

Specific diseases caused by fungi, protozoans, and helminths, along with the pathological properties of these organisms, are discussed in detail in later chapters.

* * *

In the next chapter, we will examine a group of nonspecific defenses of the host against disease. But before proceeding, examine Figure 14-8 very carefully. It summarizes some key concepts of the mechanisms of pathogenicity we have discussed in this chapter.

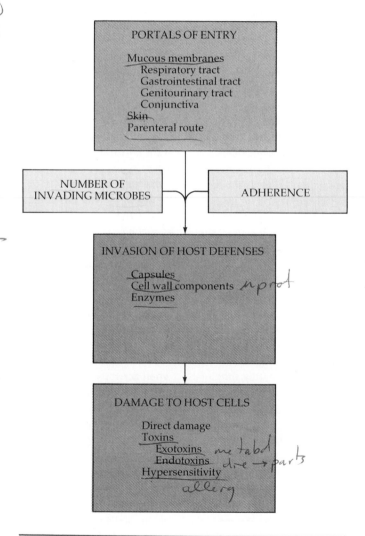

PORTALS OF ENTRY

Mucous membranes
  Respiratory tract
  Gastrointestinal tract
  Genitourinary tract
  Conjunctiva
Skin
Parenteral route

NUMBER OF INVADING MICROBES

ADHERENCE

INVASION OF HOST DEFENSES

Capsules
Cell wall components
Enzymes

DAMAGE TO HOST CELLS

Direct damage
Toxins
  Exotoxins
  Endotoxins
Hypersensitivity

Figure 14-8 Mechanisms of pathogenicity. A summary of how microorganisms cause disease.

MICROBIOLOGY HIGHLIGHTS

Unusual Pathogenesis of Colon Cancer

Investigations into the cause and development of colon cancer have begun to focus on the role played by the hundreds of species of microorganisms that flourish in the large intestine of the average person. The importance of microorganisms was first noticed when rats were fed cycasin, a substance that occurs naturally in the nut of the cycad plant. Germ-free rats were unaffected by the cycasin, but rats with the usual intestinal microflora developed colon cancer. Cycasin itself is not carcinogenic, but β-glucosidase, a bacterial enzyme, converts cycasin to methylazoxymethanol, which caused the cancer in the experimental rats.

Many substances that have been called carcinogens should more properly be called *procarcinogens* because they are not actively carcinogenic until they are chemically modified, as was cycasin in the experimental rats. Some of these procarcinogenic substances are found naturally in the environment, and others enter the body in foods or as pollutants. As in the case of cycasin, human intestinal microflora produce enzymes that convert these procarcinogens into active carcinogens.

Researchers have known since the early 1970s that populations that consume the typical North American and Western European diet—which is rich in beef, fat, and protein—are particularly prone to develop colon cancer. However, it seems now that it is not the diet itself that causes the cancer but the interaction between this diet and the microflora of the intestine. Variations in diet produce little change in the kinds of bacteria that live in the intestines, but they produce dramatic changes in the *metabolic activity* of the bacteria. It is this metabolic activity that is important.

To determine the role of intestinal bacteria in converting chemicals to carcinogens, Elena McCoy and her coworkers at New York Medical College compared the mutagenic abilities of 2-aminofluorene when activated by liver cell enzymes, intestinal cell enzymes alone, *Bacteroides fragilis* enzymes alone, and intestinal cell enzymes plus *B. fragilis* enzymes, by using the Ames test. In the Ames test, the amount of bacterial growth is an indication of the number of mutations produced (see the table). Liver enzymes, which are used in the standard Ames test, produced the highest

Mutagenic Conversion of 2-Aminofluorene in the Ames Test

Test Substance	Relative Amounts of Bacterial Growth
2-Aminofluorene (Control)	+
2-Aminofluorene Activated by:	
Liver enzymes	++++
Intestinal enzymes	++
B. fragilis enzymes	++
Intestinal enzymes plus B. fragilis enzymes	+++

conversion of this chemical to a mutagenic form. Intestinal enzymes alone and bacterial enzymes alone produced some mutagenic conversion. A combination of intestinal and bacterial enzymes seemd to have a synergistic effect almost as great as that produced by liver enzymes. These results suggest that procarcinogens reaching the intestine can be activated effectively by intestinal and bacterial enzymes working together.

Laboratory experiments have also shown that people who eat the typical Western diet have increased amounts of other bacterial enzymes in their feces. Some examples are bacterial β-glucuronidase, which retoxifies carcinogens that were previously detoxified by the liver; bacterial azoreductase, which

can produce carcinogens from azo food dyes; and bacterial nitroreductase, which can produce carcinogenic nitrosamines from amino acids and nitrates.

These findings may call into question our methods of testing the possible carcinogenic properties of many food additives and environmental chemicals. It may be necessary to test these substances in terms of their interaction with our internal ecology rather than testing them in isolation.

Study Outline

Introduction (p. 377)

1. Pathogenicity is the ability of a pathogen to produce a disease by overcoming the defenses of the host.
2. Virulence is the degree of pathogenicity.

Entry of Microorganisms into the Host (pp. 377–379)

1. The specific route by which a particular pathogen gains access to the body is called its portal of entry.

Mucous Membranes (pp. 377–378)

1. Many microorganisms can penetrate mucous membranes of the conjunctiva and the respiratory, gastrointestinal, and genitourinary tracts.
2. Microorganisms that are inhaled with droplets of moisture and dust particles gain access to the respiratory tract.
3. The respiratory tract is the most frequently used portal of entry.
4. Microorganisms enter the gastrointestinal tract via food, water, and contaminated fingers.

Skin (p. 378)

1. Most microorganisms cannot penetrate intact skin; they enter hair follicles and sweat ducts.
2. Some fungi infect the skin itself.

Parenteral Route (p. 378)

1. Some microorganisms can gain access to tissues by inoculation through the skin and mucous membranes in bites, injections, and other wounds.
2. This route of penetration is called the parenteral route.

Preferred Portal of Entry (p. 378)

1. Many microorganisms can cause infections only when they gain access through their specific portal of entry.

Numbers of Invading Microbes (p. 378)

1. Virulence can be expressed as LD_{50} (lethal dose for 50% of the inoculated hosts) or ID_{50} (infectious dose for 50% of the inoculated hosts).

Adherence (pp. 378–379)

1. Surface projections on a pathogen called adhesins (ligands) adhere to complementary receptors on the host cells.
2. Ligands can be glycoproteins or lipoproteins and are frequently associated with pili.
3. Mannose is the most common receptor.

How Pathogens Penetrate Host Defenses (pp. 379–382)

Capsules (pp. 379–380)

1. Some pathogens have capsules that prevent them from being phagocytized.

Components of the Cell Wall (p. 381)

1. Proteins in the cell wall can facilitate adherence or prevent a pathogen from being phagocytized.

Enzymes (pp. 381–382)

1. Leukocidins destroy neutrophils and macrophages.
2. Hemolysins lyse red blood cells.
3. Local infections can be protected in a fibrin clot caused by the bacterial enzyme coagulase.
4. Bacteria can spread from a focal infection by means of kinases (which destroy blood clots), hyaluronidase (which destroys a mucopolysaccharide that holds cells together), and collagenase (which hydrolyzes connective tissue collagen).

Damage to Host Cells (pp. 382–387)

Direct Damage (pp. 382–383)

1. Host cells can be destroyed when pathogens metabolize and multiply inside the host cells.
2. Some microbes can reproduce inside phagocytes or escape phagocytosis through motility.

Toxins (pp. 383–386)

1. Poisonous substances produced by microorganisms are called toxins; toxemia refers to symptoms caused by toxins in the blood.
2. The ability to produce toxins is called toxigenicity.

Exotoxins (pp. 383–384)

1. Exotoxins are produced by bacteria and released into the surrounding medium.
2. Exotoxins, not the bacteria, produce the disease symptoms.
3. Antibodies produced against exotoxins are called antitoxins.
4. Cytotoxins include diphtherotoxin (which inhibits protein synthesis) and erythrogenic toxins (which damage capillaries).
5. Neurotoxins include botulinum toxin (which prevents nerve transmission) and tetanus toxin (which prevents inhibitory nerve transmission).

6. *Vibrio* choleragen and staphylococcal enterotoxin are enterotoxins, which induce fluid and electrolyte loss from host cells.

Endotoxins (pp. 384–385)

1. Endotoxins are lipopolysaccharides—the lipid A component of the cell wall of gram-negative bacteria.
2. Bacterial cell death, antibiotics, and antibodies may cause release of endotoxins.
3. Endotoxins cause fever (by inducing release of interleukin-1) and shock (because of cachectin-induced decrease in blood pressure).

Plasmids, Lysogeny, and Pathogenicity (pp. 386–387)

1. Plasmids may carry genes for antibiotic resistance, toxins, capsules, and pili.
2. Lysogeny can result in bacteria with virulence factors, such as toxins or capsules. *phage inside!*

Pathogenic Properties of Other Microorganisms (pp. 387–390)

Viruses (pp. 387–388)

1. Signs of viral infections are called cytopathic effects (CPE).
2. Some viruses cause cytocidal effects (cell death), and others cause noncytocidal effects.
3. Cytopathic effects include the stopping of mitosis, lysis, formation of inclusion bodies, cell fusion, antigenic changes, chromosomal changes, and transformation.

Fungi, Protozoans, Helminths, and Algae (pp. 388–390)

1. Symptoms of fungal infections can be caused by capsules, toxins, and allergic responses.
2. Symptoms of protozoan and helminthic diseases can be caused by damage to host tissue or by the metabolic waste products of the parasite.

Study Questions

Review

1. List three portals of entry, and describe how microorganisms gain access through each.

2. Compare pathogenicity with virulence.

3. How would drugs that bind each of the following affect pathogenicity:
   (a) Mannose on human cell membranes
   (b) *N. gonorrhoeae* pili

4. How are capsules and cell wall components related to pathogenicity? Give specific examples.

5. Describe how hemolysins, leukocidins, coagulase, kinases, and hyaluronidase might contribute to pathogenicity.

6. Complete the following table comparing exotoxins with endotoxins.

	Exotoxin	Endotoxin
Bacterial source		
Chemistry		
Toxicity		
Pharmacology		
Example		

7. Define cytopathic effects, and give five examples.

8. Describe the factors contributing to the pathogenicity of fungi, protozoans, and helminths.

9. The ID_{50} for *Shigella* and *Salmonella* is 200 cells and 10^5 cells, respectively. Which genus is more infectious?

10. The LD_{50} for botulinum toxin is 0.000025 μg. The LD_{50} for *Salmonella* toxin is 200 μg. Which of these is the more potent toxin? How can you tell from the LD_{50} values?

11. Food poisoning can be divided into two categories, food infection and food intoxication. On the basis of toxin production by bacteria, explain the difference between these two categories.

Challenge

1. The cyanobacterium *Microcystis aeruginosa* produces a peptide that is toxic to humans. According to the graph below, when is this bacterium most toxic?

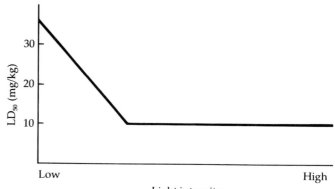

2. Explain whether each of the following examples is a food infection or intoxication. What is the probable etiological agent in each case?
   (a) Eighty-two people who ate shrimp at a dinner in Port Allen, Louisiana, developed diarrhea, cramps, weakness, nausea, chills, headache, and fever from 4 hours to 2 days after eating.
   (b) Thirty-two people in San Francisco who ate jackfish caught at Midway Island developed malaise, nausea, blurred vision, breathing difficulty, and numbness from 3 to 6 hours after eating.
3. Washwater containing *Pseudomonas* was sterilized and used to wash cardiac catheters. Three patients undergoing cardiac catheterization developed fever, chills, and hypotension. The water and catheters were sterile. Why did the patients show these reactions?

Further Reading

Brubaker, R. R. "Mechanisms of bacterial virulence." *Annual Review of Microbiology* 39:21–50, 1985. A review of nonspecific mechanisms of host defenses and bacterial virulence factors.

Crosa, J. H. "The relationship of plasmid-mediated iron transport and bacterial virulence." *Annual Review of Microbiology* 38:69–89, 1984. Discusses the actions of plasmid genes coding for virulence factors.

Habicht, G. S., G. Beck, and J. L. Benach. "Lyme disease." *Scientific American* 251(1):78–83, July 1987. Describes how investigators discovered the cause of this new disease since the first reports in 1975. Also describes the actions of bacterial LPS and interleukin-1 in causing symptoms.

Hadley, T. J., F. W. Klotz, and L. H. Miller. "Invasion of erythrocytes by malaria parasites: a cellular and molecular overview." *Annual Review of Microbiology* 40:451–477, 1986. Describes the role of ligands for attachment to red blood cells.

Middlebrook, J. L., and R. Dorland. "Bacterial toxins: cellular mechanisms of action." *Microbiology Reviews* 48:199–221, September 1984. A review of current knowledge of methods of action of many microbial toxins.

Stephen, J., and R. A. Pietrowski. *Bacterial Toxins*, 2nd ed. Washington, D.C.: American Society for Microbiology, Aspects of Microbiology, 1986. A brief treatise on important toxins.

Weiss, A. A., and E. L. Hewlett. "Virulence factors of *Bordetella pertussis*." *Annual Review of Microbiology* 40:661–686, 1986. Describes the epidemiology and diagnosis of whooping cough and illustrates that one bacterium has multiple toxins and many ways to avoid host defenses.

Nonspecific Defenses of the Host

From our discussion to this point, you can see that pathogenic microorganisms are endowed with special properties that enable them to cause disease if given the right opportunity. If microorganisms never encountered resistance from the host, we would constantly be ill and would eventually die of various diseases, but in most cases our body defenses prevent this from happening. Some of these defenses are designed to keep out microorganisms altogether. Other defenses remove the microorganisms if they do get in, and still others combat them if they remain inside. Our ability to ward off disease through our defenses is called **resistance**. Vulnerability or lack of resistance is known as **susceptibility**.

In discussing resistance, we will divide our body defenses into two general kinds, nonspecific and specific. *Nonspecific resistance* refers to defenses that protect us from *any kind* of pathogen. *Specific resistance*, or *immunity*, is the defense that the body offers against a particular microorganism. Specific defenses are based on the production of specific proteins, called antibodies, and special cells of the immune system. Chapter 16 deals with specific resistance. In this chapter, we consider nonspecific resistance.

Contributing to nonspecific resistance are the skin and mucous membranes, phagocytosis, inflammation, fever, and the production of antimicrobial substances other than antibodies. These are the main topics of this chapter. As we discuss these components of nonspecific resistance, keep in mind that nonspecific resistance provides a general response against invasion by pathogens; immunity involves a specific response against a specific microorganism.

Skin and Mucous Membranes

The skin and mucous membranes are the body's first line of defense against pathogens. This function results from both mechanical and chemical factors.

Mechanical Factors

The intact skin is one of the human body's largest organs, comprising over 15% of the body's dry weight. It consists of two distinct portions, the dermis and the epidermis (Figure 15-1). The *dermis*, the skin's inner, thicker portion, is composed of connective tissue. The *epidermis*, the outer, thinner portion, is in direct contact with the external environment. The epidermis consists of layers of continuous sheets of tightly packed epithelial cells with little or no material between the sheets. Two of the types of cells in the epidermis, Langerhans and Granstein cells, also participate in immunity by assisting helper T cells and suppressor T cells, respectively. (These cells are discussed in Chapter 16.) The top layer of epidermal cells is dead and contains a waterproofing protein called **keratin**.

If we consider the closely packed cells, continuous layering, and presence of keratin, we can see why the intact skin provides such a formidable physical barrier to the entrance of microorganisms. The intact surface of healthy epidermis is rarely, if ever, penetrated by bacteria. But when the epithelial surface is broken, a subcutaneous (below-the-skin) infection often develops. The bacteria most likely to cause such an infection are staphylococci that normally inhabit the epidermis, hair follicles, and sweat glands of the skin. Infections of the skin and underlying tissues frequently occur as a result of burns, cuts, and stab wounds or other conditions that break the skin. Moreover, when the skin is moist, as in hot, humid climates, skin infections are quite common, especially fungus infections such as athlete's foot.

Mucous membranes also consist of an epithelial layer and an underlying connective tissue layer. Mucous membranes line the entire gastrointestinal, respiratory, urinary, and reproductive tracts. The epithelial layer of a mucous membrane secretes a fluid called *mucus*, which prevents the tracts from drying out. Some pathogens that can thrive on the moist secretions of a mucous membrane are able to penetrate the membrane if the microorganism is present in sufficient numbers. *Treponema pallidum*, *Mycobacterium tuberculosis*, and *Streptococcus pneumoniae* are such pathogens. This penetration may be facilitated by toxic substances produced by the microorganism, prior injury by viral infection, or mucosal irritation. Although mucous membranes do inhibit the entrance of many microorganisms, they offer less protection than the skin.

Besides the skin and mucous membranes, several other mechanical factors help protect certain epithelial surfaces. One such mechanism that protects the eyes is the **lacrimal apparatus** (Figure 15-2), a group of structures that manufactures and drains away tears. The lacrimal glands, located toward the upper, outermost portion of each eye socket, produce the tears and pass them under the upper eyelid. From here, tears pass toward the corner of the eye near the nose and into two small holes that lead through tubes (lac-

Figure 15-1 Section through human skin. The thin layer at the top of the photo contains keratin. This layer and the darker cells beneath it make up the epidermis. The lighter material in the bottom part of the photo is the dermis. (×200)

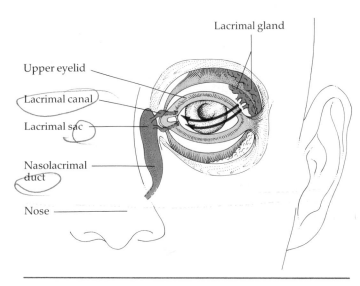

Lacrimal gland

Upper eyelid

Lacrimal canal

Lacrimal sac

Nasolacrimal duct

Nose

Figure 15-2   The lacrimal apparatus. The washing action of the tears, shown by the arrows, prevents microorganisms from settling on the surface of the eyeball. Tears produced by the lacrimal glands pass across the surface of the eyeball into two small holes that convey the tears into the lacrimal canals, lacrimal sac, and nasolacrimal duct.

rimal canals) to the nose. The tears are spread over the surface of the eyeball by blinking. Normally, the tears evaporate or pass into the nose as fast as they are produced. This continual washing action helps to keep microorganisms from settling on the surface of the eye. If an irritating substance or large numbers of microorganisms come in contact with the eye, the lacrimal glands start to secrete heavily, and the tears accumulate more rapidly than they can be carried

away. This excessive production is a protective mechanism because the excess tears dilute and wash away the irritating substance or microorganisms.

In a cleansing action very similar to that of tears, saliva is produced by the salivary glands and washes microorganisms from both the surface of the teeth and the mucous membrane of the mouth. This mechanism helps prevent colonization by microbes.

The respiratory and gastrointestinal tracts have many mechanical forms of defense. Mucus is slightly viscous (thick); thus, it traps many of the microorganisms that enter the respiratory and gastrointestinal tracts. The mucous membrane of the nose also has mucus-coated hairs that filter inhaled air and trap microorganisms, dust, and pollutants. The cells of the mucous membrane of the lower respiratory tract are covered with cilia (Figure 15-3). By moving synchronously, these cilia propel inhaled dust and microorganisms that have become trapped in mucus upward toward the throat. This so-called **ciliary escalator** keeps the mucous blanket moving toward the throat at a rate of 1 to 3 cm per hour; coughing and sneezing speed up the escalator. (Some substances in cigarette smoke are toxic to cilia and can seriously impair the functioning of the ciliary escalator.) Microorganisms are also prevented from entering the lower respiratory tract by a small lid of cartilage called the epiglottis, which covers the larynx (voicebox) during swallowing.

The cleansing of the urethra by the flow of urine is another mechanical factor that prevents microbial colonization in the urinary system. Vaginal secretions likewise move microorganisms out of the female body.

Mucus

Cilia

(a)

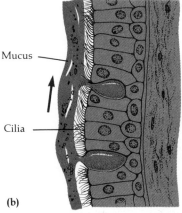

Mucus

Cilia

(b)

Figure 15-3   Mucous membrane of the trachea (windpipe). (a) Micrograph of cilia and mucus-secreting goblet cells, from which blobs of mucus are emerging (×1500). **(b)** Action of the ciliary escalator.

Chemical Factors

Mechanical factors alone do not account for the high degree of resistance of skin and mucous membranes to microbial invasion. Certain chemical factors also play important roles.

Sebaceous (oil) glands of the skin produce an oily substance called **sebum** that prevents hair from drying and becoming brittle. Sebum also forms a protective film over the surface of the skin. One of the components of sebum is unsaturated fatty acids, which inhibit the growth of certain pathogenic bacteria and fungi. The low pH of the skin, between pH 3 and 5, is caused in part by the secretion of fatty acids and lactic acid. The skin's acidity probably discourages the growth of many other microorganisms.

Bacteria that live as commensals on the skin decompose sloughed-off skin cells, and these organic molecules and the end-products of their metabolism produce body odor. As you will see in Chapter 20, certain bacteria commonly found on the skin metabolize sebum, and this metabolism forms free fatty acids that cause the inflammatory response associated with acne. Isotretinoin (Accutane), a derivative of vitamin A that prevents sebum formation, is a treatment for a very severe type of acne called cystic acne.

The sweat glands of the skin produce perspiration, which helps to maintain body temperature, eliminate certain wastes, and flush microorganisms from the surface of the skin. Perspiration also contains **lysozyme**, an enzyme capable of breaking down cell walls of gram-positive bacteria and a few gram-negative bacteria under certain conditions (see Figure 4-11). Lysozyme is also found in tears, saliva, nasal secretions, and tissue fluids, where it exhibits its antimicrobial activity. Alexander Fleming was actually studying lysozyme in 1929 when he accidentally discovered the antimicrobial effects of penicillin (Figure 1-5).

Gastric juice is produced by the glands of the stomach. It is a mixture of hydrochloric acid, enzymes, and mucus. The very high acidity of gastric juice (pH 1.2 to 3.0) is sufficient to preserve the usual sterility of the stomach. This acidity destroys bacteria and most bacterial toxins except those of *Clostridium botulinum* and *Staphylococcus aureus*. However, it is probable that many enteric pathogens are protected by food particles and thus are able to enter the intestines via the gastrointestinal tract.

The resident flora on the skin and mucous membranes usually compete successfully with other microorganisms for available nutrients and may produce metabolic end products that inhibit the growth of other microorganisms. For example, the normal lactic acid bacteria of the vagina decompose the glycogen of vaginal fluids into organic acids, creating a pH too low for the growth of most microbes. The presence of normal flora thus prevents colonization by potentially pathogenic microorganisms.

We will next discuss phagocytosis, commonly regarded as the body's second line of defense against infection.

Phagocytosis

Phagocytosis (from the Greek words for *eat* and *cell*) is the ingestion of a microorganism or any particulate matter by a cell. We have previously mentioned phagocytosis as the method of nutrition of certain protozoans. In this chapter, phagocytosis is discussed as a means by which cells in the human body counter infection. The human cells that perform this function are collectively called **phagocytes**. All are types of white blood cells or derivatives of white blood cells. Before we look at phagocytes specifically, let us consider the various components of the blood more generally.

Formed Elements in Blood

Blood consists of a fluid called **plasma**, which contains **formed elements**; that is, cells and cell fragments (Table 15-1). Of the cells listed in Table 15-1, those that concern us at present are the **leukocytes**, or white blood cells (Figure 15-4).

During many kinds of infections, especially bacterial infections, there is an increase in the total number of white blood cells; this increase is called *leukocytosis*. During the active stage of infection, the leukocyte count might double, triple, or quadruple, depending on the severity of the infection. Diseases that might cause such an elevation in the leukocyte count are meningitis, infectious mononucleosis, appendicitis, pneumococcal pneumonia, and gonorrhea. Other diseases, such as typhoid fever, tuberculosis, measles, and influenza, cause a decrease in the leukocyte count (called *leukopenia*). The source of leukocyte increase or decrease can be detected by a **differential count**, which is a calculation of the percentage of each kind of white cell in a sample of white blood cells. The percentages in a normal differential

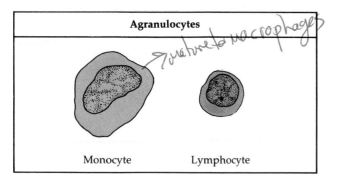

Figure 15-4   Principal kinds of leukocytes.

count of leukocytes are shown in parentheses in the first column of Table 15-1.

Leukocytes are divided into two categories, granulocytes and agranulocytes. **Granulocytes** owe their name to the presence of large granules in their cytoplasm that can be seen under a light microscope. They are differentiated into three types on the basis of how the granules stain. The granules of **neutrophils** stain red and blue with a mixture of acidic and basic dyes, respectively; those of **basophils** stain blue with the basic dye methylene blue; and those of **eosinophils** stain red with the acid dye eosin.

Neutrophils are also commonly called *polymorphonuclear leukocytes* (PMNs), or *polymorphs*. Neutrophils are highly phagocytic and motile. They have the ability to leave the blood, enter an infected tissue, and destroy microbes and foreign particles (see the box on p. 401). The details of this mechanism will be described shortly.

The role of basophils is not clear. However, they release substances, such as histamine, that are important in inflammation and allergic responses. They also release heparin, an anticoagulant.

Eosinophils are somewhat phagocytic and have the ability to leave the blood. They are believed to

TABLE 15-1   *Formed Elements in Blood*

Type of Cell	Numbers per Cubic Millimeter (mm³)	Function
Erythrocytes (Red blood cells)	4.8 to 5.4 million	Transport of O_2 and CO_2
Leukocytes (White blood cells)	5000 to 9000	
A. Granulocytes 1. Neutrophils (PMNs) (60% to 70% of leukocytes)		Phagocytosis
2. Basophils (0.5% to 1%)		Production of heparin and histamine
3. Eosinophils (2% to 4%)		Phagocytosis
B. Agranulocytes 1. Lymphocytes (20% to 25%)		Antibody production (B lymphocytes); cell-mediated immunity (T lymphocytes)*
2. Monocytes (3% to 8%)		Phagocytosis
Thrombocytes (Platelets)	250,000 to 400,000	Blood clotting

*Discussed in Chapter 16.

ingest the products of antigen–antibody reactions, and their number increases greatly during certain parasitic worm infections and hypersensitivity (allergy) reactions.

Agranulocytes are so named because they lack large numbers of granules in their cytoplasm and their granules cannot be seen under a light microscope. The two kinds of agranulocytes are **lymphocytes** and **monocytes**. Lymphocytes occur in lymphoid tissues (tonsils, lymph nodes, spleen, thymus gland, thoracic duct, bone marrow, appendix, Peyer's patches of the small intestine, and lymph nodes in the respiratory, gastrointestinal, and reproductive tracts). They also circulate in the blood (Figure 15-5). As you will see in Chapter 16, lymphocytes assume a key role in immunity. Monocytes are not actively phagocytic until they mature into macrophages during a progressing infection. In fact, the maturation and proliferation of macrophages is one factor responsible for the swelling of lymph nodes during an infection.

Kinds of Phagocytic Cells

Not all leukocytes are phagocytic. Phagocytic cells fall into two broad categories—granulocytes (sometimes called microphages) and macrophages, which include monocytes.

When an infection occurs, both granulocytes (especially neutrophils) and monocytes migrate to the infected area. During this migration, monocytes enlarge and develop into actively phagocytic cells called **macrophages** (Figure 15-6). Because these cells leave the blood and migrate through tissue to infected areas, they are called *wandering macrophages*. Some macrophages, called *fixed macrophages* or *histiocytes*, enter certain tissues and organs of the body and remain there. Fixed macrophages are found in the liver (Kupffer's cells), lungs (alveolar macrophages), nervous system (microglial cells), bronchial tissue, spleen, lymph nodes, bone marrow, and the peritoneal cavity surrounding abdominal organs. The

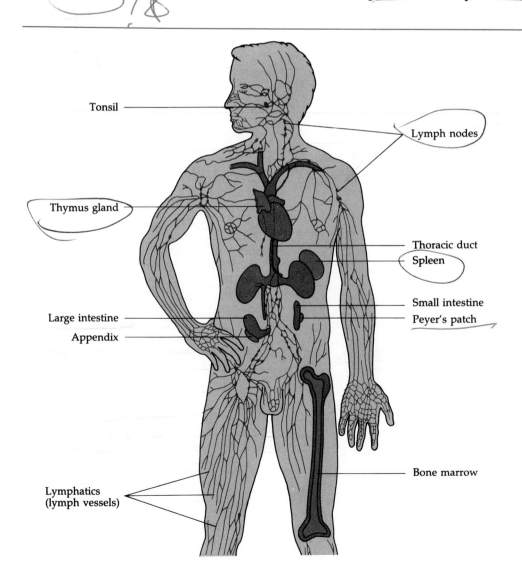

Figure 15-5 Components of the lymphatic system.

Tonsil

Lymph nodes

Thymus gland

Thoracic duct
Spleen

Large intestine

Appendix

Small intestine
Peyer's patch

Bone marrow

Lymphatics
(lymph vessels)

Lyosomes
(containing
enzymes)

Macrophage

Nucleus

Pseudopod

Yeast
cell

Figure 15-6 Electron micrograph of a macrophage engulfing a yeast cell (×5600).

various macrophages of the body constitute the **mononuclear phagocytic (reticuloendothelial) system**.

During the course of an infection, there is a shift in the type of white blood cell that predominates in the bloodstream. Granulocytes, especially neutro-phils, predominate during the initial phase of infection, at which time they are actively phagocytic; this dominance is indicated by their number in a differential count. However, as the infection progresses, the monocytes predominate; they scavenge and phagocytize remaining living bacteria and dead or dying bacteria. The increased number of monocytes will also be reflected in a differential count. As blood and lymph that contain microorganisms pass through organs with fixed macrophages, cells of the mononuclear phagocytic system remove the microorganisms by phagocytosis. The mononuclear phagocytic system also disposes of worn out blood cells.

Figure 15-7 summarizes the roles of the different types of phagocytic cells.

Mechanism of Phagocytosis

How does phagocytosis occur? For convenience of study, we will divide phagocytosis into four phases—chemotaxis, adherence, ingestion, and digestion.

Chemotaxis

Chemotaxis is the chemical attraction of phagocytes to microorganisms. (The mechanism of chemotaxis may be reviewed in Chapter 4.) Among the chemotactic chemicals that attract phagocytes are microbial products, components of white blood cells and damaged tissue cells, and peptides derived from *complement*, a system of host defense discussed later in this chapter.

Figure 15-7 Classification of phagocytes.

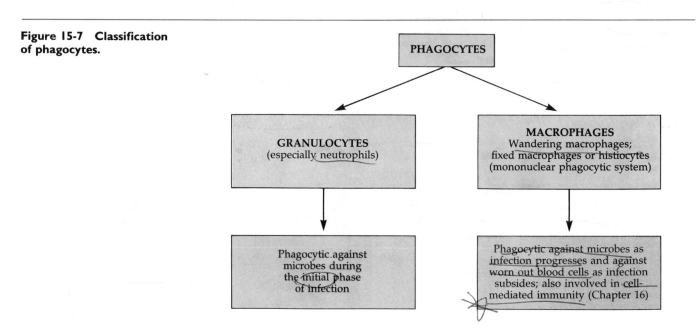

PHAGOCYTES

GRANULOCYTES
(especially neutrophils)

MACROPHAGES
Wandering macrophages;
fixed macrophages or histiocytes
(mononuclear phagocytic system)

Phagocytic against
microbes during
the initial phase
of infection

Phagocytic against microbes as
infection progresses and against
worn out blood cells as infection
subsides; also involved in cell-
mediated immunity (Chapter 16)

Adherence

As it pertains to phagocytosis, adherence is the attachment of the phagocyte plasma membrane to the surface of the microorganism or other foreign material (Figure 15-8a). In some instances, adherence occurs easily, and the microorganism is readily phagocytized. However, adherence can be hampered by the presence of large capsules and M protein. As mentioned in Chapter 14, the M protein of *Streptococcus pyogenes* inhibits the attachment of phagocytes to their surfaces and makes adherence more difficult. *Streptococcus pneumoniae* and *Klebsiella pneumoniae* also have large capsules. Such heavily encapsulated microorganisms can be phagocytized only if the phagocyte traps the microorganism against a rough surface, such as a blood vessel, blood clot, or connective tissue fibers, where the microbe cannot slide away.

Microorganisms can be more readily phagocytized if they are first coated with certain plasma proteins that promote the attachment of the microorganism to the phagocyte. This coating process is called **opsonization**. The proteins that act as *opsonins* include, among others, some components of the complement system (described later in this chapter) and antibody molecules (described in Chapter 16).

Ingestion

Following adherence, **ingestion** occurs. During the process of ingestion, the plasma membrane of the phagocyte extends projections called pseudopods that engulf the microorganism (Figure 15-8b). Once the microorganism is surrounded, the pseudopods meet and fuse, surrounding the microorganism with a sac called a *phagosome* or *phagocytic vacuole*.

MICROBIOLOGY HIGHLIGHTS

Neutrophil Defect Leads to Periodontal Disease

What happens when the body creates white blood cells that fail to function properly? Since these cells are a key line of defense against infection and disease, it is clear that malfunctioning white blood cells would leave us more vulnerable. Indeed, recent dental research indicates that an unusual periodontal disease called *localized juvenile periodontitis (LJP)* attacks children whose bodies produce faulty neutrophils. Neutrophils, which leave the blood to go to damaged tissue, normally prevent the spread of bacteria into gums and roots.

Periodontal disease is normally an adult problem in which infections of the soft tissues and bones that support the teeth cause loosening and eventual loss of the teeth. Unlike normal periodontal problems, however, LJP attacks children and adolescents. It affects the tissues that support the incisors and first molars. If the disease is left untreated, there is serious bone loss, and the affected teeth fall out by the time the victim is 20 years old.

According to the latest findings, the disease-causing agent in LJP is the gram-negative bacterium *Actinobacillus actinomycetemcomitans*, which is the only plaque bacterium that kills neutrophils. It produces a leukotoxin that binds to the neutrophil's membrane and kills it very quickly. Although *A. actinomycetemcomitans* is present in the mouths of about one-fifth of the population, it can cause the disease only in people who have reduced neutrophil function.

This reduced function is caused by depressed chemotaxis—that is, the neutrophils do not respond to the chemicals that should alert them to the presence of the bacteria, and therefore they do not approach and attack the bacteria. It appears that the neutrophils are deficient in receptors, the membrane proteins that bind them to the attracting chemicals. This deficiency seems to be hereditary.

Traditionally, LJP has been treated by killing the bacteria with antibiotics and surgically removing infected tissue. Although the affected tissues regenerate themselves, the disease often recurs. Twenty-five percent of the patients suffer a relapse within one year.

Now that the relationship of LJP to the neutrophils' chemotaxis deficiency has been recognized, a new treatment for LJP is being tested. A chemical known as Bestatin, which is purified from the bacterium *Streptomyces olivoreticuli*, is able to double the chemotactic activity of neutrophils from LJP patients and therefore is potentially useful in treating LJP. One theory is that Bestatin increases the density of chemotactic receptors. So far, Bestatin has only been tested in vitro.

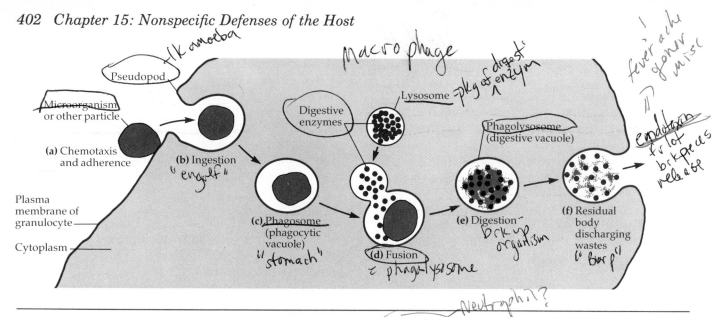

Figure 15-8 Phagocytosis. The drawing shows the mechanism within a granulocyte. **(a)** Chemotaxis and adherence. **(b)** Ingestion. **(c)** Formation of the phagosome. **(d)** Fusion of the phagosome with a lysosome forms a phagolysosome. **(e)** Digestion—destruction of the ingested microorganism. **(f)** Residual body containing indigestible material.

Digestion

In this phase of phagocytosis, the phagosome pinches off from the plasma membrane and enters the cytoplasm (Figure 15-8c). Within the cytoplasm, it contacts lysosomes that contain digestive enzymes and bactericidal substances (Chapter 4). Upon contact, the phagosome and lysosome membranes fuse (Figure 15-8d) to form a single, larger structure called a *phagolysosome*, or *digestive vacuole* (Figure 15-8e). The contents of the phagolysosome take only 10 to 30 minutes to kill many types of bacteria. It is assumed that microbial destruction occurs because of the contents of the lysosomes.

Lysosomal enzymes that attack microbial cells directly include lysozyme, which hydrolyzes peptidoglycan in bacterial cell walls. A variety of other enzymes hydrolyze other macromolecular components of microorganisms. The hydrolytic enzymes are most active at about pH 4, which is the phagolysosome's usual pH because of lactic acid produced by the phagocyte. Another kind of lysosomal enzyme, myeloperoxidase, participates in a chemical reaction that bonds chloride ions to bacteria and viruses, resulting in death of the microorganisms.

Once enzymes have digested the contents of the phagolysosome brought into the cell by ingestion, the phagolysosome moves toward the cell boundary and discharges its wastes. At this point, the phagolysosome containing indigestible material is called a *residual body* (Figure 15-8f).

Not all phagocytized microorganisms are killed by lysosomal enzymes. The toxins of some microbes, such as toxin-producing staphylococci, can actually kill the phagocytes (see the box on p. 401 for another example involving a plaque bacterium). Other microorganisms, such as *Mycobacterium tuberculosis*, can multiply within the phagolysosome and eventually destroy the phagocyte. Still others, such as the causative agents of tularemia and brucellosis, can remain dormant in phagocytes for months or years at a time.

Phagocytosis, in addition to providing a nonspecific defense of the host, also assumes a role in immunity. Macrophages help T and B lymphocytes perform vital functions in immunity. We will discuss how phagocytosis supports immunity in more detail in Chapter 16.

In the next section, we will see how phagocytosis often occurs as part of another nonspecific mechanism of resistance, inflammation.

Inflammation

Damage to the body's tissues triggers a response called inflammation. The damage can be caused by microbial infection, physical agents (such as heat, radiant energy, electricity, or sharp objects), or chemical agents (acids, bases, and gases). Inflammation is usually characterized by four fundamental symptoms—redness, pain, heat, and swelling. Sometimes a fifth symptom, loss of function, is present; its occurrence depends on the site and extent of damage. In apparent contradiction to the symptoms observed, however, the inflammatory response is actually a beneficial one. Inflammation has the following functions: (1) to destroy the injurious agent, if possible, and to remove it and its by-products from the body; (2) if destruction is not possible, to limit the effects on the body by confining or walling off the injurious agent and its by-products; and (3) to repair or replace tissue damaged by the injurious agent or its by-products.

For purposes of our discussion, we will divide the process of inflammation into three stages—vasodilation and increased permeability of blood vessels, phagocyte migration, and repair.

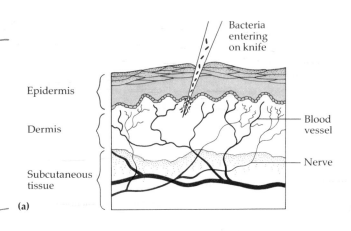

(a)

Vasodilation and Increased Permeability of Blood Vessels

Immediately following tissue damage, blood vessels dilate in the area of damage, and their permeability increases (Figure 15-9a and b). **Vasodilation** is an increase in the diameter of blood vessels. Vasodilation increases blood flow to the damaged area and is responsible for the redness (erythema) and heat associated with inflammation.

Increased permeability permits defensive substances normally retained in the blood to pass through the walls of the blood vessels and enter the injured area. The increase in permeability, which permits fluid to move from the blood into tissue spaces, is responsible for the **edema** (swelling) of inflammation. The pain of inflammation can be caused by nerve damage, irritation by toxins, or the pressure of edema.

Vasodilation and the increase in permeability of blood vessels are caused by chemicals released by damaged cells in response to injury. One such substance is **histamine**, a chemical present in many tissues of the body, especially in mast cells in connective tissue, circulating basophils, and blood platelets. Histamine is released in direct response to the injury of cells that contain it; it is also released in response to stimulation by certain components of the complement system (to be discussed later). Phagocytic granulocytes attracted to the site of injury can also produce chemicals that cause the release of histamine.

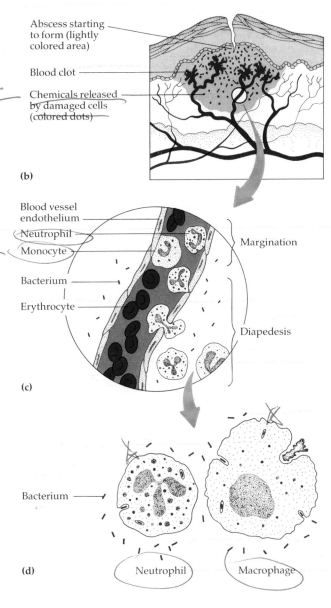

Figure 15-9   Aspects of the inflammatory response.
(a) Injury to otherwise healthy skin. **(b)** Vasodilation and increased permeability. **(c)** Phagocyte migration. **(d)** Phagocytization of bacteria and cellular debris by macrophages and neutrophils. Macrophages develop from monocytes.

Kinins are another group of substances that cause vasodilation and increase the permeability of blood vessels. These chemicals are present in blood plasma and, once activated, also attract phagocytic granulocytes—chiefly neutrophils—to the injured area. Fluid lost from capillaries into tissues puts pressure on nerve endings, which accounts for much of the pain associated with inflammation.

Prostaglandins, substances released by damaged cells, and **leukotrienes**, substances related to prostaglandins produced by mast cells and basophils, are other chemicals that cause vasodilation.

Vasodilation and the increase in permeability of blood vessels also help to deliver clotting elements of blood into the injured area. The blood clots that form around the site of activity prevent the microorganism (or its toxins) from spreading to other parts of the body. As a result, there may be a localized collection of pus in a cavity formed by the breakdown of body tissues. This focus of infection is called an **abscess** (Figure 15-9b). Common abscesses include pustules and boils.

The next stage in inflammation involves the migration of phagocytes to the injured area.

Phagocyte Migration

Generally, within an hour after the process of inflammation is initiated, phagocytes appear on the scene (Figure 15-9c). As the flow of blood gradually decreases, phagocytes (both neutrophils and monocytes) begin to stick to the inner surface of the endothelium (lining) of blood vessels. This sticking process is called **margination**. Then the collected phagocytes begin to squeeze between the endothelial cells of the blood vessel to reach the damaged area. This migration, which resembles amoeboid movement, is called **diapedesis**; the migratory process can take as little as two minutes.

As mentioned earlier, certain chemicals attract neutrophils to the site of injury (chemotaxis). These include chemicals produced by microorganisms and even other neutrophils; other chemicals are kinins and components of the complement system. The availability of a steady stream of neutrophils is ensured by the production and release of additional granulocytes from bone marrow.

As the inflammatory response continues, monocytes follow the granulocytes into the infected area. Once the monocytes are contained in the tissue, they undergo changes in biological properties and become wandering macrophages. The granulocytes predominate in the early stages of infection but tend to die off rapidly. Macrophages enter the picture during a later stage of the infection, once granulocytes have accomplished their function. They are several times more phagocytic than granulocytes and are large enough to phagocytize tissue that has been destroyed, granulocytes that have been destroyed, and invading microorganisms (Figure 15-9d).

After granulocytes or macrophages engulf large numbers of microorganisms and damaged tissue, they themselves eventually die. After a few days, an area containing dead phagocytes and damaged tissue forms in the inflamed tissue. This collection of dead cells and various body fluids is called **pus**. Pus formation usually continues until the infection subsides. At times, the pus pushes to the surface of the body or into an internal cavity for dispersal. On other occasions, the pus remains even after the infection is terminated. In this case, the pus is gradually destroyed over a period of days and is absorbed by the body.

Repair

The final stage of inflammation is tissue repair, the process by which tissues replace dead or damaged cells. Repair begins during the active phase of inflammation, but it cannot be completed until all harmful substances have been removed or neutralized at the site of injury. The ability of a tissue to regenerate, or repair itself, depends in part on the tissue involved. For example, skin has a high capacity for regeneration, whereas nervous tissue in the brain and spinal cord does not regenerate at all.

A tissue is repaired when its stroma or parenchyma produce new cells. The *stroma* is the supporting connective tissue, and the *parenchyma* is the functioning part of the tissue. For example, the capsule around the liver that encloses and protects it is part of the stroma since it is not involved in the functions of the liver; liver cells (hepatocytes) that perform functions of the liver are part of the parenchyma. If only parenchymal cells are active in repair, a perfect or near-perfect reconstruction of the tissue occurs. A familiar example of perfect reconstruction is a minor skin cut, in which parenchymal cells are more active in repair. However, if repair cells of the stroma of the skin are more active, scar tissue is formed.

The process of inflammation is summarized in Figure 15-10.

Fever

Inflammation is a local response of the body to microbial invasion. There are also systemic, or overall, responses, one of the most important being fever, an abnormally high body temperature. The most frequent cause of fever is infection from bacteria (and their toxins) or viruses.

Body temperature is controlled by a part of the brain called the hypothalamus. The hypothalamus is sometimes called the body's thermostat, and it is normally set at 37°C (98.6°F). It is believed that certain antigens affect the hypothalamus by setting it at a higher temperature. Recall from Chapter 14 that, when phagocytes ingest gram-negative bacteria, the lipopolysaccharides (LPS) of the cell wall (endotoxins) are released, causing the phagocytes to release interleukin-1 (endogenous pyrogens). Interleukin-1 causes the hypothalamus to release prostaglandins that reset the hypothalamic thermostat at a higher temperature, thereby causing fever. (See Figure 14-4.)

Assume that the body is invaded by pathogens, and the thermostat setting is increased to 39°C (102.2°F). To adjust to the new thermostat setting, the body responds with blood vessel constriction, increased rate of metabolism, and shivering, all of which raise body temperature. Even though body temperature is climbing higher than normal, the skin remains cold, and shivering occurs. This condition, called a *chill*, is a definite sign that temperature is rising. When body temperature reaches the setting of the thermostat, the chill disappears. But the body will continue to maintain its temperature at 39°C (102.2°F) until the bacterial lipopolysaccharide (LPS) or interleukin-1 is eliminated. When elimination is achieved, the thermostat is reset at 37°C. As the infection subsides, heat-losing mechanisms such as vasodilation and sweating go into operation. The skin becomes warm and the person begins to sweat. This phase of the fever, called the *crisis*, indicates that body temperature is falling.

Up to a certain point, fever is considered a defense against disease. Interleukin-1 helps step up the production of T lymphocytes. High body temperature intensifies the effect of interferon (an antiviral protein that we will discuss shortly). The high body temperature is believed to inhibit the growth of some microorganisms. Also, because the high temperature speeds up the body's reactions, it may help body tissues to repair themselves more quickly.

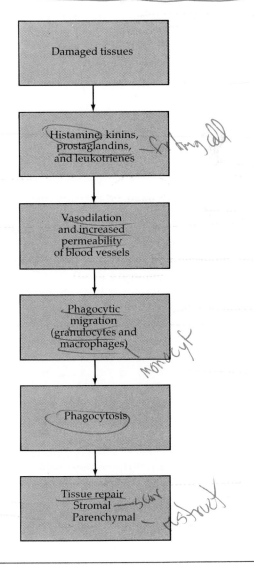

Figure 15-10 Summary of the process of inflammation.

Antimicrobial Substances

The body produces certain antimicrobial substances in addition to the chemical factors mentioned earlier. Among the most important of these are the proteins of the complement system and interferon.

Complement and Properdin

Complement and properdin comprise two related defense systems, both of which are composed of proteins. The complement system, also called the classical pathway, is the more frequently used system; the properdin system is also known as the alternate pathway.

Complement

Complement is a group of proteins found in normal blood serum. The complement system, abbreviated C, is made up of at least 20 different interacting serum proteins, designated by a complex numbering system ranging from C1 through C9. (Some of the complement proteins have subcomponents labeled with numbers and letters.) These proteins are very important to both nonspecific and specific defenses against microbial infection. The overall function of the complement system is to attack and destroy invading microorganisms.

Complement is so named because, in its "classical" role, it complements certain immune reactions involving antibodies. For this reason, many textbooks do not discuss complement until antigen–antibody reactions have been discussed. We discuss it here, however, because we regard the complement system as basically a form of nonspecific resistance. As one reason, complement has the same general role in all antigen–antibody reactions in which it participates. Also, complement itself is not specific for particular antigens and does not appear in greater amounts after immunization. Furthermore, complement participates in important defense reactions that do not involve antibodies.

Properdin

The **properdin system** is composed of three serum proteins—properdin itself, factor B, and factor D. Together, these three proteins comprise more than 10% of the proteins in serum. Proteins of the properdin system are components of the alternate pathway described below.

Pathways of Complement and Properdin Activation

The proteins of the complement and properdin systems act in an ordered sequence, or cascade, that, except for C4, follows the numerical designations. In a series of steps, the proteins *activate* one another, usually by cleaving (splitting) the next protein in the series. The fragments of a cleaved protein have new physiological or enzymatic functions. For example, one fragment might cause blood-vessel dilation while another fragment serves as part of the enzyme that cleaves the next protein in the series.

C3 plays a central role in the complement system. Its activation triggers several mechanisms that contribute to microbial destruction. As you can see in Figure 15-11, either of two pathways can activate C3.

The classical pathway is initiated by the binding of antibodies to antigen (described in Chapter 16). The antigen could be bacteria or other foreign cells. Once a pair of antibodies recognizes and attaches to the antigen, the complement protein C1 (which actually consists of three subunit proteins) binds to the two adjacent antibodies, thus activating C1 (Figure 15-12a).

Next, activated C1 in turn activates C2 and C4. It does this by functioning as a proteolytic (protein-splitting) enzyme and splitting the C2 and C4 proteins. C2 is split into fragments called C2a and C2b, and C4 into fragments called C4a and C4b (Figure 15-12b). Then C2a and C4b combine to form another proteolytic enzyme, which in turn activates C3 by splitting it into two fragments, C3a and C3b.

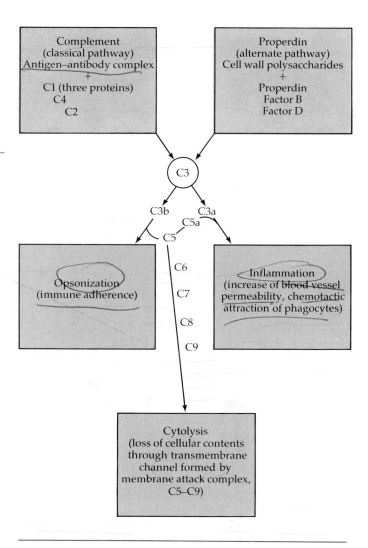

Figure 15-11   Complement activation via classical and alternative (properdin) pathways. Note the cleavage of C3 into C3a and C3b. These fragments induce three kinds of consequences destructive to microorganisms—opsonization, cytolysis, and inflammation.

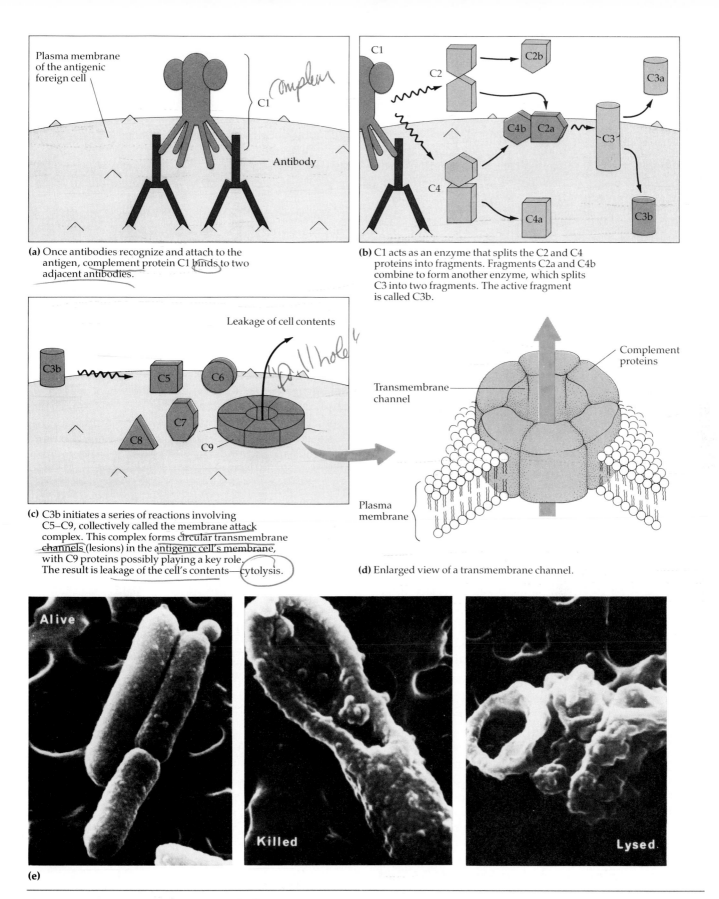

(a) Once antibodies recognize and attach to the antigen, complement protein C1 binds to two adjacent antibodies.

(b) C1 acts as an enzyme that splits the C2 and C4 proteins into fragments. Fragments C2a and C4b combine to form another enzyme, which splits C3 into two fragments. The active fragment is called C3b.

(c) C3b initiates a series of reactions involving C5–C9, collectively called the membrane attack complex. This complex forms circular transmembrane channels (lesions) in the antigenic cell's membrane, with C9 proteins possibly playing a key role. The result is leakage of the cell's contents—cytolysis.

(d) Enlarged view of a transmembrane channel.

(e)

Figure 15-12 Cytolysis caused by complement activation. This figure illustrates just a few key steps in the sequence that leads to holes in a plasma membrane. As you can see, the process is complex. Part **(e)**, ×30,000.

The alternate pathway indicated in Figure 15-11 does not involve antibodies; it is initiated by the interaction between certain polysaccharides and the proteins of the properdin system. Most of these polysaccharides are contained in the cell walls of certain bacteria and fungi, although they also include molecules on the surface of some foreign mammalian red blood cells. The properdin pathway is of particular importance in combating enteric gram-negative bacteria. The outer membrane of the bacteria's cell wall contains a lipopolysaccharide that releases an endotoxin (lipid A), triggering the alternate pathway. Note that this pathway does not involve C1, C2, or C4.

Consequences of Complement Activation

How does the complement system contribute to microbial destruction? Both the classical and alternate pathways lead to the cleavage of C3 into two fragments, C3a and C3b. These fragments induce three processes that are destructive to microorganisms—cytolysis, inflammation, and opsonization.

Cytolysis   The main function of complement is to destroy foreign cells by damaging their plasma membranes, causing the cellular contents to leak out (**cytolysis**). This is accomplished as follows (Figure 15-12c). C3b initiates a sequence of reactions involving C5, C6, C7, C8, and C9, which are known collectively as the **membrane attack complex**. The activated components of these proteins attack the invading cell's membrane and produce circular lesions, called **transmembrane channels** (Figure 15-12d), that lead to loss of ions and eventual cytolysis. The utilization of the complement components in this process is called **complement fixation**; it forms the basis of an important clinical laboratory test that will be explained in Chapter 17. Figure 15-12e illustrates the effects of complement on a microbial cell.

Inflammation   C3a, a cleavage product from C3, and C5a, a cleavage product from C5, can contribute to the development of acute inflammation (Figure 15-13). C3a and C5a bind to mast cells, basophils, and blood platelets to trigger the release of histamine, which increases blood vessel permeability. C5a also functions as a powerful chemotactic factor that attracts phagocytes to the site of complement fixation.

Opsonization   When C3b is bound to the surface of a microorganism, the C3b can interact with special receptors on phagocytes to promote phagocytosis (Figure 15-14). This phenomenon is called *opsonization* or *immune adherence*. In the process, C3b functions as an opsonin by coating the microorganism and promoting attachment of the phagocyte to the microbe.

Once complement is activated, its destructive capabilities usually cease very quickly in order to minimize destruction of the host cells. This is accomplished by the spontaneous breakdown of activated complement and by interference from inhibitors and destructive enzymes in the host's body fluids.

In addition to the importance of the complement and properdin system in defense, the system also assumes a role in causing disease as a result of inherited deficiencies. For example, deficiencies of C1, C2, or C4 cause collagen vascular disorders that result in hypersensitivity (anaphylaxis); deficiency of C3, though rare, results in increased susceptibility to bacterial infections; and C5 through C9 defects result in increased susceptibility to *Neisseria meningitidis* and *N. gonorrhoeae* infections.

Interferon (IFN)

Because viruses depend on their host cells to provide many functions of viral multiplication, it is difficult

---

Figure 15-13   Stimulation of inflammatory response by complement. (a) C3a and C5a bind to mast cells, basophils, and platelets to trigger release of histamine, which increases blood vessel permeability. **(b)** C5a functions as a chemotactic factor that attracts phagocytes to the site of complement activation.

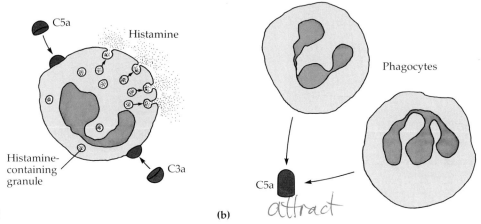

C5a

Histamine

Histamine-containing granule

C3a

Phagocytes

C5a

attract

(a)   (b)

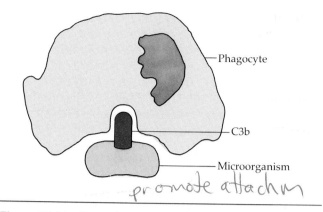

promote attachm

Figure 15-14 Opsonization by complement.

Labels on figure: Phagocyte, C3b, Microorganism

to inhibit viral multiplication without at the same time affecting the host cell itself. One way the infected host counters viral infections is with interferons. **Interferons** are a class of similar antiviral proteins produced by certain animal cells after viral stimulation. One of the principal functions of interferons is to interfere with viral multiplication. As you will see in later chapters, interferon also plays a role in immunologic activities, such as antibody production, allergies, graft rejection, and macrophage activation.

One of the most interesting features of interferons is that they are host-cell-specific but not virus-specific. This means that interferon produced by human cells protects human cells but will produce little antiviral activity for cells of other species, such as mice or chickens. However, the interferon of a species is active against a number of different viruses.

Not only do different animal species produce different interferons, but different types of cells in an animal also produce different interferons. Human interferons are of three principal types: (1) *alpha interferon* (α-IFN, formerly called leucocyte IFN), (2) *beta interferon* (β-IFN, formerly called fibroblast IFN), and (3) *gamma interferon* (γ-IFN, formerly called immune IFN). There are also various subtypes of interferon within each of the principal groups. In the human body, interferon is produced by fibroblasts in connective tissue, by lymphocytes, and by other leukocytes. The three types of interferon produced by these cells can each have a slightly different effect on the body.

All interferons are small proteins, with molecular weights between 15,000 and 30,000. They are quite stable at low pH and fairly resistant to heat.

Produced by virus-infected host cells only in very small quantities, interferon diffuses to uninfected neighboring cells. It reacts with plasma or nuclear membrane receptors, inducing the uninfected cells to manufacture mRNA for the synthesis of **antiviral proteins (AVPs)** (Figure 15-15). These proteins are enzymes that disrupt various stages of viral multiplication. For example, one AVP inhibits translation of viral mRNA by blocking initiation of protein synthesis. Another inhibits polypeptide elongation. Still another is involved in destroying viral mRNA before translation.

The low concentrations at which interferon inhibits viral multiplication are nontoxic to uninfected cells. Because of its beneficial properties, interferon would seem to be an ideal antiviral substance. But

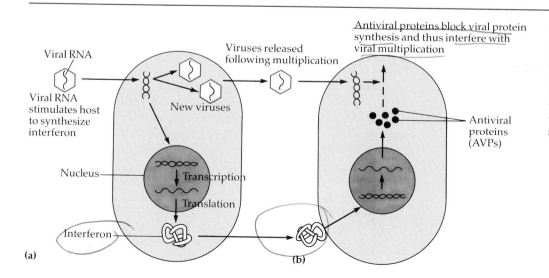

Labels on figure (a): Viral RNA; Viral RNA stimulates host to synthesize interferon; New viruses; Viruses released following multiplication; Nucleus; Transcription; Translation; Interferon

Labels on figure (b): Antiviral proteins block viral protein synthesis and thus interfere with viral multiplication; Antiviral proteins (AVPs)

Figure 15-15 Antiviral action of interferon. (a) A viral-infected cell produces interferon. **(b)** In an uninfected neighboring cell, interferon reacts with plasma or the nuclear membrane receptors and induces synthesis of antiviral proteins (AVPs).

certain problems do exist. For one thing, interferon is effective for only short periods of time. It typically plays a major role in infections that are acute and short-term, such as colds and influenza. Another problem is that it has no effect on viral multiplication in cells already infected.

The importance of interferon in protecting the body against viruses, as well as its potential as an anticancer agent, has made its production in large quantities a top public health priority. Several groups of scientists have successfully applied recombinant DNA technology in inducing certain species of bacteria to produce interferon. (This technique is described in Chapter 8.) The interferons produced with recombinant DNA techniques, called recombinant interferons (rINFs), are important for two reasons: They are pure, and they are plentiful.

Clinical trials to determine the anticancer effects of interferon were begun in 1981. At present, it appears that large doses of rINFs have only limited effects against some tumors and no effect on others. The clinical trials have also revealed that high doses of interferon can have side-effects that range from minor to quite serious, including fatigue, malaise, loss of appetite, fever, chills, pains in the joints, mental confusion, seizures, and cardiac complications.

Although glowing predictions about interferon's anticancer properties have not been justified by the results of clinical trials to date, research is continuing, for several reasons. First, there are many subtypes of interferon, and it might be possible to find the right one with definitive anticancer properties. Second, some laboratory studies indicate that interferon might work together with other chemotherapeutic agents so that the effects of interferon are enhanced. Encouraging results have been obtained by use of interferon in combination with doxorubicin (used to treat a variety of blood malignancies and solid tumors) or cimetidine (used to treat ulcers), for example. Third, it has been observed that patients who did not respond well to either specific chemotherapy or subsequent treatment with interferon showed improvement when returned to the original chemotherapy.

* * *

In Chapter 16, we will discuss specific immune responses. We will also look at the principal factors that contribute to immunity.

Study Outline

Introduction (p. 394)
1. The ability to ward off disease through body defenses is called resistance.
2. Lack of resistance is called susceptibility.
3. Nonspecific resistance refers to all body defenses that protect the body from any kind of pathogen.
4. Specific resistance refers to defenses (antibodies) against specific microorganisms.

Skin and Mucous Membranes (pp. 395–397)

Mechanical Factors (pp. 395–396)
1. The structure of intact skin and the waterproof protein keratin provide resistance to microbial invasion.
2. Some pathogens, if present in large numbers, can penetrate mucous membranes.
3. The lacrimal apparatus protects the eyes from irritating substances and microorganisms.
4. Saliva washes microorganisms from teeth and gums.
5. Mucus traps many microorganisms that enter the respiratory and gastrointestinal tracts; in the lower respiratory tract, the ciliary escalator moves mucus up and out.
6. The flow of urine moves microorganisms out of the urinary tract, and vaginal secretions move microorganisms out of the vagina.

Chemical Factors (p. 397)
1. Sebum contains unsaturated fatty acids, which inhibit the growth of pathogenic bacteria. Some bacteria commonly found on the skin can metabolize sebum and cause the inflammatory response associated with acne.
2. Perspiration washes microorganisms off the skin.

3. Lysozyme is found in tears, saliva, nasal secretions, and perspiration.
4. The high acidity (pH 1.2 to 3.0) of gastric juice prevents microbial growth in the stomach.
5. Normal flora prevent the growth of many pathogens.

microb antagonism

Phagocytosis (pp. 397–402)

1. Phagocytosis is the ingestion of microorganisms or particulate matter by a cell.
2. Phagocytosis is performed by phagocytes, certain types of white blood cells or derivatives of them.

Formed Elements of the Blood (pp. 397–399)

1. Blood consists of plasma (fluid) and formed elements (cells and cell fragments).
2. Leukocytes (white blood cells) are divided into two categories—granulocytes (neutrophils, basophils, and eosinophils) and agranulocytes (lymphocytes and monocytes).
3. During many infections, the number of leukocytes increases (leukocytosis); some infections are characterized by leukopenia (decrease in leukocytes).

Kinds of Phagocytic Cells (pp. 399–400)

1. Among the granulocytes, neutrophils are the most important phagocytes.
2. Enlarged monocytes become wandering macrophages and fixed macrophages.
3. Fixed macrophages are located in selected tissues and are part of the mononuclear phagocytic system.
4. Granulocytes predominate during the early stages of infection, whereas monocytes predominate as the infection subsides.

eat up trash leftover

Mechanism of Phagocytosis (pp. 401–402)

1. Chemotaxis is the process by which phagocytes are attracted to microorganisms.
2. The phagocyte then adheres to the microbial cells; adherence may be facilitated by opsonization—coating the microbe with plasma proteins.
3. Pseudopods of phagocytes engulf the microorganism and enclose it in a phagosome to complete ingestion.
4. Many phagocytized microorganisms are killed by lysosomal enzymes. *lysosomes*

Inflammation (pp. 402–404)

1. Inflammation is a bodily response to cell damage characterized by redness, pain, heat, swelling, and sometimes loss of function.

Vasodilation and Increased Permeability of Blood Vessels (pp. 403–404)

1. The release of histamine, kinins, and prostaglandins causes vasodilation and increased permeability of blood vessels.
2. Blood clots can form around an abscess to prevent dissemination of the infection.

Phagocyte Migration (p. 404)

1. Phagocytes have the ability to stick to the lining of the blood vessels (margination).
2. They also have the ability to squeeze through blood vessels (diapedesis).
3. Pus is the accumulation of damaged tissue and dead microbes, granuloctyes, and macrophages.

Repair (p. 404)

1. A tissue is repaired when the stroma (supporting tissue) or parenchyma (functioning tissue) produces new cells.
2. Stromal repair by fibroblasts produces scar tissue.

Fever (p. 405)

1. Fever is an abnormally high body temperature produced in response to a bacterial or viral infection.
2. Bacterial endotoxins and interleukin-1 can induce fever.
3. A chill indicates a rising body temperature; crisis (sweating) indicates that the body's temperature is falling.

Antimicrobial Substances (pp. 405–410)

Complement and Properdin (pp. 405–408)

1. Complement and properdin are serum proteins that activate one another to destroy invading microorganisms.
2. Antigen–antibody reactions eventually activate C3 protein. Certain polysaccharides and proteins activate properdin, which in turn activates C3.
3. C3 activation can result in cell lysis, inflammation, and opsonization.

Interferon (IFN) (pp. 408–410)

1. Interferons are antiviral proteins produced in response to viral infection.
2. There are three types of human interferon, α-IFN, β-IFN, and γ-IFN. Recombinant interferon has also been produced.
3. Interferon's mode of action is to induce uninfected cells to produce antiviral proteins (AVPs) that prevent viral replication.
4. Interferons are host-cell-specific but not virus-specific.

Study Questions

Review

1. Define the following terms.
   (a) Resistance
   (b) Susceptibility
   (c) Nonspecific resistance

2. Describe the mechanical factors of the skin and mucous membranes that assume a role in nonspecific resistance.

3. Describe the chemical factors of the skin and mucous membranes that assume a role in nonspecific resistance.

4. Define phagocytosis.

5. Compare the structures and functions of granulocytes and monocytes in phagocytosis.

6. How do fixed and wandering macrophages differ?

7. Define inflammation, and list its characteristics.

8. Explain how kinins could be responsible for symptoms of the common cold.

9. Why is inflammation beneficial to the body?

10. How is fever related to nonspecific defense?

11. What is the importance of chill and crisis during fever?

12. What is interferon? Discuss its role in nonspecific resistance.

13. What is complement?

14. Summarize the major outcomes of complement activation.

15. What is properdin? What are its functions in nonspecific resistance?

Challenge

1. Diagram the following processes that result in phagocytosis: margination, diapedesis, adherence, and ingestion.

2. A variety of drugs with the ability to reduce inflammation are available. Comment on the danger of misuse of these antiinflammatory drugs.

3. A hematologist often performs a differential count on a blood sample. A differential count determines the relative numbers of white blood cells. Why are these numbers important? What do you think the hematologist would find in a differential count of a patient with mononucleosis? With neutropenia? With eosinophilia?

Further Reading

Beaman, L., and B. L. Beaman. "The role of oxygen and its derivatives in microbial pathogenesis and host defense." *Annual Review of Microbiology* 38:27–48, 1984. Describes the neutrophil's "respiratory burst" using myeloperoxidase and singlet oxygen to kill microbes.

Bretscher, M. S. "How animal cells move." *Scientific American* 257(6):72–90, December 1987. Describes the movement of phagocytes, fibroblasts, and cancer cells.

Collier, H. O. J. "Kinins." *Scientific American* 207(2):111–118, August 1962. Discusses roles of chemical mediators in inflammation.

Edelson, R. L., and J. M. Fink. "The immunologic function of skin." *Scientific American* 256(6):46–53, June 1985. Discusses the role of T cells in skin.

Friedman, R. M., and N. B. Fintner, eds. *Interferon: Mechanisms of Production and Action*, Vol. 3. New York: Elsevier, 1984. Comprehensive coverage on the status of *in vivo* and *in vitro* mechanisms of action and clinical uses of interferons.

McNabb, P. C., and T. B. Tomasi. "Host defense mechanisms at mucosal surfaces." *Annual Review of Microbiology* 33:477–496, 1981. Describes various host defenses, including normal microbial flora and IgA antibodies, and how they prevent the growth of pathogens.

Old, L. J. "Tumor necrosis factor." *Scientific American* 258(5):59–75, May 1988. Discusses how bacterial (LPS) endotoxin stimulates production of tumor-destroying proteins in animals.

Ross, G. E., ed. *Immunobiology of the Complement System: An Introduction for Research and Clinical Medicine*. New York: Academic Press, 1986. A compilation of information on complement.

Trager, W. "The biochemistry of resistance to malaria." *Scientific American* 244(3):154–164, March 1981. Describes the evolution of sickle-cell anemia and how this disease offers protection from malaria.

Wood, W. B. "White blood cells versus bacteria." *Scientific American* 184(2):48–52, February 1951. Electron micrographs illustrate the action of phagocytes.

CHAPTER 16

Specific Defenses of the Host: The Immune Response

In the last chapter, we discussed several nonspecific host defenses, including the intact skin and mucous membranes, phagocytosis, inflammation, and fever. These are integral components of the body's resistance to disease. In addition to these nonspecific defenses, we have an **innate resistance** to certain illnesses. For example, all humans are resistant to many infectious animal diseases, such as canine distemper and chicken and hog cholera. However, resistance to human diseases, such as measles, can vary from person to person. Although the illness is often relatively mild for persons of European ancestry (possibly because many generations have been exposed to the measles virus), the disease caused many deaths among the Polynesians who were first exposed to measles by European explorers. A person's resistance to disease also depends on genetic factors, age, nutritional status, and general health.

In this chapter, we will study the immune system, another aspect of the body's defenses. In contrast to resistance, immunity involves a *specific* defensive response when the host is invaded by foreign organisms or other foreign substances. Invading organisms might be pathogenic bacteria, viruses, fungi, protozoans, or helminths; foreign materials include pollen, insect venom, and transplanted tissue. The body recognizes these substances as not belonging to itself, and it develops an immune response to inactivate or destroy them. We will see in this chapter and the next two chapters that the immune system is essential to our survival, but the immune system can also cause harm when its efforts are misdirected.

Learning Objectives

- Contrast the four kinds of acquired immunity.
- Define immunity, antigen, anamnestic response, lymphokines, and monoclonal antibody.
- Describe the clonal selection theory.
- Explain what an antibody is, and describe the structural and chemical characteristics of antibodies.
- Provide at least one function for each of the five classes of antibodies.
- Compare and contrast humoral and cell-mediated immunity.
- Differentiate among the types of T cells.

Types of Acquired Immunity

Acquired immunity refers to the protection an organism develops against certain types of microbes or foreign substances. Acquired immunity is developed during a person's lifetime; it is not inherited.

Immunity can be acquired either actively or passively. Immunity is acquired *actively* when a person is exposed to microorganisms or foreign substances, and the immune system responds by producing specialized lymphocytes and special proteins called **antibodies**. Organisms or substances that provoke such a response are called **antigens**; they are usually destroyed or inactivated by the specific antibodies and lymphocytes the body produces. Immunity is acquired *passively* when antibodies are transferred from one organism to another. This type of immunity is considered to be passive because the recipient does not actively synthesize the antibodies but has them provided by an outside source. Immunity in the recipient lasts only as long as the antibodies are active—in most cases, for several weeks or months.

Both actively acquired immunity and passively acquired immunity can be obtained by natural or artificial means.

Naturally Acquired Immunity

Naturally acquired active immunity is obtained when a person is exposed to antigens in the course of daily life and the immune system responds by producing antibodies and specialized lymphocytes. Immunity is life-long for some diseases, such as measles, chickenpox, and yellow fever. For other diseases, especially intestinal diseases, the immunity usually lasts for only a few years. For some diseases, such as pneumococcal pneumonia, gonorrhea, and the common cold, there appears to be only short-lived or no immunity to reinfection. Sometimes *subclinical infections* (those that produce no evident symptoms) can also confer immunity. Immunity to mumps, hepatitis, and polio, for example, is often acquired from subclinical infections.

Naturally acquired passive immunity involves the natural transfer of antibodies from a mother to her infant. An expectant mother is able to pass some of her antibodies to her fetus across the placenta. This mechanism is called *placental transfer*. If the mother is immune to such diseases as diphtheria, rubella, or polio, the newborn infant will be immune to these

diseases as well. Certain antibodies are also passed from the mother to her nursing infant in breast milk, especially in the first secretions called *colostrum*. In the infant, immunity generally lasts only as long as the transmitted antibodies are active—usually a few weeks or months.

Artificially Acquired Immunity

Artificially acquired active immunity results from vaccination. **Vaccination**, also called **immunization**, introduces specially prepared antigens called **vaccines** into the body. Vaccines consist of inactivated bacterial toxins (toxoids), killed microorganisms, or living but attenuated (weakened) microorganisms. These substances are no longer able to cause disease, but they are still able to stimulate an immune response, much as naturally acquired pathogens do. Vaccines are discussed in more detail in Chapter 17.

Unlike artificially acquired active immunity, which occurs in response to the injection of *antigens*, **artificially acquired passive immunity** involves an injection of *antibodies* into the body. These antibodies come from an animal or human who is already immune to the disease. For example, a person who is bitten by a snake might be injected with antibodies from a horse that is immune to snake venom.

The antibodies are found in the serum of the immune animal or human. *Serum* (plural, *sera*) is the fluid remaining after blood has clotted and the blood cells and clotted matter have been removed. Because most of the antibodies remain in the serum, the word **antiserum** has become a generic term for fluids containing antibodies. When serum is subjected to an electric current in the laboratory in electrophoresis (discussed in Chapter 9), the proteins within it move at different rates, as shown in Figure 16-1. One of the components that is separated out by this procedure contains most of the antibodies that were present in the original sample. This antibody-rich serum component is called **gamma globulin** or **immune serum globulin**.

When immune serum globulin is injected into the body, it gives immediate protection against a disease. Immune serum can be gathered from immune persons and used to protect others against common diseases, such as measles, rubella, hepatitis A, and rabies. However, although artificially acquired passive immunity is immediate, it is short-lived. The body produces no new antibodies. The half-life of an injected antibody is typically about three weeks.

The various types of immunity are summarized in Table 16-1.

Protein migration →

Cathode (−)   Anode (+)

Trough

γ   β   α

Globulins   Albumin

Figure 16-1   Separation of serum proteins by electrophoresis. In this procedure, serum is placed in a trough cut into a gel. In response to an electric current, the negatively charged proteins of the serum migrate through the gel from the negatively charged end (cathode) to the positively charged end (anode). Antibodies are concentrated mainly in the gamma (γ) fraction of globulins, hence the familiar term *gamma globulin*.

TABLE 16-1   *Types of Acquired Immunity*

Type of Immunity	How Acquired
Naturally acquired active immunity	Antigens enter the body; antibodies and specialized lymphocytes are produced
Naturally acquired passive immunity (placental and milk transfer)	Fetus or baby receives antibodies from an immunized mother
Artificially acquired active immunity	Prepared antigens in vaccines are injected into susceptible individual, who produces antibodies and specialized lymphocytes
Artificially acquired passive immunity	Immune serum is injected into susceptible individual, who receives preformed antibodies

The Duality of the Immune System

The human immune system, like that of all vertebrates, has two main components, the humoral immune system and the cell-mediated immune system. These components work both separately and together to protect an organism from disease.

The Humoral Immune System

The **humoral immune system** involves antibodies that are dissolved in extracellular fluids, such as blood plasma, lymph, and mucus secretions—fluids formerly known as *humors*. This system responds when specialized lymphocytes, called **B cells**, are exposed to antigens. The B cells (Figure 16-2a) produce antibodies that are specifically directed against the antigens. The humoral immune response defends mostly against bacteria, bacterial toxins, and viruses in the body's fluids.

The Cell-Mediated Immune System

The second component of the immune system is called the **cell-mediated immune system** because it directly involves specialized lymphocytes called **T cells** (Figure 16-2b). These cells are located in both the blood and the lymphoid tissues (see Figure 15-5).

(a)

(b)

Figure 16-2   B cells and T cells. These are the two major types of lymphocytes responsible for the immune response. **(a)** B cells produce humoral antibodies that are specifically directed against antigens. This transmission electron micrograph of a B cell shows its extensive endoplasmic reticulum, part of the protein-synthesizing machinery used to produce antibodies (×8900). **(b)** Transmission electron micrograph of a T cell (×11,000). T cells play a key role in cell-mediated immunity.

T cells do not secrete antibodies, as B cells do. Instead, they have antibodylike molecules, called *antigen receptors*, attached to their surfaces. These receptors allow T cells to recognize and react to many different kinds of antigens. There are several different kinds of T cells, each of which carries out a specific function.

The cell-mediated immune response is most effective against bacteria and viruses located within phagocytic or infected host cells and against fungi, protozoans, and helminths. The cell-mediated immune system also responds to transplanted tissue (such as a foreign skin graft) by mounting an immune response to reject it. This system is thought to be an important defense against cancer.

Antigens and Antibodies

Antigens and antibodies play key roles in the immune response system. As we mentioned previously, antigens (sometimes called **immunogens**) provoke a highly specific immune response in an organism. Usually, this response involves the formation of antibodies or highly specialized T cells.

The Nature of Antigens

Normally, the immune system recognizes components of the body it protects as "self" and foreign matter as "nonself." This recognition is the reason why people's defense systems do not usually produce antibodies against their own body tissues (although we will see in Chapter 18 that this sometimes occurs).

The vast majority of antigens are proteins, nucleoproteins (nucleic acid + protein), lipoproteins (lipid + protein), glycoproteins (carbohydrate + protein), or large polysaccharides. These proteins and polysaccharides are often components of invading microbes—the capsules, cell walls, flagella, pili, and toxins of bacteria, the coats of viruses, and the surfaces of many other types of cells. Nonmicrobial antigens include pollen, egg white, blood cells and serum proteins from other persons or species, and transplanted tissues and organs.

Generally, antibodies do not recognize the whole antigen. Instead, antibodies recognize and interact with specific regions on an antigen's surface, called **antigenic determinants** (Figure 16-3). The nature of this interaction depends on the size and shape of the antigenic determinant and the chemical structure of the antibody. Most antigens have many different types of determinants on their surface. Since different

determinants are recognized by different antibodies, the immune system may produce several distinct antibodies against a single antigen.

Most antigens have molecular weights of 10,000 or more. A foreign substance that has a low molecular weight is often not antigenic unless it is attached to a carrier molecule. These low-molecular-weight antigens are called **haptens** (from *haptein*, to grasp; see Figure 16-4). Once an antibody against the hapten

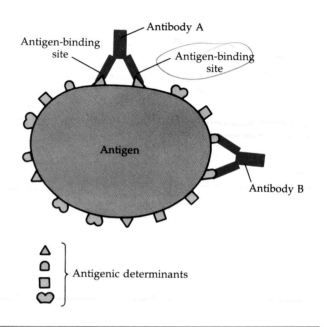

Figure 16-3 Antigenic determinants. Most antigens contain more than one antigenic determinant site, and there can be several different kinds of sites on one antigen. Each antibody molecule has at least two binding sites that can attach to a specific kind of determinant site on an antigen. It is also possible for an antibody to bind to identical sites on two different cells at the same time. This would cause neighboring antigens to aggregate if enough antibodies and antigens were present.

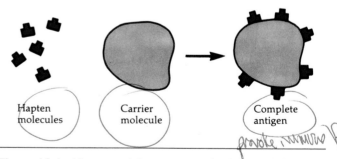

Figure 16-4 Haptens. A hapten is a molecule too small to stimulate antibody formation by itself. However, when the hapten is combined with a larger carrier molecule, usually a serum protein, the hapten and its carrier together function as an antigen and can provoke an immune response.

has been formed, the hapten will react with the antibody independent of its carrier. Penicillin is a good example of a hapten. This drug is not antigenic by itself, and relatively few people develop allergic reactions to it. However, when penicillin combines with the serum proteins of some persons, the resulting molecule initiates an immune response.

The Nature of Antibodies

Antibodies belong to a class of proteins called **immunoglobulins (Ig)**. These proteins can recognize, bind to, and help cause the destruction of antigens. Antibodies are highly specific; they can interact with only one type of antigenic determinant on an antigen.

Each antibody has at least two sites that bind to antigenic determinants. These sites are known as **antigen-binding sites**. The number of antigen-binding sites on an antibody determines the **valence** of that antibody. For example, most human antibodies have two binding sites; therefore, they are bivalent.

Antibody Structure

Because a bivalent antibody is the simplest antibody type, it is called a **monomer**. A typical antibody monomer has four protein chains—two identical *light (L) chains* and two identical *heavy (H) chains* (*light* and *heavy* refer to the relative molecular weights). The chains are joined by disulfide links (Chapter 2) and other bonds to form a Y-shaped molecule (Figure 16-5). The Y-shaped molecule is flexible and can assume a T shape (note the hinge area indicated on the antibody molecule in Figure 16-5).

Both the H and L chains have sections located at the ends of the Y's arms, called *variable (V)* regions. Together, the V regions from one L chain and one H chain form one antigen-binding site; hence, each antibody has two such sites. The amino-acid sequences of these V regions vary from one antibody to another. The resulting variations in chemical and three-dimensional structure account for the ability of different antibodies to recognize and bind with different antigens. Any single antibody molecule has only one type of L chain and one type of H chain, even if it is made up of more than one monomer. Therefore, each antibody molecule has only one type of antigen-binding site and will bind to only one type of antigenic determinant.

Figure 16-3 shows an antibody combining with antigenic determinants on an antigen. In the illustration, the antigen might be a bacterial cell with several different antigenic determinants. Note that more than one type of antibody can react with the antigen. If the two antigen-binding sites of an antibody combine with antigenic determinants on two different antigens, the antigens can aggregate into

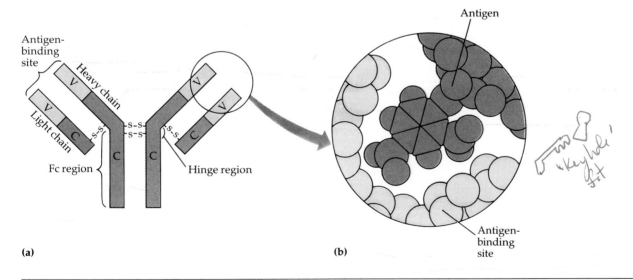

Figure 16-5 Structure of a typical antibody molecule. (a) The Y-shaped molecule is composed of two light and two heavy chains linked together by disulfide bridges (S—S). Most of the molecule is made up of constant regions (C), which are the same for all antibodies of the same class. The amino acid sequences of the variable regions (V), which form the two antigen-binding sites, differ from molecule to molecule. The stem of the Y, called the Fc region, allows the antibody to bind to the surface of certain host cells. **(b)** One antigen-binding site is shown enlarged and bound to an antigen. The three-dimensional fit between the antigen-binding site and the antigen is similar to the fit between an enzyme and its substrate.

clumps. As we will discuss in Chapter 17, this clumping can be an important factor in the diagnosis of some diseases.

The stem of the antibody monomer and the lower parts of the Y's arms are called *constant* (C) regions (Figure 16-5). The term *constant* refers to the fact that C regions of the L and H chains of different monomers are relatively invariable in their amino-acid sequence. There are five major sequences found for C regions of H chains, and there are two different sequences for L chains. Each H-chain sequence determines a different class of immunoglobulin.

The stem of the Y-shaped antibody monomer is called the *Fc* (crystallizable fragment) *region*. The antibody molecule can attach to a host cell at the Fc region. Complement (Chapter 15) can also bind to the sides of the Fc region. The two antigen-combining sites are left free to bind to compatible antigenic determinants.

Immunoglobulin Classes

The five classes of immunoglobulins (Ig) are designated IgG, IgM, IgA, IgD, and IgE, according to the properties of their heavy chains. Each class plays a different role in the immune response. The structures of IgG, IgD, and IgE molecules resemble the structure shown in Figure 16-5a. Molecules of IgM and IgA are usually two or more monomers joined together (Figure 16-6). The characteristics of the immunoglobulin classes are summarized in Table 16-2.

IgG   IgG antibodies account for about 80% to 85% of all antibodies in serum. These monomer antibodies readily cross the walls of blood vessels and enter tissue fluids. Maternal IgG antibodies, for example, can cross the placenta and confer passive immunity to the fetus. High concentrations of antigen must be present before a particular IgG is made.

IgG antibodies protect against circulating bacteria and viruses, neutralize bacterial toxins, trigger the complement system, and, when bound to antigens, enhance the effectiveness of phagocytic cells.

IgM   Antibodies of the IgM class comprise from 5% to 10% of the antibodies in serum. IgM has a pentamer structure, consisting of five Y-shaped monomers that are held in position by a *J (joining) chain* (Figure 16-6).

IgM antibodies are the first ones to appear in response to the initial exposure to an antigen. How-

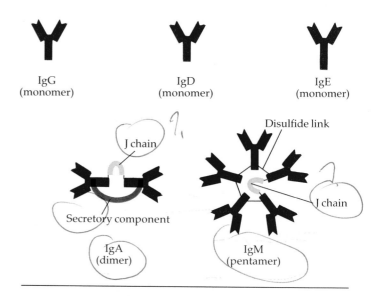

Figure 16-6   Human immunoglobulins. Structures of the five principal classes of human immunoglobulins are shown. Note that IgA and IgM are made up of two and five monomers, respectively, in this drawing. In these cases, the monomers are held together by disulfide links, and some of these are joined by a polypeptide called the J (joining) chain. IgA may also occur as a monomer, and it is usually found attached to a protein called the secretory component.

ever, the initially high IgM concentration in the blood rapidly declines, and the IgG concentration increases (Figure 16-7). A second exposure to antigen results mostly in the increased production of IgG.

The large size of the molecule may prevent IgM from moving about as freely as IgG does. IgM antibodies generally remain in the blood vessels and do not enter the surrounding tissues.

Because of its numerous antigen-binding sites, the IgM molecule is especially effective at cross-linking particulate antigens and causing their aggregation (illustrated in Figure 17-6). IgM is the predominant antibody involved in the response to the ABO blood group antigens on the surface of red blood cells (discussed in Chapter 17). It is also especially effective in reactions involving complement. IgM can enhance the ingestion of target cells by phagocytic cells, as IgG does.

The fact that IgM appears first in response to a primary infection and is relatively short lived makes it uniquely valuable in disease diagnosis. If high concentrations of IgM against a disease pathogen are detected in a sick patient, then it is likely that this disease is caused by that pathogen. Detection of IgG, which is relatively long lived, may indicate only that immunity against a particular disease was acquired in the more distant past.

TABLE 16-2 *Summary of Immunoglobulin Classes*

Characteristics	IgG	IgM	IgA	IgD	IgE
Percent of total serum antibody	80–85	5–10	15	0.2	0.002
Location	Blood, lymph, intestine	Blood, lymph, B cell surface (as monomer)	Secretions (tears, saliva, mucus, intestine, milk), blood, lymph	B cell surface, blood, lymph	Bound to mast and basophil cells throughout body, blood
Molecular weight	150,000 (monomer)	900,000 (pentamer)	320,000 (dimer)	185,000 (monomer)	200,000 (monomer)
Half-life (days) in serum*	25	5	6	3	2
Complement fixation	Yes	Yes	No**	No	No
Placental transfer	Yes	No	No	No	No
Functions known	Enhances phagocytosis, neutralizes toxins and viruses, and protects fetus and newborn	Especially effective against microorganisms and agglutinating antigens; first antibodies produced in response to initial infection	Localized protection on mucosal surfaces	Not known, but presence on B cells may indicate function in initiation of immune response	Allergic reactions; possibly expulsion or lysis of protozoan parasites

*Time required for one-half of the antibodies to disappear.
**May be "yes" via alternate pathway.

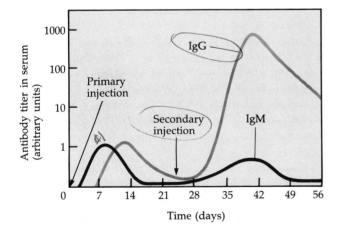

Figure 16-7 The primary and secondary immune response to an antigen. IgM appears first in response to the primary injection. IgG follows and provides longer-term immunity. The second (booster) injection results in a much faster and greater production of antibodies. The antibodies produced in response to the second injection are mostly IgG.

IgA IgA antibodies account for about 15% of the antibodies in serum. IgA can be monomers or *dimers*, two Y-shaped monomers connected by a J chain. Molecules of IgA are found in body secretions, such as saliva, sweat, and secretions of the gastrointestinal tract. IgA is also found in colostrum.

IgA becomes attached to a protein called the *secretory component* during the transport of IgA from the blood to the surface of secretory tissues. The secretory component protects the IgA from enzymatic degradation; it may also facilitate the entrance of IgA into the secretory tissues.

The main function of secretory IgA is probably to prevent the attachment of pathogens, particularly viruses and certain bacteria, to mucosal surfaces. Its presence in colostrum probably helps to protect the infant's gastrointestinal tract from infection and assists in the colonization of the tract by normal flora.

IgD IgD antibodies comprise only about 0.2% of the total serum antibodies. Their structure resembles that of the IgG molecules. IgD antibodies are found in blood and lymph and on the surfaces of B cells. These antibodies do not fix complement and cannot cross the placenta. Little is known about the function of IgD. However, the fact that IgD antibodies are found in high concentrations on the surface of B cells, especially in newborns, might indicate that these antibodies help to initiate the immune response.

IgE Antibodies of the IgE class are slightly larger than IgG molecules, but they constitute only 0.002% of the total serum antibodies. The IgE molecules bind tightly by their Fc ends to receptors on mast cells and basophils, which are specialized cells that participate in allergic reactions (Chapter 18). When an antigen such as pollen reacts with the IgE antibodies attached to a mast cell or basophil, that cell releases histamine and other chemical mediators. A person who is highly allergic or is a host to a gastrointestinal parasite often has an abnormally high serum level of IgE.

B Cells and Humoral Immunity

As we mentioned earlier, there are two basic kinds of immune responses. One of these, the humoral response, is carried out by a special group of cells called B cells. B cells are responsible for the production of antibodies.

B cells, and other cells, such as T cells and macrophages, develop from stem cells located in the bone marrow in adults and in the liver in embryos. By the time a B cell is mature, it may carry as many as 100,000 antibody molecules on its surface. Each B cell can only produce antibodies against one specific antigen; however, the total population of B cells in a person can produce antibodies against an enormous assortment of antigens.

B-Cell and Antigen Interactions

The mature B cells migrate to lymphoid organs, where they encounter antigen (Figure 16-8). When the appropriate antigen contacts the antigen receptor antibodies on a B cell, the B cell proliferates into a large clone of cells (Figure 16-9). (This phenomenon, called clonal selection, will be discussed later.) Some of these cells differentiate into **plasma cells**, which secrete antibodies against the antigen. Each plasma cell lives for only a few days but is capable of producing about 2000 humoral antibody molecules per second.

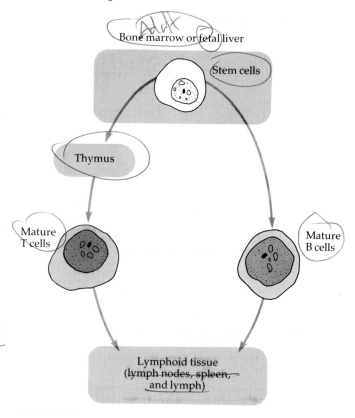

Figure 16-8 Differentiation of T cells and B cells. Both types of cells originate from stem cells in adult bone marrow or in the fetal liver. (Red blood cells, macrophages, neutrophils, and other white blood cells also originate from these same stem cells.) Some cells pass through or are influenced by the thymus and emerge as mature T cells. Other cells are influenced by the bone marrow and become B cells. Both types of cells then migrate to lymphoid tissues, such as the lymph nodes or spleen.

Stimulation of a B cell by an antigen also results in the production of a population of **memory B cells**. These cells will be discussed later in the chapter.

Sometimes, the production of antibodies by a B cell may depend on the cooperation of other cells. For example, B cells must work cooperatively with certain macrophages and T cells to produce antibodies against a type of antigen known as **T-dependent antigens**. Examples of T-dependent antigens include bacteria, foreign red blood cells, proteins, and hapten-carrier combinations that have many different antigenic determinants.

The T-dependent antigen is attacked and partly digested by macrophages or dendritic cells (cells with a highly branched appearance), which are known as *antigen-presenting cells (APC)*. (See Figure 16-10.) Portions of the antigen, usually polypeptide fragments, are taken up by an antigen-presenting cell and

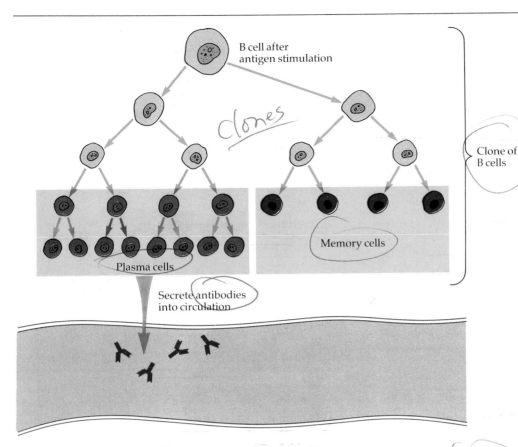

Figure 16-9 Differentiation of B cells. After a B cell has been stimulated by an antigen, the B cell divides to produce a large number of genetically identical progeny cells (a clone). Some of the progeny cells become plasma cells, which secrete antibodies. Others become memory B cells.

Figure 16-10 Antigen-presenting cells. Antigen-presenting cells are usually macrophages or dendritic cells that present antigen to helper T cells. This scanning electron micrograph shows a macrophage (top) covered with typical ridges and a dendritic cell (bottom) exhibiting the branching morphology that suggested their name ($\times 3600$). The antigen molecules being presented are presumably on the surface of the cells but are too small to be seen here.

are then positioned on the cell's surface. Specific T cells, the *helper T cells* (T_H), interact with antigen-presenting cells. T_H cells recognize the antigen fragments and certain "self" antigens (surface antigens that identify the cell as "self") that are also carried on the surface of antigen-presenting cells (Figure 16-11).

In fact, the T_H cell recognizes the foreign antigen only when it is combined with a self antigen on the APC. This encounter primes the T_H cell to activate the appropriate B cell. It is necessary that the B cell also carry self antigens; indeed, it is likely that the T-cell receptor recognizes a single complex of the antigen and the self antigen on the B-cell surface. The requirement that both types of antigens be matched in order to activate antibody production minimizes the chances of making antibodies against our own antigens. All immune responses that produce IgG, IgA, and IgE involve helper T cells.

The self antigens just mentioned are actually cell-surface proteins that are components of the **major histocompatibility complex (MHC)**. The particular collection of MHC proteins a person carries is unique to that individual. Thus, MHC proteins are well suited to serve as self antigens—signals by which the immune system distinguishes self from nonself. The MHC will be discussed further in Chapter 18.

T-independent antigens can stimulate a response from B cells without the intervention of T cells. These T-independent antigens are usually composed of repeating polysaccharide or protein subunits. Bacterial flagella, which are composed of protein, and the lipopolysaccharide layer of gram-negative are impor-

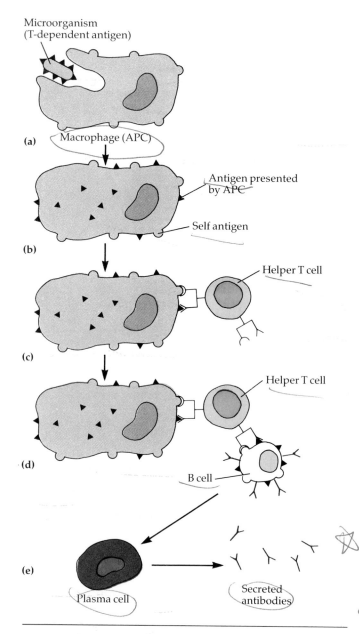

Figure 16-11   How helper T cells may activate B cells to make antibodies against T-dependent antigens. B cells require the help of antigen-presenting cells (APCs) and specialized helper T cells to produce antibodies against T-dependent antigen. The process shown is a likely one, but it is not completely understood. **(a)** The microbial antigen is attached by an APC and partially digested. **(b)** Antigen fragments are presented on the APC surface. Like all body cells, the APC also has self (MHC) antigens on the surface. **(c)** A helper T cell (T$_H$) specific for the presented antigen interacts with both that antigen and the self antigen. **(d)** The helper T cell then activates an appropriate B cell, probably one that itself has both antigens on its surface, as well as antibodies to the microbial antigen. **(e)** The interaction of T and B cells, along with direct stimulation by the microbial antigen, triggers the B cell to differentiate into a plasma cell, which secretes antibodies.

Figure 16-12   T-independent antigens. T-independent antigens have repeating units that form multiple bonds with a B cell. These antigens stimulate the B cell to make antibodies without the aid of helper T cells.

tant examples of T-independent antigens. These antigens are able to form multiple bonds with B-cell receptors, which is probably why they do not need T cell assistance (Figure 16-12). The immune response to T-independent antigens tends to be weaker than the response to T-dependent antigens. Once T-independent antigens are bound to B-cell receptors, the B cells then produce antibodies, which are almost exclusively of the IgM class.

Antibody–Antigen Binding

Antibody recognition of antigens is based on the ability of the antibody to fit and bind to an antigen. An antigen binds to an antibody in the antigen-binding site formed by the variable regions of the light and heavy chains. The resulting combination is called an **antigen–antibody complex.** Such complexes can protect a host in several ways. For example, the complexes can neutralize toxins by blocking their active sites, or they can inactivate viruses by combining with them and preventing their attachment to host cells. These complexes can also fix complement and start the lysis of invading cells.

As we will see in Chapter 18, the action of antibodies can also damage the host. For example, immune complexes of antibody, antigen, and complement can damage host tissue. Antigens combining with IgE on mast cells can initiate allergic reactions, and antibodies can react with host cells and cause autoimmune disorders.

In practice, each B-cell receptor is not always an exact fit for the antigen. When large amounts of antigen are present, some antigens attach to receptors for which they are less than a perfect fit. Because of this, some antibodies are produced that are poorer matches for the antigen than others; these are said to have less *affinity*.

Monoclonal Antibodies

Scientists have long observed that antibody-secreting B cells may become cancerous. When their proliferation is unchecked, they form tumors called myelomas. It is possible to isolate these cancer cells and propagate them in cell culture indefinitely. Cancer cells, in this sense, are "immortal," in contrast to normal B cells, which can be propagated in culture for only a few generations.

In 1975, Georges Köhler and César Milstein discovered how to fuse an "immortal" cancerous cell (a mutant that has lost the ability to form antibody) with a noncancerous antibody-secreting plasma (B) cell taken from a mouse that had been immunized with a particular antigen. The resulting hybrid cell could multiply to form a clone of cells called a **hybridoma**.

When a hybridoma is grown in culture, its genetically identical cells continue to produce the type of antibody characteristic of the ancestral B cell. The importance of the discovery is that clones of the antibody-secreting cells now can be maintained indefinitely in cell culture and can produce immense amounts of a desired pure antibody, as compared with traditional methods.

Because the antibodies are produced by a single hybridoma clone and are identical, they are called **monoclonal antibodies** (Figure 16-13). Monoclonal antibodies are extremely useful for three reasons—they are uniform, they are highly specific, and they can be readily produced in large quantities. Because of these qualities, monoclonal antibodies have assumed enormous importance as diagnostic tools. For example, kits have recently been developed that use monoclonal antibodies to recognize chlamydial and streptococcal bacteria, and pregnancy tests are also available that use monoclonal antibodies to indicate the presence of a hormone excreted only in the urine of a pregnant woman.

Figure 16-13 Production of monoclonal antibodies. A mouse injected with a specific antigen will produce antibodies against that antigen. The spleen of the mouse is removed, and B cells are isolated. The B cells are then mixed with myeloma cells, mutants that have lost the ability to form antibody. Some of the B cells and myelomas fuse to form hybrid cells. The mixture of cells is placed in a medium that allows only successfully fused cells to proliferate into clones called hybridomas. The hybridomas are then screened for production of the desired antibody. These hybridomas are then cultured to produce large amounts of monoclonal antibodies.

Antigen injected into mouse

Spleen removed

Suspension of B cells

Cultured myeloma cells (cancerous B cells), mutants that have lost the ability to form antibodies

Suspension of myeloma cells

Cells fused to form hybrid cells

Cells transferred to selective medium

Only hybrid cells grow, forming clones called hybridomas; other cells die

Hybridomas that produce a desired antibody are cultured

Desired antibody is produced in large amounts

Monoclonal antibodies are also being used to overcome unwanted effects of the immune system, such as rejection of transplanted organs (Chapter 18). In this case, antibodies are prepared that react with the T cells that are involved in rejection of the transplanted tissue; the antibodies suppress the T-cell activity.

Monoclonal antibodies can also be used to make anti-idiotypic vaccines. These vaccines, which are discussed in the next chapter, present a new approach to possible treatments for infectious diseases and cancer.

The prospect of using monoclonal antibodies to treat cancer excites the most interest. One approach is to combine monoclonal antibodies with a toxin to make an *immunotoxin.* The general idea is that such immunotoxins could be carried throughout the body by the circulatory system, specifically attach to cancer cells wherever they might be, and kill them.

The therapeutic use of monoclonal antibodies has been limited because these antibodies are currently produced by mouse cells. The immune systems of some people have reacted against the foreign mouse proteins. Monoclonal antibodies derived from human cells would probably provoke fewer reactions, and it is therefore hoped that these antibodies will eventually be commercially available.

T Cells and Cell-Mediated Immunity

Early in the history of immunological experimentation, biologists learned that some forms of immunity could be transferred between animals by transferring serum from immunized animals to nonimmunized animals. However, other forms of immunity were not transferable with blood serum. Much later it was learned that these latter forms of immunity were transferred only when certain lymphocytes were transferred; the immunity associated with these lymphocytes is now called *cell-mediated immunity.* Immunity to tuberculosis, for example, was found to be most effectively transferred by these lymphocytes.

The Components of Cell-Mediated Immunity

T cells are the key component of cell-mediated immunity. Like B cells, they derive from precursor stem cells in bone marrow, but unlike the precursors of B cells, these precursors develop and differentiate into

T cells within the thymus. After differentiation, the T cells migrate to lymphoid organs, such as the lymph nodes or the spleen. When T cells recognize foreign antigens, they do not secrete antibodies but instead differentiate into various types of **effector T cells** and into long-lived memory cells. The different classes of effector cells are responsible for attacking foreign antigens and regulating the immune response. Effector cells become active in response to antigenic stimulation.

Types of T Cells

There are three main types of effector T cells—helper T cells (T_H), suppressor T cells (T_S), and cytotoxic T cells (T_C). Mature T cells can be identified by characteristic antigens on the cell surfaces.

Helper T cells play a variety of roles. The cause of the pathogenicity of the AIDS virus was identified by using monoclonal antibodies that were programmed to recognize such antigens to show that a certain group of helper T cells was being decimated. These T cells contain a surface antigen called CD4 and are most important in a vigorous immune response. We have already seen that certain helper T cells present T-dependent antigens to B cells. Some other helper T cells help other T cells to respond to antigens. Still another type of helper T cell, called the delayed hypersensitivity T cell (T_D), is associated with certain allergic reactions, such as poison ivy, and with rejection of transplanted tissues. (T_D cells are the cells for which cell-mediated immunity was originally named.) Early investigators found that the transfer of these cells between animals also transferred immunity to tuberculosis, while the transfer of serum does not. In response to antigenic stimulation, T_D cells produce a variety of substances called *lymphokines,* which recruit defensive cells such as macrophages. T_D cells are also important in the body's defense against cancer. We will discuss these topics more fully in Chapter 18.

Suppressor T cells (T_S) are not completely understood. It is generally accepted that they inhibit the conversion of B cells into plasma cells, which might be a way of turning off the immune response when an antigen is no longer present. They apparently also suppress the activity of some other T cells so that immune responses are not developed against host antigens. Both helper T cells and suppressor T cells are also known as **regulatory T cells** because they act as regulators of the immune response.

The third main type of effector cells, **cytotoxic T cells (T_C)**, destroy target cells, such as cancer cells and transplanted tissue, upon contact. Protection against viral infections and intracellular bacterial infections is the most important function of T_C cells. Because viruses (and some bacteria) reproduce within host cells, they cannot be attacked there by humoral antibodies. T_C cells recognize viral antigens on the surface of the virus-producing host cell and cause the destruction of that cell. T_C cells cause the lysis of their target cells by coming into close contact with them and releasing a protein called *perforin*. The protein forms a pore in the target cell membrane and perforates it by a mechanism that is similar to the complement mechanism.

T_C cells will not react with free viruses but only with cells carrying both the target antigen and a class of self (MHC) antigens that are found on most body cells. After it contacts and kills a target cell, the T_C cell continues to live and can kill repeatedly. T_C cells also play at least some role in the defense against cancer and a few parasitic diseases.

Killer Cells

The body's defense system also uses other effector cells. Lymphoid cells called **natural killer (NK) cells** are capable of destroying tumor cells. In contrast to cytotoxic T cells, they do not seem to be immunologically specific; they are simply programmed to attack a broad group of cells, most particularly tumor cells. Natural killer cells are not phagocytic, and they must contact the target cell to lyse it. Another cytotoxic effector cell is the **killer (K) cell**. Killer cells are also not very specific; they will attack any target cells that are coated with antibodies. Receptors on the K-cell surface are specific for the Fc region of the antibodies. Complement activity is not involved, and the precise lethal mechanism is unknown. The functions of killer cells and the other principal cells involved in cell-mediated immunity are briefly summarized in Table 16-3.

Lymphokines

As mentioned previously, when T_D cells (and possibly other T cells) are stimulated by an antigen, they release proteins called **lymphokines**. Since it is now known that cells other than lymphoid cells produce these substances, the more general term **cytokine** is coming into use. One lymphokine produced by T_D cells is *macrophage chemotactic factor*, which attracts macrophages to the infection site. Other lymphokines are *macrophage migration inhibition factor*, which inhibits the movement of macrophages away from the infection site, and *macrophage activation factor*, which activates macrophages. Activated macrophages are much more efficient than unactivated macrophages at destroying these cellular antigens.

TABLE 16-3 *The Principal Cells in Cell-Mediated Immunity*

Cell	Function
Helper T cell (T_H)	Necessary for B-cell activation by T-dependent antigens
Delayed hypersensitivity T cell (T_D)	Provides protection against infectious agents; causes inflammation associated with tissue transplant rejection
Suppressor T cell (T_S)	Regulates immune response and helps maintain tolerance
Cytotoxic T cell (T_C)	Destroys target cells on contact
Killer cell (K)	Attacks antibody-coated target cells
Natural killer cell (NK)	Attacks and destroys target cells

TABLE 16-4 *Summary of Some T-Cell Lymphokines*

Lymphokine	Function
Macrophage chemotactic factor	Attracts macrophages to infection site
Macrophage migration inhibition factor	Prevents macrophages from leaving infection site
Macrophage activation factor	Activates macrophages to improve phagocytic activity
Interleukin-1	Promotes multiplication and activation of B cells and T cells; triggers fever response in body
Interleukin-2	Stimulates proliferation and differentiation of T cells
Lymphotoxin	Destroys nonlymphocytic target cells in vitro; function in vivo uncertain
Interferon	Inhibits viral replication
Transfer factor	Intensifies the action of sensitized T cells

Another important lymphokine is *interleukin-2* (IL-2). IL-2 is responsible for the proliferation and differentiation of T cells. The convention is to name lymphokines for their biological activity until their amino-acid sequence is known. They are then assigned an **interleukin number**.

Lymphokines are probably not involved in the contact destruction of target cells by T$_C$ cells, although a lymphokine called *lymphotoxin* can destroy nonlymphocytic target cells in vitro. The functions of these lymphokines are summarized in Table 16-4, p. 425.

The Cell-Mediated Immune Response

Like B cells, individual T cells seem to have specificity for only a single antigen, which indicates the presence of T-cell receptors to recognize the antigen. However, T cells do not respond to antigens in the same way that B cells do. For one thing, T cells cannot be stimulated by soluble antigens because the cells are unable to bind to them. But T cells do respond to antigens displayed on cell surfaces. Usually, before a T cell can respond to these antigens, the antigens must first be processed by an antigen-presenting cell (APC). As we mentioned earlier, the APC displays the processed antigens on its surface. T cells also respond to MHC antigens. In fact, an important difference between the response of a T cell to an antigen and the response of a B cell to that antigen is that the T cell will only recognize the antigen if it is in close association with an MHC antigen. This type of recognition is known as *associative recognition*.

If the T cell were triggered by the MHC antigen alone, it would mount an immune attack on the body's own tissues. However, it is likely that any T cells with receptors that react with these self antigens are eliminated in the thymus very early in life. These T-cell receptors are not the IgM and IgD molecules that serve this purpose on B cells; but their structures have been recently worked out, and they share many properties with immunoglobulins.

Once an antigen stimulates an APC cell, the cell secretes a substance called interleukin-1 (IL-1), which is a *monokine*. Monokines are biologically active substances secreted by macrophages. The monokine activates the T cell, which begins to synthesize interleukin-2. The T cell also synthesizes surface receptors for IL-2, which bind to the IL-2. When the receptors bind to the IL-2, the T cell begins to proliferate and differentiate into the different effector cells.

Only T cells that have been stimulated by an antigen have receptors for IL-2. So, although the lymphokine is nonspecific, the resulting T cells are specific to that antigen. Figure 16-14 shows the processes that produce the T cells.

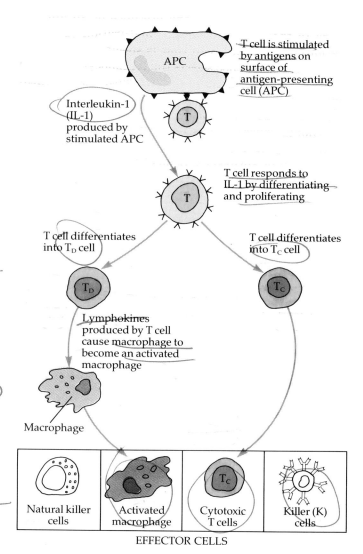

EFFECTOR CELLS

Figure 16-14   Cell-mediated immunity. The effector cells responsible for cell-mediated immunity are shown in this diagram. An activated macrophage is the indirect result of antigenic stimulation, but its ability to now kill ingested cells is not immunologically specific against that antigen. Cytotoxic T cells arise from stimulation of a T cell by an antigen and are antigenically specific against that antigen. Natural killer cells are not immunologically specific against any particular cell but attack certain foreign cells, such as tumor cells. Killer (K) cells carry determinants for the Fc region of antibody molecules. They kill antibody-coated target cells they encounter.

Clonal Selection

We have already mentioned that an individual B or T lymphocyte is specific for only a single antigen. We have also described how the presence of an antigen leads to the proliferation and maturation (differentiation) of the B or T cell that is specific for it. Thus, an antigen in a sense *selects* the lymphocyte that will multiply to form a *clone* of cells with the identical immunological specificity. This fundamental process is called **clonal selection**.

Clonal selection was first proposed in the 1950s as a theory to explain how antigens stimulate the production of antibodies. It has become a basic tenet of immunology. Furthermore, it has been proven to apply equally well to both B cells and T cells. As shown in Figure 16-15, immature B cells are directly stimulated by antigen to proliferate and mature into plasma cells that secrete antibodies against the antigen, as well as into B memory cells (which will be discussed in the next section). In a similar though more complicated process, antigen that binds to a complementary receptor on the surface of an immature T cell also triggers the cell to multiply and differentiate to produce a clone of mature T cells, as well as T memory cells.

The immature B or T cell recognizes its complementary antigen by means of special cell-surface molecules that fit with the antigen. What are these molecules? In the case of B cells, we know a simple and satisfying answer to our question: The receptor molecules are simply antibodies bound to the cell

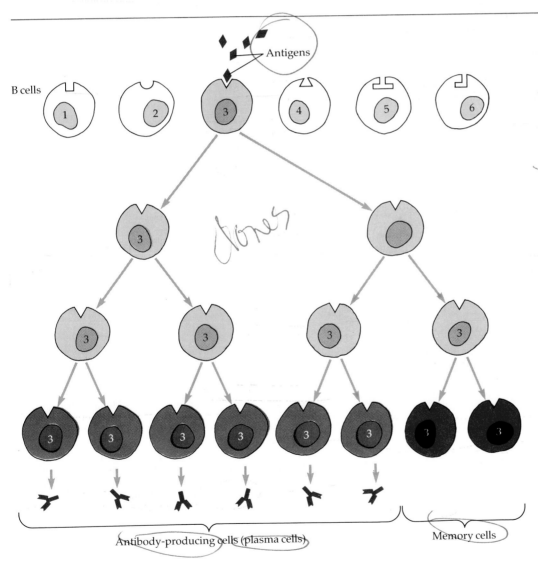

Figure 16-15 Clonal selection. The B cells and T cells of the body are capable of recognizing an almost infinite number of antigens, but each cell recognizes only one type of antigen. The presence of a particular antigen triggers proliferation of a cell that is specific for that antigen into a *clone* of cells with the same specificity. In this illustration, B cell 3 responds to the antigen by proliferating into a clone of cells (color), which then differentiate further into subclones of antibody-producing cells and memory cells. The antibodies that are produced will bind to the antigenic determinant that stimulated the B cell.

Antigens

B cells

1 2 3 4 5 6

clones

Antibody-producing cells (plasma cells)

Memory cells

membrane! T-cell receptors also share many properties with immunoglobulins—in fact, they have been drawn to resemble antibodies in some diagrams in this chapter.

The clonal selection idea raises several more important questions, as any good theory does. One is the question of how the body can possibly generate enough diversity of B and T cells to react with all the different antigens encountered during a lifetime. The answer to the "antibody diversity problem" involves shuffling (recombination) of DNA segments within the genomes of B and T cells to create genes for many

.. Microbiology in the News ....

Molecular Biology and Immunology— Nobel Prize

We have long known that the B lymphocytes of the immune system are stimulated to produce antibodies by the presence of antigens, and that each antibody molecule is a protein that combines specifically with a single antigen. But how a person (with a genome containing fewer than 100,000 genes) could inherit all the genes necessary to produce millions of different antibodies was a mystery until the late 1970s. At that time, Susumu Tonegawa discovered that genes for making millions of different antibodies are actually constructed during cell differentiation from a relatively small number of gene fragments in the original genome. For solving the mystery of antibody diversity, Tonegawa was awarded the Nobel Prize in Medicine or Physiology in 1987.

Tonegawa's discovery was revolutionary because it challenged two dogmas of traditional biology.

First, it violated the idea that the DNA coding for a polypeptide is always in one continuous piece in the genome (except for introns). Second, it is an exception to the concept that every somatic cell in the body has exactly the same DNA.

An antibody molecule consists of four polypeptide chains: two large (*heavy*) chains and two smaller (*light*) chains (see figure). Two identical heavy chains and two identical light chains are joined by disulfide bonds to form the molecule. Even before Tonegawa's discovery, scientists realized that the random combining of different heavy and light chains to form complete antibody molecules could only partially explain antibody diversity. For example, if human cells had 100 different kinds of heavy chains and 100 different kinds of light chains, the number of different ways of combining them would be 100 × 100 = 10,000.

To understand Tonegawa's discovery, we need to look more closely at the individual chains. Each chain has a *constant region*. This is a sequence of amino acids that is

identical for all antibody molecules of that type in that person. Thus, all the IgG molecules of a particular type in a given person's body have identical constant regions in their heavy chains. Two major types of constant regions can occur in light chains, and five major types can occur in heavy chains. (It is the type of constant region in the heavy chain that determines the class of the antibody. Thus, if the constant region of the heavy chains is the γ type, the antibody is an IgG; if it is the μ type, the antibody is an IgM.)

Each chain also has a *variable region*. It is the variable regions of the heavy and light chains that make the antibody react with one specific antigen and no other.

In 1976, Tonegawa and coworkers found direct experimental evidence that antibody genes are assembled by recombination events that move DNA coding for variable regions physically closer to DNA coding for constant regions. First, using myelomas (cancerous tumors that produce large amounts of single antibod-

ies), they prepared radioactive mRNA for individual light and heavy chains, and also for just the constant regions of the chains. Then they used these mRNA molecules as probes to test DNA isolated from two kinds of cells, embryonic cells that do not produce antibodies and mature B lymphocytes that do.

They found that in embryonic cells (and also in adult cells that do not produce antibodies), the DNA encoding constant regions was far away from the DNA encoding variable regions, but in cells that produce antibodies, the constant and variable coding segments were close together. Furthermore, the results showed that a nonantibody-producing cell has many separate genes for variable regions (*V-genes*) and one or two genes for constant regions (*C-genes*). Furthermore, each V-gene consists of two or three segments (called V, J, and D) that are physically separate on the embryonic genome.

As a B lymphocyte differentiates, mature genes for one type of heavy chain and one type of light chain are assembled as shown in the figure, which more kinds of antibodies than would otherwise be possible (see the box below). Thus, the resulting genomes of various B and T cells actually differ from one another—an astonishing exception to the general rule that all cells in the body have identical genomes.

Another important question is why the immune system does not react to the body's own cells and macromolecules, a phenomenon called *tolerance*. The exact answer is still a mystery, but apparently the B and T cells that interact with self antigens are somehow destroyed during fetal development.

is a simplified diagram of the process for one particular type of light chain. In this case, there are three V gene segments, one J segment, no D segments, and one C gene in the undifferentiated cell. As the cell differentiates, a long segment of DNA, from the end of one of the V segments to the beginning of a J segment, is deleted by recombination. This deletion brings a V segment (here V_2) adjacent to J and produces a gene that can be transcribed. The RNA transcript is processed as usual to remove introns (noncoding stretches of nucleic acid within genes), and the resulting mRNA is translated into a polypeptide light chain. This mature B lymphocyte ends up with a functional light-chain gene that did not exist in that form in the undifferentiated cell.

Tonegawa also played a role in the discovery of two other phenomena that contribute to antibody diversity. One is variation in the exact sites of recombination in the process just illustrated; the crossover points can vary by several nucleotides, resulting in variation in the sequence of amino acids in the polypeptide produced later (*junctional diversity*).

Another phenomenon is *somatic hypermutation*, the occurrence of single-nucleotide mutations in the fully assembled antibody genes.

Current evidence suggests that the number of different antibody gene segments in the genome of an undifferentiated cell is probably under 600. But from this relatively small collection, using the mechanisms described here, the cell can easily generate more than 200 million different antibody molecules.

Source: J. D. Watson, N. H. Hopkins, J. W. Roberts, J. A. Steitz, and A. M. Weiner, *Molecular Biology of the Gene*, 4th ed. Menlo Park, Calif.: Benjamin/Cummings, 1987, Chapter 23.

Immunological Memory

The intensity of the humoral response is reflected by the **antibody titer**, which is the amount of antibody in the serum. After the initial contact with an antigen, the exposed person's serum contains no detectable antibodies for several days. Then there is a slow rise in antibody titer; first IgM antibodies are produced, followed by IgG. Finally, a gradual decline in antibody titer occurs. This pattern is characteristic of the **primary response**. The immune responses of the host intensify after a second exposure to an antigen. This **secondary response** is also called the **memory** or **anamnestic response**, and it is explained by the clonal selection mechanism.

As we already mentioned, one component of the primary response in both humoral and cell-mediated immune responses is the production of clones of long-lived memory cells. These cells have a lifetime measured in years, or even decades. On the second or any subsequent exposure to the same antigen, these memory cells can quickly differentiate into antibody-producing plasma cells; they are thus responsible for the much faster and higher rise in antibody titer than occurred during the primary response (see Figure 16-

7). The secondary response involves mainly the production of IgG antibodies.

Probably the most familiar clinical example of the anamnestic response is seen after a tetanus toxoid vaccine is injected into someone who has received a serious cut, puncture wound, or bite. Individuals who have an established immunity to tetanus (most children acquire this in their vaccination series) display a secondary response to the booster injection of tetanus toxoid. This response renews protection against tetanus quickly enough to be effective against any toxin that might be produced if *Clostridium tetani* bacteria infect the wound. For individuals without previous immunity, the response is too slow to be effective. (For these persons, immune serum globulin can provide passive, temporary immunity.)

Memory in cellular immunity is mainly important in distinguishing between self and nonself. Certain effector T cells seem to retain a memory of antigens of the host and use it to distinguish between host cells that they tolerate and foreign cells that they attempt to eliminate.

* * *

In the next chapter, we will discuss different types of immunization and explain how they stimulate the immune system to prevent disease.

Study Outline

Introduction (p. 413)

1. An individual's genetically predetermined resistance to certain diseases is called innate resistance.
2. Individual resistance is affected by sex, age, nutritional status, and general health.
3. An individual may develop or acquire immunity after birth.
4. Immunity is the ability of the body to specifically counteract foreign organisms or substances.

Types of Acquired Immunity (pp. 414–415)

1. Acquired immunity is specific resistance to infection developed during the life of the individual.
2. Acquired immunity results from the production of antibodies and sensitized lymphocytes.

Naturally Acquired Immunity (p. 414)

1. Immunity resulting from infection is called naturally acquired active immunity; this type of immunity may be long lasting.
2. Antibodies transferred from a mother to a fetus (placental transfer) or to a newborn in colostrum results in naturally acquired passive immunity in the newborn; this type of immunity can last up to a few months.

Artificially Acquired Immunity (pp. 414–415)

1. Immunity resulting from vaccination is called artificially acquired active immunity, and can be long lasting.
2. Vaccines can be prepared from attenuated, inactivated, or killed microorganisms and toxoids.
3. Artificially acquired passive immunity refers to humoral antibodies acquired by injection; this type of immunity can last for a few weeks.

4. Antibodies made by a human or other mammal may be injected into a susceptible individual.
5. Serum containing antibodies is often called antiserum.
6. When serum is separated by electrophoresis, antibodies are found in the gamma fraction of the serum (gamma globulin or immune serum globulin).

The Duality of the Immune System   (pp. 415–416)

The Humoral Immune System   (p. 415)

1. The humoral immune system involves antibodies that are found in blood plasma and lymph.
2. Antibodies are produced by B cells in response to a specific antigen.
3. Antibodies primarily defend against bacteria, viruses, and toxins in body fluids.

The Cell-Mediated Immune System   (pp. 415–416)

1. The cell-mediated immune system depends on T cells and does not involve antibody production.
2. T lymphocytes are coated with antibodylike molecules.
3. Cellular immunity is primarily a response to intracellular viruses, multicellular parasites, transplanted tissue, and cancer cells.

Antigens and Antibodies   (pp. 416–420)

The Nature of Antigens   (pp. 416–417)

1. An antigen or immunogen is a chemical substance that causes the body to produce specific antibodies or sensitized T cells, which the antigen can then combine with.
2. As a rule, antigens are foreign substances; they are not part of the body's chemistry.
3. Most antigens are proteins, nucleoproteins, lipoproteins, glycoproteins, or large polysaccharides.
4. Generally, antigens have a molecular weight greater than 10,000.
5. Antibodies are formed against specific regions on the surface of an antigen called antigenic determinant groups.
6. Most antigens have many different determinants.
7. A hapten is a low-molecular-weight substance that combines with an antibody but cannot cause the formation of antibodies unless combined with another molecule.

The Nature of Antibodies   (pp. 417–420)

1. An antibody or immunoglobulin is a protein produced by B lymphocytes in response to the presence of an antigen and is capable of combining specifically with the antigen.
2. An antibody has at least two antigen-binding sites.
3. A single bivalent antibody unit is a monomer; multivalent antibodies are composed of monomers.
4. Most antibody monomers consist of four polypeptide chains. Two are heavy chains, and two are light chains.
5. Within each chain is a variable portion (where antigen binding occurs) and a constant portion (which serves as a basis for distinguishing the classes of antibodies).

6. An antibody monomer is Y- or T-shaped; the variable regions form the tips, and the constant regions form the base and Fc fragment.
7. The Fc region can attach to a host cell or complement.

Immunoglobulin Classes   (pp. 418–420)

1. IgG antibodies are the most prevalent in serum; they provide naturally acquired passive immunity, neutralize bacterial toxins, participate in complement fixation, and enhance phagocytosis.
2. IgM antibodies are involved in agglutination and complement fixation.
3. Secretory IgA antibodies protect mucosal surfaces from invasion by pathogens.
4. IgD antibodies may help initiate the immune response in B cells.
5. IgE antibodies bind to mast cells and basophils and are involved in allergic reactions.

B Cells and Humoral Immunity   (pp. 420–424)

1. Humoral immunity involves antibodies dissolved in extracellular fluids.
2. Bone marrow stem cells give rise to T cells, B cells, and macrophages.
3. Stem cells that mature in lymphoid organs become B cells.
4. A mature B cell has thousands of antibodies on its surface.

B-Cell and Antigen Interactions   (pp. 420–422)

1. A B cell becomes activated when an antigen reacts with antigen receptors on its surface.
2. The activated B cell produces a clone of plasma cells and memory B cells.
3. Plasma cells secrete antibody. Memory cells recognize pathogens from previous encounters.
4. T-dependent antigens are attacked by antigen-presenting cells (APC) and then presented to helper T cells. The T_H cell then activates a B cell to produce IgG, IgA, or IgE.
5. The major histocompatibility complex (MHC) consists of cell-surface proteins that are unique to each individual and provide "self" antigens.
6. T-independent antigens can directly activate a B cell and primarily elicit production of IgM.

Antibody–Antigen Binding   (pp. 422–423)

1. Antibodies react with antigens to form antigen–antibody complexes, which neutralize toxins, inactivate viruses, or lyse cells.

Monoclonal Antibodies   (pp. 423–424)

1. Hybridomas are produced in the laboratory by fusing a cancerous cell with an antibody-secreting plasma cell.
2. A hybridoma cell culture produces large quantities of the plasma cell's antibody, called monoclonal antibodies.

3. Monoclonal antibodies are used in serologic identification tests and to prevent tissue rejections. They are also an important tool in cancer research.

T Cells and Cell-Mediated Immunity (pp. 424–426)

1. Cell-mediated immunity is associated with lymphocytes.

The Components of Cell-Mediated Immunity (pp. 424–426)

1. T cells are responsible for cell-mediated immunity.
2. After processing in the thymus gland, T cells migrate to lymphoid tissue.
3. T cells differentiate into effector cells and memory cells when they are stimulated by an antigen.

Types of T Cells (pp. 424–425)

1. Helper T (T_H) cells present T-dependent antigens to B cells.
2. Delayed hypersensitivity T (T_D) cells produce lymphokines.
3. Suppressor T (T_S) cells inhibit development of plasma cells from B cells and prevent T cells from reacting to self antigens.
4. Cytotoxic T (T_C) cells lyse target cells, such as viral-infected cells and cancer cells, by release of perforin.

Killer Cells (p. 425)

1. Natural killer cells lyse tumor cells, and killer cells eliminate target cells that are coated with antibodies.

Lymphokines (pp. 425–426)

1. T_D cells release lymphokines, such as macrophage chemotactic factor, macrophage inhibition factor, and interleukins (IL-1 and IL-2).
2. IL-2 causes proliferation and differentiation of T cells.

The Cell-Mediated Immune Response (p. 426)

1. T cells recognize antigens associated with MHC (self) antigens on APCs.
2. The APC secretes IL-1 to activate the T cell, which secretes IL-2.

Clonal Selection (pp. 427–429)

1. Many different antigen-binding lymphocytes are produced during embryonic development.
2. When stimulated by an antigen, the B cell or T cell produces a clone of cells with the same antigen specificity.

Immunological Memory (p. 430)

1. The amount of antibody in serum is called the antibody titer.
2. The response of the body to the first contact with an antigen is called the primary response. It is characterized by the appearance of IgM followed by IgG.
3. Subsequent contact with the same antigen results in very high antibody titer and is called the secondary, anamnestic, or memory response. The antibodies are primarily IgG.
4. T memory cells distinguish between self and nonself.

Study Questions

Review

1. Define immunity.
2. Distinguish between the following sets of terms.
   (a) Nonspecific resistance and immunity
   (b) Humoral and cell-mediated immunity
   (c) Active and passive immunity
   (d) Innate resistance and acquired immunity
   (e) Natural and artificial immunity
   (f) T-dependent and T-independent antigens
3. Classify the following examples of immunity as naturally acquired active immunity, naturally acquired passive immunity, artificially acquired active immunity, or artificially acquired passive immunity.
   (a) Immunity following injection of diphtheria toxoid
   (b) Immunity following an infection
   (c) Newborn's immunity to yellow fever
   (d) Immunity following an injection of anti-rabies serum

4. Explain what an antigen is. Distinguish between immunogen and hapten.
5. Explain what an antibody is by describing the characteristics of antibodies. Diagram the structure of a typical antibody; label the heavy chain, light chain, and constant, variable, and Fc regions.
6. Discuss the clonal selection mechanism.
7. Define each of the following on the bases of location in the body and role in the immune response: IgG, IgM, IgA, IgD, IgE.
8. By means of a diagram, explain the role of T cells and B cells in immunity.
9. Explain a function for the following types of cells: T_C, T_D, T_H, and T_S. What is a lymphokine?

10. (a) At time *A*, the host was injected with tetanus toxoid. At time *B*, the host was given a booster dose. Explain the meaning of the areas of the curve marked *a* and *b*.

(b) Indicate in the graph the antibody response of this same individual to exposure to a new antigen at time *B*.

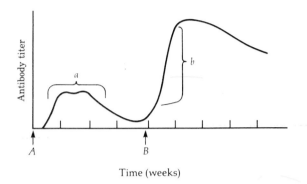

Time (weeks)

11. What effect does an antibody have on an antigen?

12. How does a T cell recognize an antigen?

13. What are natural killer cells? Killer cells?

14. How would each of the following prevent infection:
(a) antibodies against *N. gonorrhoeae* pili?
(b) antibodies against host-cell mannose?

15. How are monoclonal antibodies produced?

16. Explain why a person who recovers from a disease can attend others with the disease without fear of contracting the disease.

Challenge

1. Pooled human gamma globulin is sometimes administered to a patient after exposure to hepatitis. What is human gamma globulin? What type of immunity might this confer on the patient?

2. A patient with acquired immune deficiency syndrome (AIDS) has a low T_H/T_S cell ratio. What clinical manifestations would this have?

3. A patient with chronic diarrhea was found to lack IgA in his secretions, although he had a normal level of serum IgA. What was this patient found to be unable to produce?

4. A positive tuberculin skin test shows cell-mediated immunity to *Mycobacterium tuberculosis*. How could a person acquire this immunity?

Further Reading

Ada, G. L., and G. Nossal. "The clonal selection theory." *Scientific American* 257(2):62–69, August 1987. An excellent explanation of the development and current understanding of the clonal selection theory.

Gonda, M. A. "The natural history of AIDS." *Natural History* 95:78–81, May 1986. An interesting and scholarly discussion of the possibility that AIDS may have existed in isolated humans for centuries.

Hood, L. E., I. L. Weissman, W. B. Wood, and J. H. Wilson. *Immunology*, 2nd ed. Menlo Park, Calif.: Benjamin/Cummings, 1984. A concise textbook emphasizing the essential concepts of immunology.

Laurence, J. "The immune system in AIDS." *Scientific American* 253(6):84–93, December 1985. A well-illustrated article on the mechanism of AIDS pathogenicity.

Lerner, R. A., and A. Tramontano. "Catalytic antibodies." *Scientific American* 258(3):58–70, March 1988. Illustrates how antigen-binding sites react with antigens and suggests new approaches to vaccines.

Marrack, P., and J. Kappler. "The T cell and its receptor." *Scientific American* 254(1):36–45, January 1986. Describes the experiments that identified the T-cell receptor molecule.

Pollock, R. R., J.-L. Teillaud, and M. D. Scharff. "Monoclonal antibodies: a powerful tool for selecting and analyzing mutations in antigens and antibodies." *Annual Review of Microbiology* 38:389–417, 1984. Explains production of monoclonal antibodies and discusses their use as possible vaccines.

Tonegawa, S. "The molecules of the immune system." *Scientific American* 253(4):122–131, October 1985. Describes how billions of antibodies can be formed from a few genes.

Watson, J. D., N. H. Hopkins, J. W. Roberts, J. A. Steitz, and A. M. Weiner. *Molecular Biology of the Gene*, 4th ed. Menlo Park, Calif.: Benjamin/Cummings, 1987. Chapter 23 is an excellent up-to-date summary of the molecular and genetic aspects of immunology.

Young, J. D.-E., and Z. A. Cohn. "How killers cells kill." *Scientific American* 258(1):38–44, January 1988. Shows how killer cells cause their target cancer and virus-infected cells to leak to death.

CHAPTER 17

Practical Applications of Immunology

The science of immunology has its roots in observations about humans and disease. In this chapter, we will look at two tools that resulted from such observations—vaccines, which protect people from disease, and diagnostic tests, which tell us which microorganism is causing a disease. As we increase our understanding of the immune system at the cellular and molecular levels, scientists will be able to provide even better immunological tools.

Vaccines

Long before the invention of vaccines, the medical profession knew that people who recovered from certain diseases, such as smallpox, were immune to the disease thereafter. The Chinese may have been the first to try to exploit this phenomenon to prevent disease—they had their children inhale dried smallpox scabs. We can only speculate as to whether the results were successful or not

In the 1700s in Turkey, Lady Montague, the wife of the British ambassador, reported that it was the practice there that ". . . the old woman comes with a nutshell full of the matter of the best sort of smallpox and asks what veins you please to have opened, . . . and puts into the vein as much venom as can lie upon the head of her needle. . . ." This practice apparently often led to only a couple of days of mild illness. The success probably depended on two factors—first, the accidental choice of a different portal of entry for the infection, whose pathogen normally infects the body by entering through the respiratory system, and second, the accidental use of a less virulent form of the virus. This process, which is called **variolation**, can be a useful form of protection, but it can, on occasion, backfire and kill the recipient.

Learning Objectives

- Define serology, toxoid, attenuated, subunit vaccine, acellular vaccine, anti-idiotypic vaccine, immunodiffusion, and herd immunity.

- Explain why vaccination prevents disease.

- Differentiate between direct and indirect diagnostic tests.

- Explain how each of the following can be used in the diagnosis of a disease: precipitation, agglutination, hemagglutination-inhibition, complement fixation, neutralization, immunofluorescence, ELISA.

One of those who received this treatment, at the age of eight, was Edward Jenner. Later in life, as a physician, Jenner was intrigued by a dairy maid's assertion that she had no fear of smallpox because she had already had cowpox. Cowpox was a disease that caused lesions on cow udders; dairy maids' hands often became similarly infected during milking. Motivated by his childhood memory of variolation, Jenner began a series of experiments in 1798, in which he deliberately inoculated people with cowpox to prevent smallpox. This eventually led to the complete eradication of smallpox on Earth, the first disease for which this has been deliberately accomplished.

The development of conventional vaccines based on the model of the smallpox vaccine is the single most important application of immunology. Vaccines have greatly improved human health. Many pathogens transmitted by food or water can be controlled by sanitation or by antibiotics, if disease prevention fails. Viral diseases, however, are not readily treated once contracted, and transmission of viral pathogens by air or by direct contact is not easily prevented. Therefore, vaccination may be the only feasible method of controlling viral diseases. *Control* of a disease does not necessarily imply that everyone is immune to it. If most of the population is immune, outbreaks are limited to sporadic cases because there are not enough susceptible people to support the spread of epidemics. This is known as **herd immunity**.

Characteristics of Vaccines

A **vaccine** is a suspension of microorganisms (or some part or product of them) that will induce immunity when it is injected into a host. These microorganisms may be either *inactivated* (killed) or only *attenuated*. In the latter case, they are still living but are so weakened or altered that they are no longer virulent; yet they will still provoke an immune response. Toxoids (inactivated bacterial toxins) will also induce immunity against their active forms.

Live, attenuated virus vaccines tend to mimic an actual infection and usually provide better immunity than that provided by inactivated viruses. Examples of live vaccines are the Sabin polio vaccines and those used against yellow fever, measles, rubella, and mumps. Most attenuated virus vaccines provide lifelong immunity without booster immunizations, and an effectiveness of 95% is not unusual. This long-term effectiveness probably occurs because the attenuated viruses tend to replicate in the body, and the original

dose thereby increases considerably over time. One danger of such vaccines is that the live viruses can mutate to a virulent form, although this very rarely happens.

Viruses for vaccines may be inactivated by treatment with formalin or other chemicals. Heat is not used for this treatment because it is likely to alter the surface components of the virus and thus interfere with its ability to provoke an effective immune response. Commonly used inactivated virus vaccines include those used in humans against rabies (animals sometimes receive a live vaccine considered too hazardous for humans), influenza, and the Salk polio vaccine.

In the case of some vaccines, such as that for pneumococcal pneumonia, the antigens are the polysaccharide molecules of the bacterium's capsule. These vaccines must be readministered every few years, apparently because these antigens are less effective in stimulating antibody formation. Experience has also shown that vaccines against enteric bacterial pathogens, such as those causing cholera and typhoid, are not nearly as effective or long lived as those against viral diseases, such as measles and smallpox.

Harmful side-effects are an occasional problem with vaccines. The mass-immunization program of swine-flu vaccine during 1976 and 1977 appears to have been associated with a higher-than-normal rate of a paralytic disease called the Guillain–Barré syndrome, and there has been concern recently about the incidence of neurological side-effects caused by the pertussis vaccine.

Vaccines that are effective against bacteria (including rickettsias and mycoplasmas) and against viruses have been produced, but to date no useful vaccines against chlamydias, fungi, protozoans, or helminthic parasites in humans are in use. However, a vaccine against malaria is being developed (see the box on p. 438).

The principal vaccines used to prevent bacterial and viral diseases in the United States are listed in Tables 17-1 and 17-2. Recommendations for the administration of some of them are given in Table 17-3. Travelers who might be exposed to cholera, yellow fever, or other diseases not endemic in this country will find that current inoculation recommendations are available from the U.S. Public Health Service and local public health agencies.

TABLE 17-1 *Principal Vaccines Used in Prevention of Bacterial Diseases in Humans*

Disease	Vaccine	Recommendation	Booster
Cholera	Crude fraction of *Vibrio cholerae*	For persons who work and live in endemic areas	Every 6 months as needed
Diphtheria	Purified diphtheria toxoid	See Table 17-3	Every 10 years for adults
Meningococcal meningitis	Purified polysaccharide from *Neisseria meningitidis*	For persons with substantial risk of infection	No booster effect with additional doses
Pertussis (whooping cough)	Killed *Bordetella pertussis*	Children prior to school age; see Table 17-3	For high-risk adults
Plague	Crude fraction of *Yersinia pestis*	For persons who come in regular contact with wild rodents in endemic areas	Every 6 to 12 months as needed
Pneumococcal pneumonia	Purified polysaccharide from *Streptococcus pneumoniae*	For adults with certain chronic diseases; persons over 65	No booster effect with additional doses
Tetanus	Purified tetanus toxoid	See Table 17-3	Every 10 years for adults
Tuberculosis	BCG vaccine, an attenuated strain of *Mycobacterium bovis*	For persons who are tuberculin-negative and who are exposed for prolonged periods of time to tuberculosis	Every 3 to 4 years as needed
Typhoid fever and paratyphoid fever	Killed *Salmonella typhi, S. schottmuelleri,* and *S. paratyphi*	For persons in endemic areas or areas having outbreak	Every 3 years as needed
Typhus fever	Killed *Rickettsia prowazekii*	For scientists and medical personnel in rural areas endemic for typhus	Every 6 to 12 months as needed.

TABLE 17-2 *Principal Vaccines Used in Prevention of Viral Diseases in Humans*

Disease	Vaccine	Recommendation	Booster
Influenza	Inactivated virus	For chronically ill persons, especially with respiratory diseases, or for healthy persons over 65 years old	Annual
Measles	Attenuated virus	For infants 15 months old	*
Mumps	Attenuated virus	For infants 15 months old	*
Rubella	Attenuated virus	For infants 12 to 19 months old; for females of childbearing age who are not pregnant	*
Poliomyelitis	Attenuated or inactivated virus	For children, see Table 17-3; for adults, as risk to exposure warrants	*
Rabies	Inactivated virus	For field biologists in contact with wildlife in endemic areas; for veterinarians	Every 2 years
Yellow fever	Attenuated virus	For persons traveling to endemic areas; for military personnel	Every 10 years
Hepatitis B	Inactivated virus	Homosexual males, intravenous drug abusers, health workers exposed to blood	*

*The duration of immunity is not known because these vaccines have been in use for only a short time.

TABLE 17-3 *Recommended Schedule for Active Immunization of Children*

Recommended Age	Immunizing Agent
2 months	DPT vaccine (diphtheria toxoid, pertussis vaccine, and tetanus toxoid)
	OPV (oral poliomyelitis vaccine), trivalent preparation
4 months	DPT vaccine
	OPV, trivalent preparation
6 months	DPT vaccine
	OPV, trivalent preparation (only in high-risk areas)
15 months	DPT vaccine
	OPV, trivalent preparation
	Mumps vaccine / Measles vaccine / Rubella vaccine } or combined MMR vaccine
24 months	*Hemophilus* b polysaccharide vaccine; or at 18–23 months for high-risk children
4–6 years	DPT vaccine
	OPV, trivalent preparation
14–16 years, and each 10 years throughout life	TD vaccine (tetanus and diphtheria toxoid)

New Vaccine Development

A basic problem with developing a new vaccine is the need for sufficient quantities of the organism. In some cases, this is very difficult—for example, when the pathogen does not grow in anything but a living human. The early successful vaccines used animal cultivation—for example, the vaccinia virus for smallpox was grown on the shaved bellies of calves, and the rabies virus was grown in the central nervous system of rabbits. A vaccine against hepatitis B virus used viruses extracted from the blood of chronically infected humans because no other source was available. The vaccines produced in recent years against polio, mumps, measles, and other diseases awaited the development of successful cell culture of viruses.

Today, new techniques in molecular biology are being used to develop vaccines (see the box on p. 438).

Occasionally, harmful side-effects occur with conventional vaccines. This problem may be caused by a person's reactions to certain antigens from the injected pathogenic organism. In some cases, the troublesome antigens play little or no role in developing effective immunity. Two new techniques produce vaccines that do not contain these antigens, thus avoiding the problem. A third technique that is being developed sidesteps the problem completely by using antibodies in the vaccine instead of antigens. These techniques are discussed next.

Subunit Vaccines

A **subunit vaccine** uses only those antigenic fragments of a microorganism that are best suited to stimulating a strong immune response. The genes for these protein subunits can be introduced into the genome of a bacterium or yeast by the genetic-engineering techniques described in Chapter 8. These subunits are produced in quantity by the bacterium or yeast and are then harvested in pure form for use as a subunit vaccine. For example, hepatitis B vaccine is a subunit vaccine produced in a yeast.

Acellular Vaccines

It is also possible to fragment a conventional vaccine and collect only those portions that contain the desired antigens. Since the complete cells are not used, these vaccines are called **acellular vaccine**. A new vaccine against pertussis is of this type.

Anti-Idiotypic Vaccines

Anti-idiotypic vaccines are expected to be of importance in the near future. The basic idea of **anti-idiotypic vaccines** is that, rather than using an antigen as the vaccine, an antibody is used that mimics the shape of the antigen (and thus induces immunity) but is itself harmless. To produce such a vaccine, the first step is to make an antibody against the antigen. This antibody, called an idiotypic antibody (Ab-1), is then injected into an animal, which responds by making an antibody against the antibody (anti-idiotypic antibody, or Ab-2). The Ab-2 antibody can now function as a vaccine since it contains an antigenic determinant similar to that of the original antigen (Figure 17-1). Another animal receiving the vaccine containing Ab-2 responds by making yet a third antibody

Figure 17-1   Anti-idiotypic vaccine. (a) To make an anti-idiotypic vaccine against an antigen, the first step is to make an antibody against the antigen. This first antibody, called an idiotypic antibody (Ab-1), is then injected into an animal **(b)**, who produces an anti-idiotypic antibody (Ab-2). **(c)** The Ab-2 antibody is injected as a vaccine into a human, resulting in production of a third-generation anti-anti-idiotypic antibody (Ab-3). **(d)** This Ab-3 antibody will react with the original antigen if the antigen enters the immunized individual. Thus, immunization with Ab-3 provides protection against the original antigen. This procedure is still being developed through further research.

(anti-anti-idiotypic antibody, or Ab-3). If the immunized animal encounters the original antigen, the Ab-3 antibodies will react with the antigen, destroying or inactivating it. Thus, Ab-3 antibodies confer immunity against the original antigen.

This may seem like a complicated approach, but it has a number of advantages. First, no pathogenic organism, alive or dead, is injected. Second, such vaccines are highly specific because they are directed against the antigenic character of the pathogen best suited to stimulating an antigenic response. Third, certain nonprotein antigens are not able to stimulate immunity in newborn infants, but anti-idiotypic vaccines, being proteins, would be able to confer immunity upon the infant quickly. Fourth, this type of vaccine could be used to recognize the receptor sites on a cell and thus could be used to block the attachment of a virus to the cell. Such an approach has been

.. Microbiology in the News ....

Molecular Biology: The Latest Weapon Against Malaria

The latest microbiological and biochemical techniques of gene mapping, monoclonal-antibody production, and protein synthesis give hope that a vaccine against malaria may become available soon. Traditional techniques for creating vaccines against viruses and bacteria are ineffective against larger and more complex organisms, such as protozoans.

The need for an antimalarial vaccine has become desperate. In the early 1960s, we thought that malaria would soon disappear completely from America and other temperate-zone countries and would become relatively rare even in tropical countries. The World Health Organization planned to reduce the number of *Anopheles* mosquitoes with massive sprayings of DDT and planned to treat already infected people with a variety of drugs related to quinine.

But that battle has gone against us. The mosquitoes became resistant to DDT, and *Plasmodium* has begun to develop resistance to antimalarial drugs. About 150 million new cases of malaria occur every year throughout the world. Moreover, the number of malaria cases in the United States has increased dramatically because of Vietnam war veterans who carried the disease, increased immigration from tropical countries, and increased travel to tropical countries.

Recently researchers headed by Ruth and Victor Nussenzweig at New York University Medical Center have identified a surface antigen of the sporozoite stage of *Plasmodium*. This antigen, called circumsporozoite (CS), seems to have been involved in the temporary protection of human subjects who had been inoculated with irradiated sporozoites in a 1975 trial.

The discovery of CS has led to the production of monoclonal antibodies against it, which can be used in immunologic experiments. In 1983, the proposed as a way of developing a potential vaccine for AIDS, but the attempt has not been successful to date.

A major problem encountered in the development of anti-idiotype technology is the source of sufficient amounts of idiotypic antibody (Ab-1). One obvious source, monoclonal antibodies, awaits the development of methods of producing these antibodies with human cells rather than with mouse cells. As we mentioned in Chapter 16, when mouse cells are injected into humans, they may lead to harmful immune reactions.

Diagnostic Immunology

More than 100 years ago, when Robert Koch was trying to develop a vaccine against tuberculosis, he observed that when tubercular guinea pigs were injected with a suspension of *Mycobacterium tuberculosis*, the site of the injection became red and slightly swollen a day or two later. Many of us will recognize this symptom as essentially the same test we use today to determine whether a person has been infected by the tuberculosis pathogen. Koch, of course, had no idea of the mechanism of cell-mediated immunity that caused this phenomenon, and he had no idea about the existence of a particulate antibody.

Since then, immunology has given us many invaluable diagnostic tools, most of which are based on interactions of humoral antibodies with antigens. Because these techniques are directed at detecting the presence and amounts of antibodies in blood serum, we call this branch of immunology **serology**. In this section, we will discuss several of the most important serological techniques.

One problem our diagnostic tools must overcome is that antibodies can never be seen directly. Even at magnifications of well over 100,000×, they appear only as fuzzy, ill-defined particles. All we can do is infer their presence indirectly through a variety of reactions.

Precipitation Reactions

Precipitation reactions involve the reaction of soluble antigens with IgG or IgM antibodies (described in Chapter 16) to form large, interlocking aggregates called *lattices*. Historically, the antibodies involved in the formation of precipitates have been called *precipitins* because the lattices they produce precipitate from solution.

Precipitation reactions occur in two distinct stages. First, there is the rapid interaction between antigen and antibody to form small antigen–antibody complexes. This interaction occurs within seconds and

CS gene was cloned, making it possible to produce large quantities of the CS protein for analysis and antibody preparation.

The first human tests of a vaccine were conducted in mid-1987 at the University of Maryland School of Medicine. The next step will be to determine the vaccine's effectiveness. By mid 1989, field trials are anticipated in areas of Africa, Asia, and Latin America where the incidence of malaria is increasing.

Researchers are hopeful that the CS gene can be spliced into the genome of vaccinia viruses, which can then be scratched into the skin. This would eliminate the need for refrigeration and sterile syringes, thus making it much easier to inoculate people in the interior of underdeveloped tropical countries, where the mortality rate from malaria is very high, especially among children.

Research leading to a generally effective malaria vaccine is hindered by the fact that the life cycle of *Plasmodium* is so complex (see Figure 11-21). The vaccine currently being tested is directed against the sporozoite, which is injected into the victim by the mosquito. The immune response that is triggered by the vaccine must kill all the sporozoites within 30 minutes after the mosquito bites its victim, before the sporozoites enter the protection of the liver cells. A separate vaccine would be needed to provoke an immune response against the merozoites, which destroy red blood cells and cause the typical symptoms of malaria. Yet a third vaccine would be needed to destroy the gametocytes, which continue the life cycle within the mosquito's body. Research on vaccines to attack the merozoites and gametocytes is moving ahead at research centers worldwide.

If the experiments in developing a malaria vaccine are successful, it will be the first antiparasitic vaccine for humans and will provide a model for the development of vaccines against other parasitic diseases, such as trypanosomiasis and schistosomiasis.

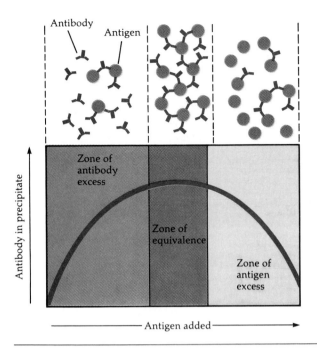

Figure 17-2   **Precipitation curve.** The curve is based on the ratio of antigen to antibody. The maximum amount of precipitate forms in the zone of equivalence, where the ratio is optimal.

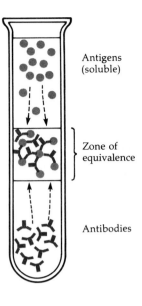

Figure 17-3   **Precipitin ring test.** (a) A diagrammatic representation of a precipitin ring test. Antigens and antibodies diffuse toward each other in a tube of small diameter and form a visible line or ring where the zone of equivalence is reached.

is followed by a slower reaction, which may take from minutes to hours, in which the antigen–antibody complexes form lattices that precipitate from solution. Precipitation reactions normally occur only when there is an optimal ratio of antigen to antibody. Figure 17-2 shows that no visible precipitate forms when there is an excess of either one. The optimal ratio is produced when separate solutions of antigen and antibody are placed adjacent to each other and allowed to diffuse together. In a **precipitin ring test** (Figure 17-3), a cloudy line of precipitation (ring) appears in the area in which the optimal ratio has been reached (the *zone of equivalence*).

Immunodiffusion tests are precipitation reactions carried out in an agar gel medium. In one such test, the **Ouchterlony test**, wells are cut into a purified agar gel in a Petri plate. A serum containing antibodies (antiserum) is added to one well, usually centrally located, and soluble test antigens are added to each surrounding well. A line of visible precipitate develops between the wells at the point where the optimal antigen–antibody ratio is reached. The Ouchterlony test can show whether the serum sample contains antibodies against more than one test antigen at a time (Figure 17-4). It may also indicate whether

(a) Line of identity

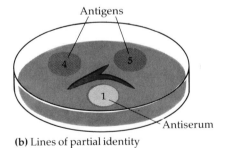
(b) Lines of partial identity

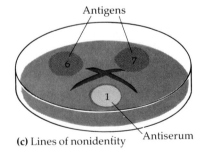
(c) Lines of nonidentity

Figure 17-4   **The Ouchterlony immunodiffusion precipitation test.** **(a)** Line of identity. Neither antigen (in wells 2 and 3) nor antibody molecules (in well 1) are able to diffuse past each other because they react and precipitate to form the dark line. Therefore, wells 2 and 3 contain the same antigen. **(b)** Lines of partial identity. The antigens are not identical, but they share many antigenic determinant sites and are therefore related. The dark spur shape indicates that some antigen did not react with antibody and diffused through the precipitation zone. **(c)** Lines of nonidentity. The antiserum contains antibodies against antigens in wells 6 and 7. The antigens are not related because they diffused across each other's zones of precipitation.

the antigens are identical, partially identical, or totally different.

Other precipitation tests do not depend entirely on the passive diffusion of antigen and antibody in a gel but instead use electrophoresis to speed up their movement. Protein mixtures can be separated rapidly, sometimes in less than an hour, with this method. A modification of the precipitation reaction combines the techniques of immunodiffusion and electrophoresis in a procedure called **immunoelectrophoresis**. This procedure is used in research in the separation of proteins in human serum and is the basis of certain diagnostic tests. The **countercurrent immunoelectrophoresis test**, also called **counterimmunoelectrophoresis (CIE)**, can be used in the diagnosis of bacterial meningitis and other diseases. A test serum known to contain certain antibodies is used to identify an unknown antigen in some body fluid. CIE is based on the fact that some antigens and antibodies have opposite charges when they are placed at opposite poles in buffers of correct ionic strength and pH. When an electrical current is applied, the antigens and antibodies move toward the opposite pole. They pass through each other to do so. A precipitation line appears within an hour if a reaction occurs (Figure 17-5).

Agglutination Reactions

Whereas precipitation reactions involve *soluble* antigens, agglutination reactions involve *particulate* antigens or soluble antigens adhering to particles. These antigens can be linked together by antibodies to form visible aggregates, a process called **agglutination** (Figure 17-6). Agglutination reactions are very sensitive, relatively easy to read, and available in great variety. Agglutination tests are classified as either *direct* or *indirect*.

Direct Agglutination Tests

Direct agglutination tests detect antibodies against relatively large cellular antigens, such as red blood cells, bacteria, and fungi. At one time, they were carried out in a series of test tubes, but now they are usually done in plastic *microtiter plates*, which have many shallow wells to take the place of individual test tubes. The amount of particulate antigen in each well is the same, but the amount of serum containing antibodies is diluted so that each successive well has half the antibodies of the previous well.

Clearly, the more antibody we start with, the more dilutions it will take to lower the amount to the point at which agglutination does not occur. This is

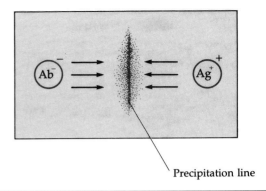

Figure 17-5 Countercurrent immunoelectrophoresis. Antigens (Ag^+) and antibodies (Ab^-) are placed in opposite wells, and the pH of the surrounding medium is adjusted so that the antigen and antibody have opposite charges. When an electrical current is applied, antigens and antibodies move in opposite directions. A line of precipitate forms where antigen and antibody meet.

Figure 17-6 Agglutination reaction. When antibodies react with antigenic determinant sites on neighboring antigens, such as bacteria or red blood cells, the particulate antigens agglutinate. IgM, the most efficient immunoglobulin for agglutination, is shown here, but IgG also participates in agglutination reactions. In indirect agglutination tests, red blood cells or other particles are coated with soluble antigens to make a particulate antigen that will agglutinate.

the measure of **titer**, or concentration of serum antibody (Figure 17-7). For infectious diseases in general, the higher the serum antibody titer, the greater the immunity to the disease. The titer alone is of limited use in diagnosing an existing illness. There is no way of knowing whether the measured antibodies were generated in response to the immediate infection or to an earlier illness. For diagnostic purposes, *a rise in titer* is significant; that is, the titer is higher later in the disease than at its onset. If it is possible to

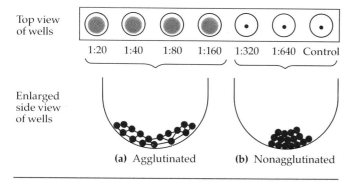

Figure 17-7 Direct agglutination test. Each well in this microtiter plate contains, from left to right, only half the concentration of serum in the preceding well. Each well contains the same concentration of particulate antigens, such as red blood cells. **(a)** In a positive (agglutinated) reaction, sufficient antibodies are present in the serum to link the antigens together, forming an antibody–antigen mat that sinks to the bottom of the well. **(b)** In a negative (nonagglutinated) reaction, insufficient antibodies are present to cause the linking of antigens. The particulate antigens roll down the sloping sides of the well, forming a pellet at the bottom. In this example, the antibody titer is 160 since the well with a 1:160 concentration is the most dilute concentration that gives a positive reaction.

demonstrate that the person's blood had no antibody titer before the illness but has a significant titer while the disease is progressing, this change in titer, called **seroconversion**, is also diagnostic.

Indirect (Passive) Agglutination Tests

Soluble antigens can respond to agglutination tests if the antigens are adsorbed onto particles, such as red blood cells, minute latex spheres, or bentonite clay. In such **indirect**, or **passive**, **agglutination tests**, the antibody reacts with the soluble antigen adhering to the particle. The particles then agglutinate with one another, much as particles do in the direct agglutination tests. Many rapid-diagnostic tests are of this type, especially to detect streptococci.

Hemagglutination

When agglutination reactions involve the clumping of red blood cells, the reaction is called **hemagglutination**. These reactions are used routinely in blood typing (Chapter 18) and in the diagnosis of infectious mononucleosis.

Certain viruses, such as those causing mumps, measles, and influenza, have the ability to agglutinate red blood cells; this process is called **viral hemagglutination**. If a person's serum contains antibodies against such viruses, these antibodies will react with the viruses and neutralize them (Figure 17-8). For example, if hemagglutination occurs in a mixture of measles virus and red blood cells but does not occur when the patient's serum is added to the mixture, this sequence indicates that the serum contains antibodies

Figure 17-8 Viral hemagglutination. (a) Viral hemagglutination is not an antigen–antibody reaction. It simply happens that certain viruses bind to red blood cells and cause their agglutination. **(b)** If serum that contains antibodies to the virus is mixed with the red blood cells and the virus, the antibodies will neutralize the virus and inhibit hemagglutination.

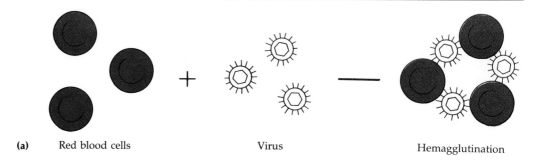

(a) Red blood cells   Virus   Hemagglutination

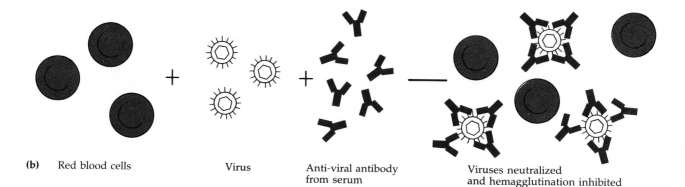

(b) Red blood cells   Virus   Anti-viral antibody from serum   Viruses neutralized and hemagglutination inhibited that have neutralized the measles virus. This **hem-agglutination-inhibition test** is widely used in the diagnosis of influenza, measles, mumps, and a number of other viral infections.

Complement-Fixation Reactions

In Chapter 15, we discussed a group of serum proteins collectively called complement. During most antigen–antibody reactions, the complement binds to the antigen–antibody complex and is used up, or fixed. This process of **complement fixation** can be used to detect even very small amounts of antibody. Antibodies that do not produce a visible reaction, such as precipitation or agglutination, can be demonstrated by the fixing of complement during the antigen–antibody reaction. Complement fixation was once used in the diagnosis of syphilis (Wasserman test) and is still used in the diagnosis of certain viral, fungal, and rickettsial diseases.

Execution of the complement-fixation test requires great care and good controls. The test is performed in two stages (Figure 17-9). First, the subject's

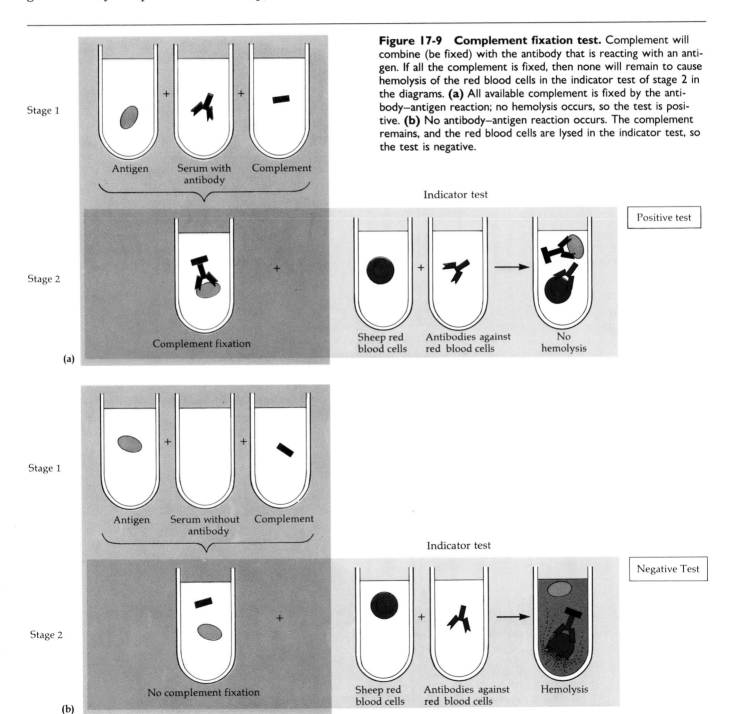

Figure 17-9 Complement fixation test. Complement will combine (be fixed) with the antibody that is reacting with an antigen. If all the complement is fixed, then none will remain to cause hemolysis of the red blood cells in the indicator test of stage 2 in the diagrams. **(a)** All available complement is fixed by the antibody–antigen reaction; no hemolysis occurs, so the test is positive. **(b)** No antibody–antigen reaction occurs. The complement remains, and the red blood cells are lysed in the indicator test, so the test is negative.

serum must be heated at 56°C for 30 minutes to inactivate its complement. The inactivated serum is then diluted and mixed with a specific amount of known antigen and fresh complement. The mixture is next incubated for about 30 minutes. An antigen–antibody reaction cannot be observed at this point. To determine whether the complement present in the mixture is fixed by an antigen–antibody reaction, which is evidence of the presence of antibodies specific for the antigen, another stage of the test must be performed.

The second stage uses an indicator system to determine whether complement is free or combined (fixed). The indicator system consists of sheep red blood cells that have specific antibodies attached to their surfaces. The exposure of these "sensitized" cells to complement causes lysis of the red blood cells (hemolysis), which changes the color of the mixture. Therefore, if complement has been fixed by an antigen–antibody reaction during the first stage, it is not available to cause blood cell lysis in the second stage; the test is positive (Figure 17-9a). But if the complement has not been fixed during the first stage, then it is available to cause blood cell lysis in the second stage (Figure 17-9b). This negative test result means that no antibodies specific for the test antigen are present in the patient's serum.

Neutralization Reactions

Neutralization reactions are antigen–antibody reactions in which the harmful effects of a bacterial exotoxin or a virus are eliminated by specific antibodies. These reactions were first described in 1890,

when investigators observed that immune serum could neutralize the toxic substances produced by *Corynebacterium diphtheriae*. Such a neutralizing substance is called an **antitoxin**. This is a specific antibody produced by a host as it responds to a bacterial exotoxin or its corresponding toxoid (inactivated toxin). The antitoxin combines with the exotoxin to neutralize it (Figure 17-10a). Antitoxins produced in an animal can be injected into humans to provide passive immunity against a toxin. Antitoxins from horses are routinely used for prevention or treatment of diphtheria and botulism; tetanus antitoxin is usually of human origin.

Neutralization tests are frequently used in the diagnosis of viral infections. The body responds to such infections by producing specific antibodies that bind to receptor sites on the viral surface. The binding of these antibodies prevents the virus from attaching to a host cell, and thus destroys the virus's infectivity (Figure 17-10b). Viruses that exhibit their cytopathic (cell-damaging) effects in cell culture or embryonic eggs can be used to detect the presence of neutralizing viral antibodies. If the serum to be tested contains antibodies against the particular virus, the antibodies will prevent that virus from infecting cells in the cell culture or eggs, and no cytopathic effects will be seen. The inability of a specific virus to cause cytopathic effects in the presence of immune serum can thus be used to determine the identity of a virus as well as the viral antibody titer.

In vitro neutralization tests are not common in modern clinical laboratories. A more frequent form of neutralization test is a skin test such as the **Schick test**, which determines the status of a person's immu-

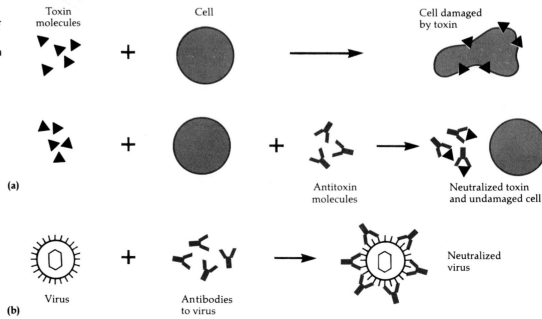

Figure 17-10 Neutralization reactions. (a) Effects of a toxin on a susceptible cell and neutralization of the toxin by antitoxin. **(b)** A specific antibody neutralizes a virus, preventing the virus from attaching to a cell.

nity to diphtheria. A small amount of diphtheria exotoxin is inoculated into the skin. If there is sufficient serum antitoxin to neutralize the exotoxin, there is no visible reaction. If antitoxin is insufficient, the exotoxin damages the tissues at the site of entry and produces a swollen, tender, reddish area that turns brown in four to five days. This reaction indicates an unsatisfactory immune response to diphtheria.

Immunofluorescence and Fluorescent-Antibody Techniques

Fluorescent-antibody (FA) techniques (Figure 17-11) can identify microorganisms in clinical specimens and detect the presence of a specific antibody in serum. These techniques use fluorescent dyes, such as *fluorescein isothiocyanate* (FITC), which are combined with antibodies to make them fluoresce when exposed to ultraviolet light. These procedures are quick, sensitive, and very specific; the fluorescent antibody test

for rabies can be done in a few hours and has an accuracy rate close to 100%.

Fluorescent-antibody tests are of two types, direct and indirect. **Direct FA tests** are usually used to identify a microorganism in a clinical specimen. During this procedure, the specimen containing the antigen to be identified is fixed onto a slide. Fluorescein-labeled antibodies are then added, and the slide is incubated briefly. The slide is washed to remove any antibody not bound to antigen and is then examined under the ultraviolet microscope for yellow-green fluorescence (see Figure 17-11a and Microview C, p. C-4).

Indirect FA tests (Figure 17-11b) are used to detect the presence of a specific antibody in serum following exposure to a microorganism. During this procedure, a known antigen is fixed onto a slide. The test serum is then added, and, if antibody that is specific to that microbe is present, it reacts with the antigen to form a bound complex. So the antigen–antibody complex can be seen, fluorescein-labeled anti-human

Figure 17-11 Fluorescent antibody techniques. (a) A direct FA test identifies group A streptococci. **(b)** In an indirect FA test, such as that used in the diagnosis of syphilis, the fluorescent dye is attached to anti-human gamma globulin, which reacts with any human immunoglobulin (such as the *Treponema pallidum*-specific antibody) that has previously reacted with antigen. **(c)** The reaction is viewed through a fluorescence microscope, and the antigen with which the dye-tagged antibody has reacted—in this case *T. pallidum*—glows in the ultraviolet illumination (×1000).

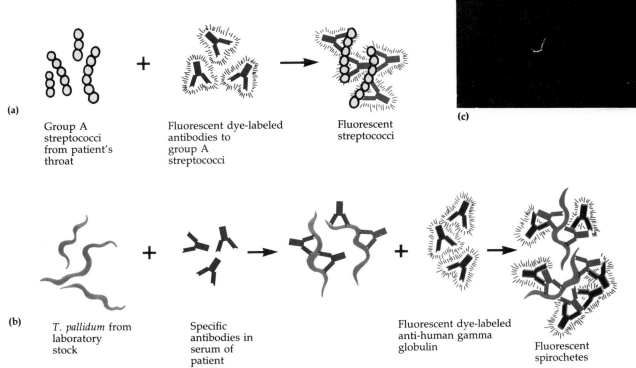

(a)

Group A streptococci from patient's throat

Fluorescent dye-labeled antibodies to group A streptococci

Fluorescent streptococci

(c)

(b)

T. pallidum from laboratory stock

Specific antibodies in serum of patient

Fluorescent dye-labeled anti-human gamma globulin

Fluorescent spirochetes

gamma globulin (anti-HGG), an antibody that reacts specifically with human antibody, is added to the slide. After the slide has been incubated and washed (to remove unbound antibody), it is examined under a fluorescence microscope. If the known antigen fixed to the slide appears fluorescent, the antibody specific to the test antigen is present.

Enzyme-Linked Immunosorbent Assay (ELISA)

The **enzyme-linked immunosorbent assay (ELISA)**, also known as the **enzyme immunoassay (EIA)**, has become a widely used serological technique. There are two basic methods. The **double antibody sandwich method** detects antigens, and the **enzyme-linked indirect immunosorbent assay** detects antibodies. A microtiter plate with numerous shallow wells is used in both procedures (see Microview C, p. C-4). The test has become highly automated in many applications, such as testing for AIDS antibodies. Variations of the test exist; for example, the reagents can be bound to tiny latex particles rather than to the surfaces of the microtiter plates.

Double Antibody Sandwich Method

In the first step of this method, antiserum is adsorbed to the surface of the wells on the microtiter plate (Figure 17-12a). A test antigen is then added to each well. If the antigen reacts specifically with the antibodies adsorbed to the well, the antigen will be retained there when the well is washed free of unbound antigen. An antibody specific for the antigen is then added. If both the antibody adsorbed to the wall of the well and the antibody known to be specific for the antigen have reacted with the antigen, a "sandwich" will have formed, with the antigen between two antibody molecules.

This reaction is visible only because the second added antibody is linked to an enzyme, such as horseradish peroxidase or alkaline phosphatase. Unbound enzyme-linked antibody is washed from the well, and then the enzyme's substrate is added to it. Enzymatic activity is indicated by a color change that can be seen by eye or read with a special spectrophotometer. The test will be positive if the antigen has reacted with adsorbed antibodies in the first step. If the test antigen was not specific for the antibody adsorbed to the wall of the well, the test will be negative because the unbound antigen will have been washed away.

Indirect Immunosorbent Assay

In the first step of this method, the antigen, rather than the antibody, is adsorbed to the walls of the shallow wells on the plate (Figure 17-12b). To see whether a serum sample contains antibodies against this antigen, the antiserum is added to the well. If the serum contains antibody specific to the antigen, the antibody will bind to the adsorbed antigen. All unreacted antiserum is washed from the well. Anti-HGG (an antibody that reacts with any human immunoglobulin) is then allowed to react with the antigen–antibody complex. The anti-HGG, which has been linked with an enzyme, reacts with the antibodies that are bound to the antigens in the well. Finally, all unbound anti-HGG is rinsed away, and the correct substrate for the enzyme is added. A colored enzymatic reaction occurs in the wells in which the bound antigen has reacted with antibody present in the serum sample. This procedure resembles that of the indirect fluorescent-antibody test, except that the anti-HGG of the indirect assay is linked with an enzyme rather than with a fluorescing dye. This assay is used to test for antibodies against the AIDS virus.

The Future of Vaccines and Diagnostic Immunology

Vaccines and diagnostic immunology are changing rapidly in response to new knowledge and new techniques. For example, genetic engineering is certain to revolutionize current vaccine production. Most of our present vaccines will soon be considered primitive. We may eventually develop new vaccines for such diseases as AIDS, malaria (see the box on p. 438), African trypanosomiasis, and leprosy, to name just a few. There will also be safer and even more effective vaccines for such diseases as cholera and pertussis and many others.

New serological tests and tests that are more sensitive, specific, rapid, and simple than those currently used will also be developed. For example, the development of monoclonal antibodies has already made it possible, for the first time, to test for chlamydial sexually transmitted diseases. The use of DNA probes, while not serological tests (see Figure 9-15), should prove increasingly useful for detecting pathogens in tissues or in foods.

It is virtually certain that rapid new developments in the practical applications of immunology can be expected in the near future, promising dramatic improvements in human health.

(a) Double antibody sandwich method
(detects antigens)

(b) Indirect immunosorbent assay
(detects antibodies)

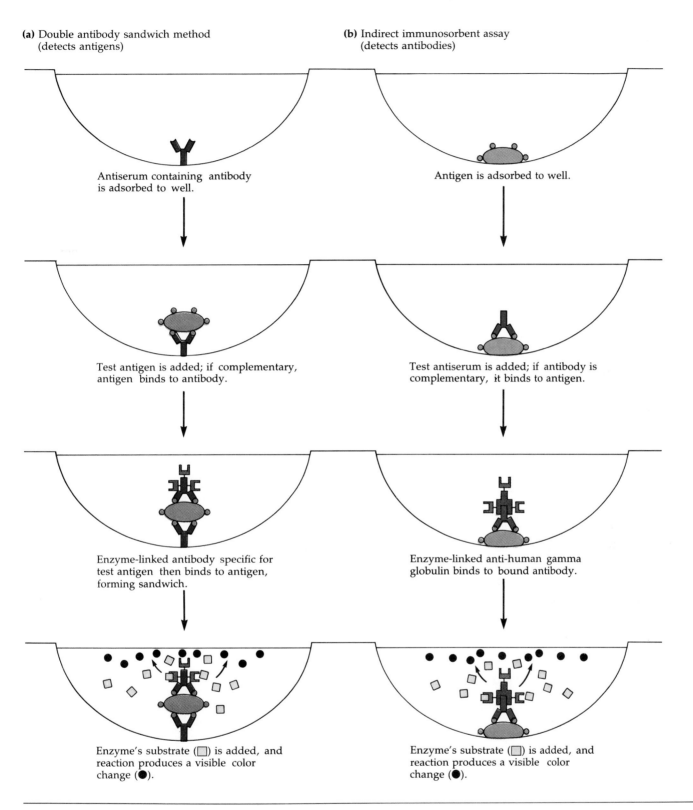

Antiserum containing antibody
is adsorbed to well.

Antigen is adsorbed to well.

Test antigen is added; if complementary,
antigen binds to antibody.

Test antiserum is added; if antibody is
complementary, it binds to antigen.

Enzyme-linked antibody specific for
test antigen then binds to antigen,
forming sandwich.

Enzyme-linked anti-human gamma
globulin binds to bound antibody.

Enzyme's substrate (▢) is added, and
reaction produces a visible color
change (●).

Enzyme's substrate (▢) is added, and
reaction produces a visible color
change (●).

Figure 17-12 The ELISA, or EIA, test. The reagents are usually contained in small wells of a
microtiter plate. **(a)** The double antibody sandwich method is used to detect antigens. **(b)** The indirect immunosorbent assay detects antibodies.

Study Outline

Vaccines (pp. 434–439)

1. The earliest vaccination procedures included inhaling dried smallpox scabs and variolation, the injection of a less virulent virus.
2. Edward Jenner developed the modern practice of vaccination when he inoculated people with cowpox virus to protect them from smallpox.
3. Herd immunity results when most of a population is immune to a disease.

Characteristics of Vaccines (pp. 435–437)

1. A vaccine is a suspension of infectious agents (or some part of them) that will induce a state of immunity.
2. Attenuated or inactivated microbes may be used.
3. Toxoids (inactivated bacterial exotoxins) may be used in vaccines.

New Vaccine Development (pp. 437–439)

1. Bacteria can be genetically engineered to produce the antigenic proteins of pathogens, and these proteins can be used for subunit vaccines.
2. Antigens separated from vaccine preparations can be used for acellular vaccines.
3. Anti-idiotypic vaccines are composed of antibodies against an antipathogen. This antibody will stimulate antibody production against the antigenic site of the pathogen.

Diagnostic Immunology (pp. 439–446)

1. Serology is the study of antibodies in serum.
2. Many serological tests can be performed quickly to determine the presence and amounts of antibodies in serum.

Precipitation Reactions (pp. 439–441)

1. The interaction of soluble antigens with IgG or IgM antibodies (precipitins) leads to precipitation reactions.
2. Precipitation reactions depend on the formation of lattices and occur best when antigen and antibody are present in optimal proportions. Excesses of either component decrease lattice formation and subsequent precipitation.
3. The precipitin ring test is performed in a small tube.
4. Immunodiffusion procedures such as the Ouchterlony test involve precipitation reactions carried out in an agar gel medium.

5. Such tests detect the presence of multiple antigens and provide information about their relatedness.
6. Immunoelectrophoresis combines electrophoresis with immunodiffusion for the analysis of serum proteins.
7. Countercurrent immunoelectrophoresis can be used to detect antigens in serum.

Agglutination Reactions (pp. 441–443)

1. The interaction of particulate antigens with antibodies leads to agglutination reactions.
2. Diseases may be diagnosed by combining the patient's serum with a known antigen in such tests as the Widal test.
3. Direct agglutination reactions can be used to determine antibody titer.
4. Diseases can be diagnosed by a rising titer or seroconversion (from no antibodies to the presence of antibodies).
5. Antibodies cause visible agglutination of soluble antigens affixed to latex spheres in indirect or passive agglutination tests.
6. Hemagglutination reactions involve agglutination reactions using red blood cells. Hemagglutination reactions are used in blood typing, diagnosis of certain diseases, and identification of viruses.
7. Antibodies against certain viruses can be detected by their ability to interfere with viral hemagglutination in hemagglutination-inhibition tests.

Complement-Fixation Reactions (pp. 443–444)

1. Complement-fixation reactions are serological tests based on the depletion of a fixed amount of complement in the presence of an antigen–antibody reaction.
2. Hemolysis or its absence is used as an indicator in evaluating complement fixation.

Neutralization Reactions (pp. 444–445)

1. In these reactions, the harmful effects of a bacterial exotoxin or virus are eliminated by a specific antibody.
2. The Schick test uses a toxin-neutralization reaction to determine the presence of antibodies to diphtheria toxin.
3. In a viral-neutralization test, the presence of antibodies against a virus can be detected by the antibodies' ability to prevent cytopathic effects of viruses in cell cultures.

Immunofluorescence and Fluorescent-Antibody Techniques (pp. 445–446)

1. Immunofluorescent techniques use antibodies labeled with fluorescent dyes.
2. Direct fluorescent-antibody tests are used to identify specific microorganisms.

3. Indirect fluorescent-antibody tests are used to demonstrate the presence of antibody in serum.

Enzyme-Linked Immunosorbent Assay (ELISA) (p. 446)

1. ELISA techniques use antibodies linked to an enzyme such as horseradish peroxidase or alkaline phosphatase.
2. Antigen–antibody reactions are detected by enzyme activity. If the indicator enzyme is present in the test well, an antigen–antibody reaction has occurred.

3. The double antibody sandwich method is used to detect antigens against a specific antibody bound in a test well.

The Future of Vaccines and Diagnostic Immunology (p. 446)

1. New and safer vaccines will be developed using genetic engineering.
2. The use of monoclonal antibodies will make serological testing faster and more accurate.

Study Questions

Review

1. Identify the following as a direct or indirect test:
   (a) Test for *Mycobacterium tuberculosis*
   (b) Test for antibodies against *M. tuberculosis*

2. Which test in Question 1 is proof of a disease state? Why doesn't the other test confirm a disease state? What is the disease?

3. Explain how an anti-idiotypic vaccine works. Why is this a desirable vaccine?

4. The following data were obtained from fluorescent-antibody tests for anti-*Legionella* in four people. What conclusions can you draw? What is the disease?

	Antibody Titer			
	Day 1	Day 7	Day 14	Day 21
Patient A	1:128	1:256	1:512	1:1024
Patient B	0	0	0	0
Patient C	1:256	1:256	1:256	1:256
Patient D	0	0	1:128	1:512

5. If a person has antibodies against streptococcal erythrogenic toxin, what results are expected in the Dick test, in which the toxin is injected into the person's skin?

6. An antigen of the hepatitis B virus is used in the HBV vaccine. Can this vaccine cause hepatitis?

7. What do the following vaccines consist of? Which could cause the disease it is supposed to prevent?
   (a) Attenuated poliomyelitis virus
   (b) Dead *Rickettsia prowazekii*
   (c) *Vibrio cholerae* toxoid

8. Influenza virus, patient's serum, and red blood cells are mixed in a tube. If the patient has antibodies against influenza, what should be observed?

9. Measles virus, patient's serum, complement, red blood cells, and anti-red blood cells are mixed in a tube. If the patient has antibodies against measles, what should be observed?

10. How would the addition of excess complement to a complement-fixation test affect the results?

11. What is a precipitin? Explain the effects of excess antigen and antibody on the precipitation reaction. How is the precipitin ring test different from an immunodiffusion test?

12. How does the antigen in an agglutination reaction differ from that in a precipitation reaction?

13. Define the following terms, and give an example of how each reaction is used diagnostically.
   (a) Viral hemagglutination
   (b) Hemagglutination inhibition
   (c) Passive agglutination

14. Explain the complement-fixation test.

15. Differentiate between a direct and indirect fluorescent-antibody test.

16. Explain the ELISA techniques.

17. Match the following serological tests to the descriptions.
   ____ Precipitation
   ____ Immunoelectrophoresis
   ____ Agglutination
   ____ Complement fixation
   ____ Neutralization
   ____ ELISA

   (a) Occurs with particulate antigens
   (b) Peroxidase activity indicates a positive test
   (c) Hemolysis is used as an indicator
   (d) Employs anti-human gamma globulin
   (e) Occurs with a free soluble antigen
   (f) Used to determine the presence of antitoxin

Challenge

1. What problems are associated with the use of live vaccines?

2. The World Health Organization has announced the complete eradication of smallpox and is working toward eradication of measles. Why would vaccination be more likely to eradicate a viral disease than a bacterial disease?

3. Many of the serologic tests require a supply of antibodies against pathogens. For example, to test for *Salmonella*, anti-*Salmonella* antibodies are mixed with the unknown bacterium. How are these antibodies obtained?

4. A test for antibodies against *Treponema pallidum* uses an antigen called cardiolipin and the patient's serum (suspected of having antibodies). Why do the antibodies react with cardiolipin? What is the disease?

Further Reading

Arnon, R. "Chemically defined antiviral vaccines." *Annual Review of Microbiology* 34:593–618, 1980. Discusses immunologic responses to viral vaccines.

Kaufman, L., and P. G. Standard. "Specific and rapid identification of medically important fungi by exoantigen detection." *Annual Review of Microbiology* 41:209–225, 1987. Discusses important rapid identification and classification tests for fungal pathogens.

Kemp, D. J., R. L. Coppel, and R. F. Anders. "Repetitive proteins and genes of malaria." *Annual Review of Microbiology* 41:181–208, 1987. Discusses how the malaria parasite evades the immune system and summarizes the possibility of development of a malaria vaccine.

Kennedy, R. C., J. L. Melnick, and G. R. Dreesman. "Anti-idiotypes and immunity." *Scientific American* 255(1):48–56, July 1986. Describes how these antibodies open the way to new manipulations of the immune system.

Langer, W. L. "Immunization against smallpox before Jenner." *Scientific American* 234(1):112–117, January 1976. An interesting history of the use of smallpox itself as a vaccination against smallpox.

Lerner, R. A. "Synthetic vaccines." *Scientific American* 248(2):66–74, February 1983. Discusses use of a synthetic viral protein to produce antibodies.

Sabin, A. B. "Evaluation of some currently available and prospective vaccines." *Journal of the American Medical Association* 246:236–241, 1981. A summary of the currently available vaccines against bacterial pneumonia, poliomyelitis, measles, rubella, and influenza, with a discussion of new vaccines under investigation.

Stites, D. P., J. D. Stobo, H. H. Fudenberg, and J. V. Wells. *Basic and Clinical Immunology*, 5th ed. Los Altos, Calif.: Lange Medical Publications, 1984. An immunology textbook with a comprehensive discussion of the serological and diagnostic aspects of immunology.

CHAPTER 18

Disorders Associated with the Immune System

Not all immune responses against an antigen produce a desirable resistance or immunity. Sometimes, the reactions of the immune system are harmful. In this chapter, we will discuss several immune responses that do the host more harm than good; we will concentrate on those involving allergy, blood group incompatibilities, autoimmunity, and immune suppression. We will see how transplanted tissue is perceived as nonself and is rejected. Similarly, cancer cells are also recognized as nonself as they arise, and they are eliminated. In this sense, we will see that contracting cancer is a failure of the immune system.

Learning Objectives

- Define hypersensitivity, desensitization, histocompatibility antigens, HLA, immunosuppression, and immunologic tolerance.
- Describe the mechanisms of anaphylaxis and contact dermatitis.
- Differentiate among the four types of hypersensitivity reactions.
- Describe the basis of human blood groups and their relationship to blood transfusions and hemolytic disease of the newborn.
- Explain how rejection of a transplant occurs and how rejection is prevented.
- Define and give a possible explanation for autoimmunity.
- Discuss one immune-complex disease.
- Discuss the causes and effects of immune deficiencies.
- Describe the immune responses to cancer and immunotherapy.

Hypersensitivity

The term **hypersensitivity** refers to sensitivity beyond what is considered normal; the term **allergy** is probably more familiar and is essentially synonymous. Hypersensitivity responses occur in people who have been previously "sensitized" by exposure to an antigen, which in this context is sometimes called an **allergen**. Once sensitized, the immune system responds to a subsequent exposure to that antigen by reacting with it in a manner that damages the host tissue.

Hypersensitivity reactions are sometimes classified as either *immediate* or *delayed*. Of course, immediate hypersensitivity reactions appear relatively faster than delayed ones, but the main difference between these reactions is in the nature of the immune responses to the antigen.

Hypersensitivity reactions are considered to be of four principal types (Table 18-1). Types I, II, and III involve humoral antibodies; type IV involves effector cells of cell-mediated immunity. Type I (anaphylaxis) reactions are the most common; they are brought

TABLE 18-1   *Types of Hypersensitivity*

Type of Reaction	Immunoglobulin or Effector Cells Involved	Examples
Type I (anaphylaxis)	IgE on mast cells or basophils	Anaphylactic shock from drug injections and insect venom; common allergic conditions, such as hay fever, asthma
Type II (cytotoxic)	IgM, IgG, complement	Transfusion reactions, Rh incompatibility; Graves' disease; myasthenia gravis
Type III (immune complex)	IgM, IgG, complement	Systemic lupus erythematosus; rheumatoid arthritis
Type IV (cell-mediated)	T cells	Rejection of transplanted tissues; Addison's disease; contact dermatitis

about by IgE antibodies bound to certain body cells. Type II (cytotoxic) reactions involve the combination of antibodies with antigens that are normal surface components of an individual's own cells. The action of complement then injures the cells. Type III (immune complex) reactions result when complexes of antibody, antigen, and complement stimulate an inflammatory response that is harmful to body tissues. Type IV (cell-mediated) reactions are brought about by sensitized T cells.

Type I (Anaphylaxis) Reactions

Type I reactions often occur within a few minutes after a person sensitized to an antigen is reexposed to that antigen. **Anaphylaxis** comes from the Greek word *phylaxis*, which means protection, plus the prefix *ana* (against), which makes the word mean unprotected. Anaphylaxis is an inclusive term for the reactions caused by the combining of certain antigens with IgE antibodies. Anaphylactic responses can be divided into *systemic reactions*, which produce shock and breathing difficulties and are sometimes fatal, and *localized reactions*, which include common allergic conditions such as hay fever, asthma, and hives (slightly raised, often itchy and reddened areas of the skin).

The IgE antibodies produced in response to an antigen such as insect venom or pollen are **cytotrophic**; that is, they bind to the surfaces of mast cells and basophils. These two cell types are similar in morphology and in their contribution to allergic reactions. **Mast cells** are especially prevalent in the connective tissue of the skin and respiratory tract and in their surrounding blood vessels. **Basophils** circulate in the bloodstream, where they constitute less than 1% of the leukocytes. Both are packed with granules containing a variety of chemicals called *mediators*. The Fc region of the IgE antibody (discussed in Chapter 16) attaches to a specific receptor site on the mast cell or the basophil, leaving two antigen-combining sites free.

These cells can have as many as 500,000 sites for IgE attachment, although not all attached IgE monomers are specific for the same antigen. When an antigen, such as plant pollen, combines with antigen-combining sites on two adjacent IgE antibodies and bridges the space between them, the mast cell or basophil is triggered to undergo *degranulation*. This releases the granules that pack the interior of these cells, thus also releasing the mediators contained in the granules (Figure 18-1).

The best-known mediator is **histamine**. The pharmacological effects of histamine are to increase the permeability and dilation of blood capillaries, resulting in edema (swelling) and erythema (redness). Other effects include increased mucus secretion (a runny nose, for example) and smooth-muscle contraction, which in the respiratory bronchi results in breathing difficulty.

Other mediators include **leukotrienes** of various types and **prostaglandins**. These mediators are not preformed and stored in the granules but are synthesized by the antigen-triggered cell. The most widely studied leukotriene, *SRS-A* (slow-reacting substances of anaphylaxis), is actually a combination of several leukotrienes. Because leukotrienes tend to cause prolonged contractions of certain smooth muscles, their action contributes to the spasms of the bronchial tubes that occur during asthmatic attacks. Prostaglandins affect smooth muscles of the respiratory system and cause increases in mucus secretions. A number of

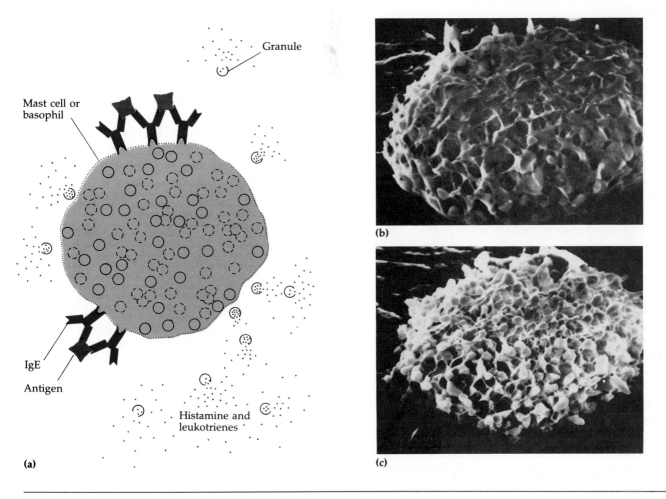

Granule

Mast cell or basophil

IgE

Antigen

Histamine and leukotrienes

(a)

(b)

(c)

Figure 18-1   Mechanism of anaphylaxis. (a) In response to an antigen, IgE antibodies are produced. These antibodies coat mast cells and basophils. When an antigen bridges the gap between two adjacent antibody molecules of the same specificity, the cell undergoes degranulation and releases mediators such as histamine and leukotrienes. **(b)** Scanning electron micrograph of a normal mast cell (×6500). **(c)** SEM of a degranulated mast cell (×6500).

other mediators are also released with the degranulation of mast cells and basophils.

Collectively, all these mediators serve as chemotactic agents that attract such cells as neutrophils and eosinophils to the mast cell site. They then activate various factors that cause inflammatory symptoms, such as dilation of the capillaries and swelling, increased secretion of mucus, and involuntary contractions of smooth muscles.

Systemic Anaphylaxis

At the turn of the century, two French biologists studied the response of dogs to the venom of the stinging jellyfish known as the Portuguese man-of-war. Large doses of the venom usually killed the dogs, but a few sometimes survived the injections. These surviving

dogs were used for repeat experiments with the venom. The results were surprising. Even a very tiny dose of the venom, one that should have been almost harmless, quickly killed the dogs. They suffered difficulty in respiration, entered shock as their cardiovascular systems collapsed, and quickly died. This phenomenon was named **anaphylactic shock**.

Systemic anaphylaxis (anaphylactic shock) can result when an individual sensitized to an injected antigen receives a subsequent injection of it. The release of mediators causes peripheral blood vessels throughout the body to undergo dilation, resulting in a drop in blood pressure (shock). This reaction can be fatal within a few minutes. The drug epinephrine (adrenaline) counteracts these effects, and kits for epinephrine self-administration are available to persons allergic to insect stings. There is very little time to act once someone develops systemic anaphylaxis.

Penicillin is a case of special interest because many of us are acquainted with people who are sensitive to this drug. In these persons, the penicillin, which is a nonimmunogenic hapten, combines with the carrier serum protein albumin and in that form is able to induce antibody formation. Penicillin allergy probably occurs in about 2% of the population. Anyone who has had an adverse reaction to penicillin that included generalized hives, swelling of throat tissues, or chest constriction should not receive the drug again. A childhood rash without hives is probably not a significant reaction. However, there is no completely reliable skin test for sensitivity to penicillin.

Localized Anaphylaxis

Whereas sensitization to injected antigens is a common cause of systemic anaphylaxis, **localized anaphylaxis** is usually associated with antigens that are ingested (foods) or inhaled (pollen). The symptoms that develop depend primarily on the route by which the antigen enters the body.

In the case of allergies involving the upper respiratory system, such as hay fever (allergic rhinitis), sensitization usually involves mast cells localized in the mucous membranes of the upper respiratory tract. Reexposure to the airborne antigen, which might be a common environmental material such as plant pollen, fungal spores, house dust, or animal dander (scaly, dried skin), results in a rapid release of large amounts of histamine, leukotrienes, and other mediators from mast cells in contact with the antigen. The typical symptoms are itchy and tearing eyes, congested nasal passages, coughing, and sneezing. Antihistamine drugs, which compete for histamine receptor sites, are often used to treat these symptoms.

Asthma is an allergic reaction that affects mainly the lower respiratory system. Such symptoms as wheezing and shortness of breath are caused by the constriction of smooth muscles in the bronchial tubes. Antihistamines are not an effective treatment for asthma since other mediators, such as leukotrienes and prostaglandins, are more important than histamine in this reaction. Treatment usually involves administration of epinephrine.

Antigens that enter the body via the gastrointestinal tract can also sensitize a host. Many of us may know someone who is allergic to a particular food. Children tend to outgrow these problems, and in fact many food allergies may not be related to hypersensitivity but are more accurately described as food

intolerances. For example, many people are unable to digest the lactose in milk because they lack the necessary enzyme. The diarrhea that results from milk ingestion is a symptom of food intolerance. Gastrointestinal upset is a common symptom of food allergies, but it can also result from many other factors. Hives are more characteristic of a true food allergy, and ingestion of large amounts of the antigen may even result in systemic anaphylaxis. Skin tests are not reliable indicators for the diagnosis of food-related allergies, and completely controlled tests for hypersensitivity to ingested foods are very difficult to perform.

Prevention of Anaphylactic Reactions

Avoiding contact with the sensitizing antigen is the most obvious way to prevent allergic reactions. Unfortunately, avoidance is not always possible. Some allergic persons may never know exactly what the antigen is. In other cases, skin tests might be of use in diagnosis (Figure 18-2). These tests involve inoculating small amounts of the suspected antigen just beneath the epidermal skin layer. Sensitivity to the antigen is indicated by a rapid inflammatory reaction that features redness, swelling, and itching at the inoculation site. This small affected area is called a *wheal*.

Once the responsible antigen has been identified, the person can either try to avoid contact with it or undertake **desensitization**. This procedure consists of a series of dosages of the antigen carefully injected beneath the skin. The objective is to cause the production of IgG antibodies rather than those of the IgE class, in the hope that the circulating IgG antibodies

Figure 18-2 Skin test to identify allergens. The allergen is inoculated into the skin. If a wheal (redness and swelling) forms at the site, this is evidence that the subject is allergic to that substance.

will act as *blocking antibodies* to intercept and neutralize the antigens before they can react with cell-bound IgE. Recent evidence indicates that desensitization might also induce the production of suppressor T cells (discussed in Chapter 16). Desensitization is not a routinely successful procedure, but it is effective in 65% to 75% of persons whose allergies are induced by inhaled antigens.

Type II (Cytotoxic) Reactions

Immunologic injury resulting from type II reactions is caused by antibodies that are directed at antigens on the host's blood cells or tissue cells. These are IgG or IgM antibodies, and their reaction with these antigens usually leads to the activation of complement. The cells involved may be red blood cells, platelets, lymphocytes, or certain organ cells.

Type II reactions damage cells through several mechanisms. One is the lysis of the host cell by plasma membrane damage caused by the antibody and complement. Other cell damage may result from the attraction of phagocytic cells or K cells to the Fc regions of antibodies attached to the host cell. These phagocytic cells, in an attempt to ingest a cell larger than themselves, may discharge lytic enzymes externally. This is a useful reaction in response to certain helminthic parasites, but not for a host tissue cell. Immunological damage may also occur when anti-

bodies block essential receptor sites and prevent their proper functions. Some researchers have suggested that this latter situation calls for a fifth type of hypersensitivity classification, but it is conventionally included in type II.

The most familiar cytotoxic hypersensitivity reactions are *transfusion reactions*, in which red blood cells are destroyed due to reaction with circulating antibodies. These involve the ABO and Rh blood group systems.

The ABO Blood Group System

In the early 1900s, it was discovered that human blood could be grouped into four principal types, which were designated A, B, AB, and O. This method of classification is called the **ABO blood group system**. Since then, at least fifteen other blood group systems have been discovered, but our discussion will be limited to two of the best known, the ABO and the Rh systems.

A person's ABO blood type depends on the presence or absence of two very similar carbohydrate antigens located on the cell membranes of red blood cells, or erythrocytes (Figure 18-3). Persons with type A blood possess a mosaic of antigens designated A on their red blood cells; persons of blood type B have antigens designated B on theirs. Persons with blood type AB have both A and B antigens on their red blood cells, and persons with blood type O lack both A and B surface antigens.

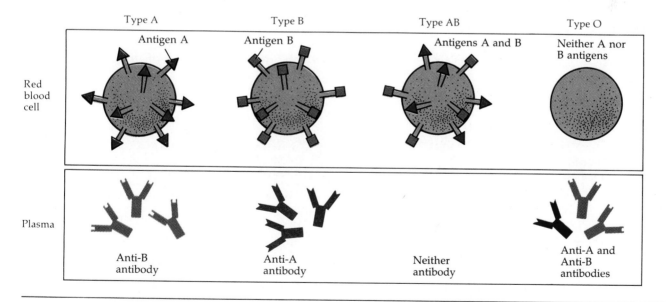

Figure 18-3 Relationships between the antigens on the surface of human red blood cells and the antibodies in the plasma.

The serum of persons with type A antigens on their red blood cells contains antibodies against type B cells (anti-B antibodies). Persons with type B blood have antibodies against type A cells (anti-A antibodies). Type O serum contains antibodies against both A and B cells, and type AB serum contains no anti-A or anti-B antibodies at all. When a transfusion is incompatible, as when type B blood is transfused into a person with type A blood, the antigens on the type B blood cells will react with the anti-B antibodies in the recipient's serum. This antigen–antibody reaction activates complement, which in turn causes lysis of the donor's red blood cells as they enter the recipient's system. The main features of the ABO blood group system are summarized in Table 18-2.

In about 80% of the population (called **secretors**), soluble antigens of the ABO type appear in saliva and other bodily fluids. In criminal investigations, it has been possible to type saliva residues from a cigarette and to type semen in cases of rape.

The Rh Blood Group System

In the 1930s, the presence of a different surface antigen on human red blood cells was discovered. It was found that when rabbits were immunized with red blood cells from Rhesus monkeys, their serum soon contained antibodies that were directed against the monkey blood cells but would also agglutinate human red blood cells. This indicated that a common antigen was present on both human and monkey red blood cells. The antigen was named the **Rh factor** (*Rh* for *Rhesus*). The roughly 85% of the population whose cells possess this antigen are called Rh⁺. Those lacking this antigen (about 15%) are Rh⁻. Antibodies that react with the Rh antigen do not occur naturally in the serum of Rh⁻ individuals, but exposure to this antigen can sensitize them to produce anti-Rh antibodies.

Blood Transfusions and Rh Incompatibility

If blood from an Rh⁺ donor is given to an Rh⁻ recipient, the donor's red blood cells stimulate the production of anti-Rh antibodies. If the recipient receives Rh⁺ red blood cells in a subsequent transfusion, a rapid hemolytic reaction will develop.

Hemolytic Disease of the Newborn

Blood transfusions are not the only way in which an Rh⁻ person can become sensitized to Rh⁺ blood. When an Rh⁻ female and an Rh⁺ male produce a child, the chances are at least 50% that the child will be Rh⁺. If the child is Rh⁺, the Rh⁻ mother can become sensitized to this antigen during birth, when the placental membranes tear and Rh⁺ fetal red blood cells enter the maternal circulation, causing the mother to produce anti-Rh antibodies. If the fetus in a later pregnancy is Rh⁺, the mother's anti-Rh antibodies will cross the placenta and destroy the fetal red blood cells

TABLE 18-2   *The ABO Blood Group System*

Characteristic	Blood Type			
	A	**B**	**AB**	**O**
Antigen present on the red blood cells	A	B	Both A and B	Neither A nor B
Antibody normally present in the plasma	anti-B	anti-A	Neither anti-A nor anti-B	Both anti-A and anti-B
Plasma causes agglutination of red blood cells of these types	B, AB	A, AB	None	A, B, AB
Percent in a mixed Caucasian population	41	10	4	45
Percent in a mixed black population	27	20	7	46

Figure 18-4 Hemolytic disease of the newborn. **(a)** Rh$^+$ father. **(b)** Rh$^-$ mother carrying her first Rh$^+$ fetus. Rh antigens from the developing fetus can enter the mother's blood across the placenta during delivery. **(c)** In response to the fetal Rh antigens, the mother will produce anti-Rh antibodies. **(d)** If the woman becomes pregnant again with an Rh$^+$ fetus, her anti-Rh antibodies will pass through the placenta into the blood of the fetus. The result is damage to fetal red blood cells.

⊕ Rh$^+$ red blood cells

⊖ Rh$^-$ red blood cells

⋎ Anti-Rh antibody

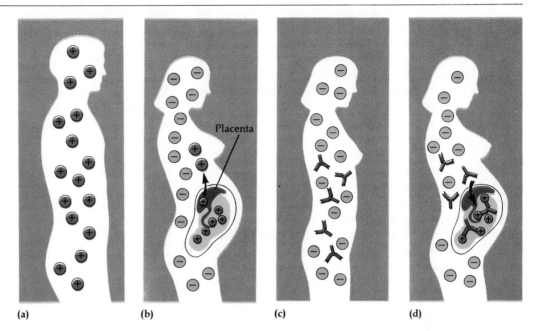

(a) (b) (c) (d)

(Figure 18-4). The fetus responds to this immune attack by producing large numbers of immature red blood cells called erythroblasts; hence the name *erythroblastosis fetalis* was once used to describe what is now called **hemolytic disease of the newborn**. Before birth, the maternal circulation removes most of the toxic by-products of fetal red cell distribution. After birth, however, the fetal blood is no longer purified by the mother, and the newborn develops jaundice and severe anemia.

Hemolytic disease of the newborn is usually prevented today by passive immunization of the Rh$^-$ mother with anti-Rh antibodies, which are available commercially. These anti-Rh antibodies combine with any fetal Rh$^+$ red blood cells that have entered the mother's circulation, so it is much less likely that she will become sensitized to the Rh antigen. If the disease is not prevented, it might be necessary that the newborn's Rh$^+$ blood, contaminated with maternal antibodies, be replaced by transfusion of uncontaminated blood.

Other Cytotoxic Diseases

Blood platelets (thrombocytes) are minute cell-like bodies that are essential to blood clotting. They are destroyed by antibodies and complement in the disease idiopathic thrombocytopenic purpura (ITP). In the situation illustrated in Figure 18-5, the platelet has become coated with drug molecules that function as haptens. Often, no drug is involved, and anti-platelet antibodies react with normal antigenic sites on the platelet. Because platelets are necessary for

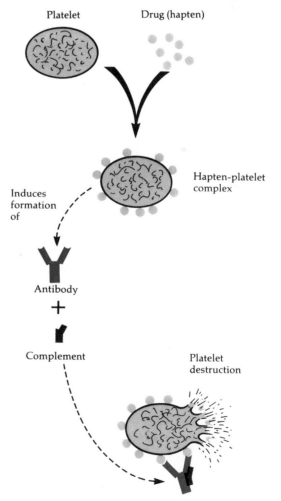

Platelet   Drug (hapten)

Induces formation of

Hapten-platelet complex

Antibody

+

Complement

Platelet destruction

Figure 18-5 Idiopathic thrombocytopenic purpura (ITP). Molecules of the drug accumulate on the surface of a platelet and stimulate an immune response that destroys the platelet.

blood clotting, their loss results in hemorrhages that appear on the skin as purple spots (purpura).

Another blood disorder of the cytotoxic type is autoimmune hemolytic anemia. This rare condition occurs when the body forms antibodies against its own red blood cells. Onset of this disease may be caused by drugs, malignancies, or autoimmune disorders.

Antibodies that block essential receptor sites on organs can cause disease. For example, myasthenia gravis is a disease in which muscle tone becomes progressively weaker. It is caused by antibodies that coat the acetylcholine receptors at the junctions at which nerve impulses reach the muscles. Eventually, the muscles controlling the diaphragm and rib cage fail to receive the necessary nerve signals, and respiratory arrest and death result. Graves' disease is caused by antibodies called long-acting thyroid stimulators. These antibodies attach to receptors on the thyroid gland that are intended for the thyroid-stimulating hormone produced by the pituitary gland. The result is that the thyroid gland is stimulated to produce increased amounts of thyroid hormones and becomes greatly enlarged. The most striking signs of the disease are goiter, which results from the enlarged thyroid, and markedly bulging, staring eyes.

Type III (Immune Complex) Reactions

Many damaging reactions can be caused by immune complexes of antigen, antibody, and complement. A primary difference between cytotoxic and immune complex reactions is that the antigens involved in the latter are not part of a host tissue cell.

Immune complexes form when certain ratios of antigen and antibody occur. The antibodies are usually of the IgM or IgG class. A significant excess of antibody (Figure 18-6a) leads to the formation of large complement-fixing complexes that are rapidly removed from the body by phagocytosis. When there is a significant excess of antigen (Figure 18-6b), soluble complexes form that do not fix complement and do not cause inflammation. However, when a certain antigen–antibody ratio exists, usually with a slight excess of antigen (Figure 18-6c), the soluble complexes that form are small and escape phagocytosis. These complexes circulate in the blood, pass between endothelial cells of the blood vessels, and become trapped in the basement membrane beneath the cells. In this

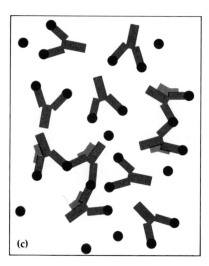

Figure 18-6 Formation of immune complexes. (a) Immune complexes form when antibody is in excess. Most of these are removed from the circulation by phagocytosis. **(b)** Immune complexes form when antigen is in excess. These generally do not fix complement or cause inflammation. **(c)** Small, soluble complexes form when there is a certain ratio of antigen to antibody; there is usually a small excess of antigen. These complexes fix complement, but are small enough to escape phagocytosis. They damage tissues when they become trapped in the basement membrane of blood vessels.

location, they may activate complement and cause a transient inflammatory reaction. Repeated introduction of the same antigen can lead to more serious inflammatory reactions.

Glomerulonephritis is an immune-complex condition resulting in inflammatory damage to the kidney glomeruli, which are sites of blood filtration. Antibodies generated in response to the M protein of streptococci are believed to be one cause of this disease (Chapter 14). It is also associated with the presence of such organisms as the malarial parasite, certain helminths that remain in the body for many years, and persistent hepatitis virus.

Another immune complex condition that often involves the kidney glomeruli (as well as other body sites) is systemic lupus erythematosus. The cause of this disease is not completely understood, but afflicted persons produce many types of antibodies directed at components of their own cells, primarily nucleic acids. The symptoms of this disease result from deposition of immune complexes in the skin, joints, heart pericardium, and kidney glomeruli.

Crippling rheumatoid arthritis is a disease in which immune complexes of IgM, IgG, and complement are deposited in the joints. The chronic inflammation caused by this deposition eventually leads to severe damage to the cartilage and bone of the joint. Cell-mediated immune responses are also thought to be involved in this disease.

Local immune-complex reactions can appear as a region of inflammation or even of necrosis (dead tissue) surrounding the site at which antigen was injected into a person already sensitized to that antigen. This is the *Arthus reaction*, caused by immune complexes deposited in blood vessels of the skin.

Type IV (Cell-Mediated) Reactions

Up to this point, we have discussed humoral types of hypersensitivity involving IgE, IgG, or IgM. Type IV reactions involve cell-mediated immune responses and are caused mainly by T cells, although macrophages may also be involved. Instead of occurring within a few minutes or hours after a sensitized individual is exposed to an antigen, these hypersensitivity reactions are often not apparent for a day or more. A major factor in the delay is the time required for the participating cells to migrate to and accumulate near the foreign antigens.

Causes of Type IV Reactions

It is thought that type IV hypersensitivity reactions occur when certain foreign antigens, particularly of a type that bind to tissue cells, are phagocytized by macrophages and then presented to receptors on the T-cell surface. Contact between the antigenic determinant sites and the appropriate T cell causes the T cell to proliferate. The T cells involved in delayed-type hypersensitivity reactions are primarily T_D cells. In some types of hypersensitivities resulting in tissue damage, T_C cells may also participate. If a person sensitized in this way is reexposed to the same antigen, a cell-mediated hypersensitivity reaction might result. A principal factor in this reaction is the release of lymphokines by T cells reacting with the target antigen. Lymphokines, as we saw in Chapter 16, contribute to the inflammatory reaction to the foreign antigen by attracting macrophages to the site and activating them.

Cell-Mediated Hypersensitivity Reactions of the Skin

We have seen that the skin is frequently the site on which hypersensitivity symptoms are displayed. One cell-mediated hypersensitivity reaction that involves the skin is the familiar skin test for tuberculosis. Because *Mycobacterium tuberculosis* is often located within macrophages, this disease can stimulate a cell-mediated immune response. As a screening test, protein components of the bacteria are injected into the skin. If the recipient has immunity from a prior exposure to tuberculosis, an inflammatory reaction to the injection of these antigens will appear on the skin in one or two days; this interval is typical of delayed hypersensitivity reactions.

Allergic contact dermatitis, another common manifestation of type IV hypersensitivity, is usually caused by haptens that combine with proteins in the skin of some persons to produce an immune response. Reactions to poison ivy (Figure 18-7), cosmetics, and the metals in jewelry (especially nickel) are familiar examples of these allergies. The identity of the offending environmental factor can usually be determined by a *patch test*. Samples of suspected materials are taped to the skin; after 48 hours, the area is examined for inflammation.

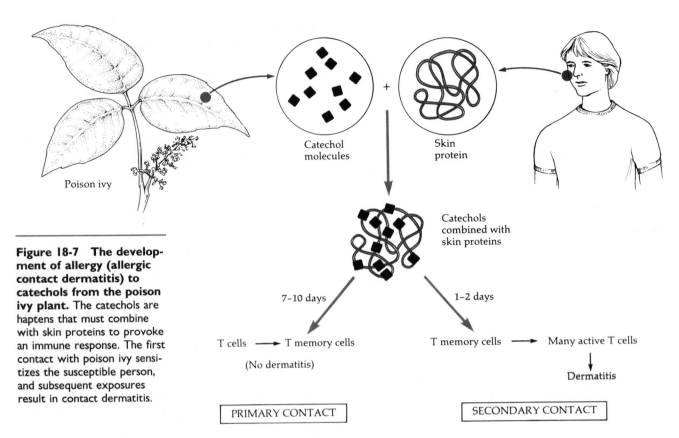

**Figure 18-7   The develop-
ment of allergy (allergic
contact dermatitis) to
catechols from the poison
ivy plant.** The catechols are
haptens that must combine
with skin proteins to provoke
an immune response. The first
contact with poison ivy sensi-
tizes the susceptible person,
and subsequent exposures
result in contact dermatitis.

Autoimmunity

Loss of Immunological Tolerance

We have mentioned that the body is normally able to
distinguish its own antigens (self) from all others
(nonself) and does not mount an immunological attack
against them. This phenomenon is called **tolerance**;
it apparently arises during fetal development. If a
fetus is exposed to an antigen, exposure to the same
antigen after birth does not stimulate the production
of antibodies or sensitized T cells. It is believed that
some clones of lymphocytes (*forbidden clones*) having
the potential to respond to self antigens may be produced during fetal life but are destroyed or suppressed
during fetal development.

Sometimes the body loses its ability to discriminate between self and nonself. This loss of tolerance
leads to **autoimmunity**, a response by antibodies or
sensitized T cells against a person's own tissue antigens. We saw in the previous section that cytotoxic

reactions are autoimmune. Autoimmunity also occurs
occasionally when cross-reacting antibodies are produced in the body's response to an infection. For
instance, the pathology of rheumatic fever might
result from the activity of antibodies formed in
response to a streptococcal infection. These antibodies
are coincidentally capable of reacting with antigenic
determinant sites on heart muscle cells.

Other autoimmune diseases specifically involve
the cell-mediated immune system. Like rheumatic
fever, some of these diseases might be related to the
immune system's response to infection. When the *lym-
phocytic choriomeningitis* virus (sometimes spread by
contact with pet hamsters) infects the membranes surrounding the brain, the body responds by producing
T cells that cause fatal neurological damage. Because
animals without a thymus are not affected by the
virus, it is believed that cell-mediated immunity is
involved in this disease.

There are a number of diseases with less obvious
etiologies, in which the cell-mediated immune system
destroys the patient's tissues. Hashimoto's thyroiditis
and Addison's disease result from T cells attacking
the thyroid and adrenal glands, respectively. Multiple
sclerosis involves inflammation and destruction of the
myelin layer coating the nerve cells; the reason for
this attack is unknown, but the disease is prevalent

only in temperate climates and is linked to geographical location during the first 15 years of life. A genetic factor is probably also important.

Many other autoimmune diseases have a genetic component. For Addison's disease, Graves' disease, rheumatic fever, and others, there is a well-known association between susceptibility to the disease and a specific determinant in the major histocompatibility complex located on one of the human chromosomes. Table 18-3 summarizes a number of autoimmune diseases.

Major Histocompatibility Complex

The inherited genetic characteristics of individuals are expressed not only in the color of their eyes and the curl of their hair but also in differences in the antigens on their cell surfaces. Some of these are called **histocompatibility antigens**. The genes controlling the production of such antigens are known as the **major histocompatibility complex (MHC)**; in humans, these genes are called the **human leukocyte antigen (HLA)** complex. We encountered these antigens in Chapter 16, where we saw that a requirement

for most antigens to stimulate an immune reaction was for them to be in association with an MHC antigen.

An important medical application of HLA typing is in transplant surgery, where the donor and the recipient must be matched. Matching of donor and recipient is done by *tissue typing*. The serological technique shown in Figure 18-8 is the one most often used. In serological tissue typing, the laboratory uses standardized antisera that are specific for particular HLA antigens. Lymphocytes from the person being tested are incubated with a selected specific antiserum. Complement and a dye, such as trypan blue, are then added. If antibodies in the antiserum have reacted specifically with the lymphocyte, the cell will lyse. The lysed cell will take up the dye (living cells will not), thus indicating that the lymphocyte possesses a certain antigen. This method is simple and rapid. Unfortunately, sera are not always available for the HLA groups most likely to cause problems with transplants between unrelated individuals.

A second method exists for identifying such HLA groups. However, it requires about six days, which is usually too long to be useful. Research to improve this situation continues, much of it aimed at developing

TABLE 18-3 *Autoimmune Diseases*

Disease	Description
Addison's disease	Destruction of adrenal gland cells that produce cortical hormones
Autoimmune hemolytic anemia	Synthesis of antibodies against own red blood cells
Diabetes mellitus, insulin dependent	Destruction of pancreatic islet cells that produce hormones and insulin
Goodpasture's syndrome	Damage to basement membranes, but most symptoms result from attack on glomeruli of kidney
Graves' disease	Binding of antibodies to receptors for thyroid-stimulating hormone leads to overstimulation of thyroid activity
Hashimoto's thyroiditis	Destruction of cells of the thyroid gland
Idiopathic thrombocytopenia purpura	Destruction of blood platelets
Myasthenia gravis	Blockage of acetylcholinesterase receptors prevents transmission of nerve signals
Rheumatoid arthritis	Inflammation of joints due to deposition of complexes of IgG and anti-IgG
Systemic lupus erythematosus	Antigen–antibody complexes (including antibodies against RNA, IgG, red blood cells, platelets, and chromosomes) are deposited in many locations, triggering inflammation; complexes containing antibodies to DNA are often deposited in kidneys

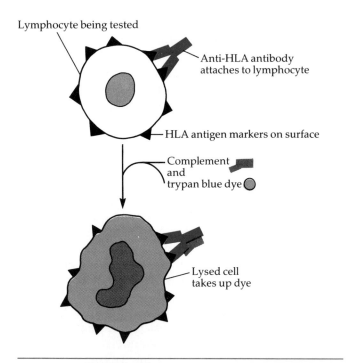

Figure 18-8   Tissue typing, serological method. Lymphocytes from the person being tested are incubated with laboratory test stocks of anti-HLA antibodies that are specific for a particular HLA antigen. If the antibodies react with a lymphocyte, then complement lyses the lymphocyte and dye can enter the cell. A positive test result indicates that the person has the particular HLA antigen being tested for.

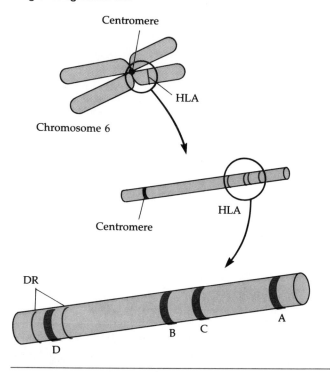

Figure 18-9   Genes controlling human leukocyte antigens (HLA). The genes that control human leukocyte antigens are located on chromosome 6. Note how HLA-A, -B, -C, -D, and -DR are located within a very short length of chromosome.

antisera for more HLA types. Eventually, it is hoped that monoclonal antibodies for all HLA types will be available.

There are two main classes of HLA antigens that are important in such matches: class I antigens (HLA-A, -B, and -C) and class II antigens (HLA-D and -DR). The genes that control production of these antigens are located close together on human chromosome 6, as shown in Figure 18-9. Matching for class I antigens has long been a standard procedure, but matching for class II antigens might well be more important, especially for tissue from a totally unrelated person (such as a transplant from a cadaver). The donor and recipient must be of the same ABO blood type, and the recipient's serum must be tested for antibodies to the donor's HLA antigens.

As we mentioned previously, certain HLA antigens are related to increased susceptibility to specific diseases, which represents another medical application of HLA typing. A few of these relationships are summarized in Table 18-4.

Transplantation

Human organ transplants have added productive years to the lives of many individuals. Since the first kidney transplant was performed in 1954, this particular type of transplant has become a nearly routine medical procedure. Other types of transplants that are now feasible include bone marrow, thymus, heart, liver, and cornea. Tissues and organs for transplant are usually taken from recently deceased individuals, although nonessential duplicates of an organ, such as a kidney, occasionally come from a living donor.

Types of Transplants

Some transplants or grafts do not stimulate an immune response. A transplanted cornea, for example, is rarely rejected, mainly because antibodies do not circulate into the anterior chamber of the eye, which is therefore considered an immunologically **privileged site.** The brain is also an immunologically privileged site, probably because it does not have lymphatic vessels and because the walls of the blood vessels in the brain differ from blood-vessel walls elsewhere in the body (the blood–brain barrier is discussed in Chapter 21). Someday, it may be possible to graft foreign nerves to replace damaged nerves in the brain and spinal cord.

TABLE 18-4 *Diseases Related to Specific HLA Antigens*

	Related HLA Antigen	Increased Risk of Occurrence as Compared with General Population	Description
Inflammatory Diseases			
Multiple sclerosis	D(DR2)	5 times	Progressive inflammatory disease affecting nervous system
Rheumatic fever	DR (antibody 883)	4 to 5 times	Autoimmune disease caused by cross-reaction with antibodies against streptococcal infection
Endocrine Diseases			
Addison's disease	D(DR3)	4 to 10 times	Deficiency in production of hormones by adrenal gland
Graves' disease	D	10 to 12 times	Antibodies attached to certain receptors in the thyroid gland cause it to enlarge and produce excessive hormones
Malignant Disease			
Hodgkin's disease	A(A1)	1.4 to 1.8 times	Cancer of lymph nodes

It is also possible to transplant **privileged tissue** that does not stimulate an immune rejection. An example is replacement of a person's damaged heart valve with a valve from a pig's heart. However, privileged sites and tissues are more the exception than the rule.

It would be interesting to understand how animals tolerate pregnancy without rejecting the fetus. It has been determined that the uterus is not a privileged site, and yet in pregnancy the tissues of two genetically different individuals are in direct contact. Current theories are that there is a barrier that masks the fetal antigens and that the maternal immune system is suppressed.

When one's own tissue is grafted to another part of the body, as is done in burn treatment or in plastic surgery, the graft is not rejected. Recent technology has made it possible to use a few cells of a burn patient's uninjured skin to culture extensive sheets of new skin. The graft of this new skin is an example of an **autograft**. Identical twins have the same genetic makeup; therefore, skin or organs such as kidneys may be transplanted between them without provoking an immune response. These transplants are called **isografts**.

Most transplants, however, are made between persons who are not identical twins; these transplants do trigger an immune response. Attempts are made to match the HLA antigens of the donor and recipient as closely as possible so that the chances of rejection are reduced. Because HLA antigens of close relatives are most likely to match, blood relatives, especially siblings, are the preferred donors. Grafts between persons who are not identical twins are called **allografts**.

Since there is a shortage of available organs, it would be helpful if **xenografts** (organs from animals) could be more successfully transplanted to humans. However, the body tends to mount an especially severe immune assault on such transplants. Perhaps the most famous xenograft recipient was Baby Fae, a two-week-old baby girl who received the heart of a baboon on October 26, 1984. The girl died 20 days later from heart and kidney problems. The various types of grafts are summarized in Table 18-5.

When bone marrow is transplanted to persons with a defective immune system, the transplanted tissue may attack the host. The purpose of the transplant is to provide such persons with the B-cell and T-cell manufacturing capability they lack. However, the result can be **graft versus host (GVH) disease**. The transplanted bone marrow contains immunocompetent cells that mount a mostly cell-mediated immune response against the tissue into which they have been transplanted. Because the recipients lack effective immunity, GVH disease is a serious complication and can even be fatal.

TABLE 18-5   *Types of Grafts*

Graft	Description
Autograft (*autos*, "self")	A graft transplanted from one site to another site on the same person—for example, the skin from the thigh grafted over a burned area on the arm
Isograft (*isos*, "equal")	A graft between genetically identical persons—identical twins
Allograft (*allos*, "other")	A graft between genetically different members of the same species
Xenograft (*xenos*, "strange")	A graft between different species—for example, a transplant of a baboon heart to a human

Immunosuppression

To keep the problem of rejection of transplanted tissue in perspective, it is useful to remember that the immune system is simply doing its job and has no way of recognizing that its attack against the transplant is not helpful. In an attempt to prevent rejection, the recipient of an allograft usually receives treatment to suppress this normal immune response against the graft. The treatments have involved drugs, antilymphocyte serum directed at T cells, and X-irradiation. Unfortunately, these treatments suppress the immune response to all antigens, by both the humoral and cell-mediated immune systems. The persons so treated then become very susceptible to infectious diseases and cancer.

The drug **cyclosporine**, derived from a mold in 1976, has revolutionized transplant medicine. Cyclosporine specifically suppresses cell-mediated immunity by acting on T cells. Its mode of action is not completely understood at this time. Apparently, the presence of cyclosporine blocks the division of T cells, probably by interfering with the synthesis by helper T cells of the T-cell growth factor interleukin-2. T-cell proliferation resumes when cyclosporine is removed. Yet, throughout the term of cyclosporine administration, the patient's humoral immunity against circulating bacteria and viruses remains essentially intact.

As a result of this discovery, the success rate of organ transplants has risen dramatically. Before cyclosporine came into general use, only about 50% of kidneys from donors who were not closely related to the patient were accepted; now as many as 90% of these transplants succeed. Liver transplants became practical only after cyclosporine became available, and heart transplants have achieved unprecedented success rates.

The greatest drawback to the use of the drug appears to be an increased risk of cancer, as that disease is combatted by the same cell-mediated immune system that rejects transplanted organs. Certain problems with liver and kidney toxicity have also necessitated careful attention to dose levels.

Recently, another approach to graft rejection has involved the use of monoclonal antibodies. A commercial preparation is available under the name *muromonab-CD3 (Orthoclone OKT3)*. It is a monoclonal IgG antibody that reacts with the CD3 surface molecule, which the T cells must have to respond to an antigen. Within minutes, the drug can clear the peripheral blood of all CD3-positive T cells and prevent their rejection of transplanted tissue. The treatment, however, is very costly; moreover, on repeated administration, an immune response to the drug itself appears.

Natural Immune Deficiencies

The drugs we have just discussed impair the immune response artificially. Occasionally, people are born with defective immune systems. A number of inherited genes can result in impaired immunity. For example, individuals with a certain recessive trait may lack a thymus gland and therefore lack cell-mediated immunity. A different recessive trait causes lowered numbers of B cells. Some immunodeficiencies have no known cause. For example, about 1 in 700 Caucasian adults produces less IgA than other adults.

A wide variety of diseases can also impair immune function. Hodgkin's disease lowers cell-mediated immunity. Many viruses are capable of infecting and killing lymphocytes, which lowers the immune response. Removal of the spleen decreases humoral immunity. Table 18-6 summarizes several of the better-known natural immune deficiency conditions, including AIDS.

TABLE 18-6   *Immune Deficiency Diseases*

Disease	Cells Affected	Comments
Acquired immune deficiency syndrome (AIDS)	T cells	Virus destroys T_H cells, allowing cancer and viral, fungal, and protozoan diseases
Common variable hypogammaglobulinemia	B, T cells	Frequent viral and bacterial infections; second most common immunodeficiency disease
Reticular dysgenesis	B, T, and stem cells	Usually fatal in early infancy
Selective dysgammaglobulinemias, may be IgA, IgG, or IgM	B cells	IgA deficiency is quite common, causing frequent mucosal infections; IgM deficiency rare; IgG deficiency causes frequent bacterial infections
Severe combined immunodeficiency	B, T, and stem cells	Occurs in several forms; causes severe viral, fungal infections; treated with transplantation of bone marrow, fetal thymus
Thymic hypoplasia (DiGeorge syndrome)	T cells	Usually fatal in infancy from *Pneumocystis* pneumonia or viral or fungal infections
Wiskott–Aldrich syndrome	B, T cells	Frequent infections by viruses, fungi, protozoans; eczema, defective blood clotting; usually causes death in childhood
X-linked infantile (Bruton's) agammaglobulinemia	B cells	Frequent extracellular bacterial infections

AIDS

Acquired immune deficiency syndrome (AIDS) is a type of immunodeficiency disease. The victim acquires the disease by a transmissible virus. This virus does not cause a disease in the same way that other viruses do. Instead, the AIDS virus destroys helper T cells (see Figure 1-1b). As the T_H cells are destroyed, the person falls victim to diseases that the immune system would normally prevent.

Origin of AIDS

In 1981, a cluster of cases of *Pneumocystis carinii* pneumonia appeared in the San Francisco area. This extremely rare disease is caused by a protozoan and is usually seen only in persons who are immunosuppressed. Investigators soon correlated the appearance of this disease with an unusual incidence of a rare form of cancer of the skin and blood vessels called Kaposi's sarcoma. The victims were all young homosexual males. By 1983, the pathogen causing the disease had been identified as a retrovirus (Chapter

12), now known as **human immunodeficiency virus (HIV)**. (It has also been called HTLV-III and LAV.)

There have been several theories about the origin of the AIDS virus. The current theory is that HIV arose by mutation of a virus that has been endemic in some areas of central Africa for many years. It is further theorized that HIV was carried to the Caribbean country of Haiti by Haitians who once lived in central Africa. From Haiti it is presumed to have moved to the United States via Haitian immigrants and American homosexual males who vacationed in Haiti.

Symptoms of AIDS

The HIV kills helper T cells known as T4 cells, so called from the CD4 antigen on their surface. The virus is also able to enter many other kinds of cells, including macrophages, B cells, endothelial cells that line blood vessels, and certain brain cells. It can remain latent in such cells for extended periods without necessarily killing them. Some of these cells, especially macrophages, may serve as vehicles of transmission.

Relatively little free (extracellular) virus is present in an infected person. This is one reason why, even though persons infected with the virus develop antibodies against the virus, the antibodies do not prevent the disease.

AIDS is actually only the end stage of an infection by HIV. Following the initial infection, the patient undergoes seroconversion—that is, shows positive results in tests that detect antibodies to HIV. The first symptoms, fever and malaise similar to those of mononucleosis (Chapter 22), may appear up to 11 years later. For a short time following this state, patients are often entirely asymptomatic. However, their body fluids are infective, and HIV can almost always be cultured from the patient. (Donations of blood, organs, or sperm should not be made by infected persons.)

After the initial stages, many infected persons exhibit **AIDS-related complex (ARC)**. Among the symptoms of ARC are swollen lymph nodes (lymphadenopathy), loss of appetite and weight, fever, rashes, night sweats, and fatigue.

It is not yet known what percentage of persons infected with HIV will develop full-blown AIDS; present indications are that the majority will develop AIDS. Almost all who develop full-blown AIDS will die within 200 to 400 days. The cause of death is often opportunistic infections such as *Pneumocystis* pneumonia and Kaposi's sarcoma. Table 18-7 lists other common infections that may cause death in AIDS victims.

Modes of Transmission

Transmission of the AIDS virus usually requires an exchange of bodily fluids, such as blood, semen, or vaginal secretions, that contain the virus or cells containing the virus. The Centers for Disease Control has established that the routes of transmission involve intimate sexual contact, contaminated needles, and blood-to-blood contact, such as in transfusions. A pregnant woman infected with the virus can transmit it to the fetus. Tears and saliva may also contain the virus, but transmission by this means has not been reliably demonstrated.

Worldwide epidemiologic studies indicate three geographic patterns of transmission: (1) Transmission is primarily among homosexual or bisexual males and intravenous drug users in North America, Western Europe, Australia, and New Zealand. (2) In African and Caribbean countries, heterosexual contact is the primary mode of transmission, and males and females are affected equally. (3) In Eastern Europe, the Middle East, and Asia, most AIDS cases have occurred among

TABLE 18-7   *Some Common Diseases Associated with AIDS*

Disease Agent	Disease
Parasitic Diseases (Protozoans)	
Pneumocystis carinii	Life-threatening pneumonia
Cryptosporidium muris	Persistent diarrhea
Toxoplasma gondii	Encephalitis
Isospora belli	Gastroenteritis
Viral Diseases	
Cytomegalovirus	Fever, encephalitis, blindness
Herpes simplex	Vesicles of skin and mucous membranes
Varicella-zoster	Chickenpox and shingles
Epstein–Barr disease	Oral hairy leukoplakia (whitish patches on mucous membranes, probably precancerous)
Bacterial Diseases	
Mycobacterium tuberculosis	Tuberculosis
Mycobacterium avium-intracellulare	May infect many organs; gastroenteritis and other highly variable symptoms
Fungal Diseases	
Cryptococcus neoformans	Life-threatening meningitis
Candida albicans	Overgrowth on mucous membranes
Cancer	
Kaposi's sarcoma	Cancer of skin and blood vessels

people who have traveled to endemic areas and who had sexual contact with infected homosexual men and female prostitutes.

AIDS is not transmitted by casual social contact. Transmission by blood transfusion is no longer likely because blood is now screened for AIDS antibodies. However, there will always be a slight risk because blood might be donated during the short interval between infection with the virus and appearance of detectable antibodies. There is no risk at all in donating blood. Anyone using intravenously administered drugs should always use a sterile, previously unused needle. Risk of sexual transmission can be minimized by the use of condoms. HIV-positive women should not become pregnant because of the probability of transmitting the virus to the fetus.

Vaccines and Treatments

There are two basic areas of AIDS research. Some researchers are searching for vaccines to prevent the disease, and others are looking for drugs that treat AIDS (see the box on p. 468).

Any AIDS vaccine will have to work against the virus-infected cells because there is relatively little free virus in circulation, even during the initial infection. There is also evidence of more than one antigenic form of HIV, which would complicate the search for a vaccine, as well as blood-testing procedures. The antigenic type presently circulating in the United States is sometimes referred to as HIV-1. Another type, HIV-2, is prevalent in Africa. Present research centers on drugs that block the reverse transcriptase enzyme, which is essential for reproduction of HIV. Because the virus infects the brain, an effective drug must cross the blood–brain barrier (a phenomenon whereby certain drugs are inhibited from entering the

central nervous system—discussed in Chapter 21). Different approaches, such as vaccines that block the receptor sites on the T cells, may be required (see the discussion of anti-idiotype vaccines in Chapter 17). A monoclonal antibody might also be used to block receptor sites. Drugs exist that are effective against viruses, and perhaps some will be found that cure or alleviate the symptoms of AIDS. However, even if an effective drug is found, its administration in the late stages of the disease would still leave the patient with a crippled immune system.

Although many experts are pessimistic about the development of an AIDS vaccine until at least the 1990s and effective chemotherapy for the disease remains elusive, it is worth noting that if AIDS had appeared only a generation ago we would not have had the tools to identify the virus or the cells that it was affecting.

The Future of AIDS

In the United States 62,638 people had been diagnosed as having AIDS by the middle of 1988; almost half of them had died. Another 150,000 Americans appear to suffer from ARC, and as many as 1.5 million may be infected with HIV. It is not certain how many of those infected with HIV will develop AIDS as clinically defined, but it is expected to be the majority. Figure 18-10 shows the number of new AIDS cases for every six months from 1981 to the middle of 1988.

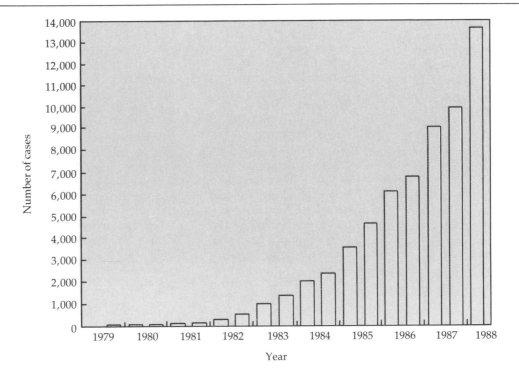

Figure 18-10 Numbers of AIDS cases. The graph shows the reported numbers of new AIDS cases in the United States for each six-month period since 1979. Source: CDC.

.. Microbiology in the News ....

New Weapons Against AIDS

Human immunodeficiency virus (HIV) has several unusual characteristics that make AIDS especially difficult to prevent and to treat.

First, it kills helper T cells, which are vitally important to the body's defenses against infection. The virus attaches its envelope glycoproteins to receptors on the T-cell membrane and then penetrates the cell. Once inside a cell, the virus can remain latent for months, during which time it is virtually immune from attack.

Second, the replication of the virus seems to be triggered by the T cell's immune response to infection. Thus, just when the T cell is most needed by the body, it is destroyed.

Third, the virus spreads from cell to cell in a unique manner. During biosynthesis, virus proteins called gp120 are incorporated into the host plasma membrane. These gp120 proteins are destined to become part of the virus's envelope when it buds out of the cell. While these proteins are in the T cell's membrane, they cause the cell to attach to other cells' receptors. This results in cell fusion and in cell-to-cell transmission of the virus.

Furthermore, HIV infects bone marrow cells from which infected cells can spread throughout the body.

To treat AIDS, we must have drugs that can cure already infected people. Vaccines against AIDS must be able to prevent initial infection. In 1987 and 1988, both in-vitro tests and human trials of several drugs and vaccines began in many laboratories around the world.

Vaccines

Human trials of several experimental vaccines have already begun, although these initial tests will not determine whether the vaccines actually prevent AIDS. The tests are intended to determine whether the vaccines produce an immune response and whether they are safe.

These vaccines attempt to create immunity against the envelope glycoproteins. Conventional vaccines prepared from inactivated whole viruses are not feasible because the virus mutates too quickly and because its oncogenes may become incorporated into the host cells. Moreover, viruses that are latent within a host cell would not be vulnerable to conventional vaccines.

The hope is that attacking the envelope glycoproteins would not only destroy free viruses but would also prevent cell fusion of T cells and budding of viruses from infected cells. In March 1987, French scientist Daniel Zagury tested a vaccine prepared from genetically engineered vaccinia virus into which the gene for gp120 had been inserted (see the figure). The vaccinia virus causes host cells to produce gp120 proteins, which bind to the cell membranes. This should then stimulate the production of antibodies against the viral proteins. Zagury, working in Zaire, noted an antibody response after injecting himself with the vaccine. In January 1988, human trials of a similar vaccine (HIVAC-1e) were begun by Oncogen, a division of Bristol-Myers.

In September 1987, MicroGeneSys, Inc., of West Haven, Connecticut, began human trials of a vaccine that uses a different envelope glycoprotein, gp160. The gp160 gene from a genetically engineered HIV-1 virus (supplied by the National Institute of Allergy and Infectious Diseases, NIAID) is inserted into baculovirus, which infects moths and butterflies. The virus is grown in cultures, and the viral protein is harvested from the cultures and used in the vaccine. After injection, the body should produce antibodies against the viral protein.

NIAID scientists are studying the effects of the vaccine in male volunteers who are healthy and free of AIDS. After eight months, one-third of the volunteers have developed anti-HIV. Cell-mediated immunity and the effectiveness of these antibodies in vivo have not been evaluated.

In England, trials of two vaccines are getting started. Two patients have been injected with anti-idiotypic antibodies, and 24 infected patients will be given a synthetic core protein (p17).

Jonas Salk has begun trials of a killed-HIV vaccine in infected volunteers at the University of Southern California. The virus is killed with gamma radiation and freeze dried. Disease progress in the vaccinated patients will be compared to unvaccinated "matched" patients.

Chemotherapy

HIV is a retrovirus. Like other retroviruses, it uses the viral enzyme reverse transcriptase to transcribe a molecule of DNA from viral RNA. This enzyme may provide a potential target for chemotherapeutic agents. In addition, several newly described genes or their products may provide potential targets. These genes include *orf*, whose product controls latency by suppressing viral replication, and *sos*, whose produce is responsible for infectivity.

Currently, only one drug is approved for treating AIDS: AZT (Retrovir®, Burroughs Wellcome Co.).

AZT works by lowering the incidence of opportunistic infections, Kaposi's sarcoma, and conversions from AIDS-related complex to AIDS. It seems to do this at least partly by increasing the number of helper T cells. However, AZT helps only certain AIDS patients. It also has side-effects that include immunosuppression and anemia.

The search for new drugs continues at a rapid pace. In 1987 and 1988, the following potentially helpful drugs were developed.

■ *Dideoxycytidine* (developed by the National Institutes for Health) inhibits infectivity in vitro. Human trials will begin soon.

■ *Ribavirin* seems to slow the progress of AIDS, according to some scientists, although it has the drawback of interfering with AZT.

■ *1-deoxynojirimycin* (*dNM*) and *castanospermine*, according to scientists at the Netherlands Cancer Institute, inhibit the action of glucosidase, which is necessary for HIV-1 biosynthesis. Both drugs have low toxicity; they are currently used in Europe to control blood-sugar levels.

■ *Ampligen*, a synthetic form of double-stranded RNA (being tested at Hahnemann University, Philadelphia) seems to stimulate the release of interferon and other antiviral proteins.

■ *D-penicillamine* (developed at George Washington University) and *3'-azido-2',3'-dideoxy-5-ethyluridine* (CS-85, developed by a consortium including the Centers for Disease Control) may inhibit viral replication. They are being tested in vitro. They have lower toxicity than AZT.

■ *Peptide T*, which combines with the receptor sites to block infection, according to the National Institute of Mental Health, has been approved for testing in AIDS patients.

■ Molecular biologists at the California Institute of Technology are looking for drugs that prevent attachment by the virus. So far, they have shown that the virus binds to *synthetic binding sites* instead of to real CD4 sites on lymphocytes in vitro.

Based on the rapid progress that scientists have made since the AIDS virus was discovered in 1984, many scientists are optimistic that AIDS will one day be as well controlled as measles and smallpox are today. Nevertheless, it will take years of continued research and testing to discover effective drugs and vaccines.

Production of AIDS vaccines using genetic engineering techniques. A gene for an HIV envelope glycoprotein is inserted into an avirulent virus. This second virus is grown in cell culture. **(a)** The viruses can be used for vaccination, or **(b)** the envelope glycoprotein produced in cell culture can be used for the vaccine.

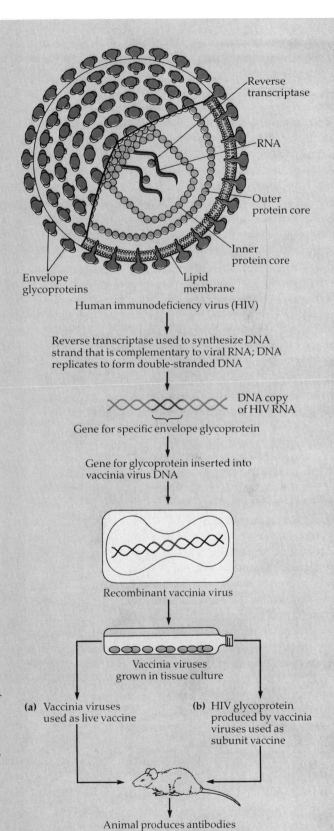

Envelope glycoproteins

Reverse transcriptase

RNA

Outer protein core

Inner protein core

Lipid membrane

Human immunodeficiency virus (HIV)

Reverse transcriptase used to synthesize DNA strand that is complementary to viral RNA; DNA replicates to form double-stranded DNA

DNA copy of HIV RNA

Gene for specific envelope glycoprotein

Gene for glycoprotein inserted into vaccinia virus DNA

Recombinant vaccinia virus

Vaccinia viruses grown in tissue culture

(a) Vaccinia viruses used as live vaccine

(b) HIV glycoprotein produced by vaccinia viruses used as subunit vaccine

Animal produces antibodies to HIV envelope glycoprotein

Immune Response to Cancer

Immunological Surveillance

The immune system's presumed responsibility for patrolling the body for cancer cells is called **immunological surveillance**. Persons who are immunosuppressed by either natural or artificial means are markedly more susceptible to cancers than the rest of the population. The conventional explanation for this susceptibility is that the immune system normally recognizes and destroys cancer cells before they become established in the body. The tendency of cancer to occur most often in the elderly, whose immune systems are thought to be less efficient, or in the very young, whose immune systems may not have developed properly, is thought to support this view.

A cell becomes cancerous when it undergoes transformation (Chapter 12) and begins to proliferate without control. Because of transformation, the surfaces of tumor cells may acquire tumor-associated antigens that mark them as nonself to the immune system (Figure 18-11). It is hoped that monoclonal antibodies will be the tool to help the immune system recognize such tumor-associated antigens and detect cancer in much earlier stages and also be useful in therapy (see the section Immunotherapy on p. 471).

Immunological Escape

Cancer sometimes occurs even in people with presumably normal immune systems. Once established, the cancer seems to become resistant to immune rejection, even though it is easy to demonstrate that cell-mediated and humoral immunity are directed against the tumor-associated surface antigens. This resistance to rejection is called **immunological escape**, and several mechanisms have been proposed to explain it. One possibility is *antigen modulation*, in which tumor cells are believed to shed their associated antigens, thus evading recognition by the immune system. Another proposed mechanism is that some cancers, such as Hodgkin's disease, release factors that eventually suppress the entire cell-mediated immune system.

An especially interesting experiment illustrating immunologic escape involves the immunization of

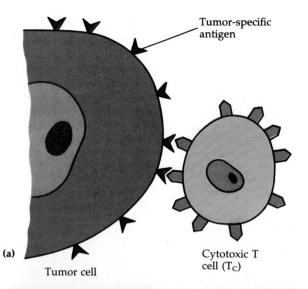

(a)

Tumor-specific antigen

Tumor cell

Cytotoxic T cell (T$_C$)

(b)

(c)

Figure 18-11 Interaction of a cytotoxic T cell (T$_c$) and a tumor cell. (a) Diagram of a cytotoxic T cell binding to a tumor-associated antigen on the surface of a tumor cell. **(b)** Scanning electron micrograph showing a cytotoxic T cell (smaller sphere at the lower left) attaching to the tumor cell after recognizing the tumor-associated antigens (×7250). **(c)** Lysis of the tumor cell is indicated by the deep folds and swellings in its surface membrane (×7250).

animals with separate cells of a certain tumor. If the same tumor is later introduced into the animal, the tumor grows readily and kills the animal. However, if the tumor is introduced into a nonimmunized animal, it is rejected in a short time. The tumor growth in the immunized animal is an example of immunologic escape by **immunological enhancement**. This effect might be due to a form of antigenic modulation. Another possible explanation is that the immunized animal forms humoral antibodies against the tumor-associated surface antigens. These antibodies then cover the tumor surface antigens so that T_C cells are unable to recognize the tumor as nonself and fail to destroy it. In the nonimmunized animal, the agents of cell-mediated immunity effectively reject the transplanted tumor.

Immunotherapy

Immunotherapy seems likely to be the most effective future approach to cancer treatment. Probably the most promising technique will involve immunotoxins. In Chapter 16, we defined an immunotoxin as a combination of a monoclonal antibody and a toxic agent, such as the drug ricin or a radioactive compound. The monoclonal antibody is targeted at a particular type of tumor antigen. The antibody selectively locates the cancer cell, and the attached toxic agent destroys the cell but causes little or no damage to healthy tissue. There is great hope that this approach will become analogous to the use of antibiotics in treating infec-

TABLE 18-8   *Principal Types of Cancers*

Type of Cancer	Description
Carcinoma	The most common form of cancer; arises from the cells forming the skin, the glands (such as the mammary, uterus, and prostate), and the membranes of the respiratory and gastrointestinal tracts; metastasizes (spreads to other parts of the body) mainly through the lymph vessels
Leukemia	Cancer of the tissues that form blood; characterized by uncontrolled multiplication and accumulation of abnormal white blood cells
Lymphoma	Cancer of the lymph nodes
Sarcoma	Another form of malignant tumor; arises from connective tissues, such as muscle, bone, cartilage, and membranes covering muscles and fat; metastasizes through the blood vessels

tious diseases. The principal types of cancers are summarized in Table 18-8.

During the final years of this century, we can expect to see more advances in the battle against cancer than in all of previous history. Almost certainly, immunology will be one of the major weapons in that fight.

Study Outline

Hypersensitivity   (pp. 415–452)

1. Hypersensitivity reactions represent immunologic responses to an antigen (allergen) that lead to tissue damage rather than immunity.
2. Hypersensitivity reactions occur only when a person has been sensitized to an antigen.
3. Hypersensitivity reactions can be divided into two main groups: Types I, II, and III are based on humoral immunity, and type IV is based on cell-mediated immunity.

Type I (Anaphylaxis) Reactions   (pp. 425–455)

1. Anaphylactic reactions involve the production of IgE antibodies that bind to mast cells and basophils to sensitize the host.
2. Binding of two adjacent IgE antibodies to an antigen causes the target cell to release chemical mediators, such as histamine, leukotrienes, and prostaglandins, which cause the observed allergic reactions.
3. Systemic anaphylaxis may develop in minutes after injection or ingestion of the antigen; this may result in circulatory collapse and death.
4. Localized anaphylaxis is exemplified by hives, hay fever, and asthma.

5. Skin testing is useful in determining sensitivity to an antigen.
6. Desensitization to an antigen can be achieved by repeated injections of the antigen, which leads to the formation of blocking (IgG) antibodies and T_S cells.

Type II (Cytotoxic) Reactions (pp. 455–458)

1. Type II reactions are mediated by IgG or IgM antibodies and complement.
2. The antibodies are directed toward host cell antigens; complement fixation may result in cell lysis or phagocytosis or K cell activity.

The ABO Blood Group System (pp. 455–456)

1. Human blood may be grouped into four principal types, designated A, B, AB, and O.
2. The presence or absence of two carbohydrate antigens designated A and B on the surface of the red blood cell determine a person's blood type.
3. Naturally occurring antibodies are present or absent in serum against the opposite AB antigen.
4. Incompatible blood transfusions lead to the complement-mediated lysis of the donor red blood cells.

The Rh Blood Group System (pp. 456–457)

1. Approximately 85% of the human population possesses another blood group antigen, designated the Rh antigen; these individuals are designated Rh⁺.
2. The absence of this antigen in certain individuals (Rh⁻) can lead to sensitization upon exposure to it.

Blood Transfusions and Rh Incompatibility (p. 456)
1. An Rh⁺ person can receive Rh⁺ or Rh⁻ blood transfusions.
2. When an Rh⁻ person receives Rh⁺ blood, that person will produce anti-Rh antibodies.
3. Subsequent exposure to Rh⁺ cells will result in a rapid hemolytic reaction.

Hemolytic Disease of the Newborn (pp. 456–457)
1. An Rh⁻ mother carrying an Rh⁺ fetus will produce anti-Rh antibodies.
2. Subsequent pregnancies involving Rh incompatibility may result in hemolytic disease of the newborn.
3. The disease may be prevented by passive immunization of the mother with anti-Rh antibodies.

Other Cytotoxic Diseases (pp. 457–458)

1. In the disease idiopathic thrombocytopenic purpura, platelets are destroyed by antibodies and complement.
2. The disease autoimmune hemolytic anemia results from antibodies against one's own red blood cells.
3. The disease myasthenia gravis is caused by antibodies reacting with neuromuscular synapses.
4. Graves' disease results in excessive thyroid hormone production when antibodies react with the thyroid gland.

Type III (Immune Complex) Reactions (pp. 458–459)
1. Immune complex diseases occur when IgM or IgG antibodies and antigen form small complexes that lodge in the basement membranes of cells.
2. Subsequent complement fixation results in inflammation.
3. Glomerulonephritis, systemic lupus erythematosus, and rheumatoid arthritis are immune complex diseases.

Type IV (Cell-Mediated) Reactions (pp. 459–460)
1. Delayed hypersensitivity responses are primarily due to T_D cell proliferation.
2. Sensitized T cells secrete lymphokines in response to the appropriate antigen.
3. Lymphokines attract and activate macrophages and initiate tissue damage.
4. The tuberculin skin test and allergic contact dermatitis are examples of delayed hypersensitivities.

Autoimmunity (pp. 460–462)

Loss of Immunological Tolerance (pp. 460–461)

1. Immunologic tolerance represents a state of unresponsiveness to a specific antigen.
2. Tolerance to self-antigens occurs under natural conditions and develops during fetal development.
3. Autoimmunity is a humoral or cell-mediated immune response against self-antigens. Autoimmune responses frequently result in disease.
4. Some antibodies produced against foreign antigens can cross-react with self-antigens (for example, rheumatic fever).

Major Histocompatibility Complex (pp. 461–462)

1. Histocompatibility antigens located on cell surfaces express genetic differences between individuals; these antigens are coded for by MHC or HLA gene complexes.
2. To prevent rejection of transplants, HLA and ABO blood group antigens are matched as closely as possible.

Transplantation (pp. 462–464)

Types of Transplants (pp. 462–463)

1. Transplantation to a privileged site (such as the cornea) or of a privileged tissue (such as pig heart valves) does not cause an immune response.
2. Four types of transplants have been defined on the basis of genetic relationships between the donor and the recipient: autografts, isografts, allografts, and xenografts.
3. Bone marrow (with immunocompetent cells) can cause graft versus host disease.

Immunosuppression (p. 464)

1. Successful transplant surgery often requires immuno-suppressive treatment, which increases the host's susceptibility to infections and cancer.
2. Cyclosporine blocks synthesis of IL-2 to inhibit T-cell growth.
3. A monoclonal antibody preparation (muromonab-CD3) destroys T cells.

Natural Immune Deficiencies (p. 464)

1. Immune deficiencies may be inherited.
2. A variety of diseases can also impair the immune response.

AIDS (pp. 465–469)

1. AIDS is caused by the retrovirus HIV, which kills cells with CD4 antigen, especially T_H cells.
2. HIV infection results in opportunistic infections that eventually cause the death of the host.

Origin of AIDS (p. 465)

1. HIV is thought to have originated in central Africa and was brought to the New World by immigration.

Symptoms of AIDS (pp. 465–466)

1. Initial symptoms may be fever and malaise.
2. AIDS-related complex includes swollen lymph nodes, loss of appetite, fever, and rashes.
3. Death is usually caused by opportunistic infections.

Modes of Transmission (pp. 466–467)

1. AIDS is transmitted by sexual contact, contaminated needles, and across the placenta.
2. Blood transfusions are not a likely source of infection because blood is tested for HIV-antibodies.

Vaccines and Treatments (pp. 467–469)

1. Drugs aimed at reverse transcriptase are being investigated.
2. Because few free viruses are circulating, vaccines that block receptor sites instead of the virus's attachment site may be useful.

The Future of AIDS (p. 467)

1. The number of AIDS cases is expected to increase during the next few years.

Immune Response to Cancer (pp. 470–471)

Immunological Surveillance (p. 470)

1. Cancer cells are normal cells that have undergone transformation, fail to exhibit contact inhibition, and possess tumor-associated antigens.
2. The response of the immune system to cancer is called immunological surveillance.

Immunological Escape (pp. 470–471)

1. The ability of tumor cells to escape immune responses against them is called immunological escape.
2. Antigen modulation and immunological enhancement represent two ways tumor cells escape detection and destruction by the immune system.

Immunotherapy (p. 471)

1. Immunotoxins are the most promising cancer treatment being researched.
2. Immunotoxins are chemical poisons with a monoclonal antibody; the antibody selectively locates the cancer cell for release of the poison.

Study Questions

Review

1. Define hypersensitivity.

2. Compare and contrast the characteristics of the four types of hypersensitivity reactions. Given an example of each type.

3. List three mediators released in anaphylactic hypersensitivities, and explain their effects.

4. Contrast systemic and localized anaphylactic reactions. Which type is more serious? Cite an example of each.

5. Explain what happens when a person develops a contact sensitivity to poison oak.
   (a) What causes the observed symptoms?
   (b) How did the sensitivity develop?
   (c) How might this person be desensitized to poison oak?

6. Discuss the roles of antibodies and antigens in an incompatible tissue transplant.

7. What happens to the recipient of an incompatible blood type?

8. Explain how hemolytic disease of the newborn develops and how this disease might be prevented.

9. Which of the following blood transfusions are compatible? Explain your answers.

Donor	Recipient
(a) AB, Rh^-	AB, Rh^+
(b) B, Rh^+	B, Rh^-
(c) A, Rh^+	O, Rh^+

10. Which type of graft (autograft, isograft, allograft, or xenograft) is most compatible? Least compatible?

11. Define immunosuppression. Why is it used, and what problems does it cause?

12. Define autoimmunity. Present a theory that could explain autoimmune responses. Discuss one autoimmune disease in relation to this theory.

13. Summarize the causes of immune deficiencies. What is the effect of an immune deficiency?

14. In what ways do tumor cells differ antigenically from normal cells? Explain how tumor cells may be destroyed by the immune system.

15. If tumor cells can be destroyed by the immune system, how does cancer develop? What does immunotherapy involve?

Challenge

1. When and how does our immune system discriminate between self and nonself antigens?

2. The first preparations used for artificially acquired passive immunity were antibodies in horse serum. A complication that resulted from the therapeutic use of horse serum was immune-complex disease. Why did this occur?

3. After working in a mushroom farm for several months, a worker develops these symptoms: hives, edema, and swelling of lymph nodes.
   (a) What do these symptoms indicate?
   (b) What mediators cause these symptoms?
   (c) How may sensitivity to a particular antigen be determined?
   (d) Other employees do not appear to have any immunological reactions. What could explain this?
   (*Note:* The allergen is conidiospores from molds growing in the mushroom farm.)

4. Physicians administering live, attenuated mumps and measles vaccines prepared in chick embryos are instructed to have epinephrine available. Since epinephrine will not treat these viral infections, what is the purpose of keeping this drug on hand?

5. What treatments are administered to a person with an immune deficiency?

Further Reading

"AIDS." *Science* 239:573–623, February 5, 1988. This special issue includes eight articles on AIDS: virus biology, epidemiology, and legal and ethical issues.

Brodt, P. "Cancer immunology—three decades in review." *Annual Review of Microbiology* 37:447–476,1983. A discussion of tumor antigens and cellular and humoral responses.

Buisseret, P. D. "Allergy." *Scientific American* 247(2):86–95, August 1982. Discusses the roles of mast cells, IgE, and antigens in hypersensitivity responses.

Cohen, I. R. "The self, the world and autoimmunity." *Scientific American* 258(4):52–60, April 1988. Describes recognition of self by T cells and suggests vaccine treatment for autoimmune diseases.

Collier, J. R., and D. A. Kaplan. "Immunotoxins." *Scientific American* 251(1):56–64, July 1984. Illustrates how cancer cells might be targeted for destruction by a monoclonal antibody carrying a toxic agent.

Høiby, N., G. Döring, and P. O. Schiøtz. "The role of immune complexes in the pathogenesis of bacterial infections." *Annual Review of Microbiology* 40:29–53, 1986. Describes immune complexes and bacterial infections (including *Pseudomonas* and leprosy) involving immune complexes.

Lockey, R. F., and S. C. Bukantz, eds. "Primer on allergic and immunologic diseases," 2nd ed. *Journal of the American Medical Association* 258(20):2829–3034, November 27, 1987. A special issue containing 26 articles on abnormal immune responses.

Old, L. J. "Cancer immunology." *Scientific American* 236(5):62–79, May 1977. An explanation of why cancer cells can escape the immune system.

Rodger, J. C., and B. L. Drake. "The enigma of the fetal graft." *American Scientist*, 75:51–57, January-February 1987. Addresses the question of why the mother doesn't reject the fetus, which is genetically foreign tissue.

Rose, N. R. "Autoimmune diseases." *Scientific American* 244(2):80–103, February 1981. Describes autoimmune diseases and suggests treatments.

"What Science Knows About AIDS." *Scientific American* 259(4), October 1988. This issue is devoted entirely to AIDS. Ten articles cover a wide range of topics, including molecular biology, epidemiology, and treatments.

CHAPTER 19

Antimicrobial Drugs

When the body's normal defenses cannot prevent or overcome a disease, it is often treated with **chemotherapy** (drugs). The term **chemotherapeutic agent** applies to any drug used for any disease, whether it be a simple headache, a "strep" throat, malaria, high blood pressure, or cancer. For example, aspirin and penicillin are both chemotherapeutic agents.

In this chapter, we focus on **antimicrobial drugs**, the class of chemotherapeutic agents used to treat infectious diseases. Like the disinfectants discussed in Chapter 7, these chemicals act by interfering with the growth of microorganisms. Unlike disinfectants, however, they must often act *within* the host. Therefore, their effects on the cells and tissues of the host are important; the ideal antimicrobial drug kills the harmful microorganism without damaging the host.

Drugs used in the chemotherapy of infectious disease are classified into two groups. Drugs that have been synthesized by chemical procedures in the laboratory are called **synthetic drugs**. Drugs produced by bacteria and fungi are called **antibiotics**.

Learning Objectives

- Define a chemotherapeutic agent, and distinguish between a synthetic drug and an antibiotic.
- Identify the contributions of Paul Ehrlich and Alexander Fleming to chemotherapy.
- List the criteria used to evaluate antimicrobial agents.
- Identify five methods of action of antimicrobial agents.
- Describe the methods of action of each of the commonly used antibacterial drugs.
- Describe the problems of chemotherapy for viral, fungal, protozoan, and helminthic infections.
- Explain the actions of currently used antiviral, antifungal, antiprotozoan, and antihelminthic drugs.
- Describe three tests for microbial susceptibility to chemotherapeutic agents.
- Describe the mechanisms of drug resistance.

Historical Development

The birth of modern chemotherapy is credited to the efforts of Paul Ehrlich in Germany during the early part of this century. While attempting to stain bacteria without staining the surrounding tissue, he speculated about some "magic bullet" that would selectively find and destroy pathogens but not harm the host. This idea provided the basis for chemotherapy, a term he coined. Recall from Chapter 1 that Ehrlich eventually discovered a chemotherapeutic agent called salvarsan, an arsenic derivative that was effective against syphilis (by the standards of the day). Prior to Ehrlich's discovery, there had been only one effective chemotherapeutic agent in the medical arsenal—quinine, a drug used for the treatment of malaria.

In the late 1930s, the discovery of sulfa drugs, a group of synthetic drugs, touched off a new interest in chemotherapy. This discovery evolved from a systematic survey of chemicals, including many synthesized derivatives of aniline dyes. The chemical prontosil, which had been first synthesized and used as a dye many years before, was found to be an effective antimicrobial agent. Oddly enough, it worked only within living animals and was completely ineffective when exposed to the same bacteria in a test tube. The active ingredient in prontosil is *sulfanilamide*, which forms as the animal metabolizes prontosil. This discovery led to the rapid development of a number of related drugs—the *sulfonamides*, or *sulfa drugs*, which are still used today.

We also mentioned in Chapter 1 that, in 1928, Alexander Fleming observed that the growth of the bacterium *Staphylococcus aureus* was inhibited in the area surrounding the colony of a mold that had contaminated a Petri plate (Figure 19-1). The mold was identified as *Penicillium notatum* (pe-ni-sil'lē-um nō-tä-tum), and its active compound, which was isolated

a short time later, was named *penicillin*. Similar inhibitory reactions between colonies on solid media are commonly observed in microbiology, and the mechanism of inhibition is called *antibiosis*. From this term comes the term *antibiotic*, meaning a substance produced by microorganisms that in small amounts will inhibit another microorganism. Therefore, the wholly synthetic sulfa drugs technically are not antibiotics.

Because microorganisms capable of antibiosis are fairly common, it is surprising that antibiotics were not developed much earlier. Actually, there is evidence in the scientific literature that penicillin had been seriously investigated in the late 1800s. But the earlier researchers had the same problem that Fleming later had—the extract of the mold was unstable and production of penicillin by the mold was unreliable, so the drug was not very useful clinically. In fact, there is no evidence that Fleming ever considered using it to treat an infection.

In 1940, a group of scientists at Oxford University headed by Howard Florey and Ernst Chain succeeded in the first clinical trials of penicillin. Intensive research in the United States then led to the isolation of especially productive *Penicillium* strains for use in mass production of the antibiotic. (The most famous of these high-producing strains was originally isolated from a cantaloupe bought at a market in Peoria, Illinois.) Penicillin is still one of our most effective antibiotics, and its enormous success led to the search for others, a search that has revolutionized medicine. You might be interested to know that penicillin has a singular toxicity for guinea pigs, and it is entirely possible that this drug would never have left the experimental laboratory under the more demanding safety requirements that must be met today in the development of drugs.

Antibiotics are actually rather easy to discover, but few are of medical or commercial value. Some are used commercially other than to treat disease, for example, as a supplement in animal feed (see the box on p. 494). Many antibiotics are toxic to humans or lack any advantage over antibiotics already in use.

More than half of our antibiotics are produced by species of *Streptomyces* (strep-tō-mī'sēs), filamentous bacteria that commonly inhabit soil. A few antibiotics are produced by bacteria of the genus *Bacillus*, and others are produced by molds, mostly of the genera *Penicillium* and *Cephalosporium* (sef-ä-lō-spô'rē-um). See Table 19-1 for the sources of many antibiotics in use today.

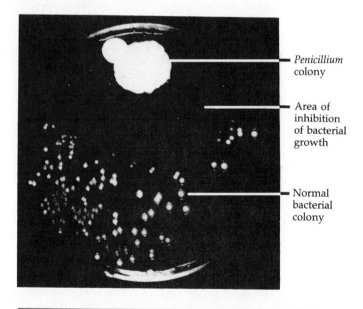

Penicillium colony

Area of inhibition of bacterial growth

Normal bacterial colony

Figure 19-1   Discovery of penicillin. Alexander Fleming took this photograph in 1929. The colony of *Penicillium* mold accidentally contaminated the plate and is inhibiting bacterial growth near it.

TABLE 19-1 *Representative Sources of Antibiotics*

Microorganism	Antibiotic
Gram-Positive Rods	
Bacillus subtilis	Bacitracin
Bacillus polymyxa	Polymyxin
Bacillus brevis	Tyrothricin
Actinomycetes	
Streptomyces parvullus	Actinomycin D
Streptomyces nodosus	Amphotericin B
Streptomyces venezuelae	Chloramphenicol
Streptomyces aureofaciens	Chlortetracycline (Aureomycin) and tetracycline
Streptomyces erythraeus	Erythromycin
Streptomyces kanamyceticus	Kanamycin
Streptomyces fradiae	Neomycin
Streptomyces noursei	Nystatin
Streptomyces rimosus	Oxytetracycline
Streptomyces griseus	Streptomycin
Micromonospora purpureae	Gentamicin
Fungi	
Cephalosporium	Cephalothin
Penicillium griseofulvum	Griseofulvin
Penicillium notatum	Penicillin

Criteria for Evaluating Antimicrobial Drugs

Antimicrobial drugs can be evaluated by a number of criteria, including the following:

1. The drug should demonstrate selective toxicity. This means that, at the optimum concentration, the drug should be toxic for the microorganism, but not for the host. With some exceptions, most chemotherapeutic agents are somewhat toxic to host cell metabolism, but they usually are not toxic enough to interfere with normal body functions.
2. The drug should not produce hypersensitivity (allergy) in most hosts (see Chapter 18). Allergic reactions, which are produced by the immune system, are not the same as toxicity. Penicillin, for example, sometimes causes allergic reactions but is one of the least toxic antibiotics for humans.
3. A drug must have solubility in body fluids so that it can rapidly penetrate body tissues. The rates at which the drug is broken down and excreted from the body must be low enough that the drug remains in the infected body tissue long enough to exert its effects. Furthermore, a good chemotherapeutic agent has a long shelf life at normal refrigeration temperatures.
4. Microorganisms should not readily become resistant to the drug. Resistance can be caused by a number of mechanisms and will be discussed later in this chapter.

Not all of these properties can be found in every chemotherapeutic agent, but developers of a drug attempt to obtain the best possible combination of properties for effective human use.

Spectrum of Activity

It is a comparatively simple task to find or develop drugs that are effective against procaryotic cells (bacteria) and do not affect the eucaryotic cells of humans. These two cell types differ substantially in many ways, such as in the presence or absence of cell walls, the fine structure of their ribosomes, and details of their metabolism. Thus, selective toxicity has numerous targets. The problem is more difficult when the pathogen is a eucaryotic cell, such as a fungus, protozoan, or helminth. At the cellular level, these organisms resemble the human cell much more closely than a bacterial cell does. We shall see that our arsenal against these types of pathogens is much more limited than our arsenal of antibacterial drugs. Viral infections are particularly difficult to treat because the pathogen is within the human host's cells, and the genetic information of the virus is directing the human cell to make viruses rather than to synthesize normal cellular materials. The more we discover about the reproduction, metabolism, and structure of pathogens, the better equipped we are to discover fresh targets for antimicrobials. No doubt the search will be never ending.

Some antimicrobial drugs have a narrow **spectrum of activity** (range of cells) they target; penicillin, for instance, affects gram-positive bacteria but very few gram-negative bacteria. Other drugs affect a large number of gram-positive or gram-negative bacteria and are therefore called *broad spectrum drugs*. Since the identity of the pathogen is not always immediately known, a broad spectrum drug would seem to have an advantage in the treatment of disease by saving valuable time. The disadvantage is that many of the normal flora of the host are destroyed by broad spectrum drugs. The normal flora ordinarily compete with and check the growth of pathogens or other microbes. If certain organisms in the normal flora are not destroyed by the antibiotic and their competitors are destroyed, the survivors may flourish and become opportunistic pathogens. An example is the overgrowth by the yeastlike fungus *Candida albicans*, which is not sensitive to bacterial antibiotics. This overgrowth is called a **superinfection**, a term that is also applied to growth of a target pathogen that has developed resistance to the antibiotic. Such an antibiotic-resistant strain replaces the original sensitive strain, and the infection continues.

Action of Antimicrobial Drugs

We will next examine the various ways in which antibiotics exert their antimicrobial activity. Antimicrobial drugs either kill microorganisms directly (**bactericidal**) or simply prevent them from growing (**bacteriostatic**). In *bacteriostasis*, the host's own defenses, such as phagocytosis and antibody production, usually destroy the microorganisms. The major modes of action are summarized in Figure 19-2.

Inhibition of Cell Wall Synthesis

Recall from Chapter 4 that the cell wall of a bacterium consists of a macromolecular network called peptidoglycan. Peptidoglycan is found only in bacterial cell walls. Penicillin and certain other antibiotics prevent the synthesis of intact peptidoglycan; consequently, the cell wall is greatly weakened, and the cell undergoes lysis (Figure 19-3). Because penicillin affects the synthesis process, only actively growing cells are

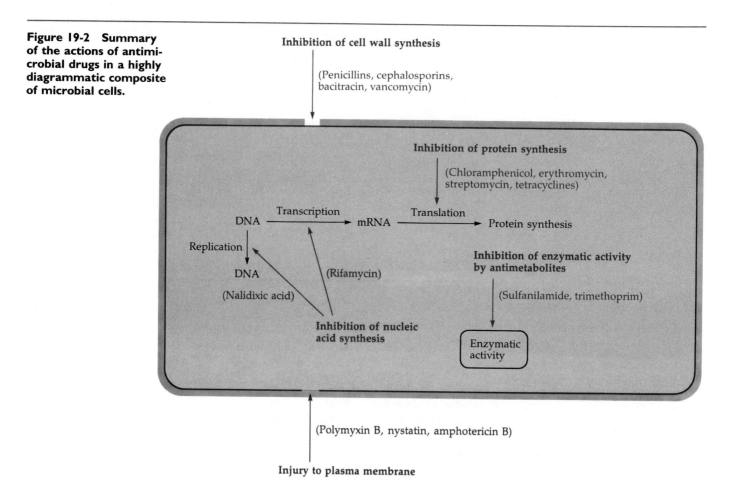

Figure 19-2 Summary of the actions of antimicrobial drugs in a highly diagrammatic composite of microbial cells.

Inhibition of cell wall synthesis

(Penicillins, cephalosporins, bacitracin, vancomycin)

Inhibition of protein synthesis

(Chloramphenicol, erythromycin, streptomycin, tetracyclines)

DNA → Transcription → mRNA → Translation → Protein synthesis

Replication

DNA

(Rifamycin)

(Nalidixic acid)

Inhibition of nucleic acid synthesis

Inhibition of enzymatic activity by antimetabolites

(Sulfanilamide, trimethoprim)

Enzymatic activity

(Polymyxin B, nystatin, amphotericin B)

Injury to plasma membrane

(a)   (b)

Figure 19-3 Effect of an antibiotic on bacterial cell walls. (a) *Staphylococcus aureus* before exposure to the antibiotic ampicillin (×24,000). **(b)** These cells of *S. aureus* are beginning to show the effects of exposure to ampicillin, which weakens cell walls (×24,000).

affected by these antibiotics. And, because human cells do not have peptidoglycan cell walls, penicillin has very little toxicity for the host cells. You may also recall that gram-negative bacteria have an outer membrane of lipopolysaccharides and other materials over a comparatively thin layer of peptidoglycan. Gram-negative bacteria are relatively resistant to penicillin because this outer membrane apparently prevents penicillin from reaching the site of peptidoglycan synthesis.

There are several ways in which cell wall synthesis can be affected. For example, cycloserine, a toxic and seldom-used antibiotic, interferes with the formation of a precursor necessary to peptidoglycan synthesis. Two commonly used antibiotics, bacitracin and vancomycin, interfere with the synthesis of the linear strands of peptidoglycan. Penicillin and cephalosporins prevent the final cross-linking of the peptidoglycans, which interferes with the construction of the macromolecular cell wall.

Inhibition of Protein Synthesis

Because protein synthesis is a common feature of all cells, both procaryotic and eucaryotic, it would seem an unlikely target for selective toxicity. One notable difference between procaryotes and eucaryotes, however, is the structure of their ribosomes. As we discussed in Chapter 4, eucaryotic cells have 80S ribosomes; procaryotic cells have 70S ribosomes. (The 70S ribosome is made up of a 50S and a 30S unit.) The difference in ribosomal structure accounts for the selective toxicity of antibiotics that affect protein synthesis. However, mitochondria (important eucaryotic organelles) also contain 70S ribosomes similar to those of bacteria. Antibiotics targeting the 70S ribosomes can therefore have adverse effects on the cells of the host. Among the antibiotics that interfere with protein synthesis are chloramphenicol, erythromycin, streptomycin, and tetracyclines (Figure 19-4).

Reacting with the 50S portion of the 70S procaryotic ribosome, chloramphenicol inhibits the formation of peptide bonds in the growing polypeptide chain. Erythromycin also reacts with the 50S portion of the 70S procaryotic ribosome. Most drugs that inhibit protein synthesis have a broad spectrum of activity; erythromycin is an exception. Because it does not penetrate the gram-negative cell wall, it affects mostly gram-positive bacteria.

Some other antibiotics react with the 30S portion of the 70S procaryotic ribosome. The tetracyclines interfere with the attachment of the tRNA carrying the amino acids to the ribosome, preventing the addition of amino acids to the growing polypeptide chain.

Tetracyclines do not interfere with mammalian ribosomes because they do not penetrate very well into intact mammalian cells. However, at least small amounts are able to enter the host cell, as is apparent from the fact that the intracellular pathogenic rickettsias and chlamydias are sensitive to tetracyclines. The selective toxicity of the drug is due to a greater sensitivity of the bacteria at the ribosomal level.

Aminoglycoside antibiotics, such as gentamicin and streptomycin, interfere with the initial steps of protein synthesis by changing the shape of the 30S portion of the 70S procaryotic ribosome. This interference causes the genetic code on the mRNA to be read incorrectly.

Figure 19-4 Inhibition of protein synthesis by antibiotics. The arrows indicate the specific points at which chloramphenicol, erythromycin, tetracyclines, and streptomycin exert their activity.

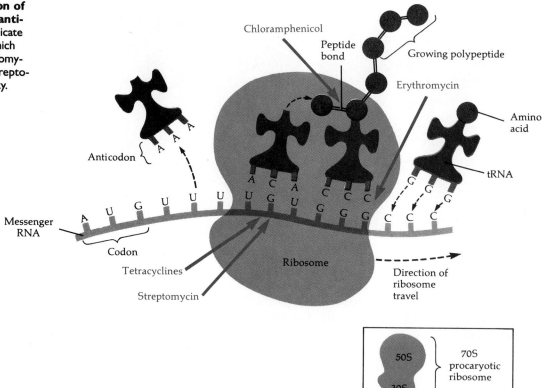

Injury to the Plasma Membrane

Certain antibiotics, especially polypeptide antibiotics, bring about changes in the permeability of the plasma membrane; these changes result in the loss of important metabolites from the microbial cell. One such antibiotic is polymyxin B. It causes disruption of the plasma membrane by attaching to the phospholipids of the membrane.

Some antifungal drugs, such as nystatin, amphotericin B, and ketoconazole, are effective against many systemic fungal diseases. Such drugs combine with sterols in the fungal plasma membrane to disrupt the membrane (Figure 19-5). Since bacterial plasma membranes generally lack sterols, these antibiotics do not act on bacteria. However, the plasma membranes of animal cells do contain sterols, and nystatin and amphotericin B can be toxic to the host. Fortunately, animal cell membranes have mostly *cholesterol*, and fungal cells have mostly *ergosterol*, against which the drug is most effective, so that the balance of the toxicity is tilted toward the fungus.

Figure 19-5 Disintegration of a yeast cell. The cell releases its cytoplasmic contents as the plasma membrane is disrupted by the antifungal drug miconazole (×12,000).

Inhibition of Nucleic Acid Synthesis

A number of antibiotics interfere with nucleic acid metabolism of microorganisms. Some drugs with this mode of action, such as the antiviral idoxuridine, have an extremely limited usefulness because they interfere with mammalian DNA and RNA synthesis as well. Others, such as rifamycin, nalidixic acid, and trimethoprim, are more widely used in chemotherapy because they are more selectively toxic.

Inhibition of Enzymatic Activity

In Chapter 5, we mentioned that an enzymatic activity of a microorganism can be *competitively inhibited* by a substance (antimetabolite) that closely resembles the normal substrate for the enzyme (see Figure 5-6). An example of competitive inhibition is the relationship between the antimetabolite sulfanilamide (a sulfa drug) and para-aminobenzoic acid (PABA). In many microorganisms, PABA is the substrate for an enzymatic reaction leading to the synthesis of folic acid, a vitamin that functions as a coenzyme for the synthesis of the purine and pyrimidine bases of nucleic acids. In the presence of sulfanilamide, the enzyme that normally converts PABA to folic acid combines with the drug instead of with PABA (Figure 19-6). This combination prevents folic acid synthesis and stops the growth of the microorganism. Because humans do not produce folic acid from PABA (they obtain it as a vitamin in ingested foods), sulfanilamide exhibits selective toxicity—it affects microorganisms that synthesize their own folic acid but does not harm the human host. Many bacteria, some fungi, and some protozoans must synthesize folic acid. Other chemotherapeutic agents that act as antimetabolites are the sulfones and trimethoprim.

Figure 19-6   Structure of sulfanilamide, a representative sulfa drug. Note the resemblance to PABA, for which it is a competitive inhibitor.

Survey of Commonly Used Antimicrobial Drugs

We will next examine the properties and activities of some commonly used antimicrobial drugs. The drugs are classified into four groups—synthetic antimicrobial drugs, antibiotics, antiviral drugs, and antiprotozoan and antihelminthic drugs. Commonly used drugs of each group are listed in Table 19-2.

Synthetic Antimicrobial Drugs

Isoniazid (INH)

Isoniazid (INH) is a very effective synthetic antimicrobial drug against *Mycobacterium tuberculosis*. It has little effect on nonmycobacteria. It is a structural analog of vitamin B_6 (pyridoxine), which serves as a coenzyme in metabolic reactions. The primary effect of INH might be to inhibit synthesis of mycolic acids, which are components of cell walls only of the mycobacteria, but the specific mode of action is not entirely understood. When used to treat tuberculosis, INH is usually administered simultaneously with other drugs, such as streptomycin, rifampin, or ethambutol.

Ethambutol

Ethambutol is effective only against mycobacteria. The drug apparently kills the microbes by inhibiting the synthesis of certain cellular metabolites. It is a comparatively weak antitubercular drug; its principal use is as the secondary drug to avoid resistance problems. For this purpose, it has essentially replaced para-aminosalicylic acid (PAS).

Sulfonamides

As we noted earlier, **sulfonamides (sulfa drugs)** were among the first synthetic antimicrobial drugs used to treat microbial diseases. Unfortunately, many pathogens have developed resistance to them. Moreover, sulfonamides cause allergic reactions in many people. For these reasons, antibiotics have tended to diminish the importance of sulfonamides in chemotherapy. The sulfonamides continue to be used to treat certain urinary tract infections and have other specialized uses, as in the combination drug **silver sulfadiazine**, used in the control of infections in burn

TABLE 19-2 *Commonly Used Antimicrobial Drugs*

Antimicrobial Drug	Effect on Microorganisms	Mode of Action	Clinical Use/Toxicity
Synthetic Drugs			
Isoniazid (INH)	Bacteriostatic	Competitive inhibition	With streptomycin and rifampin to treat tuberculosis
Ethambutol	Bacteriostatic	Competitive inhibition	With INH to treat tuberculosis
Sulfonamides	Bacteriostatic	Competitive inhibition	*Neisseria meningitidis* meningitis and some urinary tract infections
Antibiotics			
Penicillin, natural Penicillins G, V	Bactericidal	Inhibition of cell wall synthesis	Gram-positive bacteria
Penicillin, semisynthetic Ampicillin, methicillin, oxacillin, and others	Bactericidal	Inhibition of cell wall synthesis	Broad spectrum (ampicillin); resistant to penicillinase and stomach acid (methicillin, oxacillin)
Cephalosporins	Bactericidal	Inhibition of cell wall synthesis	Similar to penicillins
Carbapenems	Bactericidal	Inhibition of cell wall synthesis	Very broad spectrum
Aminoglycosides	Bactericidal	Inhibition of protein synthesis	
Streptomycin			Gram-negative bacteria
Neomycin			Topical ointment; gram-negative bacteria
Gentamicin			Gram-negative bacteria
Tetracyclines	Bacteriostatic	Inhibition of protein synthesis	Very broad spectrum; some toxic side-effects include tooth discoloration and liver and kidney damage
Chloramphenicol	Bacteriostatic	Inhibition of protein synthesis	*Salmonella*; aplastic anemia is an occasional side-effect
Macrolides Erythromycin	Bacteriostatic	Inhibition of protein synthesis	Diphtheria; legionellosis; various infections in persons allergic to penicillin
Polypeptides Bacitracin	Bactericidal	Inhibition of cell wall synthesis	Gram-positive bacteria, primarily; toxic to kidneys; usually used as topical ointment
Polymyxin B	Bactericidal	Injury to plasma membranes	Gram-negative bacteria, including *Pseudomonas*; toxic to brain and kidneys
Vancomycin	Bactericidal	Inhibition of cell wall synthesis	Staphylococci; very toxic
Rifamycins Rifampin	Bactericidal	Inhibition of RNA synthesis	Gram-positive bacteria, some gram-negatives, chlamydias, and poxviruses; often used with INH and ethambutol for tuberculosis

TABLE 19-2 (*Continued*)

Antimicrobial Drug	Effect on Microorganisms	Mode of Action	Clinical Use/Toxicity
Antifungal Drugs			
Polyenes	Fungicidal	Injury to plasma membrane	
Nystatin			*Candida* infections of skin and vagina
Amphotericin B			Histoplasmosis, coccidioidomycosis, and blastomycosis
Imidazoles	Fungicidal	Injury to plasma membranes	
Clotrimazole, miconazole			Topical use for fungal infections
Ketoconazole			Systemic fungal infections
Griseofulvin	Fungistatic	Inhibition of mitotic spindle function	Fungal infections of skin, primarily ringworm
Antiviral Drugs			
Amantadine	Antiviral	Blocks entry or uncoating	Influenza A prophylaxis and treatment
Guanine analog	Antiviral	Interferes with DNA and RNA synthesis	
Ribavirin			Viral pneumonia
Acyclovir			Herpes infections
Thymine analog Idoxuridine, Trifluridine	Antiviral	Inhibition of DNA synthesis	Herpes keratitis
Adenine analog Vidarabine	Antiviral	Inhibition of DNA and RNA synthesis	Herpes encephalitis
Interferon	Antiviral	Inhibits cell infection by virus	Experimental for virus infections and cancer
Antiprotozoan Drugs			
Chloroquine	Antiprotozoan	Inhibition of DNA synthesis	Malaria
Pentamidine isethionate	Antiprotozoan	Reacts with DNA	African trypanosomiasis, *Pneumocystis* pneumonia
Diiodohydroxyquin	Antiprotozoan	Unknown	Amoebic infections; optic nerve damage
Metronidazole	Antiprotozoan	Anaerobic metabolism products damage DNA	Ameobic infections, *Trichomonas*, anaerobic bacteria
Nifurtimox	Antiprotozoan	Unknown	Chagas' disease
Antihelminthic Drugs			
Praziquantel	Antihelminthic	Unknown	Schistosomiasis and other fluke infestations
Niclosamide	Antihelminthic	Prevents generation of ATP	Tapeworm infestations

patients. Sulfonamides are bacteriostatic; as we mentioned, their action is due to their structural similarity to para-aminobenzoic acid (PABA) (see Figure 19-6).

Probably the most widely used sulfa today is a combination of **trimethoprim** and **sulfamethoxazole (TMP–SMZ)**. This combination is an excellent example of drug *synergism* (see p. 495). When the drugs are used in combination, only about one-tenth of the concentration is needed as when each is used alone. The combination also has a broader spectrum of action and greatly reduces the emergence of resistant strains.

Figure 19-7 illustrates how the two drugs interfere with different steps of a metabolic sequence leading to the synthesis of DNA. Sulfamethoxazole inhibits the synthesis of dihydrofolic acid from PABA so that the amount made is much less than usual. Then trimethoprim, which is a structural analog of a portion of dihydrofolic acid but is not classified as a sulfa drug, blocks the conversion of any dihydrofolic acid that is present into tetrahydrofolic acid (Figure 19-8). TMP–SMZ is useful against most pathogens of the urinary tract and intestinal tract and has been the drug of choice for treatment of *Pneumocystis* pneumonia, a protozoan disease.

Other Synthetic Antimicrobial Drugs

Nitrofurans constitute a family of drugs that are active against a large number of gram-positive and gram-negative bacteria, protozoans, and fungi. *Nitrofurantoin* is widely administered orally for urinary tract infections. Like nalidixic acid (see below), it usually does not reach a high enough concentration elsewhere than in urine to be useful in systemic che-

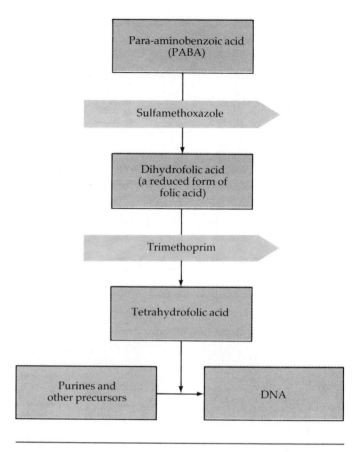

Figure 19-7 Trimethoprim and sulfamethoxazole. TMP–SMZ works by inhibiting different points in the synthesis of DNA. Together the drugs are synergistic.

motherapy. Its primary mode of action is not precisely known, but it might cause fractures of DNA polymers. Some other nitrofurans, such as *nitrofurazone* and *nifuroxime*, are used topically for skin, eye, and vagi-

Figure 19-8 Trimethoprim, a structural analog of a portion of dihydrofolic acid. Note also the presence of PABA, for which sulfamethoxzole is a structural analog, in another portion of the dihydrofolic acid molecule.

Trimethoprim

Sulfamethoxazole

Dihydrofolic acid

nal infections. *Nifurtimox* is effective against trypano-somal protozoans.

Nalidixic acid stops bacterial replication of DNA and is especially effective against gram-negative enterics. Because it accumulates in the urine, this drug is used to treat urinary tract infections in which *Escherichia, Proteus,* or *Klebsiella* species are implicated.

The *quinolones* are an emerging group of synthetic antimicrobials. They are related to nalidixic acid and may be taken orally. Their mode of action is to interfere with an aspect of DNA replication that is unique to procaryotes. Expectations are that, if toxicity problems do not appear, quinolones will be relatively inexpensive and will replace many antibiotics now used to control gram-negative infections.

The drug *5-fluorocytosine (Flucytosine)* is an antimetabolite of cytosine. It was originally synthesized as an anticancer drug, but it has found better use as an antifungal drug. Mammalian cells apparently do not convert the drug enzymatically to 5-fluorouracil, as is required for its activity, but this conversion occurs in fungi that are sensitive to the drug.

Antibiotics

Penicillin

The term **penicillin** refers to a group of chemically related antibiotics (Figure 19-9). Penicillin extracted from cultures of the mold *Penicillium* exists in several closely related forms. These are the so-called **natural penicillins** (Figure 19-9a). The prototype compound of all the penicillins is *penicillin G*. It has a narrow but useful spectrum of action and is the drug of choice against pathogenic cocci, gram-positive bacilli, and spirochetes that do not produce penicillinase. When injected intramuscularly, penicillin G is rapidly excreted from the body in three to six hours (Figure 19-10); when the drug is taken orally, the acidity of the digestive fluids in the stomach diminishes its concentration. *Procaine penicillin,* a combination of the drugs procaine and penicillin G, is retained at detectable concentrations for up to 24 hours; concentration peaks at about four hours. Still longer retention times can be achieved with *benzathine penicillin,* a combination of benzathine and penicillin G, but the concentration of penicillin in the blood is low. Although retention times of as long as four months can be obtained, the concentration of the drug is so low that the organisms must be very sensitive to the drug for it to work. Penicillin G and penicillin V, which is acid-stable and can be best taken orally, are the natural penicillins most often used.

Figure 19-9 Structure of penicillins. The portion that all penicillins have in common, containing the β-lactam ring, is shaded in color. The unshaded portions represent the side chains that distinguish one penicillin from another. **(a)** Natural penicillins. **(b)** Semisynthetic penicillins.

Figure 19-10 Penicillin G. This drug reaches different concentrations in the blood when administered by different routes and can be combined with compounds that prolong its retention in the body.

A large number of **semisynthetic penicillins** have been developed in attempts to overcome the natural penicillins' disadvantages, such as their narrow spectrum of activity and their destruction by penicillinases. In developing semisynthetic penicillins, scientists perfected a technique whereby they could either stop synthesis by *Penicillium* and obtain only the common penicillin nucleus or remove the side chains from the completed natural molecules and then chemically add other side chains. Thus the term *semisynthetic* is applied to these penicillins; part of the penicillin is produced by the mold, and part is added synthetically.

Penicillinases are enzymes produced by many bacteria, especially *Staphylococcus* spp., that cleave the β-lactam ring of the penicillin molecule (see Figure 19-9). Because of this characteristic, penicillinases are sometimes called *β-lactamases*. The type of a penicillin molecule depends on the chemical side chains attached to the β-lactam ring.

Semisynthetic penicillins have been found that are more resistant to stomach acid—and therefore are more effective when taken orally—and have a spectrum of activity that is often greater than that of the natural penicillins (see Figure 19-9b). For example, *ampicillin*, *amoxicillin*, and *carbenicillin* have a broad spectrum and are effective against a number of important gram-negative pathogens. Except for carbenicillin, they are acid-stable enough to be taken orally.

Special attention has been given to the problem of developing variants of penicillin not destroyed by penicillinases. *Methicillin* has been in use the longest, and resistance to it has appeared. *Aztreonam* avoids the action of penicillinases by having only a single ring rather than the conventional β-lactam double ring. Its spectrum of activity is remarkable for a penicillin; it affects only gram-negative bacteria, including pseudomonads. Another approach is to combine penicillins with *potassium clavulanate* (*clavulinic acid*, a product of a streptomycete), which is a noncompetitive inhibitor of penicillinase with essentially no antimicrobial activity of its own. It has been combined with broad spectrum penicillins, such as *amoxicillin* (*Augmentin®*) and *ticarcillin* (*Timentin®*).

Some newer broader spectrum variants, sometimes called *ureidopenicillins*, have also been developed. Some of these variants show antipseudomonad activity and have fewer side-effects than the available aminoglycosides currently used for these infections. Examples are *mezlocillin* and *azlocillin*.

Cephalosporins

In structure, the nuclei of **cephalosporins** resemble that of penicillin (Figure 19-11). Cephalosporins are also similar in action to penicillin; they inhibit the synthesis of cell walls. Although cephalosporins are structurally similar to penicillin, they are sufficiently different that the cephalosporins are resistant to penicillinases and are effective against more gram-negative organisms than the natural penicillins. However, the cephalosporins are susceptible to their own β-lactamases.

The number of second- and third-generation cephalosporins has proliferated in recent years. Some characteristic cephalosporins are *cephalothin*, *cefamandole*, and *cefotaxime*. An example from the most recent generation of cephalosporins, *moxalactam*, has a variation of the β-lactam ring, an oxa-β-lactam ring. Most cephalosporins are injected, but a few may be taken orally. Cephalosporins generally are more expensive than penicillins.

Carbapenems

A new class of antibiotics, the **carbapenems**, is remarkable for their extremely broad spectrum of activity. The first of this group is *Primaxin®*, a combination of a β-lactam antibiotic (*imipenem*) and *cilastatin sodium*. The cilastatin sodium prevents degradation of the combination in the kidneys. Tests have demonstrated Primaxin activity against 98% of all organisms isolated from hospital patients. Carbapenems work by inhibiting cell wall synthesis.

Cephalosporin nucleus

Penicillin nucleus

Figure 19-11 Comparison of the structure of the cephalosporin nucleus with that of penicillin. R is an abbreviation for chemical groups that make one compound different from another.

Aminoglycosides

Aminoglycosides are a group of antibiotics in which amino sugars are linked by glycoside bonds. Probably the best-known aminoglycoside is *streptomycin*, which was discovered in 1944 in a culture of *Streptomyces griseus* (gri-sē'us) taken from the throat of a chicken. Historically, streptomycin is significant as the first effective antibiotic against tuberculosis and against large numbers of gram-negative bacteria. Streptomycin is still used as an alternative drug in the treatment of tuberculosis, but rapid development of resistance and the appearance of serious toxic effects have diminished its usefulness.

Aminoglycosides include many important and widely used antibiotics. All are bactericidal and inhibit protein synthesis. However, aminoglycosides can affect hearing by causing permanent damage to the auditory nerve, and damage to the kidneys has also been reported. Neomycin is present in many topical preparations. Gentamicin (Figure 19-12) plays a very important role in treatment of many enteric gram-negative infections and *Pseudomonas aeruginosa* infections. When resistance to aminoglycosides is encountered, the semisynthetic aminoglycoside derivative *amikacin* is often prescribed.

Tetracyclines

Tetracyclines are a group of closely related broad spectrum antibiotics, produced by *Streptomyces* spp., that inhibit protein synthesis. They probably have the broadest spectrum of antibacterial activity of any known drug. Not only are they effective against gram-positive and gram-negative bacteria, they are also especially valuable against the intracellular rickettsias and chlamydias. Three of the more commonly used tetracyclines are *oxytetracycline* (*Terramycin®*), *chlortetracycline* (*Aureomycin®*), and tetracycline itself (Figure 19-13). Some newer semisynthetic tetracyclines, such as *doxycycline* and *minocycline*, are frequently used. They have the advantage of being retained longer in the body.

Tetracyclines are used for many urinary tract infections and are especially useful in treatment of mycoplasmal pneumonia, chlamydial urethritis, and rickettsial infections. They are also frequently used as alternative drugs for such diseases as syphilis and gonorrhea. Tetracyclines often suppress the normal intestinal flora because of their broad spectrum, causing gastrointestinal upsets and often leading to superinfections, particularly by *Candida albicans*. They are not advised for administration to children, who might experience a brownish discoloration of the teeth, or to pregnant women, in whom they might cause liver damage.

Figure 19-12 Structure of gentamicin, a representative aminoglycoside. Glycosidic bonds (color) are bonds between amino sugars.

Tetracycline

Figure 19-13 Structure of tetracycline. Other tetracycline-type antibiotics closely resemble tetracycline.

Chloramphenicol

Chloramphenicol is a broad spectrum bacteriostatic antibiotic that interferes with protein synthesis on the 50S ribosome (Figure 19-14). Because of its relatively simple structure, it is less expensive for the pharmaceutical industry to synthesize it chemically than to isolate it from *Streptomyces*. Its relatively small molecular size promotes its diffusion into areas of the body that are normally inaccessible to many other drugs. However, chloramphenicol has serious side-effects; most important is the suppression of bone marrow activity. This suppression affects the formation

Chloramphenicol

Figure 19-14 Structure of chloramphenicol. Note the simplicity that allows it to be synthesized inexpensively.

of blood cells. In about 1 per 40,000 users, the drug appears to cause aplastic anemia, a potentially fatal condition; the normal rate for this condition is only about 1 in 500,000 persons. Physicians have been advised not to use the drug for trivial conditions or ones for which suitable alternatives are available. It is still the drug of choice in the treatment of certain types of meningitis and typhoid fever, in spite of recent reports of resistant strains.

Macrolides

Macrolides are a group of antibiotics named for the presence of a macrocyclic lactone ring. Essentially the only macrolide in common clinical use is *erythromycin* (Figure 19-15). Its mode of action is the inhibition of protein synthesis; apparently, its site of action is similar to that of chloramphenicol. However, erythromycin is not able to penetrate the cell walls of most gram-negative bacilli. Its spectrum of activity is similar to that of penicillin G, and it is a frequent alternative drug for penicillin. Erythromycin is, however, effective against *Legionella* and some *Neisseria*—important gram-negative pathogenic genera. Because it can be administered orally, an orange-flavored preparation of erythromycin is a frequent penicillin substitute for the treatment of streptococcal and staphylococcal infections in children. Erythromycin is the drug of choice for the treatment of mycoplasmal pneumonia.

Polypeptides

A number of antibiotics are **polypeptides**, chains of amino acids linked by peptide bonds. Almost all these drugs have been isolated from the genus *Bacillus*. Two

Erythromycin

Figure 19-15   Structure of erythromycin, a representative macrolide.

well-known examples are bacitracin and polymyxin B (Figure 19-16). Both drugs are relatively toxic, and their use is mostly topical.

Bacitracin is primarily effective against gram-positive bacteria, such as staphylococci and streptococci, although it is also used against some important gram-negative pathogens, such as *Neisseria*. It inhibits the synthesis of cell walls. Its use is restricted to topical application for superficial infections.

Polymyxin B is a bactericidal antibiotic effective against gram-negative bacteria. For many years, it was one of very few drugs used against infections by the gram-negative *Pseudomonas*. The porins (Chapter 4) in the cell walls of pseudomonads are too small to admit many antimicrobials. The mode of action of polymyxin B is to injure plasma membranes. Polymyxin B is seldom used today except in topical treatment of superficial infections because of toxicity when

Figure 19-16   Structure of polymyxin B, a representative polypeptide antibiotic. The amino-acid subunits (names in color) and peptide bonds (color shading) are shown.

it is administered by injection. Both bacitracin and polymyxin B are available in nonprescription antiseptic ointments, in which the polypeptides are usually combined with neomycin, a broad spectrum aminoglycoside.

Vancomycin

Vancomycin is apparently unrelated chemically to any other antibiotic. It is a very toxic drug, is difficult to administer, and has a very narrow spectrum of activity that is based on inhibition of cell wall synthesis (peptidoglycans). The drug is important because it is probably the most effective drug in clinical use against penicillinase-producing staphylococci. Vancomycin is also used in special situations, such as treatment of streptococcal endocarditis and staphylococcus infections of such devices as prosthetic heart valves.

Rifamycins

The best-known derivative of the **rifamycin** family of antibiotics is *rifampin*. These drugs inhibit the synthesis of mRNA. By far the most important use of rifampin is against mycobacteria in the treatment of tuberculosis, but it is effective against a number of gram-positive organisms and *Neisseria*. A valuable characteristic of rifampin is its ability to penetrate tissues and reach therapeutic levels in cerebrospinal fluid and abscesses. This characteristic is probably an important factor in its antitubercular activity because the tuberculosis pathogen is usually located intracellularly in macrophages. An unusual side-effect of rifampin is the appearance of orange-red urine, feces, saliva, sweat, and even tears.

Antifungal Drugs

Polyenes

Two of the more commonly used **polyene antibiotics**, both of which are products of *Streptomyces* spp., are *nystatin* and *amphotericin B* (Figure 19-17). Both drugs are fungicidal and exert their action by damaging fungal plasma membranes; the drugs combine with the sterols in the membranes. Bacterial plasma membranes (except those of mycoplasmas) do not contain sterols, and the drugs therefore do not affect bacteria. Nystatin is used to treat local infections of the vagina and skin and can be taken orally for *Candida* infections of the intestinal tract. It is so poorly soluble that it does not cross the intestinal lining, so it does not enter the body tissue in toxic amounts. For many

Figure 19-17   Structure of amphotericin B, a representative polyene antibiotic.

years, amphotericin B has been a mainstay of clinical treatment for systemic fungal diseases, such as histoplasmosis, coccidioidomycosis, and blastomycosis. The drug's toxicity, particularly to the kidneys, is a strongly limiting factor in these uses.

Imidazoles

Imidazole antifungals, such as *clotrimazole, miconazole*, and *ketoconazole*, primarily interfere with sterol synthesis in fungi, although they probably have other antimetabolic effects. Clotrimazole and miconazole (Figure 19-18) are generally used topically, most often in treatment of cutaneous mycoses such as athlete's foot. When taken orally, ketoconazole is effective and has considerable promise as a less toxic alternative to amphotericin B for many systemic fungal infections, although occasional liver damage has been reported. Ketoconazole also has the advantage of an unusually wide spectrum of activity.

Figure 19-18   Structure of miconazole, a representative imidazole antibiotic.

Griseofulvin

Griseofulvin is an antibiotic produced by a species of *Penicillium*. It has the interesting property of being active against superficial dermatophytic fungal infections of the hair (tinea capitis, or ringworm) and nails, even though its route of administration is oral. The drug apparently binds selectively to the keratin found in the skin, hair follicles, and nails. Its mode of action is to interfere with mitosis and thereby inhibit fungal reproduction.

Other Antifungal Drugs

Tolnaftate is a common alternative to miconazole as a topical agent for the treatment of athlete's foot. Its mechanism of action is not known. *Undecylenic acid* is a fatty acid that has antifungal activity against athlete's foot, although it is not as effective as tolnaftate or the imidazoles.

Antiviral Drugs

In the developed parts of the world, it is estimated that at least 60% of infective illnesses are caused by viruses, and only about 15% are caused by bacteria. Every year, at least 90% of the population of the United States suffers from some viral disease. Yet, as of 1988, only six antiviral drugs had been approved in the United States, and these are effective against only an extremely limited group of diseases. An antiviral drug being developed is described in the box in Chapter 23, p. 584.

The first antiviral drug licensed for systemic use in the United States was *amantadine*. Although it is not known precisely how amantadine works, there is some evidence that it either prevents the virus from entering the cell or, in at least one case, from uncoating after it enters. The drug has limited usefulness in the prevention of influenza A. It has little effect on the course of the disease once it has been contracted, but in confined situations, such as institutions or nursing homes, it can limit the spread of an outbreak. Another drug with activity against influenza is *ribavirin*, an analog of guanine, a purine in DNA and RNA. Ribavirin has also been tested on infants suffering from a viral pneumonia. However, this drug must be administered by aerosol apparatus, is too expensive for general use, and is very toxic. Nevertheless, interest in ribavirin remains high because it is effective in vitro against a wide range of viruses.

Acyclovir is an analog of a guanine-containing nucleoside (Figure 19-19). It is effective only if acted on by thymidine kinase, an enzyme present in cells infected by herpesvirus. Thymidine kinase adds a phosphate to acyclovir because acyclovir resembles the nucleoside that the enzyme normally acts upon. This newly formed acyclovir triphosphate inhibits the DNA polymerase activity induced by the herpesvirus, and further synthesis of DNA is thereby terminated. Acyclovir's antiviral mechanism has little effect on host-cell DNA activity.

Acyclovir has limited usefulness in topical treatment of genital herpes. When administered intravenously, it is effective in clearing a herpes simplex infection. Intravenous use of acyclovir is especially recommended for treatment of primary genital herpes infections and herpes simplex infections in immunosuppressed persons. Oral use has been approved for genital herpes. Acyclovir is effective, and frequently used, against a number of other herpesvirus disease conditions for which it has not yet been approved.

Several other analogs of DNA nucleosides containing purine and pyrimidine have been synthesized and have limited use against herpesviruses. *Idoxuridine* and *trifluridine* are both analogs of thymidine (the thymine-containing nucleoside), and *vidarabine* is an analog of the purine adenine. All these drugs are available as topical ointments and are useful in treating herpes simplex virus infections of the eye (herpes keratitis). Vidarabine has also been approved for intravenous use in the treatment of encephalitis caused by herpesviruses and for treatment of immunosuppressed persons infected with the varicella-zoster virus (chickenpox, shingles). Zidovudine® (azidothymidine, AZT) is an analog of thymidine that has been promising in chemotherapy of AIDS patients.

A new approach to viral chemotherapy that may bear good results involves interferon, a protein secreted by certain host cells responding to viral infections (Chapter 15). Interferon inhibits cell infection by many kinds of viruses. Genetic-engineering techniques using bacteria have made sufficient amounts of interferon available for clinical studies of its possible effectiveness as a chemotherapeutic agent against cancer as well as against viruses.

Antiprotozoan and Antihelminthic Drugs

For hundreds of years, quinine from the tropical tree cinchona was the only drug known to be effective for the treatment of a parasitic infection (malaria). There are now many antiprotozoan and antihelminthic drugs, although many of them are still considered

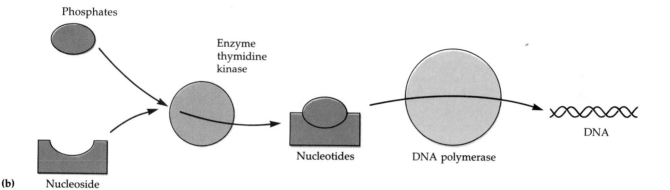

(a)

Figure 19-19 Structure and functions of acyclovir. (a) Structural resemblance between acyclovir and the nucleoside 2′-deoxyguanosine. **(b)** The enzyme thymidine kinase combines phosphates with nucleosides to form nucleotides. The nucleotides are then assembled into DNA. **(c)** Acyclovir is a structural analog of the nucleoside, and the enzyme uses it to assemble a false nucleotide that cannot be assembled into DNA.

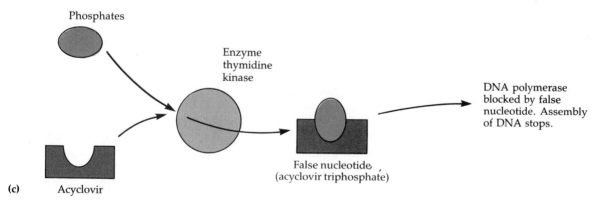

(b)

(c)

experimental. This does not preclude their use, however, by qualified physicians. The Centers for Disease Control (CDC) provide several of them on request when they are not available commercially.

Antiprotozoan Drugs

Quinine still has very limited use in the control of the protozoan disease malaria, but synthetic derivatives, such as *chloroquine*, have largely replaced it. Chloroquine fits itself between nitrogenous base pairs in DNA (called *intercalation*) and prevents the DNA from serving as a template for further DNA synthesis. *Quinacrine* functions in a very similar manner and is the drug of choice for treating giardiasis. *Diiodohydroxyquin* (*iodoquinol*) is an important drug prescribed for several intestinal amoebic diseases, but its dosage must be carefully controlled to avoid optic nerve damage. Its mode of action is yet unknown.

Metronidazole (*Flagyl®*) is one of the most widely used antiprotozoan drugs. It is unique in having activity not only against parasitic protozoans but also against obligate anaerobic bacteria. For example, as an antiprotozoan agent, it is the drug of choice for vaginitis caused by *Trichomonas vaginalis* (trik-ō-mōn′as va-jin-al′is). It is also effective in treating vaginitis caused in anaerobic conditions by the catalase-negative bacterium *Gardnerella vaginalis* (gärd-nė-rel′lä va-jin-al′is). It is also used in the treatment of giardiasis and amoebic dysentery. The mode of action is damage to DNA by products of anaerobic metabolism.

Pentamidine isethionate is used in the early stages of African trypanosomiasis, but in the United States its most familiar clinical use is in the treatment of *Pneumocystis* pneumonia, a common complication in immunocompromised persons and victims of AIDS. The drug's mode of action is unknown, but it appears to bind to DNA. *Nifurtimox* is a member of the synthetic nitrofuran drugs and is effective against the trypanosome-caused Chagas' disease. Treatment may produce the side-effects of nausea and even convulsions.

Antihelminthic Drugs

With the increased popularity of sushi, a Japanese specialty often made with raw fish, the CDC began to notice an increased incidence of tapeworm infestations. To estimate the incidence, the CDC notes requests for *niclosamide*, which is the usual first choice in treatment. The drug is effective because it inhibits ATP production under anaerobic conditions. *Praziquantel* is about equally effective for treatment of tapeworms, although its mode of action is not known. Praziquantel is highly recommended for treatment of several fluke-caused diseases, especially schistosomiasis. It causes the helminths to undergo muscular spasms, and apparently makes them susceptible to attack by the immune system. *Mebendazole* is an antihelminthic drug frequently used in the treatment of several of the most common intestinal helminths—ascariasis, *Ascaris lumbricoides* (as′kar-is lum-bri-koi′dēz); pinworms, *Enterobius vermicularis* (en-te-rō′bē-us ver-mi-kū-lar′is); and whipworm, *Trichuris trichiura* (trik′ėr-is trik-ē-yėr′a). The mode of action is disruption of the microtubules in the cytoplasm, which indirectly results in an inability to generate ATP from carbohydrates. *Piperazine* is used in the treatment of pinworms and ascariasis. It paralyzes the organisms, which soon pass out of the victim's body.

Tests for Microbial Susceptibility to Chemotherapeutic Agents

Different microbial species and strains have different degrees of susceptibility to different chemotherapeutic agents. Moreover, the susceptibility of a microorganism can change with time, even during therapy with a specific chemotherapeutic agent. Because of these factors, a physician must know the sensitivities of the disease-causing microorganism before a particular treatment can be started.

Several tests can be used to indicate which chemotherapeutic agent is most likely to combat a specific pathogen. These tests, however, are not always necessary. If the organisms have been identified—for example, as *Pseudomonas aeruginosa*, beta-hemolytic streptococci, or gonococci—certain drugs can be selected without specific testing for susceptibility. These tests are necessary only when susceptibility is not predictable, or when antibiotic resistance problems develop.

Disk-Diffusion Method

Probably the most widely used—but not necessarily the best—testing method is the **disk-diffusion method**, also known as the **Kirby–Bauer test**. A Petri plate containing an agar medium is inoculated ("seeded") uniformly over its entire surface with a standardized amount of a test organism. Next, filter-paper disks impregnated with known concentrations of chemotherapeutic agents are placed on the solidified agar surface. During incubation, the chemotherapeutic agents diffuse from the disks into the agar. The farther the agent diffuses from the disk, the lower its concentration. If the chemotherapeutic agent is effective, a *zone of inhibition* forms immediately around the disk (Figure 19-20). The diameter of the zone of inhibition can be measured. Because the size of the zone of inhibition is affected by the diffusion rate of the chemotherapeutic agent, a wider zone does not always indicate greater antimicrobial activity. The zone diameter is compared to a standard table for that drug and concentration, and the results report that the organism is *sensitive*, *intermediate*, or *resistant*. This result is often inadequate information for many clinical purposes. However, the test is simple and inexpensive and has considerable value in medical practice.

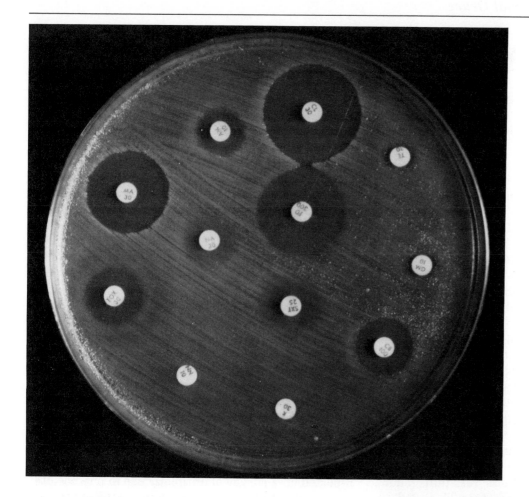

Broth Dilution Tests

A weakness of the disk diffusion method is that it does not determine if a drug is bactericidal and not just bacteriostatic. Some form of **broth dilution testing** is needed to determine the *minimal inhibitory (MIC)* and *minimal bactericidal (MBC) concentrations* of an antimicrobial. The MIC is determined by making a sequence of decreasing concentrations of the drug in a broth, which is then inoculated with the test bacteria (Figure 19-21). The lowest concentration that prevents visible growth is the MIC. The wells that do not show growth can be cultured into broth free of the antimicrobial. If growth occurs in this broth, the drug was not bactericidal, and the MBC can be determined.

Dilution tests are often highly automated. The drugs are purchased already diluted into broth in wells formed in a plastic tray. A suspension of the test organism is prepared and inoculated into all the wells simultaneously by a special inoculating device. After incubation, the turbidity may be read visually, although clinical laboratories with high work loads may read the trays with special scanners that enter the data into a computer that provides a printout of the MIC.

Control well

Doxycycline

Sulfamethoxyzole

Ethambutol

Streptomycin

Kanamycin

Decreasing concentration of drug ⟶

Figure 19-21 A microdilution MIC plate. Growth is shown by a white dot in the bottom of the well. Doxycycline had no effect at all. Kanamycin and ethambutol were about equally effective. Streptomycin was effective only at highest concentration tested. A trailing endpoint, observed with sulfamethoxyzole, is read where there is an 80% reduction in growth. The negative control well (with broth only) contained no antibiotics and no inoculum; the positive control well (not shown) contained no antibiotics but was inoculated.

Effectiveness of Chemotherapeutic Agents

Drug Resistance

Bacteria become resistant to antimicrobials in a number of different ways. Some resistance is due to the ability of a particular mutant to destroy a given antibiotic. For example, as we noted earlier, some staphylococci produce the enzyme penicillinase, which destroys natural penicillins. Resistance to tetracycline is usually related to cellular changes that decrease the cell's intake of the drug. Bacteria become resistant to sulfa drugs and antibiotics when receptor sites for the antibiotic develop less affinity for it. Resistance to trimethoprim can arise when the microbe begins synthesizing very large amounts of the enzyme against which the drug is targeted. Conversely, polyene antibiotics can become less effective when resistant organisms produce smaller amounts of the sterols with which the drug produces its effects. Of particular concern is the possibility that such *resistant mutants* will increasingly replace the susceptible normal populations (see the box below). The increasing appearance of antibiotic-resistant gonorrhea is discussed in the box in Chapter 25, p. 646.

Hereditary drug resistance is often carried by extrachromosomal genetic elements called plasmids. Some plasmids, including those called **resistance (R) factors**, can be transferred between bacterial cells in a population and between different but closely related bacterial populations (Chapter 8). R factors often contain genes for resistance to several antibiotics.

Strains of bacteria that are resistant to antibiotics are particularly common among persons who work in hospitals, where antibiotics are in constant use. *Staphylococcus aureus*, a common opportunistic pathogen that is carried in the nasal passages, develops resistance to antibiotics very frequently. Every day in hospitals, surprisingly large volumes of antibiotics are dispensed as aerosols by the universal practice of clearing syringes of air before an injection is made. As these aerosols are generally inhaled in small amounts, the staphylococci harmlessly residing in a nurse's nose, for example, can quickly develop resistance. If these resistant microbes are transferred to surgical wounds and other exposed sites, they can cause nosocomial infections (Chapter 13).

Drugs should not be used indiscriminately, and the concentrations administered should always be of optimum strength so that the chances of survival of resistant mutants are minimized. Many hospitals have special monitoring committees who review the use of antibiotics. A few years ago, one medical journal reported on a survey of antibiotic use in one hospital. The results indicated that more than 60% of the antibiotics prescribed there were either given in the wrong

.. Microbiology in the News ....

Antibiotics in Animal Feed Linked to Human Disease

Thirty years ago livestock growers began using antibiotics in the feed of closely penned animals that were being fattened for market. The drugs helped reduce bacterial infections and their spread in such ripe conditions. They also unexpectedly and inexplicably accelerated the animals' growth,

an effect that led to wider and wider use.

Meat that reaches the consumer's table is not heavily ridden with antibiotics because the Food and Drug Administration has established limits for antibiotic residues in edible tissues. In random-sample testing, only 2% of veal calves exceeded the tolerance limits in 1987. Veal would be expected to have the highest residue because the animals do not live long enough to excrete much of the antibiotics. Nevertheless, the subtherapeutic use of antibiotics has come under attack.

The Risk to Humans
Critics claim that use of antibiotics in animal feed preferentially allows the growth of strains of bacteria that are resistant to drugs commonly used to treat human infections. Such resistant bacteria could theoretically transfer their resistance genes to bacteria that infect humans by transformation or conjugation. In 1982, Dr. Thomas O'Brien of the Harvard University School of Medicine showed that bacteria that infect humans and other animals

freely share genetic information by transfer of plasmids.

Scientists also argue that some antibiotic-resistant bacteria, such as certain strains of *Salmonella* and *Listeria*, could be transferred from animals to humans directly in meat or milk. Researchers at the Centers for Disease Control (CDC) confirmed this suspicion in a study published in 1984. Dr. Scott Holmberg and his associates at CDC traced an antibiotic-resistant *Salmonella newport* infection from South Dakota beef cattle to 18 people in four states.

dosage or used in applications for which they would not be effective. Another way the development of resistant strains can be reduced is through the administration of two or more drugs simultaneously. If a strain is resistant to one of the drugs, the other might destroy it. The probability that the organism will acquire resistance to both drugs is roughly the product of the two individual probabilities.

Resistance to drugs is not the only problem with antimicrobial therapy. For example, antibiotics have no effect on toxins that might already have been released into the host's body fluids.

Effects of Combinations of Drugs

It sometimes happens that the chemotherapeutic effect of two drugs given simultaneously is greater than the effect of either given alone, a phenomenon called **synergism**. For example, in the treatment of bacterial endocarditis, penicillin and streptomycin are much more effective when taken together than when either drug is taken alone. Damage to bacterial cell walls by penicillin makes it easier for streptomycin to enter.

Other combinations of drugs can be **antagonistic**. For example, the simultaneous use of penicillin and tetracycline is often less effective than when either drug is used alone. By stopping the growth of the bacteria, the bacteriostatic drug tetracycline interferes with the action of penicillin, which requires bacterial growth.

Combinations of antimicrobial drugs should be used only for the following purposes:

1. To prevent or minimize the emergence of resistant strains.
2. To take advantage of the synergistic effect.
3. To provide optimal therapy in life-threatening situations before a diagnosis can be established with certainty.
4. To lessen the toxicity of individual drugs by reducing the dosage of each in combination.

The Future of Chemotherapeutic Agents

Antibiotics are clearly one of the greatest triumphs of medical science, although they are not without their problems and disadvantages. Our most pressing concern is with the misuse of antibiotics, which increases the likelihood of the emergence of resistant strains of pathogens.

Antiviral drugs are urgently needed. As research discovers additional aspects of the internal workings of viruses and cells, more chemicals that are effective against viruses will probably be developed. There is considerable similarity between the requirements of cancer therapy and the requirements for treatments for viral disease. It will be interesting to see whether research for one will benefit the other.

People taking antibiotics may have a higher risk of infection by resistant organisms. In an outbreak of antibiotic-resistant *S. newport* in California, people taking penicillin for respiratory infections had a higher rate of infection than others. Apparently, penicillin therapy gave the *S. newport* from beef a selective advantage over other bacteria, many of which were killed by the penicillin. In an outbreak of *S. typhimurium* linked to milk consumption in the Midwest, people taking antimicrobials were more readily infected by the bacteria. Before 1980, only one case of salmonellosis that was exacerbated by antibiotic therapy had been reported.

The number of reported cases of *Salmonella* infections of humans is increasing, and 20% to 25% of these are due to antibiotic-resistant strains of bacteria. CDC is urging public health laboratories to identify strains of *Salmonella* by studying antibiotic-resistance patterns, and by using serotype and molecular biologic techniques to trace the origins of infection.

New Techniques for Tracing Sources

Molecular biology has provided important new tools for tracing the sources of *Salmonella* infections. In 1985, nearly 1000 cases of *S. newport* resistant to multiple antibiotics were reported in Los Angeles County, California. A radioactive DNA probe was used to identify an identical plasmid carrying multiple antibiotic-resistance genes in 99% of the *S. newport* isolates. A technique called plasmid profile analysis was used to track the bacterium from the sick people back to a slaughterhouse, to a meat-deboning plant, and finally to three farms.

In spite of this evidence, cattle growers, feed manufacturers, and the drug industry still argue that the link between antibiotics in feed and human disease is sketchy, and claim that the benefits derived from the use of antibiotics outweigh the potential harmful effects. However, some farmers have reported that their sick animals are not responding to penicillin and tetracycline therapy. Thus, some limitation of antibiotic use may actually benefit the farmer.

Study Outline

Introduction   (p. 475)

1. A chemotherapeutic agent is a chemical that combats disease in the body.
2. An antimicrobial drug is a chemical substance that destroys disease-causing microorganisms with minimal damage to host tissues.
3. Antimicrobial drugs may be synthetic drugs (prepared in the laboratory) or antibiotics (produced by bacteria or fungi).

Historical Development   (pp. 475–477)

1. The first chemotherapeutic agent, discovered by Paul Ehrlich, was salvarsan, used to treat syphilis.
2. Sulfa drugs came into prominence in the late 1930s.
3. Alexander Fleming discovered the first antibiotic, penicillin, in 1929; its first clinical trials were done in 1940.
4. Antibiotics are produced by species of *Streptomyces*, *Bacillus*, *Penicillium*, and *Cephalosporium*.

Criteria for Evaluating Antimicrobial Drugs   (p. 477)

1. Antimicrobial agents should have selective toxicity for microorganisms.
2. They should not produce hypersensitivity in most patients.
3. They should be stable during storage, be soluble in body fluids, and be retained in the body long enough to be effective.
4. Microbes should not readily become resistant to the drug.

Spectrum of Activity   (pp. 477–478)

1. Antibacterial drugs affect many targets in a procaryotic cell.
2. Fungal, protozoan, and helminthic infections are more difficult to treat because these organisms have eucaryotic cells.
3. Narrow spectrum drugs affect only a select group of microbes—gram-positive cells, for example.
4. Broad spectrum drugs affect a large number of microbes.
5. Antimicrobial agents should not cause excessive harm to normal flora.
6. Superinfections occur when a pathogen develops resistance to the drug being used or when normally resistant flora multiply excessively.

Action of Antimicrobial Drugs   (pp. 478–481)

1. General action is either by directly killing microorganisms (bactericidal) or by inhibition of growth (bacteriostatic).

2. Some agents, such as penicillin, inhibit cell wall synthesis in bacteria.
3. Other agents, such as chloramphenicol, erythromycin, tetracyclines, and streptomycin, inhibit protein synthesis by acting on 70S ribosomes.
4. Agents such as polymyxin B cause injury to plasma membranes.
5. Rifamycin, nalidixic acid, and trimethoprim inhibit nucleic acid synthesis.
6. Agents such as sulfanilamide act as antimetabolites by competitively inhibiting enzyme activity.

Survey of Commonly Used Antimicrobial Drugs   (pp. 481–492)

Synthetic Antimicrobial Drugs   (pp. 481–485)

1. Isoniazid (INH) is a structural analog of vitamin B_6 and may inhibit mycolic acid synthesis in mycobacteria. INH is administered with streptomycin, rifampin, or ethambutol to treat tuberculosis.
2. The antimetabolite ethambutol is used with other drugs to treat tuberculosis.
3. Sulfonamides competitively inhibit folic acid synthesis. Sulfonamides are bacteriostatic. They are used to treat urinary tract infections and can cause allergies and drug resistance.
4. Trimethoprim and sulfamethoxazole competitively inhibit dihydrofolic acid synthesis for treatment of *Pneumocystis* pneumonia and urinary tract and intestinal infections.
5. Nitrofurans may break DNA molecules; they are used to treat urinary tract infections.
6. Nalidixic acid and quinolones interfere with DNA synthesis; they are used to treat infections caused by gram-negative bacteria.
7. The antifungal agent 5-fluorocytosine is an antimetabolite of cytosine.

Antibiotics   (pp. 458–489)

1. Penicillins have low toxicity. They inhibit peptidoglycan synthesis; cell death is by osmotic lysis.
2. Penicillin G and V are natural penicillins and are effective against cocci, gram-positive bacilli, and spirochetes.
3. Several semisynthetic penicillins such as ampicillin are resistant to stomach acid and penicillinases and are broad spectrum.
4. Ureidopenicillins such as mezlocillin are effective against *Pseudomonas* spp.
5. Cephalosporins inhibit cell wall synthesis and are used against penicillin-resistant strains.

6. Carbapenems are broad spectrum antibiotics that inhibit cell wall synthesis.
7. Aminoglycosides include streptomycin, neomycin, spectinomycin, and gentamicin; all inhibit protein synthesis and are bactericidal.
8. Tetracyclines inhibit protein synthesis and are bacteriostatic toward many bacteria, including rickettsias and chlamydias; they have the broadest spectrum of antibacterial activity.
9. Chloramphenicol inhibits protein synthesis and is bacteriostatic; a side-effect of prolonged use is aplastic anemia.
10. Macrolides, such as erythromycin, inhibit protein synthesis and are bacteriostatic; erythromycin is used to treat legionellosis and *Neisseria* infections.
11. Polypeptides include bacitracin and polymyxin B. They are applied topically to treat superficial infections.
12. Bacitracin inhibits cell wall synthesis primarily in gram-positive bacteria.
13. Polymyxin B damages plasma membranes and is effective against gram-negative bacteria.
14. Vancomycin inhibits cell wall synthesis and may be used to kill penicillinase-producing staphylococci.
15. Rifamycins are bactericidal and are especially effective against gram-positive bacteria, including mycobacteria.

Antifungal Drugs (pp. 489–490)

1. Polyenes, such as nystatin and amphotericin B, combine with plasma membrane sterols and are fungicidal.
2. Imidazoles interfere with sterol synthesis and are used to treat cutaneous and systemic mycoses.
3. Griseofulvin interferes with eucaryotic cell division and is used primarily to treat skin infections caused by fungi.

Antiviral Drugs (p. 490)

1. Amantadine blocks penetration or uncoating of influenza A virus.
2. Purine and pyrimidine analogs include ribavirin, acyclovir, idoxuridine, trifluridine, and vidarabine.
3. Interferon inhibits cell infection by viruses.

Antiprotozoan and Antihelminthic Drugs (pp. 490–492)

1. Chloroquine, quinacrine, diiodohydroxyquin, pentamidine, and metronidazole are used to treat protozoan infections.
2. Chloroquine and quinacrine stop DNA synthesis by intercalation.
3. Antihelminthic drugs include niclosamide, mebendazole, praziquantel, and piperazine.
4. Mebendazole disrupts microtubules; piperazine paralyzes intestinal roundworms.

Tests for Microbial Susceptibility to Chemotherapeutic Agents (pp. 492–493)

1. These tests are used to determine the degree of susceptibility of different microorganisms to chemotherapeutic agents.
2. They help to determine which chemotherapeutic agent is most likely to combat a specific pathogen.
3. These tests are used when susceptibility cannot be predicted or when drug resistance arises.

Disk-Diffusion Method (pp. 492–493)

1. In this test, also known as the Kirby–Bauer test, a bacterial culture is inoculated on an agar medium, and filter-paper disks impregnated with chemotherapeutic agents are overlayed on the culture.
2. After incubation, the absence of microbial growth around a disk is called a zone of inhibition.
3. The diameter of the zone of inhibition, when compared with a standardized reference table, is used to determine whether the organism is sensitive, intermediate, or resistant to the drug.

Broth Dilution Tests (p. 493)

1. In the broth dilution test, the microorganism is grown in liquid media containing different concentrations of a chemotherapeutic agent.
2. The minimum inhibitory concentration (MIC) is the lowest concentration of chemotherapeutic agent capable of preventing microbial growth.
3. The lowest concentration of chemotherapeutic agent that kills bacteria is called the minimum bactericidal concentration (MBC).

Effectiveness of Chemotherapeutic Agents (pp. 494–495)

Drug Resistance (pp. 494–495)

1. Drug resistance refers to the ability of a microorganism to resist the antimicrobial effects of a chemotherapeutic agent.
2. Resistance may be due to enzymatic destruction of a drug or cellular or metabolic changes at target areas.
3. Hereditary drug resistance is carried by plasmids called resistance (R) factors.
4. Resistance can be minimized by the discriminate use of drugs in appropriate concentrations and dosages.

Effects of Combinations of Drugs (p. 495)

1. Combinations of drugs may be used to minimize the development of resistant strains, to employ a synergistic effect, to provide therapy prior to diagnosis, and to use small concentrations of each drug to lessen toxicity.
2. Some combinations of drugs are synergistic—they are more effective when taken together.
3. Some combinations of drugs are antagonistic—when taken together, both drugs become less effective than when taken alone.

The Future of Chemotherapeutic Agents (p. 495)

1. More antiviral drugs are urgently needed. Their development might also prove to be helpful in cancer therapy.

Study Questions

Review

1. Define a chemotherapeutic agent. Distinguish between a synthetic chemotherapeutic agent and an antibiotic.

2. Paul Ehrlich discovered the first _____ . Alexander Fleming discovered _____ ; it was the first _____ .

3. List and explain five criteria used to identify an effective antimicrobial agent.

4. Fill in the following table.

Antimicrobial Agent	Synthetic or Antibiotic	Method of Action	Principal Use
Isoniazid			
Sulfonamides			
Ethambutol			
Trimethoprim			
Nitrofurans			
Penicillin, natural			
Penicillin, semisynthetic			
Cephalosporins			
Carbapenems			
Aminoglycosides			
Tetracyclines			
Chloramphenicol			
Macrolides			
Polypeptides			
Vancomycin			
Rifamycins			
Polyenes			
Griseofulvin			
Amantadine			
Acyclovir			
Chloroquine			
Niclosamide			

5. What similar problems are encountered with antiviral, antifungal, antiprotozoan, and antihelminthic drugs?

6. Identify three methods of action of antiviral drugs. Give an example of a currently used antiviral drug for each method of action.

7. Compare and contrast the tube dilution and agar dilution tests. Identify at least one advantage of each.

8. Describe the disk-diffusion test for microbial susceptibility. What information can you obtain from this test?

9. Define drug resistance. How is it produced? What measures can be taken to minimize drug resistance?

10. List the advantages of using two chemotherapeutic agents simultaneously to treat a disease. What problem can be encountered using two drugs?

Challenge

1. Why is idoxuridine effective if host cells also contain DNA?

2. The following data were obtained from a disk-diffusion test.

Antibiotic	Zone of Inhibition
A	15 mm
B	0 mm
C	7 mm
D	15 mm

   (a) Which antibiotic was most effective against the bacteria being tested?
   (b) Which antibiotic would you recommend for treatment of a disease caused by this bacterium?
   (c) Was antibiotic A bactericidal or bacteriostatic? How can you tell?

3. The following results were obtained from a tube dilution test for microbial susceptibility.

Tube Number	Antibiotic Concentration	Growth	Growth in Subculture
1	200 μg	−	−
2	100 μg	−	−
3	50 μg	−	+
4	25 μg	+	+

   (a) The MIC for this antibiotic is _____ .
   (b) The MBC for this antibiotic is _____ .

Further Reading

Abraham, E. P. "The beta-lactam antibiotics." *Scientific American* 244(6):76-86, June 1981. A discussion of semisynthetic penicillins and cephalosporins.

Baldry, P. *The Battle Against Bacteria.* Cambridge: Cambridge University Press, 1976. The history of the fight against diseases and the development of antibacterial drugs.

Bryan, L. E., ed. *Antimicrobial Drug Resistance.* New York: Academic Press, 1984. Discusses mechanisms of resistance to antibacterial, antifungal, antiviral, and antimalarial drugs.

Gauri, K. K. *Antiviral Chemotherapy: Design of Inhibitors of Viral Functions.* New York: Academic Press, 1981. Describes inhibition of viral entry, protein synthesis, and nucleic acid replication as well as the toxicology of drugs.

Hirsch, M. S. and J. C. Kaplan. "Antiviral therapy." *Scientific American* 246(4):76–85, April 1987. A review of the mechanisms by which antiviral drugs work.

Pratt, W. B., and R. Fekety. *The Antimicrobial Drugs.* New York: Oxford University Press, 1986. A very good summary of most of the commonly used antibiotics, with uses, modes of action, pharmacology, toxicity, and other information.

Sanders, C. C. "Chromosomal cephalosporinases responsible for multiple resistance to newer β-lactam antibiotics." *Annual Review of Microbiology* 41:573–593, 1987. Describes methods of action of β-lactamases and methods to detect production of these enzymes by bacteria.

Wishnow, R. M., and J. L. Steinfeld. "The conquest of the major infectious diseases in the United States: a bicentennial retrospective." *Annual Review of Microbiology* 30:427–450, 1976. A concise summary of control of diseases, including tuberculosis, cholera, malaria, and yellow fever.

When a patient is admitted to a hospital for treatment, the patient's illness poses a problem for doctors, nurses, and technicians to solve. The first step toward a solution is to identify the microorganisms responsible for the illness. The next step is to find a suitable drug and a suitable dosage to combat the pathogen. These two steps are carried out by technicians in special laboratories, utilizing powerful yet simple techniques. To solve the problem, that is, to treat the patient success-

(1)

fully, doctors depend on the results from the hospital laboratories. In the next four pages, we will tour two laboratories in the University

of California at San Francisco hospital—the microbiology lab and the immunology lab. To begin, we will visit an area where some samples are prepared for distribution to specialized labs.

(1) *Staphylococcus aureus* is a common pathogen for which rapid identification can be made. Colonies of coagulase-positive *S. aureus* are black on tellurite glycine agar. Most pathogenic strains are coagulase-positive.

Specimen-Processing Area

The specimen-processing area prepares samples for some of the specialized labs, such as the chemistry, endocrinology, and immunology labs. (The microbiology lab pro-

cesses its own samples; technicians work with clinical samples that come directly from the patient to the lab—feces, blood, urine, or anything from which bacteria can be cultured.)

(2) At UCSF hospital, the history of every patient is filed on a central computer; therefore, the first step in preparing a sample is to enter information about the sample into the patient's file and assign the sample a number. After a technician has labeled each sample with the patient's name and sample number and racked the sample in numerical order, he or she will send them to the aliquot room, part of the specimen-processing area.

(3) In the aliquot room, fluids are diluted into portions of known concentration and volume. Because the immunology lab conducts tests using serum, samples destined for the immunology lab must be centrifuged to separate the blood cells from the serum.

(2)

(3)

(4)

(4) A technician labels aliquots and sends them to the immunology lab. Fluids destined for other labs, such as spinal fluid and urine, are handled in similar ways.

Microbiology Lab

Although the microbiology lab utilizes several modern, time-saving machines, most of the work involves plating and growing cultures.

(5) Blood samples are placed in bottles containing a carbon-14 labeled (radioactive) substrate, and incubated overnight. If bacteria are present in the blood sample, they will release radioactive carbon dioxide gas as they respire. The following morning, a technician places each bottle in a machine to measure the amount of C-14 released. The results are immediately printed out on tape. The technician can quickly detect respiration levels over a certain amount. This method is much faster than the old method, in which, over a period of several days, a technician looked for cloudiness or other observable changes in each bottle, and also plated every sample. In the newer method, only suspicious samples are plated, although bottles are checked for growth each day for up to seven days.

(6)

(6) Once a suspicious sample is detected, the technician makes a smear from the specimen for immediate inspection under the microscope. The smear helps him determine which biochemical test to use later in identifying the pathogen.

(7) The technician then inoculates the sample onto blood agar. After the plates have incubated, he looks for suspicious colonies, on which he will perform specific biochemical tests.

(7)

(8)

(8) In this case, the bacterium in the blood is *Staphylococcus aureus*.

(9) Stool samples are checked for the presence of enteric pathogens. The three most common are *Campylobacter*, *Shigella*, and *Salmonella*. First, a technician inoculates a sample onto several different media. The plates are incubated, then inspected for suspicious (clear) colonies. All colored colonies, including the yellow colonies shown here, are normal.

(5)

(9)

(10)

(11)

(10) The technician then inoculates from the suspicious colony into test tubes containing three different media. Starting from the left, they are glucose broth, triple sugar iron (TSI) agar, and urea agar. A small test tube is placed in the glucose open side down; it will trap any CO_2 that may be produced as the organisms grow. When analyzing her results, the technician first notes the color of the urea agar, which tests for the enzyme *urease*. Because none of the common pathogens have urease, the absence of a color change immediately suggests the presence of a pathogen. She also notes that the slant of the TSI agar has turned red, indicating that the pathogen has fermented lactose. And, she notes that the bacterium has metabolized the glucose broth (turned it pink), as do almost all bacteria, but that it has produced no gaseous by-product, as do some pathogens. Based on

these results and the appearance of the original colony, the technician concludes that the pathogen is probably *Shigella*.

(11) These tubes show results from normal flora. Although the TSI agar looks similar to the result from *Shigella*, the magenta color of the urea agar and the gas bubble in the glucose broth indicate normal flora.

(12) To identify the pathogen conclusively, the lab technician performs a final series of biochemical tests. The UCSF microbiology lab uses an API strip containing 20 different media. Each three-letter abbreviation corresponds to the key substance on which certain species of bacteria are known to grow. The technician suspends a colony of the pathogen in a saline solution and places a sample in each well. After the strip has incubated, the technician checks for color change. These individual results are compared to a general data base, which is used to rapidly and reliably identify the bacterial pathogen.

(12)

(13) Samples taken from anaerobic parts of the body, such as from an abdominal wound, are handled in an anaerobic chamber.

(14) Once a pathogen has been identified, an appropriate drug to combat the microorganism must be selected. For any given pathogenic bacterium, several drugs may be available to combat it. A physician would like to choose the safest drug that can be used at the lowest effective concentration. This ideal concentration is called the minimum inhibitory concentration, MIC. An MIC machine can quickly test 12 sample drugs at 8 different concentrations. A technician places a liquid suspension of the organism along with different concentrations of the drugs in each well of a tray and incubates the tray overnight.

(14)

The concentration of drugs is greatest in the top row. The next day he places the tray in a special device with a mirror so that the growth in the wells can be easily seen (shown here). Cloudiness indicates growth. (The pigments and gas bubbles seen here are typical of some organisms, but do not necessarily indicate growth of the bacteria.) Each well is treated as a coordinate containing information about drug concentration and growth. An uninoculated plate is incubated to test the sterility of the drug solutions in the microtiter plate (negative control). A positive control is inoculated with reference organisms, and their susceptibility is compared to a standard table to test the effectiveness of the drug solutions being used. The physician receives a grid of the results.

(13)

Immunology Lab

The technicians in the immunology lab test clinical samples for antibodies to particular viruses. Immunologic tests are designed to detect reactions between known antigens and unknown antibodies that might be present in the patient's serum. Accuracy of procedures and reagents is ensured by running tests on known positive and known negative serum samples along with unknown patients' samples.

(15)

(15) The ELISA test is one method used to test for a specific antibody (see Chapter 17 and Figure 17-12 for a full discussion). The UCSF immunology lab uses the ELISA test to detect antibodies to rubella and hepatitis viruses in serum. A technician places beads coated with a specific antigen (one that will react to antibodies against the suspected pathogen) in wells.

(16) She then adds patient serum. If present, antibodies will adhere to the beads. The number of antibody molecules that adhere determine how many enzyme-antibody-antigen complexes will form on the bead. Each complex corresponds to the presence of an antibody in the serum.

(17) When the bead is added to a solution of substrate, the enzyme-antibody-antigen complex turns yellow, and the intensity of color indicates the amount of antibody. A spectrophotometer measures the color in the tube and gives a readout of the corresponding concentration of antibody.

(18)

(18) A fluorescent antibody test is performed using slides pre-prepared with a specific organism. In this test, the organism acts as the antigen. After serum is added to the wells, the slides are incubated. Antigen-antibody complexes form between the organism and the serum. After incubation, the technician washes off the excess, unbound patient serum. Next, she adds an antihuman fluoresceinated gamma globulin conjugate that binds to the antibody-antigen complex. Again, the slide is incubated and washed.

(19) Because the slides are specific for a given organism, the technician need look only for fluorescence under ultraviolet light. In this micrograph of a smear from an infected patient, the yellow fluorescing organism is *Chlamydia*.

(16)

(17)

(19)

PART FOUR

Microorganisms and Human Disease

CHAPTER 20

Microbial Diseases of the Skin and Eyes

Learning Objectives

- Describe the structure of the skin and the ways pathogens can invade the skin.

- Provide examples of normal skin flora, and state their locations and ecological roles.

- Differentiate between staphylococci and streptococci, and list several skin infections caused by each.

- List the etiologic agent, method of transmission, and clinical symptoms of the following skin infections: acne, warts, smallpox, chickenpox, measles, German measles, and cold sores.

- Differentiate among the types of mycoses, and provide an example of each.

- Discuss the roles of bacteria, fungi, viruses, and protozoans in conjunctivitis and keratitis.

- Describe the epidemiologies of neonatal gonorrheal ophthalmia and inclusion conjunctivitis.

In Chapter 15, we saw that the human body possesses a number of defenses that contribute to *nonspecific resistance* against many types of pathogens. These defenses include the skin and mucous membranes, antimicrobial substances (interferon, complement, and properdin), phagocytosis, the inflammatory response, and fever.

The skin, which covers and protects the body, is the body's first line of defense against pathogens. Basically, the skin is an inhospitable place for most microorganisms because the secretions of the skin are acidic and most of the skin contains little moisture. Moreover, much of the skin is exposed to radiation, which discourages microbial life. Some parts of the body, however, such as the axilla (armpit), have enough moisture to support microbial growth. Since excretions in the axillary region tend to contain more organic matter than excretions elsewhere on the body, this region can support relatively large bacterial populations, whereas other regions, such as the scalp, support rather small numbers of microorganisms. The skin is a physical as well as an ecological barrier, and it is almost impossible for pathogens to penetrate it, although some can enter through openings not readily apparent to the human eye.

Structure and Function of the Skin

The skin of an average adult occupies a surface area of about 1.9 m^2 and varies in thickness from 0.05 to 3.0 mm. As we mentioned in Chapter 15, skin consists of two principal parts, the epidermis and the dermis (Figure 20-1). The *epidermis* is the thin, outer portion, composed of several layers of epithelial cells. The outermost layer of the epidermis, the *stratum corneum*, consists of dead cells that contain a waterproofing protein called *keratin*. The epidermis, when unbroken,

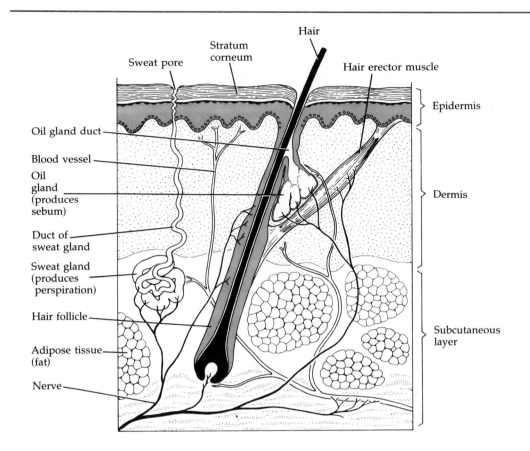

Sweat pore · Stratum corneum · Hair · Hair erector muscle

Oil gland duct

Blood vessel

Oil gland (produces sebum)

Duct of sweat gland

Sweat gland (produces perspiration)

Hair follicle

Adipose tissue (fat)

Nerve

Epidermis

Dermis

Subcutaneous layer

Figure 20-1 Structure of the skin. Note the passageways between the hair follicle and hair shaft through which microbes can penetrate the deeper tissues. Microbes can also enter the skin through sweat pores. Acne is caused by bacterial metabolism of sebum from the oil glands.

is an effective physical barrier against light, heat, microorganisms, and many chemicals.

The *dermis* is the inner, relatively thick portion of skin that is composed mainly of connective tissue. It contains numerous blood vessels, lymph vessels, nerves, hair follicles, and sweat and oil glands. A hair follicle is a small tube composed of epithelial cells in which a hair shaft grows. Sweat glands originate in the dermis and convey perspiration through ducts that terminate as sweat pores on the surface of the skin. Oil glands contain ducts that convey sebum into the hair follicles. The hair follicles, sweat-gland ducts, and oil-gland ducts provide passageways through which microorganisms can enter the skin and penetrate deeper tissues.

Perspiration is a mixture of water, salt (mostly NaCl), urea, amino acids, glucose, and lactic acid, plus small amounts of other compounds, such as the enzyme lysozyme. This mixture provides moisture and some nutrients for microbial growth. However, the salt creates a hypertonic environment that dehydrates most microorganisms, and lysozyme is capable of breaking down the cell walls of certain bacteria.

Sebum, secreted by oil glands, is a mixture of lipids (unsaturated fatty acids), proteins, and salts. It prevents hairs from drying out. Sebum passes upward through the hair follicles and spreads over the surface

of the skin, where it prevents excessive evaporation of moisture from the skin. Although the fatty acids inhibit the growth of certain pathogens, sebum, like perspiration, is nutritive for certain microorganisms, most notably the corynebacteria. When oil glands of the face become enlarged because of accumulated sebum, blackheads develop. Since sebum supports the growth of some microorganisms, pimples or boils can then develop. The color of blackheads is due to oxidized oil, not dirt.

In the linings of body cavities, such as the mouth, nasal passages, and urinary, genital, and gastrointestinal tracts, the outer protective barrier is specialized in various ways. For example, some cells are ciliated. Others secrete mucus from glands beneath this outer layer of epithelial cells—hence the name *mucous membrane* (*mucosa*). In the respiratory system, the mucous layer traps particles, including microorganisms, and the ciliary movement sweeps them upward out of the body (see Figure 15-3). Mucous membranes are frequently acidic, which tends to limit their microbial populations. The eyes are mechanically washed by tears, and the enzyme lysozyme in the tears destroys the cell walls of gram-positive bacteria and some gram-negative bacteria.

Normal Flora of the Skin

Although the skin is generally inhospitable to most microorganisms, it supports the growth of certain microbes, which are established as part of the normal flora. For example, some anaerobic bacteria of the genus *Propionibacterium* (prō-pē-on-ē-bak-ti′rē-um) live in hair follicles, where their growth is supported by secretions from oil glands. These same bacteria produce an acid (propionic acid) that helps the pH of the skin remain between 3 and 5. This acid also has a bacteriostatic effect on many other microorganisms and prevents their colonization. On superficial skin surfaces, certain aerobic bacteria can metabolize fatty acids in the presence of oxygen. This activity contributes to the skin's normal level of fatty acids, which in turn contributes to a pH that is unfavorable to more harmful microbes.

Microorganisms that secrete antimicrobial substances are part of the normal flora. Because these microorganisms can resist salt and drying, they have the advantage over pathogens that attempt to colonize the skin. Although vigorously scrubbing the skin with soap and water or disinfectants will rid it temporarily of most surface bacteria, microorganisms in hair follicles and sweat glands will soon reestablish the skin's normal floral population.

Even though the skin may be considered a single organ, its normal flora vary in different regions. For example, the flora of the face reflect those of the oropharynx (portion of the throat behind the mouth), whereas the flora of the anal region are influenced by microorganisms of the lower gastrointestinal tract.

Scanning electron micrographs show that bacteria tend to clump on the skin (or other areas) in small groups (Figure 20-2). Many of these clumps consist of fairly large, spherical, gram-positive bacteria of the genera *Staphylococcus* and *Micrococcus* (mī-krō-kok′kus). Both genera are capable of producing antimicrobial substances that prevent colonization of the skin by pathogens and help maintain the balance of skin flora. Also part of the skin's normal flora are gram-positive pleomorphic rods called *diphtheroids*. Some diphtheroids, such as *Propionibacterium acnes* (ak′nēz), are typically anaerobic and inhabit hair follicles. Others, such as *Corynebacterium xerosis* (ze-rō′sis), are aerobic and inhabit the surface of the skin. A yeast belonging to the genus *Pityrosporum* (pit-i-ros′pô-rum) is capable of growing on oily skin secretions and is a frequent inhabitant of the skin.

Figure 20-2   A section of gum showing resident bacteria (×800).

Diseases of the Skin

Many important diseases manifest themselves by such symptoms as rashes, vesicles, and other lesions that mainly affect the skin. Often the focus of infection is elsewhere in the body, but it is convenient to classify these diseases by the organ most obviously affected, the skin. Table 20-1 on p. 516 summarizes the more important diseases associated with the skin.

Our discussion is primarily concerned with bacterial, viral, and fungal skin infections. The box on p. 513 describes a skin disease, scabies, caused by a parasite.

Bacterial Diseases of the Skin

Two genera of bacteria, *Staphylococcus* and *Streptococcus* (commonly called staphylococci and streptococci), are frequent causes of skin-related diseases and merit special discussion. We will also discuss these bacteria in later chapters in relation to other organs and conditions. Superficial staphylococcal and streptococcal infections of the skin are very common. The bacteria frequently come into contact with the skin and have adapted fairly well to the physiological conditions there. Both genera also produce invasive enzymes and damaging toxins that contribute to the disease process.

Staphylococcal Skin Infections

Staphylococci are spherical gram-positive bacteria about 0.5 to 1.5 μm in diameter. They tend to form in irregular clusters like grapes because the cells divide at random points about their circumference and the daughter cells do not completely separate from each other. (See Figures 4-1c and 10-10.)

Staphylococcus aureus is the most pathogenic of the staphylococci. Typically, it forms golden yellow colonies. Almost all pathogenic strains of *S. aureus* produce **coagulase**, an enzyme that coagulates (clots) the fibrin in blood. It is possible that a fibrin clot might protect the microorganisms from phagocytosis and isolate them from other defenses of the host. There is a high correlation between the bacterium's ability to form coagulase and its production of damaging toxins, several of which may injure tissues. Staphylococci also produce *leukocidin*, which destroys phagocytic leukocytes, and *exfoliative toxin*, which is responsible for scalded skin syndrome, to be discussed shortly. Some staphylococcal toxins, *enterotoxins*, affect the gastrointestinal tract and will be discussed in Chapter 24 with diseases of the digestive system.

S. aureus is a very common problem in the hospital environment. Because *S. aureus* is carried on the skin of patients, hospital personnel, and hospital visitors, the danger of infection of surgical wounds and other breaks in the skin is very high. Moreover, such infections are difficult to treat because *S. aureus* is exposed to so many antibiotics that it quickly becomes resistant to them. At one time, this organism was almost uniformly extremely susceptible to penicillin, but now only about 10% of *S. aureus* strains are sensitive. Most of the microorganism's resistance derives from the production of penicillinases. There are some penicillinase-resistant forms of penicillin (see Chapter 19).

S. aureus often enters the body through a natural opening in the skin barrier, the hair follicle's passage through the epidermal layer. Infections of hair follicles, or **folliculitis**, often occur as **pimples**. The infected follicle of an eyelash is called a **sty**. A more serious hair follicle infection is the **furuncle (boil)**, which is a type of **abscess**, a localized region of pus surrounded by inflamed tissue. Antibiotics do not penetrate well into abscesses, which are difficult to treat. Draining pus from the abscess is frequently a preliminary step to successful treatment.

When the body fails to wall off a bacterial infection, neighboring tissue can be progressively invaded. The extensive damage is called a **carbuncle**, a hard, round, deep inflammation of tissue under the skin. At this stage of infection, the patient usually exhibits the symptoms of generalized illness with fever.

Staphylococci are the primary cause of a very troublesome problem in hospital nurseries, **impetigo of the newborn**. Symptoms of this disease are thin-walled vesicles on the skin that rupture and later crust over. To prevent outbreaks, which can reach epidemic proportions, hexachlorophene-containing skin lotions are commonly prescribed (Chapter 7). However, if these are used excessively, neurological damage may result.

Staphylococcal infections always carry the risk that the underlying tissue will become infected or that the infection will enter the bloodstream. The circulation of toxins, such as those produced by staphylococci, is called **toxemia**. One such toxin, which is produced by staphylococci lysogenized by certain phage types, causes **scalded skin syndrome**. (See Chapters 12 and 14 for discussion of lysogeny and phage conversion.) This condition is first characterized by a lesion around the nose and mouth. The lesion develops rapidly into a bright red area, and, within 48 hours, the skin of the palms and soles peels off in sheets when it is touched (Figure 20-3). The scalded skin syndrome is frequently observed in children under the age of two, especially in newborns, as a complication of staphylococcal infections. These patients are seriously ill, and vigorous antibiotic therapy is required.

Figure 20-3 Scalded skin syndrome. A staphylococcal infection has caused the skin on the foot of this newborn child to peel off in sheets.

The scalded skin syndrome is also characteristic of many cases of toxic shock syndrome (TSS). In this potentially life-threatening condition, fever, vomiting, and a sunburnlike rash are followed by shock, a sudden drop in blood pressure. The cause of TSS is most often a staphylococcal growth associated with the use of vaginal tampons; the correlation is especially high for cases in which the tampons were retained too long. Staphylococcal toxins (more than one is probably involved) enter the bloodstream from the bacterial growth site in and around the tampon, and their circulation causes the symptoms.

There is good evidence that the absorption of magnesium by the fibers of the tampon is a factor in toxic shock syndrome. As the concentration of the metal in the vagina is reduced, the staphylococci grow more slowly but are stimulated to produce toxin. (A similar mechanism involving a deficit of iron is responsible for the production of the toxin that causes diphtheria.) This syndrome has also been observed occasionally in users of contraceptive vaginal sponges and even in men with staphylococcal infections.

In the years 430 to 427 B.C. the Greek historian Thucydides reported a disease that devastated the population of Athens. He told of "men dying like sheep" and vividly described their symptoms, which included nausea, diarrhea, coughing, and fever. Until recently, no known disease seemed to fit Thucydides' description. In 1985, medical historians suggested that the symptoms and mortality rate could have been caused by influenza complicated by staphylococcal pneumonia, and they suggested that *S. aureus* was the probable pathogen. Only three months after publication of their article, ten cases of illness resembling toxic shock syndrome occurred in Minnesota in persons suffering from influenza. Six of the ten died, and *S. aureus* was isolated from their respiratory secretions, indicating staphylococcal pneumonia. This modern association between staphylococcal pneumonia and influenza is now known in medicine as the *Thucydides syndrome.*

Streptococcal Skin Infections

Like staphylococci, **streptococci** are gram-positive spherical bacteria. Unlike staphylococci, the streptococcal cells grow in chains. Prior to division, the individual cocci elongate on the axis of the chain, and then the cells divide into pairs. When the dividing pairs do not separate, chaining occurs (see Figures 4-1a and 10-11). The bridges between the individual cocci in the chain consist of cell wall material that has not cleaved.

Streptococci are classified as facultative anaerobes in *Bergey's Manual.* However, because their metabolism is strictly fermentative (they cannot use oxygen), they might better be regarded as aerotolerant anaerobes. Unlike either aerobic or facultative anaerobic bacteria, they do not produce catalase (Chapter 6).

Many nonpathogenic streptococci are commonly found inhabiting the mouth, gastrointestinal tract, and upper respiratory system, but some streptococci are responsible for skin infections. As streptococci grow, they secrete several toxins and enzymes into the growth medium. Among these toxins are **hemolysins**, which damage red blood cells. Depending on the type of destruction caused by the hemolysins, streptococci can be divided into α-hemolytic, β-hemolytic, and γ-hemolytic streptococci (Chapter 10). Of all pathogens, β-hemolytic streptococci are among the most susceptible to antimicrobial drugs. Penicillin is the drug of choice, although erythromycin is frequently used.

β-hemolytic streptococci are further differentiated into a number of immunologic groups, designated A through O, according to the antigenic carbohydrates in their cell walls. As causes of human disease, β-hemolytic group A streptococci are the most important. β-hemolytic group A streptococci can be further subdivided into over 55 immunologic types according to the antigenic properties of the *M protein* found in some strains (Figure 20-4). This protein is found external to the cell wall on hairlike fimbriae that are shorter and finer than the pili that are found on the surface of many gram-negative bacteria (Figure 4-8). The M protein has antiphagocytic properties that contribute to the strain's pathogenicity. It also appears to aid the bacteria in adhering to and colonizing on mucous membranes.

In addition to hemolysins and M proteins, a number of other substances contribute to the pathogenicity of β-hemolytic group A streptococci. These substances include *erythrogenic toxin* (responsible for the scarlet fever rash), *deoxyribonucleases* (enzymes that degrade DNA), *NADase* (an enzyme that breaks down NAD), *streptokinases* (enzymes that dissolve blood clots), *hyaluronidase* (an enzyme that dissolves hyaluronic acid, the cementing substance of connective tissue), and *leukocidins* (enzymes that kill white blood cells).

β-hemolytic group A streptococci cause a wide variety of diseases, a number of which will be discussed in later chapters. The most common species of this group, *Streptococcus pyogenes*, may be implicated in some cases of impetigo. Although impetigo of the newborn is caused by staphylococci, **impetigo** is also common in children of toddler and grade-school age and is likely in these cases to be caused by strepto-

M protein on fimbriae

Fimbriae lacking M protein

cocci. A superficial skin infection, impetigo is characterized by isolated pustules (small, round elevations containing pus) that become crusted and rupture (Figure 20-5). The disease is spread largely by contact, and the bacteria penetrate the skin through some minor abrasion or wound whose scab has been prematurely removed. Staphylococci can also be found in these lesions but are probably only secondary invaders. Fortunately, the condition is seldom serious. The drugs of choice are penicillin and erythromycin.

When streptococcal infections affect the dermis, they cause **erysipelas**. In this disease, the skin erupts into reddish patches that enlarge and thicken and swell at the margins (Figure 20-6). High fever is common. The reddening is caused by toxins produced by the streptococci as they invade new areas. Usually, the skin outbreak is preceded elsewhere in the body

Figure 20-4 Electron micrographs showing portions of group A β-hemolytic streptococci. The cell on the left carries M protein on hairlike surface fimbriae; the cell on the right lacks M protein (×90,000).

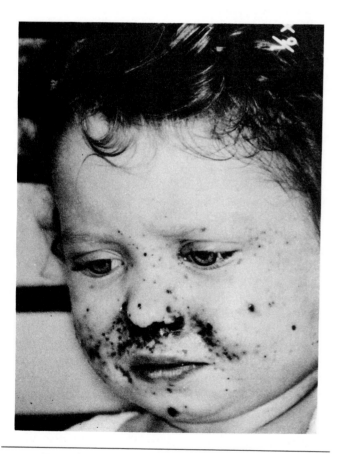

Figure 20-5 Impetigo, caused by streptococci, on the face of a three-year-old child. Note the characteristic pustules, small elevations containing pus.

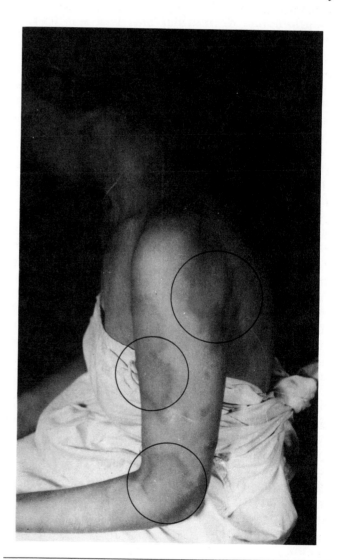

Figure 20-6 Erysipelas caused by group A β-hemolytic streptococci. The dark patches of skin circled on the photograph are areas that have been reddened by streptococcal toxins.

by a streptococcal infection, such as a streptococcal sore throat. It is not certain if the skin is infected from an external invasion or via the lymphatic system. Erysipelas is most likely to occur in the very young and the very old. The drugs of choice for the treatment of erysipelas are penicillin and erythromycin.

Infections by Pseudomonads

Pseudomonads are aerobic gram-negative rods that are widespread in soil and water. Able to survive in any moist environment, they can grow on traces of even unusual organic matter, such as soap films or cap liner adhesives, and are resistant to many antibiotics and disinfectants. The most prominent species is *Pseudomonas aeruginosa*.

Pseudomonads frequently cause outbreaks of **Pseudomonas** dermatitis. This is a self-limiting rash of about two weeks' duration often associated with swimming pools and pool-type saunas and hot tubs. When large numbers of people use these facilities, the alkalinity rises, and the chlorines become less effective; at the same time, the concentration of nutrients that support growth of pseudomonads increases. Hot water facilitates the entry of bacteria into hair follicles by causing the follicles to dilate. Competition swimmers are often troubled with **otitis externa**, or swimmer's ear, a pseudomonad infection of the external ear canal leading to the eardrum.

P. aeruginosa produces several exotoxins that account for much of its pathogenicity. It also has an endotoxin. Except for superficial skin infections and otitis externa, infection by *P. aeruginosa* is rare in healthy individuals. However, it often causes respiratory infections in persons already compromised by immunologic deficiencies (natural or drug induced) or by chronic pulmonary disease, especially cystic fibrosis. (Respiratory infections are discussed in Chapter 23.)

P. aeruginosa is also a very common and serious opportunistic pathogen in burn patients, particularly those with second- and third-degree burns. In such cases, infection may produce blue-green pus caused by the bacterial pigment *pyocyanin* (sometimes spelled *pyocyanine*). Of major concern in many hospitals is the ease with which *P. aeruginosa* is carried by flowers or plants sent by well-wishers. Because the organism can be transmitted from flowers to burn patients or individuals with fresh surgical incisions, many hospitals do not permit such patients to receive flowers.

Pseudomonads are somewhat resistant to many of the commonly used disinfectants (Chapter 7) and antibiotics (Chapter 19). Antibiotic resistance by pseudomonads is still a problem, but in recent years a number of new antibiotics have been developed, so the number of antimicrobials available for chemotherapy of these infections is not as restricted as it once was. Two antibiotics commonly used against pseudomonad infections are gentamicin and carbenicillin, often used in combination. Silver sulfadiazine is very useful in the treatment of burn infections by *P. aeruginosa*.

Acne

Acne is probably the most common skin disease of humans. More than 65% of all teenagers have the problem to some degree. In the United States, about 350,000 persons each year are estimated to suffer from severe acne, called **cystic acne**, that produces inflamed cysts and subsequent scarring of the face and upper body. Although the occurrence of acne decreases after the teen years, the scarring from these severe cases often remains.

Acne begins when channels for passage of sebum to the skin surface are blocked. The sebum accumulation leads to the formation of the familiar whiteheads and, later, blackheads. In some persons, the accumulation of sebum ruptures the lining of the hair follicle. Bacteria, especially *Propionibacterium acnes*, a diphtheroid commonly found on the skin, become involved at this stage. By metabolizing the sebum, *P. acnes* forms free fatty acids that cause an inflammatory response by the body. It is this inflammation that leads to tissue damage and subsequent acne scars. Picking or scratching the lesions—or even having tight collars or other clothing in contact with lesions—increases the incidence of scar formation. Cosmetics frequently aggravate the condition, but diet—including consumption of chocolate—has been demonstrated to have no significant effect on the disease.

Topical applications of preparations containing benzoyl peroxide are often useful, and antibiotics such as tetracycline have also been used. Severe cases should receive a dermatologist's treatment rather than household remedies. The most important recent development in treatment of severe cystic acne is isotretinoin (Accutane®). This drug, taken orally, inhibits sebum formation, and dramatic improvement follows. Isotretinoin is not recommended for use against the usual mild cases of acne because of its side-effects; for one thing, it is *teratogenic* (causes fetal damage) if taken by a pregnant woman for even a few days.

Viral Diseases of the Skin

Warts

Warts are caused by viruses that stimulate an uncontrolled but benign growth of skin cells. It was long known that warts can be transmitted from one person to another by contact, even sexually, but it was not until 1949 that viruses were identified in wart tissues. About 25 different types of papillomaviruses are now known to cause different types of warts, often of greatly differing appearance. After infection, there is an incubation period of several weeks before the warts appear.

The most common medical treatments for warts are applying extremely cold liquid nitrogen to them (cryotherapy), drying them with an electric current (electrodesiccation), or burning them with acids. Burning with acids is also the basis of some home remedies. Warts that do not respond to any other treatments, especially genital warts, can now be treated with injected interferon. An encouraging rate of success has been reported. Although warts themselves are not a form of cancer, some skin and cervical cancers seem to be associated with papillomaviruses.

Smallpox (Variola)

It is estimated that, during the Middle Ages, 80% of the population of Europe could expect to contract **smallpox** during their lives. Colonial Boston, with a total population of only 10,700, had 855 deaths from 5889 cases (14% mortality) in about 10 months during 1721 and 1722. Those who recovered from the disease carried disfiguring scars. The disease was even more devastating to Native Americans, who had no previous exposure and thus no resistance.

Smallpox is caused by a poxvirus known as the smallpox (variola) virus. There are two basic forms of this disease—**variola major**, which has a mortality rate of 20% or higher, and **variola minor**, which has a mortality rate of less than 1%. Recovery from one form of the disease produces effective immunity against both forms.

The transmission and progression of smallpox roughly parallel other viral skin diseases. Transmitted first by the respiratory route, the viruses infect many internal organs before their eventual movement into the bloodstream (**viremia**) leads to infection of the skin and the production of more recognizable

symptoms. The growth of the virus in the epidermal layers of the skin causes lesions (Figure 20-7).

Smallpox was the first disease to which immunity was artificially induced (see Chapters 1 and 17) and the first to be eradicated from the human population. The last outbreak of smallpox in the United States, which led to the mass inoculation of several million persons, was initiated by a man who succumbed in New York City in 1947. He probably acquired the disease in Mexico. It is believed that the last victim of a natural case of smallpox was one who recovered from variola minor in 1977 in Somalia, Africa.

The eradication of smallpox was possible because there are no animal host reservoirs for the disease. Once an effective vaccine became available, eradication was accomplished by a concerted vaccination effort coordinated by the World Health Organization. In some nations, such as India, rewards were given to persons who reported cases of smallpox, and all persons in contact with the infection were then vaccinated.

Routine vaccination for smallpox was discontinued in the United States in 1971, a number of years before worldwide eradication of the disease. Until

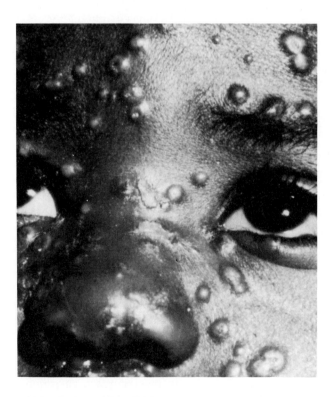

Figure 20-7 Lesions of smallpox. In some severe cases, the lesions nearly run together.

1971, several persons, usually infants, had died each year from the vaccine. The vaccination introduces live vaccinia virus, which multiplies to produce an ulcerative lesion at the site of vaccination. From there, the virus can be spread throughout the body. This risk had been acceptable as long as smallpox presented a credible threat, particularly when it could be imported by air travelers. But as the disease became less and less common in the world, the risk of vaccination could no longer be justified.

More recently, the smallpox virus collections in laboratories have been the most likely sources of new infections. The risk of such infection is not merely a hypothetical concern as there have already been several laboratory-associated infections, one of which caused death.

Chickenpox (Varicella) and Shingles (Herpes Zoster)

Chickenpox (varicella) is a relatively mild childhood disease. After gonorrhea, it is the second most common reportable infectious disease in the United States. It is probably greatly underreported, and more than 2 million cases probably occur each year in the United States. Disease summaries of the Centers for Disease Control show that about 100 deaths per year, usually from encephalitis (infection of the brain), are attributed to chickenpox.

Chickenpox is acquired by infection of the respiratory system, and the infection localizes in skin cells after about two weeks. The infected skin is vesicular for three to four days. During that time, the vesicles fill with pus, rupture, and form a scab before healing (Figure 20-8a). At any particular time, different vesicles may be in different developmental stages. The vesicular rash can also appear in the mouth and throat. When chickenpox occurs in adults—which is not frequent because the high incidence in childhood grants immunity to most persons—it is a more severe disease with a significant mortality rate.

Reye's syndrome is an occasional severe complication of chickenpox, influenza, and some other viral diseases. After the initial infection has receded, the patient persistently vomits and exhibits signs of brain dysfunction. Coma, fatty degeneration of the liver, and death can follow. Death or brain damage in survivors is from brain swelling, which prevents blood circulation. At one time, the death rate of reported cases approached 90%, but this rate has been declining with improved care and is now 10% or lower when the disease is recognized and treated in time.

(a)

(b)

Figure 20-8   Typical lesions associated with (a) chickenpox and (b) shingles.

Reye's syndrome affects children and teenagers almost exclusively. The use of aspirin to lower fevers in chickenpox and influenza increases the chances of acquiring Reye's syndrome.

The cause of chickenpox, the varicella-zoster virus, is a herpesvirus. A characteristic of herpesviruses is *latency*. Following a primary infection, the virus enters the peripheral nerves and moves to a central nerve ganglion, where it persists as viral DNA. The DNA may or may not be transcribed, but in any event it persists, latent, indefinitely. In the case of chickenpox, the location of the latent virus is in the dorsal root ganglion near the spine. Later, perhaps as long as decades later, the virus may be reactivated. The trigger can be stress or simply the lower immune competence associated with aging. The virions produced by the reactivated DNA move along the peripheral nerves to the cutaneous sensory nerves of the skin (see Figure 21-1), where they cause a new outbreak of the virus in the form of **shingles** (herpes zoster). See Figure 20-9.

In shingles, vesicles similar to those of chickenpox occur but are localized in distinctive areas. Typically, there is a distribution about the waist, although facial shingles and infections of the upper chest also occur (Figure 20-8b). The infection follows the distribution of the affected cutaneous sensory nerves and is usually limited to one side of the body at a time since these nerves are unilateral. Occasionally, such nerve infections can result in nerve damage that impairs vision or even causes paralysis. Severe pain is also frequently reported.

Shingles is simply a different expression of the virus that causes chickenpox; it expresses differently because the patient, having had chickenpox, now has partial immunity to the virus. Exposing children to cases of shingles has led to their contracting chickenpox. Shingles seldom occurs in persons under age 20, and by far the highest incidence is in the elderly population.

Immunocompromised patients are in serious danger from reactivation of the latent varicella-zoster virus—a mortality rate of 17% is common. In such cases, the antiviral drug *acyclovir* has proven helpful. A live, attenuated varicella vaccine is being tested. If finally approved, the vaccine will probably be administered at about age 15 months, with the present measles–mumps–rubella (MMR) series.

Herpes Simplex

Serologic surveys show that about 90% of the population of the United States has been infected with the herpes simplex virus. The initial infection usually occurs in infancy. Frequently, this infection is subclinical, but perhaps as many as 15% of the cases develop lesions known as **cold sores** or **fever blisters** (Figure 20-10). Usually occurring in the oral mucous membrane, these lesions heal as the infection subsides but recur when the infection strengthens again. Recurrences are usually associated with some trauma—excessive exposure to UV radiation from the sun is a frequent cause, as are the hormonal changes associated with menstrual periods or emotional upsets. These infections are caused by *herpes simplex type 1 virus*, which is transmitted primarily by oral or respiratory routes. An important complication is herpetic keratitis (discussed later in this chapter), in which the cornea of the eye becomes infected. Between recurrences, herpes simplex type 1 viruses are latent in the trigeminal nerve ganglia communicating between the face and the central nervous system.

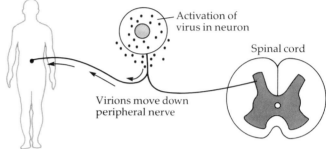

(a) Initial infection, chicken pox (varicella)

(b) Recurrence of infection, shingles (herpes zoster)

Figure 20-9 Latency of typical herpesviruses, varicella-zoster. After an initial infection of the varicella virus causes chickenpox **(a)**, the virus moves to a central nerve ganglion, where it remains latent indefinitely. Age-related immune system depression or stress can trigger a reactivation of the virus, causing shingles **(b)**.

Figure 20-10 Cold sore blisters caused by herpes simplex virus.

A different infection by a very similar virus, *herpes simplex type 2 virus*, is transmitted primarily by sexual contact. Type 2 is differentiated from type 1 by its antigens and by its effect on cells in tissue cultures. We will discuss this infection in Chapter 25, in our discussion of diseases of the urinary and genital systems. Herpes simplex type 2 viruses causing genital herpes are latent in the sacral ganglia near the base of the spine.

Very rarely, either type of the herpes simplex virus may spread to the brain, causing **herpes encephalitis**. Infections by the type 2 virus are more serious, with a fatality rate of about 70% if untreated. Only about 10% of survivors can expect to lead normal lives. However, this is one of the rare viral diseases for which a chemotherapeutic treatment is reasonably effective. Tests have shown that acyclovir is almost twice as effective as vidarabine, the previously recommended drug. Even so, the mortality in the tests was still 28%, and only 38% of the survivors escaped serious neurological damage.

Measles (Rubeola)

Measles is an extremely contagious disease that is spread by the respiratory route. Because a person with measles is infectious before symptoms appear, quarantine is not an effective measure of prevention. The causative agent is a paramyxovirus called the measles virus, which resembles the virus that causes influenza and mumps.

Humans are the only reservoir for measles in most parts of the world, although monkeys are also susceptible. Therefore, measles can potentially be eradicated, much as smallpox was. Since the licensing of a vaccine in 1963, the number of measles cases has declined from more than 400,000 cases per year to a record low of 1497 cases in 1983 (Figure 20-11). Unfortunately, the number of cases has been slowly creeping upward since then. Despite laws requiring immuni-

zation for school entry, the immunization rate is low in some socioeconomic groups, and about 40% of measles cases now occur in preschool children in these groups. In addition, although the measles vaccine is about 95% effective, cases continue to occur among those who do not develop or retain good immunity. Some of these infections are caused by contact with infected persons who come from outside the United States.

A previously developed vaccine provided poor immunity to the virus, and the practice of vaccination before the age of one year proved ineffective because of interference from inherited maternal antibodies. The new vaccine is now administered at age 15 months, frequently in combination with MMR vaccine.

The development of rubeola is similar to that of smallpox and chickenpox. Infection begins in the upper respiratory system. After an incubation period of 10 to 12 days, symptoms develop resembling those of a common cold—sore throat, headache, and cough. Shortly thereafter, a papular rash appears on the skin (Figure 20-12). Lesions of the oral cavity include the diagnostically useful *Koplik spots* (tiny red patches with central white specks) on the oral mucosa opposite the molars.

Measles is an extremely dangerous disease, especially among very young and elderly individuals. It is frequently complicated by middle ear infection or pneumonia caused by the virus itself or by secondary bacterial infection. Encephalitis strikes approximately 1 in 2000 measles victims; its survivors are often left with permanent brain damage. As many as 1 in 3000 cases is fatal, mostly in infants. The virulence of the virus seems to vary with different epi-

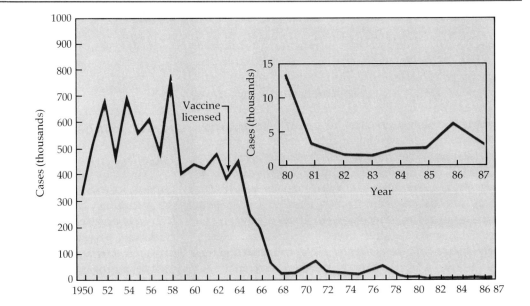

Figure 20-11 Number of reported measles cases in the United States, 1950 to 1987. Note the sharp decline in cases after introduction of the measles vaccine in 1963. Inset: The increase in measles cases in the last few years. [Source: CDC. Summary of Notifiable Diseases. *MMWR* 35:55 (1986); *MMWR* 36:52 (1/8/88).]

Figure 20-12 The rash of small raised spots typical of measles (rubeola).

demic outbreaks. Complications such as encephalitis occur, if at all, about a week after the rash appears.

Measles viruses can persist in the recovered patient. Apparently, a mutant form of the virus, which has lost the ability to complete its normal life cycle, remains in the patient's cells. The presence of this defective measles virus eventually causes **subacute sclerosing panencephalitis (SSPE)**. This disease, which usually occurs in children or young adults with a history of measles, involves a sudden rapid and fatal degeneration of the nervous system. As a result of the decline in the number of cases of measles, SSPE has almost disappeared in the United States.

Rubella

Rubella, or German measles, is caused by a togavirus (rubella virus). It is a much milder disease than rubeola (measles) and often goes undetected. A rash of small red spots and a light fever are the usual

MICROBIOLOGY HIGHLIGHTS

Scabies—An Itchy Infection by a Parasitic Arthropod

Scabies is a disease of the human skin caused by the mite *Sarcoptes scabiei*, a parasitic arthropod (see the figure). The disease is common among school children and is also found in adults. Sometimes it occurs as a nosocomial infection in hospital personnel treating patients with symptoms described as pruritic dermatitis.

The fingers, wrists, and elbows are the most frequent sites of infection. The mites burrow into the skin and fill the tunnel with their eggs and feces. The eggs hatch, and new mites mature, mate, and lay more eggs, perpetuating the life cycle. The symptoms of scabies are the result of hypersensitivity reactions to the mites. Symptoms first occur two to six weeks after the initial infection. The main symptom is itching, especially when the skin is warm (for example, when in bed at night). Red, raised lesions (erythematous papules) develop, which may become infected with bacteria through scratching. More advanced, chronic cases may result in generalized eczema.

Diagnosis is made by examination of the skin with a 10× hand lens. Burrows can sometimes be seen, and the mites can be

Female *Sarcoptes scabiei*.

picked out with a needle for microscopic examination. Alternatively, scrapings from lesions may be examined microscopically for mites. Scabies is treated by topical application of gamma benzene hexachloride (Kwell®). Clothing, bedding, and other personal objects that might contain mites must be thoroughly cleaned. Scabies is transmitted by direct contact with infected persons or by fomites carrying female mites.

symptoms (Figure 20-13). Complications are rare, especially in children, but encephalitis occurs in about 1 case in 6000, mostly in adults. Transmission is airborne, and incubation of two to three weeks is the norm. Recovery from clinical or subclinical cases appears to give a firm immunity.

The seriousness of rubella was not appreciated until 1941, when the association was made between serious birth defects and maternal infection during the first trimester (three months) of pregnancy—the **congenital rubella syndrome**. If a pregnant woman contracts the disease during this time, there is about a 35% incidence of serious fetal damage, including deafness, eye cataracts, heart defects, mental retardation, and death. Fifteen percent of babies with congenital rubella syndrome die during their first year. The last major epidemic of rubella in the United States was during 1964 and 1965. About 20,000

Figure 20-13 The rash of red spots characteristic of rubella. The spots are not raised above the surrounding skin.

severely impaired children from this epidemic are still alive.

It is very important to establish whether women of childbearing age are immune to rubella. Accurate diagnosis always requires laboratory tests; histories alone are very unreliable. Serum antibody against rubella can be assayed by hemagglutination inhibition. ELISA tests (see Chapter 17 and Microview C, p. C-4) are commercially available and provide comparably accurate results.

In 1969, a rubella vaccine was approved for use. It is important to know if the immunity conferred by the vaccine will be maintained from childhood through a woman's childbearing years. Follow-up studies indicate that there has been little decline in antibody levels in the first children to receive the vaccine, who are now entering their childbearing years. However, the question of whether vaccination grants permanent immunity will not be answered for some time.

There has been concern about the possibility of fetal damage from pregnant women receiving the rubella vaccine. In hundreds of cases in which women were vaccinated three months before or three months after their presumed dates of conception no case of congenital rubella syndrome defects has occurred. Nevertheless, the vaccine is still not recommended for pregnant women.

Fungal Diseases of the Skin

The skin is most susceptible to microorganisms that are able to resist high osmotic pressure and low moisture. It is not surprising, therefore, that fungi cause a number of skin disorders. Any fungal infection of the body is called a **mycosis**.

Superficial Mycoses

Superficial mycoses, or **piedras**, are limited to the external shafts of hair. The fungi form black or white nodules that often give the hair a gritty texture. Superficial mycoses are not very common in the United States.

Cutaneous Mycoses

Fungi that colonize the hair, nails, and the outer layer (stratum corneum) of the epidermis (see Figure 20-1) are called **dermatophytes**, and their infections are called **dermatomycoses**. Dermatophytes grow on the keratin present in those locations, causing infections called **tineas**, or **ringworms**. Some of these fungal infections are almost asymptomatic and of interest mainly for cosmetic reasons. An example is **tinea nigra**, caused by *Cladosporium werneckii* (kla-dō-spô′rē-um wėr-ne′kē-ē), in which brown or black patches appear on the palms of the hands.

The name *ringworm* arose from the ancient Greek belief that the infections, which tend to expand circularly, were caused by a worm. The Romans incorrectly associated ringworm with lice, and the term *tinea* is derived from the Latin for insect larvae. **Tinea capitis**, or ringworm of the scalp, is fairly common among elementary school children and can result in bald patches (Figure 20-14a). It is usually transmitted by contact with fomites. Dogs and cats are also frequently infected with fungi that cause ringworm in children. Ringworm of the groin, or jock itch, is known as **tinea cruris**, and ringworm of the feet, or athlete's foot, is known as **tinea pedis** (Figure 20-14b).

Three genera of fungi are involved in cutaneous mycoses. *Trichophyton* (trik-ō-fī′ton) can infect hair, skin, or nails; *Microsporum* (mī-krō-spô′rum) usually involves only the hair or skin (see Microview A, Photo 19); *Epidermophyton* (ep-i-dėr-mō-fī′ton) affects only the skin and nails. The topical drug of choice for tinea infections is usually miconazole or clotrimazole. Other topical agents found in nonprescription remedies are tolnaftate and zinc undecylenate. An oral antibiotic, griseofulvin, is often useful in these infections because it can localize in keratinized tissue.

Subcutaneous Mycoses

Subcutaneous mycoses are more serious than superficial or cutaneous mycoses. Cutaneous fungi do not seem to be able to penetrate past the dead cellular material of the stratum corneum, perhaps because they cannot obtain sufficient iron for growth in the living cells of the epidermis and the dermis. Subcutaneous mycoses usually originate in a small wound that allows common soil fungi to penetrate into the subcutaneous tissues, where they grow and form a nodule just under the skin. The nodule then ulcerates and spreads via the lymphatic system; a series of subcutaneous nodules forms along the lymphatic vessels.

In one type of subcutaneous mycosis, **chromomycosis**, the lesions are pigmented (*chromo* means color) dark brown. This mycosis is caused by one of several black molds, including *Phialophora verruscosa* (fē-ä-lō′fô-rä ver-rü-sko′sä) and *Fonsecaea pe-*

(a)

(b)

Figure 20-14 Dermatomycoses. (a) A severe case of ringworm on the side of a child's head (tinea capitis). **(b)** Ringworm of the foot, or athlete's foot (tinea pedis).

drosoi (fon-se′kē-ä pe-drō′sō-ē). It is found worldwide but mostly in the tropics. Fluctyosine, given orally, can be an effective treatment. Another subcutaneous mycosis is **maduromycosis**, which is caused by the fungus *Allescheria boydii* (al-lesh-er′ē-ä boi-dē-ē). Maduromycosis may be called a **mycetoma**, or fungal tumor. It destroys subcutaneous tissues and progresses slowly, eventually causing serious deformities. Polyenes, which have not yet been used extensively, may be of value in treating mycetomas, whose causative agent is susceptible in vitro. Surgically draining the mycetoma is also important.

TABLE 20-1   *Diseases Associated with the Skin*

	Causative Agent	Treatment
Bacterial Diseases		
Impetigo	*Staphylococcus aureus;* occasionally, *Streptococcus pyogenes*	Penicillin (for *Streptococcus* infections only); erythromycin
Erysipelas	*Streptococcus pyogenes*	Penicillin, erythromycin
Otitis externa and burn infections	*Pseudomonas aeruginosa*	Polymyxins, gentamicin, carbenicillin, and silver sulfadiazine
Acne	*Propionibacterium acnes*	Benzoyl peroxide, tetracyclines, isotretinoin
Viral Diseases		
Warts	Papillomaviruses	May be removed by liquid nitrogen cryotherapy, electrodesiccation, or acids
Smallpox (variola)	Poxvirus (smallpox virus)	None
Chickenpox (varicella)	Herpesvirus (varicella-zoster)	Vidarabine for immunocompromised patients
Shingles (zoster)	Herpesvirus (varicella-zoster)	Vidarabine for immunocompromised patients
Herpes simplex	Herpesvirus (herpes simplex type 1)	Acyclovir may modify symptoms; also see herpetic keratitis, Table 20-2
Measles (rubeola)	Paramyxovirus (measles virus)	None
German measles (rubella)	Togavirus (rubella virus)	None
Fungal Diseases		
Tinea nigra	*Cladosporium werneckii*	Griseofulvin
Ringworm (tinea)	*Microsporum, Trichophyton, Epidermophyton* species	Griseofulvin, miconazole, clotrimazole, tolnaftate
Chromomycosis	*Phialophora verruscosa, Fonsecaea pedrosoi*	Flucytosine
Maduromycosis	*Allescheria boydii*	Polyenes might be useful; surgical drainage
Candidiasis	*Candida albicans*	Topically with miconazole, clotrimazole, nystatin

Candidiasis

The bacterial flora of the mucous membranes in the genitourinary tract and mouth usually suppress the growth of such fungi as *Candida albicans* (see Figure 20-15a and b and Microview A, Photos 18 and 21). Because the fungus is not affected by antibacterial drugs, it sometimes overgrows mucosal tissue when the normal bacterial flora are suppressed by antibiotics. Changes in the normal mucosal pH may have the same effect. Such overgrowths by *C. albicans* are called **candidiasis**. Newborn infants, whose normal flora have not become established, often suffer from candidiasis of the oral cavity, called **thrush** (Figure 20-15c). *C. albicans* is also a very common cause of vaginitis (Chapter 25) and frequently is a problem on the mucosal tissues of the elderly and those debilitated by such diseases as diabetes and cancer. These superficial infections by *C. albicans* are usually treated with topical applications of miconazole, clotrimazole, or nystatin. If candidiasis becomes systemic, as can happen in immunosuppressed or immunodeficient individuals, fulminating disease and death can result.

Pseudo-hyphae

Blastospores

Chlamydo-spores

(a)

Figure 20-15 Candidiasis. (a) Photomicrograph of *Candida albicans*. Note the spherical chlamydospores (thick-walled, nonsexual spores formed from a hyphal cell), the smaller blastospores (spores produced by budding), and the pseudohyphae. *Candida* does not form filamentous hyphae but does form long cells that resemble hyphae. **(b)** Scanning electron micrograph of the yeast form of *C. albicans* (×6720). **(c)** Thrush, or oral candidiasis.

(b)

(c)

Diseases of the Eye

The epithelial cells covering the eye can be considered a continuation of the skin or mucosa. The organisms most commonly associated with the eye, *Staphylococcus epidermidis* (e-pi-dèr'mi-dis), *S. aureus*, and *diphtheroids*, usually originate from the skin and upper respiratory tract. A summary of diseases associated with the eye is presented in Table 20-2 on p. 519.

Cosmetics play a role in eye infections since microorganisms can multiply there. The cosmetics might be sterile prior to opening, but the initial application can inoculate normal flora into the cosmetics, and subsequent applications can inoculate microorganisms into the eye.

Bacterial Diseases of the Eye

Many bacteria can infect the eye, largely through the conjunctiva, the mucous membrane that lines the eyelids and covers the outer surface of the eyeball. **Conjunctivitis** is an inflammation of the conjunctiva.

Contagious Conjunctivitis (Pinkeye)

One form of conjunctivitis, which occurs primarily in children, is called **contagious conjunctivitis**, or **pinkeye**. Contagious conjunctivitis is a common disease in the Middle East, where the causative organism, *Hemophilus aegyptius* (hē-mä'fi-lus ē-jip'ti-us), is transmitted by hand contact or by flies. The disease usually can be cured by topical application of antibiotics.

Neonatal Gonorrheal Ophthalmia

Neonatal gonorrheal ophthalmia is a more serious form of conjunctivitis caused by *Neisseria gonorrhoeae* (the cause of gonorrhea). It is acquired as the infant passes through the birth canal, and infection carries a high risk of blindness. Early in this century, legislation required that the eyes of all newborn infants be treated with a 1% solution of silver nitrate,

which proved to be a very effective treatment in preventing this eye infection. Between 1906 and 1959, the percentage of admissions to schools for the blind that could be attributed to neonatal ophthalmia declined from 24% to only 0.3%. Silver nitrate use has declined in some parts of the world out of concern for occasional chemical injury. In this country, it is being replaced with antibiotics because there are frequent coinfections by the gonococcus and sexually transmitted chlamydia, and silver nitrate is not effective against chlamydia.

Inclusion Conjunctivitis

Chlamydial conjunctivitis, or **inclusion conjunctivitis**, is quite common today. It is caused by *Chlamydia trachomatis* (trä-kō'mä-tis), a bacterium that grows only as an obligate intracellular parasite. In infants, who acquire it in the birth canal, the condition tends to resolve spontaneously in a few weeks or months, but in rare cases it can lead to scarring of the cornea, much as in trachoma, discussed shortly. The current recommendation is to treat the eyes of newborn infants within an hour of birth (the timing is very important) with erythromycin or tetracyclines, which are usually effective against both the gonococcus and chlamydial organisms. Chlamydial conjunctivitis also appears to spread in the unchlorinated waters of swimming pools and is called swimming pool conjunctivitis. Tetracycline applied as an ophthalmic ointment is an effective treatment.

Trachoma

A much more serious infection caused by *Chlamydia trachomatis* is **trachoma**, the greatest single cause of blindness in the world today. Millions now have the disease, and many millions have been blinded by the infection. In the arid parts of Africa and Asia, almost all children are infected early in their lives. Trachoma also occurs occasionally in the southwestern United States, especially among Native Americans. The disease is transmitted largely by hand contact or by such objects as towels. Flies may also carry the bacteria.

The disease is a conjunctivitis that eventually leads to permanent scarring. Blindness is a result of long-term mechanical abrasion of the cornea by these scars and by turned-in eyelashes. Secondary infections by other bacterial pathogens are also a factor, although they are not often present with inclusion conjunctivitis in developed countries. Antibiotic ointments, especially tetracycline, are useful in treatment. Partial immunity is generated by recovery. The disease can be controlled through sanitary practices and health education.

Viral Disease of the Eye

Herpetic Keratitis

Herpetic keratitis is usually caused by herpes simplex type 1 virus. This is a localized infection of the cornea that can recur with an epidemiology similar to that of cold sores. It is characterized by inflammation and corneal ulcers that can be quite deep. Herpetic keratitis is one of the few viral diseases for which there is an effective chemotherapeutic treatment; trifluridine is the drug of choice.

Protozoan Disease of the Eye

Acanthamoeba Keratitis

The first case of ***Acanthamoeba* keratitis** (a-kan-thä-mē′bä kär-i-tī′tis) was reported in 1973 in a Texas rancher. Since then, over 40 cases have been diagnosed. This amoeba inhabits fresh water. All cases have been preceded by an eye injury or contact with contaminated water. Over half of the patients wear contact lenses, suggesting that contaminated lenses might transmit the infection. The symptoms resemble herpetic keratitis but are accompanied by severe pain and resistance to commonly used keratitis medications. Damage may be so severe as to require a corneal transplant to restore sight. Diagnosis is confirmed by the presence of trophozoites and cysts in corneal scrapings.

TABLE 20-2 *Diseases Associated with the Eye*

	Causative Agent	Treatment
Bacterial Diseases		
Contagious conjunctivitis (pinkeye)	*Hemophilus aegyptus*	Sulfonamides
Neonatal gonorrheal ophthalmia	*Neisseria gonorrhoeae*	Silver nitrate, tetracycline, or erythromycin for prevention; penicillin for treatment
Inclusion conjunctivitis	*Chlamydia trachomatis*	Sulfonamide, tetracycline, erythromycin
Trachoma	*Chlamydia trachomatis*	Sulfonamide, tetracycline
Viral Disease		
Herpetic keratitis	Herpesvirus (herpes simplex type 1)	Idoxuridine, vidarabine, trifluridine, acyclovir
Protozoan Disease		
Acanthamoeba keratitis	*Acanthamoeba polyphaga*	Ketoconazole, miconazole have some moderate success; surgery required to restore vision

Study Outline

Introduction (p. 502)

1. The skin is a physical and chemical barrier against microorganisms.
2. Moist areas of the skin (such as the axilla) support larger populations of bacteria than dry areas (such as the scalp).

Structure and Function of the Skin (pp. 502–503)

1. The outer portion of the skin, called the epidermis, contains keratin, a waterproof coating.
2. The inner portion of the skin, the dermis, contains hair follicles, sweat ducts, and oil glands that open to the surface.
3. Sebum and perspiration are secretions of the skin that can remove organisms from openings in the dermis.
4. Sebum and perspiration provide nutrients for some microorganisms.
5. Microorganisms can gain access to the body through the openings in the dermis.
6. Mucous membranes line the body cavities. Mucus traps microorganisms and facilitates their removal from the body.
7. The eye is mechanically washed by tears that contain lysozyme.

Normal Flora of the Skin (p. 504)

1. Members of the genus *Propionibacterium* metabolize oil from the oil glands and colonize hair follicles.
2. Microorganisms that live on human skin are resistant to desiccation and high concentrations of salt.
3. Antimicrobial substances secreted by normal flora can inhibit the growth of potential pathogens.
4. The normal flora of the face are determined by the oropharyngeal organisms, and the normal flora of the anal region are influenced by organisms in the lower gastrointestinal tract.

Diseases of the Skin (pp. 504–517)

Bacterial Diseases of the Skin (pp. 504–508)

Staphylococcal Skin Infections (pp. 505–506)

1. Staphylococci are gram-positive bacteria that often grow in clusters.
2. Almost all pathogenic strains of *S. aureus* produce coagulase.
3. Pathogenic *S. aureus* can produce enterotoxins, leukocidins, and exfoliative toxin.

4. Localized infections (sties, pimples, and carbuncles) result from *S. aureus* entering openings in the skin.
5. Many strains of *S. aureus* produce penicillinase.
6. Impetigo of the newborn is a highly contagious superficial skin infection caused by *S. aureus*.
7. Toxemia occurs when toxins enter the bloodstream; staphylococcal toxemias include scalded skin syndrome and toxic shock syndrome.

Streptococcal Skin Infections (pp. 506–508)

1. Streptococci are gram-positive cocci that often reproduce in chains. They are strictly fermentative and catalase negative.
2. Streptococci are classified according to their hemolytic enzymes and cell-wall (M) antigens.
3. Group A β-hemolytic streptococci (including *S. pyogenes*) are the pathogens most important to humans.
4. Group A β-hemolytic streptococci produce a number of virulence factors, including erythrogenic toxin, deoxyribonuclease, NADase, streptokinases, and hyaluronidase.
5. Streptococci are susceptible to penicillin.
6. Impetigo (isolated pustules) and erysipelas (reddish patches) are skin infections caused by *S. pyogenes*.

Infections by Pseudomonads (p. 508)

1. Pseudomonads are gram-negative rods. They are strict aerobes.
2. *Pseudomonas aeruginosa* is an opportunistic pathogen found primarily in soil and water. It is resistant to many disinfectants and antibiotics.
3. *P. aeruginosa* produces an endotoxin and several exotoxins.
4. Diseases caused by *P. aeruginosa* include otitis externa, burn infections, and dermatitis.
5. Infections have a characteristic blue-green pus caused by the pigment pyocyanin.
6. Carbenicillin and gentamicin are useful in treating *P. aeruginosa* infections.

Acne (p. 508)

1. *Propionibacterium acnes* can metabolize sebum trapped in hair follicles.
2. Metabolic end products (fatty acids) cause an inflammatory response known as acne.
3. Treatment with benzoyl peroxide and tetracycline is somewhat effective.

Viral Diseases of the Skin (pp. 509–514)

Warts (p. 509)

1. Papillomaviruses cause skin cells to proliferate and produce a benign growth called a wart.
2. Warts are spread by direct contact.
3. Warts may regress spontaneously or be removed chemically or physically.

Smallpox (Variola) (pp. 509–510)

1. Variola virus causes two types of skin infections, variola major and variola minor.
2. Smallpox is transmitted by the respiratory route, and the virus is moved to the skin via the bloodstream.
3. At any particular time, the vesicular or pustular lesions of the skin and mucous membranes will be in the same stage of development.
4. The only host for smallpox is humans.
5. Smallpox has been eradicated, thus routine vaccinations with live vaccinia virus are no longer made.

Chickenpox (Varicella) and Shingles (Herpes Zoster) (pp. 510–511)

1. Chickenpox is caused by varicella-zoster virus.
2. Transmitted by the respiratory route and localized in skin cells, the virus causes a vesicular rash.
3. Complications of chickenpox include encephalitis and Reye's syndrome.
4. After chickenpox, the virus can remain latent in nerve cells and subsequently activate in shingles.
5. Shingles is characterized by a vesicular rash along the affected cutaneous sensory nerves.
6. The virus can be treated with acyclovir. An attenuated live vaccine is being tested.

Herpes Simplex (pp. 511–512)

1. Herpes simplex infection of mucosal cells results in cold sores and occasionally encephalitis.
2. The virus remains latent in nerve cells, and cold sores can recur when the virus is activated.
3. Herpes simplex type 1 virus is transmitted primarily by oral and respiratory routes.
4. Acyclovir has proven successful in treating herpes encephalitis.

Measles (Rubeola) (pp. 512–513)

1. Measles is caused by measles virus (a paramyxovirus) and transmitted by the respiratory route.
2. After the virus has incubated in the upper respiratory tract, papular lesions appear on the skin, and Koplik spots appear on the oral mucosa.
3. Complications of measles include middle ear infections, pneumonia, encephalitis, and secondary bacterial infections.

4. Measles virus can persist following infection and later cause subacute sclerosing panencephalitis.
5. Vaccination with attenuated live virus provides effective long-term immunity.

Rubella (pp. 513–514)

1. The rubella virus (togavirus) is transmitted by the respiratory route.
2. A red rash and light fever might occur in an infected individual; the disease can be asymptomatic.
3. Congenital rubella syndrome can affect a fetus when a woman contracts rubella during the first trimester of her pregnancy.
4. Damage from congenital rubella syndrome includes stillbirth, deafness, eye cataracts, heart defects, and mental retardation.
5. Vaccination with live rubella virus provides immunity of unknown duration.

Fungal Diseases of the Skin (pp. 514–517)

Superficial Mycoses (p. 514)

1. Piedras are fungal infections of the external hair shaft.

Cutaneous Mycoses (pp. 514–515)

1. Tinea nigra (caused by *Cladosporium werneckii*) is a ringworm characterized by discolored blotches on the palms.
2. *Microsporum*, *Trichophyton*, and *Epidermophyton* cause dermatomycoses called ringworm, or tinea.
3. These fungi grow on keratin-containing epidermis, such as hair, skin, and nails.
4. Ringworm and athlete's foot are usually treated with topical application of antifungal chemicals.

Subcutaneous Mycoses (pp. 515–516)

1. Chromomycosis and mycetoma result from fungi that penetrate the skin through a wound.
2. The fungi grow and produce subcutaneous nodules along the lymphatic vessels.
3. Polyenes and surgical drainage are used to treat these infections.

Candidiasis (p. 517)

1. *Candida albicans* causes infections of mucous membranes and is a common cause of thrush (in oral mucosa) and vaginitis.
2. *C. albicans* is an opportunistic pathogen that may proliferate when normal bacterial flora are suppressed.
3. Topical antifungal chemicals may be used to treat candidiasis.

Diseases of the Eye (pp. 518–519)

1. *Staphylococcus epidermidis, S. aureus,* and diphtheroids are normal flora of the eye.
2. These microorganisms usually originate from the skin and upper respiratory tract.

Bacterial Diseases of the Eye (pp. 518–519)

1. Conjunctivitis is caused by a number of bacteria.
2. Contagious conjunctivitis, or pinkeye, is caused by *Hemophilus aegyptius* and occurs primarily in children.
3. Neonatal gonorrheal ophthalmia is caused by the transmission of *Neisseria gonorrhoeae* from an infected mother to an infant during its passage through the birth canal.
4. All newborn infants are treated with 1% silver nitrate or an antibiotic to prevent the growth of *Neisseria.*
5. Inclusion conjunctivitis is an infection of the conjunctiva caused by *Chlamydia trachomatis.* It is transmitted to infants during birth and is transmitted in unchlorinated swimming water.
6. In trachoma, which is caused by *C. trachomatis,* scar tissue forms on the cornea.
7. Trachoma is transmitted by hands, fomites, and perhaps flies.

Viral Disease of the Eye (p. 519)

1. Herpetic keratitis causes corneal ulcers. The etiology is herpes simplex type 1 that invades the central nervous system and can recur.
2. Trifluridine is an effective treatment for herpes keratitis.

Protozoan Disease of the Eye (p. 519)

1. *Acanthamoeba,* transmitted via water, can cause keratitis.

Study Questions

Review

1. Discuss the usual mode of entry of bacteria into the skin. Compare bacterial skin infections with those caused by fungi and viruses with respect to method of entry.

2. Compare and contrast impetigo and erysipelas.

3. A teenaged male with confirmed influenza was hospitalized when he developed respiratory distress. He had a fever, rash, and low blood pressure. *S. aureus* was isolated from his respiratory secretions. Discuss the relationship between his symptoms and the etiological agent.

4. Complete the table of epidemiology below.

Disease	Etiologic Agent	Clinical Symptoms	Method of Transmission
Acne			
Pimples			
Warts			
Smallpox			
Chickenpox			
Fever blisters			
Measles			
German measles			

5. How do these infections differ: tinea nigra, mycetoma, athlete's foot? In what ways are they similar?

6. (a) Differentiate between conjunctivitis and keratitis.
   (b) Select a bacterial and viral eye infection, and discuss the epidemiology of each.

7. A laboratory test used to determine the identity of *Staphylococcus aureus* is its growth on mannitol salt agar. The medium contains 7.5% sodium chloride (NaCl). Why is it considered a selective medium for *S. aureus*?

8. Why are people immunized against rubella since the symptoms of the disease are mild or even inapparent?

9. Explain the relationship between shingles and chickenpox.

10. Why are the eyes of all newborn infants washed with an antiseptic or antibiotic?

11. What is the leading cause of blindness in the world?

12. An opportunistic dimorphic fungus that causes skin infections is _____ .

13. Identify the following diseases based on the symptoms in the chart below.

Symptoms	Disease
Koplik spots	
Papular rash	
Vesicular rash	
Small, spotted rash	
Recurrent "blisters" on oral mucosa	
Corneal ulcer and swelling of lymph nodes	

14. What complications can occur from herpes simplex type 1 infections?

Challenge

1. You have isolated an organism from what appears to be impetigo. The organisms are gram-positive cocci in singles, pairs, and small groups. What test would help you quickly determine whether your isolate is *Staphylococcus* or *Streptococcus*? What is the result of this test if the organism is *Staphylococcus*?

2. A hospitalized patient recovering from surgery develops an infection that has blue-green pus and a grapelike odor. What is the probable etiology? How might the patient have acquired this infection?

3. Is it necessary to treat a patient for warts? Explain briefly.

4. A 12-year-old diabetic girl using continuous subcutaneous insulin infusion to manage her diabetes developed a fever (39.4°C), low blood pressure, abdominal pain, and erythroderma. She was supposed to change the needle insertion site every three days after cleaning the skin with an iodine solution. Frequently she did not change the insertion site more often than every ten days. Blood culture was negative, and abscesses at insertion sites were not cultured. What is the probable cause of her symptoms?

Further Reading

Chesney, P. J., M. S. Bergdoll, J. P. Davis, and J. M. Vergeront. "The disease spectrum, epidemiology, and etiology of toxic shock syndrome." *Annual Review of Microbiology* 38:315–338, 1984. History and recent developments related to this disease.

Cliff, A., and P. Haggert. "Island Epidemics." *Scientific American* 250(5):138–147, May 1984. An excellent review of epidemiology using measles as an example.

Henderson, D. A. "The eradication of smallpox." *Scientific American* 235(4):25–33, October 1976. A summary of the incidence of smallpox and investigation of the last endemic infection.

Langmuir, A. D., et al. "The Thucydides syndrome." *New England Journal of Medicine* 313:1027–1030, 1985. Descriptive epidemiology from translations of Thucydides' observations are used to understand an ancient epidemic.

Marples, M. J. "Life on the human skin." *Scientific American* 220(1):108–115, January 1969. A study of the microflora of the skin and the relationship between bacterial populations and hair, moisture, and temperature.

Sheagren, J. N. "*Staphylococcus aureus*, the persistent pathogen." *New England Journal of Medicine* 310:1368–1373, 1437–1442, 1984. Part One describes infections and immunity; Part Two discusses toxemias, treatment, and prevention.

Shepherd, M. G., R. T. M. Poulter, and P. A. Sullivan. "*Candida albicans*: biology, genetics, and pathogenicity." *Annual Review of Microbiology* 39:579–614, 1985. A thorough monograph on diseases caused by *C. albicans* and host defenses; includes an overview of antifungal drugs.

Tachibana, D. K. "Microbiology of the foot." *Annual Review of Microbiology* 30:351–375, 1976. An interesting report on the normal flora of the foot, foot infections, and foot odor.

See also Further Reading for Part Four at the end of Chapter 25.

Microbial Diseases of the Nervous System

Structure and Function of the Nervous System

The human nervous system consists of billions of nerve cells (neurons) and supporting cells (neuroglia) and is organized into two divisions—the central nervous system and the peripheral nervous system (Figure 21-1). The **central nervous system** consists of the brain and the spinal cord. As the control center for the entire body, it picks up sensory information from the environment, interprets the information, and sends impulses that coordinate the body's activities. The **peripheral nervous system** consists of all the nerves that branch off from the brain and the spinal cord. These nerves are the lines of communication between the central nervous system, the various parts of the body, and the external environment.

The brain is protected from physical injury by skull bones, and the spinal cord is protected by the backbone. Within these bony enclosures, both the brain and the spinal cord are covered and protected by three continuous membranes called *meninges*. These are the outermost dura mater, the middle arachnoid, and the innermost pia mater. Between the pia mater and arachnoid membranes is a space called the *subarachnoid space*, in which *cerebrospinal fluid* circulates (Figure 21-2). This fluid also circulates in spaces within the brain called *ventricles*.

A very interesting feature of the brain is the **blood–brain barrier**. Certain capillaries permit some substances to pass from the blood into the brain but restrict others. These capillaries are less permeable than others within the body and are therefore more selective in passing materials.

Figure 21-1   Organization of the nervous system.

Dura mater
Arachnoid — Cranial meninges
Pia mater

Subarachnoid space of brain

Brain
Spinal cord — Central nervous system

Cerebrum

Pons
Medulla
Cerebellum

Peripheral nervous system

Spinal cord — Dura mater
Arachnoid — Spinal meninges
Central canal — Pia mater

Subarachnoid space of spinal cord

Figure 21-2   Meninges and cerebrospinal fluid. Circulation of the fluid is indicated by arrows in the colored zones.

Drugs cannot cross the blood–brain barrier unless they are lipid soluble. (Glucose and many amino acids are not lipid soluble, but they can cross the barrier because special transport systems exist for them.) The lipid-soluble antibiotic chloramphenicol enters the brain readily. Penicillin is only slightly lipid soluble, but if it is given in very large doses, enough may cross the barrier to be effective. Inflammations of the brain tend to alter the blood–brain barrier to allow antibiotics to cross that would not be able to cross if there were no infection.

Even though the central nervous system has considerable protection, it can still be invaded by microorganisms in several ways. For example, microorganisms can gain access through trauma, such as a skull or backbone fracture, or through a medical procedure such as a lumbar puncture (spinal tap), in which a needle is inserted into the subarachnoid space of the spinal meninges to obtain a sample of fluid for diagnosis. Some microorganisms can also move along peripheral nerves. But probably the most common routes of central nervous system invasion are the bloodstream and lymphatic system (Chapter 22). Because cerebrospinal fluid communicates with the lymphatic system, invading microorganisms can enter the cerebrospinal fluid through it.

An infection of the meninges is called **meningitis**. An infection of the brain itself is called **encephalitis**. Table 21-3, p. 541, summarizes the main diseases associated with the nervous system.

Bacterial Diseases of the Nervous System

Bacterial Meningitis

Meningitis is an inflammation of the meninges. Most patients suffering from any type of meningitis complain of a headache and have symptoms of nausea and vomiting. Convulsions and coma accompany the infection in many cases. The symptoms are associated with swelling of the meninges and excess fluid. The mortality rate varies with the pathogen but is generally high for an infectious disease today. Many people who survive an attack suffer some sort of neurological damage.

Meningitis can be caused by different types of pathogens, including viruses, bacteria, fungi, and protozoans. In this section, we will discuss bacterial meningitis. There are three major types of bacterial meningitis—meningococcal meningitis caused by *Neisseria meningitidis*, pneumococcal meningitis caused by *Streptococcus pneumoniae*, and *Hemophilus influenzae* (hē-mä′fi-lus in-flü-en′zī) meningitis. These three types comprise the great majority of meningitis cases. Only meningococcal meningitis is a notifiable disease (Chapter 13). Table 21-1 presents data about these and some other types of bacterial meningitis.

Nearly 50 species of bacteria have been reported to be opportunistic pathogens that occasionally cause meningitis. These include such common organisms as *Escherichia coli*, *Pseudomonas aeruginosa*, and *Klebsiella pneumoniae*. Although the number of cases of meningitis caused by these gram-negative organisms is low, the fatality rate is in excess of 30%.

Neisseria Meningitis (Meningococcal Meningitis)

Meningococcal meningitis, or **cerebrospinal fever**, is caused by the bacterium *Neisseria meningitidis*. The neisseriae are aerobic, nonmotile, gram-negative cocci.

N. meningitidis can inhabit the portion of the throat behind the nose (nasopharynx) without causing any symptoms. This carrier state can last from several days to several months and provides the reservoir for the organism while enhancing the immunity of the carrier. If an individual without adequate immunity acquires the microorganism, usually through contact with a healthy carrier, the resulting throat infection can lead to bacteremia, which is followed by meningitis. Purplish spots may appear on the skin.

Most of the bacteria observed in cerebrospinal fluid are found in leukocytes that have entered the fluid in response to the infection. The ability of the bacteria to leave the blood to make contact with the meninges seems to be related to the concentration of microorganisms in the blood and the length of time they have circulated. Most of the symptoms associated with meningitis are thought to be caused by an endotoxin produced by the bacteria. The *N. meningitidis* endotoxin not only causes extensive damage to blood vessels but also heightens sensitivity to endotoxin. This increased sensitivity is called the *Schwartzman phenomenon*. In severe cases, death can occur within a few hours.

Meningococcal meningitis usually strikes children younger than four years of age, with the highest incidence occurring during the first year of life. Chil-

TABLE 21-1   *Types of Bacterial Meningitis*

Bacterium	Percentage of All Reported Meningitis	Comments
Neisseria meningitidis	27%	Can occur as epidemics; fatality rate of about 12%
Hemophilus influenzae	43%	Fatality rate of about 6 to 7%
Streptococcus pneumoniae	11%	Fatality rate of about 27%
Mycobacterium tuberculosis	Uncommon	Complication of tuberculosis; fatality rate of about 20%
Leptospira interrogans	Uncommon	Meningitis is a common symptom of leptospirosis
Listeria monocytogenes	Uncommon	Causes spontaneous abortions and fetal damage; meningitis in immunosuppressed persons
Nocardia asteroides	Uncommon	Meningitis is a common complication of pulmonary infections by *Nocardia*

dren are usually born with maternal immunity and become susceptible as the immunity weakens, at about six months of age. The death rate for meningococcal meningitis has been about 10% in recent years.

Explosive epidemics of meningococcal meningitis among young adults in military training camps used to be common. Apparently, the close quarters promoted the transmission of bacteria from carriers to susceptible individuals. However, a recently developed vaccine, consisting of purified capsular polysaccharide, has proved to be effective in protecting military recruits; this expression of the disease is becoming a thing of the past, although the vaccine is still not in use in the general population.

Preliminary diagnosis is based on gram-staining cerebrospinal fluid (CSF) and observing gram-negative diplococci in phagocytes. Countercurrent immunoelectrophoresis (Chapter 17) of CSF is also useful in differentiating among the several types of bacterial meningitis. The CSF can be inoculated onto blood agar for isolation of the bacteria. The bacteria will not grow on agar devoid of blood and must be incubated in an environment with 5% to 10% carbon dioxide. This concentration of carbon dioxide can be obtained by the use of a candle jar (Chapter 6) or a carbon dioxide incubator. The organism is very susceptible to desiccation and is killed if heated to 55°C for 30 minutes. Because *N. meningitidis* is very sensitive to handling once it is removed from the body, care must be taken to ensure its survival until the cultures can be used in diagnosis.

N. meningitidis, like many other pathogens, is relatively resistant to phagocytosis. Encapsulated strains are the most likely to prove virulent. At one time, sulfonamide was the drug of choice because it passes readily from the blood into cerebrospinal fluid. However, bacterial resistance to the sulfa drugs has now become common, and other drugs have come into use.

Hemophilus influenzae Meningitis

Hemophilus influenzae is a nonmotile, aerobic, gram-negative, pleomorphic bacterium (Figure 21-3). A member of the normal throat flora, it can also cause meningitis. The bacterium, which is encapsulated, is further divided into six types on the basis of its antigenic capsular carbohydrates. Only one type has real importance to medical microbiology, type b. The virulence of *H. influenzae* type b is related to its capsule; its endotoxin does not appear to play a major role in pathogenicity, unlike that of *N. meningitidis*.

Figure 21-3 Electron micrograph of *Hemophilus influenzae* type b (× 36,000). Note the thick layer of capsular material that makes it resistant to phagocytosis.

The name *Hemophilus influenzae* is somewhat misleading. The microorganism was erroneously thought to be the causative agent of the influenza pandemics of 1890 and World War I—thus the term *influenzae*. The etiologic agent of influenza is now known to be a virus. However, *H. influenzae* was probably a secondary invader during those pandemics. The term *Hemophilus* refers to the fact that the microorganism requires factors in blood for growth (*hemo* means blood). Actually, *Hemophilus influenzae* grows poorly, if at all, on blood agar, but it can be cultured on chocolate agar, which is prepared from lysed red blood cells. Lysis of the blood cell releases the X and V factors required for its growth (Chapter 10).

H. influenzae type b is by far the most common cause of bacterial meningitis; it produces an estimated 20,000 cases each year, mostly in children (Figure 21-4). It accounts for most of the cases of reported bacterial meningitis (48%) but has the lowest mortality rate of the common types of bacterial meningitis (6%). It occurs mostly in children under four years of age and is especially prevalent at about age six months. The antibodies acquired from the mother provide protection for only about two months. The carrier state in the throats of children is quite high, and transmission in daycare centers is now an increasing problem.

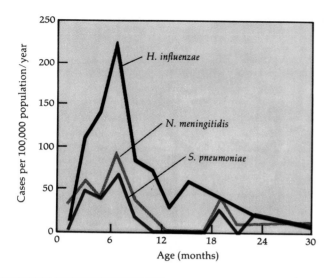

Figure 21-4 Reported cases of meningitis. The graph shows the incidence by age of the most common bacterial types of meningitis, those caused by *Hemophilus influenzae, Neisseria meningitidis*, and *Streptococcus pneumoniae.*

Invasion of the meninges by *H. influenzae* is usually preceded by a viral infection of the respiratory tract, which opens the way for the bacteria to enter the bloodstream and be transported to the meninges. A vaccine directed at the antigenic polysaccharide capsule (the Hib vaccine) has been developed but unfortunately it does not stimulate good immunity in children younger than about two years of age. Because this is the group at greatest risk, even if the vaccine were 100% effective and every person over the age of two were immunized, there would be only about a 30% reduction in cases. One approach being tried to improve this situation is to chemically bond the Hib vaccine to tetanus toxoid or diphtheria toxoid. The hope is that the immature immune system of very young children will respond better to the bonded pair of antigens.

Treatment of *H. influenzae* meningitis requires an antibiotic that will reach the CSF. Rifampin has largely replaced ampicillin, for which the bacteria rapidly developed resistance. If *H. influenzae* meningitis outbreaks occur in a daycare center, other children in the center can be given antibiotics as a preventive measure. This also has the effect of lowering the carrier rate in the group.

Laboratory diagnosis of *H. influenzae* is based on analysis of CSF and blood. If a Gram-stained smear of CSF shows gram-negative pleomorphic rods, the presence of *H. influenzae* can be assumed. If no organisms are found in the CSF, countercurrent immunoelectrophoresis can be used to differentiate among the most common bacterial types of meningitis. A passive agglutination test using latex particles has recently been developed for diagnosis of bacterial meningitis and should be more convenient to use.

Streptococcus pneumoniae Meningitis (Pneumococcal Meningitis)

Streptococcus pneumoniae, like *H. influenzae*, is a common inhabitant of the nasopharyngeal region. The capsule on *S. pneumoniae* makes it very resistant to phagocytosis and is the most important element in its pathogenicity. Most cases of pneumococcal meningitis occur among children during their first year of life. However, the hospitalized elderly are another susceptible group. For this group, some protection may be available from the vaccine for pneumococcal pneumonia, which is caused by the same organism. Penicillins remain the antibiotics of choice, much as with the other streptococcal diseases.

Although pneumococcal meningitis accounts for only about 13% of the cases of reported bacterial meningitis, the mortality rate of 26% is very high.

Listeriosis

Listeria monocytogenes (lis-te'rē-ä mo-nō-sī-tô'je-nēz) is a gram-positive rod associated with a great variety of animals. It is widely distributed in soil and water, mostly by means of animal feces. The name is derived from the proliferation of monocytes (a type of leukocyte) found in some animals infected by it. The disease **listeriosis** usually is a mild, often symptomless disease in adult humans, but it frequently results in shedding the pathogen in the feces. *L. monocytogenes* is ingested by phagocytic cells but is not destroyed and even proliferates within them.

As a pathogen, *L. monocytogenes* has two principal characteristics—it mainly affects adults who are immunosuppressed or pregnant, and it has a special affinity for growth in the central nervous system and the placenta supplying the fetus with nutrients. Growth in the central nervous system is usually expressed as meningitis. The organism seldom appears in healthy adults but is the leading cause of meningitis in cancer patients and is an important cause of meningitis in other immunosuppressed persons. When it infects a pregnant woman, the growth on the placenta leads to a high rate of spontaneous abortion or stillbirth. A surviving newborn may be acutely ill with septicemia and meningitis. The infant mortality related to this type of infection is about 60%.

In human outbreaks, the organism is mostly food-borne; dairy products are an important source. There has been some speculation that the organism's intracellular location allows it to survive pasteurization, but the current opinion is that proper pasteurization eliminates it. *L. monocytogenes* is one of the few pathogens that is capable of growth at refrigerator temperatures. In fact, it is isolated, in part, by *cold enrichment*, meaning incubation at a low temperature at which it outgrows most other bacteria.

Considerable effort is being made to develop methods by which *L. monocytogenes* can be better detected in foods, as present methods are slow and inadequate. Diagnosis in humans depends on isolation and culturing of the organism. Penicillin and ampicillin are the antibiotics of choice for treatment.

Tetanus

The causative agent of **tetanus**, *Clostridium tetani*, is an obligate, anaerobic, endospore-forming, gram-positive rod. It is especially common in soil contaminated with animal fecal wastes. In the American Civil War, it was observed that the cavalry had a higher rate of tetanus than the infantry, possibly because of animal wastes associated with the cavalry.

The symptoms of tetanus are caused by an extremely potent neurotoxin (*tetanospasmin*) that is released upon the death and lysis of the growing bacteria (Chapter 14). An amount of tetanospasmin weighing as much as the ink in one period on this page could kill 30 people. The bacteria themselves do not spread from the infection site, and there is no inflammation.

In a muscle's normal operation, a nerve impulse initiates the contraction of the muscle. At the same time, an opposing muscle receives a signal to relax so as not to oppose the contraction. The tetanus neurotoxin blocks the relaxation pathway so that both sets of muscles contract, resulting in the characteristic muscle spasms called *spastic paralysis* (Figure 21-5). The muscles of the jaw are affected early in the disease, which prevents opening of the mouth (lockjaw). Gradually, other skeletal muscles become affected, including those involved in swallowing. Death results from spasms of the respiratory muscles.

Because the organism is an obligate anaerobe, the wound through which it enters the body must provide anaerobic growth conditions. Improperly cleaned deep puncture wounds, especially those with little or no bleeding, serve very well. Of the approximately 100 tetanus cases that have occurred each year in the United States for the past 12 years (about 25% of which were fatal), many arise from trivial but fairly deep injuries that are thought to be too minor to bring to a physician. Street heroin is often diluted with quinine, which tends to make the puncture wound from injecting the drug more anaerobic. This is a contributing factor to the death rate of intravenous drug users from tetanus, which is the highest of any group.

Most persons in this country have received the DPT (diphtheria, pertussis, tetanus) immunization, which includes the tetanus toxoid that stimulates the formation of antibodies to fight tetanus toxin. Immunization is nearly 100% effective, but recovery from the clinical disease does not confer immunity because the amount of toxin usually produced is too small to be immunogenic.

When a wound is severe enough to be brought to the attention of a physician, the decision is usually made to provide protection against tetanus, and a booster of toxoid may be given. Table 21-2 summarizes the questions involved in this decision. A patient who has had the full sequence of three doses of toxoid and a booster every ten years should be immune. However,

Figure 21-5 Tetanus. Drawing of a British soldier during the Napoleonic wars dying of the final stages of tetanus. (Drawing by Charles Bell in the Royal College of Surgeons, Edinburgh.)

TABLE 21-2 *Wound Management for Prevention of Tetanus*

History of Tetanus Immunization	Minor Wounds		Extensive Wounds	
	Administer Toxoid	Administer TIG†	Administer Toxoid	Administer TIG†
Unknown or 0 to 2 doses	Yes	No	Yes	Yes
3 or more doses	No*	No	No**	No

†TIG = tetanus immune globulin
*Yes, if more than 10 years since last dose.
**Yes, if more than 5 years since last dose.

people often do not get boosters on schedule, and serological surveys show that at least 50% of the population does not have adequate protection. Even so, if toxoid has been given before, the anamnestic response (see Figure 16-7) to a booster given after the injury will be rapid enough to be protective. Tetanus immune globulin (TIG), which is prepared from the serum of immunized humans, is often given simultaneously to provide temporary immunity. It is injected at a different site, however, to avoid neutralizing the toxoid. To minimize production of more toxin, damaged tissue should be removed (debridement), and antibiotics should be administered. However, once the toxin has attached to the nerves, such therapy is of little use.

In less developed areas of the world, tetanus arising from the severed umbilical cords of infants is a major cause of death. In some cultures, the practice of dressing the cut umbilical cord with such materials as soil, clay, and cow dung is a major contributor to the development of tetanus. Worldwide, there are probably several hundred thousand cases of tetanus from all causes each year.

Botulism

Botulism, often considered to be a form of food poisoning, is caused by *Clostridium botulinum*, an obligately anaerobic, endospore-forming, gram-positive rod that is found in soil and in many freshwater sediments. Ingesting the endospores usually does no harm, as will be explained shortly. However, in anaerobic environments, such as sealed cans, the microorganism produces an exotoxin that animal assays show to be the most potent of all natural toxins. This neurotoxin is highly specific for the synaptic end of the nerve, where it blocks the release of acetylcholine, a chemical necessary for transmission of nerve impulses across synapses. Persons suffering from botulism undergo a progressive *flaccid paralysis* for one to ten days and may die from respiratory and cardiac failure. Nausea, but no fever, may precede the neurological symptoms. The initial neurological symptoms vary, but nearly all sufferers have double or blurred vision. Other symptoms include difficulty in swallowing and general weakness. Incubation time varies, but symptoms typically appear within a day or two. As with tetanus, recovery from the disease does not confer immunity because the toxin is usually not present in amounts large enough to be immunogenic.

Botulism was first described as a clinical disease in the early 1800s, when it was known as the sausage disease (*botulus* means sausage). Blood sausage was made by filling a pig stomach with blood and ground meats, tying shut all the openings, boiling it for a short time, and smoking it over a wood fire. The sausage was then stored at room temperature. This attempt at food preservation included most of the requirements for an outbreak of botulism—it killed competing bacteria but allowed the more heat-stable *C. botulinum* endospores to survive, and it provided anaerobic conditions and an incubation period for toxin production.

Most botulism results from such attempts at food preservation that fail to eliminate the *C. botulinum* endospore. Cooking preserved food is important because the botulinal toxin is very heat labile and will be destroyed by most ordinary cooking methods that bring the food to a boil. Sausage and bacon rarely cause botulism today, largely because of the addition of nitrites. It was discovered that nitrites prevent *C. botulinum* growth following germination of the endospores. However, there is considerable research aimed at minimizing or eliminating this use of nitrites because they are involved in the formation of carcinogenic (cancer-causing) nitrosamines.

Botulin toxin is not formed in foods with acidity below pH 4.7. Such foods as tomatoes therefore can be safely preserved without the use of a pressure cooker. A concern that newer strains of tomatoes did not contain enough acid for reliable preservation by boiling does not seem to be justified, but some recipes suggest addition of more acid. There have been cases of botulism from acidic foods that normally would not have supported the growth of the botulism organisms. Most of these episodes are related to mold growth, which metabolized enough acid to allow the initiation of growth of the botulism organisms.

There are several serological types of the botulinal toxin produced by different strains of the pathogen. These differ considerably in their virulence and other factors. *Type A* toxin is probably the most virulent. There have been deaths from type A toxin when the food was only tasted but not swallowed. It is even possible to absorb lethal doses through skin breaks while handling laboratory samples. In cases without treatment, there is a 60% to 70% mortality rate. The type A endospore is the most heat resistant of all *C. botulinum* strains. In the United States, it is found mainly in California, Washington, Colorado, Oregon, and New Mexico. Some eastern states have never had a type A outbreak. The type A organism is usually proteolytic (the breakdown of proteins by clostridia releases amines with unpleasant odors), but obvious spoilage odor is not always apparent in low-protein foods, such as corn or beans.

Type B toxin is responsible for most European outbreaks of botulism and is the most common type in the eastern United States. The mortality rate in cases without treatment is perhaps 25%. Type B botulism organisms occur in both proteolytic and nonproteolytic strains. The endospores of the proteolytic strains are more heat resistant and more likely to survive in food.

Type E toxin is produced by botulism organisms that are often found in marine or lake sediments. Therefore, outbreaks often involve seafood and are especially common in the Pacific Northwest, Alaska, and the Great Lakes area. The endospore of type E botulism is less heat resistant than that of other strains and is usually destroyed by boiling. Types A and B will survive much higher temperatures. Type E is nonproteolytic, so the chance of detecting spoilage in high-protein foods such as fish is minimal. It is also capable of producing a toxin at refrigerator temperatures and requires less strictly anaerobic conditions for growth.

Botulism is not a common disease. Only a few cases are reported each year, but outbreaks from restaurants occasionally involve 20 or 30 cases. Eskimos and coastal Native Americans in Alaska probably have the highest rate of botulism in the world, mostly of type E. The problem arises from their attempting to preserve food without using precious fuel for heating or cooking. The difficulty in getting prompt treatment for isolated ethnic groups such as Alaskan natives is reflected in the 40% mortality rate for type E botulism observed there between 1950 and 1973.

Botulism organisms do not seem to be able to compete successfully with the normal intestinal flora, so production of toxin by ingested bacteria almost never causes botulism in adults. However, the intestinal flora of infants are not well established, and they may suffer from **infant botulism**. First reported in 1976, an estimated 250 cases occur in the United States every year, many more than any other form of botulism. Infants have many opportunities to ingest soil and other materials contaminated with the endospores of the organism, but 30% of reported cases have been associated with honey. Endospores of *C. botulinum* are recovered with some frequency from honey, and a lethal dose may be as few as 2000 bacteria. The recommendation is not to feed honey to infants under one year of age; there is no problem with older children or adults who have normal intestinal flora.

The treatment of botulism relies heavily on supportive care. Antibiotics are of almost no use because the toxin is preformed. Antitoxins aimed at the neutralization of A, B, and E toxins are available and are usually administered together. This trivalent antitoxin will not affect the toxin already attached to the nerve ends and is probably more effective on type E than on types A and B. The toxin is quite firmly bound. Recovery requires that the nerve endings regenerate and therefore proceeds slowly. Extended respiratory assistance may be needed, and some neurological impairment may persist for months.

Botulism is diagnosed by the inoculation of mice with samples from patient serum, stool, or vomitus specimens. The toxin in food can also be identified by mouse inoculation. Different sets of mice are immunized with type A, B, or E antitoxins. All the mice are then inoculated with the test toxin; if, for example, those protected with type A antitoxin are the only survivors, then the toxin is type A.

The botulism pathogen can also grow in wounds in a manner similar to that of clostridia causing tetanus or gas gangrene (Chapter 22). Such episodes of **wound botulism** occur occasionally.

Leprosy

Mycobacterium leprae (lep'rī) is probably the only bacterium that grows mostly in the peripheral nervous system, although it can also grow in skin cells. It is an acid-fast rod closely related to the tuberculosis pathogen, *Mycobacterium tuberculosis*. The organism was first isolated and identified about 1870 by G. W. Hansen of Norway; his discovery was one of the first direct links ever made between a specific bacterium and a disease. **Leprosy** is sometimes called **Hansen's disease** in an effort to avoid using the dreaded name *leprosy*.

The organism has an optimum growth temperature of 30°C and shows a preference for the outer, cooler portions of the human body. A very slow generation time of about 12 days has been estimated. The microorganism has never been grown on artificial media, but in 1969 armadillos were experimentally infected with *M. leprae*. The armadillo body temperature is only 30° to 35°C. It is now known that armadillos can contract a leprosylike disease in the wild, although they are not considered a source of human infection. Armadillos are used in studies of the disease itself and in evaluations of the effectiveness of chemotherapeutic agents.

There are two main forms of leprosy, although intermediate forms are also recognized. The *tuberculoid (neural) form* is characterized by regions of the skin that have lost sensation and are surrounded by a border of nodules (Figure 21-6a). This form of the disease occurs in persons with effective immune reactions. Recovery sometimes occurs spontaneously. The disease can be diagnosed by detecting acid-fast bacilli in nasal secretions or by inspecting the fluids from a slit cut in a cool site, such as an earlobe. The *lepromin test* uses an extract of lepromatous tissue injected into the skin. A visible skin reaction that develops at the injection site indicates that the body has developed an immune response to the leprosy bacillus. This test is negative during the later lepromatous stage of the disease.

In the *lepromatous (progressive) form* of leprosy, skin cells are infected, and disfiguring nodules form all over the body. Patients with this type of leprosy have less effective cell-mediated immune systems, and the disease is progressive. Mucous membranes of the nose tend to become affected, and a lion-faced appearance is associated with this type of leprosy. Deformation of the hand into a clawed form and considerable necrosis of tissue can also occur (Figure 21-6b). The progression of the disease is unpredictable, and remissions may alternate with rapid deterioration.

In both forms, leprosy is spread by transfer of the bacteria in exudates (discharges) from lesions on the diseased person to minor abrasions on the skin of another. Inanimate objects, such as clothing, may transmit the disease in a similar manner. Leprosy,

(a)

(b)

Figure 21-6   Symptoms of leprosy. (a) Tuberculoid leprosy. The lesion below the shoulder on the side of a patient shows a ring of nodules around a relatively clear area in the center. **(b)** Lepromatous leprosy. Tissue necrosis is obvious on the hands of this patient.

however, is not very contagious, and transmission usually occurs only between persons in fairly intimate and prolonged contact. The time from infection to the appearance of symptoms is usually measured in years, although children can have a much shorter incubation period. Death is not usually a result of the leprosy itself but of complications, such as tuberculosis.

Much of the public's fear of leprosy can probably be attributed to biblical and historical references to the disease. In the Middle Ages, lepers were rigidly excluded from normal European society and sometimes were even given bells to wear so that people could avoid contact with them. This isolation might have contributed to the near disappearance of the disease in Europe. But patients with leprosy are no longer kept in isolation since patients can be made noncommunicable within four to five days by the administration of sulfone drugs. The National Leprosy Hospital in Carville, Louisiana, presently treats about 500 leprosy patients on an ambulatory basis. These outpatients are free to carry out their normal daily routines and return to the hospital for periodic treatment.

The number of leprosy cases in the United States has been gradually increasing. Currently, over 200 cases are reported each year, and the total number of cases is estimated at about 2600. Many of these cases are imported, for the disease is usually found in tropical climates. Millions of persons, most of them in Asia and Africa, suffer from leprosy today. Some researchers believe that unsanitary living conditions and poor nutrition, rather than climate, are responsible for the prevalence of disease in these areas.

Although the disease is still difficult to cure, prolonged treatment with sulfone drugs (such as dapsone) has been effective in arresting its progress. Dapsone has been the mainstay of treatment, but resistance to the drug is a problem. Rifampin and a fat-soluble dye, clofazimine, are the other main drugs that are used, often along with dapsone.

A vaccine for leprosy is undergoing field trials. However, because of the extremely long incubation period for leprosy, it will be many years before the vaccine's effectiveness is established. An encouraging sign that a vaccine might be effective is the fact that the BCG vaccine used to provide immunity to tuberculosis (also caused by a *Mycobacterium*) is somewhat protective against leprosy. A BCG program in Africa to control tuberculosis showed that the vaccine gave at least 50% protection against leprosy at the same time.

Viral Diseases of the Nervous System

Poliomyelitis

Poliomyelitis (polio) is best known as a cause of paralysis. However, only about 10% of infected people develop identifiable symptoms, and the paralytic form of poliomyelitis probably affects less than 1% of those infected with the poliomyelitis virus. The great majority of cases are asymptomatic, or cause mild symptoms, such as headache, sore throat, fever, and nausea, which are often interpreted as mild meningitis or influenza. Asymptomatic or mild cases of polio are most common in the very young.

In some parts of the world, because of poor sanitary conditions, most of the population contract asymptomatic poliomyelitis as infants and develop immunity. Nonetheless, the incidence of the disease in the tropical and subtropical zones is so high that there are an estimated 400,000 cases of paralytic polio each year. As sanitation has improved in the more developed regions of the world, the chance of developing the disease in infancy has decreased, particularly for those in the upper and middle socioeconomic groups. When infection has occurred in adolescence or early adulthood, the paralytic form of the disease was frequent.

Humans are the only known natural host for polioviruses. Polioviruses are more stable than most other viruses and can remain infectious for relatively long periods in water and food. The primary mode of transmission is ingestion of water that is contaminated with feces containing the virus. Thus, polio occurs mainly in the summer months in temperate regions, when people participating in water sports are readily exposed to the virus.

Because the infection is initiated by ingestion of the virus, its primary areas of multiplication are the throat and small intestine (Figure 21-7). This accounts for the sore throat and nausea. Next, the virus invades the tonsils and the lymph nodes of the neck and ileum (terminal portion of the small intestine). From the lymph nodes, the virus enters the blood (*viremia*). In most cases, the viremia is only transient, the infection does not progress past the lymphatic stage, and clinical disease does not result. If the viremia is persistent, however, the virus penetrates the capillary walls and enters the central nervous system. Once in the central nervous system, the virus displays a high affinity for nerve cells, particularly motor

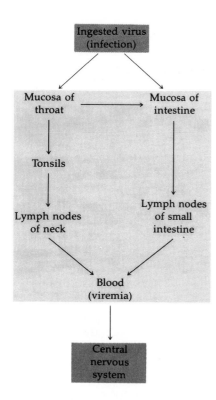

Figure 21-7 Routes taken by poliovirus in the body.

nerve cells in the upper spinal cord (anterior horn cells). The virus does not infect the peripheral nerves or the muscles. As the virus multiplies within the cytoplasm of the motor nerve cells, the cells die, and paralysis results. Death results from respiratory failure (Figure 21-8).

Diagnosis of polio is usually based on isolation of the virus from feces and throat secretions. Cell cultures can be inoculated, and cytopathic effects on the cells can be observed (see Table 14-4). Serological titrations using viral neutralization tests are also available.

The incidence of polio in the United States has decreased markedly since the availability of the polio vaccines. The last cases acquired in the United States that were attributed to a wild virus were reported in 1979. The development of the first polio vaccine was made possible by the introduction of practical techniques for cell culture, for the virus does not grow in any common laboratory animal. The polio vaccine was the prototype for the vaccines for mumps, measles, and rubella.

There are three different serotypes of the poliovirus, and immunity must be provided for all three. Two vaccines are available. The *Salk vaccine*, which was developed in 1954, uses viruses that have been inactivated by treatment with formalin. It requires a series of injections. The effectiveness of this vaccine is high, perhaps 90% against paralytic polio. The antibody levels decline with time, and booster shots are needed every few years to maintain full immunity. Using only the Salk vaccine, several European countries have almost eliminated polio from their populations.

The *Sabin vaccine*, introduced in 1963, contains three living, attenuated strains of the virus and is more popular in the United States than the Salk vaccine. It is less expensive to administer, and most people prefer taking a sip of orange-flavored drink

Figure 21-8 Iron lungs at a March of Dimes Respiratory Center in Los Angeles in the 1950s. Breathing was possible for many only with these mechanical aids. A few survivors from these epidemics still use these machines, at least part-time. Others are able to use portable respiratory aids.

containing the virus to having a series of injections. The immunity achieved with the Sabin vaccine resembles that acquired by natural infection. One disadvantage is that, on rare occasions—one in several million cases—one of the attenuated strains of the virus (type 3) seems to revert to virulence and cause the disease. A number of these cases have been secondary contacts, not the person who received the vaccine. This illustrates that recipients of the vaccine can infect contacts, leading—in most cases—to immunization.

Some medical scientists have suggested that a return to the Salk vaccine might be desirable despite its disadvantages. As the level of polio in the population declines, the cases that can be linked to the live vaccine become a more significant portion of the total cases. There is no reason why polio could not be almost eliminated in this country by an effectively administered program of immunization. Whatever the advantages of the Salk and Sabin vaccines, both have been used with remarkable effectiveness (Figure 21-9).

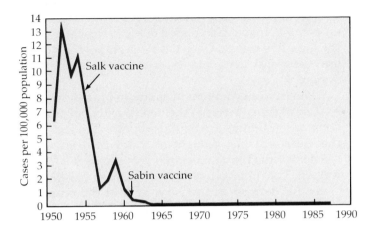

Figure 21-9 Poliomyelitis. Number of cases of poliomyelitis reported in the United States from 1951 to 1987. Note the decrease after introduction of the Salk and Sabin vaccines. In 1987, no new cases were reported. In 1986, only three new cases were reported; these were probably caused by virulent mutants in the Sabin vaccine. [Source: CDC. *Summary of Notifiable Diseases 1986, MMWR* 35(55) and 36(52), 1/8/88.]

Rabies

Rabies is an acute infectious disease that usually results in fatal encephalitis. The causative agent is a *rabies virus* (rhabdovirus) that has a characteristic bullet shape (Figure 21-10). Humans usually acquire the rabies virus from the bite of an infected animal. Even the lick of such an animal is dangerous because rabies virus in its saliva can enter minute scratches, although it does not penetrate unbroken skin. The virus can also cross the intact mucous membrane lining the eyes or body openings. Bats probably transmit the disease among themselves by inhalation of the virus, and scientists investigating bat behavior in caves have contracted rabies, presumably from rabies virus present in aerosols of bat urine in the cave atmosphere. Humans have also acquired rabies by aerosols of the virus when macerating infected tissue in a laboratory blender.

Rabies is unique, in that it usually has an incubation period long enough to allow development of immunity from postexposure vaccination. (The amount of virus introduced into the wound is usually too small to provoke adequate natural immunity in time.) Initially, the virus multiplies in skeletal muscle and connective tissue, where it remains localized for periods ranging from days to months. Then it enters and travels along the peripheral nerves to the central nervous system, where it causes encephalitis. Bites in areas rich in nerve fibers, such as the hands and face, are especially dangerous, and the resulting incuba-

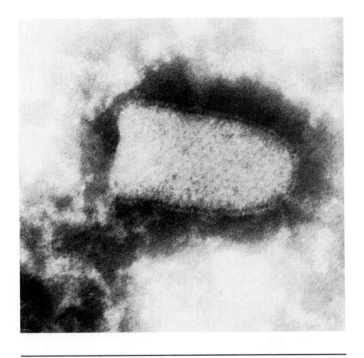

Figure 21-10 Rabies virus. Note the unusual bullet shape (×272,000).

tion period tends to be short. Bites on the hands have an average incubation period of eight weeks; bites on the face, five weeks. Once the virus enters the peripheral nerves, it is no longer accessible to the immune system.

The initial symptoms of rabies in humans include spasms of the muscles of the mouth and pharynx that occur when liquids are being swallowed. In fact, even the mere sight or thought of water can set off the spasms—thus the common name *hydrophobia* (fear of water). The final stages of the disease result from extensive damage to the nerve cells of the brain and the spinal cord. Death is almost inevitable.

Dogs with **furious rabies** are at first restless, then become highly excitable and snap at anything within reach. When paralysis sets in, the flow of saliva increases as swallowing becomes difficult, and nervous control is progressively lost. The disease is almost always fatal within a few days. Some animals suffer from **dumb (paralytic) rabies**, in which there is only minimal excitability. This form is especially common in cats. The animal remains relatively quiet and even unaware of its surroundings, but it might snap irritably if handled. There is some speculation that the two forms of the disease may be caused by slightly different forms of the virus.

Laboratory diagnosis of rabies in humans and animals is based on several findings. When the patient or animal is alive, a diagnosis can sometimes be confirmed by immunofluorescent studies, in which viral antigens are detected in saliva, serum, or cerebrospinal fluid. After death, diagnosis is confirmed by a fluorescent-antibody test performed on the brain tissue.

Any person bitten by an animal that is positive for rabies must take antirabies treatment. Another indication for antirabies treatment is any unprovoked bite by a skunk, bat, fox, coyote, bobcat, or raccoon not available for examination. Treatment after a dog or cat bite, if the animal cannot be found, is determined by the prevalence of rabies in the area.

The original Pasteur treatment, in which the virus was attenuated by drying it in the dissected spinal cords of rabies-infected rabbits, has been replaced by human diploid cell vaccine (HDCV) (Figure 21-11). The HDCV vaccine is administered in a series of injections at intervals during a 28-day period. Passive immunization is provided simultaneously by injecting human rabies immune globulin (RIG) that has been harvested from people who are immunized against rabies, such as laboratory workers.

Some problems have arisen with the HDCV vaccine. Veterinarians and laboratory workers, who must be immunized periodically, have developed allergic reactions. There have been cases of rabies in HDCV-immunized persons in Africa, causing concern that some African rabies may be caused by viruses that

---

MMWR *Centers for Disease Control*
Morbidity and Mortality Weekly Report

HUMAN RABIES IN SAN FRANCISCO

San Francisco's first case of human rabies since 1940 was reported in December 1987. The diagnosis was made postmortem in a 13-year-old Filipino immigrant boy. The patient was well until November 26, when he began experiencing painful lower back spasm. Symptoms progressed, and he presented to the emergency room on November 28, complaining of penile, buttock, and lower back pain. His neurological exam was normal, and he was afebrile, but his white blood cell (WBC) count was 20,200. He was sent home with Tylenol® for pain and Flexeril® for muscle spasm. He returned the next day, reporting priapism (penile erection) and abdominal pain causing him to scratch at his right lower abdomen. He was admitted to the hospital with a differential diagnosis of pyelonephritis or appendicitis. On admission, he had a low-grade fever (38.3°C rectal) and high blood pressure (150/90), with an elevated WBC count (17,000) and creatinine phosphokinase. An exploratory laparotomy and appendectomy were performed, but nothing abnormal was found. Over the next two days, he became increasingly agitated and combative, complaining of blurred vision and an inability to move his legs. Because no focal neurological abnormalities were identified, a psychiatric disorder was also considered in the diagnosis. Two lumbar punctures were normal. A CT scan of the head and an EEG (electroencephalogram) were unremarkable. On December 2, he experienced episodes of subnormal body temperature, abnormally low blood pressure, subnormal rate of ventilation, and slow heart action, and oral secretions increased markedly. He became comatose, was given an oxygen tub, and was then transferred to a university hospital with a discharge diagnosis of metabolic encephalopathy.

After transfer, the absence of corneal reflexes was noted. The first abnormal lumbar puncture was on December 3 (protein=52, WBC=4), but MRI and CAT scans of the head were nondiagnostic. The patient's condition continued to deteriorate, and he died on December 15, 19 days after onset of illness. At autopsy, multiple sections of brain tissue were positive by fluorescent rabies anti-

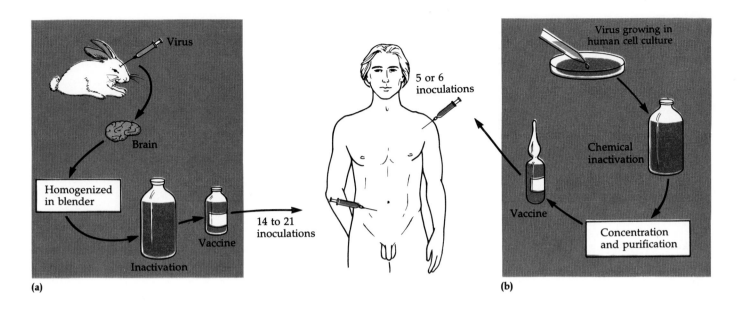

Figure 21-11   Preparation of two rabies vaccines. (a) The original Pasteur vaccine as developed in 1885. **(b)** The more recent Human Diploid Cell Vaccine (HDCV).

are antigenically different from the standard strains used to make the vaccines. Furthermore, the HDCV vaccine is much too expensive for many parts of the world; in these areas, cheaper but less effective vaccines are prepared from animal nerve tissue. How-

ever, new vaccines are being developed. One of the most promising is engineered by introducing the key rabies virus antigens into the vaccinia (smallpox vaccine) virus. This should be a much less expensive vaccine.

body test. No bite marks were recorded by the medical examiner. Serum specimens collected as late as December 9, 13 days after illness began, were negative for antibody, but a postmortem specimen had a titer of 1:65,536. No antibodies were detected in cerebrospinal fluid.

The patient had not traveled outside of California since his arrival from the Philippines in 1981. Except for a three-day camping trip to Sonoma County with 82 classmates and teachers in October 1987, he had not traveled outside of San Francisco in the previous year. No animal contacts were reported during the camping trip or while living in San Francisco. Interviews with relatives, classmates, friends, and teachers have not identified any likely animal contact, and the source of this patient's exposure to the rabies virus remains undetermined. Preliminary monoclonal antibody studies suggest the patient's isolate is compatible with viral isolates from carnivores.

Because of the exposure to the patient and his copious secretions, 12 family members and 75 of 177 health care workers received postexposure rabies prophylaxis.

Comment:

This is the first case of human rabies reported in the United States since July 1985. The last case occurred in July 1985, when a Mexican native died two months after arriving in Texas. He had no known history of exposure to rabies. Five (56%) of the nine rabies cases occurring in the United States since 1980 were acquired outside of the United States. The last California-acquired case occurred in 1969 after a two-year-old San Diego County boy was attacked by a rabid bobcat in his back yard. Including the present case, the source of rabies exposure could not be clearly determined in six (67%) of nine cases occurring in the United States since 1980. Because human rabies is so rare, the diagnosis is difficult in the absence of an exposure history. The diagnosis was made antemortem in only five (56%) of the nine cases since 1980.

Sources: *California Morbidity* 2, Jan. 22, 1988; *MMWR* 37:305–308 (5/20/88).

Figure 21-12 Reported cases of rabies in both wild and domestic animals in the United States from 1953 to 1987. The incidence of the disease has declined in domestic animals, probably because of effective vaccination programs. [Source: CDC. *Summary of Notifiable Diseases 1986, MMWR* 35:55; *MMWR* 36:52 (1/1/88).]

Rabies occurs all over the world. In the United States, 20,000 to 30,000 persons are given the rabies vaccine each year. Nearly twice that many *die* of rabies each year in India, where the vaccine is given to about 3 million persons. Australia, Great Britain, New Zealand, and Hawaii are free of the disease, a condition maintained by rigid quarantine. In North America, rabies is widespread among wildlife (Figure 21-12), the skunk is the wild animal most often reported. In a recent year, skunks accounted for 46% of animal rabies cases; bats, 19%; foxes, 10%; and raccoons, 10%. Among domestic animals, cattle accounted for 6% of the reported animal cases; dogs, 6%; and cats, 4%. Worldwide, however, dogs are the most common carriers of rabies. Rabies is seldom found in squirrels, rabbits, rats, or mice. In fact, the Centers for Disease Control records only two cases in squirrels. Rabies has long been endemic in vampire bats of South America. In Europe and North America, there are ongoing experiments to immunize wild animals with live rabies vaccine in food dropped for the animals to find. In Pennsylvania, for example, inoculated sandwiches have been used to immunize raccoons. It is too early to assess the results of these attempts. In the United States, up to 6000 cases of rabies are diagnosed in animals each year, but in recent years, only 9 cases have been diagnosed in humans (see the box on p. 536).

Arthropod-Borne Encephalitis

Encephalitis, caused by a mosquito-borne arbovirus (encephalitis virus) is rather common in the United States. Figure 21-13 shows the incidence of encephalitis in the United States from 1974 to 1987. The increase in the summer months coincides with the proliferation of adult mosquitoes during these months. A number of clinical types of this disease have been identified; all cause inflammation of the brain, spinal cord, and meninges. Active cases of these diseases are characterized by chills, headache, and fever. As the disease progresses, mental confusion and coma are observed. There is evidence that subclinical infections are commonplace.

Outbreaks of encephalitis are reported regularly when the mosquito vectors and animal reservoirs for the virus coincide. The main viral reservoir is birds, and the mosquito vector is usually a member of the genus *Culex* (kū′leks). Sentinel animals, such as caged rabbits or chickens, are tested periodically for antibodies to arboviruses. This gives health officials information on the incidence and types of virus in their area.

Horses as well as humans are frequently infected by such viruses, so there are strains called Eastern equine encephalitis (EEE) and Western equine encephalitis (WEE). St. Louis encephalitis (SLE) and California encephalitis (CE) do not infect horses at all. Most cases of CE occur among children and are caused by its La Crosse strain. This disease is endemic in a reservoir of small woodland mammals in the upper Midwest. WEE, SLE, and CE encephalitis tend to cause relatively mild diseases in humans, but there is a significant rate of serious complications. EEE is the least prevalent arboencephalitis, but it is the most severe, with a mortality rate of 25% or more and a high incidence of brain damage, deafness, and similar neurological problems in survivors.

The Far East also has endemic encephalitis. **Japanese B encephalitis** is the best known; it is a serious public health problem, especially in Japan, Korea, and China. Vaccines are used to control the disease in these countries.

Diagnosis of arthropod-borne encephalitis is usually made by serological tests, such as complement fixation. The most effective control measure is local elimination of the *Culex* mosquitoes. Persons exposed to the disease can be given immune serum to provide passive immunity, but there is no specific treatment for the disease.

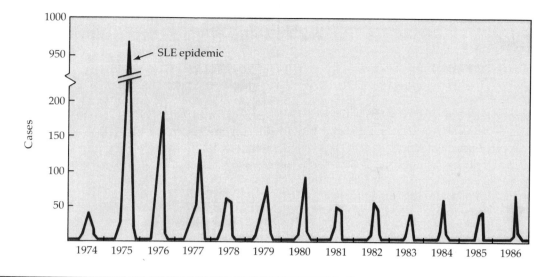

Figure 21-13 Reported encephalitis cases. Reported cases of arthropod-borne encephalitis, by month of onset, in the United States from 1974 to 1986. Note the peaks during the summer mosquito season and the exceptional outbreak of St. Louis encephalitis (SLE) in 1975. [Source: CDC. *Summary of Notifiable Diseases 1986, MMWR 35:55; MMWR 36:52 (1/1/88).*]

Fungal Disease of the Nervous System

Cryptococcus neoformans Meningitis (Cryptococcosis)

Fungi of the genus *Cryptococcus* are spherical cells resembling yeasts; they reproduce by budding and produce polysaccharide capsules, some much thicker than the cells themselves (Figure 21-14). Only one species, *Cryptococcus neoformans* (krip-tō-kok′kus nē-ō-fôr′manz), is pathogenic for humans, causing the disease called **cryptococcosis**. The organism is widely distributed in soil, especially soil enriched by pigeon droppings. It is also found in pigeon roosts and nests on the window ledges of urban buildings. Most cases of cryptococcosis occur in urban areas. It is thought to be transmitted by the inhalation of dried infected pigeon droppings.

Inhalation of *C. neoformans* initially causes infection of the lungs, frequently subclinical, and often the disease does not proceed beyond this stage. However, it can spread through the bloodstream to other parts of the body, including the brain and meninges, especially in immunosuppressed individuals and persons receiving steroid treatments for major illnesses. The disease is usually expressed as chronic meningitis, which is usually progressive and fatal if untreated.

Laboratory diagnosis of cryptococcosis is by identification of the encapsulated microorganism in an India ink wet mount and by observation of its virulence in mice. The drugs of choice for treatment are amphotericin B and flucytosine.

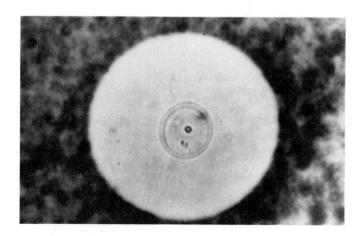

Figure 21-14 *Cryptococcus neoformans.* This yeastlike fungus has an unusually thick capsule. In this micrograph, the capsule is made visible by suspending the cells in dilute India ink (×800).

Protozoan Diseases of the Nervous System

African Trypanosomiasis

African trypanosomiasis, or sleeping sickness, is a protozoan disease that affects the nervous system (see the box in Chapter 11, p. 302). In 1907, Winston Churchill described Uganda during an epidemic of sleeping sickness as a "beautiful garden of death." Even today, the disease affects about a million people in Central and East Africa, and about 20,000 new cases are reported each year.

The disease is caused by *Trypanosoma brucei gambiense* (tri-pa-nō-sō′mä brüs′ē gam-bē-ens′) and *Trypanosoma brucei rhodesiense* (rō-dē-sē-ens′), flagellates that are injected by the bite of a tsetse fly (*Glossina*). The two organisms cause similar diseases (although *T. b. rhodesiense* is more likely to be quickly fatal), and they occupy somewhat different geographical areas. Reservoirs for the disease are domestic cattle and wild antelope that frequent the riverbank habitats of the tsetse fly in Africa.

During the early stages of the disease, a few trypanosomes can be found in the blood. During later stages, the trypanosomes move into the cerebrospinal fluid. Symptoms of the disease include decreases in physical activity and mental acuity. Untreated, the host enters a coma, and death is almost inevitable.

There are some effective chemotherapeutic agents, such as suramin and pentamidine isethionate. However, the drugs produce toxic effects, and, because they do not cross the blood–brain barrier, they are effective only when the central nervous system has not become affected. A toxic arsenical, called melarsoprol, is usually chosen when the central nervous system is affected. A vaccine is being developed, but it might be unsuccessful because the trypanosome is able to change protein coats at least 100 times and can thus evade antibodies aimed at only one or a few of the proteins. Each time the body's immune system is successful in suppressing the trypanosome, a new clone of parasites appears with a different antigenic coat (Figure 21-15).

Naegleria Microencephalitis

Naegleria fowleri (nī-gle′rē-ä fou′lėr-ē) is a protozoan (amoeba) that is known to cause the neurological disease ***Naegleria* microencephalitis**. Although cases of it are reported in most parts of the world, only a few cases per year are reported in the United States. The most common victims are children who swim in ponds or streams. The organism initially infects the nasal mucosa and later proliferates in the brain. The fatality rate is nearly 100%.

Diseases of the Nervous System Caused by Unconventional Agents

There are a number of fatal diseases of the human central nervous system thought to be caused by what have been called **slow viruses**. However, no one is absolutely sure of the causative agents, which in most cases are clearly not conventional viruses. Therefore we prefer the term **unconventional agents** (see Chapter 12). The name "slow virus" alludes to the slow course of the diseases, which are characterized by long incubation times. The damage to the central nervous system is insidious and progressive, without the fever and inflammation seen in encephalitis. In recent years, the study of slow virus diseases has been one of the most interesting areas of microbiology in medical science (Chapter 12).

A typical slow virus disease is **sheep scrapie**. The infected animal rubs against fences and walls until areas on its body are raw. During a period of several weeks or months, the animal gradually loses motor control and dies. The agent, whatever it is, can be passed to other animals, such as mice, by the injection of brain tissue from one animal to the next. The scrapie agent can give minks a disease that is indistinguishable from **transmissible mink encephalopathy**, a disease found in minks that are fed mutton at mink farms.

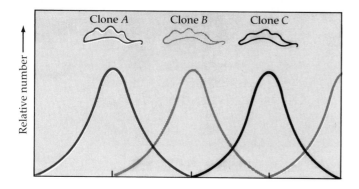

Figure 21-15 Trypanosomiasis. The population of each trypanosome clone drops nearly to zero as the immune system suppresses them, but a new clone with a different antigenic surface then replaces it.

Humans suffer from diseases similar to scrapie. **Creutzfeldt–Jakob disease (CJD)** is an example. This disease is rare (about 200 cases per year in this country) but often occurs in families, which suggests that the infection is acquired before the age of six months. There is no doubt that there is an infective agent involved since transmission via corneal transplants and accidental scalpel nicks of a surgeon during autopsy have been reported. Several cases have been traced to injection of a growth hormone made from human tissue.

Some tribes in New Guinea have suffered from a slow virus disease called **kuru**. It and CJD are sometimes so similar that it is suspected that kuru may be caused by a mutation of the CJD agent. Kuru is apparently related to the practice of ritualistic cannibalism. In some cannibalistic tribes, the women cut up the brain and packed it into bamboo cylinders, where it was steamed before being eaten. Some brain tissue was also ceremonially smeared onto the body and hair. The infection resulted when the agent was introduced into open sores or cuts. The disease is disappearing as cannibalism dies out.

As we discussed in Chapter 12, the unconventional agents that cause these diseases seem to be pure protein. If nucleic acid is present, it has not been detected to date. The exact nature of these agents and their mode of replication are still mysteries. They continue to be the subject of intense research—and contention.

TABLE 21-3 *Diseases Associated with the Nervous System*

	Causative Agent	Mode of Transmission	Treatment
Bacterial Diseases			
Meningococcal meningitis	*Neisseria meningitidis*	Contact with a healthy carrier via respiratory tract	Penicillin, chloramphenicol, rifampin
Hemophilus influenzae meningitis	*Hemophilus influenzae*	Via respiratory tract in individuals with viral infections	Rifampin
Botulism	*Clostridium botulinum*	Ingestion of toxin	Antitoxins and respiratory support
Leprosy	*Mycobacterium leprae*	Transfer of exudates from lesions or inanimate objects	Sulfone drugs (dapsone), rifampin, clofazimine
Tetanus	*Clostridium tetani*	Deep anaerobic wounds	Immune globulin and respiratory support
Fungal Disease			
Cryptococcosis	*Cryptococcus neoformans*	Inhalation of dried infected pigeon droppings	Amphotericin B, flucytosine
Viral Diseases			
Poliomyelitis	Poliovirus	Ingestion of virus	None
Rabies	Rabies virus	Bite of a rabid animal	RIG and HDCV after exposure
Arthropod-borne encephalitis	Arboviruses	*Culex* mosquitoes	Immune globulin
Unconventional Agents			
Creutzfeldt–Jakob disease	Unconventional agent	Ingestion or accidental wounds	None
Kuru	Unconventional agent	Ingestion or wound contamination	None

Study Outline

Structure and Function of the Nervous System   (pp. 524–525)

1. The central nervous system consists of the brain, which is protected by the skull bones, and the spinal cord, which is protected by the backbone.
2. The peripheral nervous system consists of the nerves that branch from the central nervous system.
3. The central nervous system is covered by three layers of membranes called meninges. Cerebrospinal fluid circulates between the inner and middle meninges and in the ventricles of the brain.
4. The blood–brain barrier normally prevents many substances, such as antibiotics, from entering the brain.
5. Microorganisms can enter the central nervous system through trauma, along peripheral nerves, and through the bloodstream and lymphatic system.
6. An infection of the meninges is called meningitis. An infection of the brain is called encephalitis.

Bacterial Diseases of the Nervous System   (pp. 526–533)

Bacterial Meningitis   (pp. 526–528)

1. Meningitis can be caused by viruses, bacteria, fungi, and protozoans.
2. Nearly 50 species of opportunistic bacteria can cause meningitis.

Neisseria Meningitis (Meningococcal Meningitis)   (pp. 526–527)

1. *Neisseria meningitidis* causes meningococcal meningitis. This bacterium is found in the throats of healthy carriers.
2. The bacteria probably gain access to the meninges through the bloodstream. The bacteria may be found in leukocytes in cerebrospinal fluid.
3. Symptoms are due to endotoxin. Purplish spots appear on the skin. The disease occurs most often in young children.
4. Military recruits are vaccinated with purified capsular polysaccharide to prevent epidemics in training camps.
5. Encapsulated *N. meningitidis* are resistant to phagocytosis.
6. Diagnosis is based on isolation and identification of the bacteria in blood or cerebrospinal fluid.
7. *N. meningitidis* must be cultured on media containing blood and incubated in an atmosphere containing 5% to 10% carbon dioxide.

Hemophilus influenzae Meningitis   (pp. 527–528)

1. *Hemophilus influenzae* is cultured on chocolate agar containing X and V factors.
2. *H. influenzae* is the most common cause of meningitis in children under four years old.
3. The disease often occurs as a secondary infection following a viral infection such as influenza.
4. A moderately successful vaccine (Hib) directed against the capsular polysaccharide antigen is available.
5. Diagnosis is based on identification of the organism in spinal fluid.

Streptococcus pneumoniae Meningitis (Pneumococcal Meningitis)   (p. 528)

1. Hospitalized patients and young children are most susceptible to *S. pneumoniae* meningitis.

Listeriosis   (pp. 528–529)

1. *Listeria monocytogenes* causes meningitis in newborns, the immunosuppressed, pregnant women, and cancer patients.
2. It is contracted by ingestion of contaminated food.
3. Listeriosis may be asymptomatic in healthy adults.
4. *L. monocytogenes* can cross the placenta and cause spontaneous abortion and stillbirth.

Tetanus   (pp. 529–530)

1. Tetanus is caused by a localized infection of a wound by *Clostridium tetani*.
2. *C. tetani* produces the neurotoxin tetanospasmin, which causes the symptoms of tetanus—spasms, contraction of muscles controlling the jaw, and death resulting from spasms of respiratory muscles.
3. *C. tetani* is an anaerobe that will grow in unclean wounds and wounds with little bleeding.
4. Acquired immunity results from DPT immunization that includes tetanus toxoid.
5. Following an injury, an immunized person may receive a booster of tetanus toxoid. An unimmunized person may receive (human) tetanus immune globulin.
6. Debridement (removal of tissue) and antibiotics may be used to control the infection.

Botulism   (pp. 530–531)

1. Botulism is caused by an exotoxin produced by *C. botulinum* growing in foods.
2. Serological types of botulinum toxin vary in virulence, with type A being the most virulent.

3. The toxin is a neurotoxin that inhibits transmission of nerve impulses, resulting in respiratory and cardiac failure.
4. Blurred vision occurs in one to two days; progressively flaccid paralysis follows for one to ten days. There can be 60% to 70% mortality in untreated cases.
5. *C. botulinum* will not grow in acidic foods or in an aerobic environment.
6. Endospores are killed by using a pressure cooker for canning. Addition of nitrites to foods inhibits outgrowth after endospore germination.
7. The toxin is heat labile and destroyed by boiling (100°C) for five minutes.
8. Wound botulism occurs when *C. botulinum* grows in anaerobic wounds.
9. Infant botulism results from the growth of *C. botulinum* in an infant's intestines.
10. For diagnosis, mice protected with antitoxin are inoculated with toxin from the patient or foods.

Leprosy (pp. 532–533)

1. *Mycobacterium leprae* causes leprosy, or Hansen's disease.
2. *M. leprae* cannot be cultured on artificial media. It can be cultured in armadillos.
3. The tuberculoid form of the disease is characterized by loss of sensation in the skin surrounded by nodules. The lepromin test is positive.
4. In the lepromatous form, disseminated nodules and tissue necrosis occur. The lepromin test is negative.
5. Leprosy is not highly contagious and is spread by prolonged contact with exudates and fomites.
6. Untreated individuals often die of secondary bacterial complications, such as tuberculosis.
7. Patients with leprosy are made noncommunicable within four or five days with sulfone drugs and then treated as outpatients.
8. The disease occurs primarily in the tropics. Many people are treated at the National Leprosy Hospital in Carville, Louisiana.
9. Laboratory diagnosis is based on observations of acid-fast rods in lesions or fluids.

Viral Diseases of the Nervous System (pp. 533–539)

Poliomyelitis (pp. 533–535)

1. The symptoms of poliomyelitis are usually headache, sore throat, fever, stiffness of the back and neck, and occasionally paralysis (less than 1% of the cases).
2. Poliovirus (picornavirus) is found only in humans and is transmitted by ingestion of water contaminated with feces.
3. Poliovirus first invades lymph nodes of the neck and small intestine. Viremia and spinal cord involvement may follow.
4. At present, outbreaks of polio in the United States are uncommon because of the use of vaccines.
5. The Salk vaccination involves injection of formalin-inactivated viruses and boosters every few years. The Sabin vaccine contains three attenuated live strains of poliovirus and is administered orally.
6. Diagnosis is based on isolation of the virus from feces and throat secretions or the presence of virus-neutralizing antibodies in the patient's serum.

Rabies (pp. 535–538)

1. Rabies virus (a rhabdovirus) causes an acute, usually fatal, encephalitis called rabies.
2. Rabies may be contracted through the bite of a rabid animal, by inhalation of aerosols, or invasion through minute skin abrasions. The virus multiplies in skeletal muscle and connective tissue.
3. Encephalitis occurs when the virus moves along peripheral nerves to the central nervous system.
4. Symptoms of rabies include spasms of mouth and throat muscles followed by extensive brain and spinal cord damage and death.
5. Laboratory diagnosis may be made by direct immunofluorescent tests of saliva, serum, and cerebrospinal fluid or brain smears.
6. Reservoirs for rabies in the United States include skunks, bats, foxes, and raccoons. Domestic cattle, dogs, and cats may get rabies. Rodents and rabbits seldom get rabies.
7. The Pasteur treatment for rabies involves multiple subcutaneous injections of rabies virus grown in rabbit brain tissue.
8. Current postexposure treatment includes administration of human rabies immune globulin (RIG) followed by multiple intramuscular injections of human diploid cell vaccine (HDCV).
9. Preexposure immunization consists of injections of HDCV.

Arthropod-Borne Encephalitis (pp. 538–539)

1. Symptoms of encephalitis are chills, headache, fever, and eventually coma.
2. Many types of arboviruses (encephalitis viruses) transmitted by *Culex* species mosquitoes cause encephalitis. The reservoir is birds.
3. Horses are frequently infected by EEE and WEE viruses.
4. Outbreaks occur primarily during the summer months when adult mosquitoes are present.
5. Diagnosis is based on serological tests.
6. Immune globulin may provide passive immunity, but no treatment is available.
7. Elimination of the vector is the most effective control measure.

Fungal Disease of the Nervous System (p. 539)

***Cryptococcus neoformans* Meningitis (Cryptococcosis)** (p. 539)

1. *Cryptococcus neoformans* is an encapsulated yeastlike fungus that causes *Cryptococcus neoformans* meningitis.
2. The disease may be contracted by inhalation of dried infected pigeon droppings.

3. The disease begins as a lung infection and then spreads to the brain and meninges.
4. Immunosuppressed individuals are most susceptible to *Cryptococcus neoformans* meningitis.
5. Diagnosis is based on observation of the fungus.

Protozoan Diseases of the Nervous System (p. 540)

African Trypanosomiasis (p. 540)

1. African trypanosomiasis is a disease caused by the protozoans *Trypanosoma brucei gambiense* or *T. b. rhodesiense* and transmitted by the bite of the tsetse fly (*Glossina*).
2. The disease affects the nervous system of the human host, causing lethargy and eventually coma. It is commonly called sleeping sickness.

Naegleria Microencephalitis (p. 540)

1. Encephalitis caused by the protozoan *N. fowleri* is almost always fatal.

Diseases of the Nervous System Caused by Unconventional Agents (pp. 540–541)

1. The diseases called slow virus diseases are of uncertain etiology. The diseases progress slowly in the host.
2. Sheep scrapie and transmissible mink encephalopathy are examples of diseases caused by unconventional agents that are transferable from one animal to another.
3. Creutzfeld–Jakob disease and kuru are human diseases similar to scrapie. Kuru occurs in isolated groups of cannibals who eat brains.
4. The unconventional agents may be pure protein, or protein associated with undetectable nucleic acid.

Study Questions

Review

1. Differentiate between meningitis and encephalitis.

2. Fill in the following table.

Causative Agent of Meningitis	Susceptible Population	Method of Transmission	Treatment
N. meningitidis			
H. influenzae			
S. pneumoniae			
L. monocytogenes			
C. neoformans			

3. Briefly explain the derivation of the name *Hemophilus influenzae*.

4. If *Clostridium tetani* is relatively sensitive to penicillin, then why doesn't penicillin cure tetanus?

5. What treatment is used against tetanus under the following conditions?
(a) Before a person suffers a deep puncture wound
(b) After a person suffers a deep puncture wound

6. Why is the following description used for wounds that are susceptible to *C. tetani* infection: ". . . Improperly cleaned deep puncture wounds . . . ones with little or no bleeding . . ."

7. List the following information for botulism: etiologic agent, suspect foods, symptoms, treatment, conditions necessary for microbial growth, basis for diagnosis, prevention.

8. Provide the following information on leprosy: etiology, method of transmission, symptoms, treatment, prevention, and susceptible population.

9. Provide the following information on poliomyelitis: etiology, method of transmission, symptoms, prevention. Why aren't Salk and Sabin vaccines considered treatments for poliomyelitis?

10. Compare and contrast the Salk and Sabin vaccines with respect to composition, advantages, and disadvantages.

11. Provide the etiology, method of transmission, reservoirs, and symptoms for rabies.

12. Outline the procedures for treating rabies after exposure. Outline the procedures for preventing rabies prior to exposure. What is the reason for the differences in the procedures?

13. Fill in the following table.

Disease	Etiology	Vector	Symptoms	Treatment
Arthropod-borne encephalitis				
African trypanosomiasis				

14. Why is the incidence of arboviral encephalitis in the United States higher in the summer months?

15. Describe the symptoms and laboratory test results that would lead to a diagnosis of slow virus disease.

16. Provide evidence that slow virus diseases are caused by transmissible agents that resemble viruses.

17. Why are meningitis and encephalitis generally difficult to treat?

Challenge

1. Most of use have been told that a "rusty nail" causes tetanus. What do you suppose is the origin of this adage?

2. A BCG vaccination will result in positive lepromin and tuberculin tests. What is the relationship between leprosy and tuberculosis?

3. Compare the Pasteur treatment with HDCV

treatment for rabies. What are the advantages of HDCV over the Pasteur treatment? Differentiate between preexposure vaccination and postexposure vaccination for rabies.

4. A one-year-old infant was lethargic and had a fever. When he was admitted to the hospital, he had multiple brain abscesses with gram-negative coccobacillary rods. Identify the disease, etiology, and treatment.

5. One week after bathing in a hot spring, a nine-year-old girl was hospitalized after a three-day history of progressive, severe headaches, nausea, lethargy, and stupor. Examination of CSF revealed ameboid organisms. Identify the disease, etiology, and treatment.

Further Reading

Bingham, R. "Outrageous ardor." *Science 81* 2:54–61, September 1981. An interesting article on slow viruses and their discoverer, Carleton Gajdusek.

Chasan, D. J. "The polio paradox." *Science 86*, 7:37–39, April 1986. A discussion about one little-used vaccine and another that is the leading cause of polio.

Gaylord, H., and P. J. Brennan. "Leprosy and the leprosy bacillus: recent developments." *Annual Review of Microbiology* 41:645–675, 1987. A review of new information about antigens and immunity to *M. leprae* from armadillo cultures.

Hall, S. S. "The La Crosse file." *Science 84* 5:54–62; July–August 1984. The epidemiology of California encephalitis virus.

John, D. T. "Primary amoebic meningoencephalitis and the biology of *Naegleria fowleri*." *Annual Review of Microbiology* 36:101–123, 1982. A good article on the natural history of the disease and the parasite.

Kaplan, C., G. S. Turner, and D. A. Warrell. *Rabies, the Facts*, 2nd ed. New York: Oxford University Press, 1986. A summary of information on rabies.

Kristensson, K., and E. Norrby. "Persistence of RNA viruses in the central nervous system." *Annual Review of Microbiology* 40:159–184, 1986. Descriptions of viruses such as measles virus that remain in the CNS for years and of the resultant host changes.

Rosen, L. "The natural history of Japanese encephalitis virus." *Annual Review of Microbiology* 40:395–414, 1986. A good lesson in natural history that describes alternate hosts and overwintering.

Schlech, W. F., et al. "Bacterial meningitis in the United States, 1978 through 1981; The National Bacterial Meningitis Study." *Journal of the American Medical Association* 253:1649–1754, 1985. Analysis and discussion of incidence and types of bacterial meningitis.

Smith, L. *Botulism: The Organism, Its Toxins, The Disease.* Springfield, Ill.: Thomas, 1977. A summary of information on botulism.

Spector, D. H., and D. Baltimore. "The molecular biology of poliovirus." *Scientific American* 232(5):24–31, May 1975. Information on viral multiplication is obtained from laboratory-grown poliovirus.

See also Further Reading for Part Four at the end of Chapter 25.

Microbial Diseases of the Cardiovascular System

The **cardiovascular system** consists of the heart, blood, and blood vessels. The **lymphatic system** consists of the lymph, lymph vessels, lymph nodes, and lymphoid organs (tonsils, appendix, spleen, and thymus gland—see Figure 15-5). Because both systems circulate various substances throughout the body, they can serve as vehicles for the spread of infection. A summary of diseases associated with the cardiovascular system is presented in Table 22-1, p. 570.

Structure and Function of the Cardiovascular System

The center of the cardiovascular system is the *heart* (Figure 22-1). The function of the heart is to circulate the blood through the body's tissues so it can deliver certain substances to cells and remove other substances from them. *Blood vessels* are the tubes that carry circulating blood throughout the body. An *artery* is a blood vessel that carries blood from the heart to different tissues of the body. When the arteries reach their destinations, they branch into smaller vessels called *arterioles*, which in turn branch into even smaller vessels called *capillaries*. Capillary walls are only one cell thick, and it is through these walls that blood and tissue cells exchange materials. The capillaries that take blood away from the tissue cells converge to form small veins called *venules*. Venules unite to form *veins*, and veins return the blood to the heart.

The *blood* itself is a mixture of cells and a liquid called plasma. The *plasma* transports dissolved nutrients to body cells and carries wastes away from the cells. The cells in blood include red blood cells, white blood cells, and platelets. Red blood cells (erythrocytes) carry oxygen and some carbon dioxide, although most of the carbon dioxide in blood is dis-

Learning Objectives

- List the symptoms of septicemia, and explain the importance of infections that develop into septicemia.

- Discuss the epidemiologies of puerperal sepsis, bacterial endocarditis, myocarditis, and Burkitt's lymphoma.

- Discuss the causes, treatments, and preventive measures of rheumatic fever and infectious mononucleosis.

- Describe the epidemiologies of tularemia, brucellosis, anthrax, and gas gangrene.

- Compare and contrast the causative agents, vectors, reservoirs, symptoms, and treatments of malaria, yellow fever, dengue, relapsing fever, typhus, Rocky Mountain spotted fever, and Lyme disease.

- List the causative agents and methods of transmission of toxoplasmosis, American trypanosomiasis, schistosomiasis, and viral hemorrhagic fevers. Also describe their worldwide effects on health.

- Identify the best control measures for vector-borne diseases.

solved in the plasma. White blood cells (leukocytes) play several roles in defending the body against infection, as was discussed in Part Three. Some white cells, in particular the neutrophils, are phagocytes. B cells and T cells, two types of lymphocytes, play key roles in immunity. Cell fragments called *platelets* are involved in blood clotting.

Structure and Function of the Lymphatic System

As part of the overall pattern of circulation, some plasma filters out of the blood capillaries and into spaces between tissue cells. These spaces are called *interstitial spaces*. The fluid circulating around and between tissue cells is called *interstitial fluid*. Micro-scopic lymphatic vessels that surround tissue cells are called *lymph capillaries*; they are larger and more permeable than blood capillaries. As the interstitial fluid moves around the tissue cells, it is picked up by the lymph capillaries; the fluid is then called *lymph* (Figure 22-1).

Because lymph capillaries are very permeable, they readily pick up microorganisms or their products. From lymph capillaries, lymph is transported into larger lymph vessels called *lymphatics*, which contain valves that keep the lymph moving toward the heart. Eventually, all the lymph is returned to the blood just before the blood enters the right atrium of the heart. As a result of this circulation, proteins and fluid that have filtered from the plasma are returned to the blood.

At various points along the lymphatic system are oval or bean-shaped structures called *lymph nodes*,

Figure 22-1 The cardiovascular and lymphatic systems. Details of circulation to head and extremities are not shown in this simplified diagram. The blood circulates from the heart through the arterial system (lighter color) to the capillaries in the lungs and other parts of the body. From these capillaries, the blood returns through the venous system (gray) to the heart. From the capillaries, some plasma filters into the surrounding tissue and enters the lymph capillaries (inset A). This fluid, now called lymph, returns to the heart through the lymphatic circulatory system (darker color). All lymph returning to the heart must pass through at least one lymph node, as shown in inset B. See also Figure 15-5.

through which lymph flows. Within the lymph nodes are cells capable of phagocytosis (fixed macrophages), which help to clear the lymph of microorganisms. At times, the number of microorganisms circulating through lymph nodes is so great that the nodes themselves become infected, and the nodes become enlarged and tender. The lymph nodes are also important to the body's immune reaction against invading microorganisms. Microorganisms entering the lymph nodes and other lymphoid tissues encounter two types of lymphocytes—B cells that are stimulated to produce humoral antibodies and T cells that differentiate into effector cells (see Figure 16-8).

Bacterial Diseases of the Cardiovascular System

Septicemia

Although blood is normally sterile, moderate numbers of microorganisms are usually not harmful. However, if the defenses of the blood and lymphatic systems fail, the microorganisms can proliferate in the blood without control. That condition is called **septicemia**, or blood poisoning. One symptom of septicemia is the appearance of **lymphangitis**, inflamed lymph vessels shown as red streaks under the skin, running along the arm or leg from the infection site. Sometimes the streaks end at a lymph node, where the lymphocytes attempt to stop the invading microorganisms. Clinically, a person suffering from septicemia exhibits the symptoms of **septic shock**. The small arteries and veins are constricted, peripheral blood circulation decreases, and tissues served by the peripheral circulation become oxygen-starved. The patient develops rapid breathing and heartbeat, a drop in blood pressure, and mental confusion. The extremities become cool and may turn blue. Septic shock may progress to coma and death.

The organisms most frequently associated with septicemia are gram-negative rods, although a few gram-positive bacteria and fungi are also implicated. Among the gram-negative rods are *Escherichia coli*, *Serratia marcescens*, *Proteus mirabilis*, *Enterobacter aerogenes*, *Pseudomonas aeruginosa*, and *Bacteroides* species. These bacteria enter the blood from a focus of infection in the body. As you may recall, the cell walls of many gram-negative bacteria contain endotoxins that are released upon the lysis of the cell. It is the endotoxin that actually causes the symptoms. Once released, the endotoxin damages blood vessels;

this damage causes the low blood pressure and subsequent shock. In some cases, antibiotics aggravate the condition by causing the lysis of large numbers of bacteria that then release damaging endotoxins. Many of these gram-negative rods are of nosocomial origin and were introduced into the bloodstream by medical procedures, such as catheters or feeding tubes, that bypass the normal barriers between the environment and the blood.

Puerperal Sepsis

Puerperal sepsis, also called **puerperal fever** and **childbirth fever**, is a nosocomial infection that frequently leads to septicemia. This begins as an infection of the uterus as a result of childbirth or abortion. *Streptococcus pyogenes*, a group A β-hemolytic streptococcus, is the most frequent cause, although other organisms may cause infections of this type. Recall from Chapter 20 that group A β-hemolytic streptococci have an M protein on their surface that increases virulence by improving adherence to the mucous membranes and by increasing resistance to phagocytosis.

Puerperal sepsis progresses from an infection of the uterus to an infection of the abdominal cavity (peritonitis) and in many cases to septicemia. At one Paris hospital between 1861 and 1864, of the 9886 women who gave birth, 1226 (12%) died of such infections. At that time, the death rate from puerperal sepsis outside such hospitals was frequently twice that. These deaths were largely unnecessary. Some 20 years before, Oliver Wendell Holmes in America and Ignaz Semmelweiss in Austria had clearly demonstrated that the disease was transmitted by the hands and instruments of the attending midwives or physicians and that disinfection of hands and instruments could prevent such transmission. Yet, in 1879, Louis Pasteur still found it necessary to lecture physicians as to the cause of the disease.

Antibiotics and modern hygienic practices have now made puerperal sepsis an uncommon complication of childbirth. In the environment of the disease, penicillin and erythromycin are effective against *S. pyogenes*. Infections related to improperly performed abortions—not strictly puerperal sepsis—are often of mixed bacterial types, many of them caused by anaerobic bacteria of the *Bacteroides* or *Clostridium* genera. Chloramphenicol and penicillin are effective in these cases.

Bacterial Endocarditis

The wall of the heart consists of three layers. The outer layer, the *pericardium*, is a sac that encloses the heart. The middle layer, the *myocardium*, is the thickest layer and consists of cardiac muscle tissue. The inner layer is a lining of epithelium and is called the *endocardium*. This layer lines the heart muscle itself and covers the valves in the heart. An inflammation of the endocardium is called **endocarditis**.

One type of bacterial endocarditis, **subacute bacterial endocarditis**, is characterized by fever, anemia, general weakness, and heart murmur. It is usually caused by α-hemolytic streptococci, although β-hemolytic streptococci or staphylococci can be involved. About 5% to 10% of the cases are caused by enterococci. The condition probably arises from a focus of infection elsewhere in the body, such as in the teeth or tonsils. Microorganisms are released by tooth extractions or tonsillectomies, enter the blood, and find their way to the heart. Normally, such bacteria would be quickly cleared from the blood by the body's defensive mechanisms. However, in persons whose heart valves are abnormal, because of either congenital heart defects or such diseases as rheumatic fever and syphilis, the bacteria lodge in the preexisting lesions. Within the lesions, the bacteria multiply and become entrapped in blood clots that protect them from phagocytes and antibodies. As multiplication progresses and the clot gets larger, pieces of the clot break off and can occlude blood vessels or lodge in the kidneys. In time, the function of the heart valves is impaired. Left untreated, subacute bacterial endocarditis is invariably fatal.

Figure 22-2 Acute bacterial endocarditis. The arrow points to an infected, ulcerated lesion on a heart valve.

Another type of bacterial endocarditis is **acute bacterial endocarditis** (Figure 22-2), which is caused by *Staphylococcus aureus* and *Streptococcus pneumoniae*. These organisms find their way from the initial site of infection to normal or abnormal heart valves; the rapid destruction of the heart valves is frequently fatal within days or weeks. Streptococci can also cause **pericarditis**, inflammation of the sac around the heart (the pericardium).

In a laboratory diagnosis of bacterial endocarditis, the bacteria are isolated from a blood culture. Penicillin is the most commonly used drug in treatment. It is good to use it prophylactically in susceptible persons, such as those suffering from rheumatic fever or undergoing tooth extractions or tonsillectomies. Erythromycin is also effective in many cases.

Rheumatic Fever

Group A β-hemolytic streptococcal infections, such as those caused by *Streptococcus pyogenes*, sometimes lead to **rheumatic fever**, which is generally considered an autoimmune type of complication (Chapter 18). The disease is usually expressed as an arthritis, especially in older persons. Another frequent form is an inflammation of the heart, which damages the valves. Other symptoms include fever and malaise. Although rheumatic fever usually does no permanent damage to the joints, there can be permanent heart damage.

The disease is usually precipitated by a streptococcal sore throat. One to five weeks later, after the original infection has probably disappeared, evidence of heart abnormalities appear. Reinfection with streptococci (for example, another sore throat) renews the attack and causes further damage to the heart. Perhaps as many as 3% of children with untreated β-hemolytic streptococcal infections contract rheumatic fever. However, many of these cases are essentially subclinical.

The exact mechanism by which streptococci produce rheumatic fever is still obscure, but some data point to an immunological reaction that is localized in the heart and joints. For example, infection with group A β-hemolytic streptococci may result in deposition of streptococcal antigens such as M proteins in the joints and heart. When the body produces antibodies against the antigens, the antigen–antibody reaction might cause the damage. It is also possible that a streptococcal antigen is cross-reactive with components of heart muscle and produces antibodies that subsequently damage the heart. Persons with symptoms of damage from rheumatic fever do have relatively high titers of antibody against streptococcal antigens. Genetic factors might also play a role in this disease. It has been demonstrated that persons of certain MHC types (Chapter 18) are about 15 times more at risk than the normal population.

The number of cases of rheumatic fever has been declining in this country because of early treatment of streptococcal infections. In the 1940s, there were an estimated 250,000 cases annually; by the 1980s, the disease had become rare. Some find this difficult to explain when group A streptococci are so frequently found in the throats of schoolchildren. One theory is that the streptococci may have mutated and lost some virulence factor, and there has been concern that another mutation might reestablish the virulence. This concern was reinforced in 1987 when a large cluster of cases appeared in a limited area of Utah. The disease remains a major health problem in much of Asia, Africa, and South America.

Penicillin administered to patients with rheumatic fever is a prophylactic. The penicillin will not alleviate the symptoms (which are treated with antiinflammatory drugs), but it will prevent subsequent infections that could cause a recurrence of rheumatic fever.

Tularemia

Tularemia is a disease caused by a small, gramnegative, facultatively anaerobic, pleomorphic, rod-shaped bacterium, *Francisella tularensis* (fran-sis-

el'lä tü-lä-ren'sis). The microorganism was named for Tulare County, California, where it was originally observed in ground squirrels in 1911.

Tularemia exhibits one of several forms, depending on whether the infection is acquired by inhalation, ingestion, bites, or, most commonly, contact through minor skin breaks. The first sign of infection is usually local inflammation and a small ulcer at the site of infection. About a week after infection, the regional lymph nodes enlarge; many will contain pockets filled with pus. If the disease is not contained by the lymphatic system, the microorganisms can produce septicemia, pneumonia, and abscesses throughout the body. Ingestion of infected, inadequately cooked meat leads to a focus of infection in the mouth and throat. The highest mortality occurs with pneumonic tularemia, which has even resulted from breathing aerosols created by handling dead but intact rabbits.

Humans most frequently acquire the infection through minor skin abrasions and by rubbing the eyes after handling small wild mammals. Probably 90% of the cases in this country are contracted from rabbits; it is estimated that 1% of all American rabbits are infected. The disease can also be spread by the bites of arthropods, such as deer flies (see Figure 11-29d), ticks, or rabbit lice, particularly in the western states; contact with small animals is a more likely cause in the central and eastern states (Figure 22-3).

F. tularensis is difficult to grow from serum samples, even on highly specialized media, because it requires a greater amount of the amino acid cysteine than is usually present in nutrient media. Colonies

Figure 22-3   Tularemia distribution. The geographical distribution, by county, of the 170 cases of tularemia reported in the United States during 1986. (Source: CDC. *Summary of Notifiable Diseases 1986, MMWR* 35:55.)

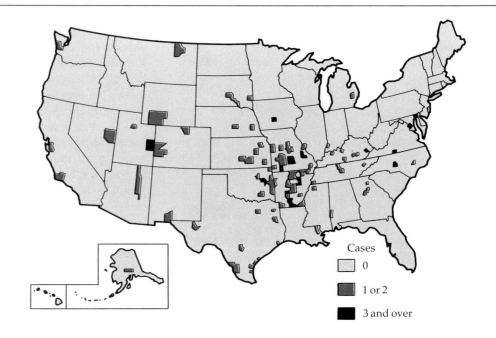

Cases

☐ 0

▨ 1 or 2

■ 3 and over

(a)

(b)

Figure 22-4 *Brucella abortus.* **(a)** *B. abortus,* the species transmitted in cow's milk (×34,200). **(b)** Phagocyte containing *B. abortus* in a vacuole. The organisms are able to survive and reproduce in phagocyte cells, where chemical and immunological defenses have difficulty reaching (×31,000).

may not appear for more than a week, but most cultures will grow within 96 hours. Although no well-defined toxins have been identified to account for its virulence, *F. tularensis* survives for long periods within body cells, including phagocytes. In fact, resistance to phagocytosis might account for the low infective dose required. Naturally acquired immunity is usually permanent, but recurrences have been reported. An attenuated live vaccine is available for high-risk laboratory workers.

Serological diagnosis can be done with an agglutination test that becomes positive after about ten days of infection. A rising titer is diagnostic. The organism is considered so dangerous to handle (because of the ease with which aerosol infections are transmitted) that such diagnosis should not be attempted without special isolation hoods.

Streptomycin is the antibiotic of choice, and prolonged administration is necessary to prevent relapses. Tetracycline is also effective, but is more likely to allow relapses. The intracellular location of the microorganism is a problem in chemotherapy.

Brucellosis (Undulant Fever)

Like the bacteria that cause tularemia, the *Brucella* bacteria that cause **brucellosis** favor intracellular growth, move through the lymphatic system, and travel to organs via the bloodstream. These micro-

organisms are very small, gram-negative, aerobic rods. They are difficult to grow on artificial media because they require that carbon dioxide be added to the atmosphere of the incubator, and they also require specially enriched media.

There are four main *Brucella* species. The most common species in the United States is *Brucella abortus* (brü-sel'lä ä-bôr'tus), most commonly found in cattle and transmitted by unpasteurized milk products (Figure 22-4). The disease caused by this species is usually mild and often self-limited. The most common form in the rest of the world is *Brucella melitensis* (me-li-ten'sis), which has a reservoir in goats and camels and is also transmitted by unpasteurized milk. The disease it causes is severe and often results in disability and death. *Brucella suis* (sü'is) is transmitted by contact with swine. Infections by *B. suis* are distinguished by the formation of destructive abscesses. There are a few cases each year of brucellosis caused by *B. canis* (kā'nis), usually from contact with stray dogs. This brucellosis resembles that caused by *B. abortus,* which causes spontaneous abortion in animals.

Brucella tend to multiply in the uterus of a susceptible animal, where their growth is favored by the presence of a carbohydrate, *mesoerythritol,* that is produced in the fetus and the membranes surrounding it. When the animal is not pregnant, it excretes the bacteria in milk. About 200 cases of brucellosis are reported in the United States each year, but there are few if any deaths. Because of pasteurization, dairy products are involved in only about 10% of brucellosis cases today; most cases occur in farmers, veterinarians, and meat packers.

The organisms apparently enter the human body by passing through minute abrasions in the skin or mucous membrane of the mouth, throat, or gastrointestinal tract. *B. melitensis* and *B. suis* are the most easily transmitted, and some think that they may even be able to penetrate intact skin. Once in the body, the microorganisms are ingested by fixed macrophages, in which they multiply and travel via the lymphatic system to the lymph nodes. From there, the microorganisms can be transported to the liver, spleen, or bone marrow. The ability of the microorganisms to grow inside the phagocytes partly accounts for their virulence and resistance to antibiotic therapy and antibodies. If any immunity is achieved, it is not reliable. A vaccine is used for cattle but not for humans.

Brucellosis is highly variable in its signs and symptoms as different organisms infect different organs. Many cases are subclinical; others are acute with fever and chills. Chronic brucellosis may be of more than one year's duration and may have very low-grade symptoms. The fever, particularly with *B. melitensis*, typically spikes to about 40°C (104°F) each evening; that is why the disease is sometimes called *undulant fever*.

Because the symptoms are difficult to interpret, serological testing is important in the diagnosis of brucellosis. Several tests are in use, the most common being an agglutination test. An ELISA test is considered more specific and sensitive, however. A simple agglutination test used on serum or milk to identify brucellosis-positive cattle has been an important factor in establishing a number of brucellosis-free areas. Culture is difficult and results in many false negative reports; it is most likely to be successful with *B. melitensis*, which tends to be present in blood samples.

Treatment must be prolonged because of the intracellular growth habit of the bacteria. Frequently, combinations of tetracycline plus streptomycin or gentamicin are used for periods of several weeks.

Anthrax

In 1877, Robert Koch isolated *Bacillus anthracis*, the bacterium that causes **anthrax** in animals (see Microview A, Photo 7). His careful work demonstrated that the rod-shaped, endospore-forming bacteria swarming in the blood of the dead animal were indeed responsible for anthrax; eventually, his methods resulted in what we now call Koch's postulates (Chapters 1 and 13). These postulates are the criteria that must be met to prove the microbial etiology of a disease.

The bacillus is a large, aerobic, gram-positive microorganism that is apparently able to grow slowly in soil types having specific moisture conditions. The endospores have survived in soil tests for up to 60 years. The organism is encapsulated, which probably improves its resistance to phagocytosis. The toxins it produces also inhibit phagocytosis.

The disease strikes primarily grazing animals, such as cattle and sheep. The *B. anthracis* endospores are ingested along with the grasses they eat. The endospores pass through the animal's stomach into its intestinal tract. There the endospores germinate, pass through the intestinal mucosa, and invade the blood, causing a fulminating, fatal septicemia. Although a reliable vaccine for the protection of farm animals is available, the sporadic occurrence of bovine anthrax fails to provide ranchers with enough incentive to vaccinate their livestock routinely. A vaccine for humans at unusual risk, such as veterinarians, is also available.

There are only a handful of human anthrax cases in the United States each year (Figure 22-5). Persons at risk are those who handle animals, hides, wool, and other animal products from certain foreign countries. Goat hair and handicrafts containing animal hides from the Middle East have been a repeated source of contamination.

If contact is made with materials containing anthrax endospores, the organism can enter the skin through a cut or abrasion and cause a malignant pustule there (Figure 22-6). This pustular infection is sometimes kept localized by the defenses of the body, but there is always danger of septicemia. Probably the most dangerous form of anthrax is pulmonary anthrax, contracted when the endospores are inhaled. *Woolsorters disease*, a dangerous form of pneumonia, results. It begins abruptly with high fever, difficulty in breathing, and chest pain. This disease eventually results in septicemia, and the mortality rate is high. Human gastrointestinal anthrax caused by ingestion of endospores has been reported in the Soviet Union but never in the United States.

Figure 22-5 **Anthrax in humans in the United States. (a)** Reported cases for five-year periods, 1920–1984. **(b)** Reported cases for each year, 1980–1987. [Source: CDC. *Summary of Notifiable Diseases 1986*, MMWR 35:55; MMWR 36:52 (1/8/88).]

Substances released from dying and dead cells provide nutrients for many bacteria. Various species of the genus *Clostridium*, which are gram-positive, endospore-forming anaerobes widely found in soil and in the intestinal tracts of humans and domesticated animals, grow readily in such conditions. *Clostridium perfringens* (pėr-frin′jens) is the species most commonly involved in gangrene, but other clostridia and several other bacteria can also grow in such wounds.

Once ischemia and the subsequent necrosis have developed, **gas gangrene** can develop, especially in muscle tissue. As the *C. perfringens* microorganisms

Figure 22-6 A cutaneous pustule of anthrax on a human arm. If this localized infection is not contained by the body's defenses, septicemia might be the next stage.

In humans, anthrax is best diagnosed by isolation of the bacterium. A number of morphological and biological tests can be used for precise identification. Penicillin is the drug of choice in human treatment. However, once septicemia is well advanced, antibiotic therapy may prove useless, probably because the exotoxins remain despite the death of the bacteria.

Gangrene

If a wound causes the blood supply to be interrupted (a condition known as *ischemia*), the wound becomes anaerobic. Ischemia leads to *necrosis*, or death of the tissue. The death of soft tissue from loss of blood supply is called **gangrene** (Figure 22-7). These conditions can also occur as a complication of diabetes.

Figure 22-7 Foot of a patient with gangrene. This disease is caused by *Clostridium perfringens* and other clostridia. The black, necrotic tissue on the toes furnishes anaerobic growth conditions for the bacteria, which then progressively destroy adjoining tissue.

grow, they ferment carbohydrates in the tissue and produce gases (carbon dioxide and hydrogen) that swell the tissue. The bacteria produce toxins that move along muscle bundles, killing cells and producing necrotic tissue that is favorable for further growth. Eventually, these toxins and bacteria enter the bloodstream and cause systemic illness. Enzymes produced by the bacteria degrade collagen and proteinaceous tissue, facilitating spread of the disease. Without treatment, the condition is uniformly fatal.

One complication of improperly performed abortions is the invasion of the uterine wall by *C. perfringens*, which resides in the genital tract of about 5% of all women. This infection can lead to gas gangrene and result in a life-threatening invasion of the bloodstream.

The surgical removal of necrotic tissue is called debridement; it and amputation are the most common medical treatments for gas gangrene. When gas gangrene occurs in such regions as the abdominal cavity, the patient can be treated in a *hyperbaric chamber*, which contains a pressurized oxygen-rich atmosphere. The oxygen saturates the infected tissues and thereby prevents the growth of the obligately anaerobic clostridia. Prompt cleaning of serious wounds and precautionary antibiotic treatment are the most effective steps in the prevention of gas gangrene. Penicillin and chloramphenicol are antibiotics with good activity against *C. perfringens*.

Systemic Diseases Caused by Animal Bites and Scratches

Animal bites can result in serious infections. Most serious bites are by domestic animals, such as dogs and cats, because they live in close contact with humans. It has been reported that 69% of dog bites become infected unless effectively cleansed. Domestic animals often harbor *Pasteurella multocida* (pas-tyèr-el'lä mul-tō'si-dä), a nonmotile, gram-negative rod similar to the *Yersinia* bacterium that causes plague. *P. multocida* is primarily a pathogen of animals; it causes septicemia, hence the name *multocida*, meaning many killing. Humans infected with *P. multocida* may show varied responses. For example, local infections with severe swelling and pain can develop at the site of the wound. Development of forms of pneumonia and septicemia is possible and life threatening. Penicillin and tetracycline are usually effective in treatment of these infections.

Clostridium species and other anaerobes, such as species of *Bacteroides* and *Fusobacterium* (fü-sō-bak-ti'rē-um), can also infect deep animal bites.

Cat Scratch Fever

Minor cat scratches sometimes result in **cat scratch fever**. An ulcerated lesion appears, soon followed by fever and painful swelling of the lymph nodes draining the area. Many other organs can become involved, and a generalized illness develops. Very recently, a minute gram-negative bacillus suspected of being the causative agent has been observed with use of special staining methods, but it has not been isolated and cultured.

Rat Bite Fever

Rat bites are typically infected by organisms of the genus *Streptobacillus* (strep-tō-bä-sil'lus) or *Spirillum* (spī-ril'lum), more commonly in Asiatic countries than in the United States. The disease, often called **rat bite fever**, is characterized by recurring fever, arthritislike symptoms (inflammation, pain, stiffness), and infections of the lymphatic vessels.

Plague

Few diseases have affected human history more dramatically than **plague**, known in the Middle Ages as the Black Death. This term comes from one of its characteristics, the blackish areas of skin caused by hemorrhages. In the fourteenth century, this disease destroyed perhaps one-fourth of the total population of Europe. Near the turn of this century, as many as 10 million people are reported to have died from plague in India in a 20-year period. The number of plague cases reported in the United States has been increasing in recent years, and there are now about 20 cases annually (Figure 22-8). A mortality rate of 25% is not unusual.

The disease is caused by a gram-negative, rod-shaped bacterium, *Yersinia pestis* (yèr-sin'ē-ä pes'tis). Normally a disease of rats, plague is transmitted from one rat to another by the rat flea, *Xenopsylla cheopis* (ze-nop-sil'lä kē-ō'pis). (See Figure 11-29c.) European rats, introduced into the United States many years ago, are the primary reservoirs of plague. In the United States, particularly in the far West and Southwest, the disease is endemic in wild rodents, especially ground squirrels, prairie dogs, and chipmunks.

If its host dies, the rat flea seeks a replacement, which may be another rodent or a human. A plague-infected flea is hungry for a meal because the growth

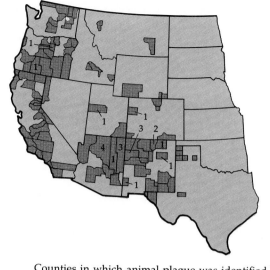

Counties in which animal plague was identified by *Yersinia pestis* isolation and/or presence of antibody in animals.

Figure 22-8 Plague distribution. Geographical distribution of human and animal plague in the United States in a typical year. Counties reporting plague in animals are shown in brown. The human cases are indicated in each county by a number.

of the bacteria blocks the flea's digestive tract, and the blood the flea ingests is quickly regurgitated. An arthropod vector is not always necessary for plague transmission. Contact from the skinning of infected animals, scratches of domestic cats, and similar contacts have been reported to cause infection.

In parts of the world where human contact with rats is common, infection from this source still prevails. From the flea bite, bacteria enter the human's bloodstream and proliferate in the lymph and blood. One factor in the virulence of the plague bacterium is its ability to survive and proliferate inside phagocytic cells rather than being destroyed by them. An increased number of highly virulent organisms eventually emerges, and an overwhelming infection results. The lymph nodes in the groin and armpit become enlarged, and fever develops as the body's defenses react to the infection. Such swellings are called *buboes*, which accounts for the name **bubonic plague** (Figure 22-9). The mortality rate of untreated bubonic plague ranges from 50% to 75%. Death from bubonic plague, if it occurs, is usually within less than a week after the appearance of symptoms.

A particularly dangerous condition arises when the bacteria are carried by the blood to the lungs and a form of the disease called **pneumonic plague** results. The mortality rate for this type of plague is nearly 100%. Even today, this disease can rarely be controlled if it is not recognized within 12 to 15 hours

of the onset of fever. Persons with pneumonic plague usually die within three days. Like influenza, pneumonic plague is easily spread by airborne droplets from human or animal cases. Great care must be taken to prevent airborne infection of persons in contact with patients. Contacts with pneumonic plague cases should receive prophylactic antibiotics.

Diagnosis of plague is most commonly done by isolating the bacterium and then identifying it by fluorescent-antibody and phage tests. Persons exposed to infection can be given prophylactic antibiotic protection. A number of antibiotics, including streptomycin and tetracycline, are effective. Recovery from the disease gives a reliable immunity. No vaccines are available, except for persons likely to come into contact with infected fleas during field operations or for laboratory personnel exposed to the organism. The long-term effectiveness of the plague vaccine is not high.

Control of rat-based plague largely involves modern sanitary procedures. Sanitary garbage disposal eliminates a source of food for rats, and building maintenance denies rats access to hiding places in attics and walls.

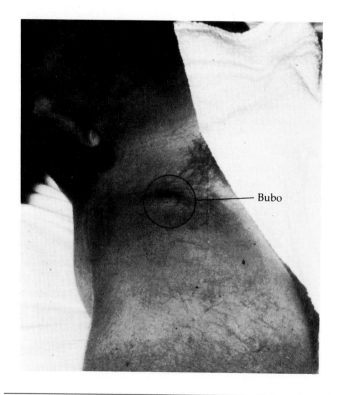

— Bubo

Figure 22-9 Bubonic plague. This photograph shows a bubo (swollen lymph node) in the groin area of the patient. Swollen lymph nodes are a common indication of systemic infection.

Relapsing Fever

Except for the species that causes Lyme disease (discussed in the next section), all members of the spirochete genus *Borrelia* (bôr-rel′ē-ä) cause **relapsing fever**. The disease is transmitted by soft ticks of the genus *Ornithodorus* (ôr-nith-ō′dô-rus) that feed on rodents. The incidence of relapsing fever increases during the summer months, when the activity of rodents and arthropods increases. The disease is characterized by a fever, sometimes in excess of 40.5°C (105°F), jaundice, and rose-colored spots. After ten days, the fever subsides. Three or four relapses may occur, each shorter and less severe than the initial fever. Diagnosis is made by observation of spirochetes in the patient's blood.

Lyme Disease

In 1975, a cluster of disease cases in young people that were first diagnosed as rheumatoid arthritis were reported near the city of Lyme, Connecticut. The seasonal occurrence (summer months), lack of contagiousness among family members, and descriptions of an unusual skin rash that appeared several weeks before the first symptoms suggested a tick-borne disease. The fact that penicillin alleviated the progression of symptoms suggested a bacterial pathogen. In 1983, a spirochete that was later named *Borrelia burgdorferi* was identified as the cause. **Lyme disease** may now be the most common tick-borne disease in this country. It is also found in Europe and Australia. In the United States, the disease is restricted to certain areas, as shown on the map (Figure 22-10a). There is some speculation that the disease may be spread to remote geographical areas by ticks attached to migratory birds. However, deer and field mice essentially maintain the spirochete in the tick population.

The tick feeds three times during its life cycle. The first and second feedings, as a larva and then a nymph, are usually on a field mouse. The third feeding, as an adult, is usually on a deer. Other animals also serve as reservoirs for the pathogen. None of these wild animals shows symptoms of being affected by the disease, although dogs and horses sometimes do.

On the Pacific Coast, the tick is the western black-legged tick *Ixodes pacificus* (iks-ō′dēs pas-i′fi-kus) (Figure 22-10b). In the rest of the country, it is mostly *Ixodes dammini* (dam′mi-nē). This latter tick is so small—the adult would fit within a small letter *o*

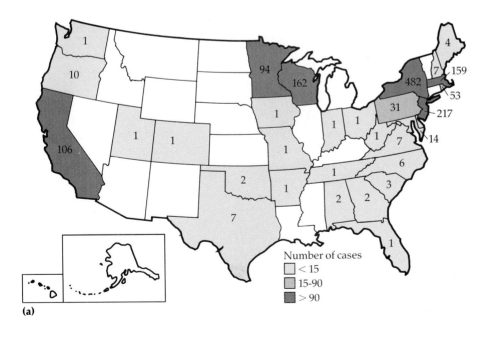

Number of cases
- < 15
- 15-90
- > 90

(a)

(b)

(c)

Figure 22-10 Lyme disease. (a) Map showing incidence of Lyme disease in the United States in 1987. The number of cases in each state is indicated. **(b)** A tick (*Ixodes pacificus*) biting a person. Ticks of this species may carry the spirochete *Borrelia burgdorferi*, which causes Lyme disease. **(c)** The classic "bull's-eye" rash of Lyme disease.

printed on this page—that it is often missed on the body. On the Atlantic Coast, almost all *Ixodes* carry the spirochete, but on the Pacific Coast few are infected.

The first symptom of Lyme disease is usually a rash that appears at the bite site. It is a red area that clears in the center as it expands to a final diameter of about 15 cm (Figure 22-10b). This distinctive rash is seen in about 75% of the cases. Flulike symptoms appear in a couple of weeks as the rash fades. Antibiotics given during this interval are very effective in limiting the disease. During a second phase, when it occurs, there is often evidence that the heart is affected. Heartbeats may become so irregular that a pacemaker is required. Neurological symptoms such as facial paralysis, meningitis, and encephalitis may be seen. Months or years later, patients may develop arthritis that may affect them for years. The knees may be so damaged that there can be difficulty walking. Immune responses to the presence of the bacteria are probably the cause of this joint damage. Many of the symptoms of long-term Lyme disease resemble those of the later stages of syphilis, also caused by a spirochete.

A number of antibiotics are effective in treatment of the disease, although in the later stages large amounts may need to be administered intravenously. Diagnosis of Lyme disease depends partly on the symptoms and the prevalence of the carrier in the geographical area. A fluorescent-antibody test has been developed that can detect even small numbers of the spirochete in blood or other fluids.

The same ticks that transmit Lyme disease can also transmit a protozoan disease much like malaria that infects red blood cells. This protozoan is *Babesia microti* (ba-bē′sē-ä mī-krō′tē), and the disease is known as **babesiosis**. The symptoms resemble those of malaria, with fever, intermittent chills, headache, and muscle pain. Most patients recover without treatment. Almost all the cases to date, about 200, have been on the Atlantic Coast.

Typhus

The various typhus diseases are caused by rickettsia bacteria, which are obligate intracellular parasites of eucaryotes. Rickettsias are spread by arthropod vectors. Rickettsias mostly infect the endothelial cells of the vascular system and multiply within them. The resulting inflammation causes local blockage of the small blood vessels. There are several related rickettsial diseases that differ mainly in their severity and in their arthropod vectors, including epidemic typhus, endemic murine typhus, and the spotted fevers (tick-borne typhus). Q fever, which is discussed in the next chapter (on respiratory diseases), is an atypical rickettsial disease caused by an atypical rickettsia.

Epidemic Typhus

Epidemic typhus (louse-borne typhus) is caused by *Rickettsia prowazekii* (ri-ket′sē-ä prou-wä-ze′kē-ē) and carried by the human body louse *Pediculus humanus corporis* (ped-ik′ū-lus hü′ma-nus cor′por-is). (See Figure 11-29b.) The pathogen grows in the gastrointestinal tract of the louse and is excreted by it. The pathogen is not transmitted directly by the bite of an infected louse but rather by the feces of the louse being rubbed into the wound when the bitten host scratches the bite. The disease can flourish only in crowded and unsanitary conditions, when lice can transfer readily from an infected host to a new host. In recent years, a number of cases of a mild form of epidemic typhus have been traced to contact with a reservoir in flying squirrels in the eastern United States. There have been no deaths. Although the vector in these cases is unknown, it is probably something other than a louse.

Epidemic typhus disease produces a high and prolonged fever for two or more weeks. Stupor and a rash of small red spots caused by subcutaneous hemorrhaging are characteristic as the rickettsias invade blood vessel linings. Mortality rates are very high when the disease is untreated.

Tetracycline- and chloramphenicol-type antibiotics are usually effective against epidemic typhus, but the elimination of conditions in which the disease can flourish is more important in its control. The most sensitive diagnostic test is an indirect fluorescent-antibody test, but a passive agglutination test using latex beads is also in use. Vaccines are available for military populations, which historically have been highly susceptible to the disease. Recovery from the disease gives a solid immunity and also renders a person immune to the related endemic murine typhus.

Endemic Murine Typhus

Endemic murine typhus occurs sporadically rather than in epidemics. The term *murine* (derived from Latin for mouse) refers to the fact that rodents, such as rats and squirrels, are the common hosts for this typhus. Endemic murine typhus is transmitted by the rat flea *Xenopsylla cheopis* (ze-nop-sil′lä kē-ō′pis; see Figure 11-29c), and the pathogen responsible for the disease is *Rickettsia typhi*, a common inhabitant of rats. Humans become involved when the flea finds a human host (Figure 22-11).

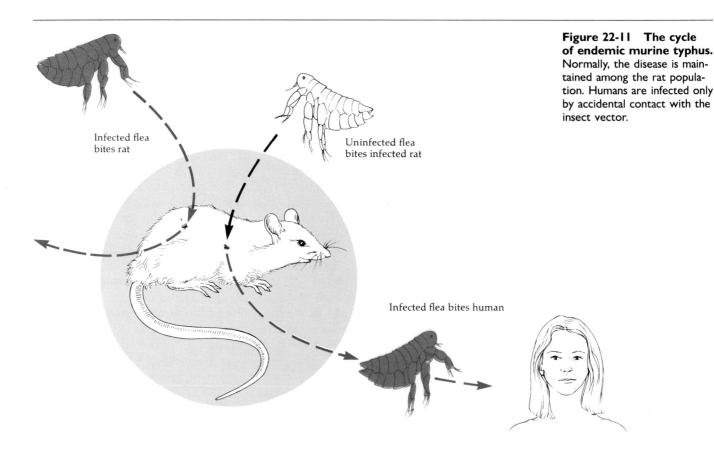

Figure 22-11   The cycle of endemic murine typhus. Normally, the disease is maintained among the rat population. Humans are infected only by accidental contact with the insect vector.

Infected flea bites rat

Uninfected flea bites infected rat

Infected flea bites human

Before World War II, there were over 5000 cases of murine typhus annually in this country. Texas has had a number of outbreaks of the disease in recent years, often associated with campaigns to eliminate rodents, which caused the rat fleas to seek a new host. With a mortality rate of less than 5%, the disease is considerably less severe than the epidemic form. Except for the reduced severity of the disease, endemic murine typhus is clinically indistinguishable from epidemic typhus.

Tetracycline and chloramphenicol are effective treatments for endemic murine typhus, and the diagnostic procedure for the disease is the same as for epidemic typhus. Rat control is the best preventive measure for the disease.

Spotted Fever

Tick-borne typhus, or **Rocky Mountain spotted fever,** is probably the best-known rickettsial disease in the United States. It is caused by *Rickettsia rickettsii.* Despite its name (it was first recognized in the Rocky Mountain area), it is most common in the southeastern United States and Appalachia (Figure 22-12). This rickettsia is a parasite of ticks and is usually passed from one generation of ticks to another through their eggs, a mechanism called *transovarian*

passage (Figure 22-13). A tick can also become infected from an animal reservoir. The rickettsia does not kill the tick. In different parts of the country, different ticks are involved—in the West, the wood tick *Dermacentor andersoni* (dėr-mä-sen′tôr an-dėr-sō′nē), in the East the dog tick *Dermacentor variabilis* (vār-ē-a′bil-is), and in the Southwest, the Lone Star tick *Amblyomma americanum.*

About a week after the tick bites, a rash develops that is sometimes mistaken for measles; however, it often appears on palms and soles, which does not occur with viral rashes. The rash is evidence of the vascular damage, resulting in leakage of blood into surrounding tissues, that is characteristic of many rickettsial infections. The rash is accompanied by fever and headache. Death, which occurs in about 3% of the approximately 1000 cases reported each year, is usually caused by kidney and heart failure. Antibiotics such as tetracycline and chloramphenicol are very effective if administered early enough.

Serological tests are available but do not become positive until late in the illness. Therapy should be started before confirmation and then be based on clinical evidence. It should be mentioned that there are a number of other tick-borne rickettsial diseases known by local names in various regions of the world.

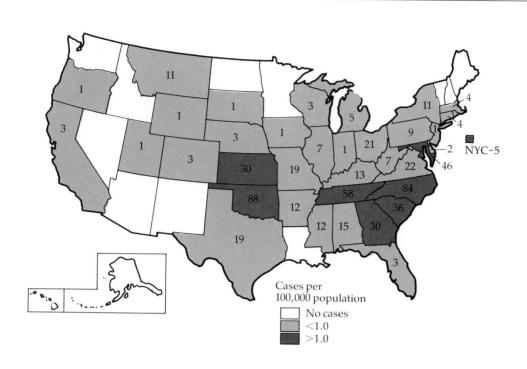

Figure 22-12 Rocky Mountain spotted fever (tick-borne typhus). The map shows the geographical distribution of the 592 cases reported in the United States in 1987. [Source: *MMWR* 36:52 (1/8/88).]

Cases per 100,000 population
- No cases
- <1.0
- >1.0

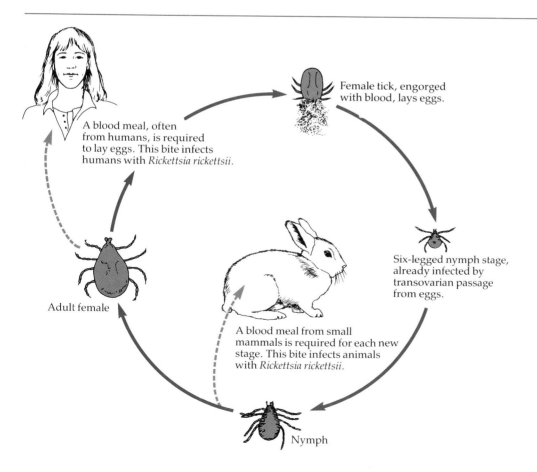

Figure 22-13 Rocky Mountain spotted fever. Mammals are not essential to maintenance of the disease in the tick population; the pathogen is passed by transovarian passage, so new ticks are infected at birth. A blood meal is required to advance to successive stages.

Female tick, engorged with blood, lays eggs.

A blood meal, often from humans, is required to lay eggs. This bite infects humans with *Rickettsia rickettsii.*

Six-legged nymph stage, already infected by transovarian passage from eggs.

A blood meal from small mammals is required for each new stage. This bite infects animals with *Rickettsia rickettsii.*

Adult female

Nymph

Viral Diseases of the Cardiovascular System

Myocarditis

The heart muscle, also called the myocardium, is sometimes subject to an inflammatory disease called **myocarditis**. Although this disease can be caused by bacteria, fungi, or protozoans, it is usually the result of a viral infection. The viruses that most frequently cause myocarditis belong to the *coxsackievirus (enterovirus)* group. After infecting the respiratory or gastrointestinal tract, the virus apparently reaches the heart via the blood or lymph. In the heart, the infecting virus damages the myocardium and leaves scar tissue. Sometimes pericarditis results from myocarditis. **Myocarditis of the newborn**, also caused by a coxsackievirus, is often fatal.

Infectious Mononucleosis

The Epstein–Barr (EB) virus (Figure 22-14) is a herpesvirus of unusual interest. In less developed parts of the world, infection with the EB virus is acquired in early childhood, and 90% of the children over four years of age have acquired antibodies. Childhood infections by the EB virus are usually asymptomatic, but if infection is delayed until young adulthood, as is often the case in the United States, the result is **infectious mononucleosis**. The peak incidence of the disease in this country occurs at about 15 to 25 years of age. College populations, particularly those from the upper socioeconomic strata, have a high incidence of the disease. About 50% of college students have no immunity, and about 15% of these students can expect to contract the disease. The symptoms are usually fever, sore throat, swollen lymph nodes in the neck, and general weakness. The spleen is often enlarged, and vigorous activity may cause it to rupture, which is the principal cause of the rare deaths from this disease. Recovery is usually complete in a few weeks, and immunity is permanent.

The usual route of infection is by transfer of saliva by kissing or, for example, by sharing drinking vessels. Reproduction of the virus appears to occur mainly in the parotid glands, which accounts for its presence in saliva. It does not spread among casual household contacts, so aerosol transmission is unlikely. The incubation period before appearance of symptoms is from four to seven weeks.

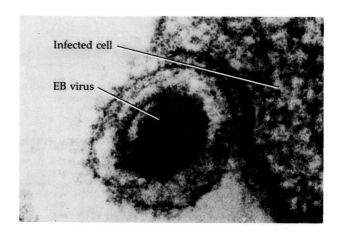

Figure 22-14   Epstein–Barr virus (EB). This virus was caught in the act of infecting a human cell (right) in culture. This particular virus particle was isolated from cells of Burkitt's lymphoma ($\times 400,000$).

Although reproduction is considered to occur in epithelial cells of the parotid glands, the virus infection is almost exclusively selective for B cells. The virus does not reproduce significantly in B cells, but the infection causes B cells to transform into plasma cells, which reproduce rapidly. These are then attacked by cytotoxic T cells of the cell-mediated immune system. The symptoms of infectious mononucleosis are associated with B-cell proliferation and the immune response. After recovery, the virus remains latent in a small number of B cells.

The disease name, *mononucleosis*, refers to lymphocytes with unusual lobed nuclei that proliferate in the blood during the acute infection. The infected B cells produce nonspecific antibodies called heterophil antibodies. Older diagnostic tests depended heavily on blood counts of the distinctive lymphocytes and on detection of the heterophil antibodies. A fluorescent-antibody test that detects IgM antibodies against the EB virus is the most specific diagnostic method.

The EB virus has been suspected as the cause of many otherwise inexplicable cases of **chronic fatigue**. People with this condition often have difficulty performing routine mental tasks and become so debilitated that they are unable to work. There have recently been several identifiable outbreaks of this condition, and many persons suffering from it have much higher than normal antibody levels for the EB virus.

Burkitt's Lymphoma

As we mentioned in Chapter 12, the EB virus has been linked to two human cancers, **Burkitt's lymphoma** and **nasopharyngeal cancer**. There is considerable evidence linking the EB virus to these human malignancies. For example, in cell cultures, EB virus will infect only human lymphoid cells. The virus then transforms some of the cells into EB-virus-carrying cells that have malignant characteristics. When inoculated into monkeys, the cells produce fatal lymphomas. There is no doubt that EB virus is potentially oncogenic.

A precondition for Burkitt's lymphoma might be infection by malaria or other tropical diseases. This form of cancer is especially prevalent in a region of Central Africa inhabited by the mosquito that transmits malaria. In peoples of most other areas of the world, the immune system seems to effectively combat tumor cells marked with EB virus antigens, and Burkitt's lymphoma is rare, although the EB virus is common. Similarly, the frequency of nasopharyngeal cancer might depend on the simultaneous presence of the EB virus and other carcinogens, such as nitrosamines found in salted fish, which constitutes a large part of certain regional diets.

Yellow Fever

Arboviruses, or arthropod-borne viruses, can reproduce either in arthropods, such as mosquitoes, or in humans. Although a number of serious human diseases are caused by such viruses, no disease is evident in the arthropods that serve as vectors. Worldwide, there are probably more than 100 arbovirus-caused diseases. **Yellow fever** is caused by an arbovirus (*yellow fever virus*) and is historically important because it was the first such virus discovered and provided the first confirmation that an insect could transmit a virus.

The yellow fever virus is injected into the skin by the mosquito. The virus then spreads to local lymph nodes, where it multiplies; from the lymph nodes, it advances to the liver, spleen, kidney, and heart, where it can persist for days. In the early stages of the disease, the person experiences fever, chills, headache, and backache, followed by nausea and vomiting. This stage is followed by jaundice, a yellowing of the skin that gave the disease its name. The jaundice is caused by the deposition of bile pigments in the skin and mucous membranes as a result of liver damage. In severe cases, the virus produces lesions in the infected organs, and hemorrhaging occurs.

Yellow fever is still endemic in many tropical areas, such as Central America, tropical South America, and Africa (Figure 22-15). Monkeys are a natural reservoir for the virus. Control of the disease depends on local control of *Aedes aegypti* (ä-e′dēz ē-jip′tē), the usual mosquito vector (see Figure 11-28a), as was done during the building of the Panama Canal many years ago. Immunization of the exposed population is also an effective control.

There is no specific treatment for yellow fever. Diagnosis is usually by clinical signs, but it can be confirmed by a rise in antibody titer or isolation of the virus from the blood. The vaccine in use is an attenuated live viral strain and yields a very effective immunity with few side-effects.

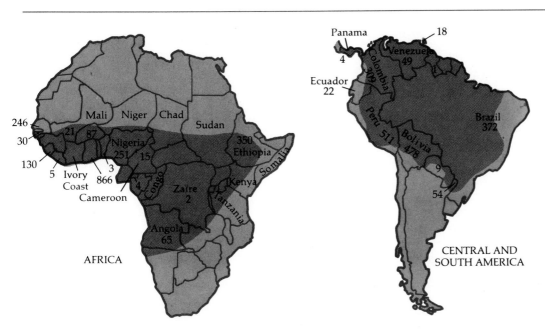

Figure 22-15 Yellow fever distribution. Zones of endemic yellow fever are shown in darker color. The number of reported cases is also indicated.

Dengue

Dengue is a similar but milder viral disease also borne by the *Aedes aegypti* mosquito. This disease, endemic in the Caribbean and other tropical environments around the world, is caused by *dengue fever virus*, an arbovirus. It is characterized by fever, severe muscle and joint pain, and rash. Except for the painful symptoms, which have led to the name **breakbone fever**, classic dengue fever is a relatively mild disease and is rarely fatal.

The countries surrounding the Caribbean are reporting an increasing number of cases of dengue— nearly 70,000 in 1985. In most years, more than 100 cases are imported into the United States, mostly by travelers from the Caribbean and South America. The mosquito vector for dengue is common in the Gulf states, and there is some concern that the virus will sooner or later be introduced into this region and become endemic. A special concern of health officials is the introduction into the United States of an Asian mosquito, *Aedes albopictus* (al-bō-pik′tus), that is an efficient carrier of the virus and an aggressive biter. It transmits the virus by transovarian passage and from person to person. The range of this mosquito can potentially cover much of the country (see the box below).

A more serious form of dengue, **dengue hemorrhagic fever**, is characterized by bleeding from the skin, gums, and gastrointestinal tract and sometimes by circulatory failure and shock. Diagnosis is usually based on symptoms, and a complement-fixation test can be used for confirmation. Control measures are directed at eliminating the *Aedes* mosquitoes. No specific treatment is effective against the virus.

Centers for Disease Control
Morbidity and Mortality Weekly Report

AEDES ALBOPICTUS—A NEW VECTOR

In August 1985, an infestation of *Aedes albopictus* (Asian tiger mosquito), an exotic mosquito known to transmit epidemic dengue fever in its native Asia, was discovered in Harris County, Texas. *Aedes albopictus* has been shown in the laboratory to be capable of efficiently transmitting dengue and several other viruses that can infect humans and are indigenous to the United States.

In the spring of 1986, efforts were begun to determine the distribution of this mosquito vector in the United States. The figure shows the present distribution of *Aedes albopictus*-positive counties.

To determine the mode of introduction and to characterize the risk of further introduction, CDC has conducted inspections of used auto and truck tires entering

Distribution of *Aedes albopictus* in the United States (1987) and confirmed cases of dengue and California encephalitis (1986). [Source: *MMWR* 36(47):771 (12/4/87) for *Aedes albopictus* data; *Mosquito and Tick-borne Surveillance*, CDC, April 1988.]

• = Counties with confirmed *Aedes albopictus* infestation (1987)

Black numbers = Number of confirmed dengue fever cases (32 in 1986)

Color numbers = Number of California encephalitis infections (64 in 1986)

Viral Hemorrhagic Fevers

In 1969, three nurses working in missionary hospitals in Nigeria contracted an unknown disease. Two died, and laboratory technicians performing virologic studies of serum from the survivor contracted the disease. The disease became known as **Lassa fever**, one of the **viral hemorrhagic fevers**. In 1967, 31 persons became ill when some African monkeys were imported into Europe. That type of viral hemorrhagic fever is now known as the **Marburg virus outbreak**. Other hemorrhagic fevers of viral origin, such as **Crimean–Congo hemorrhagic fever**, have since been reported.

The virus is transmitted very rapidly from person to person by contact with bodily fluids. Nosocomial transmission via hypodermic needles accounted for one-half of the cases of **Ebola hemorrhagic fever** in

Zaire in 1976. The fatality rates of these viral hemorrhagic fevers tend to be very high for an infectious disease. Patients suffer headache, muscle pain, skin or gastrointestinal hemorrhaging, and shock.

Protozoan Diseases of the Cardiovascular System

Toxoplasmosis

Toxoplasmosis, a disease of blood and lymphic vessels, is caused by the protozoan *Toxoplasma gondii* (toks-ō-plaz'mä gon'dē-ē), a small, crescent-shaped organism. *T. gondii* is a sporozoan, like the malarial parasite.

United States ports from Asia, the source of several million imported used tires each year. The mosquito commonly breeds in standing water in tires stored outdoors.

In October 1986, *Aedes albopictus* larvae were found in 11 tires in two cargo containers in Seattle, Washington. The two positive containers had a higher proportion of tires (91%) holding water than the negative containers did (17%). Both positive containers were shipped by the same exporter in Japan. The positive containers were fumigated with methyl bromide to kill larvae and ova.

Dengue in the Americas

Dengue has been endemic in the Americas for over 200 years. Presently, the disease appears to be following a pattern similar to that in Southeast Asia 30 years ago. Dengue fever, once a benign flulike illness, is now one of the leading causes of morbidity and mortality among Southeast Asian children. Cuba, where no dengue had been reported for 30 years, experienced a dengue epidemic in 1977, followed in 1981 by a major dengue hemorrhagic fever (DHF) epidemic caused by DEN-2, with 158 fatalities. In 1978, Mexico reported its first cases of dengue in many years, and, within six years, the disease became endemic in most coastal areas of the country. In 1984, the first DHF cases associated with epidemic dengue were reported.

Editorial Note: *Aedes albopictus*, an efficient Asian mosquito host for dengue viruses, has been shown to transmit dengue viruses both transovarially (from female mosquitoes to their offspring through infection of the eggs) and

from human to human. Laboratory studies have shown it to be a more efficient vector of dengue virus than *Aedes aegypti* and a competent vector of California encephalitis viruses and yellow fever virus. *Aedes albopictus* has not been incriminated in the spread of any viral disease in the Americas, but it represents a public health concern because of its potential to infest areas where dengue, yellow fever, or pathogenic California viruses are present. Once introduced to such areas, *Aedes albopictus* can spread these viruses to areas previously free of them.

The discontinuous distribution of *Aedes albopictus* in the southern United States suggests that the infestation may be contained through programs of surveillance, removal of breeding sites (especially tires), interruption of interstate dispersal of tires, and judicious use of insecticides in breeding sites. Studies are presently under way at CDC in collaboration with state and local agencies to determine the feasibility of these approaches. Critical features of the program include delineation of the full distribution of *Aedes albopictus*, determination of the vector's routes of spread from infested areas, and definition of the biological attributes of the mosquito that relate to control.

Sources: *MMWR* 35:51 (1/31/86); *MMWR* 35:493 (8/8/86); *MMWR* 35:649 (10/17/86); *MMWR* 35:732 (11/28/86); *MMWR* 36:450 (7/17/87); *MMWR* 36:769 (12/4/87).

Cats seem to be an essential part of the life cycle of *T. gondii* (Figure 22-16). Random tests on urban cats have shown that a large number of them are infected with the organism, which causes no apparent illness in the cat. The organism undergoes its only sexual phase in the intestinal tract of the cat. Oocysts are then shed in the cat's feces and contaminate food or water that can be ingested by other animals. The protozoan reproduces asexually in these new hosts, and the trophozoites remain in the host's tissues. A human eating the uncooked meat of such an animal, or a cat eating an infected mouse, will be infected with the parasite.

The two main modes of infection appear to be eating undercooked meats and contact with the dried feces of cats (for example, while cleaning a litter box).

Figure 22-16   Life cycle of *Toxoplasma gondii*. The domestic cat is the definitive host, in which the protozoan reproduces sexually. Oocysts are produced in the cat's intestinal tract and passed in its feces. Each oocyst contains four sporozoites, which are ingested by an intermediate host. There, in blood and other tissues, the sporozoites mature into trophozoites. Eating the undercooked meat of the intermediate host or contact with cat feces causes infection in humans.

Trophozoites in blood and other tissues; fetus infected

Feces contaminate litter box

Cats ingest mice or rats containing *Toxoplasma gondii*

T. gondii oocyst in small intestine

Oocysts in feces undergo sporogony, forming sporocysts with four sporozoites each

Feces contaminate sand box

Feces contaminate grazing soil

Humans ingest contaminated meat

Trophozoites in blood and other tissues

Trophozoites in blood and other tissues

There was a recent outbreak of toxoplasmosis in 34 persons who frequented a riding stable with a large residential population of cats. Some surveys have shown that approximately 30% of the population carries antibodies for this organism; this percentage indicates the high rate of subclinical, unrecognized infections.

The disease is a rather undefined mild illness, and the primary danger is in the congenital infection of a fetus. The actions on the fetus are drastic, including convulsions, severe brain damage, blindness, and death. The mother is probably unaware of the disease, which is transmitted across the placenta. Pregnant women should avoid cleaning cat litter boxes or having any unnecessary contact with cats; eating raw meat is also inadvisable.

Growth of the trophozoites in cells of the brain or retina cause the most damage because the cells there are killed faster than they can be replaced. Invasion of the central nervous system by *T. gondii* is one of the more serious manifestations of damage to the immune system in AIDS patients and is an important cause of AIDS-related death.

T. gondii can be isolated and grown in cell cultures. The preferred serological tests use a fluorescent-antibody method or a hemagglutination test. The latest recommendation is to use the two tests in conjunction. A rising titer in a pregnant woman may be an indication for a therapeutic abortion. Toxoplasmosis is treated with pyrimethamine in combination with either trisulfapyrimidines or sulfadiazine.

American Trypanosomiasis (Chagas' Disease)

American trypanosomiasis (Chagas' disease) is a protozoan disease of the cardiovascular system. The causative agent is *Trypanosoma cruzi* (tri-pa-nō-sō' mä kruz'ē), a flagellated protozoan (Figure 22-17). The disease occurs in the southern United States, Mexico, Central America, and into South America. Only a few cases have been reported in the United States, but the disease infects 40% to 50% of the population in some rural areas of South America.

The reservoir for *T. cruzi* includes a wide variety of wild animals, including rodents, opossums, and armadillos. The arthropod vector is the reduviid bug, often called the "kissing bug" because it often bites persons near the lips (see Figure 11-29e). The trypanosomes, which grow in the gut of the bug, are passed on if the bug defecates while feeding. The bitten human or animal often rubs the feces into the bite wound or other skin abrasions by scratching or into the eye by rubbing.

(a)

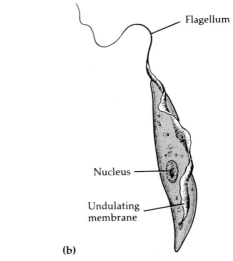

Flagellum

Nucleus

Undulating membrane

(b)

Figure 22-17 *Trypanosoma cruzi*, **the cause of American trypanosomiasis (Chagas' disease). (a)** Scanning electron micrograph (×5,600). **(b)** Trypanosomes have a single flagellum that appears as an extension of the undulating membrane.

At the site of inoculation, the trypanosomes reproduce and pass through various stages of their life cycle. A swollen lesion develops at the inoculation site (often near the eye). The organisms infect regional lymph nodes and are carried by the blood to other organs of the body, such as the liver and macrophages of the spleen, and frequently to the heart. Most damage is caused by inflammatory reactions in the host in response to the presence of the organisms. Although central nervous system involvement in adults is rare, it is not uncommon in children. One effect of nervous system involvement is loss of involuntary muscular contractions that move solids through the esophagus and gastrointestinal tract. These organs then become grossly enlarged.

Laboratory procedures for the diagnosis of American trypanosomiasis include observing the trypanosomes in blood, growing the microorganisms in blood cultures, and using fluorescent-labeled antibody on blood smears. An extraordinary form of diagnosis, which is actually done with some frequency, is *xenodiagnosis*. Reduviid bugs, reared in parasite-free conditions, are allowed to feed on the arm of a suspected patient. The trypanosome is then identified in the intestinal tract of the insect from 10 to 20 days later.

Treatment of Chagas' disease is very difficult when chronic, progressive stages have been reached. The trypanosome multiplies intracellularly and is difficult to reach chemotherapeutically. Cultivation of the parasite in tissue culture has made large-scale screening for antitrypanosomal chemicals possible. Nifurtimox was developed from such screening programs and is the drug of choice in treatment. The drug decreases the number of parasites but is not a cure.

Malaria

Malaria is characterized by chills and fever and often by vomiting and severe headache. These symptoms typically appear for one- to three-day periods that alternate with asymptomatic periods. Malaria is found wherever the mosquito vector *Anopheles* (an-of'el-ēz) is found (see Figure 11-28b) and there are human hosts for the protozoan parasite *Plasmodium*. The disease was once common in the United States, but effective mosquito control and a reduction in the number of human carriers caused the reported cases to drop below 100 by 1960 (Figure 22-18). In recent years, however, there has been an upward trend in the number of cases in the United States, reflecting a worldwide resurgence of malaria, increased travel to malarial areas, and an increase in immigration from malarial areas. Occasionally, malaria has been transmitted by unsterilized syringes used by drug addicts, but these cases cannot be spread to other humans without the mosquito. In tropical Asia, Africa, and Central and South America, malaria is still a serious problem. In fact, as the protozoan develops resistance to antimalarial drugs and as political instability continues to interfere with control measures, malaria is more often being described as a resurgent disease. Current estimates are that it affects some 300 million persons worldwide each year and causes from 2 to 4 million deaths.

The causative organisms of malaria are the spore-forming protozoans (sporozoans) of the genus *Plasmodium* (see Microview A, Photo 22). Four pathogenic species are recognized, each of which causes a distinctive form of the disease. The most dangerous species is *Plasmodium falciparum* (fal-sip'är-um), a species that is also widespread geographically. Also widely distributed is *P. vivax* (vī'vaks). *P. malariae* (mä-lā'rē-ī) and *P. ovale* (ō-vä'lē) have either a lower infection rate or cause a geographically restricted, relatively milder disease.

The mosquito carries the *sporozoite* form of the *Plasmodium* protozoan in its saliva (see Figure 11-21). The sporozoite enters the bloodstream of the bitten human and within about 30 minutes enters the liver cells. The sporozoites in the liver cells undergo reproductive schizogony by a series of stages that finally results in the release of large numbers of *merozoite* forms into the bloodstream. These merozoites infect red blood cells and undergo reproductive schizogony again. Eventually, the red blood cells rupture

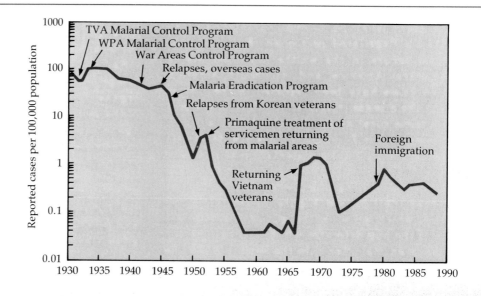

Figure 22-18 Reported cases of malaria in the United States, 1930–1987. Malaria was a common disease as recently as 1935. Its decline coincided with government control programs such as the Tennessee Valley Authority (TVA) flood control program, the Work Projects Administration (WPA) malarial control program, and war area control programs during World War II. Note the recent increases in incidence as veterans returned from Vietnam and foreign immigration increased. [Source: CDC. *Summary of Notifiable Diseases 1986, MMWR* 35:55; *MMWR* 36:52 (1/8/88).]

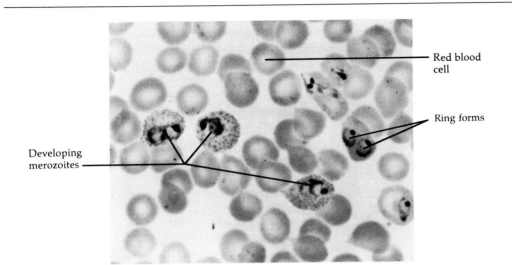

Figure 22-19 Blood smear from a malarial patient. The protozoan grows within infected red blood cells. The ring forms are an early stage. The dark spot of the ring is the nucleus, and the light central area within the circular ring is the food vacuole of the parasite. The food in the vacuole is the cytoplasm of the red blood cell. In later stages, just before lysis of the red blood cell, the cell becomes crowded with developing merozoites (×800).

and release large numbers of merozoites. This rupturing of the red blood cells is fairly simultaneous, and the release of toxic compounds is the cause of the paroxysms (recurrent intensifications of symptoms) of chills and fever that are the characteristic symptoms of malaria. The fever reaches 40°C (104°F), and a sweating stage begins as the fever subsides. Between paroxysms, the patient feels normal. Anemia results from the loss of red blood cells, and enlargement of the liver and spleen are added complications.

When the red blood cells rupture, many of the released merozoites infect other red blood cells to renew the cycle in the bloodstream. If only 1% of the red blood cells contain parasites, there are an estimated 100,000,000,000 parasites in circulation at one time in a typical malarial patient! Other merozoites become male or female *gametocytes.* When these enter the digestive tract of a feeding mosquito, they pass through a sexual cycle that produces new infective sporozoites. It took the combined labors of several generations of scientists to discover this complex life cycle of the malaria parasite.

Laboratory diagnosis of malaria is made by examination of a blood smear (Figure 22-19). Although there is no effective immunity to malaria, populations in which it has long been endemic have generally developed some resistance to it and tend to have a less severe form of the disease. Persons who have the genetic sickle-cell trait—common in many areas where malaria is endemic—are resistant to malaria.

Considerable effort is currently being expended on the search for an effective vaccine (see the box in Chapter 17, p. 438). The first step is the identification of the antigens on the surfaces of the parasite's three stages (sporozoite, merozoite, and gametocyte). Once these antigens are positively identified, then the second step—the genetic engineering of antigens for vaccines—will be attempted. The sporozoite stage is of primary interest because inhibiting its growth would prevent the initial infection from becoming well established. However, some believe that any effective vaccine would have to protect against all three stages.

Malaria was once fairly effectively treated with quinine, but newer derivatives of quinine, such as primaquine and especially chloroquine, have been the mainstays for prevention and treatment of malaria for many years. Chloroquine prevents the DNA of the parasite's merozoite stage from replicating by fitting between base pairs in the DNA. Resistance to these drugs is spreading rapidly, and current chemical alternatives are more expensive, which prohibits widespread use in the parts of the world most troubled by malaria. One approach to chemotherapy has been to use drug combinations such as Fansidar®, a combination of pyrimethamine and sulfadoxine. Both drugs are inhibitors of folic acid synthesis (see the discussion of sulfa drugs in Chapter 19); together, they have a synergistic effect. Resistance against such combinations is also emerging.

Research on antimalarial chemicals has been much aided in recent years by methods of culturing some stages of the malarial parasite. Culturing makes it possible for large numbers of chemicals to be tested. A U.S. Army research group has screened more than 250,000 such compounds.

Effective control of malaria is not in sight. It will probably require a combination of chemotherapeutic and immunological approaches. The expense and the need for an effective political organization in malarial areas are probably going to be as important in controlling the disease as advances in medical research.

Helminthic Diseases of the Cardiovascular System

Schistosomiasis

Schistosomiasis is a disease caused by a flatworm parasite, a fluke. The disease is not contracted in the United States, but more than 400,000 immigrants to the United States have the disease, and about 200 million persons in Asia, Africa, South America, and the Caribbean are also affected. Species of the genus *Schistosoma* (shis-tō-sō′mä) cause the disease; they vary with geographical location, but the characteristics of the disease are similar.

When waters become contaminated with *Schistosoma* ova excreted in human wastes (Figure 22-20), a motile, ciliated larval form of *Schistosoma* called a miracidium is released from the ova and enters certain species of snails. (The lack of a suitable host snail is the primary reason why schistosomiasis cannot be transmitted in the United States.) Eventually, the pathogen emerges from the snail in an infective form called the *cercaria*. These cercariae have forked swimming tails. When they contact the skin of a person wading or swimming in the water, they discard the tail and enzymatically penetrate the skin. They are then carried by the bloodstream to the veins of the

Figure 22-20 Life cycle of *Schistosoma*, cause of the disease schistosomiasis.

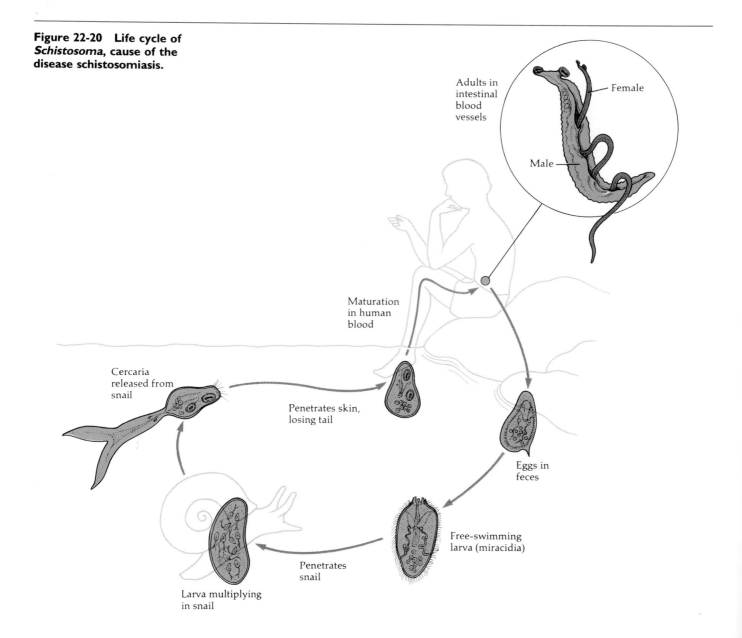

Adults in intestinal blood vessels

Female

Male

Maturation in human blood

Cercaria released from snail

Penetrates skin, losing tail

Eggs in feces

Free-swimming larva (miracidia)

Penetrates snail

Larva multiplying in snail

liver or urinary bladder, depending partly on the species of schistosome. There they mature into an adult form, in which the slender female lives in a cleft of the male (Figure 22-21). The union produces a supply of new ova, some of which cause local tissue damage (*granulomas*) from defensive body reactions (Figure 22-22). Other ova enter the water to continue the cycle.

Figure 22-21 Male and female schistosomes. The schistosomes (split-body) take their name from the life style of the adults. The female lives in a cleft on the ventral surface of the male worm, is continuously fertilized, and continuously lays eggs. The mouths by which the worms feed are visible, as is the sucker on the male by which it attaches to the host (×2700).

The adult worm appears to be unaffected by the host's immune system. Apparently, it quickly coats itself with a layer that mimics the host's tissues. However, a previous infection tends to be protective to some degree against subsequent infections, which encourages work toward development of a vaccine.

The disease often destroys the liver, but sometimes other organs, such as the lungs and the urinary system, become involved. Abscesses and ulcers are formed in response to the infestation, and the body's allergic reactions block various ducts and blood vessels.

Laboratory diagnosis consists of microscopic identification of the flukes or their ova in fecal and urine specimens, intradermal tests, and serological tests, such as complement-fixation and precipitin tests.

Praziquantel and oxiaminiquine are approved for use against schistosomes in the United States. Currently, research is under way to identify schistosome enzymes that differ from host enzymes and to develop chemotherapeutic agents that are specific for the parasite. Sanitation and elimination of the host snail are also useful forms of control.

Swimmer's Itch

Swimmers in lakes in the northern United States are sometimes troubled by **swimmer's itch**. This is a cutaneous allergic reaction to cercariae similar to that of schistosomiasis. However, these parasites mature only in wildfowl and not in humans, so infection does not progress beyond penetration of the skin and a local inflammatory response.

Figure 22-22 Granuloma from a patient with schistosomiasis. The eggs laid by the adult schistosomes lodge in tissue, and the body responds to the irritant by forming scarlike tissue surrounding it (a granuloma). This egg, surrounded by granulomatous cells, is in a section taken from a liver (×600).

TABLE 22-I   *Diseases of the Cardiovascular and Lymphatic Systems*

	Causative Agent	Mode of Transmission	Treatment
Bacterial Diseases			
Puerperal sepsis and infections related to abortions	Primarily *Streptococcus pyogenes*; *Clostridum* and *Bacteroides* species often cause postabortion infections	Unsanitary conditions in childbirth and abortion	Penicillin and erythromycin for *S. pyogenes*; chloramphenicol and penicillin for *Bacteroides* and *Clostridium* species
Endocarditis			
Subacute bacterial	Many organisms, especially α-hemolytic streptococci	Bacteremia localizing in the heart	Varies with agent; penicillin and erythromycin for α-hemolytic streptococci
Acute bacterial	*Staphylococcus aureus*, *Streptococcus pneumoniae*	Bacteremia localizing in the heart	Semisynthetic penicillin derivatives such as methicillin or oxacillin for *S. aureus*; penicillin for *S. pneumoniae*
Pericarditis	*Streptococcus pneumoniae*	Bacteremia localizing in the heart	Penicillin
Rheumatic fever	Group A β-hemolytic streptococci	Inhalation leading to streptococcal sore throat	Penicillin (for sore throat)
Tularemia	*Francisella tularensis*	Animal reservoir (rabbits); skin abrasions, ingestion, inhalation, bites	Streptomycin, tetracycline
Brucellosis	*Brucella* species	Animal reservoir (cows); ingestion in milk, direct contact with skin abrasions	Streptomycin, tetracycline
Anthrax	*Bacillus anthracis*	Reservoir is soil or animals; inhalation or ingestion of heavy spore concentrations	Penicillin, tetracycline, erythromycin, chloramphenicol
Gangrene	*Clostridium perfringens*	Contamination of open wound by clostridial endospores	Debridement, amputation, hyperbaric chamber, penicillin, clindamycin, chloramphenicol
Cat scratch fever	Unidentified bacillus	Cat scratch or bite	Tetracycline
Rat bite fever	*Spirillum minor* or *Streptobacillus moniliformis*	Rat bite	Penicillin or tetracycline
Plague	*Yersinia pestis*	*Xenopsylla cheopis* (rat flea)	Tetracycline and streptomycin
Relapsing fever	*Borrelia* species	*Ornithodorus* species (soft ticks)	Tetracycline

TABLE 22-1 (*Continued*)

	Causative Agent	Mode of Transmission	Treatment
Lyme disease	*Borrelia burgdorferi*	*Ixodes* spp. (tick)	Penicillin
Epidemic typhus	*Rickettsia prowazekii*	*Pediculus humanus corporis* body louse	Tetracycline and chloramphenicol
Endemic murine typhus	*Rickettsia typhi*	*Xenopsylla cheopis* (rat flea)	Tetracycline and chloramphenicol
Rocky Mountain spotted fever	*Rickettsia rickettsii*	*Dermacentor andersoni* and other species (tick)	Tetracycline and chloramphenicol
Viral Diseases			
Myocarditis	Several agents, especially coxsackievirus (enterovirus)	Coxsackievirus; inhalation and ingestion	None
Pericarditis	Several agents, especially coxsackievirus (enterovirus)	Complication of viral myocarditis	None
Infectious mononucleosis	Epstein–Barr virus	Oral secretions, kissing	None
Yellow fever	Arbovirus (yellow fever virus)	*Aedes aegypti* (mosquito)	None
Dengue	Arbovirus (dengue fever virus)	*Aedes aegypti* (mosquito)	None
Protozoan Diseases			
Toxoplasmosis	*Toxoplasma gondii*	Contact with cat feces or contaminated, under-cooked meats	Pyrimethamine in combination with trisulfapyrimidines or sulfadiazine
American trypanosomiasis	*Trypanosoma cruzi*	Bite of reduviid bug	Nifurtimox in early stages
Malaria	*Plasmodium* species	*Anopheles* mosquito	Chloroquine
Babesiosis	*Babesia microti*	*Ixodes* spp. (tick)	None
Helminthic Disease			
Schistosomiasis	*Schistosoma* species	Contaminated water; cercariae enter the body through the skin	Praziquantel

Study Outline

Introduction (p. 546)

1. The heart, blood, and blood vessels make up the cardiovascular system.
2. Lymph, lymph vessels, lymph nodes, and lymphoid organs constitute the lymphatic system.

Structure and Function of the Cardiovascular System (pp. 546–547)

1. The heart circulates substances to and from tissue cells.
2. Arteries and arterioles transport blood away from the heart. Veins and venules bring blood to the heart.
3. Materials are exchanged between blood and tissue cells at capillaries. Capillaries connect arterioles to venules.
4. Blood is a mixture of plasma and cells.
5. Plasma transports dissolved substances. Red blood cells carry oxygen. White blood cells are involved in the body's defense against infection. Platelets are involved in blood clotting.

Structure and Function of the Lymphatic System (pp. 547–548)

1. Fluid that filters out of capillaries into spaces between tissue cells is called interstitial fluid.
2. Interstitial fluid enters lymph capillaries and is called lymph.
3. Lymphatics return lymph fluid to the blood.
4. Lymph nodes contain fixed macrophages, B cells, and T cells.
5. Tonsils, appendix, spleen, and thymus gland are lymphoid organs.

Bacterial Diseases of the Cardiovascular System (pp. 548–559)

Septicemia (p. 548)

1. The growth of microorganisms in blood is called septicemia.
2. Symptoms include fever, decreased blood pressure, lymphangitis (inflamed lymph vessels). Septicemia can lead to septic shock.
3. Septicemia usually results from a focus of infection in the body.
4. Gram-negative rods are usually implicated. Endotoxin causes the symptoms.

Puerperal Sepsis (p. 548)

1. Puerperal sepsis begins as a uterine infection following childbirth or abortion; it can progress to peritonitis or septicemia.
2. *Streptococcus pyogenes* is the most frequent cause.
3. Oliver Wendell Holmes and Ignaz Semmelweiss demonstrated that puerperal sepsis was transmitted by the hands and instruments of midwives and physicians.
4. Puerperal sepsis is now uncommon because of modern hygenic techniques and antibiotics.
5. Anaerobic bacteria, including *Bacteroides* and *Clostridium*, can cause infections after improperly performed abortions.

Bacterial Endocarditis (p. 549)

1. The outer layer of the heart is the pericardium. The inner layer is the endocardium.
2. Subacute bacterial endocarditis is usually caused by α-hemolytic streptococci, although other gram-positive cocci can be involved.
3. The infection arises from a focus of infection, such as a tooth extraction.
4. Preexisting heart abnormalities are predisposing factors.
5. Symptoms include fever, anemia, and heart murmur.
6. Acute bacterial endocarditis is usually caused by *Staphylococcus aureus* or *Streptococcus pneumoniae*.
7. The bacteria cause rapid destruction of heart valves.
8. *Streptococcus* can also cause pericarditis.
9. Laboratory diagnosis is based on isolation and identification of the bacteria from blood.

Rheumatic Fever (pp. 549–550)

1. Rheumatic fever is an autoimmune complication of group A β-hemolytic streptococcal infections.
2. Rheumatic fever is expressed as arthritis or inflammation of the heart and can result in permanent heart damage.
3. Antibodies against *S. pyogenes* react with streptococcal antigens deposited in joints or heart valves or cross-react with the heart muscle.
4. Rheumatic fever can be a sequel to a streptococcal infection, such as streptococcal sore throat. Streptococci might not be present at the time of rheumatic fever.
5. The incidence of rheumatic fever in the United States has declined because of prompt treatment of streptococcal infections.
6. Rheumatic fever is treated with antiinflammatory drugs.

Tularemia

(pp. 550–551)

1. Tularemia is caused by *Francisella tularensis*. The reservoir is small wild mammals, especially rabbits.
2. Humans contract tularemia by handling diseased carcasses, eating undercooked meat of diseased animals, and being bitten by certain vectors (such as deer flies).
3. Symptoms include ulceration at the site of entry, followed by septicemia and pneumonia.
4. *F. tularensis* is difficult to culture in vitro. In vivo, it is resistant to phagocytosis.
5. Laboratory diagnosis is based on a slide-agglutination test on isolated bacteria.

Brucellosis (Undulant Fever)

(pp. 551–552)

1. Brucellosis can be caused by *Brucella abortus*, *B. melitensis*, *B. suis*, and *B. canis*.
2. Farm animals (cattle, pigs, goats, and dogs) comprise the reservoir.
3. The bacteria enter through minute breaks in the mucosa or skin, reproduce in macrophages, and spread via lymphatics to liver, spleen, or bone marrow.
4. Symptoms include malaise and fever that spikes each evening (undulant fever).
5. A vaccine for cattle is available.
6. Diagnosis is based on serological tests.

Anthrax

(pp. 552–553)

1. *Bacillus anthracis* causes anthrax. In soil, endospores can survive up to 60 years.
2. Grazing animals acquire an infection after ingesting the endospores.
3. Humans contract anthrax by handling hides from infected animals. The bacteria enter through cuts in the skin or through the respiratory tract.
4. Entry through the skin results in a malignant pustule that can progress to septicemia. Entry through the respiratory tract can result in pneumonia.
5. A vaccine for cattle is available.
6. Diagnosis is based on isolation and identification of the bacteria.

Gangrene

(pp. 553–554)

1. Soft tissue's death from ischemia is called gangrene.
2. Microorganisms grow on nutrients released from gangrenous cells.
3. Gangrene is especially susceptible to the growth of anaerobic bacteria such as *Clostridium perfringens*, the causative agent of gas gangrene.
4. *C. perfringens* can invade the uterine wall during improperly performed abortions.
5. Debridement, hyperbaric chambers, and amputation are used to treat gas gangrene.

Systemic Diseases Caused by Animal Bites and Scratches

(p. 554)

1. Septicemia is caused by *Pasteurella multocida* introduced by the bite of a dog or cat.
2. Anaerobic bacteria such as *Clostridium*, *Bacteroides*, and *Fusobacterium* infect deep animal bites.
3. The causative agent of cat scratch fever is unknown.
4. Rat bite fever is caused by *Streptobacillus* or *Spirillum*.

Plague

(pp. 554–555)

1. Plague is caused by *Yersinia pestis*. The vector is usually the rat flea (*Xenopsylla cheopis*).
2. Reservoirs for plague include European rats and North American rodents.
3. Symptoms of bubonic plague include bruises on the skin and enlarged lymph nodes (buboes).
4. The bacteria can enter the lungs and cause pneumonic plague.
5. Laboratory diagnosis is based on isolation and identification of the bacteria.
6. Antibiotics are effective to treat plague, but they must be administered promptly after exposure to the disease.
7. Control of the rat population is an effective deterrent to the spread and incidence of plague.

Relapsing Fever

(p. 556)

1. Relapsing fever is caused by *Borrelia* species and transmitted by soft ticks (*Ornithodorus*).
2. The reservoir for the disease is rodents.
3. Symptoms include fever, jaundice, and rose-colored spots. Symptoms recur three or four times after apparent recovery.
4. Laboratory diagnosis is based on the presence of spirochetes in the patient's blood.

Lyme Disease

(pp. 556–557)

1. Lyme disease is transmitted by a tick (*Ixodes*) and is caused by *Borrelia burgdorferi*.
2. Lyme disease is prevalent on the Atlantic Coast of the United States.
3. *Ixodes* transmits *Babesia*, a sporozoan that infects red blood cells.

Typhus

(pp. 557–558)

1. Rickettsias are obligate intracellular parasites of eucaryotic cells.

Epidemic Typhus

(p. 557)

1. The human body louse *Pediculus humanus corporis* transmits *Rickettsia prowazekii* in its feces, which are deposited while the louse is feeding.
2. Epidemic typhus is prevalent in crowded and unsanitary living conditions that allow the proliferation of lice.
3. The symptoms of typhus are rash, prolonged high fever, and stupor.
4. Tetracyclines and chloramphenicol are used to treat typhus.

Endemic Murine Typhus (pp. 557–558)

1. Endemic murine typhus is a less severe disease caused by *Rickettsia typhi* and transmitted from rodents to humans by the rat flea.

Spotted Fever (pp. 558–559)

1. *Rickettsia rickettsii* is a parasite of ticks (*Dermacentor* species) in the southeastern United States, Appalachia, and the Rocky Mountain states.
2. The rickettsia may be transmitted to humans, in whom it causes tick-borne typhus fever.
3. Chloramphenicol and tetracyclines are effective to treat the disease.
4. The complement-fixation test is used for laboratory diagnosis.

Viral Diseases of the Cardiovascular System (pp. 560–563)

Myocarditis (p. 560)

1. Inflammation of the heart muscle (myocardium) is called myocarditis.
2. Coxsackievirus (enterovirus) is the most common cause. It reaches the heart from a respiratory or gastrointestinal infection.

Infectious Mononucleosis (p. 560)

1. Infectious mononucleosis is caused by the Epstein–Barr (EB) virus (herpesvirus).
2. The virus multiplies in lymphatic tissue and causes the proliferation of atypical lymphocytes.
3. The disease is transmitted by ingestion of saliva from infected individuals.
4. Diagnosis is made by indirect fluorescent-antibody technique.

Burkitt's Lymphoma (p. 561)

1. EB virus is implicated in Burkitt's lymphoma and nasopharyngeal cancer.
2. EB virus is potentially oncogenic.

Yellow Fever (p. 561)

1. Yellow fever is caused by an arbovirus (yellow fever virus). The vector is the mosquito *Aedes aegypti*.
2. The virus multiplies in lymph nodes and then disseminates to liver, spleen, kidney, and heart.
3. Symptoms include fever, chills, headache, nausea, and jaundice.
4. Diagnosis is based on the presence of virus-neutralizing antibodies in the host.
5. No treatment is available. An attenuated, live viral vaccine is available.

Dengue (p. 562)

1. Dengue is caused by an arbovirus (dengue fever virus) and is transmitted by the mosquito *Aedes aegypti*.
2. Symptoms are fever, muscle and joint pain, and rash.
3. Dengue hemorrhagic fever involving bleeding from the skin, gums, and gastointestinal tract is a more serious form of the disease.
4. Laboratory identification is based on a complement-fixation test.

Viral Hemorrhagic Fevers (p. 563)

1. Viral hemorrhagic fevers are highly contagious, often fatal diseases.

Protozoan Diseases of the Cardiovascular System (pp. 563–567)

Toxoplasmosis (pp. 563–565)

1. Toxoplasmosis is caused by the sporozoan *Toxoplasma gondii*.
2. *T. gondii* undergoes sexual reproduction in the intestinal tract of domestic cats, and oocyts are eliminated in cat feces.
3. Humans contract the infection by ingesting undercooked meat from an infected animal or ingesting dried cat feces.
4. Subclinical infections are probably common because the disease symptoms are rather mild.
5. Congenital infections can occur. Symptoms include convulsions, brain damage, blindness, and death.
6. A rising antibody titer in a pregnant woman might indicate the need for a therapeutic abortion.

American Trypanosomiasis (Chagas' Disease) (pp. 565–566)

1. *Trypanosoma cruzi* causes Chagas' disease. The reservoir includes many wild animals. The vector is the "kissing bug."
2. The infection begins in the lymph nodes and disseminates via the bloodstream to other organs.
3. Observation of the trypanosomes in blood confirms diagnosis.

Malaria (pp. 566–567)

1. The symptoms of malaria are chills, fever, vomiting, and headache, which occur in one- to three-day cycles.
2. Malaria is transmitted by *Anopheles* species. The causative agent is any one of four species of *Plasmodium*.
3. Sporozoites reproduce in the liver and release merozoites into the bloodstream, where they infect red blood cells, which in turn produce more merozoites.
4. Laboratory diagnosis is based on microscopic observation of merozoites in red blood cells.
5. A vaccine is being developed.
6. New drugs are being developed as the protozoans develop resistance to drugs such as chloroquine.

Helminthic Diseases of the Cardiovascular System (pp. 568–569)

Schistosomiasis (pp. 568–569)

I. Species of the blood fluke *Schistosoma* cause schistosomiasis.

2. Eggs eliminated with feces hatch into larvae that infect the intermediate host. Free-swimming cercariae are released and penetrate the skin of a human.

3. The adult flukes live in the veins of the liver or urinary bladder in humans.

4. Adult flukes reproduce, and eggs are excreted or remain in the host.

5. Symptoms are from the host's defense to eggs that remain in the body.

6. Observation of eggs or flukes in feces, skin tests, or indirect serological tests may be used for diagnosis.

7. Chemotherapy is used to treat the disease; sanitation and snail eradication are used to prevent the disease.

Swimmer's Itch (p. 569)

I. Swimmer's itch is a cutaneous allergic reaction to cercariae that penetrate the skin. The definitive host for this fluke is wildfowl.

Study Questions

Review

I. What are the symptoms of septicemia?

2. How can septicemia result from a single focus of infection, such as an abscess?

3. Differentiate among endocarditis, myocarditis, and pericarditis. How are these infections contracted?

4. Complete the following table.

Disease	Frequent Causative Agent	Predisposing Condition(s)
Puerperal sepsis		
Subacute bacterial endocarditis		
Acute bacterial endocarditis		
Myocarditis		

5. Describe the probable cause of rheumatic fever. How is rheumatic fever treated? How is it prevented?

6. Complete the following table.

Disease	Causative Agent	Method of Transmission	Reservoir	Symptoms	Prevention
Tularemia					
Brucellosis					
Anthrax					
Lyme Disease					

7. Compare and contrast epidemic typhus, endemic murine typhus, and tick-borne typhus.

8. Plot the temperature of a patient with brucellosis for a one-week period.

9. List four bacterial infections that might result from animal bites.

10. Why is *Clostridium perfringens* likely to grow in gangrenous wounds?

11. List the causative agents and methods of transmission of infectious mononucleosis.

12. Cite evidence implicating EB virus in Burkitt's lymphoma. Offer an explanation as to why a large number of people who are infected with EB virus do not get Burkitt's lymphoma or nasopharyngeal cancer.

13. Complete the following table.

Disease	Causative Agent	Vector	Symptoms	Treatment
Malaria				
Yellow fever				
Dengue				
Relapsing fever				

14. What is the most effective control measure for mosquito-borne diseases?

15. Provide the following information on plague: causative agent; vector; U.S. reservoir; control; treatment; prognosis (probable outcome).

16. Differentiate between the transmission and symptoms of bubonic plague and pneumonic plague.

17. List the causative agent, method of transmission, and reservoir for schistosomiasis, toxoplasmosis, and American trypanosomiasis. Which disease are you most likely to get in the United States? Where are the other diseases endemic?

18. What is infectious mononucleosis? What is viral hemorrhagic fever?

Challenge

1. Indirect fluorescent-antibody tests on the serum of three 25-year-old women, each of whom is considering pregnancy, provided the following information. Which of these women may have toxoplasmosis? What advice might be given to each woman with regard to toxoplasmosis?

Patient	Antibody Titer		
	Day 1	Day 5	Day 12
Patient A	1:1,024	1:1,024	1:1,024
Patient B	1:1,024	1:2,048	1:3,072
Patient C	0	0	0

2. A 19-year-old man went deer hunting. While on the trail, he found a partially dismembered dead rabbit. The hunter picked up the front paws for good luck charms and gave them to another hunter in the party. The rabbit had been handled with bare hands that were bruised and scratched from the hunter's work as an automobile mechanic. Festering sores on his hands, legs, and knees were noted two days later. What infectious disease do you suspect the hunter has? How would you proceed to prove it?

3. On March 30, a 35-year-old veterinarian experienced fever, chills, and vomiting. On March 31, he was hospitalized with diarrhea, left axillary bubo, and secondary bilateral pneumonia. On March 27, he had treated a cat who had labored respiration; an X ray revealed pulmonary infiltrates. The cat died on March 28 and was disposed of. Chloramphenicol was administered to the veterinarian. On April 10, his temperature returned to normal, and on April 20, he was released from the hospital. Sixty human contacts were given tetracycline. Identify the incubation and prodromal periods for this case. Explain why the 60 contacts were treated. What was the etiologic agent? How would you identify the agent?

4. Three out of five patients who underwent aortic-valve replacement surgery developed bacteremia. The causative agent was *Enterobacter cloacae*. What were the patients' symptoms? How would you identify this bacterium? A manometer used in the operations was culture-positive for *E. cloacae*. What is the most likely source of this contaminant? Suggest a way of preventing such occurrences.

Further Reading

Bisno, A. L. "Acute rheumatic fever: forgotten but not gone." *New England Journal of Medicine* 316:421–427, 476–478, 1987. A discussion of the 1985–1986 resurgence of rheumatic fever in the intermountain area.

Freimer, E. H., and M. McCarty. "Rheumatic fever." *Scientific American* 213(6):66–74, December 1965. Discusses the link between streptococcal infection and rheumatic disease.

Gregg, C. T. *Plague: An Ancient Disease in the 20th Century.* New York: Scribner's, 1985. An interesting and factual account of plague in this century.

Habicht, G. S., G. Beck, and J. L. Benach. "Lyme disease." *Scientific American* 257(1):78–83; July 1987. Describes how investigators discovered the cause of this new disease since the first report in 1975. Also describes the actions of bacterial LPS on interleukin-1 in causing symptoms.

Henle, W., G. Henle, and E. T. Lennette. "The Epstein–Barr virus." *Scientific American* 241(1):48–59, July 1979. A summary of information on one of the most common viruses infecting humans.

Kadis, S., T. C. Montie, and S. J. Ajl. "Plague toxin." *Scientific American* 220(3):92–100, March 1969. The mode of action of *Yersinia* toxin.

McDade, J. E., and V. F. Newhouse. "Natural history of *Rickettsia rickettsii*." *Annual Review of Microbiology* 40:287–309, 1986. Discusses the ecology of these bacteria and reasons for the increasing range and incidence of the disease.

Miller, L. H., et al. "Research toward a malaria vaccine." *Science* 234:1349–1355, 1986. A recent survey of progress in the search for a vaccine against malaria.

See also Further Reading for Part Four at the end of Chapter 25.

CHAPTER 23

Microbial Diseases of the Respiratory System

The air we breathe contains microorganisms, so the upper respiratory system is a major portal of entry for pathogens. In fact, respiratory system infections are the most common type of infection—and among the most damaging. Some pathogens that enter the body via the respiratory system can infect other parts of the body—for example, the pathogens that cause measles, mumps, and rubella. Another very serious aspect of respiratory infections is the ease with which they are spread, both by *direct contact* with droplets emitted during sneezing, coughing, and talking and by *fomites* (contaminated objects). Transmission by droplets is especially common for highly communicable respiratory infections, such as pneumonic plague and influenza. Diseases associated with the respiratory system are summarized in Table 23-1, p. 598.

Structure and Function of the Respiratory System

It is convenient to think of the respiratory system as being composed of two divisions, the upper respiratory system and the lower respiratory system. The **upper respiratory system** consists of the nose, the throat, and the structures associated with them, including the middle ear and the auditory tubes (Figure 23-1). Ducts from the sinuses and the nasolacrimal ducts from the lacrimal apparatus empty into the nasal cavity (see Figure 15-2). The auditory (eustachian) tubes from the middle ear empty into the upper portion of the throat.

The upper respiratory system has several anatomical defenses against airborne pathogens. Coarse hairs in the nose filter large dust particles from the air. The nose is lined with a mucous membrane that contains numerous mucus-secreting cells and cilia. The upper portion of the throat also contains a ciliated

Learning Objectives

- Describe how microorganisms are prevented from entering the respiratory system.

- List the methods by which infections of the respiratory system are transmitted.

- Characterize the normal flora of the upper and lower respiratory systems.

- Describe the causative agent, symptoms, preferred treatment, and laboratory identification tests for ten bacterial diseases and three viral diseases of the respiratory system.

- Describe vaccines that are available to prevent respiratory system infections.

- Explain why vaccination against influenza and colds is impractical.

- List the etiologic agents, methods of transmission, preferred treatments, and laboratory identification tests for several fungal and protozoan diseases of the respiratory system.

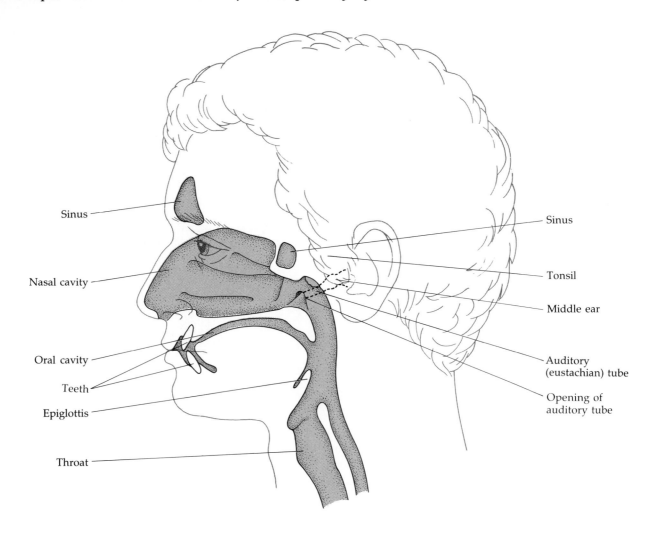

Figure 23-1 Structures of the upper respiratory system.

mucous membrane. The mucus moistens inhaled air and traps dust and microorganisms, especially particles larger than 4 to 5 μm. The cilia assist in removing these particles by moving them toward the mouth for elimination.

At the junction of the nose and throat are masses of lymphoid tissue, the tonsils and adenoids, that contribute to immunities to certain infections. Occasionally, however, these tissues become infected and help spread infection to the ears via the auditory tubes. Because the nose and throat are connected to the sinuses, nasolacrimal apparatus, and middle ear, infections commonly spread from one region to another.

The **lower respiratory system** consists of the larynx (voicebox), trachea (windpipe), bronchial tubes, and alveoli (Figure 23-2). Alveoli are the air sacs that make up the lung tissue; within them, oxygen and carbon dioxide are exchanged between the

lungs and blood. The double-layered membrane covering the lungs is the *pleura*. A ciliated mucous membrane lines the lower respiratory system down to the smaller bronchial tubes and helps to prevent microorganisms from reaching the lungs. As we discussed in Chapter 15, particles trapped in the larynx, trachea, and larger bronchial tubes are moved up toward the throat by a ciliary action called the ciliary escalator (see Figure 15-3). If microorganisms actually reach the lungs, then certain phagocytic cells—the *alveolar macrophages*, or *dust cells*—usually locate, ingest, and destroy most of them. IgA antibodies in such secretions as respiratory mucus, saliva, and tears also help to protect mucosal surfaces of the respiratory system from many pathogens. Thus, the body has several mechanisms for removing the pathogens that cause airborne infections. However, if all these mechanisms fail, then the microorganism wins the host–parasite competition, and a respiratory disease results.

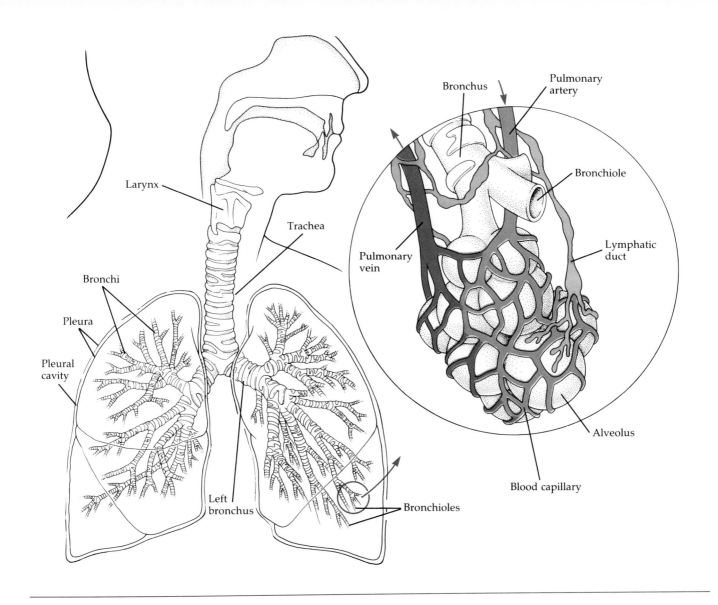

Figure 23-2   Structures of the lower respiratory system.

Normal Flora of the Respiratory System

A number of potentially pathogenic microorganisms are part of the normal flora in the upper respiratory system. The gram-positive diphtheroids, staphylococci (including *S. aureus*), micrococci, and *Bacillus* species are found in the nasal cavity. The throat near the nose might contain streptococci, including *Streptococcus pneumoniae*, as well as the gram-negative organisms *Hemophilus influenzae* and *Neisseria meningitidis*. (The normal flora of the mouth are described in Chapter 24 as part of the gastrointestinal tract.) Despite the presence of these microorganisms, they rarely cause epidemics or morbidity because certain microorganisms of the normal flora suppress their growth by competing with them for nutrients and producing inhibitory substances.

By contrast, the lower respiratory tract is usually sterile—although the trachea may contain a few bacteria—because of the normally efficient functioning of the ciliary escalator in the bronchial tubes.

Diseases of the Upper Respiratory System

As most of us know from personal experience, the respiratory system is the site of many common infections. We will soon discuss **pharyngitis**, inflammation of the mucous membranes of the throat, or sore throat. When the larynx is the site of infection, we suffer from **laryngitis**, which affects our ability to speak. This infection is caused by both bacteria, such as *H. influenzae*, and viruses, often in combination.

The organisms that cause pharyngitis also can cause inflamed tonsils, or **tonsillitis**. (The tonsils are lymphoid tissue in the rear of the nasal cavity.) The nasal sinuses are cavities in certain cranial bones that open into the nasal cavity. They have a mucous membrane lining that is continuous with that of the nasal cavity. When a sinus becomes infected with such organisms as *S. pneumoniae* or *H. influenzae*, the mucous membranes become inflamed, and there is a heavy nasal discharge of mucus. This condition is called **sinusitis**. If the opening by which the mucus leaves the sinus becomes blocked, internal pressure can cause pain or a sinus headache. These diseases are almost always *self-limiting*, meaning that recovery will usually occur even without medical intervention.

Probably the most threatening infectious disease of the upper respiratory system is **epiglottitis**, inflammation of the epiglottis. The epiglottis is a flaplike structure of cartilage that prevents ingested material from entering the larynx (see Figure 23-1). Inflammation of the epiglottis is a rapidly developing disease that can result in death within a few hours. It is caused by opportunistic pathogens, such as *H. influenzae* type b.

Bacterial Diseases of the Upper Respiratory System

Streptococcal Pharyngitis (Strep Throat)

Streptococcal pharyngitis (strep throat) is an upper respiratory infection caused by β-hemolytic group A streptococci. They are gram-positive bacteria. The most common of these is *Streptococcus pyogenes*, the same bacterium responsible for many skin and soft tissue infections, such as impetigo, erysipelas, and acute bacterial endocarditis.

The pathogenicity of group A streptococci is enhanced by their resistance to phagocytosis. They are also able to produce special enzymes that lyse fibrin clots (*streptokinase*) and DNA (*DNase*), discussed in Chapter 14. They also produce *hemolysins* that are cytotoxic to tissue cells and protective leukocytes.

Without some analysis, streptococcal pharyngitis cannot be distinguished from pharyngitis of other causes (mostly viral). Probably no more than half of so-called strep throats are actually streptococcal in origin. At one time, the diagnosis of strep throat was based on culturing of bacteria from a throat swab. Results took overnight or longer, and this test has been almost universally replaced by new indirect agglutination diagnostic tests for streptococcal pharyngitis that use microscopic latex particles that are coated with antibodies against group A streptococci. A swab of the throat is still used to collect bacteria, and the streptococcal antigens are extracted from the swab. An agglutination-type reaction is used to detect the reaction between the antigen and antibody-coated latex particles. Many of these tests take as little as ten minutes to perform.

Even when streptococci are detected, there is some controversy concerning the significance of such test results. Available tests are highly specific in detecting the presence or absence of group A streptococci. In fact, *S. pyogenes* can be detected, if care is used, in the throats of many people who are only asymptomatic carriers. A negative result may be useful because it avoids unnecessary antibiotic therapy. However, a positive result is not a certain indication that the symptoms are caused by the detected streptococci. A rise in IgM antibody titer is the best indication of a sore throat of genuinely streptococcal origin, but this is not a practical option for day-to-day use.

Streptococcal pharyngitis is characterized by inflammation and fever. Frequently, tonsillitis occurs and the lymph nodes in the neck become enlarged and tender. Another frequent complication is infection of the middle ear, called otitis media.

Nearly all β-hemolytic group A streptococci are sensitive to penicillin, the drug of choice. Erythromycin is also effective. Treatment of streptococcal pharyngitis is important to prevent such complications as rheumatic fever and glomerulonephritis, both of which are caused by an immune response to streptococcal infections.

There are more than 55 serological types of group A streptococci. Since resistance to streptococcal diseases is type-specific, a person who has recovered from infection by one type is not necessarily immune to infection by another type. A general immunity to streptococcal infections does not usually occur.

Strep throat is now most commonly transmitted by respiratory secretions, but epidemics spread by unpasteurized milk were once frequent.

Scarlet Fever

When the *S. pyogenes* strain causing streptococcal pharyngitis produces an *erythrogenic* (reddening) *toxin*, the resulting infection is called **scarlet fever**. When the strains produce this toxin, they have been lysogenized by a bacteriophage. As you may recall, this means that the genetic information of a bacterial virus has been incorporated into the chromosome of the bacterium, so the characteristics of the bacterium have been altered. The toxin causes a pinkish-red skin rash, which is probably the skin's generalized cutaneous hypersensitivity reaction to the circulating toxin, and a high fever. The tongue has a spotted, strawberrylike appearance and then, as it loses its upper membrane, becomes very red and enlarged. As the disease runs its course, the affected skin frequently peels off, as if sunburned. Complications such as deafness can occur.

Scarlet fever seems to increase and decline in severity and frequency at different times and in different locations, but the disease generally has been declining in recent years. It is a communicable disease spread mainly by inhalation of infective droplets or dust contaminated by an infected person. Classically, scarlet fever has been considered to be associated with streptococcal pharyngitis, but it might accompany a streptococcal skin infection.

Penicillin is effective in treating the original streptococcal infection. Immunity is developed to the toxin produced by the particular streptococcus that was contracted, but there are many types of streptococci, and immunity is specific to each type. Therefore, recovery from scarlet fever does not ensure immunity to streptococcal infections in general.

Diphtheria

Another bacterial infection of the upper respiratory system is **diphtheria**, once a major cause of death in children. The disease begins with a sore throat and fever, followed by general malaise and swelling of the neck. The pathogenesis of diphtheria is better understood than that of most other infectious diseases. The organism responsible is *Corynebacterium diphtheriae* (kô-rī-nē-bak-ti′rē-um dif-thi′rē-ī), which is a gram-positive, nonendospore-forming, somewhat pleomorphic rod. The dividing cells are often observed to fold together to form V- and Y-shaped figures resembling Chinese characters (Figure 23-3).

Part of the normal immunization program for children in the United States is the *DPT vaccine* (see Chapter 17). The *D* stands for diphtheria toxoid, an inactivated toxin that causes the body to produce antibodies against the diphtheria toxin. Immunity to diphtheria can be measured by the Schick test, now only rarely used, in which a standardized amount of toxin is injected into the skin. The absence of a skin reaction shows that circulating antibodies have neutralized the toxin and thus indicates that the subject is immune.

Figure 23-3 *Corynebacterium diphtheriae*, **the organism that causes diphtheria.** Note the typical clubbed shapes and the V and Y arrangements (×2000).

C. diphtheriae has adapted to a generally immunized population and may be found in the throats of symptomless carriers. The bacterium is well suited to airborne transmission and is very resistant to drying. The bacteria are usually spread by respiratory transmission, and, once established in the upper respiratory system of a susceptible person, they increase rapidly in number.

Characteristic of diphtheria is a tough (*diphtheria* means leather), grayish membrane that forms in the throat in response to the infection. It contains fibrin, dead tissue, and bacterial cells and can totally block the passage of air to the lungs.

Although the bacteria do not invade tissues, those that have been lysogenized by a phage can produce a powerful exotoxin (see Chapter 14). Circulating in the bloodstream, the toxin interferes with protein synthesis. Only 0.01 mg of this highly virulent toxin is enough to kill a 91-kg (200-pound) person. Thus, if antitoxin therapy is to be effective, it must be administered before the toxin enters the tissue cells. When such organs as the heart or kidneys are affected by the toxin, the disease can rapidly be fatal. In other cases, the nerves can be involved, and partial paralysis results.

Laboratory diagnosis can usually be made if the organisms are isolated from the primary lesion. Typical colonies grow on differential media (grayish colonies on a Loeffler slant, black colonies on tellurite agar). Once isolated, the microorganisms are tested for toxigenicity by the guinea-pig-virulence test or the gel-diffusion test. In the *guinea-pig test*, the microorganisms are injected into one side of a shaved guinea pig. After four hours, the guinea pig is injected with antitoxin. Thirty minutes later, microorganisms are injected on the opposite side. If the microorganisms are toxigenic, a characteristic lesion will form between 48 and 72 hours later at the site injected before the antitoxin was administered. In the *gel-diffusion test*, a strip of filter paper containing diphtheria antitoxin is placed on agar containing antitoxin-free calf serum. Then the microorganisms are streaked at a right angle to the filter paper. If the microorganisms are toxigenic, a visible line of antigen–antibody precipitate will form.

Even though antibiotics, such as penicillin and erythromycin, control the growth of the bacteria, they do not neutralize the diphtheria toxin. Thus, to treat diphtheria, antibiotics should only be used in conjunction with antitoxin.

The number of diphtheria cases reported in the United States each year is presently well under 100, but the death rate for respiratory cases is still from 5% to 10%, mostly among the elderly or the very young. In young children, the disease occurs mainly in groups that have not been immunized for religious or other reasons (Figure 23-4).

In the past, diphtheria was spread mainly to healthy carriers by droplet infection. Respiratory cases have been known to arise from contact with cutaneous diphtheria. With the advent of immunization, however, the carrier rate has decreased dramatically. Because humans are the only important reservoir for the diphtheria microorganism, total eradication is theoretically possible, but only if worldwide immunization becomes a reality.

Cutaneous Diphtheria

Diphtheria is also expressed as **cutaneous diphtheria**. In this form of the disease, *C. diphtheriae* infects the skin, usually at a wound or similar skin lesion, and there is minimal systemic circulation of the toxin. In cutaneous infections, the bacteria cause slow-healing ulcerations.

Cutaneous diphtheria is fairly common in tropical countries. In the United States, it is most common among Native Americans and among adults in low socioeconomic circumstances. The cutaneous form of diphtheria is responsible for most of the reported cases of this disease in persons over 30 years old (Figure 23-4). When diphtheria was more common, repeated subclinical infections reinforced the immunity, which otherwise weakens with time. Many adults now lack immunity because routine immunization was either unavailable or less common during their childhood. Because of this, when any wound or trauma requires tetanus toxoid, it is common practice to combine it with diphtheria toxoid (DT vaccine) as well.

Otitis Media

One of the more uncomfortable complications of the common cold, or of any infection of the nose or throat, is infection of the inner ear, **otitis media**, leading to earache. The inner ear can be infected directly by contaminated water from swimming pools and by some severe trauma, such as injury to the eardrum and certain skull fractures. The infecting microorganisms cause the formation of pus, which builds up

Figure 23-4 Age distribution of diphtheria cases in the United States in a typical year. Note that most cases now occur in the adult population, many of whom were not immunized in childhood or whose immunity has lapsed. Many of these cases are of cutaneous diphtheria.

pressure against the eardrum and causes it to become inflamed and painful. The condition is more common in early childhood, probably because the auditory tube connecting the middle ear to the throat is small and is more easily blocked by infection (see Figure 23-1). Enlarged adenoids might be a contributing factor for children with an unusually high frequency of infection.

A number of bacteria can cause otitis media. The one most commonly isolated is *S. pneumoniae* (about 35% of cases). Other bacteria frequently involved include *H. influenzae* (25% to 30%), *Branhamella catarrhalis* (bran-ha-mel′là ka-tär′al-is) (10% to15%), *S. pyogenes* (8% to 10%), and *S. aureus* (1% to 2%). In about 3% to 5% of cases, no bacteria can be detected. Most of these cases are probably caused by obligate anaerobic bacteria. Viruses are usually not a cause. Broad spectrum penicillins, such as amoxicillin, are usually the first choice for therapy. Penicillin and cephalosporins are also used.

Viral Disease of the Upper Respiratory System

Common Cold (Acute Coryza)

There are a number of different viruses involved in the etiology of the **common cold (acute coryza)**. About 50% of all colds are caused by *rhinoviruses*. *Coronaviruses* probably cause another 15% or 20%. About 10% of all colds are caused by one of an assortment of *adenoviruses*, *parainfluenza viruses*, and *respiratory syncytial viruses*. *Coxsackieviruses* and *echoviruses*, considered to be primarily enteric viruses, are known to cause summer colds with fever. In about 40% of colds, no agent is ever identified.

We tend to accumulate immunities against cold viruses during our lifetimes, which may be a reason why older persons tend to get fewer colds. It has been shown that young children have three or four colds per year, but adults at age 60 average less than one cold per year. Immunity is based on the ratio of IgA antibodies to single serotypes and has a reasonably high short-term effectiveness. Isolated populations may develop a group immunity, and their colds disappear until a new set of viruses is introduced. Altogether, there are probably more than 200 agents that cause colds. There are at least 113 serotypes of rhinoviruses alone, so a vaccine effective against so many different agents does not seem practical.

However, a new approach has been suggested. It has been found that most cold viruses, although antigenically different, often use as few as two different receptor sites on the mucosa. This may make it possible to approach the problem from a different direction. Instead of trying to produce antibodies against many different virus configurations, perhaps an antibody can be produced that will achieve the same results by blocking a relatively few receptor sites in the body.

The symptoms of the common cold are familiar to all of us. They include sneezing, excessive nasal secretion, and congestion. (In ancient times, one school of medical thought believed that the nasal discharges were waste products from the brain.) The infection can easily spread from the throat to the sinus cavities, the lower respiratory system, and the middle ear, leading to complications of laryngitis and otitis media. The uncomplicated cold usually is not accompanied by fever.

It is popularly thought that colds are spread by coughing and sneezing. However, research has shown that, under laboratory conditions at least, the hands are by far the most effective means of spreading colds—a contaminated surface is touched, and the virus is then deposited into the nose by the contaminated hands. A single rhinovirus deposited on the nasal mucosa is sufficient to cause a cold. The better news is that, when researchers required healthy volunteers to kiss cold sufferers for 60 to 90 seconds, only 8% of the volunteers came down with colds.

The incidence of colds definitely increases with cold weather. The rhinoviruses prefer a temperature slightly below that of normal body temperature—such as might be found in the upper respiratory system, which is open to the outside environment. However, colds will die out in isolated populations even in extremely cold climates as immunity develops in the population. Colds are also found in warm climates with little variation in temperature. No one knows exactly why the number of colds seems to increase with colder weather in temperate zones. It is not known if closer indoor contact promotes epidemic-type transmission or if physiological changes increase susceptibility.

Because colds are caused by viruses, antibiotics are of no use in treatment. The common cold will usually run its course to recovery in about a week's time. Recovery time is not affected by over-the-counter drugs presently available, although such drugs may lessen the severity of certain signs and symptoms. A chemotherapeutic drug aimed at colds is discussed in the box below.

Diseases of the Lower Respiratory System

The lower respiratory system can be infected by many of the same bacteria and viruses that infect the upper respiratory system. As the bronchi (see Figure 23-2) become involved, **bronchitis** or **bronchiolitis** develops. *Mycoplasma pneumoniae* and a number of common respiratory viruses are suspected causes. In infants, the most common agent is probably the respiratory syncytial virus, discussed later in this chapter. Diseases such as **whooping cough** are also a form of bronchitis. A severe complication of bronchitis is **pneumonia**, in which the pulmonary alveoli become involved. These often interrelated diseases are sometimes lumped under the term **croup**.

Bacterial Diseases of the Lower Respiratory System

Whooping Cough (Pertussis)

Infection by the bacterium *Bordetella pertussis* (bôr-de-tel′lä pėr-tus′sis) results in whooping cough (pertussis). *B. pertussis* is a small, nonmotile, gram-negative coccobacillus that forms a capsule when virulent. It is an obligate aerobe and produces an endotoxin as part of its cell wall; the endotoxin resembles that of other pathogenic gram-negative bacteria. In the cytoplasm, the bacteria produce an exotoxin that is apparently released along with the endotoxin when the cell autolyses upon death. The precise role of the toxins in pathogenesis has yet to be determined. The organism does not invade tissues. However, it attaches to the cilia in the trachea and impedes their action, allowing mucus to accumulate. Eventually, a toxic fragment of the bacterial cell wall, *tracheal cytotoxin*, causes loss of the ciliated cells (Figure 23-5).

.. Microbiology in the News ....

A Cure for the Common Cold?

An experimental drug has shown broad antiviral effects on a large number of the picornaviruses against which it has been tested. The tests, say scientists, while still preliminary, could yield the first drug capable of treating this family of disease agents, whose members can cause a wide variety of illnesses ranging in seriousness from polio, hepatitis A, viral meningitis, and neonatal sepsis (a

generally fatal disease affecting newborns) to those mild rhinovirus infections responsible for half of all common colds.

Based on the efficacy demonstrated thus far in animals and cell-culture experiments, the drug—known as WIN 512,711—"holds the potential for curing the common cold" and a broad spectrum of other previously untreatable infections, says Guy Diana, group leader for medicinal chemistry at the Sterling–Winthrop Research Institute in Rensselaer, New York.

The new drug halts the reproduction and spread of a virus by preventing the initial uncoating of the

(a) Normal viral reproduction

(b) Viral reproduction blocked

Bordetella pertussis   Cilia

Bordetella pertussis

Figure 23-5 Ciliated cells of the respiratory system. In this scanning electron micrograph, cells of *Bordetella pertussis* can be seen growing on the cilia; they will eventually cause the loss of the ciliated cells (×3000).

Whooping cough, which is primarily a childhood disease, can be quite severe. The initial stage, which is called the *catarrhal stage*, resembles a common cold. Prolonged sieges of coughing characterize the second, *paroxysmal stage*. When ciliary action is compromised, the mucus accumulates, and the infected person desperately attempts to cough up these mucus accumulations. The violence of the coughing can actually result in broken ribs. Gasping for air between coughs causes a whooping sound, hence the name of the disease. Coughing episodes occur several times a day for one to six weeks. The *convalescence stage*, the third stage, may last for months. The disease is of unusually long duration for a respiratory infection. Very high white blood cell counts (sometimes six to eight times normal levels) are frequently seen in per-

protein shell that encapsulates its genetic material (see the figure).

By halting viral reproduction so early in the virus-uncoating phase, the drug's mechanism of action may be unique, says Mark A. McKinlay, a virologist on the project. Exactly why it selectively binds with viruses and prevents their uncoating, however, is not fully understood.

Used therapeutically, the drug halted development of paralysis in mice that had recently been infected with polio-2 or a paralytic ECHO-9 picornavirus. Used prophylacti-cally, the drug prevented viral infection altogether when given to mice immediately before and again immediately after their exposure to various picornaviruses.

Since there are no animal models other than chimpanzees for studying rhinovirus infections, McKinlay says tests of the drug's efficacy against these cold viruses were performed using several types of cultured human cells. To date, WIN 512,711 appears effective against 34 of the 40 rhinoviruses tested. Testing is not completed, however;

there are more than 120 rhinoviruses capable of causing colds.

So far, McKinlay says, WIN 512,711 is somewhat more effective against enteroviruses—the picornavirus family's major class of nonrhinoviruses causing human disease—than against cold viruses. Diana has reported data on a close analog of the drug that showed the opposite effect—slightly better action against rhinoviruses. This leads to speculation that the endproduct of the research may not be a single anti-picornavirus drug but instead a group of related

compounds, each of whose activity has been optimized for a target class of the viruses.

Initial toxicity tests suggest the new drug is safe. The next step is clinical trials. A critical question to be answered will be whether these viral diseases are already too advanced to treat by the time symptoms appear.

Source: Reprinted with permission from *Science News, the Weekly News Magazine of Science,* copyright 1985 by Science Service, Inc. Revised January 1988.

sons with pertussis. The disease is transmitted by inhaling pathogens expelled by the coughing of infected patients.

For diagnosis, the organism is cultured from a throat swab inserted through the nose on a thin wire and held in the throat while the patient coughs. *B. pertussis* forms characteristically small, glistening, white colonies surrounded by an unusual zone of blood hemolysis on Bordet–Gengou medium. This medium was developed many years ago. Most of its additives are not nutritive but are intended to absorb substances present in most culture media that inhibit the growth of *B. pertussis*. A fluorescent-antibody test or agglutination by antiserum is used for confirmation.

Mild cases of whooping cough require no specific treatment. The disease is most severe in infants under one year of age, and nearly half the cases occur in this age group. Severe cases, especially in infants, are usually treated with erythromycin. Tetracyclines and chloramphenicol can be used. Although antibiotics might not result in rapid improvement, the drugs do render the patient noninfectious. If secondary pneumonia develops as a complication, antibiotic therapy must be continued.

After recovery, immunity is good—at least, second attacks tend to be very mild. A vaccine prepared from heat-inactivated whole bacteria is a part of the regular immunization schedule for children (the P in DPT vaccine stands for *pertussis*). Vaccination is usually recommended at about two months of age, when the titer of maternal antibodies in the baby's circulation is dropping and the child's own antibodies have not yet replaced them.

There has been considerable concern about the safety of the pertussis vaccine. It has a higher rate of adverse effects than most vaccines. The risk of serious neurological damage is about 3.2 per million doses, compared to about 1 per million doses for measles-mumps-rubella (MMR) vaccine. Vaccination has lowered the annual number of cases in the United States from more than 250,000 per year to fewer than 2000, with the number of deaths below 10. In Japan and England, adverse publicity has led to widespread avoidance of pertussis vaccination. The result was major epidemics. Because infants are less capable of coping with the effort of coughing to maintain an airway, they have a relatively high mortality rate. Authorities point to these figures as indications that vaccination should continue while research on better vaccines is intensified. An acellular vaccine, with fewer nonfunctional components, has been tested in Japan, Sweden, and limited areas of the United States. Very few neurological side-effects have been reported with this newer vaccine, but the effectiveness has been less than anticipated.

Tuberculosis

Tuberculosis is an infectious disease caused by the bacterium *Mycobacterium tuberculosis* (mī-kō-bak-ti'rē-um tü-bėr-kū-lō'sis). The microorganism is a slender rod and an obligate aerobe. The rods grow slowly (20-hour generation time), sometimes form filaments, and tend to grow in clumps. On the surface of liquid media, their growth appears moldlike, which suggested the genus name *Mycobacterium*, from the Greek *mykes*, meaning fungus (see Microview A, Photo 15).

These bacteria are relatively resistant to normal staining procedures. When stained by the Ziehl–Neelson technique of steaming in carbolfuchsin dye, they cannot be decolorized with a mixture of acid and alcohol and are therefore classified as *acid-fast*. This characteristic reflects the unusual composition of the cell wall, which contains large amounts of lipid materials. These lipids might also be responsible for the mycobacteria's resistance to environmental stresses, such as drying. In fact, these bacteria can survive for weeks in dried sputum and are very resistant to sunlight and to chemical antimicrobials used as antiseptics and disinfectants. Therefore, particular care must be taken in handling possibly infected material.

Tuberculosis is a good example of the importance of the ecological balance between host and parasite in infectious disease. Hosts are not usually aware of pathogens that invade the body and are defeated. If defenses fail, however, hosts are very much aware of the resulting disease. Several factors may affect host resistance levels—the presence of other illness and physiological and environmental factors such as malnutrition, overcrowding, and stress.

Tuberculosis is most commonly acquired by inhaling the tubercle bacillus. If early defenses fail and the bacillus reaches the lungs, it is usually phagocytized by a macrophage in the alveoli (see Figure 23-2). The macrophages of a healthy individual destroy the bacillus. If they do not, the macrophages actually protect the microbe from the chemical and immunological defenses of the body, and many of the bacilli survive and multiply within the macrophage.

If the host's macrophages fail to deal with the infection, they eventually lyse, releasing an increased number of pathogens. The tubercle bacilli released from dying macrophages form a lesion. A hypersensitivity reaction against these organisms causes formation of a *tubercle*, which effectively walls off the pathogen. These small lumps are characteristic of tuberculosis and give the disease its name. Tubercles are composed of packed masses of tissue cells and the disintegration products of bacilli and leukocytes; they usually have a necrotic center. Few bacteria are present in the tubercle.

The tubercle bacillus does not produce any injurious toxins. Tissue damage is mostly from the hypersensitivity reaction. As the reaction continues, the tubercle undergoes necrosis and eventually forms a *caseous lesion* that has a cheeselike consistency. If the caseous lesions heal, they become calcified and show clearly on X rays, where they are called *Ghon complexes* (Figure 23-6). If the disease is not arrested at this point, the caseous lesions progress to liquefaction. An air-filled *tuberculous cavity* is formed from the caseous lesion. Conditions within the cavity favor the proliferation of the tubercle bacillus, which then grows for the first time extracellularly. Bacilli soon reach very high numbers, and eventually the lesion ruptures, releasing the microorganisms into the bloodstream. This condition of rapidly spreading infection that overwhelms the body's remaining defenses is called *miliary tuberculosis* (the name is derived from the numerous millet seed-sized tubercles formed in the infected tissues).

This condition leads to a progressive disease characterized by loss of weight, coughing (often with a show of blood), and general loss of vigor. (At one time, tuberculosis was commonly known as consumption.) Even when patients are considered cured, tubercle bacilli often remain in the lung, and the disease may be reactivated. Reactivation may be precipitated by old age, poor nutrition, or immunosuppression.

The first effective antibiotic in the treatment of tuberculosis was streptomycin, although rifampin is currently preferred. Probably the drug most commonly used today is the synthetic chemical isoniazid (INH). A combination of INH and rifampin is particularly effective against phagocytized bacteria. Ethambutol and streptomycin are used in cases of intolerance or resistance to the primary drugs. To be effective, chemotherapy for tuberculosis must be continued for months—a nine-month regimen is considered a short-course treatment. Prolonged treatment is necessary, partly because the tubercle bacillus grows very slowly and many antibiotics are effective only against growing cells. Also, the bacillus is often hidden for long periods in macrophages or other locations difficult to reach with antibiotics. A major practical problem with treatment of tuberculosis is the patient's compliance with medication regimens during the long course of treatment. For preventive prophylaxis for persons exposed to tuberculosis, INH is usually given alone because there are few bacteria and development of resistance is not likely.

Ghon complexes

Figure 23-6 X ray showing Ghon complexes in a case of tuberculosis. The lower area of the right lung (left side of photo) has a large complex, and numerous smaller lesions toward the upper part of the lung are visible. The left lung is clear.

Persons infected with tuberculosis develop cell-mediated immunity against the bacterium. This form of immune response, rather than humoral immunity, is because the pathogen is mostly located within macrophages. This immunity, involving sensitized T cells, is the basis for the *tuberculin skin test.* In this test, a purified protein derivative (PPD) of the tuberculosis bacterium, derived by precipitation from broth cultures, is injected cutaneously. If the injected person has been infected with tuberculosis in the past, sensitized T cells react with these proteins, and a delayed hypersensitivity reaction appears in about 48 hours. This reaction appears as an induration (hardening) and reddening of the area around the injection site. Probably the most accurate tuberculin test is the *Mantoux test,* in which dilutions of 0.1 ml of antigen are injected and the reacting area of the skin is measured. A number of similar tests are also in common use.

A positive tuberculin test in the very young is a probable indication of an active case of tuberculosis. In older persons, it might indicate only a previous disease episode, or vaccination, not a current active case. Nonetheless, it is an indication that further examination is needed, such as a chest X ray for the detection of lung lesions and attempts to isolate the bacterium.

Confirmation of tuberculosis by isolation of the bacterium is complicated by the very slow growth of the pathogen. Eight weeks might be needed for growth of an isolate, and at least one week is required for the appearance of an exceedingly minute colony. Selective media are recommended, such as the Lowenstein–Jensen medium containing glycerol and eggs.

Another species, *Mycobacterium bovis* (bō'vis), is a pathogen mainly of cattle. It is the cause of **bovine tuberculosis**, which is transmitted to humans via contaminated milk or food. Bovine tuberculosis seldom spreads from human to human. But before the days of pasteurized milk and the development of control methods, such as tuberculin testing of cattle herds, this disease was a common form of tuberculosis in humans. *M. bovis* infections cause tuberculosis that primarily affects the bones or lymphatic system. At one time, a common manifestation of this type of tuberculosis was hunchbacked deformation of the spine.

The *BCG vaccine* is a live culture of *M. bovis* that has been made avirulent by long cultivation on artificial media. (BCG stands for the *b*acillus of *C*almette and *G*uerin, the persons who originally isolated the strain.) The vaccine is fairly effective (about 76%) in preventing tuberculosis and has been in use since the 1920s. In the United States, it is used among high-risk groups only. Apparently, the vaccine enhances cell-mediated immunity, and persons who have received the vaccine show a positive reaction to tuberculin skin tests.

There are still more than 20,000 new cases of tuberculosis and about 1700 deaths reported in the United States each year. In recent years, the mortality rate has been decreasing, probably because of more effective chemotherapy. Approximately 30% of new cases occur among foreign-born groups, especially Asian-born refugees. Certain native-born ethnic groups tend to have higher rates of tuberculosis than others—for example, blacks, Native Americans, and Eskimos (Figure 23-7). The number of cases in children is declining, an indication that person-to-person transmission is less likely to occur. The greatest increase is seen in persons about 25 to 44 years of age, probably because of long latent infections being activated. Tuberculosis, which the healthy body combats by cell-mediated immunity, is frequently a complication of AIDS, which destroys this type of immunity. Tuberculosis cases in the developing countries of the world are estimated in the tens of millions.

Bacterial Pneumonias

The term *pneumonia* is applied to many pulmonary infections. In the United States, the annual number of cases of the various forms of pneumonia probably exceeds 2 million and there are about 50,000 deaths. Most pneumonias are bacterial in origin. The more common pathogens are *S. pneumoniae*, *H. influenzae*, *S. aureus*, *Legionella pneumophila*, and *Mycoplasma pneumoniae*.

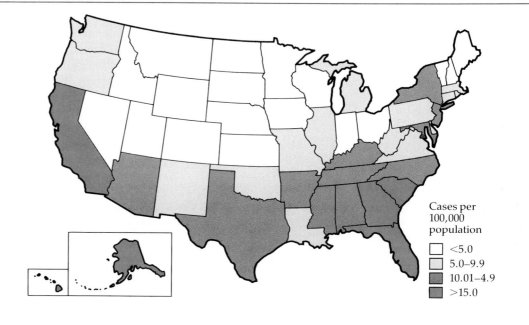

Figure 23-7   Tuberculosis distribution. The geographical distribution of reported tuberculosis cases in the United States in 1987. [Source: *MMWR* 36:52 (1/8/88.)]

Cases per 100,000 population

☐ <5.0
5.0–9.9
10.01–4.9
>15.0

Some pneumonias caused by certain gram-negative organisms are most likely to occur when the patient's defenses are naturally lowered from such conditions as diabetes or from alcoholism or drug abuse. Among such organisms that cause pneumonias are *K. pneumoniae, E. coli, P. aeruginosa, Branhamella catarrhalis*, and several species of *Enterobacter*.

Most of the organisms that cause pneumonia are normal flora of the mouth and throat that cause opportunistic infections. Viruses, fungi, and protozoans are also responsible for different pneumonias. Examples of these will be discussed later in this chapter. In practice, the causes of many pneumonia cases are never diagnosed. The physician usually makes an initial "best-guess" diagnosis based on symptoms, history, and microscopic examination of a sputum specimen. An antibiotic treatment is then directed at the suspected organisms.

Pneumococcal Pneumonia

The most common cause of pneumonia in adults is *S. pneumoniae* (the pneumococcus); the disease it causes is **pneumococcal pneumonia**. *S. pneumoniae* is a gram-positive, catalase-negative, ovoid bacterium (Figure 23-8). Because it usually forms cell pairs, it was formerly called *Diplococcus pneumoniae*. The organism produces a dense capsule that makes the pathogen resistant to phagocytosis. These capsules are also the basis of serological differentiation of pneumococci into some 83 serotypes. Before antibiotic therapy became available, antisera directed at these capsular antigens were used to treat the disease.

Pneumococcal pneumonia involves both the bronchi and the alveoli (see Figure 23-2). Symptoms include fever, difficult breathing, and chest pain. The lungs have a reddish appearance because blood vessels are dilated. In response to the infection, alveoli fill with some red blood cells, polymorphonuclear lymphocytes (PMNs), and fluid from surrounding tissues. The sputum is often rust-colored from blood coughed up from the lungs. Pneumococci can invade the bloodstream, the pleural cavity surrounding the lung, and occasionally the meninges. No bacterial toxin has been clearly related to pathogenicity.

An initial diagnosis can be made from a chest X ray because the fluid in the lungs produces recognizable shadows on the X ray. The diagnosis can be confirmed by isolation of the pneumococci from the throat, sputum, and other fluids. Pneumococci can be distinguished from other gram-positive, α-hemolytic streptococci by observing the inhibition of growth next to a disk of optochin (ethylhydrocuprein hydrochloride)

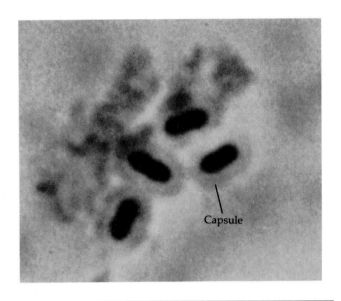

Figure 23-8 *Streptococcus pneumoniae.* Note the paired arrangement, the reason the organism was once called *Diplococcus pneumoniae* (×12,000). The capsule, important in pathogenicity, has been made more apparent here by reaction with a specific pneumonococcal antiserum (quellung reaction).

or by performing a bile-solubility test. They can also be serologically typed. The bacteria are observed under a microscope; when they react with a positive antiserum, their capsules appear to swell. This swelling is called the *quellung reaction* (from the German *Quellung*, meaning swelling).

There are many healthy carriers of the pneumococcus. Virulence of the bacteria seems to be based mainly on the carrier's resistance, which can be lowered by stress. Many illnesses of the elderly terminate in pneumococcal pneumonia. Viral infections of the respiratory system are a frequent precursor of pneumonia.

Recurrence of pneumonia is not uncommon, but the second serological type is usually different. Before chemotherapy was available, the fatality rate was as high as 25%. This has now been lowered to less than 1% for younger patients treated early in the course of their disease. Penicillin is the drug of choice, but a few strains of penicillin-resistant pneumococci have been reported. A vaccine has been developed from the purified capsular material of the 23 types of pneumococci that cause at least 80% of the pneumococcal pneumonias in the United States. This vaccine is used for the groups most susceptible to infection, the elderly and debilitated individuals.

Klebsiella pneumonia

Klebsiella pneumoniae, the causative agent of **Klebsiella pneumonia**, is a member of the enteric group of bacteria, but is not uncommon in the throat and mouth of healthy persons. The organism is an encapsulated gram-negative rod, and its virulence is strongly related to the presence of the capsule (see Microview A, Photo 5). This pneumonia is most commonly found in persons who are chronically debilitated, and malnutrition is often a contributing factor. Male alcoholics past 40 years of age are probably the most susceptible group. Probably 1% to 3% of bacterial pneumonias are of this type. Symptoms resemble those of pneumococcal pneumonia, but a prime difference is the formation of lung abscesses and permanent lung damage with *Klebsiella* pneumonia.

The fatality rate in untreated cases can be very high, perhaps exceeding 85%. Early treatment is vital but is usually not received by the most susceptible population. *K. pneumoniae*, like most gram-negative bacteria, is not sensitive to penicillin. In fact, the normal penicillin treatment for pneumococcal pneumonia may even be harmful because it suppresses the throat's normal gram-positive flora that afford some protection against invading pathogens. Quick identification is essential because delay allows the disease to progress without impedance, which is very dangerous. The antibiotics of choice are cephalosporins or gentamicin, but the bacteria are commonly resistant to these drugs.

Mycoplasmal Pneumonia (Primary Atypical Pneumonia)

Typical pneumonia has a bacterial origin. If a bacterial agent is not isolated, the pneumonia is considered *atypical*, and viral agents are usually suspected. However, there could be another source of infection—mycoplasmas. These are bacteria that do not form cell walls. The mycoplasmas do not grow under the conditions normally used to recover most bacterial pathogens. Because of this characteristic, pneumonias caused by mycoplasmas are often confused with viral pneumonias.

The bacterium *Mycoplasma pneumoniae* (mī-kō-plaz′mä) is the causative agent of **mycoplasmal pneumonia (primary atypical pneumonia)**. This type of pneumonia was first discovered when such atypical infections responded to tetracyclines, indicating that the agent was nonviral.

The disease is endemic and is a fairly common cause of pneumonia in young adults and children. The pathogens are transmitted in airborne droplets. Mycoplasmas usually infect the upper respiratory tract, and relatively few cases develop into pneumonia. The mild symptoms—low fever, cough, and headache—are the reason it is sometimes called **walking pneumonia**. The mortality rate is less than 1%.

When isolates from throat swabs and sputum grow on a medium containing horse serum and yeast extract, they form distinctive colonies of a "fried egg" appearance. The colonies are usually so small that they have to be observed with a hand lens or microscope. *M. pneumoniae* is β-hemolytic for guinea pig red blood cells; this trait differentiates it from other mycoplasmas. Conclusive identification is made by fluorescent-antibody methods. The mycoplasmas are highly varied in appearance because they lack cell walls (see Figure 10-9). Their flexibility allows them to pass through filters with pores as small as 0.2 μm in diameter, which is small enough to strain out most other bacteria.

Diagnosis based on the recovery of organisms might not be useful in treatment because as long as two weeks may be required for the slow-growing organisms to develop. A complement-fixation test, using antigens from *M. pneumoniae* cells, can be used to test for a significantly rising titer of circulating antibodies, although this procedure also takes from two to three weeks. A DNA-probe test (Figure 9-14) has recently been approved that should allow reliable diagnosis within hours. Early treatment with tetracycline and erythromycin shortens the illness.

Legionellosis

A type of pneumonia identified only recently is **legionellosis (Legionnaires' disease)**, which was discussed in Chapters 1 and 13. This disease first received public attention in 1976, when a series of deaths occurred among members of the American Legion who had attended a meeting in Philadelphia. A total of 182 persons contracted pulmonary disease, apparently at this meeting, 29 of whom died. Because no obvious bacterial cause could be found, the deaths were attributed to viral pneumonia. Close investigation, mostly with techniques directed at locating a suspected rickettsial agent, eventually identified an unknown bacterium, an aerobic gram-negative rod now known as *Legionella pneumophila* (lē-jä-nel′lä nü-mō′fi-lä). Several species of *Legionella* have now been identified.

The disease is commonly characterized by a high fever of 41°C (105°F), cough, and general symptoms of pneumonia. No person-to-person transmission seems to be involved. Recent studies have shown that the bacterium can be isolated with some frequency from natural waters. Of particular interest is the fact that the organisms can grow well in the water of air-conditioning cooling towers, which might explain some epidemics in hotels, urban business districts, and hospitals as caused by airborne transmission. The organism has also been found to inhabit the water lines of many hospitals. Most hospitals keep the temperature of hot water lines relatively low (43° to 55°C) as a safety measure, and this inadvertently maintains a good growth temperature in cooler parts of the system for this organism. The bacterium is considerably more resistant to chlorine than most other bacteria and can survive in water with a low level of chlorine for long periods of time.

Although the disease has only recently been identified, it seems now to have been fairly common. Over 700 cases are reported each year, but the actual incidence is estimated at more than 25,000. The *L. pneumophila* organism is not often enough suspected of causing disease. One hospital that had never recorded a case of nosocomial legionellosis found upon investigation that more than 14% of their nosocomial pneumonias had this cause and that most of the sites in their water-distribution system contained *L. pneumophila*. Males over 50 years of age most commonly contract the disease, especially if they are heavy smokers or alcohol abusers or have a chronic illness.

The best diagnostic method is by culture on a selective charcoal–yeast extract medium that has recently been developed. Examination of respiratory specimens by fluorescent-antibody methods (Chapter 17) may be used, and a DNA probe test has recently been approved (Figure 9-14). Erythromycin is the drug of choice for treatment.

Psittacosis (Ornithosis)

The term **psittacosis (ornithosis)** was derived from the disease's association with psittacine birds, such as parakeets and other parrots. It was later found that the disease can also be contracted from many other birds, such as pigeons, chickens, ducks, and turkeys. Therefore, the more general term *ornithosis* has come into use.

The causative organism is *Chlamydia psittaci* (kla-mi′dē-ä sit′tä-sē), a gram-negative, obligate intracellular bacterium (see Figure 10-8b). One way chlamydias differ from the rickettsias, which are also obligate intracellular bacteria, is that chlamydias

form tiny **elementary bodies** as one part of their life cycle. Unlike most rickettsias, elementary bodies are resistant to environmental stress; therefore, they can be transmitted through air and do not require a bite to transfer the infective agent directly from one host to another. The elementary bodies attach to epithelial cells of the mucous membrane of the respiratory system. They enter the cell by phagocytosis and develop into larger **reticulate bodies** that reproduce by repeated binary fission. Reticulate bodies eventually change back into infectious elementary bodies that leave the cell to infect other cells.

Psittacosis is a form of pneumonia that usually causes fever, headache, and chills. Subclinical infections are very common, and stress appears to enhance susceptibility to the disease. Disorientation and even delirium in some cases indicate that the nervous system can be involved.

The disease is seldom transmitted from one human to another but is usually spread by contact with the droppings and other exudates of fowl. One of the most common modes of transmission is inhalation of dried particles from droppings. The birds themselves usually have symptoms of diarrhea, ruffled feathers, respiratory illness, and a generally droopy appearance. The parakeets and other parrots sold commercially are usually but not always free of the disease. However, pet store employees and persons involved in raising turkeys are at greatest risk of contracting the disease.

Diagnosis is made by isolation of the bacterium in embryonated eggs or mice, or by cell culture. With use of a fluorescent-antibody staining technique, the organism can then be identified by the presence of its specific antigens. If organisms are not isolated, a complement-fixation test that detects rising serum antibody titer in the patient can be used. No vaccine is available, but tetracyclines are effective antibiotics in treating humans and animals. Effective immunity does not result from recovery, even when high titers of antibody are present in the person's serum.

Most years, fewer than 100 cases and very few deaths are reported in the United States. The main danger is in late diagnosis. Before antibiotic therapy was available, mortality was about 20%.

Recently a strain of *Chlamydia psittaci* called TWAR (Taiwan, acute respiratory) has been implicated in outbreaks of a respiratory illness resembling mycoplasmal pneumonia. This has become as common a cause of pneumonia as the influenza virus in some groups of college students. Unlike psittacosis, the disease is communicated from one human to another.

Q Fever

In Australia in the mid-1930s, a disease was identified that was characterized by a fever lasting one or two weeks, chills, chest pain, severe headache, and other evidence of a pneumonia-type infection. The disease was rarely fatal. In the absence of an obvious cause, the affliction was labeled **Q fever** (for *query*)—much as one might say "X fever." The causative agent was subsequently identified as the obligate, parasitic, intracellular bacterium *Coxiella burnetii* (käks-ē-el′lä bėr-ne′tē-ē), a rickettsia (Figure 23-9a). Most rickettsias are not resistant enough to survive airborne transmission, but this organism is an exception.

The most serious complication of Q fever is endocarditis, which occurs in about 10% of cases. Five to ten years might elapse between the initial infection and appearance of endocarditis. The organisms apparently reside in the liver during this interval.

The organism is a parasite of a number of arthropods and is transmitted among animals by direct bites. The infection in animals is usually subclinical. Transmission to humans most commonly involves cattle ticks. The cattle ticks first spread the disease among dairy herds, and the organisms are shed in the feces, milk, and urine of infected cattle. The disease is spread to humans by ingestion of unpasteurized milk and inhalation of aerosols of microbes generated in dairy barns. Inhalation of a single organism is enough to cause infection. Many dairy workers have acquired at least subclinical infections. Workers in meat- and hide-processing plants are also at risk. The pasteurization temperature of milk, which was originally aimed at eliminating tuberculosis organisms, was raised slightly in 1956 to ensure the killing of *C. burnetii*. In 1981, an endosporelike body was discovered, which may account for this heat resistance (Figure 23-9b).

Identification of the organism may be made by isolation and growth in chick embryos in eggs or in cell culture. Laboratory personnel testing for *Coxiella*-specific antibodies in the patient's serum can avoid handling the organism by using serological tests, mostly complement-fixation and agglutination types.

Most cases of Q fever in the United States are reported from the western states. The disease is endemic to California, Arizona, Oregon, and Washington. A vaccine for laboratory workers and other high-risk personnel is available. Tetracyclines are very effective in treatment.

Figure 23-9 *Coxiella burnetii.* **(a)** Note the intracellular growth of this rickettsial organism in vacuoles of the host cell (×2500). **(b)** This cell has just undergone division. Note the endosporelike body (E), which is probably responsible for the relative resistance of the organism (×103,000).

(a)

Coxiella burnetii
in vacuole

(b)

Viral Diseases of the Lower Respiratory System

Viral Pneumonia

Viral pneumonia can occur as a complication of influenza or even chickenpox. A number of enteric and other viruses have been shown to cause viral pneumonia, but viruses are isolated and identified in less than 1% of pneumonia-type infections because few laboratories are equipped to properly test clinical samples for viruses. In those cases of pneumonia for which no cause is determined, viral etiology can be assumed if mycoplasmal pneumonia has been ruled out.

Respiratory syncytial virus is probably the most common cause of viral respiratory disease among infants. The name is derived from its characteristic of causing cell fusion (syncytium formation) in cell culture. It is worldwide in occurrence. Epidemics in hospital nurseries are of particular concern. The symptoms are coughing and wheezing that last for more than one week. If an infant is ill enough to be hospitalized, the mortality rate is 1% or 2%. Approval has been given to an aerosol administration method for the antiviral drug ribavirin, which at least significantly diminishes the severity of symptoms.

Influenza (Flu)

The developed countries of the world are probably more aware of **influenza (flu)** than of any other disease, except for the common cold. The flu is characterized by chills, fever, headache, and general muscular aches. Recovery normally occurs in a few days, and coldlike symptoms appear as fever subsides. Incidentally, diarrhea is not a normal symptom of the disease, and the intestinal discomforts attributed to "stomach flu" are probably from some other cause.

The influenza virus consists of eight distinct RNA fragments of differing lengths enclosed by an inner layer of protein and an outer lipid bilayer. Embedded in the lipid bilayer are numerous projections that characterize the virus (Figure 23-10). There are two types of projections, *hemagglutinin (H) spikes* and *neuraminidase (N) spikes*, both of which are highly antigenic.

The H spikes, of which there are about 500 on each virus, allow the virus to recognize and attach to body cells before infecting them. Antibodies against the influenza virus are directed mainly at these

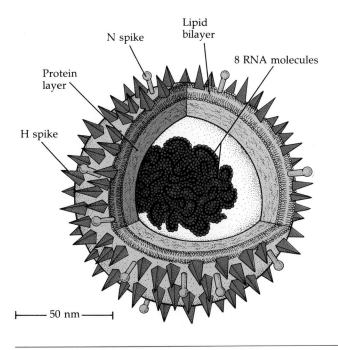

Figure 23-10 Detailed structure of the influenza virus. The envelope is composed of a protein layer, lipid layer, and two types of spikes. Because the eight separate RNA molecules can be readily rearranged, the virus can evade previous immunity by major changes in the protein composition of the H and N spikes. This is called antigenic shift. The outer spikes also undergo minor variations from time to time, which is called antigenic drift.

spikes. The term *hemagglutinin* refers to the agglutination of red blood cells that occurs when the viruses are mixed with them. This reaction is important in serological tests, such as the hemagglutination-inhibition test used to identify influenza viruses.

The N spikes, of which there are about 100 per virus, differ from the H spikes in appearance and function. Apparently, they help the virus to separate from the infected cell as the virus exits after intracellular reproduction. N spikes also stimulate formation of antibodies, but these are less important in the body's resistance to the disease than those produced in response to the H spikes.

Viral strains are identified by variation in the H and N antigens. The different forms of the antigens are assigned numbers—for example, H_1, H_2, H_3, N_1, and N_2. Each number change represents a substantial alteration in the protein makeup of the spike. The first antigenic type was isolated in 1933 and was named H_0N_1. The Asiatic flu pandemic of 1957 was of type H_2N_2, and the Hong Kong pandemic of 1968 was designated H_3N_2 (Figure 23-11). These changes are called *antigenic shifts*, and they are great enough

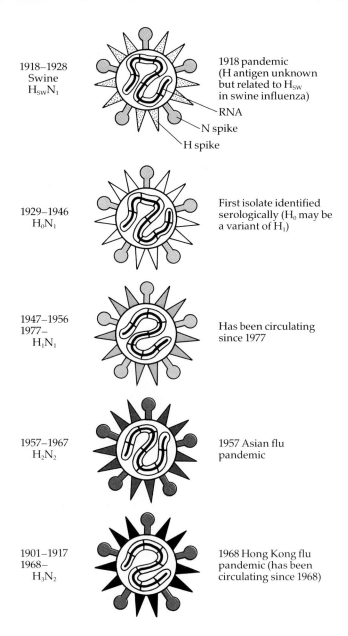

1918–1928
Swine
$H_{sw}N_1$

1918 pandemic
(H antigen unknown
but related to H_{sw}
in swine influenza)
—RNA
—N spike
—H spike

1929–1946
H_0N_1

First isolate identified
serologically (H_0 may be
a variant of H_1)

1947–1956
1977–
H_1N_1

Has been circulating
since 1977

1957–1967
H_2N_2

1957 Asian flu
pandemic

1901–1917
1968–
H_3N_2

1968 Hong Kong flu
pandemic (has been
circulating since 1968)

Figure 23-11   Differences among influenza viruses. Antigenic shifts in the influenza virus account for most of the major flu epidemics of this century.

to evade most of the immunity developed in the human population. These antigenic shifts are probably caused by a major genetic recombination. Because influenza viral RNA occurs as eight molecules, recombination is likely in infections caused by more than one strain. Recombination between the RNA of animal viral strains (found in swine, horses, and birds, for example) and the RNA of human strains might be involved. Ducks in southern Chinese farming communities have come under suspicion as the animals

most likely to be involved in genetic shifts. Wild ducks and other migratory birds then become carriers that spread the virus over large geographical areas.

Between episodes of such major antigenic shifts, there are minor annual variations in the antigenic makeup called *antigenic drift*. The virus might still be designated as H_3N_2, for example, but viral strains arise reflecting minor antigenic changes within the antigenic group. These strains are sometimes assigned names related to the locality in which they were first identified, for example, Victoria or Russia. They usually reflect an alteration of only a single amino acid in the protein makeup of the H or N spike. Such a minor, one-step mutation is probably a response to selective pressure by antibodies (usually IgA in the mucous membranes) that neutralizes all viruses except for the new mutations. Such mutations can be expected about once in each million multiplications of the virus.

The usual result of antigenic drift is that a vaccine effective against H_3, for example, will be less effective against H_3 isolates circulating ten years after the event. There will have been enough drift in that time that the virus can largely evade the antibodies originally stimulated by the earlier strain. In 1977, the Russian strain H_1N_1 circulated in the United States. Because it closely resembled a strain that had circulated in the 1950s (see Figure 23-11), it primarily affected persons under the age of about 25. This early reappearance is somewhat out of the normal character of the disease and has led to speculation that an escape from a laboratory collection might have been involved in the later outbreak.

The influenza viruses are also classified into major groups according to the antigens of their protein coats. These groups are A, B, and (rarely) C. The A-type viruses are the ones responsible for major pandemics. The B-type virus also circulates but is usually responsible for infections that are more geographically limited and milder.

To date, it has not been possible to make a vaccine for influenza that gives long-term immunity to the general population. Although it is not difficult to make a vaccine for a particular antigenic strain of virus, it has not been possible to identify each new strain of circulating virus in time for useful development and distribution of a new vaccine. Unless the strain appears to be unusually virulent, this costly process is also of questionable value. The usual practice is to limit vaccine administration to the elderly, to hospital personnel, and to similar high-risk groups. The vaccines are often *multivalent*, that is, directed at several strains in circulation at the time. At present, influenza viruses for manufacturing vaccines are grown in egg embryo cultures. The vaccines are usually 70% to 90% effective, but the duration of protection is probably no more than three years.

One new approach to the problem of antigenic variation might be to use genetic-engineering techniques. A laboratory bacterium can theoretically be made to produce very quickly a large number of antigens of the influenza virus when genes for this purpose have been inserted into the bacterial DNA.

Almost every year, epidemics of the flu spread rapidly through large populations. The disease is so readily transmissible that epidemics are quickly propagated through populations susceptible to the newly changed strain of virus. The mortality rate from the disease is not high, usually less than 1%, and these deaths are mainly among the very young and the very old. However, so many persons are infected in a major epidemic that the total number of deaths is often high. Very often, the cause of death is not the influenza virus but secondary bacterial infections. The bacterium *H. influenzae* was named under the mistaken belief that it was the principal pathogen causing influenza rather than being a secondary invader. *S. aureus* and *S. pneumoniae* are other prominent secondary bacterial invaders.

In any discussion of influenza, the great pandemic of 1918–1919 must be mentioned. Worldwide, more than 20 million people died. No one is sure why it was so unusually lethal. Usually, the very young and very old are the principal victims, but in 1918–1919, young adults had the highest mortality rate.

It is possible that the 1918 viral strain became endemic in swine in the United States. In 1976, a recruit at Fort Dix, New Jersey, died of influenza caused by a so-called swine flu strain with many of the characteristics of the 1918 pandemic strain. This case precipitated a national campaign of preventive inoculations for swine influenza. Many believe that this response was ill advised, especially because there proved to be no widespread transmission from the initial focus of infection. Another argument against such widespread vaccination is that the incidence of *Guillain–Barré syndrome* (*GBS*) was five to six times higher in persons who received flu vaccine that season than it was in unvaccinated persons. (GBS is characterized by a paralysis that is usually self-limited and reversible, although approximately 5% of the cases are fatal.) The reasons for the association between flu vaccination and GBS that year are still a mystery.

The questions raised by the 1976–1977 flu vaccination program highlight the tradeoffs that must be considered when a large-scale inoculation program is used against a new strain of flu virus. To be effective, the program must be started as soon as possible after initial infection, but this urgency does not allow enough time for the new vaccine to be thoroughly tested.

Although little can be done to alleviate a viral disease such as influenza except to treat symptoms, the bacterial complications are amenable to treatment with antibiotics. A pandemic such as that of 1918–1919 would be unlikely to have the same mortality rate today because many of those deaths were probably from secondary bacterial infections.

An antiviral drug, amantadine, has been found to significantly reduce the symptoms of influenza if administered promptly. In prophylactic use, it apparently reduces the rate of infection and illness by perhaps as much as 70%.

Fungal Diseases of the Lower Respiratory System

Histoplasmosis

Histoplasmosis superficially resembles tuberculosis. In fact, it was first recognized as a disease in the United States when X-ray surveys showed lung lesions in many persons who were tuberculin-test negative. Although the lungs are the organs most likely to be initially infected, the organisms cause lesions in almost all organs of the body. Normally, symptoms of the disease are rather poorly defined and mostly subclinical, and the disease passes for a minor respiratory infection. In a few cases, histoplasmosis becomes progressive and is a severe, generalized disease. This occurs in only a small number of infections, perhaps less than 0.1%.

The causative organism, *Histoplasma capsulatum* (his-tō-plaz′mä kap-su-lä′tum), is a dimorphic fungus; that is, it has a yeastlike morphology in tissue growth, and, in soil or artificial media, it forms a filamentous mycelium carrying reproductive conidia (Figure 23-12). In the body, the yeastlike form is found intracellularly in phagocytic cells.

Although the disease is rather widespread throughout the world, it has a limited geographical range in the United States (Figure 23-13). For example, of 513 cases reported in a recent year, 403 were

(a)

(b)

Figure 23-12 *Histoplasma capsulatum,* **a dimorphic fungus that causes histoplasmosis. (a)** Yeastlike form typical of growth in tissue (×1200). **(b)** Filamentous, spore-forming phase found in soil (×1100). The spores are usually the infectious particles.

from the states of Minnesota, Iowa, Missouri, and Alabama. In general, the disease is found in the states adjoining the Mississippi and Ohio rivers. More than 75% of the population in some of these states have antibodies against the infection. In other states—Maine, for example—a positive test is a rather rare event. Approximately 50 deaths are reported in the United States each year from histoplasmosis. The total number of infected persons may run into the millions, so the death rate is low.

Humans acquire the disease from airborne conidia produced under conditions of particular moisture and pH levels. These conditions occur particularly where droppings from birds have accumulated. Birds themselves do not carry the disease, but their droppings provide nutrients, particularly a source of nitrogen, for the fungus.

Clinical signs and history, complement-fixation tests, and, most important, isolation of the organism or its identification in tissue specimens are necessary for proper diagnosis. The organism is found in the spleen and bone marrow as well as in phagocytic cells. Currently, the most effective chemotherapy is with amphotericin B or ketoconazole.

New research might lead to a vaccine for dimorphic pathogenic fungal infections such as histoplasmosis. It involves a chemical that blocks the transition from the mycelial form of the organism to the yeast form, the one that occurs in tissue. In *H. capsulatum,* this occurs as the temperature changes from 25°C to 37°C. The blockage in the mycelial phase is permanent and is inherited by the progeny of the fungus. When injected with the mycelial phase of the fungus, mice did not develop the disease, but did develop immunity against virulent strains of the fungus.

Figure 23-13   Histoplasmosis distribution. Geographical distribution of histoplasmosis (dark area) in the United States.

Coccidioidomycosis

Another fungal pulmonary disease, also rather restricted geographically, is **coccidioidomycosis**. The causative organism is *Coccidioides immitis* (kok-sid-ē-oi'dēz im'mi-tis), a dimorphic fungus. The spores are found in dry, highly alkaline soils of the American Southwest and in similar soils of South America and northern Mexico. In tissues, the organism forms a thick-walled body filled with spores (Figure 23-14a). In soil, it forms filaments that reproduce by formation of arthrospores (Figure 23-14b). The wind carries the arthrospores to transmit the infection. Arthrospores are often so abundant that simply driving through an endemic area can result in infection, especially during a dust storm.

The symptoms of coccidioidomycosis include chest pain and perhaps fever, coughing, and loss of weight. Most infections are inapparent, and almost all patients recover in a few weeks, even without treatment. Sound immunity remains. However, in less than 1% of the infections, a progressive disease resembling tuberculosis disseminates throughout the body. The resemblance to tuberculosis is so close that isolation of the organism is necessary to properly diagnose coccidioidomycosis. A tuberculinlike skin test is used in screening for cases. Of those tested in California, more than half had positive skin tests, although most had not been aware of the infection. Farm workers are most likely to be infected, and predisposing factors, such as fatigue and poor nutrition, can lead to more severe disease.

Approximately 50 to 100 deaths each year are reported from this disease in the United States. Most cases are reported from southern California and Arizona desert regions, with only a scattering of cases reported from other states (Figure 23-15, p. 600).

A vaccine is currently being tested. If successful, it is expected to be widely used in the areas affected by coccidioidomycosis. Amphotericin B has been used to treat serious cases, but less toxic imidazole drugs, such as ketocanazole, might replace it.

Blastomycosis (North American Blastomycosis)

Blastomycosis is usually called **North American blastomycosis** to differentiate it from a similar South American blastomycosis. It is caused by the fungus *Blastomyces dermatitidis* (blas-tō-mī'sēz dėr-mä-tit'i-dis), a dimorphic fungus found most often in the Mississippi Valley, where it probably grows in soil. Approximately 30 to 60 deaths are reported each

(a)

(b)

Figure 23-14 *Coccidioides immitis.* **(a)** Scanning electron micrograph of two spherical bodies found in mouse lung tissue (×1800). Note the thick wall (SW) of the empty spherule, and the mass of spores in the other spherule. Bar length is 10 μm. **(b)** Scanning electron micrograph of arthrospores, the airborne infectious particles that spread coccidioidomycosis (×4500). The holes in the background are about 0.45 μm.

(Text continues on p. 600)

TABLE 23-1   *Diseases Associated with the Respiratory System*

	Causative Agent	Mode of Transmission	Treatment
Bacterial Diseases of the Upper Respiratory System			
Streptococcal pharyngitis (strep throat)	Streptococci, especially *Streptococcus pyogenes*	Respiratory secretions	Penicillin and erythromycin
Scarlet fever	Erythrogenic toxin-producing strains of *Streptococcus pyogenes*	Respiratory secretions	Penicillin and erythromycin
Diphtheria	*Corynebacterium diphtheriae*	Respiratory secretions, healthy carriers	Antitoxin and penicillin, tetracyclines, erythromycin
Cutaneous diphtheria	*Corynebacterium diphtheriae*	Respiratory secretions, healthy carriers	Antitoxin and penicillin, tetracyclines, erythromycin
Otitis media	Several agents, especially *Staphylococcus aureus*, *Streptococcus pneumoniae*, β-hemolytic streptococci, and *Hemophilus influenzae*	Complication of cold, or nose or throat infection	Penicillin or erythromycin
Viral Disease of the Upper Respiratory System			
Common cold	Coronaviruses, rhinoviruses	Respiratory secretions	None
Bacterial Diseases of the Lower Respiratory System			
Whooping cough	*Bordetella pertussis*	Respiratory secretions	Erythromycin, tetracyclines, chloramphenicol in severe cases, none in mild cases
Tuberculosis	*Mycobacterium tuberculosis*	Respiratory secretions; infrequently by food, especially milk	Isoniazid and rifampin
Pneumococcal pneumonia	*Streptococcus pneumoniae*	Healthy carriers; primarily a disease following viral respiratory infection or other stress	Penicillin
Klebsiella pneumonia	*Klebsiella pneumoniae*	Primarily a disease in debilitated hosts (for example, alcoholics)	Cephalosporins or gentamicin
Mycoplasmal pneumonia	*Mycoplasma pneumoniae*	Respiratory secretions	Tetracycline, erythromycin

TABLE 23-1   (*Continued*)

	Causative Agent	Mode of Transmission	Treatment
Legionellosis	*Legionella pneumophila*	Thought to be aerosols from contaminated water	Erythromycin, rifampin
Psittacosis (Ornithosis)	*Chlamydia psittaci*	Animal reservoirs; aerosols of dried droppings and other exudates of birds; person-to-person transmission is rare	Tetracyclines
Q fever	*Coxiella burnetii*	Animal reservoirs; aerosols in dairy barns; unpasteurized milk	Tetracyclines
Other bacterial pneumonias	*Pseudomonas* species; *Escherichia coli* and other bacteria	Bacteria are opportunistic	Varies with agent
Viral Diseases of the Lower Respiratory System			
Viral pneumonia	Several viruses	Aerosols; complication of other viral diseases	None
Influenza	Influenza viral strains; many serotypes	Respiratory secretions	Amantadine (prophylaxis)
Fungal Diseases of the Lower Respiratory System			
Histoplasmosis	*Histoplasma capsulatum*	Animal reservoirs; aerosols of dried bird droppings; person-to-person transmission is rare	Amphotericin B, ketoconazole
Coccidioidomycosis	*Coccidioides immitis*	Soil organisms; aerosols of dust	Amphotericin B, ketoconazole
Blastomycosis	*Blastomyces dermatitidis*	Soil organisms; aerosols of dust	Amphotericin B
Other fungal pneumonias	Species of *Aspergillus*, *Rhizopus*, *Mucor*, and other genera	Aerosols of dust containing opportunistic fungi	Amphotericin B
Protozoan Disease of the Lower Respiratory System			
Pneumocystis pneumonia	*Pneumocystis carinii*	Direct contact?	Pentamidine

Figure 23-15   Cases of coccidioidomycosis in the United States in a typical year. Because of the large number of cases that occur in the San Joaquin Valley in California, this disease is sometimes called San Joaquin Valley fever. A number of the cases shown in states outside the Southwest were probably acquired during a stay in the endemic areas, mostly California and Arizona.

year, although most infections are asymptomatic. The infection begins in the lungs and can spread rapidly. Cutaneous ulcers commonly appear, and there is extensive abscess formation and tissue destruction. The organism can be isolated from pus and biopsy specimens. Amphotericin B is usually an effective treatment.

Other Fungi Involved in Respiratory Disease

Many other opportunistic fungi may cause respiratory disease, particularly in immunosuppressed hosts or when there is exposure to massive numbers of spores. **Aspergillosis** is caused by the spores of *Aspergillus fumigatus* (a-spėr-jil'lus fü-mi-gä'tus) and other species of *Aspergillus*, which are fairly widespread in decaying vegetation. Compost piles are ideal sites for growth, and farmers and gardeners are most often exposed to infective amounts of such spores. Similar pulmonary infections sometimes result when individuals are exposed to spores of mold genera, such as *Rhizopus* or *Mucor*. Such diseases can be very dangerous, particularly invasive infections of pulmonary aspergillosis. Predisposing factors include an impaired immune system, cancer, and diabetes. As with most systemic fungal infections, there is only a limited arsenal of antifungal agents available; amphotericin B has proved the most useful.

Nocardiosis is a similar pulmonary disease, caused by the soil actinomycete (bacterium) *Nocardia*

asteroides (nō-kär'dē-ä as-tėr-oi'dēz). The disease is usually mild but can become very serious if the organism disseminates throughout the body and causes secondary infections.

Protozoan Disease of the Lower Respiratory System

Pneumocystis Pneumonia

***Pneumocystis* pneumonia** is caused by the sporozoan *Pneumocystis carinii* (nü-mō-sis'tis kär-i'nē-ī). The disease occurs throughout the world and can be endemic in hospitals. It gained attention when it was revealed to be an increasing cause of mortality among immunosuppressed patients. This group includes persons receiving immunosuppressive drugs to minimize rejection of transplanted tissue and those whose immunity is depressed because of cancer. Persons with acquired immune deficiency syndrome (AIDS) are also very susceptible to this organism.

The life cycle of this protozoan is not well known. However, outbreaks in hospitals suggest that the parasite is transmitted by direct contact between humans. *Pneumocystis* causes the alveoli to fill with a frothy exudate. Diagnosis is based on the recovery of *Pneumocystis* from the respiratory system. Trimethoprim-sulfamethoxazole is the drug of choice for treatment of this disease; an alternative is pentamidine isethionate. Untreated infections are usually fatal.

Study Outline

Introduction (p. 577)

1. Infections of the upper respiratory system are the most common type of infection.
2. Pathogens that enter the respiratory system can infect other parts of the body.
3. Respiratory infections are transmitted by direct contact and fomites.

Structure and Function of the Respiratory System (pp. 577–579)

1. The upper respiratory system consists of the nose, throat, and associated structures, such as the middle ear and auditory tubes.
2. Coarse hairs in the nose filter large particles from air entering the respiratory tract.
3. The ciliated mucous membranes of the nose and throat trap airborne particles and remove them from the body.
4. Lymphoid tissue, tonsils, and adenoids provide immunity to certain infections.
5. The lower respiratory system consists of the larynx, trachea, bronchial tubes, and alveoli.
6. The ciliary escalator of the lower respiratory system helps prevent microorganisms from reaching the lungs.
7. Microorganisms in the lungs can be phagocytized by alveolar macrophages.
8. Respiratory mucus contains IgA antibodies.

Normal Flora of the Respiratory System (p. 579)

1. Normal flora of the nasal cavity are usually gram-positive bacteria, including diphtheroids, staphylococci, micrococci, and *Bacillus* species.
2. The throat can contain *Streptococcus pneumoniae, Hemophilus influenzae,* and *Neisseria meningitidis.*
3. The lower respiratory system is usually sterile because of the action of the ciliary escalator.

Diseases of the Upper Respiratory System (pp. 579–584)

1. Specific areas of the upper respiratory system can become infected to produce pharyngitis, laryngitis, tonsillitis, sinusitis, and epiglottitis.
2. These infections may be caused by several bacteria and viruses, often in combination.
3. *H. influenzae* type b can cause epiglottitis.
4. Most respiratory tract infections are self-limiting.

Bacterial Diseases of the Upper Respiratory System (pp. 580–583)

Streptococcal Pharyngitis (Strep Throat) (p. 580)

1. This infection is caused by β-hemolytic group A streptococci, the group to which *Streptococcus pyogenes* belongs.
2. Symptoms of this infection are inflammation of the mucous membrane and fever; tonsillitis and otitis media may also occur.
3. Rapid diagnosis is made using indirect agglutination tests.
4. Penicillin is used to treat streptococcal pharyngitis.
5. Immunity to streptococcal infections does not usually develop, in part because there are many serological types.
6. Streptococcal sore throat is usually transmitted by droplets but has been associated with unpasteurized milk.

Scarlet Fever (pp. 580–581)

1. Streptococcal sore throat, caused by an erythrogenic toxin-producing *S. pyogenes*, results in scarlet fever.
2. *S. pyogenes* produces erythrogenic toxin when lysogenized by a phage.
3. Symptoms include a pink rash, high fever, and a red, enlarged tongue.

Diphtheria (pp. 581–582)

1. Diphtheria is caused by exotoxin-producing *Corynebacterium diphtheriae.*
2. Exotoxin is produced when the bacteria are lysogenized by a phage.
3. The Schick test is a skin test used to determine immunity to diphtheria.
4. A membrane, containing fibrin and dead human and bacterial cells, forms in the throat and can block the passage of air.
5. The exotoxin inhibits protein synthesis, and heart, kidney, or nerve damage may result.
6. Laboratory diagnosis is based on isolation of the bacteria and appearance of growth on differential media.
7. Toxigenicity is determined with the guinea pig test or gel-diffusion test.
8. Antitoxin must be administered to neutralize the toxin, and antibiotics can stop growth of the bacteria.
9. Routine immunization in the United States includes diphtheria toxoid in the DPT vaccine.

Cutaneous Diphtheria (p. 582)

1. Slow-healing skin ulcerations are characteristic of cutaneous diphtheria.
2. There is minimal dissemination of the exotoxin in the bloodstream.

Otitis Media   (pp. 582–583)

1. Earache, or otitis media, can occur as a complication of nose and throat infections or through direct inoculation from an external source.
2. Pus accumulation causes pressure on the eardrum.
3. Bacterial causes include *Streptococcus pneumoniae, Hemophilus influenzae, Branhamella catarrhalis, Streptococcus pyogenes,* and *Staphylococcus aureus.*

Viral Disease of the Upper Respiratory System   (pp. 583–584)

Common Cold (Acute Coryza)   (pp. 583–584)

1. Any one of approximately 200 different viruses can cause the common cold; rhinoviruses cause about 50% of all colds.
2. Symptoms include sneezing, nasal secretions, and congestion.
3. Sinus infections, lower respiratory tract infections, laryngitis, and otitis media can occur as complications of a cold.
4. Colds are most often transmitted by indirect contact.
5. Rhinoviruses prefer temperatures slightly lower than body temperature.
6. The incidence of colds possibly increases during cold weather because of increased interpersonal indoor contact or physiological changes.
7. Antibodies are produced against the specific viruses.

Diseases of the Lower Respiratory System   (pp. 584–600)

1. Many of the same microorganisms that infect the upper respiratory system also infect the lower respiratory system.
2. Diseases of the lower respiratory system include bronchitis and pneumonia, which are sometimes referrred to as croup.

Bacterial Diseases of the Lower Respiratory System   (pp. 585–592)

Whooping Cough (Pertussis)   (pp. 584–586)

1. Whooping cough is caused by *Bordetella pertussis.*
2. The initial stage of whooping cough resembles a cold and is called the catarrhal stage.
3. The accumulation of mucus in the trachea and bronchi causes deep coughs characteristic of the paroxysmal (second) stage.
4. The convalescence (third) stage can last for months.
5. Laboratory diagnosis is based on isolation of the bacteria on Bordet–Gengou medium, followed by serological tests.
6. Regular immunization for children includes dead *B. pertussis* cells as part of the DPT vaccine.

Tuberculosis   (pp. 586–587)

1. Tuberculosis is caused by *Mycobacterium tuberculosis.*
2. Large amounts of lipids in the cell wall account for the bacteria's acid-fast characteristic as well as its resistance to drying and disinfectants.
3. *M. tuberculosis* may be ingested by alveolar macrophages; if not killed, the bacteria reproduce in the macrophages.
4. Lesions formed by *M. tuberculosis* are called tubercles; necrosis results in a caseous lesion that might calcify and appear in an X ray as a Ghon complex.
5. New foci of infection can develop when a caseous lesion ruptures and releases bacteria into blood or lymph vessels; this is called miliary tuberculosis.
6. Miliary tuberculosis is characterized by weight loss, coughing, and loss of vigor.
7. The tuberculin skin test is used to determine whether a person has been exposed to tuberculosis.
8. A positive tuberculin skin test can indicate either an active case of tuberculosis or prior infection or vaccination and immunity to the disease.
9. Laboratory diagnosis is based on isolation of the bacteria and requires incubation of up to eight weeks.
10. *Mycobacterium bovis* causes bovine tuberculosis and can be transmitted to humans by unpasteurized milk.
11. *M. bovis* infections usually affect the bones or lymphatic system.
12. BCG vaccine for tuberculosis consists of a live, avirulent culture of *M. bovis.*

Bacterial Pneumonias   (pp. 588–591)

1. Most cases of pneumonia are caused by normal flora of the mouth and throat.
2. The most common etiologic agents are *S. pneumoniae, H. influenzae, S. aureus, L. pneumophila,* and *M. pneumoniae.*

Pneumococcal Pneumonia   (p. 589)

1. Pneumococcal pneumonia is caused by encapsulated *Streptococcus pneumoniae.*
2. Symptoms of this disease are fever, difficult breathing, chest pain, and rust-colored sputum.
3. Alveoli fill with red blood cells and fluids from surrounding tissues; the appearance of shadows in the X ray of lungs can be used for diagnosis.
4. The bacteria can be identified by the production of α-hemolysins, inhibition by optochin, bile solubility, and the quellung reaction.
5. A vaccine consists of purified capsular material from 23 serotypes of *S. pneumoniae.*

Klebsiella Pneumonia   (p. 590)
1. *Klebsiella pneumoniae* causes *Klebsiella* pneumonia.
2. *Klebsiella* pneumonia results in lung abscesses and permanent lung damage; mortality is 85%.

Mycoplasmal Pneumonia (Primary Atypical Pneumonia)   (p. 590)
1. *Mycoplasma pneumoniae* causes primary atypical pneumonia, or mycoplasmal pneumonia.

2. *M. pneumoniae* produces small "fried egg" colonies after two weeks' incubation on enriched media containing horse serum and yeast extract.
3. A complement-fixation test, used to diagnose the disease, is based on the rising of antibody titer. A DNA probe provides rapid identification of the bacteria.

Legionellosis   (pp. 590–591)
I. This disease is caused by the aerobic gram-negative rod *Legionella pneumophila.*
2. The bacterium can grow in water, such as air-conditioning cooling towers, and then be disseminated in the air.
3. This pneumonia does not appear to be transmitted from person to person.
4. Bacterial culture, fluorescent-antibody tests, or DNA probes are used for laboratory diagnosis.

Psittacosis (Ornithosis)   (p. 591)

I. *Chlamydia psittaci* is transmitted by contact with contaminated droppings and exudates of fowl.
2. Elementary bodies allow the bacteria to survive outside a host. Reticulate bodies develop in host cells.
3. The bacteria are isolated in embryonated eggs, mice, or cell culture; identification is based on fluorescent-antibody staining.
4. Commercial bird handlers are most susceptible to this disease.
5. The TWAR strain is transmitted from person to person.

Q Fever   (p. 592)

I. Obligately parasitic, intracellular *Coxiella burnetii* causes Q fever.
2. The disease is usually transmitted to humans through unpasteurized milk or inhalation of aerosols in dairy barns.
3. Laboratory diagnosis is made with the culture of bacteria in embryonated eggs or cell culture.

Viral Diseases of the
Lower Respiratory System   (pp. 593–595)

Viral Pneumonia   (p. 593)

I. Many pneumonia cases are thought to be caused by viral etiology.
2. A number of viruses can cause pneumonia as a complication of infections such as influenza.
3. The etiologies are not usually identified in a clinical laboratory because of the difficulty in isolating and identifying viruses.
4. Respiratory syncytial virus is a common cause of pneumonia in infants.

Influenza (Flu)   (pp. 593–595)

I. Influenza is caused by influenza virus and is characterized by chills, fever, headache, and general muscular aches.
2. Hemagglutinin (H) and neuraminidase (N) spikes project from the outer lipid bilayer of the virus.

3. Viral strains are identified by antigenic differences in the H and N spikes; they are also divided by antigenic differences in their protein coats (A, B, and C).
4. Antigenic shifts that alter the antigenic nature of the H and N spikes make natural immunity and vaccination of questionable value. Minor antigenic changes are caused by antigenic drift.
5. Deaths during an influenza epidemic are usually from secondary bacterial infections.
6. Multivalent vaccines are available for the elderly and other high-risk groups.
7. Amantadine is an effective prophylactic and curative drug.

Fungal Diseases of the
Lower Respiratory System   (pp. 595–600)
I. The following mycoses can be treated with amphotericin B.

Histoplasmosis   (pp. 595–596)

I. *Histoplasma capsulatum* causes a subclinical respiratory infection that only occasionally progresses to a severe, generalized disease.
2. The disease is acquired by inhalation of airborne conidia.
3. Isolation of the fungus or identification of the fungus in tissue samples is necessary for diagnosis.

Coccidioidomycosis   (p. 597)

I. Inhalation of the airborne arthrospores of *Coccidioidis immitis* can result in coccidioidomycosis.
2. Most cases are subclinical, but when there are predisposing factors such as fatigue and poor nutrition, a progressive disease resembling tuberculosis can result.

Blastomycosis
(North American Blastomycosis)   (pp. 597–600)

I. *Blastomyces dermatitidis* is the causative agent of blastomycosis.
2. The infection begins in the lungs and can spread to cause extensive abscesses.

Other Fungi Involved
in Respiratory Disease   (p. 600)

I. Opportunistic fungi can cause respiratory disease in immunosuppressed hosts, especially when large numbers of spores are inhaled.
2. Among these fungi are *Aspergillus, Rhizopus,* and *Mucor.*

Protozoan Disease of the
Lower Respiratory System   (p. 600)
I. Immunosuppressed patients or cancer patients are susceptible to an endemic hospital sporozoan, *Pneumocystis carinii.*

Study Questions

Review

1. Describe how microorganisms are prevented from entering the upper respiratory system. How are they prevented from causing infections in the lower respiratory system?

2. Respiratory diseases are usually transmitted by _____ . List other methods by which they can be transmitted.

3. How do the normal flora of the respiratory system illustrate microbial antagonism?

4. Complete the following table.

Disease	Causative Agent	Symptoms	Treatment
Streptococcal sore throat			
Scarlet fever			
Diphtheria			
Whooping cough			
Tuberculosis			
Pneumococcal pneumonia			
Klebsiella pneumonia			
Legionellosis			
Psittacosis			
Q fever			
Epiglottitis			
Nocardiosis			

5. How is otitis media contracted? What causes otitis media? Why was otitis media included in a chapter on diseases of the respiratory system?

6. List the causative agent, symptoms, and treatment for three viral diseases of the respiratory system. Separate the diseases according to whether they infect the upper or lower respiratory system.

7. Compare and contrast primary atypical pneumonia and viral pneumonia.

8. A patient has been diagnosed as having pneumonia. Is this sufficient information to begin treatment with antimicrobial agents? Briefly discuss why or why not.

9. List the causative agent, method of transmission, and endemic area for the fungal diseases histoplasmosis, coccidioidomycosis, and blastomycosis.

10. Under what conditions can the saprophytes *Aspergillus* and *Rhizopus* cause infections?

11. Why is the DPT vaccine routinely given to children in the United States? What does it consist of?

12. Briefly describe the procedures and positive results of the Schick test and tuberculin test, and indicate what is revealed by a positive test.

13. Discuss reasons for the increased incidences of colds and pneumonias during cold weather.

14. Match the bacteria involved in respiratory infections with the following laboratory test results.
Gram-positive cocci
Catalase-positive _____
Catalase-negative
β-hemolytic, bacitracin inhibition _____
α-hemolytic, optochin inhibition _____
Gram-positive rods
Not acid-fast _____
Acid-fast
Mycelia formed _____
Mycelia not formed _____
Gram-negative cocci _____
Gram-negative rods
Aerobes
Rods _____
Coccobacilli _____
Facultative Anaerobes
Require X and V factors _____
Capsule _____
Intracellular parasites
Form elementary bodies _____
Do not form elementary bodies _____
Wall-less _____

Challenge

1. Differentiate between the following.
(a) *S. pyogenes* causing strep throat and *S. pyogenes* causing scarlet fever.
(b) Diphtheroids and *C. diphtheriae*.

2. Why is vaccination against influenza felt to be of questionable value?

3. Why is a combination of antimicrobial drugs used to treat tuberculosis?

4. In a two-week period, eight infants in an intensive care nursery (ICN) developed pneumonia caused by respiratory syncytial virus (RSV). FA screening and viral cultures were used to diagnose infections, and

positive patients were placed in a separate room. A two-week-old girl from the newborn nursery, adjacent to ICN, also developed an RSV infection. Toward the end of this outbreak, FA tests and viral cultures were made of 15 ICN personnel. (All 15 reported having had upper respiratory illness during the previous week.) Viral cultures were negative, but the RSV-FA test was strongly positive in one nurse and weakly

positive in five others. Comment on the probable source of this outbreak. Explain the apparent discrepancy between the FA test and viral culture results. How can RSV infections in nurseries be prevented?

5. List the reasons why influenza outbreaks are fairly common in nursing homes for the elderly.

Further Reading

Al-Doory, Y. *The Epidemiology of Human Mycotic Diseases*. Springfield, Ill.: Thomas, 1975. Includes histoplasmosis, coccidioidomycosis, and blastomycosis.

"*Bordetella pertussis: pathogenesis and prevention of whooping cough.*" In L. Leive and D. Schlessinger, eds., *Microbiology—1984*. Washington, D. C.: American Society for Microbiology, pp. 157–183. A collection of articles on the epidemiology, toxins, and immunity of whooping cough.

Miller, J. A. "Clinical opportunities for plant and soil fungi." *BioScience* 36:656-658; 1986. Describes the rise in fungal infections from species previously known only as plant pathogens and soil saprophytes from new genera of fungi.

Skinner, F. A., and L. B. Quesnel. *Streptococci*. New York: Academic Press, 1978. Describes the pathogenicity, immunology, and taxonomy of the genus *Streptococcus*.

Stuart-Harris, C. "The epidemiology and prevention of influenza." *American Scientist* 69:166–172, 1981. Discusses the relationship between antigenic shift in the virus and immunity to influenza.

Thomas, G., and M. Morgan-Witts. *Anatomy of an Epidemic*. New York: Doubleday, 1982. A medical detective story about the mysterious epidemic that led to the discovery of a new bacterium (*Legionella*).

Weiss, R. "TB troubles." *Science News* 133:92–93, 1988. Discusses reasons for the recent increase in tuberculosis in the United States.

See also Further Reading for Part Four at the end of Chapter 25.

CHAPTER 24

Microbial Diseases of the Digestive System

The major sites of microbial invasion of the body are the skin and the mucous membranes of the respiratory and digestive systems. The most common routes by which pathogens exit the body are also the respiratory and digestive systems. The various respiratory diseases are the leading cause of illness in the United States, and diseases of the digestive system rank second.

Most diseases of the digestive system result from the ingestion of food or water that contains microorganisms and their toxins. When food handlers practice good sanitation methods, including proper handwashing and fly control, contamination of food is prevented. Modern methods of sewage treatment and disinfection of drinking water also help break the *fecal–oral* cycle of disease. Preservation of foods by either heat or refrigeration also minimizes contamination by disease-causing microorganisms.

Table 24-2, on p. 632, is a summary of the diseases associated with the mouth and digestive system.

Structure and Function of the Digestive System

The **digestive system** may be divided into two principal groups of organs (Figure 24-1). One group is the *gastrointestinal (GI) tract* or *alimentary canal*, essentially a tubelike structure that includes the mouth, pharynx (throat), esophagus (food tube), stomach, small intestine, and large intestine. The other group of organs, the *accessory structures*, consists of the teeth, tongue, salivary glands, liver, gallbladder, and pancreas. Except for the teeth and tongue, the accessory structures lie outside the tract and produce secretions that are conveyed by ducts into the tract.

Learning Objectives

- Describe the antimicrobial features of the digestive system.

- Describe the events that lead to dental caries and periodontal disease.

- List the etiologic agents, suspect foods, symptoms, and treatment for staphylococcal food poisoning, salmonellosis, typhoid fever, bacillary dysentery (shigellosis), cholera, and gastroenteritis.

- List the etiologic agents, methods of transmission, sites of infection, and symptoms for mumps, cytomegalovirus inclusion disease, infectious hepatitis, and serum hepatitis.

- Compare and contrast giardiasis, balantidiasis, and amoebic dysentery.

- Compare and contrast tapeworm and nematode infections.

- Discuss preventive measures for infections of the gastrointestinal system.

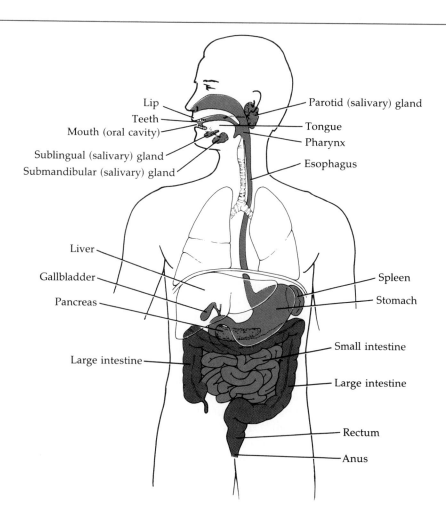

Figure 24-1 Anatomy of the human digestive system.

Lip
Teeth
Mouth (oral cavity)
Sublingual (salivary) gland
Submandibular (salivary) gland
Parotid (salivary) gland
Tongue
Pharynx
Esophagus
Liver
Gallbladder
Pancreas
Large intestine
Spleen
Stomach
Small intestine
Large intestine
Rectum
Anus

The purpose of the digestive system is to digest foods—that is, to break them down into small molecules that can be taken up and used by body cells. The combined actions of enzymes in saliva, stomach enzymes, pancreatic enzymes, intestinal enzymes, and bile from the gallbladder convert ingested nutrients into the end-products of digestion. Thus, carbohydrates are reduced to simple sugars; fats, to fatty acids and glycerol; and proteins, to amino acids. In a process called *absorption*, these end-products of digestion then pass from the small intestine into the blood or lymph for distribution to body cells.

By the time the remaining food leaves the small intestine, digestion and absorption are almost complete. As the food moves through the large intestine, water, vitamins, and nutrients are absorbed from it, and microbial cells are added to it. The resulting undigested solids, called *feces*, are stored in the rectum until they are eliminated from the body through the anus. Feces carry intestinal pathogens, and it is necessary to prevent them from contaminating food and water.

Normal Flora of the Digestive System

Bacteria heavily populate most of the digestive system. In the mouth, each milliliter of saliva can contain millions of bacteria, with various types distributed in specific areas. For example, different types of streptococci, which use sugars to form lactic acid, are found in different regions of the mouth. *Streptococcus salivarius* (sa-li-vā′rē-us) colonizes the surface of the tongue; *S. mitis* (mī′tis) inhabits mostly the of the cheek; and *S. sanguis* (san′gwis) colonizes the teeth. Spirochetes and strict anaerobes, such as *Bacteroides* (bak-tė-roi′dēz) and *Fusobacterium* (fü-sō-bak-ti′rē-um), inhabit the spaces between the gums and teeth.

The stomach and small intestine have relatively few microorganisms because of the hydrochloric acid produced by the stomach and the rapid movement of food through the small intestine. By contrast, the large intestine has enormous microbial populations, exceeding 100 billion bacteria per gram of feces. (Up to 40% of fecal mass is microbial cell material.) The population of the large intestine is comprised mostly of anaerobes of the genera *Lactobacillus* and *Bacteroides* and facultative anaerobes, such as *E. coli, Enterobacter, Klebsiella,* and *Proteus.* Most of these bacteria assist in the enzymatic breakdown of foods, and some of them synthesize useful vitamins, such as niacin, vitamins B_1, B_2, B_6, and B_{12}, folic acid, biotin, and vitamin K. The possible role of these bacteria in colon cancer is discussed in the box in Chapter 14, p. 390.

Bacterial Diseases of the Mouth

Dental Caries (Tooth Decay)

Dental caries (tooth decay) involve a gradual softening of the enamel and dentin of a tooth (Figure 24-2). If the condition remains untreated, various bacteria can invade the pulp and cause death (necrosis) of the pulp and abscess of the bone surrounding the tooth. (Such a tooth must be treated by root canal therapy.) Although dental caries are probably one of the most common infectious diseases of humans today,

you might be surprised to learn that dental caries were not common in the Western world until about the seventeenth century. In human remains from older times, only about 10% of the teeth contain caries. The introduction of table sugar, or sucrose, into the diet correlates highly with our present level of caries in the Western world. Oral bacteria convert sucrose and other carbohydrates into acid, which in turn attacks the tooth enamel.

Bacteria that convert sugar into lactic acid are commonplace in the oral cavity, but most of them are not important contributors to tooth decay. For caries to form, there must be a localization of the acid production. If a single bacterial species can be said to be *cariogenic* (causing caries), that species is *Streptococcus mutans* (mū'tans), a gram-positive coccus (Figure 24-2b and c). The organisms (some believe that it is a group of organisms) become established in the mouth when the teeth erupt from the gums. *S. mutans* then produces a sticky polysaccharide of glucose molecules called **dextran**, with which the bacteria adhere firmly to the teeth. They especially favor crevices and contact points between teeth and do not adhere to the epithelial surfaces of the mouth. This is important because about one liter of saliva is produced per day, and, if the bacteria do not adhere to some oral surface, they are washed away in the flow and swallowed.

In the production of dextran, sucrose is first hydrolyzed into its component monosaccharides, fructose

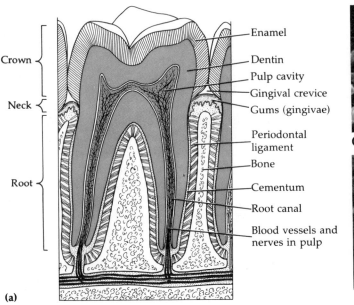

(a)

Enamel
Dentin
Pulp cavity
Gingival crevice
Gums (gingivae)
Periodontal ligament
Bone
Cementum
Root canal
Blood vessels and nerves in pulp

Crown
Neck
Root

(b)

(c)

Figure 24-2 Dental caries. (a) Structure of a normal tooth. **(b)** Scanning electron micrograph of spherical cells of *Streptococcus mutans* adhering to tooth enamel ($\times 3700$). **(c)** Scanning electron micrograph showing numerous cells of *S. mutans* in the eroded area, forming the beginning of a dental caries lesion ($\times 1300$).

and glucose. The enzyme glucosyl transferase then assembles the glucose molecules into dextran. The masses of bacterial cells and dextran adhering to the teeth are called *dental plaque* (Figure 24-3). In plaque, the bacteria produce lactic acid from sucrose, which is the dominant cause of dental caries. The primary sugar used in acid production is the fructose released by hydrolysis of sucrose.

Studies have shown that, whereas glucose and fructose are cariogenic individually, they are much more so when combined as sucrose (Figure 24-4). Because plaque is not very permeable to saliva, when lactic acid is produced it is not diluted or neutralized, and it breaks down the enamel of the teeth to which the plaque adheres. It is the localized acid-producing bacteria in the plaque that initiate dental caries.

People living on high-starch diets (starch is a polysaccharide of glucose) have a low incidence of caries unless sucrose is also part of their diet. It appears that, if sucrose is ingested only at regular meal times, the protective and repair mechanisms of the body are not overwhelmed. It is the sucrose that is ingested at

Figure 24-3 Dental plaque. After three days of not being brushed, the surface of a tooth has accumulated these organisms (×10,000).

(a)

(b)

(c)

Figure 24-4 Involvement of *S. mutans* and sucrose in dental caries. (a) Conversion of sucrose to dextran (dental plaque) and lactic acid by *S. mutans.* Dextran is a polysaccharide formed from the glucose molecule; fructose is the primary sugar fermented to lactic acid. **(b)** Scanning electron micrograph showing *S. mutans* growing in glucose broth (×3200). **(c)** Scanning electron micrograph showing *S. mutans* growing in sucrose broth (×3200). Note the large accumulation of dextran.

more frequent intervals that causes problems. In one set of experiments, it was shown that 330 grams of sucrose per day at meal times was not cariogenic, but only 120 grams per day at frequent intervals between meals caused an increase of five to seven times in the number of caries. Athough such sugars as mannitol, sorbitol, and xylitol are metabolized to acid by oral bacteria, they are not cariogenic. This is why they are used to sweeten "sugarless" candies and chewing gums.

No drugs are now available to prevent tooth decay. A vaccine, based on IgA antibodies in the saliva, is theoretically possible, and tests conducted on animals have given encouraging results. However, at present, the main strategies in the prevention of caries are minimal ingestion of sucrose; brushing, flossing, and professional cleaning to remove plaque; and the use of fluorides to provide the teeth with resistance against acidic action.

Periodontal Disease

Even persons who avoid tooth decay might, in later years, lose their teeth to **periodontal disease**. Periodontal disease is a collective term for a number of conditions characterized by inflammation and degeneration of the gums, supporting bone, periodontal ligament, and cementum (Figure 24-5). The disease is

Figure 24-5 Periodontal disease. In this X ray, note how the bone has degenerated away from the teeth. Only the tip of the root of the tooth on the right is embedded in bone.

probably initiated by bacterial growth in the gingival crevice, although the exact mechanism is unclear. One mechanism is discussed in the box in Chapter 15, p. 401. In any event, plaque deposited in the gingival crevice causes an initial inflammatory reaction. Species of *Actinomyces*, *Nocardia*, and *Corynebacterium*, among other genera, are involved in plaque formation. The formation of plaque can eventually lead to formation of abscesses, bone destruction, and general tissue necrosis.

Some periodontal disease can be controlled by frequent removal of plaque by brushing, flossing, and professional cleaning. Other forms of the disease require antibiotics, such as penicillin and tetracycline.

Bacterial Diseases of the Lower Digestive System

Most diseases of the digestive system are usually contracted when contaminated food or water is ingested. These diseases are essentially of two types, infections and intoxications.

An **infection** occurs when a pathogen enters the gastrointestinal (GI) tract and multiplies. Microorganisms can penetrate into the intestinal mucosa and grow there or can pass through to other systemic organs. Infections are characterized by a delay in the appearance of gastrointestinal disturbance while the pathogen increases in numbers or affects invaded tissue. There is also usually a fever, one of the body's general responses to an infective organism.

Some pathogens cause disease by elaborating toxins that affect the GI tract. An **intoxication** is caused by the ingestion of such a preformed toxin. Most intoxications, such as that caused by *Staphylococcus aureus*, are characterized by a very sudden appearance (usually in only a few hours) of symptoms of a GI disturbance. Fever is less often one of the symptoms.

Both infections and intoxications often cause *diarrhea*, which most of us have experienced. Severe diarrhea, accompanied by blood or mucus, is called *dysentery*. Both types of digestive system diseases are also frequently accompanied by *abdominal cramps, nausea*, and *vomiting*. The general term *gastroenteritis* is applied to diseases causing inflammation of the stomach and intestinal mucosa. Botulism (Chapter 21) is a special case of intoxication because the ingestion of the preformed toxin affects the nervous system rather than the GI tract.

In the developing countries, diarrhea is a major factor in infant mortality. Approximately one in every ten children dies of it before the age of five. It also affects the absorption of nutrients from their food and adversely affects the growth of the survivors. The cause of diarrhea may be any of several organisms. Most are not identified, but surveys in such countries as Bangladesh indicate that the three most common causes are enterotoxigenic *E. coli*, *Shigella* spp. and intestinal rotaviruses. It is estimated that mortality from childhood diarrhea could be halved by *oral rehydration therapy*. Ideally, this is a solution of sodium chloride, potassium chloride, and sodium bicarbonate. However, even a solution of a handful of table sugar and a pinch of salt in a liter of water has proved to be a very useful treatment for diarrheal diseases.

Staphylococcal Food Poisoning (Staphylococcal Enterotoxicosis)

A very common cause of gastroenteritis is **staphylococcal food poisoning**. This is an *intoxication* caused by ingestion of an *enterotoxin* produced by *S. aureus*. Staphylococci are comparatively resistant to environmental stresses, as we discussed in Chapter 10. Staphylococci have a fairly high resistance to heat; vegetative cells can tolerate 60°C for half an hour. They are also resistant to drying and radiation, which helps them survive on skin surfaces. Resistance to high osmotic pressures helps them grow in foods, such as cured ham, in which high osmotic pressure of salts inhibits the growth of competitors.

S. aureus is a common inhabitant of the nasal passages, from which it often contaminates the hands. It is also a frequent cause of skin lesions on the hands. From these sources, it can readily enter food. If the organisms are allowed to incubate in the food (a situation called *temperature abuse*), they reproduce and release enterotoxin into the food. These events, which lead to outbreaks of staphylococcal intoxication, are illustrated in Figure 24-6.

S. aureus produces several toxins that damage tissues or increase the organism's virulence. The enterotoxins causing food poisoning are classified as serological types A (which is responsible for most cases) through D. Production of this toxin is usually correlated with production of an enzyme that coagulates blood plasma. Such bacteria are described as *coagulase-positive* (Chapter 10). No direct pathogenic effect can be attributed to the enzyme, but it is useful in the tentative identification of types that are likely to be virulent. Coagulase-positive strains are considered potentially pathogenic.

Generally a population of about 1 million bacteria per gram will produce enough enterotoxin to cause

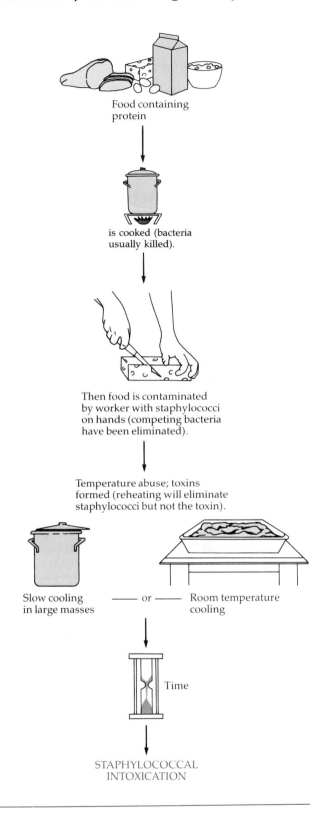

Figure 24-6 The sequence of events in a typical outbreak of staphylococcal food poisoning.

illness. The growth of the organism is facilitated if the competing organisms in the food have been eliminated—by cooking, for example. The organism is also more likely to increase if competing bacteria are inhibited by a higher-than-normal osmotic pressure or by a relatively low moisture level. *S. aureus* tends to outgrow most competing bacteria under these conditions.

Certain foods have an unusually high incidence of association with staphylococcal food poisoning. Custards and cream pies, for example, have a relatively high osmotic pressure because of their sugar content. They also have regions of low moisture that are still adequate for staphylococcal growth. If the pie is cooked during preparation, most competing organisms are eliminated. These foods can be readily inoculated with *S. aureus* by subsequent handling, and the bacteria multiply during storage.

Another high-risk food is ham, which contains many salts as preservatives or flavorings. These curing agents inhibit competing organisms more than they inhibit staphylococci. Poultry products can also harbor staphylococci if they are handled and allowed to stand at room temperatures. In fact, almost every food has been incriminated in staphylococcal food poisonings at one time or another. Hamburger, however, is rarely associated with any type of food poisoning because it has normal flora that are competitive with staphylococci, and hamburger is usually cooked immediately before consumption. Any foods that are prepared in advance and are not kept chilled are a potential source of staphylococcal food poisoning. Because contamination of foods cannot be avoided completely, the most reliable method of preventing staphylococcal food poisoning is adequate refrigeration during storage to prevent toxin formation.

The toxin itself is heat stable and can survive up to 30 minutes of boiling. So, once the toxin is formed, it is not destroyed when the food is reheated, although the bacteria will be killed.

Staphylococcal food poisoning is characterized by nausea, vomiting, and diarrhea from one to six hours after the contaminated food is ingested. The toxin affects the vagus nerve in the upper intestinal tract and triggers the brain's vomiting reflex center. The cause of the diarrheal symptoms is not as clearly identified. The symptoms usually disappear in 24 hours. Because considerable amounts of water and electrolytes are lost during episodes of vomiting and diarrhea, treatment for the disease consists of replacing the lost water and electrolytes. The disease has been identified in 25% of the outbreaks of food poisoning of known etiology in recent years.

The mortality rate of staphylococcal food poisoning is almost nil among otherwise healthy individuals, but it can be significant in weakened individuals, such as residents of nursing homes. No reliable immunity results from recovery. However, there is a great deal of variation in individual susceptibility to the toxin, and it is suspected that immunity acquired from a previous exposure might account for some of this variation.

The diagnosis of staphylococcal food poisoning is usually based on the symptoms, particularly the short incubation time characteristic of intoxication. If the food has not been reheated so that the bacteria are not killed, the organism can be recovered and grown. *S. aureus* isolates can be tested by the *phage-typing* method, which is useful in tracing the source of the contamination (Figure 9-9). These bacteria will grow well in 10% sodium chloride, so this concentration is often used in media for their isolation or identification. Pathogenic staphylococci usually ferment mannitol, produce blood hemolysins and coagulase, and form yellow colonies. They cause no obvious spoilage when growing in foods.

No simple procedure will identify the toxin in foods. Identification requires a complex extraction and a method of serological testing against antisera; these procedures are not commercially available. Food can be easily tested, however, for the presence of *thermostable nuclease*, an enzyme produced by *S. aureus* that survives heating to temperatures that will kill the cells themselves. It can be detected by a simple test, in which an extract of the food is allowed to react with DNA. The presence of detectable amounts of the enzyme usually indicates a microbial population large enough to have produced sufficient toxin to cause illness.

Salmonellosis (*Salmonella* Gastroenteritis)

As we noted earlier, in bacterial infections, disease results from microbial growth in body tissues rather than from the ingestion of food and drink already contaminated by microbial growth. Bacterial infections, such as salmonellosis, usually have longer incubation periods (from 12 hours to 2 weeks) than bacterial intoxications, reflecting the time needed for the organism to grow in the tissues of the host. Bacterial infections are also often characterized by some fever, indicating the host's response to the infection.

The *Salmonella* (sal-mōn-el′lä) bacteria (named for Daniel Salmon) are gram-negative, facultatively anaerobic, nonendospore-forming, usually motile rods that ferment glucose to produce acid and gas. Their normal habitat is the intestinal tracts of humans and many animals. All salmonellae are considered pathogenic to some degree, causing **salmonellosis**, or ***Salmonella* gastroenteritis**.

The nomenclature of the *Salmonella* organisms is confusing. According to *Bergey's Manual*, none of the present methods for naming salmonellae is satisfactory from a scientific viewpoint. Rather than recognized species, there are more than 2000 *serovars* (closely related organisms differentiated by serological testing). Only about 50 such serovars are isolated with any frequency in the United States. Some of these are named much like species—for example, *Salmonella dublin* (dub′lin) and *Salmonella eastbourne* (ēst′bôrn) were named after the sites where they were first isolated, and *Salmonella typhimurium* (tī-fi-mŭr′ē-um) causes typhoidlike disease in mice. Others are represented by the Kauffmann–White scheme (see the discussion of *Salmonella* in Chapter 10), which is generally used in clinical laboratories. This method assigns numbers and letters to different antigens: O (somatic or body), Vi (capsular), and H (flagellar).

Salmonellosis has an incubation time of about 12 to 36 hours. The salmonellae first invade the intestinal mucosa and multiply there. Sometimes they manage to pass through the intestinal mucosa to enter the cardiovascular system, and from there they spread to eventually affect many organs. The fever associated with salmonellae infections might be from endotoxins released by lysed cells, but this relationship is not certain. There is usually a moderate fever accompanied by nausea, abdominal pain and cramps, and diarrhea. As many as 1 billion salmonellae per gram can be found in the infected person's feces during the acute phase of the illness.

The mortality rate is overall very low, probably lower than 1%. However, the death rate is high in infants and among the very old; death is usually from septicemia. Individual responses to the infection vary considerably. The severity and incubation time can depend on the number of *Salmonella* ingested. Normally, recovery will be complete in a few days, but many patients will continue to shed the organisms in their feces for up to six months.

Salmonellosis is probably greatly underreported. The 20,000 to 30,000 cases reported each year are probably only from 1% to 10% of the actual total (Figure 24-7).

Meat products are particularly susceptible to contamination by *Salmonella*. If these products are mishandled, the bacteria can grow to infective numbers rather quickly. The sources of the bacteria are the intestinal tracts of many animals, and meats can be contaminated readily in processing plants. Poultry, eggs, and egg products are often contaminated by *Salmonella*.

Prevention of salmonellosis depends on good sanitation practices to deter contamination and on proper refrigeration to prevent increases in bacteria. The organisms are generally destroyed by normal cooking that heats the food to an internal temperature of at least 68°C (145°F). However, contaminated food can contaminate a surface, such as a cutting board. Although the food first prepared on the board might later be cooked and its bacteria killed, another food subsequently prepared on the board might not be cooked.

Milk has been the source of several recent outbreaks of salmonellosis. In 1985, lapses in sanitation during processing apparently introduced raw (unpasteurized) milk into already pasteurized milk, causing nearly 16,000 consumers to become ill and several deaths. In three states, unpasteurized milk may be sold as raw certified milk. Consumption of raw milk has been the source of frequent outbreaks of salmonellosis.

Figure 24-7 Incidence of salmonellosis (excluding typhoid fever). The incidence of this foodborne disease has been increasing steadily. The decrease from 1984 to 1986 is probably a slight variation and not a change in the trend. (Source: CDC. *Summary of Notifiable Diseases 1986,* MMWR 35:55.)

Diagnosis is generally made on isolation of the organisms from leftover foods or from the stools of patients. Antibiotic therapy is not useful in treating salmonellosis; treatment consists of oral rehydration therapy.

Typhoid Fever

A few serovars of *Salmonella* are much more virulent than others. They produce an intestinal disease in which they cross the intestinal wall and become invasive. The most virulent species, *Salmonella typhi*, causes the bacterial infection **typhoid fever**.

The incubation period is much longer than that of salmonellosis, normally about two weeks. The patient first suffers from a severe fever of 40°C (104°F) and continual headache. Diarrhea appears only during the second or third weeks, and the fever tends to decline. The organism becomes disseminated in the body and can be isolated from the blood, urine, and feces. In severe cases, there can be perforation of the intestinal wall. The mortality rate is now about 1% to 2%; at one time, it was about 10%. Before the days of proper sewage disposal, water treatment, and food sanitation, typhoid was an extremely common disease. It is still a frequent cause of death in parts of the world with poor sanitation.

A substantial number of recovered patients become carriers. They harbor the pathogen in the gallbladder and continue to shed the pathogen for several months. A certain number of such carriers continue to shed the organism indefinitely. Most of us are familiar with the name of Typhoid Mary. This was Mary Mallon, who worked as a cook in New York state in the early part of the century and was responsible for several outbreaks of typhoid and three deaths. Her case became well known through the attempts of the state to restrain her from working at her chosen trade.

In each recent year, there have been about 400 to 500 cases of typhoid fever in the United States. The Centers for Disease Control usually lists from 60 to 70 carriers each year. Approximately 70% of the cases in recent years were acquired during foreign travel. Cases acquired in the United States are usually from a chronic carrier. Normally, there are fewer than 10 deaths each year.

Chloramphenicol is often the drug of choice for the treatment of severe typhoid fever, despite its known toxicity. However, drug-sensitivity testing is required because of the continuing appearance of resistant strains. Ampicillin and trimethoprim-sulfamethoxazole are usually effective alternatives. Recovery from typhoid confers a life-long immunity.

Immunization for typhoid is not normally done except for high-risk laboratory and military personnel. The vaccine presently used is a killed-organism type that must be injected. In Egypt and Chile, field trials with a new oral vaccine against *S. typhi* have resulted in about 70% fewer cases of typhoid fever in the vaccinated groups. Although its effectiveness is not especially high, the oral administration makes it more attractive than the older, injected vaccines. It has been licensed in several European countries.

Bacillary Dysentery (Shigellosis)

Bacillary dysentery (shigellosis) is a severe form of diarrhea caused by a group of facultatively anaerobic gram-negative rods of the genus *Shigella* (shi-gel′lä), named for the Japanese microbiologist Kiyoshi Shiga.

There are four species of pathogenic *Shigella*: *S. sonnei* (sōn′ne-ē), *S. dysenteriae* (dis-en-te′rē-ī), *S. flexneri* (fleks′nėr-ī), and *S. boydii* (boi′de-ē). These bacteria are residents only of the human intestinal tract, and animal reservoirs are unimportant. They are closely related to the pathogenic *E. coli*. *S. sonnei*, the most common in the United States, causes a relatively mild dysentery. At the other extreme, infection with *S. dysenteriae* (sometimes called the Shiga bacillus), results in a severe dysentery and prostration. Fortunately, this is the species found least frequently in the United States.

Virulent shigellae produce an exotoxin that inhibits protein synthesis, thereby killing cells, and is probably a factor in pathogenicity. The bacteria proliferate to immense numbers in the small intestine, but the damage they cause is to the large intestine. There they produce tissue destruction in the intestinal mucosa, causing severe diarrhea with blood and mucus in the stools. Infected persons may have as many as 20 bowel movements in a day. These ulcerations in the intestinal mucosa eventually heal but form scar tissue. Additional symptoms of infection are abdominal cramps and fever. Except for *S. dysenteriae*, *Shigella* rarely invades the bloodstream. Diagnosis is usually based on recovery of the organisms from rectal swabs.

In recent years, the number of cases reported in the United States has been about 15,000 to 20,000, with 20 to 35 (about 0.2%) deaths. *S. dysenteriae* has a significant mortality rate, however, and the death rate in tropical areas can be much higher, perhaps

20%. The disease is probably more common than the reported numbers indicate. Some cases of so-called traveler's diarrhea might be mild forms of bacillary dysentery. This dysentery is especially common in institutionalized patients and in persons living on Native American reservations. Some immunity seems to result from recovery, but no satisfactory vaccine has yet been developed.

In severe cases of bacillary dysentery, antibiotic therapy and fluid and electrolyte replacement are indicated. At present, ampicillin is the drug of choice, but some resistance has appeared. Trimethoprim-sulfamethoxazole is a frequently used alternative.

Cholera

During the 1800s, the bacterial infection called Asiatic cholera crossed Europe and North America in repeated epidemics. Today, **cholera** is endemic in Asia, particularly India, and has only occasional outbreaks in Western countries. These outbreaks are caused by temporary lapses in sanitation, and the number of cases is limited.

The causative organism is *Vibrio cholerae* (vib′rē-ō kol′ér-ī), a slightly curved, gram-negative rod with a single polar flagellum (Figure 24-8). The serogroup 0:1 of *V. cholerae* causes the classically recognized epidemic form of the disease. It can be subdivided into two *biotypes* (differentiated other than by serology) as classical or *eltor* (named for the original culture isolated at a quarantine camp, El Tor, for pilgrims to Mecca). At present, the most widespread cholera pathogen is serogroup 0:1, biotype eltor. It grows in the small intestine and produces an enterotoxin that results in the secretion of chlorides, bicarbonates, and water. The excess water and mineral electrolytes are excreted, taking on the typical appearance of "rice water stools" from the masses of intestinal mucus, epithelial cells, and bacteria. The sudden loss of these fluids and electrolytes (from 12 to 20 liters of fluid might be lost in one day) causes shock, collapse, and often death. Because of the loss of fluid, the blood becomes so viscous that vital organs are unable to function properly. Violent vomiting is sometimes present. The organisms are not invasive, and a fever is usually not present. The severity of the disease varies considerably, and the number of subclinical cases might be several times the number that are reported.

In the United States, the waters of the Gulf and Pacific coasts support an indigenous population of the non-0:1 serogroup of *V. cholerae* that has been the cause of a number of outbreaks of gastroenteritis, usu-

(a)

(b)

(c)

Figure 24-8 *Vibrio cholerae*, **the pathogen that causes cholera. (a)** Note the slightly curved morphology in this scanning electron micrograph (×3300). **(b)** Normal intestinal wall of an infant rabbit (×2500). **(c)** Rabbit intestine after infection with *V. cholerae* (×2500).

ally associated with ingestion of contaminated seafood. These bacteria differ from the 0:1 serotypes in several ways, and they are more likely to invade the intestinal mucosa, causing bloody stools and fever.

The cholera bacteria can be readily isolated from the feces, partly because the organisms can grow in media that are alkaline enough to suppress the growth of many other organisms. The non-0:1 serogroup may occasionally be isolated from blood and wounds.

The most reliable control methods are based on principles of sanitation, particularly of water supplies. Recovery from the disease results in an effective immunity based on the antigenic activity of both the cells and the enterotoxin. However, because of antigenic differences among bacterial strains, the same person can have cholera more than once. Most victims in endemic areas are children. The available vaccines are protective to some extent but provide immunity of relatively short duration as compared with the immunity from a natural infection. Recent tests of attenuated vaccines have not been very encouraging.

Tetracycline or chloramphenicol is used for treatment, but chemotherapy is not very effective as replacement of the lost fluids and electrolytes. Untreated cases of cholera may have a 50% mortality rate, whereas the rate for cases having proper supportive care can be as low as 1%.

Vibrio parahaemolyticus Gastroenteritis

Vibrio parahaemolyticus (pa-rä-hē-mō-li'ti-kus) was first recognized as a pathogen in Japan in 1950. It is morphologically similar to *V. cholerae*, but it is halophilic and requires 2% to 4% sodium chloride for optimum growth. It is the most common cause of **gastroenteritis** in Japan, with thousands of cases reported annually. It is also a common cause of gastroenteritis in many nations of Southeast Asia. The organism is present in coastal waters of the continental United States and Hawaii. Crustaceans, such as shrimp and crabs, have been associated with several outbreaks caused by *V. parahaemolyticus* in the United States in recent years.

Symptoms include a burning sensation in the stomach, abdominal pain, vomiting, and watery stools, like the symptoms of cholera. The organism has also been known to cause cutaneous infections of cuts that have come in contact with contaminated clams and oysters. Incubation time is normally less

than 24 hours. Recovery usually follows in a few days, and the fatality rate is low. Tetracycline and chloramphenicol are used only in severe cases.

Because *V. parahaemolyticus* has a requirement for sodium, usually supplied as sodium chloride, isolating media containing 2% to 4% sodium chloride are used in diagnosis of the disease.

Escherichia coli Gastroenteritis (Traveler's Diarrhea)

The bacterium most familiar to microbiologists is probably *E. coli*, one of the most prolific microorganisms in the human intestinal tract. *E. coli* is normally harmless, but certain strains acquire the ability to cause gastroenteritis.

All pathogenic strains possess *adhesins* (pili, for example), by which they attach to the intestinal tract mucosa. Once in place, they produce toxins that cause diarrhea. There are several distinct pathogenic groups of *E. coli*. The *enterotoxigenic E. coli* produce an enterotoxin responsible for **traveler's diarrhea** and, in developing countries, most infant diarrhea. *Enteroinvasive E. coli* invade the epithelial lining of the large intestine and cause a dysentery very similar to that of shigellosis. The *enteropathogenic E. coli* strains cause epidemic diarrhea in young infants, and *enterohemorrhagic* strains appear to be responsible for hemorrhagic colitis (an inflammation of the colon with bleeding).

In practice, it is difficult to differentiate among the isolates of pathogenic and nonpathogenic *E. coli*. Probably 50% to 65% of traveler's diarrhea is caused by *enterotoxigenic E. coli*. Of the remaining cases, from 10% to 20% are probably caused by *Shigella*. Other enteric bacteria, such as *Salmonella* and *Campylobacter*, as well as an assortment of unidentified bacterial pathogens, viruses, and protozoan parasites, are also occasionally involved.

Traveler's diarrhea is almost impossible to avoid in parts of the world with poor sanitation practices. Prophylactic antibiotic therapy is generally not recommended, although it has been somewhat effective. An alternative preventive regimen is to take two bismuth subsalicylate (Pepto-Bismol®) tablets four times a day. One trial showed that only 14% of those following this regimen developed diarrhea during a three-week trip, whereas 40% of the control group, who took a placebo, developed traveler's diarrhea.

There has been considerable research on a vaccine, which would be very popular commercially, but results have been disappointing. In adults, the disease is usually self-limiting, and chemotherapy is not attempted. Once contracted, the best treatment is the usual oral rehydration recommended for all diarrhea. In severe cases, antimicrobials may be necessary.

In Chapter 25, *E. coli* is discussed as one of the most common causes of urinary tract infections.

Campylobacter Gastroenteritis

Campylobacter (kam-pī-lō-bak′ter) are gram-negative, spirally curved rods (Figure 24-9) that were once considered only minor causes of human gastroenteritis. Like *Salmonella*, they are part of the intestinal flora of a number of animals, especially sheep and cattle. Cows with no evidence of illness may excrete them in their milk. After selective media for their isolation and identification were developed in the late 1970s, it became clear that they were a common cause of human disease. In fact, they have replaced *Shigella* spp. as the second most common cause of diarrhea in the United States. In some states, they are even more frequently isolated in diarrhea cases than salmonellae.

Clinically, **Campylobacter gastroenteritis** is characterized by fever, cramping abdominal pain, and diarrhea with blood and mucus present in the stools. A billion organisms per gram of stool have been isolated at the peak of an infection. Normally, recovery follows within a week.

There is interesting speculation that *Campylobacter pylori* (pī′lôr-ē), may be the cause of gastric ulcers. Although some are convinced that there is a direct cause-and-effect relationship, others contend that these bacteria are simply colonizing a lesion caused by other factors. Some researchers have concluded that *C. pylori* are at least impeding healing of the ulcer, and treatment that eliminates them is likely to lead to improvement.

Yersinia Gastroenteritis

Other enteric pathogens that are being identified with increasing frequency are *Yersinia enterocolitica* (yèr-sin′ē-ä en-tér-ō-kōl-it′ik-ä) and *Yersinia pseudotuberculosis* (sū-dō-tü-ber-kū-lō′sis). These gram-negative organisms are intestinal inhabitants of many domestic animals and are often transmitted in meat and milk. Both organisms are distinctive in their ability to grow at refrigerator temperatures of 4°C (39°F). The symptoms of **yersiniosis** are diarrhea, fever, headache, and abdominal pain. The pain is often severe enough to cause a misdiagnosis of appendicitis.

Figure 24-9 *Campylobacter jejuni.* Scanning electron micrograph showing the spiral-shaped rod morphology (×8600).

Clostridium perfringens Gastroenteritis

Probably one of the more common, if underrecognized, forms of food poisoning in the United States is caused by *Clostridium perfringens*, a large, gram-positive, endospore-forming, obligately anaerobic rod. This organism is also responsible for human gas gangrene (Chapter 22).

Most outbreaks are associated with meats or stews containing meats. The organism's nutritional requirement for amino acids is met by such foods, and when the meats are cooked, the oxygen level is lowered enough for clostridial growth. The endospores survive most routine heatings, and the generation time of the vegetative bacterium is less than 20 minutes under ideal conditions. Large populations can therefore build up rapidly when foods are being held for serving or when inadequate refrigeration leads to slow cooling.

The organism grows in the intestinal tract and produces an exotoxin that causes the typical symptoms of abdominal pain and diarrhea. The toxin alters the permeability of the intestinal wall, and the resulting loss of water and electrolytes produces the symptoms of diarrhea. Most cases are mild and self-limiting and probably are never clinically diagnosed. The symptoms usually appear from 8 to 12 hours after ingestion.

Diagnosis is usually accomplished by isolation and identification of the organism in stool samples. Treatment consists of replacing lost water and electrolytes.

Bacillus cereus Gastroenteritis

Bacillus cereus (se′rē-us) is a large, gram-positive, endospore-forming organism that is very common in the environment and is usually considered harmless. It has, however, been identified as the cause of outbreaks of foodborne illness. Some cases resemble *C. perfringens* infections and are almost entirely diarrheal in nature. Other episodes involve nausea and vomiting. It is suspected that different toxins are involved in the differing symptoms. Both forms of the disease are self-limiting.

Viral Diseases of the Digestive System

Mumps

The target of the mumps virus, the parotid glands, is located just below and in front of the ear (see Figure 24-1). Since the parotids are one of the three pairs of the salivary glands of the digestive system, it is appropriate to include a discussion of mumps in this chapter. The virus itself is a *paramyxovirus*, the group to which the measles virus belongs. The mumps virus is an enveloped virus containing single-stranded, fragmented RNA (Figure 24-10).

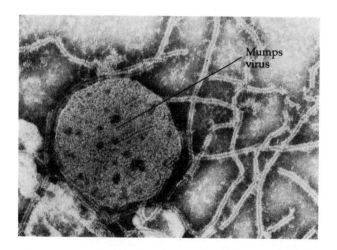

Mumps virus

Figure 24-10   Mumps virus. Transmission electron micrograph of a mumps virus, showing a single virus and filamentous helical nucleocapsids released from the virus's exterior (×45,600).

Mumps typically begins with painful swelling of one or both parotid glands from 16 to 18 days after exposure to the virus. The virus is transmitted in saliva and respiratory secretions, and its portal of entry is the respiratory tract. Once the viruses have begun to multiply in the respiratory tract and local lymph nodes in the neck, they reach the salivary glands via the blood (Figure 24-11). Viremia (viral presence in the blood) begins several days before the onset of mumps symptoms and before the virus appears in saliva. The virus is present in the blood and saliva for three to five days after the onset of the disease, and in the urine after ten days or so.

Mumps is characterized by inflammation and swelling of the parotid glands, fever, and extreme pain during swallowing. About four to seven days after the onset of symptoms, the testes can become inflamed (orchitis). This happens in about 20% to 35% of males past puberty. Sterility is a possible consequence but rarely occurs. Other complications of mumps include meningitis, inflammation of the ovaries, and pancreatitis. In children, mumps is less common than chickenpox or measles because the disease is less infectious.

An effective vaccine is available and is often administered as part of the trivalent measles–mumps–rubella (MMR) vaccine. The number of cases of mumps has dropped sharply since the introduction of the vaccine in 1968. For example, in 1968, 152,209 cases of mumps were reported, whereas in 1987 there were 12,299 cases. Second attacks are rare, and cases involving only one parotid gland, or subclinical cases, are as effective as bilateral mumps in conferring immunity.

Serological diagnosis is not usually necessary. If laboratory confirmation of a diagnosis based only on symptoms is desired, the virus can be isolated by embryonated egg or cell-culture techniques. The virus can be identified with hemagglutination-inhibition tests.

Cytomegalovirus (CMV) Inclusion Disease

The cytomegalovirus (CMV) is a herpes virus that induces a cellular swelling characterized as *cytomegaly* (large cell). These cells are known as "owl's eyes" because of their distinctive appearance (Figure 24-12). Once a person is infected with CMV, the infection persists for life. Recent research has shown that blood macrophages and T lymphocytes are the sites of latency. The virus is shed at intervals in such body secretions as saliva, urine, semen, cervical secretions, and breast milk.

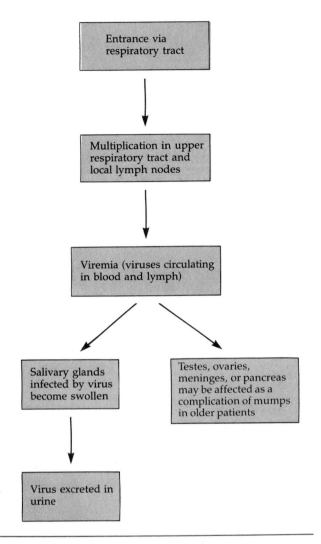

Figure 24-11 **Pathogenesis of a mumps infection.**

Some estimate that 70% of the population of the United States may carry the virus. In less developed parts of the world, infection may approach 100%. It can be spread by kissing and other personal contacts. There is a high incidence of the virus in the male homosexual population. In adults and older children, most infections are subclinical or, at worst, much like a mild case of infectious mononucleosis, However, immunosuppressed individuals may develop life-threatening pneumonia.

Infants frequently become infected at birth or by nursing; these infections are almost always subclinical. Much more serious is infection of the fetus in the uterus. There are about 36,000 new cases of such congenital cytomegalovirus infections in the United States each year. Most infants show no ill effects, although some will later have problems with hearing loss. About 10% of these infections, however, have serious consequences, known as **congenital cytomegalic inclusion (CMV) disease**. This disorder is characterized by impaired mental development, poor hearing and eyesight, and neurological problems, such as cerebral palsy and epilepsy. Many infants die within a few months. From 10% to 18% of stillborns in the United States show damage characteristic of congenital CMV disease.

To date, antiviral therapy has not been of value, although tests of some drugs have been encouraging. There is considerable interest in developing a subunit-type vaccine. An attenuated vaccine has been made, but there is concern about the safety of such a live vaccine.

The most reliable diagnostic method is isolation of the virus from bodily fluids during the first two

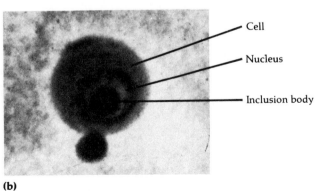

Figure 24-12 **Cytomegalovirus (CMV). (a)** Electron micrograph showing individual viral particles (VP) in tissue culture (approx. ×13,300). **(b)** Light micrograph of a single animal cell infected with cytomegalovirus (×750). The dark, intranuclear inclusion body has the typical "owl-eyed" appearance.

weeks of life; the sample is usually shipped to a central reference laboratory. Serological tests are not very specific or sensitive.

Hepatitis

Hepatitis is an inflammation of the liver. Viral hepatitis is now the second most frequently reported infectious disease in the United States. Several different viruses commonly cause hepatitis, and they can be transmitted in various ways. The virus that causes hepatitis A (infectious hepatitis) is usually transmitted by fecal–oral contamination. The virus that causes hepatitis B (serum hepatitis) is usually transmitted by parenteral transmission (other than by way of the intestines). Most often this is from blood or other bodily fluids that contain the virus. Hepatitis non-A, non-B (NANB) includes what is now the most common form of hepatitis transmitted by transfused blood in the United States. There is also a NANB hepatitis

spread by the fecal–oral route. Delta hepatitis is a particularly severe form of hepatitis. The characteristics of the various forms of viral hepatitis are summarized in Table 24-1.

Hepatitis A (Infectious Hepatitis)

The *hepatitis A virus (HAV)* is the etiological agent of **hepatitis A (infectious hepatitis)**. HAV contains single-stranded RNA and lacks an envelope. It can be grown in cell culture.

After a typical entrance via the oral route, HAV multiplies in the epithelial lining of the intestinal tract. Viremia eventually occurs, and the virus spreads to the liver, kidney, and spleen. The virus is shed in the feces and can also be detected in the blood and urine. Recent outbreaks of HAV disease indicate that the virus might also be carried in oral secretions and can be spread when fingers are put in the mouth and then are used to handle food. The amount of virus excreted is greatest before symptoms appear and then

TABLE 24-1   *Characteristics of Viral Hepatitis*

Characteristic	Hepatitis A	Hepatitis B	Delta Hepatitis	Non-A, Non-B (fecal–oral transmission)	Non-A, Non-B (parenteral transmission; may be two viruses involved)
Transmission	Fecal–oral (ingestion of contaminated food and water)	Parenteral (injection of contaminated blood or other body fluids)	Parenteral (host must be coinfected with hepatitis B)	Fecal–oral	Parenteral
Agent	Hepatitis A virus (HAV); single-stranded RNA, no envelope	Hepatitis B virus (HBV); double-stranded DNA, envelope	Delta agent (HDV); single-stranded RNA, envelope	Presumed virus of unknown characteristics	Presumed viruses of unknown characteristics
Incubation period	2 to 6 weeks	4 to 26 weeks	Uncertain	2 to 6 weeks	6 to 25 weeks
Manifestations or symptoms	Mostly subclinical; severe cases—fever, headache, malaise, jaundice	Frequently subclinical; similar to HAV, but fever, headache absent, and more likely to progress to severe liver damage	Severe liver damage, high mortality rate	Similar to HAV, but pregnant women may have high mortality rate.	Similar to HBV
Chronic liver disease	No	Yes	Yes	No	Yes
Vaccines	Under development	Two types: Heptavax B®, Recombivax HB®	None	None	None

declines rapidly. Therefore, the food handler responsible for spreading the virus might not appear to be ill. The virus is probably able to survive for several days on such surfaces as cutting boards. Contamination of food or drink by feces is aided by the resistance of HAV to disinfectants, such as chlorine, at concentrations ordinarily used in water. Mollusks, such as oysters, that live in contaminated waters are also a source of infection.

At least 50% of infections with HAV are subclinical, especially in children. In clinical cases, the initial symptoms are anorexia (loss of appetite), malaise, nausea, diarrhea, abdominal discomfort, fever, and chills. These symptoms are more likely to appear in adults; they last from 2 to 21 days. In about two-thirds of the cases, there is also jaundice, with yellowing of the skin and the whites of the eyes and the dark urine typical of liver infections. In such cases the liver becomes tender and enlarged. Liver damage is probably caused by immunological reactions.

There is no chronic form of HAV infection, and the virus is usually shed only during the acute stage of disease, although it is difficult to detect. The incubation time of two to six weeks makes epidemiological studies for the source of infections difficult. There are no animal reservoirs.

HAV is extremely widespread, and antibodies are found in a high percentage of the population in many countries—for example, Yugoslavia (97%), Taiwan (89%), France (75%), United States (45%). In the United States, the percentage is much higher among lower socioeconomic groups (72% to 88%) than among middle and upper socioeconomic groups (18% to 30%). Most cases are unreported, and the 60,000 cases reported in 1987 in the United States represent only a fraction of the actual number.

HAV antibodies can be identified with several serological tests, of which the ELISA tests are the most reliable. Acute disease is diagnosed by detection of IgM anti-HAV because these antibodies appear about four weeks after infection and disappear about three to four months after infection. IgG anti-HAV persist for many years and are indicative of prior infection but not necessarily of acute disease.

No specific treatment for the disease exists. However, the incidence of hepatitis A can be decreased by passive immunization. Persons at risk for exposure to hepatitis A can be given immune globulin (HBIG), which provides protection for several months, which is sufficient for travelers to high-risk areas. Work on a vaccine is mostly focused on an attenuated type, but none is available at present.

Hepatitis B (Serum Hepatitis)

Hepatitis B, or **serum hepatitis**, is caused by the *hepatitis B virus (HBV)*. HBV and HAV are completely different viruses. HBV is larger, its genome is double-stranded DNA, and it is enveloped. Because HBV has often been transmitted by blood transfusions, there has been great interest in studying this virus to identify contaminated blood.

The serum from patients with hepatitis B contains three distinct particles. The largest, the *Dane particle*, is probably the complete virion; it is infectious and capable of replicating. There are also smaller *spherical particles*, about half the size of a Dane particle, and *filamentous particles*, which are tubular particles with a diameter about that of the spherical particles but a length about ten times as long (Figure 24-13). The spherical and filamentous particles are probably unassembled components of Dane particles; assembly is evidently not very efficient, and large numbers of these unassembled components accumulate. There are about 1 million times more unassembled than assembled Dane particles in the blood of a hepatitis B patient. Fortunately, these numerous unassembled particles contain *hepatitis B surface antigen (HB$_S$Ag)*, which can be detected with antibodies specific against them. Although it would be difficult to detect the HB$_S$Ag on the few Dane particles in serum, even relatively insensitive tests are able to detect the numerous other particles that also carry the HB$_S$Ag. Such tests make large-scale screening of blood for HBV possible.

Physicians, nurses, dentists, medical technologists, and others who are in daily contact with blood have a considerably higher incidence of this disease than the general population. Physicians, for example, have five times, and dentists, two to three times the normal rate of infection. There have also been instances of transmission to patients by surgeons and dentists (see the box in Chapter 13, p. 368). Intravenous drug abusers often share needles and seldom concern themselves with proper sterilization; as a consequence, they also have a high incidence of hepatitis B.

In addition to transmission by virus-carrying blood, recent evidence indicates that hepatitis B can be transmitted by any secretion of bodily fluids, such as saliva, breast milk, and semen. Transmission by semen donated for artificial insemination has been

Figure 24-13   Hepatitis B virus (HBV). (a) Electron micrograph shows all three distinct types of HBV particles discussed in the text. **(b)** Diagrammatic representation.

documented, and semen transmission has been implicated in heterosexuals with multiple partners and in male homosexuals. A mother who is positive for HB$_S$Ag, especially if she is a chronic carrier, may transmit the disease to her infant, usually at birth. In most cases, this type of transmission can be prevented by giving the newborn immune globulin (HBIG) immediately after birth. These babies should also be vaccinated.

The incubation period averages about 12 weeks but ranges from 4 to 26 weeks. Because this incubation period is of uneven duration, determination of the origin of an infection can be difficult.

The CDC estimates that 200,000 or more persons, mostly young adults, are infected with hepatitis B each year. Clinically, hepatitis B varies widely. Probably about one-half of the cases are entirely asymptomatic. Symptoms are highly variable and include, in the early stages, loss of appetite, low-grade fever, and joint pains. Later, jaundice usually appears. It is difficult to distinguish between hepatitis A and B solely on clinical grounds, although there is less likely to be fever and headache with HBV infections.

Ninety percent of hepatitis B infections end in complete recovery. The mortality rate is higher than that of hepatitis A but is probably less than 1% of the generally infected population. It might be as high as 2% or 3% among hospitalized patients, many of whom are elderly. Up to 10% of patients become chronic carriers. These carriers are a reservoir for transmission of the virus, and they also have a high rate of liver disease. A special concern is the strong correlation

between the occurrence of liver cancer and the incidence of chronic hepatitis B infections. Chronic carriers are about 200 times more likely to get liver cancer than the general population. Liver cancer is the most common form of cancer in sub-Saharan Africa and the Far East, areas where hepatitis B is extremely common. Although Taiwan has a population of only 17 million it has twice the number of cases of liver cancer as the United States. Almost all these cases of liver cancer in Taiwan occur in patients with chronic hepatitis B infections.

Researchers have been unable to cultivate the HBV in cell culture. This has been the means by which vaccines for polio, mumps, measles, and rubella have been developed. However, two vaccines are available for hepatitis B. The first successful vaccine (Heptavax-B®) used HB$_S$Ag harvested from the serum of volunteer human carriers (there are as many as 2 million in the United States). In 1986, a new vaccine was licensed that used HB$_S$Ag produced by the bread yeast *Saccharomyces cerevisiae* (sak-ä-rō-mī′sēs se-ri-vis′ē-ī) called Recombivax-HB®. This is an example of recombinant DNA genetic engineering. The efficiency of the two vaccines is comparable, but the new vaccine has several advantages. It does not depend on a supply from human carriers, and any concern about transmitting viruses such as AIDS from such carriers (these concerns are unfounded) is eliminated. Vaccination is recommended for high-risk groups; these include medical personnel exposed to blood and blood products, persons undergoing hemodialysis, patients and staff at institutions for the mentally retarded, and homosexually active males. The incidence of hepatitis B is very high in this latter group and is apparently sexually transmitted.

In 1984, a campaign was begun in Taiwan to prevent mother-to-infant transmission of the disease. All infants delivered of HB_SAg-positive mothers are vaccinated, and infants of highly infectious carrier mothers also receive HBIG. In 1988, similar recommendations were implemented in the United States.

Hepatitis Non-A, Non-B (NANB)

Screening of donated blood for HB_SAg has become so effective that transmission of hepatitis B by blood transfusion has declined dramatically. But a previously unsuspected form of hepatitis appeared in the 1960s that is also caused by post-transfusion infections. This form was named, by elimination, non-A, non-B (NANB). Almost all cases of transfusion-transmitted hepatitis are now of the NANB type. It is probable that there are at least two viruses, one transmitted mainly by blood transfusions and the other by blood coagulation factor concentrates supplied to hemophiliacs. The diseases caused by these viruses clinically resemble hepatitis B and have a similar incubation period.

Another form of NANB resembles hepatitis A and is spread by fecal–oral transmission. It is known as epidemic, or enterically transmitted, NANB. It clinically resembles hepatitis B, but it has a significant mortality rate, especially among pregnant women. It bypasses immunity to hepatitis A. Epidemic NANB has occurred in large outbreaks in India and Northern Africa.

Diagnosis of NANB is made from clinical symptoms and a negative test for hepatitis A and B. Recently, a NANB virus was isolated, and an antibody test to screen donated blood was developed. Until the antibody test is in general use, blood will continue to be tested by two nonspecific tests. One test is for elevated levels of transaminase enzymes. These enzymes are indicators of damage to liver tissue and are often used in diagnosis of hepatitis. The other test measures antibodies against the core antigen of the hepatitis B virus (HB_CAg), not to be confused with the surface antigen (HB_SAg). It has been found that persons with this antigen are more likely to transmit NANB, although the association is not completely clear. There are several concerns about these two tests: (1) they will add appreciably to the cost of blood, (2) they will lead to rejection of a significant amount of good blood, and (3) they will still miss an estimated 30% to 40% of blood that is capable of transmitting NANB. Obviously, a specific test is badly needed.

Delta Hepatitis

In 1977, a new agent capable of causing hepatitis was discovered in carriers of the hepatitis B virus in Italy. Persons who carried this so-called delta antigen and were also infected with hepatitis B had a much higher incidence of severe liver damage and a much higher mortality rate than persons who had antibodies against hepatitis B alone. Since then, it has become clear that the **delta agent** is responsible for a particularly virulent form of hepatitis (with a reported mortality rate of 17%) affecting natives of the Amazon Basin and countries surrounding the Mediterranean Sea. In Europe and North America, there have also been outbreaks among intravenous drug users.

The delta agent is a defective virus, unable to cause infection on its own. It is only able to reproduce when the host is already infected with HBV, a phenomenon that is poorly understood. This coinfection causes an asymptomatic or mild, chronic HBV infection to become acutely worse. The delta agent (HDV) has an envelope identical to that of hepatitis B surrounding a protein core (HDV antigen). The nucleic acid is a single strand of RNA. The delta agent is genetically similar to certain plant viruses called viroids, and it is considered to represent a new class of animal viruses.

Viral Gastroenteritis

A number of viruses, such as the polioviruses, echoviruses, and coxsackieviruses, are transmitted by the fecal–oral route. However, despite the name *enteroviruses*, these viruses generally do not directly affect the digestive system. About 90% of cases of viral gastroenteritis are caused by rotavirus or by the Norwalk agent (see the box on p. 624).

Rotavirus (*rota* means wheel) is the most common cause of viral gastroenteritis, mostly causing disease in small children (Figure 24-14). Following an incubation period of two or three days, the patient suffers from low-grade fever, diarrhea, and vomiting, which persist for a mean of five to eight days. There may be more than one serotype of rotavirus causing disease, and immunity is only partially protective. An ELISA test can be used to detect the virus in feces.

Major epidemics of viral gastroenteritis have been caused by a virus known as the *Norwalk agent* (named after an outbreak in Norwalk, Ohio, in 1968). All age groups are affected by this virus. Following a two-day incubation period, infected persons typically suffer nausea, abdominal cramps, diarrhea, and vomiting for one to three days. Symptoms are mild, as compared

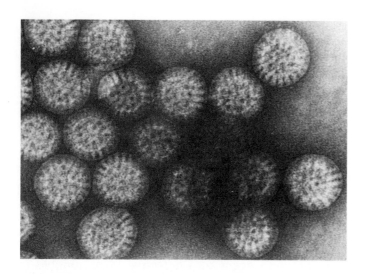

Figure 24-14 Rotavirus. The name of this virus comes from its wheel shape (×197,000).

with those of rotavirus infections. About one-half of middle-aged Americans show evidence, by serum antibodies, that they have been infected. Susceptibility to the virus varies considerably; some persons are not affected and do not even respond with an immune reaction. Others develop gastroenteritis and an immunity that might be only short term.

The only treatment for viral gastroenteritis is oral rehydration or, in exceptional cases, intravenous rehydration.

Fungal Diseases of the Digestive System

Some fungi produce toxins called **mycotoxins**, which cause blood diseases, nervous system disorders, kidney damage, and liver damage.

Ergot Poisoning

Some mycotoxins are produced by *Claviceps purpurea* (kla′vi-seps pür-pü-rē′ä), a fungus causing smut infections on crops (Figure 14-7). The mycotoxin produced by *Claviceps* causes **ergot poisoning**, which results from the ingestion of rye or other cereal grains contaminated with the fungus. Ergot poisoning was very widespread during the Middle Ages. The toxin can restrict blood flow in the extremities, and gangrene results. It may also cause hallucinogenic symptoms, producing bizarre behavior similar to that caused by LSD.

 MMWR   *Centers for Disease Control*
Morbidity and Mortality Weekly Report

OUTBREAK OF VIRAL GASTROENTERITIS— PENNSYLVANIA AND DELAWARE

Within 48 hours of the University of Pennsylvania vs. Cornell University football game in Philadelphia on September 19, 1987, 158 students with symptoms of gastrointestinal illness visited the university health service. Band members from both universities, Cornell football players, and spectators, including visiting students and university staff and faculty, had similar symptoms.

The 158 students reported nausea, vomiting, diarrhea, headache, fever, chills, and myalgia (muscle aches). The mean incubation period was 36 hours; symptoms lasted from 12 to 48 hours. Ninety-two percent of the students had purchased soda with ice from the stadium concessionaire.

None of the Pennsylvania football team members was affected until September 24 and 25, when 55 became ill. The Pennsylvania football team had used ice from a different source at the September 19 game. However, during practice on September 23, the team used ice supplied by the distributor that had provided the stadium concessionaire's ice on September 19.

On September 21, a physician notified the Delaware Department of Health of another outbreak of gastroenteritis among 750 people who attended a museum fundraiser in Wilmington, Delaware. Attendees were served food and iced drinks, most of which contained alcohol. The ice was traced to the same manufacturer that had supplied the concessionaire in Philadelphia.

Questionnaires completed by University of Pennsylvania undergraduates, football players, and band members, and by participants at the Delaware fund-raiser to examine the relationship of food, drink, and ice consumption to the development of gastrointestinal illness, revealed that ice was significantly associated with illness. The attack rate for those consuming ice in the Pennsylvania outbreak was 62%, and 10% of those not consuming ice were ill. In the Delaware outbreak, the attack rate

Aflatoxin Poisoning

Aflatoxin is a mycotoxin produced by the fungus *Aspergillus flavus* (a-spèr-jil′lus flā′vus), a common mold. Aflatoxins have been found in many foods, but they are particularly likely to be found on peanuts. Aflatoxin is highly toxic and can cause serious damage to livestock when their feed is contaminated with *A. flavus*. Although the risk to humans is unknown, there is strong evidence that aflatoxin contributes to cirrhosis of the liver and cancer of the liver in parts of the world, such as India and Africa, where food is subject to aflatoxin contamination.

Protozoan Diseases of the Digestive System

Giardiasis

Giardia lamblia (jē-är′dē-ä lam′lē-ä) is a flagellated protozoan that is able to attach firmly to a human's intestinal wall (Figure 24-15). It is the cause of a prolonged diarrheal disease in humans called **giar-**

Figure 24-15 Scanning electron micrograph of the trophozoite form of *Giardia lamblia*, the flagellated protozoan that causes giardiasis. Note the circular mark left on the intestinal wall by the ventral sucker disk the parasite uses to attach itself (×3850).

was 61% for those consuming ice and 16% for those not consuming ice.

The ice was traced to a manufacturer in southeastern Pennsylvania whose wells had been flooded by waters from Conestoga Creek following a torrential rainfall on September 8. Pennsylvania and Delaware health department laboratories found high concentrations of fecal coliforms in both the ice and the well water used to produce the ice. No source of fecal contamination of the creek has been identified.

Based on distribution records and the Pennsylvania attack rate, more than 5000 people may have become ill from consumption of the 60 to 300 tons of ice produced in the week following the flood.

Editorial Note: Identification of ice as the source in these outbreaks has raised several concerns. The high attack rates of diarrhea among people who ingested ice with alcoholic or carbonated beverages are striking because each of these beverages should have some disinfectant effect. Furthermore, since ice is not consistently con-

trolled by any state or federal agency, jurisdiction for maintaining the quality of commercially produced ice or for recalling already distributed ice is unclear.

This outbreak was characteristic of diarrheal illness caused by viruses other than rotaviruses. Stool samples contained no bacterial agents.

The Norwalk agent has been identified as the causative agent in 42% of 74 past United States diarrhea outbreaks of nonbacterial origin, on the basis of a rise in antibody titer. Other viral agents identified in United States outbreaks include the Snow Mountain and Marin County agents.

Most stool specimens from field investigations are frozen before examination. The finding of a 27-nm virus-like particle in a freshly collected, refrigerated (+4°C) stool sample indicates examination of fresh refrigerated specimens may be the key to identifying the causative agent in future investigations.

Source: *MMWR* 36:709–711 (11/6/87).

diasis. The disease sometimes persists for weeks and is characterized by malaise, nausea, weakness, weight loss, and abdominal cramps. The protozoans sometimes occupy so much of the intestinal wall that they interfere with food absorption. There are frequent outbreaks of giardiasis in the United States; it is probably the most common cause of epidemic waterborne diarrheal disease. About 7% of the population are healthy carriers and shed the cysts in their feces (see Figure 11-20).

The organism is also shed by a number of wild mammals, especially beavers, and the disease is not uncommon in backpackers who drink from wilderness waters. Most outbreaks in the United States are transmitted by contaminated water supplies. Because the cyst stage of the protozoan is relatively insensitive to chlorine, filtration of water supplies is usually necessary to eliminate the cysts from water.

Giardia are not reliably found in stools, and the *string test* is recommended for diagnosis. In this test, a gelatin capsule packed with about 140 cm of fine string is swallowed by the patient. One end of the string is taped to the cheek. The gelatin capsule dissolves in the stomach, and an enclosed weighted rubber bag attached to the other end of the string enters the upper bowel. After a few hours, the string is drawn up through the mouth and examined for parasites.

Treatment by metronidazole or quinacrine hydrochloride is effective.

Balantidiasis (Balantidial Dysentery)

Balantidium coli (bal-an-tid′ē-um kō′lē) is a ciliated protozoan that causes **balantidiasis**, or **balantidial dysentery**. The organism is the only ciliate known to be pathogenic for humans and is the largest intestinal protozoan of humans. Like *Giardia lamblia*, *B. coli* exists in a vegetative form (trophozoite) and a cyst form (Figure 24-16). *B. coli* lives in the large intestine; in rare cases, it invades the epithelial lining and causes ulceration and fatal dysentery. Typically, the disease is mild and involves abdominal pain, nausea, vomiting, diarrhea, and weight loss.

Humans acquire *B. coli* by ingesting cysts in food or water that has been contaminated by feces containing cysts. Following ingestion, the cysts descend to the colon, where the cyst walls dissolve and the released vegetative cells feed on bacteria, fecal debris, and the host's tissue cells. As feces passing through the colon are dehydrated, encystment occurs. Some cysts form after the feces are discharged. Subsequent food and water contamination initiates another cycle.

Laboratory confirmation of balantidiasis is based

(a)

(b)

Figure 24-16 *Balantidium coli.* **(a)** Ciliated trophozoite nestled in the tissue of an intestine (×2500). **(b)** Cyst of *B. coli* (×1000).

on the demonstration of vegetative cells in feces. Treatment consists of tetracyclines. Metronidazole is sometimes used as an alternative.

Amoebic Dysentery (Amoebiasis)

Amoebic dysentery (amoebiasis) is found worldwide and is spread mostly by food or water contaminated by cysts of the protozoan amoeba *Entamoeba histolytica* (en-tä-mē′bä his-tō-li′ti-kä) (Figure 11-18). Although stomach acid (HCl) can destroy vegetative cells, it does not affect the cysts. In the intestinal tract, the cyst wall is digested away, and the vegetative forms are released. The vegetative forms then multiply in the epithelial cells of the wall of the large intestine. A severe dysentery results, and the feces characteristically contain blood and mucus. The vegetative forms feed on red blood cells and destroy tissue in the gastrointestinal tract. This cycle of events is summarized in Figure 24-17.

Severe infections result if the intestinal wall is perforated. Abscesses might have to be treated surgically, and invasion of other organs, particularly the liver, is not uncommon. Perhaps 5% of the population in the United States are asymptomatic carriers of *E. histolytica*. Since 1935, 3000 to 5000 cases per year have been reported in the United States, and the mortality rate approaches 1%.

Diagnosis largely depends on recovery and identification of the organisms in feces. (Red blood cells observed within the trophozoite stage of an amoeba are considered an indication of *E. histolytica*.) There are several serological tests that can also be used for diagnosis, including immunodiffusion and fluorescent-antibody tests.

Metronidazole plus iodoquinol are the drugs of choice in treatment.

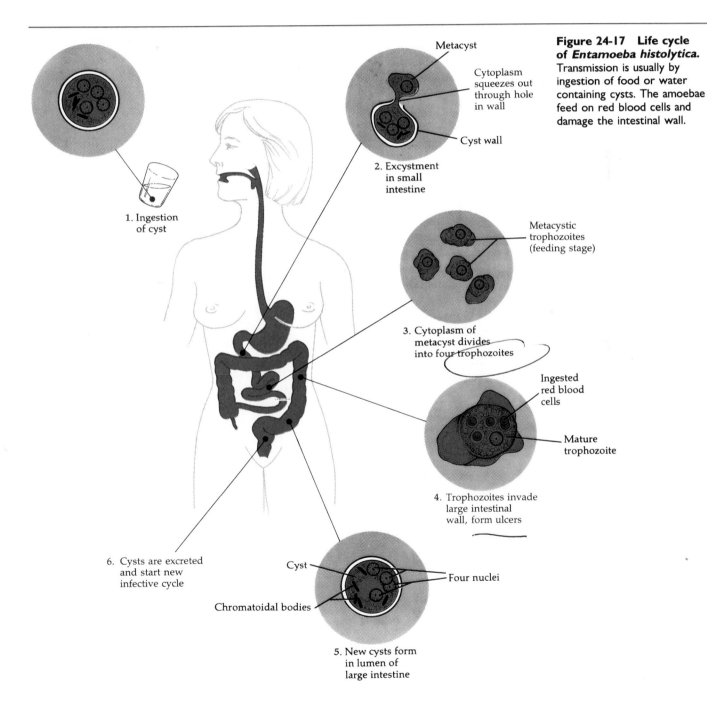

Figure 24-17 Life cycle of *Entamoeba histolytica*. Transmission is usually by ingestion of food or water containing cysts. The amoebae feed on red blood cells and damage the intestinal wall.

1. Ingestion of cyst

Metacyst
Cytoplasm squeezes out through hole in wall
Cyst wall
2. Excystment in small intestine

Metacystic trophozoites (feeding stage)
3. Cytoplasm of metacyst divides into four trophozoites

Ingested red blood cells
Mature trophozoite
4. Trophozoites invade large intestinal wall, form ulcers

6. Cysts are excreted and start new infective cycle

Cyst
Four nuclei
Chromatoidal bodies
5. New cysts form in lumen of large intestine

Cryptosporidiosis

Cryptosporidiosis is caused by the protozoan *Cryptosporidium* (krip-tō-spô-ri'dē-um), long recognized as a pathogen in calves but not known to infect humans until 1976. Since then, the number of reported cases has been increasing. This is another example of a disease with particularly serious effects on immunosuppressed individuals, including those with AIDS. In otherwise healthy individuals, the organism causes intestinal distress of short duration. Immune-deficient individuals, however, suffer from a severe life-threatening diarrhea that may last for months.

Transmission is largely fecal–oral between humans. However, a zoonoses-type transmission from calves has been observed. Diagnosis is based on microscopic examination of feces for *Cryptosporidium* oocysts. Some infections have responded to spiramycin, a macrolide antibiotic similar to erythromycin, but at present there is no reliable treatment.

Helminthic Diseases of the Digestive System

Helminthic parasites are very common in the intestinal tract of humans, especially under conditions of poor sanitation. Figure 24-18 shows the worldwide incidence of some helminthic diseases. In spite of their size and formidable appearance, they often produce few symptoms. They have become so well adapted to their human hosts, and vice versa, that when their presence is revealed it is often a surprise.

Tapeworm Infections

The life cycle of a typical **tapeworm**, *Taenia saginata* (te'nē-ä sa-ji-nä'tä), is shown in Figure 24-19. (Tapeworms were also discussed in Chapter 11.) Infestations by tapeworms begin with the consumption of undercooked beef, pork, or fish containing larval forms of the tapeworm. The larvae develop into adults that attach to the intestinal wall by suckers on the scolex. The adult beef tapeworm, *T. saginata*, can live in the human intestine for 25 years and reaches a length of six meters (18 feet) or longer. Even a worm of this size seldom causes significant symptoms

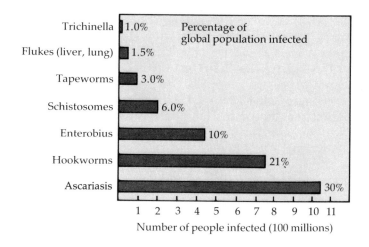

Figure 24-18   Incidence of selected intestinal helminths in the world population.

beyond a vague abdominal discomfort. There is, however, psychological distress when several feet of detached segments (proglottids) break loose and unexpectedly slip out of the anus, which happens occasionally. Eggs are shed in the feces and infect cattle by contaminating their feed. Larval forms, called *cysticeri*, then develop in the muscular and connective tissue of the animal; when undercooked meat is consumed by humans, the cycle begins again.

Taenia solium (sō'lē-um), the pork tapeworm, has a similar life cycle. An important difference is that *T. solium* may produce the larval stage in the human host, a disease condition called **cysticercosis**. In this disease, the infection arises from the tapeworm egg rather than from ingestion of undercooked meat containing cysticerci. The eggs can be ingested as a result of poor sanitary practices or possibly by autoinfection (having eggs produced by the adult tapeworm in the intestinal tract somehow enter the stomach).

Cysticerci can develop in many human organs. When this occurs in muscle tissue, the symptoms are seldom severe. Cysticerci that develop in the eyes and in the brain, however, may be more serious. Cerebral cysticercosis, which is endemic in Mexico and Central America, has become a fairly common condition in parts of the United States within Hispanic cultures. The symptoms often mimic those of a brain tumor.

The fish tapeworm, *Diphyllobothrium latum* (dī-fil-lō-bo'thrē-um lā'tum), is found in pike, trout, perch, and salmon. Recently the CDC issued a warning about the risks of fish-tapeworm infection from the increasingly popular sushi and sashimi (Japanese dishes prepared from raw fish). About ten days after eating sushi, one person developed symptoms of abdominal distention, flatulence (intestinal gas), belch-

Figure 24-19   Life cycle of the beef tapeworm, *Taenia saginata*. The adult tapeworm lives in the intestine of the human, the definitive host. Tapeworm segments and eggs are eliminated with feces and are ingested by intermediate hosts, such as grazing cattle. The tapeworm eggs hatch, and cysticerci form in the animal's muscles, which are later consumed by humans. The pork tapeworm, *Taenia solium*, has a similar life cycle, except that cysticeri may also form in human tissue.

ing, intermittent abdominal cramping, and diarrhea. Eight days later, the patient passed a 1.2-meter (four-feet) long tapeworm identified as a species of *Diphyllobothrium*.

Laboratory diagnosis of human infection consists of identification of the tapeworm eggs or segments in feces. The drug of choice for eliminating tapeworms is niclosamide.

Hydatid Disease

Not all tapeworms are large. One of the most dangerous is *Echinococcus granulosus* (ē-kīn-ō-kok′kus gra-nū-lō′sus), which is only a few millimeters in length (see Figure 11-25). Humans are not the defin-

itive hosts. The adult form lives in the intestinal tract of carnivorous animals, such as dogs and wolves. Typically, humans become infected from the feces of a dog that has become infected by eating the flesh of a sheep or deer containing the cyst form of the tapeworm. Unfortunately, humans can be an intermediate host and can develop the cyst.

Once ingested by a human, the egg of *E. granulosus* may migrate to various tissues of the body. The liver and lungs are the most common sites, but the brain and numerous other sites also may be infected. Once in place, the egg develops into a **hydatid cyst** that can grow to a diameter of 10 mm in a few months (Figure 24-20). In some locations, cysts may not be apparent for many years. Some become enormous, containing up to 15 liters (4 gallons) of fluid. Damage may arise from the size of the cyst in such areas as

Figure 24-20 Hydatid cyst. A large cyst can be seen in this X ray of an infected brain.

the brain or the interior of bones. Another factor in the pathogenicity of such cysts is that the fluid contains proteinaceous material, to which the host becomes sensitized. If the cyst suddenly ruptures, the result can be life-threatening anaphylactic shock. Treatment is limited to surgical removal, but care must be taken to avoid release of the fluid and potential anaphylactic shock. The disease is seen most frequently among people that raise sheep or hunt or trap wild animals.

Nematode Infections

The nematodes, or roundworms, cause some of the more common helminthic infections.

Pinworms

Most of us are familiar with the **pinworm**, *Enterobius vermicularis* (en-te-rō'bē-us ver-mi-kū-lar'is). (See Figure 11-26.) This tiny worm migrates out of the anus of the human host to lay its eggs, causing local itching. Whole households may become infected. Such drugs as pyrantel pamoate or mebendazole are usually effective in treatment.

Hookworms

Hookworm infections were once a very common parasitic disease in the southeastern states. In the United States, the species most commonly seen is *Necator*

americanus (ne-kā'tôr ä-me-ri-ka'nus). (See Figure 11-27.) The hookworm attaches to the intestinal wall and feeds on blood and tissue rather than on partially digested food, so large numbers of worms can lead to anemia and lethargic behavior. Large infections can also lead to a bizarre symptom, *pica*, which is a craving for peculiar foods, such as laundry starch or soil containing a certain type of clay. Pica is a symptom of iron deficiency anemia. Because the life cycle of the hookworm requires human feces to enter the soil and bare skin to contact contaminated soil, the incidence of the disease has declined greatly with improved sanitation and the practice of wearing shoes. Hookworms can be treated effectively with mebendazole.

Ascariasis

One of the most widespread helminthic diseases is **ascariasis**, caused by *Ascaris lumbricoides*. This condition is still fairly common in the United States. It does not usually cause severe symptoms, but its presence can be manifested in distressing ways. As we described in Chapter 11, diagnosis is often made when an adult worm emerges from the anus. These worms can be quite large, up to 30 cm (about 1 foot) in length. In the intestinal tract, they live on partially digested food and cause few symptoms, but ingested eggs undergo a peculiar life cycle. In the upper intestine, they hatch into small, wormlike larva that pass into the bloodstream and then into the lungs. In the lungs, they migrate into the throat and are swallowed—all this movement, just to get back to the place they started! In the lungs, they may cause some pulmonary symptoms, and extremely large numbers cause pneumonia.

The most dramatic consequences of infection with *A. lumbricoides* are from the migrations of adult worms. The worms are equipped with minute cutting teeth and can penetrate tissue. If they pass through the intestinal wall, they can infect the abdominal cavity. Worms have been known to leave the body of small children through the umbilicus (belly button) and to escape through the nostrils of a sleeping person. Once ascariasis is diagnosed, it can be effectively treated with mebendazole.

Roundworms

Most infestations by the small roundworm *Trichinella spiralis* (tri-kin-el'lä spī-ra'lis), called **trichinosis**, are insignificant. The larvae, in encysted form, are located in muscles of the host. In 1970, routine autopsies of human diaphragm muscles showed that about 4% of those tested carried this parasite. Severe cases of trichinosis can be fatal—sometimes in only a few days. The severity of the disease is generally proportional to the size of the infection. Ingestion of undercooked pork is the most common vehicle of infestation, but ingestion of the flesh of animals that feed on garbage (bears, for example) are also sources of outbreaks.

Any ground meat can be contaminated from machinery previously used to grind contaminated meats. Eating raw sausage or hamburger is a poor health habit. One person acquired trichinosis by chewing the fingernails after handling infected pork. Prolonged freezing of meats containing *Trichinella* tends to eliminate them, but freezing should not be considered a substitute for thorough cooking.

In the muscles of intermediate hosts, such as pigs, the *T. spiralis* larvae are encysted in the form of short worms about 1 mm in length. When the flesh of an infected animal is ingested by humans, the cyst wall is removed by digestive action in the intestine. The organism then matures into the adult form. The adult worms spend only about one week in the intestinal mucosa and produce larvae that invade tissue. Eventually, the encysted larvae localize in muscle (common sites include the diaphragm and eye muscles), where they are barely visible in biopsied specimens (Figure 24-21).

Symptoms of trichinosis include fever, swelling about the eyes, and gastrointestinal upset. Small hemorrhages under the fingernails are often observed. Biopsy specimens, as well as a number of serological tests, can be used in diagnosis. Recently, an ELISA serological test that detects the parasite in meats has been developed. Treatment consists of the administration of thiabendazole to kill intestinal worms and corticosteroids to reduce inflammation.

In the past ten years, the number of cases reported annually in the United States has varied from 37 to 252. Deaths are rare, and in most years none occurs.

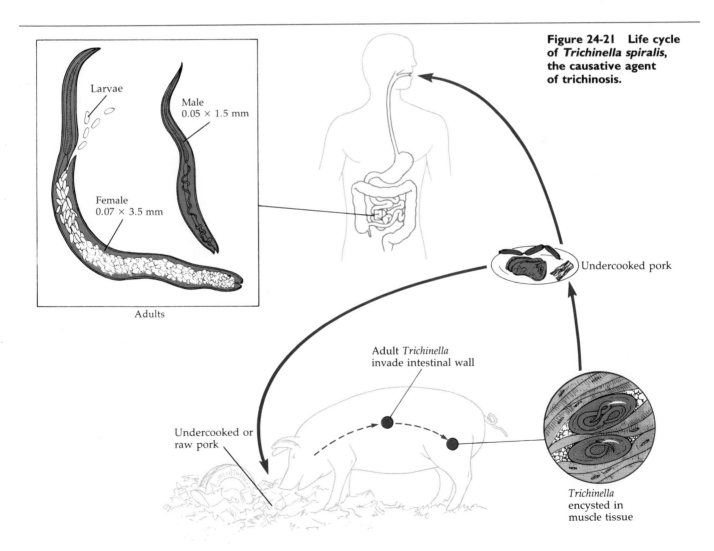

Figure 24-21   Life cycle of *Trichinella spiralis*, the causative agent of trichinosis.

Larvae

Male
0.05 × 1.5 mm

Female
0.07 × 3.5 mm

Adults

Undercooked pork

Adult *Trichinella* invade intestinal wall

Undercooked or raw pork

Trichinella encysted in muscle tissue

TABLE 24-2   *Diseases Associated with the Digestive System*

	Causative Agent	Mode of Transmission	Treatment
Bacterial Diseases of the Mouth			
Dental caries	*Streptococcus mutans*	Bacteria use sucrose to form plaque	None, but prevention includes restricting ingestion of sucrose; brushing, flossing, and professional cleaning to remove plaque; fluoridation
Periodontal disease	*Actinomyces, Nocardia, Corynebacterium*	Plaque initiates an inflammatory response	Antibiotics such as penicillin and tetracycline; cleaning teeth may prevent occurrence
Bacterial Diseases of the Lower Digestive System			
Staphylococcal food poisoning	*Staphylococcus aureus*	Ingestion of exotoxin in food, usually improperly refrigerated	Replace lost water and electrolytes
Salmonellosis	*Salmonella* spp.	Ingestion of contaminated food and drink	Replace lost water and electrolytes
Typhoid fever	*Salmonella typhi*	Ingestion of contaminated food and drink	Ampicillin and trimethoprim-sulfamethoxazole
Bacillary dysentery (shigellosis)	*Shigella* spp.	Ingestion of contaminated food and drink	Replace lost water and electrolytes; ampicillin, chloramphenicol, or trimethoprim-sulfamethoxazole
Cholera	*Vibrio cholerae*	Ingestion of contaminated food and drink	Replace lost water and electrolytes
Vibrio parahaemolyticus gastroenteritis	*Vibrio parahaemolyticus*	Often ingestion of contaminated shellfish	Tetracycline or chloramphenicol in severe cases
Escherichia coli gastroenteritis	Enterotoxigenic, enteroinvasive, and enteropathogenic strains of *Escherichia coli*	Ingestion of contaminated food and drink	Doxycycline and trimethoprim-sulfamethoxazole may help in prevention
Campylobacter gastroenteritis	*Campylobacter jejuni*	Ingestion of contaminated food and drink	Replace lost water and electrolytes
Yersinia gastroenteritis	*Yersinia enterocolitica*	Ingestion of contaminated food and drink	Replace lost water and electrolytes
Clostridium perfringens gastroenteritis	*Clostridium perfringens*	Ingestion of contaminated food (usually meat) and drink	Replace lost water and electrolytes
Bacillus cereus gastroenteritis	*Bacillus cereus*	Ingestion of contaminated food	None, self-limiting
Viral Diseases of the Digestive System			
Mumps	Mumps virus	Saliva and respiratory secretions	None, but a vaccine is available (MMR)
Cytomegalovirus (CMV) inclusion disease	Cytomegalovirus (CMV)	Placental transfer and blood transfusion	None

632

TABLE 24-2 (*Continued*)

	Causative Agent	Mode of Transmission	Treatment
Hepatitis A	Hepatitis A virus	Ingestion of contaminated food or drink	None, but passive immunization occurs
Hepatitis B	Hepatitis B virus	Blood or blood contamination by needles, transfusions; sexual contact, especially between male homosexuals	Vaccine available
Non-A, non-B hepatitis	Two unidentified viruses	Blood or blood contamination by needles, transfusions; another form by fecal–oral	None
Delta hepatitis	Delta agent	Probably same as hepatitis B	None
Viral gastroenteritis	Norwalk agent or rotavirus	Ingestion of contaminated food and water	Replace lost water and electrolytes
Fungal Diseases of the Digestive System			
Ergot poisoning	Mycotoxin produced by *Claviceps purpurea*	Ingestion of contaminated food, usually cereal grain or poisonous mushrooms	Antispasmodic drugs
Aflatoxin poisoning	Mycotoxin produced by *Aspergillus flavus*	Ingestion of contaminated food, usually peanuts	None
Protozoan Diseases of the Digestive System			
Giardiasis	*Giardia lamblia*	Ingestion of contaminated water	Metronidazole and quinacrine hydrochloride
Balantidiasis (balantidial dysentery)	*Balantidium coli*	Ingestion of contaminated food and water	Tetracycline or metronidazole
Amoebic dysentery (amoebiasis)	*Entamoeba histolytica*	Ingestion of contaminated food and water	Metronidazole plus iodoquinol
Cryptosporidiosis	*Cryptosporidium*	Uncertain, probably oral–fecal	Possibly spiramycin
Helminthic Diseases of the Digestive System			
Tapeworm infestations	*Taenia saginata* (beef tapeworm); *T. solium* (pork tapeworm); *Diphyllobothrium latum* (fish tapeworm)	Ingestion of contaminated meat	Niclosamide
Hydatid disease	*Echinococcus granulosus*	Ingestion of eggs	Surgical removal
Pinworms	*Enterobius vermicularis*	Ingestion of eggs	Pyrantel pamoate, mebendazole
Hookworms	*Necator americanus*	Larvae in soil penetrate skin	Mebendazole
Ascariasis	*Ascaris lumbricoides*	Ingestion of eggs	Mebendazole
Trichinosis	*Trichinella spiralis*	Ingestion of larvae in meat	Thiabendazole and corticosteroids

Study Outline

Introduction (p. 606)

1. Diseases of the digestive system are the second most common illnesses in the United States.
2. Diseases of the digestive system usually result from ingestion of microorganisms and their toxins in food and water.
3. The fecal–oral cycle of transmission can be broken by proper disposal of sewage and proper preparation and storage of foods.

Structure and Function of the Digestive System (pp. 606–607)

1. The gastrointestinal (GI) tract, or alimentary canal, consists of the mouth, pharynx, esophagus, stomach, small intestine, and large intestine.
2. The teeth, tongue, salivary glands, liver, gallbladder, and pancreas are accessory structures.
3. In the GI tract, with mechanical and chemical help from the accessory structures, carbohydrates are broken down to simple sugars, fats to fatty acids and glycerol, and proteins to amino acids.
4. Feces, resulting from digestion, are stored in the rectum and eliminated by defecation.

Normal Flora of the Digestive System (pp. 607–608)

1. *Streptococcus salivarius* is found on the tongue, *S. mitis*, on the cheek mucosa, and *S. sanguis*, on the teeth.
2. *Bacteroides*, *Fusobacterium*, and spirochetes are found in anaerobic niches between the teeth and gums.
3. The stomach and small intestine have few resident microorganisms.
4. The large intestine is the habitat for *Lactobacillus*, *Bacteroides*, *E. coli*, *Enterobacter*, *Citrobacter*, and *Proteus*.
5. Bacteria in the large intestine assist in degrading food and synthesizing vitamins.
6. Up to 40% of fecal mass is microbial cells.

Bacterial Diseases of the Mouth (pp. 608–610)

Dental Caries (Tooth Decay) (pp. 608–610)

1. Dental caries begin when tooth enamel and dentin are eroded and the pulp is exposed to bacterial infection.
2. *Streptococcus mutans*, found in the mouth, uses sucrose to form dextran from glucose and lactic acid from fructose.
3. Bacteria adhere to teeth with a sticky dextran capsule, forming dental plaque.

4. Acid produced during carbohydrate fermentation destroys tooth enamel at the site of the plaque.
5. Carbohydrates such as starch, mannitol, and sorbitol are not used by cariogenic bacteria to produce dextran and do not promote tooth decay.
6. Caries are prevented by restricting ingestion of sucrose and by physical removal of plaque; a vaccine against *S. mutans* is theoretically possible.

Periodontal Disease (p. 610)

1. Inflammation of the gums, bone destruction, and necrosis of tissues around the teeth can be caused by *Actinomyces*, *Nocardia*, and *Corynebacterium*; this is called periodontal disease.

Bacterial Diseases of the Lower Digestive System (pp. 610–618)

1. A gastrointestinal infection is caused by growth of a pathogen in the intestines.
2. Incubation times, the times required for bacterial cells to grow and their products to produce symptoms, range from 12 hours to 2 weeks. Symptoms generally include a fever.
3. Antibiotic therapy is useful in treatment of bacterial infections.
4. A bacterial intoxication results from ingestion of preformed bacterial toxins.
5. Symptoms appear from 1 to 48 hours after ingestion of the toxin.
6. Antibiotics are of no use because symptoms are caused by an exotoxin.
7. Gastroenteritis is an infection or intoxication that results in abdominal pain, diarrhea, or dysentery.

Staphylococcal Food Poisoning (Staphylococcal Enterotoxicosis) (pp. 611–612)

1. Staphylococcal food poisoning is caused by ingestion of an enterotoxin produced in improperly stored foods.
2. *S. aureus* is inoculated into foods during preparation. The bacteria grow and produce enterotoxin in food stored at room temperature.
3. The polypeptide exotoxin is not denatured by boiling for 30 minutes.
4. Foods with high osmotic pressure and those not cooked immediately before consumption are most often the source of staphylococcal enterotoxicosis.
5. Nausea, vomiting, and diarrhea occur from one to six hours after eating, and the symptoms last approximately 24 hours.
6. Laboratory identification of *S. aureus* isolated from foods or the presence of thermostable nuclease in foods can confirm diagnosis.

Salmonellosis (*Salmonella* Gastroenteritis) (pp. 612–614)

1. Salmonellosis, or *Salmonella* gastroenteritis, might be caused by *Salmonella* endotoxin.
2. Symptoms include nausea, abdominal pain, and diarrhea and occur from 12 to 36 hours after ingestion of large numbers of *Salmonella*. Septicemia can occur in infants and in the elderly.
3. Mortality is lower than 1%, and recovery can result in a carrier state.
4. Heating food to 68°C will usually kill *Salmonella*.
5. Laboratory diagnosis is based on isolation and identification of *Salmonella* from feces.

Typhoid Fever (p. 614)

1. A few serovars of *Salmonella* cause typhoid fever.
2. Fever and malaise occur after a two-week incubation. Symptoms last from two to three weeks.
3. *Salmonella* are harbored in the gallbladder of carriers.
4. A killed-bacteria vaccine is available for high-risk persons.

Bacillary Dysentery (Shigellosis) (pp. 614–615)

1. Bacillary dysentery is caused by four species of *Shigella*.
2. Symptoms include blood and mucus in stools, abdominal cramps, and fever. Infections by *S. dysenteriae* result in ulceration of the intestinal mucosa.
3. Isolation and identification of the bacteria from rectal swabs are used for diagnosis.

Cholera (pp. 615–616)

1. *Vibrio cholerae* produces an exotoxin that alters membrane permeability of the intestinal mucosa; vomiting, diarrhea, and loss of body fluids result.
2. The incubation period is approximately three days. The symptoms last for a few days. Untreated cholera has a 50% mortality rate.
3. Diagnosis is based on isolation of *Vibrio* from feces.

Vibrio parahaemolyticus Gastroenteritis (p. 616)

1. Gastroenteritis can be caused by the halophile *V. parahaemolyticus*.
2. Onset of symptoms begins within 24 hours after ingestion of contaminated foods. Recovery occurs within a few days.
3. The disease is contracted by ingestion of contaminated crustaceans or handling of contaminated mollusks.

Escherichia coli Gastroenteritis (Traveler's Diarrhea) (pp. 616–617)

1. Gastroenteritis may be caused by enterotoxigenic, enteroinvasive, or enteropathogenic strains of *E. coli*.
2. The disease occurs as epidemic diarrhea in nurseries, as traveler's diarrhea, as endemic diarrhea in underdeveloped countries, and as hemorrhagic colitis.
3. In adults, the disease is usually self-limiting and does not require chemotherapy. Children are treated with fluids and electrolytes.

Campylobacter Gastroenteritis (p. 617)

1. *Campylobacter* is the second most common cause of diarrhea in the United States.
2. *Campylobacter* is transmitted in cow's milk.

Yersinia Gastroenteritis (p. 617)

1. *Y. enterocolitica* and *Y. pseudotuberculosis* are transmitted in meat and milk.
2. *Yersinia* can grow at refrigeration temperatures.

Clostridium perfringens Gastroenteritis (p. 617)

1. A self-limiting gastroenteritis is caused by *C. perfringens*.
2. Endospores survive heating and germinate when foods (usually meats) are stored at room temperature.
3. Exotoxin produced when the bacteria grow in the intestines is responsible for the symptoms.
4. Diagnosis is based on isolation and identification of the bacteria in stool samples.

Bacillus cereus Gastroenteritis (p. 618)

1. Ingesting food contaminated with the soil saprophyte *Bacillus cereus* can result in diarrhea, nausea, and vomiting.

Viral Diseases of the Digestive System (pp. 618–624)

Mumps (p. 618)

1. Mumps virus (a paramyxovirus) enters and exits the body through the respiratory tract.
2. About 16 to 18 days after exposure, the virus causes inflammation of the parotid glands, fever, and pain during swallowing. About 4 to 7 days later, orchitis may occur.
3. After onset of the symptoms, the virus is found in the blood, saliva, and urine.
4. A measles–mumps–rubella (MMR) vaccine is available.
5. Diagnosis is based on symptoms or hemagglutination-inhibition tests of viruses cultured in embryonated eggs or cell culture.

Cytomegalovirus (CMV) Inclusion Disease (pp. 618–620)

1. CMV (a herpesvirus) causes intranuclear inclusion bodies and cytomegaly of host cells.
2. CMV is transmitted by saliva, urine, semen, cervical secretions, and human milk.
3. Cytomegalovirus inclusion disease can be asymptomatic, a mild disease, or progressive and fatal.
4. If the virus crosses the placenta, it can cause congenital infection of the fetus resulting in impaired mental development, neurological damage, and stillbirth.
5. Diagnosis is based on isolation of the virus; serological tests are not specific.

Hepatitis (pp. 620–623)

1. Inflammation of the liver is called hepatitis. Symptoms include loss of appetite, malaise, fever, and jaundice.
2. Viral causes of hepatitis are hepatitis A virus (infectious hepatitis), hepatitis B virus (HBV), and non-A, non-B hepatitis virus (NANB). HBV and NANB are types of serum hepatitis.

Hepatitis A (Infectious Hepatitis) (pp. 620–621)

1. Hepatitis A virus causes infectious hepatitis. At least 50% of all cases are subclinical.
2. HAV is ingested in contaminated food or water, grows in the cells of the intestinal mucosa, and spreads to the liver, kidney, and spleen in the blood.
3. The virus is eliminated with feces.
4. The incubation period is 2 to 6 weeks; the period of disease, 2 to 21 days; and recovery is complete in 4 to 6 weeks.
5. Diagnosis is based on serological tests for the virus or antibodies.
6. Passive immunization can provide temporary protection.

Hepatitis B (Serum Hepatitis) (pp. 621–623)

1. Hepatitis B virus causes serum hepatitis.
2. HBV is transmitted by blood transfusions, contaminated syringes, saliva, sweat, breast milk, and semen.
3. Dane particles (complete virion), spherical particles (HB_SAg), and filamentous particles (containing HB_SAg) are found in serum of patients with hepatitis B.
4. The average incubation period is three months; recovery is usually complete; some patients develop a chronic infection.
5. A vaccine against HB_SAg is available.

Hepatitis Non-A, Non-B (NANB) (p. 623)

1. NANB viruses presently cause most posttransfusion hepatitis.

Delta Hepatitis (p. 623)

1. Delta agent has a circular strand of RNA and uses hepatitis B surface antigens as a coat.

Viral Gastroenteritis (pp. 623–624)

1. Viral gastroenteritis is most often caused by rotavirus or the Norwalk agent.

Fungal Diseases of the Digestive System (pp. 624–625)

1. Mycotoxins are toxins produced by some fungi.
2. Mycotoxins affect the blood, nervous system, kidney, or liver.

Ergot Poisoning (p. 624)

1. Ergot poisoning is caused by the mycotoxin produced by *Claviceps purpurea.*
2. Cereal grains are most often contaminated with the *Claviceps* mycotoxin.

Aflatoxin Poisoning (p. 625)

1. Aflatoxin is a mycotoxin produced by *Aspergillus flavus.*
2. Peanuts are most often contaminated with aflatoxin.

Protozoan Diseases of the Digestive System (pp. 625–628)

Giardiasis (pp. 625–626)

1. *Giardia lamblia* grows in the intestines of humans and wild animals and is transmitted in contaminated water.
2. Symptoms of giardiasis are malaise, nausea, weakness, and abdominal cramps that persist for weeks.
3. Diagnosis is based on identification of the protozoan in the small intestine.

Balantidiasis (Balantidial Dysentery) (p. 626)

1. *Balantidium coli* causes balantidial dysentery when growing in the large intestine.
2. *B. coli* can cause ulceration of the intestinal wall and, rarely, fatal dysentery.
3. Infections are acquired by ingesting cysts in contaminated food and water.
4. Diagnosis is based on observation of trophozoites in feces.

Amoebic Dysentery (Amoebiasis) (pp. 626–627)

1. Amoebic dysentery is caused by *Entamoeba histolytica* growing in the large intestine.
2. The amoeba feeds on red blood cells and GI tract tissues. Severe infections result in abscesses.
3. Diagnosis is confirmed by observation of trophozoites in feces and several serological tests.

Cryptosporidiosis (p. 628)

1. *Cryptosporidium* causes prolonged diarrhea in immunosuppressed patients.
2. The presence of oocysts in feces confirms diagnosis.

Helminthic Diseases of the Digestive System (pp. 628–631)

Tapeworm Infections (pp. 628–629)

1. Tapeworms are contracted by the consumption of undercooked beef, pork, or fish containing encysted larvae (cysticerci).
2. The scolex attaches to the intestinal mucosa of humans (the definitive host) and matures into an adult tapeworm.
3. Eggs are shed in the feces and must be ingested by an intermediate host.
4. Adult tapeworms can be undiagnosed in a human.
5. Diagnosis is based on observation of proglottids and eggs in feces.
6. Cysticercosis in humans occurs when the pork tapeworm larvae encyst in humans.

Hydatid Disease (pp. 629–630)

1. Humans infected with the tapeworm *Echinococcus granulosus* might have hydatid cysts in their lungs or other organs.
2. Dogs and wolves are usually the definitive hosts, and sheep or deer are the intermediate hosts for *E. granulosus*.

Nematode Infections (pp. 630–631)

Pinworms (p. 630)

1. Humans are the definitive host for pinworms, *Enterobius vermicularis*.

Hookworms (p. 630)

1. Hookworm larvae bore through skin and migrate to the intestine to mature into adults.

Ascariasis (p. 630)

1. *Ascaris* adults live in human intestines.

Roundworms (pp. 630–631)

1. *Trichinella spiralis* larvae encyst in muscles of humans, swine, and other mammals to cause trichinosis.
2. The roundworm is contracted by ingesting undercooked meat containing larvae.
3. Adults mature in the intestine and lay eggs. The new larvae migrate to invade muscles.
4. Symptoms include fever, swelling around the eyes, and gastrointestinal upset.
5. Biopsy specimens and serological tests are used for diagnosis.

Study Questions

Review

1. State examples of representative normal flora, if any, in the mouth, stomach, small intestine, large intestine, and rectum.

2. What properties of *S. mutans* implicates this bacterium in the formation of dental caries? Why is sucrose, more than any other carbohydrate, responsible for the formation of dental caries?

3. List the general symptoms of gastroenteritis. Since there are many etiologies, what is the laboratory diagnosis usually based on?

4. Differentiate among the following factors of bacterial intoxication and bacterial infection: prerequisite conditions, etiologic agents, onset, duration of symptoms, and treatment.

5. Complete the following table.

Disease	Etiologic Agent	Suspect Foods	Symptoms	Treatment
Staphylococcal food poisoning				
Salmonellosis				
Bacillary dysentery				
Cholera				
Gastroenteritis				
Traveler's diarrhea				

6. Differentiate between salmonellosis and typhoid fever.

7. You probably listed *E. coli* in answer to Questions 1 and 4. Explain why this one bacterial species is both beneficial and harmful.

8. Complete the following table.

Disease	Etiologic Agent	Method of Transmission	Site of Infection	Symptoms	Prevention
Mumps					
CMV inclusion disease					
Infectious hepatitis					
Serum hepatitis					
Viral gastroenteritis					

9. What preventive treatments are currently used for infectious hepatitis? For Hepatitis B? For non-A, non-B hepatitis?

10. How is blood that is to be used for transfusions tested for HBV? For NANB?

11. Define mycotoxin. Provide an example of a mycotoxin.

12. Explain how the following diseases differ and how they are similar: giardiasis, balantidiasis, amoebic dysentery, and cryptosporidiosis.

13. Differentiate between amoebic dysentery and bacillary dysentery.

14. Diagram the life cycle for a human tapeworm.

15. Diagram the life cycle for trichinosis, and include humans in the cycle.

16. How can bacterial and protozoan infections of the GI tract be prevented?

17. Look at your diagrams for questions 14 and 15. Indicate sequences in the life cycles that could be easily broken to prevent these diseases.

Challenge

1. Why is a human infection of trichinosis considered a "dead end" for the parasite?

2. Complete the following table.

Disease	Conditions Necessary for Microbial Growth	Basis for Diagnosis	Prevention
Staphylococcal food poisoning			
Salmonellosis			

3. Twenty-eight kindergarten children and seven adults visited a certified raw milk (CRM) bottling plant, where they were given ice cream and CRM. Three to six days later, nine children and three adults developed gastroenteritis. The only foods eaten by all these children (ill and well) were in the school-provided lunches. No one else in the school became sick. What was the source of this gastroenteritis outbreak? Stool cultures showed one bacterium common to nine of the ill children and not present in samples from nine well children. This bacterium is a curved gram-negative rod; it neither ferments nor oxidizes glucose. What is the bacterium?

4. 2130 students and employees of a public school system developed diarrheal illness on April 2. The cafeteria served chicken that day. On April 1, part of the chicken was placed in water-filled pans and cooked in an oven for two hours at a dial setting of 177°C. The oven was turned off, and the chicken was left overnight in the warm oven. The remainder of the chicken was cooked for two hours in a steam cooker and then left in the device overnight at the lowest possible setting (43°C). Two serotypes of a gram-negative, oxidase-negative, lactose-negative rod were isolated from 32 patients. What is the pathogen? How could this outbreak have been prevented?

Further Reading

Betley, M. J., V. L. Miller, and J. J. Mekalanos. "Genetics of bacterial enterotoxins." *Annual Review of Microbiology* 40:577–605, 1986. Discusses the production of enterotoxins by *V. cholerae, E. coli, S. dysenteriae,* and *S. aureus.*

Centers for Disease Control. *Salmonella Surveillance Report.* Published throughout each year. Tabulates incidence and reports on current investigations.

Centers for Disease Control. *Shigella Surveillance Report.* Published throughout each year. Tabulates data and background on recent foodborne or waterborne outbreaks.

Marsh, P. *Oral Microbiology.* Washington, D.C.: American Society for Microbiology, 1980. Discusses the relationship of bacteria to dental caries and periodontal disease.

Melnick, J. L., G. R. Dreesman, and F. B. Hollinger. "Viral hepatitis." *Scientific American* 237(1):44–62, July 1977. Discusses the epidemiology and immunology of hepatitis.

Prince, A. M. "Non-A, non-B hepatitis viruses." *Annual Review of Microbiology* 37:217–232, 1983. Describes ongoing research seeking the causative agents of NANB hepatitis.

Reimann, H., and F. L. Bryan. *Food-borne Infections and Intoxications,* 2nd ed. New York: Academic Press, 1979. An authoritative discussion of nearly all foodborne diseases.

Riley, L. W. "The epidemiologic, clinical, and microbiologic features of hemorrhagic colitis." *Annual Review of Microbiology* 41:383–407, 1987. Discusses intestinal diseases caused by *E. coli* that are endemic in North America.

See also Further Reading for Part Four at the end of Chapter 25.

CHAPTER 25

Microbial Diseases of the Urinary and Reproductive Systems

The **urinary system** is comprised of organs that regulate the chemical composition and volume of the blood and excrete the waste products of metabolism. The **reproductive system** is comprised of organs that produce gametes to propagate the species and, in the female, organs that support and nourish the developing embryo. Because the urinary and reproductive systems are closely related anatomically, we will discuss them in the same chapter. In fact, some diseases that affect one system also affect the other, especially in the male. Both systems open to the external environment and thus share portals of entry for microorganisms that can cause disease. Normal flora of these and other bodily systems also cause opportunistic infections of the urinary and reproductive systems. The major microbial diseases of the urinary and reproductive systems are summarized in Table 25-2 on p. 656.

Structure and Function of the Urinary System

The urinary system consists of two *kidneys*, two *ureters*, a single *urinary bladder*, and a single *urethra* (Figure 25-1). The kidneys contain microscopic functional units called *nephrons*. As blood circulates through the kidneys, the nephrons control the concentration and volume of blood by removing and adding selected amounts of water and solutes and excreting wastes. The wastes, which include urea, uric acid, creatinine, and various salts together with water, are collectively called *urine*. The urine passes through the ureters into the urinary bladder, where it is stored prior to elimination from the body. Elimination occurs through the urethra. In the female, the urethra conveys only urine to the exterior. In the male, the urethra is a common tube for both urine and seminal fluid.

Learning Objectives

- List the normal flora of the urinary and reproductive systems and their habitats.

- Describe methods of transmission for urinary and reproductive system infections.

- List the microorganisms that cause cystitis and pyelonephritis, and name the predisposing factors for these diseases.

- Describe the cause and treatment of glomerulonephritis.

- List the etiologic agents, symptoms, methods for diagnosis, and treatments for leptospirosis, gonorrhea, syphilis, NGU, vaginitis, LGV, chancroid, granuloma inguinale, candidiasis, and trichomoniasis.

- Discuss the epidemiology of genital herpes and genital warts.

- List genital diseases that can cause congenital and neonatal infections, and explain how these infections can be prevented.

Figure 25-1   Organs of the urinary system.

The urinary tract has certain characteristics that help to prevent infection. Where the ureters enter the urinary bladder, there are valves that normally prevent the backflow of urine to the kidneys. This mech-

anism helps to shield the kidneys from lower urinary tract infections. In addition, the acidity of normal urine has some antimicrobial properties. The flushing action of urine to the exterior prevents microorganisms from setting up foci of infection.

Structure and Function of the Reproductive System

The **female reproductive system** consists of two *ovaries*, two *uterine (fallopian) tubes*, the *uterus*, the *vagina*, and *external genitals* (Figure 25-2). The ovaries produce female sex hormones and ova (eggs). When an ovum is released in the process called ovulation, it enters a uterine tube; there fertilization occurs if viable sperm is present. The fertilized ovum (zygote) descends the uterine tube and enters the uterus. It implants in the inner wall and remains there while it develops into an embryo and, later, a fetus. At birth, the infant is expelled from the uterus through the vagina. The vagina also serves as a copulatory canal. The external genitals (vulva) include the clitoris, labia, and glands that produce a lubricating secretion during copulation.

The **male reproductive system** consists of two *testes*, a system of *ducts*, *accessory glands*, and the *penis* (Figure 25-3). The testes produce male sex hormones and sperm. Newly produced sperm move into the epididymis, where they are stored until ejacula-

Figure 25-2   Female reproductive organs.

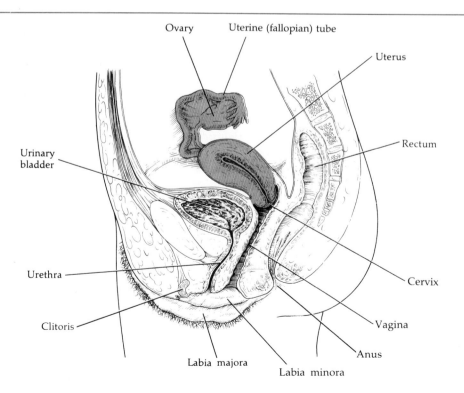

Figure 25-3 Male reproductive organs.

tion. To exit from the body, the sperm cells pass through a series of ducts—epididymis, ductus (vas) deferens, ejaculatory duct, and urethra. During ejaculation, contractions of the ductus (vas) deferens and ejaculatory duct propel the sperm toward the urethra. Along the route, the seminal vesicles, prostate gland, and bulbourethral glands secrete an alkaline fluid into seminal fluid. This alkaline fluid buffers the acidic environment of the vagina, and it is an important mechanism because sperm cells are killed in an acidic environment. During ejaculation, the seminal fluid leaves the body through the urethra. A valve system prevents the seminal fluid from entering the urinary bladder.

Normal Flora of the Urinary and Reproductive Systems

Normal urine is sterile in the urinary bladder and the organs of the upper urinary tract. The urethra, however, does contain normal resident flora that include *Streptococcus*, *Bacteroides*, *Mycobacterium*, *Neisseria*, and a few enterobacteria. Urine becomes contaminated with skin flora during its passage through the urethra.

In the female genital system, the normal flora of the vagina are greatly influenced by sex hormones. For example, within a few weeks after birth, the female infant's vagina is populated by lactobacilli. This population grows because estrogens are transferred from maternal to fetal blood and cause glycogen to accumulate in the cells lining the vagina. Lactobacilli convert the glycogen to lactic acid, and the pH of the vagina becomes acidic. This glycogen–lactic acid sequence provides the conditions under which the acid-tolerant normal flora grow in the vagina.

The physiological effects of estrogens diminish several weeks after birth, and other bacteria—including corynebacteria and a variety of cocci and bacilli—become established and dominate the flora. As a result, the pH of the vagina becomes more neutral until puberty. At puberty, estrogen levels increase, lactobacilli again dominate, and the vagina again becomes acidic. During the reproductive years, small numbers of other bacteria and yeasts become part of the flora. In the adult, a disturbance of this ecosystem by a decrease in glycogen (caused by oral contraceptives or pregnancy, for example) or elimination of the normal flora by antibiotics can lead to **vaginitis**, an infection of the vagina. When the female reaches menopause, estrogen levels again decrease, the flora return to those of childhood, and the pH again becomes neutral.

Diseases of the Urinary System

Bacterial Diseases of the Urinary System

Many infections of the urinary system appear to be opportunistic and related to a number of predisposing factors. These factors include disorders of the nervous system, toxemia associated with pregnancy, diabetes mellitus, and obstructions to the flow of urine, such as tumors and kidney stones. Because the anus, from which feces are excreted, is located close to the urethra, it is not unusual, especially in females, for the urinary tract to become contaminated with intestinal bacteria. These are usually *Escherichia coli*, *Proteus* species, and other gram-negative enteric bacteria; pseudomonads, fecal streptococci, and staphylococci are also common in such infections. These *Pseudomonas* infections are, as elsewhere, unusually troublesome to treat. The fungus *Candida albicans* is also an opportunistic agent of urinary tract infections. Moreover, a number of sexually transmitted diseases can cause inflammation in the urinary tract.

Infections usually cause inflammation of the affected tissue. Inflammation of the urethra is called *urethritis*; of the urinary bladder, *cystitis*; and of the ureters, *ureteritis*. The most significant danger from lower urinary tract infections is that they can affect the kidneys (causing *pyelonephritis*) and impair their function. The kidneys are also sometimes affected by systemic bacterial diseases, such as typhoid fever or leptospirosis; when the kidneys are thus infected, the pathogens causing these diseases can be found in excreted urine.

Many infections of the urinary tract are of nosocomial origin. In fact, about 35% of all nosocomial infections occur in the urinary tract. Operations on the urinary bladder and prostate gland and catheterization for draining the urinary bladder are procedures that can introduce bacteria into the bladder and ureters. *E. coli* causes more than half of the nosocomial infections of the urinary tract; fecal streptococci, *Proteus*, *Klebsiella*, and *Pseudomonas* also commonly cause such infections.

Treatment of diseases of the urinary tract depends on diagnostic isolation of the causative organism and determination of its antibiotic sensitivity. Normal urine contains fewer than 10,000 bacteria per milliliter. When more than 100,000 bacteria per milliliter are found, there is usually an infection. Dipslide kits of various designs are often used to estimate the bacterial content of urine. One type has a medium that is selective and differential for coliform-type bacteria on one side and a medium that selects for gram-positive bacteria on the other side.

Cystitis

Cystitis is an inflammation of the urinary bladder and is very common, especially among females. Symptoms often include *dysuria* (difficult or painful urination) and *pyuria* (the presence of leukocytes in the urine). The female urethra has many microorganisms around its opening, and it is shorter than the male urethra, so microorganisms can traverse it more readily. Sexual intercourse or careless personal hygiene facilitates such transfer. Contributing factors in females include gastrointestinal system infections and preexisting infections of the vagina, uterus, or urethra. In males, cystitis might be associated with infections of the gastrointestinal system, kidneys, or urethra. Bacteria frequently associated with cystitis in both sexes are the gram-negative rods *E. coli*, *Proteus vulgaris* (prō'tē-us vul-ga'ris), and *Pseudomonas aeruginosa*.

Treatment of cystitis depends on the bacterium that is responsible. Nitrofurantoin, which concentrates preferentially in the urine, is often the drug of choice. Broad spectrum penicillins, such as ampicillin, are also recommended.

Pyelonephritis

In 25% of untreated cases, cystitis may progress to **pyelonephritis**, an inflammation of one or both kidneys that involves the nephrons and the renal pelvis (the opening into the ureter). The disease is generally a complication of infection elsewhere in the body. In females, it is often a complication of lower urinary tract infections. The causative agent in about 75% of the cases is *E. coli*. Other bacteria associated with pyelonephritis are *Enterobacter aerogenes*, *Proteus* species, *Pseudomonas aeruginosa*, *Streptococcus pyogenes*, and staphylococci. Should pyelonephritis become chronic, scar tissue forms in the kidneys and severely impairs their function. Generally, the antibiotics effective for treatment of cystitis are also used for pyelonephritis.

Leptospirosis

The causative organism for **leptospirosis** is the spirochete *Leptospira interrogans* (lep-tō-spī'rä in-tėr'rä-ganz), shown in Figure 25-4. *Leptospira* has a characteristic shape—an exceedingly fine spiral wound so tightly that it is barely discernible under the darkfield microscope. Like other spirochetes, *L. interrogans* (so named because the hooked ends suggest a question mark) is hard to see under a normal light microscope and is therefore rarely Gram stained. It is an obligate aerobe that can be readily grown in a variety of artificial media supplemented with rabbit serum.

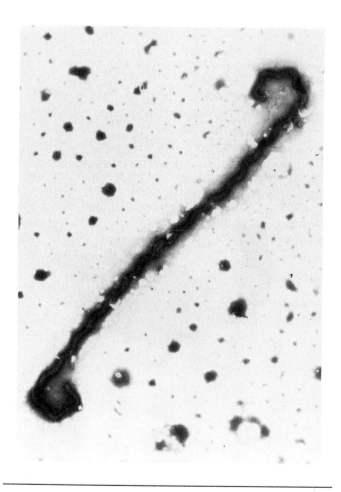

Figure 25-4 *Leptospira interrogans*, **the causative agent of leptospirosis.** The hooked ends are often seen in preparations of this organism. Remnants of the threadlike axial filament are barely visible on the hooks (×12,000).

Leptospirosis is a disease of wild mammals, and the microorganism is excreted in the urine of infected animals. Humans and animals in contact with water contaminated by such urine are most likely to contract the disease, which can be transmitted through mucous membranes. Domestic dogs are commonly immunized against the disease.

Leptospirosis is characterized by chills, fever, headache, and muscular pain. Leptospiral organisms can be found during the course of the disease in the blood, liver, cerebrospinal fluid, kidneys, and urine. Jaundice from liver damage occurs frequently. The human mortality rate is less than 10%, and kidney failure is the most common cause of death. Recovery results in solid immunity. There are about 100 cases per year in the United States.

Diagnosis by serology is complicated by an immunological overlap among the various pathogenic serotypes of *L. interrogans*. Serum antibodies usually appear during the second week of illness and reach a maximum titer during the third or fourth week. However, they are not easily identified because there are so many antigenic types. Agglutination tests, however, can be done with suspensions of killed leptospirae that have been pooled to contain the most common antigenic strains. The treatment of leptospirosis is generally unsatisfactory, partly because the disease is frequently recognized late. The drug of choice is penicillin.

Glomerulonephritis

Glomerulonephritis (Bright's disease) is an inflammation of the glomeruli in the kidneys. The glomeruli are blood capillaries that assist in filtering blood as it passes through the kidneys. Glomerulonephritis is an immune-complex disease. Most cases are a sequel to infection with β-hemolytic streptococci (*S. pyogenes*). Soluble streptococcal antigens combine with specific antibodies to form antigen–antibody complexes that interact with complement (Chapter 15). The complexes are deposited in the glomeruli, where they cause inflammation and kidney damage.

The disease is characterized by fever, high blood pressure, and the presence of protein and red blood cells in the urine. The red blood cells and protein are in the urine because the permeability of the glomeruli increases as a result of inflammation. Although a few patients die from glomerulonephritis, and some develop chronic conditions, most patients recover completely.

Diseases of the Reproductive System

Bacterial Diseases of the Reproductive System

Most diseases of the reproductive system are transmitted by sexual activity and are therefore called **sexually transmitted diseases (STDs)**. These diseases are also called **venereal diseases (VDs)**, an older term derived from the name *Venus*, the Roman goddess of love. Most of these diseases can be readily cured with antibiotics, if treated early, and can largely be prevented by the use of condoms. Nevertheless, STDs are a major public health problem in the United States. Figure 25-5 illustrates the relative frequency of the more common STDs in the United States.

Gonorrhea

By far the most common reportable communicable disease in the United States is **gonorrhea**, an STD caused by the gram-negative diplococcus *Neisseria gonorrhoeae* (see Figure 25-8 and Microview A, Photo 6). (Health care providers must report "reportable" disease cases to local, state, or federal health agencies.) An ancient disease, gonorrhea was described and given its present name by the Greek physician Galen

in A.D. 150. At present, the number of cases seems to have reached a plateau after steadily and steeply increasing for many years (Figure 25-6). About 1 million cases in the United States are reported to the Centers for Disease Control each year, and the true number of cases is probably between 3 and 4 million. More than 60% of the cases are in the 15- to 24-year-old age group.

To infect, the gonococcus must attach to the mucosal cells of the epithelial wall by means of pili. One experimental vaccine is aimed at preventing these pili from attaching. The organism does not infect the layered squamous cells characteristic of the external skin but invades the spaces separating mucosal cells. Mucosal cells are found in the oral–pharyngeal area,

(a)

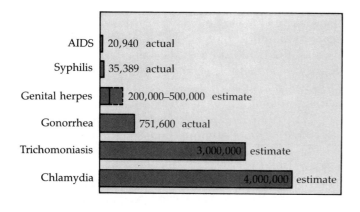

Figure 25-5 Incidence of sexually transmitted diseases. The data are estimates by the Centers for Disease Control for United States cases in 1987. They include reported cases plus the CDC's estimates of unreported cases. [Source: *MMWR* 36:52 (1/8/88).]

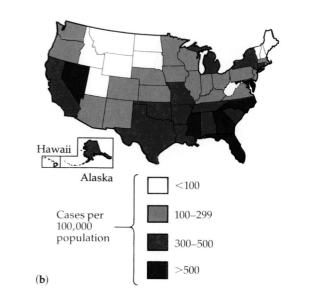
(b)

Figure 25-6 Incidence and distribution of gonorrhea.
(a) Incidence of gonorrhea in the United States since 1941.
(b) Geographical frequency of cases in 1987. [Source: CDC. *Summary of Notifiable Diseases 1986, MMWR* 36:52 (1/8/88).]

the eyes, joints, and rectum, both male and female genitals, and external genitals of prepubertal females. The invasion sets up an inflammation, and, when leukocytes move into the inflamed area, the characteristic pus formation results. In men, a single unprotected exposure results in infection with gonorrhea from 20% to 35% of the time. Women become infected from 60% to 90% of the time from a single exposure.

Males become aware of a gonorrheal infection by painful urination and discharge of pus-containing material from the urethra (Figure 25-7). About 80% of infected males show these obvious symptoms after an incubation period of only a few days; most others show symptoms in less than a week. In the days before antibiotic therapy, symptoms persisted for weeks. In untreated cases, recovery may eventually occur without complication, but, when complications do occur, they can be serious. In some cases, the urethra is scarred and partially blocked. Sterility can result when the testes become infected or when the ductus (vas) deferens, the tube carrying sperm from the testes, becomes blocked by scar tissue.

In females, the disease is more insidious. Very few women are aware of the early stages of the infection. Later in the course of the disease, there might be abdominal pain from **pelvic inflammatory disease (PID)**. Some estimate that 20% to 30% of untreated gonorrhea cases progress to PID. Probably 1 million cases occur each year, and 150,000 of these require surgery, such as removal of the infected uterine tube.

The cause of PID is not restricted to *N. gonorrhoeae*; coinfection with chlamydial bacteria is frequent, and they are also an important agent of this type of infection. *PID* is a collective term for any extensive bacterial infection of the pelvic organs, particularly the uterus, cervix, uterine tubes, and ovaries. The most serious of these is the infection of the uterine tubes (**salpingitis**). It is theorized that the bacteria frequently attach themselves to spermatozoa and are transported by them to the uterine tubes. Certainly, it has been demonstrated that women using barrier-type contraceptives, especially with spermicides, have a significantly lower rate of PID.

The result of salpingitis can be scarring that blocks passage of ova from the ovary to the uterus, resulting in sterility. Tens of thousands of women have been rendered sterile by such infections in recent years. Only about 20% of such cases of sterility can be reversed surgically. Also, implantation of the fertilized ova may take place in the uterine tube rather than in the uterus. This is an *ectopic*, or *tubal*, *pregnancy*, a life-threatening condition. The reported cases of ectopic pregnancies have been increasing steadily, which corresponds to increasing occurrence of PID. One episode of salpingitis causes infertility in

Figure 25-7 Pus-containing discharge from the urethra of a male with an acute case of gonorrhea.

10% to 15% of women, and 50% to 75% become infertile after three or more such infections.

In 1970, there were 17,800 cases of PID, and in 1983 there were 69,600 cases. The incidence of mortality is under 1%; deaths usually occur because misdiagnosis prevented treatment. The recommended treatment of PID is simultaneous administration of doxycycline (a long-acting tetracycline) and cefoxitin sodium (a cephalosporin). This combination is active against both the gonococcus and chlamydia.

In both sexes, untreated gonorrhea can become a serious, systemic infection. Complications of gonorrhea can involve the joints, heart (**gonorrheal endocarditis**), meninges (**gonorrheal meningitis**), eyes, pharynx, or other parts of the body. **Gonorrheal arthritis**, which is caused by the growth of the gonococcus in fluids in joints, occurs in about 1% of gonorrhea cases. Joints commonly affected include the wrist, knee, and ankle.

Gonorrheal eye infections occur most often in newborns. If the mother is infected, then the eyes of the infant can become infected as it passes through the birth canal. This condition, **ophthalmia neonatorum**, can result in blindness. Because of the seriousness of this condition and the difficulty of being certain that the mother is free of gonorrhea, erythromycin or silver nitrate in dilute solution is placed in the eyes of all newborn infants. If the mother is known to be infected, an intramuscular injection of penicillin is also administered to the infant. Some sort of prophylaxis is required by law in most states. Gonorrheal infections can also be transferred by hand contact from infected sites to the eyes of adults.

Gonorrheal infections can be acquired at any point of sexual contact; pharyngeal and anal gonorrhea are not uncommon. The symptoms of **pharyngeal gonorrhea** often resemble those of the usual septic sore throat. **Anal gonorrhea** can be painful and accompanied by discharges of pus. In some cases, however, the symptoms are limited to itching. The incidence of anal gonorrhea has been declining sharply since more individuals have been taking precautions against transmission of AIDS.

The increase of sexual activity with a series of partners and the fact that the disease in the female may go unrecognized have contributed considerably to the increased incidence of gonorrhea and other STDs. The widespread use of oral contraceptives has also contributed to the increase. Oral contraceptives tend to increase the moisture content and raise the pH of the vagina while increasing the susceptibility of mucosal cells. Oral contraceptives have often replaced condoms and spermicides, which helped to prevent disease transmission. All these factors predispose the female to infection and therefore have increased the overall incidence of disease. Immunity to reinfection does not result from recovery from gonorrhea.

Penicillin has been an effective treatment for gonorrhea over the years, although the dosages have had to be much increased because of the appearance of

MMWR *Centers for Disease Control*
Morbidity and Mortality Weekly Report

ANTIBIOTIC-RESISTANT GONOCOCCI AND NATURAL SELECTION

Strains of gonococci resistant to all forms of penicillin have emerged —penicillinase-producing *Neisseria gonorrhoeae* (PPNG). In 1985, for the first time since March 1976, when the first PPNG isolate was identified, PPNG was reported from all 50 states in one calendar year. Although the incidence of PPNG remains low, 16,608 cases in 1986 comprised approximately 1.8% of all gonorrhea cases reported that year. This represents a 90% increase over the PPNG cases reported in 1985.

Importation from Other Countries

A significant proportion of PPNG cases have been related to importation of infection from Southeast Asia and the Philippines. Military personnel or their dependents have been responsible for a majority of the imported cases. Over forty countries have reported cases of PPNG to the World Health Organization. The organism accounts for about 30% of all recent gonococcal isolates in the Philippines and 16% in the Republic of Singapore. (Among the factors thought to contribute to the high prevalence of PPNG in the Philippines and Singapore is the preventive use of oral penicillins, especially by prostitutes.)

Endemic Focus in the United States

In many areas of the United States, the incidence of PPNG infection is increasing. Between 1985 and 1986, the number of PPNG cases in New York City increased 154%, in Los Angeles County, 93%, and in Florida, 35%. All *N. gonorrhoeae* isolates should be tested for penicillinase production. Spectinomycin and tetracyline have been used for all PPNG infections.

Multiple-Drug Resistance

In 1983, 56 cases of penicillin-resistant gonococcal infection occurred in North Carolina. These cases represent the first reported outbreak of gonorrhea caused by bacteria that are resistant to penicillin but do not produce penicillinase, a phenomenon called chromosomally mediated resistance (CMRNG). By 1984, 400 cases of CMRNG were reported from 22 states. This is significant because the isolates are negative in a test for PPNG but are not treatable with penicillin.

The geographic distribution of tetracycline-resistant *Neisseria gonorrhoeae* (TRNG) has been widespread. In the year after TRNG was first reported (in February 1985), 79 cases were reported in Massachusetts, Georgia, Oregon, Pennsylvania, and New Hampshire.

In May 1981, a PPNG strain resistant to spectinomycin was reported from Travis Air Force Base, California. The patient had acquired the infection in the Philippines. Although few spectinomycin-resistant isolates of *N. gonorrhoeae* had been reported previously, this was the first to be penicillinase-producing also. The CDC now recommends that all isolates found to be PPNG also be tested for spectinomycin resistance. Patients who have uncomplicated anogenital infections caused by spectinomycin-resistant PPNG should be treated with ceftriaxone. Spectinomycin was used for five years before resistance was seen, which raises the question of how long the lifespan of ceftriaxone and other drugs will be.

Combinations of plasmid-mediated resistance (such as PPNG–TRNG), plasmid and chromosomally mediated resistance (such as spectinomycin-resistant PPNG), and chromosomally mediated resistance to multiple antibiotics have been isolated. Multiple-drug resistance is increasing so rapidly that as of 1987, PPNG is no longer penicillin-resistant bacteria. Penicillin resistance is usually caused by the presence of a gene for penicillinase (an enzyme that degrades penicillin). The penicillinase gene might be carried by the bacterial chromosome or by a plasmid (Chapter 8). Gonococci having such plasmids first appeared in 1976 (see the box below). Between 1985 and 1986, the incidence of resistance to antibiotics increased by more than 50%.

Spectinomycin is the favored alternative antibiotic, but resistance has now appeared against it, and the failure rate with spectinomycin treatment is increasing. Antibiotic resistance to the gonococcus has accelerated at an alarming rate in recent years. Because of frequent coinfection with chlamydia, antibiotic treatment should include an effective anti-

chlamydial antibiotic, usually of the tetracycline group.

Gonorrhea in men is diagnosed with a stained smear of pus from the urethra. The typical gram-negative diplococci within the phagocytic leukocytes are readily identified (see Figure 25-8 and Microview A, Photo 6). Gram staining of exudates is not as reliable with women. Usually, a culture is taken from within the cervix and grown on special media. Cultivation of the nutritionally fastidious bacterium requires an atmosphere containing carbon dioxide. The gonococcus is very sensitive to adverse environmental influences (desiccation and temperature) and survives poorly outside the body. It even requires special transporting media to keep it viable for short intervals

singled out for reporting. Current data reflect all drug-resistant strains. Once antibiotic-resistant gonorrhea becomes endemic, eradication is extremely difficult. Resistance can result in delays in treatment until an effective drug is found, and patients with inadequately treated gonorrhea are at high risk for complications.

Natural Selection and Antibiotic Resistance

Through natural selection, organisms best adapted to fit their environment will survive and reproduce. In an antibiotic-laden environment, the "fittest" bacterium is one that is resistant to antibiotics.

Asked to list the reasons for the emergence of drug-resistant bacterial strains, a spokesman for the National Institutes of Health blamed the unnecessary use of prophylaxis, use of antibiotics in animal feeds, the availability of over-the-counter antibiotics in many countries, and misuse by health professionals. It has been pointed out that the indiscriminate use of antibiotics may result in a vast majority of infections being caused by antibiotic-resistant bacteria.

Sources: *Summary of Notifiable Diseases 1986, MMWR 35:55; MMWR 35:304–306 (1986); MMWR 37:107–108 (1987).*

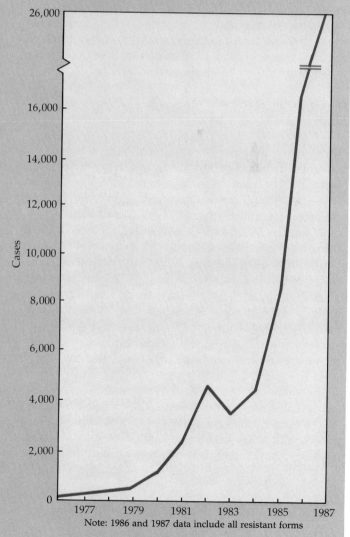

Reported penicillinase-producing *Neisseria gonorrhoeae* (PPNG) cases in the United States, 1976–1986. (Source: CDC. *Summary of Notifiable Diseases, MMWR 35:55.*)

Note: 1986 and 1987 data include all resistant forms

Figure 25-8   A smear of pus from a patient with gonorrhea. The *Neisseria gonorrhoeae* bacteria are contained within phagocytic leukocytes. They are visible as pairs of cocci and are gram-negative (×2000).

before the cultivation is under way. Diagnosis of gonorrhea has been aided recently by the development of an ELISA test that detects *N. gonorrhoeae* in urethral pus or on cervical swabs within about three hours with high accuracy.

Nongonococcal Urethritis (NGU)

Nongonococcal urethritis (NGU), also known as **nonspecific urethritis (NSU)**, refers to any inflammation of the urethra not caused by *Neisseria gonorrhoeae*. Symptoms include pain during urination and a watery discharge.

Nonmicrobial factors, such as trauma (passage of a catheter) or chemical agents (alcohol and certain chemotherapeutic agents), can cause this condition, but at least 40% of the cases of NGU are acquired sexually. In fact, NGU might be the most common sexually transmitted disease in the United States today. Although it is not a reportable disease and exact data are lacking, the Centers for Disease Control estimates that from 4 to 9 million Americans have NGU.

Because the symptoms are often mild in males, and females are usually asymptomatic, many cases go untreated. Physicians often treat NGU as a male urological disease rather than as an STD. Complications are not common but can be serious. Males may

develop inflammation of the epididymis. In females, inflammation of the uterine tubes may cause sterility by scarring the tubes. It is estimated that as many as 60% of such cases of salpingitis are from chlamydial infection.

Probably the most common pathogen associated with NGU is *Chlamydia trachomatis*. A substantial number of persons suffering from gonorrhea are coinfected with *C. trachomatis*. *C. trachomatis* is the same organism responsible for the sexually transmitted disease lymphogranuloma venereum (see the next section) and trachoma (an eye infection). Like gonorrhea, *C. trachomatis* can infect the eyes of a newborn during birth.

Chlamydias are small gram-negative bacteria that are obligately intracellular parasites (see Microview A, Photo 9). Although the symptoms caused by chlamydias are relatively mild and often asymptomatic, especially in women, they apparently have an alarming capacity to cause salpingitis and subsequent infertility.

Until recently, culture of chlamydias required cell culture lines not available in all laboratories. This diagnostic picture has changed dramatically with the availability of several new rapid tests for chlamydial infections. One test uses monoclonal antibodies specific for *C. trachomatis*. The antibodies are tagged with a fluorescein dye, and, if the antibody reacts with chlamydias on a slide smear, it is visible with fluorescence microscopy. Another rapid ELISA test is also used.

Bacteria other than *N. gonorrhoeae* and *C. trachomatis* can also be implicated in NGU. The next most common cause of urethritis and infertility is probably *Ureaplasma urealyticum* (ū-rē-ä-plas'mä ū-rē-ä-lit'i-kum). This organism is a member of the mycoplasma group (bacteria without a cell wall). Another mycoplasma, *Mycoplasma hominis* (ho'min-nis), commonly inhabits the normal vagina but can opportunistically cause salpingitis.

Both chlamydia and mycoplasma are sensitive to tetracycline-type antibiotics, such as tetracycline and doxycycline. Erythromycin is an acceptable alternative drug.

Syphilis

The earliest reports of **syphilis** date back to the end of the fifteenth century in Europe. This was coincidental with the return of Columbus from the New World and gave rise to a hypothesis that syphilis was introduced to Europe by his men. A more widely accepted hypothesis is that syphilis existed in Europe

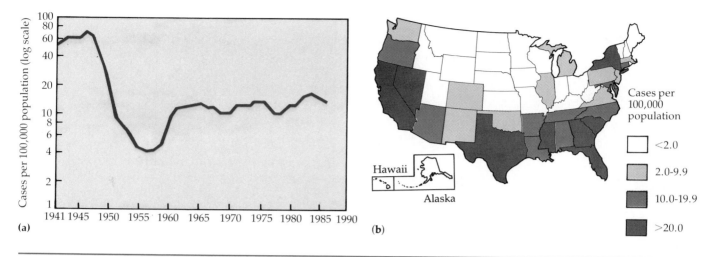

Figure 25-9   Incidence and distribution of syphilis. (a) Number of cases in the United States since 1941. **(b)** Geographical frequency of cases reported in 1987. [Source: CDC. *Summary of Notifiable Diseases 1986, MMWR* 36:52 (1/8/88).]

and Asia before the fifteenth century but was called leprosy. The early descriptions of this "leprosy" say it was transmitted by sexual contact. Although there is no way to know for certain, it appears from descriptions that in earlier times syphilis was rather more severe than it is at present. Whether this decline indicates diminished virulence of the bacterium or increased resistance in the population is unknown.

The causative agent of syphilis also causes skin diseases, such as yaws, that are endemic in the tropics. The course of infection of yaws closely resembles syphilis. Some researchers view these skin diseases and syphilis as variations of the same disease that differ in their portal of entry.

In contrast to the precipitous increase in the number of gonorrhea cases, the number of new syphilis cases in the United States has remained fairly stable (Figure 25-9). This relative stability as compared with gonorrhea (Figure 25-6) is remarkable because the epidemiology of the two diseases is quite similar, and concurrent infections are not uncommon.

The highest incidence of syphilis is in the 20- to 39-year-old age group, which is also the most sexually active. The incidence among male homosexuals is about 25 to 30 times that of the general population. Until recently, syphilis had actually become an uncommon disease in the heterosexual population. Many states discontinued the requirements for premarital syphilis tests because so few cases were detected.

The causative organism of syphilis is a gram-negative spirochete, *Treponema pallidum* (tre-pō-nē'mä pal'li-dum). The spirochete is a thin, tightly coiled helix no more than 20 μm in length (Figure 25-10). The virulent strains of the spirochetes have never been successfully cultured in vitro. Instead, they are propagated in rabbits.

Syphilis is transmitted by sexual contact of all kinds, via syphilitic infections of the genitals or other bodily parts. The incubation period averages about three weeks but can range from two weeks to several months. The disease progresses through several recognized stages.

Figure 25-10   *Treponema pallidum*, the causative agent of syphilis, as seen through a darkfield microscope. This is often the best method of diagnosis in early stages of the disease (×1000).

Figure 25-11   Characteristic lesions associated with syphilis at various stages of development. (a) Chancre of primary stage on a male. **(b)** Skin rash of secondary stage. **(c)** Gummas of tertiary stage on the back of an arm.

In the **primary stage** of the disease, the initial symptom is a small, hard-based **chancre**, or sore, which usually appears at the site of infection (Figure 25-11a). The chancre is painless, and a serous exudate is formed in the center. This fluid is highly infectious, and examination with the darkfield microscope shows many spirochetes. In a few weeks, this lesion disappears, possibly because of local immunity having developed. None of these symptoms causes any distress. In fact, many women are entirely unaware of the chancre, which is commonly on the cervix. In males, the chancre sometimes forms in the urethra and is not visible. Serological diagnostic tests are positive in about 80% of patients in the primary stage. During this stage, bacteria enter the bloodstream and lymphatic system, which distribute them widely in the body.

Several weeks after the primary stage (the exact amount of time varies), the disease enters the **secondary stage**, characterized mainly by skin rashes of varying appearance (Figure 25-11b). Other symptoms often observed are the loss of patches of hair, malaise, and mild fever. The rash is widely distributed on the skin and is also found in the mucous membranes of the mouth, throat, and cervix. At this stage, the lesions of the rash contain many spirochetes and are very infectious. Dentists or other medical workers coming into contact with fluid from these lesions can easily become infected by the spirochete entering through minute breaks in the skin. Such nonsexual transmission is possible, but the organisms do not survive long on environmental surfaces and are very unlikely to be transmitted via such objects as toilet seats. Serological tests for syphilis become almost uniformly positive at this stage.

The symptoms of secondary syphilis usually subside after a few weeks, and the disease enters a **latent period**. During this period, there are no symptoms of the disease, although serological tests are positive. After from two to four years of latency, the disease is not normally infectious, except for transmission from mother to fetus. The majority of cases do not progress beyond the latent stage, even without treatment, perhaps because immunity is developed. Some resistance to reinfection is observed at this period.

Because the symptoms of primary and secondary syphilis are not disabling, persons may enter the latent period without having received medical attention. In less than half of these cases, the disease reappears in a **tertiary stage**. This stage occurs only after an interval of many years—usually at least ten—after the latent phase is entered. It has been suggested that most of the symptoms of tertiary syphilis are caused by the body's hyperimmune reactions to surviving spirochetes. Many of the lesions are similar to those formed by the body in response to infections by *Mycobacterium tuberculosis*. These lesions, called *gummas*, are rubbery masses of tissue that appear in many organs and sometimes on the external skin (Figure 25-11c). Although many of these lesions are not very harmful, some can cause extensive tissue damage, such as deafness or blindness from lesions in the central nervous system or perforation of the palate (roof of the mouth), which interferes with speech. Few, if any, organisms are found in the lesions of the tertiary stage, and they are not considered very infectious. Today, cases of syphilis allowed to progress to this stage have become uncommon.

One of the most distressing and dangerous forms of syphilis, called **congenital syphilis**, is transmitted across the placenta to the unborn fetus. Damage to mental development and other neurological symptoms are among the more serious consequences. This type of infection is most common when pregnancy occurs during the latent stage of syphilis. A pregnancy during the primary or secondary stage is likely to produce a stillbirth.

Serological testing for primary syphilis is not completely reliable. Often, the best method for diagnosis is an examination of exudates removed from punctured lymph nodes in the affected area or from lesions. A darkfield microscope is necessary because the organisms do not stain well and are only about 0.2 μm in diameter, a size that approaches the limits of a brightfield microscope. Figure 25-10 shows the spirochete in such a preparation.

The **Venereal Disease Research Laboratory (VDRL)** slide flocculation test is probably the most widely used serological screening test. The **rapid plasma reagin (RPR)** card test, which is similar, is also in common use. Both of these test are nonspecific, in that they do not detect antibodies produced against the spirochete itself but detect *reagin-type antibodies*. Reagin-type antibodies are apparently a response to lipid materials that the body forms in indirect response to infection by the spirochete. The antigen used in such precipitation-type slide tests is thus not the syphilis spirochete but is an extract of beef heart (cardiolipin) that seems to contain lipids similar to those that stimulated the reagin-type antibody production.

These antibodies are formed from three to five weeks after infection, so the VDRL and RPR tests are of only limited usefulness in detecting primary syphilis. Each of these serological tests will detect only about 70% to 80% of primary syphilis cases, but they detect 99% of secondary syphilis cases.

Most slide tests are likely to produce a significant percentage of false positive reactions for syphilis. Therefore, positive reactions should be confirmed by a more exacting test based on spirochete-type antigens. These tests often have less than 1% false positive reactions. One such test is the **fluorescent treponemal antibody absorption (FTA–ABS) test**. This is an indirect immunofluorescence test, in which a preparation of an avirulent strain of *T. pallidum* (one that can be cultured) is allowed to react with a sample of a patient's serum on a microscope slide (see Figure 17-11c). Antibodies in the serum will combine with the spirochetes, but this reaction is not visible. To make it visible, an antibody that will combine with any human antibody (anti-human gamma globulin, or anti-HGG) is tagged with a dye that will be visible under a fluorescent microscope. The tagged anti-HGG is added to the slide and will form a combination of spirochete-antibody and tagged anti-HGG. The slide is again washed and examined with a special fluorescent-type microscope. The spirochetes are detected by their fluorescent glow when struck by ultraviolet light. A disadvantage of the FTA–ABS test is that it remains positive even after syphilis has been successfully treated. The nonspecific reagin-based tests tend to become negative upon successful treatment and are therefore more useful for determining the effect of therapy.

Benzathine penicillin, a long-acting formulation that remains effective in the body for about two weeks, is the usual antibiotic treatment of syphilis. The serum concentrations achieved by this formulation are low, but the spirochete has remained very sensitive to this treatment and shows no significant resistance. Penicillin therapy is especially effective during the primary stage, but antibiotics are generally not effective in tertiary syphilis, probably because the organisms are usually not present.

Antibiotic treatment might be prolonged. One reason for this prolonged exposure is that the spirochete grows slowly, and penicillin is effective only against growing, not dormant, organisms. For penicillin-sensitive persons, a number of other antibiotics, such as erythromycin and the tetracyclines, have also proved effective. Antibiotic therapy aimed at gonorrhea and other infections is not likely to eliminate syphilis as well, because such therapy for those other diseases is usually administered for too short a period of time. Similarly, penicillin directed at syphilis does not reach a high enough concentration to successfully treat gonorrheal infections.

Gardnerella Vaginitis

Vaginitis is most commonly caused by one of three organisms, the fungus *Candida albicans* (kan'did-ä al'bi-kans), the protozoan *Trichomonas vaginalis* (trik-ō-mon'as va-jin-al'is), or the bacterium *Gardnerella vaginalis*. Infections by *C. albicans* (see the discussion starting on page 654) are usually opportunistic, whereas infections by *T. vaginalis* and *G. vaginalis* are likely to be sexually transmitted.

Vaginal infections by *G. vaginalis* are sometimes considered a nonspecific vaginitis because the disease is an interaction between *G. vaginalis* and anaerobic bacteria in the vagina, neither of which will produce the disease alone. Probably one-third of all vaginitis cases are of this type, which occurs when the vaginal pH is above 5 (normally it would be less than 4). The condition is characterized by a pronounced fishy odor of a frothy vaginal discharge, which is usually light in volume. Diagnosis is based on the fishy odor, the level of vaginal pH, and the microscopic observation of *clue cells* in the discharge. These clue cells are sloughed-off vaginal epithelial cells covered with gram-negative rods. Treatment is primarily by metronidazole, a drug that eradicates the anaerobes essential to continuation of the disease but allows the normal lactobacilli to repopulate the vagina.

Lymphogranuloma Venereum

There are a number of STDs that are uncommon in the United States but more common in the tropical areas of the world. For example, *Chlamydia trachomatis*, the cause of trachoma (Chapter 20) and a major cause of NGU, is also responsible for **lymphogranuloma venerum**, a disease found in tropical and near-tropical regions. It is apparently caused by a strain of *C. trachomatis* that is invasive and tends to infect lymphoid tissue. In the United States, there are usually about 250 cases per year, mostly in the southeastern states.

After a latent period from 7 to 12 days, a small lesion appears at the site of infection, usually on the genitals. The lesion ruptures and heals without scarring. From one week to two months later, the microorganisms invade the lymphatic system, and the regional lymph nodes become enlarged and tender.

Suppuration (a discharge of pus) may also occur. The inflammation of the lymph nodes results in scarring that can occasionally obstruct the lymph vessels. This scarring leads to edema of the genital skin and massive enlargement of the external genitals in males. In females, rectal narrowing results from involvement of the lymph nodes in the rectal region. Enlarged nodes may be slow to subside, even after successful antibiotic therapy. These conditions can eventually require surgery.

For diagnosis, pus can be aspirated from infected lymph nodes. When infected cells are properly stained with an iodine preparation, the chlamydias can be seen as inclusions. The isolated organisms can also be grown in cell culture or in embryonated eggs. Several serological tests are available, but all are troubled with frequent false positive reactions. The drug of choice for treatment is tetracycline, and alternatives are such drugs as doxycycline or erythromycin.

Chancroid (Soft Chancre) and Granuloma Inguinale

The STDs known as chancroid and granuloma inguinale occur more frequently in tropical areas, where they are often more common than syphilis. In the United States, more than 3000 cases of chancroid and fewer than 100 cases of granuloma inguinale are reported per year. However, these diseases are seen with some frequency in sailors and overseas military personnel.

In **chancroid (soft chancre)**, a swollen, painful ulcer that forms on the genitals involves an infection of the adjacent lymph nodes. About a week elapses between infection and appearance of the symptoms. Infected lymph nodes in the groin area sometimes even break through and discharge pus to the surface. These ulcers are highly infective, as are the other genital lesions that occur. Lesions might also occur on such diverse areas as the tongue and lips. Because chancroid is so seldom seen by some physicians, they might confuse it with primary syphilis or genital herpes infections. The causative organism is *Hemophilus ducreyi* (dü-krā'ē), a small gram-negative rod that can be isolated from exudates of lesions. The recommended antibiotics are erythromycin and trimethoprim-sulfamethoxazole.

Granuloma inguinale is characterized by an initial chancre that ulcerates on or about the genitals. It can easily be confused with chancroid and primary syphilis. Without treatment, however, the ulcer spreads, and large areas of genital tissue and other tissues can be destroyed in time. The lesions of chancroid and granuloma inguinale have been suggested

Figure 25-12 Donovan bodies (rod shapes) in the cytoplasm of a cell infected with granuloma inguinale. The darkly stained nucleus is clear, but the plasma membrane around the cell is not (×1200).

as a route of transmission of the AIDS virus, thus explaining the high incidence of heterosexual transmission in Africa. The disease is most common in India, Africa, the Caribbean, and Southeast Asia.

The epidemiology of the disease is not completely understood, but it is assumed to be sexually transmitted, although it is not highly communicable. In the United States, male homosexuals are the most common sufferers of granuloma inguinale. It also comprises about 2% to 3% of the venereal disease found in newly inducted black military recruits in the United States.

The causative organism is *Calymmatobacterium granulomatis* (kal-im-mä-tō-bak-ti′rē-um gran-ū-lō′ mä-tis), a small, gram-negative, encapsulated bacillus that is very similar in many respects to *Klebsiella pneumoniae*. In smears of scrapings from lesions, the encapsulated *C. granulomatis* can often be seen with large mononuclear cells called *Donovan bodies* (Figure 25-12). The bacteria can be grown in embryonated eggs and laboratory media. Diagnosis depends on demonstration of the presence of Donovan bodies in infected lesions and tissues. Gentamicin and chloramphenicol are the drugs of choice, and tetracyclines and streptomycin are also effective.

Viral Diseases of the Reproductive System

Genital Herpes

A much publicized STD is genital herpes, caused by *herpes simplex virus* type 2. The herpes simplex virus (see Figures 12-6b and 12-8a) occurs as either type 1 or type 2. Herpes simplex type 1 is primarily respon-

sible for the common cold sore or fever blister (Chapter 20). Herpes simplex type 2, and sometimes type 1, causes a disease of sexual transmission that affects the genitals; this disease is called **genital herpes**. The lesions appear after an incubation period of about one week and cause a burning sensation. After this, vesicles appear. In both sexes, urination can be painful, and walking is quite uncomfortable; the patient is even irritated by clothing. Usually, the vesicles heal in a couple of weeks.

The vesicles contain fluid that is infectious. Condoms provide uncertain protection since the vesicles on women are usually on the external genitals (seldom on the cervix or within the vagina), and the vesicles on men are often at the base of the penis.

One of the most distressing characteristics of genital herpes is the possibility of recurrences. As in other herpes infections, such as cold sores or chickenpox—shingles, the virus enters a latent state in nerve cells. The site of latency is in nerve ganglia near the base of the spine. There is apparently considerable individual variation in latency. Some persons have frequent attacks; for others, recurrence is rare. Reactivation appears to be triggered by a number of factors, including menstruation, emotional stress or illness (especially if accompanied by fever, a factor that is also involved in appearance of cold sores), and perhaps just scratching the affected area. About 88% of patients with type 2 virus and about 50% of those with type 1 virus will have recurrences. The average patient will have recurrences every three or four months. If a person is going to experience recurrences, the first will usually appear within about six months after the primary infection.

In the United States, the incidence of genital herpes has increased so much during the past decade that it is now a very common sexually transmitted disease. Females are less likely to be aware that they have the disease. This lack of awareness can have serious effects because of *neonatal herpes*. The virus can cross the placental barrier and affect the fetus if the mother suffers from an acute infection early in pregnancy. How frequently this happens is not known, but the fetus is considered infected for practical purposes if the virus can be grown from the amniotic fluid. Infection can result in spontaneous abortion or serious damage to the fetus, such as mental retardation and defective sight and hearing.

If the fetus remains free of the virus, it is necessary to prevent infecting it as it passes through a herpes-infected birth canal. Weekly viral cultures of the mother should be taken during the final six weeks before delivery; many asymptomatic women shed the virus. If results are positive, a cesarean section is usually recommended. This operation should be done before the fetal membrane ruptures. If the woman is shedding the virus, the rupture of the fetal membrane will cause the virus to spread to the uterus within a few hours. Infected infants have a high mortality rate, and survivors almost always have severe neurological damage. Genital herpes infection also seems to be associated with a higher-than-normal rate of cervical cancer.

At present, there is no cure for genital herpes, although research on its prevention and treatment is intensive. Acyclovir, administered topically or orally, is proving to be reasonably effective in alleviating the symptoms of a primary outbreak. There is some relief of pain and other symptoms and slightly faster healing. Continuous administration for several months has been found to lower the chances of recurrence, but it has no effect on the frequency of recurrence after it is discontinued.

Considerable effort is being spent on the development of a vaccine for genital herpes. The fact that having the disease does not prevent latency or recurrence seems to indicate that this search is not likely to produce a reliable vaccine. The approach theoretically most likely to succeed is to identify the genes responsible for latency and then to target these genes with a vaccine that would prevent latency from developing. A subunit vaccine was developed and tested but was not effective. Several other vaccines are being developed for trials.

Genital Warts

Warts are an infectious disease caused by viruses. Sexual transmission of **genital warts** is common. There is considerable evidence that the papillomaviruses that cause warts are also factors in the incidence of cervical cancer. These viruses cannot be cultivated, and it is still not certain just how the transmission of the virus occurs between individuals. There are no uniformly reliable tests for genital wart infections, except for obvious morphological symptoms.

AIDS

In concluding this section on viral STDs, we should recall that AIDS is a viral disease that is frequently transmitted by sexual contact. However, its effect is on the immune system, so it was discussed in Chapter 18 and the boxes in Chapters 12 and 18, pp. 341 and 468.

Fungal Disease of the Reproductive System

Candidiasis

Candida albicans is a yeastlike fungus that often grows on mucous membranes of the mouth, intestinal tract, and genitourinary tract (see Figure 20-15). Infections are usually a result of opportunistic overgrowth when the competing microflora are suppressed by antibiotics or other factors. As we discussed in Chapter 20, *C. albicans* is the cause of thrush (oral candidiasis). It is also responsible for occasional cases of NGU in males and **vulvovaginal candidiasis**, which is the most common cause of vaginitis. About 75% of all women experience at least one episode.

The lesions of vulvovaginal candidiasis resemble those of thrush but produce more irritation, severe itching, a thick, yellow, cheesy discharge, and a yeasty odor. *C. albicans* is an opportunistic pathogen. Predisposing conditions include use of oral contraceptives and pregnancy. These cause a decrease of glycogen in the vagina (see the discussion of normal vaginal flora beginning on p. 641). Diabetes and treatment with broad spectrum antibiotics are also commonly associated with the occurrence of *C. albicans* vaginitis.

Vulvovaginal candidiasis is diagnosed by microscopic identification of the fungus in scrapings of lesions and by isolation of the fungus in culture (Figure 25-13). Treatment consists of topical application of clotrimazole or miconazole. Oral ketoconazole has also been used successfully in some difficult cases where frequent recurrences were a problem.

Figure 25-13 *Candida albicans* **in a smear prepared from a vaginal discharge (×800).**

Protozoan Disease of the Reproductive System

Trichomoniasis

The protozoan *Trichomonas vaginalis* is a fairly normal inhabitant of the vagina in females and of the urethra in many males (see Figures 25-14 and 11-19). If the normal acidity of the vagina is disturbed, the protozoan may overgrow the normal microbial population of the genital mucosa and cause **trichomoniasis**. (Males rarely have any symptoms as a result of the presence of the organism.) In response to the protozoan infection, the body accumulates leukocytes at the infection site. The resulting purulent discharge is profuse, greenish-yellow in color, and characterized by a foul odor. This discharge is accompanied by irritation and itching. This type of vaginitis is usually sexually transmitted. *T. vaginalis*, *C. albicans*, and *G. vaginalis* cause the three most common types of vaginitis. Table 25-1 summarizes the characteristics of the three types.

Diagnosis is easily made by microscopic examination and identification of the organisms in the discharge. The organism can be found in semen or urine of male carriers. Treatment is by oral metronidazole, which readily clears the infection. The organism can also be isolated and grown on laboratory media.

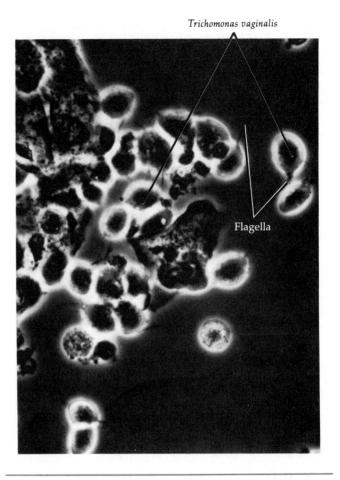

Figure 25-14 *Trichomonas vaginalis*, the flagellated protozoan that causes trichomoniasis vaginitis (×1200).

TABLE 25-1 *Characteristics of the Most Common Types of Vaginitis*

	Candida albicans	Gardnerella vaginalis	Trichomonas vaginalis
Organism	Yeastlike fungi	Coccobacillary bacteria	Flagellated protozoa
Odor	Yeasty	Fishy	Foul
Color of discharge	White	Gray-white	Greenish-yellow
Consistency of discharge	Curdy	Thin, frothy	Frothy
Amount of discharge	Varies	Copious	Varies
Appearance of vaginal mucosa	Dry, red	Pink	Tender, red
pH	Below 4	Above 5	Above 5

Source: Adapted from P. A. Gilly, "Vaginal discharge: its causes and cures." *Postgraduate Medicine* 80(8):231, December 1986.

TABLE 25-2 *Diseases Associated with the Urinary and Reproductive Systems*

	Causative Agent	Mode of Transmission	Treatment
Bacterial Diseases of the Urinary System			
Cystitis (bladder infection)	*Escherichia coli, Proteus vulgaris, Pseudomonas aeruginosa*, and others	Opportunistic infections	Sulfonamides, chloramphenicol, kanamycin, penicillin G, or polymyxin
Pyelonephritis (kidney infection)	Most common are *E. coli, Enterobacter* spp., *Klebsiella* spp., and *Proteus* spp.	From systemic bacterial infections or infections of the lower urinary tract	Antibiotic therapy depends on agent; for chronic cases, nitrofurantoin or trimethoprim–sulfamethoxazole
Leptospirosis (kidney infection)	*Leptospira interrogans*	Direct contact with infected animals, urine of infected animals, or contaminated water	Penicillin
Glomerulonephritis (kidney infection)	Autoimmune reaction to certain strains of *Streptococcus pyogenes*	Sequel to infection by particular strains of *S. pyogenes* in another part of the body	Chloramphenicol, erythromycin, lincomycin, penicillin, or vancomycin for the initial infection
Bacterial Diseases of the Reproductive System			
Gonorrhea	*Neisseria gonorrhoeae*	Direct contact, especially sexual contact	Penicillin for nonresistant strains; spectinomycin for resistant strains; tetracycline for concurrent *Chlamydia* infections
Nongonococcal urethritis (NGU)	*Chlamydias* or other bacteria, including *Mycoplasma hominis, Ureaplasma urealyticum*	Sexual contact or opportunistic infections	Tetracycline, and erythromycin
Syphilis	*Treponema pallidum*	Direct contact, especially sexual contact	Penicillin

TABLE 25-2 (*Continued*)

	Causative Agent	Mode of Transmission	Treatment
Gardnerella vaginitis	*Gardnerella vaginalis*	Opportunistic pathogen, vaginal pH of 5 to 6	Metronidazole
Lymphogranuloma venereum (LVG)	*Chlamydia trachomatis*	Direct contact, especially sexual contact	Tetracycline, doxycycline, erythromycin
Chancroid (soft chancre)	*Hemophilus ducreyi*	Direct contact, especially sexual contact	Erythromycin and trimethoprim–sulfamethoxazole
Granuloma inguinale	*Calymmatobacterium granulomatis*	Probably sexual contact	Gentamicin, chloramphenicol, tetracycline, streptomycin
Viral Diseases of the Reproductive System			
Genital herpes	Herpes simplex virus type 2, and occasionally type 1	Direct contact, especially sexual contact	Acyclovir may help symptoms
Genital warts	Papillomavirus	Direct contact	None
Fungal Disease of the Reproductive System			
Candidiasis	*Candida albicans*	Opportunistic pathogen, may be transmitted by sexual contact	Clotrimazole, miconazole
Protozoan Disease of the Reproductive System			
Trichomoniasis (usually vaginitis)	*Trichomonas vaginalis*	Usually sexually transmitted	Metronidazole

Study Outline

Introduction (p. 639)

1. The urinary system regulates the chemical composition of the blood and excretes nitrogenous waste.
2. The genital system produces gametes for reproduction and, in the female, supports the growing embryo.
3. The urinary and genital systems are closely related anatomically, and both open to the external environment.
4. Microbial diseases of these systems can result from infection from an outside source or from opportunistic infection by normal body flora.

Structure and Function of the Urinary System (pp. 639–640)

1. The nephrons of the kidneys remove urea, uric acid, creatinine, salts, and water from the blood and form urine.
2. Urine is transported from the kidneys through ureters to the urinary bladder and is eliminated through the urethra.
3. Valves prevent urine from flowing back to the urinary bladder and kidneys.
4. The flushing action of urine and the acidity of normal urine have some antimicrobial value.

Structure and Function of the Reproductive System (pp. 640–641)

1. The female genital system consists of two ovaries, two uterine tubes, the uterus, the vagina, and external genitalia.
2. The vagina functions as the birth canal and copulatory canal.
3. The male genital system consists of two testes, ducts, accessory glands, and the penis.
4. Seminal fluid leaves the male body through the urethra.

Normal Flora of the Urinary and Reproductive Systems (p. 641)

1. *Streptococcus, Bacteroides, Mycobacterium, Neisseria,* and enterobacteria are resident flora of the urethra.
2. The urinary bladder and upper urinary tract are sterile under normal conditions.
3. Lactobacilli dominate vaginal flora during reproductive years.

Diseases of the Urinary System (pp. 642–643)

Bacterial Diseases of the Urinary System (pp. 642–643)

1. Opportunistic gram-negative bacteria from the intestines often cause urinary tract infections, especially in females.
2. Predisposing factors of urinary tract infections include nervous system disorders, toxemia, diabetes mellitus, and obstructions to the flow of urine.
3. Urethritis, cystitis, and ureteritis are terms describing inflammations of tissues of the lower urinary tract.
4. Pyelonephritis can result from lower urinary tract infections or from systemic bacterial infections.
5. Diagnosis and treatment of urinary tract infections depend on the isolation and antibiotic-sensitivity testing of the etiologic agents.
6. More than 100,000 bacteria per milliliter of urine indicates an infection.
7. About 35% of all nosocomial infections occur in the urinary system. *E. coli* causes more than half of these infections.

Cystitis (p. 642)

1. Cystitis, inflammation of the urinary bladder, is common in females.
2. Microorganisms at the opening of the urethra and along the length of the urethra, careless personal hygiene, and sexual intercourse contribute to the high incidence of cystitis in females.

Pyelonephritis (pp. 642–643)

1. Inflammation of the kidneys, or pyelonephritis, is usually a complication of lower urinary tract infections.
2. About 75% of pyelonephritis cases are caused by *E. coli.*

Leptospirosis (p. 643)

1. The spirochete *Leptospira interrogans* is the cause of leptospirosis.
2. Leptospirosis is characterized by chills, fever, headache, and jaundice.
3. Serological identification is difficult because there are many serotypes of *L. interrogans.*

Glomerulonephritis (p. 643)

1. Glomerulonephritis, or Bright's disease, is an immune-complex disease occurring as a sequel to a β-hemolytic streptococcal infection.
2. Antigen–antibody complexes are deposited in the glomeruli and cause inflammation and damage.

Diseases of the Reproductive System (pp. 644–657)

Bacterial Diseases of the Reproductive System (pp. 644–653)

1. Most diseases of the genital system are sexually transmitted diseases (STDs).
2. Most STDs can be prevented by the use of condoms and are treated with antibiotics.

Gonorrhea (pp. 644–648)

1. *Neisseria gonorrhoeae* causes gonorrhea.
2. Gonorrhea is the most common reportable communicable disease in the United States.
3. *N. gonorrhoeae* attaches to mucosal cells of the oral–pharyngeal area, genitals, eyes, joints, and rectum by means of pili.
4. Symptoms in males are painful urination and pus discharge. Blockage of the urethra and sterility are complications of untreated cases.
5. Females might be asymptomatic until the infection spreads to the uterus and uterine tubes. Blockage of the uterine tubes, sterility, and pelvic inflammatory disease are complications of untreated cases.
6. Gonorrheal endocarditis, gonorrheal meningitis, and gonorrheal arthritis are complications that can affect both sexes if gonorrheal infections are untreated.
7. Ophthalmia neonatorum is an eye infection acquired by infants during passage through the birth canal of an infected mother.
8. Penicillin has been used effectively to treat gonorrhea. In 1976, penicillinase-producing strains of *N. gonorrhoeae* appeared.
9. Culturing gram-negative diplococci isolated from patients is used to confirm diagnosis.

Nongonococcal Urethritis (NGU) (p. 648)

1. NGU, or nonspecific urethritis, is any inflammation of the urethra not caused by *N. gonorrhoeae*.
2. About 40% of NGU cases are transmitted sexually.
3. Most cases of NGU are caused by *Chlamydia trachomatis*.
4. Symptoms of NGU are often mild or lacking, although salpingitis and sterility may occur.
5. *C. trachomatis* can be transmitted to infants' eyes at birth.
6. *Ureaplasma urealyticum* and *Mycoplasma hominis* also cause NGU.

Syphilis (pp. 648–652)

1. Syphilis is caused by *Treponema pallidum*, a spirochete that has not been cultured in vitro. Laboratory cultures are grown in rabbits.
2. *T. pallidum* is transmitted by direct contact and can invade intact mucous membranes or penetrate through breaks in the skin.
3. The primary lesion is a small, hard-based chancre at the site of infection. The bacteria then invade the blood and lymphatic system, and the chancre spontaneously heals.
4. The appearance of a widely disseminated rash on the skin and mucous membranes marks the secondary stage. Spirochetes are present in the lesions of the rash.
5. The patient enters a latent period after the secondary lesions spontaneously heal.
6. At least ten years after the secondary lesion, tertiary lesions called gummas can appear on many organs.
7. Congenital syphilis, resulting from *T. pallidum* crossing the placenta during the latent period, can cause neurological damage in the newborn.
8. *T. pallidum* is identifiable through darkfield microscopy of fluid from primary and secondary lesions.
9. Many serological tests, such as VDRL and FTA–ABS, can be used to detect the presence of antibodies against *T. pallidum* during any stage of the disease.

Gardnerella Vaginitis (p. 652)

1. Vaginitis can be caused by *Candida albicans*, *Trichomonas vaginalis*, or *G. vaginalis*.
2. Diagnosis of *G. vaginalis* is based on increased vaginal pH, fishy odor, and the presence of clue cells.

Lymphogranuloma Venereum (p. 652)

1. *C. trachomatis* causes LGV, which is primarily a disease of tropical and subtropical regions.
2. The initial lesion appears on the genitals and heals without scarring.
3. The bacteria are spread in the lymph system and cause enlargement of lymph nodes, obstruction of lymph vessels, and edema of the genital skin.
4. The bacteria are isolated and identified from pus taken from infected lymph nodes.

Chancroid (Soft Chancre) and Granuloma Inguinale (pp. 652–653)

1. Chancroid, a swollen, painful ulcer on the mucous membranes of the genitals or mouth, is caused by *Hemophilus ducreyi*.
2. *Calymmatobacterium granulomatis* causes granuloma inguinale.
3. Granuloma inguinale begins as an ulcer on the genitalia that ulcerates and spreads to other tissues.
4. Diagnosis is based on observation of mononuclear cells called Donovan bodies.

Viral Diseases of the Reproductive System (pp. 653–654)

Genital Herpes (pp. 653–654)

1. Herpes simplex type 2 causes an STD called genital herpes.

2. Symptoms of the infection are painful urination, genital irritation, and fluid-filled vesicles.

3. Neonatal herpes is contracted during fetal development or birth. It can result in neurological damage or infant fatalities.

4. The virus might enter a latent stage in nerve cells. Vesicles reappear following trauma and hormonal changes.

5. The disease is not contagious during the latent periods.

6. Genital herpes is associated with cervical cancer.

7. The drug acyclovir has proven effective in treating the symptoms of genital herpes; however, it does not cure the disease.

Genital Warts (p. 654)

1. Papillomaviruses cause warts.

2. The papillomaviruses that cause genital warts have been associated with cervical cancer.

AIDS (p. 654)

1. AIDS is a sexually transmitted disease of the immune system.

Fungal Disease of the Reproductive System (p. 654)

Candidiasis (p. 654)

1. *Candida albicans* causes NGU in males and vulvovaginal candidiasis in females.

2. Vulvovaginal candidiasis is characterized by lesions that produce itching and irritation.

3. Predisposing factors for candidiasis are pregnancy, diabetes, tumors, and broad spectrum antibacterial chemotherapy.

4. Diagnosis is based on observation of the fungus and its isolation from lesions.

Protozoan Disease of the Reproductive System (p. 655)

Trichomoniasis (p. 655)

1. *Trichomonas vaginalis* causes trichomoniasis when the pH of the vagina increases.

2. Diagnosis is based on observation of the protozoan in purulent discharges from the site of infection.

Study Questions

Review

1. List the normal flora of the urinary system, and show their habitats in Figure 25-1.

2. List the normal flora of the genital system, and show their habitats in Figures 25-2 and 25-3.

3. How are urinary tract infections transmitted?

4. Explain why *E. coli* is frequently implicated in cystitis in females. List some predisposing factors for cystitis.

5. Name one organism that causes pyelonephritis. What are the portals of entry for organisms that cause pyelonephritis?

6. Complete the following table.

Disease	Causative Agent	Symptoms	Method of Diagnosis	Treatment
Gardnerella vaginitis				
Gonorrhea				
Syphilis				
NGU				
LGV				
Chancroid				
Granuloma inguinale				

7. Name one fungus and one protozoan that can cause genital system infections. What symptoms would lead you to suspect these infections?

8. Leptospirosis is a kidney infection of humans and other animals. How is this disease transmitted? What types of activities would increase one's exposure to this disease? What is the etiology?

9. Describe the symptoms of genital herpes. What is the etiologic agent? When is this infection least likely to be transmitted?

10. What is glomerulonephritis? How is it transmitted? How is it treated?

11. List the genital infections that cause congenital and neonatal infections. How can transmission to a fetus or newborn be prevented?

Challenge

1. The tropical skin disease called yaws is transmitted by direct contact. Its causative agent, *Treponema pallidum pertenue*, is indistinguishable from *T. pallidum*. The appearance of syphilis in Europe coincides with the first importation of slaves. How might *T. pallidum pertenue* have evolved into *T. pallidum* in the temperate climate of Europe?

2. Why can frequent douching be a predisposing factor to bacterial vaginitis, vulvovaginal candidiasis, or trichomoniasis?

3. The chart below is a key to selected microorganisms that cause genitourinary infections. Complete this key by listing genera discussed in this chapter in the blanks that correspond to their respective characteristics.

Gram-negative bacteria
  Spirochete
    Aerobic   ————————
    Anaerobic   ————————
  Coccus
    Oxidase-positive   ————————
  Bacillus, nonmotile
    Requires X factor   ————————
    X factor not present ————————
  Obligate intracellular parasite ————————
  Lacking cell wall
    Urease-positive   ————————
    Urease-negative   ————————

Fungus
  Pseudohyphae   ————————
Protozoan
  Flagella   ————————
No organism observed/cultured from
patient   ————————

4. A previously well 19-year-old female was admitted to a hospital after two days of nausea, vomiting, headache, and neck stiffness. CSF and cervical cultures showed gram-negative diplococci in leukocytes; a blood culture was negative. What disease did she have? How was it probably acquired?

5. *Neisseria* are cultured on Thayer–Martin media, consisting of chocolate agar and nystatin, incubated in a 5% CO_2 environment. How is this selective for *Neisseria*?

Further Reading

Corey, L., and P. G. Spear. "Infections with herpes simplex viruses." *New England Journal of Medicine* 314:686–691, 749–755, 1986. Part One describes the virus; Part Two describes clinical infections and treatment.

Gunby, P. "Genital herpes research: many aim to tame maverick virus." *Journal of the American Medical Association* 250:2417–2427, 1983. A summary of prospects for control of genital herpes infections.

Holmes, K. K. "The *Chlamydia* epidemic." *Journal of the American Medical Association* 245:1718–1723, 1981. A readable summary of recent information on chlamydial bacteria and their sexually transmitted diseases.

Rapp, F. "Herpesviruses, venereal disease, and cancer." *American Scientist* 66:670–674, 1978. Differentiates between the association of a virus in cancer and the cause of the cancer.

Rosebury, T. *Microbes and Morals*. New York: Viking Press, 1971. A history of venereal diseases from Fracastor through modern antibiotics, including laboratory diagnosis, treatment, and the prospects for vaccines.

Schachter, J. "Chlamydiae." *Annual Review of Microbiology* 34:285–309, 1980. A good review of the clinical features and epidemiology of chlamydial infections.

Schoolnik, G. K., ed. *The Pathogenic Neisseriae*. Washington, D.C.: American Society for Microbiology, 1985. A collection of papers on the genetics, pathogenesis, immunology, and control of *N. gonorrhoeae* and *N. meningitidis*.

Further Reading for Part Four

The following medical microbiology textbooks discuss principles of microbiology with an emphasis on bacterial pathogens. Some texts also include viral, fungal, and parasitic diseases.

Baron, S., ed. *Medical Microbiology*, 2nd ed. Menlo Park, Calif.: Addison-Wesley, 1986.

Davis, B. D., R. Dulbecco, H. N. Eisen, and H. S. Ginsberg. *Microbiology*, 3rd ed. New York: Harper & Row, 1980.

Freeman, B. A. *Burrow's Textbook of Microbiology*, 22nd ed. Philadelphia: Saunders, 1985.

Hoeprich, P. D., ed. *Infectious Diseases*, 3rd ed. New York: Harper & Row, 1983.

Joklik, W. K., H. P. Willett, D. B. Amos, and C. M. Wilfert, eds. *Zinsser Microbiology*, 19th ed. Norwalk, Conn.: Appleton–Century–Crofts, 1988.

Mandell, G. L., R. G. Douglas, and J. E. Bennett, eds. *Principles and Practice of Infectious Disease*, 2nd ed. New York: Wiley, 1985.

Wilson, G., A. Miles, and M. T. Parker. *Topley and Wilson's Principles of Bacteriology, Virology and Immunity*, 7th ed., 4 vols. Baltimore: Williams and Wilkins, 1984.

PART FIVE

Microbiology, the Environment, and Human Affairs

Soil and Water Microbiology

Many people associate bacteria with disease or food spoilage and therefore regard microorganisms as harmful things to be avoided or controlled. Actually, harmful microorganisms are only a very small fraction of the total microbial population. Bacteria and other microorganisms are in fact essential to the maintenance of life on Earth. In previous chapters, we focused mainly on disease-causing capabilities of microorganisms. In this chapter, you will learn about many of the positive functions microbes perform in the environment.

We can find microbes, especially bacteria, in the most widely varied habitats on Earth. They are found in the frozen Antarctic and in boiling hot springs. Aircraft flying miles above the Earth can recover microbes from the thin atmosphere. Exploration of the deepest ocean sediments shows bacteria in large numbers living there, in eternal darkness and subjected to incredible pressures (see the box in Chapter 6, p. 146). Microorganisms are found in the clearest mountain streams flowing from a melting glacier and in waters nearly saturated with salts, such as those of the Dead Sea (see the box in Chapter 5, p. 120).

Some microbes are *parasites*—they derive their nutrients and their ability to reproduce from other microbes. *Bdellovibrio* bacteria, for example, prey on other bacteria (see the box in Chapter 3, p. 67). Other microorganisms, such as those that make the lactic acid of cheese, pickles, or sauerkraut, minimize their competition by forming an acidic environment that is inhospitable to other organisms.

Symbiosis, in which two organisms establish a mutually beneficial relationship, is exemplified by the association of a fungus and an alga in *lichen* (discussed later in this chapter). Some protozoans have even adapted to being hosts to single-celled algae that reside within their cytoplasm—the protozoan furnishes protection, and the algae provide additional nutrients for the host.

Commensalism, where one organism benefits from the relationship without affecting the other, is common among microbes. For example, the few bacteria that can degrade polymers (such as cellulose) release simpler compounds (such as glucose) that neighboring scavengers quickly utilize. Some organisms participate in cometabolism, which carries generosity to an extreme. The organisms degrading a particular substrate get no benefit whatsoever in terms of energy or nutrition but leave behind a product that can be used by other microbes. However, this sort of behavior is the exception—most microbes live in an intensely competitive world.

Some microbial population or another fills almost every imaginable ecological niche. Many of these niches are in soil and water. The microorganisms that inhabit soil and water are the subject of this chapter.

Soil Microbiology and Cycles of the Elements

The Components of Soil

Soil is a complex mixture of solid inorganic matter (rocks and minerals), organic matter, water, air, and living organisms.

Minerals

The weathering of rocks adds such elements as silicon, aluminum, and iron to the soil. Calcium, magnesium, potassium, sodium, phosphorus, and small amounts of other elements that are essential for life are also present in the mineral component of soil.

Organic Matter

The organic matter in soil consists of carbohydrates, proteins, lipids, and other materials. Organic matter comprises from 2% to 10% of most agriculturally important soils. Swamps and bogs, by contrast, contain a higher content of organic matter, up to 95% in some peat bog soils. The water-logged soil of swamps and bogs is an anaerobic environment, and the microbial decomposition of organic matter occurs very slowly there.

All organic matter in soil is derived from the remains of microorganisms, plants, and animals, their waste products, and the biochemical activities of various microorganisms. A great portion of the organic matter is of plant origin—mostly dead roots, wood and bark, and fallen leaves. A second source of organic matter is the vast numbers of bacteria, fungi, algae, protozoans, and viruses, which can total billions per gram of fertile soil. These organisms break down organic substances, producing and maintaining a continuous supply of inorganic substances that plants and other organisms require for growth. Much of the organic matter in soil is ultimately decomposed to inorganic substances, such as ammonia, water, carbon dioxide, and various compounds of nitrate, phosphate, and calcium.

A considerable part of the organic matter in soil occurs as **humus**. This dark material consists of partially decomposed organic matter, chiefly materials that are relatively resistant to decay.

The addition of organic matter, either completely or partially decomposed, is essential to the fertility of soil. Moreover, these spongy organic materials loosen the soil and thereby prevent the formation of heavy crusts and increase the pore spaces. This addition of pore spaces in turn increases aeration and water retention.

Water and Gases

Microbial life in soil depends on the availability of some water, which dissolves many inorganic and organic constituents in soil and makes them available to the living inhabitants. However, if the water content is too high, atmospheric oxygen finds it difficult to penetrate the soil, and this can be a factor in limiting growth efficiency and the types of organisms that flourish. Generally, because of respiration by soil organisms, soil gases contain a high proportion of carbon dioxide and a low proportion of oxygen. Most of these soil gases exist in pore spaces between soil particles or are dissolved in the water.

Organisms

Fertile soil contains a great many animals, ranging from microscopic forms, including numerous nematodes, to larger forms, including insects, millipedes, centipedes, spiders, slugs, snails, earthworms, mice, moles, gophers, and reptiles. Most of these animals are beneficial, in that they promote some mechanical movement of the soil and thereby help to keep the soil loose and open. All soil organisms also contribute to the organic matter of soils in the form of their waste products and eventual remains.

Soil also contains the root systems of higher plants and enormous numbers of microorganisms. Without the microorganisms, especially bacteria, the soil would soon become unfit to support life. Bacteria affect the soil and its ecology in many ways. Most bacteria decompose organic matter into simple products. In these reactions, nutrients are made available for use by plants, animals, and other microorganisms. Other soil bacteria are associated with the transformation of nitrogen and sulfur compounds, so that usable supplies of nitrogen and sulfur are continually provided.

Soil Microbial Flora

The soil is one of the main reservoirs of microbial life (Table 26-1). A good agricultural soil the size of a football field usually contains a microbial population of a size approaching the weight of a cow eating grass on that field. The metabolic capabilities of these vast numbers of microorganisms are probably about 100,000 times those of the grazing cow. Measurement of the carbon dioxide from soil and other evidence indicate that these microorganisms exist in near starvation and at low reproductive rates. When usable nutrients are added to soil, the microbial populations and their activity rapidly increase until the nutrients are depleted, and then the microbial activity returns to the lower levels.

The most numerous organisms in soil are bacteria. Typical garden soil has millions of bacteria in each gram. As Table 26-1 shows, the population is highest in the top few centimeters of the soil and declines rapidly with depth. The populations are usually estimated with plate counts on nutrient media, and the actual numbers are probably greatly underestimated in this method. No single nutrient medium or growth condition can possibly meet all the myriad nutrient and other requirements of soil microorganisms.

Actinomycetes are bacteria, but they are usually considered separately in enumerations of soil populations because they differ so greatly from our conventional concept of bacteria. They typically form long filaments resembling those of molds, but of bacterial dimensions. This filamentous growth habit is a considerable advantage for a microorganism in soil. Under relatively dry conditions, the filament can bridge the gap between one particle of soil and another. This type of growth also maximizes surface area for a given weight and may give the actinomycetes a nutritional advantage. These organisms are found in soil in very large numbers and produce a gaseous substance called *geosmin* that gives the soil its characteristic musty odor. The **biomass** (total mass of living organisms in a given volume) of the actinomycetes is probably about equal to the biomass of all other bacteria in soil. Interest in actinomycetes was greatly stimulated by the discovery that some genera, particularly *Streptomyces* (strep-tō-mī′sēs), produce valuable antibiotics.

Fungi are found in soil in much smaller numbers than bacteria and actinomycetes. Because plate counts of fungal colonies are based on the germination of asexual spores, the actual fungal population is probably lower than such counts indicate. It is estimated that the biomass of fungi probably equals the biomass of the bacteria and actinomycetes combined. This large biomass occurs because the dimensions of the fungal mycelium are many times greater than that of bacterial cells. Molds greatly outnumber yeasts in soil.

Algae and cyanobacteria are found even in dry desert soils. They sometimes form visible accumulations on the surfaces of moist soils. As photosynthetic organisms, they are found mainly on the soil surface, where sunlight, water, and carbon dioxide are most abundant. However, significant numbers of algae and cyanobacteria are also found more than 50 cm below the surface. The environmental contribution of these microorganisms is significant only in special cases. For example, fixation of atmospheric nitrogen (discussed shortly) by some species of cyanobacteria in grasslands, tundra regions, and after rainfall in deserts adds significantly to the fertility of soils in those regions.

TABLE 26-1 *Microorganisms per Gram of Typical Garden Soil at Various Depths*

Depth (cm)	Bacteria	Actino-mycetes	Fungi	Algae
3–8	9,750,000	2,080,000	119,000	25,000
20–25	2,179,000	245,000	50,000	5,000
35–40	570,000	49,000	14,000	500
65–75	11,000	5000	6000	100
135–145	1400	—	3000	—

Source: Adapted from M. Alexander, *Introduction to Soil Microbiology*, 2nd ed. New York: Wiley, 1977.

Protozoans are plentiful in soil. They become dormant as cysts under adverse conditions. When microbial numbers increase, many protozoans function as predators, and their numbers tend to rise and fall with the bacterial populations.

Pathogens in Soil

Human pathogens, which are mostly parasites, find the soil an alien, hostile environment. Even relatively resistant enteric pathogens, such as *Salmonella* species, have been observed to survive for only a few weeks or months when introduced into soil. Most human pathogens that can survive in soil are endospore-forming bacteria. For example, endospores of *Bacillus anthracis*, which causes anthrax in animals, can survive in certain soils for decades before finally germinating when ingested by grazing animals. Disposing of the body of an animal infected with anthrax requires considerable care so that the soil is not seeded with the endospores from the dead animal.

Clostridium tetani (the causative agent of tetanus), *Clostridium botulinum* (the causative agent of botulism), and *Clostridium perfringens* (the causative agent of gas gangrene) are also endospore-forming pathogens whose normal habitat is the soil. From the soil they are introduced into foods or wounds, where they grow and produce toxins.

Pathogens of plants are much more likely to normally inhabit soil. Most plant pathogens are fungi, mainly because fungi are capable of growing in the low moisture typical of plant surfaces. Many of the rusts, smuts, blights, and wilts affecting plants are caused by fungi that pass part of their life cycle in soil. Bacterial diseases of plants are less common, but bacterial pathogens in soil do cause plant diseases, especially rots, in which bacterial enzymes hydrolyze plant tissue.

Viruses also cause plant diseases. The mechanism by which they penetrate the thick plant cell walls is frequently via a sap-feeding insect.

Some of the microorganisms found in soil are insect pathogens, which are potentially useful for pest control. *Bacillus thuringiensis*, for example, is a soil bacterium that is pathogenic to the larvae of many insects; it is now used in their control (Chapter 10). A number of other insect pathogens, including viruses and fungi, are under investigation as biological pesticides.

Microorganisms and Biogeochemical Cycles

Perhaps the most important role of soil microorganisms is their participation in **biogeochemical cycles**—the recycling of certain chemical elements so that they can be used over and over again. Among these elements are carbon, nitrogen, sulfur, and phosphorus. Were it not for the activities of microorganisms in the various biogeochemical cycles, essential elements would become depleted, and all life would cease.

The Carbon Cycle

The first biogeochemical cycle we will consider is the carbon cycle (Figure 26-1). As you know, all organic compounds contain carbon. Most of the inorganic carbon used in the synthesis of organic compounds comes from the carbon dioxide in the atmosphere. Some carbon dioxide is also dissolved in water.

In photosynthesis—the first step in the carbon cycle—carbon dioxide is incorporated, or *fixed*, into organic compounds by such photoautotrophs as cyanobacteria, green plants, algae, and green and purple sulfur bacteria. In the next step in the cycle, chemoheterotrophs consume the organic compounds—animals eat photoautotrophs, especially green plants, and may in turn be eaten by other animals. Thus, as the organic compounds of the photoautotrophs are digested and resynthesized, the carbon atoms of carbon dioxide are transferred from organism to organism up the food chain.

Some of the organic molecules are used by chemoheterotrophs, including animals, to satisfy their energy requirements. When such energy is released through the process of respiration, carbon dioxide is released into the atmosphere. This carbon dioxide immediately becomes available to start the cycle over again. However, much of the carbon remains within the organisms until they excrete wastes or die. When the organisms die, the organic compounds of their bodies are deposited in the soil and are decomposed by microorganisms, principally by bacteria and fungi. During this decomposition, carbon dioxide is returned to the atmosphere. Although the carbon dioxide of the atmosphere comprises only about 0.03% of the atmospheric gases, it is essential for the growth of plants and algae.

Figure 26-1   The carbon cycle.

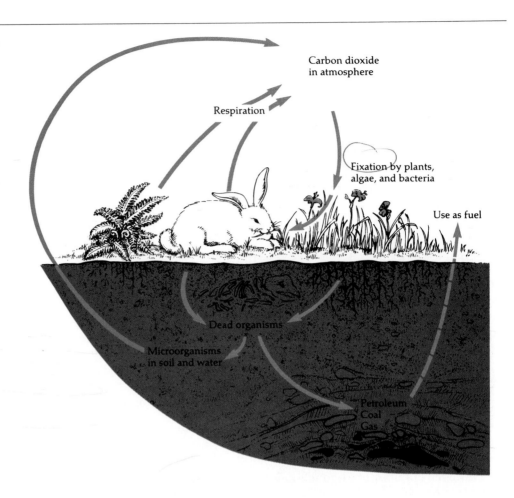

Carbon is stored in rocks, such as limestone ($CaCO_3$), and is dissolved in oceans as carbonate ions (CO_3^{2-}); it is also stored in organic mineral forms, such as coal and petroleum. Burning such fossil fuels releases carbon dioxide. There is evidence that the amount of carbon dioxide in the atmosphere has been increasing as a result of widespread use of fossil fuels. Some scientists believe that this is a potential problem because the increase of atmospheric carbon dioxide could cause a warming of the earth known as the greenhouse effect (see the box on p. 670).

The Nitrogen Cycle

The nitrogen cycle is shown in Figure 26-2. Nitrogen is needed by all organisms for their synthesis of protein, nucleic acids, and other nitrogen-containing compounds. Molecular nitrogen (N_2) comprises almost

80% of the earth's atmosphere; the atmosphere over every acre of fertile soil contains more than 30,000 tons of nitrogen. Despite the abundance of nitrogen as a molecular gas, however, no eucaryote is able to make direct use of it. Instead, the nitrogen must be fixed (combined) with other elements, such as oxygen and hydrogen. The resulting compounds, such as nitrate ion (NO_3^-) and ammonium ion (NH_4^+), are then used by autotrophic organisms. The chemical and physical forces operating in the soil, water and air, together with the activities of specific microorganisms, are important factors in the conversion of nitrogen to usable forms.

Ammonification

Almost all the nitrogen in the soil exists in organic molecules, primarily in proteins. When an organism dies, the process of microbial decomposition results in the hydrolytic breakdown of proteins into amino acids. In a process called deamination, the amino groups of amino acids are removed and converted into ammonia (NH_3). (See Figure 26-2.) This release of ammonia is

Figure 26-2   The nitrogen cycle.

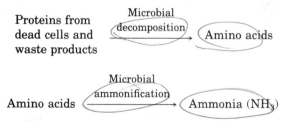

called **ammonification**. Ammonification is brought about by numerous bacteria and fungi. It can be represented as follows:

Proteins from dead cells and waste products $\xrightarrow{\text{Microbial decomposition}}$ Amino acids

Amino acids $\xrightarrow{\text{Microbial ammonification}}$ Ammonia (NH₃)

Microbial growth releases extracellular proteolytic enzymes that decompose proteins. The resulting amino acids are transported into the microbial cells, where ammonification occurs. The fate of the ammonia produced by ammonification depends on soil conditions. Because ammonia is a gas, it rapidly dis-

appears from dry soil, but in moist soil it becomes solubilized in water, and ammonium ions (NH_4^+) are formed:

$$NH_3 + H_2O \longrightarrow NH_4OH \longrightarrow NH_4^+ + OH^-$$

Ammonium ions from this sequence of reactions are used by bacteria and plants for amino-acid synthesis.

Nitrification

The next sequence of reactions in the nitrogen cycle involves the oxidation of the ammonium ion to nitrate, a process called **nitrification**. Living in the soil are autotrophic nitrifying bacteria, such as those of the genera *Nitrosomonas* (nī-trō-sō-mō′näs) and *Nitrobacter* (nī-trō-bak′tėr). These organisms obtain energy by oxidizing ammonia or nitrate. In the first stage, *Nitrosomonas* oxidizes ammonia to nitrites:

$$NH_4^+ \xrightarrow{\ Nitrosomonas\ } NO_2^-$$

Ammonium ion   Nitrite ion

In the second stage, such organisms as *Nitrobacter* oxidize nitrites to nitrates:

$$NO_2^- \xrightarrow{\ Nitrobacter\ } NO_3^-$$

Nitrite ion   Nitrate ion

Plants tend to use nitrate as their source of nitrogen for protein synthesis because nitrate is highly mobile in soil and is more likely to encounter a plant root than ammonium. Ammonium ions would actually make a more efficient source of nitrogen as they require less energy to incorporate into protein, but these positively charged ions are usually bound to negatively charged clays in the soil, whereas the negatively charged nitrate ions are not bound.

Denitrification

At various points in the cycle, atmospheric nitrogen is either added or removed. The loss of nitrogen from the cycle involves a process called **denitrification**, the conversion of nitrate to nitrogen gas. Denitrification can be represented as follows:

$$NO_3^- \longrightarrow NO_2^- \longrightarrow N_2O \longrightarrow N_2$$

Nitrate ion   Nitrite ion   Nitrous   Nitrogen
                                      oxide   gas

Pseudomonas species appear to be the most important group of bacteria in denitrification in soils. A number of other genera, including *Paracoccus* (pa-rä-kok′kus), *Thiobacillus* (thī-ō-bä-sil′lus), and *Bacillus*, also include species that are capable of carrying out denitrification reactions. Denitrifying bacteria are aerobic, but under anaerobic conditions they can use nitrate in place of oxygen as a final electron acceptor. (This process is called *anaerobic respiration*; see Chapter 5.) Thus, denitrification occurs in waterlogged soils depleted of oxygen. Because denitrifying bacteria deliver nitrogen to the atmosphere at the expense of soil nitrates, denitrification is unfavorable for soil fertility.

Nitrogen Fixation

During the initial phase of the nitrogen cycle, nitrogen is converted into ammonia in a process called **nitrogen fixation**. Only a few species of bacteria and

.. Microbiology in the News ....

Bacteria Contribute to the Greenhouse Effect

For years, scientists have been warning that the increased burning of fossil fuels is raising the carbon dioxide content of the atmosphere and that this added carbon dioxide absorbs some infrared radiation that the Earth normally radiates out into space. This absorption traps the Earth's heat, ultimately causing a greenhouse effect that can eventually warm the atmosphere enough to melt the polar ice caps and flood major coastal cities. In addition, the more pessimistic scientists have warned that the warmer air could cause continual droughts in major crop-growing areas, resulting in worldwide famine.

Other scientists, who are more optimistic, claim that nature takes care of the problem automatically. The extra carbon dioxide stimulates the growth of plants, which fix carbon dioxide into organic molecules, thus reestablishing a balance.

However, there are gases other than carbon dioxide that absorb the infrared rays leaving the Earth, thus exacerbating the greenhouse effect, and these gases, molecule for molecule, are far more active infrared absorbers than carbon dioxide. One of them is methane, which is now much more prevalent in the atmosphere than it was two or three decades ago.

One key link in the chain is *carbon dioxide fertilization*, the stimulating effect of carbon dioxide on plant growth. At first this seems to be a blessing. A Department of Energy report estimates that doubling the amount of carbon dioxide in the atmosphere could cause a 30% to 50% increase in the growth of some agricultural plants. Dr. Boyd Strain, a botanist at Duke University, suggests that much of the increase of agricultural productivity after the Industrial Revolution came not from the use of fertilizers and pesticides but from the increasing abundance of carbon dioxide in the atmosphere.

Because of the added plant growth, however, there is a corresponding increase in the amount of plant matter that decays. According to Paul Guthrie, an atmospheric scientist at the NASA Goddard Space Flight Center, the dangerous increase in the methcyanobacteria are capable of enacting this process. The nitrogenase enzyme responsible for nitrogen fixation is anaerobic, so it probably evolved early in the history of the planet, before the atmosphere contained molecular oxygen and before nitrogen-containing compounds were available from decaying organic matter. Nitrogen fixation is brought about by two types of organisms, nonsymbiotic and symbiotic.

Nonsymbiotic (free-living) *nitrogen-fixing bacteria* are found in particularly high concentrations in the *rhizosphere,* the region where the soil and roots make contact—especially in grasslands. Among the nonsymbiotic bacteria that can fix nitrogen are aerobic species such as *Azotobacter* (ä-zō-tō-bak′tėr). These aerobic organisms apparently shield the nitrogenase enzyme from oxygen by, among other things, having a very high rate of oxygen utilization that minimizes the diffusion of oxygen into the cell where the enzyme is located.

Another nonsymbiotic obligate aerobe that fixes nitrogen is *Beijerinckia* (bī-jė-rink′ē-ä). Some anaerobic bacteria, such as certain species of *Clostridium,* also fix nitrogen. The bacterium *Clostridium pasteurianum* (pas-tyėr-ē-ā′num), an obligately anaerobic, nitrogen-fixing microorganism, is a prominent example. Other nonsymbiotic nitrogen-fixing bacteria include certain species of the facultatively anaerobic *Klebsiella, Enterobacter,* and *Bacillus* and the anaerobic photoautotrophic *Rhodospirillum* (rō-dō-spī-ril′um) and *Chlorobium* (klô-rō′bē-um).

There are many species of aerobic, photosynthesizing cyanobacteria that fix nitrogen. Because their energy supply is independent of carbohydrates in soil or water, they are especially useful suppliers of nitrogen to the environment. Cyanobacteria usually carry their nitrogenase enzymes in specialized structures called *heterocysts* that provide anaerobic conditions for fixation.

Most of the nonsymbiotic nitrogen-fixing organisms are capable of fixing large amounts of nitrogen under laboratory conditions. However, in the soil there is usually a shortage of usable carbohydrates to supply the energy needed for the reduction of nitrogen to ammonia, which is then incorporated into protein. Nevertheless, these nitrogen-fixing bacteria make important contributions to the nitrogen economy of such areas as grasslands, forests, and the arctic tundra.

Symbiotic nitrogen-fixing bacteria serve an even more important role in plant growth for crop production. Members of the genera *Rhizobium* (rī-zō′bē-um) and *Bradyrhizobium* infect the roots of leguminous

. . . .

ane content of the air may largely be a result of the decay of the extra plant matter whose growth was stimulated by the increased amount of carbon dioxide. When dead plants decay in anaerobic pockets of organic debris in forests or grasslands, in the sediment of lakes and bays, or in rice paddies, bacterial fermentation breaks down the organic matter, releasing carbon dioxide and hydrogen. Anaerobic archaeobacteria known as *methanogens* (see Figure 10-19) then combine these into methane and water (see the figure).

The amount of methane in the atmosphere has doubled during the last 150 years. If it is true that a large percentage of this increase is a result of the action of methanogenic bacteria on products of organic decay, then the menace of the greenhouse effect is worse than the pessimists thought. The process that at first cheered the optimists—increased plant growth—may be a reason for even darker pessimism.

Methane is generated in the sediments of lakes and bays by the action of methanogens.

plants, such as soybeans, beans, peas, peanuts, alfalfa, and clover. (These agriculturally important plants are only a few of the thousands of known leguminous species, many of which are bushy plants or small trees found in poor soils in many parts of the world.) Rhizobia bacteria species are specifically adapted to particular leguminous plant species. The bacteria attach to the root of the host legume, usually at a root hair (Figure 26-3). In response to the bacterial infection, an indentation forms in the root hair, and an *infection thread* synthesized by the plant passes down the root hair into the root itself. The bacteria follow this infection thread and enter the cells in the root. Inside these cells, the bacteria alter their morphology into larger forms called *bacteroids* that eventually pack the plant cell. The root cells are stimulated by this infection to form a tumorlike nodule of bacteroid-packed cells

(Figure 26-4). Nitrogen is then fixed by a symbiotic process of the plant and the bacteria. The plant furnishes anaerobic conditions and growth nutrients for the bacteria, and the bacteria fix nitrogen to be incorporated into plant protein. Millions of tons of nitrogen are fixed in this way each year.

There are similar examples of symbiotic nitrogen fixation in nonleguminous plants, such as alder trees. These trees are among the first to appear in forests after fires or glaciation. The alder tree is symbiotically infected with an actinomycete (*Frankia*) and forms nitrogen-fixing root nodules. About 50 kg of nitrogen can be fixed each year by the growth of one acre of alder trees; the trees thus make a valuable addition to the forest economy.

Another important contribution to the nitrogen economy of forests is made by **lichens**, which are a

Figure 26-3 Formation of a root nodule.

(a)

(b)

Figure 26-4   Bacterial nitrogen fixation. (a) Soybean root nodules. **(b)** Bacteroids (sliced for this transmission electron micrograph) of the nitrogen-fixing bacterium *Rhizobium japonicum* in vacuoles in an infected cell of a soybean root nodule (×12,000).

symbiosis between a fungus and an alga or cyanobacterium. When one symbiont is a nitrogen-fixing cyanobacterium, the product is fixed nitrogen that eventually enriches the forest soil. Free-living cyanobacteria can fix significant amounts of nitrogen in desert soils after rains and on the surface of arctic tundra soils. Rice paddies can accumulate heavy growths of such nitrogen-fixing organisms. The cyanobacteria also form a symbiosis with a small floating fern, *Azolla*, which grows thickly in rice paddy waters. So much nitrogen is fixed by these organisms that other nitrogenous fertilizers are often unnecessary for rice cultivation.

The Role of Mycorrhizae

A very important contribution to plant growth is made by **mycorrhizae** (*myco* means fungus; *rhiza* means root). There are two types of these fungi: **endomycorrhiza**, also known as **vesicular-arbuscular mycorrhizae**, and **ectomycorrhizae**. Both types function as root hairs on plants—that is, they extend the surface area through which the plant can absorb nutrients, especially phosphorus, which is not very mobile in soil. Vesicular-arbuscular mycorrhizae (Figure 26-5a) form large spores that can be isolated easily from soil by sieving. The hyphae from these germinating spores penetrate into the plant root and form two types of structures, vesicles and arbuscules. **Vesicles** are smooth oval bodies that are most likely storage structures. **Arbuscules** (Figure 26-5b) are formed inside plant cells. Nutrients travel from the soil through fungal hyphae to these arbuscules, which gradually break down and release the nutrients to the plants. Most grasses and other plants are surprisingly dependent on these fungi for proper growth, and their presence is nearly universal in the plant world.

Ectomycorrhizae mainly infect trees such as pine and oak. The fungus forms a mycelial *mantle* over the smaller roots of the tree (Figure 26-5c). Ectomycorrhizae do not form vesicles or arbuscules. Managers of commercial pine tree farms must take care to see that seedlings are inoculated with soil containing effective mycorrhizae. Truffles, known as a culinary delicacy, are ectomycorrhizae. In France, trained pigs are often used to find them by smell and root them up.

Other Biogeochemical Cycles

Microorganisms are also important in cycles involving other elements, such as sulfur. In addition, there are important microbial transformations of potassium, iron, manganese, mercury, selenium, zinc, and other minerals. The various chemical reactions in these cycles are often essential to make the minerals available to plants in soluble form for their metabolism.

(a)

(b)

Figure 26-5 Scanning electron micrographs of mycorrhizae. (a) Endomycorrhiza (*Glomus fasciculatus*) on a soybean root. Note the mycorrhiza's spores and finely branched lateral hyphae surrounding the root (×140). These hyphae increase the plant's absorption of nutrients. **(b)** A fully developed arbuscule of an endomycorrhiza in a plant cell (×900). (The term *arbuscule* means little bush.) As the arbuscule decomposes, it releases accumulated nutrients for the plant. **(c)** The mycelial mantle of a typical ectomycorrhizal fungus surrounding a eucalyptus tree root (×100).

(c)

Degradation of Synthetic Chemicals in the Soil

We seem to take for granted that soil microorganisms will degrade materials entering the soil. Natural organic matter, such as falling leaves or animal residues, are in fact readily degraded. However, in this industrial age, there are many chemicals that do not occur in nature—such as agricultural pesticides and plastics—that enter the soil in large amounts. Many of these synthetic chemicals are highly resistant to degradation by microbial attack.

A well-known example is the insecticide DDT. When DDT was first introduced, its property of *recalcitrance* (resistance to degradation) was considered quite beneficial because one application remained effective in the soil for an extended time. But it was soon found that the chemical tended to accumulate and concentrate in parts of the food chain because of its solubility in fat. Eagles and other predatory birds accumulated DDT from contaminated food and suffered impaired reproductive ability—that is, their eggs had soft shells and broke during incubation.

Not all synthetic chemicals are as recalcitrant as DDT. Some are made up of chemical bonds and subunits that are subject to attack by bacterial enzymes. Small differences in chemical structure can make large differences in biodegradability. The classic example is that of two herbicides—2,4-D, the common chemical used to kill lawn weeds, and 2,4,5-T, which is used to kill shrubs. The addition of a single chlorine atom to the structure of 2,4-D extends its life in soil from a few days to an indefinite period (Figure 26-6).

A growing problem is the leaching into groundwaters of toxic materials that are not biodegradable or that degrade very slowly. The sources of these materials may include landfills, illegal industrial dumps, or pesticides applied to agricultural crops. Once groundwater becomes contaminated, the economic damage can be devastating.

(a) 2,4-D 2,4,5-T

(b)

Figure 26-6 Slight structural differences often affect biodegradability. (a) The structures of herbicides 2,4-D and 2,4,5-T. **(b)** The rates of microbial decomposition of 2,4-D and 2,4,5-T.

Aquatic Microbiology and Sewage Treatment

Aquatic Microorganisms

Aquatic microbiology refers to the study of microorganisms and their activities in natural waters, such as lakes, ponds, streams, rivers, estuaries, and the sea. Large numbers of microorganisms in a body of water generally indicate high nutrient levels in the water. Water contaminated by inflows from sewage systems or from biodegradable industrial organic wastes is relatively high in bacterial counts. Similarly, ocean estuaries (fed by rivers) have higher nutrient levels and therefore have higher microbial counts than other shoreline waters.

In water, particularly in water with low nutrient concentrations, microorganisms tend to grow on stationary surfaces and on particulate matter. In this way, the microorganism has contact with more nutrients than if it were randomly suspended and floating freely with the current. Many bacteria whose main habitat is water have appendages and holdfasts that attach to various surfaces. One example is *Caulobacter* (kô-lō-bak′tėr). (See Figure 10-16a.) Some bacteria also have gas vesicles that they can fill and empty to adjust buoyancy.

Freshwater Microbial Flora

Figure 26-7 shows a typical lake or pond that serves as an example to represent the various zones and the kinds of microbial flora found in a body of fresh water. The **littoral zone** along the shore has considerable rooted vegetation, and light penetrates throughout it. The **limnetic zone** consists of the surface of the open water area away from the shore. The **profundal zone** is the deeper water under the limnetic zone. The **benthic zone** contains the sediment at the bottom.

Microbial populations of freshwater bodies tend to be affected mainly by the availability of oxygen and light. Light is, in many ways, the more important resource because photosynthetic algae are the main source of organic matter, and hence of energy, for the

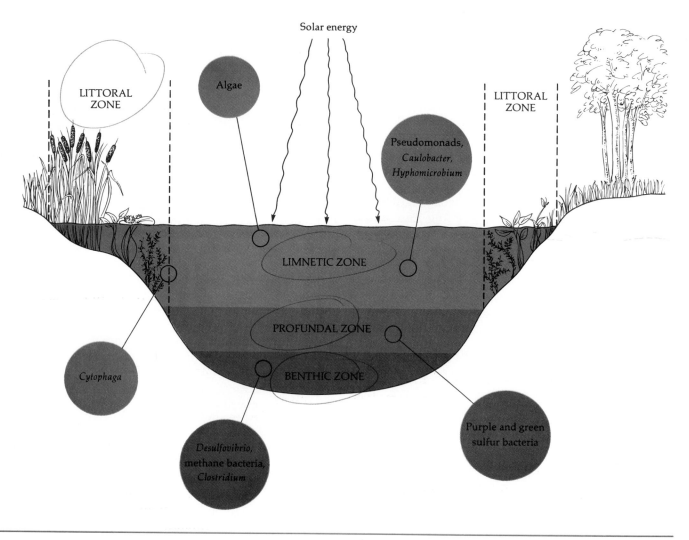

Figure 26-7   The zones of a typical lake or pond and some representative microorganisms of each zone. The microorganisms fill niches that vary in light, nutrients, and oxygen availability.

lake. These organisms are the *primary producers* of a lake that supports a population of bacteria, protozoans, fish, and other aquatic life. Photosynthetic algae are located in the limnetic zone.

Areas of the limnetic zone with sufficient oxygen contain pseudomonads and species of *Cytophaga* (sī-tăf′äg-ä), *Caulobacter*, and *Hyphomicrobium* (hī-fõ-mī-krō′bē-um). Oxygen does not diffuse into water very well, as any aquarium owner knows. Microorganisms growing on nutrients in stagnant water quickly use up the dissolved oxygen in the water. In the oxygenless water, fish die, and odors (from hydrogen sulfide and organic acids, for example) are produced from anaerobic activity. Wave action in shallow

layers, or water movement in rivers, tends to increase the amount of oxygen throughout the water and aid in the growth of aerobic populations of bacteria. Movement thus improves the quality of water and aids in the degradation of polluting nutrients.

Deeper waters of the profundal zone have low oxygen concentrations and less light. Algal growth near the surface often filters the light, and it is not unusual for photosynthetic microorganisms in deeper zones to use different wavelengths of light from those used by surface-layer photosynthesizers (Figure 11-10). The purple and green sulfur bacteria are found in these deeper zones. These bacteria are anaerobic photosynthetic organisms that metabolize hydrogen sulfide to sulfur and sulfate in the bottom sediments of the benthic zone. The sediment also includes bacteria such as *Desulfovibrio* (dē-sul-fō-vib′rē-ō) that use sulfate (SO_4^{2-}) as an electron acceptor and reduce it to hydrogen sulfide (H_2S), which is responsible for the rotten-egg odor of many lake muds.

Methane-producing bacteria are also part of these anaerobic benthic populations. In swamps, marshes, or bottom sediments, they produce methane gas (see the box on p. 670). *Clostridium* species are common in bottom sediments and may include botulism organisms, particularly those causing outbreaks of botulism in water fowl.

Seawater Microbial Flora

The open ocean is relatively high in osmotic pressure, low in nutrients, and quite cold at great depths. The pH also tends to be higher than is optimal for most microorganisms. Bacterial populations in such waters tend to be much lower than in estuaries and in small freshwater bodies fed by rivers and streams. Much of the microscopic life of the ocean is composed of photosynthetic diatoms and other algae. Largely independent of preformed organic nutrient sources, these microbes use energy from photosynthesis and atmospheric carbon dioxide for carbon. These organisms constitute the marine **phytoplankton** community, the basis of the oceanic food chain. Ocean bacteria benefit from the eventual death and decomposition of phytoplankton and also attach to their living bodies. Protozoans in turn feed on bacteria and the smaller phytoplankton. Krill—shrimplike crustaceans—feed on the phytoplankton and in turn are an important food supply of larger sea life. Many fish and whales are also able to feed directly on the phytoplankton.

Microbial luminescence is an interesting minor aspect of deep-sea life. Certain dinoflagellates (see Figure 11-11a) produce light flashes when they are agitated by wave action or by a boat's wake. Many bacteria are also luminescent (see Microview A, Photos 12 and 13), and some have established symbiotic relationships with benthic-dwelling fish. These fish sometimes use the glow of their resident bacteria as an aid in attracting and capturing prey in the complete darkness of the ocean depths. These bioluminescent organisms have an enzyme called luciferase that picks up electrons from flavoproteins in the electron transport chain and then emits some of the electron's energy as a photon of light.

The Role of Microorganisms in Water Quality

Water in nature is seldom totally pure. Even rainfall is contaminated as it falls to Earth. For example, combustion of fossil fuels puts sulfur compounds in the air. These encounter the falling rain and form sulfurous acid—the notorious acid rain. Water quality is measured by the degree of its contamination by soluble chemicals, suspended colloids such as clay particles (turbidity), and microbes.

Water Pollution

The form of water pollution that will be our primary interest is microbial pollution, especially by pathogenic organisms.

Transmission of Infectious Diseases

Water that moves below the ground's surface undergoes a filtering that tends to remove microorganisms. For this reason, water from springs and deep wells is generally of good quality. The most dangerous form of water pollution occurs when human feces enter the water supply. Many diseases are perpetuated by the fecal–oral route of transmission, in which a pathogen shed in human feces contaminates water or food and is later ingested (Chapter 24). Typical of these diseases are typhoid fever and cholera, which are caused by bacteria, and hepatitis A, which is caused by a virus. An especially troublesome waterborne disease is diarrhea caused by ingestion of cysts of the protozoan *Giardia lamblia*. This disease is a problem not only in municipal water systems but also in mountain streams, which are contaminated by beavers and other animals. Mollusks that feed by filtering tend to concentrate viruses and bacteria from water in their tissues, so mollusks from polluted water may be dangerous to eat.

Not all pathogens must be ingested to cause disease. For example, the helminthic disease schistosomiasis is spread among persons who swim or wade in waters contaminated by human wastes. The pathogens are not usually ingested but are in the form of swimming cercaria that bore through the skin.

As good sanitation practices have nearly eliminated a few diseases, such as typhoid fever and cholera in this country, attention has focused on other waterborne diseases. Note in Table 26-2 that most outbreaks of waterborne bacterial diseases are now caused by *Shigella* species, which cause shigellosis

TABLE 26-2 *Outbreaks of Waterborne Diseases in the United States, 1961–1983*

Etiologic Agent	Number of Outbreaks	Number of Cases	Number of Deaths
Bacteria			
Shigella spp.	52	7462	6
Salmonella spp.	37	19,286	3
Campylobacter spp.	5	4773	0
Toxigenic *E. coli*	5	1188	4
Vibrio spp.	1	17	0
Yersinia spp.	1	16	0
Viruses			
Hepatitis A	51	1626	1
Norwalk	16	3973	0
Rotavirus	1	1761	0
Protozoans			
Giardia lamblia	84	22,897	0
Entamoeba histolytica	3	39	2
Chemical			
Inorganic (metals, nitrate)	29	891	0
Organic (pesticides, herbicides)	21	2725	7
Unidentified Agents	266	86,740	0
Total	572	153,394	23

Source: Data adapted from G. F. Craun, "An overview of statistics on acute and chronic water contamination problems." In *Fourth Domestic Water Quality Symposium: Point-of-Use Treatment and Its Implications.* Lisle, Ill.: Water Quality Association, 1985, pp. 5–15; and E. C. Lippy and S. C. Waltrip. "Waterborne disease outbreaks, 1946–1980: a thirty-five year perspective." *American Water Works Association Journal* 76:60–67, 1984.

(bacterial dysentery). There is some anxiety about pathogens such as *Legionella*, the cause of a pneumonia often spread by inhalation of water aerosols. *Mycobacteria* spp., *Pseudomonas* spp., and other opportunistic pathogens that are dangerous to immunosuppressed individuals are being found more and more often in tap water.

Chemical Pollution

Preventing chemical contamination of water is a difficult problem. Industrial and agricultural chemicals leached from the land enter water in great amounts and in forms that are resistant to biodegradation. Many of these chemicals become biologically concentrated in some of the organisms in the food chain.

A striking example of industrial water pollution involved mercury, used in the manufacture of paper. The metallic mercury was allowed to flow into waterways as waste. It was assumed that the mercury was inert and would remain segregated in the sediments. However, bacteria in the sediments converted the mercury into a soluble chemical compound, methyl mercury, which was then taken up by fish and invertebrates in the waters. When such seafood is a substantial part of the human diet, the mercury concentrations can accumulate with devastating effects on the nervous system.

Another example of chemical pollution is the synthetic detergents developed immediately after World War II. These rapidly replaced many of the soaps then in use. Because these new detergents were not biodegradable, they rapidly accumulated in the waterways. In some rivers, large rafts of detergent suds could be seen traveling downstream; in some cities, a small head of bubbles might appear on the surface of a glass of water. These detergents were replaced in 1964 by new biodegradable formulations.

The new detergents, however, brought new problems. Substantial amounts of phosphates were added to many of these detergents to improve their effectiveness. Unfortunately, phosphates pass virtually unchanged through most sewage treatment systems and can cause **eutrophication** (*eu* means well; *troph* means nourish) of lakes and streams—an overabundance of nutrients resulting in overgrowth of algae or cyanobacteria and the eventual death of other organisms.

To understand how phosphate pollution leads to entrophication, we must first consider the nutritional requirements of algae and cyanobacteria. These organisms get their energy from sunlight and their carbon from carbon dioxide dissolved in the water. Most waters contain adequate amounts of the minerals required for algal growth, except for nitrogen and phosphorus. Both of these nutrients can enter water from domestic, farm, and industrial wastes when waste treatment is absent or inefficient. These additional nutrients cause dense aquatic growths called **blooms**. Since many cyanobacteria are able to fix nitrogen from the atmosphere, these photosynthesizing organisms require only traces of phosphorus to initiate blooms. Once eutrophication results in blooms of algae or cyanobacteria, the eventual effect is the same as the addition of equivalent amounts of organic matter. But in this case, blooms generate additional organic matter by photosynthesis and nitrogen fixation from the atmosphere. In the short run, these algae and cyanobacteria produce oxygen. However, they eventually die and are degraded by bacteria. During the degradation process, the oxygen in the water is used up, which may kill the fish. Undegraded remnants of organic matter settle to the bottom and hasten the filling of the lake.

Municipal wastes containing detergent are likely to be the main source of phosphates in lakes and streams because soil phosphorus is relatively insoluble and tends to be retained in soil. Because of this, phosphate-containing detergents are banned in some localities.

Coal-mining wastes, particularly in the eastern United States, are very high in sulfur content, mostly iron sulfide (FeS_2). In the process of obtaining energy from the oxidation of the ferrous ion (Fe^{2+}), bacteria such as *Thiobacillus ferrooxidans* (fer-rō-oks′i-danz) convert the sulfide into sulfate. (See the box in Chapter 27, p. 708.) The sulfate enters streams as sulfuric acid, which lowers the pH of the water and damages aquatic life. The low pH also promotes the formation of insoluble iron hydroxides, which form the yellow precipitates often seen clouding such polluted waters.

Tests for Water Purity

Historically, most of our concern about water purity has been related to the transmission of disease. Therefore, tests have been developed to determine the safety of water; many of these tests are also applicable to foods.

It is not practical, however, to look only for pathogens in water supplies. For one thing, if we were to find the pathogen causing typhoid or cholera in the water system, the discovery would already be too late to prevent an outbreak of the disease. Moreover, such pathogens would probably be present only in small numbers and might not be included in tested samples.

The tests for water safety in use today are aimed instead at detecting particular **indicator organisms**.

There are several criteria for an indicator organism. The most important criterion is that the organism be consistently present in human feces in substantial numbers so that its detection will be a good indication that human wastes are entering the water. The indicator organisms should also survive in the water at least as well as the pathogenic organisms would. The indicator organisms must also be detectable by simple tests that can be carried out by persons with relatively little training in microbiology.

In the United States, the usual indicator organisms are the *coliform* bacteria. Coliforms are defined as aerobic or facultatively anaerobic, gram-negative, nonendospore-forming, rod-shaped bacteria that ferment lactose to form gas within 48 hours of being placed in lactose broth at 35°C. Because some coliforms are not solely enteric bacteria but are more commonly found in plant and soil samples, many standards for food and water specify the determination of *fecal coliforms*. The predominant fecal coliform is *E. coli*, which constitutes a large proportion of the human intestinal population. There are specialized tests to distinguish between fecal coliforms and nonfecal coliforms. It is important to note that coliforms are not themselves pathogenic under normal conditions, although they can cause diarrhea and opportunistic urinary tract infections.

The United States Environmental Protection Agency Drinking Water Standards specify the minimum number of water samples to be examined each month and the maximum number of coliform organisms permitted in each 100 ml of water. Figure 6-16 shows the most probable number (MPN) method for determining the number of bacteria in a 100-ml sample; Appendix D provides the MPN table used for calculations. For detection and enumeration of coliforms, selective and differential media are required, and one of two specialized methods is used. These are summarized in Figure 26-8.

Coliforms have been very useful as indicator organisms in water sanitation, but they have limitations. One problem has been the growth of coliform bacteria in biofilms on the inner surfaces of water pipes. Their presence in tap water has led to a number of community orders to boil water even though these coliforms are not necessarily related to fecal contamination and there is no evidence that they have a significant effect on public health. A more serious problem is that several pathogens are even more resistant than coliforms to chemical disinfection—in their order of increasing resistance to chlorination,

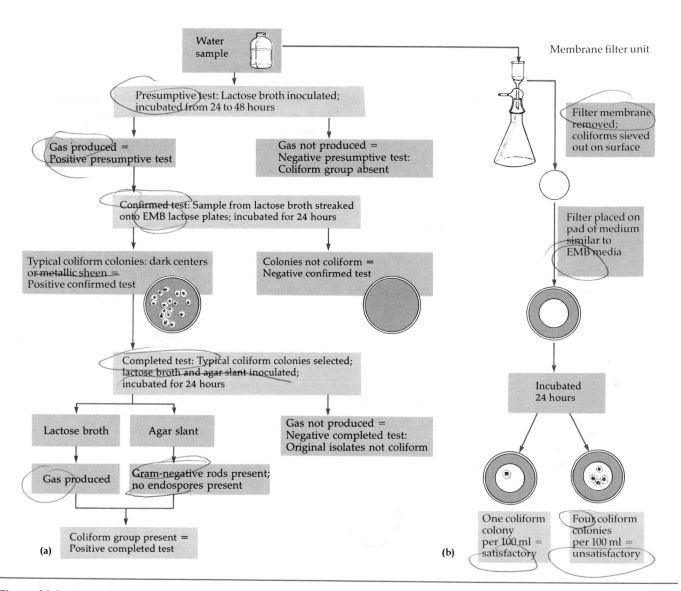

Figure 26-8   Analysis of drinking water for coliforms. (a) Note how the series of tests uses the characteristics associated with coliforms—the presence of gas from lactose fermentation and gram-negative rods. Coliform numbers are estimated by inoculating a number of tubes of lactose broth. The percentage of tubes positive for coliforms is a statistical indication of the number of coliforms in the sample. **(b)** Most laboratories now use the membrane filtration method for coliforms. The coliform colonies have a distinctive appearance (dark centers or metallic sheen).

coliforms, viruses, and protozoan cysts. Using sophisticated methods of detecting viruses, it has been found that chemically disinfected water samples that are free of coliforms are often contaminated with enteric viruses. The cysts of *G. lamblia* are so resistant to chlorination that eliminating them by this method is probably impractical. Mechanical methods are probably necessary, such as filtration and flocculation to remove colloidal particles.

Concern for these problems has led to recommendations for revision of drinking water regulations. These call for a dual barrier treatment of all drinking water. Surface waters should be treated first by flocculation and then by sand or diatomaceous earth filtration to remove cysts and turbidity; these processes would remove most bacteria and viruses as well. This barrier treatment would be followed by chemical disinfection to inactivate any remaining pathogens. These purification methods are described in the next section.

Water Treatment

When water is obtained from uncontaminated reservoirs fed by clear mountain streams or from deep wells, it requires minimal treatment to make it safe to drink. Many cities, however, obtain their water from badly polluted sources, such as rivers that have received municipal and industrial wastes upstream. Very turbid (cloudy) water is allowed to stand in a holding reservoir for a period of time to allow as much particulate suspended matter as possible to settle out (Figure 26-9). The water then undergoes **flocculation treatment**, that is, removal of colloidal materials such as clay, which would otherwise remain in suspension indefinitely. A flocculant chemical, such as aluminum potassium sulfate (alum), forms aggregations of fine suspended particles called floc. As these aggregations slowly settle out, they entrap colloidal material and carry it to the bottom. Large numbers of viruses and bacteria are also removed by this treatment. Alum was used to clear muddy river water during the first half of the nineteenth century in the military forts of the American West, long before the germ theory of disease was developed. There were also observations at the time in European cities that people who used water filtered through sand to remove turbidity did not have the same incidence of cholera during outbreaks.

After flocculation treatment, water is passed through beds of sand or diatomaceous earth to accomplish **sand filtration**. As we mentioned previously, some protozoan cysts, such as those of *G. lamblia*, appear to be removed from water only by such filtration treatment. The microorganisms are trapped mostly by surface adsorption in the sand beds. Bacteria do not penetrate the tortuous routing of the sand beds, even though the openings might be larger than the organisms that are filtered out. These sand filters are periodically backflushed to clear them of accumulations. Water systems of cities that have an exceptional concern for toxic chemicals supplement sand filtration with filters of activated charcoal (carbon). Charcoal has the advantage of removing not only particulate matter but also some organic chemical pollutants dissolved in solution.

Before entering the municipal distribution system, the filtered water is chlorinated. Because organic matter neutralizes chlorine, the plant operators must pay constant attention to maintaining effective levels of chlorine. There has been some concern that chlorine itself might be a health hazard, that it might react with organic contaminants of the water to form carcinogenic compounds. At present, this possibility is considered minor as compared with the proven usefulness of chlorination of water.

As we noted in Chapter 7, one substitute for chlorination is ozone treatment. Ozone (O_3) is a highly reactive form of oxygen that is formed by electrical spark discharges and ultraviolet light. (The fresh odor of air following an electrical storm or around an ultraviolet light bulb is from ozone.) Ozone for water treatment is generated electrically at the site of treatment. Use of ultraviolet light is also a possible alternative to chemical disinfection. Arrays of ultraviolet tube lamps are arranged in quartz tubes so that water flows close to the lamps. This is necessary because of the low penetrating power of ultraviolet radiation.

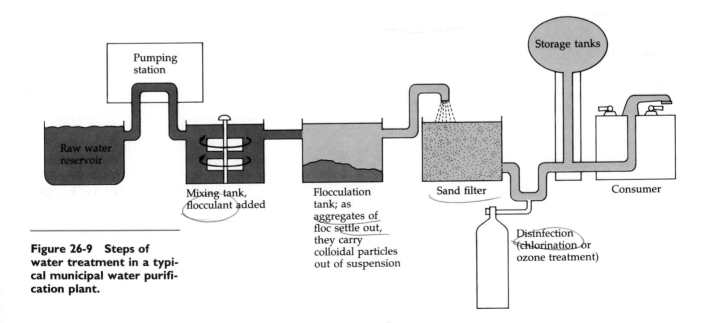

Figure 26-9 Steps of water treatment in a typical municipal water purification plant.

Pumping station

Storage tanks

Raw water reservoir

Mixing tank, flocculant added

Flocculation tank; as aggregates of floc settle out, they carry colloidal particles out of suspension

Sand filter

Consumer

Disinfection (chlorination or ozone treatment)

Sewage Treatment

After water has been used, it becomes sewage. Sewage includes all the water from a household that is used for washing, as well as toilet wastes. Rainwater flowing into street drains and some industrial wastes enter the sewage systems in some cities. Sewage is mostly water and contains little particulate matter, perhaps only about 0.03%. Even so, in large cities, this solid portion of sewage can total more than 1000 tons of solid material per day.

Until environmental awareness intensified, a surprising number of large cities in this country had only rudimentary sewage treatment systems or no system at all. Raw sewage, untreated or nearly so, was simply discharged into rivers or oceans. A flowing, well-aerated stream is capable of considerable self-purification. Therefore, until increases in populations and their wastes exceeded this capability, casual treatment of municipal wastes caused little complaint. In the United States, most methods of simple discharge have been improved. However, at this time, 90% of the 125 largest cities on the Mediterranean Sea release all sewage into this body of water with no treatment at all.

Primary Treatment

The usual first step in sewage treatment is called **primary treatment** (Figure 26-10). In this process, incoming sewage receives preliminary treatment—large floating materials are screened out, the sewage is allowed to flow through settling chambers so that

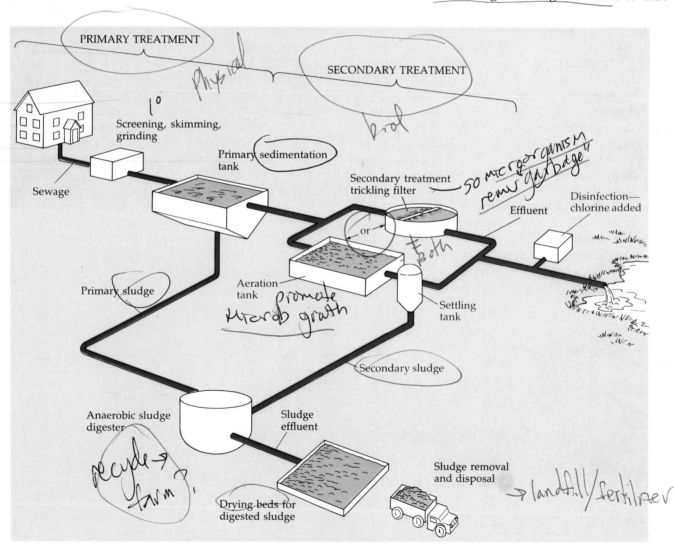

Figure 26-10 Steps of typical sewage waste treatment. A particular system would use either activated sludge aeration tanks (see also Figure 26-11) or trickling filters (see also Figure 26-13), not both, as shown in this figure. The sludge (see also Figure 26-14) is disposed of in landfills or agricultural land.

sand and similarly gritty material can be removed, skimmers remove floating oil and grease, and floating debris is shredded and ground. After this step, the sewage passes through sedimentation tanks, where solid matter settles out. (The design of these primary settling tanks varies.) Sewage solids collecting on the bottom are called **sludge**; sludge at this stage is called primary sludge. From 40% to 60% of suspended solids are removed from sewage by this settling treatment, and flocculating chemicals that increase the removal of solids are sometimes added at this stage. Biological activity is not particularly important in primary treatment, although some digestion of sludge and dissolved organic matter can occur during long holding times. The sludge is removed on either a continuous or intermittent basis, and the effluent (the liquid flowing out) then undergoes secondary treatment.

Biochemical Oxygen Demand

Primary treatment removes approximately 25% to 35% of the **biochemical oxygen demand (BOD)** of the sewage. An important concept in sewage treatment and in the general ecology of waste treatment, BOD is a measure of the biologically degradable organic matter in water. BOD is determined by the amount of oxygen required by bacteria to metabolize the organic matter. The classic method of measurement is to use special bottles with airtight stoppers. Each bottle is first filled with the test water or dilutions of the test water. The water is initially aerated to provide a relatively high level of dissolved oxygen and is seeded with bacteria if necessary. The filled bottles are then incubated in the dark for five days at 20°C, and the decrease in dissolved oxygen is determined by a chemical or electronic testing method. The more oxygen that is used up as the bacteria degrade the organic matter in the sample, the greater the BOD—which is usually expressed in milligrams of oxygen per liter of water. The amount of oxygen that normally can be dissolved in water is only about 10 mg/l. Typical BOD values of waste water may be twenty times this amount. If this waste water enters a lake, for example, bacteria in the lake begin to consume the organic matter responsible for the high BOD, rapidly depleting the oxygen in the lake water.

Secondary Treatment

After primary treatment, the greater part of the BOD remaining in the sewage is in the form of dissolved organic matter. **Secondary treatment**, which is primarily biological, is designed to remove most of this organic matter and reduce the BOD. In this process, the sewage undergoes strong aeration to encourage the growth of aerobic bacteria and other microorganisms that oxidize the dissolved organic matter to carbon dioxide and water. Two commonly used methods of secondary treatment are activated sludge systems and trickling filters.

In the aeration tanks of the **activated sludge system**, air or pure oxygen is added to the effluent from primary treatment (Figure 26-11). The sludge in the effluent contains large numbers of metabolizing bacteria, together with yeasts, molds, and protozoans. An especially important ingredient of the sludge are species of *Zoogloea* (zō-ō-glē'ä) bacteria, which form

(a)

(b)

Figure 26-11 Activated sludge system of secondary treatment. (a) An aeration tank. Note that the surface is frothing from aeration. **(b)** Diagram of an activated sludge system.

Figure 26-12 ***Zoogloea*, the organism that contributes to floc formation in activated sludge systems. (a)** *Zoogloea ramigera* cell with single polar flagellum (×12,000). **(b)** Particle of floc formed by *Z. ramigera* (×560). Note the cells embedded in gelatinous matter synthesized by the bacteria.

flocculant masses (floc) in the aeration tanks (Figure 26-12). The activity of these aerobic microorganisms oxidizes much of the effluent's organic matter into carbon dioxide and water. When the aeration phase is completed, the floc (secondary sludge) is allowed to settle to the bottom, just as the primary sludge settles in primary treatment.

Soluble organic matter in the sewage is adsorbed onto the floc and is incorporated into microorganisms in the floc. As the floc settles out, this organic matter is removed with the floc and is subsequently treated in an anaerobic sludge digester. More organic matter is probably removed by this process than by the relatively short-term aerobic oxidation.

Most of the settled sludge is removed for treatment in an anaerobic sludge digester; some of the sludge is recycled to the activated sludge tanks as a starter culture for the next sewage batch. The effluent water is sent on for final treatment. Occasionally, when aeration is stopped, the sludge will float rather than settle out; this phenomenon is called *bulking*. When this happens, the organic matter in the floc flows out with the discharged effluent and often causes serious problems of local pollution. A considerable amount of research has been devoted to the causes of bulking and its possible prevention. It is apparently caused by the growth of filamentous bacteria of various types; the sheathed bacterium *Sphaerotilus natans* (sfe-rä′ti-lus nā′tans) is often mentioned as the primary offender. Activated sludge systems are quite efficient: they remove from 75% to 95% of the BOD from sewage.

Trickling filters are the other commonly used method of secondary treatment. In this method, the sewage is sprayed over a bed of rocks or molded plastic (Figure 26-13). The rocks or other components of the bed must be large enough so that air penetrates to the bottom but small enough to maximize the surface area available for microbial activity. Actually, there is no filtering action. A slimy, gelatinous film of aerobic microorganisms grows on the rocks' surfaces. In many ways, this film is functionally similar to the organisms found in activated sludge systems. Because air circulates throughout the rock bed, these aerobic microorganisms in the slime layer are able to oxidize much of the organic matter trickling over the surfaces into carbon dioxide and water. Trickling filters remove from 80% to 85% of the BOD, so they are less efficient than activated sludge systems. However, trickling filters are usually less troublesome to operate and have fewer problems from overloads or toxic sewage.

Treated sewage is disinfected, usually by chlorination, before being discharged. Not all microbes are necessarily killed by chlorination. The number of coliforms allowed in the effluent depends on where it is discharged. The discharge is usually into an ocean or into flowing streams, although spray irrigation fields are sometimes used to avoid phosphorus and bacterial contamination of waterways. For ocean discharge, 240 coliforms per 100 ml of water may be allowed, but in a smaller body of water coliforms might have to be reduced to 23 per 100 ml.

In cities with limited freshwater supplies, there is already some recycling of sewage into drinking water. This is safe and practical and will probably become commonplace in the future.

Sludge Digestion

Primary sludge accumulates in primary sedimentation tanks; sludge also accumulates in activated sludge and in trickling filter secondary treatments. For further treatment, these sludges are often pumped

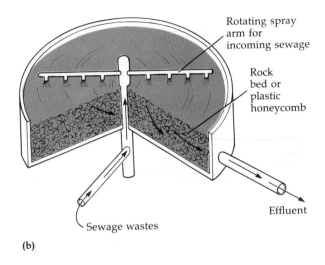

(a)

(b)

Figure 26-13 Trickling filter method of secondary treatment. (a) The sewage is sprayed from the system of rotating pipes onto a bed of about 6 meters of plastic honeycomb designed to have a maximum surface area and allow oxygen to penetrate deeply into the bed. Microorganisms grow on the enormous surface area of the plastic filling (some systems use rocks) and aerobically metabolize the organic matter in the sewage trickling down through the bed. **(b)** Diagram of a trickling filter system.

to **anaerobic sludge digesters** (Figure 26-14). The process of sludge digestion is carried out in large tanks, from which oxygen is almost completely excluded. In secondary treatment, emphasis is placed on the maintenance of aerobic conditions so that organic matter is converted to carbon dioxide, water, and solids that can settle out. By comparison, anaerobic fer-

mentations are metabolically inefficient, and the microorganisms leave large amounts of only partially digested organic materials in the form of fatty acids, alcohols, and similar products that retain much of their original BOD.

An anaerobic sludge digester is designed to encourage the growth of anaerobic bacteria, especially

(a)

Figure 26-14 Sludge digestion. (a) An anaerobic sludge digester at a California treatment plant. Much of a typical digester is below ground level. In colder climates, the above-ground part of the tank is insulated by mounded earth. Methane from such a digester is often used to run pumps or heaters in the treatment plant. **(b)** Diagram of a sludge digester. The scum layer and supernatant layers are low in solids and are recirculated through secondary treatment.

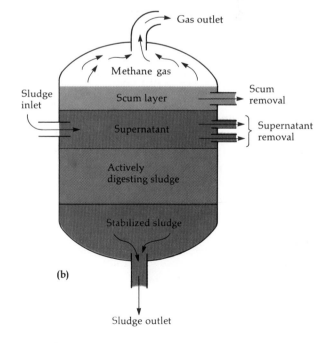

(b)

methane-producing bacteria that decrease organic solids by degrading them to soluble substances and gases, mostly methane (60% to 70%) and carbon dioxide (20% to 30%). The methane is routinely used as a fuel for heating the digester and is also frequently used to run power equipment in the plant.

There are essentially three stages in the activity of an anaerobic sludge digester. The first stage is the production of carbon dioxide and organic acids from anaerobic fermentation of the sludge by various anaerobic and facultatively anaerobic microorganisms. In the second stage, the organic acids are metabolized to form considerable hydrogen and carbon dioxide as well as organic acids, such as acetate. These products are the raw materials for a third stage, in which the methane-producing bacteria produce methane (CH_4). Most of the methane is derived from the energy-yielding reduction of carbon dioxide by hydrogen gas:

$$CO_2 + 4H_2 \longrightarrow CH_4 + 2H_2O$$

Other methane-producing microbes split acetic acid (CH_3COOH) to yield methane and carbon dioxide:

$$CH_3COOH \longrightarrow CH_4 + CO_2$$

Methane and carbon dioxide are relatively innocuous end-products, comparable to the carbon dioxide and water from aerobic treatment. However, considerable amounts of undigested sludge still remain, although it is relatively stable and inert. To reduce its volume, this sludge is pumped to shallow drying beds or filters. It is then carried away for disposal. It can be used for landfill or as a soil conditioner, or it can be incinerated. Sludge has about one-fifth of the value of normal commercial lawn fertilizers but has desirable soil-conditioning qualities, much as soil humus does.

Septic Tanks

Homes and businesses in areas of low population density that are not connected to municipal sewage systems often use a **septic tank** (Figure 26-15), a device whose operation is similar in principle to primary treatment. Sewage enters a holding tank, and suspended solids settle out. The sludge in the tank must be pumped out periodically and disposed of. The effluent flows through a system of perforated piping into a leaching (soil-drainage) field. The effluent entering the soil is decomposed by soil microorganisms. These systems work well when not overloaded and when the drainage system is properly sized to the load and soil type. Heavy clay soils require extensive drainage systems because of the soil's poor permeability. The high porosity of sandy soils can allow chemical or bacterial pollution of nearby water supplies.

Oxidation Ponds

Many small communities and many industries use **oxidation ponds**, also called **lagoons** or **stabilization ponds**, for water treatment. These are inexpensive to build and operate but require considerable land. Designs vary, but most incorporate two stages. The first stage is analogous to primary treatment—the sewage pond is deep enough that conditions are almost entirely anaerobic. Sludge settles out in this stage. In the second stage, which corresponds to secondary treatment, effluent is pumped into an adjoining pond or system of ponds that are shallow enough to be aerated by wave action. Because it is difficult to maintain aerobic conditions for bacterial growth in ponds with so much organic matter, the growth of algae is encouraged to produce oxygen. Bacterial action in decomposing the organic matter in the

(a)

(b)

Figure 26-15 Septic tank system. (a) Overall plan. Solids settle out in the septic tank as sludge—most soluble organic matter—is disposed of by percolation into soil. **(b)** The inside of a septic tank. The sludge is pumped out at periodic intervals.

wastes generates carbon dioxide. Algae, which use carbon dioxide in their photosynthetic metabolism, grow and produce oxygen, which in turn encourages activity of aerobic microorganisms in the sewage. Considerable amounts of organic matter in the form of algae accumulate, but this is not a problem because the oxidation pond, unlike a lake, already has a large nutrient load.

Some small sewage-producing operations, such as isolated campgrounds or highway rest stop areas, use an **oxidation ditch** for sewage treatment. In this method, a small oval channel in the shape of a race track is filled with sewage water. A paddle wheel similar to that on a Mississippi steamboat propels the water in a self-contained flowing stream aerated enough to oxidize the wastes.

Tertiary Treatment

As we have seen, primary and secondary treatments of sewage do not remove all the biologically degradable organic matter. Amounts of organic matter that are not excessive can be added to a flowing stream without causing a serious problem. Eventually, however, the pressures of increased population might increase wastes beyond a body of water's carrying capacity, and additional treatments might be required. Even now, primary and secondary treatments are inadequate in certain situations, such as when the effluent is discharged into small streams or recreational lakes. Some communities have therefore developed tertiary treatment plants. Lake Tahoe in the Sierra Nevada, surrounded by extensive development, is the site of one of the best-known tertiary systems.

The effluent from secondary treatment plants contains some residual BOD. It also contains about 50% of the original nitrogen and 70% of the original phosphorus—which can greatly affect a lake's ecosystem. Tertiary treatment is designed to remove essentially all the BOD, nitrogen, and phosphorus. Tertiary treatment depends less on biological treatment than on physical and chemical treatments. Some systems encourage denitrifying bacteria to form volatile nitrogen gas. Nitrogen is converted to ammonia and evaporates into the air in stripping towers. Phosphorus is precipitated out by combining with such chemicals as lime, alum, and ferric chloride. Filters of fine sands and activated charcoal remove small particulate matter and dissolved chemicals. Finally, chlorine is added to the purified water to kill or inhibit any remaining microorganisms and to oxidize any remaining odor-producing substances.

Tertiary treatment provides water that is suitable for drinking, but the process is extremely costly. Secondary treatment is less costly, but water that has undergone only secondary treatment contains many nutrients that pollute water. Much work is being done at present to design secondary treatment plants in which the effluent can be used for irrigation. This design would eliminate a source of water pollution, provide nutrients for plant growth, and reduce the demand on already scarce water supplies. The soil would act as a trickling filter to remove chemicals and microorganisms before the water reaches groundwater and surface water supplies. Even now, waste water with coliform counts below 2.2/100 ml is being used to irrigate food crops, orchards, and pastures; water with coliform counts below 23/100 ml is being used to irrigate landscaping and recreational areas.

Study Outline

Water and Gases   (p. 665)

1. Microbial life in soil depends on the availability of water and oxygen.

Organisms   (pp. 665–667)

1. Microorganisms in the soil decompose organic matter and transform nitrogen- and sulfur-containing compounds into usable forms.
2. Bacteria are the most numerous organisms in the soil.
3. The characteristic musty odor of soil is from geosmin produced by actinomycetes.
4. Soil is not a reservoir for human pathogens, except for some endospore-forming bacteria.
5. Insect and plant pathogens are found in soil; human pathogens find soil to be a hostile environment.

Microorganisms and Biogeochemical Cycles   (pp. 667–674)

1. In biogeochemical cycles, certain chemical elements are recycled.
2. Microorganisms are essential to the continuation of biogeochemical cycles.

The Carbon Cycle   (pp. 667–668)

1. CO_2 is fixed into organic compounds by photoautotrophs.
2. These organic compounds provide nutrients for chemoheterotrophs.
3. Chemoheterotrophs release CO_2 that is then used by photoautotrophs.

The Nitrogen Cycle   (pp. 668–673)

1. Microorganisms decompose proteins from dead cells and release amino acids.
2. Ammonia is liberated by microbial ammonification of the amino acids.
3. Ammonia is oxidized to nitrates for energy by nitrifying bacteria.
4. Denitrifying bacteria reduce nitrates to molecular nitrogen (N_2).
5. N_2 is converted into ammonia by nitrogen-fixing bacteria.
6. Nitrogen-fixing bacteria include free-living genera such as *Azotobacter*, cyanobacteria, and the symbiotic bacteria *Rhizobium* and *Frankia*.
7. Ammonium and nitrate are used by bacteria and plants to synthesize amino acids.
8. Symbiotic fungi called mycorrhizae live in and on plant roots.
9. Mycorrhizae increase the surface area and nutrient absorption of the plant.

Other Biogeochemical Cycles   (p. 673)

1. Microorganisms are also involved in the transformation of other elements, including sulfur, potassium, iron, manganese, mercury, zinc, and selenium.

2. These reactions make minerals available in soluble form to plants for their metabolism.

Degradation of Synthetic Chemicals in the Soil   (pp. 674–675)

1. Many synthetic chemicals, such as pesticides and plastics, are recalcitrant (resistant to degradation).
2. Recalcitrance is based on the nature of the chemical bonding.

Aquatic Microbiology and Sewage Treatment   (pp. 675–687)

Aquatic Microorganisms   (p. 675)

1. The study of microorganisms and their activities in natural waters is called aquatic microbiology.
2. Natural waters include lakes, ponds, streams, rivers, estuaries, and the sea.
3. The concentration of bacteria in water is proportional to the amount of organic material in the water.
4. Most aquatic bacteria tend to grow on surfaces rather than in a free-floating state.

Freshwater Microbial Flora   (pp. 675–677)

1. Numbers and locations of microbial flora depend on the availability of oxygen and light.
2. Photosynthetic algae are the primary producers of a lake. They are found in the limnetic zone.
3. Pseudomonads, *Cytophaga*, *Caulobacter*, and *Hyphomicrobium* are found in the limnetic zone, where oxygen is abundant.
4. Microbial growth in stagnant water uses available oxygen and can cause odors and the death of fish.
5. The amount of dissolved oxygen is increased by wave action.
6. Purple and green sulfur bacteria are found in the benthic zone, which contains light and H_2S but no oxygen.
7. *Desulfovibrio* reduces SO_4^{2-} to H_2S in benthic mud.
8. Methane-producing bacteria are also found in the benthic zone.

Seawater Microbial Flora   (p. 677)

1. The open ocean is not a favorable environment for most microorganisms because of its high osmotic pressure, low nutrients, and high pH.
2. Phytoplankton, consisting mainly of diatoms, are the primary producers of the open ocean.
3. Some algae and bacteria are bioluminescent. They possess the enzyme luciferase, which can emit light.

The Role of Microorganisms in Water Quality (pp. 677–687)

Water Pollution (pp. 677–679)

1. Microorganisms are filtered from water that percolates into groundwater supplies.
2. Some pathogenic microorganisms are transmitted to humans in water supplies.
3. Recalcitrant chemical pollutants may be concentrated in animals in an aquatic food chain.
4. Mercury is metabolized by certain bacteria into a soluble compound that is concentrated in animals.
5. Nutrients such as phosphates cause eutrophication of aquatic ecosystems.
6. Eutrophication means well nourished. It is the result of the addition of pollutants or natural nutrients.
7. The metabolic activities of *T. ferrooxidans* can be used to recover uranium ore.

Tests for Water Purity (pp. 679–680)

1. Tests for the bacteriological quality of water are based on the presence of indicator organisms.
2. The most common indicator organisms are coliforms.
3. Coliforms are aerobic or facultatively anaerobic, gram-negative, nonendospore-forming rods that ferment lactose with the production of acid and gas within 48 hours of being placed in a medium at 35°C.
4. Fecal coliforms, predominantly *E. coli*, are used to indicate the presence of human feces.

Water Treatment (p. 681)

1. Drinking water is held in a holding reservoir long enough that suspended matter settles.
2. Flocculation treatment uses a chemical such as alum to coalesce and then settle colloidal material.
3. Sand filtration removes bacteria, viruses, and protozoan cysts.
4. Drinking water is disinfected with chlorine to kill remaining pathogenic bacteria.

Sewage Treatment (pp. 682–687)

1. Domestic wastewater is called sewage.
2. It includes household water, toilet wastes, industrial wastes, and rainwater.

Primary Treatment (pp. 682–683)

1. Primary sewage treatment is the removal of solid matter called sludge.
2. Biological activity is not very important in primary treatment.

Biochemical Oxygen Demand (p. 683)

1. Primary treatment removes approximately 25% to 35% of the biochemical oxygen demand (BOD) of the sewage.
2. BOD is a measure of the biologically degradable organic matter in water.
3. It is determined by the amount of oxygen bacteria require to degrade the organic matter.

Secondary Treatment (pp. 683–684)

1. Secondary treatment is the biological degradation of organic matter in sewage after primary treatment.
2. Activating sludge and trickling filters are methods of secondary treatment.
3. Microorganisms degrade the organic matter aerobically.
4. Secondary treatment removes up to 95% of the BOD.

Sludge Digestion (pp. 684–686)

1. Sludge is placed in an anaerobic sludge digester; bacteria degrade organic matter and produce simpler organic compounds, methane, and CO_2.
2. The methane produced in the digester is used to heat the digester and to operate other equipment.
3. Excess sludge is periodically removed from the digester, dried, and disposed of as landfill or as soil conditioner or incinerated.

Septic Tanks (p. 686)

1. Septic tanks can be used in rural areas to provide primary treatment of sewage.
2. They require a large leaching field for the effluent.

Oxidation Ponds (pp. 686–687)

1. Small communities can use oxidation ponds for secondary treatment.
2. These require a large area in which to build an artificial lake.

Tertiary Treatment (p. 687)

1. Tertiary treatment employs physical filtration and chemical precipitation to remove all the BOD, nitrogen, and phosphorus from water.
2. Tertiary treatment provides drinkable water, whereas secondary treatment provides water usable only for irrigation.

Study Questions

Review

1. Write a one-sentence description of each of the following soil constituents: solid inorganic matter, water, gases, organic matter.

2. The precursor to coal is peat found in bogs. Why does peat accumulate in bogs?

3. The metabolic activities of microorganisms often produce acids. Why would microbial growth increase the solid inorganic matter in soil?

4. Compare and contrast humus and recalcitrant chemicals.

5. Diagram the carbon cycle in the presence and absence of oxygen. Name at least one microorganism that is involved at each step.

6. Fill in the following table with the information provided below.

Process	Chemical Reactions	Microorganisms
Ammonification		
Nitrification		
Denitrification		
Nitrogen fixation		

Choices:

$$NO_3^- \longrightarrow N_2$$
$$N_2 \longrightarrow NH_3$$
$$-NH_2 \longrightarrow NH_3$$
$$NH_3 \longrightarrow NO_2^-$$
$$NO_2^- \longrightarrow NO_3^-$$

Bacillus
Rhizobium
Nitrosomonas
Azotobacter
Nitrobacter
Proteolytic bacteria

7. The following organisms have important roles as symbionts with plants and fungi. Describe the symbiotic relationship of each organism with its host: Cyanobacteria, mycorrhizae, *Rhizobium*, *Frankia*.

8. Compare and contrast the physical conditions of the ocean with those of fresh water.

9. Matching.

____ $CO_2 + H_2S \xrightarrow{\text{light}} C_6H_{12}O_6 + S^0$
____ $SO_4^{2-} + H^+ \longrightarrow H_2S$
____ $CO_2 + H_2 \longrightarrow CH_4$

(a) Methane-producing bacteria
(b) *Desulfovibrio*
(c) Photosynthetic bacteria

10. Indicate which reactions in Question 9 require atmospheric oxygen (O_2).

11. Outline the treatment process for drinking water.

12. What is the purpose of a coliform count on water?

13. If coliforms are not normally pathogenic, why are they used as indicator organisms for bacteriological quality?

14. The following processes are used in wastewater treatment. Match the type of treatment with the processes. Each choice can be used once, more than once, or not at all.

Processes	Types of treatment
____ Leaching field	(a) Primary
____ Removal of solids	(b) Secondary
____ Biological degradation	(c) Tertiary
____ Activated sludge	
____ Chemical precipitation of phosphorus	
____ Trickling filter	
____ Results in drinking water	
____ Effluent can be used for irrigation	
____ Produces methane	

15. Define BOD.

16. Why is activated sludge a more efficient means of removing BOD than a sludge digester?

17. Why are septic tanks and oxidation ponds not feasible for large municipalities?

18. Explain the effect of dumping untreated sewage into a pond on the eutrophication of the pond. The effect of sewage that has primary treatment? The effect of sewage that has secondary treatment? Contrast your previous answers with the effect of each type of sewage on a fast-moving river.

Challenge

1. A sewage treatment plant in Santee, California, is a tertiary treatment plant. Outline the flow of water and solids through this plant. What is the final quality of the water with respect to BOD, nitrogen, and phosphorus?

2. Here are the formulas of two detergents that have been manufactured:

$$C—C—C—C—C—C—C—C—C—C— \ldots$$

```
              C
              |
              C       C
              |       |
C — C — C — C — C — C — ...
              |
              C
```

Which of these would be recalcitrant, and which would be readily degraded by microorganisms? (*Hint*: Refer to the degradation of fatty acids in Chapter 5.)

3. What is the MPN of a water sample with these multiple-tube fermentation test results: 10 ml portions: 4 positive for acid and gas; 1 ml portions: 3 positive for acid and gas; 0.1 ml portions: 1 positive. (See MPN table, Appendix D.)

4. Flooding after two weeks of heavy rainfall in Tooele, Utah, preceded the high rate of diarrheal illness shown below. *G. lamblia* was isolated from 26% of the patients. A comparison study of a town 65 miles away revealed that there was diarrheal illness in 2.9% of the 103 persons interviewed. Tooele has a municipal water system and a municipal sewage treatment plant. Explain the probable cause of this epidemic and method(s) of stopping it.

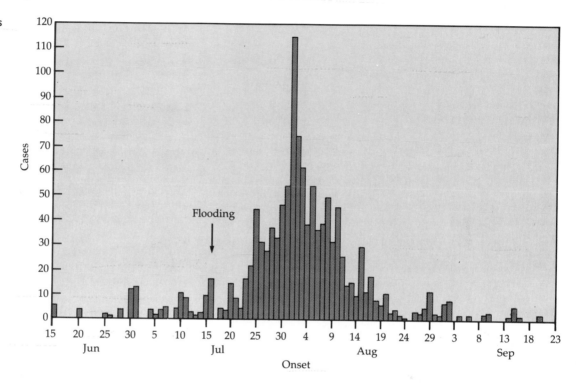

Distribution of cases of diarrheal illness.

Further Reading

Atlas, R. M., and R. Bartha. *Microbial Ecology: Fundamentals and Applications*, 2nd ed. Menlo Park, Calif.: Benjamin/Cummings, 1987. The foremost textbook on the subject of microbial ecology.

AWWA Organisms in Water Committee. "Committee report: microbiological considerations for drinking water regulation revisions." *Journal of the American Water Works Association* 79:81–88, 1987. A discussion of the current status of the microbiology of drinking water and suggested changes for the future.

Brock, T. D. *Thermophilic Microorganisms and Life at High Temperatures*. New York: Springer-Verlag, 1978. Discusses the microorganisms, their metabolism, and growth requirements of the unique environment of the hot springs in Yellowstone National Park.

Childress, J. J., H. Felbeck, and G. N. Somero. "Symbiosis in the deep sea." *Scientific American* 256(5):114–120, May 1987. Describes the chemoautotrophic bacterial symbionts in worms living at hydrothermal vents.

Cloud, P. "The biosphere." *Scientific American* 249(3):176–189, September 1983. A discussion of how life has been shaped and sustained on Earth in soil, water, and the atmosphere.

Dowling, D. N., and W. J. Broughton. "Competition for nodulation of legumes." *Annual Review of Microbiology* 40:131–157, 1986. Describes the effects of the environment, other organisms, and host genes on root-nodule formation.

Erlich, H. L. *Geomicrobiology*. New York: Marcel Dekker, 1981. A textbook on the role of microorganisms in transformations of carbon, phosphorus, mercury, manganese, and other elements.

Margulis, L., D. Chase, and R. Guerero. "Microbial communities." *BioScience* 36:160–170, 1986. A discussion about bacteria in their natural environment. Photographs of living systems show relationships that are not apparent in vitro.

Padwa, D., ed. "Benefits to agriculturists from the studies of microorganisms." *Developments in Industrial Microbiology* 24:19–67, 1983. Papers on technological uses of nitrogen-fixing bacteria, mycorrhizae, yeasts, and agrobacteria.

Shapiro, J. A. "Bacteria as multicellular organisms." *Scientific American* 258(6):82–89, June 1988. Describes colonies of bacterial cells that function as multicellular organisms, such as predatory *Myxococcus* colonies.

Sieburth, J. M. *Sea Microbes*. New York: Oxford University Press, 1979. Discusses microorganisms and their adaptations for living in ocean waters.

Whitely, H. R., and H. E. Schnepf. "The molecular biology of parasporal crystal body formation in *Bacillus thuringiensis*." *Annual Review of Microbiology* 40:549–576, 1985. Describes the types of BT toxins, their insecticidal properties, and insect specificity.

CHAPTER 27

Applied Microbiology and Biotechnology

We will now turn our attention from soil and aquatic microbiology to food and industrial microbiology. This chapter will discuss food spoilage and preservation, foodborne infections and food poisoning, and food production by microorganisms. We will also look at some important aspects of industrial microbiology and the impact of genetic engineering on industrial microbiology.

Food Microbiology

Food Preservation and Spoilage

Modern civilization and its large populations could not be supported without effective methods of food preservation. In fact, civilization arose only after agriculture produced a year-round stable food supply in a single site so that people were able to give up the nomadic hunting cultures. Many of the methods of food preservation used today were probably discovered by chance in centuries past. Primitive people observed that dried meat and salted fish resisted decay. Nomads no doubt observed that soured animal milk resisted further decomposition and was still palatable. Moreover, if the curd of the soured milk was pressed to remove moisture and allowed to ripen (in effect, cheese-making), it was even more effectively preserved and tasted better. Farmers soon learned that if grains were kept dry, they did not become moldy.

All such phenomena are readily understood by anyone familiar with the factors that control the growth of microorganisms (Chapters 6 and 7). Bacteria and fungi require a minimum amount of avail-

Learning Objectives

- Provide a brief history of the development of food preservation.

- Explain why sterilization of canned foods is important.

- Describe thermophilic anaerobic spoilage and flat sour spoilage by mesophilic bacteria.

- Describe low-temperature preservation and aseptic packaging.

- Define pasteurization, and explain why dairy products are pasteurized.

- Provide examples of chemical food preservatives, and explain why they are used.

- Outline at least four beneficial activities of microorganisms in food production.

- Describe the role of microorganisms in the production of single-cell protein (SCP), alternative energy sources, industrial chemicals, and chemotherapeutic agents.

able moisture for growth. Drying food and adding salt or sugar lowers the available moisture and thus prevents spoilage. Acidity created by the natural fermentation of milk and vegetable juices, as in sauerkraut, also prevents the growth of many spoilage bacteria.

These methods of preservation are still in use today. However, a tour of any supermarket will demonstrate that heat sterilization, pasteurization, refrigeration, and freezing are now much more popular methods for the control of food spoilage. These modern methods preserve food nearer its natural state and palatability than the methods that were used in the past.

Canning

Preservation by heat originated in the early 1800s, before microorganisms were known to cause human disease and food spoilage. The technology of canning was promoted by the military. When armies were small professional organizations, they could be supported by the land, with supplements of portable rations of wine, cheese, and dried grains. But with the rise of large armies of citizen soldiers, a development of the French Revolution, this strategy was no longer viable. Thus, the French government offered a prize to anyone who could devise a method of preserving food, particularly meat. In 1810, the prize was won by a confectioner, Nicholas Appert, who showed that food could be preserved if it was sealed in tightly

stoppered containers and boiled for specified periods of time. He developed detailed tables of the boiling times required for different foods and container sizes. As we know today, there are many endospore-forming bacteria that will survive hours of such boiling. Failures were manageably few, however, and were usually attributed to faulty sealing.

The concept of heat preservation was followed quickly by the invention of the metal can. Temperatures above the boiling point of water were introduced into processing by oil or salt baths. The use of pressurized steam in closed containers, which is used today, was introduced late in the nineteenth century, when reliable pressure controls and safety valves were developed. By then, of course, there was more theoretical understanding of microbe control.

Commercial Sterilization

Preserving foods by heating a properly sealed container is not difficult. The problem is to determine and to use the minimum amount of heat necessary to kill spoilage organisms and dangerous microorganisms, such as the endospore-forming *Clostridium botulinum*. Because heating degrades the quality of food, much research was needed to determine the exact heat treatment that would succeed in sterilizing the food but minimize these quality changes.

Canned goods today undergo what is called **commercial sterilization** (Figure 27-1), which is not as rigorous as true sterilization. Commercial sterilization is intended to destroy *C. botulinum* endospores.

Figure 27-1 Industrial canning. (1) The food to be canned is washed, sorted, and blanched. Blanching is a treatment with hot water or live steam that softens the product so that the can will fill better. It also destroys enzymes that might alter the color, flavor, or texture of the product and lowers the microbial count. (2) Cans are filled to capacity, leaving as little dead space as possible. (3) Cans are placed in a steam box, where the heat exhausts (drives out) dissolved air. (4) Sealing. (5) Sterilization is done in a large retort with steam under pressure. (6) Cooling is done with a water spray or submergence. (7) Cans are then labeled, stored, and delivered.

If this is accomplished, then any other significant spoilage or pathogenic bacterium will also be destroyed. To ensure complete sterilization, enough heat is applied for the **12D treatment** (12-decimal reductions), by which a theoretical population of botulism endospores would be decreased by 12 logarithmic cycles. What this means is that if there were 10^{12} (1,000,000,000,000) botulism endospores in a can, after treatment there would be only one survivor. Because 10^{12} is an improbably large population, this treatment is considered quite safe. Certain thermophilic endospore-forming bacteria have endospores that are more resistant to heat treatment than *C. botulinum*. However, these bacteria are obligate thermophiles and generally remain dormant at temperatures lower than about 45°C (112°F). Therefore, they are not a problem at normal storage temperatures.

Spoilage of Canned Food

If canned foods are incubated at high temperatures, such as exist in a truck in the hot sun or next to a steam radiator, the thermophilic bacteria that often survive commercial sterilization can germinate and grow. **Thermophilic anaerobic spoilage** is a fairly common cause of spoilage in low-acid canned foods. The can usually swells from gas, and the contents have a lowered pH and a sour odor. A number of thermophilic species of *Clostridium* can cause this type of spoilage. When thermophilic spoilage occurs but the can is not swollen by gas production, the spoilage is termed **flat sour spoilage**. This type of spoilage is caused by thermophilic organisms such as *Bacillus stearothermophilus* (ste-rō-thèr-mä′fil-us), which is found in the starch and sugars used in food preparation. Many industries have standards for the numbers of such thermophilic organisms permitted in raw materials. Both types of spoilage occur only when the cans are stored at higher-than-normal temperatures, which permits the growth of bacteria whose endospores are not destroyed by normal processing.

Mesophilic bacteria can spoil canned foods if the food is underprocessed or if the can leaks. Normally, these bacteria are killed by proper processing procedures. Underprocessing is more likely to result in spoilage by endospore-formers; the presence of nonendospore-formers strongly suggests that the can leaks. Leaking cans are often contaminated during the cooling of cans after processing by heat. The hot cans are sprayed with cooling water or passed through a trough filled with water. As the can cools, a vacuum is formed inside, and external water can be sucked through a leak past the heat-softened sealant in the crimped lid (Figure 27-2). Contaminating bacteria in the cooling water are drawn into the can with the water. Spoilage from underprocessing or can leakage is likely to produce odors of putrefaction, at least in high-protein foods, and occurs at normal storage temperatures. In such types of spoilage, there is always the potential of botulism bacteria being present.

Some acidic foods, such as tomatoes or preserved fruits, are preserved by heat of 100°C or below. As a rule, the only important spoilage organisms in acidic foods are molds, yeasts, and occasional species of acid-

(a) Formation of side seam

Sealing compound

(b) Formation of double seam for top or bottom

Figure 27-2 Construction of a metal can. Note the seam construction. During cooling after sterilization, the vacuum formed in the can may actually force the entry of contaminating organisms along with water.

tolerant, nonendospore-forming bacteria. These organisms are capable of growing at the normal pH of these foods but are easily killed by temperatures of less than 100°C. Occasional problems in acidic foods develop from a few organisms that are both heat-resistant and acid-tolerant. Examples of heat-resistant fungi are the mold *Byssochlamys fulva* (bis-sō-klam′is fül′va), which produces a *heat-resistant ascospore*, and a few molds, especially species of *Aspergillus*, that sometimes produce specialized resistant bodies called *sclerotia*. A spore-forming bacterium, *Bacillus coagulans* (kō-āg′ū-lans), is unusual in that it is capable of growth at a pH of almost 4.0. A summary of the types of canned food spoilage is presented in Table 27-1.

Home canning is an important means of preserving food. Because of the possibility of botulism food poisoning resulting from improper canning methods, persons involved in home canning should obtain reliable directions and follow them exactly.

Pasteurization

As we discussed in Chapter 1, Louis Pasteur began his career as a microbiologist when the French brewing industry commissioned him in 1871 to investigate the causes of spoilage in beer and wine. He concluded that spoilage was caused by the growth of microorganisms, and he worked out a method of eliminating

TABLE 27-1 *Types of Canned Food Spoilage*

Type of Spoilage	Indications of Spoilage	
	Appearance of Can	Contents of Can
Low- and Medium-Acid Foods (pH above 4.5)		
Flat sour (*Bacillus stearothermophilus*)	Possible loss of vacuum on storage	Appearance not usually altered; pH markedly lowered; sour; may have slightly abnormal odor; sometimes cloudy liquid
Thermophilic anaerobic (*Clostridium thermosaccharolyticum*)	Can swells	Fermented, sour, cheesy, or butyric odor
Sulfide spoilage (*Desulfotomaculum nigrificans*)	Can flat	Usually blackened; "rotten egg" odor
Putrefactive anaerobic (*Clostridium sporogenes*)	Can swells	May be partially digested; pH slightly above normal; typical putrid odor
Aerobic endospore-formers (*Bacillus* spp.)	Usually no swelling, except in cured meats when nitrate and sugar are present	Coagulated evaporated milk; black beets
High-Acid Foods (pH below 4.5)		
Flat sour (*Bacillus coagulans*)	Can flat, little change in vacuum	Slight pH change; off odor and flavor
Butyric anaerobic (*Clostridium butyricum*)	Can swells	Fermented, butyric odor
Nonendospore-formers (mostly lactic-acid bacteria)	Can swells	Acid odor
Yeasts	Can swells, may burst	Fermented; yeasty odor
Molds	Can flat	Surface growth; musty odor

Source: Data from the National Canners Association, 1950 6th Street, Berkeley, Calif., 94710.

them with mild heating. Too much heat would alter the beverage's characteristics to an unacceptable degree, but killing all the microorganisms was found to be unnecessary. The principle of the selective killing of microorganisms, now called pasteurization, was later applied to other foods, milk in particular (Chapter 7). Other products, such as ice cream, yogurt, and beer, all have individual pasteurization times and temperatures, which often differ considerably.

Aseptic Packaging

A recent development in food preservation is the increasing use of **aseptic packaging**. Packages are usually made of some material that cannot tolerate conventional heat treatment, such as laminated paper or plastic. The packaging materials come in continuous rolls that are fed into a machine that sterilizes the material with a hot hydrogen peroxide solution, sometimes aided by ultraviolet light, which enhances hydrogen peroxide activity. If the containers are metal, sterilization can be done with superheated steam or other high-temperature methods. High-energy electron beams can also be used to sterilize the packaging materials. While still in the sterile environment, the material is formed into packages, which are then filled with liquid foods that have been conventionally sterilized by heat. The filled package is not sterilized after it is sealed.

Low-Temperature Preservation

Low temperatures increase the reproduction time of microorganisms. Even so, some molds and bacteria can grow at a significant rate at temperatures below the freezing point of water, 0°C. (Food typically does not freeze solid until it is several degrees below this point.) Growth of microorganisms in foods a few degrees below 0°C is common, and growth at even −18°C has been reported. A properly set refrigerator maintains a temperature range from 0°C to 7°C. Many microorganisms grow slowly at these temperatures and will alter the taste and appearance of foods that are stored for too long.

Pathogenic bacteria, with a few exceptions, will not grow at these temperatures. Among the exceptions are the clostridiae that cause type E botulism, *Yersinia enterocolitica* (yèr-sin′e-ä en-tèr-ō-kōl-it′ik-ä), and *Listeria monocytogenes* (lis-te′rē-ä mo-nō-sī-tô′je-nēz). Freezing does not immediately kill significant numbers of bacteria, but frozen bacterial populations are dormant and decline slowly with time. Some parasites, such as the roundworms that cause trichinosis, are killed by several days of freezing. Some important temperatures associated with microorganisms and food spoilage are shown in Figure 27-3.

Anyone responsible for preparing food in large amounts should be aware of the spoilage that can arise

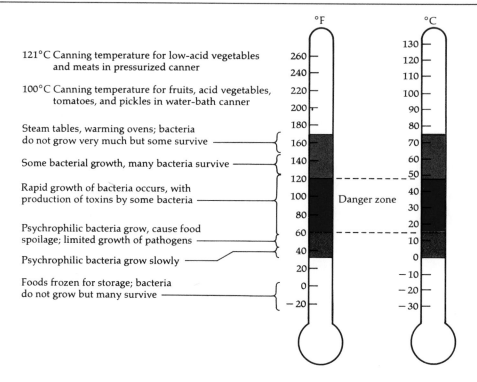

Figure 27-3   Spoilage temperatures.

121°C Canning temperature for low-acid vegetables and meats in pressurized canner

100°C Canning temperature for fruits, acid vegetables, tomatoes, and pickles in water-bath canner

Steam tables, warming ovens; bacteria do not grow very much but some survive

Some bacterial growth, many bacteria survive

Rapid growth of bacteria occurs, with production of toxins by some bacteria

Psychrophilic bacteria grow, cause food spoilage; limited growth of pathogens

Psychrophilic bacteria grow slowly

Foods frozen for storage; bacteria do not grow but many survive

Danger zone

Figure 27-4 The effect of a food's mass on its cooling rate. Note that the two-inch-deep rice cooled through the incubation temperature of the *Bacillus cereus* in about an hour, whereas the six-inch-deep rice remained in this temperature range for more than five hours.

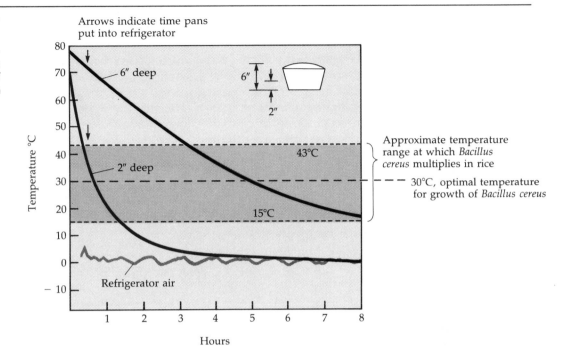

Arrows indicate time pans put into refrigerator

6″ deep

2″ deep

Refrigerator air

Temperature °C

Hours

Approximate temperature range at which *Bacillus cereus* multiplies in rice

43°C

15°C

30°C, optimal temperature for growth of *Bacillus cereus*

from the slow cooling of large amounts of hot food. Hot food placed in a refrigerator cools more slowly than most persons realize, and the food can spend a considerable time at the incubation temperatures of pathogenic or spoilage bacteria before the temperature of the food drops to that of the refrigerator (Figure 27-4).

Radiation and Food Preservation

There has been considerable research into the use of ionizing radiation for food preservation, especially for military applications. It is possible to sterilize food by radiation, but it is unlikely that this procedure will soon replace conventional heat sterilization for most purposes. Taste and appearance of irradiated food have been much improved in recent years. This improvement has been accomplished mainly by freezing the food in liquid nitrogen and exhausting oxygen from the package before it is irradiated.

However, a major factor affecting the adoption of this type of preservation is that the public is apprehensive about foods exposed to radiation, even though there is no evidence that harmful effects result from the amount and types of radiation used. Irradiation is considered to be another form of food additive. As such, it must be reported on the label and must meet the stringent requirements for safety testing required of chemical food additives. These are expensive and

can only be justified where a considerable market is expected.

Pasteurizationlike radiation treatments that fall short of attempts at total sterilization may prove to be commercially practical. The Food and Drug Administration has allowed irradiation to decontaminate spices for several years. More recently, they have given approval for the irradiation in moderate doses of certain foods to eliminate insects and to prevent sprouting or other undesirable maturation of certain fruits and vegetables. Pork may also be irradiated to eliminate the parasitic worms that cause trichinosis. Foods treated in this manner must be labeled with a logo (Figure 27-5) and a statement, such as "Treated with gamma radiation to extend shelf life." After a few years, to allow for familiarization, only the logo will be required.

Figure 27-5 Irradiation logo. This logo indicates that a food has received irradiation treatment.

Many European countries and Japan have allowed the use of irradiation on foods for some time. In the United States, the greatest commercial interest is in irradiation treatment of spices and pet foods. The preferred method for irradiation is gamma rays produced by Cobalt-60, when deep penetration is a requirement. However, this type of treatment requires several hours of exposure in isolation behind protective walls. High-energy electron accelerators are much faster and sterilize in a few seconds, but this treatment has low penetrating power and is suitable only for sliced meats, bacon, or similar thin products.

Microwaves, such as those used for cooking, kill bacteria only by heating foods and have little or no direct killing effect. Bacteria can survive (and are easily isolated) on the interior walls of microwave ovens.

Chemical Preservatives

Chemical preservatives are frequently added to foods to retard spoilage. Among the more common additives are sodium benzoate, sorbic acid, and calcium propionate. These chemicals are simple organic acids, or salts of organic acids, which the body readily metabolizes and which are generally judged to be safe in foods. **Sorbic acid**, or its more soluble salt *potassium sorbate*, and **sodium benzoate** prevent molds from growing in certain acidic foods, such as cheese and soft drinks. Such foods, usually with a pH of 5.5 or less, are most susceptible to mold-type spoilage. **Calcium propionate** is an effective fungistat used in bread. It prevents the growth of surface molds and the *Bacillus* species of bacteria that causes ropy bread. These organic acids inhibit mold growth, not by affecting the pH but by interfering with the mold's metabolism or the integrity of the plasma membrane.

Sodium nitrate and **sodium nitrite** are added to many meat products, such as ham, bacon, wieners, and sausage. The active ingredient is sodium nitrite, which certain bacteria in the meats can also produce from sodium nitrate. These bacteria use nitrate as a substitute for oxygen under anaerobic conditions, much as it is used in denitrification in soil. The nitrite has two main functions—to preserve the pleasing red color of the meat by reacting with blood components in the meat and to prevent the germination and growth of any botulism endospores that might be present. There has been some concern that the reaction

of nitrites with amino acids can form certain carcinogenic products known as **nitrosamines**, and the amount of nitrites added to foods has generally been reduced recently for this reason. However, use of nitrites continues because of their established value in preventing botulism. Since nitrosamines are formed in the body from other sources, the added risk posed by limited use of nitrates and nitrites in meats might be lower than was once thought.

Foodborne Infections and Microbial Intoxications

We mentioned in Chapter 24 that illness resulting from microbial growth in food is associated with two principal mechanisms. In the mechanism known as *foodborne infection*, the contaminating microorganism infects the person who ingests contaminated food. As the pathogen grows in this host, it produces damaging toxins. Diseases caused by this mechanism include gastroenteritis, typhoid fever, and dysentery. In the other mechanism, called *microbial intoxication*, the toxin is formed in the food by microbial growth and then is ingested with the food. Diseases associated with this mechanism include botulism, staphylococcal food poisoning, and mycotoxicoses (intoxications caused by fungal toxins, such as ergot or aflatoxin). Several microbial intoxication diseases are described in Table 27-2.

Dairy products, which are often consumed without having been cooked, are particularly likely to transmit food-related diseases. Standards for sanitation in the dairy industry are therefore very stringent. Because most dairy milk is drawn mechanically and is promptly put into cooled holding tanks, most of the bacteria in milk today are gram-negative psychrophiles. Grade A pasteurized cultured products, such as buttermilk and cultured sour cream, have high counts of lactic acid bacteria as a natural result of their method of manufacture. Therefore, the standards require only that they contain fewer than 10 coliforms per milliliter. Pasteurized Grade A milk is required to have a standard plate count of fewer than 20,000 bacteria and not more than 10 coliforms per milliliter. Grade A dry milk products are required to have a standard plate count of fewer than 30,000 bacteria per gram or a coliform count of fewer than 10 per gram.

Some people prefer raw milk because of the lack of additives or because they believe it has not been altered chemically. (Pasteurization can denature enzymes found in milk.) Unpasteurized (raw) dairy products are available commercially in about half the states in this country. Three states have boards that certify raw milk from dairies meeting the required standards for cleanliness; the dairy herds are also inspected for disease. The standard for certified raw milk is a maximum bacterial count of 10,000 per milliliter and 10 coliforms per milliliter. The microbial standards guaranteed by certification do not prevent transmission of disease in raw milk; in fact, certified raw milk regularly is a source of outbreaks of salmonellosis.

TABLE 27-2 *Foodborne Microbial Intoxications*

Disease	Foods Involved	Prevention	Clinical Features	Onset and Duration
Aflatoxin	Moldy grains, peanuts	Avoid contaminated grains, peanut products	Low doses may induce liver cancer; high doses cause general liver damage	
Bacillus cereus intoxication	Custard, cereal, starchy foods	Refrigeration of foods	Cramps, diarrhea, nausea, vomiting	Onset: 8–16 hours Duration: Less than 1 day
Botulism	Canned foods	Proper canning procedures; boiling food prior to consumption	Difficulty swallowing, double vision, respiratory paralysis	Onset: 2 hours–6 days Duration: Weeks
Ciguatera	Carnivorous tropical fish in which dinoflagellate toxin is concentrated	Avoid eating large fish	Tingling rash, fever, breathing discomfort	Onset: 2–6 hours Duration: To 2 weeks
Ergotism	Moldy grains	Avoid contaminated grains	Burning abdominal pain, hallucinations	Onset: 1–2 hours Duration: Months
Methyl mercury poisoning	Freshwater or ocean fish	Stop dumping mercury into waters	Blurred vision, numbness, apathy, coma	Onset: 1 week Duration: May be chronic
Mushroom poisoning	*Amanita* species	Avoid poisonous mushrooms	Vomiting, liver necrosis, neurotoxic effects	Onset: Less than 1 day Duration: Less than 10 days
Paralytic mollusk poisoning	Bivalve mollusks during red tide (dinoflagellate blooms)	Avoid eating mollusks during red tide	Tingling, rash, fever, respiratory paralysis	Onset: Less than 1 hour Duration: Less than 12 hours
Scombroid poisoning	Histaminelike substance produced by *Proteus* growing on certain fish, such as tuna	Refrigeration of fish	Headache, cramps, hives, shock (rare)	Onset: Several minutes–1 hour
Staphylococcal intoxication	High-osmotic-pressure foods not cooked before eating	Refrigeration of foods	Nausea, vomiting, diarrhea	Onset: Less than 1 day Duration: Less than 3 days

The Role of Microorganisms in Food Production

In the latter part of the nineteenth century, microbes used in food production were grown in pure culture for the first time. This development quickly led to improved understanding of the relationships between specific microbes and their products and activities. This period can be considered as the beginning of industrial food microbiology. For example, once it was understood that a certain yeast grown under certain conditions produced beer and that certain bacteria could spoil the beer, brewers were better able to control the quality of their products. Specific industries became active in microbiological research and selected certain microbes for their special qualities. For example, the brewing industry extensively investigated the isolation and identification of yeasts and selected those that were able to produce more alcohol.

As our knowledge of microbial genetics increases, we will continue to improve our ability to select and use desirable strains of microorganisms. For example, scientists are developing *Zymomonas* that are able to produce alcohol using cellulose from agricultural wastes rather than from carbohydrates. The selecting of ethanol-producing microorganisms with an improved tolerance for ethanol might allow the microorganisms to produce the alcohol more efficiently in higher concentrations. Tremendous advances in molecular biology have produced techniques for altering microorganisms that are increasingly replacing the relatively hit-or-miss approaches of the past. The application of these new techniques is known as *genetic engineering*, which we will discuss in the final section of this chapter. In this section, we will discuss the role of microorganisms in the production of several common foods.

Cheese

The United States leads the world in the manufacture of cheese, producing more than 1.5 million tons each year. Although there are many types of cheeses, all require the formation of a *curd*, which can then be separated from the main liquid fraction, or *whey* (Figure 27-6). The curd is made up of a protein, *casein*, and is usually formed by the action of an enzyme,

Figure 27-6 Cheese-making. In this Swiss plant, curd is drawn from whey in a porous cheesecloth.

rennin, which is aided by acidic conditions provided by certain lactic-acid-producing bacteria. These inoculated lactic-acid bacteria also provide the characteristic flavors and aromas of fermented dairy products during the ripening process. The curd undergoes a microbial ripening process, except for a few unripened cheeses, such as ricotta and cottage cheese.

Cheeses are generally classified by their hardness, which is produced in the ripening process. The more moisture lost from the curd, and the more the curd is compressed, the harder the cheese. Romano and Parmesan cheeses, for example, are classified as *very hard* cheeses; Cheddar and Swiss are *hard* cheeses. Limburger, blue, or Roquefort cheeses are classified as *semisoft*; Camembert is an example of a *soft* cheese. The hard Cheddar and Swiss cheeses are ripened by lactic-acid bacteria growing anaerobically in the interior. Such hard, interior-ripened cheeses can be quite large.

The longer the incubation time, the higher the acidity and the sharper the taste of the cheese. Semisoft cheeses, such as Limburger, are ripened by bacteria and other contaminating organisms growing on the surface. Blue and Roquefort cheeses are ripened by *Penicillium* molds inoculated into the cheese. The texture of the cheese is loose enough that adequate oxygen can reach the aerobic molds. The growth of the *Penicillium* molds is visible as blue-green clumps in the cheese. Camembert cheese is ripened in small packets so that the enzymes of *Penicillium* mold growing aerobically on the surface will diffuse into the cheese for ripening. A *Propionibacterium* (prō-pē-on-ē-bak-ti′rē-um) species produces carbon dioxide that forms the holes in Swiss cheese.

Other Dairy Products

Butter is made by churning cream until the fatty globules of butter separate from the liquid buttermilk. The typical flavor and aroma of butter and buttermilk are from diacetyls, a combination of two acetic acid molecules that is a metabolic end-product of the fermentation by some lactic-acid bacteria. Today buttermilk is usually not a by-product of butter making but is made by inoculating skim milk with bacteria that form lactic acid and the diacetyls. The inoculum is allowed to grow for 12 or more hours before the buttermilk is cooled and packaged. Sour cream is made from cream inoculated with organisms similar to those used to make buttermilk.

A wide variety of slightly acidic dairy products—probably a heritage of a nomadic past—are found around the world. Many of them are part of the daily diet in the Balkans, Eastern Europe, and Russia. One such product is yogurt, which is also popular in the United States. Commercial yogurt is made from low-fat milk, from which much of the water has been evaporated in a vacuum pan. The resulting thickened milk is inoculated with a mixed culture of *Streptococcus thermophilus*, primarily for acid production, and *Lactobacillus bulgaricus* (bul-ga′ri-kus), to contribute flavor and aroma. The temperature of the fermentation is about 45°C for several hours, during which time *S. thermophilus* outgrows *L. bulgaricus*. Maintaining the proper balance between the flavor-producing and the acid-producing organisms is the secret of good yogurt.

Kefir and kumiss are popular beverages in Eastern Europe. The usual lactic-acid-producing bacteria are supplemented with a lactose-fermenting yeast to give these drinks an alcohol content of 1% or 2%.

Nondairy Fermentations

Microorganisms are also used in baking, especially for bread. The sugars in bread dough are fermented, much as they are in the fermentation of alcoholic beverages (discussed in the next section). Anaerobic conditions for production of ethanol by the yeasts are mandatory for production of alcoholic beverages. In baking, carbon dioxide forms the typical bubbles of leavened bread. Aerobic conditions favor carbon dioxide production and are encouraged as much as possible. This is the reason the bread dough is kneaded repeatedly. Whatever ethanol is produced evaporates during baking. In some breads, such as rye or sourdough, the growth of lactic-acid bacteria gives the typical tart flavor (see the box in Chapter 1, p. 6).

Fermentation is also used in the production of such foods as sauerkraut, pickles, and olives. In Asia, extremely large amounts of soy sauce are produced by molds that form starch-degrading enzymes to produce fermentable sugars. This principle is used in making other Asian fermented foods, including sake, the Japanese rice wine. In soy sauce production, molds such as *Aspergillus oryzae* (a-spėr-jil′lus ô-rī′zē) are grown on wheat bran and then are allowed to act along with lactic-acid bacteria on cooked soybean and crushed wheat mixtures. After this process has produced fermentable carbohydrates, a prolonged fermentation results in soy sauce. Table 27-3 lists many fermented foods.

Alcoholic Beverages and Vinegar

Microorganisms are involved in the production of almost all alcoholic beverages. Beer and ale are products of grain starches fermented by yeast (Table 27-4). Because yeasts are unable to use starch directly, the starch from grain must be converted to glucose and maltose, which the yeasts can ferment into ethanol and carbon dioxide. In this conversion, called **malting**, starch-containing grains, such as malting barley, are allowed to sprout, then are dried and ground. This product, called **malt**, contains starch-degrading enzymes (amylases) that convert cereal starches into carbohydrates that can be fermented by yeasts. For **distilled spirits**, such as whiskey, vodka, and rum, carbohydrates from cereal grains, potatoes, and molasses are fermented to alcohol. The alcohol is then distilled to make a concentrated alcoholic beverage.

Wines are typically made from grapes, which contain sugars that can be used directly by yeasts for fermentation; malting is unnecessary in wine-

TABLE 27-3 *Some Fermented Foods and Related Products*

Foods and Products	Raw Ingredients	Fermenting Organisms	Location Produced
Dairy Products			
Cheeses (ripened)	Milk curd	*Streptococcus* spp., *Leuconostoc* spp.	Worldwide
Kefir	Milk	*Streptococcus lactis, Lactobacillus bulgaricus, Candida* spp.	Southwestern Asia
Kumiss	Raw mare's milk	*L. bulgaricus, Lactobacillus leichmannii, Candida* spp.	Russia
Taette	Milk	*S. lactis* var. *taette*	Scandinavia
Yogurt	Milk, milk solids	*Streptococcus thermophilus, L. bulgaricus*	Worldwide
Meat and Fish Products			
Country-cured hams	Pork hams	*Aspergillus, Penicillium* spp.	Southern United States
Dry sausages	Pork, beef	*Pediococcus cervisiae*	Europe, United States
Fish sauces	Small fish	Halophilic *Bacillus* spp.	Southeast Asia
Izushi	Fresh fish, rice, vegetables	*Lactobacillus* spp.	Japan
Nonbeverage Plant Products			
Cocoa beans	Cacao fruits (pods)	*Candida krusei, Geotrichum* spp.	Africa, South America
Coffee beans	Coffee cherries	*Erwinia dissolvens, Saccharomyces* spp.	Brazil, Congo, Hawaii, India

making. Grapes usually need no additional sugars, but other fruits might be supplemented with sugars to ensure enough alcohol production. The steps of wine-making are shown in Figure 27-7. Lactic-acid bacteria are important when wine is made from grapes that are especially acidic from high concentrations of malic acid. These bacteria convert the malic acid to the weaker lactic acid in a process called **malolactic fermentation**. The result is a less acidic, better-tasting wine than would otherwise be produced.

Winemakers who allowed wine to be exposed to air found that it soured from the growth of aerobic bacteria that converted the ethanol in the wine to acetic acid. The result was vinegar (*vin* means wine; *aigre* means sour). The process is now used deliberately to make vinegar. Ethanol is first produced by anaerobic fermentation of carbohydrates by yeasts. The ethanol is then aerobically oxidized to acetic acid

by acetic-acid-producing bacteria of the genera *Acetobacter* and *Gluconobacter*. The basic process of vinegar production, which requires vigorous aeration, is shown in Figure 27-8.

Microorganisms as a Food Source

In the future, as human populations increase, if the amount of arable land does not correspondingly increase, microorganisms may become more important to the food requirements of the world. Protein is in particularly short supply. Microorganisms, which can usually double their weight in a few hours and often in less than an hour, may help solve this problem. When used as a food source, microorganisms are referred to as **single-cell protein (SCP)**. Microorganisms can use as substrates many carbonaceous materials, such as cellulose, methanol, or petroleum hydrocarbons. Petroleum yields the most food for its

TABLE 27-3 *(Continued)*

Foods and Products	Raw Ingredients	Fermenting Organisms	Location Produced
Kimchi	Cabbage and other vegetables	Lactic-acid bacteria	Korea
Miso	Soybeans	*Aspergillus oryzae, Saccharomyces rouxii*	Japan
Olives	Green olives	*Leuconostoc mesenteroides, Lactobacillus plantarum*	Worldwide
Poi	Taro roots	Lactic-acid bacteria	Hawaii
Sauerkraut	Cabbage	*L. mesenteroides, L. plantarum*	Worldwide
Soy sauce (shoyu)	Soybeans	*A. oryzae* or *Aspergillus soyae*; *S. rouxii, Lactobacillus delbrueckii*	Japan
Tempeh	Soybeans	*Rhizopus oligosporus*; *Rhizopus oryzae*	Indonesia, New Guinea, Surinam
Breads			
Idli	Rice and bean flour	*Leuconostoc mesenteroides*	Southern India
Rolls, cakes, etc.	Wheat flours	*Saccharomyces cerevisiae*	Worldwide
San Francisco sourdough bread	Wheat flour	*Saccharomyces exiguus, Lactobacillus sanfrancisco*	Northern California
Sour pumpernickel	Wheat flour	*L. mesenteroides*	Switzerland, other areas

Source: J. M. Jay. *Modern Food Microbiology*, 3rd ed. New York: Van Nostrand Reinhold, 1986.

weight, but its demand for use as a fuel makes it expensive to use for food production. Wastes from meat packing, paper production, agriculture, and forestry might also serve as substrates. Cellulose products from wood or other plants are the most obvious candidates, but relatively few microorganisms can use cellulose or the lignin often associated with it. Considerable effort is being directed in research to economically convert cellulose into other products that can be readily metabolized, such as glucose.

Among all the microorganisms from which single-cell protein can be made, bacteria are economically the most attractive. They have shorter generation times than yeasts and molds, and, even more important, their protein content is as high as 80%, as compared with 45% protein in soybeans and 50% protein in yeasts. Photosynthetic organisms are of special interest because light is an inexhaustible energy source. However, algae and photosynthetic bacteria tend to make relatively unpalatable end-products. An

exception is the cyanobacterium *Spirulina* (spī-rü-lī′na), which for centuries has been used in dried form from a lake in Chad, Africa. *Spirulina* was also eaten by the Aztecs in Mexico. Experiments in eating *Spirulina* have been conducted in the United States, France, and Mexico, and, as a result, the United States and Mexico permit the sale of *Spirulina*.

Yeasts such as *Saccharomyces cerevisiae* (sak-ä-rō-mī′sēs se-ri-vis′ē-ī) or *Candida utilis* (kan′did-ä ū′til-is) can be grown in large quantities, but they lead to adverse gastrointestinal reactions in humans. Also, their high nucleic-acid content exceeds the human body's capacity to metabolize them and may precipitate ailments such as gout. Yeasts are therefore most likely to be used as an animal food supplement.

In England, the bacterium *Methylophilus methylotrophus* (meth-i-lo′fi-lus meth-i-lō-trōf-us) is used to produce protein supplements for animal feeds. As much as half a gram of cells, which contain about 70%

Figure 27-7 Basic steps in making red wine. For white wines, the pressing precedes fermentation so that the color is not extracted from the solid matter.

Figure 27-8 The classic method of vinegar production. This type of vinegar generator is still in use, but most vinegar producers now use a suspension of organisms agitated by air bubbles.

crude protein, can be produced from a gram of methanol. Methanol might seem to be an unlikely substrate, but it has the advantages of being free of impurities, soluble in water, and readily separated from the end-product.

Industrial Microbiology

Much of industrial microbiology—the use of microbes to make industrial products—has centered on the lactic-acid and ethanol fermentations that are the basis of the major food industries we just discussed. Although the lactic acid and ethanol that are produced are incidental to food fermentation, they have industrial uses. Microbes are used to make a variety of products, many of which are not related to foods. During World War I, microbes were used to make chemical compounds such as glycerol and acetone. In the future, as the petroleum products from which many chemicals are synthesized become scarce, the world may come to depend on microbial fermentations of renewable substrates to produce these chemicals. The potential shortage of petroleum has also led to interest in microbes as alternative energy producers. In the coming years, there will no doubt be a revolution in the application of genetically engineered microorganisms in industry—what is now known as **biotechnology**.

In this section, we will first discuss some of the more important commercial microbial products, then the growing alternative energy industry, and finally the industrial applications of genetic engineering.

Industrial Products

Amino Acids

Amino acids have become a major industrial product from microorganisms. For example, over 165,000 tons of glutamic acid, used to make the flavor enhancer monosodium glutamate, are produced every year. Certain amino acids, such as lysine, cannot be synthesized by animals and are present only at low levels in the normal diet. Therefore, the commercial synthesis of lysine and some of the other essential amino acids as food supplements is an important industry. More than 35,000 tons of lysine are produced every year. In nature, microbes rarely produce amino acids in excess of their own needs because feedback inhibition (see Chapter 8) prevents wasteful production of primary metabolites. Commercial microbial production of amino acids such as lysine depends on specially selected mutants and sometimes on ingenious manipulations of metabolic pathways.

Lysine is produced by a bacterium, *Corynebacterium glutamicum* (kôr-rī-nē-bak-ti'rē-um glü-tam'i-kum), which normally produces both threonine and lysine. If both accumulate, feedback inhibition prevents further production of both amino acids, which share a common enzyme in their synthetic pathway. A mutant bacterium that lacks the enzymes necessary to make threonine is used to produce lysine in large amounts. Enough threonine is added to the medium to allow this mutant organism to grow, but not enough is added to trigger feedback inhibition. The organism then continues to produce lysine because lysine alone does not cause feedback inhibition.

Glutamic acid is produced by the same bacterium that produces lysine. The bacterium is induced to excrete glutamic acid when it is provided with only a minimal amount of the vitamin biotin. The plasma membrane is weakened by the vitamin deficiency, and the glutamic acid leaks into the medium. This technique of inducing the cell to form a leaky membrane is also sometimes used in antibiotic production.

About the only amino acids that are not produced by microbial fermentation are methionine, alanine, glycine, and cysteine. These are chemically synthesized. Individual amino acids made by microbial fermentation are produced in amounts of at least 1000 tons per year.

Citric Acid

Citric acid is a constituent of citrus fruits, such as oranges and lemons, and at one time these were its only industrial source. However, over a hundred years ago, citric acid was identified as a product of mold metabolism. It can be chemically synthesized, but not cheaply enough to replace production by mold fermentation. Citric acid has an extraordinary range of uses beyond the obvious ones of giving tartness and flavor to foods. It is an antioxidant and pH adjuster in many foods, and in dairy products it often serves as an emulsifier. It has even found uses in the making of detergents and in the removal of sulfur from power station stack gases. Most of the uses are, however, in foods, and only about 30% of its applications are industrial and pharmaceutical. Well over 300,000 tons of citric acid are produced every year. Much of it is produced by a mold, *Aspergillus niger* (nī'jėr), using molasses as a substrate. The mold excretes citric acid when provided with only a limited supply of iron and manganese.

Enzymes

Enzymes are widely used in different industries. For example, amylases are used in the production of syrups from corn starch, in the production of paper sizing (a coating for smoothness, as on this page), and in the production of glucose from starch. Glucose isomerase is an important enzyme; it converts the glucose that amylases form from starches into fructose, which is used in place of sucrose as a sweetener in many foods. Probably half of the bread baked in this country is made with proteases, which adjust the amounts of glutens (protein) in wheat so that baked goods are

TABLE 27-4   *Production of Alcoholic Beverages by Yeasts*

Beverage	Yeast	Method of Preparation	Function of Yeast
Root beer (pre-1905 and homemade)	*Saccharomyces cerevisiae*	Molasses, sassafras bark, wintergreen bark, and sarsaparilla root added for flavor; add yeast; incubate aerobically.	Converts sugar into carbon dioxide; by aerobic metabolism; 0.03% alcohol.
Beer	*S. carlsbergensis* (bottom yeast)	Germinated barley releases starches and amylase enzymes (malting). Enzymes in malt hydrolyze starch to fermentable sugars (mashing). Liquid (wort) sterilized. Hops (flowers) added for flavor. Yeast added, incubated at 37°–49°C.	Converts sugar into alcohol and carbon dioxide; 4.0% alcohol. Yeast grow on bottom of fermenting vessel.
Ale	*S. cerevisiae* (top yeast)	As in beer; incubated at 50°–70°C.	Converts sugar into alcohol; 6% alcohol. Yeast grow at top of fermentation vessel.
Sake	*S. cerevisiae*	*Aspergillus oryzae* converts starch in steamed rice into sugar; yeast added; incubated at 20°C.	Converts sugar into alcohol; 14%–16% alcohol.
Wine, natural	*S. cerevisiae*	Strain of grape provides various flavors and sugar concentrations. Grapes crushed into must; sulfur dioxide added to inhibit wild yeast; yeast added. Red wines: incubated at 25°C. Aged in oak for 3–5 years and in bottle for 5–15 years. White wines: incubated at 10°–15°C. Aged 2–3 years in bottle.	Converts grape-sugar into alcohol; 14% or less alcohol.
Wine, sherry	*S. cerevisiae* and *S. beticus* or *S. bayanus*	As natural wine with additional surface growth (flor) at 27°C. Alcohol added to 18%–21%.	*S. beticus* grows as surface film producing aldehydes from alcohol.
Wine, sparkling (champagne)	*S. cerevisiae*	As natural wine with secondary fermentation in bottle. 2.5% sugar and yeast added to bottled wine; incubated at 15°C; bottle inverted to collect yeast in neck.	In secondary fermentation produce carbon dioxide; yeast settle quickly.
Distilled Beverages			
Rum, Jamaica	Wild yeast	Cane molasses inoculated from previous fermentation. Oak-aging adds color. Distilled to concentrate.	Converts sugar to alcohol; 50%–95% alcohol.
Brandy	*S. cerevisiae*	Fruits pressed; yeast added. Distilled to concentrate alcohol, blended with other brandies.	Converts sugar into alcohol; 40%–43% alcohol.
Whiskey	*S. cerevisiae*	Wort (see beer) is fermented by yeast. Distilled to concentrate alcohol; aged in charred oak barrels.	Converts sugar to alcohol; 50%–95% alcohol.

improved or made uniform. Other proteolytic enzymes are used as meat tenderizers or in detergents as an additive to remove proteinaceous stains. Several enzymes produced commercially by microorganisms are listed in Table 27-5.

Vitamins

Vitamins are sold in large quantities combined in tablet form. They are used as individual food supplements. Microbes can provide an inexpensive source of some vitamins. **Vitamin B$_{12}$** is produced by *Pseudomonas* spp. and *Propionibacterium* spp. Cobalt is a part of the molecule, and cobalt chloride must be provided in the medium. **Riboflavin** is another vitamin produced by fermentation, mostly by fungi such as *Ashbya gossypii* (ash'bē-ä gos-sip'ē'ē).

Pharmaceuticals

Modern pharmaceutical microbiology developed after World War II, with the production of antibiotics. To increase the number of random mutations, antibiotic-producing organisms were subjected to ultraviolet radiation in the hope that some mutations would be useful. These and other techniques soon led to strains that produce antibiotics thousands of times faster than the original isolates did. All **antibiotics** were originally the products of microbial metabolism. Many antibiotics are still produced by microbial fermentations, and work continues on selection of more productive mutants and nutritional and genetic manipulations. Antibiotics are typically made industrially by inoculating a solution of growth medium with spores of the appropriate mold or streptomycete

TABLE 27-5 *Some Microbial Enzymes Produced Commercially*

Enzyme	Microorganism	Use of Enzyme
Amylase	*Aspergillus niger* *Aspergillus oryzae* *Bacillus subtilus*	Baking: Flour supplement Brewing: Mashing Food: Precooked foods, syrup Pharmaceuticals: Digestive aids Starch: Cold-water laundry Textiles: Desizing agent
Cellulase	*Aspergillus niger*	Food: Liquid coffee concentrate
Dextransucrase	*Leuconostoc mesenteroides*	Pharmaceuticals: Dextran
Glucose oxidase	*A. niger*	Food: Glucose removal from egg solids Pharmaceuticals: Test papers
Invertase	*Saccharomyces cerevisiae*	Candy: Prevents granulation in soft center Food: Artificial honey
Lactase	*Saccharomyces fragilis*	Dairy: Prevents crystallization of lactose in ice cream and concentrated milk
Lipase	*A. niger*	Dairy: Flavor production in cheese
Pectinase	*A. niger*	Wine and juice: Clarification
Penicillinase	*B. subtilis*	Pharmaceuticals: Diagnostic agent
Protease	*A. oryzae*	Brewing: Beer stabilizer Baking: Bread-making Food: Meat tenderizer Pharmaceuticals: Digestive aid Textiles: Desizing
Streptodornase	*Streptococcus pyogenes*	Pharmaceuticals: Reagent, wound debridement

Source: J. R. Porter, "Microbiology and the food and energy crisis." *ASM News* 40(11):822, November 1974.

and vigorously aerating it. After the antibiotic reaches a satisfactory concentration, it is extracted by solution, precipitation, and other standard industrial procedures. Many of the antibiotics produced in the United States are fed to animals to increase their growth (see the box in Chapter 19, p. 494). The search for new and modified antibiotics is accelerating as worldwide microbial resistance to antibiotics continues to increase alarmingly (see the box in Chapter 25, p. 646).

Vaccines are a product of industrial microbiology.

Many antiviral vaccines are mass produced in chicken eggs or cell cultures. To produce vaccines against bacterial diseases usually requires the growth of large amounts of the bacteria. Genetic engineering techniques are increasingly important in the area of vaccine development and production. For instance, genetic engineering is being used to develop an AIDS vaccine (see the box in Chapter 18, p. 468).

Steroids (Figure 27-9) are a very important group of chemicals that include cortisone, used as an antiinflammatory drug, and estrogens and progesterone,

MICROBIOLOGY HIGHLIGHTS

Microbial Miner

For thousands of years, refiners have been extracting copper from low-grade copper ore by a process called *leaching*. Until 1957, however, nobody realized that the extraction would be impossible without the bacterium *Thiobacillus ferrooxidans*, which thrives in acidic environments such as mining sites. Thanks to this microbe, millions of tons of copper are recovered from billions of tons of low-grade ore every year.

Thiobacillus ferrooxidans is one of several species of

chemolithotrophs, "rock-eating" bacteria, which are becoming increasingly important in the extraction of a variety of metals. The chemolithotrophs obtain their energy from the oxidation of inorganic substances (iron and sulfur, in the case of *Thiobacillus ferrooxidans*). This allows them to stimulate reactions involving minerals that are unaffected by, and may even be poisonous to, other bacteria. Moreover, chemolithotrophic bacteria thrive in the absence of organic matter because they extract their carbon directly from the carbon dioxide in the atmosphere.

The bacteria are naturally present in the leaching solution prepared at the mining site. The refiners grind the copper-bearing rock and heap it into a pit, called a *dump* (see photo). Then they pour water mixed with sulfuric acid (the *leach solution*) through the dump. When the leach solution runs out through the bottom of the dump, it

Figure 27-9 Production of steroids. Conversion of a precursor compound such as a sterol into a steroid by *Streptomyces*. These changes are extremely difficult to obtain by synthetic chemistry techniques, so microorganisms are often essential for economic production of pharmaceutical steroids.

Sterol → *Streptomyces* species → Steroid

contains copper sulfate ($CuSO_4$) in solution. The refiners then add metallic iron to the leach solution. The iron reacts with the copper sulfate to release metallic copper (see the equations below, and Figure 27-10).

Thiobacillus frees copper from the ore in a variety of ways. Most commonly, the bacillus reacts with iron and sulfur compounds that are in the rock. The name *ferrooxidans* means "oxidizing iron compounds."

The rock is likely to contain compounds of iron and sulfur, such as iron pyrite (FeS_2), also known as "fool's gold." Note that the pyrite molecule contains a divalent *ferrous* ion (Fe^{2+}) and two monovalent sulfur ions (S^-). As the leach solution trickles down through the ore-bearing rock, *Thiobacillus ferrooxi-*

dans oxidizes the Fe^{2+} (ferrous) ion by removing an electron from it. This turns the ferrous ion into a trivalent *ferric* (Fe^{3+}) ion.

The sulfur that was in the FeS_2 molecule can now combine with hydrogen ions and oxygen molecules to form sulfuric acid (H_2SO_4).

The equation describing the chemical reaction looks like this:

FeS_2 + $2H^+$ + $2O_2$

Iron pyrite / Hydrogen ions / Oxygen

⟶ Fe^{3+} + H_2SO_4 + S^0

Ferric ion / Sulfuric acid / Sulfur

In addition to the iron pyrite, the ore contains compounds of copper and sulfur, such as copper sulfide (CuS). The ferric ion (Fe^{3+}) oxidizes the monovalent copper ion (Cu^+) to create a divalent copper ion (Cu^{2+}), which then combines with the *sulfate* ion (SO_4^{2-}), contributed by the sulfuric acid, to form

copper sulfate ($CuSO_4$), as follows:

CuS + Fe^{3+} + H_2SO_4

Copper sulfide / Ferric ion / Sulfuric acid

⟶ $CuSO_4$ + Fe^{2+} +

Copper sulfate / Ferrous ion

$2H^+$ + S^0

Hydrogen ions / Sulfur

Now that the leach solution has copper sulfate dissolved in it, the refiners add metallic iron (in the form of iron filings or scrap iron). The iron reacts with the copper sulfate to form ferrous sulfate and metallic copper, as in the following equation:

$CuSO_4$ + $2Fe^0$ + H_2SO_4

Copper sulfate / Iron / Sulfuric acid

⟶ $2FeSO_4$ + Cu^0

Ferrous sulfate / Copper

+ $2H^+$

Hydrogen

Thiobacillus can now reoxidize the ferrous ion to ferric ion.

The mining industry has begun to use chemolithotropic bacteria also to recover selenium, vanadium, manganese, molybdenum, and uranium, all of which are present in low concentrations in ores. Furthermore, microbial mining offers a possible solution to another problem that threatens modern industrial societies—pollution of the soil and groundwater by unextracted metallic ions in waste water. Chemolithotrophic bacteria that can take up and accumulate toxic metal ions in their cells may be useful in removing metal pollutants from waste water.

both used in birth control pills. It is difficult to recover steroids from animal sources, but microorganisms can synthesize steroids from sterols or from related compounds derived from animals and plants. Certain strains of molds and bacteria (particularly strepto-mycetes) can produce steroids by converting chemical groups on sterols and similar molecules. Genetic engineering will probably revolutionize the microbial production of medical pharmaceuticals.

Uranium and Copper

Thiobacillus ferrooxidans (thī-ō-bä-sil′lus fer-rō-oks′i-danz) is used in the recovery of otherwise unprofitable grades of uranium and copper ores (see Figure 27-10 and the box on p. 708). For example, when solutions containing ferric ion (Fe^{3+}) are washed through deposits of insoluble copper compounds, the copper is oxidized to soluble compounds. In this process, Fe^{3+} is reduced to Fe^{2+}. The Fe^{2+} can be reoxidized by *T. ferrooxidans*. The soluble copper then moves out of the ores and is reclaimed.

Microorganisms

Microorganisms themselves sometimes constitute an industrial product. A simple example is the sale of yeast cultures for making wines at home. Baker's yeast is produced in large aerated fermentation tanks.

At the end of the fermentation, the contents of the tank are about 4% yeast solids. The concentration can be much higher, but only at a cost in efficiency. The cells are harvested by continuous centrifuges and are pressed into the familiar yeast cakes sold in the supermarket for home baking. Wholesale bakers purchase yeast in 50-lb. boxes.

Other important microbes that are sold industrially are the symbiotic nitrogen-fixing bacteria *Rhizobium* and *Bradyrhizobium*. These organisms are usually mixed with peat moss to preserve moisture; the farmer mixes the peat moss and bacterial inoculum with the seeds of legumes to assure infection of the plants with efficient nitrogen-fixing strains (Chapter 26). Gardeners for many years have used the insect pathogen *Bacillus thuringiensis* to control leaf-eating insect larvae. Commercial preparations containing toxic crystals and endospores of this organism are available at almost any gardening supply store.

Alternative Energy Sources Using Microorganisms

Each year, the United States produces many hundreds of tons of organic wastes from crops, forests, and municipalities. This waste is called biomass, and so far it has been disposed of by burning or placing in a dump. Conversion of biomass into alternative fuel

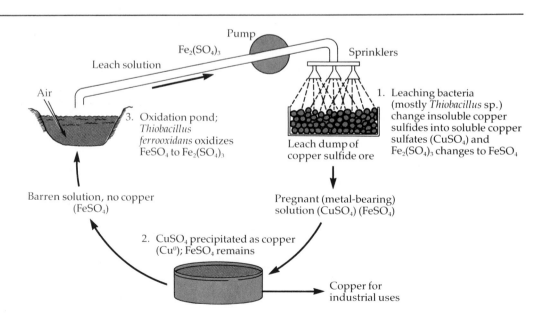

Figure 27-10 Biological leaching of copper ores. The chemistry of the process is considerably more complicated than shown here. The *Thiobacillus* bacteria essentially derive their energy from oxidation of sulfides to sulfates.

Pump

$Fe_2(SO_4)_3$

Leach solution

Sprinklers

Air

3. Oxidation pond; *Thiobacillus ferrooxidans* oxidizes $FeSO_4$ to $Fe_2(SO_4)_3$

Leach dump of copper sulfide ore

1. Leaching bacteria (mostly *Thiobacillus* sp.) change insoluble copper sulfides into soluble copper sulfates ($CuSO_4$) and $Fe_2(SO_4)_3$ changes to $FeSO_4$

Barren solution, no copper ($FeSO_4$)

Pregnant (metal-bearing) solution ($CuSO_4$) ($FeSO_4$)

2. $CuSO_4$ precipitated as copper (Cu^0); $FeSO_4$ remains

Copper for industrial uses

sources, called **bioconversion**, would help meet our energy needs as well as help dispose of troublesome accumulations of solid waste. The bioconversion of organic materials into alternative fuels by microorganisms is a developing industry.

Probably one of the most convenient energy sources produced from biomass conversion is **methane**. Methane was discussed in Chapter 26 as a product of the anaerobic treatment of sewage sludge. New York City recently opened a plant that will produce 10 million cubic feet of methane per day from wastes at one of its landfill sites. Large cattle-feeding lots must dispose of immense amounts of animal manure, and considerable effort has been devoted to devising practical methods for producing methane from these wastes. A major problem with any scheme for large-scale methane production is the need to economically concentrate the widespread biomass material. If it could be concentrated, the animal and human wastes in the United States could supply much of our energy now supplied by fossil fuels and natural gas.

An interesting aspect of methane production from animal wastes is the inhibition of the process by certain antibiotics fed to cattle. The purpose of these antibiotics is to suppress methane production from the largely cellulose foods in the animal's rumen. It is desirable that as much of the nutrients in these foods as possible be converted into meat or milk, but certain anaerobic bacteria convert some of these nutrients into methane gas. This gas tends to be lost to the atmosphere and represents an inefficient use of nutrients. The antibiotics used to inhibit these bacteria are excreted in the animals' wastes, and, unfortunately, they inhibit desired methane production from the wastes.

The agricultural industry has encouraged the production of ethanol from agricultural products. *Gasohol* (90% gasoline + 10% ethanol) is available in many parts of the United States and is used in automobiles worldwide. Corn is presently the most frequently used substrate, but eventually agricultural waste products may be used. A by-product of ethanol production is large quantities of ethanol-producing yeasts. The Food and Drug Administration recently approved use of such yeasts as a food additive, although they are most likely to be used as an animal food supplement. In the future, we will see more efforts to recover the energy now lost in industrial, municipal, and rural wastes.

Industrial Applications of Genetic Engineering

As we saw in Chapter 8, the basic technique of genetic engineering is the recombination of DNA molecules. In this technique, DNA is enzymatically cut up, and those fragments identified as containing useful genetic characteristics are spliced into a vector or carrier molecule, such as a plasmid (usually) or a bacterial virus, that can replicate the recombinant DNA. The plasmid can then be tranferred to a host cell by any of various methods. Once in the cell, the plasmid is capable of replicating independently of the chromosome. Bacteria normally carry only a few plasmids, but by a process of *gene amplification* the plasmids can be encouraged to replicate, and a single cell can be made to contain several hundred or even a few thousand plasmids.

Eucaryotic cells such as yeasts may also carry plasmids, and they can be more useful in some ways than bacteria. For example, eucaryotic cells are more likely to continuously excrete the desired product. A bacterium such as *E. coli* usually must be disrupted to recover the product. The ability of an organism to continuously excrete the product is a very valuable industrial characteristic.

Special techniques are available to produce the DNA segments to be used in genetic engineering. DNA sequences containing a particular gene can be synthesized, for example, with the messenger RNA isolated from the cell of a microorganism or an animal. From the messenger RNA, an enzyme, reverse transcriptase, can synthesize a complementary section of DNA that contains the gene. Short sequences of DNA can even be synthesized by machine (Figure 27-11). A keyboard on the machine is used to enter the sequence of bases desired, much as one enters the letters to compose a sentence. A microprocessor controls the synthesis of the DNA from stored supplies of nucleotides and the other necessary reagents. A chain of about 40 nucleotides can be synthesized by this method.

Protoplast fusion is a technique that increases the rate of recombination; it can be used to recombine DNA of bacterial species that normally do not exchange genetic information. The cell walls are enzymatically removed, and spherical *protoplasts* form. Two protoplasts are caused to fuse together; they then synthesize the normal cell wall, resulting in a hybrid organism with new combinations of genes. This technique is often used on antibiotic-producing streptomycetes to increase the frequency of isolation of useful

(a)   (b)

Figure 27-12 Fusion of two plant protoplasts. This process was induced with polyethylene glycol. **(a)** Mesophyll cell protoplast of potato is on the left, and protoplast of a tomato leaf cell is on the right ($\times 180$). **(b)** Nearly completed fusion after 15 minutes ($\times 180$). This technique combines genetic information in different plants much more rapidly and efficiently than conventional plant breeding.

Figure 27-11 A DNA synthesis machine. Short sequences of DNA can be synthesized by instruments such as this one. The necessary chemical components are placed in containers in the machine.

mutants. Another important use is the transfer of genetic information between plant species (Figure 27-12).

A different approach constructs new antibiotics by isolating a mutant microorganism that synthesizes only a portion of the complete antibiotic. The incomplete antibiotic lacks an essential chemical constituent. Adding the missing chemical constituent to the medium allows the mutant to form the normal antibiotic. However, adding substitute chemicals can result in the synthesis of new, more useful antibiotics.

Herbicides are not particularly selective in their activity and may kill crops as well as weeds. An approach to this problem is to use recombinant DNA techniques to make a crop plant resistant to the herbicide. Glyphosate herbicides kill plants by suppressing the production of amino acids needed for growth of the plants. Tobacco plants have been made resistant to glyphosates by inserting into their genome an enzyme-coding gene from *Salmonella* bacteria that allows the plant to continue producing these necessary amino acids. A plasmid of *Agrobacterium tumefaciens* can be used to change a plant's genome (see the box in Chapter 8, p. 224).

Gene transfers as therapy for inherited disease is a potential application of genetic engineering. As early as 1977, it was demonstrated that a cultured mouse cell that was deficient in the enzyme thymidine

Figure 27-13 Production of an insect toxin by genetic-engineering techniques. Root-dwelling *Pseudomonas fluorescens* bacteria have been genetically engineered to produce the insecticidal toxin normally produced by *Bacillus thuringiensis* (see Figure 10-12b). The crystal of toxin can be seen in the cell of *P. fluorescens*, which is capable of producing the toxin in much larger amounts than *B. thuringiensis* can. When root-eating insect larvae ingest the engineered bacteria, they are killed by the toxin. The use of the engineered bacteria is several times more effective than the conventional method of applying *B. thuringiensis* spores and toxin to the plant leaves ($\times 67,200$).

kinase could have the deficiency corrected by inserting a gene for this enzyme from a herpes simplex virus into the cell. Generations later, the progeny still produced the enzyme, indicating that the gene had been permanently inserted into the mouse cell's DNA.

The list of products and applications for genetic engineering based on the techniques of recombinant DNA is long and is expanding almost daily. As these products come into use in medicine and in the marketplace, they will affect our lives and well-being in ways that can only be guessed at today. Table 27-6 lists a representative group of such products now being manufactured or under final development. Figure 27-13 shows a product of genetic-engineering techniques.

TABLE 27-6 *Some Products of Genetic Engineering*

Product	Comments
Pharmaceutical Products	
Tissue plasminogen activator (t-PA)	Dissolves the fibrin of blood clots; therapy for heart attacks
Erythropoietin (EPO)	Treatment of anemia
Insulin	Produced by *E. coli*; therapy for diabetics; this human insulin is better tolerated than insulin extracted from animals
Interleukin-2 (IL-2)	Produced by *E. coli*; possible treatment for cancer; stimulates the immune system
Gamma-interferon	Produced by *E. coli*, *Saccharomyces cerevisiae*; possible cancer, virus-disease therapy
Tumor necrosis factor (TNF)	Causes disintegration of tumor cells
Human growth hormone (HGH)	Produced by *E. coli*; corrects growth deficiencies in children
Epidermal growth factor (EGF)	Heals wounds, burns, ulcers
Monoclonal antibodies	Fusion of cancer cell and antibody-producing body cell; possible therapy for cancer, transplant rejection; diagnostic tests
Hepatitis B vaccine	Plasmid with hepatitis gene inserted into vaccinia, which is readily cultivated, whereas the hepatitis virus is not
Agricultural Products	
Ice-minus bacterium, *Pseudomonas syringae*	Lacks normal protein product that initiates undesirable ice formation on plants
Pseudomonas fluorescens bacterium	Has toxin-producing gene from insect pathogen *Bacillus thuringiensis*; toxin kills root-eating insects that ingest bacteria (Figure 27-13)
Rhizobium meliloti bacterium	Modified for enhanced nitrogen fixation
Animal Husbandry Products	
Porcine growth hormone (PGH)	Improves weight gain in swine
Bovine growth hormone (BGH)	Improves weight gain in cattle
Food Production Products	
Rennin	Produced by *Aspergillus niger*; formation of milk curds for dairy products
Recreation Products	
Snomax®	*Pseudomonas syringae* makes a protein that causes water to freeze at higher-than-normal temperatures; used for making snow at ski resorts

Study Outline

Food Microbiology (pp. 692–704)

Food Preservation and Spoilage (pp. 692–698)

1. The earliest methods of preserving foods were drying, the addition of salt or sugar, and fermentation.

Canning (pp. 693–695)

1. In 1810, Nicholas Appert devised a canning process that involved keeping air out of sealed cans and heating food in the cans.
2. In the nineteenth century, steam under pressure (pressure cooking) was used in canning.

Commercial Sterilization (pp. 693–694)

1. Today, commercial sterilization heats canned foods to the minimum temperature necessary to destroy *C. botulinum* endospores while minimizing alteration of the food.
2. The commercial sterilization process uses sufficient heat to reduce a population of *C. botulinum* by 12 logarithmic cycles (12D treatment).
3. Endospores of thermophiles can survive commercial sterilization.

Spoilage of Canned Food (pp. 694–695)

1. Canned foods stored above 45°C can be spoiled by thermophilic anaerobes.
2. Thermophilic anaerobic spoilage might be accompanied by gas production; if no gas is formed, the spoilage is called flat sour spoilage.
3. Spoilage by mesophilic bacteria is usually from improper heating procedures or leakage.
4. Acidic foods can be preserved by heat of 100°C because microorganisms that survive are not capable of growth in a low pH.
5. *Byssochlamys*, *Aspergillus*, and *Bacillus coagulans* are acid-tolerant and heat-resistant organisms that can spoil acidic foods.

Pasteurization (pp. 695–696)

1. Louis Pasteur determined that, because spoilage was caused by only certain microorganisms, it was not necessary to kill all microorganisms to prevent spoilage.
2. Pasteurization is the application of heat that selectively kills microorganisms without greatly altering the quality of the food.

Aseptic Packaging (p. 696)

1. Presterilized materials are assembled into packages and aseptically filled with heat-sterilized liquid foods.

Low-Temperature Preservation (pp. 696–697)

1. Many microorganisms, although few pathogens, will grow slowly at refrigeration temperatures (0°C to 7°C).
2. *Clostridium botulinum* type E, *Yersinia enterocolitica*, and *Listeria monocytogenes* are pathogens that will grow at low temperatures.
3. Freezing might not kill bacteria, but it will prevent their growth.

Radiation and Food Preservation (pp. 697–698)

1. Gamma radiation can be used to sterilize food, kill insects and parasitic worms, and prevent sprouting of fruits and vegetables.
2. Microwaves kill bacteria by heating the food.

Chemical Preservatives (p. 698)

1. Sodium benzoate, sorbic acid or potassium sorbate, or calcium propionate are added to foods to inhibit the growth of molds.
2. These chemicals are metabolized by humans.
3. Sodium nitrate and sodium nitrite are added to meats to preserve the red color and to prevent the germination of *C. botulinum* endospores.
4. Nitrates are converted to nitrites by some bacteria using anaerobic respiration; nitrites are the active ingredient.

Foodborne Infections and Microbial Intoxications (pp. 698–699)

1. In a foodborne infection, the pathogen grows in the infected host.
2. Gastroenteritis, typhoid fever, and dysentery are foodborne infections.
3. Microbial intoxication results from ingestion of a toxin produced by microbes growing in food.
4. Botulism, staphylococcal food poisoning, and mycotoxicoses are examples of microbial intoxications.
5. Pasteurized dairy milk might contain gram-negative psychrophiles.
6. Pasteurized products of cultured milk (such as buttermilk and sour cream) contain large numbers of lactic-acid bacteria.

7. Public health service standards for Grade A pasteurized milk are a standard plate count of fewer than 20,000 bacteria per milliliter and not more than 10 coliforms per milliliter.

8. Dry milk products must have a standard plate count of fewer than 30,000 bacteria per gram or a coliform count fewer than 10 per gram.

9. The species of bacteria in unpasteurized (raw) milk are not controlled, and raw milk can transmit disease.

The Role of Microorganisms in Food Production (pp. 700–704)

Cheese (pp. 700–701)

1. The milk protein casein curdles because of the action by lactic-acid bacteria or the enzyme rennin.

2. Cheese is the curd separated from the liquid portion of milk, called whey.

3. Hard cheeses are produced by lactic-acid bacteria growing in the interior of the curd.

4. The growth of microorganisms in cheeses is called ripening.

5. Semisoft cheeses are ripened by bacteria growing on the surface.

6. Soft cheeses are ripened by *Penicillium* growing on the surface.

Other Dairy Products (p. 701)

1. Old-fashioned buttermilk was produced by lactic-acid bacteria growing during the butter-making process.

2. Commercial buttermilk is made by letting lactic-acid bacteria grow in skim milk for 12 hours.

3. Sour cream, yogurt, kefir, and kumiss are produced by lactobacilli, streptococci, or yeast growing in low-fat milk.

Nondairy Fermentations (p. 701)

1. Sugars in bread dough are fermented by yeast to ethanol and CO_2; the CO_2 causes the bread to rise.

2. Sauerkraut, pickles, olives, and soy sauce are the products of microbial fermentations.

Alcoholic Beverages and Vinegar (pp. 701–702)

1. Carbohydrates obtained from grains, potatoes, or molasses are fermented by yeast to produce ethanol in the production of beer, ale, and distilled spirits.

2. The sugars in fruits such as grapes are fermented by yeast to produce wines.

3. In wine-making, lactic-acid bacteria convert malic acid into lactic acid in malolactic fermentation.

4. *Acetobacter* and *Gluconobacter* oxidize ethanol in wine to acetic acid (vinegar).

Microorganisms as a Food Source (pp. 702–704)

1. Microorganisms can grow and produce proteinaceous cell material from otherwise unusable substrates such as cellulose.

2. The microbial cells can be made into a food called single-cell protein (SCP).

3. *Spirulina*, yeasts, and *Methylophilus* are cultured and used as SCP.

Industrial Microbiology (pp. 704–713)

1. Microorganisms produce alcohol, lactic acid, glycerol, and antibiotics that are used in food manufacturing and other industrial processes.

Industrial Products (pp. 705–710)

1. Most amino acids used in foods and medicine are produced by bacteria.

2. Lysine and glutamic acid are produced by *Corynebacterium glutamicum*.

3. Citric acid, used in foods, is produced by *Aspergillus niger*.

4. Enzymes used in manufacturing foods, medicines, and other goods are produced by microbes.

5. Some vitamins used as food supplements are made by microorganisms.

6. Vaccines, antibiotics, and steroids are products of microbial growth.

7. Some microbes (*Rhizobium*, *Bradyrhizobium*, and *Bacillus thuringiensis*) are grown for agricultural use.

8. Coal-mining wastes contain sulfides that are converted into sulfuric acid by *Thiobacillus ferrooxidans*.

Alternative Energy Sources Using Microorganisms (pp. 710–711)

1. Organic waste, called biomass, can be converted by microorganisms into alternative fuels, a process called bioconversion.

2. Fuels produced by microbial fermentation are methane and ethanol.

Industrial Applications of Genetic Engineering (pp. 711–713)

1. Selected genes can be inserted into bacteria and yeasts by genetic engineering so these cells will produce large quantities of desired hormones, vaccines, and antibiotics.

2. Genes from bacteria can also be inserted into plants (by bacteria or other vectors) to improve plant growth and crop productivity.

Study Questions

Review

1. Why are foods preserved?

2. List five methods used to preserve foods.

3. Matching.

   ____ Sorbic acid   (a) Sterilization
   ____ 5°C   (b) Pasteurization
   ____ 72°C for 15 seconds   (c) Low-temperature
   ____ 121°C for 15   preservation
         minutes   (d) Chemical
   ____ 12D treatment   preservation

4. Since many mesophilic bacteria can grow at low temperatures, what value is it to store food in a refrigerator?

5. Define pasteurization and aseptic packaging.

6. Pasteurization does not kill all microorganisms, so food can still spoil. Why then are dairy products pasteurized?

7. State one advantage and one disadvantage of the addition of nitrites to foods.

8. Outline the steps in the production of cheese, and compare the production of hard and soft cheeses.

9. Outline the steps in the process of wine production.

10. List at least two advantages and two disadvantages of SCP.

11. How can microorganisms provide energy sources? What metabolic processes can result in fuels?

12. Nicholas Appert's invention was the forerunner of hermetically sealed cans. Why do hermetically sealed cans prevent food spoilage?

13. Why is a can of blackberries preserved by commercial sterilization typically heated to 100°C instead of at least 116°C?

14. Under what conditions would you expect to find thermophiles spoiling foods? Psychrophiles?

15. Discuss the importance of industrial microbiology.

16. How will advances in genetic engineering improve production of expensive chemicals? Outline the steps taken to engineer a cell to produce a selected protein.

Challenge

1. What bacteria seem to be most frequently used in the production of food? Can you guess why?

2. Why do the following processes preserve food?
   (a) Fermentation
   (b) Salting
   (c) Drying

3. Explain the processes involved in the production of sourdough bread.

4. Three to five days after eating Thanksgiving dinner at a restaurant, 112 people developed fever and gastroenteritis. All the food had been consumed except for five "doggie" bags. Bacterial analysis of the mixed contents of the bags (containing roast turkey, giblet gravy, and mashed potatoes) showed the same bacterium that was isolated from the patients. The gravy had been prepared from giblets of 43 turkeys that had been refrigerated for three days prior to preparation. The uncooked giblets were ground in a blender and added to a thickened, hot stock mixture. The gravy was not reboiled and was stored at room temperature throughout Thanksgiving Day. What was the source of the illness? What was the most likely etiologic agent? Was this an infection or an intoxication? (*Hint*: Refer to Chapter 24.)

5. *Methylophilus methylotrophus* can convert methane (CH_4) into proteins. Amino acids are represented by

$$H_2N-\underset{\underset{H}{|}}{\overset{\overset{R}{|}}{C}}-\underset{OH}{\overset{O}{\underset{\diagdown}{\overset{\diagup\diagup}{C}}}}$$

Diagram a pathway illustrating the production of at least one amino acid.

Further Reading

Biotechnology. *Science* 219:611ff, 1983. A special issue on biotechnology, including discussions of monoclonal antibodies and applications of genetic engineering to industrial microbiology.

Brierley, C. L. "Microbiological mining." *Scientific American* 247(2):44–53, August 1982. Discusses the role of *Thiobacillus* in leaching copper from ore.

Hesseltine, C. W. "Microbiology of oriental fermented foods." *Annual Review of Microbiology* 37:575–601, 1983. An interesting monograph on the bacteria and fungi used to prepare traditional Asian foods.

Industrial microbiology. *Scientific American* 245(3), September 1981. An issue devoted entirely to the topic of industrial microbiology and genetic engineering.

Jay, J. M. *Modern Food Microbiology*, 3rd ed. New York: Van Nostrand Reinhold, 1986. A current textbook on food microbiology.

Kosikowski, F. V. "Cheese." *Scientific American* 252(5):88–99, May 1985. Describes how types of milk, species of microbes, and ripening times are used to produce 2000 cheese varieties.

Microbes for hire. *Science 85* 6(6):30–46, July/August 1985. A special issue on the use of microbes in biotechnology.

Reed, G. *Prescott and Dunn's Industrial Microbiology*, 4th ed. Westport, Conn.: AVI, 1983. A standard reference book in industrial microbiology.

Rose, A. H. ed. *Economic Microbiology*, 8 vols. New York: Academic Press. Published over several years, these volumes cover food, energy, and other aspects of applied microbiology.

Sawyer, C. A. D., and J. J. Pestka. "Foodservice systems: presence of injured bacteria in foods during food product flow." *Annual Review of Microbiology* 39:51–67, 1985. An analysis of microbiological quality of foods that are prepared and stored, both hot and cold, before serving.

Society for Industrial Microbiology. *Developments in Industrial Microbiology*. Arlington, Virginia: Society for Industrial Microbiology. Proceedings of the general meeting of the Society for Industrial Microbiology, published annually. Each volume contains discussions of current advances in food, dairy, and pharmaceutical microbiology.

Weaver, R. F. "Changing life's genetic blueprint." *National Geographic Magazine* 166(6):818–847, December 1984. Well-illustrated with diagrams and photos explaining the basics of genetic engineering.

Zeikus, J. G. "Chemical and fuel production by anaerobic bacteria." *Annual Review of Microbiology* 34:423–464, 1980. A summary of the technology and practicality of producing hydrocarbon fuels and organic solvents from agricultural and municipal wastes.

APPENDIX A

Classification of Bacteria According to Bergey's Manual of Systematic Bacteriology

Kingdom Procaryotae

Division I. **Gracilicutes** Procaryotes with thin cell walls, implying a gram-negative type of cell wall

Division II. **Firmicutes** Procaryotes with thick and strong skin, indicating a gram-positive type of cell wall

Division III. **Tenericutes** Procaryotes of a pliable and soft nature, indicating the lack of a rigid cell wall

Division IV. **Mendosicutes** Procaryotes with faulty cell walls, suggesting the lack of conventional peptidoglycan

VOLUME I

Section 1 The Spirochetes

Order I. Spirochaetales
    Family I. Spirochaetaceae
        Genera
          I. *Spirochaeta*   III. *Treponema*
          II. *Cristispira*   IV. *Borrelia*
    Family II. Leptospiraceae
        Genus I. *Leptospira*
Other organisms: Hindgut spirochetes of termites and *Cryptocercus punctulatus* (wood-eating cockroach).

Section 2 Aerobic/Microaerophilic, Motile, Helical/Vibrioid Gram-Negative Bacteria

    Genera
    Aquaspirillum   *Campylobacter*
    Spirillum   *Bdellovibrio*
    Azospirillum   *Vampirovibrio*
    Oceanospirillum

Section 3 Nonmotile (or Rarely Motile), Gram-Negative Curved Bacteria

    Family I. Spirosomaceae
        Genera
          I. *Spirosoma*
          II. *Runella*
          III. *Flectobacillus*
    Other genera
        Microcyclus
        Meniscus
        Brachyarcus
        Pelosigma

Section 4 Gram-negative Aerobic Rods and Cocci

    Family I. Pseudomonadaceae
        Genera
          I. *Pseudomonas*   III. *Frateuria*
          II. *Xanthomonas*   IV. *Zoogloea*
    Family II. Azotobacteraceae
        Genera
          I. *Azotobacter*
          II. *Azomonas*
    Family III. Rhizobiaceae
        Genera
          I. *Rhizobium*   III. *Agrobacterium*
          II. *Bradyrhizobium*   IV. *Phyllobacterium*
    Family IV. Methylococcaceae
        Genera
          I. *Methylococcus*
          II. *Methylomonas*
    Family V. Halobacteriaceae
        Genera
          I. *Halobacterium**
          II. *Halococcus**
    Family VI. Acetobacteraceae
        Genera
          I. *Acetobacter*
          II. *Gluconobacter*
    Family VII. Legionellaceae
        Genus I. *Legionella*
    Family VIII. Neisseriaceae
        Genera
          I. *Neisseria*   III. *Acinetobacter*
          II. *Moraxella*   IV. *Kingella*

Other genera

Beijerinckia	*Alcaligenes*
Derxia	*Serpens*
Xanthobacter	*Janthinobacterium*
Thermus	*Brucella*
Thermomicrobium	*Bordetella*
Halomonas	*Francisella*
Alteromonas	*Paracoccus*
Flavobacterium	*Lampropedia*

Section 5 Facultatively Anaerobic Gram-Negative Rods

Family I. Enterobacteriaceae
Genera

I. *Escherichia*	VIII. *Serratia*
II. *Shigella*	IX. *Hafnia*
III. *Salmonella*	X. *Edwardsiella*
IV. *Citrobacter*	XI. *Proteus*
V. *Klebsiella*	XII. *Providencia*
VI. *Enterobacter*	XIII. *Morganella*
VII. *Erwinia*	XIV. *Yersinia*

Other genera of the family Enterobacteriaceae

Obesumbacterium	*Rahnella*
Xenorhabdus	*Cedecea*
Kluyvera	*Tatumella*

Family II. Vibrionaceae
Genera

I. *Vibrio*	III. *Aeromonas*
II. *Photobacterium*	IV. *Plesiomonas*

Family III. Pasteurellaceae
Genera
   I. *Pasteurella*
   II. *Haemophilus*
   III. *Actinobacillus*
Other genera

Zymomonas	*Gardnerella**
Chromobacterium	*Eikenella*
Cardiobacterium	*Streptobacillus*
Calymmatobacterium	

Section 6 Anaerobic Gram-Negative Straight, Curved, and Helical Rods

Family I. Bacteroidaceae
Genera

I. *Bacteroides*	VIII. *Wolinella*
II. *Fusobacterium*	IX. *Selenomonas*
III. *Leptotrichia*	X. *Anaerovibrio*
IV. *Butyrivibrio**	XI. *Pectinatus*
V. *Succinimonas*	XII. *Acetivibrio*
VI. *Succinivibrio*	XIII. *Lachnospira**
VII. *Anaerobiospirillum*	

Section 7 Dissimilatory Sulfate- or Sulfur-Reducing Bacteria

Genera

Desulfuromonas	*Desulfobacter*
Desulfovibrio	*Desulfobulbus*
Desulfomonas	*Desulfosarcina*
Desulfococcus	

Section 8 Anaerobic Gram-Negative Cocci

Family I. Veillonellaceae
Genera
   I. *Veillonella*
   II. *Acidaminococcus*
   III. *Megasphaera*

Section 9 The Rickettsias and Chlamydias

Order I. Rickettsiales
Family I. Rickettsiaceae
Tribe I. Rickettsieae
Genera
   I. *Rickettsia*
   II. *Rochalimaea*
   III. *Coxiella*
Tribe II. Ehrlichieae
Genera
   IV. *Ehrlichia*
   V. *Cowdria*
   VI. *Neorickettsia*
Tribe III. Wolbachieae
Genera
   VII. *Wolbachia*
   VIII. *Rickettsiella*
Family II. Bartonellaceae
Genera
   I. *Bartonella*
   II. *Grahamella*
Family III. Anaplasmataceae

I. *Anaplasma*	III. *Haemobartonella*
II. *Aegyptianella*	IV. *Eperythrozoon*

Order II. Chlamydiales
Family I. Chlamydiaceae
Genus I. *Chlamydia*

Section 10 The Mycoplasmas

Order I. Mycoplasmatales
Family I. Mycoplasmataceae
Genera
   I. *Mycoplasma*
   II. *Ureaplasma*
Family II. Acholeplasmataceae
Genus I. *Acholeplasma*
Family III. Spiroplasmataceae
Genus I. *Spiroplasma*
Other genera
   Anaeroplasma
   Thermoplasma*
   Mycoplasmalike organisms of plants and invertebrates

Section 11 Endosymbionts

A. Endosymbionts of Protozoa
Genera

I. *Holospora*	IV. *Lyticum*
II. *Caedibacter*	V. *Tectibacter*
III. *Pseudocaedibacter*	

"Stigonemataceae"
Genera
Chlorogloeopsis
Fischerella
Geitleria
Prochlorales
Prochloraceae
Genus *Prochloron*

Section 20 Aerobic Chemolithotrophic Bacteria and Associated Organisms

Nitrifying Bacteria
Family I. Nitrobacteraceae
Genera

I. *Nitrobacter*	VI. *Nitrosococcus*
II. *Nitrospina*	VII. *Nitrosospira*
III. *Nitrococcus*	VIII. *Nitrosolobus*
IV. *Nitrospira*	IX. "*Nitrosovibrio*"
V. *Nitrosomonas*	

Colorless Sulfur Bacteria
Genera

Thiobacterium	*Thiomicrospira*
Macromonas	*Thiosphaera*
Thiospira	*Acidiphilium*
Thiovulum	*Thermothrix*
Thiobacillus	

Obligate Chemolithotrophic Hydrogen Oxidizers
Genus *Hydrogenbacter*
"Siderocapsaceae"
"Siderocapsa"
"Naumaniella"
"Siderococcus"
"Ochrobium"
The Magnetotactic Bacteria
Genera
Aquaspirillum (*A. magnetotacticum*)
"Bilophococcus"

Section 21 Budding and/or Appendaged Bacteria

Prosthecate Bacteria
Budding bacteria
Genera

Hyphomicrobium	*Ancalomicrobium*
Hyphomonas	*Prosthecomicrobium*
Pedomicrobium	*Stella*

Nonbudding bacteria
Genera

Caulobacter	*Prosthecobacter*
Asticcacaulis	"*Thiodendron*"

Nonprosthecate Bacteria
Budding bacteria
Genera

Planctomyces	*Angulomicrobium*
Isophaera	*Gemmiger*
Blastobacter	*Ensifer*

Nonbudding bacteria
Genera
Gallionella
Nevskia

Morphologically unusual budding bacteria (involved in iron and manganese deposition)
Genera

Seliberia	"*Caulococcus*"
"*Metallogenium*"	"*Kuznezovia*"

Others
Spinate bacteria

Section 22 Sheathed Bacteria

Genera

Sphaerotilus	*Phragmidiothrix*
Leptothrix	*Crenothrix*
Haliscominobacter	*Clonothrix*
Lieskeella	

Section 23 Nonphotosynthetic, Nonfruiting, Gliding Bacteria

Order I. Cytophagales
Family I. Cytophagaceae
Genera

I. *Cytophaga*	V. *Flexibacter*
II. *Sporocytophaga*	VI. *Microscilla*
III. *Capnocytophaga*	VII. *Saprospira*
IV. *Flexithrix*	VIII. *Herpetosiphon*

Order II. Lysobacterales
Family I. Lysobacteraceae
Genus *Lysobacter*
Order III. Beggiatoales
Family I. Beggiatoaceae
Genera

I. *Beggiatoa*	III. *Thiospirillopsis*
II. *Thioploca*	IV. *Thiothrix*

Others
Simonsiellaceae
Genera
I. *Simonsiella*
II. *Alysiella*
Pelonemataceae
Genera

"*Pelonema*"	"*Achroonema*"
"*Peloploca*"	"*Desmanthus*"

Families and genera incertae sedis
Genera

Chitinophaga	*Vitreoscilla*
"*Toxothrix*"	*Desulfonema*
Leucothrix	*Agitococcus*

Section 24 Gliding, Fruiting Bacteria

Order I. Myxobacterales
Family I. Myxococcaceae
Genus *Myxococcus*
Family II. Archangiaceae
Genus *Archangium*
Family III. Cystobacteraceae
Genera
I. *Cystobacter*
II. *Melittangium*
III. *Stigmatella*

Family IV. Polyangiaceae
Genera
    I. *Polyangium*
    II. *Nannocystis*
    III. *Chondromyces*

Section 25 Archaeobacteria

Order I. Methanobacteriales
  Family I. Methanobacteriaceae
    Genera
      I. *Methanobacterium*
      II. *Methanobrevibacter*
  Family II. Methanothermaceae
    Genus I. *Methanothermus*
Order II. Methanococcales
  Family I. Methanococcaceae
    Genus I. *Methanococcus*
Order III. Methanomicrobiales
  Family I. Methanomicrobiaceae
    Genera
      I. *Methanomicrobium*
      II. *Methanogenium*
      III. *Methanospirillum*
  Family II. Methanosarcinaceae
    Genera
      I. *Methanosarcina*   IV. *Methanothrix*
      II. *Methanococcoides* V. *Halomethanococcus*
      III. *Methanolobus*
  Family III. Methanoplanaceae
    Genus I. *Methanoplanus*
  Others
    Methanosphaera
Order IV. Thermoproteales
  Family I. Thermoproteaceae
    Genera
      I. *Thermoproteus*
      II. *Thermophilum*
  Family II. Desulfurococcaceae
    Genera
      I. *Desulfurococcus*   III. *Thermodiscus*
      II. *Thermococcus*   IV. *Pyrodictium*
Order V. "Thermococcales"
  Family I. "Thermococcaceae"
    Genera
      I. *"Thermococcus"*
      II. *"Pyrococcus"*
Order VI. "Sulfolobales"
  Family I. "Sulfolobaceae"
    Genus I. *"Sulfolobus"*
Order VII. "Thermoplasmales"
  Family I. "Thermoplasmaceae"
    Genus I. *Thermoplasma*
Order VIII. "Halobacteriales"
  Family I. Halobacteriaceae
    Genera
      I. *Halococcus*   IV. *Haloarcula*
      II. *Halobacterium* V. *Natronococcus*
      III. *Haloferax*   VI. *Natronobacterium*

VOLUME 4

Section 26 Nocardioform Actinomycetes
  Genera
    I. *Nocardia*   VII. *Faenia*
    II. *Rhodococcus*   (*Micropolyspora*)
    III. *Nocardioides* VIII. *Promicromonospora*
    IV. *Pseudonocardia* IX. *Intrasporangium*
    V. *Oerskovia*   X. *Actinopolyspora*
    VI. *Saccharopolyspora* XI. *Saccharomonospora*

Section 27 Actinomycetes with Multi-Locular Sporangia
  Genera
    I. *Geodermatophilus*
    II. *Dermatophilus*
    III. *Frankia*

Section 28 Actinoplanetes
  Genera
    I. *Actinoplanes*   IV. *Dactylosporangium*
    II. *Ampullariella* V. *Micromonospora*
    III. *Pilimelia*

Section 29 *Streptomyces* and Related Genera
  Genera
    I. *Streptomyces*   III. *Kineosporia*
    II. *Streptoverticillium* IV. *Sporichthya*

Section 30 Maduromycetes
  Genera
    I. *Actinomadura*   V. *Planomonospora*
    II. *Microbispora*   VI. *Spirillospora*
    III. *Microtetraspora* VII. *Streptosporangium*
    IV. *Planobispora*

Section 31 *Thermomonospora* and Related Genera
  Genera
    I. *Thermomonospora* III. *Nocardiopsis*
    II. *Actinosynnema*   IV. *Streptoalloteichus*

Section 32 Thermoactinomycetes
  Genus I. *Thermoactinomyces*

Section 33 Other Genera
  Genera
    Glyomyces   *Kitasatospora*
    Kibdelosporangium *Saccharothrix*

---

* Genera that have been placed in more than one volume: *Gardnerella*, *Lachnospira*, and *Butyrivibrio* have thin, gram-positive walls but stain as gram-negatives. *Halobacterium*, *Halococcus*, and *Thermoplasma* are *Archaeobacteria* that are also included among the Gram-Negative Rods and Cocci or the Mycoplasmas. Nocardioforms are included with Gram-Positive Eubacteria and Actinomycetes because their relatedness to either group is uncertain.

Pronunciation of Scientific Names

RULES OF PRONUNCIATION

There are rules for the pronunciation of scientific names which depend, in part, on the derivation of the root word and its vowel sounds. We have provided some general guidelines below. Pronunciations frequently do not follow the rules because a common usage has become "accepted" or the derivation of the name cannot be determined. Furthermore, for many scientific names there are alternative correct pronunciations.

Vowels. Pronounce all the vowels in scientific names. Two vowels written together and pronounced as one sound are called a **diphthong** (for example, the "ou" in "sound"). A special comment is needed about the pronunciation of the vowel endings *-i* and *-ae*: There are two alternative ways to pronounce each of these. In this book, we usually give the pronunciation of a long e ("ē") to the *-i* ending and a long i ("ī") to the *-ae* ending. However, the reverse pronunciations are also correct and in some cases are preferred. For example, *coli* is usually pronounced kō′lī.

Consonants. When *c* or *g* is followed by ae, e, oe, i, or y, it has a soft sound. When *c* or *g* is followed by a, o, oi, or u, it has a hard sound. When a double *c* is followed by e, i, or y, it is pronounced as ks (i.e., cocci).

Accent. The accented syllable is usually the next to last or third to last syllable.

1. The accent is on the next to last syllable:
   - When the name contains only two syllables, for example, péstis.
   - When the next to last syllable is a diphthong, for example, *Amoéba.*
   - When the vowel of the next to last syllable is long, for example, *Treponéma.* The vowel in the next to last syllable is long in words ending in these suffixes:

Suffix	Example
-ales	orders such as Eubacteriáles
-ina	*Sarcína*
-anus, -anum	*pasteuriánum*
-uta	*diminúta*

   - When the word ends in one of these suffixes:

Suffix	Example
-atus, -atum	*caudátum*
-ella	*Salmonélla*

2. The accent is on the third to last syllable in family names. Families end in -aceae, which is always pronounced -ā′sē-ē.

PRONUNCIATION OF ORGANISMS IN THIS TEXT

Pronunciation key:

a	hat	ē	see	o	hot	th	thin
ā	age	ė	term	ō	go	u	cup
ã	care	g	go	ô	order	ù	put
ä	father	i	sit	oi	oil	ü	rule
ch	child	ī	ice	ou	out	ū	use
e	let	ng	long	sh	she	zh	seizure

Acanthamoeba polyphaga   a-kan-thä-mē′bä pol-ē-fā′gä
Acetobacter   ä-sē-tō-bak′tėr
Acinetobacter   ā-si-ne′tō-bak-tėr
Actinobacillus actinomycemcomitans   ak-tin-ō-bä-sil′lus ak-tin-ō-mī-sem-kō′mi-tanz
Actinomyces israelii   ak-tin-ō-mī′sēs is-rā′lē-ē
Aedes aegypti   ä-e′dēz ē-jip′tē
A. albopictus   al-bō-pik′tus
Aeromonas hydrophilia   ār-ō-mō′nas hī-dro′fil-ä
Agaricus augustus   ä-gär′i-kus au-gus′tus
Agrobacterium tumefaciens   ag-rō-bak-ti′rē-um tü-me-fā′shenz
Ajellomyces   ä-jel-lō-mī′sēs
Allescheria boydii   al-lesh-er′ē-ä boi′dē-ē
Amanita phalloides   am-an-ī′ta fal-loi′dēz
Amblyomma americanum   am-blē-ōm′mä ä-mãr-i-kā′num
Anabaena   an-ä-bē′nä
Anopheles   an-of′el-ēz
Aquaspirillum bengal   a-kwä-spī-ril′lum ben′gal
A. graniferum   gra-ni′fir-um
A. magnetotacticum   mag-ne-tō-tak′ti-kum
Arcula   är-kū′lä
Arthroderma   är-thrō-dėr′mä
Ascaris lumbricoides   as′kar-is lum-bri-koi′dēz
Ashbya gossypii   ash′bē-ä gos-sip′ē′ē
Aspergillus flavus   a-spėr-jil′lus flā′vus
A. fumigatus   fü-mi-gä′tus
A. niger   nī′jėr
A. oryzae   ô-rī′zē
A. soyae   soi′ī

A. tamarii   ta-mär'ē-ē

Azolla   ā-zō'lä

Azomonas   ā-zō-mō'nas

Azospirillum   ā-zō-spī-ril'lum

Azotobacter   ä-zō-tō-bak'tėr

Babesia microti   ba-bē'sē-ä mī-krō'tē

Bacillus anthracis   bä-sil'lus an-thrā'sis

B. brevis   bre'vis

B. cereus   se'rē-us

B. coagulans   kō-ag'ū-lanz

B. licheniformis   lī-ken-i-fôr'mis

B. megaterium   meg-ä-tėr'ē-um

B. polymyxa   po-lē-miks'ä

B. sphaericus   sfe'ri-kus

B. stearothermophilus   ste-rō-thėr-mä'fil-us

B. subtilis   su'til-us

B. thuringiensis   thür-in-jē-en'sis

Bacteroides fragilis   bak-tė-roi'dēz fra'jil-is

B. hypermegas   hī'pėr-meg-äs

Balantidium coli   bal-an-tid'ē-um kō'lī (or kō'lē)

Bdellovibrio   del-lō-vib'rē-ō

Beauveria   bō-vär'ē-a

Beggiatoa   bej-jē-ä-tō'ä

Beijerinckia   bī-jė-rink'ē-ä

Bifidobacterium globosum   bī-fi-dō-bak-ti'rē-um glob-ō'sum

B. pseudolongum   sū-dō-lông'um

Blastomyces dermatitidis   blas-tō-mī'sēz dėr-mä-tit'i-dis

Blattabacterium   blat-tä-bak-ti'rē-um

Boletus edulis   bō-lē'tus e'dū-lis

Bordetella pertussis   bôr-de-tel'lä pėr-tus'sis

Borrelia burgdorferi   bôr-rel'ē-ä burg-dôr'fėr-ē

Bradyrhizobium   brad-ē-rī-zō'bē-um

Branhamella catarrhalis   bran-ha-mel'lä ka-tär'al-is

Brucella abortus   brü-sel'lä ä-bôr'tus

B. canis   kā'nis

B. melitensis   me-li-ten'sis

B. suis   sü'is

Byssochlamys fulva   bis-sō-klam'is fül'vä

Calvulina cristata   kal-vū-lī'nä kris'tä-tä

Calymmatobacterium granulomatis   kal-im-mä-tō-bak-ti'rē-um gran-ū-lō'mä-tis

Campylobacter fetus   kam-pī-lō-bak'tėr fē'tus

C. jejuni   jē-jū'nē

C. pylori   pī'lôr-ē

Candida albicans   kan'did-ä al'bi-kanz

C. krusei   krūs'ä-ē

C. utilis   ū'til-is

Canis familiaris   kā'nis fa-mil'yär-is

Carpenteles   kär-pen'tel-ēz

Caulobacter bacteroides   ko-lō-bak'tėr bak-tėr-oi'dēz

Cephalosporium   sef-ä-lō-spô'rē-um

Ceratocystis ulmi   sē-rä-tō-sis'tis ul'mē

Chlamydia psittaci   kla-mi'dē-ä sit'tä-sē

C. trachomatis   trä-kō'mä-tis

Chlamydomonas   klam-i-dō-mō'näs

Chlorobium   klô-rō'bē-um

Chloroflexus   klô-rō-flex'us

Chromatium vinosum   krō-mä'tē-um vi-nō'sum

Chroococcus turgidus   krō-ō-kok'kus tėr'gi-dus

Chrysops   krī'sops

Citrobacter   sit'rō-bak-tėr

Cladosporium werneckii   kla-dō-spô'rē-um wėr-ne'kē-ē

Claviceps purpurea   kla'vi-seps pùr-pù-rē'ä

Clonorchis sinensis   klo-nôr'kis si-nen'sis

Clostridium acetobutylicum   klô-stri'dē-um a-sē-tō-bū-ti'li-kum

C. bifermentens   bī-fėr-men'tans

C. botulinum   bo-tū-lī'num

C. butyricum   bü-ti'ri-kum

C. histolytica   his-tō-lit'i-kä

C. novyi   nō'vi-ē

C. pasteurianum   pas-tyėr-ē-ä'num

C. perfringens   pėr-frin'jens

C. septicum   sep'ti-kum

C. sporogenes   spô-rä'jen-ēz

C. subterminale   sub-tėr-mi-na'lē

C. tetani   te'tan-ē

C. thermosaccharolyticum   thėr-mō-sak-kär-ō-li'ti-kum

Coccidioides immitis   kok-sid-ē-oi'dēz im'mi-tis

Corynebacterium diphtheriae   kô-rī-nē-bak-ti'rē-um dif-thi'rē-ī

C. glutamicum   glü-tam'i-kum

C. xerosis   ze-rō'sis

Coxiella burnetii   käks-ē-el'lä bėr-ne'tē-ē

Cristispira pectinis   kris-tē-spī'rä pek'tin-is

Cryptococcus neoformans   krip-tō-kok'kus nē-ō-fôr'manz

Cryptosporidium muris   krip-tō-spô-ri'dē-um mū'ris

Culex   kū'leks

Cytophaga   sī-tō'fä-gä

Daptobacter   dap'to-bak-tėr

Dermacentor andersoni   dėr-mä-sen'tôr an-dėr-sō'nē

D. variabilis   vär-ē-a'bil-is

Desulfotomaculum   dē-sul-fō-to-ma'kū-lum

Desulfovibrio   dē-sul-fō-vib'rē-ō

Didinium nasutum   dī-di'nē-um nä-sūt'um

Diphyllobothrium latum   dī-fil-lō-bo'thrē-um lā'tum

Dryas octopetala   drī'äs ok-tō-pet'äl-ä

Echinococcus granulosus   ē-kīn-ō-kok'kus gra-nū-lō'sus

Emericella   em-ėr-ē-sel'lä

Emmonsiella   em-mon-sē-el'lä

Endothia parasitica   en-dō'thē-ä par-ä-si'ti-kä

Entamoeba histolytica   en-tä-mē'bä his-tō-li'ti-kä

Enterobacter aerogenes   en-te-rō-bak'tėr ā-rä'jen-ēz

E. agglomerans   ag-glom'ėr-anz

E. cloacae   klō-ā'kī

Enterobius vermicularis   en-te-rō'bē-us ver-mi-kū-lar'is

Epidermophyton   ep-i-dėr-mō-fī'ton

Erwinia carotovora   ėr-wi'nē-ä ka-ro-tä'vô-rä

E. dissolvens   dis-solv'ens

E. herbicola   hėr-bik'ō-lä

Erysipelothrix rhusiopathiae ār-i-si-pel′ō-thrix rus-ē-ō-path′ē-ī

Erysiphe grammis ār′i-sīf gram′mis

Escherichia coli esh-ėr-i′kē-ä kō′lī (or kō′lē)

Euglena ū-glē′nä

Eurotium yėr-ō′tē-um

Filobasidiella fi-lō-ba-si-dē-el′lä

Fonsecaea pedrosoi fon-se′kē-ä pe-drō′sō-ē

Francisella tularensis fran-sis-el′lä tü-lä-ren′sis

Frankia frank′ē-ä

Fusobacterium fü-sō-bak-ti′rē-um

Gambierdiscus toxicus gam′bėr-dis-kus toks′i-kus

Gardnerella vaginalis gärd-nė-rel′lä va-jin-al′is

Gelidium jel-id′ē-um

Geotrichum gē-ō-trik′um

Giardia lamblia jē-är′dē-ä lam′lē-ä

Gloeocapsa glē-ō-kap′sä

Glomus fasciculatus glo′mus fa-sik-ū-lä′tus

Glossina gläs-sē′nä

Gluconobacter glü-kon-ō-bak′tėr

Gonyaulax gä-nē-ä′luks

Gymnoascus jim-nō-as′kus

Halobacterium halobium hal-ō-bak-ti′rē-um hal-ō′bē-um

Halococcus hal-ō-kok′kus

Hebeloma alpinum hē-bel-ō′mä al-pin′um

Hemophilus aegyptius hē-mä′fi-lus ē-jip′ti-us

H. ducreyi dü-krā′ē

H. influenzae in-flü-en′zī

H. parainfluenzae pa-ra-in-flü-en′zī

Histoplasma capsulatum his-tō-plaz′mä kap-su-lä′tum

Holospora hō-lo′spô-rä

Homo sapiens hō′mō sā′pē-ens

Hydrogenomonas hī-drō-je-nō-mō′näs

Hyphomicrobium hī-fō-mī-krō′be-um

Isospora belli ī-so′spô-rä bel′lē

Ixodes dammini iks-ō′dēs dam′mi-nē

I. pacificus pas-i′fi-kus

Klebsiella pneumoniae kleb-sē-el′lä nü-mō′nē-ī

Lactobacillus acidophilus lak-tō-bä-sil′lus a-si-dä′fi-lus

L. bulgaricus bul-gā′ri-kus

L. delbrueckii del-brük′ē-ē

L. leichmannii līk-man′nē-ē

L. plantarum plan-tä′rum

L. sanfrancisco sän-fran-sis′kō

Laminaria lam-i-nä′rē-ä

Legionella pneumophila lē-jä-nel′lä nü-mō′fi-lä

Leptospira interrogans icterohaemorrhagiae lep-tō-spī′rä in-tėr′rä-ganz ik-tėr-ō-hem-ôr-raj′ē-ī

Leuconostoc mesenteroides lü-kō-nos′tok mes-en-ter-oi′dēz

Listeria monocytogenes lis-te′rē-ä mo-nō-sī-tô′je-nēz

Lycoperdon perlatum lī-kō-pėr′don pėr-lä′tum

Meniscus mē-nis′kus

Methanobacterium meth-a-nō-bak-ti′rē-um

Methylophilus methylotrophus meth-i-lo′fi-lus meth-i-lō-trôf′us

Microcladia mī-krō kläd′ē-ä

Micrococcus luteus mī-krō-kok′kus lū′tē-us

Microcystis aeruginosa mī-krō-sis′tis ä-rü-ji-nō′sä

Micromonospora purpureae mī-krō-mo-nä′spô-rä pùr-pù-rē′ī

Microsporum gypseum mī-krō-spô′rum jip′sē-um

Mixotricha mix-ō-trik′ä

Moraxella lacunata mô-raks-el′lä la-kü-nä′tä

Mucor mū-kôr

Mycobacterium avium-intracellulare mī-kō-bak-ti′rē-um ā′vē-um in-trä-cel-ū-lär′e

M. leprae lep′rī

M. smegmatis smeg-ma′tis

M. tuberculosis tü-bėr-kū-lō′sis

Mycoplasma hominis mī-kō-plaz′mä ho′mi-nis

M. pneumoniae nu-mō′nē-ī

Naegleria fowleri nī-gle′rē-ä fou′lėr-ē

Nannizia nan-nī′zē-ä

Necator americanus ne-kā′tôr ä-me-ri-ka′nus

Neisseria gonorrhoeae nī-se′rē-ä go-nôr-rē′ī

N. meningitidis me-nin-ji′ti-dis

Nereocystis nē-rē-ō-sis′tis

Nicotiana tabacum ni-kō-tē′ä-nä ta-ba′kum

N. sylvestris sil-ves′tris

Nitrobacter nī-trō-bak′tėr

Nitrosomonas nī-trō-sō-mō′näs

Nocardia asteroides nō-kär′dē-ä as-tėr-oi′dēz

Oocystis ō-ō-sis′tis

Ornithodorus ôr-nith-ō′dô-rus

Paracoccidioides brasiliensis pär-ä-kok-sid′ē-oi-dēz bra-sil-ē-en′sis

Paracoccus denitrificans pär-ä-kok′kus dē-nī-tri′fi-kanz

Paragonimus westermanni pär-ä-gōn′e-mus we-ster-man′nē

Paramecium multimicronucleatum pär-ä-mē′sē-um mul-tē-mī-krō-nü-klē-ä′tum

Pasteurella haemolytica pas-tyėr-el′lä hēm-ō-lit′i-kä

P. multocida mul-tō′si-dä

Pediculus humanus ped-ik′ū-lus hü′ma-nus

Pediococcus pe-dē-ō-kok′kus

Penicillium chrysogenum pen-i-sil′lē-um krī-so′gen-um

P. griseofulvum gri-sē-ō-fül′vum

P. marneffei mar-nif′fē-ē

P. notatum nō-tä′tum

Peridinium per-i-din′ē-um

Petriellidium pet-rē-el-li′dē-um

Phialophora verruscosa fē-ä-lō′fô-rä ver-rü-skō′sä

Phormidium luridum fôr-mi′dē-um ler′i-dum

Phytophthora infestans fī-tof′thô-rä in-fes′tans

Pichia pik′ē-ä

Pityrosporum pit-i-ros′pô-rum

Plasmodium falciparum plaz-mō′dē-um fal-sip′är-um

P. malariae mä-lā′rē-ī

P. ovale ō-vä′lē

P. vivax vī′vaks

Plesiomonas shigelloides ple-sē-ō-mō′nas shi-gel-loi′des

Pneumocystis carinii nü-mō-sis′tis kär-i′nē-ī (or kär-i′nē-ē)

Propionibacterium acnes prō-pē-on-ē-bak-ti′rē-um ak′nēz
P. freudenreichii froi-den-rīk′ē-ē
Proteus mirabilis prō′tē-us mi-ra′bi-lis
P. vulgaris vul-ga′ris
Prototheca prō-tō-thā′kä
Pseudomonas aeruginosa sū-dō-mō′nas ā-rü-ji-nō′sä
P. cepacia se-pā′sē-ä
P. diminuta di-mi-nü′tä
P. fluorescens flôr-es′ens
P. syringae sėr-in′gī
Quercus kwer′kus
Rhizobium japonicum rī-zō′bē-um jap-on′i-kum
R. meliloti mel-i-lot′ē
Rhizopus nigricans rī-zō′pùs nī′gri-kans
R. oligosporus ol-i-gō-spôr′us
R. oryzae ôr′ī-zī
Rhodomicrobium vannielii rō-dō-mī-krō′bē-um van-niel′ē-ē
Rhodopseudomonas rō-dō-sū-dō-mō′nas
Rhodospirillum rubrum rō-dō-spī-ril′lum rūb′rum
Rickettsia prowazekii ri-ket′sē-ä prou-wä-ze′kē-ē
R. rickettsii ri-ket′sē-ē
R. typhi tī′fē
Riftia rift′ē-ä
Rosa multiflora rō′sä mul-ti-flô′rä
Saccharomyces carlsbergensis sak-ä-rō-mī′sēs kärls-bėr′gen′sis
S. cerevisiae se-ri-vis′ē-ī
S. ellipsoideus ē-lip-soi′dē-us
S. exiguus egz-ij′ū-us
S. fragilis fra′jil-is
S. rouxii rō-ē′ē
Salmonella cholerae-suis sal-mōn-el′lä kol-ėr-ä-sü′is
S. dublin dub′lin
S. eastbourne ēst′bôrn
S. enteritidis en-tėr-ī′ti-dis
S. paratyphi pa-rä-tī′fē
S. typhi tī′fē
S. typhimurium tī-fi-mùr′ē-um
Sargassum sär-gas′sum
Sartorya sär-tô′rē-ä
Schistosoma shis-tō-sō′mä or skis-tō-sō′mä
Serratia marcescens ser-rä′tē-ä mär-ses′sens
Shigella boydii shi-gel′lä boi′dē-ē
S. dysenteriae dis-en-te′rē-ī
S. flexneri fleks′nėr-ī
S. sonnei son′ne-ē
Sphaerotilus natans sfe-rä′ti-lus nä′tans
Spirillum ostreae spī-ril′lum ost′rä-ī
S. volutans vō′lū-tans
Spiroplasma spī-ro-plaz′mä
Spirosoma spī-rō-sō′mä
Spirulina spī-rü-lī′nä
Sporosarcina ureae spô-rō-sär-sī′nä yė′rē-ī
Sporothrix schenkii spô-rō′thriks shen′kē-ē
Staphylococcus aureus staf-i-lō-kok′kus ô′rē-us
S. epidermidis e-pi-dėr′mi-dis
Stella stel′lä
Stigmatella aurantiaca stig-mä′tel′lä ô-rän-tē′ä-kä

Streptobacillus strep-tō-bä-sil′lus
Streptococcus faecalis strep-tō-kok′kus fē-kā′lis
S. mitis mī′tis
S. mutans mū′tans
S. pneumoniae nü-mō′nē-ī
S. pyogenes pī-äj′en-ēz
S. salivarius sa-li-vā′rē-us
S. sanguis san′gwis
S. thermophilus thėr-mo′fil-us
Streptomyces aureofaciens strep-tō-mī′sēs ô-rē-ō-fa′si-ens
S. erythraeus ā-rith′rē-us
S. fradiae frā′dē-ī
S. griseus gri′sē-us
S. kanamyceticus kan-ä-mī-sē′ti-kus
S. nodosus nō-dō′sus
S. noursei nôr′sē-ī
S. olivoreticuli ō-liv-ō-re-tik′ū-lē
S. parvullus pär-vū′lus
S. rimosus ri-mō′sus
S. venezuelae ve-ne-zü-e′lī
Sulfolobus sul-fō-lō′bus
Taenia saginata te′nē-ä sa-ji-nä′tä
T. solium sō′lē-um
Talaromyces ta-lä-rō-mī′sēs
Thermoplasma thėr-mō-plaz′mä
Thiobacillus ferrooxidans thī-ō-bä-sil′lus fer-rō-oks′i-danz
T. thiooxidans thī-ō-oks′i-danz
Thiocapsa floridana thī-ō-kap′sä flôr-i′dä-nä
Toxoplasma gondii toks-ō-plaz′mä gon′dē-ē
Trachelomonas trä-kel-ō-mōn′as
Treponema pallidum trē-pō-nē′mä pal′li-dum
Triatoma trī-ä-tō′ma
Trichinella spiralis trik-in-el′lä spī-ra′lis
Trichomonas vaginalis trik-ō-mōn′as va-jin-al′is
Trichonympha sphaerica trik-ō-nimf′ä sfe′ri-kä
Trichophyton trik-ō-fī′ton
Trichosporon trik-ō-spôr′on
Trichuris trichiura trik′ėr-is trik-ē-yėr′a
Tridacna trī-dak′nä
Trypanosoma brucei gambiense tri-pa-nō-sō′mä brüs′ē gam-bē-ens′
T. brucei rhodesiense rō-dē-sē-ens′
T. cruzi kruz′ē
Ureaplasma urealyticum ū-rē-ä-plaz′mä ū-rē-ä-lit′i-kum
Veillonella vi-lo-nel′lä
Vibrio cholerae vib′rē-ō kol′ėr-ī
V. fischeri fish′ėr-ē
V. parahaemolyticus pa-rä-hē-mō-li′ti-kus
Wuchereria bancrofti vū-kėr-är′ē-ä ban-krof′tē
Xenopsylla cheopis ze-nop-sil′lä kē-ō′pis
Yersinia enterocolitica yėr-sin′ē-ä en-tėr-ō-kōl-it′ik-ä
Y. pestis pes′tis
Y. pseudotuberculosis sū-dō-tü-bėr-kū-lō′sis
Zoogloea ramigera zō-ō-glē′ä ram-i′ger-ä
Zymomonas zī-mō-mō′näs

Word Roots Used in Microbiology

The Latin rules of grammar pertain to singular and plural forms of scientific names.

	Gender		
	Feminine	Masculine	Neuter
Singular	-a	-us	-um
Plural	-ae	-i	-a
Examples	alga,	fungus,	bacterium,
	algae	fungi	bacteria

a-, an- absence, lack. Examples: abiotic, in the absence of life; anaerobic, in the absence of air.

-able able to, capable of. Example: viable, having the ability to live or exist.

actino- ray. Example: actinomycetes, bacteria that form star-shaped (with rays) colonies.

aer- air. Examples: aerobic, in the presence of air; *aerate*, to add air.

albo- white. Example: *Streptomyces albus* produces white colonies.

ameb- change. Example: ameboid, movement involving changing shapes.

amphi- around. Example: amphitrichous, tufts of flagella at both ends of a cell.

amyl- starch. Example: amylase, an enzyme that degrades starch.

ana- up. Example: anabolism, building up.

ant-, anti- opposed to, preventing. Example: antimicrobial, a substance that prevents microbial growth.

archae- ancient. Example: archaeobacteria, "ancient" bacteria, thought to be like the first form of life.

asco- bag. Example: ascus, a baglike structure holding spores.

aur- gold. Example: *Staphylococcus aureus*, gold pigmented colonies.

aut-, auto- self. Example: autotroph, self-feeder.

bacillo- a little stick. Example: bacillus, rod-shaped.

basid- base, pedestal. Example: basidium, a cell that bears spores.

bdell- leech. Example: *Bdellovibrio*, a predatory bacterium.

bio- life. Example: biology, the study of life and living organisms.

blast- bud. Example: blastospore, spores formed by budding.

bovi- cattle. Example: *Mycobacterium bovis*, a bacterium found in cattle.

brevi- short. Example: *Lactobacillus brevis*, a bacterium with short cells.

butyr- butter. Example: butyric acid, formed in butter, responsible for rancid odor.

campylo- curved. Example: *Campylobacter*, curved rod.

carcin- cancer. Example: carcinogen, a cancer-causing agent.

-caryo, -karyo a nut. Example: eucaryote, a cell with a membrane-enclosed nucleus.

caseo- cheese. Example: caseous, cheeselike.

caul- a stalk. Example: *Caulobacter*, appendaged or stalked bacteria.

cerato- horn. Example: keratin, the horny substance making up skin and nails.

chlamydo- covering. Example: chlamydospores, spores formed inside hypha.

chloro- green. Example: chlorophyll, green-pigmented molecule.

chrom- color. Examples: chromosome, readily stained structure; metachromatic, intracellular colored granules.

chryso- golden. Example: *Streptomyces chryseus*, golden colonies.

-cide killing. Example: bactericide, an agent that kills bacteria.

cili- eyelash. Example: cilia, a hairlike organelle.

cleisto- closed. Example: cleistothecium, completely closed ascus.

co-, con- together. Example: concentric, having a common center, together in the center.

cocci- a berry. Example: coccus, a spherical-shaped cell.

coeno- shared. Example: coenocyte, a cell with many nuclei not separated by septa.

col-, colo- colon. Examples: colon, large intestine; *Escherichia coli*, a bacterium found in the large intestine.

conidio- dust. Example: conidia, spores developed at the end of aerial hypha, never enclosed.

coryne- club. Example: *Corynebacterium*, club-shaped cells.

-cul small form. Example: particle, a small part.

-cut the skin. Example: *Firmicutes*, bacteria with a firm cell wall, gram-positive.

cyano- blue. Example: cyanobacteria, blue-green pigmented organisms.

cyst- bladder. Example: cystitis, inflammation of the urinary bladder.

cyt- cell. Example: cytology, the study of cells.

de- undoing, reversal, loss, removal. Example: deactivation, becoming inactive.

di-, diplo- twice, double. Example: diplococci, pairs of cocci.

dia- through, between. Example: diaphragm, the wall through or between two areas.

dys- difficult, faulty, painful. Example: dysfunction, disturbed function.

ec-, ex-, ecto- out, outside, away from. Example: excrete, to remove materials from the body.

en-, em- in, inside. Example: encysted, enclosed in a cyst.

entero- intestine. Example: *Enterobacter*, a bacterium found in the intestine.

epi- upon, over. Example: epidemic, a disease affecting all the people.

erythro- red. Example: erythma, redness of the skin.

eu- well, proper. Example: eucaryote, a proper cell.

exo- outside, outer layer. Example: exogenous, from outside the body.

extra- outside, beyond. Example: extracellular, outside the cells of an organism.

firmi- strong. Example: *Bacillus firmus* forms resistant endospores.

flagell- a whip. Example: flagellum, a projection from a cell; in eucaryotic cells, it pulls cells in a whiplike fashion.

flav- yellow. Example: *Flavobacterium* cells produce yellow pigment.

fruct- fruit. Example: fructose, fruit sugar.

-fy to make. Example: magnify, to make larger.

galacto- milk. Example: galactose, monosaccharide from milk sugar.

gamet- to marry. Example: gamete, a reproductive cell.

gastr- stomach. Example: gastritis, inflammation of the stomach.

gel- to stiffen. Example: gel, a solidified colloid.

-gen an agent that initiates. Example: pathogen, any agent that produces disease.

-genesis formation. Example: pathogenesis, production of disease.

germ, germin- bud. Example: germ, part of an organism capable of developing.

-gony reproduction. Example: schizogony, multiple fission producing many new cells.

gracili- thin. Example: *Aquaspirillum gracile*, a thin cell.

halo- salt. Example: halophile, an organism that can live in high salt concentrations.

haplo- one, single. Example: haploid, half the number of chromosomes or one set.

hema-, hemato-, hemo- blood. Example: *Hemophilus*, a bacterium that requires nutrients from red blood cells.

hepat- liver. Example: hepatitis, inflammation of the liver.

herpes creeping. Example: herpes, or shingles, lesions appear to creep along the skin.

hetero- different, other. Example: heterotroph, obtains organic nutrients from other organisms; other feeder.

hist- tissue. Example: histology, the study of tissues.

hom-, homo- same. Example: homofermenter, an organism that produces only lactic acid from fermentation of a carbohydrate.

hydr-, hydro- water. Example: dehydration, loss of body water.

hyper- excess. Example: hypertonic, having a greater osmotic pressure in comparison with another.

hypo- below, deficient. Example: hypotonic, having a lesser osmotic pressure in comparison with another.

im- not, in. Example: impermeable, not permitting passage.

inter- between. Example: intercellular, between the cells.

intra- within, inside. Example: intracellular, inside the cell.

io- violet. Example: iodine, a chemical element that produces a violet vapor.

iso- equal, same. Example: isotonic, having the same osmotic pressure when compared with another.

-itis inflammation of. Example: colitis, inflammation of the large intestine.

kin- movement. Example: streptokinase, an enzyme that lyses or moves fibrin.

lacti- milk. Example: lactose, the sugar in milk.

leuko- whiteness. Example: leukocyte, a white blood cell.

lip-, lipo- fat, lipid. Example: lipase, an enzyme that breaks down fats.

-logy the study of. Example: pathology, the study of changes in structure and function brought on by disease.

lopho- tuft. Example: lophotrichous, having a group of flagella on one side of a cell.

luc-, luci- light. Example: luciferin, a substance in certain organisms that emits light when acted upon by the enzyme luciferase.

lute-, luteo- yellow. Example: *Micrococcus luteus,* yellow colonies.

-lysis loosening, to break down. Example: hydrolysis, chemical decomposition of a compound into other compounds as a result of taking up water.

macro- largeness. Example: macromolecules, large molecules.

mendosi- faulty. Example: Mendosicutes, archaeobacteria lacking peptidoglycan.

meningo- membrane. Example: meningitis, inflammation of the membranes of the brain.

meso- middle. Example: mesophile, an organism whose optimum temperature is in the middle range.

meta- beyond, between, transition. Example: metabolism, chemical changes occurring within a living organism.

micro- smallness. Example: microscope, an instrument used to make small objects appear larger.

-mnesia memory. Examples: amnesia, loss of memory; anamnesia, return of memory.

molli- soft. Example: Mollicutes, a class of wall-less eubacteria.

-monas a unit. Example: *Methylomonas*, a unit (bacterium) that utilizes methane as its carbon source.

mono- singleness. Example: monotrichous, having one flagellum.

morpho- form. Example: morphology, the study of the form and structure of organisms.

multi- many. Example: multinuclear, having several nuclei.

mur- wall. Example: murein, a component of bacterial cell walls.

mus-, muri- mouse. Example: murine typhus, a form of typhus endemic in mice.

mut- to change. Example: mutation, a sudden change in characteristics.

myco-, -mycetoma, -myces a fungus. Example: *Saccharomyces*, sugar fungus, a genus of yeast.

myxo- slime, mucus. Example: Myxobacteriales, an order of slime-producing bacteria.

necro- a corpse. Example: necrosis, cell death or death of a portion of tissue.

-nema a thread. Example: *Treponema* has long, thread-like cells.

nigr- black. Example: *Aspergillus niger,* a fungus that produces black conidia.

ob- before, against. Example: obstruction, impeding or blocking up.

oculo- eye. Example: monocular, pertaining to one eye.

-oecium, -ecium a house. Examples: perithecium, an ascus with an opening that encloses spores; ecology, the study of the relationships between organisms and between an organism and its environment (household).

-oid like, resembling. Example: coccoid, resembling a coccus.

-oma tumor. Example: lymphoma, a tumor of the lymphatic tissues.

-ont being, existing. Example: schizont, a cell existing as a result of schizogony.

ortho- straight, direct. Example: orthomyxovirus, a virus with a straight, tubular capsid.

-osis, -sis condition of. Examples: lysis, the condition of loosening; symbiosis, the condition of living together.

pan- all, universal. Example: pandemic, an epidemic affecting a large region.

para- beside, near. Example: parasite, an organism that "feeds beside" another.

peri- around. Example: peritrichous, projections from all sides.

phaeo- brown. Example: *Phaeophyta*, brown algae.

phago- eat. Example: phagocyte, a cell that engulfs and digests particles or cells.

philo-, -phil liking, preferring. Example: thermophile, an organism that prefers high temperatures.

-phore bears, carries. Example: conidiophore, a hypha that bears conidia.

-phyll leaf. Example: chlorophyll, the green pigment in leaves.

-phyte plant. Example: saprophyte, a plant that obtains nutrients from decomposing organic matter.

pil- a hair. Example: pilus, a hairlike projection from a cell.

plano- wandering, roaming. Example: plankton, organisms drifting or wandering in water.

plast- formed. Example: plastid, a formed body within a cell.

-pnoea breathing. Example: dyspnea, difficulty in breathing.

pod- foot. Example: pseudopod, a footlike structure.

poly- many. Example: polymorphism, many forms.

post- after, behind. Example: posterior, places behind (a specific) part.

pre-, pro- before, ahead of. Examples: procaryote, a cell with the first nucleus; pregnant, before birth.

pseudo- false. Example: pseudopod, false foot.

psychro- cold. Example: psychrophile, an organism that grows best at cold temperatures.

-ptera wing. Example: Diptera, the order of true flies, insects with two wings.

pyo- pus. Example: pyogenic, pus-forming.

rhabdo- stick, rod. Example: Rhabdovirus, an elongated, bullet-shaped virus.

rhin- nose. Example: rhinitis, inflammation of mucous membranes in the nose.

rhizo- root. Examples: *Rhizobium*, a bacterium that grows in plant roots; mycorrhiza, mutualism between a fungus and the roots of a plant.

rhodo- red. Example: *Rhodospirillum*, a red-pigmented, spiral-shaped bacterium.

rod- gnaw. Example: rodents, the class of mammals with gnawing teeth.

rubri- red. Example: *Clostridium rubrum*, red-pigmented colonies.

rumin- throat. Example: *Ruminococcus*, a bacterium associated with a rumen (modified esophagus).

saccharo- sugar. Example: disaccharide, a sugar consisting of two simple sugars.

sapr- rotten. Example: *Saprolegnia*, a fungus that lives on dead animals.

sarco- flesh. Example: sarcoma, a tumor of muscle or connective tissues.

schizo- split. Example: schizomycetes, organisms that reproduce by splitting, an early name for bacteria.

scolec- worm. Example: scolex, the head of a tapeworm.

-scope, -scopic watcher. Example: microscope, an instrument used to watch small things.

semi- half. Example: semicircular, having the form of half a circle.

sept- rotting. Example: aseptic, free from bacteria that could cause decomposition.

septo- partition. Example: septum, a cross-wall in a fungal hypha.

serr- notched. Example: serrate, with a notched edge.

sidero- iron. Example: *Siderococcus*, a bacterium capable of oxidizing iron.

siphon- tube. Example: Siphonaptera, the order of fleas, insects with tubular mouths.

soma- body. Example: somatic cells, cells of the body other than gametes.

speci- particular thing. Examples: species, the smallest group of organisms with similar properties; specify, to indicate exactly.

spiro- coil. Example: spirochete, a bacterium with a coiled cell.

sporo- spore. Example: sporangium, a structure that holds spores.

staphylo- grapelike cluster. Example: *Staphylococcus*, a bacterium that forms clusters of cells.

-stasis arrest, fixation. Example: bacteriostasis, cessation of bacterial growth.

strepto- twisted. Example: *Streptococcus*, a bacterium that forms twisted chains of cells.

sub- beneath, under. Example: subcutaneous, just under the skin.

super- above, upon. Example: superior, the quality or state of being above others.

sym-, syn- together, with. Examples: synapse, the region of communication between two neurons; synthesis, putting together.

-taxi to touch. Example: chemotaxis, response to the presence (touch) of chemicals.

taxis- orderly arrangement. Example: taxonomy, the science dealing with arranging organisms into groups.

tener- tender. Example: Tenericutes, the phylum containing wall-less eubacteria.

thallo- plant body. Example: thallus, an entire macroscopic fungus.

therm- heat. Example: *Thermus*, a bacterium that grows in hot springs (to 75°C).

thio- sulfur. Example: *Thiobacillus,* a bacterium capable of oxidizing sulfur-containing compounds.

-thrix. See **trich-**

-tome, -tomy to cut. Example: appendectomy, surgical removal of the appendix.

-tone, -tonic strength. Example: hypotonic, having less strength (osmotic pressure).

tox- poison. Example: antitoxin, effective against poison.

trans- across, through. Example: transport, movement of substances.

tri- three. Example: trimester, three-month period.

trich- a hair. Example: peritrichous, hairlike projections from cells.

-trope turning. Example: geotropic, turning toward the earth (pull of gravity).

-troph food, nourishment. Example: trophic, pertaining to nutrition.

-ty condition of, state. Example: immunity, the condition of being resistant to disease or infection.

undul- wavy. Example: undulating, rising and falling, presenting a wavy appearance.

uni- one. Example: unicellular, pertaining to one cell.

vaccin- cow. Example: vaccination, injection of a vaccine (originally pertained to cows).

vacu- empty. Example: vacuoles, an intracellular space that appears empty.

vesic- bladder. Example: vesicle, a bubble.

vitr- glass. Example: in vitro, in culture media in a glass (or plastic) container.

-vorous eat. Example: carnivore, an animal that eats other animals.

xantho- yellow. Example: *Xanthomonas*, produces yellow colonies.

xeno- strange. Example: axenic, sterile, free of strange organisms.

xero- dry. Example: xerophyte, any plant that tolerates dry conditions.

xylo- wood. Example: xylose, a sugar obtained from wood.

zoo- animal. Example: zoology, the study of animals.

zygo- yoke, joining. Example: zygospore, a spore formed from the fusion of two cells.

-zyme ferment. Example: enzyme, any protein in living cells that catalyzes chemical reactions.

APPENDIX D

Most Probable Numbers (MPN) Table

MPN Index for Various Combinations of Positive and Negative Results When Five 10-ml Portions, Five 1-ml Portions, and Five 0.1-ml Portions Are Used (see pp. 164–165)

No. of Tubes Giving Positive Reaction out of			MPN Index per 100 ml	No. of Tubes Giving Positive Reaction out of			MPN Index per 100 ml
5 of 10 ml Each	5 of 1 ml Each	5 of 0.1 ml Each		5 of 10 ml Each	5 of 1 ml Each	5 of 0.1 ml Each	
0	0	0	<2	4	3	1	33
0	0	1	2	4	4	0	34
0	1	0	2	5	0	0	23
0	2	0	4	5	0	1	30
1	0	0	2	5	0	2	40
1	0	1	4	5	1	0	30
1	1	0	4	5	1	1	50
1	1	1	6	5	1	2	60
1	2	0	6	5	2	0	50
2	0	0	4	5	2	1	70
2	0	1	7	5	2	2	90
2	1	0	7	5	3	0	80
2	1	1	9	5	3	1	110
2	2	0	9	5	3	2	140
2	3	0	12	5	3	3	170
3	0	0	8	5	4	0	130
3	0	1	11	5	4	1	170
3	1	0	11	5	4	2	220
3	1	1	14	5	4	3	280
3	2	0	14	5	4	4	350
3	2	1	17	5	5	0	240
4	0	0	13	5	5	1	350
4	0	1	17	5	5	2	500
4	1	0	17	5	5	3	900
4	1	1	21	5	5	4	1600
4	1	2	26	5	5	5	$\geq$2400
4	2	0	22				
4	2	1	26				
4	3	0	27				

Source: *Standard Methods for the Examination of Water and Wastewater*, 16th ed., New York: American Public Health Association, 1985.

Methods for Taking Clinical Samples

To diagnose a disease, it is often necessary to obtain a sample of material that may contain the disease-causing organism. Samples must be taken aseptically. The sample container should be labeled with the patient's name, room number (if hospitalized), date, time, and medications being taken. Samples must be transported to the laboratory immediately for culture. Delay in transport may result in growth of some organisms, and their toxic products may kill other organisms. Pathogens tend to be fastidious and die without their optimum environmental conditions.

In the laboratory, samples from infected tissues are cultured on differential and selective media in an attempt to isolate and identify any pathogens or organisms that are not normally found in association with that tissue.

UNIVERSAL PRECAUTIONS

The following procedures should be used by all health care workers, including students, whose activities involve contact with patients or with blood or other body fluids. These procedures were developed to minimize the risk of transmitting AIDS in a health care environment, but adherence to these guidelines will minimize transmission of *all* nosocomial infections.

1. Gloves should be worn when touching blood and body fluid, mucous membranes, and nonintact skin and when handling items or surfaces soiled with blood or body fluids. Gloves should be changed after contact with each patient.
2. Hands and other skin surfaces should be washed immediately and thoroughly if contaminated with blood or other body fluids. Hands should be washed immediately after gloves are removed.
3. Masks and protective eyewear or face shields should be worn during procedures that are likely to generate droplets of blood or other body fluids.
4. Gowns or aprons should be worn during procedures that are likely to generate splashes of blood or other body fluids.
5. To prevent needlestick injuries, needles should not be recapped, purposely bent or broken, or otherwise manipulated by hand. After disposable syringes and needles, scalpel blades, and other sharp items are used, they should be placed in puncture-resistant containers for disposal.
6. Although saliva has not been implicated in HIV transmission, mouthpieces, resuscitation bags, or other ventilation devices should be available for use in areas in which the need for resuscitation is predictable. Emergency mouth-to-mouth resuscitation should be minimized.
7. Health care workers who have exudative lesions or weeping dermatitis should refrain from all direct patient care and from handling patient-care equipment.
8. Pregnant health care workers are not known to be at greater risk of contracting HIV infection than health care workers who are not pregnant; however, if a health care worker develops HIV infection during pregnancy, the infant is at risk of infection. Because of this risk, pregnant health care workers should be especially familiar with and strictly adhere to precautions to minimize the risk of HIV transmission.

Source: Centers for Disease Control. "Recommendations for prevention of HIV transmission in health-care settings." *MMWR* 36(S2), 1987.

INSTRUCTIONS FOR SPECIFIC PROCEDURES

Wound or Abscess Culture
1. Cleanse the area with a sterile swab moistened in sterile saline.
2. Disinfect the area with 70% ethanol or iodine solution.
3. If the abscess has not ruptured spontaneously, a physician will open it with a sterile scalpel.
4. Wipe the first pus away.
5. Touch a sterile swab to the pus, taking care not to contaminate the surrounding tissue.
6. Replace the swab in its container, and properly label the container.

Ear Culture
1. Clean the skin and auditory canal with 1% tincture of iodine.
2. Touch the infected area with a sterile cotton swab.
3. Replace the swab in its container.

Eye Culture
This procedure is usually performed by an ophthalmologist.
1. Anesthetize the eye with topical application of a sterile anesthetic solution.
2. Wash the eye with sterile saline solution.
3. Collect material from the infected area with a sterile cotton swab. Return the swab to its container.

Blood Culture

1. Close room windows to avoid contamination.
2. Clean skin around selected vein with 2% tincture of iodine on a cotton swab.
3. Remove dried iodine with gauze moistened with 80% isopropyl alcohol.
4. Draw a few milliliters of venous blood.
5. Aseptically bandage the puncture.

Urine Culture

1. Provide the patient with a sterile container.
2. Instruct the patient to collect a mid-stream sample. This is obtained by voiding a small volume from the bladder before collection. This washes away extraneous skin flora.
3. A urine sample may be stored under refrigeration (4°–6°C) for up to 24 hours.

Fecal Culture

For bacteriological examination, only a small sample is needed. This may be obtained by inserting a sterile swab into the rectum or feces. The swab is then placed in a tube of sterile enrichment broth for transport to the laboratory. For examination for parasites, a small sample may be taken from a morning stool. The sample is placed in a preservative (polyvinyl alcohol, buffered glycerol, saline, or formalin) for microscopic examination for eggs and adult parasites.

Sputum Culture

1. A morning sample is best as microorganisms will have accumulated while the patient is sleeping.
2. Patient should rinse his or her mouth thoroughly to remove food and normal flora.
3. Patient should cough deeply from the lungs and expectorate into a sterile glass wide-mouth jar.
4. Care should be taken to avoid contamination of personnel.
5. In cases such as tuberculosis where there is little sputum, stomach aspiration may be necessary.
6. Infants and children tend to swallow sputum. A fecal sample may be of some value in these cases.

APPENDIX F

Biochemical Pathways

Figure F-I The Calvin–Benson Cycle for Photosynthetic Carbon Metabolism. Steps 1–3 provide for the initial fixation and reduction of carbon, generating the three-carbon compounds glyceraldehyde 3-phosphate and dihydroxyacetone phosphate, which are interconvertible (step 4). On the average, 2 out of every 12 three-carbon molecules are used in the synthesis of glucose (steps A–D). Ten out of every 12 three-carbon molecules are used to generate ribulose 5-phosphate by a complex series of reactions (step 5). The ribulose 5-phosphate is then phosphorylated at the expense of ATP (step 6), forming ribulose 1,5-diphosphate, the acceptor molecule with which the sequence began. (See Figure 5-11 on p. 118 for a simplified version of the Calvin–Benson cycle.)

Figure F-2   Glycolysis (Embden–Meyerhof Pathway). Each of the ten steps of glycolysis is catalyzed by a specific enzyme, which is named under each step number. (See Figure 5-13 on p. 123 for a simplified version of glycolysis.)

Step 1   Glucose enters the cell and is phosphorylated by the enzyme hexokinase, which transfers a phosphate group from ATP to the number-six carbon of the sugar. The product of the reaction is glucose 6-phosphate. The electrical charge of the phosphate group traps the sugar in the cell because of the impermeability of the plasma membrane to ions. Phosphorylation of glucose also makes the molecule more chemically reactive. Although glycolysis is supposed to *produce* ATP, in step 1, ATP is actually consumed—an energy investment that will be repaid with dividends later in glycolysis.

Step 2   Glucose 6-phosphate is rearranged to convert it to its isomer, fructose 6-phosphate. Isomers have the same number and types of atoms but in different structural arrangements.

Step 3   In this step, still another molecule of ATP is invested in glycolysis. An enzyme transfers a phosphate group from ATP to the sugar, producing fructose 1,6-diphosphate.

Step 4   This is the reaction from which glycolysis gets its name ("sugar splitting"). An enzyme cleaves fructose 1,6-diphosphate into two different three-carbon sugars: glyceraldehyde 3-phosphate and dihydroxyacetone phosphate. These two sugars are isomers of one another.

Step 5   The enzyme isomerase interconverts the three-carbon sugars. The next enzyme in glycolysis uses only glyceraldehyde 3-phosphate as its substrate. This pulls the equilibrium between the two three-carbon sugars in the direction of glyceraldehyde 3-phosphate, which is removed as fast as it forms.

Step 6 An enzyme now catalyzes two sequential reactions while it holds glyceraldehyde 3-phosphate in its active site. First, the sugar is oxidized at the number one carbon and NAD$^+$ is reduced, resulting in the formation of NADH + H$^+$. Second, the enzyme couples this reaction to the creation of a high-energy phosphate bond at the number one carbon of the oxidized substrate. The source of the phosphate is inorganic phosphate, which is always present in the cell. As products, the enzyme releases NADH + H$^+$ and 1,3-diphosphoglyceric acid. Notice in the figure that the new phosphate bond is symbolized with a squiggle (~), which indicates that the bond is at least as energetic as the phosphate bonds of ATP.

Step 7 At this step, glycolysis produces ATP. The phosphate group, with its high-energy bond, is transferred form 1,3-diphosphoglyceric acid to ADP. For each glucose molecule that began glycolysis, step 7 produces two molecules of ATP, because every product after the sugar-splitting step (step 4) is doubled. Of course, two ATPs were invested to get sugar ready for splitting. The ATP ledger now stands at zero. By the end of step 7, glucose has been converted to two molecules of 3-phosphoglyceric acid.

Step 8 Next, an enzyme relocates the remaining phosphate group of 3-phosphoglyceric acid to form 2-phosphoglyceric acid. This prepares the substrate for the next reaction.

Step 9 An enzyme forms a double bond in the substrate by extracting a water molecule from 2-phosphoglyceric acid to form phosphoenolpyruvic acid. This results in the electrons of the substrate being rearranged in such a way that the remaining phosphate bond becomes very unstable.

Step 10 The last reaction of glycolysis produces another molecule of ATP by transferring the phosphate group from phosphoenolpyruvic acid to ADP. Since this step occurs twice for each glucose molecule, the ATP ledger now shows a net gain of two ATPs. Thus, the glycolysis of one molecule of glucose results in two molecules of pyruvic acid, two molecules of NADH + H$^+$, and two molecules of ATP. Each molecule of pyruvic acid can now undergo respiration or fermentation.

Figure F-3   The Pentose Phosphate Pathway. This pathway, which operates simultaneously with glycolysis, provides an alternate route for the oxidation of glucose and plays a role in the synthesis of biological molecules, depending on the needs of the cell. Possible fates of the various intermediates are shown in color. (See p. 124 for a discussion of the pentose phosphate pathway.)

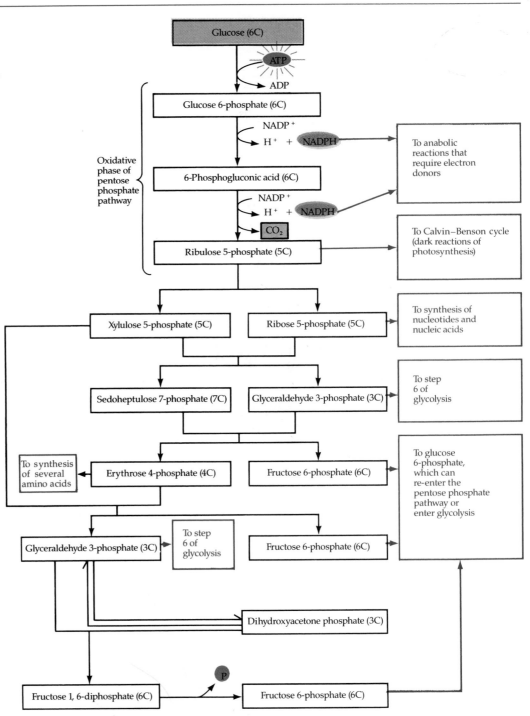

Figure F-4   The Entner–Doudoroff Pathway. This pathway is an alternate to glycolysis for the oxidation of glucose to pyruvic acid. (See Figure 5-14 on p. 125 for a simplified version of the Entner–Doudoroff pathway.)

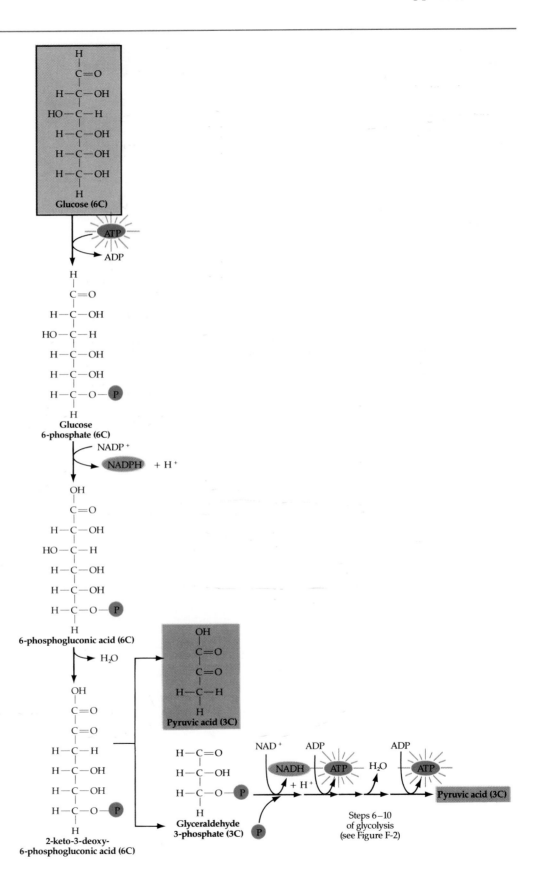

Figure F-5 The Krebs Cycle. (See Figure 5-16 on p. 126 for a simplified version.)

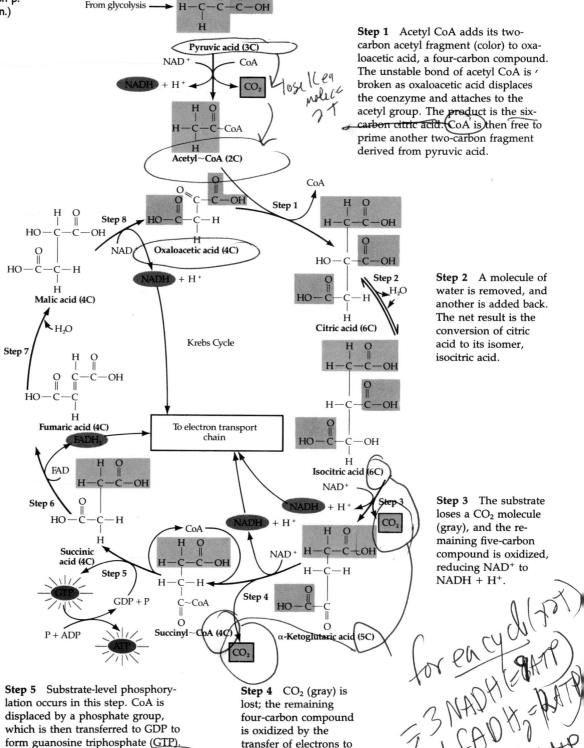

Step 1 Acetyl CoA adds its two-carbon acetyl fragment (color) to oxaloacetic acid, a four-carbon compound. The unstable bond of acetyl CoA is broken as oxaloacetic acid displaces the coenzyme and attaches to the acetyl group. The product is the six-carbon citric acid. CoA is then free to prime another two-carbon fragment derived from pyruvic acid.

Step 8 The last oxidative step produces another molecule of NADH + H$^+$ and regenerates oxaloacetic acid, which accepts a two-carbon fragment from acetyl CoA for another turn of the cycle.

Step 2 A molecule of water is removed, and another is added back. The net result is the conversion of citric acid to its isomer, isocitric acid.

Step 7 Bonds in the substrate are rearranged in this step by the addition of a water molecule.

Step 3 The substrate loses a CO$_2$ molecule (gray), and the remaining five-carbon compound is oxidized, reducing NAD$^+$ to NADH + H$^+$.

Step 6 In another oxidative step, two hydrogens are transferred to FAD to form FADH$_2$. The function of this coenzyme is similar to that of NADH + H$^+$, but FADH$_2$ stores less energy.

Step 5 Substrate-level phosphorylation occurs in this step. CoA is displaced by a phosphate group, which is then transferred to GDP to form guanosine triphosphate (GTP). GTP is similar to ATP, which is formed when GTP donates a phosphate group to ADP.

Step 4 CO$_2$ (gray) is lost; the remaining four-carbon compound is oxidized by the transfer of electrons to NAD$^+$ to form NADH + H$^+$ and is then attached to CoA by an unstable bond.

Exponents, Exponential Notation, and Logarithms

Very large and very small numbers—such as 4,650,000,000 and 0.00000032—are cumbersome to work with. It is more convenient to express such numbers in exponential notation, that is, as a power of 10. For example, 4.65×10^9 is in **standard exponential notation**, or **scientific notation**. 4.65 is the **coefficient**, and 9 is the power or **exponent**. In standard exponential notation, the coefficient is always a number between 1 and 10, and the exponent can be positive or negative.

To change a number into exponential notation, follow two steps. First, determine the coefficient by moving the decimal point so there is only one nonzero digit to the left of it. For example,

$$0.0000003\ 2$$

The coefficient is 3.2. Second, determine the exponent by counting the number of places you moved the decimal point. If you moved it to the left, the exponent is positive. If you moved it to the right, the exponent is negative. In the example, you moved the decimal point 7 places to the right, so the exponent is −7. Thus

$$0.00000032 = 3.2 \times 10^{-7}$$

Now suppose we are working with a large number instead of a very small number. The same rules apply, but our exponential value will be positive rather than negative. For example,

$$4,650,000,000 = 4.65 \times 10^{+9}$$
$$= 4.65 \times 10^9$$

To multiply numbers written in exponential notation, multiply the coefficients and *add* the exponents. For example,

$$(3 \times 10^4) \times (2 \times 10^3) =$$
$$(3 \times 2) \times 10^{4+3} = 6 \times 10^7$$

To divide, divide the coefficient and *subtract* the exponents. For example,

$$\frac{3 \times 10^4}{2 \times 10^3} = \frac{3}{2} \times 10^{4-3} = 1.5 \times 10^1$$

Microbiologists use exponential notation in many situations. For instance, exponential notation is used to describe the number of microorganisms in a population. Such numbers are often very large (Chapter 6). Another application of exponential notation is to express concentrations of chemicals in a solution—chemicals such as media components (Chapter 6), disinfectants (Chapter 7), or antibiotics (Chapter 19). Such numbers are often very small. Converting from one unit of measurement to another in the metric system requires multiplying or dividing by a power of 10, which is easiest to carry out in exponential notation.

A **logarithm (log)** is the power to which a base number is raised to produce a given number. Usually we work with logarithms to the base 10, abbreviated $\log_{10}$. The first step in finding the $\log_{10}$ of a number is to write the number in standard exponential notation. If the coefficient is exactly 1, the $\log_{10}$ is simply equal to the exponent. For example,

$$\log_{10} 0.00001 = \log_{10}(1 \times 10^{-5})$$
$$= -5$$

If the coefficient is not 1, as is often the case, a logarithm table or calculator must be used to determine the logarithm.

Microbiologists use logs for calculating pH levels and for graphing the growths of microbial populations in culture (Chapter 6).

GLOSSARY

Abscess   A localized accumulation of pus.

Acetyl coenzyme A (acetyl CoA)   A substance composed of an acetyl group and a carrier molecule called coenzyme A (CoA); provides the means for pyruvic acid to enter the Krebs cycle. *See* Krebs cycle.

Acetyl group

$$CH_3-\overset{\displaystyle O}{\overset{\displaystyle \|}{C}}-$$

Acid   A substance that dissociates into one or more hydrogen ions and one or more negative ions.

Acid-fast stain   A differential stain used to identify bacteria that are not decolorized by acid-alcohol.

Acidic dye   A salt in which the color is in the negative ion; used for negative staining.

Acquired immunity   The ability, obtained during the life of the individual, to produce specific antibodies.

Actinomycetes   Gram-positive bacteria that tend to form branching filaments; may form true mycelia; may produce conidiospores; see *Bergey's Manual*, Vols. 2 and 4.

Activated sludge   Aerobic digestion used in secondary sewage treatment.

Activation energy   The minimum collision energy required for a chemical reaction to occur.

Actively acquired immunity   Production of antibodies by an individual in response to an antigen.

Active site   Region on an enzyme that interacts with the substrate.

Active transport   Net movement of a substance across a membrane against a concentration gradient; requires energy.

Acute disease   A disease in which symptoms develop rapidly but last for only a short time.

Adenine   A purine nucleic acid base that pairs with thymine in DNA and uracil in RNA.

Adenocarcinoma   Cancer of glandular tissue.

Adenosine diphosphate (ADP)   The substance formed when ATP is split and energy is released.

Adenosine diphosphoglucose (ADPG)   Glucose activated by ATP; precursor for glycogen synthesis.

Adenosine triphosphatase   The enzyme that catalyzes the following reactions: ADP + ℗ $\longrightarrow$ ATP and ATP $\longrightarrow$ ADP + ℗.

Adenosine triphosphate (ATP)   An important intracellular energy source.

Adherence   Attachment of a microbe to a host cell or other surface.

Adhesin   *See* Ligands.

Aerial mycelium   A mycelium composed of fungal hyphae that project above the surface of the growth medium and produce asexual spores.

Aerobe   An organism requiring oxygen (O_2) to grow.

Aerobic respiration   Respiration in which the final electron acceptor in the electron transport chain is oxygen (O_2).

Aerotolerant anaerobe   An organism that does not use oxygen (O_2) but is not affected by its presence.

Aflatoxin $C_{17}H_{10}O_6$, a carcinogenic toxin produced by *Aspergillus flavus*.

Agar   A complex polysaccharide derived from a marine alga and used as a solidifying agent in culture media.

Agglutination   A joining together or clumping of cells.

Agranulocyte   A leukocyte without apparent granules in the cytoplasm; includes lymphocytes and monocytes.

Airborne transmission   The spread of pathogens farther than 1 meter in air from reservoir to susceptible host.

Alcohol   An organic molecule with the functional group —OH.

Aldehyde   An organic molecule with the functional group $-\overset{\displaystyle \|}{\underset{\displaystyle O}{C}}-H$.

Algae   A group of photosynthetic eucaryotes; some are included in the Kingdom Protista, and some in the Kingdom Plantae.

Alkaline   Having more OH^- ions than H^+ ions; pH is greater than 7.

Allergen   An antigen that evokes a hypersensitivity response.

Allergy   *See* Hypersensitivity.

Allograft   A graft between persons who aren't identical twins.

Allosteric inhibition   The process in which an enzyme's activity is changed because of binding on the allosteric site.

Allosteric site   The site on an enzyme at which an inhibitor binds.

Alpha-amino acid   An amino acid with —COOH and —NH_2 attached to the same carbon atom.

Alum   Aluminum sulfate [$Al_2(SO_4)_3$].

Amanitin   Polypeptide mushroom toxin that causes liver and nerve damage.

Ames test   A procedure using bacteria to identify potential carcinogens.

Amination   The addition of an amino group.

Amino acid   An organic acid containing an amino group and a carboxyl group.

Aminoglycoside   An antibiotic consisting of amino sugars and an aminocyclitol ring; for example, streptomycin.

Amino group —NH$_2$

Ammonification Removal of amino groups from amino acids to form ammonia.

Amoeba An organism belonging to the Kingdom Protista that moves by means of pseudopods.

Amphibolic pathway A pathway that is anabolic and catabolic.

Amphitrichous Having tufts of flagella at both ends of a cell.

Anabolism All synthesis reactions in a living organism.

Anaerobe An organism that does not require oxygen (O$_2$) for growth.

Anaerobic respiration Respiration in which the final electron acceptor in the electron transport chain is an inorganic molecule other than oxygen (O$_2$); for example, a nitrate or sulfate.

Anaerobic sludge digester Anaerobic digestion used in secondary sewage treatment.

Anal pore A site in certain protozoans for elimination of waste.

Analytical epidemiology Comparison of a diseased group and a healthy group to determine the cause of the disease.

Anamnestic response A rapid rise in antibody titer following exposure to an antigen after the primary response to that antigen.

Anaphylaxis A hypersensitivity reaction involving IgE antibodies, mast cells, and basophils.

Angstrom (Å) A unit of measurement equal to 10^{-10}m, 10$^{-4}\mu$m, and 10^{-1}nm; no longer an official unit.

Animalia A kingdom composed of multicellular eucaryotes lacking cell walls.

Animal virus A virus that multiplies in animal tissues.

Anion An ion with a negative charge.

Anoxygenic Not producing molecular oxygen; typical of bacterial photosynthesis.

Antagonism Active opposition; for example, between two drugs or two microbes.

Antibiotic An antimicrobial agent produced naturally by a bacterium or fungus.

Antibody A protein produced by the body in response to an antigen and capable of combining specifically with that antigen.

Antibody titer The amount of antibody in serum.

Anticodon The three nucleotides by which a transfer RNA recognizes an RNA codon.

Antigen Any substance that, when introduced into the body, causes antibody formation and reacts only with its specific antibody.

Antigen-binding sites Sites on an antibody that bind to antigenic determinants.

Antigenic determinant A specific region on the surface of an antigen against which antibodies are formed.

Antigenic drift Minor variations in the antigenic makeup of a virus that occur with time.

Antigenic shift Major genetic changes in influenza viruses causing changes in H and N antigens.

Antigen modulation The process whereby tumor cells shed their specific antigens, thus avoiding the host's immune response.

Antigen receptors Antibodylike molecules on T cells.

Antihuman gamma globulin Antibodies that react specifically with human antibodies.

Anti-idiotypic antibody Antibody that mimics the shape of an antigen to induce immunity.

Antimetabolite Any substance that interferes with metabolism by competitive inhibition of an enzyme.

Antimicrobial agent A chemical that destroys pathogens without damaging body tissues.

Antisense strand (−strand) Viral RNA that cannot act as mRNA.

Antiseptic A chemical for disinfection of the skin, mucous membranes, or other living tissues.

Antiserum A solution containing antibodies.

Antitoxin A specific antibody produced by the body in response to a bacterial exotoxin or its toxoid.

Antiviral protein (AVP) Protein made in response to interferon that blocks viral multiplication.

Apoenzyme The protein portion of an enzyme, which requires activation by a coenzyme.

Arbuscule Fungal mycelia in plant root cells.

Archaeobacteria Procaryotic organisms lacking peptidoglycan.

Arthropod An animal phylum characterized by an exoskeleton and jointed legs; includes insects and ticks.

Arthrospore An asexual fungal spore formed by fragmentation of a septate hypha.

Artificially acquired active immunity The production of antibodies by the body in response to a vaccination.

Artificially acquired passive immunity The transfer of humoral antibodies formed by one individual to a susceptible individual, accomplished by injection of antiserum.

Ascospore A sexual fungal spore produced in an ascus, formed by the Ascomycetes class of fungi.

Ascus A saclike structure containing ascospores.

Asepsis The absence of contamination by unwanted organisms.

Asexual reproduction Reproduction without opposite mating strains.

Atom The smallest unit of matter that can enter into a chemical reaction.

Atomic number The number of protons in the nucleus of an atom.

Atomic weight The total number of protons and neutrons in the nucleus of an atom.

Attenuation Lessening of virulence of a microorganism. Also, a regulatory mechanism for protein synthesis.

Autoclave Equipment for sterilization by steam under pressure, usually operated at 15 psi and 121°C.

Autograph A tissue graft from one's self.

Autoimmunity An immunologic response against a person's own tissue antigens.

Autotroph An organism that uses carbon dioxide (CO$_2$) as its principal carbon source.

Auxotroph A mutant microorganism with a nutritional requirement not possessed by the parent.

Axial filament The structure for motility found in spirochetes.

Bacillus Any rod-shaped bacterium; when written as a genus, refers to rod-shaped, endospore-forming, facultatively anaerobic, gram-positive bacteria.

Bacteremia A condition of bacteria in the blood.

Bacteria All living organisms with procaryotic cells.

Bacterial growth curve A graph indicating the growth of a bacterial population over time.

Bactericidal Capable of killing bacteria.

Bacteriochlorophyll The light-absorbing pigment found in green sulfur and purple sulfur bacteria.

Bacteriocinogenic plasmid Plasmid containing genes for the synthesis of bacteriocins.

Bacteriocin Toxic protein produced by bacteria that kill other bacteria.

Bacteriophage (phage) A virus that multiplies in bacterial cells.

Bacteriorhodopsin The light-absorbing purple pigment in *Halobacterium*'s cell membrane.

Bacteriostatic Capable of inhibiting bacterial growth.

Bacteroid Enlarged *Rhizobium* cells found in root nodules.

Basal body A structure that anchors flagella to the cell wall and plasma membrane.

Base A substance that accepts hydrogen ions and is capable of uniting with water to form an acid.

Base analog A chemical that is structurally similar to the normal nitrogenous bases in nucleic acids but with altered basepairing properties.

Base pairs The arrangement of nitrogenous bases in nucleic acids based on hydrogen bonding; in DNA, base pairs are A–T and G–C; in RNA, base pairs are A–U and G–C.

Base substitution The replacement of a single base in DNA by another base, causing a mutation.

Basic dye A salt in which the color is in the positive ion; used for bacterial stains.

Basidiospore A sexual fungal spore produced in a basidium, characteristic of the Basidiomycetes.

Basidium A pedestal that produces basidiospores; found in Basidiomycetes.

Basophil A granulocyte that readily takes up basic dye.

B cells Stem cells that differentiate into plasma cells that secrete antibodies.

BCG vaccine A live, attenuated strain of *Mycobacterium bovis* used to provide immunity to tuberculosis.

Benign tumor A noncancerous tumor.

Benthic zone The sediment at the bottom of a body of water.

Bergey's Manual The standard taxonomic reference on bacteria.

Beta oxidation The removal of two carbon units of a long chain of fatty acid to form acetyl-CoA.

Binary fission Bacterial reproduction by division into two daughter cells.

Binomial nomenclature The system of having two names (genus and specific epithet) for each organism.

Biochemical oxygen demand (BOD) A measure of the biologically degradable organic matter in water.

Biochemical pathway A sequence of enzymatically catalyzed reactions occurring in a cell.

Biochemistry The science of chemical processes in living organisms.

Bioconversion Changes in organic matter brought about by the growth of microorganisms.

Biogenesis The concept that living cells can only arise from preexisting cells.

Biogeochemical cycles The recycling of chemical elements by microorganisms for use by other organisms.

Biological transmission Transmission of a pathogen from one host to another when the pathogen reproduces in the vector.

Bioluminescence The emission of light from the electron transport chain of certain living organisms.

Biomass Organic matter produced by living organisms and measured by weight.

Biotechnology The use of microbes in industry.

Biovar A subgroup of a *Salmonella* serovar based on biochemical testing.

Bladder A bag. In brown algae, an air sac.

Blastospore An asexual fungal spore produced by budding from the parent cell.

Blocking antibody An IgG antibody that reacts with an allergen to prevent a hypersensitivity reaction.

Blood–brain barrier Cell membranes that allow some substances to pass from the blood to the brain but restrict others.

Bloom (algal) Abundant growth of microscopic algae, producing visible colonies in nature.

Brightfield microscope A microscope that uses visible light for illumination; the specimens are viewed against a white background.

Broad spectrum antimicrobial agent A chemical that has antimicrobial activity against many infectious microorganisms.

Brownian movement The movement of particles, including microorganisms, in a suspension due to bombardment by the moving molecules in the suspension.

Bubo An enlarged lymph node caused by inflammation.

Budding Asexual reproduction beginning as a protuberance from the parent cell that grows to become a daughter cell; also, release of an enveloped virus through the plasma membrane of an animal cell.

Buffer A substance that tends to stabilize the pH of a solution.

Burst size The number of newly synthesized bacteriophage particles released from a single cell.

Burst time The time required from bacteriophage adsorption to release.

Cachectin A polypeptide released by phagocytes in response to bacterial endotoxins; induces shock.

Calvin–Benson cycle Conversion of CO_2 into reduced organic compounds.

Cancer   A malignant, invasive cellular tumor that has the capability of spreading throughout the body or body parts.

Candle jar   A sealed jar containing a lighted candle, used to incubate bacterial cultures in high CO_2 atmosphere.

Capsid   The protein coat of a virus that surrounds the nucleic acid.

Capsomere   A protein subunit of a capsid.

Capsule   An outer, viscous covering on some bacteria composed of a polysaccharide or polypeptide.

Carbohydrates   Organic compounds composed of carbon, hydrogen, and oxygen, with the hydrogen and oxygen present in a 2:1 ratio; includes starches, sugars, and cellulose.

Carbon cycle   The series of processes that converts carbon dioxide (CO_2) to organic substances and back to carbon dioxide in nature.

Carbon fixation   *See* Calvin–Benson cycle.

Carbon skeleton   The basic chain or ring of carbon atoms in a molecule; $-C-C-C-$, for example.

Carboxyl group $-COOH$.

Carboxysome   Procaryotic inclusion containing ribulose 1,5-diphosphate carboxylase.

Carbuncle   An inflammation of the skin and subcutaneous tissue due to dissemination of a furuncle.

Carcinogen   Any cancer-producing substance.

Cardiolipin   A beef heart extract used in the venereal disease research laboratory (VDRL) slide test to detect antibodies against syphilis.

Carrier   An individual who harbors a pathogen but exhibits no signs of illness.

Casein   Milk protein.

Catabolic repression   Inhibition of decomposition of alternate carbon sources by glucose.

Catabolism   All decomposition reactions in a living organism.

Catalase   An enzyme that catalyzes the breakdown of hydrogen peroxide to water and oxygen.

Catalyst   A substance that affects the rate of a chemical reaction, usually increasing the rate, but isn't changed in the reaction.

Cation   A positively charged ion.

cDNA (complementary DNA)   DNA made from mRNA in vitro.

Cell   The basic microscopic unit of structure and function of all living organisms.

Cell culture   Animal or plant cells grown in vitro.

Cell-mediated immunity   An immune response that involves the binding and elimination of antigens by T cell lymphocytes.

Cell theory   The principle that all living things are composed of cells.

Cellulose   A glucose polysaccharide that is the main component of plant cell walls.

Cell wall   The outer covering of most bacterial, fungal, algal, and plant cells: in eubacteria, it consists of peptidoglycan.

Centrioles   Paired, cylindrical structures found in the centrosome of eucaryotic cells.

Cephalosporin   An antibiotic produced by the fungus *Cephalosporium* that inhibits the synthesis of gram-positive bacterial cell walls.

Cercaria   A free-swimming larva of trematodes.

Chancre   A hard sore, the center of which ulcerates.

Chemical bond   Attractive force between atoms forming a molecule.

Chemical element   A fundamental substance composed of atoms that have the same atomic number and behave the same way chemically.

Chemical energy   The energy of a chemical reaction.

Chemically defined medium   A culture medium in which the exact chemical composition is known.

Chemical reaction   The process of making or breaking bonds between atoms.

Chemiosmosis   A proton gradient across a cytoplasmic membrane; can be used to generate ATP.

Chemistry   The science of the interactions of atoms and molecules.

Chemoautotroph   An organism that uses an inorganic chemical as an energy source and carbon dioxide (CO_2) as a carbon source.

Chemoheterotroph   An organism that uses organic molecules as a source of carbon and energy.

Chemostat   An apparatus to keep a culture in log phase indefinitely.

Chemotaxis   A response to the presence of a chemical.

Chemotherapy   Treatment of a disease with chemical substances.

Chemotroph   An organism that uses oxidation-reduction reactions as its primary energy source.

Chitin   A glucosamine polysaccharide that is the main component of fungal cell walls and arthropod skeletons.

Chlamydospore   An asexual fungal spore formed within a hypha.

Chloramphenicol   A broad spectrum bacteriostatic chemical.

Chlorobium vesicle   *See* Chlorosome.

Chlorophyll *a*   The light-absorbing pigment in cyanobacteria, algae, and plants.

Chloroplast   The organelle that performs photosynthesis in photoautotrophic eucaryotes.

Chlorosome   Plasma membrane folds in green sulfur bacteria containing bacteriochlorophylls.

Choleragen   A *Vibrio* enterotoxin.

Chromatin   Threadlike, uncondensed DNA in an interphase eucaryotic cell.

Chromatophore   An infolding in the plasma membrane where bacteriochlorophyll is located in photoautotrophic bacteria.

Chromosome   The structure that carries hereditary information.

Chronic disease   An illness that develops slowly and is likely to continue or recur for long periods of time.

Cilia   Relatively short cellular projections that move in a wavelike manner.

Ciliate A member of the protozoan Phylum Ciliata that uses cilia for locomotion.

Cisternae Stacked elements of the Golgi complex.

Class A taxonomic ranking between phylum and order.

Clone A population of cells that are identical to the parent cell.

Coagulase A bacterial enzyme that causes blood plasma to clot.

Coccobacillus A bacterium that is an oval-shaped rod.

Coccus A spherical or ovoid bacterium.

Codon A group of three nucleotides in mRNA that specifies the insertion of an amino acid into a protein.

Coenocytic hyphae Fungal filaments that are not divided into uninucleate cell-like units because they lack septa.

Coenzyme A nonprotein substance that is associated with and that activates an enzyme.

Coenzyme A A coenzyme that functions in decarboxylation.

Cofactor The nonprotein component of an enzyme.

Coliforms Aerobic or facultatively anaerobic, gram-negative, nonspore-forming, rod-shaped bacteria that ferment lactose with acid and gas formation within 48 hours at 35°C.

Collagen The main structural protein of muscles.

Collision theory The principle that chemical reactions occur because energy is gained as particles collide.

Colony A clone of bacterial cells on a solid medium that is visible to the naked eye.

Commensalism A system of interaction in which two organisms live in association and one is benefited while the other is neither benefited nor harmed.

Commercial sterilization A process of treating canned goods aimed at destroying the endospores of *Clostridium botulinum*.

Common vehicle transmission Transmission of a pathogen to a large number of people by an inanimate reservoir.

Communicable disease Any disease that can be spread from one host to another.

Competence The physiological state in which a recipient cell can take up and incorporate a large piece of donor DNA.

Competitive inhibition The process by which a chemical competes with the normal substrate for the active site of an enzyme.

Complement (C) A group of serum proteins involved in phagocytosis and lysis of bacteria.

Complement fixation The process in which complement combines with an antigen–antibody complex.

Complete digestive system A digestive system with a mouth and an anus.

Completed test The final test for detection of coliforms in the multiple-tube fermentation test.

Complex medium A culture medium in which the exact chemical composition is not known.

Complex virus A virus with a complicated structure, such as a bacteriophage.

Compound A substance composed of two or more different chemical elements.

Compound light microscope An instrument with two sets of lenses that uses visible light as the source of illumination.

Compromised host A person whose resistance to infection is impaired.

Condensation reaction A chemical reaction in which a molecule of water is released.

Condenser A lens system located below the microscope stage that directs light rays through the specimen.

Confirmed test The second stage of the multiple-tube fermentation test, used to identify coliforms on solid differential media.

Congenital disease A disease present at birth as a result of some condition that occurred in utero.

Conidiophore An aerial hypha bearing conidiospores.

Conidiospore An asexual spore produced in a chain from a conidiophore.

Conjugation The transfer of genetic material from one cell to another involving cell-to-cell contact.

Conjugative plasmid A plasmid with genes for carrying out conjugation.

Constitutive enzyme An enzyme that is produced regardless of how much substrate is present.

Contact inhibition The cessation of animal cell movement and division from contact with other cells.

Contact transmission Spread of disease by direct contact, indirect contact, or droplet.

Contagious disease A disease that is easily spread from one person to another.

Continuous cell line Animal cells that can be maintained through an indefinite number of generations in vitro.

Convalescent period The period of recovery from a disease.

Corepressor The molecule (end-product) that brings about repression of a repressible enzyme.

Countercurrent immunoelectrophoresis (CIE) The movement of antigen and antibody toward each other through an electric field. *See also* Electrophoresis.

Counterstain A stain used to give contrast in a differential stain.

Covalent bond A chemical bond in which the electrons of one atom are shared with another atom.

Cresol A mixture of isomers from petroleum.

Crisis The phase of a fever characterized by vasodilation and sweating.

Cristae Foldings of the inner membrane of a mitochondrion.

Crossing over Process by which a portion of a chromosome is exchanged with a portion of another chromosome.

Culture Microorganisms that grow and multiply in a container of culture medium.

Culture medium The nutrient material prepared for growth of microorganisms in a laboratory.

Curd The solid part of milk that separates from the liquid in the making of cheese, for example.

Cutaneous mycosis A fungal infection of the epidermis, nails, and hair.

Cuticle Nonliving outer covering of helminths.

Cyanobacteria Oxygen-producing photoautotrophic procaryotes; formerly called blue-green algae.

Cyclic photophosphorylation Movement of an electron from chlorophyll through a series of electron acceptors and back to chlorophyll; purple and green bacterial photophosphorylation.

Cyclosporine A drug that suppresses the T-cell response.

Cyst A sac with a distinct wall containing fluid or other material; also, a protective capsule of some protozoans.

Cysticercus Encysted tapeworm larva.

Cystitis Inflammation of the urinary bladder.

Cytochrome oxidase An enzyme that oxidizes cytochrome *c*.

Cytochromes Proteins that function as electron carriers in respiration and photosynthesis.

Cytocidal Resulting in cell death.

Cytokines *See* Lymphokines.

Cytomegaly Enlarged cells.

Cytopathic effects (CPE) Tissue deterioration caused by viruses.

Cytoplasm In a procaryote, everything inside the plasma membrane; in a eucaryote, everything inside the plasma membrane and external to the nucleus.

Cytoplasmic streaming The flowing of cytoplasm in a eucaryotic cell.

Cytosine A pyrimidine nucleic acid base that pairs with guanine.

Cytoskeleton Microfilaments and microtubules that provide support and movement for eucaryotic cytoplasm.

Cytostome The mouthlike opening in some protozoans.

Cytotoxic T cells (T$_C$) Cells that destroy antigens.

Cytotrophic Binding to cells; for example, IgE antibodies bind to target cells.

D- Prefix describing a stereoisomer.

Dane particle A structure in the serum of a patient with hepatitis B having hepatitis B virus (HBV) surface antigens.

Darkfield microscope A microscope that has a device to scatter light from the illuminator so that the specimen appears white against a black background.

Dark reaction *See* Calvin–Benson cycle.

Deamination Removal of an amino group.

Death phase Period of logarithmic decrease in a bacterial population.

Debridement Surgical removal of necrotic tissue.

Decarboxylation Removal of carbon dioxide (CO$_2$) from an amino acid.

Decimal reduction time The time (in minutes) required to kill 90% of a bacterial population at a given temperature.

Decolorization The process of removing a stain.

Decomposition reaction A chemical reaction in which bonds are broken to produce smaller parts from a large molecule.

Definitive host An organism that harbors the adult, sexually mature form of a helminthic parasite.

Degeneracy (of the genetic code) Redundancy of genetic code; that is, most amino acids are signaled by several codons.

Degranulation Release of contents of secretory granules.

Dehydration The removal of water.

Dehydrogenation The loss of hydrogen atoms from a substrate.

Delayed hypersensitivity Cell-mediated hypersensitivity.

Delayed hypersensitivity T cells (T$_D$) Cells that produce lymphokines in type IV hypersensitivities.

Denaturation A change in the molecular structure of a protein.

Denitrification The reduction of nitrates to nitrites or nitrogen gas.

Dental plaque A combination of bacterial cells, dextran, and debris adhering to the teeth.

Deoxyribonucleic acid (DNA) The nucleic acid of genetic material.

Deoxyribose A five-carbon sugar contained in DNA nucleotides.

Dermatophyte A fungus that causes a cutaneous mycosis.

Dermis The inner portion of the skin.

Descriptive epidemiology Analysis of all data regarding the occurrence of a disease to determine the cause of the disease.

Desensitization The prevention of allergic inflammatory responses.

Desiccation The absence of water.

Detergent Any substance that reduces the surface tension of water.

Dextran A polymer of glucose.

Diacetyl CH$_3$COCOCH$_3$ produced from carbohydrate fermentation; chiefly responsible for the odor of dairy products.

Diagnosis Identification of a disease.

Diapedesis The process by which phagocytes move out of blood vessels.

Differential count The number of each kind of leukocyte in a sample of 100 leukocytes.

Differential medium A solid culture medium that makes it easier to distinguish colonies of the desired organism.

Differential stain A stain that distinguishes objects on the basis of reactions to the staining procedure.

Diffusion The net movement of molecules or ions from an area of higher concentration to an area of lower concentration.

Digestion The process of breaking down substances physically and chemically.

Digestive vacuole An organelle in which substrates are broken down enzymatically.

Dimorphism The property of having two growth forms.

Dioecious Referring to organisms in which organs of different sexes are located in different individuals.

Diphtheroid A gram-positive pleomorphic rod.

Diplobacilli Rods that divide and remain attached in pairs.

Diplococci Cocci that divide and remain attached in pairs.

Diploid Having two sets of chromosomes; normal state of a eucaryotic cell.

Dipocolinic acid Chemical substance found in bacterial endospores and not in vegetative cells.

Direct contact A method of spreading infection from one host to another through some kind of close association of the hosts.

Direct FA test A fluorescent-antibody test to detect the presence of an antigen.

Direct microscopic count Enumeration of cells by observation through a microscope.

Disaccharide A sugar consisting of two monosaccharides.

Disease Any change from a state of health.

Disinfectant Any substance used on inanimate objects to kill or inhibit the growth of microorganisms.

Disk-diffusion test *See* Kirby–Bauer test.

Dissimilation plasmids Plasmids containing genes coding for the production of enzymes that catalyze the catabolism of certain unusual sugars and hydrocarbons.

Dissociation Transformation of a compound into positive and negative ions in solution.

Disulfide bond Two atoms of sulfur held together by a covalent bond (S—S).

Division A phylum; used in botany and microbiology.

DNA probe A labeled piece of DNA used to identify the presence of bacteria in a sample.

Donor cell A cell that gives DNA to a recipient cell in recombination.

Donovan bodies Inclusion bodies in large mononuclear cells infected with *Calymmatobacterium granulomatis*.

DPT vaccine A combined vaccine used to provide active immunity, containing diphtheria and tetanus toxoids and killed *Bordetella pertussis* cells.

Droplet infection The transmission of infection by small liquid droplets carrying microorganisms.

Dysentery A disease characterized by frequent, watery stools with blood and mucus.

Eclipse period The time during viral multiplication when complete, infective virions are not present.

Edema An abnormal accumulation of fluid in body parts or tissues, causing swelling.

Effector T cells *See* Cytotoxic T cells, Delayed hypersensitivity T cells, Helper T cells, Killer cells, Memory cells, Natural killer cells, Regulatory T cells, Suppressor T cells.

Electron A negatively charged particle in motion around the nucleus of an atom.

Electron acceptor An ion that picks up an electron that has been lost from another atom.

Electron donor An ion that gives up an electron to another atom.

Electronic configuration The arrangement of electrons in shells or energy levels in an atom.

Electron microscope A microscope that uses a flow of electrons instead of light to produce an image.

Electron shells Regions of an atom corresponding to different energy levels.

Electron transport chain A series of compounds that transfer electrons from one compound to another, generating ATP by oxidative phosphorylation.

Electrophoresis The separation of substances (for example, serum proteins) by their rate of movement through an electric field.

Elementary body An infectious form of *Chlamydia*.

ELISA (enzyme-linked immunosorbent assay) A group of serologic tests that use enzyme reactions as indicators.

Embden–Meyerhof pathway *See* Glycolysis.

Emulsify To mix two liquids that do not dissolve in each other.

Encephalitis Inflammation of the brain.

Encystment Formation of a cyst.

Endemic disease A disease that is constantly present in a certain population.

Endergonic reaction A chemical reaction that requires energy.

Endocarditis Inflammation of the lining of the heart (endocardium).

Endocytosis The process of moving material into a eucaryotic cell.

Endoplasmic reticulum A membrane network in eucaryotic cells connecting the plasma membrane with the nuclear membrane.

Endospore A resting structure formed inside some bacteria.

Endotoxin Part of the outer portion of the cell wall of most gram-negative bacteria; lipid A.

Energy level The energy of an electron and its position relative to the atomic nucleus.

Enrichment culture A culture medium used for preliminary isolation that favors the growth of a particular microorganism.

Enterics The common name for bacteria in the family Enterobacteriaceae.

Enterotoxin An exotoxin that causes gastroenteritis; produced by *Staphylococcus*, *Vibrio*, and *Escherichia*.

Entner–Doudoroff pathway An alternate pathway for the oxidation of glucose to pyruvic acid.

Envelope An outer covering surrounding the capsid of some viruses.

Enzyme A protein that catalyzes chemical reactions in a living organism.

Enzyme induction The process by which a substance can cause the synthesis of an enzyme.

Enzyme repression The process by which a substance can stop the synthesis of an enzyme.

Enzyme-substrate complex A temporary union of an enzyme and its substrate.

Eosinophil A granulocyte whose granules take up the stain eosin.

Epidemic disease A disease acquired by many people in a given area in a short time.

Epidemiology The science dealing with when and where diseases occur and how they are transmitted.

Epidermis The outer portion of the skin.

Equilibrium The point of even distribution.

Ergot A substance produced in sclerotia by the fungus *Claviceps purpurea* that causes contraction of arteries and uterine muscle.

Erythrogenic toxin A substance produced by some streptococci that causes erythema.

Ester linkage The bonding between two organic molecules (R) as R— C —O—R.

Ethambutol A synthetic antimicrobial agent that interferes with the synthesis of RNA.

Etiology The study of the cause of a disease.

Eubacteria Procaryotic organisms; possess peptidoglycan cell walls.

Eucaryote A cell with DNA enclosed within a distinct membrane-bounded nucleus.

Eutrophication The addition of organic matter and subsequent removal of oxygen from a body of water.

Exchange reaction A chemical reaction that has both synthesis and decomposition components.

Exergonic reaction A chemical reaction that releases energy.

Exon A region of a eucaryotic chromosome that codes for a protein.

Exonuclease An enzyme that cuts DNA.

Exotoxins Protein toxins released from bacterial cells into the surrounding medium.

Experimental epidemiology The study of a disease using controlled experiments.

Extracellular enzyme An enzyme released from a cell to break down large molecules.

Extracellular polymeric substance (EPS) *See* Glycocalyx.

Extreme halophile An organism that requires a high salt concentration for growth.

Facilitated diffusion The transfer of a substance across a plasma membrane from an area of higher concentration to an area of lower concentration mediated by carrier proteins (permeases).

Facultative anaerobe An organism that can grow with or without oxygen (O_2).

Facultative halophile An organism capable of growth in, but not requiring, 1% to 2% salt.

Family A taxonomic group between order and genus.

Fat An organic compound consisting of glycerol and fatty acids.

Fatty acids Long hydrocarbon chains ending in a carboxyl group.

Feedback inhibition Inhibition of an enzyme in a particular pathway by the accumulation of end-product from the pathway.

Fermentation The enzymatic degradation of carbohydrates in which the final electron acceptor is an organic molecule, ATP is synthesized by substrate-level phosphorylation, and oxygen (O_2) is not required.

Fever An abnormally high body temperature.

F factor Fertility factor; a plasmid found in the donor cell in bacterial conjugation.

Fibrinolysin A kinase produced by streptococci.

Filamentous form A structure in the serum of a patient with hepatitis B having hepatitis B virus (HBV) surface antigens.

Filtration Passage of a liquid or gas through a screenlike material.

Fimbria *See* Pilus.

Fixed macrophage Macrophage that is located in a certain organ or tissue—for example, in liver, lungs, spleen, or lymph nodes.

Fixing (in slide preparation) The process of attaching the specimen to the slide.

Flagella Thin appendages that arise from one or more locations on the surface of a cell and are used for cellular locomotion.

Flagellate A member of the protozoan Phylum Mastigophora that uses flagella for locomotion.

Flaming The process of sterilizing an inoculating loop by holding it in an open flame.

Flat sour Thermophilic spoilage of canned goods not accompanied by gas production.

Flatworm An animal belonging to the Phylum Platyhelminthes.

Flavoprotein Protein with flavin coenzyme that functions as an electron carrier in respiration.

Flocculation Removal of colloidal material by addition of a chemical that causes the colloidal particles to coalesce.

Flora The microbial population of an area, such as of human skin.

Fluid mosaic model A way of describing the dynamic arrangement of phospholipids and proteins comprising the plasma membrane.

Fluke A flatworm belonging to the Class Trematoda.

Fluorescence The ability to give off light of one color when exposed to light of another color.

Fluorescent-antibody technique A diagnostic tool using antibodies labeled with fluorochromes and viewed through a fluorescent microscope.

Fluorescent microscope A microscope that uses an ultraviolet light source to illuminate specimens that will fluoresce.

Fluorochromes Dyes used to stain bacteria that fluoresce when illuminated with ultraviolet light.

Focal infection A systemic infection that began as an infection in one place.

Folliculitis Infection of hair follicles.

Fomite A nonliving object that can spread infection.

Forespore A structure consisting of chromosome, cytoplasm, and endospore membrane inside a bacterial cell.

Frameshift mutation A mutation caused by the addition or deletion of one or more bases in DNA.

Free radical A highly reactive particle with an unpaired electron; designated X·.

Freeze drying *See* Lyophilization.

FTA-ABS test An indirect fluorescent-antibody test used to detect syphilis.

Functional groups Arrangement of elements in organic molecules that are responsible for most of the chemical properties of those molecules.

Fungi Organisms that belong to the Kingdom Fungi; eucaryotic chemoheterotrophs.

Furuncle An infection of a hair follicle.

Gamete A male or female reproductive cell.

Gametocyte A male or female protozoan cell.

Gamma globulin *See* Immune serum globulin.

Gangrene Tissue death from loss of blood supply.

Gastroenteritis Inflammation of the stomach and intestine.

Gas vacuole A procaryotic inclusion for buoyancy compensation.

Gene A segment of DNA or a sequence of nucleotides in DNA that codes for a functional product.

Gene amplification A mechanism that causes a gene to be replicated many times.

Generalized transduction Transfer of bacterial chromosome fragments from one cell to another by a bacteriophage.

Generation time The time required for a cell or population to double in number.

Genetic engineering Manufacturing and manipulating genetic material in vitro.

Genetics The science of heredity.

Genotype The genetic makeup of an organism.

Genus The first name of the scientific name (binomial); the taxon between family and species.

Geosmin An alcohol produced by actinomycetes that has an earthy odor.

Germ Part of an organism capable of developing.

Germicidal Capable of killing microorganisms.

Germicidal lamp An ultraviolet light (wavelength = 260 nm) capable of killing bacteria.

Germination The process of starting to grow from a spore or endospore.

Germ theory The principle that microorganisms cause disease.

Globulin The protein type to which antibodies belong.

Glomerulonephritis Inflammation of the glomeruli of the kidneys, but not a result of kidney infection.

Glucan A polysaccharide component of yeast cell walls.

Glycerol An alcohol; $C_3H_5(OH)_3$.

Glycocalyx A gelatinous polymer surrounding a procaryotic cell wall.

Glycogen A polysaccharide stored by some cells.

Glycolysis The main pathway for the oxidation of glucose to pyruvic acid.

Golgi complex An organelle involved in the secretion of certain proteins.

Graft-versus-host (GVH) disease A condition that occurs when a transplanted tissue has an immune response to the tissue recipient.

Gram-negative cell wall A peptidoglycan layer surrounded by a lipopolysaccharide outer membrane.

Gram-positive cell wall Composed of peptidoglycan.

Gram stain A differential stain that divides bacteria into two groups, gram-positive and gram-negative.

Granulocyte A leukocyte with visible granules in the cytoplasm; includes neutrophils, basophils, and eosinophils.

Griseofulvin A fungistatic antibiotic.

Group translocation In procaryotes, active transport in which a substance is chemically altered during transport across the plasma membrane.

Guanine A purine nucleic acid base that pairs with cytosine.

Guillain–Barré syndrome A neurological syndrome, of unknown etiology, marked by muscular weakness and paralysis.

Gumma A rubbery mass of tissue characteristic of tertiary syphilis and tuberculosis.

Halogen One of the following elements: fluorine, chlorine, bromine, iodine, or astadine.

Halophile An organism that grows in high concentrations of salt.

Haploid A eucaryotic cell or organism with one of each type of chromosome.

Hapten An antigen that has reactivity and not immunogenicity.

Heavy metals Certain elements with specific gravity greater than 4 that are used as antimicrobial agents; for example, silver (Ag), copper (Cu), and mercury (Hg).

Helminth A parasitic roundworm or flatworm.

Helper T cells (T_H) Cells that interact with an antigen before B cells interact with the antigen.

Hemagglutination Clumping of red blood cells.

Hemagglutination-inhibition A process whereby an antibody inhibits viral hemagglutination.

Hemoflagellate A parasitic flagellate found in the circulatory system of its host.

Hemolysins Enzymes that lyse red blood cells.

Herd immunity The presence of immunity in most of a population.

Hermaphroditic Having both male and female reproductive capacities.

Heterocyst A large cell in certain cyanobacteria; site for nitrogen-fixation.

Heterolactic Any organism producing lactic acid and other acids or alcohols as end-products of fermentation (for example, *Escherichia*).

Heterophilic antibody An antibody that reacts with distantly related antigens.

Heterotroph An organism that requires an organic carbon source.

Hexachlorophene A chlorinated phenol used as an antiseptic.

Hexose monophosphate shunt *See* Pentose phosphate pathway.

Hfr   A bacterial cell in which the F factor has become integrated into the chromosome.

High-temperature short-time (HTST) pasteurization   Pasteurizing at 72°C for 15 seconds.

Histamine   A substance released by tissue cells that causes vasodilation.

Histocompatibility antigens   Antigens on the surface of human cells.

Histones   Proteins associated with DNA in eucaryotic chromosomes.

Holdfast   The branched base of an algal stipe.

Holoenzyme   An enzyme consisting of an apoenzyme and a cofactor.

Homolactic   An organism producing only lactic acid from fermentation (for example, *Streptococcus*).

Homologous chromosome   A chromosome that has the same base sequence as another. In a diploid cell, one of a pair of chromosomes.

Host   An organism infected by a pathogen.

Hot-air sterilization   The use of an oven at 170°C for approximately 2 hours.

H (hemagglutination) spikes   Antigenic projections from the outer lipid bilayer of influenza virus.

Humoral immunity   Immunity produced by antibodies dissolved in body fluids, mediated by B cells.

Humus   Organic matter that remains in soil following partial decomposition.

Hyaluronic acid   A mucopolysaccharide that holds together certain cells of the body.

Hybridoma   A cell made by fusing an antibody-producing B cell with a cancer cell.

Hydrogen bond   A bond between a hydrogen atom covalently bonded to oxygen or nitrogen and another covalently bonded oxygen or nitrogen.

Hydrogen ion (H$^+$)   A proton.

Hydrolysis   A decomposition reaction in which chemicals react with the H$^+$ and OH$^-$ of a water molecule.

Hydroxy group   —OH.

Hydroxyl ion   OH$^-$.

Hyperbaric chamber   An apparatus to hold gases at pressures greater than 1 atmosphere.

Hypersensitivity   Altered, enhanced immune reactions leading to pathologic changes.

Hypertonic   Describing a solution that has a higher concentration of solutes than an isotonic solution.

Hyphae   Long filaments of cells in fungi or actinomycetes.

Hypotonic   Describing a solution that has a lower concentration of solutes than an isotonic solution.

Icosahedron   A polyhedron with 20 triangular faces and 12 corners.

ID$_{50}$ The bacterial concentration required to produce a demonstrable infection in 50% of the test host population.

IgA   The class of antibodies found in secretions.

IgD   Antibodies found on B cells.

IgE   The class of antibodies involved in hypersensitivities.

IgG   The most abundant antibodies found in serum.

IgM   The first antibodies to appear after exposure to an antigen.

Illuminator   A light source.

Imidazoles   Antifungal drugs that interfere with sterol synthesis.

Immediate hypersensitivity   Allergic reactions involving humoral antibodies.

Immune complex   A circulating antigen–antibody aggregate capable of fixing complement.

Immune serum globulin   The serum fraction containing immunoglobulins (antibodies).

Immunity   The body's defense against a particular microorganism.

Immunization   A process that produces immunity.

Immunoassay   Detection of an antigen or antibody by serologic methods.

Immunodiffusion test   A test consisting of precipitation reactions carried out in an agar gel medium.

Immunoelectrophoresis   Identification of proteins by electrophoretic separation and then serologic testing.

Immunofluorescence   Procedures using the fluorescent-antibody technique.

Immunogen   *See* Antigen.

Immunoglobulin (Ig)   An antibody.

Immunological enhancement   The binding of antibodies to tumor antigens.

Immunological escape   Resistance of cancer cells to immune rejection.

Immunological surveillance   The body's immune response to cancer.

Immunosuppression   Inhibition of the immune response.

Immunotherapy   Treatment using antibodies.

Immunotoxin   An immunotherapeutic agent consisting of a poison bound to a monoclonal antibody.

IMViC   Biochemical tests used to identify enterics, including indole, methyl red, Voges–Proskauer, and citrate.

Inapparent infection   *See* Subclinical infection.

Incidence   The fraction of the population that contracts a disease during a particular length of time.

Inclusion   Material inside a cell.

Incomplete digestive system   A digestive system with one opening (mouth) for intake of food and elimination of waste.

Incubation period   The time interval between the actual infection and first appearance of any signs or symptoms of disease.

Indirect (passive) agglutination test   Indirect agglutination test using soluble antigens attached to latex or other small particles.

Indirect contact   Transmission of pathogens by agents such as food and water.

Indirect FA test   A fluorescent-antibody test to detect the presence of specific antibodies.

Inducer   A substrate that brings about an increased amount of an enzyme.

Inducible enzyme   *See* Enzyme induction.

Inert   Inactive.

Infection   Growth of microorganisms in the body.

Infection thread   An invagination in a root hair that allows *Rhizobium* to infect the root.

Infectious disease   A disease caused by pathogens.

Inflammation   A host response to tissue damage characterized by reddening, pain, heat, and swelling.

Innate resistance   Resistance of an individual to diseases that affect other species and other individuals of the same species.

Inoculate   To introduce microorganisms into a culture medium or host.

Inoculating loop (needle)   An instrument used to transfer bacteria from one culture medium to another.

Inorganic compounds   Small molecules not usually containing carbon.

Insertion sequence   *See* Transposon.

Interferon   An antiviral protein produced by certain animal cells in response to a viral infection.

Interleukins   Small proteins released from human cells in response to bacterial infection; directly or indirectly may induce fever, pain, or T-cell proliferation.

Intermediate host   An organism that harbors the larval stage of a helminth.

Intoxication   Poisoning.

Intron   A region of a eucaryotic chromosome that does not code for a protein.

Invasiveness   The ability of microorganisms to establish residence in a host.

In vitro   "In glass"; not in a living organism.

In vivo   Within a living organism.

Iodophor   A complex of iodine and a detergent.

Ion   A negatively or positively charged atom or group of atoms.

Ionic bond   A chemical bond formed when atoms gain or lose electrons in the outer energy levels.

Ionization   Separation of a molecule into groups of atoms with electrical charges.

Ionizing radiation   High-energy radiation that causes ionization; for example, X rays and gamma rays; wavelengths less than 1 nm.

Ischemia   Local loss of blood supply.

Isoelectric focusing   Electrophoresis of a protein until it reaches its isoelectric point.

Isoelectric point   The pH at which a protein no longer migrates in electrophoresis.

Isograft   Tissue graft from an identical twin.

Isomers   Two molecules with the same chemical formula but different structures.

Isoniazid (INH)   A bacteriostatic agent used to treat tuberculosis.

Isotonic   Referring to a solution in which osmotic pressure is equal across a membrane.

Isotope   A form of a chemical element in which the number of neutrons in the nucleus is different from the other forms of that element.

Keratin   A protein found in epidermis, hair, and nails.

Killer (K) cells   Effector cells that kill antibody-coated target cells.

Kinases   (1) Bacterial enzymes that break down fibrin (blood clots); (2) Enzymes that remove at ⓟ from ATP and attach it to another molecule.

Kingdom   The highest category in the taxonomic hierarchy of classification.

Kinins   Substances released from tissue cells that cause vasodilation.

Kirby–Bauer test   An agar-diffusion test to determine microbial susceptibility to chemotherapeutic agents.

Koch's postulates   Criteria used to determine the causative agent of infectious diseases.

Krebs cycle   A pathway that converts two-carbon compounds to carbon dioxide (CO_2), transferring electrons to NAD^+ and other carriers.

L-   Prefix describing a stereoisomer; L-amino acids are more commonly found in proteins.

Lagging strand   During DNA replication, the daughter strand synthesized discontinuously.

Lag phase   The time interval in a bacterial growth curve with no growth.

Larva   The sexually immature stage of a helminth or arthropod.

Latent disease   A disease characterized by a period of no symptoms when the pathogen is inactive.

Latent viral infection   A condition in which a virus remains in the host without producing disease for long periods of time.

LD$_{50}$ The lethal dose for 50% of the inoculated hosts within a given period of time.

Leading strand   During DNA replication, the daughter strand synthesized continuously.

Lepromin test   A skin test to determine the presence of antibodies to *Mycobacterium leprae*.

Leukemias   Cancers characterized by abnormally high numbers of leukocytes.

Leukocidins   Substances produced by some bacteria that can destroy neutrophils and macrophages.

Leukocyte   White blood cell.

Leukocytic pyrogen   A protein produced by PMNs that causes an increase in body temperature. *See* Polymorphonuclear leukocyte.

Leukocytosis-promoting factor   A substance released by inflamed tissues that increases the production of granulocytes.

Leukopenia   A condition in which the number of leukocytes is less than normal.

Leukotrienes   Mediators of anaphylaxis.

L form   A natural bacterium with a defective cell wall.

Lichen   A symbiosis between a fungus and an alga or cyanobacterium.

Ligands   Projections on procaryotic cells for adherence.

Ligase   An enzyme that joins together pieces of DNA.

Light reaction   *See* Photophosphorylation.

Limnetic zone   The surface zone of an inland body of water away from the shore.

Lipase An exoenzyme that breaks down fats into their component fatty acids and glycerol.

Lipid A molecule composed of glycerol and fatty acid.

Lipid A A component of gram-negative outer membrane; endotoxin.

Lipopolysaccharide (LPS) A molecule consisting of a lipid and a polysaccharide, forming the outer layer of gram-negative cell walls.

Lipoprotein A molecule consisting of a lipid and protein.

Liposome A fatty globule that may be used to administer chemotherapeutic agents.

Lithotroph *See* Autotroph.

Littoral zone The region along the shore of an inland body of water where there is considerable vegetation and where light penetrates to the bottom.

Local infection An infection in which pathogens are limited to a small area of the body.

Localized reaction An anaphylaxis-type reaction, such as hayfever, asthma, and hives.

Log phase Period of bacterial growth or logarithmic increase in a cell number.

Lophotrichous Having two or more flagella at one end of a cell.

Luciferase An enzyme that accepts electrons from flavoproteins and emits a photon of light.

Lymphangitis Inflammation of the lymph vessels.

Lymphocyte An agranulocyte involved in antibody production.

Lymphokines Proteins released by T cells and macrophages.

Lymphoma Cancer of lymphoid tissue.

Lymphotoxin A lymphokine that can destroy target cells in vitro.

Lyophilization Freeze-drying; freezing a substance and sublimating the ice in a vacuum.

Lysis Disruption of the plasma membrane. In disease, a gradual period of decline.

Lysogeny A state in which phage DNA is incorporated into the host cell without lysis.

Lysosome An organelle containing digestive enzymes.

Lysozyme An enzyme capable of lysing bacterial cell walls.

Lytic cycle A sequence for replication of phages that results in host cell lysis.

Macrolides Antibiotics that inhibit protein synthesis; for example, erythromycin.

Macromolecules Large organic molecules.

Macrophage A phagocytic cell; an enlarged monocyte.

Macrophage activation factor Increases macrophages' efficiency at destroying ingested cells.

Macrophage chemotactic factor Attracts macrophages to infection site.

Macrophage migration-inhibiting factor Prevents macrophages from leaving infection site.

Macular rash Small red spots appearing on the skin.

Major histocompatibility complex (MHC) The genes that code for the histocompatibility antigens; also known as human leukocyte antigens (HLA).

Malaise A feeling of general discomfort.

Malignant Cancerous.

Malolactic fermentation Conversion of malic acid to lactic acid by lactic-acid bacteria.

Malt Germinated barley grains containing maltose and amylase.

Malting Germination of starchy grains resulting in glucose and maltose production.

Mannan A polysaccharide component of yeast cell walls.

Mantoux test A tuberculin skin test.

Margination The process by which phagocytes stick to the lining of blood vessels.

Mast cell A type of cell found throughout the body that contains histamine and other substances that stimulate vasodilation.

Maximum growth temperature The highest temperature at which a species can grow.

Mechanical transmission The process by which arthropods transmit infections by carrying pathogens on their feet and other body parts.

Meiosis The process that leads to the formation of haploid gametes in a diploid organism.

Membrane filter A screenlike material with pores small enough to retain microorganisms.

Memory cells Long-lived B or T cells responsible for anamnestic response.

Meningitis Inflammation of the meninges covering the central nervous system.

Merozoite A trophozoite of *Plasmodium* found in infected red blood cells or liver cells.

Mesophile An organism that grows between 25°C and 40°C.

Mesosome An irregular fold in the plasma membrane of a procaryotic cell.

Messenger RNA (mRNA) The type of RNA molecule that directs the incorporation of amino acids into proteins.

Metabolism The sum of all the chemical reactions that occur in a living cell.

Metacercaria The encysted stage of a fluke in its final intermediate host.

Metachromatic granule The intracellular volutin stored by some bacteria.

Metalloprotein A protein conjugated with a metal atom.

Metastasis The spread of cancer from a primary tumor to other parts of the body.

Meter The standard unit of length in the metric system; one ten-millionth of the distance from the equator to the pole.

Microaerophile An organism that grows best in an environment with less oxygen (O_2) than is found in air.

Microbiota *See* Flora.

Micrometer (μm) A unit of measure equal to 10^{-6} m.

Microorganism A living organism too small to be seen with the naked eye; includes bacteria, fungi, protozoans, and microscopic algae; also includes viruses.

Microtubule The structure of the proteins comprising eucaryotic flagella and cilia.

Microwave An electromagnetic wave with wavelength between 10^{-1} and 10^{-3} m.

Minimal bactericidal concentration (MBC) The lowest concentration of a chemotherapeutic agent that will result in no growth of the test microorganisms.

Minimal inhibitory concentration (MIC) The lowest concentration of a chemotherapeutic agent that will prevent growth of the test microorganism.

Minimum growth temperature The lowest temperature at which a species will grow.

Miracidium The free-swimming, ciliated larva of a fluke that hatches from the egg.

Missense mutation A mutation that results in substitution of an amino acid in a protein.

Mitochondria Organelles containing the respiratory ATP-synthesizing enzymes.

Mitosis The division of the cell nucleus, often followed by division of the cytoplasm of the cell.

Mixed culture A culture containing more than one kind of microorganism.

MMWR *Morbidity and Mortality Weekly Report*; weekly publication of the Centers for Disease Control containing data on notifiable diseases and topics of special interest.

Molds Fungi that form mycelia and appear as cottony tufts.

Mole An amount of a chemical equal to the atomic weights of all the atoms in a molecule of the chemical.

Molecular biology The science dealing with DNA and protein synthesis of living organisms.

Molecular weight The sum of the atomic weights of all atoms making up a molecule.

Molecule A combination of atoms forming a specific chemical compound.

Monera The kingdom to which all procaryotic organisms belong.

Monoclonal antibodies Specific antibodies produced by in vitro clones of B cells hybridized with cancerous cells.

Monocyte A phagocytic agranulocyte.

Monolayer A single layer of cells due to the cessation of cell division by contact inhibition.

Monomers The units that combine to form polymers.

Mononuclear phagocytic system A system of fixed macrophages located in the spleen, liver, lymph nodes, and bone marrow.

Monotrichous Having a single flagellum.

Morbidity The incidence of a specific notifiable disease.

Mordant A substance added to a staining solution to make it stain more intensely.

Morphology The external appearance.

Mortality The deaths from a specific notifiable disease.

Most probable number (MPN) A statistical determination of the number of coliforms per 100 ml of water or food.

Motility The ability of an organism to move by itself.

M protein A heat- and acid-resistant protein of streptococcal cell walls.

Multiple-tube fermentation test A method of detecting the presence of coliforms.

Mutagen An agent in the environment that brings about mutations.

Mutation Any change in the base sequence of DNA.

Mutation rate The probability of a gene mutating each time a cell divides.

Mutualism A symbiosis in which both organisms are benefited.

Mycelium A mass of long filaments of cells that branch and intertwine, typically found in molds.

Mycetoma A chronic infection caused by certain fungi and *Nocardia* characterized by a tumorlike appearance.

Mycology The science dealing with fungi.

Mycorrhiza A fungus growing in symbiosis with plant root hairs.

Mycosis A fungal infection.

Mycotoxin A toxin produced by a fungus.

Myocarditis Inflammation of the heart muscle.

Naked virus A virus without an envelope.

Nanometer (nm) A unit of measurement equal to 10^{-9} m, $10^{-3}\mu$m, and 10 Å.

Natural killer (NK) cells Lymphoid cells that destroy tumor cells.

Naturally acquired active immunity Antibody production in response to an infectious disease.

Naturally acquired passive immunity The natural transfer of humoral antibodies; transplacental transfer.

Necrosis Tissue death.

Negative (indirect) selection *See* Replica plating.

Negative stain A procedure that results in colorless bacteria against a stained background.

Neurotoxin A chemical that is poisonous to the nervous system.

Neutralizing antibody An antibody that inactivates a bacterial exotoxin or virus.

Neutron An uncharged particle in the nucleus of an atom.

Neutrophil Also called polymorphonuclear leukocyte; a highly phagocytic granulocyte.

Nicotinamide adenine dinucleotide (NAD$^+$) A coenzyme that functions in the removal and transfer of H$^+$ and electrons from substrate molecules.

Nicotinamide adenine dinucleotide phosphate (NADP$^+$) A coenzyme similar to NAD$^+$.

Nitrification The oxidation of nitrogen from ammonia to nitrites and nitrates.

Nitrofuran A synthetic antimicrobial drug.

Nitrogen cycle The series of processes that converts nitrogen (N_2) to organic substances and back to nitrogen in nature.

Nitrogen fixation The conversion of nitrogen (N_2) into ammonia.

Nitrosamine A carcinogen formed by the combination of nitrite and amino acids; nitroso-: $-N=O$.

Nomenclature The system of naming things.

Noncommunicable disease A disease that is not transmitted from one person to another.

Noncompetitive inhibitor *See* Allosteric inhibition.

Noncyclic photophosphorylation Movement of an electron from chlorophyll to NAD^+; plant and cyanobacterial photophosphorylation.

Nonionizing radiation Radiation that does not cause ionization; for example, ultraviolet radiation.

Nonsense codon A special terminator codon that does not code for any amino acid.

Nonsense mutation A base substitution in DNA that results in a nonsense codon.

Nonspecific resistance Host defenses that tend to afford protection from any kind of pathogen.

Normal flora Microorganisms that colonize an animal without causing disease.

Nosocomial infection An infection that develops during the course of a hospital stay and was not present at the time the patient was admitted.

Notifiable disease A disease that physicians must report to the public health service.

N (neuraminidase) spikes Antigenic projections from the outer lipid bilayer of influenza virus.

Nuclear envelope The double membrane that separates the nucleus from the cytoplasm in a eucaryotic cell.

Nucleic acid A macromolecule consisting of nucleotides; for example, RNA and DNA.

Nucleic acid hybridization The process of combining single complementary strands of DNA.

Nucleoid The region in a bacterial cell containing the chromosome.

Nucleoli Areas in a eucaryotic nucleus where rRNA is synthesized.

Nucleoplasm The gel-like fluid within the nuclear envelope.

Nucleoprotein A macromolecule consisting of protein and nucleic acid.

Nucleoside A compound consisting of a purine or pyrimidine and pentose sugar.

Nucleotide A compound consisting of a purine or pyrimidine base, a five-carbon sugar, and a phosphate.

Nucleus The part of a eucaryotic cell that contains the genetic material; also, the part of an atom consisting of the protons and neutrons.

Numerical taxonomy A method of comparing organisms on the basis of many characteristics.

Nutrient broth (agar) A complex medium made of beef extracts.

Objective lens In a compound light microscope, the lens closest to the specimen.

Obligate anaerobe An organism that is unable to use oxygen.

Ocular lens In a compound light microscope, the lens closest to the viewer.

Oligodynamic action The ability of small amounts of a heavy metal compound to exert antimicrobial activity.

Oncogene A gene that can bring about malignant transformation.

Oncogenic virus A virus that is capable of producing tumors.

Operator The region of DNA adjacent to structural genes that controls their transcription.

Operon The operator site and structural genes it controls.

Opportunistic pathogen An organism that does not ordinarily cause a disease but can become pathogenic under certain circumstances.

Opsonization The enhancement of phagocytosis by coating microorganisms with certain serum proteins (opsonins).

Optimum growth temperature The temperature at which a species grows best.

Oral groove On some protozoans, the site at which nutrients are taken in.

Order A taxonomic classification between class and family.

Organelles Membrane-bounded structures within eucaryotic cells.

Organic compounds Molecules that contain carbon and hydrogen.

Organic growth factor An essential organic compound that an organism is unable to synthesize.

Organotroph *See* Heterotroph.

Osmosis The net movement of solvent molecules across a selectively permeable membrane from an area of higher concentration to an area of lower concentration.

Osmotic pressure The force with which a solvent moves from a solution of lower solute concentration to a solution of higher solute concentration.

Ouchterlony test An immunodiffusion test.

Outer membrane The outer layer of a gram-negative cell wall consisting of lipoproteins, lipopolysaccharides, and phospholipids.

Oxidase test A diagnostic test for the presence of cytochrome c using *p*-aminodimethylaniline.

Oxidation The removal of electrons from a molecule or the addition of oxygen to a molecule.

Oxidation pond A method of secondary sewage treatment.

Oxidation-reduction (redox) reaction A coupled reaction in which one substance is oxidized and one is reduced.

Oxidative phosphorylation The synthesis of ATP coupled with electron transport.

Oxygen cycle The processes that convert molecular oxygen (O_2) to oxides, water, and organic compounds and back to O_2.

Oxygenic Producing oxygen, as in plant and cyanobacterial photosynthesis.

PABA Para-aminobenzoic acid; a precursor for folic acid synthesis.

PAGE Polyacrylamide gel electrophoresis. *See also* Electrophoresis.

Pandemic disease An epidemic that occurs worldwide.

Papular rash A skin rash characterized by raised spots.

Parasite An organism that derives nutrients from a living host.

Parasitism A symbiosis in which one organism (the parasite) exploits another (the host) without providing any benefit in return.

Parenteral route Deposition directly into tissues beneath the skin and mucous membranes.

Passively acquired immunity Immunity acquired when antibodies produced by another source are transferred to the individual who needs them.

Pasteurization The process of mild heating to kill particular spoilage organisms or pathogens.

Pathogen A disease-causing organism.

Pathology Study of the causes and development of disease.

Pellicle The flexible covering of some protozoans.

Penicillins A group of antibiotics produced either by *Penicillium* (natural penicillins) or by adding side chains to the β-lactam ring (semisynthetic penicillins).

Pentose phosphate pathway A metabolic pathway that can occur simultaneously with glycolysis to produce pentoses and NADH without ATP production.

Peptide A chain of two (di-), three (tri-), or more (poly-) amino acids.

Peptide bond A bond joining the amino group of one amino acid to the carboxyl group of a second amino acid with the loss of a water molecule.

Peptidoglycan The structural molecule of bacterial cell walls consisting of the molecules N-acetylglucosamine, N-acetylmuramic acid, tetra peptide side chain, and peptide side chain.

Peptone Short chains of amino acids produced by the action of acids or enzymes on proteins.

Pericarditis Inflammation of the sac around the heart (pericardium).

Periplasmic space The region between the outer membrane and the cytoplasmic membrane.

Peritrichous Having flagella distributed over the entire cell.

Permease A carrier protein in the plasma membrane.

Peroxidase An enzyme that breaks down hydrogen peroxide; $H_2O_2 + NADH + H^+ \longrightarrow 2H_2O + NAD^+$.

Peroxide An oxygen oxide consisting of two atoms of oxygen.

pH The symbol for hydrogen ion concentration; a measure of the relative acidity of a solution.

Phage *See* Bacteriophage.

Phage typing A method of identifying bacteria using specific strains of bacteriophages.

Phagocyte A cell capable of engulfing and digesting particles that are harmful to the body.

Phagocytosis The ingestion of solids by cells.

Phagolysosome A digestive vacuole.

Phalloidin *See* Amanitin.

Phase-contrast microscope A compound light microscope that allows examination of structures inside cells through the use of a special condenser.

Phenol C_6H_5OH; carbolic acid.

Phenol coefficient A standard of comparison for the effectiveness of disinfectants; the disinfecting action of a chemical is compared with phenol for the same length of time on the same organism under identical conditions.

Phenolic A synthetic derivative of phenol.

Phenotype The external manifestations of the genetic makeup of an organism.

Phosphate group A portion of a phosphoric acid molecule attached to some other molecule:

$$-O-\overset{\overset{\textstyle O}{\|}}{\underset{\underset{\textstyle OH}{|}}{P}}-O-$$

Phospholipid A complex lipid composed of glycerol, two fatty acids, and a phosphate group.

Phosphorylation The addition of a phosphate group to an organic molecule.

Photoautotroph An organism that uses light for its energy source and carbon dioxide (CO_2) as its carbon source.

Photoheterotroph An organism that uses light for its energy source and an organic carbon source.

Photophosphorylation The production of ATP by photosynthesis.

Photosynthesis The light-driven synthesis of carbohydrate from carbon dioxide (CO_2).

Phototaxis The response to the presence of light.

Phototroph An organism that uses light as its primary energy source.

Phylogeny The evolutionary history of a group of organisms.

Phylum A taxonomic classification between kingdom and class.

Phytoplankton Free-floating photoautotrophs.

Pilus An appendage on a bacterial cell for attachment and conjugation.

Pinocytosis The engulfing of small molecules by infolding the plasma membrane.

Plankton Free-floating organisms.

Plantae A kingdom composed of multicellular eucaryotes with cellulose cell walls.

Plant virus A virus that multiplies in plant tissues.

Plaque A clearing in a confluent growth of bacteria due to lysis by phages. *See also* Dental plaque.

Plaque-forming units Visible plaques counted, perhaps due to more than one phage.

Plasma The liquid portion of blood in which the formed elements are suspended.

Plasma cells Cells produced by B lymphocytes that manufacture specific antibodies.

Plasma membrane The selectively permeable membrane enclosing the cytoplasm of a cell; the outer layer in animal cells, internal to the cell wall in other organisms.

Plasmid A small cyclic DNA molecule in bacteria replicating independent of the chromosome.

Plasmodium A multinucleated mass of protoplasm; when written as a genus, refers to the etiology of malaria.

Plasmolysis Loss of water from a cell in a hypertonic environment.

Plate count A method of determining the number of bacteria in a sample by counting the number of colony-forming units on a solid culture medium.

Pleomorphic Having many shapes.

Pneumonia Inflammation of the lungs.

Point mutation *See* Base substitution.

Polar molecule A molecule with an unequal distribution of charges.

Poly-β-hydroxybutyric acid A fatty acid storage material unique to bacteria.

Polyene An antimicrobial agent that alters sterols in eucaryotic plasma membranes and contains more than four carbon atoms and at least two double bonds.

Polyhedron A many-sided solid.

Polykaryocyte A multinucleated giant cell.

Polymer A molecule consisting of a sequence of similar units or monomers.

Polymerase An enzyme that synthesizes specific polymers.

Polymorphonuclear (PMN) leukocyte A neutrophil.

Polypeptide A chain of amino acids. Also, a group of antibiotics.

Polyribosome An mRNA strand with several ribosomes attached to it.

Porins Proteins in the outer membrane of gram-negative cell walls that allow passage of small molecules.

Portal of entry The avenue by which a pathogen gains access to the body.

Portal of exit The route by which a pathogen leaves the body.

Positive (direct) selection A procedure for picking out mutant cells by growing them.

Potential energy Energy that is stored.

Pour plate method A method of inoculating a solid nutrient medium by mixing bacteria in the melted medium and pouring the medium into a Petri plate to solidify.

PPD (Purified Protein Derivative) Proteins extracted from a boiled culture of *M. tuberculosis*; the antigen used in the tuberculin skin test.

Precipitation reaction A reaction between soluble antigens and multivalent antibodies to form aggregates.

Precipitin ring test A precipitation test performed in a capillary tube.

Predisposing factor Anything that makes the body more susceptible to a disease or alters the course of a disease.

Presumptive test The first step in the detection of coliforms.

Prevalence The fraction of a population having a specific disease at a given time.

Primary cell line Human tissue cells that grow only a few generations in vitro.

Primary infection An acute infection that causes the initial illness.

Primary producer Photoautotroph that converts light energy into chemical energy.

Primary response Antibody production to the first contact with an antigen.

Primary treatment Physical removal of solid matter from waste water.

Prion Infectious protein.

Privileged tissue Body tissue to which there is no immune response.

Procaryote A cell whose genetic material is not enclosed in a nuclear envelope.

Prodromal period The time following incubation when the first symptoms of illness appear.

Product The substance formed in a chemical reaction.

Profundal zone The deeper water under the limnetic zone in an inland body of water.

Proglottid A body segment of a tapeworm containing male and female organs.

Promoter site The starting point on DNA for transcription of RNA by RNA polymerase.

Properdin system Three serum proteins that function with complement to kill bacteria.

Prophage Phage DNA inserted into the host cell's DNA.

Prostaglandins Hormonelike substances that are synthesized in many tissues and circulate in the blood, the exact function of which is unknown.

Prosthetic group A coenzyme that is bound tightly to its apoenzyme.

Protein A large molecule containing carbon, hydrogen, oxygen, and nitrogen (and sulfur); some have a globular structure and others are pleated sheets.

Proteolytic An enzyme capable of hydrolyzing a protein.

Protista The kingdom to which protozoans belong; unicellular, eucaryotic organisms.

Proton A positively charged particle in the nucleus of an atom.

Protooncogene An oncogene that is functioning normally.

Protoplasmic streaming Movement of protoplasm in a plasmodial slime mold.

Protoplast A gram-positive bacterium without a cell wall; a plant cell lacking a cell wall.

Protozoa Unicellular eucaryotic organisms belonging to the Kingdom Protista.

Provirus Viral DNA that is integrated into the host cell's DNA.

Pseudohypha A short chain of cells that results from the lack of separation of daughter cells after budding.

Pseudopods Extensions of a cell that aid in locomotion and feeding.

Psychrophile An organism that grows best at 15°C and does not grow above 20°C.

Psychrotroph An organism that is capable of growth at 4°C and above 20°C.

Purines The class of nucleic acid bases that includes adenine and guanine.

Pus An accumulation of dead phagocytes, dead bacterial cells, and fluid.

Pustules Small elevations containing pus.

Pyocin A bacteriocin produced by *Pseudomonas aeruginosa*.

Pyocyanin Blue-green pigment produced by *Pseudomonas aeruginosa*.

Pyogenic Pus-forming.

Pyrimidines The class of nucleic acid bases that includes uracil, thymine, and cytosine.

Quaternary ammonium compound (quat) A cationic detergent with four organic groups attached to a central nitrogen atom; used as a disinfectant.

Quellung reaction Apparent swelling of a bacterial capsule in the presence of a specific antibody.

Quinine An antimalarial drug derived from the cinchona tree and effective against sporozoites in red blood cells.

Rapid plasma reagin A serologic test for syphilis.

Reactants Substances that are combined in a chemical reaction.

Reactivity The ability of an antigen to combine with an antibody.

Reagin IgE antibodies made in response to a treponemal infection characterized by their ability to combine with lipids; reagin–lipid complex will fix complement.

Recalcitrance Being resistant to degradation.

Receptor An attachment for a pathogen on a host cell.

Recipient cell A cell that receives DNA from a donor cell in recombination.

Recombinant DNA A DNA molecule produced by recombination.

Recombinant DNA techniques *See* Genetic engineering.

Recombinant RNA technology Techniques used to make RNA molecules.

Recombination The process of joining pieces of DNA from different sources.

Redia A trematode larval stage that may reproduce asexually one or two times before developing into a cercaria.

Red tide A bloom of planktonic dinoflagellates.

Reducing medium A culture medium containing ingredients that will remove dissolved oxygen from the medium to allow the growth of anaerobes.

Reduction The addition of electrons to a molecule; the gain of hydrogen atoms.

Refractive index The relative velocity with which light passes through a substance.

Regulator gene The gene that codes for a repressor protein.

Regulatory T cells Helper T and suppressor T cells that regulate the immune response.

Rennin A proteolytic enzyme obtained from a calf's stomach that forms curd from milk.

Replica plating A method of inoculating a number of solid minimal culture media from an original plate of complete medium; mutant colonies that don't grow on the minimal media can be selected from the original plate.

Replication fork The point where DNA separates, and new strands of DNA will be synthesized.

Replicative form A double-stranded RNA molecule produced during the multiplication of certain RNA viruses.

Repressible enzyme *See* Enzyme repression.

Repressor A protein that binds to the operator site to prevent transcription.

Reservoir of infection A continual source of infection.

Resistance The ability to ward off diseases through nonspecific and specific defenses.

Resistance (R) factor A bacterial plasmid carrying genes that determine resistance to antibiotics.

Resistance transfer factor (RTF) A group of genes for replication and conjugation on the R factor.

Resolution The ability to distinguish fine detail with a magnifying instrument.

Respiration An ATP-generation process in which chemical compounds are oxidized and the final electron acceptor is usually an inorganic molecule; also, the process by which living organisms produce carbon dioxide (CO_2).

Restriction enzymes Enzymes that cut double-stranded DNA leaving staggered ends.

Reticulate body An intracellular stage of *Chlamydia*.

Reverse transcriptase RNA-dependent DNA polymerase; an enzyme that synthesizes a complementary DNA from an RNA template.

Reversible reaction A chemical reaction in which the end products can readily revert to the original molecules.

R genes Genes carried on the R factor that code for enzymes that inactivate certain drugs.

Rhizosphere The region in soil where the soil and roots make contact.

Riboflavin A B vitamin that functions as a flavoprotein.

Ribonucleic acid (RNA) The class of nucleic acids that comprises messenger RNA, ribosomal RNA, and transfer RNA.

Ribose A five-carbon sugar that is part of ribonucleotide molecules and RNA.

Ribosomal RNA (rRNA) The RNA molecules that form the ribosomes.

Ribosomes The sites of protein synthesis in a cell, composed of RNA and protein.

Rifamycin An antibiotic that inhibits bacterial RNA synthesis.

Ring stage A young *Plasmodium* trophozoite that looks like a ring in a red blood cell.

RNA-dependent RNA polymerase An enzyme that synthesizes a complementary RNA from an RNA template.

RNA primer A short strand of RNA used to start synthesis of the lagging strand of DNA.

Root nodule A tumorlike growth on the roots of certain plants containing a symbiotic nitrogen-fixing bacterium.

Roundworms Animals belonging to the phylum Aschelminthes.

S (Svedberg unit) Notes the relative rate of sedimentation during ultra-high-speed centrifugation.

Sabin vaccine A preparation containing three attenuated strains of polio virus administered orally.

Saccharide   Sugar; general formula $(CH_2O)_n$.

Salk vaccine   A preparation of a formalin-inactivated polio virus that is injected.

Salt   A substance that dissolves in water to cations and anions, neither of which is H^+ or OH^-.

Sanitation   Removal of microbes from eating utensils and food-preparation areas.

Saprophyte   An organism that obtains its nutrients from dead organic matter.

Sarcina   A group of eight bacteria that remain in a packet after dividing.

Sarcoma   A cancer of fleshy, nonepithelial tissue or connective tissue.

Saturation   The condition in which the active site on an enzyme is occupied by the substrate or product at all times.

Saxitoxin   Neurotoxin produced by some dinoflagellates.

Scanning electron microscope   An electron microscope that provides three-dimensional views of the specimen magnified about 10,000 times.

Schaeffer–Fulton stain   An endospore stain that uses malachite green to stain the endospores and safranin as a counterstain.

Schick test   A skin test to detect the presence of antibodies to diphtheria.

Schizogony   The process of multiple fission, where one organism divides to produce many daughter cells.

Sclerotia   The reddish hardened ovaries of a grain that are filled with mycelia of the fungus *Claviceps purpurea*.

Scolex   The head of a tapeworm, containing suckers and possibly hooks.

Secondary infection   An infection caused by an opportunistic pathogen after a primary infection has weakened the host's defenses.

Secondary treatment   Biological degradation of the organic matter in waste water, following primary treatment.

Secretion   The production and release of fluid containing a variety of substances from a cell.

Secretory granule   A vesicle containing protein produced from the Golgi complex.

Selective medium   A culture medium designed to suppress the growth of unwanted bacteria and encourage the growth of desired microorganisms.

Selective permeability   The property of a plasma membrane to allow certain molecules and ions to move through the membrane while restricting others.

Selective toxicity   The property of some antimicrobial agents to be toxic for a microorganism and nontoxic for the host.

Semiconservative replication   The process of DNA replication in which each double-stranded molecule of DNA contains one original strand and one new strand.

Sense codon   A codon that codes for an amino acid.

Sense strand (+strand)   Viral RNA that can act as mRNA.

Sepsis   The presence of unwanted bacteria.

Septate hypha   A hypha consisting of uninucleate cell-like units.

Septicemia   A condition characterized by the multiplication of bacteria in the blood.

Septic tank   A tank, built into the ground, in which waste water is treated by primary treatment.

Septum   A crosswall dividing two parts.

Serial dilution   The process of diluting a sample several times.

Seroconversion   Change in a person's response to an antigen in a serological test.

Serology   The branch of immunology concerned with the study of antigen–antibody reactions in vitro.

Serovar   A subgroup of *Salmonella* based on the presence of antigens.

Serum   The liquid remaining after blood plasma is clotted, which contains immunoglobulins.

Sewage   Domestic waste water.

Sex pilus   A pilus used for the transfer of genetic material during bacterial conjugation.

Sexual dimorphism   The distinctly different appearance of adult male and female organisms.

Sexual reproduction   Reproduction that requires two opposite mating strains, usually designated male and female.

Signs   Changes due to a disease that a physician can observe and measure.

Simple stain   A method of staining microorganisms with a single basic dye.

Single-cell protein (SCP)   A food substitute consisting of microbial cells.

Singlet oxygen   Highly reactive O_2.

Skin test   The intradermal injection of an antigen or antibody to determine susceptibility to an antigen.

Slide agglutination test   A method of identifying an antigen by combining it with a specific antibody in a slide.

Slime layer   *See* Capsule.

Slime mold   Funguslike protists.

Slow-reacting substance of anaphylaxis (SRS-A)   Leukotrienes released by target cells after being bound by IgE antibodies.

Slow virus infection   A disease process that occurs gradually over a long period of time, caused by viruses or unconventional agents. *See also* Prion, Virino, Viroid.

Sludge   Solid matter obtained from sewage.

Smear   A thin film of material on a slide.

Soap   A surface-active agent made from animal fats and lye (NaOH).

Soil   A mixture of solid inorganic matter, water, air, organic matter, and living organisms.

Solubility   The ability to be dissolved, usually in water.

Solute   A substance dissolved in another substance.

Solvent   A dissolving medium.

Somatic   Relating to the cell itself.

Specialized transduction   The process of transferring a piece of cell DNA adjacent to a prophage to another cell.

Species   (1) The most specific level in the taxonomic hierarchy. **Bacterial species:** A population of cells with similar characteristics. (2) The second name in scientific binomial nomenclature.

Specific channel proteins Proteins in the outer membrane of gram-negative cell walls that allow passage of specific molecules.

Specific epithet The second name in a scientific binomial.

Specific resistance *See* Immunity.

Spherical particle A particle in the serum of a patient with hepatitis B having hepatitis B virus (HBV) surface antigens.

Spheroplast A gram-negative bacterium lacking a complete cell wall.

Spike A carbohydrate–protein complex that projects from the surface of certain enveloped viruses.

Spirillum A spiral or corkscrew-shaped bacterium.

Spirochete A corkscrew-shaped bacterium with an axial filament.

Spontaneous generation The idea that life could arise spontaneously from nonliving matter.

Spontaneous mutation A mutation that occurs without a mutagen.

Sporadic disease A disease that occurs occasionally in a population.

Sporangiophore Aerial hypha supporting a sporangium.

Sporangiospore An asexual fungal spore formed within a sporangium.

Sporangium A sac containing one or more spores.

Spore A reproductive structure formed by fungi and actinomycetes.

Sporogenesis The process of spore and endospore formation.

Sporozoite A trophozoite of *Plasmodium* found in mosquitoes, infective for humans.

Stability The condition of not deteriorating with time.

Staphylococci A broad sheet of spherical cells.

Stationary phase The period in a bacterial growth curve when the number of cells dividing equals the number dying.

STD Sexually transmitted disease.

Stem cells Fetal cells that give rise to bone marrow, blood cells, and B and T cells.

Stereoisomers Two molecules consisting of the same atoms, arranged in the same manner but differing in their relative positions; mirror images.

Sterile Free of microorganisms.

Steroids A specific group of chemical substances, including cholesterol and hormones.

Sterol A lipid alcohol found in the plasma membranes of fungi and *Mycoplasma*.

Stipe A stemlike supporting structure of multicellular algae and basidiomycetes.

Strain A group of cells all derived from a single cell.

Streak plate method A method of isolating a culture by spreading microorganisms over the surface of a solid culture medium.

Streptobacilli Rods that remain attached in chains after cell division.

Streptococci Cocci that remain attached in chains after cell division.

Structural gene A gene that codes for an enzyme.

Sty An infection of an eyelash follicle.

Subacute disease A disease with symptoms between acute and chronic.

Subclinical infection An infection that does not cause a noticeable illness.

Subcutaneous mycosis A fungal infection of tissue beneath the skin.

Substrate Any compound with which an enzyme reacts.

Substrate-level phosphorylation The synthesis of ATP by direct transfer of a high-energy phosphate group from an intermediate metabolic compound to ADP.

Sulfa drugs Any synthetic chemotherapeutic agent containing sulfur and nitrogen. *See also* Sulfonamides.

Sulfhydral group —SH.

Sulfonamides Bacteriostatic compounds that interfere with folic acid synthesis by competitive inhibition.

Superficial mycosis A fungal infection localized in surface epidermal cells and along hair shafts.

Superinfection Growth of the target pathogen that has developed resistance to the antimicrobial drug being used; growth of an opportunistic pathogen.

Superoxide dismutase Enzyme that destroys O_2^-; $O_2^- + O_2^- + 2H^+ \longrightarrow H_2O_2 + O_2$.

Superoxide free radical O_2^-.

Suppressor T cells (T_S) Cells that inhibit an immune response.

Surface-active agent (surfactant) Any compound that decreases the tension between molecules lying on the surface of a liquid.

Susceptibility The lack of resistance to a disease.

Sylvatic Belonging in the woods; a wild animal.

Symbiosis The living together of two different organisms.

Symptom A change in body function that is felt by a patient due to a disease.

Syndrome A specific group of signs or symptoms accompanying a particular disease.

Synergistic effect The principle whereby the effectiveness of two drugs used simultaneously is greater than either drug used alone.

Synthesis reaction A chemical reaction in which two or more atoms combine to form a new, larger molecule.

Synthetic chemotherapeutic agent An antimicrobial agent that is prepared in a laboratory.

Synthetic drug A chemotherapeutic agent that is prepared from chemicals in a laboratory.

Systemic anaphylaxis Hypersensitivity reaction causing vasodilation and resulting in shock.

Systemic (generalized) infection An infection throughout the body.

Systemic mycosis A fungal infection in deep tissues.

T antigen An antigen in the nucleus of a tumor cell.

Tapeworm A flatworm belonging to the Class Cestoda.

Target cell A body cell to which antibodies bind.

Taxon A taxonomic category.

Taxonomy The science of classification.

T cell Stem cell processed in the thymus gland that is responsible for cellular immunity.

Teichoic acid A polysaccharide found in gram-positive cell walls.

Temperate phage A bacteriophage existing in lysogeny with a host cell.

Terminator site The site on DNA at which transcription ends.

Tertiary treatment Physical and chemical treatment of waste water to remove all BOD, nitrogen, and phosphorus, following secondary treatment.

Tetracyclines Broad spectrum antibiotics that interfere with protein synthesis.

Tetrad A group of four cocci.

Tetrahedron A four-sided solid structure.

T-even bacteriophage A complex virus with double-stranded DNA that infects *E. coli*; for example, T2, T4, T6.

Thallus The entire vegetative structure or body of a fungus, lichen, or alga.

Thermal death point (TDP) The temperature required to kill all the bacteria in a liquid culture in 10 minutes at pH 7.

Thermal death time (TDT) The length of time required to kill all bacteria in a liquid culture at a given temperature.

Thermoduric Heat resistant.

Thermophile An organism whose optimum growth temperature is between 50°C and 60°C.

Thermophilic anaerobic spoilage Spoilage of canned foods due to the growth of thermophilic bacteria.

Thermostable nuclease A heat-stable enzyme produced by *Staphylococcus aureus*.

Thylakoid A chlorophyll-containing membrane in a chloroplast.

Thymine A pyrimidine nucleic acid base in DNA that pairs with adenine.

Tincture An alcoholic or aqueous solution.

Tinea A cutaneous fungal infection; ringworm.

Tissue culture *See* Cell culture.

Titer Reciprocal of a dilution.

Tolerance A state of immunological nonresponsiveness to self antigens.

Total magnification Magnification of a specimen, determined by multiplying the ocular lens magnification by the objective lens magnification.

Toxemia Symptoms due to toxins in the blood.

Toxigenicity The capacity of a microorganism to produce a toxin.

Toxin Any poisonous substance produced by a microorganism.

Toxoid An inactivated toxin.

Trace element A chemical element required in small amounts for growth.

Transamination The transfer of an amino group from an amino acid to an organic acid.

Transcription The process of synthesizing RNA from a DNA template.

Transduction Transfer of DNA from one cell to another by a bacteriophage. *See also* Generalized transduction, Specialized transduction.

Transfer RNA (tRNA) The class of molecules that brings amino acids to the site where they are incorporated into proteins.

Transformation The process in which genes are transferred from one bacterium to another as "naked" DNA in solution; also, the changing of a normal cell into a cancerous cell.

Transient flora Microorganisms that are present on an animal for a short time without causing a disease.

Translation The use of RNA as a template in the synthesis of protein.

Translocation Movement of a gene from one chromosomal locus to another.

Transmission electron microscope An electron microscope that provides high magnifications of thin sections of a specimen.

Transposon A small piece of DNA that can move from one region of the DNA molecule to another.

Transverse fission *See* Binary fission.

Transverse septum A crosswall that separates genetic material into two daughter cells in binary fission.

Trickling filter A method of secondary sewage treatment.

Trophozoite The vegetative form of a protozoan.

Tube-agglutination test A method for determining antibody titer using serial dilutions of serum mixed with antigen.

Tube-dilution test A method of determining the MIC using serial dilutions of an antimicrobial drug.

Tuberculin test A skin test used to detect the presence of antibodies to *Mycobacterium tuberculosis*.

Tumor Excessive tissue caused by uncontrolled cell growth.

Tumor-specific transplantation antigen (TSTA) A viral antigen on the surface of a transformed cell.

Turbidity The cloudiness of a suspension.

Turnover number The number of substrate molecules metabolized per enzyme molecule per second.

12D treatment A sterilization process that results in a decrease of the bacterial population by 12 logarithmic cycles.

Ubiquinones Low molecular weight, nonprotein carriers in the electron transport chain.

UDP-N-acetyl glucosamine (UDPNAG) A compound necessary for the biosynthesis of peptidoglycan.

Ultrastructure Fine detail not seen with a compound light microscope.

Ultraviolet (UV) radiation Radiation from 10 to 390 nm.

Uncoating The separation of viral nucleic acid from its protein coat.

Undulating membrane A highly modified flagellum on some protozoans.

Uracil A pyrimidine nucleic acid base in RNA that pairs with adenine.

Uridine diphosphoglucose (UDPG) Precursor for synthesis of glycogen.

Use-dilution test   A method of determining effectiveness of a disinfectant using serial dilutions.

Vaccination   The process of conferring immunity using a vaccine.

Vaccine   A preparation of killed, inactivated, or attenuated microorganisms or toxoids to induce artificially acquired active immunity.

Vacuole   An intracellular inclusion, in eucaryotic cells, surrounded by a plasma membrane containing raw food; in procaryotic cells, surrounded by a proteinaceous membrane containing gas.

Valence   The combining capacity of an atom or molecule.

Vancomycin   An antibiotic that inhibits cell-wall synthesis.

Vasodilation   Dilation or enlargement of blood vessels.

VDRL test   A rapid screening test to detect the presence of antibodies against *Treponema pallidum.*

Vector   An arthropod that carries disease-causing organisms from one host to another.

Vegetative cells   Cells involved with obtaining nutrients, as opposed to reproduction or resting.

Venereal disease   A sexually transmitted disease.

Vesicle   The expanded, terminal area of the Golgi complex; also, a fluid-filled blister. In a procaryote, a protein-covered hollow cylinder in a gas vacuole.

V factor NAD^+ or $NADP^+$.

Vibrio   A curved or comma-shaped bacterium.

Viral hemagglutination   The ability of certain viruses to cause agglutination of red blood cells.

Viremia   The presence of viruses in the blood.

Virino   An infectious agent composed of a small piece of nucleic acid and a protein from the host.

Virion   A fully developed complete viral particle.

Viroid   An infectious piece of "naked" RNA.

Virulence   The degree of pathogenicity of a microorganism.

Virus   A submicroscopic, parasitic, filterable agent consisting of a nucleic acid surrounded by a protein coat.

Visible light   Radiation from 400 to 700 nm, which the human eye can see.

Volutin   Stored phosphate in a procaryotic cell.

Wandering macrophage   A macrophage that leaves the blood and migrates to infected tissue.

Wassermann test   A complement fixation test used to diagnose syphilis.

Wheal   An area of edema of the skin resulting from a skin test.

Whey   The fluid portion of milk that separates from the curd.

Widal test   An agglutination test used to detect typhoid fever.

Xenograft   A tissue graft from another species.

X factor   A precursor necessary to synthesize cytochromes.

Yeast   A unicellular fungus belonging to the Phylum Ascomycetes.

Zone of inhibition   The area of no bacterial growth around an antimicrobial agent in the disk-diffusion test.

Zoonosis   A disease that occurs primarily in wild and domestic animals but can be transmitted to humans.

Zygospore   A sexual fungal spore characteristic of the Zygomycetes.

Zygote   A diploid cell produced by the fusion of two haploid gametes.

PHOTOGRAPH ACKNOWLEDGMENTS

Note: CDC = Centers for Disease Control, Public Health Service, U.S. Department of Health and Human Services, Atlanta, Georgia. BPS = Biological Photo Service, Moss Beach, California.

Chapter 1   Opener: CDC. 1-1a: National Cancer Institute. 1-1b: CDC. 1-2a and 1-6: C.L. Case, Skyline College. 1-7a: H.S. Pankratz, Michigan State University/BPS. 1-7b: G.T. Cole, University of Texas at Austin/BPS. 1-7c: H.S. Wessenberg and G.A. Antipa. *J. Protozoology* 17:250–270, 1970. 1-7d: J.R. Waaland, University of Washington/BPS. 1-7e: S.C. Holt, University of Texas Health Science Center, San Antonio/BPS. 1-8: Z. Skobe/BPS.

Chapter 2   Opener: N.L. Max, University of California/BPS.

Chapter 3   Opener: CDC. 3-1a: Cambridge Instruments, Buffalo, N.Y. 3-3a-c: J.R. Waaland, University of Washington/BPS. 3-5: P.W. Johnson and J.M. Sieburth, University of Rhode Island/BPS. 3-6b: CDC. 3-8a: R. Rodewald, University of Virginia/BPS. 3-8b: E. Golub, Purdue University. 3-10a: S.C. Holt, University of Texas Health Science Center, San Antonio/BPS. 3-10b: L.J. LeBeau, University of Illinois Hospital/BPS. 3-10c: CDC. Box, p. 67: Jeffrey C. Burnham, Ph.D.

Chapter 4   Opener: Z. Skobe/BPS. 4-1a top: Cheun-mo To and Charles C. Brinton, Jr. 4-1a bottom: Z. Skobe/BPS. 4-1b: T.J. Beveridge, University of Guelph/BPS. 4-1c top: T.J. Beveridge and C. Forsberg, University of Guelph/BPS. 4-1c bottom: A.E. McKee, Naval Medical Research Institute, Bethesda, Md. 4-2a: J.J. Cardamone, Jr., University of Pittsburgh/BPS. 4-2b and c: Z. Skobe/BPS. 4-3a: G.T. Cole, University of Texas at Austin/BPS. 4-3b: J.P. Burans, Naval Medical Research Institute, Bethesda, Md. 4-3c: N.S. Hayes et al., University of North Carolina, Chapel Hill. 4-4a: Courtesy of Dr. Heinz Schlesner, Institut fur Allgemeine Mikrobiologie, Christian-Albrechts University, Kiel, W. Germany. 4-4b: Courtesy Walther Stoeckenius, from *J. Bacteriology* 148:352, 1981. 4-6a top and lower left: P.W. Johnson and J. M. Sieburth, University of Rhode Island/BPS. 4-6a lower middle: S.C. Holt, University of Texas Health Science Center, San Antonio/BPS. 4-6a lower right: L.E. Sieburth, Rutgers University, New Brunswick, N.J. 4-7a: S.C. Holt, University of Texas Health Science Center, San Antonio/BPS. 4-8: S. Abraham and E.H. Beachey, V.A. Medical Center, Memphis, Tenn. 4-12a: T.J. Beveridge, Uni-

versity of Guelph/BPS. 4-13: H.S. Pankratz, Michigan State University/BPS. 4-14: T. J. Beveridge, University of Guelph/BPS. 4-18b: S.C. Holt, University of Texas Health Science Center, San Antonio/BPS. 4-18c left: H.S. Pankratz, Michigan State University/BPS. 4-18c middle and right: T. J. Beveridge, University of Guelph/BPS. 4-18b: S.C. Holt, University of Texas Health Science Center, San Antonio/BPS. 4-18c left: H.S. Pankratz, Michigan State University/BPS. 4-18c middle and right: T.J. Beveridge, University of Guelph/BPS. 4-20a: E.B. Small, unpublished. 4-20b: H.S. Wessenberg and G.A. Antipa, 1970, "Capture and Ingestion of Paramecium by Didinium Nasutum," *J. Protozoology*, 17: 250–270. 4-21a: R. Rodewald, University of Virginia/BPS. 4-22a and 4-23a: G.E. Palade, Yale University Medical School. 4-24a: Courtesy of Keith R. Porter. 4-25: E.H. Newcomb and T.D. Pugh, University of Wisconsin/BPS. Box, p. 81: D. Balkavill and D. Maratea.

Chapter 5   Opener: T.J. Beveridge, University of Guelph/PBS.

Chapter 6   Opener: CDC. 6-7: L.J. Le Beau, University of Illinois Hospital/BPS. 6-8a, b, d: CDC. 6-8c: R.J. Hawley, Georgetown University School of Dentistry. 6-9: CDC. 6-10b: L.E. Simon, Rutgers University 6-15a: H.W. Jannasch, Woods Hole Oceanographic Institution. 6-15b: Courtesy of Nalge Company. Box, p. 147: K. Crane, Woods Hole Oceanographic Institution.

Chapter 7   Opener: C.L. Case, Skyline College. 7-5: B.R. Funke, North Dakota State University 7-8: R. Humbert, Stanford University/BPS.

Chapter 8   Opener: L. Caro and R. Curtiss. 8-1a: Dr. Jack Griffith. 8-1b: G.F. Bahr, Armed Forces Institute of Pathology. 8-5a: J. Cairns, Imperial Cancer Research Fund Laboratory, Mill Hill, London. 8-7b: O.L. Miller, Jr., B.A. Hamkalo, and C.A. Thomas, Jr., *Science* 169:392, July 24, 1970. Copyright 1970 by the American Association for the Advancement of Science. 8-23: L.G. Caro and R. Curtiss. 8-27a: R. Welch, Medical School, University of Wisconsin. Box, p. 224: USDA Photo.

Chapter 9   Opener: T.J. Beveridge, University of Guelph, and G.D. Sprott, National Research Council of Canada, Ottawa/BPS. 9-4a, b: S.W. Schopf, Earth and Space Sciences, University of California, Los Angeles. 9-4c: S. M. Awramik,

University of California/BPS. 9-8: J.J. Le Beau, University of Illinois Hospital/BPS. 9-9: B.D. Davis et al., *Microbiology*, 3rd ed. New York: Harper & Row, 1980. 9-11: Biavatti et al., *Int. J. Systematic Bacteriology* 32:358–373, 1982. 9-12d, e: Courtesy of Steffen Kjelleberg. From Jouper-Jaan et al., *Applied and Environmental Microbiology* 1986, 53:1419–1421. Box, p. 250: © Stephen Leatherwood, Hubbs Marine Research Center.

Chapter 10 Opener: R.P. Burchard, University of Maryland Baltimore County. 10-1b: M.A. Listgarten and S.S. Socransky, "Electron Microscopy of Axial Fibrils, Outer Envelope and Cell Division of Certain Oral Spirochetes," *J. Bacteriology* 88:1087, 1964, © American Society for Microbiology. 10-1c: N.S. Hayes et al., University of North Carolina, Chapel Hill. 10-2a: T.J. Beveridge, University of Guelph/BPS. 10-2b: J.C. Burnham, T. Hashimoto, and S.F. Conti, "Electron Microscopic Observations on the Penetration of *Bdellovibrio bacteriovorus* into Gram-negative Bacterial Hosts," *J. Bacteriology* 96:1366–1381, 1968, © American Society for Microbiology. 10-3: John A. Fuerst, from *Bergey's Manual of Systematic Bacteriology* (Williams & Wilkins). 10-4: Cheun-mo To and Charles C. Brinton, Jr. 10-5: Judith F.M. Hoeniger, *J. Gen. Microbiology*, 40:29–42, 1965. Reproduced with permission of the Society for General Microbiology. 10-6: G.T. Cole, University of Texas at Austin/BPS. 10-7a: CDC. 10-7b: Courtesy of Sydney M. Finegold. 10-8a: W. Burgdorfer, Rocky Mountain Laboratory, Hamilton, Mont. 10-8b: R.C. Cutlip, National Animal Disease Center, Ames, Ia. 10-9: M.G. Gabridge, Cytographics, Inc. 10-10: A.E. McKee, Naval Medical Research Institute, Bethesda, Md. 10-11: Lin Tao, J.M. Tanzer, T.J. Mac Alister, *J. Bacteriology* 169:2543, 1987, © American Society for Microbiology. 10-12a: R.E. Strange and J.R. Hunter, in G.W. Gould and A. Hurst, eds., *The Bacterial Spore*. New York: Academic Press, 1969, p. 461, Fig. 4. 10-12b: C.L. Hannay and P. Fitz-James, *Canadian J. Microbiology* 1:694, 1955. 1012c: Courtesy of H.M. Solomon, FDA. From "Botulism" by D.A. Kautter and R.K. Lynt, Jr., in *FDA Papers*, November 1971. 10-13 and 10-14: CDC. 10-15: R.J. Hawley, Georgetown University School of Dentistry. 10-16a: J.S. Poindexter, *Bacteriology Review* 28:231–295, 1964. Reprinted with permission of the American Society for Microbiology. 10-16b: R.P. Burchard, University of Maryland Baltimore County. 10-16c: K. Stephens, Stanford University/BPS. 10-17: R.L. Moore/BPS. 10-18: Courtesy of J.L. Stokes. 10-19: T.J. Beveridge, University of Guelph/BPS. 10-20a: S.C. Holt, University of Texas Health Science Center, San Antonio/BPS. 10-20b: Courtesy of Norbert Pfennig. 10-21a: P.W. Johnson and J. M. Sieburth, University of Rhode Island/BPS. 10-21b: J.R. Waaland, University of Washington/BPS. 10-21c: Courtesy of Jeffrey C. Burnham, Ph.D. 10-22: From J. Burgess, "Morphological and Physiological Differentiation in *Streptomyces*," in R. Losick and L. Shapiro, eds., *Microbial Development*, Cold Spring Harbor, New York, 1984.

Chapter 11 Opener: G.T. Cole, University of Texas at Austin/BPS. 11-2: C.L. Case, Skyline College. 11-3: C. Robinow,

University of Western Ontario. 11-4 and 11-7: C.L. Case, Skyline College. 11-8: CDC. 11-9a: R. Humbert/BPS. 11-9b, c: J.R. Waaland, University of Washington/BPS. 11-11d: C.L. Case, Skyline College. 11-12a: C.M. Pringle, University of California, Berkeley/BPS. ll-13: C.L. Case, Skyline College. 11-15 and 11-17: Carolina Biological Supply. 11-18 and 11-28a: CDC. Box, p. 302: CDC.

Chapter 12 Opener: R.C. Valentine and H.G. Pereira. *J. Molecular Biology* 13:13–20, 1965. 12-3b: Dr. Jack Griffith 12-4b: R.C. Valentine and H.G. Pereira. *J. Molecular Biology* 13:13–20, 1965. 12-5b: F.A. Murphy, Viral Pathology Branch, CDC. 12-6b: A.K. Harrison, Viral Pathology Branch, CDC. 12-7a and c: Robley Williams. 12-8a: Courtesy of Bernard Roizman. 12-8b: C. Garon and J. Rose, National Institute of Allergy and Infectious Diseases. 12-9a: A.K. Harrison, Division of Viral Diseases, CDC. 12-9b: J.J. Cardamone, Jr., University of Pittsburgh/BPS. 12-9c: G.H. Smith, National Cancer Institute. 12-10: R. Humbert, Stanford University/BPS. 12-13: G. Wertz, School of Medicine, University of North Carolina, Chapel Hill. 12-18: C. Morgan, H.M. Rose, and B. Mednis. *J. Virology* 2:507–516, 1968. 12-23b: D.R. Brown et al. *J. Virology* 10:524–536, 1972. 12-24: A.K. Harrison, Division of Viral Diseases, CDC. 12-25: T.O. Diener, U.S. Department of Agriculture.

Chapter 13 Opener: Z. Skobe/BPS. 13-1a: Courtesy of Dwayne C. Savage, University of Illinois. 13-1b: Z. Skobe/BPS. 13-1c: Courtesy of Dwayne C. Savage, University of Illinois. 13-6: M.W. Jennison and the Department of Biology, Syracuse University.

Chapter 14 Opener: S. Murphree, Auburn University/BPS. 14-1a: L.R. Inman and J.R. Cantey, *J. Clinical Investigation* 71:1–8, 1983. 14-1b: S. Murphree, Auburn University/BPS. 14-2: S.C. Holt, University of Texas Health Science Center, San Antonio/BPS. 14-5: F.A. Murphy, Viral Pathology Branch, CDC. 14-6: J.P. Bader, National Cancer Institute.

Chapter 15 Opener: C.L. Sanders/BPS. 15-1: L. Winograd, Stanford University/BPS. 15-3a: K.E. Muse, Duke University Medical Center/BPS. 15-6: C.L. Sanders/BPS. 15-12e: Schreiber et al. *J. Experimental Medicine* 149:870–882, 1979.

Chapter 16 Opener: R.M. Steinman et al. *J. Experimental Medicine* 149:1–16, 1979. © Rockefeller University Press. 16-2a: From D. Zucker-Franklin, M.D., et al., *Atlas of Blood Cells: Function and Pathology*. Reprinted by permission of Edi Ermes, s. r. l., Italy. 16-2b: Courtesy of D. Zucker-Franklin, M.D., from *Seminars in Hematology* 6:4, 1969. Figure 10A. Reprinted with permission of Grune & Stratton, Inc. 16-10: R.M. Steinman et al. *J. Experimental Medicine* 149:1–16, 1979. © Rockefeller University Press.

Chapter 17 Opener: J.J. Cardamone, Jr., and S. Salvin, University of Pittsburgh/BPS. 17-11c: CDC.

Chapter 18 Opener: J.J. Cardamone, Jr., and S. Salvin, University of Pittsburgh/BPS. 18-1b, c: Courtesy of J.C. Forman, University College London. 18-2: L.J. Le Beau, University of Illinois Hospital/BPS. 18-11b, c: A. Liepins.

Chapter 19 Opener: L.J. LeBeau, University of Illinois Hospital/BPS. 19-1: A. Fleming. *British J. of Experimental Pathology* 10:226–236, 1929. 19-3: D. Greenwood and F. O'Grady, *Science* 163:1076, 1969. Copyright 1969 by the American Association for the Advancement of Science. 19-5: M. Bastide, S. Jouvert, and J.-M. Bastide. *Canadian J. Microbiology* 28:1119–1126, 1982. 19-20: L.J. Le Beau, University of Illinois Hospital/BPS. 19-21: Courtesy of Donald Nash, from *J. Clinical Microbiology*, Dec. 1986, p. 977. Reprinted with permission of the American Society for Microbiology.

Chapter 20 Opener: G.T. Cole, University of Texas at Austin/BPS. 20-2: Dr. Jean-Paul Revel, California Institute of Technology. 20-3: From J.C. Sherris, ed., *Medical Microbiology: An Introduction to Infectious Diseases*, Elsevier. 20-4: P.P. Cleary, University of Minnesota School of Medicine/BPS. 20-5: L.J. LeBeau, University of Illinois Hospital/BPS. 20-6: Armed Forces Institute of Pathology, Neg. No. 58-6180-6. 20-7 and 20-8a: World Health Organization, Geneva, Switzerland. 20-8b: Courtesy of H. Mensing. 20-10: P. Weary, University of Virginia School of Medicine. 20-12, 20-13, 20-14, and 20-15a: CDC. 20-15b: G.T. Cole, University of Texas at Austin/BPS. 20-15c: N. Goodman, University of Kentucky Medical College.

Chapter 21 Opener: P.A. Merz, New York State Institute for Basic Research. 21-3: F. L. A. Buckmire, The Medical College of Wisconsin. 21-5: Bettmann Archive. 21-6: CDC. 21-8: Courtesy of Warren E. Collins, Inc. 21-11: A.K. Harrison, CDC. 21-14: Edward J. Bottone, The Mount Sinai Hospital.

Chapter 22 Opener: S.C. Holt, University of Texas Health Science Center, San Antonio/BPS. 22-2: Armed Forces Institute of Pathology, Neg. No. N-64686. 22-4a: J.J. Cardamone, Jr., University of Pittsburgh/BPS. 22-4b: Marilyn J. Tufte. 22-6: CDC. 22-7: Armed Forces Institute of Pathology, Slide No. 79-18280-2. 22-9: CDC. 22-10b: © Edward S. Ross. 22-10c: CDC. 22-14: Dr. Jack Griffith. 22-17a: S.G. Baum, Albert Einstein College of Medicine/BPS. 22-19: CDC. 22-21: Wilmar Jansma, University of Southern Illinois; Harry Schafer, NIH; Bruce Wetzel, NIH. 22-22: Armed Forces Institute of Pathology, Neg. No. 70-1483.

Chapter 23 Opener: G.T. Cole, University of Texas at Austin/BPS. 23-3: CDC. 23-5: K.E. Muse, Duke University Medical Center/BPS. 23-6: R.B. Morrison, Austin, Tex. 23-8: G.L. Gooodhart, V.A. Hospital, Philadelphia. 23-9a: J.A. Stuekemann and D. Paretsky, University of Kansas/BPS. 23-9b: Thomas F. McCaul and Jim C. Williams, "Developmental Cycle of *Coxiella burnetii*: Structure and Morphogenesis of Vegetative and Sporogenic Differentiations," *J. Bacteriology* Sept. 1981, p. 1068, © American Society for Microbiology. 23-12a: G.D. Roberts, Mayo Clinic, Rochester, Minn. 23-12b: G.T. Cole, University of Texas at Austin/BPS. 23-14a: David J. Drutz and Milton Huppert, *J. Infectious Diseases* 147:379. © 1983 University of Chicago. 23-14b: G.T. Cole, University of Texas at Austin/BPS.

Chapter 24 Opener: Reprinted by permission of the publisher from "Ultrastructural Observations on Giardiasis in a Murine Model," by Robert L. Owen, Paulina C. Nemanic, and David P. Stevens, *Gastroenterology* 76:757–769. Copyright 1979 by the American Gastroenterological Association. 24-2b, 24-2c, 24-3: Z. Skobe/BPS. 24-4b, c: Hutton D. Slade, "Biology, Immunology, and Cariogenicity of *Streptococcus mutans*," *Microbiology Review* 44(2):331–384, June 1980. Reprinted with permission of the American Society for Microbiology. 24-5: M.A. Listgarten, University of Pennsylvania/BPS. 24-8a: G.T. Cole, University of Texas at Austin/BPS. 24-8b, c: Reprinted from *Food Technology* 36(3):93–96, 1982. Copyright © by Institute of Food Technologists. 24-9: L.-K. Ng, R. Sherburne, D.E. Taylor, and M.E. Stiles, "Morphological Forms and Viability of *Campylobacter* Species Studied by Electron Microscopy," *J. Bacteriology* 164:339. © American Society for Microbiology. 24-10: F.A. Murphy, Viral Pathology Branch, CDC. 24-12a: A.K. Harrison, Division of Viral Diseases, CDC. 24-12b: CDC. 24-13a: Dr. Jack Griffith. 24-14: Taken by E. Calomeni and provided by M. Estes, Baylor College of Medicine, Houston, Texas. 24-15: Reprinted by permission of the publisher from "Ultrastructural Observations on Giardiasis in a Murine Model," by Robert L. Owen, Paulina C. Nemanic, and David P. Stevens, *Gastroenterology* 76:757–769. Copyright 1979 by the American Gastroenterological Association. 24-16a: A.E. McKee, Naval Medical Research Institute, Bethesda, Md. 24-16b: CDC. 24-20: Armed Forces Institute of Pathology, Neg. No. 68-2740.

Chapter 25 Opener: CDC. 25-4: D. Bromley, West Virginia University Medical Center. 25-7, 25-8, 25-10, 25-11, 25-12, 25-13, and 25-14: CDC.

Chapter 26 Opener: M.F. Brown, University of Missouri. 26-4a: R. Toja, Charles R. Kettering Laboratory. 26-4b: E.H. Newcomb and S.R. Tandon, University of Wisconsin-Madison/BPS. 26-5a, b: M.F. Brown, University of Missouri. 26-5c: R.L. Peterson, University of Guelph. 26-11a: C.W. May/BPS. 26-12: R.F. Unz. *Int. J. Systematic Bacteriology* 21:91–99, 1971. 26-13a and 26-14a: C.W. May/BPS.

Chapter 27 Opener: © 1986 Mycogen Corporation, used by permission. 27-6: Switzerland Cheese Association, New York, N.Y. 27-11: Courtesy of Biosystems, Inc. 27-12: J. Shepard, D. Bidney, T. Barsby, and R. Kemble, *Science* 219:684. © 1983 American Association for the Advancement of Science. 27-13: © 1986 Mycogen Corporation, used by permission.

Part Openers One: Z. Skobe/BPS. Two: S. Dales, University of Western Ontario. Three: G.T. Cole, University of Texas at Austin/BPS. Four: Z. Scobe/BPS. Five: E.H. Newcomb and S.R. Tandon, University of Wisconsin-Madison/BPS.

Microview A   1, 2: L.J. LeBeau, University of Illinois Hospital/BPS. 3: CDC. 4, 5: L.J. LeBeau, University of Illinois Hospital/BPS. 6, 7, 8, 9,: CDC. 10, 11, 14: C.L Case, Skyline College. 12, 13, 16, 17, 18: C.L. Case, Skyline College. 15, 19: CDC. 20: Michael H. Nachtigall, San Mateo County Department of Health Services, San Mateo, CA. 21, 22, 23, 24, 25: CDC. 26, 27, 28, 29: C.L. Case, Skyline College. 30: R. Humbert/BPS. 31, 32: C.L. Case, Skyline College.

Microview B   1, 2, 3, 4: Genentech, Inc. 5: The DuPont Company. 6, 7, 8, 9: © Nita Winter, 1985/Cetus Corporation. 10: J.C. Hunter-Cevera, Cetus Corporation. 11: © Nita Winter, 1985/Cetus Corporation. 12: © Chuck O'Rear, 1985/Cetus Corporation. 13, 14, 15, 16: © Nita Winter, 1985/Cetus Corporation. 17, 18: Genentech, Inc. 19, 20: Chiron Corporation. 21: Cetus Corporation.

Microview C   1: C.L. Case, Skyline College. 2, 3, 4, 5, 6, 7: © David Powers, 1985/University of California, Davis. 8: L.J. LeBeau, University of Illinois Hospital/BPS. 9, 10, 11, 12, 13, 14, 15, 16, 17, 18: © David Powers, 1985/ University of California, Davis. 19: Michael H. Nachtigall, San Mateo County Department of Health Services, San Mateo, Calif.

INDEX

Amphotericin B
action of, 480
blastomycosis treated with, 600
coccidioidomycosis treatment with, 597
cryptococcosis treatment with, 539
histoplasmosis treatment with, 596
properties of, *483*, 489, *489*
source of, *477*
Ampicillin
bacillary dysentery treated with, 615
bacterial cell walls, effect on, *479*
cystitis and pyelonephritis treated with, 642–643
H. influenzae meningitis treatment with, 528
properties of, *482*
spectrum of activity of, 486
structure of, *485*
typhoid fever treatment with, 614
Ampligen, AIDS treatment with, 469
Amylases, commercial production of, 705, *707*
Anabaena, *261*
Anabolic reactions, 106–107
Anabolism, 106–107
biochemical pathways of, 134–137
integration with catabolism, 137–139
Anaerobic gram-negative cocci, characteristics of, *260*, 269, *269*
Anaerobic growth media and methods, 154–155, *154*, *155*
Anaerobic respiration, 121, 125, 130, *133*, 150, 264, 670
Anaerobic sludge digesters, 685, *685*
Anal gonorrhea, 646
Anal pore, 300
Analytical epidemiology, 370–371
Anamnestic response, 430
Anaphylactic shock, 453, 630
Anaphylaxis, 451–455, *452*, *453*
complement system and, 408
localized, 452, 454
prevention of, 454–455
systemic, 452–454
Anemia
autoimmune hemolytic, 458, 461
sickle-cell, 45
Animal bites, 554, *570*
Animalcules, 6–7
Animal feed supplements
antibiotic use as, 476, 494–495
microbial production of, 229
M. methylotrophus as, 703–704
Animalia, 16, 237, *238*, *240*

Animals, virus inoculation in, 328
Animal viruses
characteristics of, 319
classification of, 324, *326–327*
multiplication of, 334–340
Anions, 28
Anopheles mosquitoes, *313*
DDT resistance of, 438
malaria transmitted by, *312*, 566, *571*
Plasmodium transmission by, 303–304, *304*, 313
Anoxygenic photosynthetic bacteria, *261*
Anoxygenic process, 119
Antagonistic drugs, 495
Anthrax, 9, *10*
agent of, 274
characteristics of, 552–553, *553*, *570*
gastrointestinal, 552
pulmonary, 552
reported cases of, *553*
transmission of, *362*
Anti-anti-idiotypic antibodies, 438, *438*
Antibiosis, 476
Antibiotic-resistance genes, 225–226, *225*
Antibiotics, 12–13
aminoglycoside, action of, 479
animal feed, use as supplement in, 476, 494–495
bacterial destruction by, 85
endotoxin release by, 384
genetic engineering in production of, 711–712
historical development of, 475–476
microbial production of, 707–710
outer membrane barrier to, 83
properties of, *482*, 485–489
protein synthesis inhibition by, 91
recombinant DNA production of, 228–229
resistance to, 13
R factors and resistance to, 224, *225*
sources of, *477*
Antibodies, *10*, *11*, *15*, 43. *See also* Monoclonal antibodies.
affinity of, 423
AIDS, 446
anti-anti-idiotypic, 438, *438*
anti-idiotypic, 437, *438*, 468
fluorescent, 61
genetics of production of, *11*
heterophil, 560
idiotypic, 437, *438*, 439
immune system production of, 414
nature of, 417–420

reagin-type, 651
serology, in, 245–246
structure of, 417–418, *417*, 428
valence of, 417
Antibody-antigen binding, 422–423
Antibody diversity problem, 428–429
Antibody titer, 430, 441–442, 444
microbial production of, 229
Anticaterpillar toxin
Anticodons, 203–205, *205*
Antifungal drugs
action of, 480
properties of, *483*, 489–490
Antigen-antibody complexes
functions of, 422
in precipitation reactions, 439–440
Antigen-antibody reactions, chemical bonding in, 28
Antigen-binding sites, 417–418, *417*
Antigenic changes, virally induced, 388
Antigenic determinants, 415, *415*
Antigenic drift, *593*, 594
Antigenic shifts, 593–595, *593*, *594*
Antigen modulation, 470
Antigen-presenting cells (APC), 420–421, *421*, *422*, 426
Antigen receptors, 416, 422
Antigens
ABO blood group, 418
bacterial relatedness and, 246
B cell interactions with, 420–422
clumping of, 417–418
hepatitis B, 368, 621–623
histocompatibility, 461
HLA, *463*
immune response to, 414, 418, *419*
nature of, 416–417
particulate, 441, *442*
self, 421
soluble, 439, 441
types A and B, 455–456, *455*, *456*
Antihelminthic drugs, properties of, *483*, 490, 492
Anti-HGG. *See* Anti-human gamma globulin.
Antihistamine drugs, 454
Anti-human gamma globulin (anti-HGG), *445*, 446, 651
Anti-idiotypic antibodies, 437, *438*
in AIDS virus development, 468
Anti-idiotypic vaccines, 424, 437–439, *438*
Antimetabolites, action of, 481
Antimicrobial agents, plasma membrane destruction by, 87–88

Branhamella catarrhalis
  otitis media caused by, 583
  pneumonia caused by, 589
Breakbone fever, 562
Breed count method, 164
Brightfield illumination, 59, *59*, *60*, *64*
Bright's disease, 643
Broad spectrum drugs, 478
Brome grass mosaic virus, *348*
5-Bromouracil, 213, *213*
Bronchitis, agents of, 268
Brood capsules, 308
Broth dilution tests, 493
Brown algae, 292, *293*, *294*, *295*, *295*
Brown kelp, 292
Brucella abortus, 551
Brucella canis, 551
Brucella melitensis
  brucellosis caused by, 551–552
  portal of entry of, *380*
Brucella spp.
  characteristics of, *260*, 265
  infections by, 551–552
  transmission of, *362*
Brucella suis, 551–552
Brucellosis, *362*, 551–552, *570*
  agent of, 265; portal of entry of, *380*
Bruton's agammaglobulinemia, *465*
"Bubble boy," 345, 358
Bubonic plague, *362*, 555, *555*
  agent of, 267
Bud, 159
Budding
  bacteria, *261*, 276, *276*
  enveloped viruses, 342, *342*
  yeasts, 284, *285*
Buffers, 147
Bugs, *312*
Bulking, of sludge, 684
Burkitt's lymphoma, 561
  agent of, *326*, 345
  oncogene translocation in, 346
Burn patients, susceptibility of, 366
Burst size, 332
Burst time, 332
Butter, production of, 701
Buttermilk, production of, 701
Butyric anaerobic spoilage, *695*
B vitamins, 5

Cachectin, 386
Calcium hypochlorite, 184
Calcium propionate, 186, *189*, 698
California encephalitis, 538
  cases of in the United States, *562*
Calvin-Benson cycle, 118, *118*, 134, *735*
Calvulina cristata, *291*

Calymmatobacterium granulomatis, 653, *657*
Campylobacter fetus, 262
Campylobacter gastroenteritis, 617, *632*
Campylobacter jejuni, 262, *617*
Campylobacter pylori, 617
Campylobacter spp.
  characteristics of, *260*
  diarrhea caused by, 617
  waterborne, outbreaks of, *678*
Cancer
  cervical, 654
  colon, 390
  immune response to, 470–471
  liver, 389, 622
  monoclonal antibodies in treatment of, 424
  principal types of, *471*
  viruses and, 343–346
Cancer-causing viruses, 334
Candida
  infections, nystatin used for, *483*, 489
  in normal flora, 357
  reclassification of, *290*
Candida albicans, 286, *517*
  AIDS-associated, *466*
  overgrowths by, 517
  superinfection by, 478, 487
  urinary tract contamination with, 642
  vaginitis caused by, 356, 652, 655, *655*
Candida utilis, as food source, 703
Candidiasis, 292, *516*, 517, *517*
Candle jars, 155–156, *155*
Canine demyelinating encephalo-myelitis, *347*
Canned food, spoilage of, 694–695, *695*
Canning, 693–695
  industrial, *693*
CAP (catabolite activator protein), *210*
Capillaries, 546
Capsids, 321–323, *321*, *322*
Capsomeres, 321–322, *321*, *322*
Capsules
  adherence, effect on, 401
  characteristics of, 77
  genetic transformation and formation of, 218–219, *218*
  host defense resistance by, 379–380
  of *K. aerogenes*, *379*
  negative staining of, 68–69, *68*, *69*, 77
  in vaccine development, 435

Carbapenems, properties of, *482*, 486
Carbenicillin
  pseudomonad infections, use in, 508
  spectrum of activity of, 486
  structure of, *485*
Carbohydrates, 39–40
  catabolism of, 122–134
Carbolfuchsin, 68–69, *69*
Carbolic acid, 182. *See also* Phenol.
Carbon
  atomic structure of, *25*
  chemical bonding of, 29, *30*, 37–38, *37*
  electronic configuration of, 26, *27*, 34
  microbial growth requirements for, 148
Carbon cycle, 19, 667–668, *668*
Carbon dioxide fertilization, 670
Carbon dioxide incubators, 155
Carbon fixation, 117–118
Carbon skeleton, 37
Carbonyl group, *38*
Carboxyl group, *38*
Carboxysomes, 92
Carbuncle, 505
Carcinogens
  chemical, identification of, 217
  mutagens as, 213, 217
Carcinoma, *471*
Cardiovascular system
  bacterial diseases of, 548–559, *570–571*
  helminthic diseases of, 568–569, *571*
  protozoan diseases of, 563–567, *571*
  structure and function of, 546–547, *547*
  viral diseases of, 560–563, *571*
Carotenoids, 136
Carpenteles, *290*
Carrageen, 296
Carrier proteins, 43, 88, *88*
Carriers
  disease transmission by, 361
  of *E. histolytica*, 627
  of giardiasis, 626
  of hepatitis B, 622–623
  of typhoid, 614
Cascade, 406
Case control method, 370
Casein, 700
Caseous lesions, 587
Case reporting, 371–372
  of AIDS cases, 371–372
Castanospermine, AIDS treatment with, 469
Catabolic reactions, 106–107

Phase halo, *59*

Phase-contrast microscopy, *59*, 60–61, *60*, *64*

Phenol, *182*
  disinfectant use of, 179, 182, *188*
  in infection control, 9

Phenol coefficient, 179–182, *188*

Phenolics, *182*
  as disinfectants, 182–183, *188*

Phenotype, 195
  altered, selecting or testing for, 215–216

Phenylalanine, *47*

O-Phenylphenol, *182*, 183, *188*

Phialophora verruscosa, chromomycosis caused by, 515–516, *516* 516, *516*

Phormidium luridum, *279*

Phosphates, eutrophication caused by, 678–679

Phosphoenolpyruvate (PEP), 90

Phospholipase C, 49

Phospholipid bilayer, 42, *43*, 85–86, *86*

Phospholipids, 41–42, *41*
  biosynthesis of, 136

Phosphorus
  electronic configuration of, *27*
  microbial growth requirements for, 149

Phosphorylation, types of, 115–116

Photoautotrophs, 117–119, *117*, 667
  carbon cycle, role in , 667
  carbon sources of, 148

Photoheterotrophs, *117*, 119

Photophosphorylation, 115–117, *116*, 118
  in *H. halobium*, 120–121

Photosynthesis, 117–119, 667
  in algae, 18
  anoxygenic, 278
  coenzymes in, 109
  microorganisms and, 5
  sites of, 87, *87*

Photosynthetic bacteria, 117–119, *117*
  inclusions of, 92

Phototaxis, 79

Phototrophic bacteria, characteristics of, 278

Phototrophs, 117

Phylogenetic hierarchy, 242

Phylogenetic relationships, 237–240, 247–248

Phylum, 242

Phytophthora infestans, 292

Phytoplankton, 677

Pica, 630

Pichia, *290*

Picornaviruses
  antiviral drug for, 584

multiplication of, 336–337
plant viruses similar to, *348*
portal of entry of, *381*

PID. *See* Pelvic inflammatory disease.

Piedras, 514

Pili, 76, 79–80, *80*

Pilin, 79

Pimples, 503, 505

Pinkeye, 518, *519*

Pinocytosis, plasma membrane in, 96

Pinworms, 310, *310*, *311*, 492, 630, *633*

Piperazine, 492

Pityrosporum
  in normal flora, 357
  skin, growth on, 504

Placental transfer, 414, 418, *419*

Plague, 554–555, *570*. *See also* Bubonic plague.
  agent of, portal of entry of, *381*
  distribution in the United States, *555*
  vaccine for, *436*, 555
  vector of, *365*

Planktonic algae, 293, 296

Plantae, 16, 237, *238*, *241*

Plant diseases, 667
  fungal, *11*

Plant-microbe associations, 229

Plant pathogens, 267, 667

Plants
  carbon cycle, role in, 667
  herbicide-resistant, 712
  leguminous, 671–672
  nutritional patterns of, 117–119, *117*, *119*

Plant viruses, 319, 324, 348
  classification of, *348*

Plaque, dental, *355*, 609–610, *609*

Plaque-forming units (pfu), 325

Plaque method, 325

Plaques, 247, *247*, 325
  bacteriophage, *325*

Plasma, 397

Plasma cells, 420, *421*, *422*

Plasma membrane
  antibiotic effects on, *478*, 480, *480*
  antimicrobial effects on, 173
  electron transport carriers and ATPase in, 117
  eucaryotic, 96
  procaryotic, 81, 85–88, *86*, 96, *101*
  saturation and fluidity of, 42–43

Plasmids, 91, 223–224
  in conjugation, 220, *221*, 223
  in genetic engineering, 711
  pathogenicity and, 386–387
  recombinant, 226–228, *227*

Plasmodial slime molds, 298–299

Plasmodium falciparum, 566

Plasmodium malariae, 566

Plasmodium ovale, 566

Plasmodium spp., 298–299, *299*, 313
  characteristics of, 302–304, *303*, *305*
  malaria caused by, 566
  portal of entry of, *381*
  resistance of, 438
  transmission of, *363*
  vector of, *365*

Plasmodium vivax
  life cycle of, *304*
  malaria caused by, 566

Plasmolysis, 147–148, *148*

Plate counts, 162, *163*

Platelets, 547
  destruction of in ITP, 457–458, *457*
  numbers and function of, *398*

Platyhelminthes, 306–308, *310*

Pleated sheets, 45, *48*

Pleomorphic bacteria, 76

Plesiomonas shigelloides, isolated from marine animals, *251*

PMNs. *See* Polymorphonuclear leukocytes.

Pneumococcal pneumonia, 589, *598*
  agent of, 272; portal of entry of, *380*
  immunity in, 414
  leukocyte count in, 397
  vaccine for, 435, *436*, 589

Pneumococci, capsular antigens of, 589

Pneumocystis carinii, *305*
  in AIDS, 341

Pneumocystis carinii pneumonia, *305* 599, 600
  AIDS-associated, 2, 356–357, 360, 465–466, *466*, 600
  pentamidine isothionate used for, *483*, 492
  treatment of, *483*, 484

Pneumonias
  AIDS-associated, 356–357, 360, 465–466, *466*, 600
  atypical, 590
  bacterial, 588–592
  B. catarrhalis, 589
  H. influenzae, 588
  influenza-associated, 506
  Klebsiella, 590, *598*
  Legionella, 4, 588
  mycoplasmal, 588, 590, *598*
  P. carinii, 2, *305*, *483*, 484, 492, *599*, 600
  pneumococcal, 272, 380, 397, 414, 435, 589, *598*
  primary atypical ("walking"), 271

Subacute sclerosing panencephalitis (SSPE), 347, *347*, 513
Subclinical infection, 360
  immunity conferred by, 414
Subcutaneous infection, 395
Subcutaneous mycoses, 291, 515–516
Substrate-level phosphorylation, 115
Substrates, 33, 108, 110
  concentration of, enzymatic activity and, *111*, 112
Subunit vaccines, 437
Subunits, 38
Sucrase, *111*
Sucrose, 39, *39*
  conversion to dextran, *609*
  dental caries, correlation with, 608–610
Sugar-phosphate backbone, 195, 198, *198*
  covalent bond breaking in, 214
Sugars, 39–40
Sulfadiazine, toxoplasmosis treated with, 565
Sulfadoxine, malaria treatment with, 567
Sulfa drugs, 12, 476, 481
  resistance to, 494, 527
Sulfamethoxazole
  action of, 484, *484*
  typhoid fever treatment with, 614
Sulfanilamide, *112*, 476, 481
  enzyme inhibition by, 112
  structure of, *481*
Sulfhydryl groups, *38*, 45
Sulfide spoilage, *695*
Sulfolobus, characteristics of, *261*, 277
Sulfonamides, 476
  bacterial susceptibility to, *84*
  conjunctivitis treatment with, *519*
  gram-positive bacteria and, 66
  meningococcal meningitis treated with, 527
  properties of, 481–484, *482*
Sulfones
  action of, 481
  leprosy treatment with, 533, *541*
Sulfur
  electronic configuration of, 27
  microbial growth requirements for, 149
Sulfur bacteria, 92
Sulfur cycle, *10, 11*
  bacteria in, 277
Sulfur granules, 92
Sulfur-reducing bacteria, characteristics of, *260*, 268–269
Sunlight, antimicrobial effect of, 179
Superficial mycoses, 291, 514
Superinfection, 478, 487

Superoxide dismutase, 150–151
Superoxide free radicals, 150–151
Suppressor T cells, 424–425, *425*
  in desensitization, 455
Suramin, trypanosomiasis treatment with, 540
Surface-active agents, antimicrobial action of, *189*
Susceptibility, 394
  to viruses, individual differences in, 335
Sushi, 628
Sweat glands, 397
Swimmer's ear, 508
Swimmer's itch, 569
Swine influenza, 595
  vaccine for, 435, 595
Symbiosis, 149, 356, 664
  in termites, 262–263
Symptomatology, classification by, 324
Symptoms, 359, 369
Synapsis, 217
Syndrome, 359
Synergism, of drugs, 484, 495, 567
Synthesis reactions, 32
Synthetic chemicals, degradation of in soil, 674–675
Synthetic drugs, 12, 475
Synthetic reactions, 106
Syphilis, *10, 11*, 648–652, *657*
  agent of, 259; portal of entry of, *381*
  diagnostic tests for, 443, 445
  incidence and distribution of, 3, *649*
  latent period of, 650
  lesions of, 650
  stages of, 650–651, *650*
  tetracyclines used for, 487
  treatment of, 12, 475–476
Systemic infection, 360
Systemic lupus erythematosus, 459, *461*
Systemic mycoses, 291

2,4,5-T, microbial decomposition of, 675, *675*
2,4-D, microbial decomposition of, 675, *675*
Taenia saginata, 308, *310*, 628, *629*
  transmission of, *363*
Taenia solium, 628, *629*
Taette, *702*
Talaromyces, 290
Tampons, toxic shock syndrome and, 4, *371*
T antigen, 344–345

Tapeworms, 308, *308*, *310*, *363*
  antihelminthic drugs used for, *483*, 492
  infections with, 628–629, *633*
  life cycle of, 628, *629*
Taxa, 236
Taxis, 79
Taxonomic hierarchy, 241–242
Taxonomy, 236
TCA. *See* Tricarboxylic acid cycle.
T-cell receptors, immunoglobulins, similarity to, 426, 428
T cells, 415–416, *415*
  cell-mediated immunity, role in, 424–426
  differentiation of, *420*, 427
  in hypersensitivity reactions, *452*, 459
  types of, 424, *425*
T-dependent antigens, 420, *422*
T-DNA, 224–225
TDP. *See* Thermal death point.
TDT. *See* Thermal death time.
TD vaccine, immunization schedule for, *437*
Tears, 395–396, *396*
Teichoic acids, 82–83, *83, 84*
Temperate phages, 332–333
Temperature
  body, in fever, 405
  enzymatic activity and, 111–112, *111*
  microbial growth inhibition by, 172, 177, *180*
  microbial growth ranges of, 145–146, *145*
  reaction rate and, 33–34
  spoilage, *696*
  sterilization, typical values of, *171*
Temperature abuse, 611
Termination factors, *206*
Termites, symbiotic microorganisms in, 262–263
Terramycin, 487
Tertiary treatment, of sewage, 687
Tetanospasmin, 384, 386, 529
Tetanus, *10*, 529–530, *529, 541*
  agent of, 274; portal of entry of, *381*
  exotoxins and, 384, *385*
  vaccine for, *436*
Tetanus antitoxin, 444
Tetanus immune globulin (TIG), 530, *530*
Tetanus neurotoxin, 384
Tetanus toxin
  characteristics of, 384
  lethality of, 383
Tetanus toxoid
  boosters of, 529–530, *530*